D1749922

The BOTANY of BEER

Giuseppe Caruso

The BOTANY of BEER

An Illustrated Guide to More Than 500 Plants Used in Brewing

Foreword by Marika Josephson
Translated by Kosmos, Reggio Emilia, Italy

Columbia University Press
New York

This book was translated thanks to a grant awarded by the Italian Ministry of Foreign Affairs and International Cooperation.

COLUMBIA UNIVERSITY PRESS

Columbia University Press gratefully acknowledges the generous support for this book provided by Publisher's Circle member the Knapp Family Foundation.

Columbia University Press
Publishers Since 1893
New York   Chichester, West Sussex
cup.columbia.edu

Translation copyright © 2022 Columbia University Press
Originally published as *La botanica della birra. Caratteristiche e proprietà di oltre 500 specie vegetali usate nel brassaggio*, by Giuseppe Caruso.
Texts and illustrations by Giuseppe Caruso
© 2019 Slow Food Editore
Via Audisio, 5—12042 Bra (Cn)
www.slowfoodeditore.it

All rights reserved
"An Infinite World" excerpted from T. Musso and M. Drago, *Baladin. La birra artigianale è tutta colpa di Teo* (*Craft Beer Is All Teo's Fault*) (Milan: Feltrinelli, 2013). Authorized reproduction © Giangiacomo Feltrinelli Editore, Milan. First edition in Serie Bianca, May 2013.

∞
Columbia University Press books are printed on permanent and durable acid-free paper.
Printed in the United States of America

Cover design: Noah Arlow
Cover image: Giuseppe Caruso

To Aurora Iolanda Pellegrino, my mother,
to whom I owe the determination
that writing this book required.

An Infinite World

Teo Musso

I would be there like a madman, trying to capture every little nuance, every little change in flavor, in smell. I would put the ingredients in cold water, then I would try an infusion at 95 degrees Celsius for half an hour, everything to try out nuances . . . I began with the ten spices that are usually used, only to soon realize that that world is infinite: spices, herbs, petals, sticks, leaves, really everything is good for creating flavor. There's no end to it. I experimented with a hundred types of petals, thirty varieties of coriander, gentian root sourced from twenty-five different origins. I proceeded obsession by obsession: one month submerged in mountain herbs, one month in something else. My most impressive fixation was resins. All things considered, the world of petals is more familiar, while the world of resins is unexplored—and therefore even more fascinating.

Really awesome. We have very limited knowledge of resins: not many people know that the incense burned in churches is a resin from the myrrh family. This makes us automatically associate resins with something exclusively aromatic, and above all we are mostly familiar with burnt resins. Instead, I started making infusions and discovered a fantastic universe of flavors and scents.

It is a very complicated world because we are talking about the products of the earth, subject to changes and variables. If you base your work on things like flavors and fragrances, you put yourself completely in the hands of your suppliers. When you change suppliers, you have to restart smelling everything because sometimes the differences are remarkable. For me, this is the most beautiful thing in the world.

The master brewer Teo Musso is the founder of the innovative Baladin brewery and Birra Baladin.

Contents

An Infinite World	vii
Foreword	xi
Acknowledgments	xv
Introduction	1
In Praise of G-locality	9
About Plants and Beer Making	11
A Botanical Beer-Making Compendium	13
The Botanical Beer-Making Profile	15
Botanical Beer-Making Profiles	33
Other Beer-Making Plant Species	572
Glossary	575
Bibliography	589
Index of Common Names	613

Foreword

In the summer of 2009, my partner (now husband) and I packed up a U-Haul with the modest belongings acquired by a couple of twenty-somethings who had been bouncing between tiny apartments in New York City. Nick had been accepted into a doctoral program at Southern Illinois University, Carbondale, and I was finishing my dissertation at the New School for Social Research, mobile and working from home.

We were two philosophers ready for a new adventure as we pulled into the spacious duplex that would become our home, smack dab in the middle of the Shawnee National Forest. In two days we had traded New York City, population 9 million, for Makanda, Illinois, population 350.

Steps outside our condo were wild persimmons, elderflowers, nettle, juniper, rose, sassafras, and a densely packed oak and hickory forest peppered by the hooting of great horned owls. But all I could name when we settled down that summer were the dandelions that grew abundantly in the grassy lawn between the homes in our small, unpaved cul-de-sac. I had no idea what was in store for me in the next ten years.

Almost immediately I began brewing beer at home. I had been surrounded by beer throughout my childhood. Carboys lining the side of the garage of our San Diego home are some of my earliest memories, and my father never missed an opportunity to stop at a new brewery while we were on vacation or had a free weekend or were simply on the way to the grocery store. He never really needed a reason. Throughout the 1990s, when craft beer in Southern California was in its infancy, I remember having root beer with my mother and her parents at the tables of the Karl Strauss brewpub in downtown San Diego and buying bottles of Stone Brewing Company's Arrogant Bastard Ale for uncles visiting from Canada and North Dakota.

So it was little surprise that when we arrived in Makanda with a full kitchen and a back deck and plenty of storage space for our hobbies, I jumped right into making beer. My family and friends equipped me with books and kettles as a kind of housewarming gift. During the day I worked on my dissertation, and on the weekends I brewed beer.

True to the spirit of that time, I started a blog for family and friends to follow my progress and to connect with others in the area who were exploring the now burgeoning craft beer scene. In no time I found a small and passionate group of beer lovers and home brewers who lived in the southern delta of Illinois, a huge expanse that is defined on the west by the Mississippi River, and on the south and east by the Ohio and Wabash Rivers. We gathered once a week at a now-defunct store and restaurant, where we shared beer and stories, including the beer we brewed at home. These were people with an intimate knowledge of the region—a life hunting, fishing, gardening, and foraging in the woods for morel mushrooms—all with a passion for the ingredients and processes that made good beer. I loved them immediately and found it easy to call this new place home.

Around that time, some major books and films were being released nationally that took a deep dive into our food systems. Michael Pollan's *The Omnivore's Dilemma* was the first I remember reading, followed by the powerful documentary *Food, Inc.* At home, Nick and I made an ethical commitment to buying local food as much as possible and eating meat only if we knew where it was from—practices we still adhere to. We fit the vision of what Mark Bittman called "less-meat-atarians," confusing most of our family who to this day don't know if we're vegetarian, vegan, or just don't like meat.

I was at the farmer's market in Carbondale every weekend, and it was inevitable for me, an adventurous home brewer, that those ingredients would end up in my beer. I made a pale ale with chiles and sage. I made beer with juniper branches and dandelions. I found all the books I could that talked about brewing with ingredients other than malt, water, yeast, and hops. There were approximately two.

I was also beginning to sense a deep irony in the craft beer market, which had grown wildly since I had experienced it in the eighties and nineties in San Diego. New breweries were popping up in all major cities and many small towns, asking customers on bumper stickers to "support your local brewer," but I was becoming more and more uncomfortable with the fact that virtually none of the ingredients in my "local" beer were grown locally. Hops were from the Pacific Northwest or Germany. Grain was from Idaho or England. Yeast came from a lab on a coast. Yes, brewers were part of a local network of businesses the same way a laundromat was—but how deeply did they support a regional economy? How did they cut down on their carbon footprint? How did they convey a unique sense of place—of terroir? These were questions being asked of the global food system but for some reason not of beer.

Right around this time, I received an email out of the blue. The subject line read simply, "hi." If not for the fact that the name looked believable I may have deleted it, suspecting it to be spam.

It wasn't spam.

"My name is Aaron Kleidon," it began. "I found your blog through the Co-Op website. I have recently moved back to Ava, after spending 5 years in Colorado." Aaron had seen that I was part of a group of home brewers that met weekly and was eager to share his family's interest in plants and some of the beverages he'd been making with them. He continued, "It's nice to see someone with a common interest in regional food and beverage. I have several beers which were made last year which I feel are good. One is an octoberfest brewed with sassafrass. The other one is a persimmon saison fermented on American oak. I was curious if I could meet you and your husband and share some of my drinks as well as ideas that might help to develop a regional beverage market."

I knew immediately that we'd have much to share. Aaron brought his beer the next week and it was probably one of the most important beers I have had in my life. I still remember it distinctly. I remember the flavor of the persimmon, stone-fruit-like, lightly tannic, and finishing of baking spice. I remember the distinctive, sweet flavor of sassafras and how it blended beautifully with the malt. Between the two of us, combining our experiments with ingredients we'd found in the farmer's market and in our backyard, we knew this was the beer we wanted to make and share with the world. Within months we were working with a third partner to start a brewery.

There was no road map for a brewery of this nature. In fact, there was no road map for *beer* of this nature. We consulted the great resource *Sacred and Herbal Healing Beers* by Stephen Harrod Buhner as a guide for many plants, which included in-depth information about historical and medicinal uses of plants in beer. But his book was an encyclopedia of sorts, of recipes written in their own time and in their own context (some recipes were calibrated for hogshead barrels), not a practical guide for modern brewing techniques. And of course it couldn't possibly include all of the plants we had in our woods in the Shawnee. We quickly branched out to resources beyond beer, reading about Native American uses of plants, food preparations, syrup- and jam-making techniques, historical liqueur manufacturing—anything that would give us insight into how plants had been used in human consumption. The magic then lay in how to translate those ideas to beer.

We brewed together weekly for almost a year and a half before opening Scratch in March 2013. During that time we had settled on building a brewery in the woods. Nearly everyone was worried about our being in such a remote location, but the space was a natural fit. Many of our weekly brews were made down a dirt road from where we were now building our brewery, in a little clearing that Aaron's family used for hunting and fishing. A small camper was parked there with running water, and we harvested plants just steps from our kettles. Aaron had grown up in these woods and never stopped showing me flowers, trees, and shrubs and telling me about their uses or lore. I was learning about the woods, and together we were learning how to incorporate the woods into beer. Most importantly, we were encouraged by the results: fascinating flavors that melded well with traditional styles to give subtle background notes and unique character.

FOREWORD

When we finally opened to the public, we were brewing on a small, forty-nine-gallon system, which allowed us to start up on a shoestring budget. It also meant that to keep up we had to brew a couple times a week. Each brew day we made a different beer and used a new ingredient. In those first years we never brewed the same beer twice, experimenting with everything we could find: leaves, roots, bark, seeds, flowers, nuts, fruit, even mushrooms. The vast majority of those plants we'd never seen or tasted in beer before. We had never read a written recipe for those plants or encountered a resource that suggested how to use them. In our first year alone we made over a hundred different beers and put all of that knowledge in our pocket to take into the next year and the next seasons. Our beer evolved over the ensuing years, along with our knowledge and skills, until eventually we felt like we were making beer that was entirely our own.

The book you have in your hands is the book I wish we'd had ten years ago. Dr. Caruso's work is an essential collection that documents the vast life of plants in the creation of beer, beyond simply hops and grain. This isn't a *how-to*, per se; rather it is a vital record that shows precisely which plant parts can be (and have been) used in beer, chemical compounds of each that convey flavor and aroma, and both historical and contemporary instances of use in beer manufacturing. This is a road map filled with inspiration and practical advice to aid any brewer who wants to explore new ingredients.

That alone would be enough, but this book does more. Learning about the plants contained within these pages and incorporating them into the brewhouse may counterintuitively help conserve plants even as we use them as a resource. It helps us better appreciate biodiversity in order to ensure that we nurture that diversity for this and future generations. Indeed, this is a work that prods us to become simultaneously both more humble and more *creative* as brewers—as human beings—that are part of a connected world. The sheer abundance of plants included here should make any brewer, at least once, set aside the pineapple and vanilla beans and find inspiration closer to home. In so doing this work conveys the underlying philosophy we aspire to at Scratch that puts regional agriculture and terroir at the forefront of what it means to make "local" beer. Such an approach has the groundbreaking power to do in beer what we've been doing in food now for the last couple of decades. Armed with an appreciation of local flora and agriculture, we are better able to counteract the effects of monoculture farming, reduce our collective carbon footprint, and convey the ineffable sense of place through beer.

Marika Josephson
Owner and brewer at Scratch Brewing Company
Ava, Illinois

Acknowledgments

Thanks to Anna Alfieri, Carlo Bogliotti, LeeAnn Bortolussi, Chiara Cauda, Eraldo Corti, Kara Cowan, Jennifer Crewe, Nicola Ferrentino, Marco Ferrini, Roberto Fidale, Michele Filippo Fontefrancesco, Zachary Friedman, Michael Haskell, Meredith Howard, Marika Josephson, Ben Kolstad, Emanuela Laratta, Sheniqua Larkin, Nicolò Lo Conte, Bianca Minerdo, Giulia Montepaone, Pietro Mungo, Teo Musso, Grazia Novellini, Andrea Pieroni, Antonio Procopio, Stella Ricciardelli, Roberto Sia, Eugenio Signoroni, and Anselmo Verrino. Each of these people, sometimes even unknowingly, has contributed to the realization of this project.

Nomina si nescis, perit et cognitio rerum. [If you don't know the names of things, knowledge of them also dies.]
—Carl Linnaeus (1707–1778)

It will be well for brewers to be prepared for all eventualities.
—Ezra Meeker (1830–1928)

The BOTANY of BEER

Introduction

Beer has always had an atavistic, complex, and dialectical relationship with wine. Wine has been seen as a noble and prestigious beverage since ancient times, the beverage of wisdom, of stability, of the wealthy classes—in irreconcilable opposition to beer, the drink of the common folk but also of experimentation and freedom. Since wine is made exclusively by pressing grapes, the dichotomy seems to have deepened even more. Between the two worlds, apart from a group of proud fans of this or that product, when not solemn admirers of both, according to the circumstances, there is however a fundamental difference: wine is produced with just one ingredient, grapes, whereas beer is produced with many ingredients whose choice and combination, when not strictly bound to local availability, is up to the brewer.

In objection to this statement, which I recognize as being grossly simplified, it could be argued that there are countless cultivars of wine-making grapes (and, having done my thesis in agricultural sciences in viticulture, how could I forget!) and that each, in the specific environmental conditions of a given place, is capable of expressing a different terroir. This is true beyond any doubt. However, it should be pointed out, grapes are only one botanical species, despite being articulated in infinite genotypic and phenotypic variations. Wine, in short, even with all its interpretations, is indissolubly connected to the cultivation of the grapevine, a species with rather specific bioclimatic needs and, consequently, a limited range. Numerous attempts to artificially enlarge the range of grapevine cultivation—even extending it to tropical and subtropical areas of the globe—have often clashed with insufficient ecological plasticity and have rarely produced qualitatively significant results. In addition, the pulp of the grape is 70–80 percent water, a characteristic that makes grapes easily perishable and therefore not very transportable, suitable at most for local trade. Distances between production and processing areas must be short, a couple of hours by truck at most, and the costs involved in transporting grapes over long distances are prohibitive. In short, wine, for reasons directly linked to the intrinsic characteristics of its unique raw material, is mostly anchored to a specific territory. We should also note that this wine–territory link has been consolidated in the popular imagination by wise marketing and shrewd storytelling.

According to the *Reinheitsgebot*, the Bavarian purity edict of 1516, beer should be made only with water, barley, and hops. The successive processes of unification of the German lands, which included the edict in the legislation of the new geopolitical entity each time, caused the extinction of an unspecified number of endemic beer styles in Germany and induced a surreal flattening of German beer production, with very few and little-appreciated exceptions. Today we could call it a law against creativity, in light of the effects, and definitely consigned to history. A law from five hundred years ago that had negative influences even at a global level, causing a surprising push toward the homogenization of consumers' taste that still endures.

The production of beer, when not shackled by the obtuseness of censorship and regulations that depress creativity, is not necessarily tied to a territory or, least of all, to a single raw material. In fact, many ingredients can be used in the production of beer, and many contain little water (or can be treated to optimize their shelf life), thus proving to be easily transportable. It is precisely this wide availability of ingredients, thanks to globalization, that makes beer suitable for the most experimental and creative souls. But not only that. According to some, beer exists in that mysterious place where art and science meet. The fact that the creativity of the brewer benefits from the availability of many ingredients does not necessarily mean they should not characterize their beer production by establishing a profitable bond with their territory of origin. In some cases, it is enough to simply look to the past, when the availability of exotic beer-making materials was scarce. Some breweries in Northern Europe rediscovered ancient ingredients that way: an example

INTRODUCTION

of such an ingredient is *Calluna vulgaris*, heather, which produces surprising heather ales, yet another beer-making style that was until recently at risk of extinction. Renewing tradition while remaining faithful to it is not an easy task, but it is possible. Alternatively, when freed from ties with a cumbersome beer-making past, it is possible for creativity to wink at other more deeply rooted traditions. This is how in Italy an innovative combination was created, in which beer was brought together with the country's great wine-making tradition to create an entirely new style now recognized by the Beer Judge Certification Program: Italian Grape Ale (IGA). Malt wort and grape must ferment together. Who would ever have thought that beer and wine could resolve their ancient quarrel and form this unpredictable alliance? However, despite all the knowledge that we imagined having at a certain point, one need only read the book *Ancient Brews*, by Patrick E. McGovern, to learn that as far back as nine thousand years ago, in Neolithic China, the most ancient alcoholic beverage in recorded history was made of cereals and grapes, along with many other ingredients. History, even the history of beer, is full of interesting occurrences and recurrences. We speak of "beer" but perhaps should refer to "beers," to paraphrase Lorenzo "Kuaska" Dabove, a great beer expert and tireless communicator. Dabove is considered by many a sort of putative father of Italian craft beer who, with a simple plural, admirably summarizes the extraordinary kaleidoscope of the world of beer as a vast and varied family of alcoholic beverages produced from a molecule that plants use as a reserve substance: starch. Plants accumulate starch by polymerizing photosynthetic glucose in underground organs (tubers, bulbs, rhizomes) or, more frequently, seeds. Brewing has always had global dimensions, given that practically every society with access to some source of starch, from one end of the world to the other, produces beer. In Africa, the preferred source of starch is millet. In South America, it is corn and cassava. In Asia, rice is the plant of choice, whereas in Europe, wheat and, above all, barley are the main ones. Elsewhere, sorghum, rye, oats, agave, palms, and many other species give their starch contribution to the world of beer.

The problem with starch is its structure, made of glucose chains that are strongly branched and impossible for common yeasts to metabolize. To reduce grains' starch into a form useful to feed yeast metabolism, malting is required. This process activates the endogenous enzymes of the seed, those that normally supply energy to the embryo during germination. A similar process of amylolysis is done on some rice beers by means of the enzymes of certain mushrooms, whereas some South American fermented beverages take advantage of human saliva, a known source of amylolytic enzymes.

Once the problem of properly feeding yeast is solved, a crucial step in obtaining ethyl alcohol, numerous other plants are used to give the finished product fermentable sugars, flavors, aromas, bitterness, and preservative properties according to its individual needs. In pre-medieval times, dozens if not hundreds of different plants were used to make beer. The plants used in brewing were the hallmark of the region where each beer came from, and a mix of wild or cultivated herbs used before the advent of hops was called *gruit*, which in Old German means "herb." Some believe that the use of the term should be extended to any beers brewed without hops. In medieval Germany, the archbishop of Cologne held a monopoly on gruit production and had an interest in promoting this style of beer, as opposed to the emerging hopped beers. German states and the Church competed for supremacy in beer production for a long time. Today, it appears that in the long-lasting gruit–hops debate, the latter eventually prevailed. But is that really the case? With the craft brewing revolution now a global phenomenon, most successful breweries produce hopped, even heavily hopped, beers that many craft beer consumers continue to appreciate. But creating yet another India Pale Ale (IPA) is less and less frequently the motivation that drives so many to make beer. Increasingly, research is going in exactly the opposite direction, that of gruit, the ancient hopless ale. Ever less rare are the breweries that regularly produce beers with herbs other than hops to give bitterness, flavor, and aroma, and ever more are the brewers who consider hops a sort of gilded cage, inevitably lacking the vivacity and stimuli that can be provided by hundreds of other plant species. According

INTRODUCTION

to these gruit lovers, it is not worth wasting one's life trying to dole out hops when one has so many other ingredients to explore.

On the other hand, beer has an incredibly long history. Some archaeologists are even willing to swear that humankind first stumbled upon the process of fermentation something like thirty thousand years ago. And if from that point on we know that many plant species have been used by various populations in brewing, mainly because of the species being locally available, we can assume that hops were just one of them. There are many thousands of years of ascertained beer making, and most of those millennia are without or nearly without hops. With hops having so many useful qualities for brewing in a single species, it was frankly inevitable that sooner or later hopping would become a systematic practice, notwithstanding strong territorial differences throughout Europe. Over the centuries, "gruit" was first a singular term and then became plural (a pool of plant species) until it fell into disuse following the legal sanctions placed on it by the *Reinheitsgebot*. Huge economic interests connected to the control of herbs always lurked behind such reversals. In seventeenth-century Britain, one of the countries most resistant to the increasing use of hops, the terms "ale" and "beer" both referred to beverages made of fermented grains, but the first was made without hops, the second with. Only in the eighteenth century, with the almost universal adoption of hops, did the terms became synonymous.

I like to imagine, as suggested by many other authors, that the classic tribal brewery of the past was essentially composed of a group of women sitting in a circle, intently chewing cereals and spitting the bolus mixed with saliva in a common pot to form a fermentable mass (thanks to ptyalin, salivary amylase) that was soon colonized by wild yeasts floating in the air, starting its transformation into beer. This task, like many others, is presumed to have been women's work for many generations (for example, by the so-called brewsters or alewives of the Anglo-Saxon world), providing beverages and intoxication first for the family, then in small family taverns. The step from this practice to becoming healers—and maybe even witches—must have seemed a small one, not least because women often came into conflict with organized forms of religion and medicine, as well as with shrewd businessmen who saw an appealing business in the production of beer. With the Protestant Reformation, an epochal change took place for these fierce competitors between the sixteenth and seventeenth centuries. The stimulating herbs contained in gruit—to tell the truth, sometimes these were toxic concoctions with psychotropic effects—were banned as dangerous adulterants, thus encouraging the use of the depressant hops.

It was the peaceful revolution beginning in the 1980s on the American beer scene, opposed to the homogenization of consumers' taste toward a single brewing style, that slowly spread throughout the world. And the revolution continues, setting new growth records, but above all exploring new ways of brewing. An emerging trend is *localization*, the reduced use—in some cases drastic—of beer ingredients found in places physically distant from the brewery. Faced with a seemingly endless availability of products and beer ingredients and the effects of *globalization*, there are those who believe it is more ethical, more functional, and even more convenient to use local products. On the other hand, given the choice, no one would choose to make beer with immature hops, picked and shipped in unknown thermal conditions for an unspecified period of time, instead of freshly harvested, fully ripe hops from a producer close to home. Taste and aroma work in every plant just as in hops: harvesting, storage conditions, transportation, and seasonality greatly influence the quality of products and of the beer that can be obtained from them.

Plato wrote that the health of his republic was based on the contributions of individuals, and long ago identified farmers—producers of food for the community—as fundamental members of society. Farmers, however, not only produce food but also care for the soil, forming an interface between nature and human society. Preserving soil fertility preserves the land on which we all depend to provide healthy, nutrient-rich food to communities. The health of the soil is transferred to plants, animals, individuals, families, rural communities, cities, cultures, societies, and economies.

INTRODUCTION

For millennia, much of human society has been organized into small rural communities directly linked to the soil in which the food that kept them alive was grown. Today, society is aggregating into ever larger and more crowded cities, increasingly detached from farmers and the land. The growing drain of young human resources from rural communities to urban areas has been matched by a growing concentration of agriculture in regions far beyond the reach of individual farmers, if not beyond the borders of individual nations. In many countries, crop diversity has been drastically reduced to the benefit of just a few species, whose production processes have been almost totally industrialized; the others are mostly imported, often not at the right point of maturity and therefore nutritionally poor. The health of people and local communities inevitably suffers; consequently, so does that of entire nations. For these reasons, a return to local agriculture should be hoped for and promoted. And while the technical details of a modern approach to sustainable agriculture—including from a social perspective—are learned in specialized schools, militant agriculture, like other professions, is learned in the field. It is precisely for these reasons that institutes like that where I work are practicing an increasing amount of laboratory work that integrates theoretical study with direct experience in the countryside.

It is based on this new awareness, which often has significant economic implications, that experiences collectively referred to as "civic agriculture" are being created: economic models that pursue the common good as the sum of the welfare of individuals in a community, through cultivating plants and raising livestock for food. In these models, the involvement of local communities and citizens; innovative production and marketing systems; a vision of society based on sustainable social, economic, and environmental practices; ethics; and a sense of responsibility and reciprocity are all fundamental. Owing to the moral nature of civic agriculture, its impact should be assessed through environmental and social multipliers, as well as economic ones. The productive reference point for civic agriculture is usually the medium-small farm, strongly integrated into the local production system, actively interconnected with the community and the natural resources of the area. In addition to food of proven quality and healthfulness, civic agriculture ensures that citizens have access to natural infrastructure (landscapes, management of natural resources, and biodiversity) and social infrastructure (knowledge of the rural world, the identity and vitality of the community, human well-being, and socio-educational and welfare services). It is no coincidence that forms of civic agriculture such as community-supported agriculture and forms of social agriculture practiced by farms, cooperatives, and community gardens, are increasingly successful. These are all practices of recognized economic, environmental, and social quality that do not exhaust their role in simple mercantile exchanges but instead become bearers of lasting and continuous relational values. Thus, brewers who take the concept of local to the extreme and refuse to use products sourced from more than a hundred kilometers from their brewery are becoming less rare. One wonders how far this trend will go. What is the ultimate expression of the concept of "local"? Well, the number of brewers who have chosen to cultivate a small vegetable garden seems to be increasing. These gardens might be set up right behind the brewery so that some of its products can be used in the brewing process. Could this be the best of the "zero kilometer" concept?

Actually, much more is possible in this courageous direction. One of the problems of modern industrial agriculture concerns the cultivars used. Without wishing to venture into the age-old question of patents on living organisms, an aberration against which Slow Food has long fought, it should be remembered that modern cultivars are appreciated above all for the remarkable shelf life of their products, often to the detriment of their flavor and organoleptic qualities. Choosing traditional varieties, particularly if they are used for brewing, means relying on organisms that have adapted to a certain bioclimatic context over the course of generations, proving capable of overcoming the most common adversities and providing products with extraordinary sensory and nutritional qualities. Initiatives to protect and spread the seeds of traditional cultivars are multiplying everywhere, and certainly not by chance. Let us think about what it means for our health and the

INTRODUCTION

environment to bring all these agricultural practices together in highly sustainable forms such as organic farming, biodynamic agriculture, and permaculture.

But is it possible to go even further? It would seem so. The new frontier taking sustainability and localism to another level is *foraging*. The ancient practice of "going for herbs," collecting edible wild plants, has regained popularity in recent decades. It is then a short step from ethnobotany to economic botany, from the plate to the brewery.

Foraging is the knowledge, understanding, and awareness of plants and their ecology, so practicing it implies getting deeply in tune with the surrounding environment. It is only when we feel comfortable and relaxed in a place that we begin to notice things and accumulate knowledge. Getting to know the forest, the prairie, the mountains, the desert, the waterways, and the coasts, however, takes time. Frequenting these environments in all seasons, taking notes, reading guides, listening, smelling, and enjoying the natural environments takes time and commitment. At first everything looks the same, but with time and experience, differences emerge. That appreciation eventually begets collection. For some it is just a passing fad; for others it is truly a way of life. If this practice has been vital to all cultures, it may be vital to ours as well, much to the chagrin of the urban lifestyle adopted by most of us. And if the learning process is slow, be aware that it could go on indefinitely without the practice ever being fully understood, which makes it both mysterious and exciting.

However, foraging is not necessarily an individual practice: it can easily take on the connotations of a rewarding social experience, for example by leading you to meet people who know the plants of the area. For lack of a better way, you can visit the local farmers' market and observe what wild species are sold there. Talk to farmers, to gatherers. It is not uncommon for people with a deep knowledge of the natural world to seek an opportunity to share that knowledge. Then there are parks, botanical gardens, and environmental associations that have naturalist guides who can often provide useful information. In any case, it is always a good idea to start with species that are easy to recognize and then gradually acquire a systematic knowledge of the flora of the area. Once the identity of a certain species has been ascertained, researching the plant and its uses becomes much easier. The main way to do this is to consult a botanist, either an independent professional or an academic. More and more breweries are using the expertise of professional botanists and foragers with specific floristic and applied botanical skills, who are willing to explore natural areas on behalf of the brewery to collect roots, flowers, leaves, fruits, barks, and branches. This knowledge, and even these collaborations, pave the way for the production of beer with plants gathered in nature, so-called wildcrafting.

Once a plant is definitively identified, good rules for collection should be followed. First, never harvest near roads: that causes the pollution produced by cars to end up on our plates or in our mugs. Sustainability, then. We need to think of wild plants in terms of sharing. Other foragers and wildlife may need them. Never significantly reduce (by more than a quarter) the local population of a certain species unless it is an invasive alien species. All the more reason to pay attention to plants protected by laws, and to endangered plants, as well as to endemic species, which by definition are quite rare. Moreover, brewing with wild plants involves a series of additional problems. Wort containing gruit is very different from hopped wort. Since it does not provide the level of antimicrobial protection provided by hops, it is particularly vulnerable, although it is plausible that many beer-making herbs have antibacterial properties. The problems of sanitizing spaces and equipment are multiplied by the absence of hops. The inclusion of these plants, often chopped into small pieces, during the various phases of the brewing process can also cause some inconvenience. However, clogging can easily be avoided by putting the herbs in nylon mesh bags and dropping them into the boiling wort at the right time. The predictability of the effects of a certain herb on fermentation and on the finished product is rather limited. But isn't the discovery of intriguing new combinations of ingredients one of the joys of brewing? Only experimentation, at the moment performed by individual brewers, will give answers to this.

According to some brewers, the first step in the process of getting rid of hops is reinterpreting classic beer styles without them, after which it should be possible to proceed to new interpretations. According to others, the use of many species will open totally new stylistic scenarios. It will be interesting to see how the situation evolves. In the meantime, I like to think that, when the new trends in the production of craft beer are eventually consolidated, hops will be a sort of *primus inter pares*, and the new (or old) beer-making species will conquer spaces in the recipes of those increasingly numerous brewers who intend to showcase the properties of ingredients available in their local territories instead of vacuum-packed stuff shipped from the other side of the world. In addition to the excellent beers they produce, the communities in which these forward-thinking brewers operate are also beginning to appreciate the economic benefits of reinvesting beer-making profits on a local scale, by supporting local farmers, hop growers, maltsters, botanists, and foragers. One of the most functional approaches to the valorization of a territory's beer-making resources seems to be the one developed by Marika Josephson, Aaron Kleidon, and Ryan Tockstein. In *The Homebrewer's Almanac*, their original combination of foraging, sustainable cultivation, and strictly local market supply is cleverly organized according to the four seasons. Each period of the year appears incredibly full of fruits, seeds, leaves, flowers, buds, saps, and bark, ingredients as unlikely as they are extraordinary, capable of earning a place in the surprising beer and food offerings of Scratch Brewing Company, which appears to have found plenty of success in parts of southern Illinois.

It is hard to say whether this self-sufficient brewing movement is really pulling the plug on the corporations that still dominate the world's beer scene, but they are certainly working to re-establish vital relationships with both the natural world and the people who provide the raw materials and drink the beers.

Perhaps the most advanced point of the movement, capable of harmoniously combining the systematic sensory screening of the flora of a territory in search of potential beer-making resources with the rigor of botanical science, is represented by Pascal Baudar. In his splendidly illustrated *The Wildcrafting Brewer*, he demonstrates a remarkable ability to exploit the plant resources of each ecosystem at hand in a surprising way, producing extraordinary beers that encapsulate the sensorial essence of the place, his so-called forest beers. Bottling the sensations offered by a natural ecosystem is in itself something revolutionary and perhaps deserves the more appropriate collective name of "ecosystem beers," but being able to make a different one for every season really means taking brewing to a higher level: entering a dimension of deep cognitive, sensorial, and emotional harmony with the place. As if all this were not enough to make this visionary's work something out of the ordinary, here he is raising the bar even higher. Why limit yourself to describing your experiences with ecosystems, plants, and wild fermentations when you can lay out the guidelines of a real working method to govern what we could call "ecosystem wildcrafting"? Why not make this approach to beer research of *genius loci* universal? Was this search for a sense of place taken for granted? Was it perhaps easy to bring senses and science together? No, indeed it was not. And yet he succeeded.

But there is no shortage of surprising perspectives on craft brewing, such as the undoubtedly super-qualified Sam Calagione, co-founder of Dogfish Head Craft Brewery and one of the world's most creative and successful brewers. In *Project Extreme Brewing*, the international brewing superstar states:

There are really two components of brewing for me that are existentially rewarding. One: the opportunity to take risks and try ingredient combinations and techniques that may have never been done before. And two: the opportunity to meet, learn from, and be inspired by so many amazing people who have informed my journey as a brewer.

I imagine there can be no better *viaticum* than this, nor any equally authoritative, pointing the way to embarking on a path in brewing that is both informed and experimental, steeped in creativity, and perhaps even destined for success.

Brewing is an activity completely dependent on plants. Cereals (or other vegetable sources of sugars) and herbs that bitter, preserve, and give flavor to beer. Yeast itself can be considered a plant ingredient to some extent. And plants are very sensitive and aware organisms. They have a precise sense of what

INTRODUCTION

is happening to them, as well as what is happening around them. They can modify their chemistry in response to stress to fight adversity, reproduce, and communicate with their peers and other organisms. It is exactly this flexibility, after all, that makes the so-called terroir. I envision a day, if it is not already happening while I am writing this, when the senses of producers and consumers will get to the point of recognizing the subtle differences given by the various terroirs to the ingredients of beer, barley, and hops in particular.

Waiting for such an awareness, this book intends to cast a light on a substantial selection of the huge variety of plant ingredients available to creative brewers. It was certainly necessary, if not urgent, to organize the babel of nomenclature adopted all over the world to identify the plants used in making beer. It was necessary to provide the tools—for both professionals and amateurs—to unambiguously associate a Latin binomial with a specific species and to connect morphological information useful for plant recognition to a plant's biogeographical origin, ecology, cultivation area, etc., without neglecting references to its most important beer-making characteristics. This is the book I would have liked to read as a beer lover—a universal (or almost-universal) compendium of plants used in making beer—but I had to admit it didn't exist until now. And that is why I started my research, during the process of which I also came across attempts to describe people's experiences of brewing as local, radical, or extreme, just because of the ingredients used. Each of these attempts, while well written and rich in insights, did not go beyond a third of the species described here, and more importantly lacked the necessary rigor of botanical science. Almost four years of research pushed the database to five hundred species: the limit I agreed upon with the publisher. A respectable number. The actual number of possible beer-making species, however, is probably much higher.

I must confess that if I had known from the beginning how much time and effort the project would have taken, I probably would have given up. If I had known then that I would have had to draw all these plants, I certainly would have. Today, however, I am happy to have invested so much energy that I can now offer the world of beer, in all its facets, as rigorous and comprehensive a guide as possible on beer botanicals and their properties. I see it as if not the repayment of a debt, at least the return of a courtesy as a botanist to those who have taken the trouble to brew the great variety of beers that have enlivened our evenings with original and unusual aromas and flavors. My hope is that it will be these extraordinary people, these magical elves who wander in that intangible place between art and science, who will get the most out of reading this book, promoting the development of new projects and new perspectives for the sector. Traditions to be recovered, original combinations of ingredients with which to experiment, sustainable agriculture, foraging, and wildcrafting promise sublime sensory experiences and vast new frontiers of freedom. It is my unshakable belief that a deep knowledge of the ingredients of beer is an indispensable tool with which to govern such freedom.

Catanzaro Lido, May 18, 2019

In Praise of G-locality

Andrea Pieroni and Michele F. Fontefrancesco

Asking about the botany of beer today is not simple curiosity. Rather, it addresses the central theme of the recent transformation of a beverage whose history spans at least eight thousand years, and the process by which it evolved to become a sort of liquid synthesis of the biodiversity of territories. At the center of this transformation is the theme of embeddedness, the creation of a strong link between a product and its place of origin. This is motivated by a new sensitivity in the market, especially in the West, that ascribes increasing value to products expressing a local character, a close relationship with the community and the country in which they are made.

Over the past twenty years, even outside the high-end market, consumers' attention has become increasingly drawn toward not only the distinctiveness of products' sensory characteristics but also their territoriality and the ability of ingredients and production methods to express the biodiversity and cultural specificity of their place of origin. The consumer becomes informed, then recognizes and seeks out this link through guides, brands, websites, and social media. The gourmet traveler, an explorer of rural and urban space, is interested in tracing the connections that link the "biographies" of food products to particular areas and communities. In this sense, we can conclude that today as never before, food demand has developed in search not only of food but also of territories in which to let one's gastronomic imagination take root—we could even say that, without a specific place to evoke, a product remains in some way "tasteless."

Identifying a food with a territory is not a recent phenomenon. We can find stories from long ago that link the creation of strong local identities to the expansion of the urban world and the transformation of the rural setting. These stories have their roots in the early modern era and have become firmly established, especially in Italy, and are now a shared cultural concept capable of creating widespread awareness of the territorial bond that underpins a food and defines its character. Wine perhaps best expresses this synthesis between a product and its local geography, since it responds to the demand for a sense of place in terms of both ingredients and production. On the one hand it is possible to identify the origin of particular grapes and yeasts with great precision, reaching levels of minute topographical detail; on the other, it is just as easy to define the place of production, often located just a few kilometers from the rows of vines. This makes it almost a truism that wine is a "local" product, one strongly linked to its territory; however, this is less the case for beer.

Beer's contemporary image is bound to its industrial and mass production; to a product that is a skillful blend of malts, hops, and yeasts purchased on the international market, whose precise origins and history are a mystery to most consumers, and in many cases to the producers themselves. Despite the constant growth that the beer market has experienced in recent years, it is clear that a product so anonymously global can hardly meet the demand for local character, land, people, and landscapes. The major brands in the sector have tried to meet this demand through the creation of "super-premium" lines but have not always been successful. Other brewers have been able to better interpret this need, as shown by an annual growth rate about four times that of industrial producers and a rapid, robust increase in the number and revenue of craft breweries.

Frequently working in interstitial and marginal areas, often without an ancient tradition but interpreting the productive knowledge of the master brewers as a sign of innovation, the protagonists of the craft revolution play a fundamental role today, not only in understanding the product's market but above all in understanding how beer has created its own embeddedness, even in areas where it was not historically part of the foodscape. Today, that embeddedness is pursued at various levels, particularly by creating short, localized production chains; by centering the role of water, indigenous cereals, and local yeasts; and by experimenting with process and product innovations capable of expanding the repertoire of additional ingredients. Craft breweries, more than others, have in fact extended the cereal–hop pairing that has defined the production of modern beer by experimenting with new combinations of herbs, roots, and fruits. Where the addition of cocoa and coffee, vanilla and exotic spices can take us far away to distant horizons, the use of blends of grapes, wild herbs, and local fruits tell us about nearby territories. This emerging botany of beer, which the book you are leafing through helps to clarify, becomes a tool for making beer a unique object. Even more than wine, beer then becomes the synthesis of a meticulous interpretation of human geographies and natural resources from near and far, and a source of inspiration for new recipes that better express a territorial identity for their products.

In this sense and from a contemporary perspective, Giuseppe Caruso's book, written with extraordinary botanical rigor and passionate attention to every possible ingredient in beer, becomes a valuable resource to help us fully understand the phenomenon and appreciate the botanical and sensory diversity that beer can express. This approach and understanding come together in this book to present a new concept of beer as a fully "g-local"—that is, global and local—product.

Andrea Pieroni,
Professor of ethnobotany,
Magnificent rector of the University of Gastronomic Sciences

Michele F. Fontefrancesco,
Social anthropologist and researcher at the University of Gastronomic Sciences

About Plants and Beer Making

The purpose of this book is to offer a scientifically correct and comprehensive overview of vascular plants (*Spermatophytes* and *Pteridophytes*) used in the production of beer, both today and in the past. In fact, some plant species that were once very much in vogue in beer technology have now almost completely fallen into disuse. In some cases, only small enclaves of brewers (and consumers), courageously going against the trend, have admirably decided to take on the task of keeping ancient brewing traditions alive, although they may seem rather démodé in a sector undergoing such rapid transformation. It would certainly be shortsighted to disregard these beer-making plants merely because they are being neglected today. It is likely that one day they will, once again, find their place among beer ingredients, providing interesting, new, and creative ideas for master brewers and rewarding sensory experiences for beer lovers. In some cases, this is already taking place.

If this volume were not limited to vascular plants—those with specialized tissues through which water and nutrients circulate—it would have been necessary to include information on other groups of organisms that perform important functions in making beer. First, of course, there are yeasts (*Saccharomyces* spp., *Brettanomyces* spp., etc.), fungi responsible for the fermentation process and numerous properties of the final product. At the same time, it would have been impossible to avoid giving an account of the numerous bacteria that play a recognized role in the production of beer and in some variations of the product (*Acetobacter* spp., *Kloeckera* spp., *Lactobacillus* spp., *Pediococcus* spp., *Torulaspora* spp., etc.). Space would also need to be devoted to algae such as *Chondrus crispus*, *Macrocystis* spp., *Nereocystis* spp., *Palmaria palmata*, and *Pyropia* spp. that are sometimes used in brewing technology. In the past, in addition to plants in the strict sense of the term, botany included the study of a range of other organisms: algae (in truth, algae, although studied by specialists as a specific branch of science, from the taxonomic point of view still belong to the *Plantae* kingdom), bacteria, and fungi, among others.

Over time, the accumulation of enormous amounts of information resulting from scientific research, as well as the clearer definition of phylogenetic relationships between large groups of "plant" organisms, has contributed to marking the end of botany as a generalist science and opening the door to inevitable specializations. Microbiology, virology, algology, and mycology today represent only some of the new branches of what was once broadly considered botanical science. And since experts, as well as botanical matter, tend to specialize, it would have been equally imprudent to venture into the unfamiliar intricacies of other biological disciplines.

Most of the useful information making up the compendium of beer-making plant species included in this volume comes from sources in Italian and English. Since I do not have immediate access to other languages, some doubt remains as to how much information I may have excluded from this list, which is already very long, since that information may be available only in other languages. The problem has been solved, I believe, at least for the most part, through the extensive exploration of an online database in English concerning beer-making ingredients from all over the world.

Although every effort has been undertaken to make the book as complete as possible, it cannot realistically include everything. One reason for this is the rapid evolution of the brewing industry worldwide. It is objectively very difficult to learn about a new plant ingredient used in making beer if information about it is not shared in a way that is widely accessible. In addition, several other plant species, along with those listed in the book, have been used (or at least tested) in some form of brewing. I have not included some of these species here because exceeding the five hundred botanical beer-making profiles (and related iconography) could have compromised the sustainability of the whole project owing to the increased

ABOUT PLANTS AND BEER MAKING

size of the volume. Others have been omitted simply because their suitability as an ingredient in beer became known only after the book was set to go to print. This contingent of species, currently excluded from the compendium of plants used in making beer, could perhaps form the nucleus of a second volume.

In its original configuration, to introduce the topic to people unfamiliar with the brewing process, this volume would have included short but meaningful chapters such as "The History of Beer," "The Brewing Process," and "Beer Styles." Unfortunately, the compendium of plants used in making beer has grown so much that it reached and then exceeded five hundred, so it was necessary to limit the information to the specific subject of the book. Brief inserts on these topics, have, however, been included to provide the necessary information to readers. Books and websites are now widely available in many languages so that anyone who wants to learn the secrets of making beer can do so rather easily.

A Botanical Beer-Making Compendium

The challenge was to establish, as completely as possible, a botanical beer-making compendium from the point of view of the botanist. The goal was to assemble currently available knowledge on the vascular plants used, for whatever reason, in the beer-making process. For the first time, an attempt has been made to take stock of the wide variety of plant species currently or previously used in the production of beer. This compendium, however, should not be understood merely as a sterile exercise in style. On the contrary, it is intended to provide the reader with useful, applicable information that can be concretely used in experimentation and—why not—in the production of beer. This book is designed primarily to help improve brewers' awareness of the characteristics and origins of the plant products they already use in brewing. This book can also serve as a guide for anyone retracing paths already traveled by others, as well as anyone rediscovering plant ingredients that have fallen into disuse but are perhaps closer to the tastes, sensibilities, and mentality of today's consumers. It is also aimed at encouraging readers to look for new uses for these plants in the production of beer and maybe even to try new and original combinations of these "ingredients" to create innovative beer products. Another hoped-for outcome may be to induce readers to experiment with new plant species, different from those listed here but perhaps taxonomically related to some of them. The reader might even experiment with endemic species (without risking their extinction), which, by their very nature, are linked to rather limited and exclusive territories. Thanks to the characteristics of these unique plants, one could create a commercially valued new product in a creative way that highlights a distinct taste associated with a specific territory. Of course, this would be possible only after making the necessary checks (lack of toxicity and poisonousness, dosage, and technique).

As you can easily imagine, the compendium does not provide creativity; that is something each person must look for within themselves. The volume is limited to providing ideas and technical insights so that you might gain a few new perspectives about a world, the beer world, that is very much in tumultuous evolution.

Creating such a compendium, starting from nothing, is a complicated, demanding, nerve-racking, long, systematic, and often tedious process. As is well known, the difficulty in creating a compendium from scratch lies mainly in the rate of dispersion of information. In this case, the dispersion rate was undoubtedly quite high. The list of plant names, the basis for any subsequent in-depth study, was essentially derived from many, many books about making beer. Other species were added thanks to continual and fruitful discussions with master brewers, traders, beer lovers, experts in the field, and simple enthusiasts. A good deal of information came from the web. Scanning hundreds of websites, managed by just as many craft and other breweries from all over the world; reading thousands of profiles of just as many types of breweries; and exploring dozens of thematic sites provided an essential contribution regarding the number of species included here. Despite efforts to make this "checklist" as complete as possible, new species continued to emerge until the production of this book. The fact that even beer makers were unaware of some of these species is an indication of the probability and inevitability that this list is incomplete. Moreover, even systematic investigation of the web is not necessarily a guarantee of access to all useful information. Suffice it to say that many breweries, especially smaller ones (which often have a higher rate of innovation, as well as a highly localized distribution), do not have detailed websites. Since the production of beer is a global phenomenon, the internet research inevitably clashed with a babel of languages, most of which are unknown to me.

A very useful website, https://www.brewersfriend.com, came to the rescue, as a large section of it is dedicated to "Other Ingredients," referring to the ingredients primarily used in beer production (water, barley malt, hops, etc.). This impressive database, apparently containing every type of material used in the production of beer, played a crucial role in enhancing this botanical beer-making guide. Despite this critical contribution, some things were not included. At a certain point, it became imperative to put time and quantitative limits on my research. However, the species listed and described here represent a respectable body of information that certainly makes it possible to undertake numerous projects and experiments in the production of beer. But I hope that readers who are aware of plant species not listed here that are used today or were used in the past in the production of beer will share this valuable information with me. This will earn my eternal gratitude and actively contribute to the compendium, thus allowing the optimization of any subsequent editions.

The Botanical Beer-Making Profile

The structure of each of the five hundred or so botanical profiles in this book, given the specificity of the theme and the eminently applicative aims of the work, is based on two fundamental aspects: a botanical section (including nomenclature, morphological description, phytogeographic classification, ecology, etc.) and a beer-making section (including parts of the plant that can be used in making beer, their chemical composition, possible beer-making applications, etc.).

Botanical Section

Scientific Name—Although the general public usually have some aversion to Latin plant names, using the scientific name minimizes ambiguities and is far more reliable than using the common name (or, more often, names). Therefore, the decision to use Latin names in this volume should not be considered a mere affectation aimed at hard-core botanists devoted to an incomprehensible nomenclatural orthodoxy but rather the best system currently available for the communication of the names of plant organisms for enthusiasts, botanists, and botanophiles worldwide. The scientific name consists of three elements: the first indicates the **genus**, the grouping to which a certain species belongs (e.g., *Hordeum*). The genus name is written in Latin, and the first letter is capitalized. The second element is the name of the species (its specific epithet), also in Latin but entirely lowercase (e.g., *vulgare*). **Species** is the basic unit of plant classification. Several related species are grouped in the same genus. The third element of the scientific name is the **name of the author** who first described that species. In the scientific name of many plant species (including those described in this book), the letter "L" appears, capitalized, with a period ("L.") immediately after genus and species. This refers to the great Swedish taxonomist Carl von Linné, better known in English as Linnaeus (from the Latinized name Carolus Linnaeus), author of the first descriptions of many plant species. In fact, having established the rules of the binomial nomenclature and the first modern classification structure, he was also the first to describe a large number of living organisms, including numerous species of plants. Based on these necessary premises, the full scientific name of the example taken here is therefore *Hordeum vulgare* L. In some cases, a fourth element may be necessary to complete the name of a species: the year of publication of the work in which the first description (protologue) of the species is recorded.

Family—Today's plant classification framework provides for the inclusion of several other taxonomic categories. A taxonomic category of great importance is the **family**, which groups similar genera. Similar **families** form an **order**, more orders form a **class**, and so on until **division**, **kingdom**, and **domain**. As an example, here is the classification of barley:

Taxonomic rank	Taxon name
Domain	*Eukaryota*
Kingdom	*Plantae*
Subkingdom	*Tracheobionta*
Superdivision	*Spermatophyta*
Division	*Magnoliophyta*
Class	*Liliopsida*
Subclass	*Commelinidae*
Order	*Poales*
Family	*Poaceae*
Subfamily	*Pooideae*
Tribe	*Hordeae*
Subtribe	*Hordeinae*
Genus	*Hordeum*
Species	***Hordeum vulgare***
Hordeum vulgare L., 1753	

In addition to the most important taxonomic categories, highlighted in bold, there are several other categories. Many not shown here are commonly used by taxonomists.

THE BOTANICAL BEER-MAKING PROFILE

BEER-MAKING SECTION

Beer-Making Parts (and Possible Uses)—Rarely is a plant species used in its entirety in beer making. This applies both to basic plant ingredients (barley, wheat, hops, etc.) and to less frequently used plants. It is therefore useful to provide accurate information on the parts of each species used in the production process. Beer production is thus divided into several phases. Each plant species is mostly used in only one of these phases or can contribute to one or more characteristics of the final product (aroma, taste, bitterness, etc.) through the contribution of different substances (sugars, carbohydrates, essential oils, etc.). Nevertheless, it can happen that the same species can be used, with different methods and purposes, in the production of different beers. But it may also happen that the plant species may be involved in several stages of the production process, with different purposes in some cases. Including clear and concise information on the uses and applications of individual plant species, when available, is entirely consistent with the theme and purpose of this book.

Toxicity or Poisonousness—Since beer is a food product, the need to provide the reader with information that in no way endangers human health is obvious. Some plant species used to make beer, mostly in the past, are now recognized as toxic, if not poisonous. Nevertheless, some were used in the past to adulterate the final product, giving it greater ability to induce cheap inebriation (while ethanol was considered costly because it was produced from an expensive raw material, malt). It should be remembered that even species commonly used for food (and therefore possibly also for making beer) can sometimes be dangerous, depending on numerous factors such as the growth conditions of the plant, the dosages used, the length of use, the conditions of the consumer (pathological states, pregnancy, allergies, immunosuppression, etc.), and possible interactions with other foods or substances (e.g., drugs, other ingredients in beer), to name just a few. Surely it would seem irresponsible to say that my research has highlighted every possible aspect of the potential hazards of these plants. There could certainly be other potentially dangerous aspects that have escaped me and that deserve investigation and study. Moreover, it is virtually impossible to exclude the possibility that in the future, some of the species included in this book may present one or more potential risks to human health that are unknown or insufficiently studied today. Therefore, any plant species must be used with the utmost caution in making beer to avoid risk to the health and safety of the consumer. The use of plants in the beer-making process must not be approached lightly; on the contrary, it must be considered carefully and shrewdly, with brewers conducting in-depth studies, researching beyond the information contained in this book, continually updating their knowledge in step with scientific research, and corroborating every choice with the essential trials. Some notoriously dangerous species have been included in this book to ensure completeness of information, and their toxicity has been highlighted. The author and publisher disclaim all civil and criminal liability, in Italy and abroad, arising from the application of information contained in this book.

Chemistry and Chemical Composition—The botanical profile also includes an information box about the chemical composition of the plant portion of specific beer-making interest (per 100 g of the edible part unless otherwise specified). In some cases, several technological considerations related to plant parts used indirectly in brewing (e.g., to produce raw materials such as wood for malt smoking) have been included. The botanical profile includes the chemical composition of the parts of each plant with possible application in producing beer to give the reader an idea, at least in principle, of the type of nutritional and aromatic-tasting-sensory contribution that a certain plant (or plant part) may give to the finished product. It is necessary to be aware that beer is the result of complex chemical and biochemical interactions, some not yet fully understood. These reactions concern the modifications incurred by the main beer ingredients, but they can also involve the biochemical components of additional ingredients and even be released from some types of containers. The qualitative results of one or more additions to the basic beer ingredients are therefore unpredictable on the sole basis of prior knowledge of the chemical composition of the ingredients used. The results depend on the creativity and desire of the home brewer, the craft brewer, or the industrial brewery to experiment with new plausible combinations and then evaluate and select those that can realistically aspire to have enough admirers to make it a viable product, perhaps a niche product but still an economically sustainable one. For some species, complete information on the chemical composition of the parts in use in beer technology was unavailable. In some cases, unfortunately, no information at all was found. To this concern must be added the great heterogeneity of the information available, often deriving from diverse bibliographic sources on plants frequently studied with purposes different from those of this book. Finally, many chemical characteristics of plants, particularly those related to secondary metabolism, are highly dependent on the environmental conditions in which the plant grows and develops. Secondary metabolism is a set of endocellular processes that, biochemically, are not essential for the growth and survival of the cell. Secondary metabolism is accompanied by primary metabolism and introduces, for example, new biosynthetic pathways that lead to the formation of molecules called secondary metabolites, which assist the survival of the plant cell and/or the organism. For each species, for reasons of space, I have included information on chemical composition with reference to only one location. The information on chemical composition should therefore be considered, when available, only as roughly indicative, since it could vary significantly compared to what has been recorded in different locations.

Characteristics of Wood—Some of the plants described in this book have a use in beer making that can be defined as indirect. These are not ingredients introduced directly into the production cycle of beer but materials that accompany some phases of brewing, playing a significant role in the achievement of the final product. In particular, I thought it useful to provide some information on the main technological characteristics (e.g., specific weight, porosity) of the trees that provide the wood used for the construction of the barrels or for smoking the malt.

Beer Style (or Type of Brewing)—Each beer style involves the use of different ingredients. Each plant, therefore, lends itself to use in the production of one or more beer styles. The existence of so many types of beer, especially since the practice of home brewing and the production of craft beers have become global phenomena, discounts almost any type of classification. Nevertheless, an interesting and ambitious categorization experiment has been developed by the Beer Judge Certification Program (BJCP), an American organization that has already released two versions of its guidelines, the BJCP Style Guidelines (in 2008 and 2015). Today these guidelines represent a valuable standard to which many people in the world refer. Nevertheless, the BJCP itself has recognized that categorization, designed to facilitate judging in beer-producing competitions, by grouping beers with similar perceptive characteristics may not be the best way to fully understand beer styles. However, the BJCP Style Guidelines have adopted a beer-making taxonomy based on three taxa: category, subcategory, and style. Categories are large and arbitrary groupings of beer styles, generally with similar characteristics. Each subcategory is defined by the most significant label of a certain beer style and, similarly to the concept of style, identifies the main characteristics of a type of beer. In fact, some subcategories are not related to others belonging to the same category. Given that the purpose of the BJCP Style Guidelines is to group beer styles for the sole purpose of facilitating judgement during competitions, it follows that it would be misleading to try to derive additional meanings or geographic or historical evidence from such groupings.

The nomenclature adopted in the guidelines consists of simple identifiers chosen to best represent the styles and groupings described. Styles have been given names, and these have been grouped by similar characteristics. Therefore, names have been given to groups. However, many styles may have different names in different places. By proposing a selection of names (commonly used or descriptive of a style) that can form a common nomenclature for simple reference, the guidelines attempt to prevent the misuse of several names at once. Some of the names used are protected. This does not mean that they should not be respected or that their use should be extended to all commercial breweries but rather that these names have been considered the most appropriate to describe their respective styles, avoiding the need to indicate "style" each time. Reference to a country or region of origin is often included to differentiate styles with the same name. Considering the global diffusion of many styles, geographic references should be considered merely descriptors, without implication that a certain style constitutes exclusivity of a certain area. If anything, country or region of origin refers only to where a beer originated or is produced.

Correct use of the guidelines implies coherence with their intended purpose: providing a standard set of descriptions of beer styles to be used in beer competitions. However, especially in countries with an emerging craft beer market, the definitions of the guidelines have frequently been adopted to describe beers in general. To avoid misuse of the indications provided, the BJCP points out that the BJCP Style Guidelines are only guidelines, not technical specifications—suggestions, not rigid limits—useful to describe the most common types of beer and broad enough to allow the flexibility (and creativity) necessary to enhance beer production. The guidelines have also been drafted in detail to facilitate the process of structured evaluation of beer and to sometimes draw a clear dividing line between styles, avoiding deleterious overlaps. In fact, not all commercial beers fall within the styles (or other taxa) proposed by the guidelines, which forgo (for various reasons) defining every possible beer style. In addition, in applying the guidelines, one cannot fail to notice factors such as the change of styles over time (often without a corresponding change in nomenclature) and the change of ingredients. An example of this is the large variety of hops available today, as well as the tendency of breweries to adopt (and often just as quickly abandon) different ingredients to differentiate themselves.

Each category or subcategory is identified by a series of descriptors (overall impression; appearance, aroma, taste, and mouthfeel; history; characteristic ingredients; basic statistics; and commercial examples) collected in a special format. Since this is a subject of increasing complexity, and one subject to differences in classification across the world, it was useful to try to propose here a common basis of communication that facilitates the understanding of the plant–style link, minimizing possible ambiguities. Therefore, while recognizing the risk connected to the adoption of a specific stylistic classification, especially if it is unmistakably Americentric, like this one, it was considered that the possible disadvantages were compensated for by greater communicative clarity. The beer taxonomy proposed by the BJCP Style Guidelines is therefore summarized next, including its numbering system and original names. The original and complete version of this courageous document can be found, together with other interesting systematic information about the beer-making universe, on the organization's website (https://www.bjcp.org).

1. Standard American Beer
Typical American beers, with great appeal. Including both ales and lagers, beers in this category are generally simple, light tasting, and accessible. The ales tend to have the qualities of lagers or are designed for the mass market of lager drinkers as crossover beers.
01A. Lite American Lager
01B. American Lager
01C. Cream Ale
01D. American Wheat Beer

2. International Lager
These are mass-produced premium lagers in most countries of the world. Developed from American or European styles, they tend to have a fairly uniform character and are widely marketed. Freely derived from the original pilsner-type lagers, with color variations, they have an additional malt flavor while maintaining a wide appeal to most palates.
02A. International Pale Lager
02B. International Amber Lager
02C. International Dark Lager

3. Czech Lager

These are generally divided by gravity class (draft, lager, special) and color (pale, amber, dark). The Czech names of gravity classes are *výčepní* (draft, 7–10°P), *ležák* (lager, 11–12°P) and *speciální* (special, 13°P+). The color categories are *světlé* (pale), *polotmavé* (amber), and *tmavé* (dark). *Pivo* is the Czech word for beer. The division into gravity classes is similar to the German groupings of *schankbier*, *vollbier*, and *starkbier*, although with different gravity ranges. Within the classes, Czech beers are often simply referred to by their gravity. There are often variations within the gravity-color groupings, especially in the *speciální* class. The style guidelines combine some of these classes, whereas other existing beers on the Czech market are not described (like the strong Czech porter). This does not imply that the underlying categories cover all Czech beers: they are simply a way to group some of the most common types for judging purposes. The Czech lagers generally differ from German lagers and other Western lagers in that the Germans are almost always completely attenuated, whereas the Czech lagers may have a small amount of nonfermented extract remaining in the finished beer. This helps to provide a slightly higher final gravity (and thus a slightly lower apparent attenuation), a slightly fuller body and mouthfeel, and a richer flavor profile, slightly more complex in beers that are equivalent in color and strength. German lagers tend to have a cleaner fermentation profile, whereas Czech lagers are often fermented at cooler temperatures (7–10°C) for a longer time and may have a slight, barely detectable (near-threshold) amount of diacetyl that is often perceived more as rounded body than openly as aroma and flavor (diacetyl in significant amounts is a flaw). Czech lager yeast strains are not always as clean and attenuating as German strains, which helps to achieve a higher final gravity (along with lower-temperature fermentation and mashing methods). Czech lagers are traditionally produced with decoction mash (often double decoction), even with modern malts, whereas most modern German lagers are produced with infusion or step-infusion mash. These differences lead to the richness, mouthfeel, and aromatic profile that distinguish Czech lagers.

03A. Czech Pale Lager
03B. Czech Premium Pale Lager
03C. Czech Amber Lager
03D. Czech Dark Lager

4. Pale Malty European Lager

This category includes German pale malty lagers, produced with pilsner malt in the gravity range from *vollbier* to *starkbier*. Although malted, they are still clean, well-attenuated lagers like most German beers.

04A. Munich Helles
04B. Festbier
04C. Helles Bock

5. Pale Bitter European Beer

Beers of German origin that are pale and have a bitter balance, with a bland to moderately strong hoppy character typical of classic German hops. They are generally bottom fermented or lagered to provide a smooth profile and are well attenuated like most German beers.

05A. German Leichtbier
05B. Kölsch
05C. German Exportbier
05D. German Pils

6. Amber Malty European Lager

Amber-colored lagers, of German origin, bottom fermented, with a malty balance and a strength from *vollbier* to *starkbier*.

06A. Märzen
06B. Rauchbier
06C. Dunkels Bock

7. Amber Bitter European Beer

Amber beers, balanced like bitter beers of German or Austrian origin.

07A. Vienna Lager
07B. Altbier
07C. Kellerbier
07C1. Kellerbier: Munich Kellerbier
07C2. Kellerbier: Franconian Kellerbier

THE BOTANICAL BEER-MAKING PROFILE

8. Dark European Lager
This category brings together the German lager *vollbiers* darker than the dark amber color.
08A. Munich Dunkel
08B. Schwarzbier

9. Strong European Lager
Lagers from Germany and the Baltic region with a strong aroma and high alcohol content. Most are dark, but light versions are also known.
09A1. Doppelbock (Pale)
09A2. Doppelbock (Amber)
09B. Eisbock
09C. Baltic Porter

10. German Wheat Beer
German wheat beers of *vollbier* and *starkbier* strength without sourness, light to dark colors.
10A. Weizen/Weissbier
10B. Dunkels Weissbier
10C. Weizenbock

11. English Pale Ale (British Bitter)
The English bitter family was derived from the English pale ales as draft beer in the late nineteenth century. The use of crystal malt in bitters became more widespread after the First World War. Traditionally served very cool and unpressurized (by gravity or hand pump) at cellar temperature (e.g., real ale). Most bottled or kegged bitters produced in the United Kingdom are more alcoholic and carbonated versions of the cask product for export and have a different character and balance compared to their British draft counterparts (which are often sweeter and less hoppy than the cask versions). These guidelines reflect the real ale style, not the export formulation of commercial products. There are several regional variations of bitters, ranging from the darker and sweeter served with nearly no head to the brighter, hopped, light versions, with a sizeable layer of foam, and everything in between. The caramel component of these styles should not be overemphasized. Exported bitters can be oxidized, which increases the caramel flavor (as well as more negative flavors). The aromas derived from oxidation should not be considered traditional or required for the style.

11A. Ordinary Bitter
11B. Best Bitter
11C. Strong Bitter

12. Pale English Beer (Pale Commonwealth Beer)
This category contains light, moderately strong, hoppy bitter ales from countries formerly part of the British Empire.
12A. English Golden Ale
12B. Australian Sparkling Ale
12C. English IPA

13. English Brown Beer
While dark mild, brown ale, and English porter have long histories, these guidelines describe only the modern versions. They are grouped together for judging purposes only because they have similar flavor and balance, not because of a common, implicit ancestry. Common characteristics are low to moderate strength, dark color, a generally malty balance, and British ancestry. These styles do not have a historical relationship with each other; in particular, no one evolved into one of the others or was a component of another. The category name was not used historically to describe this grouping of beers. Brown beer was a distinct historical product and has no relation to the category name.
13A. Dark Mild
13B. English Brown Ale
13C. English Brown Porter

14. Scottish Ale
The original meaning of "shilling (/-) ale" has been wrongly described for years. A single style of beer has never been designated as 60/-, 70/- or 80/-. "Shillings" referred only to the cost of a barrel of beer, meaning there were 54/- stouts, 86/- IPAs, and so on. The Scottish ales in question were called light, heavy, and export; altogether they covered the cost spectrum from about 60/- to 90/- and were simply dark ale, based on malt. The 120/- ale falls outside this range, as the stronger Scotch ales (also known as wee heavies) do. The guidelines for Scottish light, heavy, and export are similar for each style of beer. As gravity increases, so does the character of the beers

in question. Historically, the three types of beer were parti-gyled to different strengths and represented an adaptation of English pale ale with reduced strength and hopping rates and darker colors (often from added caramel). More modern versions (at least post–World War II ones) tend to use more complex grists.
14A. Scottish Light
14B. Scottish Heavy
14C. Scottish Export

15. Irish Beer
The traditional beers of Ireland included in this category are amber to dark beers, top-fermented beers, moderate to slightly strong, and often widely misinterpreted owing to differences in export versions or overly focusing on the specific attributes of beer produced by well-known, large-volume breweries. Each style in this group has a wider assortment than is commonly believed.
15A. Irish Red Ale
15B. Irish Stout
15C. Irish Extra Stout

16. Dark British Beer
Modern British and Irish stouts, from medium to strong and from bitter to sweet, originated in England, although some are now widely associated with Ireland. In this case, "British" has the broader meaning of "British Isles."
16A. Sweet Stout
16B. Oatmeal Stout
16C. Tropical Stout
16D. Foreign Extra Stout

17. Strong British Ale
This category contains the strongest nonroasty beers of the British Isles.
17A. English Strong Ale
17B. Old Ale
17C. Wee Heavy
17D. English Barleywine

18. Pale American Beer
Modern American ales of medium strength and light color, from moderately malty to moderately bitter.
18A. Blonde Ale
18B. American Pale Ale

19. Dark American Beer (Amber and Brown American Beer)
Modern American beers of amber to dark color, warm fermented, with a standard strength from balanced to bitter.
19A. American Amber Ale
19B. California Common
19C. American Brown Ale

20. American Porter and Stout
All these beers have evolved from their English namesakes to be completely transformed by American craft brewers. In general, they are bigger, stronger, more toasted, and more hop-centric than their English cousins. These styles are grouped together because of a shared history and aromatic profile.
20A. American Porter
20B. American Stout
20C. Imperial Stout

21. IPA
The category is intended for modern American and American-derived IPAs. This does not imply that English IPAs are not properly IPAs or that there is no relationship between them. It is simply a method of grouping similar styles for competitions. English IPAs are grouped with other English-derived beers, and the stronger Double IPAs are grouped with the stronger American beers. "IPA" is used intentionally over "India Pale Ale" because historically, none of these beers had anything to do with India, and the origins of many are unclear. However, the term has ended up referring to a defined-balance style of modern craft beer.
21A. American IPA
21B. Specialty IPA
21B1. Specialty IPA: Belgian IPA
21B2. Specialty IPA: Black IPA
21B3. Specialty IPA: Brown IPA
21B4. Specialty IPA: Red IPA
21B5. Specialty IPA: Rye IPA
21B6. Specialty IPA: White IPA

22. Strong American Ale
Modern, strong American ales, with a variable balance of malt and hops. The category is mostly defined by the alcoholic strength and the absence of toasting.
22A. Double IPA
22B. American Strong Ale
22C. American Barleywine
22D. Wheatwine

23. European Sour Ale
This category contains the traditional sour beer styles of Europe, many of which (but not all) are still produced with a wheat component. Most have a bitterness content, with an acidity that balances the contribution from hops. Some are sweetened or flavored in the brewery or at the time of consumption.
23A. Berliner Weisse
23B. Flanders Red Ale
23C. Oud Bruin
23D. Lambic
23E. Gueuze
23F. Fruit Lambic

24. Belgian Ale
Belgian and French ales, from more malted to balanced, more strongly flavored.
24A. Witbier
24B. Belgian Pale Ale
24C. Bière de Garde

25. Strong Belgian Ale
Light beers, well attenuated, from balanced to bitter, often dominated more by yeast character than malt aroma, generally with a higher alcohol content (although there is a range of styles within the category).
25A. Belgian Blond Ale
25B. Saison (Standard)
25C. Belgian Golden Strong Ale

26. Trappist Ale
"Trappist" is a protected legal name that may be used commercially only by Trappist monasteries that produce beer. However, it may be used to describe types or styles produced by those breweries and those who make beers of a similar style. Trappist beers are characterized by high attenuation, high carbonation through bottle conditioning, and an interesting (often aggressive) yeast character.
26A. Belgian Single
26B. Belgian Dubbel
26C. Belgian Tripel
26D. Belgian Dark Strong Ale

27. Historical Beer
This category groups styles that are now extinct or that were more popular in the past and are now known in the form of re-creations. It also includes traditional or indigenous beers of cultural importance in certain countries. The inclusion of a beer in this category does not imply that it is no longer brewed but that it represents a minor style or is being rediscovered by artisan breweries.
27A. Historical Beer: Gose
27B. Historical Beer: Kentucky Common
27C. Historical Beer: Lichtenhainer
27D. Historical Beer: London Brown Ale
27E. Historical Beer: Pivo Grodziskie
27F. Historical Beer: Pre-Prohibition Lager
27G. Historical Beer: Pre-Prohibition Porter
27H. Historical Beer: Roggenbier
27I. Historical Beer: Sahti

28. American Wild Ale
The name "American wild ale" is in common use among craft brewers and home brewers. However, the word "wild" does not imply that these beers are necessarily naturally fermented; rather, it indicates that they are influenced by microorganisms other than traditional yeasts. The category includes a broad spectrum of beers that cannot be defined as traditional European sour beers or wild styles. All styles in this category are essentially specialty beers for which many creative interpretations are possible, and styles are defined only by specific fermentation profiles and ingredients. Within the category, the term "Brett" is used as an abbreviation of *Brettanomyces*. Many craft brewers and home brewers use this term in conversation, if not in formal communication.
28A. Brett Beer
28B. Mixed Fermentation Sour Beer
28C. Soured Fruit Beer

29. Fruit Beer

Beers made with any fruit or combination of fruits. According to the culinary (nonbotanical) definition adopted here, a fruit is made from the fleshy structures of the plant—sweet or sour and edible raw—associated with the seed. Examples are pome fruits (e.g., apple, pear, quince), stone fruits (e.g., cherry, plum, peach, apricot), berries (any fruit whose name includes the word "berry"), currants, citrus fruits, dried fruits (e.g., date, plum, raisin), tropical fruits (e.g., banana, mango, pineapple, guava, papaya), figs, pomegranates, and prickly pears. Spices, herbs, and vegetables as defined in category 30 are excluded, especially botanical fruits treated as culinary vegetables. If you have to justify the meaning of "fruit" using the word "technically," then it is not included in this category.

29A. Fruit Beer
29B. Fruit and Spice Beer
29C. Specialty Fruit Beer

30. Spice Beer

Here the common or culinary definitions of spices, herbs, and vegetables are used, not the botanical or scientific ones. In general, spices are the dried seeds, pods, fruits, roots, or barks of plants used to flavor food. Aromatics are leafy plants or parts of them (leaves, flowers, petals, stems) used to flavor food. Vegetables are salty, or less sweet, edible plant products, primarily cooked, sometimes raw; they may include some fruits (in botanical terms). This category includes all spices, herbs, and kitchen vegetables, as well as nuts (any vegetable with "nut" in its name, including coconut), chili, coffee, chocolate, spruce tips, rose hips, hibiscus, fruit peels or zest (not the juice), rhubarb, and the like. It does not include culinary fruits and cereals. Sugars and aromatic fermentable syrups (e.g., agave nectar, maple syrup, molasses, sorghum, honey) may be included only in combination with other eligible ingredients and should not have a dominant character. Any combination of eligible ingredients may be included. See category 29 for a definition and examples of fruits.

30A. Spice, Herb, or Vegetable Beer
30B. Autumn Seasonal Beer
30C. Winter Seasonal Beer

31. Alternative Fermentable Beer

Specialty beers with a few additional ingredients (grains or sugars) that add a distinctive character.
31A. Alternative Grain Beer
31B. Alternative Sugar Beer

32. Smoked Beer

Specialty beers with a smoky character.
32A. Classic Style Smoked Beer
32B. Specialty Smoked Beer

33. Wood Beer

Specialty beers with an aging character of wood, with or without an adjunctive character of alcohol.
33A. Wood-Aged Beer
33B. Specialty Wood-Aged Beer

34. Specialty Beer

While the guidelines include many specialty beers, the Specialty Beer category is intended for those beers that do not fit into any other category. Since the number of specialty categories is quite large, most beers are expected to fit somewhere else, unless there is something unique about them. It is a good idea to check the other specialty categories before deciding to include a beer in one of these styles.

34A. Clone Beer
34B. Mixed-Style Beer
34C. Experimental Beer

35. Appendix B: Local Styles

This appendix contains styles, presented by local BJCP members, that are not yet well codified and stabilized but are important for the individual country. They are not included in the main guidelines but are available for use by those who wish.

35A. Dorada Pampeana
35B. Pampas Golden Ale
35C. IPA Especialidad: IPA Argenta
35D. Argentine IPA
35E. Italian Grape Ale

When available, I have included the "stylistic" information. In addition to the BJCP styles, in some of the botanical beer–making profiles you may find the term "gruit." Gruit (also known as ogruyt, gruut, grut, and grjut) is a variable mix of plants used, mostly in the past, to give beer bitterness, flavor, and resistance to alterations, properties conferred by hops. Ancient brewing practices using gruit have been preserved, and gruit continues to be used in the production of interesting niche products. In many ways, the practice is experiencing a revival in various brewing districts.

Beer—One or more examples of beers made with a specific plant species are given as reference to the possible application of a plant species in brewing. Especially with regard to the species most commonly used in brewing, it would be impossible not to exclude many beers available on the market. In these cases, the choice to include particular beers was made based mainly on the ease of access to relevant information.

Sources of Information—Because of the difficulty of compiling this complex and detailed compendium, it seemed useful to highlight the origin of the references to the use or uses of a certain species for making beer. The sources are so numerous for some plants that it was decided not to include all of them. Conversely, there are several species for which there are few or no references to their use in making beer.

Iconography—A large part of this book consists of the iconographic collection related to the species described. Each botanical profile is associated with an iconographic table, a useful tool to facilitate plant recognition. I created these botanical iconographies using live samples of each species when possible. When fresh samples were not available, I turned to herbarium specimens. When this type of material was also unavailable, particularly for exotic and tropical species, I used photographs and illustrations from various bibliographical sources and the internet. In the creation of each iconography, I have attempted to include every element useful to recognize the taxon, as well as a synthesis of the main organs of the plant, with particular attention paid to the reproductive apparatus. The iconographies have not all been made at the same scale; therefore, to have a more precise idea of the size of each organ of a given species, I suggest referring to the morphological description contained in the botanical beer-making profile. In addition to the iconographic tables related to the morphology of each species, a detailed iconography has sometimes been added to illustrate aspects of particular beer-making interest.

Icons—To make each species quickly classifiable in terms of its beer-making potential, I thought it useful to prepare a system of icons to be associated with each species, thus highlighting its prevailing beer-making characteristics. However, the possible contribution of many species described in this book to beer making goes far beyond the brief description that an icon system can show.

Statistics—The following information is provided to facilitate an understanding of how much a family (or a genus) is represented within this compendium. Synoptic data on the geographic origin of the plants, preferred habitats, the parts most used in making beer, and other elements are included to contribute to an overall view of the compendium. Let us go in order.

The pool of plants includes 501 species, divided among 98 families, for an average value of 5.1 species per family. The 24 most represented families (5–48 species) include 380 species (75.8 percent of the total); of the remaining 74 families (121 species; 24.2 percent of the total), 28 include 2–4 species, for a total of 75 (15 percent of the total), and 46 families are represented by only one species (9.1 percent of the total). As table 1 shows, the most represented family is Rosaceae, with 48 species. This is followed by Lamiaceae (37), Asteraceae (35), Fabaceae and Fagaceae (27), Poaceae (23), Rutaceae (22), Ericaceae (19), Apiaceae (16), Solanaceae and Pinaceae (13), Myrtaceae (12), Amaranthaceae (10), and so on. The compendium includes 312 genera, of which 242 are represented by only one species, 56 including 2–4 species, 10 including 5–10 species, and only 4 genera represented by more than 10 species (*Prunus*, 11; *Rubus*, 13; *Citrus*, 19; *Quercus*, 23). Table 2 provides the number of species for the 32 most represented genera in the compendium.

Icon Legend

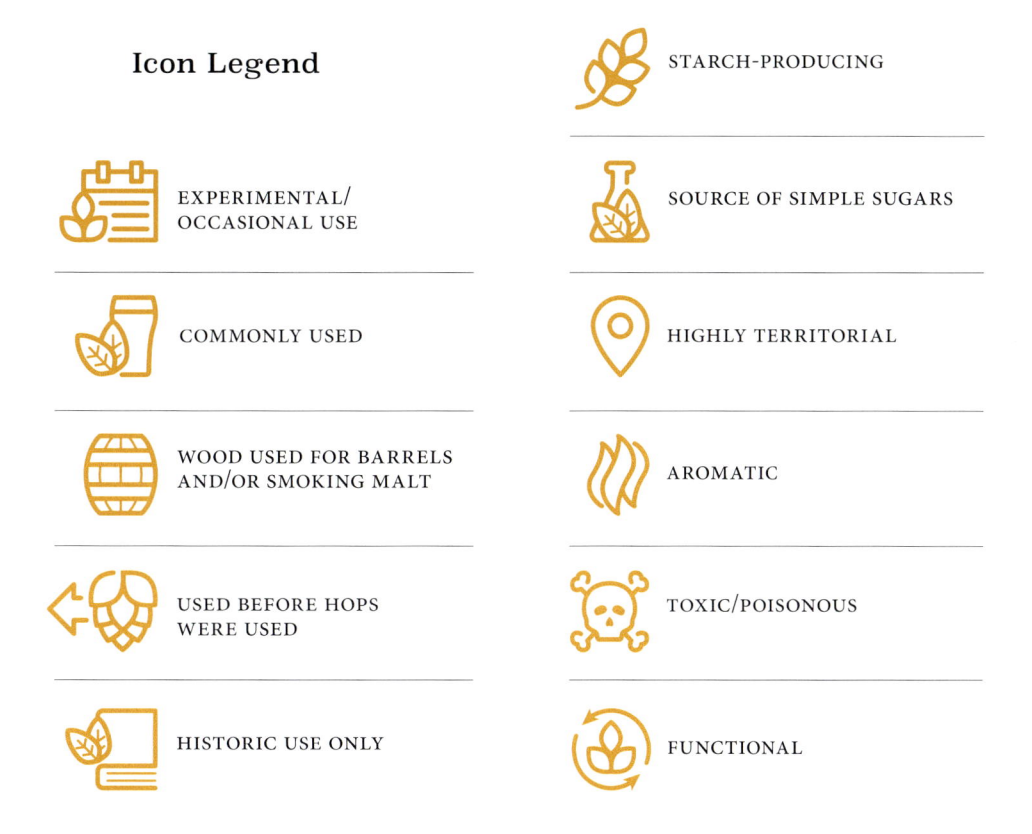

- **EXPERIMENTAL/OCCASIONAL USE**
- **COMMONLY USED**
- **WOOD USED FOR BARRELS AND/OR SMOKING MALT**
- **USED BEFORE HOPS WERE USED**
- **HISTORIC USE ONLY**
- **STARCH-PRODUCING**
- **SOURCE OF SIMPLE SUGARS**
- **HIGHLY TERRITORIAL**
- **AROMATIC**
- **TOXIC/POISONOUS**
- **FUNCTIONAL**

Table 3 provides information on geographic origin. Considering the large continental and subcontinental areas of the planet and considering that each species can be found on different continents, species of Asian origin are most prevalent, with 175 species (27.2 percent of the total). Europe follows with 156 (24.2 percent), North America with 125 (19.4 percent), South America with 64 (9.9 percent), Africa with 44 (6.8 percent), India with 21 (3.3 percent), Australia with 11 (1.7 percent), and New Zealand with 4 (0.6 percent). Other species have been identified without a clear geographic origin and/or with wide distribution. Among these, the cultivated species include 22 entities selected by humans that do not have a correspondent in nature. The 19 cosmopolitan and subcosmopolitan species are widespread, and 3 have unknown origin.

Different parts of each species included in the compendium can be used in brewing (table 4). The total beer-making parts amount to 755 units. Considering that 501 species are included, each provides on average 1.5 plant organs (or parts) for beer making. Fruit is the most used plant organ (22.0 percent), followed by the leaf (17 percent), seed (12.1 percent), wood (11.1 percent), and stem (10.6 percent).

GLOBAL DISTRIBUTION OF PLANTS USED IN BEER MAKING
(NUMBER OF SPECIES)

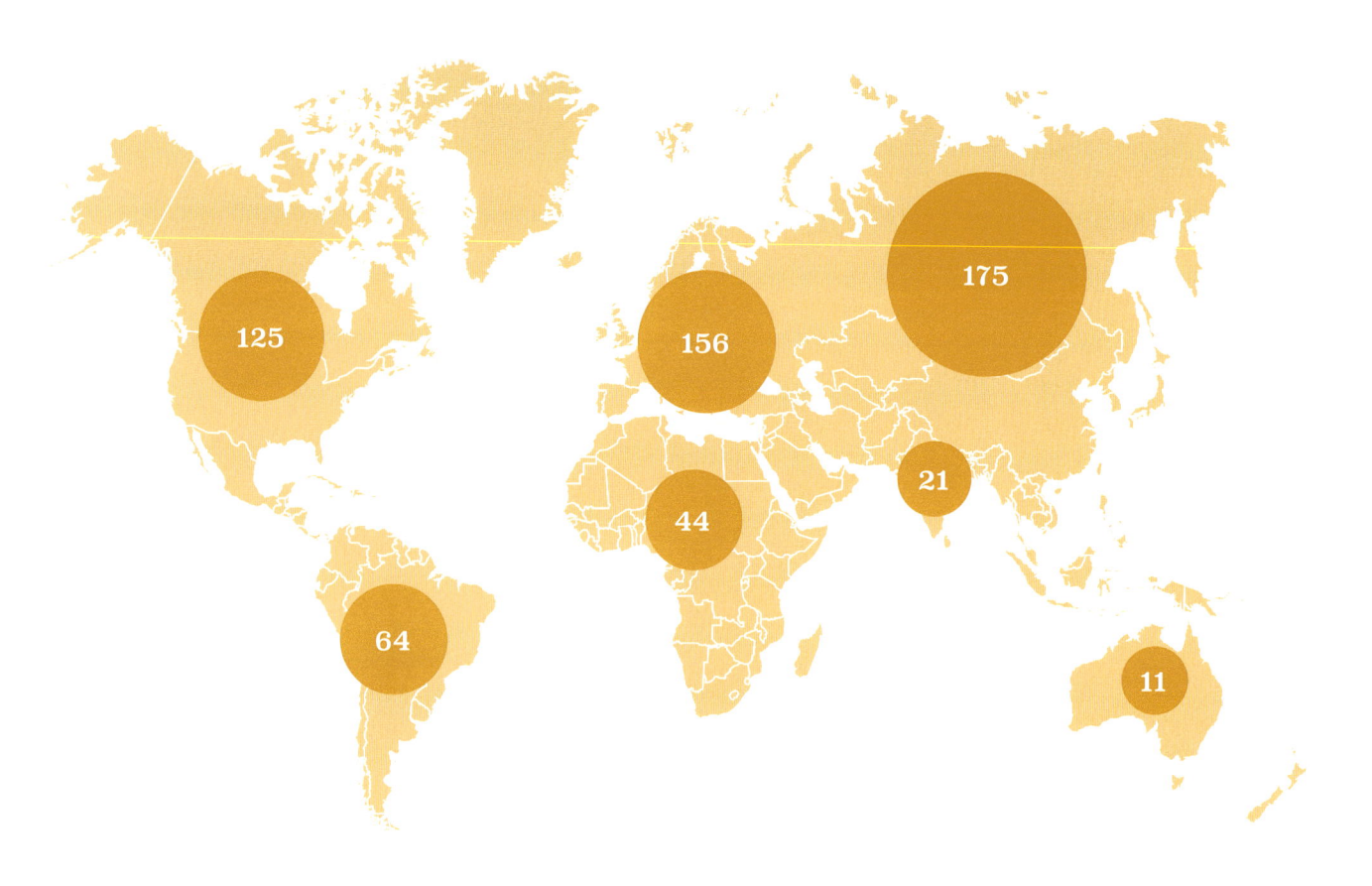

- 125 (North America)
- 156 (Europe)
- 175 (Northern Asia)
- 21 (South Asia)
- 44 (Africa)
- 64 (South America)
- 11 (Australia)

Table 1. Beer-making species by family

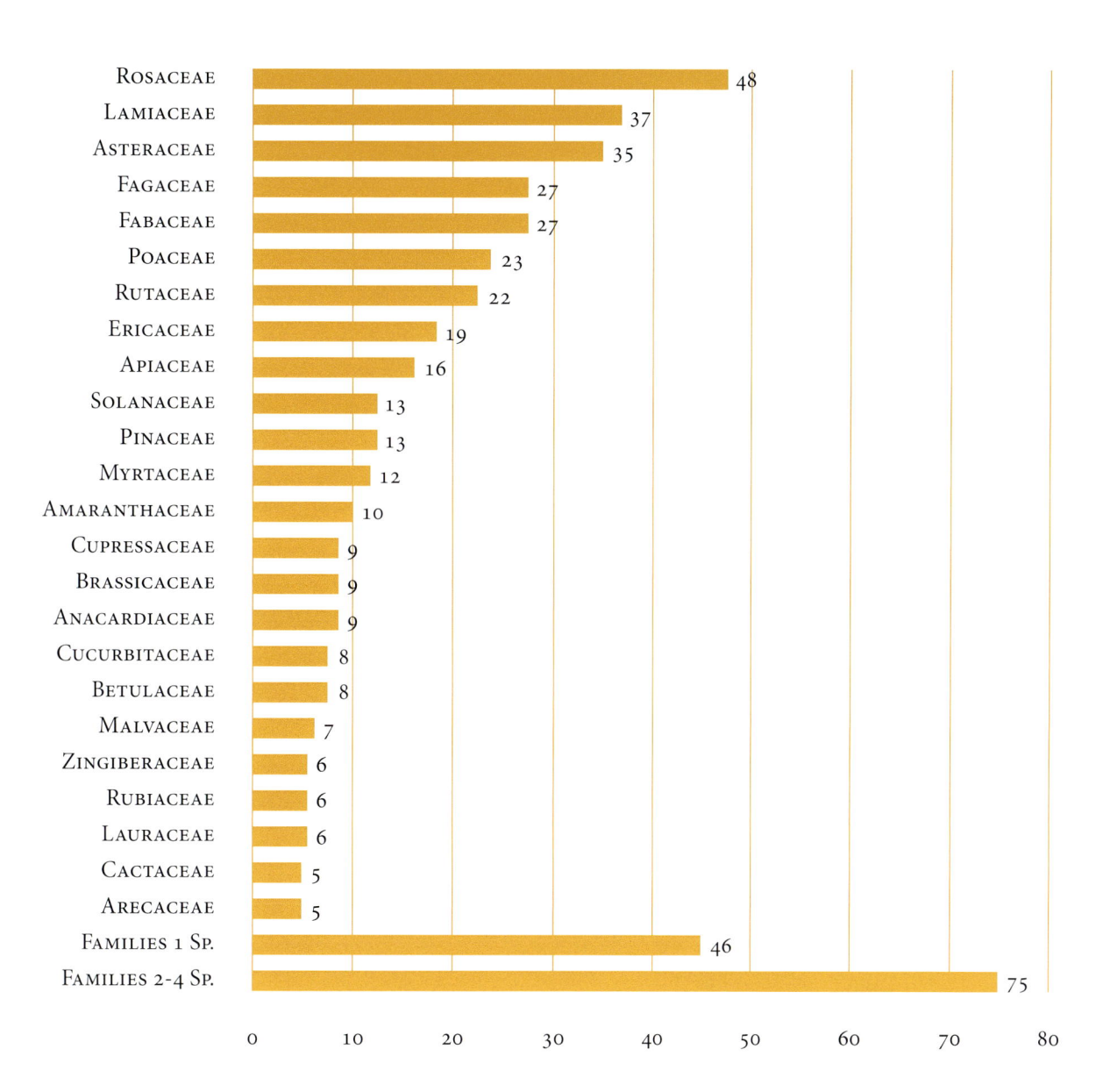

Family	Species
Rosaceae	48
Lamiaceae	37
Asteraceae	35
Fagaceae	27
Fabaceae	27
Poaceae	23
Rutaceae	22
Ericaceae	19
Apiaceae	16
Solanaceae	13
Pinaceae	13
Myrtaceae	12
Amaranthaceae	10
Cupressaceae	9
Brassicaceae	9
Anacardiaceae	9
Cucurbitaceae	8
Betulaceae	8
Malvaceae	7
Zingiberaceae	6
Rubiaceae	6
Lauraceae	6
Cactaceae	5
Arecaceae	5
Families 1 Sp.	46
Families 2-4 Sp.	75

Table 2. Number of species for the
32 most-represented genera

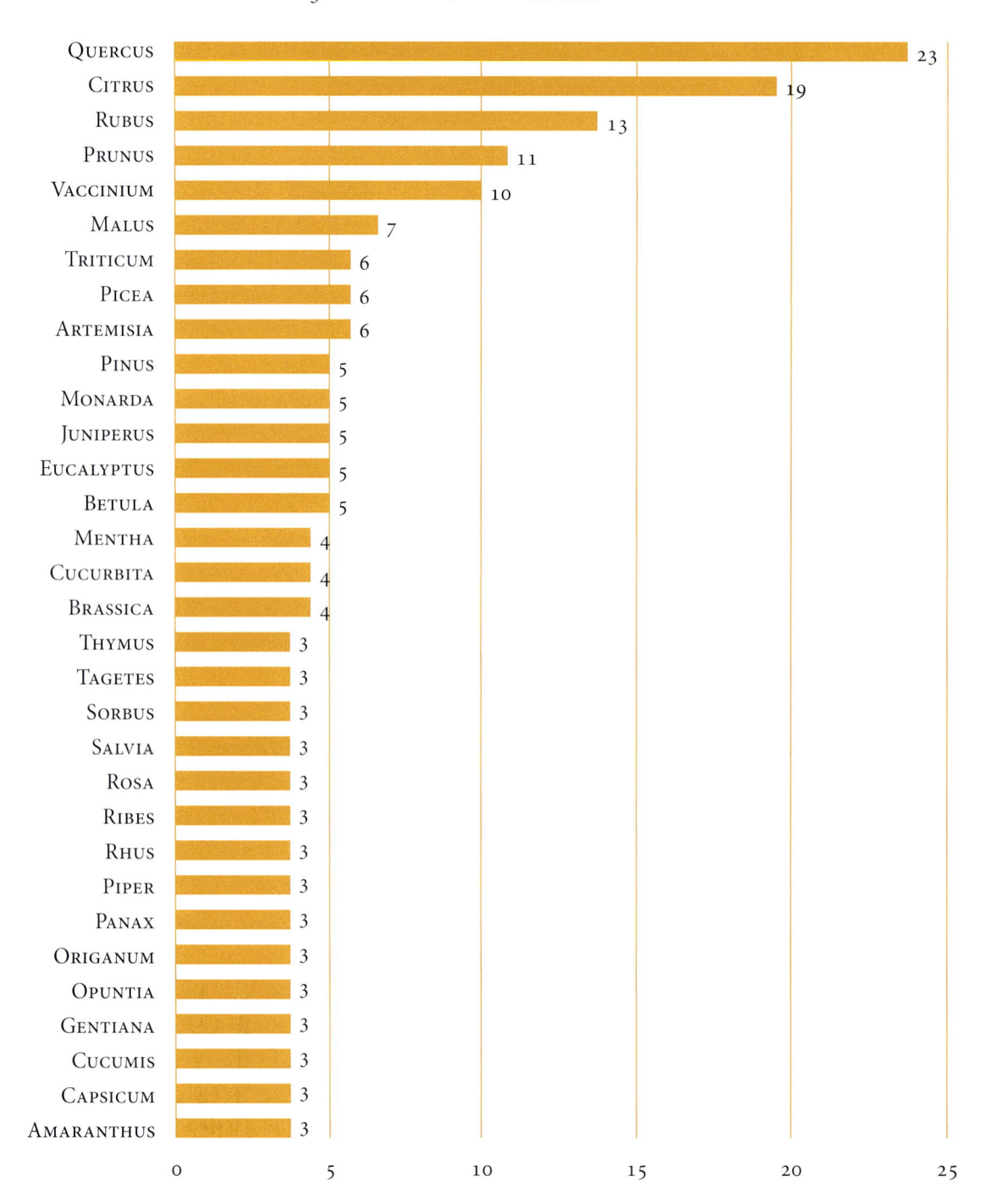

Genus	Number of species
Quercus	23
Citrus	19
Rubus	13
Prunus	11
Vaccinium	10
Malus	7
Triticum	6
Picea	6
Artemisia	6
Pinus	5
Monarda	5
Juniperus	5
Eucalyptus	5
Betula	5
Mentha	4
Cucurbita	4
Brassica	4
Thymus	3
Tagetes	3
Sorbus	3
Salvia	3
Rosa	3
Ribes	3
Rhus	3
Piper	3
Panax	3
Origanum	3
Opuntia	3
Gentiana	3
Cucumis	3
Capsicum	3
Amaranthus	3

TABLE 3. SPECIES BY REGION		
REGION	**No.**	**%**
Asia	175	27.2
Europe	156	24.2
North America	125	19.4
South America	64	9.9
Africa	44	6.8
Cultivated species	22	3.4
India	21	3.3
Cosmopolitan species	19	3.0
Australia	11	1.7
New Zealand	4	0.6
Unknown origin	3	0.5
TOTAL	644	100.0

TABLE 4. PLANT ORGANS USED IN BEER MAKING		
PLANT ORGAN (OR PART OF IT)	**No.**	**%**
Fruit (or parts) + false fruits	166	22.0
Leaf (or parts) + galls	128	17.0
Seed (or parts)	91	12.1
Wood	84	11.1
Stem (parts; +/- young)	80	10.6
Root + hypogeal organs	56	7.4
Flower (or parts, +/- mature)	53	7.0
Inflorescences (flowering buds)	42	5.6
Bark	22	2.9
Sap + latex	15	2.0
Resin	7	0.9
Entire plant	6	0.8
Oil (seed and/or essential)	5	0.7
TOTAL	755	100.0

BOTANICAL BEER-MAKING PROFILES

Abies alba Mill.
PINACEAE

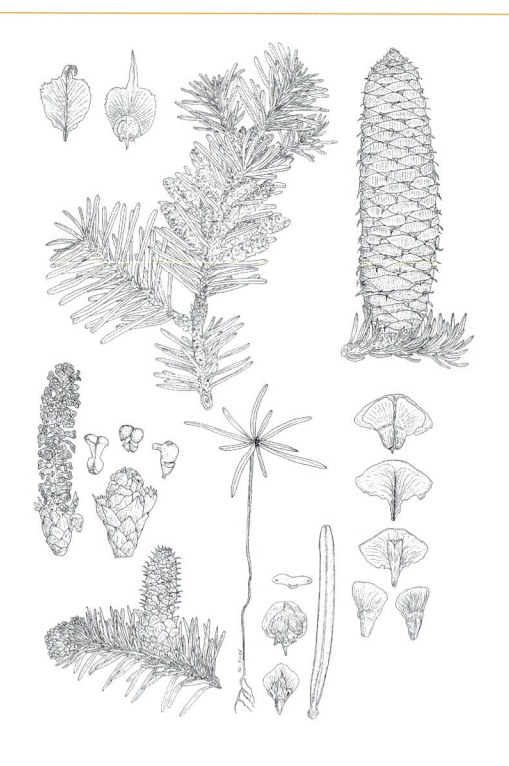

Synonyms *Abies argentea* Chambray, *Abies baldensis* (Zuccagni) Zucc. ex Nyman, *Abies candicans* Fisch. ex Endl., *Abies chlorocarpa* Purk. ex Nyman, *Abies duplex* Hormuz. ex Beissn., *Abies metensis* Gordon, *Abies miniata* Knight ex Gordon, *Abies minor* Gilib., *Abies nobilis* A.Dietr., *Abies pardei* Gaussen, *Abies rinzii* K.Koch, *Abies taxifolia* Desf., *Abies vulgaris* Poir.

Common Names Silver fir.

Description Tree (10–45 m). Bark grayish-white or pale gray, flaky with thin scales. Branches dark brown, young reddish and pubescent. Leaves linear-flat (1.5–2 × 10–20 mm) all around the branches but facing the same side, obtuse at the apex, grooved along the midrib, with 2 white longitudinal lines below. Pinecones upright, up to 4–9 cm.

Subspecies The species includes several taxa of subspecific rank, some even recently described (e.g., *Abies alba* Mill. subsp. *apennina* Brullo, Scelsi & Spamp.). However, the taxonomic interpretation of these entities is not univocal.

Related Species *Abies* sp. div. (information regarding use in making beer, based on the common name, relates to the genus, not just to the species *A. alba*, which is given here as an example; it is conceivable that other species belonging to the genus have similar potential use); the use of *Abies balsamea* (L.) Mill. and *Abies concolor* (Gordon) Lindl. ex Hildebr. in making beer is ascertained.

Geographic Distribution Orophyte S Europe.
Habitat Mountain forests in the beech forest belt.
Range Tree widely cultivated even outside the original geographic distribution for forestry purposes.

Beer-Making Parts (and Possible Uses)
Young apices (to confer aroma/flavor).

Chemistry Essential oil, wood: 4-hydroxy-4-methyl-2-pentanone 73.36%, α-cedrol 10.08%, 2,6-dimethyl-1,3,6-heptatriene 7.35%, 4-terpineol 3.25%, α-phellandrene 2.96%. Essential oil, bark: di(2-ethylhexyl) phthalate 59.83%, methyl cyclopentane 16.63%, 13-epimanool 6.31%, methyl cyclohexane 3.73%, 3-methylhexane 3.29%.

Style British Golden Ale, Mum.

Source Hieronymus 2010; https://www.brewersfriend.com/other/sapin/; https://www.honest-food.net/2016/05/05/gruit-herbal-beer/.

Acacia melanoxylon R.Br.
FABACEAE

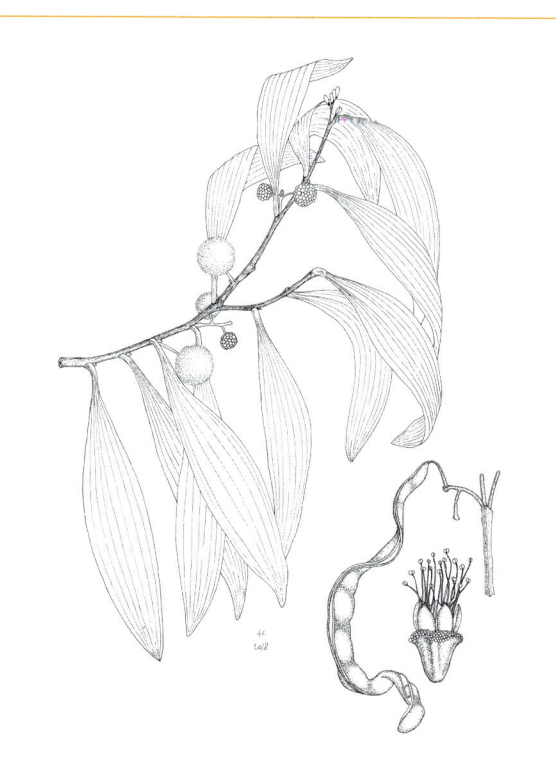

Synonyms *Acacia arcuata* Spreng., *Mimosa melanoxylon* (R.Br.) Poir., *Racosperma melanoxylon* (R.Br.) C.Mart., *Racosperma melanoxylon* (R.Br.) Pedley.

Common Names Blackwood, Australian acacia, black acacia, blackwood acacia.

Description Tree or shrub (up to 35–40 m). Trunk cylindrical (Ø up to 1.5 m), bark rough, brownish-gray to dark gray, longitudinally furrowed and scaly; ovoid foliage. Leaves in the young specimens alternate and bipinnate-compound, soon replaced by phyllodes, strictly elliptical to lanceolate, usually curved (4–16 × 0.5–2.5 cm), leathery, with reticulated veins. Inflorescence capitulum (Ø about 6 mm), organized in racemes with 30–55 inflorescences: peduncle 5–13 mm. Flowers bisexual, regular, pentamerous, small, creamy to pale yellow, almost sessile; cup-shaped calyx, with small lobes; corolla about 2 mm, half lobed; numerous stamens, free, about 4 mm; ovary ellipsoid, sessile, slender stylus, tubular stigma. Fruit legume flat, twisted, coiled (6–15 × 0.4–0.8 cm), glabrous, brown, with 6–10 seeds. Elliptical seeds, compressed, 3–5 mm, glossy black.

Related Species The *Acacia* genus encompasses about 1,500 taxa.

Geographic Distribution E Australia (Queensland, New South Wales, Tasmania, Victoria). **Habitat** Rain forests with rich soils up to the mountain zones; generally, undergrowth tree in the forests of *Eucalyptus* sp. **Range** Australia (New South Wales, Tasmania, Victoria), locally naturalized in SW Europe.

Beer-Making Parts (and Possible Uses) Wood tested for wine containers with little success (uncertain use in beer production).

Wood Differentiated, with yellowish-white sapwood and reddish-brown or reddish heartwood, mostly variegated (WSG 0.67–0.75), long lasting, fairly easy to work, used for fine furniture and interior decoration, veneers, rifle stocks.

Style None verified.

Source Cantwell and Bouckaert 2016; Cantwell and Bouckaert 2018.

Acer pseudoplatanus L.
SAPINDACEAE

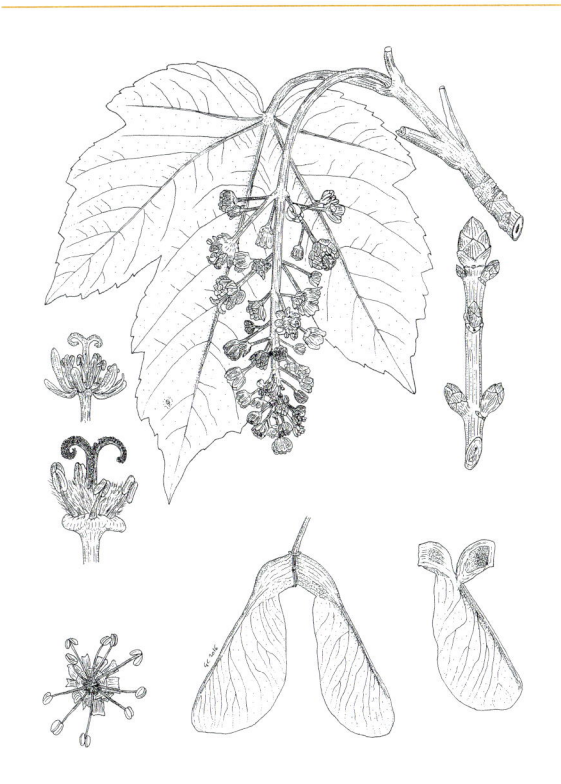

Synonyms *Acer abchasicum* Rupr., *Acer atropurpureum* Dippel, *Acer dittrichii* Ortm., *Acer erythrocarpum* Dippel, *Acer euchlorum* Dippel, *Acer hybridum* Bosc, *Acer majus* Gray, *Acer melliodorum* Opiz, *Acer opizii* Ortmann ex Opiz, *Acer procerum* Salisb., *Acer purpureum* Dippel, *Acer rafinesquianum* Dippel, *Acer villosum* C.Presl, *Acer wondracekii* Opiz, *Acer worleei* Dippel.

Common Names Maple.

Description Tree (5–40 m), deciduous. Trunk (Ø up to 3.5 m) single, foliage globular and broad bark initially gray or yellowish, then reddish and detaching into large plates; opposite buds, like all maples. Leaves long petiolate (1–1.5 times the lamina), simple (14–22...), opposite, deciduous, 5-lobed with small, acute lobes, margin lightly dentate, dark green above, glaucous below, foliar scars that do not touch. Flowers, appearing after the leaves, on 1 cm peduncles, sepals and petals 0.5 × 3 mm, 8 mm stamens with dark anthers, united in apical catkins 1–2 dm long. Fruit samara with divergence wings about 90°.

Related Species *Acer* sp. div. (details on use in making beer, based on the common name, are related to the genus, not only to the species *A. pseudoplatanus*, which is given here as an example; it is conceivable that other species belonging to the genus have similar potential use); the genus *Acer* includes more than 230 recognized taxa.

Geographic Distribution CW Europe, Mediterranean, Asia Minor, Caucasus. **Habitat** In Italy, the tree is found in mountainous areas, but it also appears at low altitudes; for example, it is present sporadically in the high plains at the foot of the Alps and is common in the Prealps. It can be found up to an altitude of 1,500–1,900 m. It is most commonly found in maple-ash and beech forests. In Italy's Po Valley it is almost completely replaced by hedge maple. In Central Europe it is present at gradually lower altitudes as you move northward. **Range** Eurasia.

Beer-Making Parts (and Possible Uses) Sap (sweetener), wood (to smoke the malt).

Wood Compact and homogeneous, with fine texture and a pleasant ivory color; WSG 0.62–0.70; easy to work, used for furniture both in massive pieces in veneers and for panel coating, for small objects, billiard cues, backs and necks of stringed musical instruments (in this use the marbling is highly prized).

Style American Amber Ale, American Brown Ale, American Pale Ale, American Porter, Robust Porter, Smoked Beer.

Source https://www.bjcp.org (32A. Classic Style Smoked Beer); https://www.brewersfriend.com/other/maple/.

Acer saccharum Marshall
Sapindaceae

Synonyms *Acer hispidum* Schwer., *Acer palmifolium* Borkh., *Acer saccharophorum* K.Koch, *Acer subglaucum* Bush, *Acer treleaseanum* Bush.

Common Names Sugar maple.

Description Tree (25–40 m) with typical expanded foliage; bark gray or grayish-brown, smooth or furrowed by light vertical lines, dark and fissured with age; branches young and glabrous. Leaves (up to 13 cm), long petiolate, opposite, deciduous, with deep veins, 3- to 5-lobed, lobes slightly incised and separated by obtuse angles, margins roughly dentate, dark green color, in autumn golden yellow to orange, up to crimson and scarlet, especially in the northern part of the geographic distribution; dark petiole. Flowers small (about 5 mm), yellow-green color, grouped (8–14) in sessile corymbs, hanging from long branches that form at the same time as or just before the leaves. The species is dioecious (sometimes even monoecious). Fruit samara (2.5 cm), with slightly divergent wings, brown color in autumn.

Subspecies Different forms of this taxon have been recognized at different taxonomic ranks (variety, subspecies, species). Two of these are now considered autonomous species: *Acer barbatum* Michx., including *Acer floridanum* (Chapm.) Pax and *Acer leucoderme* Small. The taxonomic position of *Acer nigrum* F.Michx. (subspecies or species) is instead still controversial. According to some authors, the entities of subspecific rank of this taxon could be *Acer saccharum* Marshall subsp. *saccharum*, *Acer saccharum* Marshall subsp. *floridanum* (Chapm.) Desmarais, *Acer saccharum* Marshall subsp. *grandidentatum* (Torr. & A.Gray) Desmarais, *Acer saccharum* Marshall subsp. *leucoderme* (Small) Desmarais, *Acer saccharum* Marshall subsp. *nigrum* (F.Michx.) Desmarais, *Acer saccharum* Marshall var. *rugelii* (Pax) Rehder, *Acer saccharum* Marshall var. *schneckii* Rehder, *Acer saccharum* Marshall var. *sinuosum* (Rehder) Sarg., *Acer saccharum* Marshall subsp. *skutchii* (Rehder) A.E.Murray.

Cultivar Among the most popular cultivars of this species are Apollo, Arrowhead, Astis (Steeple), Bonfire, Caddo, Columnare (Newton Sentry), Fall Fiesta, Goldspire, Green Mountain, Inferno, Legacy, Lord Selkirk, Monumentale, Sweet Shadow, Temple's Upright, and Unity.

Related Species *Acer saccharinum* L., *Acer negundo* L., *Acer rubrum* L.; the genus *Acer* includes more than 230 recognized taxa; the beer-making use of *Acer negundo* L., *Acer nigrum* F.Michx., and *Acer rubrum* L. is ascertained.

Geographic Distribution SE Canada, NE United States. **Habitat** Fertile, deep, cool soils; cold continental climates with a marked thermal difference between summer and winter. **Range** Superimposable to the area.

Beer-Making Parts (and Possible Uses) Sap extracted and transformed into syrup by hot concentration (volume ratio 32:1) as a sweetener; sap used as is instead of brewing water; wood for barrels (uncertain use for beer making) and for smoking malt, bark (tested in brewing), branches, buds, leaves.

CHEMISTRY Maple syrup: energy 261 kcal, sucrose 67%, water 33%.

WOOD Pinkish sapwood, reddish-brown heartwood, WSG 0.68–0.70, used for furniture, models, and shoe forms, coils and lathe work, bodywork, carpentry, handles, but above all prized for floorboards.

STYLE American Amber Ale, American Brown Ale, American Pale Ale, American Porter, Maple Beer, Robust Porter, Scotch Ale.

BEER Maple Nipple, Lawson's Finest Liquids (Vermont, United States); Équinox du Printemps, Dieu du Ciel (Quebec, Canada).

SOURCE Calagione et al. 2018; Cantwell and Bouckaert 2016; Cantwell and Bouckaert 2018; Daniels and Larson 2000; Giaccone and Signoroni, 2017; Josephson et al. 2016; McGovern 2017; https://www.bjcp.org (30. Spiced Beer; 30B. Autumn Seasonal Beer); https://www.omnipollo.com/beer/8/; https://www.omnipollo.com/beer/agamemnon/; https://www.brewersfriend.com/other/maple/.

Maple Beers: A North American Tradition

One of the best-known products of *Acer saccharum*, like various other taxonomically related species, is its sap in its natural form, instead of in the mashing water or as maple syrup, as a total or partial replacement for malts. Maple beer, widespread in much of New England and Quebec, uses maple syrup as the main and often only source of saccharides. In fact, in many U.S. states the definition of beer is linked exclusively to the alcohol content of the drink, without any reference to the ingredients used (whereas under Italian law, beer must be produced with wort obtained from at least 60% barley malt and must contain hops). In these beers, classically flavored with hops and almost always brewed using high fermentation, unlike what you might think, the sweetness is low because most of the sugar is transformed into alcohol. The contribution of the syrup is mainly aromatic and linked to caramel notes. Sometimes, to increase the perception of the maple syrup, the beer stays for some time in barrels in which it was previously aged. The addition of sap instead has a less decisive impact on the organoleptic profile of the beer.

Achillea filipendulina Lam.
ASTERACEAE

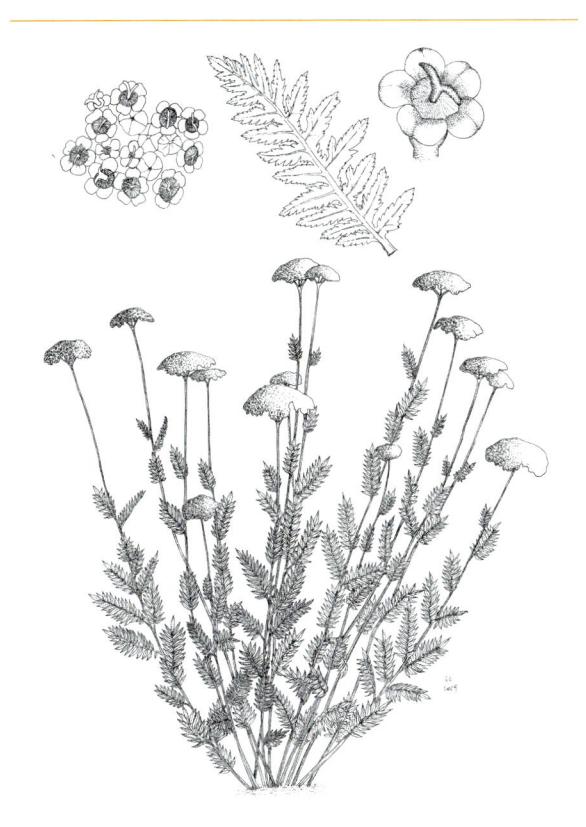

SYNONYMS *Achillea eupatorium* M.Bieb., *Achillea filicifolia* M.Bieb., *Tanacetum angulatum* Willd.

COMMON NAMES Yarrow, fernleaf yarrow, fern leaf yarrow, milfoil, nosebleed.

DESCRIPTION Herbaceous perennial (up to 1 m). Root woody (Ø 5–10 mm). Stems with short hairs, closely clustered to more or less patent, densely leafy. Leaves green, with densely affixed hairs and scattered glandular hairs; low basal and petiolate cauline leaves (0–5 cm) enlarged at the base, oblong-lanceolate lamina (10–20 × 3–7 cm), oblong ± incised and acute pinnatipartite segments (10–20 × 5–10 mm); upper cauline leaves pinnatifide, with short obtuse elements. Inflorescences capitula (30–50 or rarely more), with bracts 2–8 mm, in dense corymbs of unequal height (Ø 4–10 cm); obovoid-oblong or obovoid-cylindrical involucre (3.5–5 × 2.5–3.5 mm), basally attenuated, phillaries pale to whitish-green, externally pubescent and dotted with glands, externally linear-lanceolate, acute, internally triangular-lanceolate, ± acute

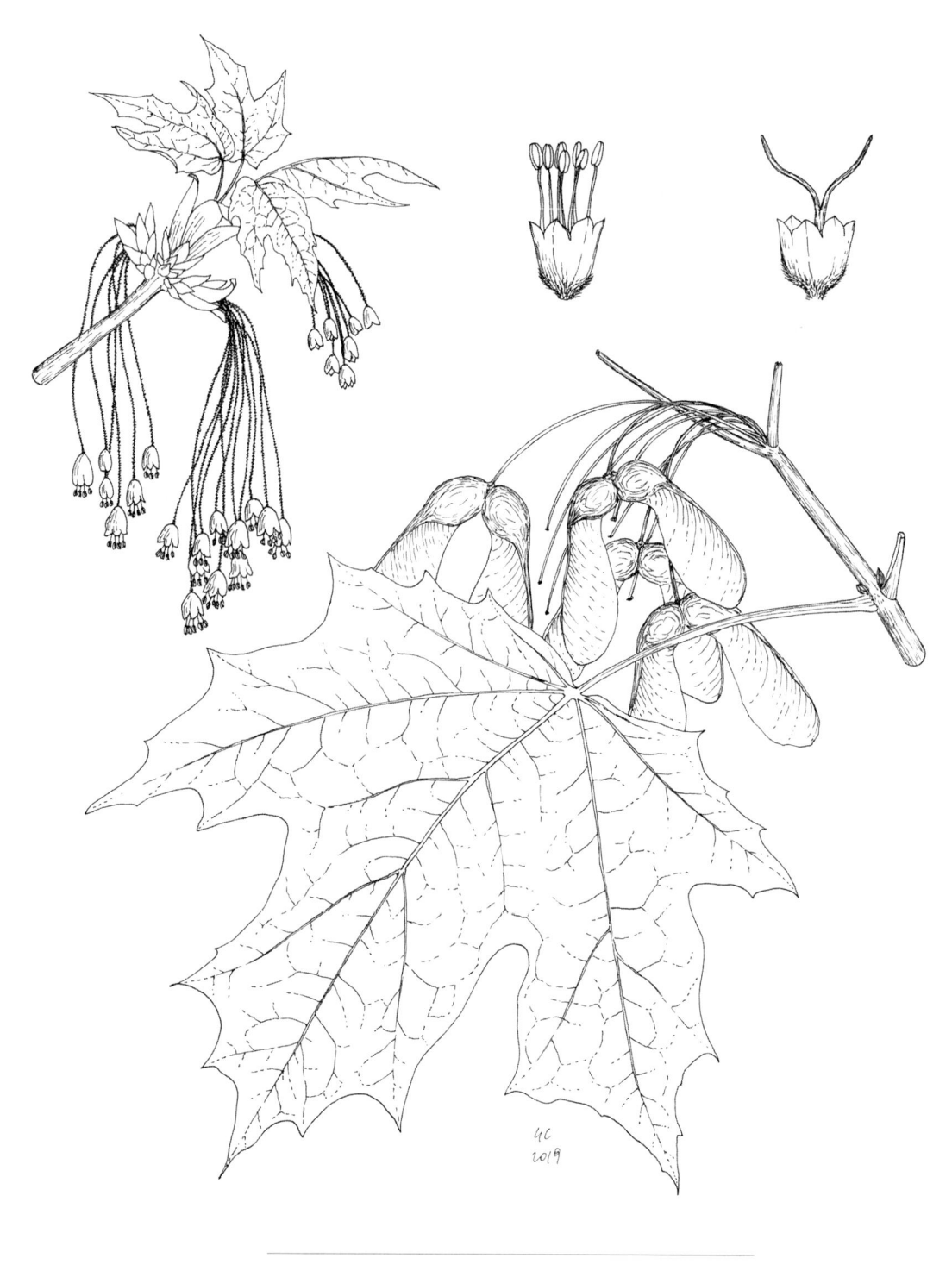

Acer saccharum Marshall
SAPINDACEAE

to obtuse; oblong-lanceolate scales, shorter than flowers, membranous, glabrescent, acute; flowers ligulate (2–4), with a gold-colored ligula (0.7–1 × 1.3–1.5 mm), 3-lobed; flowers tubular (15–30), with a corolline tube (2–2.5 mm), yellow. Fruit oblong cypsela (1.5–1.75 mm), light brown color.

CULTIVAR Some cultivars of this species are Cloth of Gold, Coronation Gold, Credo, Gold Plate, Heidi, Helios, Hella Glashof, King Edward, Lachsschönheit, Lucky Break, Martina, Mondpagode, Moonshine, Parker's Variety, and Summerwine.

RELATED SPECIES The *Achillea* genus includes more than 180 taxa.

GEOGRAPHIC DISTRIBUTION Turkey, Iraq, Iran, C Asia, Afghanistan, Pakistan. **HABITAT** Semi-arid environments. **RANGE** Cultivated as an ornamental plant and frequently naturalized.

BEER-MAKING PARTS (AND POSSIBLE USES) Flowering tops (*Achillea millefolium* substitute).

CHEMISTRY Essential oil: trans-2,7-dimethyl-4,6-octadien-2-ol 27.93%, borneol 21.44%, santolina triene 7.13%, camphor 7.08%, α-pinene 1.58%, camphene 1.52%, pinocarveol 1.33%, bornyl acetate 1.33%, yomogi alcohol 3.34%, α-terpineol 1.39%, carene 1%.

STYLE Herb Beer, Spice Beer.

SOURCE Fisher and Fisher 2017; Josephson et al. 2016; McGovern 2017.

Achillea millefolium L.
ASTERACEAE

SYNONYMS *Achillea albida* Willd., *Achillea ambigua* Boiss., *Achillea ambigua* Pollini, *Achillea anethifolia* Fisch. ex Herder, *Achillea angustissima* Rydb., *Achillea arenicola* A.Heller, *Achillea bicolor* Wender., *Achillea borealis* Bong., *Achillea californica* Pollard, *Achillea ceretanica* Sennen, *Achillea compacta* Lam., *Achillea coronopifolia* Willd., *Achillea crassifolia* Colla, *Achillea cristata* Hort. ex DC., *Achillea dentifera* Rchb., *Achillea eradiata* Piper, *Achillea fusca* Rydb., *Achillea gigantea* Pollard, *Achillea gracilis* Raf., *Achillea haenkeana* Tausch, *Achillea intermedia* Schleich., *Achillea lanata* Lam., *Achillea lanulosa* Nutt., *Achillea laxiflora* A.Nelson, *Achillea laxiflora* Pollard & Cockerell, *Achillea magna* L., *Achillea marginata* Turcz. ex Ledeb., *Achillea megacephala* Raup, *Achillea nabelekii* Heimerl, *Achillea occidentalis* (DC.) Raf. ex Rydb., *Achillea ochroleuca* Eichw., *Achillea ossica* K.Koch, *Achillea pacifica* Rydb., *Achillea palmeri* Rydb., *Achillea pecten-veneris* Pollard, *Achillea pratensis* Saukel & R.Länger, *Achillea pseudo-tanacetifolia* Wierzb. ex Rchb., *Achillea puberula* Rydb., *Achillea pumila* Schur, *Achillea rosea* Desf., *Achillea setacea* Schwein., *Achillea sordida* (W.D.J.Koch) Dalla Torre & Sarnth., *Achillea subalpina* Greene, *Achillea submillefolium* Klokov & Krytzka, *Achillea sylvatica* Becker, *Achillea tanacetifolia* Mill., *Achillea tenuis* Schur, *Achillea virgata* Hort. ex DC., *Achillios millefoliatus* St.-Lag., *Alitubus millefolium* (L.) Dulac, *Alitubus tomentosus* Dulac, *Chamaemelum millefolium* (L.) E.H.L.Krause, *Chamaemelum tanacetifolium* (All.) E.H.L.Krause.

COMMON NAMES Yarrow.

DESCRIPTION Herbaceous perennial (3–8 dm), aromatic, rhizome creeping and branched with secondary roots and stolons. Aerial stem erect, striped, pubescent. Leaves lanceolate 2 (3) pennatosette; basal (1.5–3 × 10–20 cm) petiolate, medians sessile (1 × 5–10 cm). Flowers of 2 types: external (4–6) female (2 mm), ligulate (ligula wider than long, 3-lobed, with smaller central lobe), white or pink; core tubulous (2 mm) with 5 yellowish-white hermaphrodite petals; flowers grouped in capitula (up to 8 mm), involucre ovoid (up to 5 mm) composed of ovate scales with membranous margin; capitula grouped in corymbs. Fruit achene (1.7–2 mm) without pulp.

SUBSPECIES In addition to the typical subspecies, *Achillea millefolium* L. subsp. *millefolium*, there are four other recognized subspecies: *Achillea millefolium* L. subsp. *alpestris* (Wimm & Grab.) Gremli, *Achillea millefolium* L. subsp. *ceretanica* (Sennen) Sennen, *Achillea millefolium* L. subsp. *stricta* (Schleich. ex Heimerl) Hyl., and *Achillea millefolium* L. subsp. *sudetica* (Opiz) Oborny.

RELATED SPECIES The genus *Achillea* includes about 184 species.

GEOGRAPHIC DISTRIBUTION Euro-Siberian. **HABITAT** Dry meadows, mainly mountain and subalpine. **RANGE** Cosmopolitan plant (mostly harvested in natural or seminatural environments, rarely cultivated).

(4–8 cm), erect or slightly curved, carrying on the whole length numerous tiny greenish-yellow hermaphrodite flowers (1.8–2 mm). Flower with perigonium formed by 6 oblong tepals (1 × 2.5 mm), hulled, membranous; stamens 6, free; ovary superior with 6 loggias; stigma 1. Fruit capsule (Ø 1.5–2 cm) oblong, bacciform, reddened at maturity, containing 1 or more ovate seeds.

Subspecies In the *Acorus calamus* species, there are three recognized varieties: *Acorus calamus* L. var. *americanus* Raf., *Acorus calamus* L. var. *angustatus* Besser, and *Acorus calamus* L. var. *calamus*.

Related Species The *Acorus* genus includes two species: *Acorus calamus* L. and *Acorus gramineus* Aiton.

Geographic Distribution E Asia circumboreal region. **Habitat** Channel banks, swamps, ponds. **Range** Circumboreal.

Beer-Making Parts (and Possible Uses) Rhizomes.

Toxicity Hallucinogenic effects.

Chemistry Essential oil, root (21 constituents, 76.6% of total oil): β-gurjunene 28.0%, (Z)-asarone 13.7%, aristolene 13.4%, (E)-asarone 7.9% (composition of root oil quite different from that of *A. calamus* leaves).

Style English Barleywine, Pale Ale.

Source Heilshorn 2017; https://www.brewersfriend.com/other/calamus/.

🍺 In *The Art of Brewing*, published in 1826, David Booth cites *Acorus calamus* among the plants used to add a bitter and warm note to beer. According to Booth, the dried root of this plant has long been used throughout the United Kingdom to intensify bitter flavors.

Actinidia deliciosa (A.Chev.) C.F.Liang & A.R.Ferguson
ACTINIDIACEAE

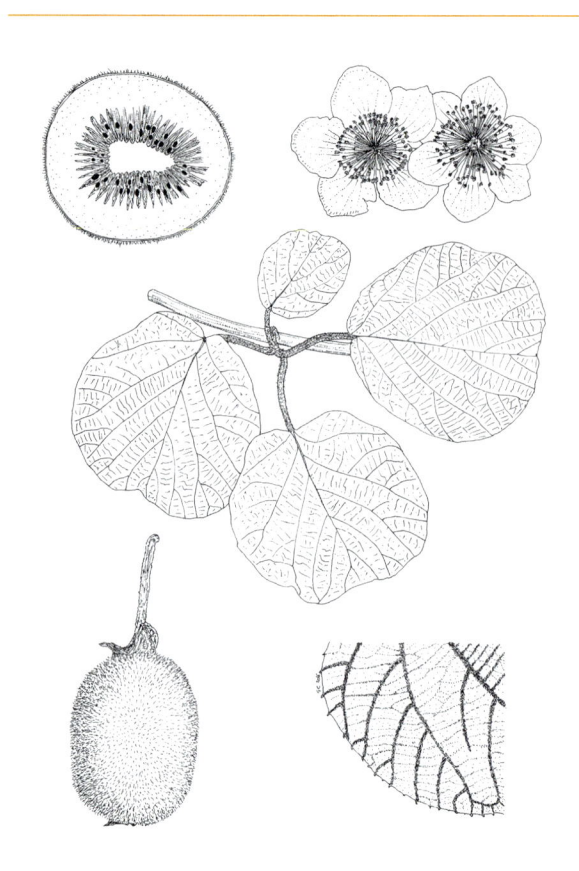

Synonyms None
Common Names Fuzzy kiwifruit.

Description Climbing tree or shrub (up to 9 m). Leaves alternate, long petiolate, deciduous, oval to almost circular (7.5–12.5 cm), cordate at the base; young ones covered with reddish hairs, mature ones dark green and glabrous on the upper surface, white-woolly with light and prominent veins underneath. Flowers fragrant, dioecious or unisexual, 1–3 at the foliar axilla; 5–6 petals, white at the beginning, then brown, Ø 2.5–5 cm, both sexes have a central tuft consisting of many stamens, although that of female flowers is without vital pollen; nectar absent. Fruit oblong (up to 6.5 cm), rust-colored epicarp densely covered with short stiff brown hairs; flesh bright green (or yellow or brown) except for white succulent center, from which radiate many fine clear lines; flesh hard until fully ripe, shiny, juicy, delicious; taste subacidic to rather acidic.

Subspecies *Actinidia deliciosa* var. *chlorocarpa* (A.Chev.) C.F.Liang & A.R.Ferguson, *Actinidia deliciosa* var. *deliciosa* (A.Chev.) C.F.Liang & A.R.Ferguson.

Cultivar *Actinidia deliciosa* cultivar Hayward, developed by Hayward Wright in Avondale, New Zealand, in 1924, is the most cultivated cultivar in the world.

Related Species *Actinidia arguta* (Siebold & Zucc.) Planch. ex Miq., *Actinidia chinensis* Planch., *Actinidia kolomikta* (Rupr. & Maxim.) Maxim.

Geographic Distribution S China. **Habitat** Unknown. **Range** Italy, New Zealand, Chile, France, Greece, Japan, United States.

Beer-Making Parts (and Possible Uses) Fruits.

Chemistry Energy 51.80 kcal, water 87 g, protein 1 g, carbohydrates 9.12 g, fiber 2.12 g, lipids 0.8 g, calcium 34.11 mg, iron 0.37 mg, iodine 0.33 mg, magnesium 14.93 mg, zinc 0.16 mg, selenium 0.60 μg, sodium 4 mg, potassium 290 mg, vitamin B1 thiamine 0.02 mg, vitamin B2 riboflavin 0.04 mg, niacin 0.6 mg, vitamin B6 pyridoxine 0.12 mg, folic acid 26.83 μg, vitamin C ascorbic acid 43.14 mg, carotenoids (β-carotene) 37.25 μg, vitamin A retinol 6.22 μg, palmitic acid 0.05 g, stearic acid 0.04 g, oleic acid 0.10 g, linoleic acid 0.34 g, linolenic acid 0.06 g, tryptophan 18 mg, glucose 4.32 g, fructose 4.60 g, sucrose 0.21 g, available organic acids 1.49 g, oxalic acid 1 g, citric acid 1 g, malic acid 0.5 g, soluble polysaccharides 0.59 g, insoluble polysaccharides (noncellulose) 0.42 g, cellulose 0.83 g, lignin 0.28 g.

Style American IPA, American Pale Ale, American Wheat or Rye Beer, California Common Beer, English IPA, Fruit Beer, Mixed-Fermentation Sour Beer, Saison, Specialty Fruit Beer, Specialty IPA (Black IPA), Weissbier, Weizenbock, Witbier.

Beer 41° Parallelo, Birrificio Pontino (Latina, Lazio, Italy).

Source https://www.brewersfriend.com/other/kiwi/.

Aframomum melegueta K.Schum.
ZINGIBERACEAE

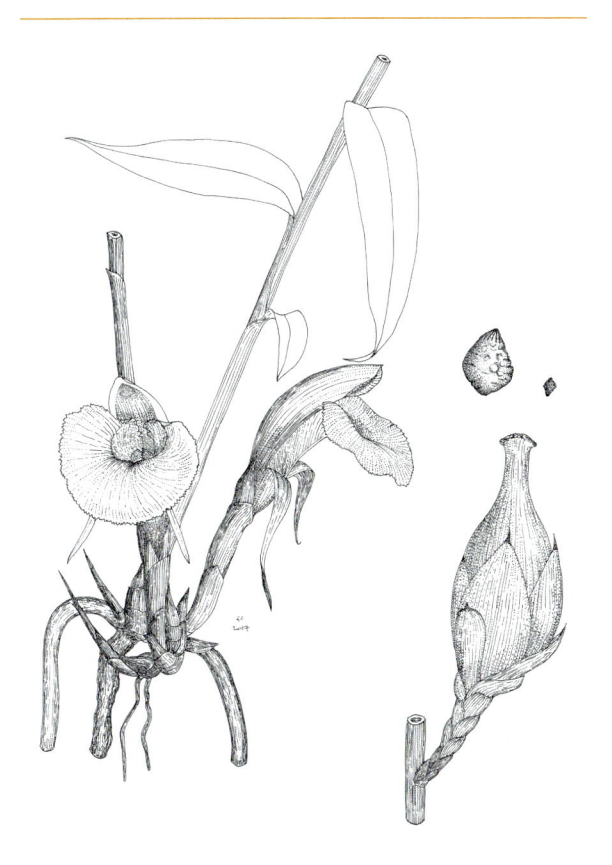

Synonyms *Aframomum grana-paradisi* (L.) K.Schum., *Aframomum meleguetella* K.Schum., *Alpinia grana-paradisi* (L.) Moon, *Amomum grana-paradisi* L., *Amomum grandiflorum* Sm., *Cardamomum grana-paradisi* (L.) Kuntze, *Cardamomum grandiflorum* (Sm.) Kuntze.

Common Names Grains of paradise, ossame, melegueta pepper, alligator pepper, guinea grains, fom wisa, guinea pepper.

Description Herbaceous perennial (up to 2 m), rhizomatous. Stem with leaves. Leaves sessile or subsessile, basal sheath surmounted by a ligula trunk (1 mm); lamina lanceolate to linear-lanceolate (22 × 2.5 cm), sharp apex, glabrous. Inflorescence panicle at the base of the stems. Flowers trimerous, hermaphrodite, pink to pale pink color, calyx tubular (4.5 cm), laterally divided above; corolla tubular at the base, 3-lobed, labellum obovate (Ø 10 cm); 1 fertile stamen. Fruit capsule (about 5 cm), trivalve, bright red color, emerges at the base of the stem at ground level, ovoid and tapered on top. Seeds small (Ø about 3.5 mm), abundant, subglobose, rough surface.

RELATED SPECIES The genus *Aframomum* has 56 recognized species; the beer-making use of *Aframomum citratum* (C.Pereira) K.Schum., *Aframomum daniellii* (Hook.f.) K.Schum., and *Aframomum exscapum* (Sims) Hepper is also established.

GEOGRAPHIC DISTRIBUTION W Africa (coasts from Senegal to Cameroon): Angola, Benin, Burundi, Cameroon, Central African Republic; Democratic Republic of Congo, Equatorial Guinea, Ethiopia, Gabon, Ghana, Guinea, Ivory Coast, Kenya, Liberia, Madagascar, Malawi, Mozambique, Nigeria, Rwanda; Sao Tome and Principe, Seychelles, Sierra Leone, Sudan, Tanzania, United Republic of Togo, Uganda, Zambia, Zimbabwe. **HABITAT** Tropical forests. **RANGE** Cultivated throughout tropical and subtropical Africa (Mascarenes, Madagascar, tropical Africa from Mozambique to Sudan and Côte d'Ivoire).

BEER-MAKING PARTS (AND POSSIBLE USES) Seeds.

CHEMISTRY Essential oil, leaves (19 compounds): myrtenyl acetate 29.06%, iso-limonene 19.43%, γ-elemene 8.84%, germacrene-D 7.83%, β-guaiene 6.39%, elemene 4.56%, longiborneol 3.82%. Essential oil, stem (26 compounds): caryophyllene oxide 19.7%, myrtenyl acetate 14.7%, β-eudesmene 10.83%, β-caryophyllene 10.13%, β-chamigrene 4%. Essential oil, root (25 compounds): myrtenyl acetate 22.7%, pinocarvil acetate 11.5%, cyperene 8.96%, caryophyllene oxide 5.97%, myrtenol 5.29%, β-caryophyllene 4.4%, elixene 4.44%. Essential oil, seeds (9 compounds): α-caryophyllene 48.78%, β-caryophyllene 32.5%, linalool 5.4%, E-nerolidol 4.33%.

STYLE Double Wit, Summer Ale.

BEER Summer Ale, Samuel Adams (Massachusetts, United States).

SOURCE Calagione et al. 2018; Heilshorn 2017; Hieronymus 2010; Markowski 2015; https://boulevard.com/BoulevardBeers/two-jokers-double-wit/.

Agastache foeniculum (Pursh) Kuntze
LAMIACEAE

SYNONYMS *Agastache anethiodora* (Nutt.) Britton & A.Br., *Hyptis marathrosma* (Spreng.) Benth., *Hyssopus anethiodorus* Nutt., *Hyssopus anisatus* Nutt., *Hyssopus discolor* Desf., *Hyssopus foeniculum* (Pursh) Spreng., *Lophanthus anisatus* (Nutt.) Benth., *Lophanthus foeniculum* (Pursh) E.Mey., *Perilla marathrosma* Spreng., *Stachys foeniculum* Pursh, *Vleckia albescens* Raf., *Vleckia anethiodora* (Nutt.) Greene, *Vleckia anisata* (Nutt.) Raf., *Vleckia bracteata* Raf., *Vleckia bracteosa* Raf., *Vleckia discolor* Raf., *Vleckia foeniculum* (Pursh) MacMill., *Vleckia incarnata* Raf.

COMMON NAMES Blue giant hyssop, anise hyssop, anise mint, fennel, fennel giant hyssop, fragrant giant hyssop, giant hyssop, lavender giant hyssop, licorice mint.

DESCRIPTION Herbaceous perennial (60–120 cm). Taproot. Stems tetragonal, facing upward, forming a group (Ø about 30 cm). Leaves oval to triangular, margin dentate, underside whitish color. Flowers lavender colored, grouped in showy verticillasters, or false spirals, occasionally branching at the apex.

CULTIVAR There are several cultivars of this species, including Alabaster, Album, Blue Blazes, Blue Fortune, Blue Fountaine, Camphorata, and Senior.

Related Species The genus *Agastache* includes about 28 taxa.

Geographic Distribution North America. **Habitat** Grasslands, forest areas of arid highlands, plains and fields. **Range** North America.

Beer-Making Parts (and Possible Uses) Leaves, flowers.

Chemistry Essential oil (total 99.266%): 1-octen-3-ol 0.461%, 3-octanone 0.407%, 1,8-cineol 3.334%, otten-3-yl-acetate 0.386%, estragole 94.003%, α-copaene 0.029%, β-bourbonene 0.084%, E-caryophyllene 0.058%, germacrene D 0.430%, bicyclogermacrene 0.020%, spathulenol 0.039%, β-eudesmol 0.015%.

Style Porter, Stout.

Beer Ale from de Old Vale, Oldvale Farm and Brewing Company (Minnesota, United States).

Source Fisher and Fisher 2017; Hieronymus 2016b.

Agathis robusta (C.Moore ex F.Muell.) F.M.Bailey
ARAUCARIACEAE

Synonyms *Agathis palmerstonii* (F.Muell.) F.M.Bailey, *Dammara bidwillii* Guilfoyle, *Dammara palmerstonii* F.Muell., *Dammara robusta* C.Moore ex F.Muell.

Common Names Kauri, kauri pine, Queensland kauri, Queensland kauri pine, smooth-bark kauri, smooth-barked kauri.

Description Tree (30–50 m), evergreen, monoecious. Trunk with smooth to slightly wrinkled bark, grayish-brown or orange-brown color; young branches glabrous, smooth, green or bluish-green color. Leaves narrowly oblong to narrowly elliptical (5–13 × 1–5 cm), entire margin, obtuse or acute apex, glabrous, upper side bright dark green, paler and duller than the lower side, without prominent central vein; alternate leaves on the larger branches, opposite to subopposite on the lateral branches; petioles 2–10 mm, leathery. Male cones subsessile (12 mm peduncle), axillary, surrounded at the base by 7–13 small overlapping bracts, cylindrical (4–10 × 0.7–1.5 cm), consisting of hundreds of microsporophylls, initially green, then brown or reddish-brown; female cones globular to ovoid (9–15 × 8–10.5 cm), formed by 300–500 macrosporophylls (3.5–4 × 4–4.5 cm) in the shape of densely packed scales, each carrying a seed at the base, green then brown. Seeds strictly ovoid (10–12 × 5–10 mm), brown in color, with a large wing on one side (up to 20 mm).

Subspecies Besides the typical subspecies, the taxon *Agathis robusta* (C.Moore ex F.Muell.) F.M.Bailey subsp. *nesophila* Whitmore has been described.

Related Species The genus *Agathis* includes 19 taxa.

Geographic Distribution Australia (N Queensland, SE Queensland), Papua New Guinea (Port Moresby, the subspecies *Agathis robusta nesophila* in Owen Stanley Range and New Britain). **Habitat** Riparian environments, closed forests, open woodlands. **Range** Grown as an ornamental species in the original range and as a curiosity in botanical gardens and gardens around the world.

Beer-Making Parts (and Possible Uses) Wood tested for wine containers with little success (uncertain use in beer making).

Wood Little differentiated, with yellow sapwood and slightly darker heartwood (WSG 0.48–0.60).

Fine and regular texture, straight grain, rather slow drying with modest shrinkage, not very sensitive to changes in ambient humidity. Easy sawing, shearing, flaking, cleaning; suitable for carpentry, furniture, cabinet making, plywood, matches, maritime construction or weatherproofing, chemical containers.

STYLE None verified.

SOURCE Cantwell and Bouckaert 2016; Cantwell and Bouckaert 2018.

Agave americana L.
ASPARAGACEAE

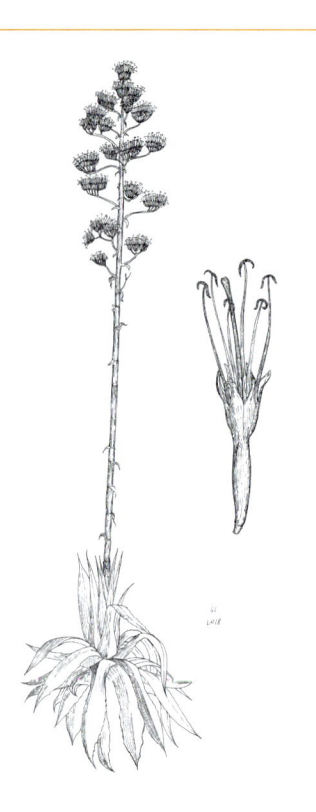

SYNONYMS *Agave altissima* Zumagl., *Agave communis* Gaterau, *Agave complicata* Trel. ex Ochot., *Agave cordillerensis* Lodé & Pino, *Agave felina* Trel., *Agave fuerstenbergii* Jacobi, *Agave gracilispina* (Rol.-Goss.) Engelm. ex Trel., *Agave ingens* A.Berger, *Agave melliflua* Trel., *Agave milleri* Haw., *Agave ornata* Jacobi, *Agave picta* Salm-Dyck, *Agave rasconensis* Trel., *Agave subtilis* Trel., *Agave subzonata* Trel., *Agave theometel* Zuccagni, *Agave zonata* Trel., *Aloe americana* (L.) Crantz.

COMMON NAMES Agave, wild century plant, century plant.

DESCRIPTION Acaulescent plant or with short stem, commonly suckering. Stem up to 2 m; rosettes noncaespitose (10–20 × 20–37 dm). Leaves upright, spread to ascending, occasionally reflected, 80–200 × 15–25 cm; leaf blade light green to green or glaucous-gray, sometimes variegated, closely to broadly lanceolate, smooth, rigid; margin almost straight or wavy to crenate, prickly, single teeth, 5–10 mm, 1–4 cm apart; apical spine dark brown to grayish, conical or subulate, 2–6 cm. Scape 5–9 m. Inflorescence paniculate, nonbulbiferous; bracts persistent, triangular, 5–15 cm; branches lateral, 15–35, horizontal to slightly ascending, including 1/3–1/2 distal inflorescence, longer than 10 cm. Flowers erect, 7–10, 5 cm; perianth yellow in color, tube ± cylindrical (8–20 × 12–20 mm), lobes of petals erect, sub-equal, 20–35 mm; stamens long exerted; filaments inserted above the middle of the perianth tube, erect, yellow, 6–9 cm; anthers yellow, 25–35 mm; ovary 3–4.5 cm, narrow neck, 3–8 mm. Fruit capsule briefly pediculate, oblong, 3.5–8 cm, apex with beak. Seeds 6–8 mm.

SUBSPECIES *Agave americana* L. presents a typical variety (var. *americana*) and the variety *Agave americana* L. var. *expansa* (Jacobi) Gentry. According to other authors, the taxonomic situation is more complex: *Agave americana* L. subsp. *americana* (leaves 100–200 cm, 6–10 times longer than wide; capsules 4–8 cm; SW United States, Mexico) var. *americana* (leaf apical spine 3–5 cm, leaves frequently reflected; Texas), *Agave americana* L. subsp. *americana* var. *expansa* (Jacobi) Gentry (leaf apical spine 2–3 cm, leaves upright; Arizona), *Agave americana* L. subsp. *protamericana* Gentry (leaves 80–135 cm, 4–6 times longer than wide; capsules 3.5–4 cm; Texas; NE Mexico); however, the taxonomic schemes mentioned do not seem to include the great morphological and chromatic diversification of this species that, according to other classifications, would include other varieties (e.g., var. *latifolia*, var. *marginata*, var. *medio-picta*, var. *medio-picta alba*, var. *oaxacensis*, var. *picta*, var. *striata*, var. *theometel*), some of which are of ornamental interest.

CULTIVAR The great variability of this species represents a considerable challenge for any attempt at classification. There are numerous botanical varieties and ornamental cultivars.

RELATED SPECIES There are about 233 species belonging to the genus *Agave*, some of which have possible beer-making applications.

Geographic Distribution SW United States, Mexico. **Habitat** Desert sandy places, grasslands (200–1,300 m above sea level [ASL]). **Range** Used as ornamental species in hot climate areas around the world and frequently naturalized.

Beer-Making Parts (and Possible Uses) Sap, nectar.

Toxicity American agave sap can cause pain and dermatitis in humans if it touches the skin. The species appears in the Poisonous Plant Database of the U.S. FDA.

Chemistry Foliar powder: fiber 38.40%, sugars 45.83%, protein 35.33%, ash 5.94%, lipids 2.03%.

Style Gluten-free Beer.

Source Heilshorn 2017; Hieronymus 2016b (the term "variety" in this work should be interpreted as "species"); McGovern, 2017.

Agave tequilana F.A.C.Weber
ASPARAGACEAE

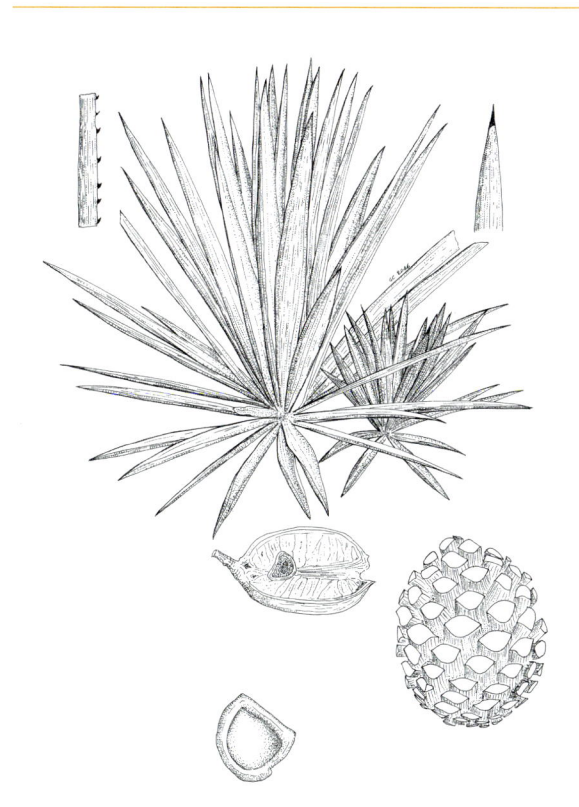

Synonyms *Agave palmaris* Trel., *Agave pedrosana* Trel., *Agave pesmulae* Trel., *Agave pseudotequilana* Trel. Several sources report the binomial *Agave tequilensis* (Blue Moon®); however, this appears to be a trade name only and is not scientifically recognized.

Common Names Blue agave.

Description Perennial succulent, suckering perennial plant (ability exploited to asexually multiply the plant to obtain new individuals for cultivation). Ripe stem short, thick, (30–50 cm). Leaf rosette radiating, 1.2–1.8 m. Leaves lanceolate (90–120 × 8–12 cm), ascending to horizontal, strongly fibrous, rigid, narrow and thick at the base, wider in the middle, sharp, generally glaucous, bluish or gray-green color, linear, wavy or slightly wavy margin; margin teeth light to dark brown, generally regular in size (3–6 mm) and distance (1–2 cm), with thin cusps curved by low pyramidal bases; terminal spine flattened or openly grooved above, generally short (1–2 cm), rarely longer, dark brown, decurrent or not. Inflorescence 5–6 m, densely branched panicle; 20–25 individual units per inflorescence, each large, dense, composed several times. Flower 68–75 mm, ovary 32–38 mm, short neck without shrinkage; tepals green, tube funnel shaped (10 mm), lobes subequal (25–28 mm).

Cultivar *Agave tequilana* is considered a domestic species derived from *Agave vivipara* L., distinguished by wider leaves, thicker stems, wider and more widespread inflorescence, and relatively large flowers with rather short tubes. These morphological differences, however, do not appear so sharp. The economic and cultural importance of *A. tequilana* has contributed to the taxonomic distinction. The name *Agave tequilana* var. *azul* includes the only species of *Agave* usable, according to Mexican law, for the production of tequila. The sap is the main ingredient that defines every form of the alcoholic beverage. Although many botanists recognize that the classification of *A. tequilana* as a separate species does not have sufficient taxonomic evidence, a reluctance to synonymize it with wild species is justified by the value of separation for the tequila industry. This industry counts on the current classification for the commercial distinction between tequila and other mezcal (products mostly with *A. vivipara*).

Related Species In addition to *A. tequilana*, *Agave salmiana* Otto ex Salm-Dyck is one of the 233 species belonging to the genus *Agave* with possible beer-making applications.

Geographic Distribution Mexico.
Habitat Grown in arid tropical environments.
Range Mexico.

Beer-Making Parts (and Possible Uses) The sap (trade name: nectar or agave syrup) of *Agave tequilana* is sweet (47–56% fructose, 16–20% glucose) and edible, with a consistency and taste similar to honey. The agave grown to 7–10 years of age is deprived of its leaves, revealing the plant's stem (called the pina), weighing 25–75 kg. To produce the agave nectar, the sap is extracted from the pina, filtered, and heated at low temperature, a treatment that promotes the transformation of complex carbohydrates into simple sugars.

Chemistry Water 24–37%, fructose 47–56%, glucose 16–20%.

Style Double IPA, Spiced Beer.

Beer Perfect Circle, Crak (Campodarsego, Veneto, Italy).

Source https://www.allaboutagave.com/; https://www.bjcp.org (30. Spiced Beer); https://www.eticamente.net/24977/birra-quali-sostanze-sono-contenute-al-suo-interno.html; https://www.nchomebrewing.com/brewing-beer-with-agave-syrup/.

Agrimonia eupatoria L.
ROSACEAE

Synonyms None
Common Names Agrimony.

Description Herbaceous perennial (3–6 dm). Stem erect, cylindrical, with short and long mixed hairs. Leaves imparipinnate with oblanceolate contour (6–8 × 10–15 cm), with 4–5 pairs of main segments (up to 2.5–5 cm) and short segments (5–15 mm) interspersed; leaf underside paler; usually leaves cauline, smaller than internodes. Inflorescence elongated spiciform raceme (1–3 dm); flower peduncles 2 mm, bent on the fruit; sepals 1.5 mm; petals yellow (2 × 3.5 mm); stamens 10–20. Fruit clubbed (3 × 7 cm) at the axilla of a bract divided into 5 laciniae, in the upper half with a ring of hooked spines.

Subspecies Besides the typical subspecies, scientists also recognize *Agrimonia eupatoria* L. subsp. *asiatica* (Juz.) Skalický and *Agrimonia eupatoria* L. subsp. *grandis* (Andrz. ex C.A.Mey.) Bornm.

Related Species The genus *Agrimonia* includes 20 taxa, but only *A. eupatoria* is used in making beer.

Geographic Distribution Subcosmopolitan. **Habitat** Arid uncultivated meadows. **Range** Subcosmopolitan.

Beer-Making Parts (and Possible Uses) Flower tops, leaves.

Chemistry Flavonoids, catechol tannins, essential oil, bitter substance quercetin, ursolic, silicic, citric, malic, nicotinic and ascorbic acid, prohormones, vitamins K, B1.

Style California Common Beer.

Source https://www.brewersfriend.com/other/agrimony/.

Allium cepa L.
AMARYLLIDACEAE

Synonyms *Allium cepaeum* St.-Lag., *Allium commune* Noronha, *Allium esculentum* Salisb., *Allium napus* Pall. ex Kunth, *Allium pauciflorum* Willd. ex Ledeb., *Allium salota* Dostál, *Ascalonicum sativum* P.Renault, *Cepa alba* P.Renault, *Cepa esculenta* Gray, *Cepa pallens* P.Renault, *Cepa rubra* P.Renault, *Kepa esculenta* Raf., *Porrum cepa* (L.) Rchb.

Common Names Onion.

Description Herbaceous perennial (6–15 dm), bulbous. Bulb first oblong, then pyriform or rounded-flat (Ø 5–10 cm) with generally pinkish tunics. Stem tubular scape (Ø 1–3 cm) with maximum width in the middle. Leaves up to 15 mm wide, all basal. Inflorescence spherical (Ø 5–10 cm), dense, spathe with 2–4 short, bent valves. Flowers on 15–40 mm peduncles, tepals whitish (4.5 mm), obtuse, briefly mucronated; stem filaments protruding beyond the perigonium, alternately dentate.

Cultivar The wide diffusion and long cultivation of the plant have produced a great variety of cultivars in every country. As it is beyond the aims of this work to list all those globally available, I report a selection of the major Italian cultivars as examples. These are usually named after the cultivation area, shape, color, bulb size, earliness, or, more generally, the color of the outer tunics (white, golden yellow, or red). The most common traditional cultivars in Italy include red onion from Tropea (Calabria), red onion from Valle Alifana (Campania), red onion from Acquaviva delle Fonti (Apulia), red onion from Suasa (Marche), red onion from Certaldo (Tuscany), copper onion from Montoro (Campania), red onion from Cavasso Nuovo (Friuli Venezia Giulia), red onion from Vatolla (Campania), onion of Sermide (Lombardy), onion Borettana (Emilia-Romagna), onion of Brunate (Lombardy), onion of Cannara (Umbria), sweet onion, golden onion of Voghera (Lombardy), onion of Banari (Sardinia), Paglina onion of Castrofilippo (Sicily), onion of Isernia (Molise), and blonde onion of Cureggio and Fontaneto (Piedmont).

Related Species The genus *Allium* includes about 1,000 taxa among which, apart from those mentioned here, it is likely that some may be useful in making beer.

Geographic Distribution W Asia. **Habitat** Cultivated and often subspontaneous. **Range** Extensively cultivated.

Beer-Making Parts (and Possible Uses) Bulbs.

Chemistry Energy 26 kcal, water 92.1 g, protein 1.0 g, lipids 0.1 g, carbohydrates 5.7 g, starch 0 g, soluble sugars 5.7 g, dietary fiber 1.0 g, sodium 10 mg, potassium 140 mg, iron 0.4 mg, calcium 25 mg, phosphorus 35 mg, vitamin B1 (thiamine) 0.02 mg, vitamin B2 (riboflavin) 0.03 mg, vitamin PP (niacin) 0.5 mg, vitamin A (retinol) 3 µg, vitamin C 5 mg.

STYLE Mild/Brown Ale, Barleywine.

BEER Onion Mild, Maui Brewing Company (Hawaii, United States).

SOURCE Calagione et al. 2018; https://www.ratebeer.com/beer/maui-brewing-onion-mild/145647/.

Allium sativum L.
AMARYLLIDACEAE

SYNONYMS *Allium controversum* Schrad. ex Willd., *Allium longicuspis* Regel, *Allium ophioscorodon* Link, *Allium pekinense* Prokh., *Porrum ophioscorodon* (Link) Rchb.

COMMON NAMES Garlic.

DESCRIPTION Herbaceous perennial (5–8 dm), bulbous. Bulb (2–4 cm) often proliferous, with oblong or globular bulbs and white papyraceous tunics; scape cylindrical (Ø 1–1.5 cm), wrapped in leaf sheaths up to half. Leaves linear, 6–12 mm wide, smooth. Inflorescence subspherical, rich (Ø 6–10 cm) or almost completely sterile and bulbiferous. Spathe univalve, rostrate long, longer than the umbel. Tepals whitish or greenish, 3 mm; stamens embedded.

CULTIVAR This species has long been cultivated in many countries. Over time, numerous cultivars have been selected. Examples from Italy include garlic from Caraglio (Piedmont), white garlic of Piacenza, garlic from Voghiera (Emilia-Romagna), red garlic from Sulmona (Abruzzo), garlic from Vessalico (Liguria), red garlic from Nubia (Sicily), garlic from Resia (Friuli Venezia Giulia), and red garlic from Proceno (Lazio).

RELATED SPECIES The genus *Allium* includes about 1,000 taxa among which, apart from those mentioned here, it is likely that some may be useful in making beer.

GEOGRAPHIC DISTRIBUTION Perhaps native to CW Asia. **HABITAT** Cultivated and sometimes subspontaneous. **RANGE** Cultivated everywhere.

BEER-MAKING PARTS (AND POSSIBLE USES) Bulbs.

CHEMISTRY Energy 623 kJ (149 kcal), carbohydrates 33.06 g, sugars 1 g, fiber 2.1 g, lipids 0.5 g, protein 6.36 g, thiamine (B1) 0.2 mg, riboflavin (B2) 0.11 mg, niacin (B3) 0.7 mg, pantothenic acid (B5) 0.596 mg, vitamin B6 1.235 mg, folate (B9) 3 µg, vitamin C 31.2 mg, calcium 181 mg, iron 1.7 mg, magnesium 25 mg, manganese 1.672 mg, phosphorus 153 mg, potassium 401 mg, sodium 17 mg, zinc 1.16 mg, selenium 14.2 µg.

STYLE Amber Ale, American IPA, American Pale Ale, American Porter.

BEER Black Garlic Beer, the Garlic Farm (Newchurch, United Kingdom).

SOURCE Calagione et al. 2018; https://www.birradegliamici.com/news-birraie/birre-strane/; https://www.brewersfriend.com/other/garlic/; https://www.thegarlicfarm.co.uk/product/black-garlic-beer.

The well-known organoleptic and sensory characteristics of Allium sativum make it especially difficult to manipulate, manage, and dose in the beer-making process. Nevertheless, some brewers have experimented with the use of garlic in its fermented form, which, because of its appearance, is called black garlic. Unlike the natural version, it has soft, subtle, toffee-like, and slightly acidic notes.

Alnus glutinosa (L.) Gaertn.
BETULACEAE

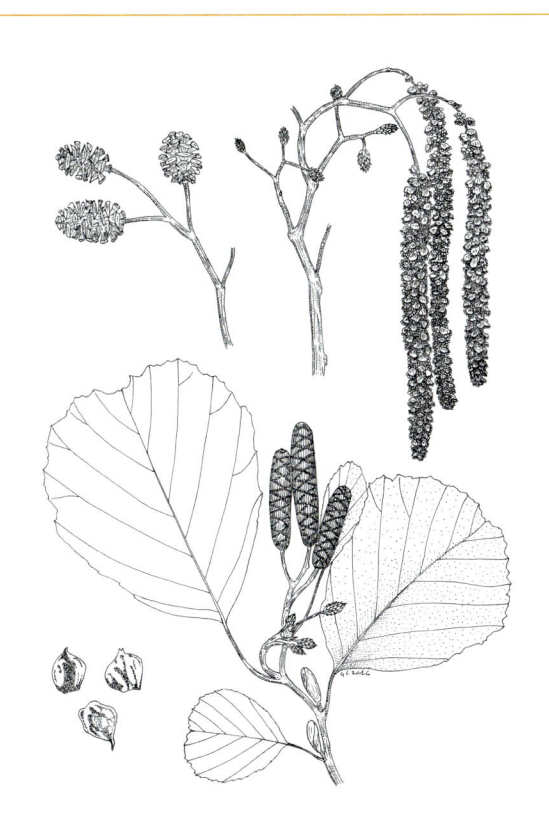

Synonyms *Alnus aurea* K.Koch, *Alnus cerifera* Hartig ex Regel, *Alnus emarginata* Krock., *Alnus imperialis* Dippel, *Alnus incisa* Steud., *Alnus macrocarpa* Lodd. ex Loudon, *Alnus morisiana* Bertol., *Alnus nitens* K.Koch, *Alnus oxyacanthifolia* Lodd., *Alnus prunifolia* K.Koch, *Alnus quercifolia* Willd., *Alnus rotundifolia* Stokes, *Betula glutinosa* (L.) Lam.

Common Names Black alder.

Description Tree (2–25 m). Bark greenish-brown, shiny, with transverse lenticels 1–3 mm long; buds pedunculate. Branches young and leaves sticky. Leaves with 1–2 cm petiole and obovate lamina (4–6 × 5–7 cm) or orbicular (Ø 10 cm) lamina, coarsely dentate, with truncated base, apex truncated or bidentate; veins 6–8 on each side. Inflorescences completely formed in winter, without basal leaves; catkins ovoid (1–3 cm).

Subspecies Besides the typical subspecies, the following taxa are recognized: *Alnus glutinosa* (L.) Gaertn. subsp. *barbata* (C.A.Mey.) Yalt., *Alnus glutinosa* (L.) Gaertn. var. *incisa* (Willd.) Regel, and *Alnus glutinosa* (L.) Gaertn. var. *laciniata* (Willd.) Regel.

Related Species The genus *Alnus* includes 57 taxa. The English-speaking literature generally refers to a generic "alder" rather than a precise species for beer making. Useful species for brewing could therefore also include *Alnus incana* (L.) Moench., while *Alnus alnobetula* (Ehrh.) K.Koch, *Alnus incana* (L.) Moench, and *Alnus rubra* Bong are certainly used.

Geographic Distribution Paleotemperate. **Habitat** Woods and bushes on the banks of watercourses and in peaty soils. **Range** Eurasia, N Africa, Middle East.

Beer-Making Parts (and Possible Uses) Wood (to smoke the malt), young shoots (gruit), twigs (to flavor; to filter the wort of Sahti, a traditional Finnish beer).

Chemistry (3R,5S)-1,7-bis-(3,4-dihydroxyphenyl)-3-hydroxyheptano-5-O-β-D-xylopyranoside, rubranoside A, rubranoside B, mapleside VII, 3(R)-1,7-di(3,4-dihydroxyphenyl)-3-O-β-D-[6-(E-3,4-dimethoxycinnamoil glucopyranosyl)]heptane, 3(R)-1,7-di(3,4-dihydroxyphenyl)-5-O-β-D-[6-(Z-3,4-dimethoxycinnamoil glucopyranosyl)] heptane, 3(R)-1,7-di(3,4-dihydroxyphenyl)-5-O-β-D-[6-(E-3,4,5-trimethoxycinnamoilglucopyranosyl)]heptane, hirsutanonol, 5(S)-O-methylhirsutanonol, alnuside A, alnuside B, platifillonolo-5-O-β-D-xylopyranoside, oregonine, 5(S)-hirsutanonolo-5-O-β-D-glucopyranoside, platifilloside, (5S)-1-(4-hydroxyphenyl)-7-(3,4-diidrossi-fenil)-5-O-β-D glycopyranosyl-eptane-3-one, 5(S)-1,7-di(4-idroxifenil)-5-O-β-D-[6-(E-p-coumaroilglucopiranosil)] heptan-3-one, 5(S)-1,7-di(4-idrossifenil)-5-O-β-D-[6-(E-3,4-dimethoxycinnamoyl glucopyranosyl)] heptan-3-one, 5(S)-1-(4-hydroxyphenyl)-7-(3,4-hydroxyphenyl)-5-O-β-D-[6-(E-3,4-dimethoxycinnamoyl glucopyranosyl)] heptane-3-one, 5(S)-1-(3,4-dihydroxyphenyl)-7-(4-hydroxyphenyl)-5-O-β-D-[6-(Z-3,4-dimethoxycinnamoylglucopyranosil)] heptan-3-one, 5(S)-1,7-di(3,4-dihydroxyphenyl)-5-O-β-D-[6-(E-p-cumaroylglucopyranosil)] heptan-3-one, 5(S)-1,7-di(3,4-dihydroxyphenyl)-5-O-β-D-[6-(Z-3,4-dimethoxycinnamoil glucopyranosyl)] heptane-3-one, 5(S)-1,7-di(3,4-dihydroxyphenyl)-5-O-β-D-[6-(E-3,4,5-trimethoxycinnamoil glucopyranosyl)] heptane-3-one, 5(S)-1,7-di(3,4-dihydroxyphenyl)-5-O-β-D-[6-(E-3,

4-dimethoxycinnamoil glucopyranosyl)]]heptane-3-one, hirsutene, praecoxin-D, genkwanin, quercetin-3-soforoside, rhododendrin, taraxerone, β-amyrin, ursolic acid, uvaol, hop, betulin, betulin aldehyde, phenylacetate, lupenone, simiarenol, β-sitosterol, brassinolide, castasterone.

Wood Undifferentiated, takes its characteristic reddish color immediately after cutting; reddish-brown with drying; soft, easy to work, WSG 0.42–0.64, not very resistant to alterations but durable if permanently immersed in water; suitable for lathe work, reels, spindles, clogs, plywood; valuable for piles in submerged land and agricultural poles; good fuel.

Style Classic Style Smoked Beer, Sahti.

Beer Sahti, Finlandia Sahti (Suodenniemi, Finland).

Source Cantwell and Bouckaert 2016; Cantwell and Bouckaert 2018; Jackson 1991; https://www.bjcp.org (32A. Classic Style Smoked Beer).

🍺 Sahti, a Centuries-Old Finnish Tradition
Sahti is a kind of beer traditionally brewed in Finland. Although neither its history nor its characteristics are certain, what has survived to this day is a fermented drink made from malted and nonmalted cereals (barley, oats, spelt) and juniper berries, filtered using branches of juniper and black alder (*Alnus glutinosa*) inside a container, called a *kurnaa*, obtained from the trunk of the same tree. The result is an amber-colored drink, gas free, and with a lot of body. The taste tends to be sweet, with hints of banana (linked to fermentation) and light balsamic notes given by juniper and wood.

Aloysia citriodora Palau
VERBENACEAE

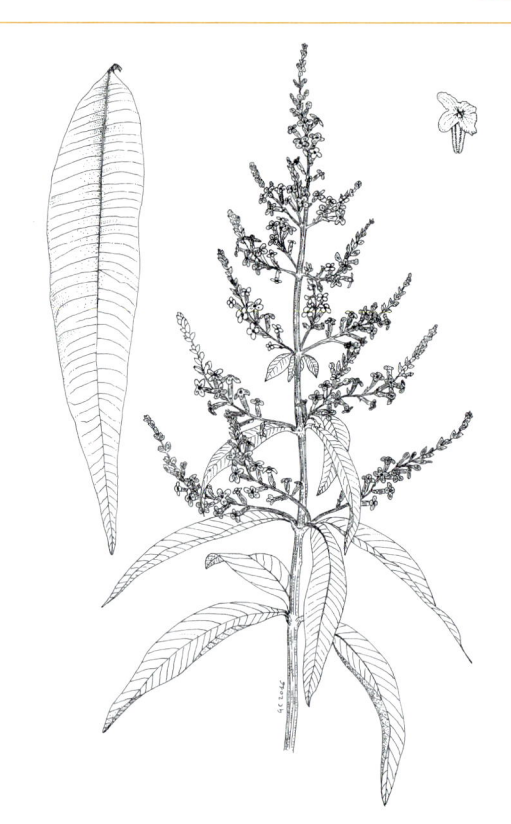

Synonyms *Aloysia sleumeri* Moldenke, *Aloysia triphylla* (L'Hér.) Britton, *Aloysia triphylla* Royle, *Cordia microcephala* Willd. ex Roem. & Schult., *Lippia citriodora* (Palau) Kunth, *Lippia triphylla* (L'Hér.) Kuntze, *Verbena citriodora* (Palau) Cav., *Verbena triphylla* L'Hér., *Zappania citriodora* (Palau) Lam.

Common Names Lemon verbena, lemon beebrush.

Description Shrub or bush (2–3 m), deciduous. Branches upright, woody. Leaves lanceolate (1.5–3 × 7–10 cm), whole, verticillate to 3, with intense lemon scent when rubbed. Inflorescence elongated spikelet forming an apical pyramidal panicle with opposite branches. Flower with light purple corolla (and calyx) (4 mm).

Related Species The genus *Aloysia* includes 46 taxa among which none, other than *A. citriodora*, seems to have beer-making applications.

Geographic Distribution Peru, Chile, Argentina. **Habitat** Unknown. **Range** Cultivated for ornamental and food purposes in all temperate areas of the world.

Beer-Making Parts (and Possible Uses) Leaves.

Chemistry Essential oil: trans-caryophyllene oxide 13.52%, β-spatulenol 13.27%, Ar-curcumene 11.47%, black 10.02%, cis-verbenol 7.78%, ς-cadinene 3.89%, isoledene 3.48%, β-caryophyllene 3.11%, alloaromadendrene oxide-(l) 3.06%, trans-nuciferol 2.31%, ς-elemene 2.3%, cis-nuciferol 2.17%, 1,8-cineol 1.97%, nerol 1.85%, ledene oxide-(I) 1.75%, 2-carene 1.71%, alloaromadendrene 1.66%, α-copaene 1.25%, tricycle [5,2,2,0(1,6)] undecan-3-ol, 2-methylene-6,8,8-trimethyl 0.97%, limonene 0.71%, 1,3,4-trimethyl-3-cyclohexene-1-carboxaldehyde 0.7%, Di-epi-α-cedrene 0.67%, geranyl-acetate 0.66%, trans-verbenol 0.64%, α-humulene 0.6%, campholene aldehyde 0.58%, myrtenol 0.58%, trans-2-caren-4-ol 0.53%, tricycle[5,2,2,0(1,6)]undecan-3-ol, 2-methylene-6,8-trimethyl 0.49%, murolan-3,9(11)-diene-10-peroxy 0.36%, ledene oxide-(II) 0.35%, ς-himachalene 0.34%, γ-terpinene 0.33%, cisp-mentha-2,8-dien-1-ol 0.33%, verbenyl ethyl ether 0.31%, eudesma-4,11-dien-2-ol 0.31%, trans-sabinene-hydrate 0.29%, germacrene-D 0.28%, alloaromadendrene oxide-(2) 0.28%, D-carvone 0.27%, 2,6,6-trimethyl-1-cyclohexene-1-acetaldehyde 0.27%, α-pinene 0.23%, eugenol 0.19%, trans-p-mentha-2,8-dienol 0.18%, γ-cadinene 0.17%, cis-limonene dioxide 0.16%, piperitone 0.15%, β-pinene 0.14%, cis-pmentha-1(7), 8-dien-2-ol 0.13%, α-terpinene 0.12%, 4,4,11,11-tetramethyl-7-tetracyclo [6,2,1,0(3,8)0(3,9)]undecanol 0.1%, α-thujene 0.09%, trans-α-bergamotene 0.09%, calamenene 0.09%, 6-canphenol 0.08%, trans-carveyl acetate 0.07%, cis-carveol 0.06%, aromadendrene 0.06%, para-cymene-7-ol 0.05%, 9-isopropyl-1-methyl-2-methylene-5-oxatricycle[5,4,0,0(3,8)]undecane 0.05%, sabinene 0.04%, trans-β-o-cymene 0.04%, (Z,Z,Z)-8,11,14-icosatrienoic acid 0.04%, 6-methyl-5-epten-2-one 0.03%, β-myrcene 0.03%, perillic aldehyde 0.03%, 1-(1,3-dimethyl-buta-1,3-dienyl)-3,7,7-trimethyl-2-oxabicycl [3,2,0] ept-3-ene 0.03%, paracimene 0.02%, cis-β-o-cymene 0.02%, β-guaiene 0.01%.

Style California Common Beer, Saison.

Beer Lips of Faith, New Belgium (Colorado, United States).

Source Heilshorn 2017; Hieronymus 2016a; Hieronymus 2016b; https://www.brewersfriend.com/other/cedron/.

Alpinia galanga (L.) Willd.
ZINGIBERACEAE

Synonyms *Alpinia alba* (Retz.) Roscoe, *Alpinia bifida* Warb., *Alpinia carnea* Griff., *Alpinia pyramidata* Blume, *Alpinia rheedei* Wight, *Alpinia viridiflora* Griff., *Amomum galanga* (L.) Lour., *Amomum medium* Lour., *Galanga officinalis* Salisb., *Hellenia alba* (Retz.) Willd., *Heritiera alba* Retz., *Languas galanga* (L.) Stuntz, *Languas pyramidata* (Blume) Merr., *Languas vulgare* J.Koenig, *Maranta galanga* L., *Zingiber galanga* (L.) Stokes, *Zingiber medium* Stokes, *Zingiber sylvestre* Gaertn.

Common Names Galangal, greater galangal.

Description Herbaceous perennial (1–2 m). Rhizome tuberous and slightly aromatic. Leaves oblong-lanceolate, acute, glabrous, green above, paler below, with slightly callous white margins; sheaths long, glabrous, short, and rounded. Flowers greenish-white, gathered in dense panicles (30 cm); bracts ovate-lanceolate; calyx tubular, irregularly 3-dentate; lobes corollary, oblong, greenish claw, white blade, streaked with red (more than 1 cm long), coarsely elliptical, briefly 2-lobed at the apex, with a pair of subulate glands at the base of the claw. Fruit orange-red, the size of a small cherry.

Related Species The genus *Alpinia* includes about 247 taxa.

Geographic Distribution SC China, Cambodia, Myanmar, Thailand, Vietnam, Borneo, Java, Malaysia, Philippines, Sumatra. **Habitat** Wet tropical forests. **Range** India, Sri Lanka, Bangladesh (+ geographic distribution).

Beer-Making Parts (and Possible Uses) Rhizome.

Chemistry Essential oil: 3-phenyl-2-butanone 23.2%, 4,5-dihydro-2-phenyl-1H-imidazole 16.84%, propanal benzene 11.68%, 5-hydroxy-1,7-diphenyl-3-heptanone 9.05%, 1-(4-hydroxyphenyl)-1-methoxy-4 phenylbutane 5.03%, nortrachelogenin 3.36%, 2-methyl-3-nitrophenyl β-phenylpropionate 2.55%, α-3-cyclohexene-1-methanol 2.27%, naphthalene 1.65%, α-farnesene 0.78%, hexadecanoic acid methyl ester 0.59%, caryophyllene 0.55%, 9,12 octodecanoic acid 0.48%, cubenol 0.34%, eucalyptol 0.25%.

Style Braggot, Experimental Beer, Mild, Saison, Schwarzbier, Special/Best/Premium Bitter, Spice/Herb/Vegetable Beer, Tripel, Witbier.

Beer Elysian He Said, 21st Amendment Brewery (California, United States).

Source Fisher and Fisher 2017; Heilshorn 2017; https://www.brewersfriend.com/other/galangal/; https://www.brewersfriend.com/other/galingale/.

Althaea officinalis L.
MALVACEAE

Synonyms *Althaea kragujevacensis* Pančić, *Althaea micrantha* Borbás, *Althaea multiflora* Rchb. ex Regel, *Althaea sublobata* Stokes, *Althaea taurinensis* DC., *Althaea vulgaris* Bubani, *Malva althaea* E.H.L.Krause, *Malva maritima* Salisb., *Malva officinalis* (L.) Schimp. & Spenn. ex Schimp. & Spenn.

Common Names Marshmallow.

Description Herbaceous perennial (up to 1.5 m). Root fleshy, fusiform, cylindrical, whitish inside, yellowish outside. Stems tomentose, erect, and numerous, simple or little branched. Leaves briefly petiolate, stipulated, alternate, palminerved, with oval lamina with acute lobes and crenate margin, tomentose on both sides, whole or 3- to 5-lobed, major central lobe. Flowers (Ø 3–5 cm) with axillary or terminal peduncles; calyx with 5–8 bracts attached to the base, calyx with 5 divisions, corolla twisted prefoliate with 5 pink petals, alternating with the divisions of the calyx. Fruits densely pubescent, formed by numerous achenes arranged circularly one next to the other, with 10–20 seeds, reniform.

Related Species The *Althaea* genus has 17 taxa of which none, other than *A. officinalis*, seems to have applications in making beer.

Geographic Distribution SE Europe, S Siberia. **Habitat** Widespread in wetlands, along the banks of ditches, even in the presence of brackish water (0–1,200 m ASL). **Range** Does not appear to be cultivated.

Beer-Making Parts (and Possible Uses) All parts of the plant, dried roots.

Chemistry Root (without rhizodermal): polysaccharide mucilage 6.2–11.6% (galacturonorhamnans, arabinans, glucarides, arabinogalactans), carbohydrates 25–35%, flavonoids, glycosides, sugars (10% sucrose); amines (up to 12% asparagine), pectin, lipids 1.7%; calcium oxalate, coumarins, phenolic acid, sterols, scopoletin, quercetin, kaempferol, chlorogenic acid, caffeic acid, p-coumaric acid, aluminum, iron, magnesium, selenium, tin, calcium.

Style American Light Lager, American Porter, American Stout, Cream Ale, Herb/Vegetable Beer, Holiday/Winter Special Spiced Beer, Imperial IPA, Imperial Stout, Oatmeal Stout, Russian Imperial Stout, Spice, Sweet Stout.

Beer Imperial Puft, Tiny Rebel (Rogerston, Wales).

Source Heilshorn 2017; Hieronymus 2016b; https://www.brewersfriend.com/other/marshmallow.

Althaea officinalis is well known, especially in the United States, where in the past people used its roots to make marshmallows. It offers brewers various possibilities of use. As Stan Hieronymus describes it in *Brewing Local* (2016), all parts of the plant are edible, and all are quite sticky (especially the roots). The root parts, if boiled, can contribute a sweet taste, while the seeds, if roasted, give the same notes as coffee but without adding caffeine.

Amaranthus caudatus L.
AMARANTHACEAE

Synonyms *Amaranthus abyssinicus* L.H.Bailey, *Amaranthus alopecurus* Hochst. ex A.Br. & C.D.Bouché, *Amaranthus cararu* Moq., *Amaranthus dussii* Sprenger, *Amaranthus edulis* Speg., *Amaranthus leucocarpus* S.Watson, *Amaranthus mantegazzianus* Pass., *Amaranthus maximus* Mill., *Amaranthus pendulinus* Moq., *Amaranthus pendulus* Moq., *Euxolus arvensis* Rojas Acosta.

Common Names Amaranth, love lies bleeding, pendant amaranth, tassel flower, velvet flower, foxtail amaranth.

Description Herb, annual (3–8 dm). Stem and inflorescence mostly reddened. Leaves rhombic-oval, acuminate. Inflorescence linear spikelet, terminal elongated, pendulous. Flowers with 5 tepals, oblanceolate-spatulate, as long as the fruit.

Cultivar According to some authors, the remarkable intraspecific variety of these taxa is partly a result of natural diversification (many taxa of subspecific rank have been included in the synonyms) and of long and widespread cultivation; therefore, it is reasonable to confirm the existence of numerous cultivars.

Related Species The genus *Amaranthus* includes 113 taxa, many of which are edible, but only the 3 mentioned here have use in making beer.

Geographic Distribution Paleotropical. **Habitat** Cultivated for food or ornament and often naturalized. **Range** Cultivated in many countries.

Beer-Making Parts (and Possible Uses) Seeds.

Chemistry Protein 14%, lipids 10%, ash 2.5%, starch 64%, fiber 8–16%, lysine 5.2–6.0 g/16 g N, leucine, valine, threonine, tannins 0.3%.

Style American Wheat Beer, Rye Beer, Saison.

Beer Saison Z, Jolly Pumpkin Artisan Ale (Michigan, United States).

Source https://www.brewersfriend.com/other/amaranto/.

Amaranthus cruentus L.
AMARANTHACEAE

Synonyms *Amaranthus anacardana* Hook.f., *Amaranthus arardhanus* Sweet, *Amaranthus carneus* Moq., *Amaranthus chlorostachys* Moq., *Amaranthus esculentus* Besser ex Moq., *Amaranthus farinaceus* Roxb. ex Moq., *Amaranthus guadeloupensis* Voss, *Amaranthus guadelupensis* Moq., *Amaranthus incarnatus* Moq., *Amaranthus montevidensis* Moq., *Amaranthus paniculatus* L., *Amaranthus purgans* Moq., *Amaranthus rubescens* Moq., *Amaranthus sanguineus* L., *Amaranthus sanguinolentus* Schrad. ex Moq., *Amaranthus speciosus* Sims, *Amaranthus spicatus* Wirzén, *Amaranthus strictus* Willd.

Common Names Amaranth, blood amaranth, red amaranth, purple amaranth, prince's feather, Mexican grain amaranth.

Description Herb, annual (2–20 dm). Stem branched from the base, with arched branches, furrowed by longitudinal, green or reddish stripes, glabrous or covered with sparse fluff in the apical part. Leaves green or reddish, petiolate, ovate-oblong to oval (1.5–12 × 1–6 cm), attenuated to wedges at the base, with evident reddish veins. Inflorescence green or deep red, the longest terminal of the sides (about 45 cm). Flowers unisexual pentamerous, wrapped by bracts (2–4 mm) briefly aristate, erect, ± as much as tepals (1–1.5 times); ovate-lanceolate tepals (1.5–2.5 mm) with obtuse, muticate, or briefly mucronate apex and yellow-brown median rib (1–2 mm); female flower bracts generally not exceeding the stigma but with short awn (1.3–1.5 times perianth); male flowers with 5 stamens. Fruit capsule compressed (pyxidium), ellipsoidal, longer than perigonium, dehiscent horizontally. Compressed seeds whitish, yellowish, or blackish (1 mm).

Cultivar There are many cultivars of this species, cultivated in North and Central America for more than 5,000 years, both as grain and ornamental.

Related Species The genus *Amaranthus* includes 113 taxa, many of which are edible, but only the three listed here have use in making beer.

Geographic Distribution Neotropical. **Habitat** Ruins and wastelands near homes. **Range** Cultivated in many countries as subsistence food (Africa) or as health food (North America, Europe).

Beer-Making Parts (and Possible Uses) Seeds.

Chemistry Flour (g/100 g): protein 16.6 g, water 10.3 g, ash 3.35 g, fiber 9.8 g, lipids 8.77 g.

Style American Wheat Beer, Rye Beer.

Source https://www.brewersfriend.com/other/amaranto/.

Amaranthus hypochondriacus L.
AMARANTHACEAE

Synonyms *Amaranthus anardana* Buch.-Ham. ex Moq., *Amaranthus atrosanguineus* Moq., *Amaranthus aureus* Besser, *Amaranthus bernhardii* Moq., *Amaranthus flavus* L., *Amaranthus frumentaceus* Buch.-Ham. ex Roxb., *Amaranthus macrostachyus* Mérat ex Moq., *Amaranthus monstrosus* Moq.

Common Names Amaranth, Prince-of-Wales feather.

Description Herb (5–12 dm), annual, monoecious. Stem furrowed, smooth. Leaves alternate, long petiolate, oblong-lanceolate (10–20 cm), light green with purple marks. Flowers unisexual on the same plant; 3 bracts bright purple-red in color, 3–5 sepals same color as bracts. Inflorescence long, dense, upright spikelet.

Subspecies Besides the typical variety, the taxon *Amaranthus hypochondriacus* L. var. *powellii* (S.Watson) Pedersen is accepted.

Cultivar Burgundy (Burgundy Elite), Golden, Golden Giant, Opopeo.

Related Species The genus *Amaranthus* includes 113 taxa, many of which are edible, but only the three listed here have beer-making use.

Geographic Distribution Neotropical (Mexico?). **Habitat** Ruins and fallow land near homes. **Range** Cultivated in many countries.

Beer-Making Parts (and Possible Uses) Seeds.

Chemistry Seeds: water 5.22%, ether extract 5.9%, protein 16.2%, fiber 2.0%, ash 3.0%, carbohydrates 67.1%.

Style American Wheat Beer, Rye Beer.

Source Fisher and Fisher 2017; https://www.brewersfriend.com/other/amaranto/.

Amburana cearensis (Allemao) A.C.Sm.
FABACEAE

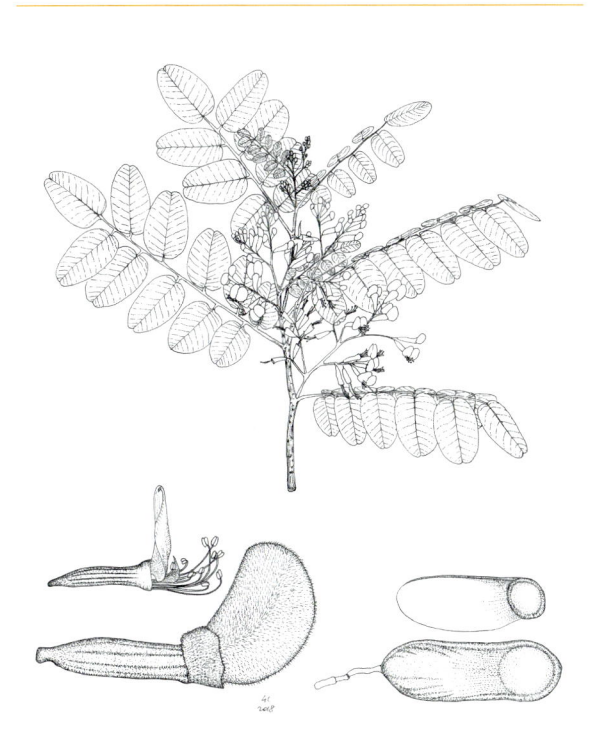

Synonyms *Amburana claudii* Schwacke & Taub., *Torresea cearensis* Allemao.

Common Names Amburana.

Description Tree (15–35 m), deciduous. Stem straight (Ø 50–130 cm), bark orange-brown, smooth, detaching in papillary scales; foliage flattened, dark green. Leaves alternate, pinnate (8–22 cm), with 7–12 elliptical, alternate (3–5 × 2–3 cm) leaves. Inflorescence raceme terminal or axillary; flowers with round petals, white in color. Fruit oblong legume (5–8 × 1–2 cm), hard, dark brown. Seed ovoid (4–7 cm), brown in color, widely winged.

Related Species Besides *Amburana cearensis* (Allemao) A.C.Sm., the genus includes the species *Amburana acreana* (Ducke) A.C.Sm.

Geographic Distribution South America (N Argentina, Brazil, Paraguay, Bolivia, Peru). **Habitat** Arid tropical forests, often in deep, well-drained soils; in rain forests, usually found only in calcareous and rocky areas. **Range** South America (N Argentina, Brazil, Paraguay, Bolivia, Peru).

Beer-Making Parts (and Possible Uses) Wood for barrels (first they contain cachaça, then beer for aging).

Wood Yellowish sapwood, very similar to heartwood, yellowish or brownish with pink or orange tones, strongly marked by veins and with streaks of parenchyma; oily appearance; fragrant of coumarin, with a slight taste of vanilla. Coarse but uniform texture, irregular grain. Not difficult to work, but the sawn surfaces are quite cottony; holds quite firm and considered to be of medium durability (WSG 0.55–0.65). Used for construction, weatherproofing, carpentry, joinery, fixtures, staves.

Style Barleywine.

Beer Amburana Barley Wine, Way Beer (Pinais, Brazil).

Source Cantwell and Bouckaert 2016; Cantwell and Bouckaert 2018.

Anacardium occidentale L.
ANACARDIACEAE

Synonyms *Acajuba occidentalis* (L.) Gaertn., *Anacardium microcarpum* Ducke, *Cassuvium pomiferum* Lam., *Cassuvium reniforme* Blanco, *Cassuvium solitarium* Stokes.

Common Names Cashew tree, cajù.

Description Tree (10–12 m). Stem short, often irregularly shaped. Leaves spiral, leathery, elliptical to obovate (4–22 × 2–15 cm), with entire margins. Inflorescence panicle or corymb (up to 26 cm). Flowers small, light green, then turning reddish, with 5 slender, acute (7–15 mm) petals. Fruit consists of a pseudocarp (false fruit); what appears to be the fruit is an oval to pyriform structure (hypocarp, which develops from the peduncle and flower receptacle) called a cashew apple (*marañón* in Spanish), which ripens in a yellow or red structure (about 5–11 cm long), edible and with a strong, sweet taste and smell; juicy flesh and fragile exocarp make it unsuitable for transport (in Latin America, it is used for the preparation of a refreshing fruit drink). The real fruit is a kidney-shaped drupe that grows at the distal end of the cashew apple; inside is a single seed (often considered a nut in a culinary, not botanical, sense). Seed surrounded by a double shell containing an allergenic phenolic resin (cashew acid), a powerful skin irritant chemically similar to urushiol, a *Hedera helix* toxin; roasting destroys the toxin but should be done outdoors because the smoke contains urushiol, which, if inhaled, can cause dangerous (even fatal) allergic lung reactions.

CULTIVAR Various studies confirm the existence of numerous cultivars of this species, even well outside the original area.

RELATED SPECIES The genus *Anacardium* includes 20 species, but only *A. occidentale* is known to be used in making beer.

GEOGRAPHIC DISTRIBUTION Brazil. **HABITAT** Unknown. **RANGE** Vietnam, Nigeria, India, Ivory Coast, Benin.

BEER-MAKING PARTS (AND POSSIBLE USES) Seeds.

CHEMISTRY Energy 2,310 kJ (553 kcal), water 5.2 g, carbohydrates 30.19 g, starch 0.74 g, sugars 5.91 g, fiber 3.3 g, lipids 43.85 g, saturated 7.783 g, monounsaturated 23.797 g, polyunsaturated 7.845 g, protein 18.22 g, thiamine (B1) 0.423 mg, riboflavin (B2) 0.058 mg, niacin (B3) 1.062 mg, pantothenic acid (B5) 0.86 mg, vitamin B6 0.417 mg, folate (B9) 25 µg, vitamin C 0.5 mg, vitamin E 0.9 mg, vitamin K 34.1 µg, calcium 37 mg, iron 6.68 mg, magnesium 292 mg, manganese 1.66 mg, phosphorus 593 mg, potassium 660 mg, sodium 12 mg, zinc 5.78 mg.

STYLE Northern English Brown Ale, Robust Porter.

BEER Cashew Mountain Brown, Founders (Michigan, United States).

SOURCE https://www.brewersfriend.com/other/cashews/.

Anamirta cocculus (L.) Wight & Arn.
MENISPERMACEAE

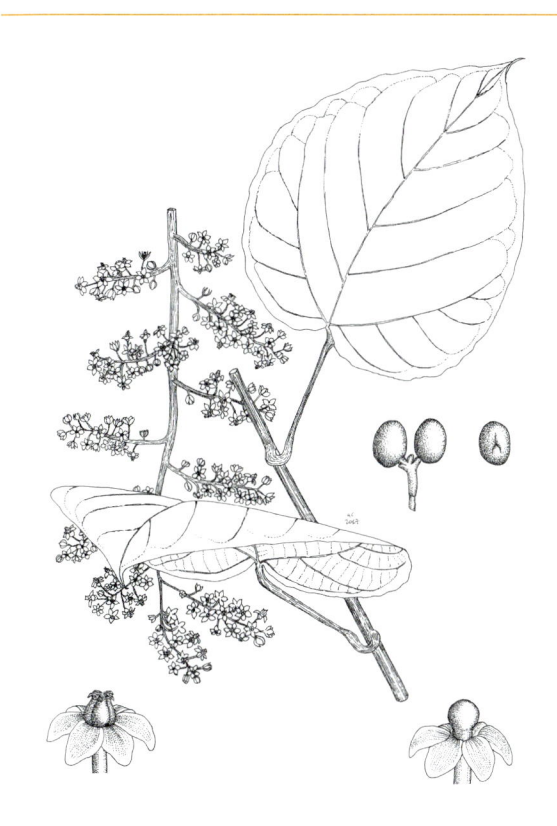

SYNONYMS *Anamirta baueriana* Endl., *Anamirta jucunda* Miers, *Anamirta paniculata* Colebr., *Anamirta populifolia* (DC.) Miers, *Anamirta racemosa* Colebr. ex Steud., *Anamirta toxifera* Miers, *Cocculus indicus* Royle, *Cocculus lacunosus* DC., *Cocculus suberosus* DC., *Menispermum cocculus* L.

COMMON NAMES Poison berry, cocculus, cocculus indicus.

DESCRIPTION Climbing creeper shrub. Stem woody (Ø 10 cm) exuding sap, white if cut; bark thick, spongy, deeply fissured. Leaves subtly leathery, long stalked (5–12 cm), leaves cordate or wide ovate (10–16 cm), basal veins 3–5. Flowers green, briefly pedunculate, unisexual, strongly fragrant, no petals, sepals 6; female 4- to 5-carpellar. Inflorescence long panicle cauliflorus erect or hanging. Fruit drupe (Ø 1.2 cm), usually 2, dark purple or black. Seeds cup shaped.

RELATED SPECIES *Anamirta* is a monotypical genus.

GEOGRAPHIC DISTRIBUTION E Asia (India, Thailand, Indonesia, Philippines, New Guinea). **HABITAT**

armoracia L., *Cochlearia lancifolia* Stokes, *Cochlearia variifolia* Salisb., *Crucifera armoracia* E.H.L.Krause, *Nasturtium armoracia* (L.) Fr., *Rorippa armoracia* (L.) Hitchc., *Rorippa rusticana* (G. Gaertn., B. Mey. & Scherb.) Godr.

Common Names Horseradish.

Description Herbaceous perennial (4–7 dm), rhizomatous. Rhizome whitish, with pungent aroma. Stem erect, glabrous, striped. Basal leaves with grooved petiole (2–3 dm) and lanceolate lamina (up to 10 × 25 cm), with thick rounded teeth on edges; cauline leaves sessile, pinnatifide. Inflorescences axillary racemes on the upper leaves. Flowers on peduncles 10–15 mm; sepals green, ovate (3 mm); petals white (6 × 10 mm). Fruits silique ovoid or subspherical (4–6 mm).

Cultivar Numerous.

Related Species The genus *Armoracia* includes three species: *Armoracia macrocarpa* (Waldst. & Kit.) Kit. ex Baumg., *Armoracia rusticana* P. Gaertn., B. Mey. & Scherb., and *Armoracia sisymbrioides* (DC.) N.Busch ex Ganesh.

Geographic Distribution E Europe. **Habitat** Cultivated as a condiment and semiwild in vegetable gardens. **Range** CE Europe.

Beer-Making Parts (and Possible Uses) Rhizomes.

Chemistry 48 kcal, high vitamin C content, low sodium content, folic acid, fiber, allyl isothiocyanate.

Style California Common Beer, Saison, Stout.

Beer Quadro, Horseradish (Folgaria, Trentino-Alto Adige, Italy).

Source Calagione et al. 2018; Hieronymus 2016b; Josephson et al. 2016; https://www.brewersfriend.com/other/horseradish/.

Artemisia absinthium L.
ASTERACEAE

Synonyms *Absinthium officinale* Brot., *Absinthium vulgare* (L.) Lam., *Artemisia absinthia* St.-Lag., *Artemisia baldaccii* Degen, *Artemisia doonense* Royle, *Artemisia inodora* Mill., *Artemisia kulbadica* Boiss. & Buhse, *Artemisia pendula* Salisb., *Artemisia rehan* Chiov., *Artemisia rhaetica* Brügger.

Common Names Wormwood.

Description Bush (4–12 dm), white-tinted with hair, fragrant vermouth smell, bitter taste. Stalk erect, woody at the base. Basal leaves with 5–10 cm petiole, 3-pennatosette with 3–4 mm wide segments, rounded at the apex; minor cauline leaves (3–4 cm), subsessile. Inflorescences capitula 30–60, hemispherical (3 × 3 mm) in leaf terminal panicle, ± unilateral. Flowers yellowish-brown (2 mm).

Cultivar *Artemisia absinthium* is available in a variety of cultivated forms: Annua, Chinese Absinthe, Lambrook Silver, Major, Shrub, and probably others.

Related Species The genus *Artemisia* includes about 530 species of which some are of alimentary use and those mentioned here can be used in making beer.

Geographic Distribution E Mediterranean, then subcosmopolitan (origin not yet identified, perhaps

Near East). **HABITAT** Dry fallow, hedges, walls. **RANGE** Cultivated everywhere.

BEER-MAKING PARTS (AND POSSIBLE USES) Leaves, flowers.

TOXICITY The monoterpene thujone and its metabolites, as well as distillates with high alcohol content (50–70% ABV), widely consumed in France until the last century, have long been banned everywhere because they are considered harmful to health. People dependent on this drink develop absinthism, a form of alcoholism. The thujone contained in *Artemisia absinthium* essential oil at high doses (80–100 g) can cause epileptic seizures, delirium tremens, and death. However, such doses are practically impossible to consume, as absinthe normally contains 30–40 mg of thujone per kg. In addition, the application of the correct manufacturing technique results in the loss of most of the thujone in two distinct moments of processing: the drying of *Artemisia absinthium*, which involves the loss of 70–80% of the highly volatile thujone, and the "head cut" of the distillate. Anethole and phenethol, substances no less toxic than thujone, and also considered dangerous in the past, are present in plants of daily use (parsley, laurel, rosemary, nutmeg, etc.) and in various alcoholic beverages. Thujone, in contrast, was exclusive to beverages such as absinthe, vermouth, and genepì (though the last two were never criminalized as absinthe was). Recent studies show that in correctly distilled absinthe, even those made with traditional recipes and procedures, only a minimum quantity of thujone remains. The European Union and the United States have removed the ban on the production of *A. absinthium*–based beverages.

CHEMISTRY Flowering leaves, stems, and tops contain active ingredients such as thujone, absinthin, anabsinthin, artabasin, anabasine, and anabasinine.

STYLE American IPA, American Pale Ale, American Wheat or Rye Beer, Belgian Dark Strong Ale, Brown Porter, California Common Beer, English Barleywine, Extra Special/Strong Bitter (ESB), Gruit, Holiday/Winter Special Spiced Beer, Mild, Old Ale, Roggenbier (German Rye Beer), Saison, Specialty IPA (Belgian IPA, Rye IPA), Spice/Herb/Vegetable Beer, Wee Heavy, Witbier, Wood-Aged Beer.

BEER Beermouth, Baladin (Piozzo, Piedmont, Italy); The Last Word, Perennial Artisan Ale (Missouri, United States).

SOURCE Hieronymus 2012; Hieronymus 2016b; Josephson et al. 2016; Steele 2012; https://www.brewersfriend.com/other/malurt/; https://www.brewersfriend.com/other/wormwod/; https://www.gruitale.com/bot_wormwood.htm.

Artemisia absinthium is the main aromatic essence of vermouth. This spiced wine, created at the end of the eighteenth century in Piedmont, Italy, may in fact contain various herbs depending on the house of production, but one is essential: absinth. When Teo Musso thought of creating a beer-based vermouth–Beermouth–he created the recipe from this plant with such an intense and recognizable aromatic contribution. Alongside *Artemisia absinthium*, the beermouth recipe includes 12 other botanicals whose essences are extracted using modern distillation systems that use ultrasound or vacuum to keep the fragrances as intact as possible.

Artemisia douglasiana Besser ex Besser
ASTERACEAE

SYNONYMS *Artemisia heterophylla* Besser.

COMMON NAMES California mugwort, Douglas's sagewort, dream plant.

DESCRIPTION Herbaceous perennial (0.5–2.5 m). Stems erect, develop from strong rhizomes. Leaves gray-green, spaced, elliptical, 3- to 5-lobed at the end (lobes ± defined). Inflorescences with bell-shaped capitula carrying 5–9 pistillate flowers and 6–25 tubular flowers.

RELATED SPECIES The genus *Artemisia* includes about 530 species of which some are of alimentary use and those mentioned here can be used in making beer.

GEOGRAPHIC DISTRIBUTION W United States. **HABITAT** Pitches, riverbanks. **RANGE** Cultivated in the area of origin for phytotherapeutic uses and environmental reconstruction.

BEER-MAKING PARTS (AND POSSIBLE USES) Leaves.

CHEMISTRY Leaves contain thujone and cineol.

STYLE Gruit.

SOURCE https://honest-food.net/2016/05/05/gruit-herbal-beer/.

Artemisia dracunculus L.
ASTERACEAE

SYNONYMS *Achillea dracunculus* Hort. ex Steud., *Artemisia aromatica* A.Nelson, *Artemisia cernua* Nutt., *Artemisia changaica* Krasch., *Artemisia dracunculoi des* Pursh, *Artemisia glauca* Pall. ex Willd., *Artemisia inodora* Hook. & Arn., *Artemisia inodora* Willd., *Artemisia nuttalliana* Besser, *Artemisia redowskyi* Ledeb., *Draconia dracunculus* (L.) Soják, *Oligosporus dracunculiformis* (Krasch.) Poljakov, *Oligosporus dracunculus* (L.) Poljakov, *Oligosporus glaucus* (Pall. ex Willd.) Poljakov.

COMMON NAMES Tarragon, estragon.

DESCRIPTION Glabrous bush (6–15 dm). Rhizome often used for multiplication. Branched stem, slender branches, erect. Leaves glossy green, margin entire, the lower ones trifid, the others linear-lanceolate (2–10 × 20–80 mm), glossy. Inflorescences hanging capitula (Ø 2–4 mm), each containing up to 40 small yellow or greenish flowers, the outer ones pistilled and fertile, the inner ones sterile; capitula grouped in leafy pyramidal panicle. Achene seeds (1.5 mm) sterile in some cultivars, fertile in others.

SUBSPECIES Besides the typical variety (var. *dracunculus*), *Artemisia dracunculus* L. var. *glauca* (Pall. ex Willd.) H.M.Hall & Clem. is known.

CULTIVAR There are several chemotypes of this species.

RELATED SPECIES The genus *Artemisia* includes about 530 species of which many are of alimentary use and those mentioned here, like *Artemisia pontica* L., can be used in making beer.

GEOGRAPHIC DISTRIBUTION Sarmatic-Siberian. **HABITAT** Cultivated and subspontaneous in gardens. **RANGE** Cultivated.

BEER-MAKING PARTS (AND POSSIBLE USES) Leaves.

TOXICITY The potential carcinogenicity of estragole and methyleugenol has led to restrictions on the use of herbs containing these components, including *Artemisia dracunculus*.

CHEMISTRY Pentahydroxy-methoxy-methyl-flavone, estragoniside, glycopyranoside dihydroxy-pyranoside, pinocembrin, glucopyranoside, luteolin, quercetin, rutin, kaempferol, naringenin, 5,7-dihydroxypyranoside, naringenin, trihydroxy-methoxy-flavanone, trihydroxy-dimethoxy-flavanone, dihydroxy-methoxydihydrocalcone, davidigenin, sakuranetin, chicoric acid, hydroxybenzoic acid, hydroxy-methoxycinnamic acid, chlorogenic acid, caffeic acid, caffeoylquinic acid, herniarin, methoxy-dihydroartemisinin, epoxy-artemidine, dracumerine, hydroxyacetamidine, isovalerate capillarin, methylenedioxy-methoxycoumarin, di-methylenedioxy-methoxycoumarin, dimethyl allyl ether of esculetin, scopoletin, scoparone, daphnetin-methyl-ether, daphnetin methyl ether, artemidiol, hydroxycoumarin, artemidine, artemidinal, artemidinol, aesculine, capillarin, hydroxic-pillarin, hydroxyartemidine, dimethoxy capillaris in, neopellitorine A, neopellitorine B.

Style American Pale Ale, American Wheat or Rye Beer, Baltic Porter, Belgian Tripel, Bière de Garde, Brown Porter, English Barleywine, Experimental Beer, Gueuze, Mild, Saison, Specialty Beer, Spice/Herb/Vegetable Beer, Witbier.

Beer Drago'nCella, Vale La Pena (Rome, Lazio, Italy).

Source Heilshorn 2017; https://www.brewersfriend.com/other/estragon/; https://www.brewersfriend.com/other/tarragon/.

Artemisia genipi Weber ex Stechm.
ASTERACEAE

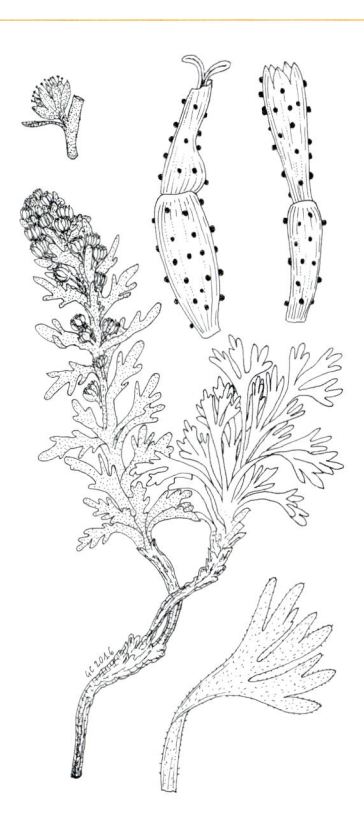

Synonyms *Absinthium tanacetifolium* (L.) Gaertn., *Artemisia bocconei* All., *Artemisia macrophylla* Fisch. ex Besser, *Artemisia mertensiana* Wallr., *Artemisia mirabilis* Rouy, *Artemisia orthobotrys* Kitag., *Artemisia racemosa* Miégev., *Artemisia rupestris* Vill., *Artemisia serreana* Pamp., *Artemisia spicata* (Baumg.) Wulfen ex Jacq., *Artemisia sylvatica* Ledeb., *Artemisia tanacetifolia* All.

Common Names Genepi.

Description Low shrub (5–12 cm), white and hairy, with aromatic smell. Woody stems at the base, with ascending branches, simple, axillary to rosettes. Basal leaves with petiole 10–25 mm and lamina 2–3 times triforked; terminal laciniae of 1.5 × 5 cm. Inflorescence capitulum ± sessile, ovoid (2 × 3 mm), with scales usually edged in black.

Subspecies At least two subspecies of this taxon are known: the typical one, *Artemisia genipi* Weber ex Stechm. subsp. *genipi*, usually silicicolous, and *Artemisia genipi* Weber ex Stechm. subsp. *foliosa* Giac. et Pign., calcicole.

Related Species The genus *Artemisia* includes about 530 species of which some are of alimentary use and those mentioned here can be used in making beer; very similar to *Artemsia genipi* are *Artemisia umbelliformis* Lam. and *Artemisia glacialis* L.

Geographic Distribution Endemic alpine species. **Habitat** Cliffs and gravel in alpine and nival areas. **Range** Alps (?).

Beer-Making Parts (and Possible Uses) Young apexes, leaves.

Chemistry Essential oil, leaves: α-pinene 1.2%, β-pinene 17.9%, p-cymene 3.1%, γ-terpinene 2.8%, α-thujone 26.0%, β-thujone 6.8%, terpinen-4-ol 12.2%, sabinil-isobutyrate 1.0%, sabinil-isovalerate 2.6%, sabinil n-valerate 1.5%, spatulenol 1.7%, valeranone 3.3%.

Style Herb Beer.

Beer GNP, La Bière du Grand Saint Bernard (Gignod, Aosta Valley, Italy).

Source https://www.giornaledellabirra.it/homebrewing/home-beer-aromatizzata-al-genepy/.

Artemisia tridentata Nutt.
ASTERACEAE

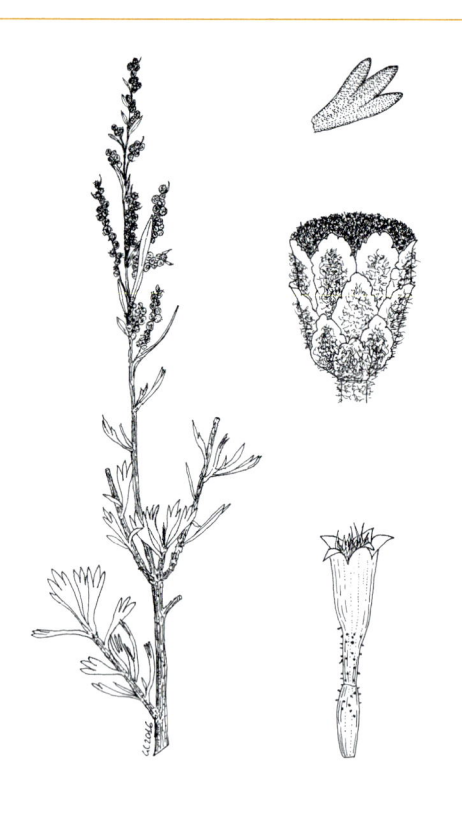

Synonyms *Artemisia angusta* Rydb., *Seriphidium tridentatum* (Nutt.) W.A.Weber.

Common Names Sagebrush, big sagebrush, Great Basin sagebrush.

Description Shrub (0.4–4.5 m), evergreen, long-lived (more than 100 years), erect, light gray color, with a strong and pungent odor, more intense when wet. Deep taproot (1–4 m) with superficial lateral roots. Stems branched, ascending, dark brown or black bark; distinct vegetative and floriferous branches. Leaves partially deciduous, wedge shaped, spatula-obovate or oblanceolate (1–6.5 × 0.2–2 cm), margin smooth, covered with fine silvery-gray hairs; the large and external ones usually end with 3 lobes (2–4), hence the specific epithet; apical internodes so short that the leaves often form dense rosettes at the ends of the branches. Flowers yellow in color, grouped in capitula and these in long clusters. Fruits edible achenes.

Subspecies Three subspecies are recognized: *Artemisia tridentata* Nutt. subsp. *tridentata*, *Artemisia tridentata* Nutt. subsp. *vaseyana* (Rydb.) Beetle, and *Artemisia tridentata* Nutt. subsp. *wyomingensis* Beetle & A.L.Young.

Cultivar There are at least a dozen.

Related Species The genus *Artemisia* includes about 530 species of which some are of alimentary use and those mentioned here can be used in making beer.

Geographic Distribution SW United States (Great Basin). **Habitat** Arid, semi-arid. **Range** SW United States.

Beer-Making Parts (and Possible Uses) Achenes.

Toxicity Highly allergenic by contact (dermatitis) especially in sensitive individuals; essential oils of the plant metabolized in the liver into toxic compounds causing blood clotting and formation of micro–blood clots in the liver and digestive tract; symptoms of intoxication usually appear 24 to 48 hours after ingestion.

Chemistry Essential oil: L-camphor 40%, pinene 20%, cineol 7%, methacrolein 5%, terpinene-D-camphor-esquiterpenoids 12%.

Style American Pale Ale, Pils, Specialty Beer.

Beer Sagebrush Classic Pils, Deschutes (Oregon, United States).

Source Heilshorn 2017; https://www.brewersfriend.com/other/sagebrush/.

Artemisia vulgaris L.
ASTERACEAE

Synonyms *Absinthium spicatum* (Wulfen ex Jacq.) Baumg., *Artemisia affinis* Hassk., *Artemisia apetala* hort.pest. ex Steud., *Artemisia cannabifolia* H.Lév., *Artemisia coarctata* Forselles, *Artemisia discolor* Douglas ex DC., *Artemisia eriophora* Ledeb., *Artemisia flodmanii* Rydb., *Artemisia glabrata* DC., *Artemisia heribaudii* (Sennen) Sennen, *Artemisia heyneana* Wall., *Artemisia hispanica* Stechm. ex Besser, *Artemisia javanica* Pamp., *Artemisia leptophylla* D.Don, *Artemisia leptostachya* D.Don, *Artemisia longiflora* Pamp., *Artemisia michauxii* Besser, *Artemisia officinalis* Gateau, *Artemisia opulenta* Pamp., *Artemisia paniculiformis* DC., *Artemisia parviflora* Wight, *Artemisia rubriflora* Turcz. ex Besser, *Artemisia ruderalis* Salisb., *Artemisia samamisica* Besser,

Artemisia selengensis Turcz. ex Besser, *Artemisia superba* Pamp., *Artemisia tongtchouanensis* H.Lév., *Artemisia violacea* Desf., *Artemisia virens* Moench, *Artemisia wallichiana* Besser.

Common Names Mugwort, common wormwood.

Description Herbaceous perennial (5–20 dm), faint vermouth smell. Stem erect, striped, very coppery, without stolons. Leaves pinnatisect, almost glabrous and dark above, white and woolly below, lower leaves (8–10 × 9–12 cm) with 2–4 dentate laciniae on each side, semi-clasping, with 4–6 mm wide blade in the apical portion, reduced to the rachis in the basal portion; upper leaves reduced and ± linear. Flowers tubular, reddish-yellow to brown (2–3 mm), with glandular corolla, grouped in ovoid subsessile capitula (1–1.5 × 3 mm) with an involucre consisting of grayish or lanate scales; capitula in turn forming a large, leafy pyramidal panicle. Fruit glabrous, oblong, pointed, achene smooth and without pappus.

Related Species The genus *Artemisia* includes about 530 species of which some are of alimentary use and those mentioned here can be used in making beer.

Geographic Distribution Circumboreal. **Habitat** Fallow, rubble; plant generally synanthropic. **Range** Circumboreal.

Beer-Making Parts (and Possible Uses)
Leaves, tops.

Chemistry Essential oil: hexane 17.619%, menthol 9.717%, β-eudesmol 8.297%, spatulenol 4.582%, p-menton 3.265%, 1,8-cineol 2.855%, β-pinene 2.122%, borneol 2.106%, trans-caryophyllene 2.097%, vulgarola 1.974%, camphano 1.856%, hedycaryol 1.512%, caryophyllene oxide 1.437%, ethanol 1.381%, germacrene D 1.309%, pinocarveol 1.212%, trans-β-terpineol 1.104%, unknown compound 1.020%, α-pinene 23.56%, α-campholene aldehyde 0.965%, cis-sabinene hydrate 0.792%, pulegone 0.664%, l-limonene 0.567%, camphene 0.552%, α-caryophyllene 0.433%, myrtenol 0.42%, δ-3-carene 0.418%, alcanfor 0.407%, veridiflorol 0.389%, trans-carveol 0.343%, cymene 0.294%, p-menthan-3-one 0.257%, aromadendrene 0.25%, p-menth-3-ene 0.247%, α-gurjunene 0.242%, inesol 0.233%, bicyclogermacrene 0.232%, α-copaene 0.2%, γ-terpinene 0.191%, 2,4-heptadiene 0.188%, piperitone 0.188%, verbenone 0.182%, menthofuran 0.181%, β-selinene 0.165%, 1-octen-3-ol 0.159%, decanal 0.146%, β-elemene 0.145%, β-farnesene 0.141%, endo-bornyl acetate 0.139%, α-guaiene 0.13%, 2-hexenal 0.128%, phenolic aldehyde 0.126%, 4-thujanol 0.123%, α-citral 0.107%, verbenene 0.077%, geranyl acetone 0.056%, β-o-cymene Y 0.048%, pentanal 0.029%, 2-methyl butanal 0.027%, octene 0.022%, hexanal 0.018%, methylcyclopentane 0.011%, 1-octene 0.011%.

Style American Amber Ale, American Brown Ale, American Pale Ale, American Stout, Baltic Porter, Belgian Specialty Ale, Belgian Tripel, Bière de Garde, Brown Porter, California Common Beer, English Barleywine, Experimental Beer, Extra Special/Stong Bitter (ESB), Flanders Brown Ale/Oud Bruin, Gruit, Holiday/Winter Special Spiced Beer, Irish Red Ale, Northern English Brown Ale, Robust Porter, Specialty Beer, Spice/Herb/Vegetable Beer, Strong Scotch Ale, Sweet Stout, Wee Heavy, Wild Specialty Beer, Witbier.

Beer Prove It Gruit, Spiteful Brewing (Illinois, United States).

Source Heilshorn 2017; Hieronymus 2016b; Markowski 2015; McGovern 2017; Steele 2012; https://www.actaplantarum.org/floraitaliae/mod_viewtopic.php?t=38475; https://www.brewersfriend.com/other/mugwort/; https://www.gruitale.com/bot_mugwort.htm.

Artocarpus heterophyllus Lam.
MORACEAE

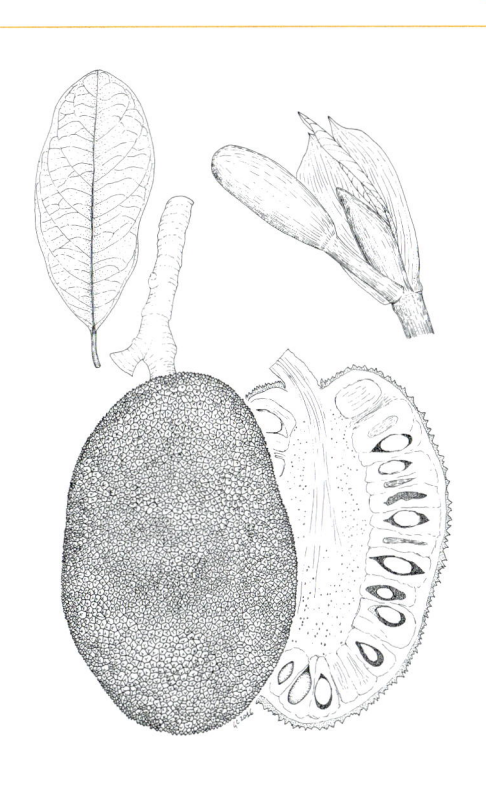

Artemisia vulgaris L.
ASTERACEAE

Synonyms *Artocarpus brasiliensis* Ortega, *Artocarpus maximus* Blanco, *Artocarpus philippensis* Lam.

Common Names Jackfruit.

Description Tree (8–25 m) cauliflorous, monoecious. Main root is a long taproot. Stem sturdy (Ø 30–80 cm), branched near the base, with dense domed or pyramidal foliage; bark rough, irregular, slightly scaly, gray-brown in color, inner one thick, ochre in color; all parts appear smooth, glabrous, or with tiny white hairs (0.5 mm), with brittle tips, which give the twigs and leaves a slightly rough appearance; if wounded, all live parts of the plant exude a copious white, rubbery sap. Leaves larger in the median or distal portion (4–25 × 2–12 cm), with nibs (5–12 pairs of veins), leathery, glossy, usually glabrous, dark green on the upper side, light green on the underside, alternate or spiral; those on floriferous branches obovate or oblong, on young branches oblong, narrow; margin entire at maturity, young ones 2- to 3-lobed; apex obtuse, short, and pointed; base wedge shaped or pointed; main veins light green–yellowish in color; at the knots, fused stipules around the stem leave a circular scar when they fall. Flowers on elongated axis forming racemoid inflorescences; male spikelets (3–10 × 1–5 cm) smooth, whitish-green, yellowish, rough when ripe, oblong, cylindrical, clavate, ellipsoid, distal ends with a 1.5–2.5 mm wide, slightly hairy ring; peduncle (0.4–0.5 × 1.5–3.5 cm) carrying many fertile flowers (tubular perianth, 2-lobed) or sterile flowers (full perianth, compact); female spikelets (5–15 cm) solitary or coupled, oblong or cylindrical, with wrinkled surface, light to dark green in color, 8–9 mm thick peduncle, and base with green ring 3–4 mm wide. Multiple fruit (20–100 × 15–50 cm), heavy (4.5–5 kg), oval, oblong or ellipsoid, light or dark green when young, greenish yellow-brown when ripe, formed by more than 500 achenes (syncarp), each indehiscent and monosperm; exterior covered by hard pyramidal thorns, pointed or blunt; true fruits are in the interior (4–11 × 2–4 cm; 6–53 g), consisting of the seed (2–4.5 × 1–3.7 cm; 2.5–14 g) and a fleshy aril, waxy, compact or soft, yellow-orange in color, sweet, aromatic (2–6.5 × 0.1–0.7 mm; 5–42 g); peduncle green (2–10 × 1–3.5 cm).

Cultivar Numerous.

Related Species The genus *Artocarpus* includes 68 species.

Geographic Distribution India.
Habitat Himalayan mountain slopes. **Range** SE Asia, N Australia, Brazil, tropical Africa.

Beer-Making Parts (and Possible Uses) Fruits.

Chemistry Energy 95 kcal, water 74%, carbohydrates 23%, protein 2%, lipids 1%, vitamin B6, vitamin C, potassium.

Style American IPA, American Light Lager, Blonde Ale, Fruit IPA.

Beer Jackfruit IPA, Ookapow Brewing Co. (Florida, United States).

Source https://www.brewersfriend.com/other/jackfruit/.

Asimina triloba (L.) Dunal
ANNONACEAE

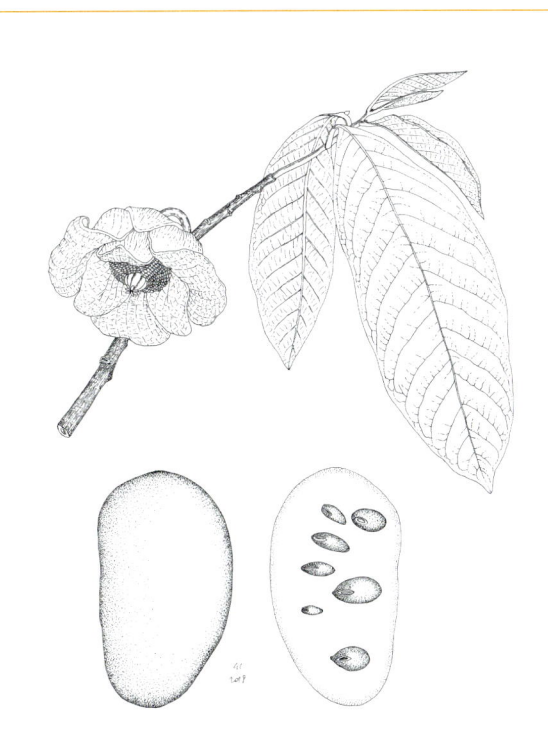

Synonyms *Annona pendula* Salisb., *Annona triloba* L., *Asimina campaniflora* Spach, *Asimina conoidea* Spach, *Asimina glabra* K.Koch, *Asimina virginiana* Poit. & Turpin, *Porcelia triloba* (L.) Pers., *Uvaria triloba* (L.) Torr. & A.Gray.

Common Names Pawpaw, dog banana, Indian banana.

Description Shrub or tree (1.5–14 m). Trunk slender (Ø up to 20–30 cm); bark not deeply furrowed in larger specimens, branches spreading and rising, slender; new sprouts moderately to copiously hairy, brown in color, hairless with age. Leaves with 5–10 mm petiole; lamina oblong-obovate to oblanceolate, 15–30 cm, membranous, strictly wedge-shaped base, margins poorly or not at all revolute, apex acute to sharp; underside densely hairy, then sparse on veins; upper side sparse appressed-pubescent on veins, then glabrous. Inflorescence develops on the bud of the previous year or during the emergence of new leaves; peduncle inclined, 1–2.5 cm, densely hairy, dark brown to red-brown hairs; bracts 1–2, basal, usually ovate-triangular, rarely more than 2–3 mm, hairy. Flowers brown, fetid, Ø 2–5 cm; triangular-deltate sepals, 8–12 mm, densely hairy underneath; outer petals curved outward, oblong-elliptical, 1.5–2.5 cm, underside puberulent on the veins; inner petals elliptical, 1/3–1/2 the length of the outer petals, base saccate, apex curved, glabrous lower surface, veins engraved on top, distinct area of corrugated nectars; pistils 3–12. Fruit berries (5–15 cm) yellow-green in color. Seeds (1.5–2.5 cm) brown to light brown.

Cultivar Some of the many cultivars of this species are 3–21, 10–35, Allegheny, Davis, G9-108, Jeremy's Gold, Mango, Mitchell, NC-1, Overleese, PA Golden, Potomac, Prima 1216, Prolific, Rappahannock, Rebecca's Gold, Shenandoah, Sunflower, Susquehanna, Taytwo, Wabash, Wells, and Wilson.

Related Species The genus *Asimina* includes 11 taxa.

Geographic Distribution CE United States, SE Canada (Ontario). **Habitat** Mesic woodlands, alluvial sites, fallow areas near fences (0–1,500 m ASL). **Range** Expanding cultivation in temperate zones on all continents.

Beer-Making Parts (and Possible Uses) Fruits (only edible part of the plant; everything else is poisonous, seeds included).

Toxicity All parts of the plant (except the pulp of ripe fruits) are toxic owing to the presence of acetogenins. Substances with anti-mitotic activity (inhibition of cell replication) are being studied for the treatment of tumors.

Chemistry Fruits (with peel): carbohydrates 18.8 g, fiber 2.6 g, lipids 1.2 g, protein 1.2 g, vitamin A 87 µg, thiamine (B1) 0.01 mg, riboflavin (B2) 0.09 mg, niacin (B3) 1.1 mg, vitamin C 18.3 mg, calcium 63 mg, iron 7 mg, magnesium 113 mg, manganese 2.6 mg, phosphorus 47 mg, potassium 345 mg, zinc 0.9 mg.

Style Belgian Abbey Ale, Fruit Beer, Wild Ale.

Beer Wild Paw Paw Ale, Fullsteam (North Carolina, United States).

Source Cantwell and Bouckaert 2016; Cantwell and Bouckaert 2018; Hieronymus 2016b; Josephson et al. 2016.

Aspalathus linearis (Burm.f.) R.Dahlgren
FABACEAE

Synonyms *Aspalathus contaminata* (Thunb.) Druce, *Aspalathus corymbosa* E.Mey., *Psoralea linearis* Burm. f.

Common Names Roiboos, red bush.

Description Shrub (up to 2 m) ± erect, very variable. Young stems often reddish in color. Leaves green, needle shaped (15–60 × 1 mm), sessile, in dense whorls. Flowers yellow, solitary or in dense groups at the tips of branches. Fruits small, lanceolate legumes, usually containing 1–2 hard seeds.

SUBSPECIES In addition to the typical subspecies, *Aspalathus linearis* (Burm.f.) R.Dahlgren subsp. *linearis*, *Aspalathus linearis* (Burm.f.) R.Dahlgren subsp. *pinifolia* (Marloth) R.Dahlgren is recognized as a taxon of a subspecific rank.

CULTIVAR This species has a fair variety of ecotypes.

RELATED SPECIES The genus *Aspalathus* includes more than 300 species.

GEOGRAPHIC DISTRIBUTION Cederberg (South Africa). **HABITAT** From Vanrhynsdorp in the north to Cape Peninsula and Betty's Bay in the south, on sandy soils in climates characterized by winter rainfall (300–350 mm/year), cold and humid winters, dry and hot summers. **RANGE** Species endemic to a small area of the west coast of the Western Cape Province in South Africa; the symbiotic relationship with local microorganisms is so strong that attempts at cultivation in the United States, Australia, and China have failed until recently.

BEER-MAKING PARTS (AND POSSIBLE USES) Chopped and oxidized leaves (process often referred to as fermentation, similar to what happens in tea production), then infused.

CHEMISTRY Fresh leaves have a high vitamin C content; oxidized leaves contain polyphenols (flavonols, flavones, dihydrochalcones, aspalathin, nothofagin), benzoic acid, cinnamic acid, and relatively few tannins (compared to black or green tea).

STYLE American Amber Ale, American IPA, American Pale Ale, American Wheat Beer, Blonde Ale, California Common Beer, Experimental Beer, Holiday/Winter Special Spiced Beer, Irish Red Ale, Specialty Fruit Beer, Spice/Herb/Vegetable Beer, Standard/Ordinary Bitter.

BEER Jackie O's Elaborative #5, Hill Farmstead (Vermont, United States).

SOURCE https://www.brewersfriend.com/other/rooibos/.

Asparagus officinalis L.
ASPARAGACEAE

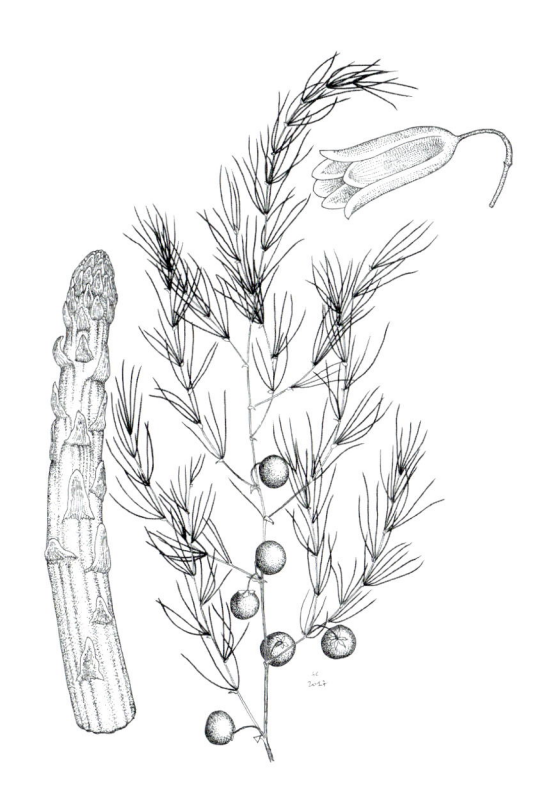

SYNONYMS *Asparagus altilis* (L.) Asch., *Asparagus caspius* Schult. & Schult.f., *Asparagus esculentus* Salisb., *Asparagus fiori* Sennen, *Asparagus hedecarpus* Andrews ex Baker, *Asparagus hortensis* Mill. ex Baker, *Asparagus littoralis* Steven, *Asparagus oxycarpus* Steven, *Asparagus paragus* Gueldenst. ex Ledeb., *Asparagus polyphyllus* Steven, *Asparagus polyphyllus* Steven ex Ledeb., *Asparagus sativus* Mill., *Asparagus setiformis* Krylov, *Asparagus vulgaris* Gueldenst. ex Ledeb.

COMMON NAMES Asparagus, garden asparagus.

DESCRIPTION Herbaceous perennial (up to 1.5 m). Rhizome horizontal, creeping, roots fasciculate. Stems green, erect, glabrous, branchy. Leaves replaced by silky-subulate cladodes (10–25 mm), in bundles (3–6); true leaves reduced to tiny scales (3 × 6 mm), membranous, triangular, acute, at the base of branches, slightly spurred at the base. Flowers hermaphrodite or unisexual (monoecious or dioecious plants), actinomorphic of small size, solitary or paired (rarely 3–4) at the base of the branches, bell-shaped perigonium, 6 whitish subpatent tepals striped with green on capillary peduncles, curved, and up to 1 cm long;

anthers (1.5 mm) about as long as the filament, usually with 6 stamens (replaced by staminodes in female flowers); ovary with 3 carpels, usually monolocular. Fruit spherical berry, globular, scarlet red color (Ø 11–14 mm). Seeds black (1–4). The edible part of the plant is the young spring bud (shoot).

Subspecies In addition to the typical subspecies *Asparagus officinalis* L. subsp. *officinalis*, there is another subspecies, *Asparagus officinalis* L. subsp. *prostratus* (Dumort.) Corb.

Cultivar There are hundreds. Some examples are Apollo, Atlas, Avalim, Cumulus, Ercole, Ginjlim, Guelph Millennium, H 666, HP 149, Jersey Deluxe, Jersey Gem, Jersey Giant, Jersey Supreme, Mo 2/12, Mondeo, NJ 857, NJ 931, NJ 953, Pacific 2000, Pacific Challenger, Ramires, Rapsody, UC 157, Victor, J855, and White Angel.

Related Species The genus *Asparagus* includes about 220 species, some of which are used for food purposes (e.g., *Asparagus acutifolius* L., *Asparagus albus* L., *Asparagus horridus* L.), but only *A. officinalis* is currently used for beer-making.

Geographic Distribution Euro-Mediterranean. **Habitat** Cultivated in vegetable gardens, prefers full sun but adapts well to half shade, grows well in both cold and heat, soil must be well drained with an acidity not exceeding 6. It can be naturalized in wet, sandy, marshy fallow soils, near rivers and/or forest edges (0–600 m ASL). **Range** China, Peru, Mexico, United States (California, Michigan, Washington), Germany.

Beer-Making Parts (and Possible Uses) Shoots.

Chemistry Shoots: energy 85 kJ (20 kcal), carbohydrates 3.88 g, sugars 1.88 g, fiber 2.1 g, lipids 0.12 g, proteins 2.2 g, vitamin A 38 µg, β-carotene 449 µg, lutein zeaxanthin 710 µg, thiamine (B1) 0.143 mg, riboflavin (B2) 0.141 mg, niacin (B3) 0.978 mg, pantothenic acid (B5) 0.274 mg, vitamin B6 0.091 mg, folate (B9) 52 µg, choline 16 mg, vitamin C 5.6 mg, vitamin E 1.1 mg, vitamin K 41.6 µg, calcium 24 mg, iron 2.14 mg, magnesium 14 mg, manganese 0.158 mg, phosphorus 52 mg, potassium 202 mg, sodium 2 mg, zinc 0.54 mg.

Style IPA.

Beer Asparagus IPA, Stone (California, United States).

Source https://thefullpint.com/four-brewers/four-brewers-asparagus-ipa-beancurdturtle-brewing/.

Atropa belladonna L.
SOLANACEAE

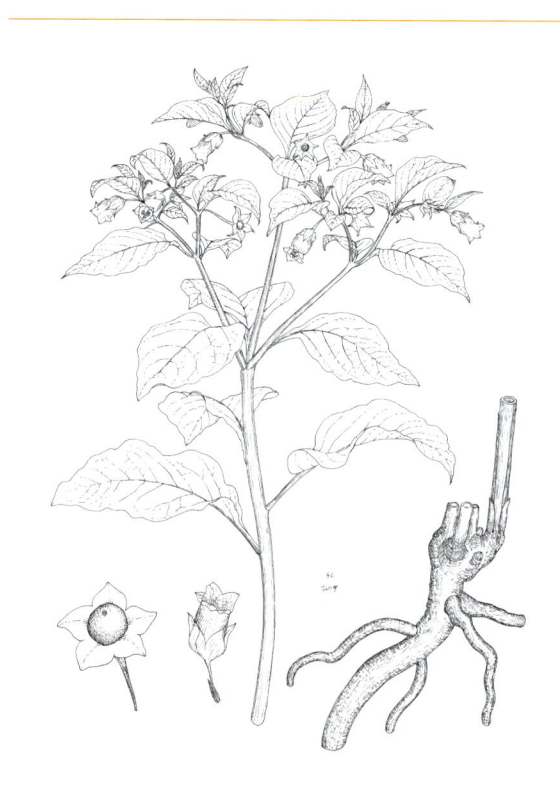

Synonyms *Atropa borealis* Kreyer ex Pascher, *Atropa cordata* Pascher, *Atropa digitaloides* Pascher, *Atropa lethalis* Salisb., *Atropa lutescens* Jacquem. ex C.B.Clarke, *Atropa mediterranea* Kreyer ex Pascher, *Belladonna baccifera* Lam., *Belladonna trichotoma* Scop., *Boberella belladonna* (L.) E.H.L.Krause.

Common Names Belladonna, deadly nightshade.

Description Herbaceous perennial (5–20 dm), fetid, viscous. Root enlarged. Stem glabrous with enlarged branches. Leaves alternate with 1–2 cm petiole and oval-lanceolate (5–9 × 10–15 cm), whole to irregularly crenate or lobed, ± rounded at the apex. Flowers 1–3 at the axilla of upper leaves, peduncle 10–18 mm, usually arched; calyx with tube 5–6 mm and teeth 4–5 mm; corolla 15–30 mm, violet-brown, yellowish inside with dark veins. Fruit glossy black berry, spherical (Ø 15–20 mm).

Subspecies In addition to the typical subspecies, *Atropa belladonna* L. subsp. *belladonna*, the subspecies *Atropa belladonna* L. subsp. *caucasia* (Kreyer) Avet. is known.

Related Species The genus *Atropa* includes 4 species: *Atropa acuminata* Royle ex Lindl., *Atropa baetica* Willk., *Atropa belladonna* L., and *Atropa pallidiflora* Schönb.-Tem.

Geographic Distribution Mediterranean mountain areas. **Habitat** Humid glades, coppices, woodland clearings. **Range** Mediterranean mountain areas.

Beer-Making Parts (and Possible Uses) The whole plant.

Toxicity The whole plant contains very toxic alkaloids and is subject to legal restrictions.

Chemistry Alkaloids: josciamine, atropine, scopolamine.

Style Findings in the Neolithic settlement of Skara Brae (Orkney Islands, Scotland, United Kingdom).

Source https://www.nationalgeographic.it/food/2015/01/09/news/short_history_of_drinks-2862950/.

Avena sativa L.
POACEAE

Synonyms *Avena algeriensis* Trab., *Avena distans* Schur, *Avena georgica* Zuccagni, *Avena glabrata* Hausm., *Avena grandis* Nevski, *Avena heteromalla* Haller, *Avena macrantha* (Hack.) Malzev, *Avena macrantha* (Hack.) Nevski, *Avena* × *mutata* Samp., *Avena mutica* Krock., *Avena nodipilosa* (Malzev) Malzev, *Avena orientalis* Schreb., *Avena polonica* Schwägr. ex Schmalh., *Avena praecocioides* Litv., *Avena praecoqua* Litv., *Avena praegravis* (E.L.Krause) Roshev., *Avena pseudosativa* (Thell.) Herter, *Avena racemosa* Thuill., *Avena rubra* Zuccagni, *Avena sexflora* Larrañaga, *Avena shatilowiana* Litv., *Avena tatarica* Ard., *Avena thellungii* Nevski, *Avena trabutiana* Thell., *Avena trisperma* Roem. & Schult., *Avena verna* Heuze.

Common Names Oat.

Description Herb (5–12 dm), annual. Culms cylindrical, erect, ascending, plant generally glaucous and glabrescent. Leaves 8–15 mm wide. Panicle rich and broad; spikelets (2–3 flowers) not articulated on the rachis and therefore persistent inside the glumes until the rachis breaks; glume 20–30 mm; lemma 20 mm, glabrous or with few basal hairs, with remains of 3–5 cm. Fruit caryopsis.

Subspecies In addition to the typical variety, the following subspecific taxa are recognized: *Avena sativa* L. var. *cinerea* (Körn.) Vascon., *Avena sativa* L. var. *leiantha* (Malzev) E. Morren, *Avena sativa* L. var. *subpilosa* (Thell.) E. Morren, and *Avena sativa* L. var. *subuniflora* (Thell.) Thell.

Cultivar There are numerous cultivars of *Avena sativa* around the world. North Carolina State University has released some, including Brooks, Caballo, Gerard 224, Gerard 229, NC09-4503N, Rodgers, Southern States 76–30, and SS 76–50.

Related Species The genus *Avena* includes about 35 species.

Geographic Distribution Cultivated plant. **Habitat** Fields, fallow land, ruderal areas. **Range** Cultivated in areas with a temperate climate with wet summers: Russia, Canada, Poland, Finland, Australia, United States, Spain, Sweden, Germany, United Kingdom.

Beer-Making Parts (and Possible Uses) Caryopsis, as it is or malted.

Chemistry Caryopsis: energy 1,628 kJ (389 kcal), carbohydrates 66.3 g, fiber 10.6 g, β-glucan (soluble fiber) 4 g, lipids 6.9 g, protein 16.9 g, thiamine (B1) 0.763 mg, riboflavin (B2) 0.139 mg, niacin (B3) 0.961 mg, pantothenic acid (B5) 1.349 mg, vitamin B6 0.12 mg, folate (B9) 56 µg, calcium 54 mg, iron 5 mg, magnesium 177 mg, manganese 4.9 mg, phosphorus 523 mg, potassium 429 mg, sodium 2 mg, zinc 4 mg.

Style American Amber Ale, American IPA, American Stout, Berliner Weisse, Bière de Garde, Blonde Ale, Dry Stout, Foreign Extra Stout, Mild, Oatmeal Stout, Oktoberfest/Märzen, Robust Porter, Russian Imperial Stout, Saison, Specialty Beer, Traditional Bock.

Beer English Oatmeal Stout, Samuel Smith (Tadcaster, United Kingdom).

Source Calagione et al. 2018; Dabove 2015; Fisher and Fisher 2017; Heilshorn 2017; Hieronymus 2010; Hieronymus 2016a; Hieronymus 2016b; Jackson 1991; Markowski 2015; McGovern 2017; Sparrow 2015; Steele 2012; https://www.bjcp.org (31A. Alternative Grain Beer); https://www.brewersfriend.com/other/oat/; https://www.brewersfriend.com/other/oatmeal/; https://www.brewersfriend.com/other/oats/; https://www.eticamente.net/24977/birra-quali-sostanze-sono-contenute-al-suo-interno.html; https://www.omnipollo.com/beer/fatamorgana/.

🍺 Oat has accompanied the history of beer at least since the sixteenth century, when this cereal was frequently used, although not always with positive results mainly owing to its bitterness, which brewers did not always appreciate. Its use became more widespread with the invention of oat malt by the Scottish brewer James Mclay in 1895. Today *Avena sativa* is often used in the production of British-style Stout. This ingredient gives the beer softness, creaminess, and structure.

Averrhoa carambola L.
OXALIDACEAE

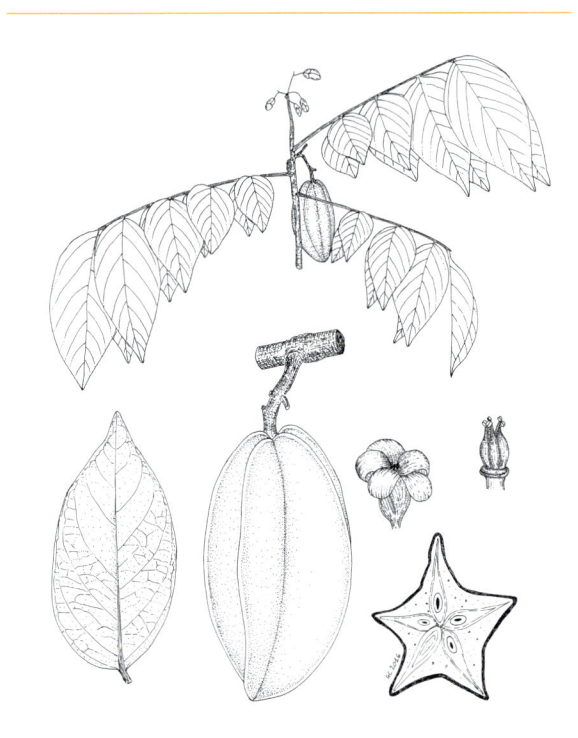

Synonyms *Averrhoa acutangula* Stokes, *Sarcotheca philippica* (Villar) Hallier f.

Common Names Starfruit.

Description Evergreen tree (5–12 m). Trunk short, branches drooping, forming a rounded crown. Leaves (15–20 cm) soft, alternate, compound, paripinnate, with 5–11 leaflets (4–9 cm) opposite and one terminal leaflet, ovate or ovate-oblong, upper side smooth and green, underside finely hairy and whitish. Flowers downy, bell shaped, perfect, at the axilla of the leaves at the end of the branches, gathered in loose branched panicles attached to the stem by a red peduncle, at the end of the small branches or on larger stems or on the trunk; single flower about 6 mm wide, with 5 petals with curved ends, pink, lilac, red, or streaked with violet, and whitish margins. Fruits oblong (6.35–15 × up to 9 cm), cross-section forming a 5- to 6-pointed star, thin, waxy, yellow-orange skin; juicy, crunchy, yellow when ripe, acid and sweet taste. Seeds (up to 12/fruit) flat (6–13 mm), dark in color; some cultivars produce seedless fruit.

Cultivar Numerous and far from equivalent in quality. Some are Arkin, B-2, B-10, B-16, B-17, Cheng Chui, Dah Pon, Demak, Erlin, Fwang Tung, Golden Star, Hew-1, Kajang, Kary, Lara, Maha, Miss, Newcomb, Pasi, Sri Kembangan, Star King, Team Ma, Thayer, Waiwei, and Wubento.

Related Species The genus *Averrhoa* includes 5 species: *Averrhoa bilimbi* L., *Averrhoa carambola* L., *Averrhoa dolichocarpa* Rugayah & Sunarti, *Averrhoa leucopetala* Rugayah & Sunarti, and *Averrhoa microphylla* Tardieu.

Geographic Distribution Indonesia, Malaysia. **Habitat** Hot and humid tropical climate, with reduced tolerance to low temperatures, in deep and well-drained but also sandy and clayey soils. **Range** In addition to the areas of origin, cultivated in Ghana, Guyana, French Polynesia, Guam, Hawaii, Australia, Brazil, Cambodia, China, Colombia, Dominican Republic, India, Israel, Japan, Laos, Myanmar, Philippines, Singapore, Taiwan, Thailand, Trinidad and Tobago, Uganda, United States, Vietnam.

Beer-Making Parts (and Possible Uses) Fruits.

Toxicity Fruit rich in oxalic acid that can be toxic in high concentrations. Some people on dialysis show signs of intoxication because of the neurotoxin caramboxin contained in the fruit. Normally neutralized in the kidneys, this toxin can be fatal in people with kidney problems.

Avena sativa L.
POACEAE

Chemistry Mature fruits: water 89.96 g, protein 0.45 g, lipids 0.32 g, fiber 0.96 g, sugars 5.60 g, pectin 1.02 g, starch 1.04 g, organic acids 0.36 g, ascorbic acid 23.4 mg, tannins 0.14 mg, ash 351 mg, iron 0.49 mg, calcium 4.83 mg, phosphorus 19.24 mg.

Style American Pale Ale, Berliner Weisse, Flanders Red Ale, Saison.

Beer Star Fruit Berliner Weisse, Funky Buddha (Florida, United States).

Source https://www.brewersfriend.com/other/starfruit/.

Bambusa vulgaris Schrad.
POACEAE

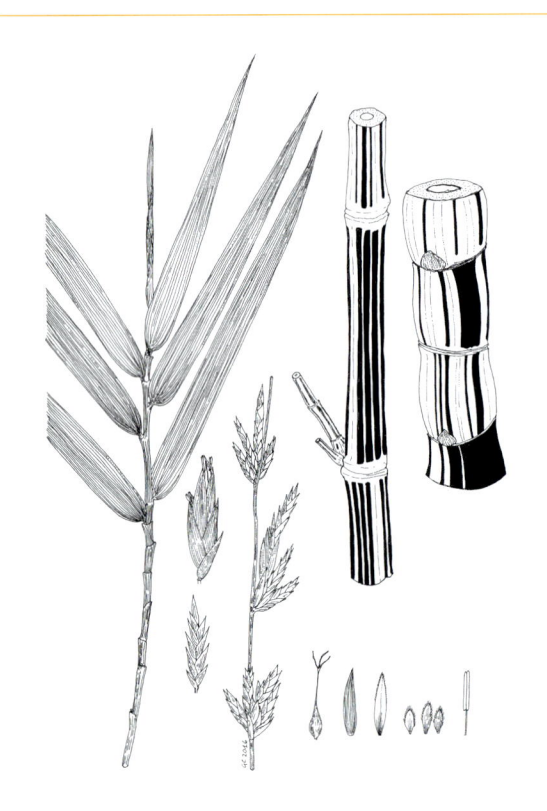

Synonyms *Arundarbor blancoi* (Steud.) Kuntze, *Arundarbor fera* (Oken) Kuntze, *Arundarbor monogyna* (Blanco) Kuntze, *Arundarbor striata* (Lindl.) Kuntze, *Arundo fera* Oken, *Bambusa auriculata* Kurz, *Bambusa blancoi* Steud., *Bambusa fera* (Oken) Miq., *Bambusa monogyna* Blanco, *Bambusa nguyenii* Ohrnb., *Bambusa sieberi* Griseb., *Bambusa striata* Lodd. ex Lindl., *Bambusa surinamensis* Rupr., *Bambusa thouarsii* Kunth, *Gigantochloa auriculata* (Kurz) Kurz, *Leleba vulgaris* (Schrad. ex J.C.Wendl.) Nakai, *Nastus thouarsii* (Kunth) Raspail, *Nastus viparus* Raspail, *Oxytenanthera auriculata* (Kurz) Prain, *Phyllostachys striata* (Lodd. ex Lindl.) Nakai.

Common Names Bamboo, common bamboo.

Description Herbaceous perennial (10–20 m), rhizomatous. Culm stems (Ø 4–10 cm) lemon yellow in color with green streaks, rigid, thick walled, slightly bulging knots, internodes (20–45 cm) straight or flexuous at the base, drooping at the end, branching from median knots upward. Leaves dark green, deciduous, narrowly lanceolate, with dense pubescence. Flowering uncommon; at intervals of several decades, the entire population blooms. Rare fruits and seeds because of poor pollen function caused by irregular meiosis. Vegetative propagation from rhizome that stretches underground and then lets new shoots emerge.

Cultivar There are three main groups within this taxon: bamboo with green stems; golden bamboo, with yellow stems and green streaks of different intensity (*Bambusa vulgaris* var. *striata*); and Buddha's belly bamboo, with green stems, about 3 m high, Ø 1–3 cm, internodes 4–10 cm long (*B. vulgaris* var. *wamin*). The most common varieties are *Bambusa vulgaris* Schrad. var. *aureovariegata* Beadle (the most popular), *Bambusa vulgaris* Schrad. var. *striata* (Lodd. ex Lindl.) Gamble, *Bambusa vulgaris* Schrad. f. *waminii* T.H.Wen, and *Bambusa vulgaris* Schrad. f. *vittata* (Rivière & C.Rivière) McClure, and also Kimmei, Maculata, and Wamin Striata.

Related Species The genus *Bambusa* includes 138 species.

Geographic Distribution S China, Asia. **Habitat** Riverbanks, roadsides, wastelands, open land, wetlands. **Range** Cultivated in areas with tropical and subtropical climates.

Beer-Making Parts (and Possible Uses)
Sprouts, culm wood (to smoke malt).

Toxicity Contains taxiphyllin, a cyanogenic glycoside extremely toxic to humans (50–60 mg lethal dose) because it acts as an enzyme inhibitor; degrades in boiling water.

Chemistry Energy 27 kcal, cellulose 41–44%, pentosanes 21–23%, lignin 26–28%, ash 1.7–1.9%, silica 0.6–0.7%, B vitamins (thiamine, riboflavin, niacin, vitamin B-6 or pyridoxine), pantothenic acid, manganese, copper, potassium 533 mg.

Style Ale, Premium Lager.

Beer Bambusa Cholula, Cholula de Rivadabia (Puebla, Mexico).

Source Daniels and Larson 2000; https://bamboo-beer.ca/index.html.

Beta vulgaris L. subsp. *vulgaris* convar. *vulgaris* var. *altissima*
AMARANTHACEAE

Synonyms *Beta vulgaris* L. subsp. *vulgaris* convar. *saccharifera* Alef.

Common Names Sugar beet.

Description Herb, biennial (up to 1 m). Taproot edible, white, thickened, conical or conical-elongated, often branched, with rough surface; thick white flesh. Basal leaves (20–30) forming a rosette, dark green, wavy lamina; petiole long, green; cauline leaves small, fleshy, alternating, whole, lanceolate, almost sessile, smooth or wavy. Stem herbaceous, erect, very branched, emerging in the plant's second year of life. Flowers hermaphrodite, pentamerous, with or without 3 bracts, small, green or whitish in color, single or in groups of 2–5 or more, which subsequently form glomerules of seeds united in a long infructescence; bracts linear-lanceolate; calyx perianth, pentamerous, concrescent at the base with semi-inferior gynoecium; stigma 2- to 3-lobate, almost sessile; stamens 5 on a fleshy disc around the gynoecium, filaments as long as perianth or shorter. Fruits covered by the withered and suberified remains of the perianth, which, concrescent with the gynaeceum, has a fleshy pericarp that gradually hardens; fruits collected in rather small seed glomerules, yellow, yellow-brown, or brown in color; at maturity easily divisible. Seeds flat, glabrous, with a rostrum, 2–8 seeds per glomerule (there are varieties with 1 seed per glomerule). One thousand glomerules weigh 18–50 g; 1,000 seeds weigh 1.5–6 g.

Subspecies According to some sources, this taxon includes three entities of subspecific rank: *Beta vulgaris* L. subsp. *adanensis* (Pamukç.) Ford-Lloyd & J.T.Williams, *Beta vulgaris* L. subsp. *vulgaris*, and *Beta vulgaris* L. var. *trojana* (Pamukç.) Ford-Lloyd & J.T.Williams.

Cultivar According to other sources, *Beta vulgaris* includes three subspecies: *Beta vulgaris* L. subsp. *adanensis* (Pamukç. ex Aellen) Ford-Lloyd & J.T.Williams (= *Beta adanensis* Pamukç. ex Aellen), found in the disturbed habitats and steppes of SE-Europe (Greece) and W-Asia (Cyprus, Israel, Syria, Turkey); *Beta vulgaris* L. subsp. *maritima* (L.) Thell., wild ancestor of all cultivated beets, present from the coasts of W-Europe and the Mediterranean Sea to the Middle East; and *Beta vulgaris* L. subsp. *vulgaris* [= *Beta vulgaris* L. subsp. *cicla* (L.) Arcang., *Beta vulgaris* L. subsp. *rapacea* (Koch) Döll]. All cultivated beets belong to this subspecies; they are divided into five groups: *Altissima*, sugar beet (= *Beta vulgaris* L. subsp. *vulgaris* convar. *vulgaris* var. *altissima*); *Cicla*, leaf beet (= *Beta vulgaris* L. subsp. *vulgaris* convar. *cicla* var. *cicla*); *Conditiva*, red turnip (= *Beta vulgaris* L. subsp. *vulgaris* convar. *vulgaris* var. *vulgaris*); *Crassa*, fodder beet (= *Beta vulgaris* L. subsp. *vulgaris* convar. *vulgaris* var. *crassa*); and *Flavescens*, beet (= *Beta vulgaris* L. subsp. *vulgaris* convar. *cicla* var. *flavescens*).

Related Species The genus *Beta* includes nine species: *Beta corolliflora* Zosimovic ex Buttler, *Beta lomatogona* Fisch. & C.A.Mey., *Beta macrocarpa* Guss., *Beta macrorhiza* Steven, *Beta nana* Boiss. & Heldr., *Beta palonga* R.K.Basu & K.K.Mukh., *Beta patula* Aiton, *Beta trigyna* Waldst. & Kit. and, of course, *Beta vulgaris* L.

Geographic Distribution Cultivated plant.
Habitat Exists in the wild but as a result of human selection. **Range** Russia, France, United States, Germany, Turkey, China, Ukraine, Poland, Egypt, United Kingdom.

Beer-Making Parts (and Possible Uses)
Taproots from which sucrose is extracted.

Chemistry Taproot: sucrose 17–19%.

Style Alternative Sugar Beer, American Amber Ale, American Barleywine, American Brown Ale, American IPA, American Light Lager, American Pale Ale, American Porter, American Stout, American Wheat or Rye Beer, Belgian Blond Ale, Belgian Dark Strong Ale, Belgian Dubbel, Belgian Golden Strong Ale, Belgian Pale Ale, Belgian Specialty Ale, Belgian Tripel, Berliner Weisse, Blonde Ale, Brown

Porter, California Common Beer, Cream Ale, Double IPA, Dry Stout, Dunkelweizen, Düsseldorf Altbier, English Barleywine, English IPA, Experimental Beer, Extra Special/Strong bitter (ESB), Foreign Extra Stout, Fruit Beer, Holiday/Winter Special Spiced Beer, Imperial IPA, Irish Red Ale, Light American Lager, Mild, Munich Dunkel, North German Altbier, Northern English Brown Ale, Old Ale, Premium American Lager, Roggenbier (German Rye Beer), Russian Imperial Stout, Saison, Scottish Export 80/-, Specialty Beer, Spice/Herb/Vegetable Beer, Trappist Single, Wheatwine, Witbier, Wood-Aged Beer.

Source Calagione et al. 2018; Hieronymus 2016a; Hieronymus 2016b; Markowski 2015; Steele 2012; https://www.brewersfriend.com/other/sugar/.

Beta vulgaris L. subsp. *vulgaris* convar. *vulgaris* var. *vulgaris*
AMARANTHACEAE

Synonyms *Beta vulgaris* L. subsp. *vulgaris* gr. *Conditiva*, *Beta vulgaris* L. var. *conditiva* Alef.

Common Names Beetroot.

Description Morphology similar to *Beta vulgaris* L. var. *altissima* but with taproot shape ± roundish and color yellowish-white to intense violet.

Subspecies Three entities of subspecific rank of this taxon are known: *Beta vulgaris* L. subsp. *adanensis* (Pamukç.) Ford-Lloyd & J.T.Williams, *Beta vulgaris* L. subsp. *vulgaris*, and *Beta vulgaris* L. var. *trojana* (Pamukç.) Ford-Lloyd & J.T.Williams.

Cultivar Among the most common are Albino, Bull's Blood, Chioggia, Crosby's Egyptian, Cylindra, Detroit Dark Red Medium Top, Early Wonder, Formanova, Golden Beet/Burpee's Golden, Perfected Detroit, Red Ace Hybrid, Ruby Queen, and Touchstone Gold.

Related Species The genus *Beta* includes 9 species: *Beta corolliflora* Zosimovic ex Buttler, *Beta lomatogona* Fisch. & C.A.Mey., *Beta macrocarpa* Guss., *Beta macrorhiza* Steven, *Beta nana* Boiss. & Heldr., *Beta palonga* R.K.Basu & K.K.Mukh., *Beta patula* Aiton, *Beta trigyna* Waldst. & Kit. and, of course, *Beta vulgaris* L.

Geographic Distribution Cultivated plant. **Habitat** It doesn't exist in the wild but is a result of human selection. **Range** NE Europe, Australia, India, United States, Hungary, Poland, Germany, Serbia.

Beer-Making Parts (and Possible Uses) Taproot, used as food or as a source of nonanthocyanic pigments of the betalaine group (betanine, isobetanine, probetanine, and neobetanine, collectively known as betacyanine); also a source of indicaxanthin and vulgaxanthin (yellow-orange pigments known as betaxanthin).

Toxicity The presence of oxalic acid can cause kidney problems.

Chemistry Energy 180 kJ (43 kcal), water 87.58g, carbohydrates 9.56 g, sugars 6.76 g, fiber 2.8 g, lipids 0.17 g, protein 1.61 g, beta-carotene 2–20 µg, thiamine (B1) 0.031 mg, riboflavin (B2) 0.04 mg, niacin (B3) 0.334 mg, pantothenic acid (B5) 0.155 mg, vitamin B6 0.067 mg, folate (B9) 109 µg, vitamin C 4.9 mg, calcium 16 mg, iron 0.8 mg, magnesium 23 mg, manganese 0.329 mg, phosphorus 40 mg, potassium 325 mg, sodium 78 mg, zinc 0.35 mg.

Style American Amber Ale, American IPA, American Wheat Beer, American Wheat or Rye Beer, Berliner Weisse, Bière de Garde, Blonde Ale, California Common Beer, English Barleywine, English IPA, Irish Red Ale, Mild, Saison, Specialty Beer, Specialty IPA (Red IPA), Spice/Herb/Vegetable Beer.

Beer Beet Weiss, Crane Brewing (Kansas, United States).

Source Heilshorn 2017; Josephson et al. 2016; https://www.brewersfriend.com/other/beetroot/; https://www.brewersfriend.com/other/beets/; https://www.brewersfriend.com/other/remolacha/.

Although appreciated for the color it gives to the beer, *Beta vulgaris* is a complicated ingredient. If its aesthetic contribution is objectively pleasing, the same cannot be said of its aromatic component, which gives earth, vegetal, and soil notes. Certainly, the practice of peeling the beet before use can lead to a better result, but as this is not always sufficient, it is necessary to precisely calibrate the quantity and the most suitable time to introduce the taproot pulp during the brewing phases.

Beta vulgaris L. subsp. *vulgaris* convar. *vulgaris* var. *altissima*
Beta vulgaris L. subsp. *vulgaris* convar. *vulgaris* var. *vulgaris*

Betula alleghaniensis Britton
BETULACEAE

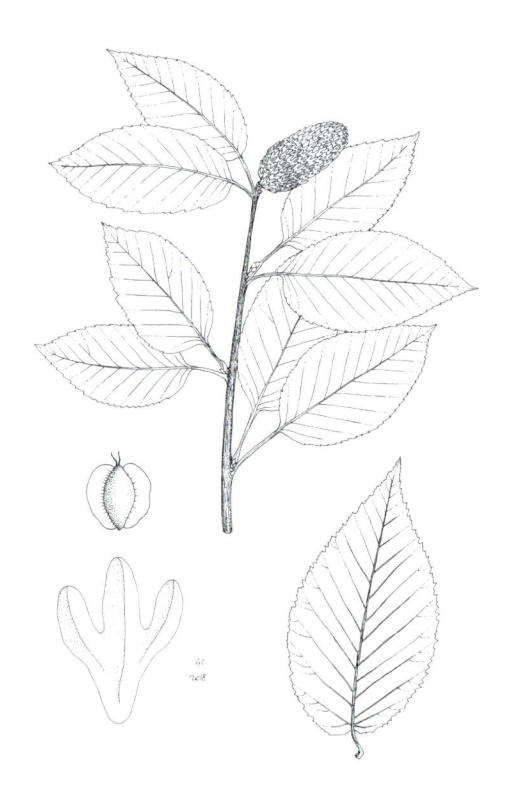

Synonyms None
Common Names Yellow birch.

Description Tree (up to 30 m). Trunk straight, foliage closely rounded; bark of young trunks and branches dark reddish-brown, mature bark brown-reddish, yellowish, or grayish, shiny, smooth, irregularly exfoliating or, sometimes, dark and closed; lenticels dark, expanded horizontally; small branches with smell and taste of *Gaultheria procumbens* if crushed, glabrous to sparsely pubescent, usually covered with small resin glands. Leaves with narrowly ovate to ovate-oblong lamina (6–10 × 3–5.5 cm) with 9–18 pairs of lateral veins, rounded to wedge or cordate base, margins doubly serrate, coarse, rather irregular teeth, sharp apex; underside mostly moderately pubescent, especially along the major veins and at the axilla of the veins, often with scattered, tiny resin glands. Infructescence erect, ovoid (1.5–3 × 1–2.5 cm), generally intact when the fruit is released; scales from sparsely to moderately pubescent, lobes diverging proximal to medial; central lobes tapered to a narrow tip, lateral lobes ascending or partially extended, wider, rounded. Fruit samara with wings narrower than the body, wider near the top, not or only slightly extended apically beyond the body.

Related Species The genus *Betula* includes about 135 taxa.

Geographic Distribution SE Canada, NE United States. **Habitat** Riverbanks, marshy forests, rich and humid forest slopes (0–500 m ASL). **Range** SE Canada, NE United States.

Beer-Making Parts (and Possible Uses) Young twigs, sap.

Chemistry Essential oil: methyl salicylate 97%. Abundant sap but less sugary than maple sap.

Style Birch Beer.

Source Fisher and Fisher 2017; Hieronymus 2016b. (Hieronymus mentions *Betula* species of interest in brewing including two European species naturalized in American territory, presumably *Betula pendula* and *Betula pubescens*, both mentioned in this book, and five native American species widely known among foragers but not better specified. Considering that *Flora of North America* mentions as many as 16 native species, not 5, as belonging to this genus, as a precautionary measure only a few North American *Betula* species, including this one, identified through other sources are mentioned here)

Betula lenta L.
BETULACEAE

Synonyms *Betula carpinifolia* Ehrh.

Common Names Sweet birch, cherry birch, black birch.

Description Tree (up to 20 m). Trunk tall, straight, crown narrow; trunk bark and mature branches light grayish-brown to dark brown or almost black, smooth, furrowed and fragmented into shallow scales with age; twigs with smell and taste of *Gaultheria procumbens* L. when crushed, glabrous to sparsely pubescent, usually covered with small resin glands. Leaves ovate to oblong-ovate (5–10 × 3–6 cm), with 12–18 pairs of lateral veins, rounded to cordate base, margins finely and sharply serrate or slightly doubly serrate, fine, sharp teeth, sharp apex; underside mostly glabrous, except scattered pubescence along main veins and axil of veins, often with scattered, tiny resin glands. Infructescence erect, ovoid to almost globular (1.5–4 × 1.5–2.5 cm), usually intact after fruit release; scales mostly glabrous, lobes diverging from proximal end to middle, central lobe short, wedged, lateral lobes extended to slightly ascending, longer and wider than central lobe. Fruit samara with wings narrower than the body, wider in the center, not extended apically beyond the body.

Subspecies The two subspecies of this taxon are the typical *Betula lenta* L. subsp. *lenta* and *Betula lenta* L. subsp. *uber* (Ashe) A.E.Murray.

Related Species The genus *Betula* includes about 135 taxa.

Geographic Distribution E Canada, E United States. **Habitat** Mesophilic, rich, humid forests, especially along protected slopes, but also in more exposed rocky sites (0–1,500 m ASL). **Range** E Canada, E United States.

Beer-Making Parts (and Possible Uses) Young twigs, sap.

Toxicity Very toxic orally; methyl salicylate absorbed by the skin can be fatal, even in small doses.

Chemistry Leaves (× 100 g dry weight): protein 28.1 g, lipids 8.6 g, carbohydrates 55.6 g, fiber 16.9 g, ashes 7.7 g, monotropitoside (salicylic acid primvercoside, gaultherin) 3%, essential oil 0.23–0.6% (97–99% of which is methyl salicylate). Buds: 4–6% essential oil containing betulin.

Style Birch Beer.

Source Fisher and Fisher 2017; Hieronymus 2016b (Hieronymus mentions *Betula* species of interest in brewing including two European *Betula* species naturalized in American territory, presumably *Betula pendula* and *Betula pubescens*, both mentioned in this book, and five native American species widely known among foragers but not better specified. Considering that *Flora of North America* mentions as many as 16 native species, not 5, as belonging to this genus, as a precautionary measure only a few North American *Betula* species, including this one, identified through other sources are mentioned here); Josephson et al. 2016 (this author generically uses the common name "birch" without further specifying the taxon).

Betula papyrifera Marshall
BETULACEAE

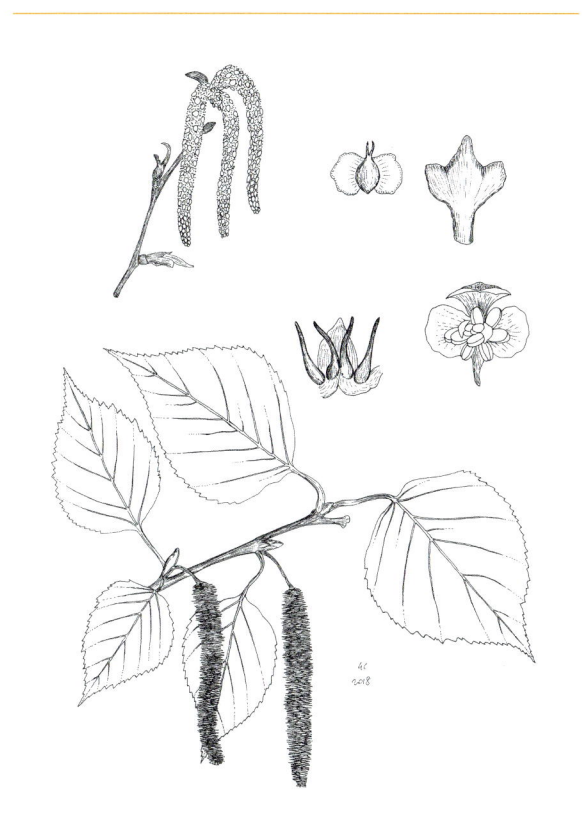

Synonyms *Betula excelsa* Aiton, *Betula grandis* Schrad., *Betula latifolia* Tausch, *Betula lyalliana* (Koehne) Bean, *Betula montanensis* Rydb. ex B.T.Butler, *Betula papyracea* Aiton, *Betula pirifolia* K.Koch, *Betula subcordata* Rydb. ex B.T.Butler.

Common Names Paper birch, canoe birch, white birch.

Description Tree (20 m, occasionally up to 30 m). Trunk single or sometimes multiple; mature canopy tightly rounded; bark of young stems reddish-dark brown, smooth, at maturity creamy to white-chalk or light to dark brown (rare), exfoliating in paper-like sheets; lenticels light, horizontal, dark and larger at maturity; twigs devoid of smell and taste of *Gaultheria procumbens* L., slightly to moderately pubescent, rarely scattered small resin glands. Leaf with ovate lamina (5–12 × 4–7 cm), with 9 fewer pairs of lateral veins, base rounded, wedged, or truncated, margins sharp to coarse or irregularly doubly serrate or dentate-serrated, apex sharp to short sharp; underside sparsely to moderately pubescent, often velvety along the major veins and at the axils of the veins, covered with tiny resin glands. Fruits pendulous, cylindrical (2.5–5 × 0.6–1.2 cm), swiftly crushed when the fruit is released; scales pubescent to glabrous, lobes diverging from the proximal end to the center, central lobe tightly elongated, obtuse, lateral lobes about as long as the central lobe but several times larger, strongly divergent, wide. Fruit samara with wings as wide as the body or just larger, extended apically just beyond the body.

Subspecies *Betula papyrifera* Marshall var. *cordifolia* (Regel) Regel is recognized in addition to the typical variety.

Related Species The genus *Betula* includes about 135 taxa.

Geographic Distribution N North America. **Habitat** Humid forests of the highlands, ± open, especially on rocky slopes, sometimes in marshy forests (300–900 m ASL). **Range** N North America.

Beer-Making Parts (and Possible Uses) Sap added to the wort to produce beer.

Chemistry No data available.

Style Birch Beer.

Source Fisher and Fisher 2017; Hieronymus 2016b. (Hieronymus mentions *Betula* species of interest in brewing including two European *Betula* species naturalized in American territory, presumably *Betula pendula* and *Betula pubescens*, both mentioned in this book, and five native American species widely known among foragers but not better specified. Considering that *Flora of North America* mentions as many as 16 native species, not 5, as belonging to this genus, as a precautionary measure only a few North American *Betula* species, including this one, identified through other sources are mentioned here.)

🍺 Birch Beer

Typical of Pennsylvania, birch beer is traditionally a low-alcohol drink (2–3%) produced from the bark of the younger branches of *Betula papyrifera*. During the Prohibition era, the bark was boiled in water for a few hours to release its essential oils and some sugar; the resulting liquid was fermented. Today the process is simplified, with the essential oils extracted directly.

Betula pendula Roth
BETULACEAE

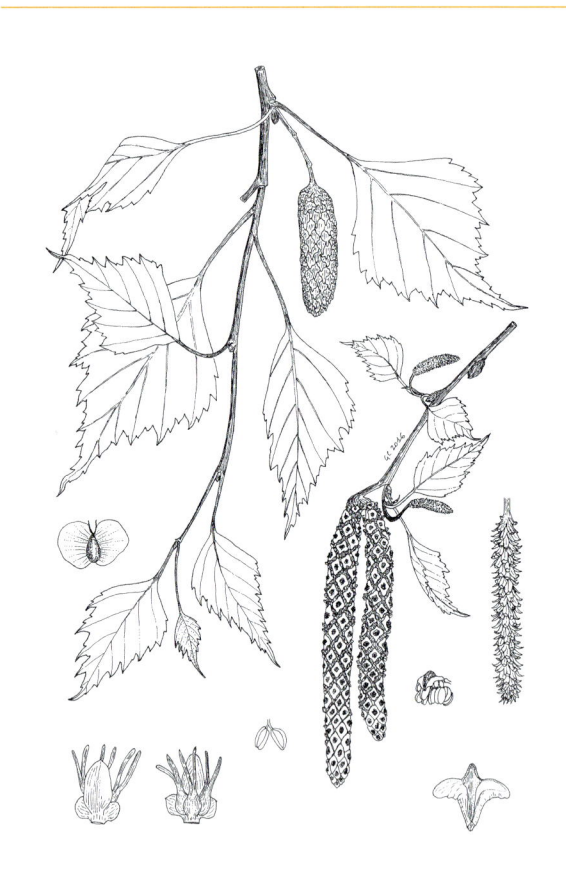

Synonyms *Betula brachylepis* V.N.Vassil., *Betula cajanderi* Sukaczev, *Betula demetrii* I.V.Vassil., *Betula ellipticifolia* V.N.Vassil., *Betula gummifera* Bertol., *Betula hippolyti* Sukaczev, *Betula hybrida* Blom,

Betula insularis V.N.Vassil., *Betula kossogolica* V.N. Vassil., *Betula lobulata* Kit., *Betula ludmilae* V.N. Vassil., *Betula microlepis* I.V.Vassil., *Betula mongolica* V.N.Vassil., *Betula montana* V.N.Vassil., *Betula palmata* Borkh., *Betula platyphylloides* V.N.Vassil., *Betula pseudopendula* V.N.Vassil., *Betula talassica* Poljakov, *Betula tiulinae* V.N.Vassil., *Betula transbaicalensis* V.N.Vassil., *Betula urticifolia* (Spach) Regel, *Betula verrucosa* Ehrh., *Betula virgultosa* Fr. ex Regel, *Betula vladimirii* V.N.Vassil.

COMMON NAMES Birch, silver birch.

DESCRIPTION Tree (25–30 m), not very long lasting, with expanded foliage, especially vertically. Main trunk erect and slender, ascending primary branches, young branches hanging to form a sparse and light crown; bark thin and smooth, first golden brown, then papyraceous-white, with large blackish horizontal slits toward the base, peels into thin silvery horizontal strips; young glabrous branches with abundant glandular-resinous warts. Leaves (up to 6 cm) deciduous, simple, alternate, petiolate, triangular to rhomboidal with sharp apex, widely wedged at the base, margin very thin, doubly dentate (primary teeth sharp and prominent), at first viscous and downy, then glabrous with a gray-green reflection below, intense green color above. Flowers unisexual, male flowers with brown bracts, bipartite stamens up to the base, yellow anthers, united in yellowish catkins (3–10 cm), sessile, pendulous; the female ones are light green with dark red stigmas, grouped in shorter catkins (1–2 cm), pedunculate, lateral, thin, first erect, then patent, finally hanging; female flowers wrapped by 3-lobed leathery scales, caducous together with the fruits, with lateral lobes folded down. Fruits small ovoid achenes (achene-cones), compressed, glabrous, with 2 small wings that facilitate their dispersion, enclosed in hanging cones that disintegrate when ripe.

SUBSPECIES In addition to the typical variety, *Betula pendula* Roth var. *fontqueri* (Rothm.) G.Moreno & Peinado and *Betula pendula* Roth var. *oycowiensis* (Besser) Dippel are also recognized.

RELATED SPECIES The genus *Betula* includes 135 taxa, 20 of which are probably of hybrid origin. Certainly, other species besides *B. pendula* are used in brewing (*Betula lenta* L., *Betula alleghaniensis* Britton, etc.).

GEOGRAPHIC DISTRIBUTION Euro-Siberian.
HABITAT Species heliophilous, hygrophilous, frugal, pioneering, consolidating in clearings and in denuded soils; it also forms pure woods, more often sporadic or in small groups in the sparse mountain woods of broad-leaved trees or conifers and in subalpine bushes. **RANGE** Euro-Siberian.

BEER-MAKING PARTS (AND POSSIBLE USES) Sap; syrup (similar use to *Acer* sp. syrup); vegetative apexes, twigs (to flavor; to filter the wort of Sahti, a traditional Finnish beer); wood for barrels (uncertain use for beers) and to smoke the malt.

CHEMISTRY Freshly extracted sap has a sugar content of 0.5–2%, depending on the *Betula* species, location, weather, and season; the syrup has a sugar content > 66% (fructose 42–54%, glucose 45%, sucrose, and traces of galactose).

WOOD Undifferentiated, whitish, of very fine texture and mostly straight (except in briarwood); modest shrinkage, easy to work but quickly altered by fungi (WSG 0.58–0.70). It can be used for furniture, small objects, carving work, and plywood.

STYLE Brown Porter, Holiday/Winter Special Spiced Beer, Mum, Sahti.

SOURCE Cantwell and Bouckaert 2016; Cantwell and Bouckaert 2018; Daniels and Larson 2000; Fisher and Fisher 2017; Heilshorn 2017; Hieronymus 2010; Hieronymus 2016b; McGovern 2017; https://www.brewersfriend.com/other/birch/.

Style Ancient Ale.

Beer Theobroma, Dogfish Head (Delaware, United States).

Source McGovern 2017; en.wikipedia.org/wiki/Dogfish_Head_Brewery.

Boesenbergia rotunda (L.) Mansf.
ZINGIBERACEAE

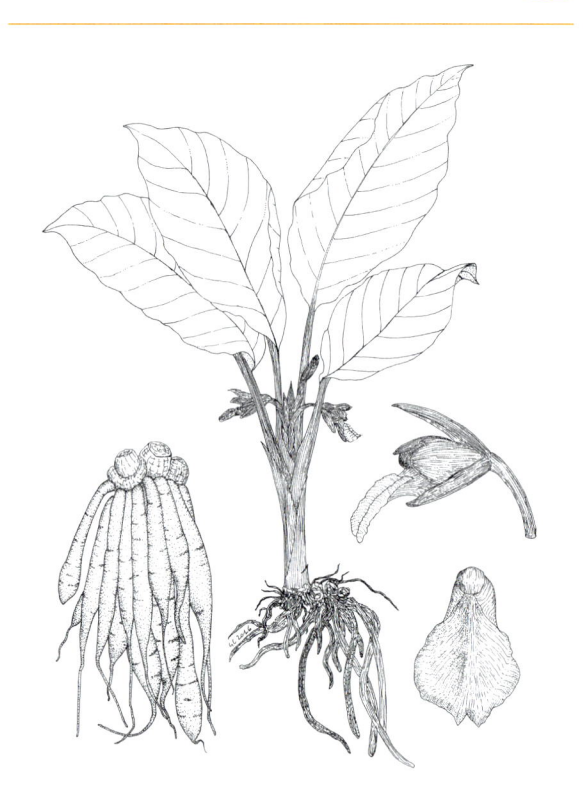

Synonyms *Boesenbergia cochinchinensis* (Gagnep.) Loes., *Boesenbergia pandurata* (Roxb.) Schltr., *Curcuma rotunda* L., *Gastrochilus panduratus* (Roxb.) Ridl., *Gastrochilus rotundus* (L.) Alston, *Kaempferia cochinchinensis* Gagnep., *Kaempferia ovata* Roscoe, *Kaempferia pandurata* Roxb.

Common Names Fingerroot, Chinese keys, Chinese ginger, lesser galangal.

Description Herb, perennial (15–50 cm), erect. Rhizome small, globular (Ø 1.5–2 cm), central, from which several long tubers (1–1.5 × 5–10 cm) branch off, bright yellow in color (varieties with red or black tubers also known), strongly aromatic, similar to the fingers of a hand. Leaves usually 3–5 to 7–9 × 10–20 cm, entire, elliptical to ovate-oblong, bright green color, red sheath. Flowers tubular, pink and red, aromatic, grouped in terminal inflorescences.

Cultivar Numerous cultivars of this species are known.

Related Species The genus *Boesenbergia* includes 69 species.

Geographic Distribution From S Yunnan (China) to W Malaysia. **Habitat** Dense forests. **Range** SE Asia.

Beer-Making Parts (and Possible Uses) Rhizomes.

Chemistry Active compounds: 1,3-tetradecadiene, 1,8-cineol, 2,4-dihydroxy-6-phenethylbenzoic acid methyl ester, 2,6-dihydroxy-4-methoxydihydrocalcone, 2′,4′-dihydroxy-3′-(-geranyl)-6′-methoxicalcone, 2′-hydroxy-4,4′,6′-trimethoxycalcone, 2′-hydroxy-4,4′,6′-dimethoxycalcone, 2-cyclohexylacetate, 2-isopropyl-4,5-dimethyl oxazole, 2-N-pyroyl-4,5-dimethylthiazole, 3,5,7,3′,4′-pentamethoxyflavone, 3,5,7,4′-tetramethoxyflavone, 3,5,7-trimethoxyflavone, 3′,4′-dimethoxytophenone, 3-carene, 4-hydroxypropanduratin A, 5,6-dehydrokawain, 5,7,4′-trimethoxyflavone, 5,7-dihydroxy-8-geranylflavanone, 5,7-dihydroxyflavanone, 5,7-dihydroxyflavone, 6-geranylpinocembrin, allo-ocimene, alpinetin, bicyclo (2.2.1) heptan-2-ol, boesenbergin A, boesenbergin B, borneol, caffeic acid, camphene, camphor, cardamonin, chlorogenic acid, cinnamyl cinnamate, cis-linalool oxide, citronellol, coumaric acid, cyclohexyl-n-propionate, dihydro-5,6-dehydrokawain, elemene, ethyl benzoate, farnesene, flavokawain C, geranial, geranyl benzoate, geranyl format, geranyl propionate, guaiacol, helichrysum, hesperidin, isoborneol, isopanduratin A, isopanduratin A1, isopanduratin A2, kaempferol, krachaizine A, krachaizine B, linalool, methyl 3-phenylpropionate, methyl cinnamate, methyl-n-nonanoate, myrcene, myristicin, naringin, n-butyl-n-pentadecanoate, n-eicosane, neral, nerolidol, neryl acetate, N-hexanal, N-hexyl angelate, nicolaioidesin B, ocimene, panduratin A, panduratin B1, panduratin B2, panduratin C, panduratin D, panduratin E, panduratin F, panduratin G, panduratin H, panduratin I, phellandrene, pinene, pinocembrin, pinocembrin chalcone, pinostrobin, pinostrobin chalcone, prenylcalcone, propiophenone, quercetin, rosafenone, rotundaflavone Ia, rotundaflavone Ib, rubranine, sabinene, sakuranetin, sesquiterpene hydrocarbons, tectochrysin, terpinen-4-ol, terpinene, terpineol, terpinolene, terpinyl valerate, thujaplicin, thujene, trans-2-hexanil-n-propionate, trans-caryophyllene, trans-geraniol, trans-ocimene, tricyclene, uvangoletin.

Style Robust Porter.

Source https://www.brewersfriend.com/other/krachai/.

Borago officinalis L.
BORAGINACEAE

Synonyms *Borago hortensis* L.

Common Names Borage, starflower, bee bread.

Description Herb, annual (up to 70 cm), entirely covered with long, white, subspiny bristles, which make it shaggy. Stems erect, branchy at the top, often veined red. Lower leaves, long petiolate, ovate-lanceolate lamina, toothed, wavy margin, veined; cauline lanceolate leaves, briefly petiolate or amplexicaul. Flowers pedunculate, pendulous in full bloom and ephemeral; calyx 5 narrow sepals, lanceolate, merged only at the base; corolla with short tube, azure-blue color, rarely white, 5-lobed, stamens 5, anthers deriving from the union of the stamens violet in color. Terminal inflorescences. Fruits tetrachene, light brown in color, oval, hard, containing several small seeds.

Related Species The genus *Borago* includes five species: *Borago longifolia* Poir., *Borago morisiana* Bigazzi & Ricceri, *Borago officinalis* L., *Borago pygmaea* (DC.) Chater & Greuter, and *Borago trabutii* Maire.

Geographic Distribution Euro-Mediterranean, Steno-Mediterranean. **Habitat** Synanthropic and ruderal; nitrophilic; humid, sandy, or clayey environments. **Range** Mediterranean.

Beer-Making Parts (and Possible Uses) Aerial parts (flowers, leaves).

Toxicity Use (alimentary and therapeutic) not recommended because of the presence of pyrrolizidine alkaloids in the aerial parts of the plant, which are hepatotoxic, genotoxic, and carcinogenic and not removed by cooking or fermentation.

Chemistry Mucilage, tannins, flavonoids, saponins, essential minerals (calcium, potassium), palmitic acid and essential fatty acids, omega-6.

Style Spice/Herb/Vegetable Beer.

Beer Four Flower, Beau's (Ontario, Canada).

Source Fisher and Fisher 2017; Hieronymus 2016b; https://www.brewersfriend.com/other/borage/.

Borassus flabellifer L.
ARECACEAE

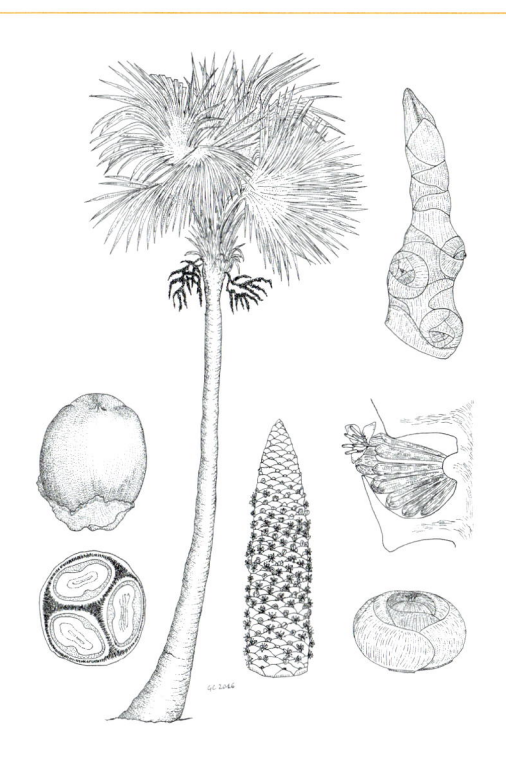

Synonyms *Borassus flabelliformis* L., *Borassus flabelliformis* Roxb., *Borassus sundaicus* Becc., *Borassus tunicatus* Lour., *Pholidocarpus tunicatus* (Lour.) H.Wendl., *Thrinax tunicata* (Lour.) Rollisson.

Common Names Palmyra palm, doub palm, tala palm, toddy palm, wine palm, sugar palm.

Description Evergreen tree (up to 30 m). Trunk grayish, sturdy (Ø up to 70 cm), covered with foliar scars; old leaves persistent for several years before finally falling off. Leaves bluish-green, costapalmate, up to about 3 m wide and cut for about half of their length into rigid segments; petiole sturdy, black at the base, about 1.3 cm long and irregularly toothed at the margins. Dioecious species (male and female flowers on different plants). Male flowers less than 1 cm long, forming semicircular groups, hidden under scaly bracts inside inflorescences similar to those of catkins; solitary female flowers, the size of a golf ball, placed along the axis of an inflorescence about 1 m long. Fruits fleshy (Ø 15–25 cm; about 2 kg), brown to black when ripe, sweet flesh, fibrous, translucent, pale white. Seeds 1–4 per fruit, each enclosed in a woody endocarp.

Related Species *Borassus aethiopum* Mart., *Borassus akeassii* Bayton, Ouédr. & Guinko, *Borassus heineanus* Becc., *Borassus madagascariensis* (Jum. & H.Perrier) Bojer ex Jum. & H.Perrier.

Geographic Distribution India, Bangladesh, Nepal, Sri Lanka, Myanmar, Cambodia, Vietnam, Laos, Thailand, Indonesia, Malaysia, Papua New Guinea, Philippines. **Habitat** Sparse forests, along coasts, along watercourses mainly in sandy soils. **Range** Beyond the original geographic distribution, naturalized in Pakistan, Socotra, China.

Beer-Making Parts (and Possible Uses) Sugar obtained from sap (at harvest the sap contains 14% sugar; concentrated in syrup through boiling; after cooling solidifies, forming lumps of sugar).

Chemistry Sap: protein 0.35 g, sugar 10.93 g, reduced sugar 0.96 g, ash 0.54 g, calcium traces, phosphorus 0.14 g, iron 0.4 g, vitamin C 13.25 mg, vitamin B1 3.9 mg. Sugar: energy 398 kcal, protein 0.24 g, lipids 0.37 g, minerals 0.50 g, carbohydrates 98.6 g, calcium 0.08 g, phosphorus 0.064 g, iron 30 mg, nicotinic acid 4.02 mg, riboflavin 229 mg.

Style Saison.

Source https://www.ratebeer.com/forums/lemon-grass-keffir-lime-and-galangal-ginger_150182.html.

Boswellia frereana Birdw.
BURSERACEAE

Synonyms None
Common Names Yagar, yagcar, maidi, meydi.

Description Tree (up to 8 m). Trunk swollen at the base and often disc shaped; pale yellow or brown bark with outer layers flaking and papery and an inner layer often reddish-brown; copious, milky, dried yellowish resin; strong branches. Leaves oblanceolate (10–30 cm), subglabrous, usually with tiny glandular hairs and a few longer nonglandular hairs on the rachis, petiole 1–5 cm, composed of 9–15 leaflets or occasionally simple; leaflets wavy, whole, or sinuous or with some irregular crenature on each side, truncated to cordate base, apex obtuse, generally ovate-orbicular (5 × 3 cm or rarely larger), less frequently elliptical-oblong, with a network of veins slightly prominent below. Inflorescence thyrsus (raceme like), 10–30 cm, glabrous to pubescent, half peduncle, lateral ramifications rarely more than 5 mm; peduncles 1–4 mm; bracts 1.5–6 mm, lower ones similar to leaves. Flower with calyx (about 2 mm) glabrous to pubescent; petals reddish or greenish red (3.5–5 × 1.5–2.5 mm); glabrous filaments (about 1.5 mm), linear but widened; plate-shaped disc, yellowish, greenish, or purplish color. Fruit with 5–8 pods (5.5–9 × 3–7 mm), pyriform, glabrous; trigonal kernel, narrow at both ends, upper part often slightly winged.

RELATED SPECIES The genus *Boswellia* includes 28 species.

GEOGRAPHIC DISTRIBUTION N Somalia. **HABITAT** Rocky slopes and gorges (0–1,000 m ASL), on cliffs and limestone ridges; characteristic of the species is the basal part of the trunk, which widens like a sucker sticking firmly to the rock, even on almost vertical walls; often this basal part appears whitish because it is covered by the excess of resin running along the trunk. **RANGE** Somalia, Yemen (?).

BEER-MAKING PARTS (AND POSSIBLE USES) Resin. (Incense is a gum resin that exudes from the bark of *Boswellia* plants. Harvested by producing oval decortications on the branches with a tool called a *menghaf*, a kind of chisel sharp on one side to debark the branches and dull on the other to collect the resin. The species that produce incense of the best quality are *Boswellia frereana*, *Boswellia papyrifera*, and *Boswellia sacra*. Resin is selected in 4–5 degrees of quality depending on the size of the dried grains, color, and purity. Quality also depends on the harvest period and the environment where the plants grow.)

CHEMISTRY Essential oil: α-thujene 8.1%, α-pinene 38%, p-cymene 11%, limonene 2.4%, sabinene 2.6%, trans-verbenol 4.2%, bornyl acetate 2.8%, isomers of α-phellandrene dimers.

STYLE Spice Beer.

SOURCE Dabove 2015.

Brassica napus L. var. *napobrassica* (L.) Rchb.
BRASSICACEAE

SYNONYMS *Brassica napus* L. subsp. *rapifera* Metzg.

COMMON NAMES Rutabaga, swede, Swedish turnip.

DESCRIPTION Herb, biennial (30–70 cm). Taproot enlarged, in longitudinal section, elliptical, circular, obovate, square or oblong, whitish-yellow to orange-reddish color, upper part green, bronze, or purple; flesh white to yellow in color, compact, fine grained. Pseudo-stem short to long, purple or light purple color between the leaf scars. Leaves with lamina entire to lobed (lobed leaf variety with few to many lobes and terminal lobe from short to long), bluish-green color, smooth, waxiness of variable intensity, petiole thin to thick, erect to patent. Flowers with or without pollen production.

SUBSPECIES This species has a complex taxonomic–nomenclature history and was first reported by the Swedish botanist Gaspard Bauhin in 1620. It was validly described as *Brassica oleracea* var. *napobrassica* by Linnaeus in 1753; later it was moved to other taxa as a variety, subspecies, or species (*Brassica napobrassica* Mill.). With a chromosomal number of

2n = 38, this taxon is the result of a cross of *Brassica rapa* × *oleracea*. Its number of chromosomes was later duplicated, thus becoming an allopolyploid.

Cultivar There are numerous cultivars of rutabaga. Those listed here constitute the type for the main characteristics used to describe the cultivars themselves: Acme, Airlie, Angus, Aubigny Green Top, Blanc Hors Terre, Brora, Champion, Doon Major, Dryden, Excelsior, Frise Gele, Harrietfield, Heinkenborsteler, Helena, Jaune à Collet Rouge, Jaune à Collet Verte, Joan, Kenmore, Laurentian, Lizzy, Magres, Marian, Melfort, Mella, Merrick, Niko, Ruby, Ruta Otofte, Sator Otofte, Seedfelder, Teviotdale, Tweed, Vittoria, Vogesa, and Wilhelmsburger.

Related Species The genus *Brassica* has about 53 taxa of which several are edible and some, listed here, are used in making beer.

Geographic Distribution Scandinavia, Russia. **Habitat** Cultivated plant derived from human-made selection and crossbreeding, not naturally occurring in the wild. **Range** Cultivated in temperate regions.

Beer-Making Parts (and Possible Uses) Taproots.

Chemistry Energy 157 kJ (38 kcal), carbohydrates 8.62 g, sugars 4.46 g, fiber 2.3 g, lipids 0.16 g, protein 1.08 g, thiamine (B1) 0.09 mg, riboflavin (B2) 0.04 mg, niacin (B3) 0.7 mg, pantothenic acid (B5) 0.16 mg, vitamin B6 0.1 mg, folate (B9) 21 µg, vitamin C 25 mg, calcium 43 mg, iron 0.44 mg, magnesium 20 mg, manganese 0.131 mg, phosphorus 53 mg, potassium 305 mg, zinc 0.24 mg.

Style Saison.

Source https://www.brewersfriend.com/other/rutabagas/.

Brassica nigra (L.) K.Koch
BRASSICACEAE

Synonyms *Brassica brachycarpa* P.Candargy, *Brassica bracteolata* Fisch. & C.A.Mey., *Brassica persoonii* Rouy & Foucaud, *Brassica sinapioides* Roth, *Brassica sinapioides* Roth ex Mert. & W.D.J.Koch, *Brassica sinapis* Noulet, *Crucifera sinapis* (L.) E.H.L.Krause, *Melanosinapis communis* K.F.Schimp. & Spenn., *Melanosinapis nigra* (L.) Calest., *Mutarda nigra* (L.) Bernh., *Raphanus sinapis-officinalis* Crantz, *Sinapis erysimoides* Roxb., *Sinapis nigra* L., *Sinapis tetraedra* J.Presl & C.Presl, *Sinapis torulosa* Pers., *Sisymbrium nigrum* (L.) Prantl.

Common Names Black mustard.

Description Herb, annual (3–15 dm). Stem erect, ramose, shaggy. Leaves shaggy, rarely glabrous, lower ones dark green, lyrate-pinnatisect (up to 5 × 15 cm) with 1–3 pairs of lateral segments; leaves cauline, ± whole. Inflorescences elongated racemes. Flowers with erect-patent sepals (3–4.5 mm), yellow petals (2–2.5 × 7–9 mm). Fruit siliqua (1.5–2 × 15–20 mm), cross-section ± rectangular, erect, affixed to the axis; peduncle 2–3 mm, beak 2 mm.

Cultivar The cultivars of this species are Barn, California, English, and Trieste.

Related Species The genus *Brassica* has about 53 taxa of which several are edible and some, mentioned here, are used in making beer.

Geographic Distribution Mediterranean (?).
Habitat Grain fields, fallow, farmyards.
Range Plant cultivated in temperate climates.

Beer-Making Parts (and Possible Uses) Seeds.

Toxicity Seeds contain an alkaloid (sinapine) and a glucoside (sinigrin) that may be toxic.

Chemistry Leaves (100 g): energy 31 kcal, water 89.5 g, protein 3.0 g, lipids 0.5 g, carbohydrates 5.6 g, fiber 1.1 g, ash 1.4 g, calcium 183 mg, phosphorus 50 mg, iron 3.0 mg, sodium 32 mg, potassium 377 mg, b-carotene 4200 mg, thiamine 0.11 mg, riboflavin 0.22 mg, niacin 0.8 mg, ascorbic acid 97 mg. Seeds (100 g): water 7.6 g, protein 29.1 g, lipids 28.2 g, carbohydrates 30.2 g, fiber 11.0 g, ash 0.5 g. Essential oil: Di-(9-octadecenoyl)-glycerol 42.16%, (Z,Z)-9,12-octadecadienoyl chloride 41.4%, hexadecanoic acid, 1-(hydroxymethyl)-1,2-ethanedyl ester 13.2%.

Style Spice/Herb/Vegetable Beer.

Source Heilshorn 2017; https://www.brewersfriend.com/other/sinapinsiemen/.

Brassica oleracea L. subsp. *oleracea* convar. *acephala* DC. cv. *sabellica* L.
BRASSICACEAE

Synonyms None
Common Names Borecole, kale.

Description Herb, annual (3–7 dm). Leaves numerous, lanceolate (up to 5–8 dm), rippled surface for emerging blisters between the ribs of all orders, margin revolute, bluish-green color. Apical bud absent.

Related Species The genus *Brassica* has about 53 taxa of which several are edible and some, mentioned here, are used in making beer.

Geographic Distribution C Italy. **Habitat** Cultivated plant. **Range** Cultivated mainly in Italy.

Beer-Making Parts (and Possible Uses) Leaves.

Chemistry Carbohydrates, protein 3.3%, lipids 0.7%, fiber 2%, water 84.5%, ash 1.5%, calcium, iron, magnesium, phosphorus, potassium, sodium, zinc, copper, manganese, sulfur, selenium, fluorine, vitamin A, vitamin B (B1, B2, B3, B6), vitamin C, vitamin K, beta-carotene, lutein, zeaxanthin.

Style American IPA, English IPA, Spice/Herb/Vegetable Beer.

Source https://www.brewersfriend.com/other/kale/.

Brassica oleracea L. subsp. *oleracea* convar. *botrytis* L. cv. *italica* Pleuk
BRASSICACEAE

Synonyms *Brassica oleracea* L. convar. *botrytis* (L) Alef. var. *cymosa* Duch.

Common Names Broccoli.

Description Herb (3–10 dm), annual, biennial, or perennial. Stem ± woody at the base. Leaves few, basal ones (up to 3 dm) lyrate-pinnatisect, crenate; those cauline lyrate ± crenate or pinnatisect, petiolate, wrapping the inflorescence. Inflorescence compact, contracted, capituliform, gathered on axis and branches shortened, fleshy (probably malformation or monstrosity caused by cultivation). Flower with yellow petals (15–20 mm). Fruit siliqua (2–3 × 40–80 mm), beak 5–15 mm.

Cultivar Among the many cultivars are Amadeus, Apollo, Arcadia, Asteroid, Avenger, Belstar, Blue Wind, Calabrese, Centennial, Constellation, De Cicco, Diplomat, Early Purple Sprouting, Emerald Crown, Express, Fiesta, Gemini, Gypsy, Happy Rich, Imperial, Kailaan, Lieutenant, Marathon, Paraiso, Romanesco, Suiho, Tahoe RZ, and Waltham 29.

Related Species The genus *Brassica* has about 53 taxa of which several are edible and some, listed here, are used in making beer; the beer-making use of *Brassica oleracea* L. var. *capitata* L. and *Brassica rapa* L. has been verified.

Geographic Distribution S Italy. **Habitat** Cultivated plant. **Range** Widely grown, especially in China, India, Spain, Mexico, Italy.

BEER-MAKING PARTS (AND POSSIBLE USES)
Unripe inflorescences.

CHEMISTRY Energy 141 kJ (34 kcal), water 89.3 g, carbohydrates 6.64 g, sugars 1.7 g, fiber 2.6 g, lipids 0.37 g, protein 2.82 g, beta-carotene 31 µg, lutein 361 µg, zeaxanthin 1403 µg, thiamine (B1) 0.071 mg, riboflavin (B2) 0.117 mg, niacin (B3) 0.639 mg, pantothenic acid 0.573 mg, vitamin B6 0.175 mg, folate (B9) 63 µg, vitamin C 89.2 mg, vitamin E 0.78 mg, vitamin K 101.6 µg, calcium 47 mg, iron 0.73 mg, magnesium 21 mg, manganese 0.21 mg, phosphorus 66 mg, potassium 316 mg, sodium 33 mg, zinc 0.41 mg.

STYLE Kölsch.

SOURCE https://www.brewersfriend.com/other/broccoli/.

Bulnesia sarmientoi Lorentz ex Griseb.
ZYGOPHYLLACEAE

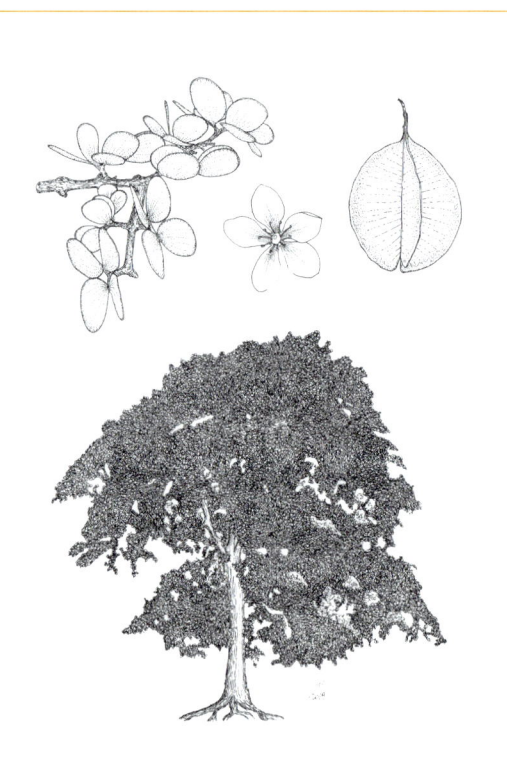

SYNONYMS None

COMMON NAMES Paraguayan palo santo, ironwood (wood marketed under the names Argentine Lignum Vitae, Paraguay Lignum Vitae, Vera, and Verawood [the last two may also refer to the related *Bulnesia arborea*], and under the rather ambiguous names Aura Palo Santo and Palo Santo).

DESCRIPTION Tree (7–20 m), deciduous, long-lived, heliophilious, mesoxerophilic. Foliage expanded, very branched, with falling branches. Trunk straight, short (3–4 m), and cylindrical (50–80 cm). Bark grayish-brown, not very thick, with shallow fissures and small irregular plates. Leaves bifoliate. Flowers hermaphrodite, 1–2, yellowish-white color. Fruit pods 3-winged, dark green to brown.

RELATED SPECIES In addition to *Bulnesia sarmientoi* Lorentz ex Griseb., the genus *Bulnesia* comprises nine recognized taxa: *Bulnesia arborea* (Jacq.) Engl., *Bulnesia bonariensis* Griseb., *Bulnesia carrapo* Killip & Dugand, *Bulnesia chilensis* Gay, *Bulnesia foliosa* Griseb., *Bulnesia retamo* (Gillies ex Hook. & Arn.) Griseb., *Bulnesia rivas-martinezii* G.Navarro, *Bulnesia schichendanzii* Hieron. ex Griseb., and *Bulnesia schickendantzii* Hieron.

GEOGRAPHIC DISTRIBUTION Paraguay, Argentina, Colombia, E Brazil (according to the IUCN, the species is endangered, and its use should be avoided, if not through sustainable processes, to avoid contributing to the decline of the population). **HABITAT** Deciduous, calcareous, open and arid forests periodically flooded, in very saline soils with deep humidity. **RANGE** Paraguay, Argentina, Colombia, E Brazil.

BEER-MAKING PARTS (AND POSSIBLE USES) Wood (an essential oil used in cosmetics is also obtained from the wood, whose use in making beer for barrels and fermenters has never been recorded).

CHEMISTRY Essential oil: α-pinene 0.66–0.70%, limonene 34.16–62.88%, menthofuran 0.70–6.07%, terminene-4-ol 0.54–0.60%, α-terpineol 19.67–23.53%, carvone 3.68–4.05%, sesquiterpene 3.95–25.53%, acid acetic 0.1360%, guaiaretic acid 0.007%.

WOOD Duramen of green to brown-black color, with streaks; thin sapwood light yellow in color. Very heavy wood (WSG 0.92–1.1), hard, resistant, durable (because of calcium oxalate crystals and resin between the fibers) even if exposed, pleasantly aromatic, with fine and homogeneous texture. Used where considerable resistance to abrasion is required, for engraving work, lathe work, making durable poles, and the production of a good charcoal; high-quality wood that is easily ignited even though

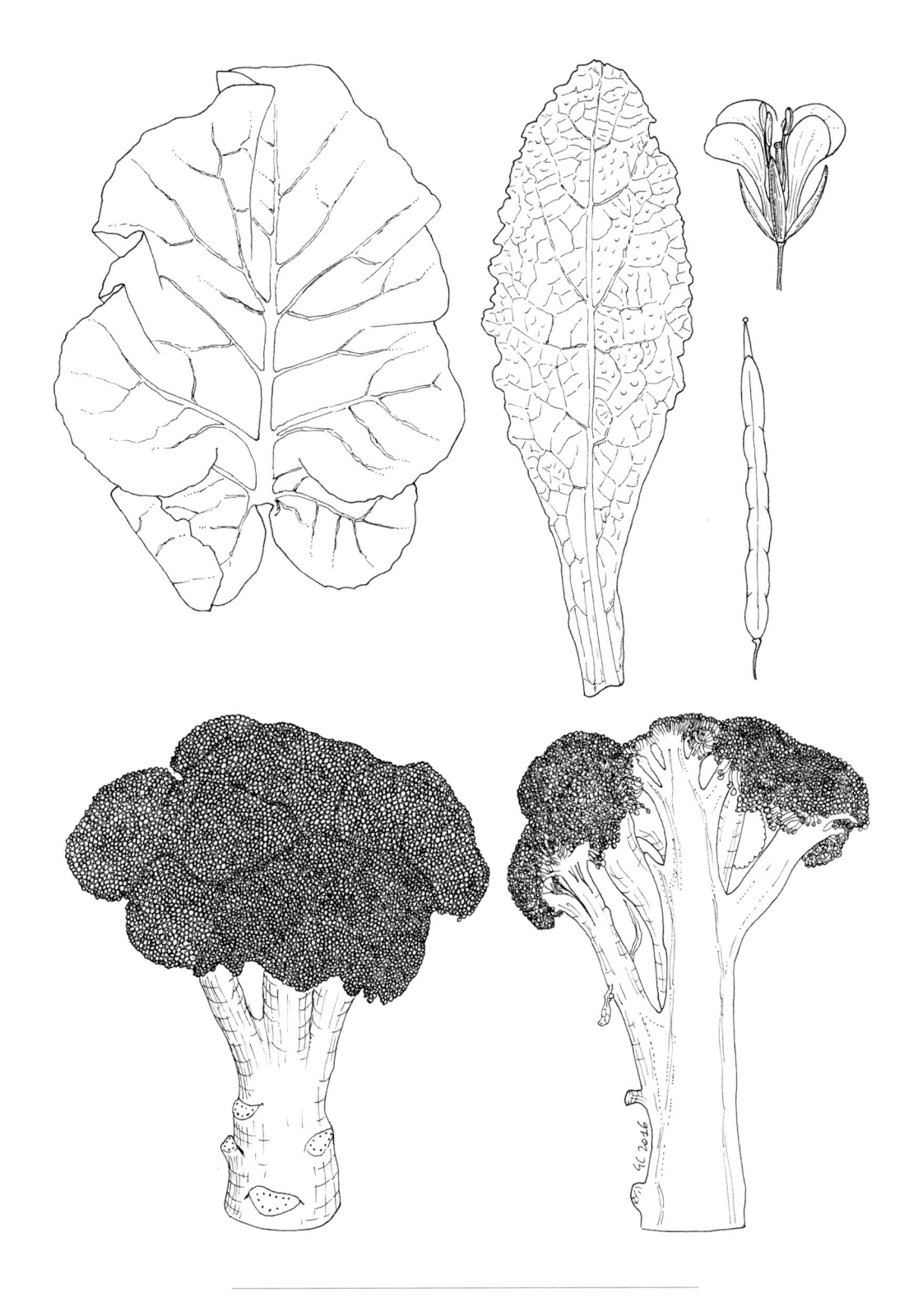

Brassica oleracea L. subsp. *oleracea* convar. *acephala* DC. cv. *sabellica* L.
Brassica oleracea L. subsp. *oleracea* convar. *botrytis* L. cv. *italica* Pleuk

it is quite dense and releases a fragrant smoke (small pieces are used instead of incense). An oil known as guaiac oil (or guayacol), produced from the wood used for distillation, is an ingredient in soaps and perfumes, appreciated for its skin-healing properties. The resin, extracted by means of organic solvents, is used to make varnishes.

STYLE Brown Ale, Wood-Aged.

BEER Palo Santo Marron, Dogfish Head (Delaware, United States).

SOURCE https://www.dogfish.com/brewery/beer/palo-santo-marron; https://www.newyorker.com/magazine/2008/11/24/a-better-brew. (The confusion between *Bulnesia sarmientoi* and *Bursera graveolens* is a typical example of the ambiguity that the use of common names in beer-making botanicals can generate. For example, the common name "palo santo," indicating both species, is the basis of two different reports converging on a single botanical species used in making brewing vessels of considerable size owing to the aromatic characteristics of the wood; given this confusion, both species have been included.)

Sam Calagione's Palo Santo Marron
In a 2008 *New Yorker* article, Burkhard Bilger told the story of a special beer produced in Delaware and aged in a huge barrel built with one of the hardest woods in the world. The beer in question was called Palo Santo Marron, in honor of the wood in which it is refined and that lent it part of its aroma, and the brewery was Dogfish Head, the creation of one of the most famous and eclectic beer makers in the world: Sam Calagione. The story reads like something halfway between history and legend: it says that this barleywine was born from a bright idea from John Gasparine, the owner of a flooring company that worked with sustainable wood. In Paraguay for business, Gasparine encountered some artifacts in *Bulsenia sarmientoi*, whose wood is often used by the locals to produce bowls, containers, and, occasionally, barrels. The intense scent of this species struck Gasparine to such an extent that he wrote to his friend Calagione and suggested he make a beer intense enough to be aged in such an aromatic wood. The Dogfish Head owner not only enthusiastically welcomed the idea but also asked his friend to get the wood as soon as possible to build a large barrel. It is in that very container that Palo Santo Marron, a barleywine with rich and elegantly roasted dried fruit notes, still ages today.

Bursera graveolens (Kunth) Triana & Planch.
BURSERACEAE

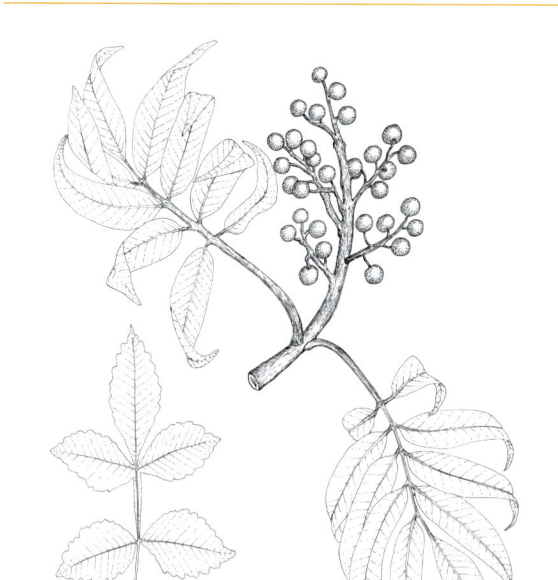

SYNONYMS *Amyris caranifera* Willd. ex Engl., *Amyris graveolens* Spreng, *Bursera anderssonii* B.L.Rob., *Bursera pilosa* (Engl.) L.Riley, *Bursera tatamaco* (Tul.) Triana & Planch., *Elaphrium graveolens* Kunth, *Elaphrium pilosum* (Engl.) Rose, *Spondias edmonstonei* Hook. f., *Terebinthus graveolens* (Kunth) Rose, *Terebinthus pilosa* (Engl.) Rose.

COMMON NAMES Palo santo.

DESCRIPTION Tree (up to 10 m). Trunk smooth, gray in color, branches ferruginous, tomentose toward the apex, wrinkled by leaf scars that form transversal streaks. Leaves pinnately compound (27.5 × 13 cm), ± tomentose, grouped at the ends of branches, ± thickly glandular for glands stipitate with white to orange color, winged rachis, wing 1.5–3 mm (up to 5–6 mm); leaflets 5–9, ovate to round, rarely ovate-lanceolate, obtuse to acute, base wedge shaped to rounded, coarsely crenate, apex sharp to widely attenuated, sessile or briefly pedunculate leaves, opposite or subopposite, lateral pairs slightly unequal (2–6 × 1–4 cm) and different from the terminal (3.5–7 × 1–3 cm). Inflorescence longer than the leaves (9–15 cm), minutely glandular (orange stipitate glands), puberulent; linear-lanceolate bracts, ciliate, puberulent (up to 2 mm). Flowers tetramerous; sepals ovate or ovate-triangular, puberulent, ciliate, glandular for stipitate glands (up to 1.5 mm), persistent; petals oblong, thick, minutely puberulent, with stipitate glands, orange color (4 × 1–1.5 mm), deciduous in pistillate flowers; linear-subulate filaments; anthers yellow; ovary ovoid (Ø about 1.5 mm), with stipitate glands; stylus about 1 mm, persistent; stigma 2-lobed. Fruit obovoid drupe (9–12 × 7–9 mm), both ends acute, brown when ripe, rough, minutely glandular, 2-valve, fleshy endocarp. Seeds black in color, with a whitish-pubescent, fleshy basal aril.

Subspecies Besides the typical *Bursera graveolens* (Kunth) Triana & Planch. subsp. *graveolens*, a second subspecies, *Bursera graveolens* (Kunth) Triana & Planch. subsp. *malacophylla* (B.L.Rob.) A.Weeks & Tye, endemic to the Galápagos, is recognized.

Related Species The genus *Bursera* includes about 120 taxa.

Geographic Distribution Mexico, Yucatan, Peru, Venezuela, N-Argentina, Paraguay, Bolivia, Brazil, Costa Rica, El Salvador, Guatemala, Honduras, Colombia, Ecuador, Galapagos Islands. **Habitat** Pacific coastal dry tropical forests. **Range** Beyond the natural geographic distribution, introduced to Cuba.

Beer-Making Parts (and Possible Uses) Wood (used after 4–10 years of aging after natural tree death) for casks or in cubes (or blocks) to be introduced into the beer during the aging process.

Toxicity The essential oil, widely used in aromatherapy, appears to be slightly toxic when inhaled or used externally in large quantities.

Chemistry The essential oil extracted from the wood of this species has the following major constituents: limonene 58.6%, α-terpineol 10.9%, menthofuran 6.6%, carvone 2.0%, germacrene D 1.7%, γ-murolene 1.2%, trans-carveol 1.1%, pulegone 1.1%.

Wood Little differentiated, yellowish color, irregular grain; used for small carpentry work but especially as a phytochemical for many diseases.

Style Barleywine.

Source Cantwell and Bouckaert 2016; Cantwell and Bouckaert 2018; Giaccone and Signoroni 2017; https://en.wikipedia.org/wiki/Dogfish_Head_Brewery. (The confusion between *Bulnesia sarmientoi* and *Bursera graveolens* is a typical example of the ambiguity that the use of common names in beer-making botanicals can generate. For example, the common name "palo santo," indicating both species, is the basis of two different reports converging on a single botanical species, used in making brewing vessels of considerable size owing to the aromatic characteristics of the wood; given this confusion, both species have been included.)

Buxus sempervirens L.
BUXACEAE

Synonyms *Buxus angustifolia* Mill., *Buxus arborescens* Mill., *Buxus argentea* Steud., *Buxus aurea* Steud., *Buxus caucasica* K.Koch, *Buxus colchica* Pojark., *Buxus crispa* K.Koch, *Buxus cucullata* K.Koch, *Buxus elegantissima* K.Koch, *Buxus fruticosa* Borkh., *Buxus handsworthii* K.Koch, *Buxus marginata* Steud., *Buxus mucronata* Baill., *Buxus myrtifolia* Lam., *Buxus rosmarinifolia* Baill., *Buxus salicifolia* K.Koch, *Buxus suffruticosa* Mill., *Buxus tenuifolia* Baill., *Buxus variegata* Steud., *Buxus vulgaris* Bubani.

Common Names Box, European box, boxwood.

Geographic Distribution Amphi-Atlantic, Europe. **Habitat** Acid soils and siliceous soils of moors, undergrowth, and pastures. **Range** Widely cultivated as an ornamental species in temperate-oceanic climates and in acidic soils.

Beer-Making Parts (and Possible Uses)
Flower stems.

Chemistry 5,7-dihydroxychromone-7-ß-D-glucoside, apigenin-7-(2-acetyl-6-methylglucuronide), apigenin exuronolactone, herbacetin, kaempferol, myricetin, quercetin, isoscutellarin, isorhamnetin-3-O-galactoside, kempferol-3-O-arabinoside, kaempferol-3-O-diacetylarabinohexoside, kaempferol-3-O-(6-p-coumaroyl)-β-D-glucoside (tiliroside), kaempferol-3-O-galactoside, kaemp-ferol-3-O-glucoside, kaempferol-3-[2,3,5-triacetin-α-L-arabinofuranosyl-(1→6)-ß-D-glucoside], kaempferol-3-[2,3,4-triacetyl-α-L-arabinopyranosyl-(1→6)-ß-D-glucoside], 3-methoxy-5,7-dihydroxyflavone-7-O-glucoside (galangin-3-methylethyltherate-7-glucoside; 3-methylgalangin), 3,5,7,8,4-pentahydroxyflavone-8-O-gentiobioside, 3,5,7,8,4-pentahydroxyflavone-4-O-ß-D glucoside, quercetin-3-O-arabinoside (f), quercetin-3-O-arabinoside (p), quercetin-3-O-diacetylarabinohexoside, quercetin-3-O-galactoside, quercetin-3-O-monoacetin therabinohexoside, quercetin-3-O-tetra acetylarabinohexoside, quercetin-3-[2,3,5-triacetyl-α-L-arabinofuranosyl-(1→6)-ß-D-glucoside], quercetin-3-[2,3,4-triacetyl-α-L-arabinopyranosil-(1→6)-ß-D-galactoside], quercetin-3-[2,3,4-triacetil-α-L-arabinopyranosyl-(1→6)-ß-D-glucoside], dihydroherbacetin, 3,5,7,8,4-pentahydroxy flavanone-8-(2-acetilglucoside), 3,5,7,3,4-pentahydroxy flavanone-3-O-glucoside, 3,5,7,8,4-pentahydroxy flavanone-8-g lucoside (callunine), 5,7,8,4-tetrahydroxy-flavanone-8-glucoside (3-desoxychallunine), (+)-catechin, (-)-epicatechin, procyanidin A2, procyanidin B1, procyanidin B2, procyanidin B3, procyanidin B4, procyanidin B5, procyanidin C1, procyanidin D1, cyanidin 3-O-glucoside, caffeic acid, 5-O-caffeoylquinic acid (chlorogenic acid), 3-O-caffeoylquinic acid, p-coumaric acid, p-coumaroylquinic acid, ferulic acid, isochlorogenic acid, 5-hydroxysalicylic acid (gentisic acid), protocatechuic acid, syringe acid, vanillic acid, 3,5-dihydroxytoluene (orcinol), hydroquinone, hydroquinone-ß-D-glucopyranoside (arbutin), 1-[4-(β-D-glucopyranosyloxy)-2-hydroxy-6-methoxyphenyl]-ethanol (rhodiolinozide), orcinol-ß-D-glucoside (sakakin), myristic acid, n-nonacosane, octadecenoic acid, octadecatrienoic acid, octadecadienoic acid, palmitic acid, stearic acid, α-amyrin, ß-sitosterol, ursolic acid, ascorbic acid, mucilage.

Style American Amber Ale, American Pale Ale, Belgian Dark Strong Ale, Braggot, California Common Beer, Experimental Beer, Gruit, Heather Ale, Northern English Brown, Scottish Export 80/-, Scottish Light 60/-, Specialty Beer, Spice/Herb/Vegetable Beer, Strong Scotch Ale.

Beer Fraoch Heather Ale, Williams Brother (Kelliebank, Scotland, United Kingdom).

Source Fisher and Fisher 2017; Heilshorn 2017; Hieronymus 2016b; Josephson et al. 2016; Markowski 2015; Sparrow 2015; https://www.brewersfriend.com/other/heather/; https://www.gruitale.com/bot_heather.htm;https://www.williamsbrosbrew.com/beer-board/bottles/fraoch-heather-ale.

🍺 The Ancient Tradition of Heather Ale
If we were to take Robert Louis Stevenson's word for it, there would be no doubt that using heather to flavor beer was a Scottish practice already known in ancient times. "Heather Ale" is in fact the title of a poem, written by the author of *The Strange Case of Dr. Jekyll and Mr. Hyde* in 1891, in which the genesis of this drink "much sweeter than honey and much stronger than wine" is attributed to the people of Scotland. It seems, however, that flavoring beer with *Calluna vulgaris* flowers was a widespread practice throughout the British Isles, particularly in Ireland. Several texts describe its use: heather buds were placed on the bottom of a vat used as a filter after mashing. In this way, the hot wort was impregnated with the scent of the flowers without the need to resort to infusions or other more complicated practices.

Calluna vulgaris (L.) Hull
ERICACEAE

Calocedrus decurrens (Torr.) Florin
CUPRESSACEAE

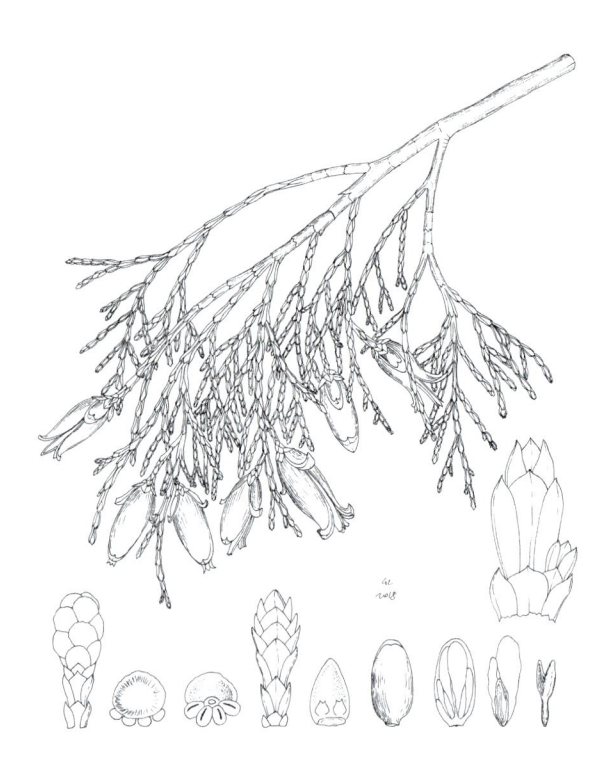

Synonyms *Abies cupressoides* Poir., *Heyderia decurrens* (Torr.) K.Koch, *Libocedrus decurrens* Torr., *Thuja craigana* A.Murray bis, *Thuja decurrens* (Torr.) Voss.

Common Names Incense cedar, incense-cedar, California incense-cedar.

Description Tree (up to 57 m). Trunk single (Ø up to 3.6 m). Bark brown, fibrous, furrowed and rippled. Branches composed of segments 2 to several times longer than wide, which widen distally. Leaves 3–14 mm, including long decurrent base, rounded underneath, acute apex (often suddenly), usually mucronate. Male cones red-brown to light brown in color. Female cones oblong-ovate when closed, red-brown to golden brown in color, proximal scales often reflected to mature cone, median scales, then widely open to curved, distal scales erect. Seeds 4 or fewer per cone, 14–25 mm (including wings), light brown in color.

Related Species In addition to *Calocedrus decurrens*, the genus *Calocedrus* includes three species: *Calocedrus formosana* (Florin) Florin, *Calocedrus macrolepis* Kurz, and *Calocedrus rupestris* Aver., T.H.Nguyên & P.K.Lôc.

Geographic Distribution W North America (United States [California, Nevada, Oregon], Mexico [Baja California]). **Habitat** Mountain forests (300–2,800 m ASL). **Range** Widely cultivated as an ornamental or forest species even outside the geographic distribution.

Beer-Making Parts (and Possible Uses) Sprigs.

Chemistry Essential oil, leaves (total 93.8% composition): Δ-3-carene 31.3%, myrcene 13.4%, α-pinene 11.2%, terpinolene 6.9%, limonene 6.4%, pin-2-en-8-ol 4.2%, methyl pin-2-en-8-oate 3%, α-terpinyl acetate 2.4%, methyl myrtenate 2.3%, β-phellandrene 1.7%, pin-2-en-8-ale 1.6%, terpinen-4-ol 1%, 1-(2-methylene cyclopropyl) cyclopentene 0.9%, α-phenchene 0.7%, (E)-dec-4-enal 0.7%, pin-2-en-8-acetate 0.6%, sylvestrene 0.5%, β-pinene 0.4%, bornyl acetate 0.4%, p-cymene 0.3%, γ-terpinene 0.3%, p-cymenene 0.3%, verbenone 0.3%, piperitone 0.3%, camphene 0.2%, α-terpinene 0.2%, linalool 0.2%, p-cymen-8-ol 0.2%, α-terpineol 0.2%, methil campholenate 0.2%, (E,E)-deca-2.4-dienale 0.2%, cedrene 0.2%, tricyclene 0.1%, α-thujene 0.1%, sabinene 0.1%, α-phellandrene 0.1%, camphor 0.1%, trans-verbenol 0.1%, camphene hydrate 0.1%, borneol 0.1%, p-methylacetophenone 0.1%, α-campholenol 0.1%, carvone 0.1%, traces of fenchone and myrtenyl-methyl ether. Essential oil, young branches (total 89.9% of compounds): α-pinene 22.3%, Δ-3-carene 11.1%, pin-2-en-8-ol 10.3%, myrcene 9.6%, limonene 5.5%, methyl pin-2-en-8-oate 3.6%, methyl myrtenate 3.5%, terpinolene 3.2%, α-terpinyl acetate 2.1%, pin-2-en-8-ale 2%, β-phellandrene 1.4%, pin-2-en-8-acetate 1.4%, terpinen-4-ol 1.1%, cedrol 1.1%, bornil acetate 0.7%, (E)-dec-4-enal 0.6%, (E)-dec-4-enal 0.6%, β-pinene 0.6%, verbenone 0.6%, piperitone 0.5%, methyl campholenate 0.5%, α-campholenol 0.5%, carvone 0.5%, camphene hydrate 0.4%, β-elemene 0.3%, thujopsene 0.3%, β-selinene 0.3%, 1-(2-methylene cyclopropyl) cyclopentene 0.3%, p-cymen-8-ol 0.3%, α-terpineol 0.3%, camphor 0.3%, borneol 0.3%, myrtenyl methyl ether 0.2%, (E)-β-caryophyllene 0.2%, α-selinene 0.2%, p-bisabolene 0.2%, eudesm-11-en-4a-ol 0.2%, α-phenchene 0.2%, sylvestrene 0.2%, p-cymene 0.2%, γ-terpinene 0.2%, p-cymenene 0.2%, camphene 0.2%, α-terpinene 0.2%, linalool 0.2%, (E,E)-deca-2,4-dienal 0.2%, tricyclene 0.2%, sabinene 0.2%, fenchone 0.1%, methyl phenylethylether 0.1%, prezizaene

0.1%, γ-cadinene 0.1%, caryophyllene dioxide 0.1%, (E)-biformene 0.1%, abietatriene 0.1%, sandaracopimarinal 0.1%, dehydroabietal 0.1%, α-thujene 0.1%, α-phellandrene 0.1%, trans-verbenol 0.1%, traces of the following compounds: perillene, trans-p-menth-2-en-1-ol, trans-pinocamphone, isopinocamphone, myrtenol, thymolmethyl ether, carvacrol methyl ether, (Z)-dec-4-en-1-ol, myrtenyl acetate, α-humulene, selin-4,11-diene,γ-cuprenene, β-elemol, γ-cadinol, β-eudesmol,α-eudesmol, (Z)-eptadec-8-ene, manool oxide. Essential oil, branches (total 99.8% of compounds): α-pinene 56.6%, myrcene 8.4%, Δ-3-carene 5.2%, limonene 5.1%, pin-2-en-8-ol 4.5%, methyl pin-2-en-8-oate 1.8%, methyl myrtenate 1.6%, terpinolene 1.5%, β-phellandrene 1.2%, β-pinene 1.2%, α-terpinyl acetate 1%, pin-2-en-8-ale 0.8%, terpinen-4-ol 0.6%, pin-2-en-8-yl acetate 0.6%, cedrene 0.6%, β-elemene 0.5%, bornyl acetate 0.5%, camphene 0.5%, β-selinene 0.4%, verbenone 0.4%, α-selinene 0.3%, abietatriene 0.3%, dehydroabietal 0.3%, piperitone 0.3%, α-terpineol 0.3%, tricyclene 0.3%, sabinene 0.3%, camphene hydrate 0.3%, α-campholenol 0.3%, carvone 0.3%, myrtenyl methyl ether 0.2%, thujopsene 0.2%, (E)-biformene 0.2%, sandaracopimarinal 0.2%, α-fenchene 0.2%, (E)-dec-4-enale 0.2%, p-cimene 0.2%, γ-terpinene 0.2%, p-cymen-8-ol 0.2%, methyl campholenate 0.2%, camphre 0.2%, fenchone 0.1%, methyl phenylethyl ether 0.1%, trans-pinocarveol 0.1%, (E)-β-caryophyllene 0.1%, prezizaene 0.1%, p-bisabolene 0.1%, β-elemol 0.1%, β-eudesmol 0.1%, α-eudesmol 0.1%, manool oxide 0.1%, 1-(2-methylene cyclopropyl) cyclopentene 0.1%, sylvestrene 0.1%, p-cymenene 0.1%, α-terpinene 0.1%, α-thujene 0.1%, α-phellandrene 0.1%, traces of the following compounds: 2-pentylfuran, 1,3,8-p-menthatriene, p-methylanisole, perillene, α-humulene, seline-4,11-diene, γ-cadinene, caryophyllene oxide, γ-eudesmol, linalool, trans-verbenol.

Style Spruce Beer (?).

Source Hieronymus 2016b.

Camellia sinensis (L.) Kuntze
THEACEAE

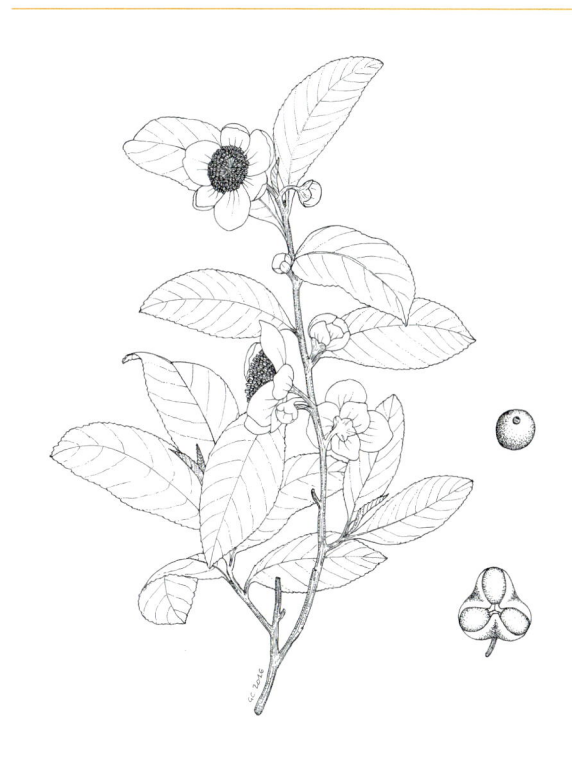

Synonyms *Camellia arborescens* Hung T.Chang & F.L.Yu, *Camellia bohea* (L.) Sweet, *Camellia chinensis* (Sims) Kuntze, *Camellia viridis* Sweet, *Thea bohea* L., *Thea cantoniensis* Lour., *Thea chinensis* Sims, *Thea cochinchinensis* Lour., *Thea grandifolia* Salisb., *Thea olearia* Lour. ex Gomes, *Thea oleosa* Lour., *Thea parvifolia* Salisb., *Thea sinensis* L., *Thea viridis* L., *Theaphylla cantonensis* (Lour.) Raf.

Common Names Tea.

Description Upright shrub (over 2 m), evergreen. Leaves ovate-acuminate (4–15 × 2–5 cm), toothed margin, bright light green. Flowers simple (Ø 4 cm), 7–8 white or yellow petals, stamens golden yellow.

Subspecies Within *Camellia sinensis*, at least three entities of varietal rank are recognized in addition to the typical *Camellia sinensis* (L.) Kuntze var. *sinensis*: *Camellia sinensis* (L.) Kuntze var. *assamica* (J.W. Mast.) Kitam., *Camellia sinensis* (L.) Kuntze var. *dehungensis* (H.T.Chang & B.H.Chen) T.L.Ming, and *Camellia sinensis* (L.) Kuntze var. *waldenae* (S.Y.Hu) H.T.Chang. The entity *Camellia sinensis* (L.) Kuntze subsp. *buisanensis* (Sasaki) S.Y.Lu & Y.P.Yang has also been described, but its taxonomic classification remains doubtful.

Cultivar Among the numerous cultivars are Benifuuki, Fushun, Kanayamidori, Meiryoku, Okumidori, Saemidori, and Yabukita.

Related Species The genus *Camellia* has 280 taxa, but only *C. sinensis* is known to be used to make beer.

Geographic Distribution S and SE Asia. **Habitat** Cultivated worldwide, especially in tropical and subtropical climate regions. **Range** Cambodia, China, India, Laos, Myanmar, Thailand, Vietnam.

Beer-Making Parts (and Possible Uses) Leaves.

Chemistry Fresh leaves: caffeine 4%. Active ingredients: essential oil caffeine 3%, polyphenols (epicatechins), aromatic and aliphatic heterosides, tannins 5.27%, proteins 17%, sugars 5%, vitamin C, vitamins B group, calcium, iron, fluorine, nickel, theobromine 0.05%, traces of theophylline.

Style American IPA, American Pale Ale, American Wheat or Rye Beer, British Golden Ale, Irish Red Ale, Lambic, Metheglin, Southern English Brown, Specialty Beer, Specialty Fruit Beer, Spice/Herb/Vegetable Beer.

Beer Yu Lu, Siren (Finchampstead, England, United Kingdom).

Source Calagione et al. 2018; Dabove 2015; Heilshorn 2017; Jackson 1991; McGovern 2017; https://www.beverfood.com/dreher-radler-mix-birra-dreher-infuso-te-succo-limone_zwd_63775/; https://www.brewersfriend.com/other/tea/; https://www.mixerplanet.com/dreher-te-un-mix-tra-birra-e-te_102537/.

Cannabis sativa L. ssp. *indica* (Lam) E.Small & Cronq. var. *indica* Wehmer
CANNABACEAE

Synonyms *Cannabis indica* Lam., *Cannabis indica* Lam. f. *afghanica* (Vavilov) Vavilov, *Cannabis indica* Lam. var. *kafiristanica* Vavilov, *Cannabis sativa* L. ssp. *indica* (Lam) E.Small & Cronq. var. *kafiristanica* Vavilov.

Common Names Cannabis, hashish.

Description Herb, annual (5–18 dm), dioecious. Very similar to *Cannabis sativa* ssp. *sativa*. Leaves palmate-divided with lanceolate segments.

Subspecies *Cannabis sativa* L. has two subspecies: *Cannabis sativa* L. ssp. *sativa* (L.) Small & Cronquist and *Cannabis sativa* L. ssp. *indica* (Lam) E.Small & Cronq. This differentiation, based on distinct morphologies and biochemical peculiarities (e.g., content of tetrahydrocannabinol [THC] and other alkaloids), is not universally accepted. According to some, the two should be traced back to varietal rank or form, or even completely synonymized.

Cultivar The varietal picture of this entity is quite wide and varied, depending on geographical area, productivity, relative alkaloid content (especially all THC and cannabidiol [CBD]) and aromatic profile. Among the most popular cultivars are Bubble Kush, Cheese Auto, Critical, Medical Mass, Northern Light Auto, Royal AK Auto, Royal Highness, Royal Moby, Royal White Widow Auto, and Skunk XL.

Related Species *Cannabis* is a monotypic genus.

Geographic Distribution C China (?). **Habitat** Cultivated. **Range** Cultivated.

Beer-Making Parts (and Possible Uses) Resin extracted from female inflorescences.

Toxicity THC, the major psychoactive constituent of marijuana has low toxicity, but at high doses it can be very dangerous. A total of 483 other compounds in the plant are known, including at least 65 other cannabinoids.

Chemistry Biochemical components of the genus *Cannabis*: cannabigerolic acid, cannabigerolic monomethylether acid, cannabigerol, cannabigerol monomethylether, cannabigerovarinic acid, cannabigerovarine, cannabichromenic acid, cannabichrome, cannabicromene-varinic acid, cannabichromene-varine, cannabidiolic acid, cannabidiol, monomethylether cannabidiol, cannabidiol-C4, acido-cannabidivarinic acid, cannabidivarin, cannabidiorcol, delta-9-tetrahydrocannabinolic acid A, delta-9-tetrahydrocannabinolic acid B, delta-9-tetrahydrocannabinol, delta-9-tetrahydrocannabinolic acid C4, delta-9-tetrahydrocannabivarinic acid, delta-9-tetrahydrocannabivarinic acid, delta-9-tetrahydrocannabiorcolic acid, delta-9-tetrahydrocannabiorcolic acid, delta-7-cis-tetrahydrocannabivarin, delta-8-tetrahydrocannabinol, cannabicylolic acid, cannabicylol, cannabicyclovarin, cannabielsoic acid A, cannabielsoic acid B, cannabielsoin, acid cannabinol, cannabinol,

cannabinol methylether, cannabinol-C4, cannabivarin, cannabinol-C2, cannabiorcol, cannabinodiol, cannabinovarin, cannabitriol, 10-ethoxy-9-hydroxy-delta-6a-tetrahydrocannabinol, 8,9-dihydroxy-delta-6a-tetrahydrocannabinol, cannabitriovarin, ethoxycannabitriovarin, dehydrocannabifuran, cannabifuran, cannabichromanone, cannabicitran, 10-oxo-delta-6a-tetrahydrocannabinol, delta-9-cis-tetrahydrocannabinol, 3,4,5,6-tetrahydro-7-hydroxy-alpha-alpha-2-trimethyl-9-n-propyl-2,6-methane-2H-1-benzoxocin-5-methanol, cannabiripsol, trihydroxy-delta-9-tetrahydrocannabinol, myrcene, limonene, linalool, trans-ocimene, beta-pinene, alpha-pinene, beta-caryophyllene, delta-3-carene, trans-gamma-bisabolene, trans-alpha-farnesene, beta-phenol, beta-phellandrene, alpha-umulene, guaiacol, alpha-eudesmol, terpinolene, alpha-selinene, alpha-terpineol, fencon, camphene, cis-sebinenehydrate, cis-ocimene, beta-eudasmol, beta-selinene, alpha-trans-bergamotene, gamma-eudasmol, borneol, cis-beta-farnesene, gammacurcumene, cis-gamma-bisabolene, alpha-thujene, epi-alpha-bisabolol, ipsdienol, alpha-ylangene, beta-elemene, alpha-cis-bergamotene, gamma-murolene, alpha-adenine, alpha-longipinene, caryophyllene oxide, cannabisativine, anhydrocannabisativine, N-trans-feruloyltyramine, N-p-cumaroil tyramine, N-trans-captopril tyramine, grossamide, cannabis-A, cannabis-B, cannabis-C, cannabis-D, apigenin, luteolin, kaempferol, quercetin, orientin, vitexin, cannflavin A, cannflavin B, linoleic acid, alpha-linolenic acid, oleic acid, cannabis perane, isocannabispirane, cannabistilbene-I, cannabistilbene-II, cannitrene-I, cannitrene-II.

Style American IPA, American Pale Ale, California Common Beer, Fruit Beer, Imperial IPA, International Pale Lager, Old Ale, Porter.

Beer George Washington's Secret Stash, Dad and Dude's Breweria (Colorado, United States) [no longer operating].

Source Daniels 2016; https://www.brewersfriend.com/other/hash/; https://www.brewersfriend.com/other/marijuana/.

🍺 The American Boom of Beers Made with Cannabis

Cannabis sativa belongs to the same family as *Humulus lupulus*. The two species not only share some morphological aspects but also have a rather similar aromatic profile. It is therefore not surprising that one of the first consequences of the legalization of marijuana was the production of beers using cannabis as a substitute, or partial substitute, for hops. California and Colorado, among the first states to make the sale of marijuana legal, are also those states whose breweries were the first to start producing beer using this ingredient.

Cannabis sativa L. ssp. *sativa* (L.) Small & Cronquist
CANNABACEAE

Synonyms *Cannabis americana* Pharm. ex Wehmer, *Cannabis chinensis* Delile, *Cannabis erratica* Siev., *Cannabis foetens* Gilib., *Cannabis generalis* E.H.L.Krause, *Cannabis gigantea* Crevost, *Cannabis intersita* Soják, *Cannabis lupulus* Scop., *Cannabis macrosperma* Stokes, *Cannabis ruderalis* Janisch. (according to some authors, this last one should be considered a distinct entity from *Cannabis sativa*).

Common Names Hemp.

Description Herb, annual (5–25 dm), dioecious. Stem erect, rough. Opposite lower leaves, the upper ones mostly alternate, with linear stipules, petiole 2–5 cm and palmate-divided into 5–11 lanceolate-acuminated segments (the larger ones 1–3 × 5–15 cm). Male flowers yellow-greenish (5 mm) in panicles; female flowers paired at the axilla of the upper leaves. Fruit nucule (3–5 mm) grayish, glossy.

Subspecies *Cannabis sativa* L. includes two subspecies: *Cannabis sativa* L. ssp. *sativa* (L.) Small & Cronquist and *Cannabis sativa* L. ssp. *indica* (Lam) E.Small & Cronq. This differentiation, based on distinct morphologies and biochemical peculiarities (e.g., content of THC and other alkaloids), is not universally accepted. According to some, the two should be traced back to varietal rank or form, or even completely synonymized.

Cultivar The varietal distinction within this taxon is made on a morphological basis (e.g., size and vigor of plants, entity of cauline ramifications), but especially on a sexual basis (monoecious versus dioecious). Among the dioecious cultivars are Armanca, Asso, Bergnaturhanf Ladir, Bernabeo, Bredemann, Bundy Gem, Chameleon, CanMa, Cannakomp, Carmagnola, Carmen, CFX-1, CFX-2, Crag, CRS-1, CS, Dioica 88, Eletta Campana, Ermakovskaya Mestnaya, ESTA-1, Fibranova, Fibridia, Fibrimor, Finola,

Fleischmann Hemp, Havelländer, Helvetica, Kompolti, Kompolti sargaszárú, Kuban Kuhnow, Lovrin 110, Novosadska Konomplja, Petera, Rastslaviska, Red petiole, Schurig, Silvana, Suomi, Superfibra, Tiborszallasi, X59, Yunma 1–6, and Zenica. Among the hybrid cultivars with a prevalence of female individuals are Alyssa, Fedora 17-19-74, Fédrina 74, Felina 32–34, Fibriko, Futura 75–77, Hohenthürmer Gleichzeitig Reifender, Kompolti TC Hybrid, Lipko, and UNIKO-B. Monoecious cultivars include Anka, Antal Avorio, Bialobrzeskie, Beniko, Canda, Carma, Carmono, Codimono, Dacia Secuieni, Delores, Delta-405, Delta-I-losa, Deni, Denise, Diana, Dneprovskaya Odnodomnaya 6–14, Epsilon 68, Ermes, Fasamo, Férimon, Fibrimon (21, 24, and 56), Fibrol, Irene, Joey, Jutta, KC Dóra, KC Virtus, KC Zuzana, Major, Marcello, Markant, Monoica, Rajan Santhica (23, 27, and 70), Schurigs, Secuieni 1, Secueni Jubileu, Silesia, Szarvasi, Tisza, Tygra, UC-RGM, Wojko, Yuso (–14, –16, and –31), Yvonne, Zenit, and Zolotonoshskaya (11, 13, and 15). There are also ornamental cultivars such as Ermes and Panorama.

Related Species *Cannabis* is a monotypic genus.

Geographic Distribution C-Asia (widely cultivated). **Habitat** Cultivated. **Range** Cultivated in all tropical and subtropical areas of the world, especially China and India.

Beer-Making Parts (and Possible Uses) Apical leaves, infructescences, seeds; stems (burned for smoking malt).

Chemistry Essential oil, seeds: linoleic acid 50–70%, alpha-linolenic acid 15–25%, gamma-linolenic acid 1–6%, oleic acid 10–16%, palmitic acid 6–9%, stearic acid 2–3%, eicosanoic acid 0.79–0.81%, eicosenoic acid 0.39–0.41%, eicosadienoic acid 0.0–0.09%, cannabidiol 10 mg/kg, delta-9-tetrahydrocannabinol 50 mg/kg, myrcene 160 mg/L, beta-caryophyllene 740 mg/L, beta-sitosterol 100–148 g/L, alpha-tocopherol 7–80 ppm, gamma-tocopherol 710–870 ppm, methyl salicylate.

Style California Common Beer, Specialty IPA (Black IPA).

Beer Futura, Blandino (Strongoli, Calabria, Italy).

Source Dabove 2015; Daniels and Larson 2000; McGovern 2017; https://www.absinth.cz/product/cannabis-beer/cannabis-beer; https://www.dodimalto.it/birra/futura-del-birrificio-blandino/.

Capsicum annuum L.
SOLANACEAE

Synonyms *Capsicum abyssinicum* A.Rich., *Capsicum angulosum* Mill., *Capsicum axi* Vell., *Capsicum bauhinii* Dunal, *Capsicum caerulescens* Besser, *Capsicum cerasiforme* Mill., *Capsicum ceratocarpum* Fingerh., *Capsicum cereolum* Bertol., *Capsicum comarim* Vell., *Capsicum conicum* G.Mey., *Capsicum conicum* Lam., *Capsicum conoide* Mill., *Capsicum conoides* Roem. & Schult., *Capsicum conoideum* Mill., *Capsicum cordiforme* Mill., *Capsicum crispum* Dunal, *Capsicum cydoniforme* Roem. & Schult., *Capsicum dulce* Dunal, *Capsicum fasciculatum* Sturtev., *Capsicum fastigiatum* Blume, *Capsicum frutescens* L., *Capsicum globiferum* G.Mey., *Capsicum globosum* Besser, *Capsicum grossum* L., *Capsicum indicum* auct., *Capsicum longum* DC., *Capsicum milleri* Roem. & Schult., *Capsicum minimum* Mill., *Capsicum odoratum* Steud., *Capsicum odoriferum* Vell., *Capsicum oliviforme* Mill., *Capsicum ovatum* DC., *Capsicum petenense* Standl., *Capsicum pomiferum* Mart. ex Steud., *Capsicum purpureum* Roxb., *Capsicum purpureum* Vahl ex Hornem., *Capsicum pyramidale* Mill., *Capsicum quitense* Willd. ex Roem. & Schult., *Capsicum silvestre* Vell., *Capsicum sphaerium* Willd., *Capsicum tetragonum* Mill., *Capsicum tomatiforme* Fingerh. ex Steud., *Capsicum torulosum* Hornem., *Capsicum tournefortii* Besser, *Capsicum ustulatum* Paxton.

Common Names Chili.

Description Herb, annual (4–10 dm), glabrous. Stem erect, striated, nodes enlarged, very branchy. Leaf with oval-acuminated lamina (3–8 × 7–13 cm) entire or irregularly crenated, progressively narrowed at the base; small groove above, as long as the lamina. Flowers isolated at the axilla of the upper leaves, peduncle curved (1 cm), enlarged at the end and up to 3 cm long; calyx at flowering 3–4 mm with obtuse teeth, later increasing; corolla yellowish-white with laciniae of 5–6 × 10–15 mm. Fruit edible berry (bell pepper or chili) of very different size (2–15 cm), shape, and color (green, yellow, red).

Subspecies Besides the typical variety, *Capsicum annuum* L. var. *glabriusculum* (Dunal) Heiser & Pickersgill is recognized.

Cultivar A large representation of cultivars of *Capsicum annuum*: Abbraccio, Aci Sivri, Acrata, Agrifoglio, Alba, Aleppo, Alma Paprika, Amando, Anaheim,

Cannabis sativa L. ssp. *indica* (Lam.) E.Small & Cronq. var. *indica* Wehmer Cannabaceae
Cannabis sativa L. ssp. *sativa* (L.) Small & Cronquist CANNABACEAE

Ancho, Apache, Arlecchino, Assam, Aurora, Azteco, Bacio di Satana, Banana Pepper, Bassotto, Bell, Bhavnagari Long, Black Namaquakand, Black Olive, Black Pearl, Black Prince, Blu di Prussia, Blueberry, Bolivian Rainbow, Bulgarian Carrot, Calico, Calusa Indian Mound, Candlelight, Capriglio, Carricillo, Cascabel, Catarina, Cayenna Golden, Cayenna Red, Cedrino, Cerasella di Calabria, Charleston, Cherry, Chiara, Chicanna Purple, Chile de Arbol, Chile de Onza, Chilly Chili, Chiltepin, Chimayo, Chinese Five Color, Chipotle, Chispas, Ciliegino, Corno di Toro, Costeño Amarillo, Cubanella, Czechoslovakian Black, Diavolicchio Diamante, Dob, Dutch Hot, Ecuadorian Purple, Elephant Trunk, Ethiopian Brown, Etna, Eureka, Explosive Ember, Filius Blu, Fips (or Fiesta), Firecracker, Fish, Fresno, Frigitello, Golden Nugget, Guajillo, Guindilla, Hatch, Holiday Cheer, Hot Portugal, Hungarian Black, Jalapeño, Jalapeño Purple, Jaloro, Jamaican Mushroom, Joe's Long Cayenne, Joe's Round, Jwala Finger, Krakatoa, Lingua di Fuoco, Loco, Macarena, Madagascar, Marbles, Mariachi, Masquerade, Maule's Red, Media Noche, Medusa, Mirasol, Naso di Cane, Nepali, NuMex Bailey, NuMex Big Jim, NuMex Centennial, NuMex Easter, NuMex Heritage 6–4, NuMex Joe E. Parker, NuMex Mirasol, NuMex Pinata, NuMex R. Naky, NuMex Sandia, NuMex Twilight, Orozco, Paprika, Pasilla, Peppa Orangina, Pequin, Peter Pepper, Piment d'Anglet, Pimento d'Espelette, Pimento del Sahara, Pimiento de Padron, Pimiento del Piquillo, Pimiento di Guernica, Poblano, Poinsettia, Prairie Fire, Pretty Purple, Purple Flash, Ring of Fire, Riot, Sangria, Santa Fe Grande, Santaka, Scozzese, Serrano, Shishito, Sigaretta di Bergamo, Siling Mahaba, Spagna, Sparkler, Stromboli, Tabiche, Tasmanian Black, Tianjin, Tomato, Trifetti, Venezuelan Purple, Vezena Piperka, Vietato, Violetta, Violetto Fuoco Nero, Yatsufusa, Zimbabwe Black.

Related Species There are 42 species belonging to the genus *Capsicum*; however, *Capsicum annuum* is the most cultivated and includes the most widespread varieties (sweet peppers, chili, cayenne bell pepper, Mexican jalapeño).

Geographic Distribution S America.
Habitat Cultivated everywhere. **Range** Cultivated everywhere.

Beer-Making Parts (and Possible Uses) Fresh, chipotle, dried, and smoked fruits (merkén).

Toxicity Pure capsaicin is a toxic substance, and its intake in large quantities can have lethal effects.

Chemistry Fruits (Adorno cultivar): 4-hydroxy-4-methyl-2-pentanone 0.98%, 1-hexadecene 0.06%, 2-methylpentadecane 0.04%, hexadecane 0.04%, heptadecane 0.15%, pentadecanoic acid 0.25%, 1-heptadecanol acetate 0.10%, octadecane 0.21%, alloaromadendrene 0.52%, oleic acid 0.14%, 2-methyl-3,13-octadecadien 0.10%, 5-eicosene 0.16%, cycloeicosane 0.04%, 1-octadecanamine 5.28%, 7,11-hexadecadiene 0.16%, 9-octadecenamide 0.12%, 3,6-dimethyl-2,3,3a,4,5,7a-hexa-hydrobenzofuran 0.07%, heicosane 0.04%, 2-hydroxycyclopentadecanone 0.16%, heneicosane 0.36%, nonivamide 7.66%, nordihydrocapsaicin 0.62%, capsaicin 37.22%, dihydrocapsaicin 28.68%, N-vanillildecanamide 1.39%, docosane 0.52%, homo-capsaicin 0.74%, homo-capsaicin II 0.42%, hemodihydrocapsaicin 1.85%, hemodihydrocapsaicin II 0.43%, 2-methyltricosane 0.08%, squalene 0.14%, tetracosane 0.23%, pentacosane 0.58%, hexacosane 0.14%, heptacosane 0.76%, vitamin E 2.1%, ergost-5-en-3-ol 0.3%, 11-decyldocosane 0.25%, p-sitosterol 0.59%, α-amyrin 0.41%, β-amyrin 0.46%, methyl 3-hydroxycholestenoate 0.21%.

Style American Amber Ale, American Brown Ale, American IPA, American Lager, American Pale Ale, American Strong Ale, American Wheat Beer, Blonde Ale, Cream Ale, Dry Stout, Experimental Beer, Fruit Beer, Holiday/Winter Special Spiced Beer, Imperial Ipa, Kölsch, Oktoberfest/Märzen, Other Smoked Beer, Premium American Lager, Robust Porter, Saison, Special/Best/Premium Bitter, Spice/Herb/Vegetable Beer, Sweet Stout.

Beer Helles Diablo, Malt Cross (Trecate, Piedmont, Italy).

Source Calagione et al. 2018; Dabove 2015; Giaccone and Signoroni 2017; Heilshorn 2017; Hieronymus 2016b; Josephson et al. 2016; McGovern 2017; https://www.birraaltaquota.it/prodotti/birre-estrose/20-chicano.html; https://www.bjcp.org (28C. Wild Specialty Beer; 30. Spiced Beer); https://www.brewer-sfriend.com/other/poblano/; https://www.chilibeer.com/chilibeer/Photo_History.html#-grid; https://en.wikipedia.org/wiki/Chili_pepper; https://it.wikipedia.org/wiki/Capsicum_annuum chipotle.

🍺 The Difficulty of Making Beers with Chili Pepper

Chili pepper is one of the most intriguing ingredients for brewers around the world. The aromatic and spicy characteristics of this fruit, if well dosed, can give the beer fascinating new notes, but finding that correct dose can be problematic. Alessio Selvaggio, cofounder with Federico Casari of the Croce di Malto brewery in Trecate (Novara, Italy), knows this well. In their Helles Diablo, Alessio and Federico used three hot peppers (Bhut Jolokia, Cornetto Calabrese, and Rocoto) selected more for their aromatic contribution than for their spiciness. To evaluate them, several tests were made by infusing the hot peppers in boiling water for five minutes. After choosing the varieties, Alessio and Federico had to decide in which phase of production to use them. After repeated tests, they opted for boiling: the mix of hot peppers is added in two stages of ten and five minutes. In this way, the drink remains marked by a slight spicy tone that goes well with the slightly smoky and honeyed notes of this light beer.

Capsicum baccatum L.
SOLANACEAE

Synonyms *Capsicum cerasiflorum* Link, *Capsicum chamaecerasus* Nees, *Capsicum ciliare* Willd., *Capsicum conicum* Vell., *Capsicum microcarpum* Cav., *Capsicum microphyllum* Dunal, *Capsicum pulchellum* Salisb., *Capsicum umbilicatum* Vell.

Common Names Peppadew, ají.

Description Shrub (up to 2 m in the original area), similar to a small tree. Stem with an erect position. Leaves lanceolate, green in color. Flowers single, axillary, one per node; corolla formed by 5–7 petals, white with yellow-green to brown spots, stamens yellow, anthers yellow to brown. Fruit berry of quite variable shape so that many cultivars have ornamental use, green at the beginning, white, yellow, orange, red, purple when ripe.

Subspecies Besides the typical variety, *Capsicum baccatum* L. var. *baccatum*, two other varieties are recognized: *Capsicum baccatum* L. var. *pendulum* (Willd.) Eshbaugh and *Capsicum baccatum* L. var. *praetermissum* (Heiser & P.G.Sm) Hunz.

Cultivar Many, are used both for food and ornamental purposes. Some of the most common are Aji Amarillo, Ají Brazilian Red Pumpkin, Ají Criolla Sella, Ají Crystal, Aji Habanero, Ají Omnicolor, Ají Pineapple, Atomic Starfish, Bird Pepper, Bishop Crown, Brazilian Starfish, Christmas Bell, Lemon Drop, Piquanté Pepper, Praetermissum, Rain Forest, Sugar Rush Peach, and Wild Baccatum.

Related Species There are 42 species belonging to the genus *Capsicum*; many are used in food preparations.

Geographic Distribution Bolivia, Peru. **Habitat** Tropical and humid subtropical areas or areas with an arid season, but also arid mountain valleys at high altitudes and on extremely variable soils. **Range** Argentina, Brazil, Bolivia, Colombia, Ecuador, Peru.

Beer-Making Parts (and Possible Uses) Fruits.

Toxicity Pure capsaicin is toxic, and its intake in large quantities can have lethal effects.

Chemistry Capsaicinoids (total 197–239 μg/g): nordihydrocapsaicin 16–19 μg/g, capsaicin 100–124 μg/g, dihydrocapsaicin 60–70 μg/g, homocapsaicin 24–28 μg/g.

Style California Common Beer.

Beer Peppadew Ale, Ass Clown Brewery (North Carolina, United States).

Source https://www.brewersfriend.com/other/peppadew/.

Capsicum chinense Jacq.
SOLANACEAE

Synonyms *Capsicum sinense* Murray, *Capsicum toxicarium* Poepp. ex Fingerh.

Common Names Habanero, yellow lantern chili.

Description Slow-growing perennial bush (up to 5 dm). Roots shallow. Leaves lanceolate, sharp, fleshy, wrinkled, ± slightly hairy on the underside. Flowers bell shaped, white or greenish in color, petals 5, purple *jimenezii* Bertoni, *Carica mamaya* Vell., *Carica peltata* Hook. & Arn., *Carica pinnatifida* Heilborn, *Carica portoricensis* Urb., *Carica posopora* L., *Carica pyriformis* Willd., *Carica rochefortii* Solms, *Carica sativa* Tussac, *Papaya carica* Gaertn., *Papaya cimarrona* Sint. ex Kuntze, *Papaya citriformis* (Jacq.) A. DC., *Papaya communis* Noronha, *Papaya cubensis* (Solms) Kuntze, *Papaya cucumerina* Noronha, *Papaya edulis* Bojer, *Papaya hermaphrodita* Blanco, *Papaya peltata* (Hook. & Arn.) Kuntze, *Papaya rochefortii* (Solms) Kuntze, *Papaya sativa* Tuss., *Papaya vulgaris* A. DC., *Vasconcellea peltata* (Hook. & Arn.) A. DC.

Common Names Papaya, papaw, pawpaw.

Description Single-trunk tree (5–10 m), usually without branches. Leaves distributed in a spiral along the trunk, confined mostly at the apex; lower part of trunk covered with scars of leaves and fruits of previous years; leaves wide (Ø 50–70 cm), deeply palmate-lobed, with 7 lobes. Flowers fragrant, with 5 petals (2.5–5.1 cm long), creamy white to yellow-orange in color, pale green stigma, hanging yellow stamens. The presence or absence of functional stamens and stigma and ovary determine different types of flower: female flowers, generally large and rounded at the base, have stigma but no stamens; to produce fruit, these flowers must receive pollen from the outside brought by insects or wind; male flowers are thin, tubular, and perfect (they contain male and female organs), but the small vestigial ovary is not functional; hermaphrodite flowers have characters intermediate between male and female and, having both organs, usually self-pollinate. The flowers emerge at the axilla of the leaves. Fruit berry rather large when ripe (15–45 × 10–30 cm), soft when ripe and with epicarp from amber to orange.

Cultivar Two main groups: yellow flesh and red flesh.

Related Species The genus *Carica* includes 22 species; however, only *C. papaya* is used in making beer.

Geographic Distribution CS America.
Habitat Cultivated. **Range** India, Brazil, Indonesia, Nigeria, Mexico.

Beer-Making Parts (and Possible Uses) Fruits.

Chemistry Energy 179 kJ (43 kcal), carbohydrates 10.82 g, sugars 7.82 g, fiber 1.7 g, lipids 0.26 g, protein 0.47 g, vitamin A equivalent 47 µg, beta-carotene 274 µg, lutein zeaxanthin 89 µg, thiamine (B1) 0.023 mg, riboflavin (B2) 0.027 mg, niacin (B3) 0.357 mg, pantothenic acid (B5) 0.191 mg, folate (B9) 38 µg, vitamin C 62 mg, vitamin E 0.3 mg, vitamin K 2.6 µg, calcium 20 mg, iron 0.25 mg, magnesium 21 mg, manganese 0.04 mg, phosphorus 10 mg, potassium 182 mg, sodium 8 mg, zinc 0.08 mg, lycopene 1828 µg.

Style American IPA, American Wheat Beer, Belgian Blond Ale, Blonde Ale, Fruit Lambic, Gose, Imperial IPA.

Beer Papaya Sour Ale, Jungle Juice & MC77 (Rome, Lazio, Italy).

Source https://www.bjcp.org (29. Fruit Beer); https://www.brewersfriend.com/other/papaya/; https://www.craftbeer.com/news/beer-release/oregon-fruit-products-collaborates-with-breakside-brewery.

Cariniana legalis (Mart.) Kuntze
LECYTHIDACEAE

Synonyms *Cariniana brasiliensis* Casar., *Couratari legalis* Mart.

Common Names Legal cariniana, jequitiba-rosa, jequitibá, jequitibá branco, jequitibá vermelho, jequitiba rosa, pau carga.

Description Tree (up to 25 m). Trunks glabrous, sparsely lenticulated when young. Leaves with oblong lamina (3–6.5 × 1.5–3.2 cm), margins slightly serrated-crenulate, rarely crenate, apex sharp (1–5 mm), base straight, wedge shaped, petiole decurrent and decurrent portion characteristically covered in dry material; central rib under-prominent, glabrous; secondary ribs 9–11 pairs, anastomosed near the margins; petiole 3–5 mm, winged upper part. Inflorescence panicle or raceme poorly branched, terminal and subterminal, rachis and puberulous branches; pedicels 0.5–1 mm, sparsely puberulous. Flower with calyx 2.5–3 mm, dark lobes, rounded, sparsely puberulous outside; petals oblong-ovate, about 4 mm, white in color; androecium Ø 3 mm at the base, with about 50 stamens attached mostly at the apex with few free filaments from the base of the androecium, reddish-purple color; very short stylus. Fruit pyxidium 4.5–6.5 × 1.5–3 cm, indistinct calyx ring, 0.4–0.8 cm below apex, pericarp 2–3 mm thick at apex, smooth, not toothed opercular dehiscence line; operculum Ø 1–1.5 cm.

Related Species In addition to *Cariniana legalis* (Mart.) Kuntze, the genus *Cariniana* includes eight species: *Cariniana domestica* (Mart.) Miers, *Cariniana estrellensis* (Raddi) Kuntze, *Cariniana ianeirensis* R.Knuth, *Cariniana micrantha* Ducke, *Cariniana parvifolia* S.A.Mori, Prance & Menandro, *Cariniana penduliflora* Prance, *Cariniana pyriformis* Miers, and *Cariniana rubra* Gardner ex Miers.

Geographic Distribution Brazil (Mata Atlântica). **Habitat** Riparian environments. **Range** Brazil.

Beer-Making Parts (and Possible Uses) Wood for barrels (first they contain cachaça, then beer for aging).

Wood Differentiated, with pale sapwood and reddish heartwood, sometimes variegated, medium texture, straight grain, not always easy to work; WSG 0.50–0.70; used for construction and shipbuilding, plywood, furniture, barrels, floors, shoe heels, etc.

Style Brazilian Beer.

Source Cantwell and Bouckaert 2016; Cantwell and Bouckaert 2018.

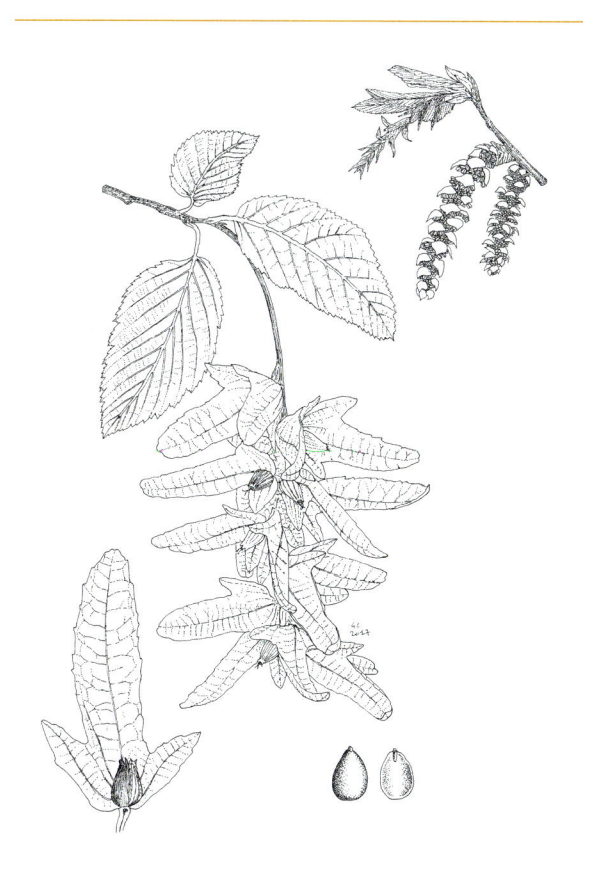

Carpinus betulus L.
BETULACEAE

Synonyms *Carpinus carpinizza* Kil., *Carpinus caucasica* Grossh., *Carpinus intermedia* Wierzb. ex Rchb., *Carpinus nervata* Dulac, *Carpinus quercifolia* Desf. ex Steud., *Carpinus sepium* Lam., *Carpinus ulmifolia* St.-Lag, *Carpinus ulmoides* Gray, *Carpinus vulgaris* Mill.

Common Names European hornbeam, common hornbeam.

Description Tree (15–25 m), deciduous. Roots wide but not deep, lateral ones robust. Trunk straight, irregular section with grooves; bark ash-gray with whitish spots, smooth; branches apical ascending, foliage dense and oval; branches of the year reddish-green, slender, pubescent; buds alternate, fusiform (5–7 mm), attached to the sprig, pubescent apically. Leaves alternate, distichous, petiolate (1 cm), oblong-ovate (4–10 × 2.5–5 cm), margin double serrate, base truncated or cordate, apex acute, secondary veins 10–15. Male flowers without bracts and perianth (4-6-12-stamens); female flowers consist of long, pointed, ciliated bracts. Inflorescences unisexual catkins, the male ones on the lateral branches, cylindrical (2–5 cm), pendulous, the female ones shorter

(1–3 cm) on the main branches, at the base carrying 2 flowers with 2 styles each and 6 basal bracts that, after fertilization, grow into a characteristic trilobed bract, with a median lobe of 3–5 cm. Fruit ovoid achene (7–10 mm) compressed on one face, furrowed, hard, greenish to brown in color (anemochorous dissemination together with the bract).

RELATED SPECIES The genus *Carpinus* consists of 54 recognized species. Some may have use in beer making similar to *C. betulus*.

GEOGRAPHIC DISTRIBUTION European-Caucasian. **HABITAT** Relatively sciaphilous species, prefers loose, deep, well-humidified soils; from subacidic to calcareous; it is a soil improver and therefore also preparatory for more demanding species. **RANGE** Tree widely cultivated for ornamental purposes (especially the Fastigiata cultivar).

BEER-MAKING PARTS (AND POSSIBLE USES) Wood (for toasting and smoking malt).

WOOD Undifferentiated, whitish or light brown-grayish color, very fine texture, not very regular grain, especially depending on the bad shape of the trunks often crooked and noncircular section (ribbed); hard, compact, WSG 0.75–0.85, poor drying, during which cracks, deformations, and collapse often occur. Despite the difficulty of processing, it is valuable for shoe lasts, textile or mechanical industry tools, cart work, and road paving. Excellent fuel both in its natural state and when converted to charcoal.

STYLE Brown Ale.

SOURCE Daniels 2016.

Carum carvi L.
APIACEAE

SYNONYMS *Bunium carvi* (L.) M.Bieb., *Carum aromaticum* Salisb., *Carum gracile* Lindl., *Carum officinale* Gray, *Carum rosellum* Woronow, *Carum velenovskyi* Rohlena, *Carvi careum* Bubani, *Falcaria carvifolia* C.A.Mey., *Foeniculum carvi* (L.) Link, *Karos carvi* (L.) Nieuwl. & Lunell, *Ligusticum carvi* (L.) Roth, *Pimpinella carvi* (L.) Jess., *Selinum carvi* (L.) E.H.L.Krause, *Seseli carvi* (L.) Spreng, *Sium carvi* (L.) Bernh.

COMMON NAMES Caraway, Persian cumin.

DESCRIPTION Biennial herbaceous plant (30–100 cm). Root tapered. Stems erect, striped, glaucous, glabrous, hollow, very branchy and leafy. In the first year, the root produces a rosette of basal leaves with a petiole that widens at the base in a membranous and striated sheath; leaves oblong 2-3-pinnatisect, filiform; in the second year, the stem develops with sessile cauline leaves, reduced to a few filiform laciniae, with membranous amplexicaul sheath reddened at least at the margin. Inflorescences umbel at the apex of the branches, 6–16 unequal rays, rarely 1–3 involucral bracts (deciduous), umbels with 5–15 rays usually without bracts. Flowers with calyx consisting of 5 inconspicuous teeth, corolla consisting of 5 white or reddish-pink petals, deeply toothed at the apex; peripheral flowers generally longer; stamens arranged between the petals. Fruits polyachenes (diachenes) with 2 mericarps (3–7 mm long), oblong, narrow at the extremities and slightly curved, aromatic, dark brown in color, with 5 ribs in relief, paler.

RELATED SPECIES The genus *Carum* has 31 species among which only *C. carvi* has use in making beer.

GEOGRAPHIC DISTRIBUTION Paleotemperate-Eurasian. **HABITAT** Mountain meadows, pastures, paths; usually found from 800–2,250 m ASL but sometimes found from 200–300 m ASL. **RANGE** Europe, United States.

BEER-MAKING PARTS (AND POSSIBLE USES) Fruits, seeds.

CHEMISTRY Energy 1.567 kJ (375 kcal), water 8.06 g, carbohydrates 44.24 g, sugars 2.25 g, fiber 10.5 g, lipids 22.27 g (saturated 1.535 g, monounsaturated 14.04 g, polyunsaturated 3.279 g), protein 17.81 g, vitamin A equivalent 64 μg, thiamine (B1) 0.628 mg, riboflavin (B2) 0.327 mg, niacin (B3) 4.579 mg, vitamin B6 0.435 mg, folate (B9) 10 μg, vitamin C 7.7 mg, vitamin E 3.33 mg, vitamin K 5.4 μg, calcium 931 mg, iron 66.36 mg, magnesium 366 mg, phosphorus 499 mg, potassium 1788 mg, sodium 168 mg, zinc 4.8 mg.

STYLE Berliner Weisse, Gruit, Spice/Herb/Vegetable Beer, Witbier.

BEER Põhjala Cellar Series, Laugas (Tallinn, Estonia).

SOURCE Calagione et al. 2018; Cantwell and Bouckaert 2016, 2018; Heilshorn 2017; Hieronymus 2010; Hieronymus 2012; Hieronymus 2016b; Josephson et al. 2016; Serna-Saldivar 2010; https://www.brewersfriend.com/other/cominho/.

Carya illinoinensis (Wangenh.) K.Koch
JUGLANDACEAE

SYNONYMS *Carya angustifolia* Sweet, *Carya diguetii* Dode, *Carya oliviformis* (F.Michx.) Nutt., *Carya pecan* (Marshall) Engl. & Graebn., *Carya pecan* (Marshall) Nutt., *Carya tetraptera* Liebm., *Hicoria pecan* (Marshall) Britton, *Hicorius oliviformis* Nutt., *Juglans illinoinensis* Wangenh., *Juglans oliviformis* Marshall, *Juglans pecan* Marshall.

COMMON NAMES Pecan.

DESCRIPTION Tree (30–40 m), deciduous. Bark grayish-brown or light brown, with ripples and grooves only superficial. Leaves alternate, pinnate-compound, 25–50 cm long, with 11–17 leaflets 10–20 cm long. Flowers unisexual, both sexes in separate groups on the same tree. Fruit nut with thin casing, 4-winged from base to apex, in groups of 3–12; a relief is formed where the two halves of the outer casing join; dark brown fruit covered with yellow scales; shell thin and fragile, often persistent on branches in winter after the fall of the nut; nut dark red, pointed at both ends.

CULTIVAR In the United States, about 100 varieties are known. In Italy, Kiowa, Shoshoni, and Wichita are cultivated.

Related Species The genus *Carya* has 31 taxa of which 9 are of hybrid origin.

Geographic Distribution SE United States. **Habitat** Riparian environments, floodplains, well-drained soils. **Range** Cultivated in the United States (Texas, Georgia, Alabama, Louisiana, Oklahoma), Brazil, Australia, Israel; small cultivations in Italy (Sicily, Apulia).

Beer-Making Parts (and Possible Uses) Wood (to smoke the malt), pecans (kernels added to the wort), bark.

Chemistry Fruits: energy 2,889 kJ (690 kcal), water 3.52 g, carbohydrates 13.86 g, starch 0.46 g, sugars 3.97 g, fiber 9.6 g, lipids 71.97 g (saturated 6.18 g, monounsaturated 40.8 g, polyunsaturated 21.614 g), protein 9.17 g, beta-carotene 29 µg, lutein zeaxanthin 17 µg, vitamin A 56 IU, thiamine (B1) 0.66 mg, riboflavin (B2) 0.13 mg, niacin (B3) 1.17 mg, pantothenic acid (B5) 0.863 mg, vitamin B6 0.21 mg, folate (B9) 22 µg, vitamin C 1.1 mg, vitamin E 1.4 mg, vitamin K 3.5 µg, calcium 70 mg, iron 2.53 mg, magnesium 121 mg, manganese 4.5 mg, phosphorus 277 mg, potassium 410 mg, zinc 4.53 mg.

Wood Differentiated, with white sapwood and brown heartwood, sometimes tending to reddish or variegated, medium or coarse texture, straight grain; easily worked and holds nails but not ideal for gluing; nonresistant to alterations and insects (WSG 0.70–0.85). High mechanical resistance, therefore sought after for vehicles, sports equipment, handles.

Style American Amber Ale, American Brown Ale, American Light Lager, American Porter, Baltic Porter, Brown Porter, Dry Stout, Mild, Robust Porter, Spice/Herb/Vegetable Beer, Strong Scotch Ale, Sweet Stout.

Beer Pecan Porter, (512) Brewing Company (Texas, United States).

Source Cantwell and Bouckaert 2016, 2018; Daniels and Larson 2000; Hieronymus 2016b; Josephson et al. 2016; https://www.bjcp.org (32A. Classic Style Smoked Beer); https://www.brewersfriend.com/other/pacanes/; https://www.brewersfriend.com/other/pecan/; https://www.omnipollo.com/beer/90000/; https://www.omnipollo.com/beer/noa-double-barrel/.

Carya ovata (Mill.) K.Koch
JUGLANDACEAE

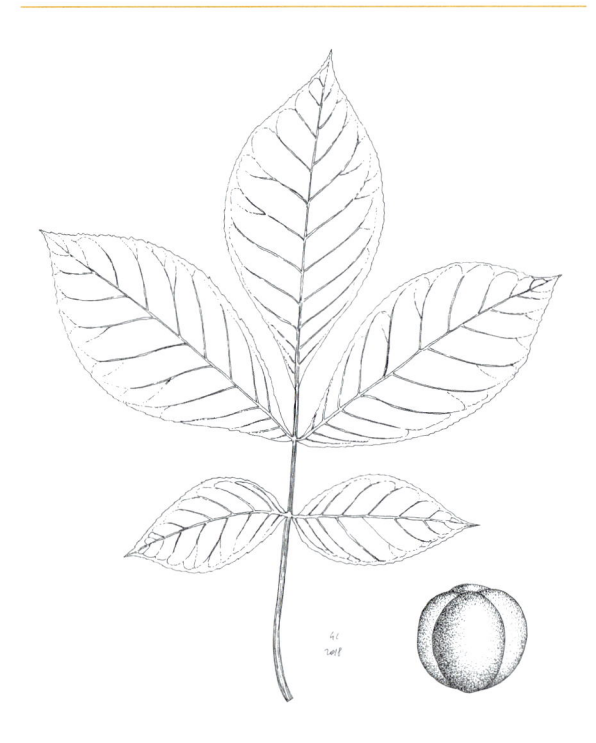

Synonyms *Carya borealis* (Ashe) C.K.Schneid., *Hicoria borealis* Ashe, *Hicorius borealis* Ashe, *Scoria ovata* (Mill.) Macmill.

Common Names Shagbark hickory, shellbark hickory.

Description Tree (up to 46 m). Trunk with light gray bark, fissured or exfoliating in long strips or wide plates that often curl on the trunk; greenish, reddish, or dark orange branches that hold their color or become black, dry, rigid or flexible, hirsute or glabrous; dark brown to black terminal buds, ovoid, 6–18 mm, tomentose or almost glabrous; perules imbricate; axillary buds protected by bracts fused in a sort of shield. Leaves 30–60 cm; petiole (4–13 cm) and rachis hirsute or mostly glabrous; leaflets (3–7), lateral petioles 0–1 mm, terminal petioles 3–17 mm; ovate, obovate, or elliptical, lamina nonfalcated (4–26 × 1–14 cm), margins finely to coarsely tightened, with tufts of hair in axillae of proximal ribs, apex from acute to sharp; lower surface hirsute for unicellular hairs and with 2–4 rays, occasionally restricted to the central and main ribs or substantially hairless, with a few to many large, irregular and small, round, irregular 4-lobed hairy scales. Male inflorescence pedunculated catkin, up to 13 cm, bracts and axis glabrous, anthers hirsute. Fruit color brown to reddish-brown, spherical to subspherical (2.5–4 × 2.5–4 cm); husk wrinkled, 4–15 mm thick, dehiscent at the base, smooth sutures; nut color light brown, ovoid, obovoid, or ellipsoid, compressed, 4-angled, wrinkled; thick casing. Seed sweet.

Subspecies The following taxa of varietal rank are recognized: *Carya ovata* (Mill.) K.Koch var. *ovata*, *Carya ovata* (Mill.) K.Koch var. *australis* (Ashe) Little, and *Carya ovata* (Mill.) K.Koch var. *mexicana* (Engelm. ex Hemsl.) W.E.Manning.

Cultivar Cultivars of this species exist as ornamental plants; some have been selected to improve the technological characteristics of the fruit.

Related Species The genus *Carya* includes 31 taxa of which there are at least a dozen recognized hybrids; another species whose use in making beer is known is *Carya cordiformis* (Wangenh.) K.Koch.

Geographic Distribution North America (Alaska to Texas). **Habitat** Wet basins, rocky hillsides, limestone outcrops. **Range** North America.

Beer-Making Parts (and Possible Uses) Wood (tested in making beer and for smoking malt), bark (tested in making beer), leaves, drupes (unripe and ripe).

Wood Differentiated, with white sapwood and brown heartwood, sometimes tending to reddish or variegated, medium or coarse texture, straight grain; easily worked and holds nails but not ideal for gluing; nonresistant to alterations and insects (WSG 0.70–0.85). High mechanical resistance, therefore sought after for vehicles, sports equipment, handles.

Style None known.

Source Cantwell and Bouckaert 2016; Cantwell and Bouckaert 2018; Hieronymus 2016b; Josephson et al. 2016.

Castanea dentata (Marshall) Borkh.
FAGACEAE

Synonyms *Castanea americana* (Michx.) Raf., *Fagus dentata* Marshall.

Common Names American chestnut.

Description Tree, often imposing in the past (up to 30 m), now found mostly as shoots (up to 5–10 m) owing to extensive destruction caused by a fungal disease. Bark gray, smooth when young, furrowed with age. Branches hairless. Leaves with petiole 8–40 mm, lamina strictly obovate to oblanceolate (9–30 × 3–10 cm), base wedge shaped, margins sharply tightened, each triangular tooth gradually thinning into a thick beard more than 2 mm long, apex acute or sharp; underneath often free of starry trichomes, with glabrous appearance but covered with evenly distributed, multicellular small hairs, glands sunk between the ribs and scattered, straight, simple trichomes concentrated on the ribs. Flowers stem with conspicuous pistils, straight whitish or yellowish hairs in the center of the flower. Pistillate flowers 3 per cupule. Fruits dome, 4-valve, including 3 flowers/fruits, irregularly dehiscent valves along 4 sutures at maturity, cupule thorns essentially glabrous, with a few scattered simple trichomes; nuts 3 per cupule, obovate, 18–25 × 18–25 mm, flattened on one or both sides, beak up to 8 mm excluding stylus.

Related Species The genus *Castanea* includes nine taxa of specific rank.

Geographic Distribution Canada (Ontario), W United States. **Habitat** In the past, common in rich mixed deciduous forests, particularly with *Quercus* species (0–1,200 m ASL). **Range** After 1930, many populations of *Castanea dentata* were almost destroyed by ascomycetes infection from *Cryphonectria parasitica* (Murrill) M.E.Barr (= *Endothia parasitica* [Murrill] P.J.Anderson & H.W.Anderson). Although the species still exists, in many places the plants are usually rejected and almost never produce fertile seeds or are hybrids with *Castanea sativa* Miller, *Castanea mollissima* Blume, or *Castanea crenata* Siebold & Zuccarini, the result of artificial crossbreeding aimed at creating disease-resistant *C. dentata* clones.

Beer-Making Parts (and Possible Uses) Seeds for starch (similar to Munich malt).

Chemistry Seeds: carbohydrates 86.26% (starch, sucrose, cellulose), ashes (calcium, magnesium, potassium).

Style Chestnut Ale.

Beer Winged Nut, Urban Chestnut (Missouri, United States).

Source Hieronymus 2016b.

Castanea sativa Mill.
FAGACEAE

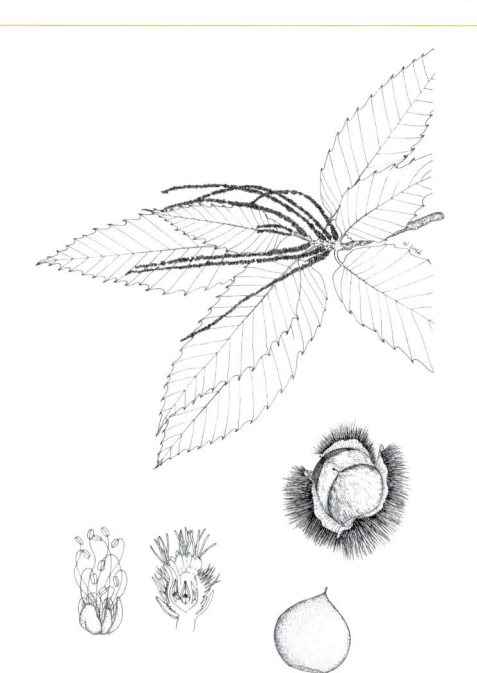

Synonyms *Castanea prolifera* (K.Koch) Hickel, *Castanea vesca* Gaertn., *Fagus castanea* L.

Common Names Chestnut, Spanish chestnut.

Description Tree (30–35 m), deciduous, long-lived (more than 500 years). Trunk remarkable (Ø 4–6 m); bark young and smooth, olive in color, then gray-brown retreat with long vertical grooves; branches of the year cylindrical or angled, bark smooth, red-brown in color. Leaves simple, alternate, spiral shaped, elliptical-lanceolate with serrated margin, wedge-rounded base; petiole 1.5–2.5 cm long with early deciduous stipules; upper side of summer leaves smooth, glossy, deep green with prominent ribs, underside light. Monoecious plant with mixed or even only male catkin inflorescences; male inflorescence consisting of flowers united in axillary glomerules (7 flowers) or in buds (40 per catkin), erect and up to 15 cm long; male flower with hexamerous perianth (6 parts) and 6–12 (20) stamina; shorter, more complex mixed inflorescences develop at the apex of the branch, consisting of about 20 axillary buds; at the base of the inflorescence are 1–4 female buds, each composed of 2–3 flowers enclosed by a dome; the following ones are formed by male flowers from 3–7 per tip (the apicals only 2); female flowers formed by a hexamerous perianth with an inferior ovary with 6–9 carpels and as many rigid, hairy stylus at the base; after fertilization, the scaly dome turns into a husk. Pollination mainly anemophilous. Husk (Ø 5–10 cm) thorny, normally containing 3 fruits (2–7); fruits achene with smooth, leathery brown pericarp, at the base with light scar (hilum) and at the apex remains of the stylus; cotyledons large, formed by a hard ivory-colored pulp, protected by a light brown membrane film (episperm).

Cultivar There are hundreds (e.g., Brown Chestnut of Cuneo, Garfagnana Chestnut): often they are quite homogeneous populations, genetically propagated by grafting, ± adapted to specific territories and environmental conditions. The modern selections are more aimed at improving resistance to fungi (*Chyphonectria parasitica*, *Phytophthora cambivora*, *Phytophthora cinnamomi*) and insects (*Dryocosmus kuriphilus*) through hybridization with *Castanea* sp. of eastern origin (*Castanea crenata* Siebold & Zuccarini, *Castanea dentata* Borkhausen, *Castanea henryi* [Skan] Rehder & E.H.Wilson, *Castanea mollissima* Blume, *Castanea pumila* Miller *Castanea seguinii* Dode).

Related Species Besides *Castanea sativa* Mill., the genus *Castanea* includes eight taxa: *Castanea crenata* Siebold & Zucc., *Castanea dentata* (Marshall) Borkh., *Castanea henryi* (Skan) Rehder & E.H.Wilson, *Castanea mollissima* Blume, *Castanea × neglecta* Dode, *Castanea ozarkensis* Ashe, *Castanea pumila* (L.) Mill., and *Castanea seguinii* Dode.

Geographic Distribution SE Europe. **Habitat** Widely cultivated, moderately thermophilic and heliophilic species, mesophilic, acidophilic, prefers fresh and rich soils. **Range** South Korea, China, Italy, Turkey, Japan, Portugal, Spain, France, North Korea.

Beer-Making Parts (and Possible Uses) Seeds (raw, boiled, or roasted), wood (for barrels).

Chemistry Seeds: energy 891 kJ (213 kcal), water 48.65 g, carbohydrates 45.54 g, fiber 8.1 g, lipids 2.26 g (saturated 0.425 g, monounsaturated 0.780 g, polyunsaturated 0.894 g), protein 2.42 g (tryptophan 0.027 g, threonine 0.086 g, isoleucine 0.095 g, leucine 0.143 g, lysine 0.143 g, methionine 0.057 g, cystine 0.077 g, phenylalanine 0.102 g, tyrosine 0.067 g, valine 0.135 g, arginine 0.173 g, histidine 0.067 g, alanine 0.161 g, aspartic acid 0.417 g, glutamic acid 0.312 g, glycine 0.124 g, proline 0.127 g, serine 0.121 g), vitamin A 1 µg, thiamine (B1) 0.238 mg, riboflavin (B2) 0.168 mg, niacin (B3) 1.179 mg, pantothenic acid (B5) 0.509 mg, vitamin B6 0.376 mg, folate (B9) 62 µg, vitamin C 43 mg, calcium 27 mg, iron 1.01 mg, magnesium 32 mg, manganese 0.952 mg, phosphorus 93 mg, potassium 518 mg, sodium 3 mg, zinc 0.52 mg.

Wood Whitish or yellow-brown in the sapwood, more or less dark brown in the heartwood, with vascular system evident, making the texture appear coarse; grain normally straight; WSG 0.58–0.70. Drying and processing not particularly difficult. Has the advantage of being a very durable, solid wood. While coppice shoots are valuable for poles, weaving, and staves; the wood of tall trees is used for fixtures, furniture, beams, and veneers. Not much valued as a fuel, but its high tannin content makes it valuable for the leather tanning industry as a source of tannic extracts.

Style Barleywine, California Common Beer, Chestnut Ale, Chestnut Lager, Northern English Brown Ale.

Beer Bastarda Rossa, Amiata (Arcidosso, Tuscany, Italy).

Source Calagione et al. 2018; Cantwell and Bouckaert, 2016; Cantwell and Bouckaert 2018; Dabove 2015; Giaccone and Signoroni 2017; Sparrow 2015; https://www.brewersfriend.com/other/castagne/; https://www.sangabriel.it/it/birre/le-stagionali/birra-dell-apostolo.

The Chestnut: Italian Artisans' First Flagship

Looking back to the end of the nineties and to the first stirrings of the craft beer movement in Italy, one notices how many producers saw the chestnut as the ingredient that could best present their Italianness to the world. *Castanea sativa* is found throughout Italy in many varieties and methods of use. In Liguria, for example, the fruits are dried and smoked, whereas in Piedmont and Lombardy it is traditional to roast them in a fire, and in Central Italy it is the flour that is used. Processing methods that give rise to aromas, perfumes, and sensations with distinctive characteristics allowed Italian breweries to distinguish themselves at a time when they were still little known and poorly trusted. Some, like Amiata in Tuscany, have made the interesting choice to devote almost all their range to the many nuances of chestnuts. Today the scene has changed, and, especially owing to complications in the production phase, very few brewers still offer chestnut beers; nevertheless, this fruit continues to be a symbol of uniquely Italian craft beer.

Cedrela odorata L.
MELIACEAE

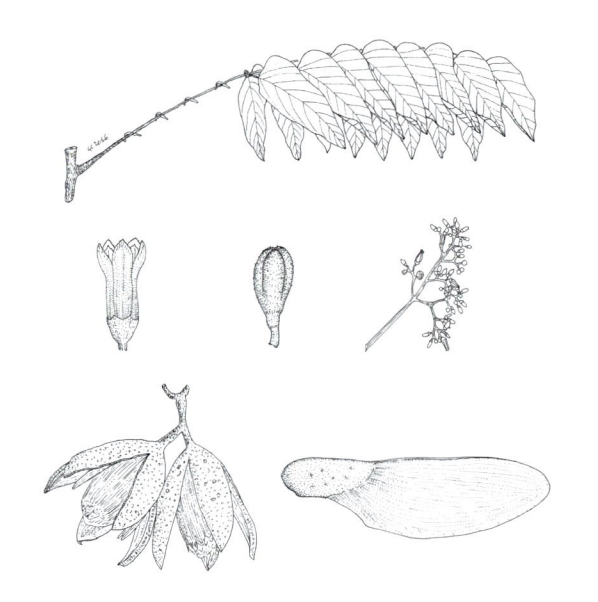

Synonyms *Cedrela brachystachya* (C.DC.) C.DC., *Cedrela brownii* Loefl. ex Kuntze, *Cedrela caldasana* C.DC., *Cedrela cedro* Loefl., *Cedrela cubensis* Bisse, *Cedrela glaziovii* C.DC., *Cedrela guianensis* A.Juss., *Cedrela hassleri* (C.DC.) C.DC., *Cedrela huberi* Ducke, *Cedrela imparipinnata* C.DC., *Cedrela longipes* S.F.Blake, *Cedrela mexicana* M.Roem., *Cedrela mourae* C.DC., *Cedrela occidentalis* C.DC. & Rose, *Cedrela odorata* Vell., *Cedrela palustris* Handro, *Cedrela paraguariensis* Mart., *Cedrela rotunda* S.F.Blake, *Cedrela sintenisii* C.DC., *Cedrela vellozoana* M.Roem., *Cedrela whitfordii* S.F.Blake, *Cedrela yucatana* S.F.Blake, *Cedrus odorata* Mill., *Surenus glaziovii* (C.DC.) Kuntze, *Surenus guianensis* (A.Juss.) Kuntze, *Surenus mexicana* (M.Roem) Kuntze, *Surenus paraguariensis* (Mart.) Kuntze, *Surenus vellozoana* (M.Roem) Kuntze.

Common Names Spanish cedar, Cuban cedar, cigar-box wood, red cedar.

Description Tree (40–60 m), monoecious, deciduous. Trunk straight, cylindrical (Ø 120–300 cm), without branches up to about 25 m; buttresses absent or modest up to 2 m; bark wrinkled and fissured, red-brown at the base of the trunk, grayish above; inner bark pink or purple-red; young branches ± densely lenticulate. Leaves alternate, paripinnate with 5–15 pairs of leaflets; leaflets from opposite to alternate, entire, ovate to oblong-lanceolate, 5–16 cm long, usually glabrous, oblique base, apex sharp to shortly acuminate. Inflorescence in terminal panicles. Flowers unisexual but with well-developed vestiges of the opposite sex, actinomorphic, pentamerous, greenish-white, subsessile, 6–9 mm long, with garlic odor; calyx cup shaped, divided on one side, from superficially to deeply toothed; petals free, imbricate and adorned for 1/3 of their length, white or creamy red shaded toward the margin; stamens 5, free, dorsifixed anthers; ovaries 5-locular, pubescent, each loculus with 10–14 ovules, stylus short, stigma discoid. Fruit capsule oblong-ellipsoid to obovoid (1.5–4 cm long), reddish-brown, pendulous, with 5 thin woody valves. Seeds winged.

Related Species There are 17 species belonging to the genus *Cedrela*.

Geographic Distribution West Indies, Trinidad and Tobago, CS America. **Habitat** Heliophilous and intolerant of water stagnation, typical species of primary and secondary evergreen and semideciduous rain forests of lowlands and low mountains. **Range** Cultivated in tropical climate areas around the world (Costa Rica, Fiji, Indonesia, Kenya, Madagascar, Malaysia, Nigeria, Philippines, Samoa, Singapore, Solomon Islands, South Africa, Tanzania, Thailand, Uganda, United States).

Beer-Making Parts (and Possible Uses) Wood for barrels (first they contain cachaça, then beer for aging).

Wood Relatively light and soft (density 410–525 kg/m^3); heartwood cream-white when freshly cut, then pink-brown, clearly distinguished by a narrow band of sapwood; the first is valuable and durable, unlike the second. Grain generally interconnected, straight or woolly, with the presence of tension wood; moderately fine to moderately coarse texture; when freshly cut it has a distinctive onion smell that disappears in two to three days. Easy to work and finish, although internal tensions can sometimes produce cracks; many applications in carpentry, among which perhaps the best known is the making of cigar boxes; attracting increasing interest for the production of barrels.

Style Wood-Aged Beer.

Source Cantwell and Bouckaert 2016; Cantwell and Bouckaert 2018; https://www.bjcp.org (33A. Wood-Aged Beer).

Centaurea benedicta (L.) L.
ASTERACEAE

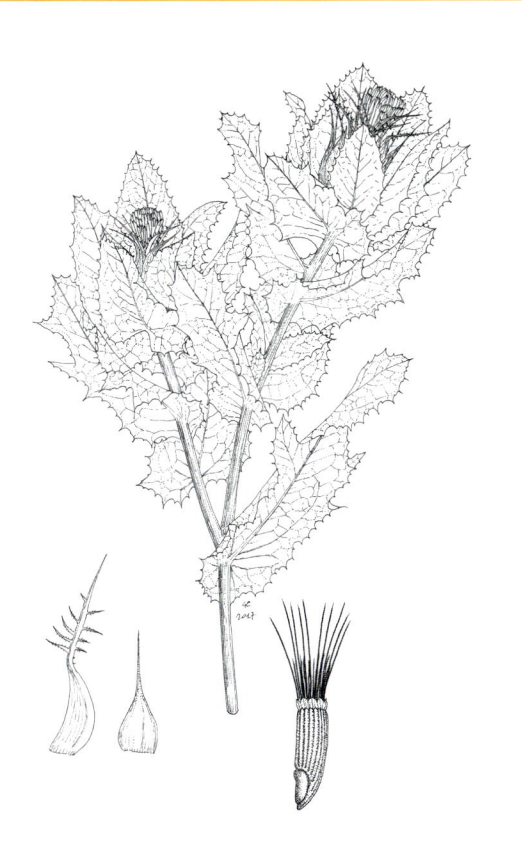

Synonyms *Benedicta officinalis* Bernh., *Calcitrapa benedicta* (L.) Sweet, *Calcitrapa lanuginosa* Lam., *Carbeni benedicta* (L.) Adans., *Carbeni benedicta* (L.) Arcang, *Cardosanctus officinalis* Bubani, *Carduus benedictus* Auct. ex Steud., *Carduus benedictus* (L.) Garsault, *Centaurea centriflora* Friv. ex Gugler, *Centaurea pseudobenedicta* (Asch.) E.H.L.Krause, *Cirsium horridum* (Adams) Petr., *Cnicus benedictus* L., *Cnicus bulgaricus* Panov, *Cnicus kotschyi* Sch.Bip., *Cnicus microcephalus* Boiss., *Cnicus pseudo-benedictus* hort.dorpat. ex Asch., *Epitrachys microcephala* K.Koch.

Common Names Blessed thistle.

Description Herb, annual (20–60 cm). Taproot white. Stem hairy-lush, purple-reddish in color, erect, spreading-ramose, leafy up to the apex. Leaves pale green, glandulous-wooly, alternate, sinuate-pinnatifid, rachis white ribbed; basal leaves slightly petiolate, up to 30 cm long, cauline leaves semiamplexicaul or briefly flowing on the stem, all with triangular toothed spiny segments. Inflorescences of solitary capitula (Ø 2–3 cm), apical, surrounded by a set of large leafy bracts, inserted under the capitulum itself; involucre globose-cobwebby and covered with variable scales, ovate-acute, red-yellowish, toothed-spiny, ending with a long pinnate thorny appendage. Flower with yellow corolla, purple-greenish veins; flowers all tubular, sterile exteriors, hermaphrodite interiors. Fruits cylindrical achene (8–11 mm) slightly curved, with about 20 longitudinal ribs, surmounted by a pappus composed of a membranous crown and 2 sets of hairs (short inner bristles, middle stiff bristles).

Related Species There are about 900 taxa (species and subspecies) in the genus *Centaurea*.

Geographic Distribution W Mediterranean. **Habitat** Uncultivated, ruins, abandoned fields (0–800 m ASL). **Range** Cultivated as a medicinal plant.

Beer-Making Parts (and Possible Uses) Heads, leaves, stems (used for bitterness).

Chemistry Cnicin 0.2–0.7%, polyacetylene, absithin, salonitenolide, artemisiifoline, α-amyrenone, α-amyrin acetate, α-amyrin, multiflorenol acetate, trachelogenin, nortracheloside, arctigenin, apigenin-7-O-glucoside, luteolin, astragalin, tannins 8%, essential oils 0.3% (n-nonane, n-undecane, n-tridecane, dodeca-1,11-dien-3,5,7,9-tetrain, p-cymene, fenchone, citral, cinnamaldehyde, phenolic compounds, saponins, alkaloids, starch, glycosides, triperpenes, coumarin.

Style Bitter, Mum.

Beer Blessed Thistle, Cairngorm (Aviemore, Scotland, United Kingdom).

Source Fisher and Fisher 2017; Heilshorn 2017; Hieronymus 2010; Hieronymus 2016b.

Beer-Making Parts (and Possible Uses)
Flowering tops, flowers (early in boiling for bittering, late for flavoring).

Chemistry Essential oil (major components): chamazulene 27.8%, β-pinene 7.93%, 1.8-cineole 7.51%, α-pinene 5.94%, α-bisabolol 5.76%.

Style Blanche.

Beer MCDXCII—Eataly Birreria Roma (Rome, Lazio, Italy).

Source Fisher and Fisher 2017; Hieronymus 2016b; https://www.effettofood.com/la-nuova-birreria-di-eataly-roma-atmosfera-accogliente-e-birre-di-qualita/.

Chenopodium album L.
AMARANTHACEAE

Synonyms *Anserina candidans* (Lam) Montandon, *Atriplex alba* (L.) Crantz, *Atriplex viridis* (L.) Crantz, *Blitum viride* (L.) Moench, *Botrys alba* (L.) Nieuwl., *Botrys pagana* (Rchb.) Lunell, *Chenopodium agreste* E.H.L.Krause, *Chenopodium bernburgense* (Murr) Druce, *Chenopodium bicolor* Bojer ex Moq., *Chenopodium borbasiforme* (Murr) Druce, *Chenopodium borbasii* F.Murr, *Chenopodium browneanum* Schult., *Chenopodium candicans* Lam., *Chenopodium catenulatum* Schleich. ex Steud., *Chenopodium concatenatum* Willd., *Chenopodium* × *densifoliatum* (Ludw. & Aellen) F.Dvořák, *Chenopodium elatum* Shuttlew. ex Moq., *Chenopodium glomerulosum* Rchb., *Chenopodium laciniatum* Roxb., *Chenopodium lanceolatum* Muhl. ex Willd., *Chenopodium leiospermum* DC., *Chenopodium lobatum* (Prodán) F.Dvořák, *Chenopodium missouriense* Aellen, *Chenopodium neglectum* Dumort., *Chenopodium neoalbum* F.Dvořák, *Chenopodium opulaceum* Neck., *Chenopodium ovalifolium* (Aellen) F.Dvořák, *Chenopodium paganum* Rchb., *Chenopodium paucidentatum* (Aellen) F.Dvořák, *Chenopodium pedunculare* Bertol., *Chenopodium probstii* Aellen, *Chenopodium riparium* Boenn. ex Moq., *Chenopodium subaphyllum* Phil., *Chenopodium superalbum* F.Dvořák, *Chenopodium viride* L., *Chenopodium viridescens* (St.-Amans) Dalla Torre & Sarnth., *Chenopodium vulgare* Gueldenst. ex Ledeb., *Chenopodium zobelli* A. Ludw. & Aellen.

Common Names Lamb's quarter, melde, white goosefoot, manure weed, fat-hen.

Description Herb, annual (10–200 cm). Stem erect, irregularly branched from the base, ± glaucous, ribbed, with vermillion streaks at the axils of the branches and floury at the top. Leaves polymorphous, petiolate (1–2.5 cm), ± farinose, faces concolorous, alternate, lamina rhombic-ovate to lanceolate (3–6 × 2.5–8 cm), usually 1.5 times as long as wide, progressively narrowed at base and subobtuse at apex; margin entire or sinuously toothed (3–6 pairs of teeth per side) or serrated; minor leaves 3–4 cm lanceolate, subentire with powdery lamina below; young leaves may show redness near the petiole junction or throughout the underside and in the leaf margin. Inflorescence spiciform and leafy, formed by compact glomerules, subglobose and interrupted (3–4 mm), placed at the axil of the primary and secondary branches; the spikes of the upper part are more elongated. Flowers hermaphrodite, greenish in color (Ø 1.5 mm); perianth segments 5 ovate-elliptic, keeled and with membranaceous margin, covering the fruit to maturity, stamens 5; stigmas 2 (0.2–0.3 mm); pericarp adnate. Fruit utricle depressed-ovoid (1.5 mm), enveloping 1 obovate seed (1.2–1.6 mm), black, round, lenticular, horizontally arranged, with flattened to acute margin and slightly rugose-crested surface.

Subspecies The taxonomic situation within the taxon *Chenopodium album* L. is rather controversial. According to some authors, the following taxa of subspecific rank should be recognized: *Chenopodium album* L. subsp. *iranicum* Aellen and *Chenopodium album* L. subsp. *pseudostriatum* Zschacke, but also two elements of varietal rank as *Chenopodium album* L. var. *boscianum* (Moq.) A.Gray and *Chenopodium album* L. var. *reticulatum* (Aellen) Uotila. However, this classification is not universally accepted.

Related Species The genus *Chenopodium* has 164 taxa, including some suitable for human consumption (both wild and cultivated species).

Geographic Distribution Cosmopolitan.
Habitat Weeded crops, spring cereals, roadsides, ruins, paths, nitrogen-rich and ± sunny soils (0–1,500 m ASL). **Range** Noncultivated cosmopolitan species.

Beer-Making Parts (and Possible Uses) Leaves, stems, seeds.

Chemistry Essential oil, leaves (0.64% volume/weight of extract; aromatic compounds 60.1%): p-cymene 40.9%, ascaridole 15.5%, pinan-2-ol 9.9%, α-pinene 7.0%, β-pinene 6.2%, α-terpineol 6.2%. Composition: total phenols 224.99–304.98 mg, simple phenols 72.50–101.007 mg, tannins 152.49–203.91 mg, saponins 0.043–0.867 g, phytic acid 238.3–268.33 mg, phytate phosphorus 67.16–75.62 mg, alkaloids 1.27–1.53 mg, flavonoids 220.0–406.67 mg, oxalates 394.19–477.08 mg, oils, proteins, trace elements, other bioactive components.

Style Herb Beer.

Source Heilshorn 2017; Hieronymus 2016b.

Chenopodium quinoa Willd.
AMARANTHACEAE

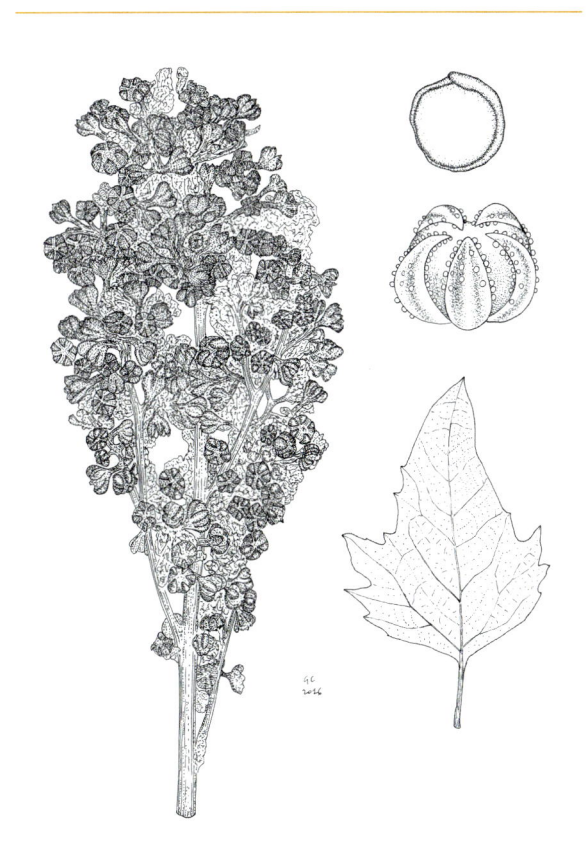

Synonyms *Chenopodium canihua* Cook, *Chenopodium ccoyto* Toro Torrico, *Chenopodium ccuchi-huila* Toro Torrico, *Chenopodium guinoa* Krock., *Chenopodium nuttalliae* Saff.

Common Names Quinoa.

Description Herb, annual (0.3–3 m). Central stem woody and erect, ± branched (varietal character), color green, red, or purple. Leaves alternate, lamina broad, usually pubescent, powdery, margin smooth, toothed or lobed. Flowers hermaphrodite or female, perigonium composed of 5 green tepals, 5 stamens, feathery pistil with 2–3 stigmas and superior unilocular ovary. Inflorescence terminal panicle from which emerge secondary axes carrying flowers (amaranthiform) or tertiary axes with flowers (glomeruliform). Fruits (Ø about 2 mm) variable in color (white, red, black).

Cultivar There are more than 200 varieties of quinoa; the one most used for its low saponin content is Real; other cultivars are Bear, Cherry Vanilla, Cochabamba, Dave 407, Faro, Gossi, Isluga, Kaslatla,

Kcoito, Linares, Rainbow, Red Faro, Red Head, and Temuco.

Related Species The genus *Chenopodium* has 164 taxa, some of which are suitable for human consumption (both wild and cultivated species).

Geographic Distribution Peruvian Andes. **Habitat** Mountainous environments. **Range** Peru, Bolivia, Ecuador.

Beer-Making Parts (and Possible Uses) Seeds.

Toxicity The seed coating contains saponins (toxic glycosides) that impart an unpleasant bitter taste and can cause irritation to the respiratory and gastrointestinal systems and have hemolytic effects if applied directly to the blood; the coating must therefore be removed from any marketed seeds.

Chemistry Energy 1,539 kJ (368 kcal), water 13.3 g, carbohydrates 64.2 g, fiber 7 g, lipids 6.1 g (monounsaturated 1.6 g, polyunsaturated 3.3 g), protein 14.1 g, vitamin A 1 µg, thiamine (B1) 0.36 mg, riboflavin (B2) 0.32 mg, niacin (B3) 1.52 mg, vitamin B6 0.49 mg, folate (B9) 184 µg, choline 70 mg, vitamin E 2.4 mg, calcium 47 mg, iron 4.6 mg, magnesium 197 mg, manganese 2 mg, phosphorus 457 mg, potassium 563 mg, sodium 5 mg, zinc 3.1 mg.

Style Kölsch, Kosher (in 2013, quinoa was declared kosher, therefore suitable for consumption even during Passover), Stout.

Beer Survival, Hub (Oregon, United States).

Source Fisher and Fisher 2017; https://www.brewersfriend.com/other/quinoa/.

Chrysanthemum indicum L.
ASTERACEAE

Synonyms *Arctotis elegans* Thunb., *Bidens marginata* DC., *Chrysanthemum japonicum* Thunb., *Chrysanthemum koraiense* Nakai, *Chrysanthemum lushanense* Kitam., *Chrysanthemum nankingense* Hand.-Mazz., *Chrysanthemum procumbens* Lour., *Chrysanthemum purpureum* Pers., *Chrysanthemum tripartitum* Sweet, *Collaea procumbens* (Rich.) Spreng, *Dendranthema indicum* (L.) Des Moul., *Dendranthema nankingense* (Hand.-Mazz.) X.D.Cui, *Matricaria indica* (L.) Desr., *Pyrethrum indicum* (L.) Cass., *Tanacetum indicum* (L.) Sch.Bip.

Common Names Chrysanthemum.

Description Herb, perennial (2.5–10 dm). Rhizome superficial. Leaves petiolate (1–2 cm), lamina oval or elliptic-oval (3–7 × 2–4 cm), pubescent, pinnatifid, pinnate lobed or slightly lobed. Inflorescence heads in loose terminal cymes, each wrapped in an involucre formed by 5 rows of bracts (phyllaries), scarious. Flowers of 2 types: peripheral female, fertile, bearing a corolla with a tube prolonged laterally in a tongue (ligula), yellow in color (1–1.3 cm), apex tridentate; central hermaphrodite, with tubular corolla, yellow in color, apex 5-dentate. Fruits achenes (1.5–1.8 mm), subterete to obovoid, without pappus.

Cultivar Among those mostly used in China are Huángshān Gòngjú (or Gòngjú), Hángbáijú (or Hángjú), Chújú, and Bójú.

Related Species There are 42 taxa in the genus *Chrysanthemum*.

Geographic Distribution E Asia (E China, CS Japan). **Habitat** Upland grasslands, thickets, wet areas near rivers, fields, roadsides, salt marshes near the sea, under shrubby vegetation (100–2,900 m ASL). **Range** Asia.

Beer-Making Parts (and Possible Uses) Capitula, flowers.

Chemistry Essential oil, capitula: α-pinene 14.63%, 1,8-cineole 10.71%, germacrene D 5.25%, (-)-sinularene 3.95%, β-bisabolene 3.95%, bornyl acetate 3.64%, β-elemene 3.18%, borneol 3.02%.

Style Mum, Saison.

Source McGovern 2017; https://www.beeradvocate.com/beer/profile/31894/147138/ (species chosen as an example of *Chrysanthemum* sp.); https://www.mumbeer.be/en.

Chrysanthemum morifolium Ramat.
ASTERACEAE

Synonyms *Anthemis artemisifolia* Willd., *Anthemis grandiflora* Ramat., *Anthemis stipulacea* Moench, *Chrysanthemum hortorum* W.Mill., *Chrysanthemum maximoviczianum* Ling, *Chrysanthemum procumbens* Blume, *Chrysanthemum sabini* Lindl., *Chrysanthemum sinense* Sabine, *Chrysanthemum sinense* Sabine ex Sweet, *Dendranthema sinensis* (Sabine) Des Moul., *Matricaria morifolia* (Ramat.) Ramat., *Pyrethrum sinense* (Sabine) DC., *Tanacetum sinense* (Sabine) Sch.Bip.

Common Names Chrysanthemum, florist's daisy.

Description Herb (30–90 cm), annual or perennial. Leaves alternate, simple, base truncate, cordate, or sagittate, margin lobed and toothed, leaf apex obtuse, rounded. Inflorescence capitulum (Ø 30–60 mm) pedunculated (about 50 mm), enveloped by cycles of lanceolate-oblong or ovate bracts; inner disk flowers tubular, yellow, ray flowers ligulate, yellow or orange. Fruit achene, pappus absent or small (0.5 mm).

Cultivar There are dozens.

Related Species There are 42 taxa in the genus *Chrysanthemum*.

Geographic Distribution *Chrysanthemum morifolium* is a species of probable hybridogenic origin between *Chrysanthemum indicum* L. and *Chrysanthemum japonicum* (Maxim.) Makino (= *Dendranthema japonicum* [Makino] Kitam.). **Habitat** Anthropized environments. **Range** Widely cultivated in China for phytotherapeutic purposes and elsewhere as an ornamental plant.

Beer-Making Parts (and Possible Uses) Flowers.

Chemistry Essential oil, flowers (13 components; 97.5% of the oil): cis-chrysanthenyl acetate 21.6%, octadecanoic acid 19.5%, borneol 15.5%. Essential oil, leaves (21 components; 93.7% of the oil): borneol 20.5%, 1,8-cineole 15.2%, trans-α-bergamotene 14%, camphor 10.6%, α-pinene 9.3%.

Style Mum, Saison.

Source https://www.beeradvocate.com/beer/profile/31894/147138/ (species chosen as an example of *Chrysanthemum* sp.); https://www.mumbeer.be/en.

Cichorium intybus L.
ASTERACEAE

Synonyms *Cichorium balearicum* Porta, *Cichorium byzantinum* Clem., *Cichorium byzantinum* Clementi, *Cichorium cicorea* Dumort., *Cichorium commune* Pall., *Cichorium divaricatum* Heldr. ex Nyman, *Cichorium glabratum* C.Presl, *Cichorium glaucum* Hoffmanns. & Link, *Cichorium hirsutum* Gren., *Cichorium illyricum* Borb., *Cichorium officinale* Gueldenst. ex Ledeb.

Common Names Radicchio, Italian chicory.

Description Herb, biennial or perennial (20–150 cm). Taproot cylindrical or conical, long and branched, when severed releases a bitter-tasting white sap. Stem erect, branched with stiff, spreading branches, hollow, angular, shaggy from downward-facing hairs. Leaves united in a basal rosette, petiolate, irregularly pinnate-partite with acute triangular segments, the primordial ones can also be nontoothed, undivided, hairy (in dry places) or glabrous (in grassy places), dark green, often suffused with red, especially on the central vein; alternate and sessile cauline leaves, the lower ones lobed and hairy above, the upper ones oblong and lanceolate, sheathing. Flowers all ligulate, with 5-toothed ligule, united in heads of 2–3 elements carried by short peduncles, intense blue color, rarely white or pink. Inflorescences with involucre arranged in 2 rows; bracts of the involucre ciliate, the outer ones short and oval, the inner ones oblong, lanceolate, and straight; inflorescences close in the afternoon and in bad weather. Fruits sand-colored achenes with 5 bristly sides on the rim, surmounted by pappus with very short flakes; pappus 1/10–1/8 the length of the achene.

Cultivar Red Treviso radicchio is one of the several cultivars typical of the Veneto region, obtained by selection from the species *Cichorium intybus*; another, often used in Roman cuisine, is catalogna, locally known as puntarelle; there are many others.

Related Species The genus *Cichorium* has 10 species: *Cichorium alatum* Hochst. & Steud., *Cichorium bottae* Deflers, *Cichorium callosum* Pomel, *Cichorium calvum* Sch.Bip. ex Asch., *Cichorium dubium* E.H.L.Krause, *Cichorium endivia* L., *Cichorium hybridum* Halácsy, *Cichorium intybus* L., *Cichorium pumilum* Jacq., and *Cichorium spinosum* L.

Geographic Distribution Cosmopolitan.

Habitat Grassy areas and uncultivated fields, along the edges of roads. **Range** The production area of Treviso radicchio, authorized by the Protected Geographical Indication (PGI) specification, includes municipalities in the province of Treviso (Carbonera, Casale sul Sile, Casier, Istrana, Mogliano Veneto, Morgano, Paese, Ponzano Veneto, Preganziol, Quinto di Treviso, Silea, Spresiano, Trevignano, Treviso, Vedelago, Villorba, Zero Branco), in the province of Padua (Piombino Dese, Trebaseleghe), and in the province of Venice (Martellago, Mirano, Noale, Salzano, Scorzè); other varieties are cultivated everywhere.

Beer-Making Parts (and Possible Uses) Leaves.

Chemistry Low in calories, water 92–94%, rich in antioxidants, vitamins A, B1, B2, calcium, iron, anthocyanins, tryptophan.

Style American Stout, Brown Porter, Imperial IPA, Northern English Brown Ale, Old Ale, Robust Porter, Specialty Beer, Sweet Stout.

Beer Radicchio, Ivan Borsato (Camalò, Veneto, Italy).

Source Calagione et al. 2018; Dabove 2015; Giaccone and Signoroni 2017; Hieronymus 2016b; Josephson et al. 2016; Steele 2012; https://www.birradegliamici.com/news-birraie/birre-strane/; https://www.brewersfriend.com/other/chicory/; https://www.marcadoc.it/gustare/Birra-Ambra-Rossa-San-Gabriel-la-birra-al-Radicchio-Rosso-di-Treviso.htm.

Cichorium intybus L.
ASTERACEAE

135

Cinchona calisaya Wedd.
RUBIACEAE

Synonyms *Cinchona amygdalifolia* Wedd., *Cinchona australis* Wedd., *Cinchona carabayensis* Wedd., *Cinchona delondriana* Wedd., *Cinchona euneura* Miq., *Cinchona forbesiana* Howard ex Wedd., *Cinchona gammiana* King, *Cinchona gironensis* Mutis, *Cinchona hasskarliana* Miq., *Cinchona ledgeriana* (Howard) Bern. Moens ex Trimen, *Cinchona pahudiana* Howard, *Cinchona peruviana* Howard, *Cinchona thwaitesii* King, *Cinchona weddelliana* Kuntze, *Quinquina calisaya* (Wedd.) Kuntze, *Quinquina carabayensis* (Wedd.) Kuntze, *Quinquina ledgeriana* (Howard) Kuntze.

Common Names Calysaia, bark, Bolivian bark, yellow bark.

Description Shrub or sapling (3–5 m). Leaves opposite, whole, broadly ovate-rounded or elliptic (3.5–4 × 10 cm), obtuse, lamina above glabrous, below pubescent; petiole short. Flowers pleasant smelling, corolla tubular, 5-lobed, white to pink in color, externally densely pubescent, long hairy laciniae at margins; anthers emerging from fauces; style short (about 1/2 of corolla tube); calyx with triangular, acute teeth. Inflorescence panicle triramified. Fruit capsule ovoid or subcylindrical (4–6 mm), ribbed, opening from base. Seeds numerous, winged on both sides, apex bifid, margin dentate-fimbriate.

Related Species The genus *Cinchona* has 25 species of which only two, *Cinchona officinalis* L. and *Cinchona calisaya* Wedd., are used for making beer.

Geographic Distribution S America, Peru, Bolivia. **Habitat** Mountain slopes of the Andes. **Range** W South America (Bolivia, Peru), India, Java.

Beer-Making Parts (and Possible Uses) Bark.

Chemistry Most important alkaloids: quinine, cinchonine, quinidine.

Style Russian Imperial Stout.

Beer Marchè 'L Re, LoverBeer (Marentino, Piedmont, Italy).

Source https://www.brewersfriend.com/other/kinabark/.

Cinchona officinalis L.
RUBIACEAE

Synonyms *Cascarilla officinalis* (L.) Ruiz, *Cinchona academica* Guibourt, *Cinchona chahuarguera* Pav., *Cinchona coccinea* Pav. ex DC., *Cinchona colorata* Lambert, *Cinchona condaminea* Bonpl., *Cinchona crispa* Tafalla ex Howard, *Cinchona legitima* Ruiz ex Lamb., *Cinchona palton* Pav., *Cinchona peruviana*

Mutis, *Cinchona suberosa* Pav. ex Howard, *Cinchona uritusinga* Pav. ex Howard, *Hindsia subandina* Krause, *Quinquina officinalis* (L.) Kuntze, *Quinquina palton* (Pav.) Kuntze.

Common Names Chinabark, quinine tree, lojabark, quinine, red cinchona, cinchona bark, Jesuitŏs bark, loxa bark, Jesuitŏs powder, countess powder, Peruvian bark.

Description Shrub or small tree (4–6 m). Bark wrinkled, branches covered with minute hairs. Leaves lanceolate to elliptic or ovate (10 × 3.5–4 cm), acute, acuminate, base attenuate, leathery, above glabrous and often shiny, below glabrous or puberulent or short hairs, especially on the veins. Stipules lanceolate or oblong, acute or obtuse, glabrous. Inflorescences terminal panicles, multiflorous. Flower with hypanthium with short, coarse hairs; calyx reddish, glabrous or nearly so, with triangular lobes; corolla pink or red, silky, with ovate, acute lobes, coralline tube about 1 cm long; 5 stamens included in corolla; ovary 2-locular, stigma 2-lobed. Fruit capsule oblong (1.5–2 cm), nearly glabrous.

Related Species The genus *Cinchona* has 25 species of which only two, *Cinchona officinalis* L. and *Cinchona calisaya* Wedd., have use in beer making.

Geographic Distribution W South America (Colombia, Ecuador, Peru, Bolivia). **Habitat** Rain forest. **Range** Cultivated in the original area and in tropical areas.

Beer-Making Parts (and Possible Uses) Bark.

Chemistry Most important alkaloids: quinine, cinchonine, quinidine.

Style American Barleywine.

Beer Quirks and Quinine, De Molen (Bodegraven, Netherlands).

Source https://www.brewersfriend.com/other/quinine/.

Cinnamomum verum J.Presl
LAURACEAE

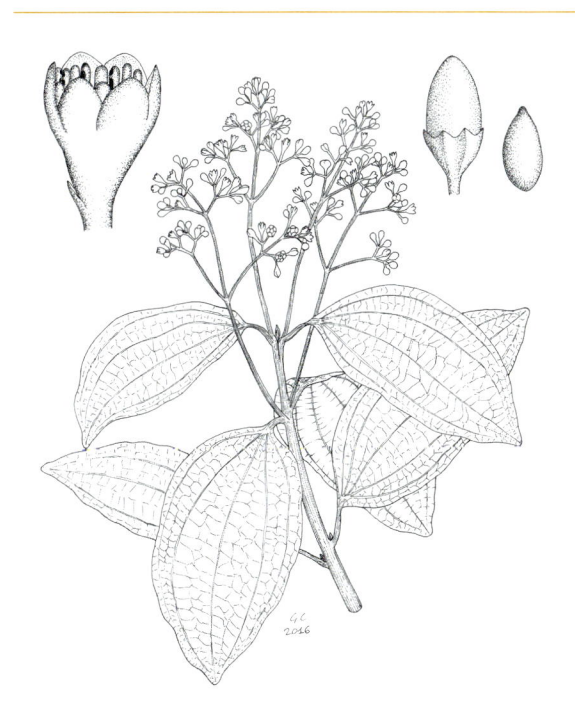

Synonyms *Camphorina cinnamomum* (L.) Farw., *Cinnamomum alexei* Kosterm., *Cinnamomum aromaticum* J.Graham, *Cinnamomum barthii* Lukman., *Cinnamomum bengalense* Lukman., *Cinnamomum biafranum* Lukman., *Cinnamomum bonplandii* Lukman., *Cinnamomum boutonii* Lukman., *Cinnamomum capense* Lukman., *Cinnamomum cayennense* Lukman., *Cinnamomum commersonii* Lukman., *Cinnamomum cordifolium* Lukman., *Cinnamomum decandollei* Lukman., *Cinnamomum delessertii* Lukman., *Cinnamomum ellipticum* Lukman., *Cinnamomum erectum* Lukman., *Cinnamomum humboldtii* Lukman., *Cinnamomum karrouwa* Lukman., *Cinnamomum leptopus* A.C.Sm., *Cinnamomum leschenaultii* Lukman., *Cinnamomum madrassicum* Lukman., *Cinnamomum maheanum* Lukman., *Cinnamomum mauritianum* Lukman., *Cinnamomum meissneri* Lukman., *Cinnamomum ovatum* Lukman., *Cinnamomum pallasii* Lukman., *Cinnamomum pleei* Lukman., *Cinnamomum pourretii* Lukman., *Cinnamomum regelii* Lukman., *Cinnamomum roxburghii* Lukman., *Cinnamomum sieberi* Lukman., *Cinnamomum sonneratii* Lukman., *Cinnamomum vaillantii* Lukman., *Cinnamomum variabile* Lukman., *Cinnamomum wolkensteinii* Lukman., *Cinnamomum zeylanicum* Blume, *Cinnamomum zollingeri* Lukman., *Laurus cinnamomum* L.

Common Names True cinnamon tree, Ceylon cinnamon tree.

Description Tree, evergreen (8–17 m). Trunk (Ø 30–60 cm) solid, bark thick and gray, branches downward. Leaves stiff, stipulate, opposite, somewhat variable in shape and size; petiole 1–2 cm long, grooved on the upper surface; lamina usually ovate or elliptic (5–18 × 3–10 cm), with more or less rounded base and rather sharp apex; 3–5 conspicuous longitudinal veins starting from the base of the lamina and almost reaching the apex; young leaves reddish in color, then dark green above, lighter veins, glaucous and pale on the underside. Flowers very small (Ø 3 mm), pale yellow, with fetid odor, each subtended by a small ovate bract; calyx bell shaped and pubescent with 6 sharply pointed segments; corolla absent. Flowers collected in loose axillary and terminal panicles at the ends of the branches; peduncles creamy-white, slightly hairy, 5–7 cm long. Fruit ovoid, fleshy, black drupe, 1.5–2 cm long when ripe, with the enlarged calyx persistent at the base.

Cultivar They are divided into types: type 1 (Pani Kurundu, Pat Kurundu, Mapat Kurundu), type 2 (Naga Kurundu), type 3 (Pani Miris Kurundu), type 4 (Weli Kurundu), type 5 (Sewala Kurundu), type 6 (Kahata Kurundu), and type 7 (Pieris Kurundu).

Related Species The genus *Cinnamomum* includes 344 species.

Geographic Distribution India, Sri Lanka. **Habitat** Countries with a tropical climate. **Range** Cultivated in countries with a tropical climate (India, Sri Lanka, Madagascar, Malaysia, Vietnam, Sumatra, Indonesia).

Beer-Making Parts (and Possible Uses) Bark.

Toxicity The potential toxicity of cinnamon comes from its coumarin content, known to cause liver and kidney damage. The European Food Safety Authority recommends a maximum daily dose of 0.1 mg of coumarin per kg of body weight. The coumarin content of *Cinnamomum verum* is rather low (< 0.10 mg/g), whereas it is much higher and more variable (0.10–12.18 mg/g) in *Cinnamomum cassia* (L.) J.Presl, often sold as cinnamon just like *Cinnamomum verum*.

Chemistry The aroma of cinnamon comes from the essential oil contained in the dried bark (0.5–1%), composed of about 90% of cinnamaldehyde, which by oxidation forms resins; other components are ethyl cinnamate, eugenol, beta-caryophyllene, linalool, and methyl caviculus.

Style American Amber Ale, American Brown Ale, American IPA, American Light Lager, American Stout, Belgian Dark Strong Ale, Belgian Tripel, Brown Porter, California Common Beer, Dunkelweizen, Gruit, Holiday/Winter Special Spiced Beer, Metheglin, Mild, Oatmeal Stout, Old Ale, Robust Porter, Saison, Special/Best/Premium Bitter, Specialty Beer, Spice/Herb/Vegetable Beer, Sweet Stout, Vienna lager.

Beer Felina, Menaresta (Carate Brianza, Lombardy, Italy).

Source Calagione et al. 2018; Dabove 2015; Heilshorn 2017; Hieronymus 2010; McGovern 2017; https://www.beeradvocate.com/beer/style/72/; https://www.bjcp.org (30B. Autumn Seasonal Beer); https://www.fermentobirra.com/homebrewing/birre-speziali/; https://www.omnipollo.com/beer/symzonia/.

🌿 Spicy Christmas Beers

Christmas beers are an ancient European tradition, but their origin is unclear. Some link them to beverages produced in Roman times to celebrate Saturnalia; for others, their creation is linked to traditional Northern European celebrations of Yule, the winter solstice. No matter their history, the term "Christmas beers" identifies a wide group of beverages spread across many European countries and characterized by a high ABV. Especially in Belgium, Kerstbier beers have an additional peculiarity: their intense spiciness. The main ingredient of Christmas beers is often *Cinnamomum verum*; however, many other spices are often used, such as star anise, clove, pepper, and raisin, which are also used in the sweets of this period.

Citrullus lanatus (Thunb.) Matsum. & Nakai
CUCURBITACEAE

Synonyms *Anguria citrullus* Mill., *Citrullus amarus* Schrad., *Citrullus anguria* (Duchesne) H.Hara, *Citrullus aquosus* Schur, *Citrullus battich* Forssk., *Citrullus caffer* Schrad., *Citrullus caffrorum* Schrad., *Citrullus chodospermus* Falc. & Dunal, *Citrullus citrullus* H.Karst., *Citrullus citrullus* Small, *Citrullus edulis* Spach, *Citrullus mucosospermus* (Fursa) Fursa, *Citrullus pasteca* Sageret, *Citrullus vulgaris* Schrad., *Colocynthis amarissima* Schltdl., *Colocynthis citrullus* Fritsch, *Colocynthis citrullus* (L.) Kuntze, *Cucumis amarissimus* Schrad., *Cucumis citrullus* (L.) Ser., *Cucumis dissectus* Decne., *Cucumis laciniosus* Eckl. ex Schrad., *Cucumis laciniosus* Eckl. ex Steud., *Cucumis vulgaris* (Schrad.) E.H.L.Krause, *Cucurbita anguria* Duchesne, *Cucurbita caffra* Eckl. & Zeyh., *Cucurbita citrullus* L., *Cucurbita gigantea* Salisb., *Cucurbita pinnatifida* Schrank, *Momordica lanata* Thunb.

Common Names Watermelon.

Description Herb, annual (1–4 m). Stem shaggy, enlarged (Ø 1 cm). Leaves 1-pinnate (5–12 × 8–18 cm) with rounded lobes; tendrils mostly branched. Flower with yellow corolla (15 mm) with ± rounded lobes. Fruit spherical to ovoid pepo (2–4 dm or more) with juicy red, rarely white, flesh. Seeds oval and flattened (6 × 8 mm), dark brown and shiny.

Subspecies The subdivision of the species into two varieties, *Citrullus lanatus* (Thunb.) Matsum. & Nakai var. *lanatus* and *Citrullus lanatus* (Thunb.) Matsum. & Nakai var. *citroides* (L.H.Bailey) Mansf. (= *Citrullus caffer* Schrad.), originated from the erroneous synonymization of *Citrullus lanatus* (Thunb.) Matsum. & Nakai and *Citrullus vulgaris* Schrad. Although recent molecular investigations have shown that the two taxa are not closely related, given the large number of articles since 1930 that have erroneously applied the name *Citrullus lanatus* (Thunb.) Matsum. & Nakai to denote watermelon, it has been proposed that this meaning be retained.

Cultivar The approximately 1,200 known cultivars are divided into the following groups: Citroides (*C. lanatus* subsp. *lanatus* var. *citroides*) with yellow, sweet flesh, grown for fodder or for pectin extraction; Lanatus (*C. lanatus* var. *caffer*; *C. caffer* Schrad.; *C. amarus* Schrad.) with white flesh, an important food for travelers in the Kalahari Desert; and Vulgaris, including watermelon in the strict sense, cultivated by humans for millennia and having the closest wild relative in *Citrullus lanatus* (Thunb.) Matsum. & Nakai subsp. *mucospermus* (Fursa) Fursa from W-Africa, cultivated for forage.

Related Species The genus *Citrullus* has four species: *Citrullus colocynthis* (L.) Schrad., *Citrullus ecirrhosus* Cogn., *Citrullus lanatus* (Thunb.) Matsum. & Nakai, and *Citrullus rehmii* De Winter.

Geographic Distribution Paleotropical (Kalahari Desert?). **Habitat** Desert, sandy plains. **Range** China, Turkey, Iran, Brazil, Egypt.

Beer-Making Parts (and Possible Uses) Fruits.

Chemistry Energy 127 kJ (30 kcal), water 91.45 g, carbohydrates 7.55 g (sugars 6.2 g), fiber 0.4 g, lipids 0.15 g, protein 0.61 g, vitamin A 28 μg, beta-carotene 303 μg, thiamin (B1) 0.033 mg, riboflavin (B2) 0.021 mg, niacin (B3) 0.178 mg, pantothenic acid (B5) 0.221 mg, vitamin B6 0.045 mg, choline 4.1 mg, vitamin C 8.1 mg, calcium 7 mg, iron 0.24 mg, magnesium 10 mg, manganese 0.038 mg, phosphorus 11 mg, potassium 112 mg, sodium 1 mg, zinc 0.1 mg, lycopene 4,532 μg.

Style American Amber Ale, American IPA, American Light Lager, American Pale Ale, American Wheat Beer, American Wheat or Rye Beer, Berliner Weisse, Blonde Ale, California Common Beer, Cream Ale, Fruit Beer, Gueuze, Piwo Grodziskie, Saison, Specialty Fruit Beer, Weizen/Weissbier, Witbier.

Beer Sargeniska, B94 (Lecce, Apulia, Italy).

Source Calagione et al. 2018; Hieronymus 2016b; https://www.brewersfriend.com/other/watermelon/; https://www.craftbeer.com/brewers_banter/sippin-summer-10-must-try-watermelon-beers.

Making Beer with Watermelon: Raffaele Longo's Experience

In the summer of 2018, Raffaele Longo, brewer of the Lecce-based company B94, produced one of the few Italian beers that include *Citrullus lanatus* among the ingredients. The idea was to use a fruit that has always evoked freshness and refreshment. A fruit that in the countryside of Nardò, in Salento, is widely cultivated and in different varieties, above all Melania. Raffaele also wanted to send a political message with this product, given that the harvest of watermelons is sometimes tainted by the sad phenomenon of *caporalato* (the illegal recruitment of workers for very low wages) and that, because of the low price they fetch on the market, watermelons are often left to rot in the fields. Producing a beer dedicated to watermelon was a way to honor the fruit and to highlight both its merits and problems. But which type should be produced with such a unique aromatizer? Sargeniska—named for *sargeniscu*, the local dialect word for "watermelon"—is a saison 5.5% ABV, pale with light pinkish reflections. Just like the fruit it is named for, it is meant to be consumed in summertime, for refreshment on the hottest days. For the first production of Sargeniska, four quintals of watermelons were used. Their juice was added to the fermentation in a quantity corresponding to about 40% of the wort.

Citrus × aurantium L.
RUTACEAE

Synonyms *Aurantium × acre* Mill., *Aurantium × bigarella* Poit. & Turpin, *Aurantium × corniculatum* Mill., *Aurantium × corniculatum* Poit. & Turpin, *Aurantium × coronatum* Poit. & Turpin, *Aurantium × distortum* Mill., *Aurantium × humile* Mill., *Aurantium × myrtifolium* Descourt., *Aurantium × orientale* Mill., *Aurantium × silvestre* Pritz., *Aurantium × sinense* (L.) Mill., *Aurantium × variegatum* Barb. Rodr., *Aurantium × vulgare* (Risso) M.Gómez, *Citrus × amara* Link, *Citrus × aurata* Risso, *Citrus × benikoji* Yu.Tanaka, *Citrus × bigaradia* Loisel., *Citrus bigaradia* Risso & Poit., *Citrus × calot* Lag, *Citrus × canaliculata* Yu.Tanaka, *Citrus × changshan-huyou* Y.B.Chang, *Citrus × communis* Poit. & Turpin, *Citrus × dulcimedulla* Pritz., *Citrus × dulcis* Pers., *Citrus × florida* Salisb., *Citrus × funadoko* Yu.Tanaka, *Citrus × fusca* Lour., *Citrus × glaberrima* Yu.Tanaka, *Citrus humilis* (Mill.) Poir., *Citrus × humilis* (Mill.) Poir., *Citrus × intermedia* Yu.Tanaka, *Citrus × iwaikan* Yu.Tanaka, *Citrus × karna* Raf., *Citrus × kotokan* Hayata, *Citrus × medioglobosa* Yu.Tanaka, *Citrus × mitsuharu* Yu.Tanaka, *Citrus × myrtifolia* (Ker Gawl.) Raf., *Citrus × natsudaidai* (Yu.Tanaka) Hayata, *Citrus × omikanto* Yu.Tanaka, *Citrus × pseudogulgul* Shirai, *Citrus × reshni* (Engl.) Yu.Tanaka, *Citrus × rokugatsu* Yu.Tanaka, *Citrus × rumphii* Risso, *Citrus × sub-compressa* (Tanaka) Yu.Tanaka, *Citrus × taiwanica* Yu.Tanaka & Shimada, *Citrus × taiwanica* Tanaka & Shimada, *Citrus × tangelo* J.W.Ingram & H.E.Moore, *Citrus × tosa-asahi* Yu.Tanaka, *Citrus × truncata* Yu.Tanaka, *Citrus × vulgaris* Risso, *Citrus × yatsushiro* Yu.Tanaka, *Citrus × yuge-hyokan* Yu.Tanaka.

Common Names Sour orange, bitter orange, Seville orange, bigarade.

Description Small tree (2–6 m), evergreen. Leaves with narrowly winged petiole and more or less ovate lamina, rounded at the base, almost without aroma. White flowers in short racemes. Fruit subspherical or slightly ovoid (5–7 × 6–9 cm) with peel attached to the segments.

Cultivar There are many cultivars of this species, among which the Bizzarria that has chimera traits (it bears bitter orange and citron fruits at the same time, as well as fruits with characteristics intermediate between the two species). Caniculata (Canaliculata) has flattened fruits, an orange color, and an irregular peel. In the Salicefolia (Salicifolia), the leaves are narrow and long (similar to those of willow), but the fruits are the same as those of the classic bitter orange. Corniculata is an ancient Italian cultivar whose fruits have wrinkled peels with protuberances resembling small horns. Other cultivars are Consolei, Crispifolia, Dulcis, Fasciata, Foetifera, and Listata.

Related Species Notwithstanding an extremely complex taxonomic picture, some sources report 33 species for the genus *Citrus*, many of which have food and beer-making applications.

Geographic Distribution SE Asia. **Habitat** Cultivated. **Range** Spain, Italy.

Beer-Making Parts (and Possible Uses) Dried fruit peels.

Toxicity Wide range of constituents including flavonic glycosides (hesperidin), coumarin, polymethoxyflavones, aldehydes, amines, and monoterpenes; active principles octopamine and synephrine, mainly contained in the peel and in the pulp of the fruit. In addition to its historical uses (culinary, medicinal), today it is also used in the preparation of dietary supplements for weight loss instead of *Ephedra sinica* Stapf (Ephedraceae). These food supplements provide 10–40 mg of synephrine per dose; Italian regulations (Circ. no. 3 of July 18, 2002, Official Journal 188 of August 12, 2002) state that the daily dose must not exceed 30 mg (about 800 mg of *C. × aurantium* at 4% synephrine); individuals with cardiovascular pathologies and/or hypertension must consult a doctor because it causes vasoconstriction and increases blood pressure; consumption during pregnancy or lactation and by those under 12 years of age is not recommended. These products (*C. × aurantium*, synephrine) are banned in Canada; in the United States, the species is listed in the Poisonous Plant Database, and the FDA has proposed a ban on the sale of products containing synephrine, although the species is allowed as a dietary supplement.

Chemistry In the essential oil, 33 compounds (99% of the total oil) have been identified, the most important of which are limonene 27.5%, E-nerolidol 17.5%, α-terpineol 14%, α-terpinyl acetate 11.7%, and E-farnesol 8%.

Style American Brown Ale, American Pale Ale, American Strong Ale, Baltic Porter, Belgian Blond Ale, Belgian Dark Strong Ale, Belgian Tripel, Blonde Ale, California Common Beer, Holiday/Winter Special Spiced Beer, Old Ale, Southern English Brown, Specialty IPA (Brown IPA, White IPA), Witbier.

Beer Friska, Barley (Maracalagonis, Sardinia, Italy).

Source Calagione et al. 2018; Giaccone and Signoroni 2017; Hieronymus 2010; Markowski 2015; https://www.brewersfriend.com/other/pomerans/; https://www.fermentobirra.com/homebrewing/birre-speziali/.

Citrus × aurantium L. subsp. *currassuviensis*
RUTACEAE

Synonyms None
Common Names Curaçao, laraha.

Description Small tree (up to 6 m). Foliage intricate, branches armed with long spines (2.5–7.6 cm). Leaves dark green. Flowers white, sweetly fragrant. Fruits with wrinkled peel and vivid color; flesh suffused orange color, seeds numerous.

Related Species Notwithstanding an extremely complex taxonomic picture, some sources report 33 species for the genus *Citrus*, many of which have food and beer-making applications.

Geographic Distribution Curaçao Island: an island off the coast of Venezuela inhabited by the Arawak people when the first European, Alonso de Ojeda, at the head of a Spanish expeditionary force, set foot there in 1499. The Spanish considered the island uninteresting, so much so that they did not begin to colonize it until 1526. As was common practice at the time, the colonists tried to import European cultivations, among which the bitter oranges of Seville (*C. × aurantium*, also used for ornamental purposes in the city of Seville). However, the plant did not adapt to the arid climate of the island, producing only green shriveled fruits with a bitter, fibrous, and inedible pulp; not even goats could eat them. This ended up being fortunate because the abandoned cultivations disseminated, producing new individuals that survived the hard selection created by the climate and poor soils. In 1634, the Dutch took possession of the island, establishing a flourishing commercial base for the Dutch West India Company, which traded all kinds of merchandise, slaves included. The Dutch, or, according to some sources, the Portuguese, rediscovered that strange wild orange, calling it *lahara* (the etymology of the term is uncertain, but it seems connected to the Portuguese *laranja* for "orange" and to *papamientu*, a term used in a Spanish Creole language with additions of Portuguese and Dutch, spoken in the Caribbean islands of Aruba, Bonaire, and Curaçao). The name of the island, Curaçao ("cured" in Portuguese) would become connected to the consumption of the fruit, which, although terribly bitter, had a curative effect on sailors at risk of scurvy after long Atlantic crossings. From its consumption as such, the step to producing a liqueur was pretty short: the dried peels of the citrus fruit, strongly aromatic, soon proved useful for the production of various alcoholic beverages such as the famous Blue Curaçao, whose vivid color is however obtained with an artificial coloring, E133 Brilliant Blue. **Habitat** Cultivated and spontaneous species. **Range** Island of Curaçao.

Beer-Making Parts (and Possible Uses) Dried fruit peels.

Chemistry No information available.

Beer Blanche de Namur, Brasserie du Bocq (Purnode, Belgium).

Style American IPA, American Pale Ale, American Wheat Beer, American Wheat or Rye Beer, Belgian Blond Ale, Belgian Dark Strong Ale, Belgian Golden Strong Ale, Belgian Pale Ale, Belgian Specialty Ale, Belgian Tripel, Bière de Garde, Blonde Ale, California Common Beer, Saison, Special/Best/Premium Bitter, Specialty Beer, Specialty IPA (Brown IPA, White IPA), Weizen/Weissbier, Witbier.

Source Hieronymus 2010; Hieronymus 2016; Jackson 1991; Markowski 2015; Stewart 2013; https://www.brewersfriend.com/other/curacao/; https://www.guidabirreartigianali.it/birre-blanche-o-bianche.html.

The Citrus of Blanche

Curaçao is one of the plant species best known to brewers around the world. But it was not always so. For a long time, the peel of this citrus fruit was used almost exclusively as one of the ingredients of a typical style from the Belgian town of Hoegaarden, in Flanders, the main production center of blanche wheat beers (sometimes made with more than 50% unmalted cereal), spiced with coriander seeds (*Coriandrum sativum*) and dried curaçao peel. In 1956, however, the history of this style was interrupted. In that year, in Hoegaarden, the last producer of blanche ceased his activity, and for ten years nobody bothered to resume brewing the fragrant and refreshing beer that had quenched the thirst of generations. It was Pierre Celis, the son of a milkman who had worked for some years in the city's brewery, who started to reproduce the recipe in his garage. The success was immediate, and soon his business was taken over by a large industrial group that still owns it today. Blanche beers quickly spread and became one of the most renowned and appreciated styles in the world.

Citrus × bergamia Risso & Poit.
RUTACEAE

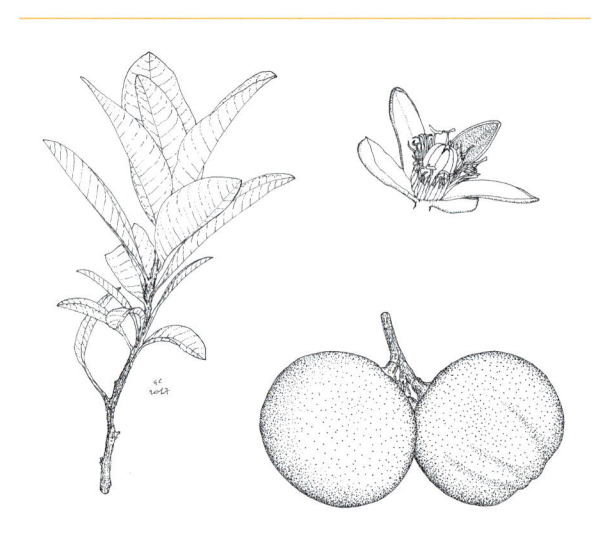

Synonyms *Citrus bergamia* Risso.

Common Names Bergamot.

Description Tree (up to 12 m; in cultivation pruned to 4–5 m), erect, much branched, thornless; trunk Ø up to 25 cm. Leaves alternate, simple, glandular, aromatic when rubbed; petiole about 13 mm, moderately winged, articulated close to the lamina; lamina lanceolate (12 × 6 cm), weakly toothed in the upper third. Inflorescence terminal, racemose, multiflorous; peduncle up to 8 mm long; flowers bisexual, 4- to 10-merous, fragrant; calyx cup shaped with short lobes, yellow-green; corolla Ø 3.8 cm, with 5 petals, narrow and elongated, white without any purple color; stamens 13–28, in 2–6 groups, sometimes petaloid; nectar disc; pistil with subglobose ovary, short and thick style, stigma distinct to indistinct. Fruit hesperidium subglobose, slightly flattened to pyriform (6.5–7 × 6–7.5 cm), often with

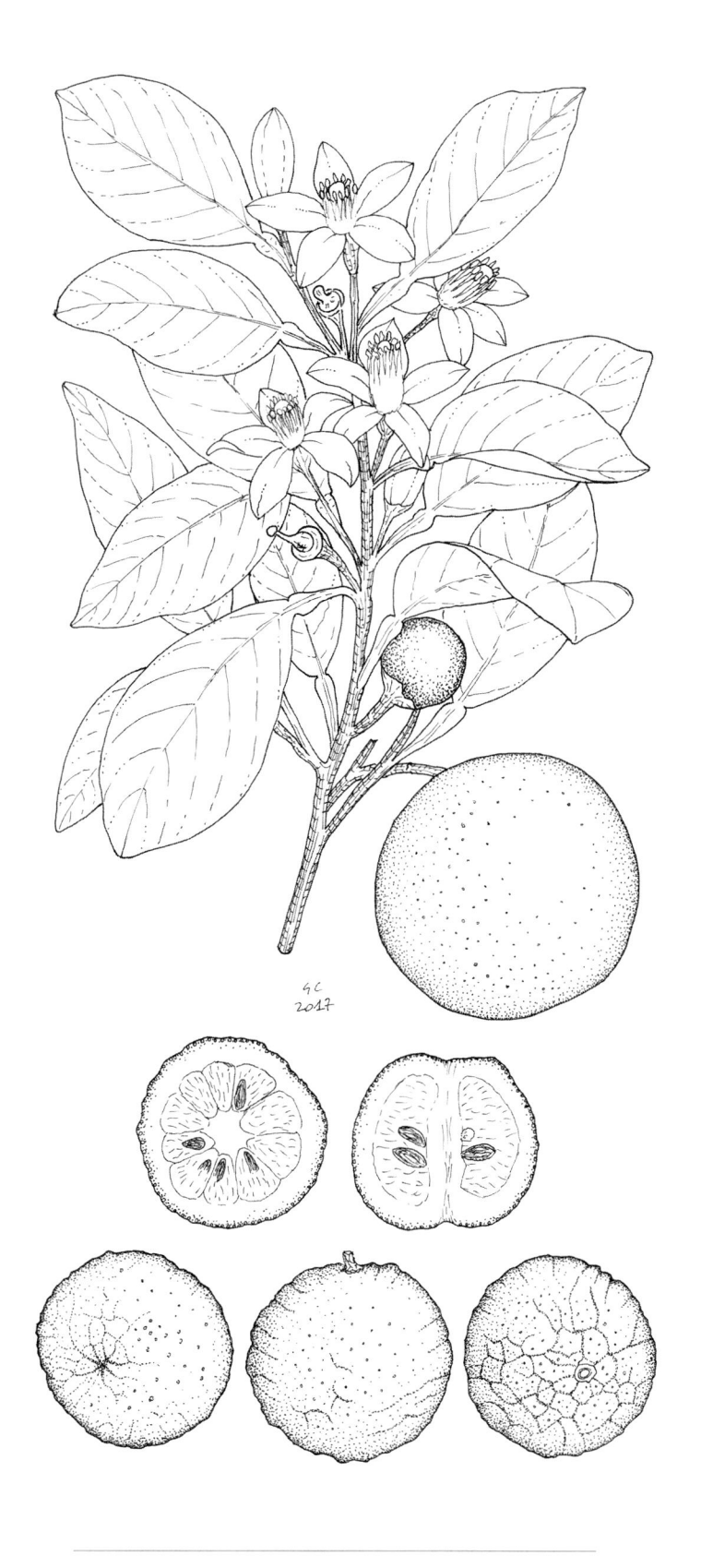

Citrus × aurantium L. RUTACEAE
Citrus × aurantium L. subsp. currassuviensis RUTACEAE

a small umbilicus and persistent style; peel 6–7 mm thick, with numerous glands, hard, smooth to wrinkled, sometimes ridged, adherent, bright green, then yellow when ripe; flesh yellowish, firm, very acidic and bitter, divided into 8–14 segments. Seeds 0–13 per fruit, flattened (11 × 6 × 4.4 mm), pale yellow, usually monoembryonic.

Cultivar There are four well-defined groups: Common Bergamot, Melarosa (rather flat fruit), Torulosa (wrinkled fruit), and Piccola (dwarf cultivar). Only common bergamot is cultivated for commercial purposes for its essential oil, and there are three cultivars: Castagnaro, Femminello, and Inserto. Previously Femminello and Castagnaro constituted practically all the commercial plantations of the world, but they have largely been replaced by the cultivar Inserto (Fantastico), a hybrid between Femminello and Castagnaro. Femminello is less vigorous and smaller than Castagnaro but is earlier and more regular in production; the fruit is spherical or nearly so, the smooth rind more aromatic and therefore preferred. Castagnaro is more erect and vigorous, reaching a larger size than Femminello, but is less fruitful; the fruit, roundish, often has a short neck with obovate control and sometimes ribs; the rind is usually rougher and the oil less aromatic. Inserto is a fairly vigorous tree that produces well and has only a slight tendency to alternate production; the fruit is medium-sized (about 130 g), with a rough rind.

Related Species Notwithstanding an extremely complex taxonomic framework, some sources report 33 species for the genus *Citrus*, many of which have food and beer-making applications.

Geographic Distribution *C.* × *bergamia* is most likely of hybridogenic origin. Some believe it is a hybrid between *Citrus* × *aurantium* and *Citrus limon* or a mutation of the latter. Others think it is a hybrid between *Citrus* × *aurantiun* and *Citrus aurantifolia* (Christm. & Panzer) Swingle (lime). Bergamot is not known in its natural state but only in cultivation. **Habitat** Cultivated. **Range** Italy (the Ionian coast of the province of Reggio Calabria accounts for about 80% of the world production), Ivory Coast, Turkey (Antalya), Mauritius.

Beer-Making Parts (and Possible Uses) Dried fruit peels, pulp (?).

Toxicity Some components of the essential oil have recognized negative effects on sun-exposed skin (erythematous, photosensitizing, phototoxic); some sources also report the essential oil as having carcinogenic effects.

Chemistry Essential oil: volatile components, limonene 37.2%, linalyl acetate 30.1%, linalool 8.8%, γ-terpinene 6.8%, β-pinene 2.8%, other/ignoids 14.3%; nonvolatile components (g/L), bergamottin 21.42 g, citroptene 2.58 g, bergaptene 2.37 g, 5-geranyloxyi-7-methoxycoumarin 1.12 g. Juice (g/L or mg/L): isocitric acid 0.36–0.55 g, citric acid 92–140 g, L-malic acid 0.97–2.25 g, D-lactic acid 48–57 g, L-lactic acid 43–49 mg, ascorbic acid 380–440 mg, sucrose 9.6–24.1 g, glucose 9.1–15.1 g, fructose 9.9–14.8 g, pectin 560–740 mg, sodium 11–25 mg, potassium 1,150–1,500 mg, calcium 65–85 mg, magnesium 67–116 mg, phosphate 201–370, protein 4.87–5.63 g, naringin 480–600 mg, neoeriocitrin 67–135 mg, narirutin 10–30 mg, naringin 71–118 mg, neohesperidin 45–85 mg.

Style Blanche, Spiced beer, Summer Session Ale.

Beer Seta Special, Birrificio Rurale (Desio, Lombardy, Italy).

Source Dabove 2015; https://www.birraofelia.it/shop/birre-stagionali/beergamotta/; https://www.fermento-birra.com/homebrewing/birre-speziali/.

Citrus × *clementina* hort. ex Tan.
RUTACEAE

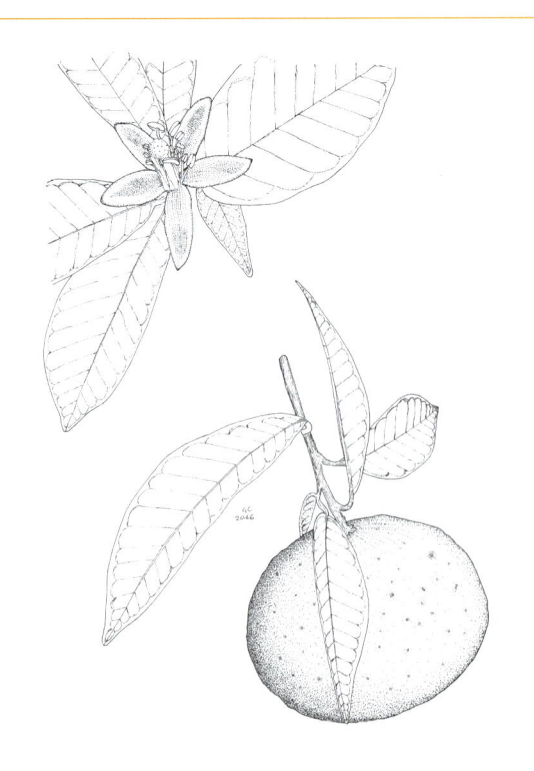

Synonyms None
Common Names Clementine.

Description Shrub or small tree (up to 5 m), with morphological characteristics ± intermediate to parental species (see description of *Citrus × sinensis* and *Citrus reticulata*). Fruit hesperidium, usually seedless (in absence of cross-pollination), thin exocarp easy to remove (peel).

Subspecies May be native to Algeria, owing to a casual crossing in the 1940s by Father Clément Rodier (of the Missergin convent in Oran, Algeria); may be the result of a previous crossing by the priest Pierre Clément; a more recent hypothesis is of an ancient hybrid (from China or Japan) introduced to the Mediterranean by an unknown Algerian religious. After the first hybridizations of the twentieth century, it was evident that the plant was a new species of *Citrus*, as the characteristics remained unchanged over time and reproduction presented no problems.

Cultivar The high quality of clementines makes them one of the most important cultivated *Citrus*. As in sweet oranges and satsumas, the genetic variability within the species, analyzed through molecular markers, is minimal because the existing varieties have been obtained not by hybridization but by the selection of spontaneous mutants for characters of agronomic interest. Another possible explanation for the large agronomic diversity compared to the low variability of molecular markers is the failure of the markers used to focus on the molecular modification characteristic of new clementine cultivars. More refined analyses have compared the variability of as many as 24 varieties of clementines, essentially distinguishing only 2, corresponding to two distinct geographical groups: North Africa and Spain.

Related Species Notwithstanding an extremely complex taxonomic picture, some sources report 33 species for the genus *Citrus*, many of which have food and beer-making applications.

Geographic Distribution Algeria (?), China (?). **Habitat** Cultivated in Mediterranean and subtropical climates. **Range** Tunisia, Algeria, Morocco, Spain, Italy (Calabria, Apulia, Sicily), United States (Florida).

Beer-Making Parts (and Possible Uses) Fruits.

Chemistry Energy 198 kJ (47 kcal), water 86.58 g, carbohydrates 12.02 g, sugars 9.18 g, fiber 1.7 g, lipids 0.15 g, protein 0.85 g, thiamin (B1) 0.086 mg, riboflavin (B2) 0.030 mg, niacin (B3) 0.636 mg, pantothenic acid (B5) 0.151 mg, vitamin B6 0.075 mg, folate (B9) 24 µg, choline 14 mg, vitamin C 48.8 mg, vitamin E 0.20 mg, calcium 30 mg, iron 0.14 mg, magnesium 10 mg, manganese 0.023 mg, phosphorus 21 mg, potassium 177 mg, sodium 1 mg, zinc 0.06 mg; the oil contains mostly limonene, myrcene, linalool, α-pinene, and many aromatic complexes.

Style American IPA, American Wheat or Rye Beer, Belgian Dark Strong Ale, Belgian Tripel, Fruit Beer, Russian Imperial Stout, Saison, Witbier.

Beer Clementine, Jolly Pumpkin Artisan Ale (Michigan, United States).

Source https://www.brewersfriend.com/other/clementine/.

Citrus × floridana (J.Ingram & H.Moore) Mabb.
RUTACEAE

Synonyms *Citrofortunella floridana* J.W.Ingram & H.E.Moore, *Citrus floridana* (J.Ingram & H.E. Moore) Mabberley, *Citrus japonica* Thunb. × *Citrus auraitfolia* (Christm.) Swingle.

Common Names Limequat.

Description Shrub or sapling. *Citrofortunella* hybrid resulting from a cross between lime (*Citrus × aurantifolia*) and kumquat (*Fortunella* sp.) obtained by Walter Tennyson Swingle, a noted farmer-botanist, on behalf of the USDA in 1909. Leaves and other morphological characteristics typical of *Citrus*. Fruits produced in abundance from a young age, small, oval, yellow-greenish, containing seeds; peel (epicarp) with a sweet taste, pulp with a taste similar to lime.

Cultivar Three cultivars of limequat are known, all obtained in Florida and therefore carrying the names of various locations in the American state: Eustis (*Citrus japonica × Citrus aurantiifolia*), Lakeland (*Citrus japonica × Citrus aurantiifolia*), and Tavares (*Citrus japonica Margarita × Citrus aurantiifolia*).

Related Species Notwithstanding an extremely complex taxonomic framework, some sources report 33 species for the genus *Citrus*, many of which have food and beer-making applications.

Geographic Distribution Artificial hybrid obtained in Florida. **Habitat** Cultivated. **Range** Japan, Israel, Spain, Malaysia, South Africa, Armenia, United Kingdom, United States (California, Florida, Texas).

Beer-Making Parts (and Possible Uses) Fruits.

Chemistry Juice: vitamin C. Volatile components, peel: linalyl acetate 0.18%, neryl acetate 0.15%, octyl acetate 0.10%, aromandendrene 0.56%, (-)-β-pinene 0.43%, camphene 0.04%, caryophyllene 0.37%, cymenene 0.93%, D-germacrene 0.30%, pinene 0.82%, terpinolene 0.56%, D-limonene 88.72%, α-phellandrene 0.99%, α-cubebene 0.08%, β-myrcene 2.53%, β-phellandrene 0.75%, δ-cadinene 0.14%, γ-terpinene 0.93%, α-trans-bergamotene 0.30%, β-bisabolene 0.82%, β-elemene 0.08%, n-hexanal 0.39%, pent-1-en-3-one 0.25%.

Style Gose, Witbier.

Beer Limequat Gose, Cigar City (Florida, United States).

Source https://www.brewersfriend.com/other/limequats/.

Citrus × microcarpa Bunge
RUTACEAE

Synonyms × *Citrofortunella microcarpa* (Bunge) Wijnands, × *Citrofortunella mitis* (Blanco) J.W.Ingram & H.E.Moore, *Citrofortunella mitis* (Blanco) J.Ingram & H.E.Moore, *Citrus × mitis* Blanco.

Common Names Calamansi, calamondin.

Description Shrub or small tree (3–6 m), medium vigor, upright and columnar, nearly thornless. Leaves small, broadly oval, mandarin-like. Flower tangerine-like. Fruit hesperidium very small, oblate to spherical, apex flat or depressed; exocarp orange to orange-red, very thin, smooth, finely glandular, easily separable only at maturity, sweet and edible; segments about 9, axis small and semi-empty; flesh orange, tender, juicy, and acidic. Seeds few, small, round, polyembryonic, with green cotyledons.

Related Species Notwithstanding an extremely complex taxonomic framework, some sources report 33 species for the genus *Citrus*, many of which have food and beer-making applications.

Geographic Distribution China. **Habitat** Cultivated. **Range** Widely cultivated in Asia (China,

Taiwan, Philippines, Japan, Java, Indonesia, India, Ceylon, etc.); in the United States cultivated as an ornamental.

Beer-Making Parts (and Possible Uses)
Fruits.

Chemistry Volatile compounds, peel (Malaysia) (99% of compounds present): α-pinene 34.9 ppm, camphene 0.55 ppm, β-pinene 41.2 ppm, sabinene 18.0 ppm, δ-3-carene 2.67 ppm, β-myrcene 174 ppm, limonene 8640 ppm, β-phellandrene 29.1 ppm, trans-β-ocimene 0.41 ppm, terpinene 0.36 ppm, cis-β-ocimene 3.82 ppm, ρ-cymene 0.52 ppm, terpinolene 0.81 ppm, trace allo-ocimene, α-cubebene 0.56 ppm, δ-elemene 13.8 ppm, bicycloelemene 1.39 ppm, α-copaene 0.33 ppm, β-bourbonene 0.48 ppm, β-elemene 3.64 ppm, β-caryophyllene 1.12 ppm, γ-elemene 1.51 ppm, β-cubebene 0.58 ppm, β-arnesene 1.03 ppm, α-humulene 1.79 ppm, germacrene D 146 ppm, β-selinene 6.18 ppm, α-selinene 6.50 ppm, bicyclogermacrene 14.5 ppm, α-farnesene 3.08 ppm, germacrene B 2.89 ppm, hexanol 0.24 ppm, cis-3-hexanol 6.12 ppm, linalool 32.6 ppm, sabinene hydrate 0.72 ppm, 1-octanol 12.7 ppm, 1-nonanol 3.81 ppm, α-terpineol 14 ppm, citronello 1.96 ppm, trans-2-decenol 0.9 ppm, trace carveol, perilla alcohol 2.43 ppm, trans-nerol 2.43 ppm, elemol 15 ppm, γ-eudesmol 4.13 ppm, α-eudesmol 6.97 ppm, β-eudesmol 17.9 ppm, phytol 11.4 ppm, octanal 3.80 ppm, nonanal 5.42 ppm, decanal 10.1 ppm, undecanal 4.15 ppm, trans-2-decenal 2.51 ppm, neral 1.29 ppm, geranial 1.94 ppm, trans, cis-2-4-decadienal 1.87 ppm, perilla aldehyde 2.23 ppm, trans, trans-2-4-decadienal 2.16 ppm, trans-2-dodecenal 1.18 ppm, trans cis-2-6-dodecadienal 1.3 ppm, cis-3-hexenyl acetate 6.12 ppm, heptyl acetate 0.55 ppm, ethyl octanoate trace, octyl acetate 0.73 ppm, citronellyl acetate 1.03 ppm, decyl acetate 1.31 ppm, geranyl acetate 34.4 ppm, methyl salicylate 2.77 ppm, geranyl propionate 1.64 ppm, dodecyl acetate 0.99 ppm, perillyl acetate 2.07 ppm, methyl-N-methyl anthranilate 3.27 ppm, acetic acid 0.19 ppm, octanoic acid 4, 09 ppm, nonanoic acid 3.97 ppm, decanoic acid 2.04 ppm, myristic acid 0.39 ppm, palmitic acid 24.4 ppm, stearic acid 21.1 ppm, linoleic acid 0.98 ppm, limonene oxide 0.28 ppm, camphor 0.35 ppm, isopiperitenone 1.12 ppm, carvone 9430 ppm.

Style Bière de Garde, California Common Beer, Sour Beer, Stout, Witbier.

Beer Barrel Project 18.02, Kees (Middelburgh, Netherlands).

Source https://www.brewersfriend.com/other/calamansi/.

Citrus × myrtifolia (Ker Gawl.) Raf.
RUTACEAE

Synonyms *Citrus × aurantium* L. var. *myrtifolia* Ker Gawl.

Common Names Myrtle-leaved orange tree.

Description Small tree (up to 3 m), compact, slow-growing, thornless. Leaves small, elliptical, pointed, leathery, bright green, resembling those of myrtle (hence the Latin name). Flowers small, white, very fragrant, solitary or gathered in groups, in axillary or apical position, with appreciated decorative functions (plant also cultivated in pots). Fruits hesperidium, intense orange color, produced after 3–4 years of life, smaller than those of mandarin and flattened at the poles; pulp divided into 8–10 segments, bitter and acidic.

Cultivar Chinotto of Savona.

Related Species Notwithstanding an extremely complex taxonomic framework, some sources report 33 species for the genus *Citrus*, many of which have food and beer-making applications.

Geographic Distribution China (?). **Habitat** Cultivated in Mediterranean climates. **Range** Malta, Libya, France, Italy (Liguria, Tuscany, Sicily, Calabria).

Beer-Making Parts (and Possible Uses) Fruits (juice?).

Chemistry Juice: ascorbic acid 895 g/L, isocitric acid 0.136 g/L, D-lactic acid 0.23 g/L, L-lactic acid 0.15 g/L, glucose 23.1 g/L, fructose 26.0 g/L, sucrose 31.5 g/L, pectin 232 mg/L, eriocitrin 33 mg/L, neo-eriocitrin 467 mg/L, naringin 800 mg/L, neo-hexperidin 711 mg/L.

Style Fruit beer, Witbier.

Beer N°8, Scarampola (Millesimo, Liguria, Italy).

Source Dabove 2015; https://www.birramoretti.it/le-birre-di-casa/birra-moretti-chinotto/.

Citrus × paradisi Macfad.
RUTACEAE

Synonyms None
Common Names Grapefruit.

Description Tree (5–6 m, although in the wild it can reach 13–15 m), evergreen. Leaves dark green, glossy, up to 15 cm long, thin. Flowers Ø 5 cm, with 4 white petals. Fruit with yellow-orange peel, usually spheroidal (Ø 10–15 cm), slightly flattened at the poles; pulp segmented, acidic, variable in color according to cultivar (white, pink, red).

Cultivar Cultivars are distinguished according to the color of their pulp: white (Duncan, Jaffa Marsh, Melogold, Oroblanco Sweetie, Sweetie), pink (Foster, Henderson, Marsh Pink Thompson, Ray Ruby, Redblush Ruby Red, Shambar), and red (Flame, Jaffa Sunrise, Rio Red, Star Ruby Sunrise).

Related Species Notwithstanding an extremely complex taxonomic picture, some sources report 33 species for the genus *Citrus*, many of which have food and beer-making applications.

Geographic Distribution Barbados (Caribbean). **Habitat** Cultivated. **Range** China, United States, Mexico, Thailand, South Africa, Israel, Turkey, Argentina, India, Sudan, Ghana.

Beer-Making Parts (and Possible Uses) Fruits, peels, wood (for barrels).

Chemistry Pulp: energy 138 kJ (33 kcal), water 90.48 g, carbohydrates 8.41 g, sugars 7.31 g, fiber 1.1 g, lipids 0.10 g, protein 0.69 g, thiamine (B1) 0.037 mg, riboflavin (B2) 0.020 mg, niacin (B3) 0.269 mg, pantothenic acid (B5) 0.283 mg, vitamin B6 0.043 mg, folate (B9) 10 µg, choline 7.7 mg, vitamin C 33.3 mg, vitamin E 0.13 mg, calcium 12 mg, iron 0.06 mg, magnesium 9 mg, manganese 0.013 mg, phosphorus 8 mg, potassium 148 mg, zinc 0.07 mg. Essential oil, peel: α-pinene 1.19%, sabinene 0.52%, β-myrcene 3.51%, γ-terpinene 0.91%, limonene 50.8%, linalool oxide 2.29%, linalool 1.07%, trans-p-2,8-mentadien-1-ol 0.63%, limonene oxide 0.87%, citronellal 0.83%, 4-terpineol 0.78%, 4-carvon menthenol 0.88%, α-terpineol 0.75%, α-terpinenol 1.33%, decanal 2.29%, Z-carveol 1.50%, citronellol 1.78%, E-carveol 1.87%, isophorone 1.29%, d-carvone 0.64%, geraniol 0.44%, citral 1.22%, 4-vinyl guaiacol 0.33%, carvil acetate 0.66%, eugenol 1.34%, geranyl butyrate 0.16%, geranyl acetate 0.55%, α-copaene 0.79%, β-cubebene 0.81%, β-caryophyllene 1.73%, germacarene D 0.83%, valencene 3.36%, α-panasinsene 0.39%, δ-cadinene 1.25%, β-gurjunene 0.65%, γ-gurjunene 0.67%, dodecanal 0.31%, elemol 0.52%, nerolidol 0.32%, β-caryophyllene oxide 0.56%, β-sinensal 0.51%, farnesol 3.84%, α-sinensal 0.19%, santolin epoxide 0.42%, nootkatone 8.47%.

Wood Similar characteristics to those of *Citrus limon*.

Style American Amber Ale, American IPA, American Pale Ale, Belgian Blond Ale, Belgian Pale Ale, Belgian Specialty Ale, Belgian Tripel, Berliner Weisse, Blonde Ale, British Golden Ale, California Common Beer, Experimental Beer, Fruit Beer, Imperial IPA, Saison, Specialty Beer, Specialty IPA (Belgian IPA, Rye IPA, White IPA), Spice/Herb/Vegetable Beer, Weissbier, Witbier.

Beer Ipé Grapefruit, Birrificio San Paolo (Torino, Piedmont, Italy).

Source Calagione et al. 2018; Dabove 2015; Giaccone and Signoroni 2017; Hieronymus 2010, 2016b; Josephson et al. 2016; https://www.brewersfriend.com/other/toronja/; https://dreher.it/prodotti/dreher-pompelmo/; https://www.oregonfruit.com/fruit-brewing/category/products.

🍺 Hops and Citrus Fruits: Great Teamwork

One of the reasons for the success of American IPAs—the style that has most defined the craft revolution—is the use of hops coming from the Yakima and Willamette Valley areas, in the states of Washington and Oregon. Characterized by intense notes of citrus and tropical fruit, since the early years of the movement, they have been one of the key elements of what a new generation of brewers had in mind to stand out in the industry that had dominated the world scene for years. To make the sensations even more intense and pronounced, starting from the late nineties, some producers began adding citrus fruit peel or tropical fruit pulp to their already aromatic IPAs. One of the first to do this experimentation was the Californian brewery Ballast Point (San Diego), which added grapefruit zest to its iconic beer Sculpin. This double IPA, hopped with Amarillo and Simcoe, has been the brewery's flagship product since its first releases thanks to an unexpected drinkability and a great aromaticity given by the hops, which give an overall "juicy" and fresh sensation. Pink grapefruit, which reinforces one of the main notes given by Amarillo, adds freshness. The Ipé Grapefruit of the Turin brewery San Paolo is dedicated to Sculpin.

Citrus × sinensis (L.) Osbeck
RUTACEAE

Synonyms *Aurantium × acre* Mill., *Aurantium × bigarella* Poit. & Turpin, *Aurantium × corniculatum* Mill., *Aurantium × corniculatum* Poit. & Turpin, *Aurantium × coronatum* Poit. & Turpin, *Aurantium × distortum* Mill., *Aurantium × humile* Mill., *Aurantium × myrtifolium* Descourt., *Aurantium × orientale* Mill., *Aurantium × silvestre* Pritz., *Aurantium × sinense* (L.) Mill., *Aurantium × variegatum* Barb. Rodr., *Aurantium × vulgare* (Risso) M.Gómez, *Citrus × amara* Link, *Citrus × aurata* Risso, *Citrus × benikoji* Yu.Tanaka, *Citrus × bigaradia* Loisel., *Citrus bigaradia* Risso & Poit., *Citrus × calot* Lag, *Citrus × canaliculata* Yu.Tanaka, *Citrus × changshan-huyou* Y.B.Chang, *Citrus × communis* Poit. & Turpin, *Citrus × dulcimedulla* Pritz., *Citrus × dulcis* Pers., *Citrus × florida* Salisb., *Citrus × funadoko* Yu.Tanaka, *Citrus × fusca* Lour., *Citrus × glaberrima* Yu.Tanaka, *Citrus humilis* (Mill.) Poir., *Citrus × humilis* (Mill.) Poir., *Citrus × intermedia* Yu.Tanaka, *Citrus × iwaikan* Yu.Tanaka, *Citrus × karna* Raf., *Citrus × kotokan* Hayata, *Citrus × medioglobosa* Yu.Tanaka, *Citrus × mitsuharu* Yu.Tanaka, *Citrus × myrtifolia* (Ker Gawl.) Raf., *Citrus × natsudaidai* (Yu.Tanaka) Hayata, *Citrus × omikanto* Yu.Tanaka, *Citrus × pseudogulgul* Shirai, *Citrus × reshni* (Engl.) Yu.Tanaka, *Citrus × rokugatsu* Yu.Tanaka, *Citrus × rumphii* Risso, *Citrus ×*

subcompressa (Tanaka) Yu.Tanaka, *Citrus taiwanica* Yu.Tanaka & Shimada, *Citrus × taiwanica* Tanaka & Shimada, *Citrus × tangelo* J.W.Ingram & H.E. Moore, *Citrus × tosa-asahi* Yu.Tanaka, *Citrus × truncata* Yu.Tanaka, *Citrus × vulgaris* Risso, *Citrus × yatsu-shiro* Yu.Tanaka, *Citrus × yuge-hyokan* Yu.Tanaka (some of the taxa associated with *Citrus × sinensis* [L.] Osbeck perhaps deserve greater nomenclatural and taxonomic autonomy; the intricate phylogeny of the genus *Citrus* is still the basis of heated taxonomic controversies that genetic techniques may help settle in the near future).

COMMON NAMES Orange tree, sweet orange tree.

DESCRIPTION Tree or small tree (6–13 m), evergreen, roots surface, crown closed, branches spiny. Branches angled when young, often with spines. Leaves smooth, oval, 5–15 × 2–8 cm, dark green above, glossy, with a distinctive scent similar to the fruit, petiole winged. Flowers small, waxy greenish-white, fragrant; calyx broad, saucer-shaped, 5 petals white, elliptic, 1.3–2.2 cm long. Fruit hesperidium (orange), red-green to yellow-green, round (Ø 4–12 cm), consisting of a leathery peel (average 6 mm thick), tightly adhered, protecting the juicy inner pulp, divided into segments that may or may not contain seeds depending on the cultivar.

CULTIVAR The number of cultivars of orange is indefinite. Cultivars are distinguished by characteristics such as shape, pulp color, juice acidity, and presence or absence of navel. Main groups and varietal names include Valencia (Belladonna, Berna, Cadanera, Calabrese [Calabrese Ovale], Carvalhal, Castellana, Cherry Orange, Clanor, Common Blonde, Curly Blonde, Dom João, Fukuhara, Gardner, Hamlin, Hart's Tardiff Valencia, Homosassa, Jaffa Orange, Jincheng, Joppa, Khettmali, Kona, Lue Gim Gong, Macetera, Malta, Maltaise Blonde, Maltaise Ovale, Marrs, Midsweet, Moro Tarocco, Mosambi, Narinja, Parson Brown, Pera, Pera Coroa, Pera Natal, Pera Rio, Pineapple, Premier, Queen, Rhode Red, Roble, Salustiana, Sathgudi, Seleta, Shamouti Masry, Sunstar, Tomango, Verna, Vicieda, Westin, Xã Đoài orange), Navel (Bahianinha [Bahia], Dream Navel, Late Navel, Red Navel, Washington or California Navel), Red (Maltese, Moro, Sanguina, Sanguinella, Scarlet Navel, Tarocco), and Nonsour (or Sweet).

RELATED SPECIES Notwithstanding an extremely complex taxonomic framework, some sources report 33 species for the genus *Citrus*, many of which have food and beer-making applications.

GEOGRAPHIC DISTRIBUTION China, India, Vietnam. **HABITAT** Cultivated. **RANGE** Brazil, United States, China, India, Mexico, Spain, Egypt, Turkey, Italy, South Africa.

BEER-MAKING PARTS (AND POSSIBLE USES) Fresh and dried fruit peels, flowers (orange blossoms), fruit juice, wood (for smoking malt).

CHEMISTRY Energy 197 kJ (47 kcal), water 86.75 g, carbohydrates 11.75 g, sugars 9.35 g, fiber 2.4 g, lipids 0.12 g, protein 0.94 g, vitamin A 11 µg, thiamine (B1) 0.087 mg, riboflavin (B2) 0.04 mg, niacin (B3) 0.282 mg, pantothenic acid (B5) 0.25 mg, vitamin B6 0.06 mg, folate (B9) 30 µg, choline 8.4 mg, vitamin C 53.2 mg, vitamin E 0.18 mg, calcium 40 mg, iron 0.1 mg, magnesium 10 mg, manganese 0.025 mg, phosphorus 14 mg, potassium 181 mg, zinc 0.07 mg.

STYLE American Amber Ale, American IPA, American Light Lager, American Pale Ale, American Wheat or Rye Beer, Belgian Blond Ale, Belgian Dark Strong Ale, Belgian Golden Strong Ale, Belgian Pale Ale, Belgian Specialty Ale, Belgian Tripel, Blonde Ale, California Common Beer, Cream Ale, Dark American Lager, Double IPA, English IPA, Extra Special/Strong bitter (ESB), Foreign Extra Stout, Fruit Beer, Holiday/Winter Special Spiced Beer, Imperial IPA, Metheglin, Northern English Brown Ale, Old Ale, Robust Porter, Russian Imperial Stout, Saison, Specialty beer, Specialty IPA (Rye IPA, White IPA), Spice/Herb/Vegetable Beer, Standard American Lager, Vienna Lager, Weissbier, White IPA, Witbier.

BEER Trupija, 'A Magara (Nocera Terinese, Calabria, Italy).

SOURCE Calagione et al. 2018; Dabove 2015; Daniels and Larson 2000; Heilshorn 2017; Hieronymus 2010; Hieronymus 2016; Hieronymus 2016b; Josephson et al. 2016; Markowski 2015; Steele 2012; https://www.brewersfriend.com/other/apelsinjuice/; https://www.brewersfriend.com/other/laranja/; https://www.brewersfriend.com/other/orange/; https://www.fermentobirra.com/homebrewing/birre-speziali/.

Citrus × *tangelo* J.W.Ingram & H.E.Moore
RUTACEAE

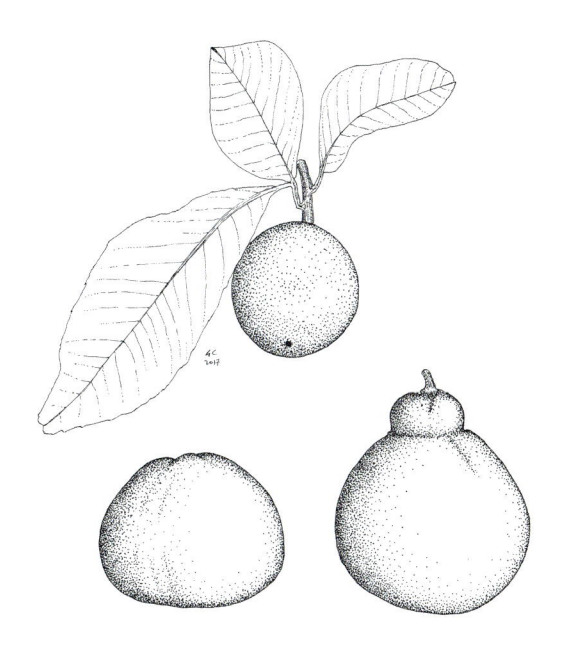

Synonyms *Citrus* × *aurantium* L. (?).

Common Names Tangelo.

Description Tree of medium-strong structure, globular, extended with branches that tend to bend downward, thorns present only on the strongest branches. Other morphological characteristics more or less similar to those of *Citrus*. Fruit hesperidium of analogous shape to that of grapefruit but slightly smaller (Ø 7.5–9 cm), American cultivars ± umbonate at the proximal extremity; exocarp (peel) thin, bright green color, yellow when mature; pulp uniform yellow-orange color, taste pleasantly sour with notes of mandarin.

Cultivar Group of mandarin × grapefruit citrus hybrids. Mapo (Havana mandarin × Duncan grapefruit) obtained in 1950 at the Research Center for Citrus and Mediterranean Crops in Acireale, Sicily, and released for cultivation in 1972; the first to experiment with the cross, in the 1970s, was the Sicilian agronomist Francesco Russo. Minneola (tangerine [*Citrus* × *tangerine*] Dancy × Duncan grapefruit] probably obtained by Walter Tennyson Swingle in 1911, released in 1931 by the USDA Horticultural Research Station in Orlando, Florida. Ugli, native to Jamaica, exported in the 1930s mainly to France.

Related Species Notwithstanding an extremely complex taxonomic framework, some sources report for the genus *Citrus* 33 species, many of which have food and beer-making applications.

Geographic Distribution Artificial hybrid absent in nature. **Habitat** Warm climates, rich and well-drained soils. **Range** Grown in warm climates.

Beer-Making Parts (and Possible Uses) Fruits.

Chemistry Water 88 g, total solids 11.9 g, sugars 9 g, protein 0.66, citric acid 1.83, ash 0.42 g, vitamin A, vitamin B, vitamin C, folic acid, potassium.

Style American Pale Ale, Fruit Beer, IPA, Wiezen.

Beer Citrus IPA, Greenport Harbour (New York, United States).

Source https://www.brewersfriend.com/other/tangelo/.

Citrus aurantiifolia (Christm.) Swingle
RUTACEAE

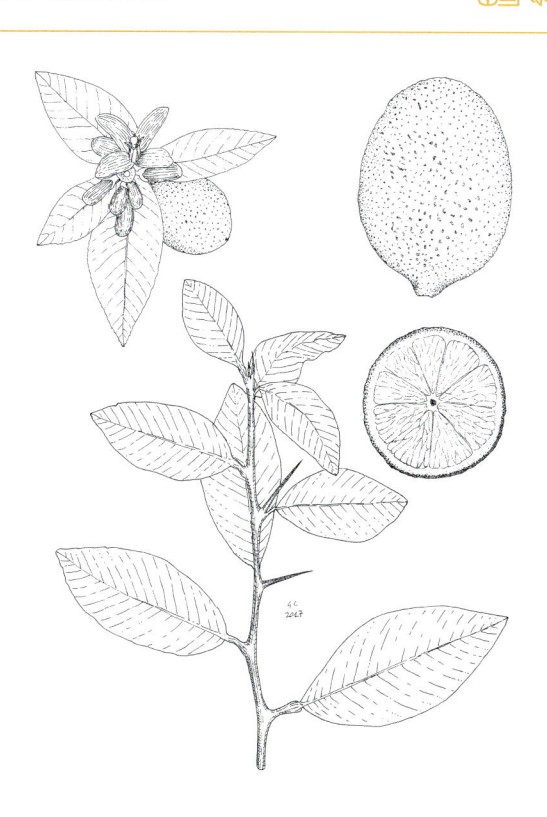

Synonyms *Citrus* × *acida* Pers., *Citrus* × *davaoensis* (Wester) Yu.Tanaka, *Citrus* × *excelsa* Wester, *Citrus* × *javanica* Blume, *Citrus* × *lima* Macfad., *Citrus* ×

limettioides Yu.Tanaka, *Citrus* × *limonellus* Hassk., *Citrus* × *macrophylla* Wester, *Citrus* × *montana* (Western) Yu.Tanaka, *Citrus* × *nipis* Michel, *Citrus* × *notissima* Blanco, *Citrus* × *papaya* Hassk., *Citrus* × *pseudolimonum* Wester, *Citrus* × *spinosissima* G.Mey., *Citrus* × *voangasay* (Bory) Bojer, *Limonia* × *aurantiifolia* Christm.

COMMON NAMES Key lime, West Indian lime, bartender's lime, Omani lime, Mexican lime.

DESCRIPTION Shrub or sapling (up to 5 m), with many thorns. Stem rarely straight, much branched, often from below. Leaves ovate, 2.5–9 cm long, similar to orange leaves (as suggested by the specific epithet). Flowers Ø 2.5 cm, yellowish-white with a slight purple color at the margins. Fruit roundish hesperidium (Ø 3–6 cm) with persistent stylus.

RELATED SPECIES Notwithstanding an extremely complex taxonomic picture, some sources report 33 species for the genus *Citrus*, many of which have food and beer-making applications.

GEOGRAPHIC DISTRIBUTION SE Asia. **HABITAT** Cultivated in tropical climates. **RANGE** SE Asia (India), Mexico, China, South America (Argentina, Brazil), Caribbean.

BEER-MAKING PARTS (AND POSSIBLE USES) Fruits.

CHEMISTRY Essential oil: β-pinene 12.6%, limonene 53.8%, γ-terpinene 16.5%, terpinolene 0.6%, α-terpneol 0.4%, citral 2.5%. Juice: energy 126 kJ (30 kcal), water 88.3 g, carbohydrates 10.5 g, sugars 1.7 g, fiber 2.8 g, lipids 0.2 g, protein 0.7 g, thiamine (B1) 0.03 mg, riboflavin (B2) 0.02 mg, niacin (B3) 0.2 mg, pantothenic acid (B5) 0.217 mg, vitamin B6 0.046 mg, folate (B9) 8 µg, vitamin C 29.1 mg, calcium 33 mg, iron 0.6 mg, magnesium 6 mg, phosphorus 18 mg, potassium 102 mg, sodium 2 mg.

STYLE American IPA, American Pale Ale, American Wheat or Rye Beer, Baltic Porter, Belgian Blond Ale, Belgian Golden Strong Ale, Belgian Tripel, Berliner Weisse, Blonde Ale, California Common Beer, Classic American Pilsner, Clone Beer, Doppelbock, Fruit Beer, International Pale Lager, Maibock/Helles Bock, Mixed-Style Beer, Premium American Lager, Saison, Specialty Beer, Spice/Herb/Vegetable Beer, Strong Scotch Ale, Weissbier, Witbier.

BEER Honey Ginger Lime, Cascade (Oregon, United States).

SOURCE Calagione et al. 2018; Heilshorn 2017; https://www.brewersfriend.com/other/lime/; https://www.omnipollo.com/beer/42/; https://www.omnipollo.com/beer/karl-framboise/; https://www.omnipollo.com/beer/stolen-fruit/.

Citrus hystrix DC.
RUTACEAE

SYNONYMS *Citrus auraria* Michel, *Citrus balincolong* (Yu.Tanaka) Yu.Tanaka, *Citrus boholensis* (Wester) Yu.Tanaka, *Citrus celebica* Koord., *Citrus combara* Raf., *Citrus hyalopulpa* Yu.Tanaka, *Citrus kerrii* (Swingle) Yu.Tanaka, *Citrus kerrii* (Swingle) Tanaka, *Citrus latipes* Hook.f. & Thomson ex Hook.f., *Citrus macroptera* Montrouz., *Citrus micrantha* Wester, *Citrus papeda* Miq., *Citrus papuana* F.M.Bailey, *Citrus southwickii* Wester, *Citrus torosa* Blanco, *Citrus tuberoides* J.W.Benn., *Citrus ventricosa* Michel, *Citrus vitiensis* Yu.Tanaka, *Citrus westeri* Yu.Tanaka, *Fortunella sagittifolia* K.M.Feng & P.Y.Mao, *Papeda rumphii* Hassk.

COMMON NAMES Kaffir lime, makrut lime, magrood lime, Mauritius papeda.

Description Shrub or small tree (2–10 m), very thorny (hence the specific epithet). Thick and luxuriant foliage. Leaves up to 12 cm long, prickly, with a central narrowing (a distinctive feature compared to other species of the genus *Citrus*); "double" leaves made up of a normal leaf blade to which a laminar petiole similar to a second leaf blade is attached. Flowers strongly odorous. Fruit hesperidium roundish to oval (up to 4 cm long), sometimes elongated at the junction like a fig; epicarp (peel) strongly rippled and irregularly rounded, color from bright green to dark green; pulp extremely sour, therefore fruit not consumed fresh.

Related Species Notwithstanding an extremely complex taxonomic picture, some sources report 33 species for the genus *Citrus*, many of which have food and beer-making applications.

Geographic Distribution SE Asia (India, Nepal, Bangladesh, Thailand, Indonesia, Malaysia, Philippines). **Habitat** Cultivated in tropical climates. **Range** Thailand, Vietnam, Laos, Cambodia, Indonesia, Malaysia, Reunion Island, Madagascar.

Beer-Making Parts (and Possible Uses) Fruits, leaves (?).

Chemistry Different essential oils from different organs (peel, leaves, petioles, juice). Peel (38 constituents, 89% of essential oil): monoterpenes 87% (β-pinene 10.9–25.93%, sabinene 20.36%, limonene 4.7%, terpinen-4-ol 13%, α-terpineol 7.6%, 1.8-cineole 6.4%, citronellol 6%, citronellal 23.64%). Leaves (57 constituents): citronellal 81%. Petioles: citronellal 78.64%. Juice: β-pinene 39.5%, terpinene-4-ol 17.55%. Other unusual components of the oils of this species are 2,6-dimethyl-5-heptenal, citronellal acid, and safrole.

Style Saison.

Source https://www.brewersfriend.com/other/keffirlime/.

Citrus japonica Thunb.
RUTACEAE

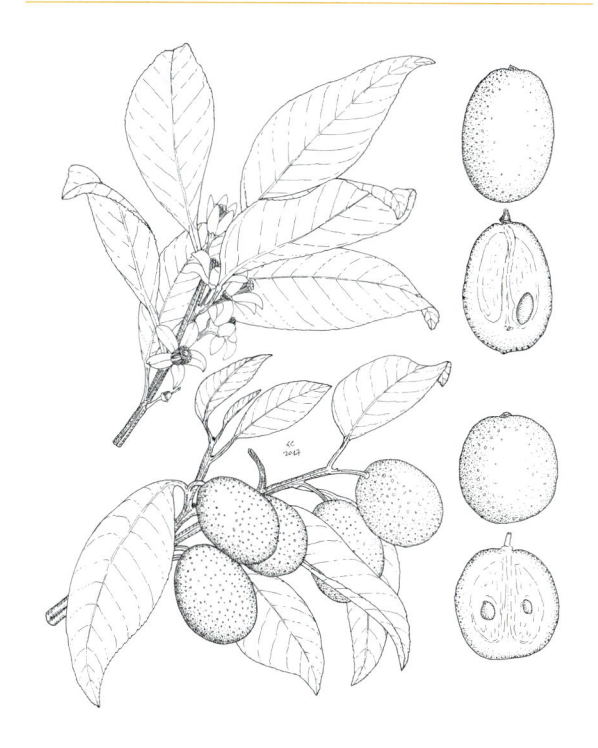

Synonyms *Atalantia hindsii* (Champ. ex Benth.) Oliv., × *Citrofortunella madurensis* (Lour.) D.Rivera & al., *Citrus erythrocarpa* Hayata, *Citrus hindsii* (Champ. ex Benth.) Govaerts, *Citrus inermis* Roxb., *Citrus kinokuni* Yu.Tanaka, *Citrus madurensis* Lour., *Citrus margarita* Lour., *Citrus microcarpa* Bunge, *Fortunella bawangica* C.C.Huang, *Fortunella chintou* (Swingle) C.C. Huang, *Fortunella crassifolia* Swingle, *Fortunella hindsii* (Champ. ex Benth.) Swingle, *Fortunella japonica* (Thunb.) Swingle, *Fortunella margarita* (Lour.) Swingle, *Fortunella obovata* Tanaka, *Sclerostylis hindsii* Champ. ex Benth., *Sclerostylis venosa* Champ. ex Benth.

Common Names Kumquat, cumquat.

Description Shrub or small tree (2.5–4.5 m), slow-growing, with dense branches, sometimes bearing small thorns. Leaves dark green, shiny. White flowers, similar to others of the genus *Citrus*, single or grouped at the leaf axil. Fruit (hundreds or thousands per year for each tree) hesperidium, oval to roundish (varietal character), usually small (Ø 2–4 cm).

Cultivar There are several entities within this species, now mostly considered varieties but in the past considered taxa in their own right. Round Kumquat or Marumi or Morgani (*Citrus japonica s.s.*), evergreen tree that usually produces roundish fruits (sometimes oval), golden yellow color, usually consumed cooked or in preserves. Kumquat Ovale or Nagami (*Citrus margarita* Lour.), orange fruits, edible whole raw. Kumquat Jiangsu or Fukushu (*Citrus obovata* Raf.), round or bell-shaped fruits, intense orange in color, edible raw or as preserves; this group is distinguished from the other kumquats by their round leaves. Kumquat Centennial Variegated, thornless tree spontaneously derived from Nagami; produces round fruits with thin peel, recognizable for the characteristic variegation given by alternated longitudinal green and yellow stripes.

Related Species In an extremely complex taxonomic picture, some sources report 33 species for the genus *Citrus*, many of which have food and beer-making applications.

Geographic Distribution China. **Habitat** Cultivated. **Range** China, Chile, Japan, Korea, Taiwan, SE Asia, Nepal, Pakistan, Iran, Middle East, Europe (Corfu, Greece), S United States (Florida, Alabama, Louisiana, California, Hawaii, Nevada, Arizona, other states).

Beer-Making Parts (and Possible Uses) Fruits.

Chemistry Essential oil: limonene 93%, α-pinene 0.34%, α-bergamotene 0.021%, caryophyllene 0.18%, α-humulene 0.07%, α-muurolene 0.06%, isopropyl propanoate 1.8%, terpenyl acetate 1.26%, carvone 0.175%, citronellal 0.6%, 2-methylundecanal, nerol 0.22%, trans-linalool oxide 0.15%. Whole fruit: energy 296 kJ (71 kcal), carbohydrates 15.9 g, sugars 9.36 g, fiber 6.5 g, lipids 0.86 g, protein 1.88 g, vitamin A 15 µg, lutein zeaxanthin 129 µg, thiamin (B1) 0.037 mg, riboflavin (B2) 0.09 mg, niacin (B3) 0.429 mg, pantothenic acid (B5) 0.208 mg, vitamin B6 0.036 mg, folate (B9) 17 µg, choline 8.4 mg, vitamin C 43.9 mg, vitamin E 0.15 mg, calcium 62 mg, iron 0.86 mg, magnesium 20 mg, manganese 0.135 mg, phosphorus 19 mg, potassium 186 mg, sodium 10 mg, zinc 0.17 mg.

Style American Pale Ale, American Wheat Ale, Belgian Blond Ale, Belgian Specialty Ale, California Common Beer, English IPA, Fruit Lambic, Witbier.

Beer Summer Wheat Ale, 't IJ (Amsterdam, Netherlands).

Source Hieronymus 2016b; Josephson et al. 2016; Steele 2012; https://www.brewersfriend.com/other/kumkvat/.

Citrus junos Siebold ex Tanaka
RUTACEAE

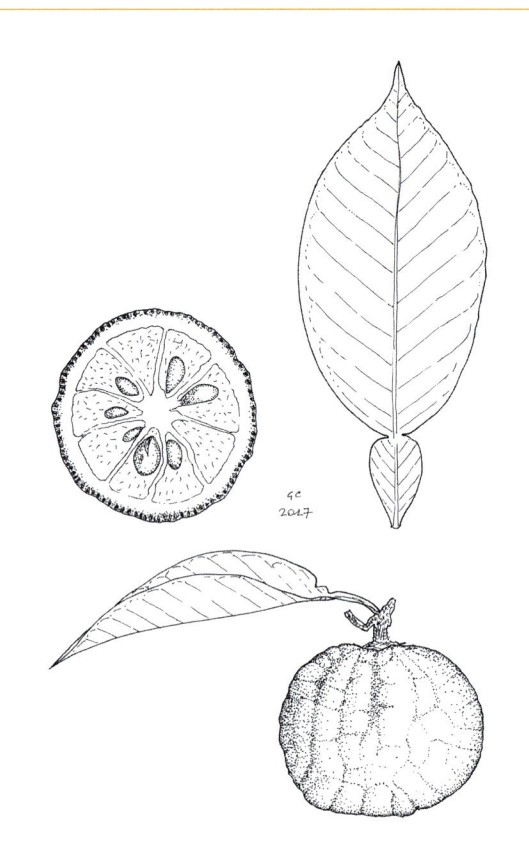

Synonyms *Citrus ichangensis* Swingle × *Citrus reticulata* Blanc var. *austera* Swingle (?).

Common Names Yuzu.

Description Shrub or small tree (3–6 m). Stem usually has large thorns. Leaves characterized by a wide petiole (similar to that *of Citrus hystrix*, but with a greater ratio of lamina:petiole in this species), strongly aromatic. Fruit (similar to that of *Citrus sudachi* Yu.Tanaka) hesperidium, very aromatic, orange color when ripe, vaguely similar to a grapefruit with irregular peel (Ø 5.5–10 cm).

Cultivar A few cultivars are known, both for ornamental and productive use: Hana Yuzu, Shishi Yuzu, and Yuku.

Related Species Notwithstanding an extremely complex taxonomic picture, some sources report 33 species for the genus *Citrus*, many of which have food and beer-making applications.

Geographic Distribution According to some sources, it is a hybrid, *Citrus ichangensis* Swingle × *Citrus reticulata* Blanc var. *austera* Swingle, native to China and Tibet, where it grows wild.

Habitat One of the few *Citrus* able to bear rather harsh winter temperatures (down to −9°C).

Range China, Japan, Korea.

Beer-Making Parts (and Possible Uses) Fruits.

Chemistry Volatile compounds, peel: α-pinene 2.7%, camphene traces, β-pinene 1.1%, sabinene 0.5%, δ-3-carene traces, myrcene 3.2%, α-phellandrene 0.9%, pseudolimonene traces, α-terpinene 0.3%, limonene 63.2%, β-phellandrene 5.4%, (Z)-β-ocimene traces, γ-terpinene 12.5%, (E)-β-ocimene traces, ρ-cymene 0.6%, terpinolene 0.7%, octanal traces, tetradecane traces, nonanal traces, α-ρ-dimethyl styrene traces, 1,3,8-ρ-menthatriene traces, (Z)-limonene oxide traces, α-cubebene traces, trans-sabinene hydrate 0.1%, δ-elemene 0.2%, (E)-linalool oxide traces, bicyclohexane traces, (-)-α-copaene 0.1%, decanal traces, β-cubebene 0.1%, linalool 2.8%, cis-sabinene hydrate 0.1%, (E)-α-bergamotene traces, β-elemene 0.1%, β-caryophyllene 0.3%, terpinen-4-ol 0.1%, aromadendrene traces, caryophyllene traces, γ-elemene traces, (E)-2-decenal traces, (Z)-ρ-farnesene traces, (E)-ρ-farnesene 1.3%, α-humulene 0.1%, α-terpineol 0.3%, dodecanal traces, germacrene D 0.4%, guaiene traces, α-muurolene traces, piperitone traces, bicyclogermacrene 2%, α-farnesene traces, 5-cadinene 0.1%, citronellol traces, β-sesquiphellandrene 0.1%, perillaldehyde traces, nerol traces, germacrene B 0.2%, β-ionone traces, perillyl alcohol traces, (E)-nerolidol traces, germacrene D-4-ol 0.4%, elemol traces, spatulenol traces, eugenol traces, thymol 0.3%, α-cadinol traces, (E,E)-farnesyl acetate traces, β-eudesmol traces.

Style American IPA, American Pale Ale, American Wheat or Rye Beer, California Common Beer, Saison, Witbier.

Beer Saison du Japon, Hitachino Nest Beer (Ibaraki, Japan).

Source https://www.brewersfriend.com/other/yuzu/.

Citrus limon (L.) Osbeck
RUTACEAE

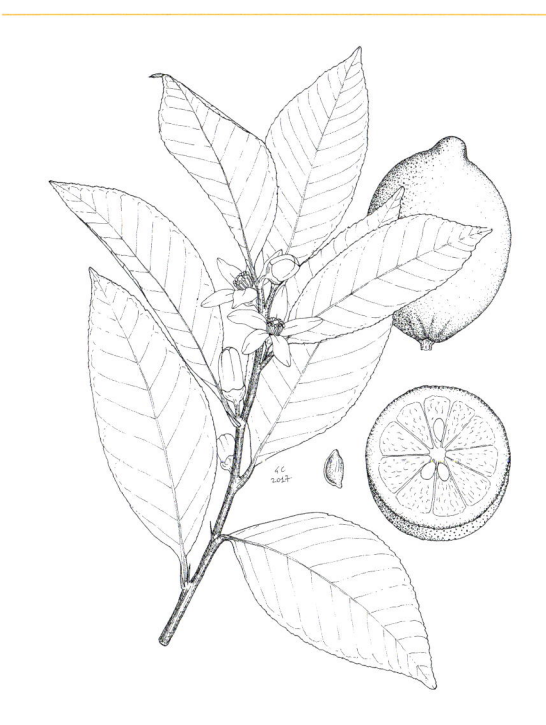

Synonyms *Citrus* × *limodulcis* D.Rivera, Obón & F.Méndez, *Citrus* × *limonelloides* Hayata, *Citrus* × *limonia* Osbeck, *Citrus* × *limonum* Risso, *Citrus* × *mellarosa* Risso, *Citrus* × *meyeri* Yu.Tanaka, *Citrus* × *vulgaris* Ferrarius ex Mill., *Limon* × *vulgaris* Ferrarius ex Miller.

Common Names Lemon tree.

Description Small tree (2–5 m), evergreen. Leaves broadly elliptical (5–10 cm). Flowers solitary or in pauciflorous racemes; corolla white inside, suffused with purplish-red outside. Fruit yellow, fusiform (4–5 × 5–7 cm), with about 10 cloves and short apical conical appendage.

Subspecies A related taxon, whose origin is uncertain, is *Citrus limon* (L.) Osbeck var. *pompia* Camarda, native of a small area in the province of Nuoro (Sardinia) and now one of Slow Food Presidia (Slow Food, the international association for sustainable, healthy and quality food, put some rare food products, often under extinction risk, called Presidia, under special attention and protection, in order to promote customer interest, production, and the long-term conservation).

Cultivar There are hundreds, some of which have an exclusively local distribution because they are not very durable once the fruits are ripe (e.g., Sorrento lemon).

lemon scented, ovate-lanceolate or ovate-elliptic. Flower buds large, white or purplish. Fragrant flowers with 4–5 pinkish or purplish petals with 30–60 stamens. Fruit fragrant, oblong or oval, highly variable even on the same branch; peel yellow, usually wrinkled and very thick; flesh light yellow or greenish, divided into 14–15 segments, firm, not very juicy, sour or sweet, containing numerous seeds.

Subspecies The taxonomy of the genus *Citrus* is very complex and still partly controversial. The classification based on morphology seems insufficient to explain the complexity of the genus, in part because hybrids often have very different characters compared to both parents. According to a recent classification based on genetics, the four ancestral *Citrus*, from which all the other forms derived by mutation, selection, and hybridization, are *Citrus hystrix* (papeda), *Citrus maxima* (pomelo), *Citrus medica* (citron), and *Citrus reticulata* (mandarin).

Cultivar There are many varieties classified according to the characteristics of the pulp. Those with acid pulp include Fiorentina and Diamante (Italy), Greca (Corfu), and Balady (Israel). Sweet pulp varieties include Corsa (Corsica) and Marocchina (Morocco). The pulpless varieties are the hand-shaped ones (*Citrus medica* L. var. *sarcodactylis* or Buddha's Hand) and the Yemenites (Yemen). *Citrus medica* is also the basis of several hybrids: Ponderosa Lemon (*Citrus limon* × *medica*), Lumia (*Citrus* × *lumia* Risso & Poit.), and Rhobs el Arsa (presumed to be *Citrus medica* × *aurantium*).

Related Species Within an extremely complex taxonomic framework, some sources report 33 species for the genus *Citrus*, many of which have food and beer-making applications.

Geographic Distribution India (Himalaya, Western Ghats). **Habitat** Cultivated. **Range** Italy, Greece, Yemen, Israel, China, Japan.

Beer-Making Parts (and Possible Uses) Fruit peels.

Chemistry Essential oil, peel: thujene 1.2%, α-pinene 2.5%, camphene 0.9%, sabinene 0.4%, β-pinene 2.6%, β-myrcene 2.1%, α-phellandrene 0.2%, α-terpinene 1.2%, ρ-cymene traces, limonene 35.4%, (E)-β-ocimene 2.4%, γ-terpinene 24.5%, cis-sabinene hydrate traces, terpinolene 1.5%, linalool 0.6%, n-nonanal 0.2%, citronellal 0.3%, terpinen-4-ol 1.5%, α-terpineol 1.1%, n-decanal 0.1%, nerol 1%, neral 4.4%, geraniol 0.3%, perillaldehyde 0.2%, geranial 5.5%, citronellyl acetate 0.2%, neryl acetate 0.5%, geranyl acetate 0.4%, β-elemene 0.1%, β-cubebene 0.1%, α-bergamotene 0.1%, β-caryophyllene 0.3%, trans-β-farnesene 0.5%, β-bisabolene 1.2%, α-humulene 0.2%, γ-cadinene traces, δ-cadinene 0.2%, tetradecanal traces, β-bisabolol 0.4%, eicosane 0.1%, docosane traces, citroptene 0.9%.

Style Lambic, Farmhouse Ale, Pilsner (flavored), Saison.

Beer Struiselensis, De Struise (Lo-Reninge, Belgium).

Source Dabove 2015; https://www.fermentobirra.com/struiselensis-de-struise-brouwerij/; https://www.gustidicorsica.com/it/9-7/producteur/paisolu-dilutina.html; https://untappd.com/b/l-inconsistent-brewery-bustese-beer-at-cedar/384690.

Citrus reticulata Blanco
RUTACEAE

Synonyms *Citrus chrysocarpa* Lush., *Citrus crenatifolia* Lush., *Citrus daoxianensis* S.W.He & G.F.Liu, *Citrus deliciosa* Ten., *Citrus depressa* Hayata, *Citrus erythrosa* Yu.Tanaka, *Citrus himekitsu* Yu.Tanaka, *Citrus koozi* (Sieb. ex Yu.Tanaka) Yu.Tanaka, *Citrus lycopersiciformis* (Lush.) Yu.Tanaka, *Citrus mangshanensis* S.W.He & G.F.Liu, *Citrus nippokoreana* Yu.Tanaka, *Citrus otachihana* Yu.Tanaka, *Citrus papillaris* Blanco, *Citrus ponki* Yu.Tanaka, *Citrus poonensis* Yu.Tanaka, *Citrus succosa* Yu.Tanaka, *Citrus succosa* Tanaka, *Citrus suhuiensis* Hayata, *Citrus sunki* (Hayata) Yu.Tanaka, *Citrus tachibana* (Makino) Yu.Tanaka, *Citrus tangerina* Yu.Tanaka, *Citrus tankan* Hayata, *Citrus vangasy* Bojer.

Common Names Mandarin, tangerine.

Description Shrub or small tree (2–4 m). Leaves lanceolate, small (3.2–3.8 × 7.5–9 cm), fragrant. Flowers white in color, petals lanceolate (4.7 × 11 mm). Fruit hesperidium, spheroidal, slightly flattened at the poles (oblate) (5–6.2 × 4.4–5 cm), peel color orange, thin (about 2 mm), fragrant, with albedo very rarefied and grainy, which allows for easy peeling; juicy flesh, sweet, light orange color, divided into segments easily separable, within which are numerous seeds.

Subspecies The taxonomy of the genus *Citrus* is very complex and still partly controversial. The classification based on morphology seems insufficient to explain the complexity of the genus, in part because hybrids often have very different characters compared to both parents. According to a recent classification based on genetics, the four ancestral *Citrus*, from which all other forms derived by mutation, selection, and hybridization, are *Citrus hystrix* (papeda), *Citrus maxima* (pomelo), *Citrus medica* (citron), and *Citrus reticulata* (mandarin).

Cultivar Given the strong tendency to hybridization, in the absence of genetic data, it is not easy to distinguish "pure" mandarins from hybrids and related taxa such as *Citrus unshiu* and *Citrus tangerina*. Some authors distinguish two types of pure mandarin: edible mandarins (Bang Mot, Dancy, Nanfengmiju) and sour mandarins, too sour to be eaten fresh (Cleopatra, Shekwasha, Sunki).

Related Species Notwithstanding an extremely complex taxonomic framework, some sources report 33 species for the genus *Citrus*, many of which have food and beer-making applications.

Geographic Distribution Vietnam, S China, Japan. **Habitat** Cultivated. **Range** China, Spain, Brazil, Japan, Turkey, Italy, Egypt, Iran, Morocco, South Korea, United States, Pakistan, Mexico, Argentina, Thailand, Peru, Algeria, Taiwan, Nepal, Maldives (aggregated production of mandarins, clementines, tangerines, satsumas).

Beer-Making Parts (and Possible Uses) Fruits.

Chemistry Pulp: energy 223 kJ (53 kcal), carbohydrates 13.34 g, sugars 10.58 g, fiber 1.8 g, lipids 0.31 g, protein 0.81 g, vitamin A 34 µg, beta-carotene 155 µg, thiamin (B1) 0.058 mg, riboflavin (B2) 0.036 mg, niacin (B3) 0.376 mg, pantothenic aicdo (B5) 0.216 mg, vitamin B6 0.078 mg, folate (B9) 16 µg, choline 10.2 mg, vitamin C 26.7 mg, vitamin E 0.2 mg, calcium 37 mg, iron 0.15 mg, magnesium 12 mg, manganese 0.039 mg, phosphorus 20 mg, potassium 166 mg, sodium 2 mg, zinc 0.07 mg. Essential oil, peel: α-pinene 1.27%, sabinene 0.49%, β-pinene 0.40%, β-myrcene 3.27%, limonene 69.9%, Z-β-ocimene 0.28%, γ-terpinene 0.23%, octanal 1.06%, 1-octanol 0.95%, linalool oxide 0.60%, linalool 1.1%, menthadien-1-ol 0.42%, trans-p-1,8-dienol 0.52%, citronellal 0.78%, α-terpineol 1.1%, 4-carvon menthenol 0.95%, α-terpineneol 1.51%, decanal 2.33%, Z-carveol 1.29%, citronellol 0.80%, carvone 0.47%, perillaldehyde 1.64%, isopropyl cresol 1.36%, 4-vinyl guaiacol 2.32%, α-cubebene 0.48%, copaene 0.89%, allyl isovalerate 0.36%, β-cubebene 0.84%, β-caryophyllene 1.39%, germacarene 1.07%, α-farnesene 0.48%, γ-munrolene 1.16%, δ-cadinene 0.88%, dodecanal 0.42%, elemol 0.35%, γ-eudesmol 1.08%, α-cadinol 0.46%, β-sinensal 0.67%, farnesol 1.14%, α-sinensal 0.93%, nootkatone 3.95%.

Style American IPA, American Pale Ale, American Stout, American Wheat Beer, American Wheat or Rye Beer, Belgian Pale Ale, Belgian Specialty Ale, Bière de Garde, Blonde Ale, California Common Beer, Fruit Beer, Holiday/Winter Special Spiced Beer, Old Ale, Saison, Specialty Beer, Standard/Ordinary Bitter, Weissbier, Witbier.

Beer Bianca, Bruno Ribadi (Cinisi, Sicily, Italy).

Source Hieronymus 2016b; https://www.brewersfriend.com/other/mandarin/; https://www.brewersfriend.com/other/tangerine/.

Citrus unshiu (Yu.Tanaka ex Swingle) Marcow.
RUTACEAE

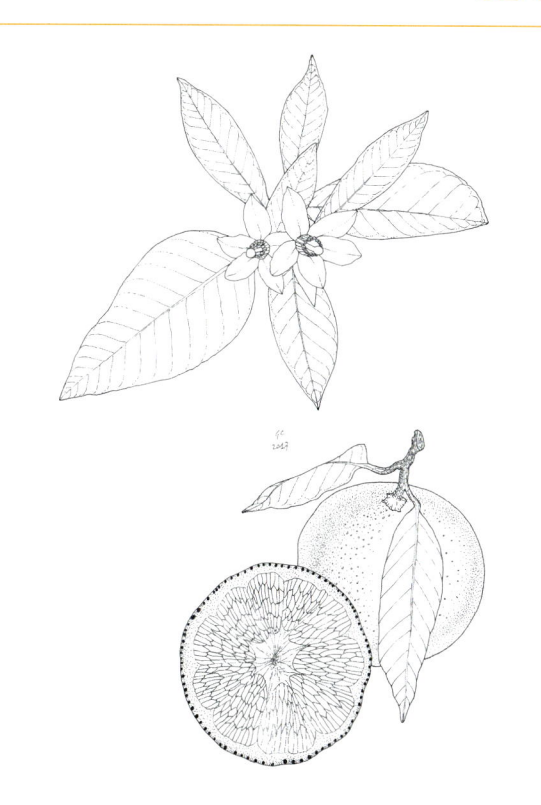

Synonyms According to some sources, *Citrus reticulata* Blanco.

Common Names Unshu mikan, cold hardy mandarin, satsuma mandarin, satsuma orange, Christmas orange, tangerine, naartjie.

Description Shrub or small tree (2–4 m). Foliage dense. Leaves lanceolate, broad. Fruit similar to *C. reticulata*, round, slightly flattened at the poles (oblate), yellow or deep green peel even when ripe, very thin, leathery, punctate with large and prominent oil glands, not very adherent to the fruit, which is thus easy to peel; orange flesh, extremely sweet, with delicate structure, not very resistant to handling, usually seedless.

Cultivar Some examples: Citrus Unshiu, Egan nr. 2, Guoqing nr. 1, Guoqing nr. 4 Satsuma Mandarin, Kawano Wase Satsuma, Mapo, Miyagawa, Miyagawa Wase, Okitsu, Satsuma, Satsuma Mandarin, Satsuma Miyagawa Wase, Satsuma Owari, Satsuma Unshiu Wase, Satsuma Wase, Sochi nr. 1, Ueda Unshiu.

Related Species Notwithstanding an extremely complex taxonomic picture, some sources report 33 species for the genus *Citrus*, many of which have food and beer-making applications; in addition to the numerous species described here, *Citrus sudachi* Yu.Tanaka also has recognized beer-making uses.

Geographic Distribution China. **Habitat** Cultivated. **Range** Japan, Spain, C China, Korea, United States, South Africa, South America, Black Sea, Croatia.

Beer-Making Parts (and Possible Uses) Fruits.

Chemistry Essential oil, peel: β-myrcene 0.91%, γ-terpinene 4.66%, L-limonene 88.11%, O-cymene 0.85%, β-linalol 0.97%, 2,4-diisopropenyl-1-methyl-1-vinyl cyclohexane 1.82%, α-farnesene 0.91%, diethyl phthalate 1.02%.

Style American IPA.

Source https://www.brewersfriend.com/other/naartjie/.

Clinopodium nepeta (L.) Kuntze subsp. *nepeta*
LAMIACEAE

Synonyms *Acinos transsilvanica* Schur, *Calamintha acinifolia* Sennen, *Calamintha alboi* Sennen, *Calamintha athonica* Rchb., *Calamintha barolesii* Sennen, *Calamintha bonanovae* Sennen, *Calamintha bonanovae* Sennen & Pau, *Calamintha brevisepala* Sennen, *Calamintha caballeroi* Sennen & Pau, *Calamintha cacuminiglabra* Sennen, *Calamintha cantabrica* Sennen & Elias, *Calamintha dilatata* Schrad., *Calamintha dufourii* Sennen, *Calamintha enriquei* Sennen & Pau, *Calamintha eriocaulis* Sennen, *Calamintha ferreri* Sennen, *Calamintha guillesii* Sennen, *Calamintha josephi* Sennen, *Calamintha largiflora* Klokov, *Calamintha litardierei* Sennen, *Calamintha longiracemosa* Sennen, *Calamintha mollis* Jord. ex Lamotte, *Calamintha nepeta* (L.) Savi, *Calamintha nepetoides* Jord., *Calamintha obliqua* Host, *Calamintha peniciliata* Sennen, *Calamintha rotundifolia* Host, *Calamintha sennenii* Cadevall, *Calamintha suavis* Sennen, *Calamintha thessala* Hausskn., *Calamintha transsilvanica* (Jáv.) Soó, *Calamintha trichotoma* Moench, *Calamintha vulgaris* Clairv., *Melissa aetheos* Benth., *Melissa nepeta* L., *Melissa obtusifolia* Pers., *Micromeria byzantina* Walp., *Satureja mollis* (Jord. ex Lamotte) E.Perrier, *Satureja nepeta* (L.)

Citrus reticulata Blanco
RUTACEAE

Scheele, *Satureja nepetoides* (Jord.) Fritsch, *Thymus athonicus* Bernh. ex Rchb., *Thymus minor* Trevir., *Thymus nepeta* (L.) Sm.

Common Names Lesser calamint.

Description Herb, perennial or shrub (2–8 dm), mint scented. Stems ascending, woody below, with inclined hairs. Leaves ovate, acute, with revolute, entire or weakly toothed margins. Inflorescence top 5–20 flowered, on peduncles up to 2 cm long, leafy; calyx humped, shaggy, with 3–4 mm tube, 0.5–2 mm teeth and hairs protruding from the fauces; corolla pubescent, with violet or pale tube (8–10 mm) and upper labellum 2 mm, lower labellum 4.5 mm, tri-lobed, pale in the middle with 2 violet blotches.

Subspecies In the Italian flora, there are also the subspecies *Clinopodium nepeta* (L.) Kuntze ssp. *sylvaticum* (Bromf.) Peruzzi & F.Conti (calyx 6–10 mm, 2 lower teeth 2–4 mm, distinctly longer than the upper 3, regularly edged with hairs up to 0.5 mm long, fauces without protruding hairs; corolla 15–22 mm, bright pink in color; flower pedicels up to 15 mm long; leaves 3–6 cm larger, triangular-acuminate with 6–10 regular, acute teeth per side) and *Clinopodium nepeta* (L.) Kuntze ssp. *spruneri* (Boiss.) Bartolucci & F.Conti (calyx with abundant spherical glands and ± equal teeth).

Related Species The genus *Clinopodium* includes 177 species, some of which, including the one mentioned here, have aromatic properties.

Geographic Distribution Euro-Mediterranean.
Habitat Arid meadows, uncultivated, walls. **Range** Mediterranean.

Beer-Making Parts (and Possible Uses) Inflorescences.

Chemistry Essential oil: menthol, borneol, other terpenic components (giving an aroma similar to mint but more intense and camphorated).

Style Specialty IPA (Belgian IPA).

Source https://www.brewersfriend.com/other/neboda/.

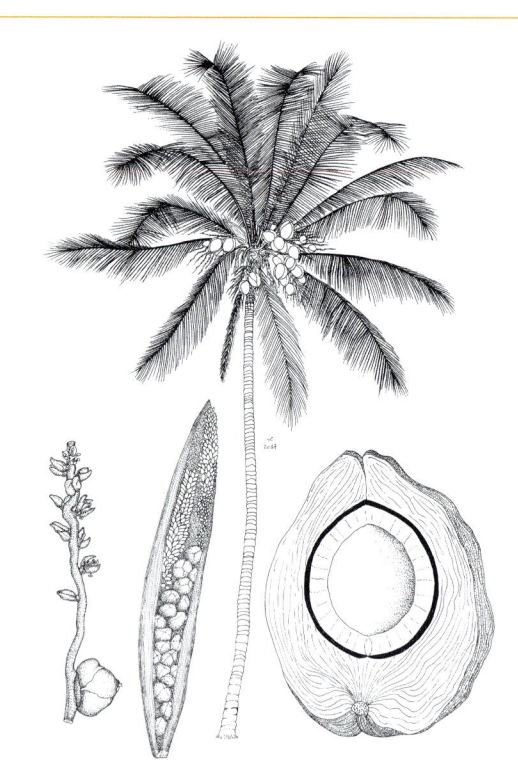

Cocos nucifera L.
ARECACEAE

Synonyms *Calappa nucifera* (L.) Kuntze, *Cocos indica* Royle, *Cocos nana* Griff.

Common Names Coconut tree, coconut palm.

Description Tree (20–40 m). Trunk columnar, slender (Ø 50–70 cm at base, 25–35 cm above), surface gray in color, marked by leaf scars, topped by a crown of leaves. Roots fasciculate, cylindrical, Ø uniform, short-lived, soon replaced by adventitious roots at plant collar. Leaves paripinnate (4–5 m), erect in first 2 years of life, then drooping, each consisting of light leaflets, elongate and briefly petiolate, divided in 2, striate and poorly arched; leaf petiole basally dilated into a broad sheath. Inflorescences (6–12 per year per plant, uneven-aged) spadix-like, branched (± 50 branching), protected by a concave spathe. Monoecious plant with diclinous flowers, male and female on the same individual. Flowers small, yellowish color, unisexual, often with rudimentary organs of the other sex; female flowers (± 5) on basal branches, perigonium 6 tepals (2 imbricate whorls of 3 elements), stamens rudimentary, ovary 3-carpellar, only 1 will be fertilized; male flowers numerous (about 300/ramification spadix), perigonium 6 tepals, 6 stamens, 1 rudimentary pistil (apically 3 teeth with nectar glands to attract pollinators). Mixed pollination (anemophilous and entomophilous). Fruits voluminous drupes (± 1 kg), erroneously called coconuts, exocarp (skin) smooth, thin, reddish-brownish color, mesocarp fibrous and light at maturity, closely united to the endocarp (shell), woody and very hard, with 3 pores of lesser thickness at the base. Seed, consisting of the pulp of the "nut," with very thin brown integument, tightly adherent to the endosperm rich in fats (copra), formed by a 1–3 cm thick layer that forms a cavity containing a milky liquid (coconut water); embryo wrapped by the endosperm at one end of the fruit.

Cultivar More than 80 varieties of this species have been described, grouped into two major categories: tall and dwarf. Tall varieties are allogamous, long-lived (more than 80 years), bear fruit late (after about 10 years), and produce larger fruits. Dwarf varieties are autogamous, with a maximum height shorter than tall varieties, have smaller fruits and bear fruit early (at four years). In this second group, the most famous variety is Malayan Dwarf, which has three subtypes (eburnea, pumila, regia) based on the color of the drupe. There are also varieties with intermediate characteristics, usually of little commercial value mainly in the limited production areas.

Related Species *Cocos* is a monotypic genus.

Geographic Distribution Indonesian archipelago (?), South America (?). **Habitat** Coastal and subcoastal areas of tropical regions (0–600 m ASL), with rainfall 1,300–2,000 mm/year. **Range** Current distribution of the species between the 22nd parallel north and the 22nd parallel south.

Beer-Making Parts (and Possible Uses) Seeds, sap (transformed into solid sugar by dehydration), oil, nut shells (to smoke malt).

Chemistry Seed pulp: energy 354 kcal, lipids 33 g (saturated fatty acids 30 g, polyunsaturated fatty acids 0.4 g, monounsaturated fatty acids 1.4 g), sodium 20 mg, potassium 356 mg, carbohydrates 15 g, dietary fiber 9 g, sugars 6 g, protein 3.3 g, vitamin C 3.3 mg, calcium 14 mg, iron 2.4 mg, vitamin B6 0.1 mg, magnesium 32 mg. Coconut water: energy 19 kcal, lipids 0.2 g (saturated fatty acids 0.2 g), sodium 105 mg, potassium 250 mg, carbohydrates 3.7 g, dietary fiber 1.1 g, sugars 2.6 g, protein 0.7 g, vitamin C 2.4 mg, calcium 24 mg, iron 0.3 mg, magnesium 25 mg.

Style American Stout, Foreign Extra Stout, Fruit Beer, Oatmeal Stout, Saison, Spice/Herb/Vegetable Beer.

Beer Monster Tones, Modern Times (California, United States).

Source Calagione et al. 2018; Daniels and Larson 2000; Heilshorn 2017; Jackson 1991; https://www.brewersfriend.com/other/cocco/.

Coix lacryma-jobi L.
POACEAE

Synonyms *Coix agrestis* Lour., *Coix arundinacea* Lam., *Coix exaltata* Jacq. ex Spreng, *Coix ouwehandii* Koord., *Coix palustris* Koord., *Coix pumila* Roxb., *Coix stigmatosa* K.Koch & Bouché, *Lithagrostis lacryma-jobi* (L.) Gaertn.

Common Names Job's tears, Job's-tears, coixseed, tear grass.

Description Herb (0.7–4 m), annual, perennial in frost-free areas, culm erect, with supporting roots from lower nodes, internodes full. Leaves not aggregated at the base and not auriculate; leaf blade lanceolate and broad, 30–70 mm wide, cordate, flat, and persistent, without cross-veins. Monoecious plant. Male inflorescence in pairs or triads, several per raceme; spikelet rachilla terminating with a male flower, male spikelets having proximal glumes and flowers more or less incomplete (flowers below sterile or male); female inflorescence in which spikelets fall together with glumes, the rachilla terminating with a fertile female flower; glumes of fertile multiveined spikelets. Fruit globular (6–12 mm long). Oval seeds yellow, purple, white, or brown in color.

Subspecies Four varieties of this species are recognized: *Coix lacryma-jobi* L. var. *lacryma-jobi* (throughout the Asian subcontinent to Malaysia and Taiwan but also naturalized elsewhere), *Coix lacryma-jobi* L. var. *ma-yuen* (Rom.Caill.) Stapf. (from S-China to Malaysia and the Philippines), *Coix lacryma-jobi* L. var. *puellarum* (Balansa) A.Camus. (from Assam to Yunnan and Indochina), and *Coix lacryma-jobi* L. var. *stenocarpa* Oliv. (from E-Himalaya to Indochina).

Cultivar There are numerous cultivars, including Guangxi Longlin #1, Guangxi Longlin #2, Guangxi Xilin #1, Pinzhong #0, Pinzhong #1, Pinzhong #2, Pinzhong #10, Pinzhong #11, Qianyin #1, XYYBT, XYYH12-1, Xingrenxiaobaike, Yuenanhuake mi, and ZYYB12.

Related Species The genus *Coix* has 4 species: *Coix aquatica* Roxb., *Coix gasteenii* B.K.Simon, *Coix lacryma-jobi* L., and *Coix puellarum* Balansa.

Geographic Distribution Asia. **Habitat** Cultivated. **Range** Asia.

Beer-Making Parts (and Possible Uses) Seeds.

Chemistry Water 10.8 g, protein 13.6 g, lipids 6 g, carbohydrates 58.5 g, fiber 8.4 g, ash 2.6 g.

Style Ancient Brew (grain used in the earliest known beer in China).

Source McGovern 2017; https://www.fermentobirra.com/la-birra-preistorica-cina/.

Ancient Chinese Beer

One of the earliest breweries in human history may have sprung up in China between 3500 and 2900 BCE. Archaeological excavation led to the discovery of a series of objects that may have been containers used to produce a sort of beer. Confirming this hypothesis are the organic residues found in the containers, among which barley, millet, and seeds of Job's tears stand out, as well as a number of spices.

Coffea arabica L.
RUBIACEAE

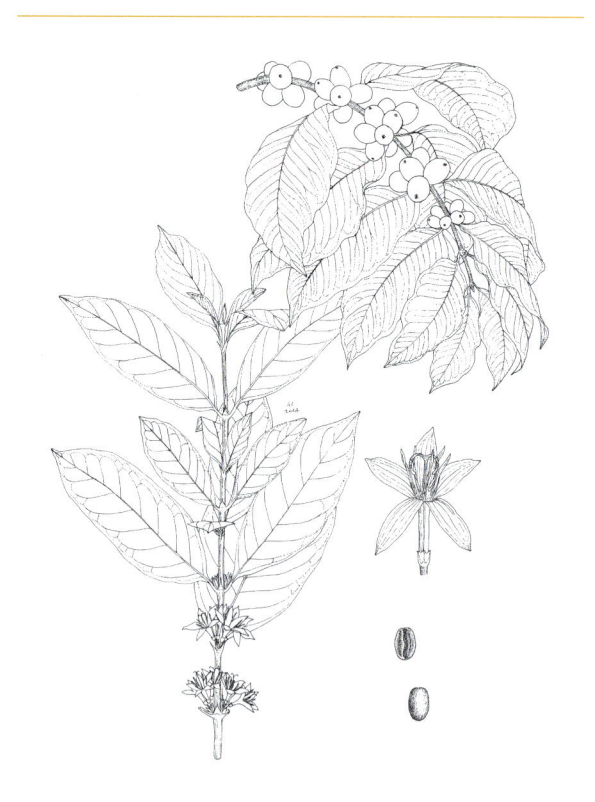

Synonyms *Coffea corymbulosa* Bertol., *Coffea laurifolia* Salisb., *Coffea moka* Heynh., *Coffea sundana* Miq., *Coffea vulgaris* Moench.

Common Names Coffee shrub of Arabia, mountain coffee, arabica coffee.

Description Shrub or small tree (up to 5 m), evergreen, glabrous. Leaves opposite, dark green, glossy, elliptical (5–20 × 1.5–7.5 cm; usually 10–15 × 6 cm), sharply pointed, briefly petiolate. Flowers white, fragrant, in axillary clusters; corolla tubular, 1 cm long, 5-lobed; calyx small, cup shaped. Fruit drupe, about 1.5 cm long, oval-elliptical in shape, green when unripe, yellow and then crimson when ripe, black after drying; seeds usually 2, ellipsoidal, 8.5–12.5 mm long, with deeply grooved inner surface, consisting mainly of green horny endosperm and small embryos.

Subspecies According to some sources, *Coffea arabica* L. var. *angustifolia* (Roxb.) Miq. ex A.Froehner would be a taxon of subspecific rank of this species.

Cultivar Innumerable varieties of *Coffea sp.* are known; among the cultivated ones are *Arabica* varieties (Arusha, Bergendal, Blue Mountain, Bourbon, Brutte,

Catuai, Caturra, Charrieriana, Colombian, Ethiopian Harar, Ethiopian Sidamo, Ethiopian Yirgacheffe, French Mission, Gesha, Geisha T.2722, Guadeloupe Bonifieur, Hawaiian Kona, K7, Maragaturra, Maragogipe, Mayagüez, Mocha, Mundo Novo, Orange, Pacamara, Pacas, Pache Colis, Pache Comum, Ruiru 11, S795, Santos, Sidikalang, SL28, SL34, Sulawesi Toraja Kalossi, Sumatra Lintong, Sumatra Mandheling, Typica, Yellow Bourbon), *Arabica* × *Robusta* inter-specific hybrids (Arabusta, Catimor, Java, Sarchimor, Timor, Uganda), *Robusta* varieties (Kapéng Alamid, Kopi Luwak, Robusta), and *Liberica* varieties (Barako Coffee, Café Baraco, Kapeng Barako).

RELATED SPECIES Of the 128 species known for the genus *Coffea*, some are used in making beer, including *Coffea canephora* Pierre ex A.Froehner (= *Coffea robusta* L.Linden), *Coffea liberica* Hiern, and *Coffea liberica* Hiern var. *dewevrei* (De Wild. & T.Durand) Lebrun (= *Coffea excelsa* A.Chev.).

GEOGRAPHIC DISTRIBUTION Ethiopia, Yemen.
HABITAT Plant cultivated in tropical climates.
RANGE Africa, South America, SE-Asia, China, Caribbean Islands, Pacific Islands.

BEER-MAKING PARTS (AND POSSIBLE USES) Seeds, powder (also instant decaffeinated).

CHEMISTRY Energy 1,201 kJ (287 kcal), water 4.1 g, protein 10.4 g, lipids 15.4 g, carbohydrates 28.5 g, sodium 74 mg, potassium 2,020 mg, iron 4.1 mg, calcium 130 mg, phosphorus 160 mg, riboflavin 0.2 mg, niacin 10 mg.

STYLE American Stout, Baltic Porter, Belgian Specialty Ale, Brown Porter, Coffee Beer, Dry Stout, Foreign Extra Stout, Imperial Stout, Irish Extra Stout, Oatmeal Stout, Robust Porter, Russian Imperial Stout, Special/Best/Premium Bitter, Sweet Stout.

BEER Coffee Brett, Carrobiolo (Monza, Lombardy, Italy).

SOURCE Calagione et al. 2018; Cantwell and Bouckaert 2016; Cantwell and Bouckaert 2018; Giaccone and Signoroni 2017; Heilshorn 2017; https://www.beeradvocate.com/beer/profile/18134/55/; https://www.beeradvocate.com/beer/profile/287/2010/; https://www.bjcp.org (30. Spiced Beer); https://en.wikipedia.org/wiki/Caff%C3%A8; https://www.hort.purdue.edu/newcrop/duke_energy/Coffea_arabica.html; https://www.omnipollo.com/beer/symzonia/.

Coffee Ales

The use of coffee in brewing became popular, especially in the United States, between the late 1990s and the early 2000s, beginning with porters and stouts. Coffee, particularly when roasted, blends very well with the roasted notes of dark malts giving nuances and depth that cereal alone cannot achieve. As time has passed, the use of coffee has become more refined. Today, many producers are experimenting with this ingredient by varying the method of extraction (from espresso to mocha to cold brew), the type (many varieties, different levels of roasting, including the use of green coffee), and the grind, or when it is added. In recent years, the range of styles enriched by coffee notes has greatly expanded. Today it is not rare to find a coffee IPA or a blonde ale with a light sensation derived from this splendid plant.

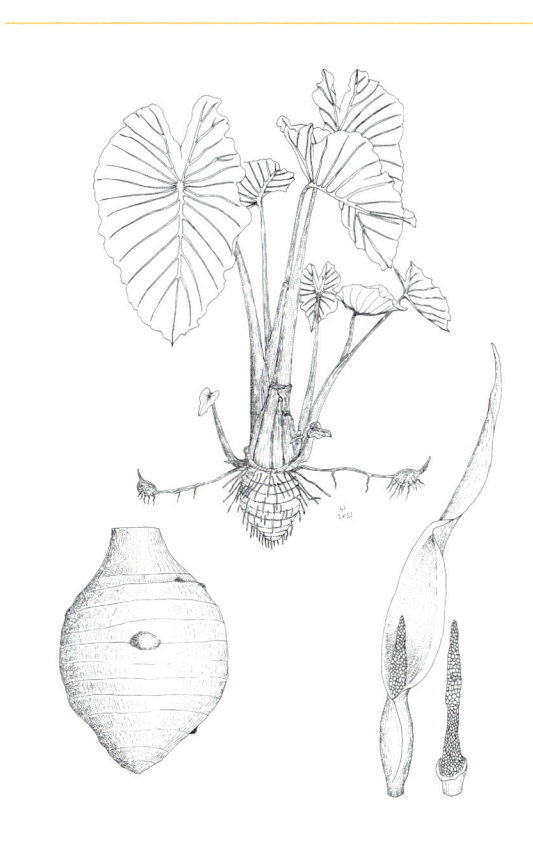

Colocasia esculenta (L.) Schott
ARACEAE

SYNONYMS *Alocasia dussii* Dammer, *Alocasia illustris* W.Bull, *Aron colocasium* (L.) St.-Lag, *Arum chinense* L., *Arum colocasia* L., *Arum colocasioides* Desf., *Arum esculentum* L., *Arum lividum* Salisb., *Arum nymphaeifolium* (Vent.) Roxb., *Arum peltatum* Lam., *Caladium acre* R.Br., *Caladium colocasioides*

(Desf.) Brongn., *Caladium esculentum* (L.) Vent., *Caladium glycyrrhizum* Fraser, *Caladium nymphaeifolium* Vent., *Caladium violaceum* Desf., *Caladium violaceum* Engl., *Calla gaby* Blanco, *Calla virosa* Roxb., *Colocasia acris* (R.Br.) Schott, *Colocasia aegyptiaca* Samp., *Colocasia euchlora* K.Koch & Linden, *Colocasia formosana* Hayata, *Colocasia gracilis* Engl., *Colocasia himalensis* Royle, *Colocasia konishii* Hayata, *Colocasia neocaledonica* Van Houtte, *Colocasia nymphaeifolia* (Vent.) Kunth, *Colocasia peltata* (Lam) Samp., *Colocasia vera* Hassk., *Colocasia violacea* (Desf.) auct., *Colocasia virosa* (Roxb.) Kunth, *Colocasia vulgaris* Raf., *Leucocasia esculenta* (L.) Nakai, *Steudnera virosa* (Roxb.) Prain, *Zantedeschia virosa* (Roxb.) K.Koch.

Common Names Taro.

Description Herb, perennial (0.8–1.8 m), robust. Underground tuberous stem, subglobose (6 × 4 cm), starchy, outer dark, inner white; pseudostem (10–17 × 1 cm), usually short in wild forms but occasionally subarborescent, strongly stoloniferous, green, carmine, or purple in color. Leaves numerous, petiole 0.8–1.8 m, lamina ovate-sagittate (25–90 × 20–70 cm), color greenish to dark green, apex acute. Inflorescences at leaf axil; peduncle (± 70 cm) curved downward at fruiting; spathe (± 30 cm) green to yellow-orange in color, restricted at the level of the sterile zone of the spadix; spadix stipitate (± 13.5 cm), female zone (3 cm) bearing pistils with sparse staminodes, ovary globose (Ø 0.15 cm), green, with many ovules, stigma 1-lobed, with rounded lobes, pale yellow on a short (0.05 cm) style; sterile zone (1 cm), about 6 synandrous whorls (concrescence of sterile staminodes) ± rhombus-hexagonal; male zone (4 cm), yellow, with irregularly rounded or rhombus-hexagonal synandria (concrescence of fertile stamens) (Ø 0.1 cm). Fruit berry (± 0.4 cm), green, with thick coating and few seeds.

Subspecies *Colocasia esculenta* (L.) Schott var. *aquatilis* Hassk. (Kimberley in Western Australia) is a taxon of subspecific rank not unanimously recognized.

Cultivar Some authors recognize five cultivars: Msaanga (plant up to 0.75 m, petiole with red stripes at the base), Msale (plant up to the height of a person, petiole and lamina entirely dark green), Msaru (light green petiole, ribs with margins of laminar tissue on the underside), Mshele (plant up to 1 m, dark red petiole, dark green lamina), and Mujasa (very tall plant, light green petiole, tuber not used as food).

Related Species The genus *Colocasia* includes eight species: *Colocasia affinis* Schott, *Colocasia antiquorum* Schott, *Colocasia esculenta* (L.) Schott, *Colocasia fallax* Schott, *Colocasia gigantea* (Blume) Hook.f., *Colocasia mannii* Hook.f., *Colocasia menglaensis* J.T.Yin, H.Li & Z.F.Xu, and *Colocasia oresbia* A.Hay.

Geographic Distribution Malaysia (?), S India (?), S Asia (?). **Habitat** In warm tropical and subtropical climates, along riverbanks, in open swamps, in the immediate vicinity of waterfalls; occasionally in undergrowth; tends to naturalize at roadsides, in wet fields, and in disturbed areas. **Range** Major producing countries: Nigeria, China, Cameroon, Ghana, Papua New Guinea (others: EW Africa, Oceania, S India, E India, Nepal, Bangladesh, SE Asia, E Asia, Pacific Islands, Egypt, E Mediterranean, Caribbean, Americas).

Beer-Making Parts (and Possible Uses) Tubers, leaves (?).

Toxicity Toxic root owing to the presence of calcium oxalate (in the form of cellular raphides, a type of needle-like crystals); if eaten raw, contact causes strong burning to the pharyngeal mucosa; cooking (boiling) eliminates the toxic principles; after having handled the raw pulp, avoid touching the eyes to avoid irritation.

Chemistry Tuber (cooked): energy 594 kJ (142 kcal), carbohydrates 34.6 g, sugars 0.49 g, fiber 5.1 g, lipids 0.11 g, protein 0.52 g, thiamine (B1) 0.107 mg, riboflavin (B2) 0.028 mg, niacin (B3) 0.51 mg, pantothenic acid (B5) 0.336 mg, vitamin B6 0.331 mg, folate (B9) 19 μg, vitamin C 5 mg, vitamin E 2.93 mg, calcium 18 mg, iron 0.72 mg, magnesium 30 mg, manganese 0.449 mg, phosphorus 76 mg, potassium 484 mg, zinc 0.27 mg.

Style Robust Porter.

Beer Pia Taro, Brasserie de Tahiti (Polynesia).

Source https://www.brewersfriend.com/other/taros/.

Commiphora myrrha (Nees) Engl.
BURSERACEAE

Synonyms *Balsamea myrrha* Baill., *Balsamea myrrha* (T.Nees) Oken, *Balsamea playfairii* Engl., *Balsamodendrum myrrha* T.Nees, *Commiphora coriacea* Engl., *Commiphora cuspidata* Chiov., *Commiphora molmol* (Engl.) Engl. ex Tschirch, *Commiphora rivae* Engl.

Common Names Myrrh, African myrrh, herabol myrrh, Somali myrrhor, common myrrh, gum myrrh.

Description Shrub or small tree (up to 4 m), glabrous, stout, spiny, usually with a well-defined short trunk. Outer bark silvery, whitish-gray, or bluish, with small or large papery flakes detaching from the greenish sub-bark; exudate unscented, slimy, produces a hard, translucent, yellowish gum resin; all branches have spines dorsally and at nodes. Leaves trifoliate, papery, grayish green or glaucous, highly variable in shape and size; petiole 1–10 mm long; some lateral leaflets, sometimes very minute, may be found on short, long shoots; leaves elliptic, spatulate, or lanceolate (6–44 × 3–20 mm), attenuate, cuneate, rounded or truncate at base, apex rounded or acute, with 3–4 rather weak main veins, margin entire or 6 teeth on each side. Male flowers usually early, 2–4 in dichasial cymes, 3–4 mm long, often sparsely glandular; bracts light brown (0.5–0.7 × 0.5–0.7 mm), often slightly attached at base forming a fragile detachable collar; receptacle, petals oblong, thinner and curved at tip, 4.5 mm long, 1.5 mm wide, glass shaped; filaments 1.4 and 1.2 mm, anthers 1.2 and 1.0 mm long. Fruits 1–2 on articulated stems, ovoid, flattened, beak 2–4 mm long. Seed smooth with slight swellings.

Related Species The genus *Commiphora* has 22 species, some of which produce aromatic resins.

Geographic Distribution Ethiopia, Kenya, Oman, Saudi Arabia, Somalia. **Habitat** Altitude 250–1,300 m, rainfall 230–300 mm/year, calcareous substrates. **Range** Cultivated in the original area.

Beer-Making Parts (and Possible Uses) Resin (from dehydrated sap).

Chemistry Volatile oils up to 17% (m-cresol, limonene, eugenol, formic acid, acetic acid, heerabolone, furano-sesquiterpene, pinene), resins up to 40% (camphoric acid, mycolic acid, myrrhine), gums and bitter principles up to 60%, ashes, salts, sulfates, benzoates, malate, potassium acetate.

Style American Barleywine, California Common Beer.

Beer Nora, Baladin (Piozzo, Piedmont, Italy).

Source https://www.baladin.it/it/productdisplay/nora; https://www.brewersfriend.com/other/myrrh/.

Common Names Freijo, laurel blanco.

Description Tree (up to 16 m), thornless, sub-deciduous, heliophilous, monoecious. Trunk (Ø up to 45 cm), gray-brown bark, fissured and narrowly furrowed. Leaves simple, alternate, shortly petiolate, ovate to elliptic, margin entire or slightly toothed (10–18 × 6–10 cm). Inflorescence corymb (8–15 cm), axillary. Flowers pentamerous, actinomorphic, hermaphrodite, fragrant, (Ø 6–8 mm); calyx short, persistent in fruit; corolla campanulate, whitish-cream in color. Fruits globose (Ø 10–15 mm), yellow and succulent when ripe, endocarp hard, with irregular, yellowish surface, containing a seed.

Related Species The genus *Cordia* includes about 410 taxa.

Geographic Distribution Argentina, Paraguay, Brazil. **Habitat** Typical arboreal species of Cerrado, semideciduous deciduous forest. **Range** Argentina, Paraguay, Brazil.

Beer-Making Parts (and Possible Uses) Wood for barrels (first they contain cachaça, then used to age beer).

Wood Differentiated, with white sapwood and dark brown variegated heartwood (WSG 0.40–0.78), of good durability and easy to work; used for construction, carpentry, furniture, doors, windows, frames, tables, musical instruments, decorative pieces.

Style Brazilian Beer (?).

Source Cantwell and Bouckaert 2016; Cantwell and Bouckaert 2018 (although these sources mention *C. sellowiana*, it is likely that other species of the same genus are more frequently used to make barrels, such as *Cordia alliodora* [Ruiz and Pav.] Oken and *Cordia goeldiana* Huber; the wood characteristics included here refer to the former).

Coriandrum sativum L.
APIACEAE

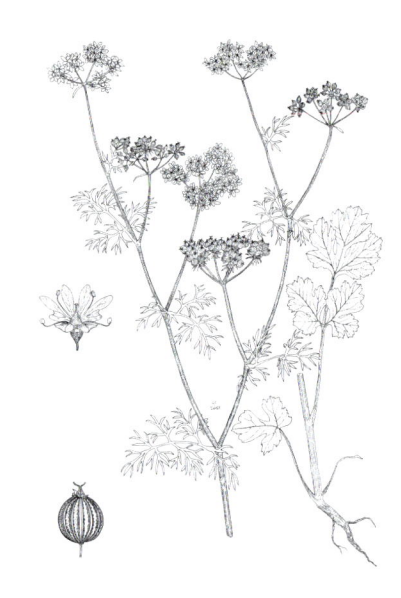

Synonyms *Bifora loureiroi* Kostel., *Coriandropsis syriaca* H.Wolff, *Coriandrum globosum* Salisb., *Coriandrum majus* Gouan, *Selinum coriandrum* Krause.

Common Names Coriander, cilantro, Chinese parsley.

Description Herb, annual (2–5 dm). Stem erect, cylindrical, smooth, with fetid odor. Leaves with tenuous, 1- to 3-pinnate lamina (1–2 dm); last segments acute, 0.5–1 mm wide. Umbels with 4–6 rays; bracts absent, rarely 1; involucre of 3 linear bracts; petals white or slightly pinkish, those of peripheral flowers elongate to 3–4 mm. Fruit subspherical (Ø 3–5 mm), with barely noticeable ribs.

Cultivar Numerous.

Related Species The genus *Coriandrum* has two species: *Coriandrum sativum* L., mentioned here, and *Coriandrum tordylium* (Fenzl) Bornm.

Geographic Distribution SW Mediterranean. **Habitat** Cultivated and grown wild, invasive in cereal crops. **Range** Cultivated throughout the world.

Beer-Making Parts (and Possible Uses) Leaves, fruits.

Chemistry Leaves (raw): energy 95 kJ (23 kcal), water 92.21 g, carbohydrates 3.67 g, sugars 0.87 g, fiber 2.8 g, lipids 0.52 g, protein 2.13 g, vitamin A 337 µg, β-carotene 3930 µg, lutein zeaxanthin 865 µg, thiamine (B1) 0.067 mg, riboflavin (B2) 0.162 mg, niacin (B3) 1.114 mg, pantothenic acid (B5) 0.57 mg, vitamin B6 0.149 mg, folate (B9) 62 µg, vitamin C 27 mg, vitamin E 2.5 mg, vitamin K 310 µg, calcium 67 mg, iron 1.77 mg, magnesium 26 mg, manganese 0.426 mg, phosphorus 48 mg, potassium 521 mg, sodium 46 mg, zinc 0.5 mg. Essential oil (29 compounds): decanol 38.3%, 2-decenol 24.9%, (E)-2-decenal 6.9%, 2-undecenal 2.7%, dodecenal 2.2%, decanal 6.6%, undecenal 2.2%, 1-undecanol 1.9%, undecenol 3.0%, (E)-2-undecanal 2.9%.

Style American Amber Ale, American IPA, American Lager, American Light Lager, American Pale Ale, American Strong Ale, American Wheat or Rye Beer, Belgian Blond Ale, Belgian Dark Strong Ale, Belgian Golden Strong Ale, Belgian Pale Ale, Belgian Specialty Ale, Belgian Tripel, Berliner Weisse, Bière de Garde, Blonde Ale, Brown Porter, California Common Beer, Classic Rauchbier, Experimental Beer, Foreign Extra Stout, Holiday/Winter Special Spiced Beer, Irish Red Ale, Kölsch, Maibock/Helles Bock, Old Ale, Saison, Specialty Beer, Specialty IPA (Black IPA, White IPA), Spice/Herb/Vegetable Beer, Strong Scotch Ale, Weizen/Weissbier, Witbier.

Beer Irie, Almond '22 (Loreto Aprutino, Abruzzo, Italy); Orval, Orval (Villers-devant-Orval, Belgium).

Source Calagione et al. 2018; Dabove 2015; Fisher and Fisher 2017; Giaccone and Signoroni 2017; Hamilton 2014; Heilshorn 2017; Hieronymus 2010; Hieronymus 2012; Hieronymus 2016; Jackson 1991; Josephson et al 2016; Markowski 2015; McGovern 2017; Serna-Saldivar 2010; Steele 2012; https://www.brewersfriend.com/other/cilantro/; https://www.brewersfriend.com/other/koriander/; https://www.fermentobirra.com/homebrewing/birre-speziali/; https://unabirralgiorno.blogspot.co.uk/2013/10/traquair-jacobite-ale.html.

🍺 Not Only in Blanche

Although mainly known for their use in blanche beers, *Coriandrum sativum* seeds are actually among the most frequently used spices by brewers. The citric note they bring enriches many traditional and contemporary styles, giving lightness and depth to the drink.

Cornus mas L.
CORNACEAE

Synonyms *Cornus erythrocarpa* St.-Lag, *Cornus flava* Steud., *Cornus homerica* Bubani, *Cornus mascula* L., *Cornus nudiflora* Dumort., *Cornus praecox* Stokes, *Cornus vernalis* Salisb., *Eukrania mascula* (L.) Merr., *Macrocarpium mas* (L.) Nakai.

Common Names Cornelian cherry.

Description Shrub or small tree (6–8 m), deciduous. Trunk erect, often twisted, much branched above, with quadrangular branchlets, bark peeling, gray with reddish cracks, branches short, erect-patent. Buds enveloped by 4 acute, pubescent scales, arranged 2 × 2 (so as to overlap for better thermal insulation of the floral bud), but when fully open they are arranged on the same plane. Leaves, opposite and acuminate, with short hairy petiole, lamina oval, 3–5 veins converging toward the apex, light green in color, underside hairy, upper side almost glabrous. Flowers small, yellow, blooming before the leaves, exuding a faint honey odor; calyx with 4 acute sepals, greenish color; corolla 4 acute petals, glabrous, golden-yellow in color; stamens alternate with petals, inserted around an epigynous nectariferous disc; ovary inferior, locular with only 1 ovule per locule; stylus 1 with a headed stigma. Inflorescences axillary umbels on stout peduncles, forming before leaves, enveloped by 4 acuminate bracts, greenish tinged with red. Fruits (dogwoods) drupes ovoid, pendulous, edible, fleshy, dark red, contain a hard, bispermous kernel.

Related Species The genus *Cornus* includes about 67 taxa.

Geographic Distribution Pontic, SE Europe, Steno-Mediterranean. **Habitat** Stream banks, edges of deciduous woods, shrubs; thermophilous, xerophilous, calciphilous species (0–1,500 m ASL). **Range** Grown sporadically as an ornamental plant and for the fruit.

Beer-Making Parts (and Possible Uses) Fruits.

Chemistry Fruit (mesocarp): total dry matter 18.26–33.39%, soluble solid content 17.40–32.37%, total acids 1.62–3.75%, total sugars 11.77–26.30%, reducing sugars 9.50–24.07%, sucrose 0.38–3.25%, Ca-pectates 0.32–2.44%, vitamin C 14.56–39.22 mg/100 g, protein 0.20–2.71%, cellulose 0.43–0.95%, anthocyanins 35.63–126.53 mg/100 g, tannins 0.56–1.47%.

Style Fruit Beer.

Beer Scarlet, Ofelia (Sovizzo, Veneto, Italy).

Source Giaccone and Signoroni 2017; https://www.birraofelia.it/shop/birre/scarlet/.

Corylus avellana L.
BETULACEAE

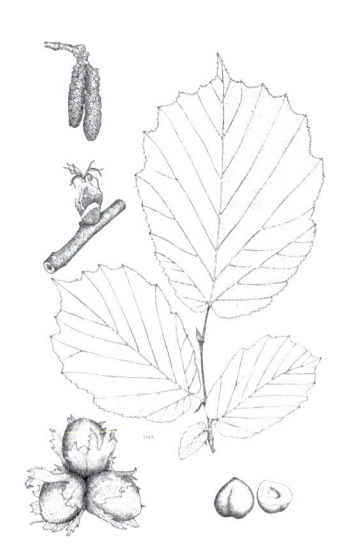

Synonyms *Corylus alba* Aiton ex Steud., *Corylus arborea* Steud., *Corylus ardua* Poit. and Turpin, *Corylus filicifolia* A.DC., *Corylus grandis* Aiton, *Corylus hispanica* Mill. ex D.Rivera and al., *Corylus laciniata* A.DC., *Corylus ovata* Lam. ex Steud., *Corylus pumila* Lodd. ex Loudon, *Corylus serenyiana* Pluskal, *Corylus urticifolia* Dippel.

Common Names Common hazel.

Description Shrub or small tree (1–5 m), branched at the base, with shiny gray-brown bark, at the end with long longitudinal fractures, lenticels initially small (1 mm), longitudinal, then transverse (3–7 mm); buds elliptic, glabrous, green; young branches pubescent. Leaves short (1.5 cm), hirsute, glandular petiole, lamina elliptic (4–5 × 6–7 cm) or subrounded (Ø 9–13 cm), heart-shaped base, acute apex, margin doubly toothed. Male inflorescences aments, in autumn pinkish (6 × 30–50 mm), at anthesis (late winter) golden-yellow and 6–10 cm long, pendulous; female inflorescences aments similar to buds (3 × 6 mm), with a tuft of purple stigmas (2 mm). Fruits (nuts) clustered 2–5, almost completely enveloped by 2 pubescent, fringed, leafy bracts.

Subspecies The variety *Corylus avellana* L. var. *pontica* (K.Koch) H.J.P.Winkl. is recognized.

Cultivar Among the best known are Barcelona, Butler, Casina, Clark Cosford, Daviana, Delle Langhe, England, Ennis, Fillbert, Halls Giant, Jemtegaard, Kent Cob, Lewis, Tokolyi, Tonda Gentile, Tonda di Giffoni, Tonda Romana, Wanliss Pride, and Willamette.

Related Species In addition to the taxa included here, another 20 entities are recognized within the genus *Corylus*, and some produce edible seeded fruits.

Geographic Distribution European-Caucasian. **Habitat** Undergrowth of deciduous and coniferous forests. **Range** Turkey, Italy, United States, Azerbaijan, Georgia, China, Iran, Spain, France, Poland.

Beer-Making Parts (and Possible Uses) Seeds, branches (used in wort filtration in Scandinavian countries), branches (for smoking malt).

Chemistry Seeds: energy 2,629 kJ (628 kcal), water 5.31 g, carbohydrates 16.7 g, sugars 4.34 g, fiber 9.7 g, lipids 60.75 g, protein 14.95 g, vitamin A 1 µg, β-carotene 11 µg, lutein zeaxanthin 92 µg, thiamine (B1) 0.643 mg, riboflavin (B2) 0.113 mg, niacin (B3) 1.8 mg, pantothenic acid (B5) 0.918 mg, vitamin B6 0.563 mg, folate (B9) 113 µg, vitamin C 6.3 mg, vitamin E 15.03 mg, vitamin K 14.2 µg, calcium 114 mg, iron 4.7 mg, magnesium 163 mg, manganese 6.175 mg, phosphorus 290 mg, potassium 680 mg, zinc 2.45 mg.

Wood Sticks, barrel hoops, and charcoal (once used as drawing charcoal) can be produced from the shoots of this species.

Style American Brown Ale, Brown Porter, California Common Beer, Dunkelweizen, Düsseldorf Altbier, Irish Stout, Märzen, Mild, Northern English Brown Ale, Robust Porter, Southern English Brown, Sweet Stout, Winter Seasonal Beer.

Beer Hazelnut Brown Ale, Rogue (Oregon, United States).

Source Calagione et al. 2018; Cantwell and Bouckaert 2016; Cantwell and Bouckaert 2018; Daniels and Larson 2000; Hieronymus 2016b; McGovern 2017; https://www.brewersfriend.com/other/hazelnut/.

Crataegus aestivalis (Walter) Torr. & A.Gray
ROSACEAE

Synonyms *Anthomeles aestivalis* (Walter) M.Roem., *Crataegus elliptica* Elliott, *Crataegus lucida* Elliott, *Crataegus luculenta* Sarg, *Crataegus maloides* Sarg, *Mespilus aestivalis* Walter.

Common Names Mayhaw, may hawthorn, eastern may hawthorn, apple may hawthorn, summer haw.

Description Small tree (up to 12 m), deciduous. Crown rounded. Young branches brown to gray in color, spiny. Leaves alternate, simple, narrow, broader above or at the midpoint, margin serrate, rarely lobed or entire, bright dark green. Flowers white in color, single or in groups of 2–3. Fruit fleshy pome, red in color, wider above the middle or rounded.

Related Species Within the genus *Crataegus*, in addition to the nominal species, *Crataegus aestivalis* (Walter) Torr. & A.Gray, the Section *Coccineae* Series *Aestivales* (Sarg. ex C.K.Schneid.) Rehder includes *Crataegus opaca* Hook. & Arn. ex Hook. and *Crataegus rufula* Sarg.

Geographic Distribution United States (North Carolina to Mississippi). **Habitat** Outer coastal plains, in seasonally flooded depressions, on floodplains or in mountainous areas in riparian swamps, in reservoirs, along riverbanks. **Range** SE United States.

Beer-Making Parts (and Possible Uses) Fruits.

Chemistry Fruit rich in vitamin C, β-carotene, and minerals; however, because of its sour taste, it is rarely eaten fresh or unsweetened; mostly used in the preparation of jellies, jams, and syrups that find numerous applications in cooking.

Style Specialty Fruit Beer.

Beer Come What Mayhaw, Great Raft Brewing (Louisiana, United States).

Source https://www.brewersfriend.com/other/mayhaw/.

Crataegus monogyna Jacq.
ROSACEAE

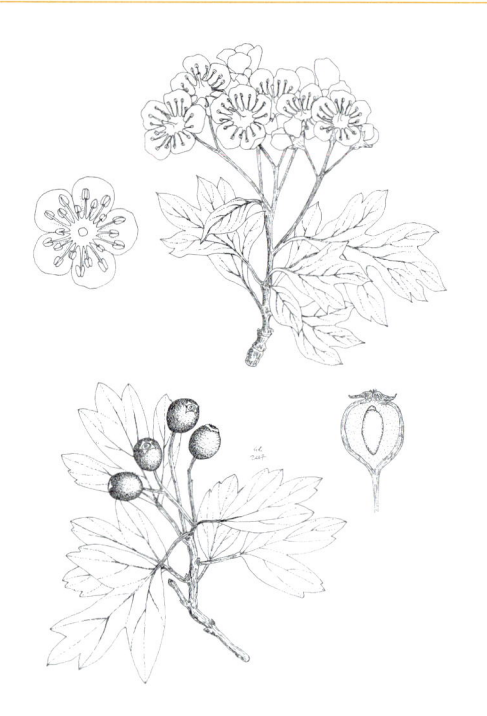

SYNONYMS *Crataegus aegeica* Pojark., *Crataegus alemanniensis* Cinovskis, *Crataegus apiifolia* Medik., *Crataegus azarella* Griseb., *Crataegus* × *borealoides* R.Doll, *Crataegus bracteolaris* Gand., *Crataegus brevispina* Kunze, *Crataegus chlorocarpa* Gand., *Crataegus cuneata* Halcsy, *Crataegus* × *curvisepaloides* R.Doll, *Crataegus dissecta* Borkh., *Crataegus floribunda* Gand., *Crataegus granatensis* Boiss., *Crataegus hirsuta* Schur, *Crataegus* × *integerrima* R.Doll, *Crataegus intermedia* Fuss, *Crataegus intermedia* Schur, *Crataegus* × *krima* R.Doll, *Crataegus krumbholzii* R.Doll, *Crataegus leiomonogyna* Klokov, *Crataegus lipskyi* Klokov, *Crataegus maroccana* Pers., *Crataegus maura* L.f., *Crataegus oligacantha* Gand., *Crataegus orientobaltica* Cinovskis, *Crataegus parvifolia* Lojac., *Crataegus petiolulata* Gand., *Crataegus polyacantha* (Jan ex Guss.) Nyman, *Crataegus praearmata* Klokov, *Crataegus pulchella* Gand., *Crataegus septempartita* Pojark., *Crataegus stevenii* Pojark., *Crataegus subborealis* Cinovskis, *Crataegus subintegriloba* Pojark., *Crataegus sublucens* Gand., *Crataegus thyrsoidea* Gand., *Crataegus transalpina* A.Kern. ex Hayek, *Crataegus xeromorpha* Pojark., *Mespilus crataegus* Borkh., *Mespilus diversifolia* (Pers.) Poir., *Mespilus elegans* Poir., *Mespilus fissa* Poir., *Mespilus insegnae* Tineo ex Guss., *Mespilus maroccana* (Pers.) Poir., *Mespilus maura* (L.f.) Poir., *Mespilus monogyna* (Jacq.) All., *Mespilus oliveriana* Poir., *Mespilus polyacantha* Jan ex Guss., *Oxyacantha apiifolia* (Medik.) M.Roem., *Oxyacantha azarella* (Griseb.) M.Roem., *Oxyacantha elegans* (Poir.) M.Roem., *Oxyacantha granatensis* (Boiss.) M.Roem., *Oxyacantha monogyna* (Jacq.) M.Roem.

COMMON NAMES Hawthorn.

DESCRIPTION Shrub or small tree (2–12 m), long-lived (up to 500 years), deciduous, bushy. Root fasciculate. Canopy globular or elongated, trunk sinuous, often branched from the base; bark compact, smooth, light gray in young plants, flaking in plaques, brown or red-ochraceous in old specimens; branches reddish-brown, the lateral ones with sharp (up to 2 cm), dark thorns. Buds alternate, spiraled, bright reddish in color; straight spines beneath lateral buds. Leaves deciduous, with fluted petiole, alternate, simple, bright green, shiny on upper side, glaucous green below, glabrous, rhomboidal or oval, toothed margin, divided into 3–7 very deep lobes with entire margin, toothed only apically; stipules toothed and glandular. Flowers white or pinkish, calyx with 5 triangular-ovate laciniae; corolla with 5 subrounded petals, purple stamens in multiple numbers to the petals (15–20), inserted on the margin of a brownish-green receptacle with a glabrous monocarpellar ovary and only 1 greenish-white stylus with flattened stigma (rarely 3-styled flowers). Inflorescence simple or compound corymb, erect, borne by villous peduncle. Fruits pome (Ø 7–10 mm), fleshy, rose-colored when ripe, crowned at apex by remnants of calyceal laciniae; fruits clustered in dense clusters. Seed 1 per fruit, yellow-brown in color.

SUBSPECIES The considerable variability within this species is evidenced by the number of varieties and forms described. The following are accepted by most scholars: *Crataegus monogyna* Jacq. f. *pendula* (Lodd. ex Loudon) Rehder, *Crataegus monogyna* Jacq. f. *pteridifolia* (Lodd. ex Loudon) Rehder, *Crataegus monogyna* Jacq. f. *semperflorens* (Andr.) C.K.Schneid., *Crataegus monogyna* Jacq. f. *stricta* (Loudon) Zabel, and *Crataegus monogyna* Jacq. var. *lasiocarpa* (Lange) K.I.Chr. However, there are many other forms whose status has not yet been clarified.

RELATED SPECIES The genus *Crataegus* includes about 420 taxa, many of which have edible fruits.

Geographic Distribution Eurasian. **Habitat** Xerophilous woods, hedges, thickets, shrubs, scrubs, forest edges, grassy slopes, with preference for calcareous soils (0–1,600 m ASL). **Range** Eurasia.

Beer-Making Parts (and Possible Uses)
Fruits.

Chemistry Flavonoids (hyperoside, luteolin, apigenin, luteolin-3,7-diglucoside and quercetin), oligomer procyanidins, bioflavonoids and flavoglucosides, rutin, triterpene complexes, ascorbic acid, essential oils, tannins, crategin, chlorogenic acid and sapogenins, vitamin C.

Style Blanche, Bohemian Pilsner.

Beer Monflowers, Civale (Alessandria, Piedmont, Italy).

Source Heilshorn 2017; Jackson 1991; https://www.brewersfriend.com/other/biancospino/.

Crithmum maritimum L.
APIACEAE

Synonyms None
Common Names Seafennel, samphire, rock samphire.

Description Herb, perennial (20–50 cm), lignified at the base, caespitose, glabrous, glaucous. Root rhizomatous. Stem woody, branched, with ascending, zigzagging herbaceous scapes, glaucous green in color, longitudinally striated, with flexuous, prostrate habit. Leaves alternate, persistent, glabrous, triangularly outlined, bi- or tri-pinnate, lanceolate segments fleshy, succulent, long petiole widens forming a sheath that wraps around the base of the stem; cauline leaves simpler, upper ones monopinnate, inserted on the sheath amplexicaule. Inflorescence terminal umbel, 8–36 rays, each with involucre and whorls of pendulous lanceolate bracts. Flowers with 5 entire petals, suborbicular, tip folded toward center, greenish-white color. Fruit schizocarp (polyachene), 5–6 mm, ovoid, yellowish or reddish color, glabrous with surface traversed by enlarged longitudinal ribs. Seeds floating (hydrocoria) in spongy pericarp tissue.

Related Species *Crithmum* is a monospecific genus.

Geographic Distribution Euro-Mediterranean, Steno-Mediterranean. **Habitat** Environments directly influenced by saltiness (beaches, piers, cliffs, rocks). **Range** Plant collected in the wild, rarely cultivated.

Beer-Making Parts (and Possible Uses)
Young branches (?), fruits (?).

Chemistry Leaves and fruits: mineral salts (calcium, magnesium, manganese, sodium, zinc), oils, essential oil, iodine, vitamin C, β-carotene, protein. Essential oil: γ-terpinene 32.4%, β-phellandrene 22.3%, sabinene 9.1%, methyl thymol 8.6%, ρ-cymene 7.6%, cis-β-ocimene 3.2%, α-pinene 1.8%, β-myrcene 1.4%, α-phellandrene 0.7%, α-thujene 0.4%, β-pinene 0.2%, α-terpinene 0.2%, trans-β-ocimene 0.2%, terpinen-4-ol 0.3%, α-terpineol 0.2%, methyl thymol isomer 0.2%, germacrene-D 0.3%, terpinolene 0.1%, allo-ocimene 0.1%, Δ3-carene 0.1%, camphene 0.1%, limonene traces, dillapiol traces.

Style Sour Ale.

Source https://www.brewersfriend.com/other/samphire/.

Crocus sativus L.
IRIDACEAE

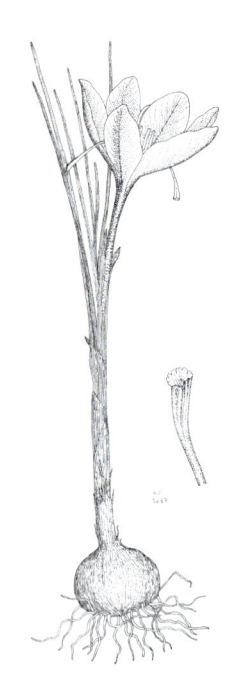

Synonyms *Crocus officinalis* (L.) Honck., *Crocus orsinii* Parl., *Crocus pendulus* Stokes, *Crocus setifolius* Stokes, *Geanthus autumnalis* Raf., *Safran officinarum* Medik.

Common Names Saffron.

Description Herb, perennial (10–20 cm), bulbous. Bulb subspherical (Ø 1.5–2 cm) with fine brown fibers, not reticulate. Lower leaves (3–4) reduced to sheath and present already at flowering; definitive leaves present at anthesis, 4–6 mm × 2–3 dm, ciliate. Flower single, inodorous; 2 spathes, funnel shaped, perigonium purplish, with fauces more light purplish, pubescent; outer laciniae 4–5 cm, with darker veins; inner laciniae smaller, anthers twice as long as filaments; stigmas entire clavate, 25–27 mm long, orange-red, fragrant, projecting beyond perigonial laciniae.

Related Species The genus *Crocus* includes about 127 taxa, some of which have edible parts.

Geographic Distribution W Asia (the species, unknown in the wild if not by naturalization following anthropic introduction in cultivation, would have derived from the mutation of *Crocus cartwrightianus* Herb., native to the island of Crete and Greece; it is a triploid, sterile species—multiplication through bulbs—whose differentiation has occurred by artificial selection of the individuals with elongated and bigger stigmas). **Habitat** Cultivated as condiment, rarely wild. **Range** Iran (90–93%), Greece, Morocco, India, Afghanistan, Azerbaijan, Italy, Austria, Germany, Switzerland, Australia, China, Egypt, Great Britain, France, Israel, Mexico, New Zealand, Sweden, Turkey, California, C-Africa.

Beer-Making Parts (and Possible Uses) Pistils.

Chemistry Dried pistils: energy 1,298 kJ (310 kcal), water 11.9 g, carbohydrates 65.37 g, fiber 3.9 g, lipids 5.85 g (saturated 1.586 g, monounsaturated 0.429 g, polyunsaturated 2.067 g, protein 11.43 g, vitamin A 530 IU, thiamine (B1) 0.115 mg, riboflavin (B2) 0.267 mg, niacin (B3) 1.460 mg, vitamin B6 1.01 mg, folate (B9) 93 µg, vitamin C 80.8 mg, calcium 111 mg, iron 11.1 mg, magnesium 264 mg, phosphorus 252 mg, potassium 1,724 mg, sodium 148 mg, zinc 1.09 mg, selenium 5.6 µg. Volatile and nonvolatile aromatic component: more than 150 volatile aromatic compounds identified, but also many nonvolatile active components such as carotenoids (zeaxanthin, lycopene, various α- and β-carotenes); the yellow-orange color depends mainly on α-crocin (crocins are a group of hydrophilic carotenoids), a carotenoid pigment that can constitute up to 10% of the dry weight of saffron. The bitter glycoside picrocrocin (up to 4% by weight) is responsible for flavor. When saffron is dried after harvest, heat and enzymatic action split the picrocrocin into D-glucose and a free molecule of safranal (a volatile oil that imparts flavor).

Style American Barleywine, American IPA, American Pale Ale, Belgian Dubbel, Belgian Golden Strong Ale, Belgian Specialty Ale, Belgian Tripel, Blonde Ale, California Common Beer, Clone Beer, Cream Ale, Double IPA, Dry Stout, English IPA, Extra Special/Strong Bitter (ESB), Foreign Extra Stout, Fruit Beer, Gruit, Holiday/Winter Special Spiced Beer, Imperial Stout, Metheglin, Northern English Brown Ale, Oatmeal Stout, Old Ale, Russian Imperial Stout, Saison, Specialty Beer, Specialty Smoked Beer, Spice/Herb/Vegetable Beer, Standard American Lager, Winter Seasonal Beer, Witbier.

Beer Crocus, Amiata (Arcidosso, Tuscany, Italy).

Source Giaccone and Signoroni 2017; McGovern 2017; https://www.birraaltaquota.it/prodotti/birre-estrose/62-croco.html; https://www.brewersfriend.com/other/saffron/; https://www.fermentobirra.com/homebrewing/birre-speziali/.

Cryptomeria japonica (Thunb. ex L.f.) D.Don
CUPRESSACEAE

Synonyms *Cryptomeria araucarioides* Henkel & W.Hochst., *Cryptomeria compacta* Beissn., *Cryptomeria elegans* Jacob-Makoy, *Cryptomeria fortunei* Hooibr. ex Billain, *Cryptomeria generalis* E.H.L.Krause, *Cryptomeria kawaii* Hayata, *Cryptomeria lobbiana* Billain, *Cryptomeria mairei* (H.Lév.) Nakai, *Cryptomeria mucronata* Beissn., *Cryptomeria nigricans* Carrière, *Cryptomeria pungens* Beissn., *Cryptomeria variegata* Beissn., *Cryptomeria viridis* Beissn., *Cupressus japonica* Thunb. ex L.f., *Cupressus mairei* H.Lév., *Schubertia japonica* (Thunb. ex L.f.) Jacques, *Taxodium japonicum* (Thunb. ex L.f.) Brongn.

Common Names Cryptomeria, Japanese cedar.

Description Tree (up to 40 m). Trunk (Ø at least 2 m), bark reddish-brown, fibrous, peeling off in strips; crown pyramidal; branches in verticils, horizontal or slightly pendulous; pendulous branches of the first year green. Leaves on sterile branches at 15°–45° to axis, those on fertile branches at 30°–55°, subulate to linear, ± straight or strongly incurved (0.4–2 cm × 0.8–1.2 mm), rigid, stomatal bands with 2–8 rows of stomata on each side. Inflorescences cones formed from fifth year; male cones in racemes of 6–35, ovoid or ovoid-ellipsoid (2–8 × 1.3–4 mm), each (except basal and apical) subtended by a leaf; female cones in clusters of 1–6, globose or subglobose, (0.9–2.5 × 1–2.5 cm); cone scales 20–30. Two proximal margins often convex, or all 4 margins ± concave, median part with or without shoulders at widest point, apex usually curved, umbo-rhombic, distally with 4–7 serrations (1–3.5 mm). Seeds 2–5 per scale, brown or dark brown, irregularly ellipsoid or multiangular and ± compressed (4–6.5 × 2–3.5 mm); wings 0.2–0.25 mm wide.

Subspecies In the taxon *Cryptomeria japonica* (Thunb. ex L.f.) D.Don, according to some authors, two varieties should be distinguished: *Cryptomeria japonica* (Thunb. ex L.f.) D.Don var. *sinensis* Miq. in Siebold and Zuccarini and *Cryptomeria japonica* (Thunb. ex L.f.) D.Don var. *japonica* Miq.

Related Species *Cryptomeria* is a monotypic genus.

Geographic Distribution China (Fujian, Jiangxi, Sichuan, Yunnan, Zhejiang). **Habitat** Forests in well-drained soils, subject to hot, humid conditions. **Range** Introduced elsewhere in China (Anhui, Fujian, Gansu, Guangdong, Guangxi, Guizhou, Henan, Hubei, Hunan, Jiangsu, Jiangxi, Shandong, Sichuan, Taiwan, Yunnan, Zhejiang), Japan, Europe, on the American mainland for wood and as an ornamental species.

Beer-Making Parts (and Possible Uses) Wood (for barrels).

Wood Pleasant resinous scent, with narrow whitish sapwood and red to brown heartwood; fine texture and mostly straight grain; WSG 0.29–0.44; used for construction, bridges, ships, shingles, joinery, fine lacquerware, furniture, tools, paper.

Style Japanese Ale.

Source Cantwell and Bouckaert 2016; Cantwell and Bouckaert 2018.

Cucumis melo L.
CUCURBITACEAE

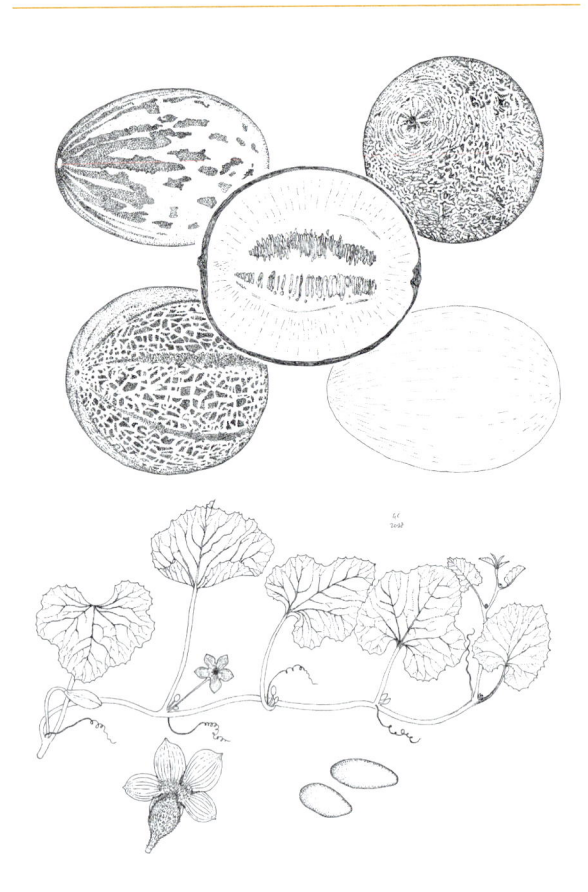

Synonyms *Cucumis acidus* Jacq., *Cucumis alba* Nakai, *Cucumis arenarius* Schumach. & Thonn., *Cucumis aromaticus* Royle, *Cucumis bisexualis* A.M. Lu & G.C.Wang, *Cucumis callosus* (Rottler) Cogn., *Cucumis campechianus* Kunth, *Cucumis cantalupo* Rchb., *Cucumis chate* Hasselq., *Cucumis chate* L., *Cucumis chinensis* (Pangalo) Pangalo, *Cucumis chito* C.Morren, *Cucumis cicatrisatus* Stocks, *Cucumis conomon* Thunb., *Cucumis cubensis* Schrad., *Cucumis dudaim* L., *Cucumis eriocarpus* Boiss. & Noë, *Cucumis flexuosus* L., *Cucumis jamaicensis* Bertero ex Spreng, *Cucumis jucunda* F.Muell., *Cucumis laevigatus* Chiov., *Cucumis maculatus* Willd., *Cucumis microcarpus* (Alef.) Pangalo, *Cucumis microsperma* Nakai, *Cucumis microspermus* Nakai, *Cucumis momordica* Roxb., *Cucumis officinarum-melo* Crantz, *Cucumis orientalis* Kudr., *Cucumis pancherianus* Naudin, *Cucumis pedatifidus* Schrad., *Cucumis persicodorus* Seitz, *Cucumis persicus* (Sarg.) M.Roem., *Cucumis pictus* Jacq., *Cucumis princeps* Wender., *Cucumis pseudocolocynthis* Royle, *Cucumis pseudocolocynthis* Wender., *Cucumis pubescens* Willd., *Cucumis reginae* Schrad., *Cucumis schraderianus* M.Roem., *Cucumis serotinus* Haberle ex Seitz, *Cucumis trigonus* Roxb., *Cucumis turbinatus* Roxb., *Cucumis utilissimus* Roxb., *Ecballium lamber-tianum* M.Roem., *Melo adana* (Pangalo) Pangalo, *Melo adzhur* Pangalo, *Melo agrestis* (Naudin) Pangalo, *Melo* × *ambiguua* Pangalo, *Melo ameri* Pangalo, *Melo cantalupensis* (Naudin) Pangalo, *Melo cassaba* Pangalo, *Melo chandalak* Pangalo, *Melo chinensis* Pangalo, *Melo conomon* Pangalo, *Melo dudaim* (L.) Sageret, *Melo figari* Pangalo, *Melo flexuosus* (L.) Pangalo, *Melo microcarpus* (Alef.) Pangalo, *Melo monoclinus* Pangalo, *Melo orientalis* (Kudr.) Nabiev, *Melo persicus* Sageret, *Melo sativus* Sageret, *Melo vulgaris* Moench ex Cogn., *Melo zard* Pangalo.

Common Names Muskmelon, cantaloupe, rockmelon.

Description Herb, annual (3–5 m), creeping or climbing. Roots fibrous, extensive (more than 150 cm). Stem flexuous, branched, striate, pubescent, bearing simple cirri. Leaves simple, alternate, opposite the cirri, long petiolate (4–12 cm), lamina lobed (3–7 lobes palmate ± evident) or rounded (Ø 6–20 cm), cordate at the base, margin wavy-dentate, with hairiness on both faces and palmate vein. Flowers axillary, pedunculate (0.5–4 cm), mostly unisexual (gynoecious species, with only female flowers, or monoecious species, with separate sexes on distinct flowers borne by the same plant), but also hermaphrodite; calyx campanulate (0.3–0.8 cm) with 5 linear lobes (0.2–0.4 cm), pubescent; corolla 5-partite, yellow, lobes obtuse apically (0.3–2.4 × 0.2–2 cm); hypanthium (0.7–0.8 cm) broader at the apex; flowers staminate solitary or fasciculate, with 3 stamens; pistillate flowers solitary, bearing staminodes; ovary (0.4–1.1 cm) inferior, ellipsoid, densely pubescent, surmounted by a short stylus (1–2 mm) and a 3-lobed stigma (2–2.4 mm). Fruit pepo (0.4–2.2 kg) with extremely variable characteristics (varietal characters); oval or roundish shape, rind (epicarp) color white, pale yellow, deep yellow, orange, green, epicarp surface hairy during the juvenile stages of development, when ripe smooth, reticulated, wrinkled (variable sculpture), ribbed; flesh color white, yellow, orange, pink, juicy, very fragrant; central cavity fibrous, hosts numerous seeds. Seeds smooth, elliptical, flattened (0.5–1.2 × 0.2–0.7 cm).

Subspecies Most authors do not recognize the intraspecific variability of this taxon as expressing entities of subspecific rank in botanical terms but advocate a horticultural classification.

Cultivar The numerous varieties belonging to this polymorphic species are subdivided based on harvest period with respect to the stage of maturity (ripe fruit or unripe vegetable) and into further subgroups based on the characteristics of the epicarp and the fruit pulp. The group of melons as fruit (harvested when ripe) includes the subgroups Cantalupensis or Cantalupio (fruits of medium size, smooth surface, yellow-orange pulp, so called because they were brought by Asian missionaries to the castle of Cantalupo in Sabina, in the hills of Rieti, Lazio, in Italy), Reticulatus (or netted melons, of medium size, white or yellow-green pulp, with reticulated surface), and Inodorus (winter melons, whitish or pinkish pulp with smooth rind, with a taste between pear and melon, typical Christmas dish in Sicilian and Italian traditions). In the group of melons as vegetables (harvested before ripening) are the subgroups Flexuosus (snake melon or tortarello, used raw in the same way as cucumber) and Momordica (bitter melon, mainly used as a medicinal plant because it is rich in vitamins A, C, and E).

Related Species The genus *Cucumis* includes about 54 taxa, some of which produce edible fruits.

Geographic Distribution E Africa. **Habitat** Species mostly cultivated. **Range** China, Turkey, United States, Spain, Morocco, Romania, Iran, Israel, Egypt, India.

Beer-Making Parts (and Possible Uses) Fruits (pulp).

Chemistry Cantaloupe melon: energy 34 kcal, lipids 0.2 g (saturated fatty acids 0.1 g, polyunsaturated fatty acids 0.1 g), carbohydrates 8 g, sugars 8 g, fiber 0.9 g, proteins 0.8 g, vitamin A 3,382 IU, vitamin C 36.7 mg, vitamin B6 0.1 mg, sodium 16 mg, potassium 267 mg, calcium 9 mg, iron 0.2 mg, magnesium 12 mg.

Style American IPA, American Pale Ale, Weissbier.

Source Heilshorn 2017; https://www.brewersfriend.com/other/cantaloupe/; https://www.brewersfriend.com/other/galiamelon/.

Cucumis metuliferus E.Mey. ex Naudin
CUCURBITACEAE

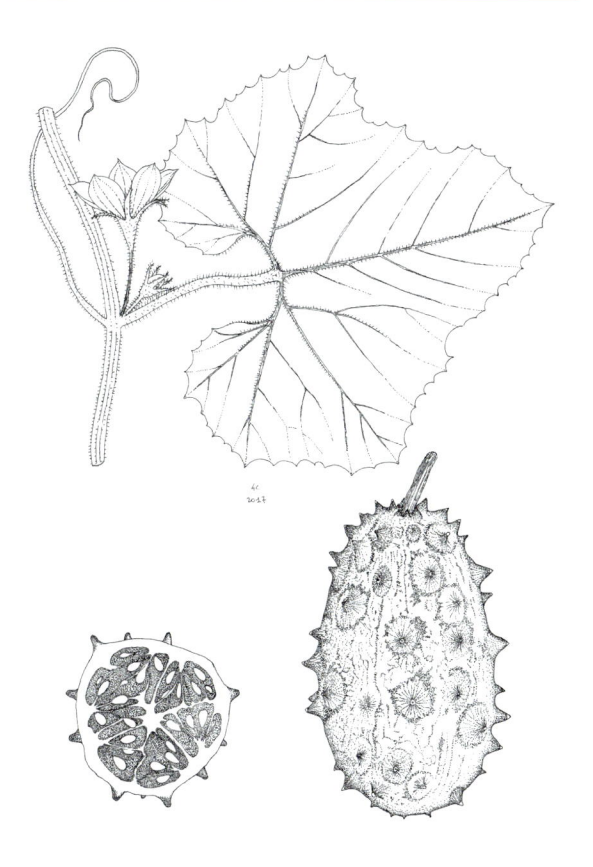

Synonyms *Cucumis tinneanus* Kotschy & Peyr.

Common Names Kiwano, horned melon, African cucumber, horned cucumber.

Description Herb, annual (several m), vigorous, prostrate or climbing. Root robust, fibrous. Stems elongate, fluted, with long spreading stiff hairs and tendrils (4–10.5 cm) simple, solitary. Leaves alternate, simple; petioles (3–12 cm) bristly; lamina ovate or pentagonal (3.5–14 cm × 3.5–13.5 cm), superficially palmate, 3- to 5-lobed, bristly and silky especially on the veins of the underside, which is scabro-punctate. Flowers unisexual, regular, pentamerous; sepals filiform (2–4 mm); petals (0.5–1.5 cm) united at base, yellow; male flowers in clusters of 1–10, on peduncle 2–18 mm, stamens 3; female flowers solitary, peduncle 5–35 mm, ovary inferior, ellipsoid (1–2.5 cm), covered with large soft spines, stigma 3-lobed. Fruit pepo oblong-cylindrical (6–16 × 3–9 cm), on 2–7 cm peduncle, rounded at ends and bristling with spiny protuberances (1–1.5 cm) with broad base, mottled dark green, yellow-orange when ripe, with many seeds. Seeds narrowly ovoid (5–8 mm), compressed, with rounded margins and silky hairs.

Cultivar In various countries (Kenya, New Zealand, France, Israel), improved cultivars of this species have been selected and are now cultivated for export.

Related Species
The genus *Cucumis* includes about 54 taxa, some of which produce edible fruits.

Geographic Distribution
Tropical and subtropical sub-Saharan regions of Africa (from Senegal to Somalia to South Africa); naturalized in Yemen, Kenya, New Zealand, France, Israel, Australia, Croatia. **Habitat** From semi-evergreen riparian forests to semi-arid highlands to Kalahari sands (0–1,800 m ASL) in a wide variety of soils. **Range** Cultivated in S Africa (Zimbabwe), New Zealand, Japan, United States (California), Israel, Kenya, S France.

Beer-Making Parts (and Possible Uses)
Fruits (unripe or ripe).

Toxicity Among wild plants (not cultivated), it is possible to find some with bitter fruits (that local fauna avoid) and others with nonbitter fruits; these types are morphologically indistinguishable. Plants with bitter fruits can cause poisoning (the bitterness increases as the fruit ripens) because of the presence of cucurbitacins (cucurbitacin B), bitter toxic compounds responsible for great discomfort and even death owing to their powerful laxative action.

Chemistry Fruit (composition changes according to level of ripeness): water 91 g, energy 134 kJ (32 kcal), proteins 1.1 g, lipids 0.7 g, carbohydrates 5.2 g, fiber 1.1 g, calcium 11.9 mg, magnesium 22.3 mg, phosphorus 25.5 mg, iron 0.53 mg, thiamine 0.04 mg, riboflavin 0.02 mg, niacin 0.55 mg, vitamin C 19 mg.

Style Blonde Ale.

Source https://www.brewersfriend.com/other/kiwano/.

Cucumis sativus L.
CUCURBITACEAE

Synonyms *Cucumis esculentus* Salisb., *Cucumis hardwickii* Royle, *Cucumis muricatus* Willd., *Cucumis rumphii* Hassk., *Cucumis setosus* Cogn., *Cucumis sphaerocarpus* Gabaev, *Cucumis vilmorinii* Spreng.

Common Names Cucumber.

Description Herb, annual (1–3 m). Stem climbing, shaggy. Leaves pentagonal in outline (7–18 cm), with 5 lobes up to 1/3 as deep as the lamina. Flowers with golden yellow corolla (Ø 2–3 cm), wavy at the margin. Fruit pepo green, cylindrical (2–4 × 10–20 cm), with white flesh.

Cultivar Among the most common varieties are Apple, Beit Alpha, Dosakai, East Asian, English, Kekiri, Lebanese, Persian, and Schälgurken.

Related Species The genus *Cucumis* includes about 54 taxa, some of which produce edible fruits.

Geographic Distribution India.
Habitat Cultivated, rarely wild. **Range** China, Turkey, Iran, Russia, Ukraine.

Beer-Making Parts (and Possible Uses)
Unripe fruits.

Chemistry Unripe fruit: energy 16 kcal, water 94–96 g, carbohydrates 3.6 g, fiber 0.5 g, sugars 1.7 g, protein 0.7 g, lipids 0.1 g, vitamin A 105 IU, vitamins C 2.8 mg, potassium 147 mg, calcium 16 mg, iron 0.3 mg, magnesium 13 mg, sodium 2 mg.

Style American IPA, American Pale Ale, Belgian Golden Strong Ale, Berliner Weisse, Blonde Ale, California Common Beer, Experimental Beer, Fruit and Spice Beer, Kölsch, Maibock/Helles Bock, Old Ale, Premium American Lager, Saison, Specialty Beer, Spice/Herb/Vegetable Beer, Weissbier, Weizenbock, Witbier.

Beer Kill Green, Argo (Collecchio, Emilia-Romagna, Italy).

Source Calagione et al. 2018; Hieronymus 2016b; Josephson et al. 2016; https://www.brewersfriend.com/other/cucumber/; https://unabirralgiorno.blogspot.it/2014/02/alesmith-speedway-stout.html.

Cucurbita ficifolia Bouché
CUCURBITACEAE

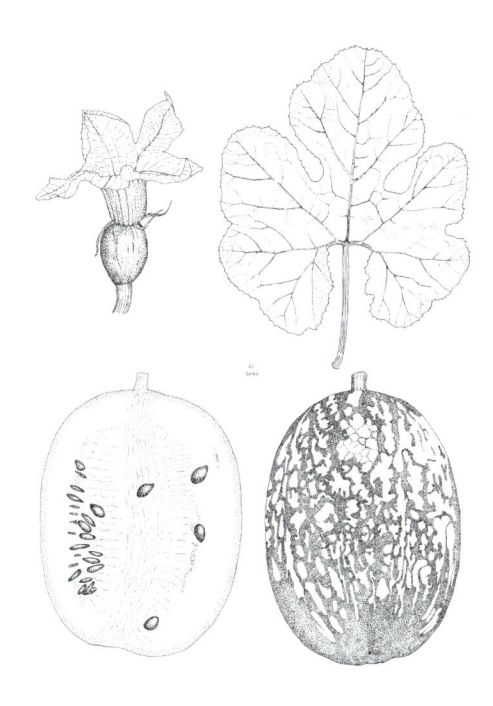

Synonyms *Cucurbita melanosperma* A.Braun ex Gasp., *Cucurbita mexicana* Dammann, *Pepo malabaricus* Sageret.

Common Names Fig-leaf gourd, Malabar gourd, black-seeded gourd.

Description Herb (a few meters), perennial, short-lived, climbing by long branching tendrils, herbaceous at first, then somewhat woody. Taproot up to 2 m, lateral roots forming a shallow network below the soil surface. Stems numerous, elongate, vaguely pentagonal or rounded in section, prickly or spiny, often rooting at nodes. Leaves alternate, simple, lamina circular-ovate to nearly reniform, sinuate to obtusely lobed (Ø 18–25 cm), margin toothed to entire. Flowers solitary, unisexual, regular, pentamerous, up to Ø 7.5 cm, yellow to pale orange in color; calyx and corolla campanulate with a short tube; male flowers with short, thick columnar androecium, filaments with trichomes more than 1 mm long; female flowers on short, wrinkled peduncles, ovary inferior. Fruit large pepo, globose to cylindrical (15–50 cm long), green with white stripes and spots; peel smooth and hard, flesh of ripe fruit white, coarse, firm, fibrous, and rather dry, with many seeds; fruit peduncle little or no widening at apex. Seeds oblong-ellipsoid (1.5–2.5 cm), flat, usually black in color.

Cultivar There are several cultivated varieties of this species.

Related Species The genus *Cucurbita* includes 22 taxa, some of which produce edible fruits and have beer-making applications.

Geographic Distribution Cultigen originated in the mountainous regions of Latin America (from Mexico to Chile), where it is still widely cultivated. **Habitat** Mountainous areas, in the tropics at altitudes higher than 1,000 m, in Latin America at altitudes of 1,000–3,000 m ASL. **Range** Besides the original range (S America), the species is cultivated in tropical Africa, in the mountains of Ethiopia, Kenya, Tanzania, Angola; in Asia, it is cultivated in India, Japan, Korea, China, Philippines.

Beer-Making Parts (and Possible Uses)
Leaves, unripe fruits.

Toxicity All Cucurbitaceae contain triterpene glycosides called cucurbitacins. These compounds are present in all organs of the plant at different concentrations; if concentrated in the edible parts, they give a bitter taste. Although having recognized medicinal properties, cucurbitacins in high doses can be poisonous.

Chemistry No data are available on the composition of *Cucurbita ficifolia*; however, it is likely to be similar to that of other Cucurbitaceae. *Cucurbita* spp. leaves: energy 105 kJ (25 kcal), water 89 g, protein 4 g, lipids

0.2 g, carbohydrates 2 g, fiber 2.4 g, calcium 475 mg, phosphorus 135 mg, iron 0.8 mg, β-carotene 1 mg, thiamin 0.08 mg, riboflavin 0.06 mg, niacin 0.3 mg, ascorbic acid 80 mg. The composition of unripe fruits is comparable to that of *Cucurbita pepo*: energy 17 kcal, carbohydrates 3.11 g, protein 1.21 g, lipids 0.32 g, fiber 1 g, folate 24 µg, niacin 0.451 mg, pantothenic acid 0.204 mg, pyridoxine 0.163 mg, riboflavin 0.094 mg, thiamine 0.045 mg, vitamin A 200 IU, vitamin C 17, 9 mg, vitamin E 0.12 mg, vitamin K 4.3 µg, sodium 8 mg, potassium 261 mg, calcium 16 mg, iron 0.37 mg, magnesium 18 mg, manganese 0.177 mg, phosphorus 38 mg, selenium 0.2 µg, zinc 0.32 mg, ß-carotene 120 µg, lutein-zeaxanthin 2,125 µg.

Style Spice/Herb/Vegetable Beer.

Source https://www.brewersfriend.com/other/chilacayote/.

Cucurbita maxima Duchesne
CUCURBITACEAE

Synonyms *Cucumis rapallito* Carrière, *Cucumis zapallito* Carrière, *Cucurbita farinae* Mozz. ex Naudin, *Cucurbita pileiformis* M.Roem., *Cucurbita rapallito* Carrière, *Cucurbita sulcata* Blanco, *Cucurbita turbaniformis* M.Roem., *Cucurbita zapallito* Carrière.

Common Names Pumpkin, winter squash.

Description Herb, annual (2–6 m). Stems creeping or climbing, hirsute, with branching cirri. Leaves with heart-shaped lamina (2–3 dm), rough, obscurely toothed. Female flowers with unenlarged peduncles; corolla campanulate (5–10 cm), orange-yellow. Fruit pepo usually ellipsoid, flattened at the poles (Ø 3–6 dm), with orange-yellow flesh; seeds white, oval-flattened (1 × 1.5–2 cm). Flesh and seeds are edible.

Subspecies Among the many varieties of *Cucurbita maxima* Duchesne ssp. *maxima* are Amphora (*C. maxima* Duchesne ssp. *maxima* cv. *maxima* 'Amphora'), Atlantic Giant (*C. maxima* Duchesne 'Atlantic Giant'), Aurantiaca-Alba (*C. maxima* Duchesne ssp. *maxima* cv. *turbaniformis* [M.Roem] Alef. var. *turbaniformis* 'Aurantiaca-Alba'), Australian Butter (*C. maxima* Duchesne 'Australian Butter'), Big Max (*C. maxima* Duchesne ssp. *maxima* cv. *maxima* 'Big Max'), Big Moon (*C. maxima* Duchesne ssp. *maxima* cv. *hubbardiana* Grebensc. 'Big Moon'), Bleu de Hongrie (*C. maxima* Duchesne ssp. *maxima* cv. *maxima* 'Bleu de Hongrie'), Blue Banana (*C. maxima* Duchesne ssp. *maxima* cv. *bananina* Grebensc. 'Blue Banana'), Buttercup Squash (*C. maxima*), Ebisu (*C. maxima* Duchesne 'Ebisu'), Flat White Boer A (*C. maxima* Duchesne ssp. *maxima* cv. *maxima* 'Flat White Boer A'), Galeux d'Eysines (*C. maxima* Duchesne 'Galeux d'Eysines'), Gelber Zentner (*C. maxima* Duchesne ssp. *maxima* cv. *maxima* 'Gelber Zentner'), Giraumon Turban (*C. maxima* Duchesne ssp. *maxima* cv. *turbaniformis* [M.Roem] Alef. var. *turbaniformis* 'Giraumon Turban'), Golden Delicious (*C. maxima* Duchesne ssp. *maxima* cv. *hubbardiana* Grebensc. 'Golden Delicious'), Golias (*C. maxima* Duchesne 'Golias'), Green Hubbard (*C. maxima* Duchesne ssp. *maxima* cv. *hubbardiana* Grebensc. 'Green Hubbard'), Green Hubbard, Chicago Warted (*C. maxima* Duchesne ssp. *maxima* cv. *hubbardiana* Grebensc. 'Green Hubbard, Chicago Warted'), Grosser Gelber ssp. (*C. maxima* Duchesne ssp. *maxima* cv. *maxima* 'Grosser Gelber Zentner'), Grosser Gruener (*C. maxima* Duchesne ssp. *maxima* cv. *maxima* 'Grosser Gruener'), Hokkaido (*C. maxima* Duchesne ssp. *maxima* cv. *maxima* 'Hokkaido'), Hubbard Squash (*C. maxima* Duchesne ssp. *maxima* cv. *hubbardiana* Grebensc. 'Hubbard Squash'), Kiszombori (*C. maxima* Duchesne ssp. *maxima* cv. *maxima* 'Kiszombori'), Mammut (*C. maxima* Duchesne ssp. *maxima* cv. *maxima* 'Mammut'), Marina di Chioggia (*C. maxima* Duchesne ssp. *maxima* cv. *hubbardiana* Grebensc. 'Marina di Chioggia'), Massimo di Chioggia (*C. maxima* Duchesne ssp. *maxima* cv. *maxima*

'Massimo di Chioggia'), Muscat de Provence (*C. maxima* Duchesne 'Muscat de Provence'), Orange Früchte (*C. maxima* Duchesne ssp. *maxima* cv. *turbaniformis* [M.Roem] Alef. 'Orange Früchte'), Peruaner Kürbis (*C. maxima* Duchesne ssp. *maxima* cv. *maxima* 'Peruaner Kürbis'), Potimarron (*C. maxima* Duchesne ssp. *maxima* cv. *hubbardiana* Grebensc. 'Potimarron'), Queensland Blue (*C. maxima* Duchesne 'Queensland Blue'), Red Hokkaido (*C. maxima* Duchesne ssp. *maxima* cv. *maxima* 'Red Hokkaido'), Red Kuri (*C. maxima* Duchesne ssp. *maxima* cv. *maxima* 'Red Kuri'), Roter Zentner (*C. maxima* Duchesne ssp. *maxima* cv. *maxima* 'Roter Zentner'), Sweet Meat (*C. maxima* Duchesne ssp. *maxima* cv. *maxima* 'Sweet Meat'), Tondo di Nizza (*C. maxima* Duchesne ssp. *maxima* cv. *hubbardiana* Grebensc. 'Tondo di Nizza'), Triambelkürbis/Triamble (*C. maxima* Duchesne ssp. *maxima* 'Triambelkürbis/Triamble'), Trombone (*C. maxima* Duchesne ssp. *maxima* cv. *bananina* Grebensc. 'Trombone'), Turban squash (*C. maxima* Duchesne ssp. *maxima* cv. *turbaniformis* [M.Roem] Alef. var. *turbaniformis* 'Turban-Kürbis'), Valenciano (*C. maxima* Duchesne 'Valenciano'), Veltruska (*C. maxima* Duchesne 'Veltruska'), Viridi-Alba (*C. maxima* Duchesne ssp. *maxima* cv. *turbaniformis* [M.Roem] Alef. var. *turbaniformis* 'Viridi-Alba'), Wildsippe Argentinien (*C. maxima* Duchesne ssp. *andreana* [Naudin] Filov 'Wildsippe Argentinien'), and Yellow Hubbard (*C. maxima* Duchesne ssp. *maxima* cv. *hubbardiana* Grebensc. 'Yellow Hubbard').

CULTIVAR The names of the many cultivars of this species often incorporate one of the many English terms for this plant: calabash, gourd, head, marrow, pumpkin (to which the classic Halloween pumpkin refers), and squash. Among the best-known cultivars are Arikara squash (traditional cultivar of the Arikara, a Lakota tribe, with 2–5 kg fruit, teardrop or round shape, orange and mottled green color), Banana squash (elongated shape, skin color blue, pink, and orange, bright orange flesh), Boston marrow (sweet flavor, narrow at one end and bulbous at the other), Buttercup squash (turban-shaped winter variety, 1.3–2.2 kg, with dense yellow-orange flesh), Candy Roaster (landrace developed by the Cherokee in the southern Appalachians, quite variable in fruit shape, color, and weight, orange flesh), Hubbard squash (teardrop shape), Jarrahdale pumpkin (gray skin, very similar to the Queensland Blue and Sweet Meat varieties), Kabocha (Japanese variety), Lakota squash (American variety), Nanticoke squash (rare turban-shaped traditional cultivar of the Nanticokes of Delaware and eastern Maryland), and Turk's turban (traditional cultivar similar to Buttercup squash).

RELATED SPECIES The genus *Cucurbita* includes 22 taxa, some of which produce edible fruits and have applications in making beer.

GEOGRAPHIC DISTRIBUTION C America. **HABITAT** Cultivated, often grows wild near gardens or other anthropogenic environments, disturbed environments, meadows and fields, banks of lakes and rivers. **RANGE** Cultivated everywhere.

BEER-MAKING PARTS (AND POSSIBLE USES) Fresh fruits, roasted seeds.

TOXICITY All Cucurbitaceae contain triterpene glycosides called cucurbitacins. These compounds are present in all organs of the plant at different concentrations; if concentrated in the edible parts, they give a bitter taste. Although having recognized medicinal properties, cucurbitacins in high doses can be poisonous.

CHEMISTRY Energy 77 kJ (18 kcal), water 94.6 g, protein 1.1 g (lysine 54 mg, histidine 16 mg, arginine 54 mg, aspartic acid 102 mg, threonine 29 mg, serine 44 mg, glutamic acid 184 mg, proline 26 mg, glycine 27 mg, alanine 28 mg, cystine 3 mg, valine 35 mg, methionine 11 mg, isoleucine 31 mg, leucine 46 mg, tyrosine 42 mg, phenylalanine 32 mg, tryptophan 12 mg), carbohydrates 3.5 g (soluble sugars 2.5 g, starch 0.9 g), lipids 0.1 g (saturated 0.05 g, monounsaturated 0.01 g, polyunsaturated 0.5 g), fiber 0.5 g, sodium 1 mg, thiamin (B1) 0.03 mg, riboflavin (B2) 0.02 mg, niacin (B3) 0.5 mg, vitamin A 599 μg, vitamin C 9 mg, vitamin E 1.06 mg, calcium 20 mg, iron 0.9 mg, phosphorus 40 mg, magnesium 12 mg, potassium 340 mg, copper 0.12 mg, selenium 0.3 mg, zinc 0.32 mg.

STYLE American Amber Ale, American Barleywine, American Brown Ale, American IPA, American Light Lager, American Pale Ale, American Porter, American Stout, American Strong Ale, American Wheat or Rye Beer, Autumn Seasonal Beer, Baltic Porter, Belgian Dark Strong Ale, Belgian Dubbel, Belgian Pale Ale, Belgian Specialty Ale, Bière de Garde, Blonde Ale, Brown Porter, California Common Beer, Clone Beer, Cream Ale, Dark American Lager, Dunkelweizen, Experimental Beer, Extra Special/Strong Bitter (ESB),

Description Herb, annual (up to 3 m). Root taproot or fibrous. Stems creeping or climbing, forming adventitious roots at nodes; bristly with strong persistent hairs, pustulate, and an underlying hispidulo-hirsute layer; tendrils 2- to 7-ramified, 1–5 cm above base, hispidulous to hirsute, needle-like. Leaves petiole 5–24 cm, pustulate-hispidulous and hispidulo-hirsutulous; lamina, sometimes spotted white at intersections of veins, broadly ovate-cordate to triangular-cordate or reniform, ± deeply palmate, 3- to 7-lobed (20–30 × 20–35 cm), usually wider than long or with the 2 sizes equivalent, base cordate, lobes ovate-deltate to obovate or obovate-rhombic, midrib of leaf lobes not distinctly elongate, margins denticulate to serrate-denticulate, surfaces hirsute, hirsute-strigose, villous-strigose, or hispidulo-scabrous, needle-like; Peduncles in fruit 5-costulute, ± gradually expanded at point of insertion with fruit, hardened, woody. Flowers with campanulate hypanthium (8–12 mm); sepals linear to subulate-linear (8–25 mm); corolla yellow to golden yellow or orange in color, tubular-campanulate (4–10 cm); filaments of anthers glabrous; ovary villous. Fruit pepo, uniform color light to dark green, green with white stripes, or with minute cream, yellow, orange, or bicolor green-yellow speckles, globose or depressed-globose to ovoid, obovoid, ellipsoid-ovoid, broadly ellipsoid, slightly pyriform, cushion-shaped, or cylindrical (5–25 cm), usually smooth or ribbed, rarely warty; flesh color whitish to yellowish or light orange, soft, not bitter. Seeds whitish to cream or reddish in color, narrowly to broadly elliptical or obovate, rarely orbiculate (7–26 mm), margins raised-thickened, surface ± smooth.

Subspecies Within the taxon *Cucurbita pepo* L., the typical subspecies (subsp. *pepo*), the subspecies *ovifera* (L.) D.S.Decker, and the variety *texana* (Scheele) D.S.Decker are recognized.

Cultivar Those with edible fruits have been subdivided into eight groups, mainly on the basis of the morphology of the pepo, which has many differences (size, shape, durability, color, fibrousness, pulp bitterness) compared to those of wild lines.

Related Species The genus *Cucurbita* includes 22 taxa, some of which produce edible fruits and are used in making beer.

Geographic Distribution Mexico. **Habitat** Plant cultivated in temperate areas. **Range** Besides widespread cultivation, this species tends to form nonperennial formations wherever cultivated, especially in temperate and tropical areas.

Beer-Making Parts (and Possible Uses) Fruits.

Chemistry Fruit, pulp: protein 1.35%, lipids 1.36–1.81%, copper 1.17 ppm, magnesium 4.02 ppm, zinc 1.32 ppm, iron 2.87 ppm, sodium 1.89 ppm, potassium 6.86 ppm, calcium 23.82 ppm. Fruit, peel: protein 8.03%, lipids 1.51–2.06%. Seeds: protein 9.53–19.73%, lipids 23.65–37.46%.

Style Pumpkin Ale, Pumpkin Porter.

Source Calagione et al. 2018; Heilshorn 2017; Hieronymus 2016b; Josephson et al. 2016.

Cuminum cyminum L.
APIACEAE

Synonyms *Cuminia cyminum* J.F.Gmel., *Cuminum aegyptiacum* Mérat ex DC., *Cuminum hispanicum* Mérat ex DC., *Cuminum odorum* Salisb., *Cuminum sativum* J.Sm., *Cyminon longeinvolucellatum* St.-Lag.

Common Names Cumin.

Description Herb, annual (2–5 dm). Stems erect, glabrous. Leaves completely divided into 1 × 20–50 mm filiform laciniae. Inflorescences 3- to 5-rayed umbels; bracts often trifid; bracts linear; petals white or ± reddened. Fruit ellipsoid or clavate (4–5 mm), at apex with persistent lesiniform (1–2 mm) calcine teeth, in dorsum with 5 bristly-sharp ribs.

Cultivar In some countries (e.g., Sri Lanka) where this spice is widely used, there are different commercial types with specific names (e.g., Duru, Maduru, Suduru), although it is not clear if these products are obtained from different plants (cultivars?) or from different methods of processing the same plant.

Related Species Besides *C. cyminum* L., the genus *Cuminum* includes the species *Cuminum borszczowii* (Regel and Schmalh.) Koso-Pol., *Cuminum setifolium* (Boiss.) Koso-Pol., and *Cuminum sudanense* H.Wolff.

Geographic Distribution C Asia. **Habitat** Arid grasslands. **Range** India, Syria, Iran, Turkey.

Beer-Making Parts (and Possible Uses) Seeds.

Chemistry Energy 375 kcal, lipids 22 g (saturated fatty acids 1.5 g, polyunsaturated fatty acids 3.3 g, monounsaturated fatty acids 14 g), carbohydrates 44 g, fiber 11 g, sugar 2.3 g, protein 18 g, vitamin A 1,270 IU, vitamin C 7.7 mg, vitamin B6 0.4 mg, sodium 168 mg, potassium 1788 mg, magnesium 366 mg, calcium 931 mg, iron 66.4 mg.

Style American Barleywine, American Porter, American Stout, Belgian Blond Ale, Belgian Dark Strong Ale, Belgian Dubbel, Belgian Golden Strong Ale, Belgian Tripel, California Common Beer, Gruit, Holiday/Winter Special Spiced Beer, Other Smoked Beer, Saison, Specialty IPA (Belgian IPA), Spice/Herb/Vegetable Beer, Strong Scotch Ale, Weizen/Weissbier, Witbier.

Beer Ommegang Abbey Ale, Ommegang (New York, United States).

Source Hieronymus 2010; Hieronymus 2012; Jackson 1991; Markowski 2015; McGovern 2017; https://www.brewersfriend.com/other/cumin/.

Curcuma longa L.
ZINGIBERACEAE

Synonyms *Amomum curcuma* Jacq., *Curcuma brog* Valeton, *Curcuma domestica* Valeton, *Curcuma ochrorhiza* Valeton, *Curcuma soloensis* Valeton, *Curcuma tinctoria* Guibourt, *Stissera curcuma* Giseke.

Common Names Turmeric, tumeric.

Description Herb, perennial (up to 1 m). True root departs from a large, cylindrical, branched underground rhizome, yellow or orange in color, strongly aromatic, most commercially relevant part of the plant. Leaves large (20–45 cm), arranged in 2 rows, each divided into sheaths, petiole (50–115 cm) and lamina oblong-elliptic (76–230 × 38–45 cm), narrowed at the apex; the sheaths together form a false stem. Flowers hermaphrodite, zygomorphic, and trimerous; 3 sepals (0.8–1.2 cm) fused, unequal calyx teeth, white in color, with a soft down; 3 petals, bright yellow, fused into a coralline tube (3 cm) with 3 lobes (1–1.5 cm), triangular (middle lobe, labellum, larger than lateral), with upper termination softly spiny, labellum obovate (1–2 cm), yellowish, with a yellow stria in the center; androecium consisting of fertile stamens (inner circle) and sterile staminodes (outer) shorter than labellum; 3 carpels form a 3-lobed, sparsely hairy ovary. Flowers collected in terminal pseudo-inflorescence (12–20 cm) on the pseudostem; this with large green bracts below, white or purplish above; green ovate-oblong bracts (3–5 cm) form a series of pockets that host large yellow flowers (with possible orange hues); at the apex of the pseudo-inflorescence the tapering colored bracts do not have flowers. Fruit capsule trilocular.

Cultivar As for other plants of agricultural interest, different species, varieties, and cultivars (about 30) have also been developed in Pakistan (Bannu, Faisalabad, Kasur) and India (Alleppey Finger, Amalapuram, Armour, Balaga, Bilaspur, BRS1, Cuddapah, Dindigam, Duggirala, Erode, GL Purm I, GL Purm II, Jangir, Kasturi, Krishna, Kodur, Lakadong, Moovattupuzha, Mudaga, Nizamabad Bulb, P317, Raigarth, Rajapore, Rajendra Sonia, RH2, RH10, Roma, Salem, Sangli, Shillong, Sudarshana, Sugandham, Suguna, Suvarna, Tekurpet, Vontimitra, Wyanad).

Related Species The genus *Curcuma* includes more than 90 taxa, some of which have edible parts.

Geographic Distribution S Asia (from India to Malaysia). **Habitat** Areas with a tropical climate, with temperatures of 20–35°C and high rainfall. **Range** Widely cultivated in tropical and subtropical areas, in particular in Asia and Africa.

Beer-Making Parts (and Possible Uses) Rhizomes (fresh or boiled, dried and powdered).

Chemistry Rhizome (powder): energy 354 kcal, carbohydrates 65 g, protein 8 g, lipids 10 g (saturated fatty acids 3.1 g, polyunsaturated fatty acids 2.2 g, monounsaturated fatty acids 1.7 g), fiber 21 g, sugar 3.2 g, vitamin C 25.9 mg, vitamin B6 1.8 mg, sodium 38 mg, potassium 2.52 g, calcium 183 mg, iron 41.4 mg, magnesium 193 mg. Active components: curcuminoids (3–5%), mixtures of cinnamoylmethane derivatives (curcumin, demethoxycurcumin, bis-demethoxycurcumin), volatile fraction (3–5%), containing characteristic terpene compounds (zingiberene, curcumol, β-turmerone), smaller amounts of some arabinogalactans (ukonan acid A, B, C, D).

Style Belgian Dark Strong Ale, Belgian Specialty Ale, Belgian Tripel, Classic American Pilsner.

Beer Shangrila, Troll (Vernante, Piedmont, Italy).

Source Josephson et al. 2016; https://www.brewersfriend.com/other/tumeric/; https://www.brewersfriend.com/other/turmeric/.

Cydonia oblonga Mill.
ROSACEAE

Synonyms *Cydonia communis* Loisel., *Cydonia cydonia* (L.) Pers., *Cydonia europaea* Savi, *Cydonia maliformis* Beck, *Cydonia sumboshia* Buch.-Ham. ex D.Don, *Cydonia vulgaris* Pers., *Cydonia vulgaris* Pers. var. *oblonga* (Mill.) DC., *Pyrus cydonia* L., *Pyrus oblonga* (Mill.) Steud., *Sorbus cydonia* Crantz.

Common Names Quince.

Description Shrub or tree (2–5 m). Leaves with ovate lamina (3–5 × 5–9 cm), acute, whole. Flowers solitary; peduncles tomentose; petals pinkish (15 mm). Fruit pome (3–12 cm in cultivation) vaguely apple-like but characteristically shaped, with fragrant, spongy flesh.

Subspecies *Cydonia oblonga* Mill. f. *lusitanica* (Mill.) Rehder, *Cydonia oblonga* Mill. var. *maliformis* (Mill.) Rehder, *Cydonia oblonga* Mill. f. *marmorata* (Dippel) C.K.Schneid., *Cydonia oblonga* Mill. f. *pyramidalis* (Dippel) Rehder, *Cydonia oblonga* Mill. f. *pyriformis* (Dierb.) Rehder.

Cultivar Cultivars of this species include Aromatnaya, Bereczcki, Champion, Cooke's Jumbo (Jumbo), Dwarf Orange, Gamboa, Iranian, Isfahan, Le Bourgeaut, Lescovacz, Ludovic, Maliformis, Meeches Prolific, Morava, Orange (Apple Quince), Perfume,

Pineapple, Portugal (Lusitanica), Shams, Siebosa, Smyrna, Van Deman, and Vrajna (Bereczcki).

Related Species Apart from the variability within *C. oblonga* Mill., the genus *Cydonia* is considered monospecific.

Geographic Distribution SW Asia. **Habitat** Plant cultivated in the Mediterranean, often subspontaneous. **Range** Turkey, China, Uzbekistan, Morocco, Iran, Argentina, Azerbaijan, Spain, Serbia, Algeria.

Beer-Making Parts (and Possible Uses) Fruits, also in jam (quince paste).

Chemistry Energy 238 kJ (57 kcal), carbohydrates 15.3 g, fiber 1.9 g, lipids 0.1 g, protein 0.4 g, thiamin (B1) 0.02 mg, riboflavin (B2) 0.03 mg, niacin (B3) 0.2 mg, pantothenic acid (B5) 0.081 mg, vitamin B6 0.04 mg, folate (B9) 3 µg, vitamin C 15 mg, calcium 11 mg, iron 0.7 mg, magnesium 8 mg, phosphorus 17 mg, potassium 197 mg, sodium 4 mg, zinc.

Style American Amber Ale, American Pale Ale, Clone Beer, Witbier.

Beer Malagrika, B94 (Lecce, Apulia, Italy).

Source Calagione et al. 2018; Giaccone and Signoroni 2017; https://www.bjcp.org (29. Fruit Beer); https://www.brewersfriend.com/other/birsalma/; https://www.brewersfriend.com/other/quince/.

Cymbopogon citratus (DC.) Stapf
POACEAE

Synonyms *Andropogon cerifer* Hack., *Andropogon ceriferus* Hack., *Andropogon citratus* DC., *Andropogon fragrans* C.Cordem., *Andropogon roxburghii* Nees ex Steud.

Common Names Lemon grass, lemongrass, oil grass.

Description Herb, perennial (up to 2 m), aromatic. Leaves linear (1 m × 2 cm) tapering toward the sheath; smooth and hairless, white on the upper side, green below; ligule (appendage between sheath and leaf blade) less than 2 mm long, rounded or truncated (abrupt end as if from a sharp cut). Inflorescence panicle, lax and bent, about 60 cm, reddish to rust color; spikelet pedicels purple.

Related Species The genus *Cymbopogon* includes about 54 taxa, some of which are edible.

Geographic Distribution Indonesia. **Habitat** Cultivated in the tropics. **Range** Philippines, Indonesia, Brazil, India, Nepal, Sri Lanka, South Africa.

Beer-Making Parts (and Possible Uses) Leaves.

Chemistry Essential oil: geranial 45.2%, neral 32.4%, geraniol 5.5–40%, myrcene 10.2–18%, geranyl acetate 1.2%, α-terpineol 0.9%, limonene 0.4%, citronellol 0.3%, citronellal 0.2%, camphene, camphor, α-camphorene, Δ-3-carene, caryophyllene, caryophyllene oxide, 1,8-cineole, citronellal, citronellol, n-decyldehyde, α,β-dihydropseudoionone, dipentene, β-elemene, elemol, farnesal, farnesol, fenchone, furfural, iso-pulegol, iso-valeraldehyde, limonene, linalyl acetate, menthol, menthone, methyl heptenol, ocimene, α-oxobisabolene, β-phellandrene, α-pinene, β-pinene, terpineol, terpinolene, 2-undecanone, neral, nerolic acid, geranic acid.

Style American Amber Ale, American IPA, American Light Lager, American Pale Ale, American Stout, American Wheat Beer, American Wheat or Rye Beer, Belgian Blond Ale, Belgian Pale Ale, Belgian Specialty Ale, Bière de Garde, Blonde Ale, British Golden Ale, California Common Beer, Classic American Pilsner, English Barleywine, Experimental Beer, Festbier, Imperial IPA, Kölsch, Russian Imperial Stout, Saison, Schwarzbier, Specialty Beer, Specialty IPA (Belgian IPA, White IPA), Spice/Herb/Vegetable Beer, Weissbier, Witbier.

Beer Whistlin' Beers, Suarez Family (New York, United States).

Source Dabove 2015; Giaccone and Signoroni 2017; Heilshorn 2017; Hieronymus 2016a; Hieronymus 2016b; Josephson et al. 2016; Steele 2012; https://www.brewersfriend.com/other/citroengras/; https://www.brewersfriend.com/other/lemongrass/; https://www.craftbeer.com/styles/herb-and-spice-beer/; https://en.wikipedia.org/wiki/Cymbopogon_citratus.

Cynara scolymus L.
ASTERACEAE

Synonyms None
Common Names Artichoke, globe artichoke.

Description Herb, perennial (2–15 dm). Stem stout, erect, usually simple. Leaves 1–2-pinnate or ± entire with violet or no thorns, to 1 m, arcuate-patent. Flowers blue or purplish, gathered in large capitula (Ø 8–15 cm), pyriform, scales oval, ± spiny depending on cultivar. Fruit achene ellipsoidal or more or less prismatic with long-haired pappus.

Cultivar There are two varietal groups divided according to their system of propagation. The group of traditional cultivars propagated vegetatively is divided into many subgroups: Large Green (Camus de Bretagne, Castel [France], Green Globe [United States, South Africa], Vert de Laon [France]), Medium Green (Argentina, Bayrampasha [Turkey], Blanc d'Oran [Algeria], Blanca de Tudela [Spain], Española [Chile], Sakiz, Verde Palermo [Sicily, Italy]), Large Violet (C3 [Italy], Romanesco), Viola Medie (Baladi [Egypt], Brindisino, Catanese, Ñato [Argentina], Niscemese [Sicily, Italy], Violet d'Algérie [Algeria], Violet de Provence [France], Violetta di Chioggia [Italy]), Spinose (Criolla [Peru], Ingauno [Liguria], Spinoso Sardo [Sardinia]), and Bianche (very rare). The group of cultivars propagated by seed is divided into three subgroups: cultivars for industry (A-106, Imperial Star, Lorca, Madrigal), Verdi (Harmony, Symphony), and Viola (Concerto, Opal, Tempo).

RELATED SPECIES In addition to *C. scolymus*, the genus *Cynara* includes 11 recognized taxa: *Cynara algarbiensis* Mariz, *Cynara aurantiaca* Post, *Cynara auranitica* Post, *Cynara baetica* (Spreng.) Pau, *Cynara cardunculus* L., *Cynara cardunculus* L. subsp. *flavescens* Wiklund, *Cynara cornigera* Lindl., *Cynara cyrenaica* Maire & Weiller, *Cynara humilis* L., *Cynara syriaca* Boiss., and *Cynara tournefortii* Boiss. & Reut.

GEOGRAPHIC DISTRIBUTION Sicily (genetic and historical research suggest that the domestication of the artichoke [*Cynara scolymus*] from its wild progenitor [*Cynara cardunculus*] may have occurred in CW Sicily, where cultivars with an appearance and genetic characteristics intermediate between wild and cultivated forms are still grown today). **HABITAT** Cultivated. **RANGE** Italy, Egypt, Spain, Argentina, Peru, Algeria, China, Morocco, United States, France, Turkey, Tunisia.

BEER-MAKING PARTS (AND POSSIBLE USES) Green heads.

CHEMISTRY Energy 47 kcal, lipids 0.2 g (polyunsaturated fatty acids 0.1 g), sodium 94 mg, potassium 370 mg, carbohydrates 11 g, fiber 5 g, sugars 1 g, protein 3.3 g, vitamin A 13 IU, vitamin C 11.7 mg, calcium 44 mg, iron 1.3 mg, vitamin B6 0.1 mg, magnesium 60 mg.

STYLE Spice/Herb/Vegetable Beer.

SOURCE https://www.birradegliamici.com/news-birraie/birre-strane/.

Cynodon dactylon (L.) Pers.
POACEAE

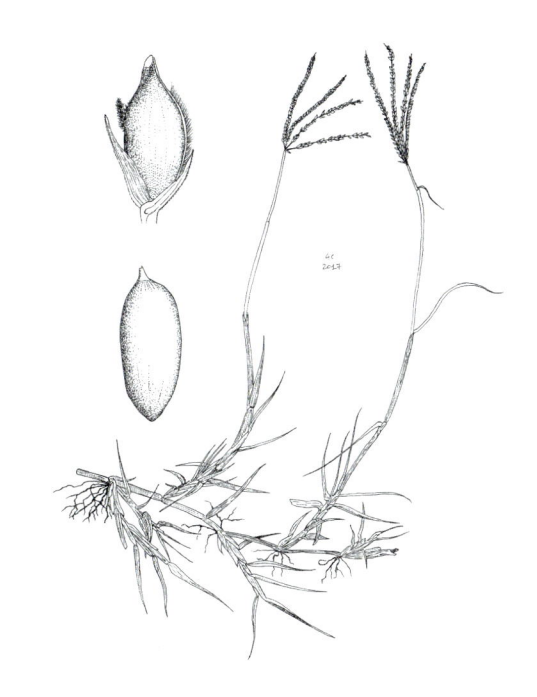

SYNONYMS *Agrostis linearis* Retz., *Agrostis stellata* Willd., *Capriola dactylon* (L.) Kuntze, *Chloris maritima* Trin., *Chloris paytensis* Steud., *Cynodon affinis* Caro & E.A.Sánchez, *Cynodon aristiglumis* Caro & E.A.Sánchez, *Cynodon aristulatus* Caro & E.A.Sánchez, *Cynodon decipiens* Caro & E.A.Sánchez, *Cynodon distichloides* Caro & E.A.Sánchez, *Cynodon erectus* J.Presl, *Cynodon glabratus* Steud., *Cynodon hirsutissimus* (Litard. & Maire) Caro & E.A.Sánchez, *Cynodon iraquensis* Caro, *Cynodon laeviglumis* Caro & E.A.Sánchez, *Cynodon linearis* Willd., *Cynodon maritimus* Kunth, *Cynodon mucronatus* Caro & E.A.Sánchez, *Cynodon nitidus* Caro & E.A.Sánchez, *Cynodon pascuus* Nees, *Cynodon pedicellatus* Caro, *Cynodon polevansii* Stent, *Cynodon scabrifolius* Caro, *Cynodon stellatus* Willd., *Cynodon tenuis* Trin., *Cynodon umbellatus* (Lam.) Caro, *Cynosurus dactylon* (L.) Pers., *Cynosurus uniflorus* Walter, *Digitaria dactylon* (L.) Scop., *Digitaria glumaepatula* (Steud.) Miq., *Digitaria glumipatula* (Steud.) Miq., *Digitaria linearis* (L.) Pers., *Digitaria maritima* (Kunth) Spreng., *Fibichia dactylon* (L.) Beck, *Milium dactylon* (L.) Moench, *Panicum ambiguum* (DC.) Le Turq., *Panicum dactylon* L., *Panicum glumipatulum* Steud., *Panicum lineare* L., *Paspalum ambiguum* DC., *Paspalum dactylon* (L.) Lam., *Paspalum umbellatum* Lam., *Phleum dactylon* (L.) Georgi, *Syntherisma linearis* (L.) Nash, *Vilfa linearis* (Retz.) P.Beauv., *Vilfa stellata* (Willd.) P.Beauv.

Common Names Bermuda grass, dog's tooth grass, Bahama grass, devil's grass, couch grass, wiregrass, scutch grass.

Description Herb, perennial (3–4 dm). Rhizome tenacious, creeping, and rooting at nodes, with distichous leaves. Culms ascending, wrapped in leaf sheaths to inflorescence. Leaves with lamina 3–3.5 mm wide, those of sterile shoots short, lanceolate, the others 3–5 cm long; lamina canaliculate, rigid, bristling with 1.5 mm patent hairs. Spikes digitate to 3–5, thin, usually violet; spikelets 1-flowered (with the rudiment of an abortive upper flower); 2 glumes, 0.7 mm and 2.6 mm, respectively; lemma 2.3 mm.

Cultivar Several cultivars of this species are known in the United States: common Bermuda grass (*C. dactylon* var. *dactylon*, a tetraploid [2n = 36], which originated in the Near East and is a common cropland weed), coastal Bermuda grass (cross *C. dactylon* var. *dactylon* × *C. dactylon* var. *elegans*, bigger than the previous one and therefore particularly suitable for hay production), Tiff Bermuda (found in cotton fields near Tifton in Georgia, has long decumbent culms and is suitable for hay production and grazing), St. Lucia Bermuda (plant without rhizomes, propagates by stolons in sandy soils rich in organic matter on the southern coast of Florida), Alicia (quick to expand but does not produce quality hay), Callie (adapted to conditions in the area between North Carolina and Texas, well digestible), Oklan (good digestibility), Suwannee Bermuda (adapted to low-fertility soils), coast cross-1 (adapted to the S-United States), Midland (Oklahoma, high productivity even in dry conditions), Hardie (adapted to deep, rich soils with high yields), and Greenfield (excellent erosion control, grows well and fast in thin, eroded soils).

Related Species Besides *Cynodon dactylon*, the genus includes 12 taxa: *Cynodon aethiopicus* Clayton & Harlan, *Cynodon barberi* Rang. & Tadul., *Cynodon coursii* A.Camus, *Cynodon flexicaulis* (Schwägr.) Steud., *Cynodon inclinatus* (Hedw.) Brid, *Cynodon incompletus* Nees, *Cynodon × magennisii* Hurcombe, *Cynodon nlemfuensis* Vanderyst, *Cynodon parviglumis* Ohwi, *Cynodon plectostachyus* (K.Schum) Pilg, *Cynodon radiatus* Roth, and *Cynodon transvaalensis* Burtt Davy.

Geographic Distribution Thermocosmopolitan.

Habitat Crops, weeds in cultivated fields (0–2,300 m ASL; 650–1,750 mm rainfall/year). **Range** Crops, weeds in cultivated fields in hot-arid climates.

Beer-Making Parts (and Possible Uses) Rhizomes (as an addition or as a single product for an inexpensive and refreshing beer).

Toxicity Nontoxic species but has high concentrations of oxalic acid, and there have been occasional cases of hydrocyanic acid poisoning.

Chemistry Active ingredients: fructan, triticin, mannitol, inositol, malic acid, mucilage, saponin, vanilligloside, amygdalin, agropyrene in essential oil, silicic acid, potassium and iron salts, vitamins A, B.

Style Spice/Herb/Vegetable Beer.

Source https://www.erbemedicinali.netsons.org/index.php?option=com_content&view=article&id=254:-gramigna&catid=29:piante&Itemid=29.

Cyphomandra betacea (Cav.) Sendtn.
SOLANACEAE

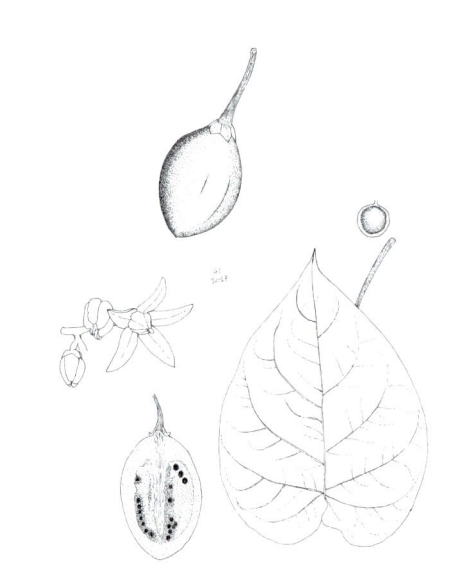

Synonyms *Pionandra betacea* (Cav.) Miers, *Solanum insigne* Lowe.

Common Names Tamarillo, tree tomato, tamamoro.

Description Tree, small (2–5 m), perennial, fast-growing, with soft wood. Roots shallow and not very pronounced, may be damaged by high winds. Stem single, flowers and fruit on side branches. Leaves large (15–30 × 10–20 cm), broadly cordate-ovate, usually confined to distal end of branches. Flowers small, pink in color, in small clusters (10–50 flower units) at or near the leaf axil, near the distal end of branches. Fruits ovate berries (4–10 cm), in groups (1–6), red-purple or orange-yellow color. Seeds more numerous than in common tomato.

Cultivar In New Zealand three main varietal groups are recognized according to the color of the berry: red (excellent consumed raw or cooked or used for decoration of other foods, red skin and dark red color around the seeds, provide antioxidants, vitamins, and minerals), amber (sweeter fruits, smaller size, golden skin streaked with red), and gold (intermediate sweetness, versatile in cooking as ingredient or condiment).

Related Species Besides the species reported here, only *Cyphomandra tegorea* (Aubl.) Hemsl. is recognized as belonging to the genus *Cyphomandra*. However, many names associated with this genus are awaiting taxonomic and nomenclatural revision.

Geographic Distribution Andes (Ecuador, Colombia, Peru, Chile, Bolivia). **Habitat** Areas with a subtropical climate (rainfall 600–4,000 mm/year; temperatures 15–20°C); species intolerant to cold (below −2°C) and to aridity stress; areas similar to those used to cultivate species of the genus *Citrus sp.* (e.g., Mediterranean climate) also provide good conditions for tamarillo. **Range** Cultivated in subtropical areas of Ecuador, Colombia, Peru, Chile, Bolivia, Rwanda, South Africa, India, Nepal, Hong Kong, China, United States, Australia, Butan, New Zealand, Malaysia, Philippines, Puerto Rico.

Beer-Making Parts (and Possible Uses) Fruits.

Chemistry Energy 40 kcal/fruit, water 81–87 g, protein 1.5–2.5 g, lipids 0.05–1.28 g, fiber 1.4–6 g, vitamin A 0.32–1.48 mg, vitamin C 19.7–57.8 mg, calcium 3.9–11.3 mg, magnesium 19.7–22.3 mg, iron 0.4–0.94 mg.

Style Witbier.

Beer Portmarillo, Dogfish Head (Delaware, United States).

Source https://www.brewersfriend.com/other/tamarillo/.

Cytisus scoparius (L.) Link
FABACEAE

Synonyms *Sarothamnus bourgaei* Boiss., *Sarothamnus oxyphyllus* Boiss., *Sarothamnus scoparius* (L.) W.D.J.Koch, *Sarothamnus vulgaris* Wimm, *Spartium scoparium* L.

Common Names Broom, Scotch broom.

Description Shrub (1–3 m), deciduous, heavily branched, sparsely leafy at fruiting. Stem erect, glabrous, striate, and angular with 5 longitudinal sharp ribs, branches young, green, straight. Lower leaves on flattened petioles, divided into 3 obovate-oblong leaflets, slightly pubescent, upper ones simple, lanceolate, subsessile (1–2 cm). Flowers numerous, hermaphrodite, deep golden yellow, isolated or paired at the axil of the leaves of the stems of the previous year, united in raceme inflorescences, carried by glabrous peduncles, whitish, at least 2 times longer than the calyx; calyx bilabiate, glabrous (6–7 mm), corolla deciduous, papilionaceous (hull obtuse, vexillum emarginated at apex and rounded at base, wings oblong and glabrous, all ± the same length, 16–24 mm); androecium of 10 stamens (4 long, 1 medium, 5 short), diadelphous (filaments joined together in 2 groups) with basifixed and dorsifixed anthers; ovary with ciliated margins and glabrous or hairy style. Fruit legume elliptic, flattened (20–50 × 7–12 mm), ciliate at margins, green, then blackish-brown at maturity. Seeds (± 13 per legume) ovoid or elliptical and flat (2–4 × 2–3 mm), brown or green with white-yellowish strophiole.

Subspecies While only the nominal subspecies is present in Italy, *Citysus scoparius* (L.) Link subsp. *maritimus* (Rouy) Heywood grows on the maritime cliffs of NW Europe, with procumbent branching and leaves and young twigs densely covered with silky hairs.

Related Species The genus *Cytisus* includes about 90 taxa.

Geographic Distribution Europe. **Habitat** Woods, heaths, and clearings of hilly areas, at path edges, in every type of soil although it prefers siliceous soils (0–1,400 m ASL in Italy). **Range** Widely cultivated in the area and elsewhere for ornamental purposes and for the consolidation of slopes and stony inclines but also to obtain various products (fibers for fabrics, paper, baskets, brooms, pigments, repellents, elastic wood used to make crossbows); it can become invasive.

Beer-Making Parts (and Possible Uses) Young flowering shoots.

Toxicity Plant contains toxic glycosides whose ingestion causes agitation, nausea, vomiting, hypotension, hallucinations.

Chemistry Among the alkaloids contained are tyramine, dopamine, N-methyl-dopamine, epinine, and L-sparteine.

Style Scottish Export, Specialty Beer.

Source Heilshorn 2017; https://www.brewersfriend.com/other/broom/.

Dacrycarpus dacrydioides (A.Rich.) de Laub.
PODOCARPACEAE

Synonyms *Dacrydium ferrugineum* Van Houtte ex Gordon, *Nageia dacrydioides* (A.Rich.) F.Muell., *Nageia excesla* (D.Don) Kuntze, *Podocarpus dacrydioides* A.Rich., *Podocarpus thujoides* R.Br.

Common Names Kahikatea, New Zealand white pine.

Description Tree (more than 50 m). Trunk (Ø 1.5 m) often grooved and reinforced at the base, free of branches for a considerable height; crown conical in young individuals; bark dark gray, irregularly covered with small bumps (about 1–2 × 1–2 mm); branches slender, drooping. Plantlets bear juvenile foliage to 1–2 m, then adult leaves appear; young, semi-adult, adult leaves may be present simultaneously in the same individual; young leaves subdense, subpatent, narrow-linear, subfalcate, acuminate, decurved (3–7 × 0.5–1 mm); adult ones (1–2 mm), imbricate, appressed, subtrigonous, lanceolate-subulate to acuminate, with broader base. Male cones terminal, orange in color, to 1 cm, with bisporangiate sporophylls, apex acute. Seeds (4–5 mm) broadly ovoid, black, solitary, terminal on short twigs, the upper 2–3 leaves forming a distinct receptacle, red, swollen, and succulent when mature.

Related Species The genus *Dacrycarpus* includes 11 taxa.

Geographic Distribution New Zealand. **Habitat** Coastal to subalpine forests (0–600 m ASL) on both main islands. **Range** Cultivated as a forest species in the native range and as a curiosity in botanical gardens and orchards worldwide.

Beer-Making Parts (and Possible Uses) Wood tested for making beer.

Wood Broad whitish sapwood and yellowish heartwood, fine texture, regular grain, mostly free of defects but not very resistant to alteration and insects (WSG 0.42–0.48). Used for packaging, joinery, and interior panels.

Style None verified.

Source Cantwell and Bouckaert 2016; Cantwell and Bouckaert 2018.

Daucus carota L.
APIACEAE

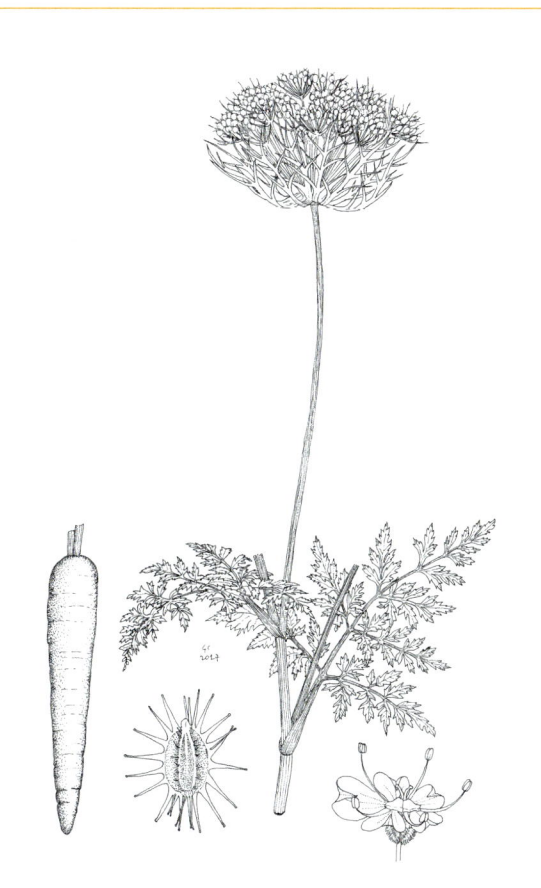

Synonyms *Carota sylvestris* (Mill.) Rupr., *Caucalis carnosa* Roth, *Caucalis carota* (L.) Crantz, *Caucalis daucus* Crantz, *Daucus alatus* Poir., *Daucus allionii* Link, *Daucus blanchei* Reut., *Daucus brevicaulis* Raf., *Daucus communis* Rouy & E.G.Camus, *Daucus dentatus* Bertol., *Daucus esculentus* Salisb., *Daucus exiguus* Steud., *Daucus gingidium* Georgi, *Daucus glaber* Opiz ex Čelak., *Daucus heterophylus* Raf., *Daucus kotovii* M.Hiroe, *Daucus levis* Raf., *Daucus marcidus* Timb.-Lagr., *Daucus montanus* Schmidt ex Nyman, *Daucus neglectus* Lowe, *Daucus nudicaulis* Raf., *Daucus officinalis* Gueldenst. ex Ledeb., *Daucus scariosus* Raf., *Daucus sciadophylus* Raf., *Daucus strigosus* Raf., *Daucus sylvestris* Mill., *Daucus vulgaris* Neck.

Common Names Carrot, wild carrot, Queen Anne's lace.

Description Herb, annual or biennial (up to 80 cm), ± bristly. Long taproot yellowish (orange in most cultivated varieties), fleshy, fusiform, branched. Stems erect, hairy, rarely glabrous, sometimes striped, simple, and slightly branched in the upper part; bearing quite variable, from compact to slender to expanded. Leaves highly variable, lower leaves with incised-toothed oval segments, upper leaves divided into linear laciniae. Inflorescence umbels (5–7 cm or +) with 20–40 rays, involucre formed by 7–10 laciniate bracts, straight at flowering, folding into a ball at fruiting. Flowers small, petals pinkish-white, the central one blackish-purple. Fruits elliptical achenes, with bristly main ribs and secondary ribs with generally simple pointed spines.

Subspecies In addition to the nominal subspecies (*Daucus carota* L. subsp. *carota*), in Italy: *Daucus carota* L. subsp. *commutatus* (Paol.) Thell. (leaves with divisions of each order, inserted at right angles on the rachis), *Daucus carota* L. subsp. *drepanensis* (Arcang.) Heywood (small plant, less than 20 cm, umbels without a central flower, contracted at fruiting), *Daucus carota* L. subsp. *hispanicus* (Gouan) Thell. (densely bristly, bracts with membranous margins, umbels often contracted), *Daucus carota* L. subsp. *major* (Vis.) Arcang. (bristly with erect stems, umbels Ø 6–10 cm, single or double hooked barbs), *Daucus carota* L. subsp. *maritimus* (Lam.) Batt. (subglabrous, stem suberect, leaves with deeply pinnate-partite linear segments, umbels Ø 3–5 cm, barbs with generally star-shaped hooks), *Daucus carota* L. subsp. *maximus* (Desf.) Ball (sparsely bristly, ascending stem, umbels Ø 12–30 cm, hooked barbs generally star shaped), *Daucus carota* L. subsp. *rupestris* (Guss.) Heywood (small, not more than 10 cm, densely hairy with fleshy bipinnate oval perimeter leaves, umbels Ø 5 cm or so, not contracted at fruiting), *Daucus carota* L. subsp. *sativus* (Hoffm.) Schübl. & G.Martens (cultivated type). Some authors recognize within *Daucus carota* L. subsp. *sativus* (Hoffm.) Arcangeli two varieties, from which all the numerous existing cultivars would derive: *Daucus carota* L. subsp. *sativus* (Hoffm.) Arcangeli var. *sativus* (western carrot), with orange, yellow, or white reserve organs, yellowish-green leaves, highly dissected, last segments linear-lanceolate, relatively hairless (fewer than 50 hairs/mm^2), cultivated everywhere and predominant variety except in Asia; and *Daucus carota* L. subsp. *sativus* (Hoffm.) Arcangeli var. *atrorubens* (Oriental carrot), with purple and/or yellow (rarely yellowish-orange) reserve organs, leaves grayish-green (glaucous), sparsely dissected, last segments lanceolate to ovate, relatively pubescent (more than 50 hairs/mm^2), although introduced in other places common only in Asia because it does not conserve well. Many hybrids between the two varieties (*D. carota* var. *sativus* × var. *atrorubens*) have been obtained, cultivated and consumed mainly in Asia (e.g., *D. carota* f. *kintoki*, also of commercial interest in the West).

Cultivar Nantes varieties (Bolero, Cosmic Purple Kaleidoscope Mix, Merida, Napa, Nelson, Parano, Purple Dragon, Scarlet Nantes, Touchon, White Satin, Yaya) are very common and easy to grow; they produce cylindrical carrots (15–18 cm), are sweet and crisp, and have a rounded tip. Imperator varieties (Atomic Red, Autumn King) are the classic long, tapered carrots. Chantenay varieties (Hercules, Red-Cored Chantenay) are instead short and robust (4–8 × 15 cm), with a wide crown that tapers abruptly. Mini varieties (Babette, Romeo) are very small and of different shapes.

Related Species The genus *Daucus* includes about 41 taxa.

Geographic Distribution Paleotemperate subcosmopolitan. **Habitat** Grasslands, meadows, roadsides, arid environments (0–1,400 m ASL in Italy). **Range** China, Russia, United States, Uzbekistan, Ukraine, Poland, Turkey, Morocco, United Kingdom, Japan.

Beer-Making Parts (and Possible Uses)
Flowers, roots (taproots), leaves, seeds.

Chemistry Taproot: water 91.6 g, protein 1.1 g, lipids 0.2 g, carbohydrates 7.6 g, fiber 3.1 g, energy 147 kJ (35 kcal), sodium 95 mg, potassium 220 mg, iron 0.7 mg, calcium 44 mg, phosphorus 37 mg, magnesium 11 mg, zinc 2.92 mg, copper 0.19 mg, selenium 1 µg, thiamin 0.04 mg, riboflavin 0.04 mg, niacin 0.7 mg, vitamin A 1148 µg, vitamin C 4 mg. Essential oil, root: α-terpinolene 26.2–56.3%, β-pinene 4.1–8.2%, γ-terpinene 3.3–5.6%, p-cymene 2.7–7.4%, α-pinene 2.7–5.3%, E-bisabolene 0.8–2.0%, sabinene 5.9%, limonene 5.5%, myristicin 4.9%, myrcene 3%, p-cymene-8-ol 2.2%, germacrene D 2.2%, caryophyllene oxide 1.6%, octanal 1.2%, dehydro-p-cymene 1.2%, Z-β-ocimene 1%, Z-γ-bisabolene 1%, bornylacetate 0.9%, β-caryophyllene 0.8%, γ-himachalane 0.8%, trans-2-nonenal 0.7%, methylacetophenone 0.7%, terpinen-4-ol 0.6%, (E,E)-2,4-decadienal 0.6%, β-elemene 0.6%, spathulenol 0.6%, viridiflorol 0.6%, α-phellandrene 0.5%, 2E-decenal 0.5%, trans-β-farnesene 0.5%, α-terpinene 0.3%, undecane 0.3%, α-humulene 0.3%, α-selinene 0.3%, heptanal 0.2%, α-thujene 0.2%, camphene 0.2%, 1,3,8-p-menthatriene 0.2%, thymolmethylether 0.2%. Essential oil, leaves: α-pinene 20.9–44.8%, sabinene 11.3–14.2%, limonene 11.3–12.7%, myrcene 7.4–11.2%, germacrene D 4.9–14.0%, β-pinene 1.3–5.9%, caryophyllene oxide 3.7%, β-caryophyllene 2.3%, α-bisabolol 1.2%, camphene 1%, γ-terpinene 0.9%, terpinen-4-ol 0.8%, Z-β-ocimene, 0.7%, trans-β-farnesene 0.7%, spathulenol 0.7%, β-elemene 0.6%, bicyclogermacrene 0.6%, p-cymene 0.5%, β-bourbonene 0.5%, α-terpinene 0.4%, α-terpinolene 0.4%, α-humulene 0.4%, α-selinene 0.4%, β-bisabolene 0.4%, myristicin 0.4%, α-thujene 0.3%, E-β-ocimene 0.3%, nonanal 0.3%, α-copaene 0.3%, germacrene A 0.3%. Essential oil, fruits: α-pinene 28.5%, sabinene 19.5–46.6%, geranyl acetate 3.9–28.1%, β-pinene 3.0–13.1%, limonene 2.3–3.9%, trans-sabinene hydrate 0–1.2%, terpinen-4-ol 0.8–2.2%, linalool 0.5–1.2%, caryophyllene oxide 0.4–1.2%, α-thujene 0.3–8.8%, germacrene D 0.3–6.5%, γ-terpinene 0.3–4.1%, camphor 0.3–4.1%, β-caryophyllene 0.3–1.2%, α-terpineol 0.3–0.7%, cis-sabinene hydrate 0.2–1.8%, α-terpinene 0.2–1.1%, Z-β-ocimene 0.2–1.0%, α-terpinolene 0.2–0.8%, myrcene 4%, α-bisabolol 1.9%, camphene 1.4%, trans-β-farnesene 1.1%, bicyclogermacrene 0.6%, p-cymene 0.4%, bornylacetate 0.4%, α-copaene 0.4%, spathulenol 0.4%, E-β-ocimene 0.3%, α-terpinylacetate 0.3%, (E,E)-α-farnesene 0.3%, γ-cadinene 0.3%, carotol 0.3%, α-canfolenal 0.2%, trans-2-nonenal 0.2%, β-cubebene 0.2%, β-elemene 0.2%, α-humulene 0.2%, trans-verbenol 0.1%, β-bourbonene 0.1%.

Style American Amber Ale, British Golden Ale, California Common Beer, Holiday/Winter Special Spiced Beer.

Source Heilshorn 2017; Hieronymus 2016b; Josephson et al. 2016; Steele 2012; https://www.brewersfriend.com/other/carrot/.

Dioscorea alata L.
DIOSCOREACEAE

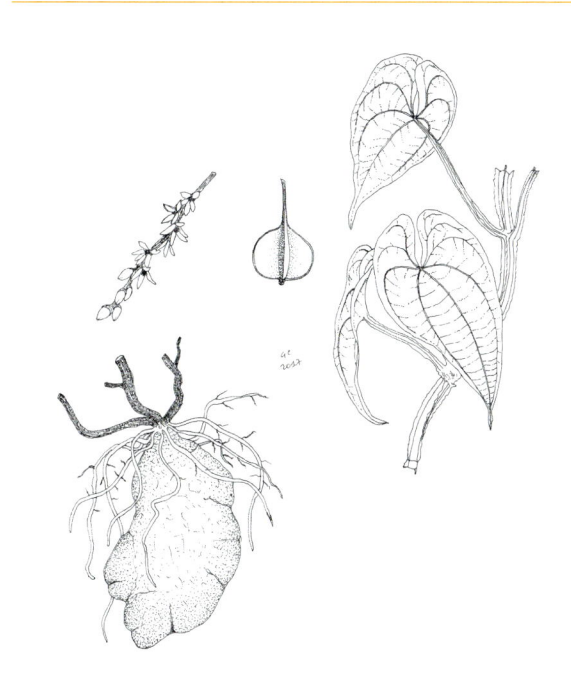

Synonyms *Dioscorea atropurpurea* Roxb., *Dioscorea colocasiifolia* Pax, *Dioscorea eburina* Lour., *Dioscorea eburnea* Lour., *Dioscorea globosa* Roxb., *Dioscorea javanica* Queva, *Dioscorea purpurea* Roxb., *Dioscorea rubella* Roxb., *Dioscorea sapinii* De Wild., *Dioscorea vulgaris* Miq., *Elephantodon eburnea* (Lour.) Salisb., *Polynome alata* (L.) Salisb.

Common Names Yam, purple yam, greater yam, Guyana arrowroot, ten-months yam, water yam, white yam, winged yam.

Description Herb, perennial, climbing (10–15 m), nonwoody, glabrous. Roots fibrous, shallow, mostly confined to the first meter of soil. Tubers usually single, cylindrical or clavate ± curved (1.5 m deep) or globose, stout and short, pyriform, often lobed or digitate; skin dark or black, flesh white, cream, or purplish. Stems quadrangular by 4 longitudinal wings, wavy, green to reddish in color; mature stems at base cylindrical and spiny. Leaves opposite, sometimes alternate on rapidly growing stems, leathery, broadly ovate (10–30 × 5–18 cm), 5- to 7-ribbed, apex acute or acuminate, sometimes reflexed, base cordiform; upper leaf dark green, shiny, under leaf pale green, with prominent veins; petioles (4–12 cm) 4-winged, forming an auriculate sheath at the base, with 2 pseudostipules embracing the stem; bulbils elongate (up to 15 cm), pendulous, produced when the leaves begin to wilt. Inflorescences axillary, unisexual, pendulous; those staminate paniculate (5–15 cm), with numerous lateral flexuous spikes bearing numerous male flowers; those pistillate racemose, with few flowers; perianth 1–1.5 mm in staminate flowers, 2–2.8 mm in pistillate flowers. Fruit capsule trilocular (2–3 cm), each locule flattened like a wing, with 2 seeds. Seeds orbicular, winged all around.

Cultivar A satisfactory classification system has not yet been found for the hundreds cv. of this species. According to some authors, as many as 15 groups should be considered (out of a total of 235 cultivars studied). The greatest variability was found in material from New Guinea, Indonesia, and the Philippines. The most significant varietal groups from SE-Asia are Purple Compact (plants with high anthocyanin content, short, pigmented tubers; Philippines [Kinampay, Morado, Vino Violeta]); Primitive Purple (plants with high anthocyanin content in leaves, unbranched, pigmented tubers; New Guinea [Aupik, Buster, Kiubu]); Primitive Green (green leaves, long tuber with prominent collar, unpigmented; New Guinea, Indonesia [Lanswa, Masi, Suabab]); Compact (anthocyanin-pigmented leaves, short, broad tubers, poorly pigmented; New Guinea, Philippines (Kabusak, Toma, Uhbisi)); and Poor White (solitary tubers, pigmented exterior, whitish interior; New Guinea, India [Baron, Goarmago, Toa]).

Related Species The genus *Dioscorea* includes more than 600 species. Those important for food include *Dioscorea bulbifera* L., *Dioscorea cayennensis* Lam., *Dioscorea communis* (L.) Caddick & Wilkin,

Dioscorea convolvulacea Cham. & Schltdl., *Dioscorea dumetorum* (Kunth) Pax, *Dioscorea esculenta* (Lour.) Burkill, *Dioscorea hispida* Dennst., *Dioscorea japonica* Thunb., *Dioscorea nummularia* Lam., *Dioscorea oppositifolia* L., *Dioscorea pentaphylla* L., *Dioscorea polystachya* Turcz., *Dioscorea rotundata* Poir., *Dioscorea trifida* L.f., and *Dioscorea villosa* L.

Geographic Distribution SE Asia (Indochina, Philippines, Indonesia, etc.) and adjacent areas (Taiwan, Ryukyu Islands, Assam, Nepal, New Guinea, Christmas Island). **Habitat** Mild temperatures (15–35°C) and high rainfall (>1000 mm/year), low to medium elevation, partially shaded habitats. **Range** Cultivated and naturalized in various areas of China, Africa, Madagascar, Western Hemisphere and Indian and Pacific Ocean islands, United States (Louisiana, Georgia, Alabama, Florida, Puerto Rico, U.S. Virgin Islands) so much so as to be considered an alien invasive species.

Beer-Making Parts (and Possible Uses) Tubers.

Toxicity Oxalates, alkaloids (diosgenin), malic acid, ferulic acid are contained in small quantities (higher in other *Dioscorea* sp.); most of these substances are inactivated by cooking and rinsing with water.

Chemistry Energy 118 kcal, water 69.6 g, carbohydrates 27.89 g, protein 1.53 g, lipids 0.17 g, fiber 4.1 g, vitamin C 17.1 mg, thiamine (B1) 0.11 mg, riboflavin (B2) 0.03 mg, niacin (B3) 0.55 mg, pantothenic acid 0.31 mg, pyridoxine (B6) 0.29 mg, folate 23 µg, calcium 17 mg, magnesium 21 mg, phosphorus 55 mg, sodium 9 mg, potassium 816 mg, iron 0.54 mg, zinc 0.24 mg, copper 0.18 mg, selenium 0.7 µg.

Style American Brown Ale, American Stout, Baltic Porter, Bière de Garde, Robust Porter, Russian Imperial Stout, Spice/Herb/Vegetable Beer.

Beer Shiretoko Draft, Abashiri Beer (Abashiri, Japan).

Source Josephson et al. 2016; McGovern 2017; https://www.brewersfriend.com/other/yam/; https://www.brewersfriend.com/other/yams/.

Diospyros kaki L.f.
EBENACEAE

Synonyms *Diospyros amara* Perrier, *Diospyros argyi* H.Lév., *Diospyros bertii* André, *Diospyros costata* Carrière, *Diospyros kaempferi* Naudin, *Diospyros lycopersicon* Carrière, *Diospyros mazelii* E.Morren, *Diospyros roxburghii* Carrière, *Diospyros schi-tse* Bunge, *Diospyros sinensis* Naudin, *Diospyros sphenophylla* Hiern, *Diospyros trichocarpa* R.H.Miao, *Diospyros wieseneri* Carrière, *Embryopteris kaki* (Thunb.) G.Don.

Common Names Persimmon, Japanese persimmon, kaki, keg fig, date plum.

Description Deciduous tree (12–27 m). Young branches densely pubescent to glabrous, sometimes with reddish-brown lenticels. Winter buds small, blackish. Leaves alternate, lamina lanceolate to elliptic or ovate to obovate (5–18 × 2.6–9 cm), papery, pubescent when young, adaxially often glabrescent when mature and paler with darker veins, base cuneate, subtruncate or rarely cordate, apex usually acuminate, lateral veins 5–7 per side, secondary veins reticulate clearly defined, flat and dark; petiole 0.8–2 cm. Plant generally monoecious (male and female flowers on different plants), sometimes dioecious (male and female flowers on the same individual). Flowers yellowish-white, rather small (about 2 cm); male ones in cymes of 3–5, calyx ± as long as the corolla, hairy on both sides, 4 lobes, corolla color white to yellowish to red (6–10 mm), stamens 14–24; female flowers solitary, calyx Ø 3 cm or more, 4 lobes, corolla color usually yellowish-white, bell shaped (0.9–1.6 cm), lobes ovate recurved, staminodes (stamens sterile owing to absence of anthers and pollen) 8–16; some varieties require pollination for fruit development whereas others develop seedless fruits without pollination; ovary glabrous or pubescent. Fruit berry, color yellow to orange, flattened-globose to conical or ovoid (Ø 2–8.5 cm), 8-locular, glabrescent. Seeds dark brown (13–16 × 7.5–9 × 4–5 mm).

Cultivar From the commercial point of view, there are two varietal groups characterized by fruit astringency: astringent and nonastringent. Astringent cultivars (Dōjō hachiya, Fuji, Gionbō, Hachiya, Hongsi, Kōshū hyakume, Ormond, Saijō, Sheng) contain high levels of soluble tannins that make fruits inedible before they are completely ripe, whereas the nonastringent cultivars (Dan gam, Fuyū, Hanagosho, Izu, Jirō, Sōshū, Taishū, Vanilla), even though they are not free from tannins, are much less astringent and can be consumed when they are still firm. There is a rare third group of varieties (Goma, Hyakume, Maru, Tsurunoko) whose pulp is dark and has particular aromas (vanilla, chocolate, brown sugar, etc.).

Related Species The genus *Diospyros* includes more than 740 taxa, some with edible fruits; *Diospyros virginiana* L. is certainly used in making beer.

Geographic Distribution *Diospyros kaki* is not found in the wild but only in cultivation, and it is believed to have been selected from the wild species *Diospyros roxburghii* Carrière, an entity that, according to some authors, would be synonymous with *Diospyros kaki*. **Habitat** Cultivated. **Range** China, Korea, Japan, Brazil, Azerbaijan.

Beer-Making Parts (and Possible Uses) Fruits.

Chemistry Energy 293 kJ (70 kcal), carbohydrates 18.59 g, sugars 12.53 g, fiber 3.6 g, lipids 0.19 g, protein 0.58 g, vitamin A 81 µg, β-carotene 253 µg, lutein zeaxanthin 834 µg, thiamine (B1) 0.03 mg, riboflavin (B2) 0.02 mg, niacin (B3) 0.1 mg, vitamin B6 0.1 mg, folate (B9) 8 µg, choline 7.6 mg, vitamin C 7.5 mg, vitamin E 0.73 mg, vitamin K 2.6 µg, calcium 8 mg, iron 0.15 mg, magnesium 9 mg, manganese 0.355 mg, phosphorus 17 mg, potassium 161 mg, sodium 1 mg, zinc 0.11 mg.

Style Belgian Dark Strong Ale, Belgian Specialty Ale, Gose, Sour Ale.

Beer Upland Persimmon Beers, Upland Brewery (Indiana, United States).

Source Cantwell and Bouckaert 2016; Cantwell and Bouckaert 2018; Hieronymus 2016b; McGovern 2017; https://www.brewersfriend.com/other/percimmon/.

Drosera rotundifolia L.
DROSERACEAE

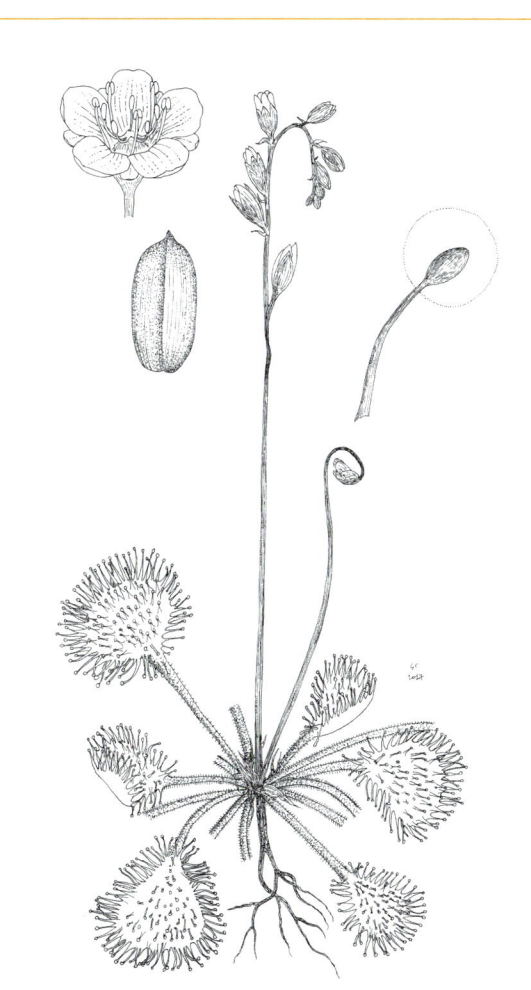

Synonyms *Drosera corsica* (Maire) A.W.Hill, *Drosera septentrionalis* (Scop.) Stokes, *Rorella rotundifolia* (L.) All., *Rossolis rotundifolia* (L.) Moench, *Rossolis septentrionalis* Scop.

Common Names Round-leaved sundew, common sundew.

Description Herb, perennial (5–12 cm), carnivorous, delicate, brittle, reddish color. Leaves spoon shaped, patent, all in basal rosette, ± adherent to the ground, petiolate (10–15 cm), lamina orbicular, covered with long purple tentacular cilia, viscous, sensitive and very thin, ending with small globular glands secreting a viscous, sugary, shiny liquid (small insects attracted by the sugary secretion get trapped by viscosity, tentacles and leaf fold in on them, and they are digested by special enzymes; when the process is complete, the leaf redistends, reactivating the trap). Flower scape erect, aphyllous, at the center of the rosette with small pauciflorous raceme, often branched. Flowers with corolla consisting of 5 spatulate petals (4–6 mm), with 5 acute, persistent laciniae, white in color, slightly longer than calyx, 5 stamens, 3 styles. Fruit smooth oval capsule.

Related Species The genus *Drosera* consists of more than 200 species, some of which may have properties (therapeutic or for making beer?) similar to *D. rotundifolia*.

Geographic Distribution Circumboreal-Eurosiberian. **Habitat** Bogs, swamps, marshes, in acidic waters, among sphagnums and mosses (0–2,000 m ASL). **Range** Species sometimes cultivated for ornamental purposes or as a botanical curiosity.

Beer-Making Parts (and Possible Uses) Leaves.

Chemistry The naphthoquinones (especially plumbagin, 7-methyl-juglone, flavonoids) secreted by the plant are considered the major agents of its many therapeutic actions (antiviral, antifungal, antibacterial, aphrodisiac, antispasmodic, anti-leprosy, anti-sclerosis).

Style Mum.

Source Hieronymus 2010.

Dysphania ambrosioides (L.) Mosyakin & Clemants
AMARANTHACEAE

Synonyms *Ambrina ambrosioides* (L.) Spach, *Ambrina anthelmintica* (L.) Spach, *Ambrina incisa* Moq., *Ambrina parvula* Phil., *Ambrina spathulata* Moq., *Atriplex ambrosioides* (L.) Crantz, *Atriplex anthelmintica* (L.) Crantz, *Blitum ambrosioides* (L.) Beck, *Botrys ambrosioides* (L.) Nieuwl., *Botrys anthelmintica* (L.) Nieuwl., *Chenopodium amboanum* (Murr) Aellen, *Chenopodium ambrosioides* L., *Chenopodium angustifolium* Pav. ex Moq., *Chenopodium anthelminticum* L., *Chenopodium citriodorum* Steud., *Chenopodium cuneifolium* Vent. ex Moq., *Chenopodium integrifolium* Vorosch., *Chenopodium querciforme* Murr, *Chenopodium sancta-maria* Vell., *Chenopodium santamaria* Vell., *Chenopodium spathulatum* (Moq.) Sieber ex Moq., *Chenopodium suffruticosum* Willd., *Chenopodium vagans* Standl., *Chenopodium variegatum* Gouan, *Dysphania anthelmintica* (L.) Mosyakin & Clemants, *Orthosporum ambrosioides* (L.) Kostel., *Orthosporum suffruticosum* Kostel., *Roubieva anthelmintica* (L.) Hook. & Arn., *Teloxys ambrosioides* (L.) W.A.Weber, *Vulvaria ambrosioides* (L.) Bubani.

Common Names Epazote, wormseed, Jesuit's tea, Mexican-tea.

Description Herb (3–8 dm), annual or perennial, with aromatic scent. Stems erect, coppiced, reddened, striated, bristly from short bristles and with sessile glands. Leaves lanceolate (1–2.5 × 5–12 cm), sinuate-dentate on edges. Inflorescence panicle branched, leafy; glomerules partially reddened, numerous, with short bract leaves at axil; perianth greenish; stamens 5, exserted, with yellow anthers.

Related Species The genus *Dysphania* includes 38 accepted taxa.

Geographic Distribution Neotropical (CS America) turned into cosmopolitan. **Habitat** Ruins, rubble. **Range** Invasive species in many territories other than its area of origin.

Beer-Making Parts (and Possible Uses) Flowering tops (substitute for the leaves of *Coriandrum sativum*).

Chemistry Essential oil: α-terpinene 61.04%, p-cymene 12.94%, limonene 0.79%, δ-3-carene 0.58%, cis-verbenol 0.49%, ascaridole 0.4%, 4-carene 13.55%, thujanol 0.77%, thymol 2.19%, carvacrol 1.4%, limonene dioxide 0.42%, β-ionone 0.2%, 1-isopropenyl-3-propenylcyclopentane 0.18%.

Style American Amber Ale, California Common Beer, Holiday/Winter Special Spiced Beer.

Source Fisher and Fisher 2017; https://www.brewersfriend.com/other/epazote/.

Echinacea purpurea (L.) Moench
ASTERACEAE

Synonyms *Brauneria purpurea* (L.) Britton, *Echinacea intermedia* Lindl. ex Paxton, *Echinacea serotina* (Sweet) D.Don ex G.Don f., *Echinacea speciosa* (Wender.) Paxton, *Helichroa purpurea* (L.) Raf., *Rudbeckia purpurea* L.

Common Names Eastern purple coneflower, purple coneflower.

Description Herb, perennial (up to 1.8 m) growing from a rhizome. Epigeous stems rough, often branching at the upper end and covered with short fluffy hairs. Leaves (15 × 10 cm) cover stems from their base, margin coarsely and irregularly toothed, upper ones tapering toward the base. Inflorescence head (Ø up to 15 cm) with central disc of small hermaphrodite flowers orange in color and outer sterile flowers (3–8 cm) red-purple in color with green ends. Fruit achene.

Cultivar Among the many varieties of *Echinacea purpurea* are Coconut Lime, Doppelganger, Magnus, PowWow White, PowWow® Wild Berry, Rubinstern, Ruby Giant, and White Swan.

Related Species Besides *Echinacea purpurea*, the genus includes 11 species: *Echinacea angustifolia* DC., *Echinacea atrorubens* (Nutt.) Nutt., *Echinacea* *laevigata* (F.E.Boynton & Beadle ex C.L.Boynton & Beadle) S.F.Blake, *Echinacea pallida* (Nutt.) Nutt., *Echinacea paradoxa* (Norton) Britton, *Echinacea paradoxa* (Norton) Britton var. *neglecta* McGregor, *Echinacea sanguinea* Nutt., *Echinacea serotina* (Nutt.) DC., *Echinacea simulata* McGregor, and *Echinacea tennesseensis* (Beadle) Small.

Geographic Distribution United States (Georgia, Louisiana, Oklahoma, Virginia, Ohio, Michigan, Illinois, Iowa, Missouri, Kansas, Florida, North Carolina, Texas, New Mexico, New York, West Virginia, Kentucky, Tennesse, Alabama, Arkansas, etc.). **Habitat** Rocky environments, open woodlands, prairies, sometimes in wetter soils adjacent to rivers and streams. **Range** Widely cultivated in temperate climate areas for ornamental or therapeutic purposes (rhizome).

Beer-Making Parts (and Possible Uses) Flowers, leaves, rhizomes.

Chemistry Phenolic composition, leaves: 4-hydroxy benzoic acid 0.18 mg, ferulic acid 0.01 mg, caffeic acid 1.35 mg, cichoric acid 0.09 mg, quercetin 0.01 mg, rutin 0.12 mg.

Style American IPA.

Source Heilshorn 2017; Hieronymus 2016b; https://www.brewersfriend.com/other/echinacea/.

Elaeagnus commutata Bernh. ex Rydb.
ELAEAGNACEAE

Synonyms None

Common Names Silverberry, wolf-willow, American silverberry.

Description Shrub or rarely small tree (1–4 m), strongly rhizomatous and stoloniferous, sometimes forming ± dense thickets. Branches spineless, reddish-brown, sparsely to densely covered with silvery scales. Leaves deciduous, simple, alternate, ovate-oblong or ovate-lanceolate (2–10 cm), base wedge shaped, briefly petiolate, both surfaces covered with minute silvery scales, sometimes with scattered brown scales beneath. Flowers hermaphrodite or unisexual, sweet smelling, found in lateral cymes of 1–3 from a short peduncle on twigs of the current year; petals absent, sepals forming a 4-lobed tube (12–15 mm) from the top of the developing fruit (inferior ovary), yellowish inside, silver outside. Fruit ovate to ellipsoid (8–10 mm), silver in color, with dry floury pulp covering a single ellipsoid seed.

Related Species The genus *Elaeagnus* includes about 100 species.

Geographic Distribution North America (Canada, United States [including Alaska]). **Habitat** Riparian and other wetland sites but also arid woodlands, from prairies to boreal coniferous forests. **Range** Naturalized plant in some U.S. states in addition to the original ones; elsewhere cultivated for ornamental purposes.

Beer-Making Parts (and Possible Uses) Fruits.

Chemistry Organic acids, fatty acids, minerals, vitamin A, vitamin C, vitamin E.

Style Kölsch, Saison.

Beer Silverberry Sour, Hermit Thrush (Vermont, United States).

Source https://www.brewersfriend.com/other/silverberries/.

Elaeagnus rhamnoides (L.) A.Nelson
ELAEAGNACEAE

Synonyms *Argussiera rhamnoides* (L.) Bubani, *Hippophae rhamnoides* L., *Hippophaes rhamnoideum* (L.) St.-Lag., *Osyris rhamnoides* (L.) Scop., *Rhamnoides hippophae* Moench.

Common Names Common sea buckthorn, seaberry.

Description Shrub or small tree (up to 9 m), deciduous, dioecious. Well-developed root. Stem woody, much branched with stiff branches, some of them aphyllous with strong, cauline, prickly spines; bark whitish-gray, silvery in young branches with brown peltate scales and star hairs; buds bare, subrounded. Leaves alternate-spiral, stipule free, subsessile, linear-lanceolate (3–5 × 40–60 mm), gray-green and whitish tomentum caducous on the upper leaf, silver-green underneath with dense silvery-white peltate hairs mixed with ferrugineous hairs. Flowers small (2.5–3.5 mm), apetalous, emerging with the first leaves on the previous year's branches; male sessile, united in short amentiform racemes, with perigonium ovate-orbicular, concave, formed by 2 tepals joined for 1/4 of their length, covered with brownish-white peltate hairs, stamens 4; female pedicellate, solitary or in small pauciflorous racemes, perianth tubular, bilobed at apex and covered with brownish peltate hairs, ovary semi-inferior unilocular with 1 exserted stylus to clavate stigma. Fruit pseudo-drupe yellow-orange in color, ovoid or subglobose (4–8 × 2–6 mm), enveloped by fleshy, juicy mesocarp. Seed single, ovoid, brown in color with hard, bony shell.

Related Species The genus *Elaeagnus* includes about 100 species; *Elaeagnus umbellata* Thunb. is known for its use in making beer.

Geographic Distribution Eurasian. **Habitat** Stony environments, riverbanks, landslides, gullies, preferably in a calcareous substrate (50–1,700 m ASL). **Range** Widely cultivated in temperate climate areas for ornamentation and edible fruits.

Beer-Making Parts (and Possible Uses) Fruits.

Chemistry Antioxidants: vitamin C 50–300 mg, vitamin A 5–10 mg, vitamin E 3 mg, flavonoids 150–300 mg, thiamine (B1), riboflavin (B2), niacin (B3), pyridoxine (B6), biotin, folic acid, vitamin K. Essential oil, seeds: linoleic acid 34%, α-linolenic acid 25%, oleic acid 19%. Essential oil, fruit pulp: palmitic acid 33%, oleic 26%, palmitoleic 25%.

Style American Wheat or Rye Beer, Berliner Weisse, Saison, Sour Ale, Spice/Herb/Vegetable Beer.

Beer Seaberry Sour, Big Axe (New Brunswick, Canada).

Source https://www.brewersfriend.com/other/havtorn/; https://www.brewersfriend.com/other/seabuckthorn/.

Melderis, and *Elymus hispidus* (Opiz) Melderis subsp. *varnensis* (Velen.) Melderis.

Cultivar Studies conducted in North America by the Land Institute (a U.S. nonprofit organization committed to research for sustainable agriculture) have contributed to the development of an agricultural system based on perennial species (to give ecological stability to the American prairies and ensure grain production comparable to that of annual plants) that led to the registration and marketing of the perennial Kernza® "wheat", a bristly crabgrass that can be used as a multifunctional crop, producing at the same time raw materials (caryopses) and ecosystem services (protection against soil erosion, formation of organic matter, CO^2 sequestration, preservation of soil microbiomes, etc.), potentially converting agriculture into a soil-forming activity similar to natural ecosystems.

Related Species The genus *Elymus* includes about 250 species.

Geographic Distribution Eurasia, India.
Habitat Open spaces, plains, abandoned fields, riverbanks, riverbeds, in calcareous or marly-arenaceous soils (up to 2,000 m ASL). **Range** Eurasia, North America (naturalized).

Beer-Making Parts (and Possible Uses) Caryopsis.

Chemistry Caryoxides (g/100 g flour): protein 18.2–24.8 g, lipids 2.2–4.0 g, ash 2.4–3.0 g, insoluble fiber 19.5–20.1 g, soluble fiber 1.9–5.4 g, starch 43.0–53.0 g, amylose 19.5–27.3 g/100 g starch, ferulic acid 794–1325 μg/g flour, p-coumaric acid 21.0–105.8 μg/g flour, sinapic acid 73–117 μg/g flour, lutein 1.9–16.0 μg/g flour, zeaxanthin 0.6–2.9 μg/g flour.

Style Kipa (Kernza IPA), Northwest-Style Pale Ale, Pacific Ale, Wit.

Beer Long Root Pale Ale, Patagonia Provisions (Oregon, United States).

Source McMillan and Cornett 2018; https://hopworksbeer.com/2016/10/03/long-root-ale-a-partnership-between-patagonia-provisions-and-hopworks/; https://www.patagoniaprovisions.com/pages/long-root-ale; https://www.bangbrewing.com/

Empetrum nigrum L.
ERICACEAE

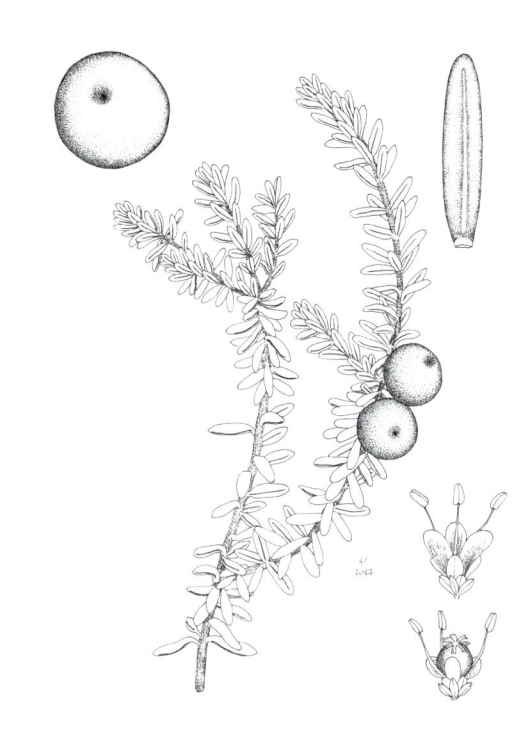

Synonyms *Chamaetaxus nigra* (L.) Bubani, *Empetrum crassifolium* Raf., *Empetrum medium* Carmich., *Empetrum procumbens* Gilib., *Empetrum scoticum* auct.

Common Names Crowberry, black crowberry, blackberry (W Alaska).

Description Shrub (5–12 dm). Stems woody, elongated, often stout and erect, with bark first reddish, then brown in color. Leaves leathery, evergreen, lanceolate (± 1.5 × 4–5 mm), lamina glossy green above, margins revolute, whitish in color, not recessed below. Flowers isolated at axil of upper leaves; corolla greenish or ± mottled red (1.5 mm). Fruit drupe, subspherical (Ø 5–10 mm). Seeds light brown in color (1.5–3 mm).

Subspecies Eight subspecies of *Empetrum nigrum*: *Empetrum nigrum* L. subsp. *albidum* (V.N.Vassil.) Kuvaev, *Empetrum nigrum* L. subsp. *androgynum* (V.N.Vassil.) Kuvaev, *Empetrum nigrum* L. subsp. *asiaticum* (Nakai ex H.Ito) Kuvaev, *Empetrum nigrum* L. subsp. *caucasicum* (Juz.) Kuvaev, *Empetrum nigrum* L. subsp. *hermaphroditum* (Hagerup) Böcher, *Empetrum nigrum* L. subsp. *kardakovii* (V.N.Vassil.) Kuvaev, *Empetrum nigrum* L. subsp. *nigrum*, *Empetrum nigrum* L. subsp.

sibiricum (V.N.Vassil.) Kuvaev, *Empetrum nigrum* L. subsp. *subholarcticum* (V.N.Vassil.) Kuvaev.

Related Species The genus *Empetrum* includes three species: *Empetrum eamesii* Fernald & Wiegand (two subspecies) and *Empetrum nigrum* L., *Empetrum rubrum* Vahl ex Willd.

Geographic Distribution Circum-Arctic-Alpine. **Habitat** Sphagnum bogs, acidic bogs. **Range** Circumboreal.

Beer-Making Parts (and Possible Uses) Fruits.

Chemistry Fruits (used as a natural coloring or as part of functional foods owing to remarkable antioxidant action and other beneficial effects [antiinflammatory, antibacterial, antifungal, hypocholesterolemic action]): flavonols (quercetin, rutin, myricetin, naringenin, morin, kaempferol), anthocyanins.

Style Fruit/Vegetable Beer, Stout.

Beer Kreklingøl, Haandbryggeriet (Drammen, Norway).

Source https://www.beeradvocate.com/beer/profile/15711/76615/; https://www.facebook.com/northernbrewers/?hc_ref=Pages_Timeline&fref=nf.

Epilobium angustifolium L.
ONAGRACEAE

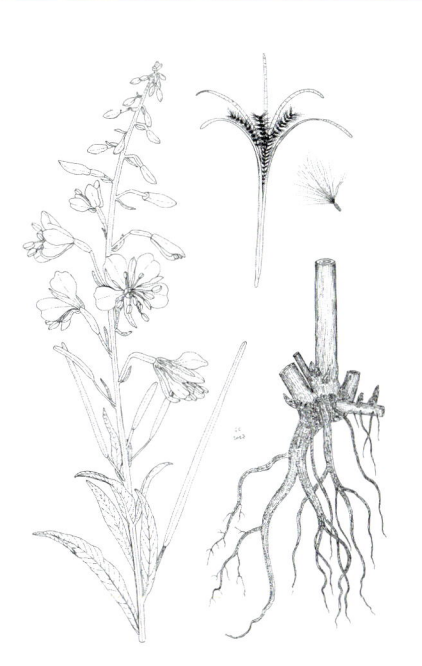

Synonyms *Chamaenerion angustifolium* (L.) Schur, *Chamaenerion angustifolium* (L.) Scop., *Chamaenerion denticulatum* Schur, *Chamaenerion spicatum* (Lam) Gray, *Chamerion angustifolium* (L.) Holub, *Epilobium antonianum* Pers., *Epilobium elatum* Munro ex Hausskn., *Epilobium gesneri* Vill., *Epilobium gracile* Brügger, *Epilobium leiostylon* Peterm., *Epilobium macrocarpum* Steph., *Epilobium persicifolium* Vill., *Epilobium rubrum* Lucé, *Epilobium salicifolium* Stokes, *Epilobium spicatum* Lam., *Epilobium variabile* Lucé, *Epilobium verticillatum* Ten.

Common Names Fireweed, great willowherb, rosebay willowherb.

Description Herb, perennial (5–22 dm), rhizomatous. Stems angularly erect or sparsely branched, glabrous. Leaves sessile, alternate, lanceolate, narrowed at the base, pointed at the apex, with entire margin, glaucescent below, with prominent lateral veins, at the top of the stem becoming bracts. Inflorescence long, dense terminal raceme. Flowers stalked; calyx reddish, downy, sepals reddish, lanceolate-linear; corolla consisting of 4 pinkish-purple petals, the upper 2 larger than the lower; style curved downward. Fruit oblong capsule, very downy, 4-valved, containing tiny seeds with feathery filaments for anemochorous dissemination.

Subspecies Two are known, the typical *Epilobium angustifolium* L. subsp. *angustifolium* and *Epilobium angustifolium* L. subsp. *circumvagum* Mosquin.

Related Species The genus *Epilobium* includes about 240 taxa.

Geographic Distribution Circumboreal. **Habitat** Disturbed humid montane areas, in large colonies at forest edges, in moist, stony, or gravelly places (600–2,500 m ASL). **Range** Circumboreal.

Beer-Making Parts (and Possible Uses) Various edible parts (young shoots, dried leaves, rhizome ends).

Chemistry Extract (mg/g of dry matter): L-ascorbic acid 17.66 mg, gallic acid 4.46 mg, protocatechuic acid 3.22 mg, hyperoside 3.53 mg, ellagic acid 1.18 mg, octyl gallate 0.08 mg.

Style Gose.

Source https://www.brewersfriend.com/other/fireweed/.

Epimedium grandiflorum C.Morren
BERBERIDACEAE

Geographic Distribution China, Japan, Korea. **Habitat** Undergrowth, shady places. **Range** Species widely cultivated for ornamental and phytotherapeutic purposes.

Beer-Making Parts (and Possible Uses) Inflorescences.

Toxicity In Italy, *Epimedium grandiflorum* is included in the list of plant extracts not allowed in food supplements.

Chemistry Some of the chemical compounds identified in *Epimedium*: flavonoids, glucides (19–31%), vitamin C, icariin (2.7%, active principle, which, according to traditional Chinese medicine, has aphrodisiac properties; however, there is a lack of scientific evidence).

Style California Common Beer.

Source https://www.brewersfriend.com/other/epimedium/.

Erica herbacea L.
ERICACEAE

Synonyms *Endoplectris tricolor* Raf., *Epimedium macranthum* C.Morren & Decne., *Epimedium pumilum* Baker.

Common Names Large flowered barrenwort, bishop's hat, longspur barrenwort.

Description Herb, perennial (up to 30 cm), deciduous, rhizomatous. Stem red in color. Leaves leathery, heart shaped, green, coppery when young, slightly hairy, spiny at the margins. Flowers with 4 long spurred petals, pink, white, yellow, or purple.

Subspecies In addition to the typical variety, *Epimedium grandiflorum* C.Morren var. *grandiflorum*, a second variety, *Epimedium grandiflorum* C.Morren var. *thunbergianum* (Miq.) Nakai, is accepted.

Cultivar Cultivars include Nanum, Orion, Purple Prince, Red Beauty, Rose Queen, Shiho, and White Queen.

Related Species The genus *Epimedium* includes about 70 taxa.

Synonyms *Erica carnea* L., *Erica lugubris* Salisb., *Erica mediterranea* L., *Erica mediterranea* Rchb. ex Nyman, *Erica purpurascens* L., *Erica saxatilis* Salisb.,

Ericoides herbaceum Kuntze, *Gypsocallis carnea* D.Don, *Gypsocallis mediterranea* D.Don, *Gypsocallis purpurascens* D.Don.

Common Names Winter heath, winter-flowering heather, spring heath, alpine heath.

Description Shrub (2–4 dm). Stem woody, creeping, glabrous. Leaves needle shaped (0.5 × 5–8 mm), patent or reflexed, shiny above, with revolute margin completely covering the leaf underside. Flowers in unilateral terminal racemes on 3 mm peduncles; sepals pinkish, acute, triveined (1 × 3 mm); corolla pinkish-red, 5–5.5 mm; anthers brown, projecting (1 mm), without appendages; style not projecting beyond anthers.

Cultivar There are numerous cultivars of this species with wide ornamental applications. Many are derived from forms that are part of the natural variability of the species: typical (Adrienne Duncan, Challenger, Loughrigg, Myretoun Ruby, Nathalie, Pink Spangles, Rosalie, Vivellii); "alba," with white flowers (Golden Starlet, Ice Princess, Isabell, Springwood White); and "aureifolia," with golden leaves (Foxhollow, Sunshine Rambler, Westwood Yellow). There are also many cultivars obtained by hybridization with other *Erica* species.

Related Species The genus *Erica* includes about 940 taxa.

Geographic Distribution Orophyte S Europe. **Habitat** Heaths, coniferous forests, dry calcareous meadows and pastures (0–2,650 m ASL). **Range** Orophyte S Europe.

Beer-Making Parts (and Possible Uses) Flowering branches (*Calluna vulgaris* substitute).

Chemistry Leaves and flowers of the plant contain phenols 120 mg (expressed in mg of gallic acid/g of extract) and flavonoids 26.90 mg (expressed in mg of rutin/g of extract) that exert an important antioxidant and antibacterial action.

Style Gruit, Heath Beer.

Source Fisher and Fisher 2017.

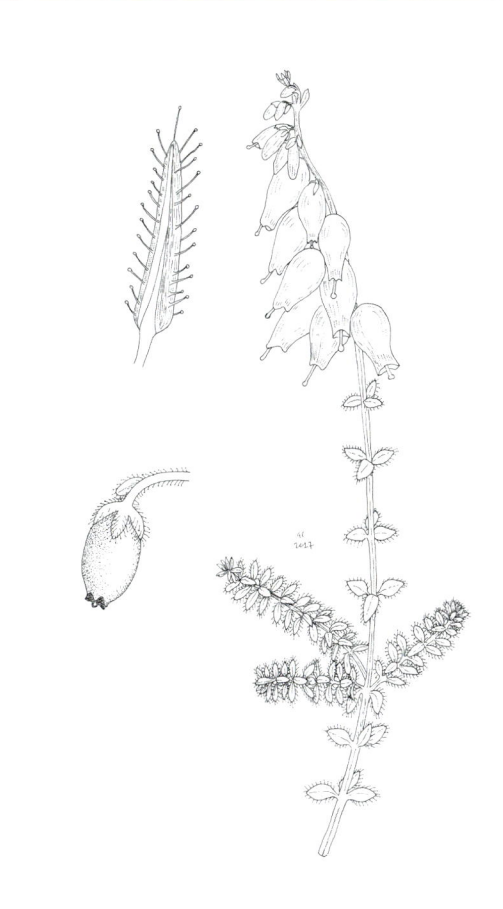

Erica tetralix L.
ERICACEAE

Synonyms *Eremocallis glomerata* Gray, *Erica botuliformis* Salisb., *Erica calycinades* J.Forbes, *Erica glomerata* Salisb., *Erica martinesii* Lag. ex Benth., *Erica rubella* Ker Gawl., *Ericoides glomeratum* (Andrews) Kuntze, *Ericoides mackeyi* Kuntze, *Ericoides tetralix* (L.) Kuntze.

Common Names Bog heather, cross-leaved heath.

Description Shrub (30–80 cm), stems twisted, pubescent or often hairy-glandular. Leaves in whorls, 4.4 mm long, linear-oblong, white underneath, bristling with stiff cilia. Flowers pink, clustered (5–12) in a terminal umbel inflorescence; peduncles shorter than flowers; calyx with lanceolate lobes, long ciliate, 3–4 times shorter than the campanulate-oval corolla (5–7 mm), with short, reflexed teeth, anthers included, stylus slightly projecting with stigma. Silky-hairy fruit capsule.

Cultivar There are several cultivars of this species, of which Alba Mollis and Con Underwood are two.

Related Species The genus *Erica* includes about 940 taxa.

Geographic Distribution WC Europe. **Habitat** Acidic siliceous wetlands and grasslands. **Range** WC Europe.

Beer-Making Parts (and Possible Uses) Leaves.

Chemistry Little information available. Aerial parts: waxes 1.36–1.42%, ursolic acid 1.5%.

Style Gruit.

Source https://www.gruitale.com/bot_bell_heather.htm.

Ericameria nauseosa (Pall. ex Pursh) G.L.Nesom & G.I.Baird
ASTERACEAE

Synonyms *Bigelowia nauseosa* M.E.Jones, *Chondrophora nauseosa* (Pall. ex Pursh) Britton, *Chrysocoma nauseosa* Pall. ex Pursh, *Chrysothamnus collinus* Greene, *Chrysothamnus concolor* (A.Nelson) Rydb., *Chrysothamnus pallidus* A.Nelson, *Chrysothamnus plattensis* (Greene) Greene.

Common Names Rabbitbrush, rubber rabbitbrush.

Description Shrub (10–250 cm). Stems erect or ascending, white to green in color, strongly branched, intricate, tomentose. Leaves (usually crowded) ascending or spreading, alternate, simple, grayish-green, lamina filiform to narrowly lanceolate (10–70 × 0.3–10 mm), leaf upper side furrowed to concave, midrib evident, apex acute, faces glabrous or tomentose, often punctate with glands; axillary fascicles absent. Inflorescence terminal top paniculate, roundish or with flattened apex (up to 12 cm), consisting of tubular flowers only (ligulate flowers absent); peduncles 1–20 mm (bracts absent or sometimes 1–5), reduced, squamiform; involucre obconic to subcylindric (6–16 × 2–4 mm), phyllaries 10–31, in 3–5 series (often in vertical ranks), brownish color, ovate to lanceolate (1.5–14 × 0.7–1.5 mm), highly unequal, often papery and keeled, midrib prominent for almost their entire length, apically expanded, apices acute to obtuse, abaxial face resinous. Disc flowers 4–6, yellow, corolla 6–12 mm, 5-lobed. Fruit cypsela, light brown, turbinate to cylindrical or oblanceoloid (3–8 mm), glabrous or hairy (often ± hairy or silky); pappus consisting of numerous capillary bristles (3–13 mm), whitish in color.

Subspecies *Ericameria nauseosa* includes considerable subspecific variability as demonstrated by the large number of varieties (24) in addition to the typical one (var. *nauseosa*): *Ericameria nauseosa* (Pall. ex Pursh) G.L.Nesom & G.I.Baird var. *ammophila* L.C.Anderson, *Ericameria nauseosa* (Pall. ex Pursh) G.L.Nesom & G.I.Baird var. *arenaria* (L.C.Anderson) G.L.Nesom & G.I.Baird, *Ericameria nauseosa* (Pall. ex Pursh) G.L. Nesom & G.I.Baird var. *arta* (A.Nelson) G.L.Nesom & G.I.Baird, *Ericameria nauseosa* (Pall. ex Pursh) G.L.Nesom & G.I.Baird var. *bernardina* (H.M.Hall) G.L.Nesom & G.I.Baird, *Ericameria nauseosa* (Pall. ex Pursh) G.L.Nesom & G.I.Baird var. *bigelovii* (A.Gray) G.L.Nesom & G.I.Baird, *Ericameria nauseosa* (Pall. ex Pursh) G.L.Nesom & G.I.Baird var. *ceruminosa* (Durand & Hilg.) G.L.Nesom & G.I.Baird, *Ericameria nauseosa* (Pall. ex Pursh) G.L.Nesom & G.I.Baird subsp. *consimilis* (Greene) G.L.Nesom & G.I.Baird, *Ericameria nauseosa* (Pall. ex Pursh) G.L. Nesom & G.I.Baird var. *graveolens* (Nutt.) Reveal & Schuyler, *Ericameria nauseosa* (Pall. ex Pursh) G.L.Nesom & G.I.Baird var. *hololeuca* (A.Gray) G.L.Nesom & G.I.Baird, *Ericameria nauseosa* (Pall. ex Pursh) G.L.Nesom & G.I.Baird var. *iridis* (L.C.Anderson) G.L.Nesom & G.I.Baird, *Ericameria nauseosa* (Pall. ex Pursh) G.L.Nesom & G.I.Baird var. *juncea* (Greene) G.L.Nesom & G.I.Baird, *Ericameria* *nauseosa* (Pall. ex Pursh) G.L.Nesom & G.I.Baird var. *latisquamea* (A.Gray) G.L.Nesom & G.I.Baird, *Ericameria nauseosa* (Pall. ex Pursh) G.L.Nesom & G.I.Baird var. *leiosperma* (A.Gray) G.L.Nesom & G.I.Baird, *Ericameria nauseosa* (Pall. ex Pursh) G.L.Nesom & G.I.Baird var. *mohavensis* (Greene) G.L.Nesom & G.I.Baird, *Ericameria nauseosa* (Pall. ex Pursh) G.L.Nesom & G.I.Baird var. *nana* (Cronquist) G.L.Nesom & G.I.Baird, *Ericameria nauseosa* (Pall. ex Pursh) G.L.Nesom & G.I.Baird var. *nitida* (L.C.Anderson) G.L.Nesom & G.I.Baird, *Ericameria nauseosa* (Pall. ex Pursh) G.L.Nesom & G.I.Baird var. *oreophila* (A.Nelson) G.L.Nesom & G.I.Baird, *Ericameria nauseosa* (Pall. ex Pursh) G.L. Nesom & G.I.Baird var. *psilocarpa* (S.F.Blake) G.L.Nesom & G.I.Baird, *Ericameria nauseosa* (Pall. ex Pursh) G.L. Nesom & G.I.Baird var. *salicifolia* (Rydb.) G.L.Nesom & G.I.Baird, *Ericameria nauseosa* (Pall. ex Pursh) G.L.Nesom & G.I.Baird var. *speciosa* (Nutt.) G.L. Nesom & G.I.Baird, *Ericameria nauseosa* (Pall. ex Pursh) G.L.Nesom & G.I.Baird var. *texensis* (L.C.Anderson) G.L.Nesom & G.I.Baird, *Ericameria nauseosa* (Pall. ex Pursh) G.L.Nesom & G.I.Baird var. *turbinata* (M.E.Jones) G.L.Nesom & G.I.Baird, and *Ericameria nauseosa* (Pall. ex Pursh) G.L.Nesom & G.I.Baird var. *washoensis* (L.C.Anderson) G.L.Nesom & G.I.Baird.

Related Species The genus *Ericameria* includes 78 taxa.

Geographic Distribution W North America (United States, Canada), NW Mexico. **Habitat** Plains, hills, valleys, bare alkaline or gypsiferous soils, arid calcareous cliffs, rock fissures, mesas, sandy gravels of dry riverbeds, dunes and deep sands, other arid, semi-arid, cold habitats (50–3,500 m ASL). **Range** Overlapping the geographic distribution.

Beer-Making Parts (and Possible Uses) Flowers, roots, leaves.

Toxicity The toxicity of this plant to grazing animals has been reported in some cases.

Chemistry The great taxonomic variability of this taxon, as well as the breadth of habitat, is reflected in an equally wide phytochemical variability. Leaves (the following list includes all compounds found in the subspecies of this taxon and other related taxa): α-pinene, β-pinene, myrcene, α-phellandrene, β-phellandrene, limonene, Δ-3-carene, 4,7-methaneazulene-1,2,3,4,5,6,7,8-octahydro-1,4,9,9-tetramethyl, β-caryophyllene, α-humulene, β-cubebene, cyclohexane, β-elemene, γ-muurolene, Δ-cadinene, cyclohexanomethanol, nerolidol, α-elemene, elemol, decanoic acid, methyl-11,14-eicosadienoate, 1-eicosine, eicosane, various naphthalenic acid derivatives, tetradecanal, petacosane, agatolic-like acid, 2-bromo-octadecanal, lycopersene, nonacosane, entriacontane, olean-12-ene, other compounds.

Style None verified (?).

Source Hieronymus 2016b.

Eriobotrya japonica (Thunb.) Lindl.
ROSACEAE

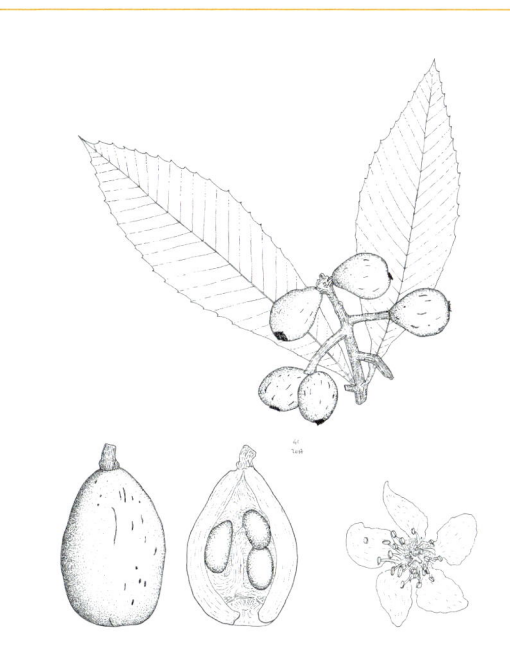

Synonyms *Crataegus bibas* Lour., *Mespilus japonica* Thunb., *Photinia japonica* (Thunb.) Franch. & Sav.

Common Names Loquat.

Description Tree or shrub (1–8 m). Branches tomentose. Leaves evergreen, narrowly elliptic (3–6 × 12–25 cm), margin toothed, dark green above, reddish tomentum underneath. Inflorescence panicle dense, pyramidal. Flowers with yellowish-white petals (4 mm). Fruit ellipsoid (3–6 cm), orange-yellow in color, flesh edible.

Cultivar In Asia there are more than 800.

Related Species The genus *Eriobotrya* includes 12 taxa.

Geographic Distribution China. **Habitat** Mainly cultivated plant. **Range** Japan, Israel, Brazil, Spain (cultivated and naturalized in many countries: Japan, Armenia, Afghanistan, Australia, Azerbaijan, Bermuda, Chile, Kenya, India, Iran, Iraq, South Africa, Mediterranean basin, Pakistan, New Zealand, Réunion, Tonga, Central America, Mexico, South America, warmest states of United States [Hawaii, California, Texas, Louisiana, Alabama, Florida, Georgia, South Carolina]).

Beer-Making Parts (and Possible Uses) Fruits.

Chemistry Energy 197 kJ (47 kcal), carbohydrates 12.14 g, fiber 1.7 g, lipids 0.2 g, protein 0.43 g, vitamin A 76 μg, thiamine (B1) 0.019 mg, riboflavin (B2) 0.024 mg, niacin (B3) 0.18 mg, vitamin B6 0.1 mg, folate (B9) 14 μg, vitamin C 1 mg, calcium 16 mg, iron 0.28 mg, magnesium 13 mg, manganese 0.148 mg, phosphorus 27 mg, potassium 266 mg, sodium 1 mg, zinc 0.05 mg.

Style American Pale Ale.

Source https://www.brewersfriend.com/other/loquat/.

Eryngium foetidum L.
APIACEAE

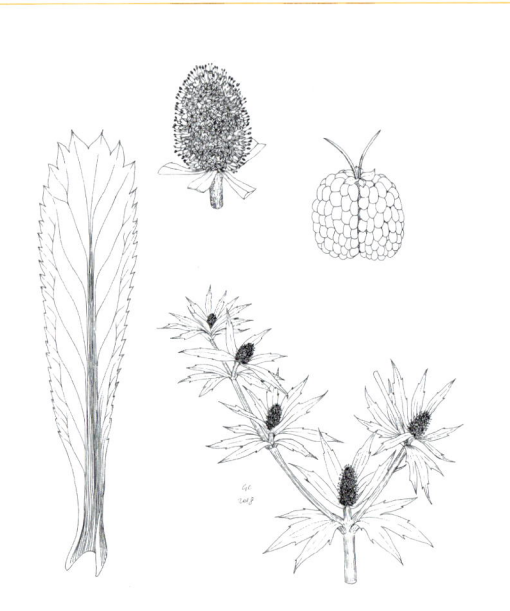

Synonyms *Eryngium antihystericum* Rottler.

Common Names Culantro, Mexican coriander, long coriander, sawtooth coriander, fitweed.

Description Herb, perennial (up to 8 dm). Taproot fusiform, often branched. Stem erect, glabrous, light to dark green, fluted, elongated before flowering, repeatedly branched to dichasium at the end. Leaves simple, subsessile to petiolate, those in basal rosette with fetid odor when crushed (rosettes disappear in older plants); lamina lanceolate-obovate to spatulate-oblong (5–32 × 1–4 cm), base narrowly involucral, margin spiny serrate. Inflorescence terminal capitulum spike shaped (actually extremely contracted umbel), eventually combined into widely branched corymb, with 2 bracts at each branching; bracts rigid (1–6 cm), deeply incised, aculeate-dentate, veins evident, the lower ones resembling normal leaves; pedunculate to 1 cm; involucral bracts 5–7, lanceolate (1–3 cm × 3–7 mm), with some spiny teeth; rays absent; capitulum cylindrical (1–2 cm × 3–5 mm), multifloral, terminal on a branch or short peduncle in the middle of the dichotomy; involucral bracts linear to lanceolate (2–3 mm), below the sessile flower; capitula tubular (1 mm), with 5 small, erect, triangular, persistent teeth; petals 5, elliptic-oblong (0.5–0.75 × 0.25 mm), greenish-white in color, apex incurved; stamens 5, larger than corolla, filaments white in color; pistil with 2 filiform styles. Fruit schizocarp ovoid-obovoid (up to 1.5 × 0.75 mm), densely tuberculate, opening into 2 semiglobose mericarps.

Related Species The genus *Eryngium* includes more than 270 taxa.

Geographic Distribution Mexico, South America, Caribbean. **Habitat** Partly shady, fertile, well-watered environments; meadows, plantations, dumps, roadsides, forest edges (0–1,700 m ASL). **Range** India, Vietnam, Australia, SE Asia, Africa, South America.

Beer-Making Parts (and Possible Uses) Leaves.

Chemistry Leaves: flavonoids, tannins, saponins, tri-terpenoids; Essential oil, leaves (94.8% of total): (E)-2-dodecenal 59.7%, 2,3,6-trimethylbenzaldehyde 9.6%, dodecanal 6.7%, (E)-2-tridecenal 4.6%, 2-formyl-1,1,5-trimethylcyclohexa-2,4-dien-6-ol 3.5%, 2-formyl-1,1,5-trimethylcyclohexa-2,5-dien-4-ol (ferulol) 2.1%, 2,3,4-trimethylbenzaldehyde 1.8%, decanal 1.7%, 2-methylchrotonic acid 1.3%, tetradecanal 0.7%, phenylacetaldehyde 0.6%, (E)-2-undecenal 0.4%, 3,4-dimethylbenzaldehyde 0.4%, 2-formyl-1,1,5-trimethylcyclohexa-2,4-dien-6-one 0.4%, γ-decalactone 0.2%, (E)-2-decenal 0.1%, hexanal 0.1%,

3,5,5-trimethyl-3-cyclohexen-1-one 0.1%, (Z)-3-hexenol 0.1%, nonanal 0.1%, 1,2,4-trimethylbenzene 0.1%, 2-octanone 0.1%, 3,5,5-trimethyl-2-cyclohexen-1-one (isophorone) 0.1%, 2,6,6-trimethyl-1,3-cyclohexadiene-1-carboxyaldehyde (safranal) 0.1%, terpinen-4-ol 0.1%, octanal, (E)-2-heptenal, hexanol, benzaldehyde, heptanal, p-cymene, 1-octen-3-ol, (E,E)-2,4-heptadienal, 2-octylfuran, (E)-2-nonenal, 1,2,3-trimethylbenzene, (E,E)-2,4-dodecadienal, vanillin.

Style American Pale Ale, American Wheat or Rye Beer, Irish Red Ale.

Source https://www.brewersfriend.com/other/culantro/.

Erythroxylum coca Lam.
ERYTHROXYLACEAE

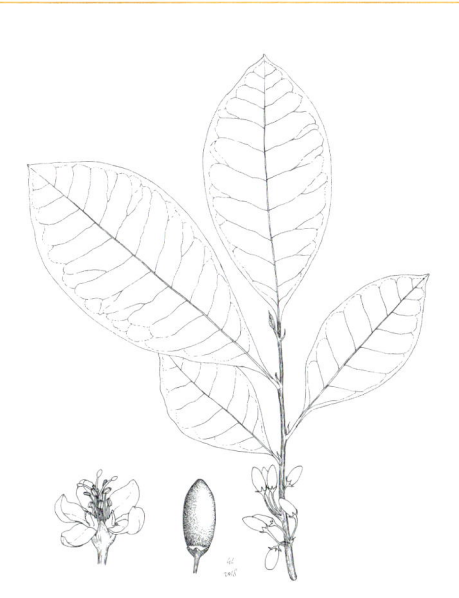

Synonyms *Erythroxylum bolivianum* Burck, *Erythroxylum chilpei* E.Machado.

Common Names Coca-bush.

Description Shrub or small tree (1–3 m). Stem much branched, bark reddish color. Leaves alternate, deep green, oval, entire, isolated. Flowers small, pentamerous, yellowish-white, actinomorphic, isolated or grouped at the leaf axil; 10 stamens concrescent at the base; 3 carpels forming a trilocular ovary that at maturity presents only 1 locule. Fruit small reddish drupe containing a single seed.

Subspecies Subspecific entities of this taxon include *Erythroxylum coca* Lam. var. *coca*, *Erythroxylum coca* Lam. var. *novogranatense* D.Morris, and *Erythroxylum coca* Lam. var. *spruceanum* Burck. According to some authors, simple cultivars should more properly be taken into consideration, whereas according to others, at least some would have their own autonomy.

Related Species The genus *Erythroxylum* includes about 260 species. With the exception of *E. coca*, no use in making beer is known.

Geographic Distribution Bolivia, Peru. **Habitat** Tropical environments. **Range** Bolivia, Peru, Chile, Colombia, Brazil, China, Indonesia, Madagascar.

Beer-Making Parts (and Possible Uses) Leaves.

Toxicity Although there are doubts about their toxicity, in much of the world the use of the leaves is illegal, as is the purified form of the drug extracted from the leaves by a complex chemical process. This extract, cocaine, has powerful euphoric effects, is addictive, and can cause serious damage to the central nervous system. The legal use of the leaves is limited to certain countries (or regions) and some indigenous ethnic groups (Arhuaco, Aymara, Kogi, Wiwa) where traditional consumption is well established.

Chemistry Leaves: cocaine (0.3–1.5%, average 0.8%), benzoylecgonine, benzoyltropine, cinnamilcocaine, cuscoigrin, dihydroxytropane, igrin, igroline, methylcocaine, methylecgonidine, nicotine, tropacocaine, A-truxillin, B-truxillin. According to other sources, only four alkaloids are present in the leaves: cocaine, norcocaine, cis-cinnamyl cocaine, trans-cinnamyl cocaine (these sources state that the other compounds are substances produced during analysis because of the analytical techniques used); coca leaves also contain several nutrients.

Style Foreign Extra Stout, Oatmeal Stout, Russian Imperial Stout, Sweet Stout.

Source https://www.brewersfriend.com/other/coca/.

Eucalyptus camaldulensis Dehnh.
MYRTACEAE

Synonyms *Eucalyptus acuminata* Hook., *Eucalyptus longirostris* F.Muell. ex Miq., *Eucalyptus mcintyrensis* Maiden.

Common Names Red gum, red river rum, gum, Murray red gum, river red eucalyptus, river red, Australian kino, Australian red rum, Botany Bay kino, eucalyptus kino, kino australiensis.

Description Tree (up to 20 m, occasionally 45 m). Trunk (1–1.5 m diameter) covered with smooth bark, white, gray, brown, or red in color. Juvenile leaves ovate to broadly lanceolate, green, gray-green, or blue-green in color; adult leaves lanceolate to narrowly lanceolate, acuminate, moderately thick; lamina 8–30 × 0.7–2 cm, green or gray-green in color; lateral veins at 40–50 degrees; intramarginal veins up to 2 mm from margin; petiole terete or grooved, 12–15 mm. Inflorescence umbel with 7–11 flowers, peduncle elongate, terete, or quadrangular, 6–15 mm; pedicels slender, 5–12 mm. Flower (floral bud) globular-rostrate or ovoid-conical; operculum hemispherical, rostrate or conical, obtuse, 4–6 × 3–6 mm; hypanthium hemispherical, 2–3 × 3–6 mm. Fruit capsule hemispherical or ovoid, 5–8 × 5–8 mm, disc broad, ascending, valves 3–5. Seeds yellow.

Subspecies According to some sources, *Eucalyptus camaldulensis* Dehnh. subsp. *camaldulensis* and *Eucalyptus camaldulensis* Dehnh. subsp. *obtusa* (Blakely) Brooker & M.W.McDonald should be the only two taxa of subspecific rank of this entity. According to a recent revision, there should be seven subspecies: *Eucalyptus camaldulensis* Dehnh. subsp. *acuta* Brooker & M.W.McDonald, *Eucalyptus camaldulensis* Dehnh. subsp. *arida* Brooker & M.W.McDonald, *Eucalyptus camaldulensis* Dehnh. subsp. *camaldulensis*, *Eucalyptus camaldulensis* Dehnh. subsp. *minima* Brooker & M.W.McDonald, *Eucalyptus camaldulensis* Dehnh. subsp. *obtusa* (Blakely) Brooker & M.W.McDonald, *Eucalyptus camaldulensis* Dehnh. subsp. *refulgens* Brooker & M.W.McDonald, and *Eucalyptus camaldulensis* Dehnh. subsp. *simulata* Brooker & Kleinig.

Related Species The genus *Eucalyptus* comprises more than 850 taxa.

Geographic Distribution Australia. **Habitat** Permanent or seasonal riparian environments, near or immediately adjacent to streams; also in drier environments. **Range** Australia, Africa (Algeria, Egypt, Ethiopia, Benin, Botswana, Cameroon, Canary Islands, Kenya, Libya, Madagascar, Malawi, Morocco, Mozambique, Niger, Rwanda, Senegal, Somalia, South Africa, Sudan, Tanzania, Tunisia, Zambia, Zimbabwe, Uganda), Asia (Caucasus, China, India, Iran, Iraq, Israel, Oman, Saudi Arabia, Seychelles, Taiwan, Yemen), Europe (Cyprus, France, Greece, Italy, Mediterranean, Portugal, Spain, Turkey), America (Brazil, Costa Rica, Ecuador, Guyana, Mexico, Nicaragua, Panama, United States, Venezuela).

Beer-Making Parts (and Possible Uses) Wood tested for wine vessels with little success (use uncertain for making beer); inflorescences.

Wood More or less dark red in color, medium texture, various or intertwined grain (WSG 0.67–0.87), with strong shrinkage that makes it difficult to obtain good seasoning without cracks, deformations, and collapse phenomena; the material coming from adult trees behaves better; considerable internal tensions in the stems occur upon cutting or immediately afterward. Despite its hardness, it has good workability. Used for structural elements, constructions, railway sleepers, parquet, and road paving.

Style Sour Ale.

Beer Brae, Edge Brewing Project (Melbourne, Australia).

Source Calagione et al. 2018; Cantwell and Bouckaert 2016; Cantwell and Bouckaert 2018.

> 🍺 **"Zero Kilometer" Eucalyptus**
> Brewers sometimes collaborate to brew special products, combining ingredients and inspirations from different territories and cultures. Increasingly, some are opening the doors of their facilities not to colleagues but to restaurateurs to create beverages that can be paired with the restaurant's specialties. In Melbourne, the Edge Brewing Project partnered with the restaurant Brae. The result is a beer that uses honey and *Eucalyptus camaldulensis* flowers collected from the garden behind the restaurant and added during fermentation to provide a floral and aromatic note.

Eucalyptus diversicolor F.Muell.
MYRTACEAE

Synonyms *Eucalyptus colossea* F.Muell.

Common Names Karri.

Description Tree (up to 90 m). Adult leaves broadly lanceolate, acuminate; lamina 9–12 × 2–3.2 cm; lateral veins conspicuous, 35–50°; intramarginal veins up to 1 mm from margin; petiole fluted, 10–20 mm. Inflorescence with flattened or angular peduncle, 18–28 mm; pedicels 3–6 mm; operculum conical, 5–7 × 5–7 mm; hypanthium cylindrical to obconic, 7–8 × 5–7 mm. Fruit capsule 8–12 × 7–10 mm.

Related Species The genus *Eucalyptus* includes more than 850 taxa.

Geographic Distribution SW Western Australia. **Habitat** High open forests in hilly environment. **Range** Species introduced for forestry purposes in various locations outside Australia. In some cases (e.g., South Africa) it tends to become invasive.

Beer-Making Parts (and Possible Uses)
Wood tested for wine vessels with little success (uncertain use in making beer).

Wood Variable in color from light brown to reddish, braided grain (WSG 0.85), compact, with high mechanical strength but not durable against soil and termites; tends to split during seasoning, not easy to work. Suitable for construction, flooring, tools, packaging, plywood.

Style None verified.

Source Cantwell and Bouckaert 2016; Cantwell and Bouckaert 2018.

Eucalyptus regnans F.Muell.
MYRTACEAE

SYNONYMS None
COMMON NAMES Mountain ash.

DESCRIPTION Tree (usually up to 75 m, specimens recorded up to 100 m). Trunk with rough bark to about 15 m, smooth, white, or gray-green above, often detaching in strips. Adult leaves lanceolate or broadly lanceolate, lamina 9–14 × 1.6–2.7 cm; petiole fluted, 12–22 mm. Inflorescence from angular peduncle, 5–13 mm; pedicels 2–4 mm; operculum conical, 2–3 × 3–4 mm; hypanthium obconic about 3 × 3–4 mm. Fruit 5–9 × 4–7 mm.

RELATED SPECIES The genus *Eucalyptus* includes more than 850 taxa.

GEOGRAPHIC DISTRIBUTION Australia (S Victoria, Tasmania). **HABITAT** High open forests, often in pure formations. **RANGE** Australia (S Victoria, Tasmania).

BEER-MAKING PARTS (AND POSSIBLE USES)
Wood tested for wine vessels with little success (uncertain use in making beer).

WOOD Rosy brown in color, medium texture (WSG 0.70); suitable for furniture, carpentry, flooring, veneers, packaging, wood wool, mechanical pulp, cellulose.

STYLE None verified.

SOURCE Cantwell and Bouckaert 2016; Cantwell and Bouckaert 2018.

Eugenia uniflora L.
MYRTACEAE

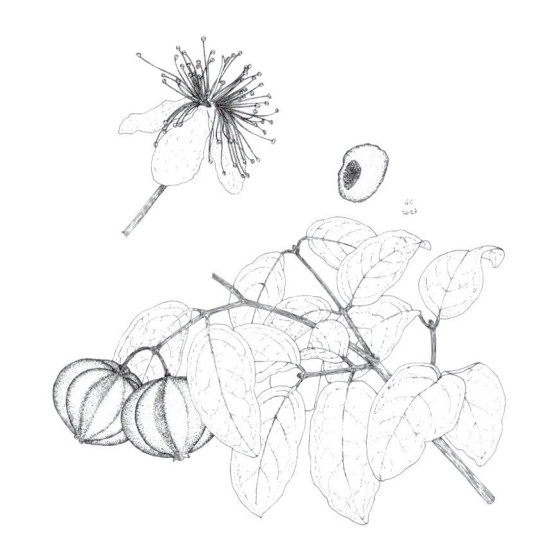

SYNONYMS *Eugenia arechavaletae* Herter, *Eugenia costata* Cambess., *Eugenia dasyblasta* (O.Berg) Nied., *Eugenia decidua* Merr., *Eugenia indica* Nicheli, *Eugenia lacustris* Barb. Rodr., *Eugenia michelii* Lam., *Eugenia microphylla* Barb. Rodr., *Eugenia myrtifolia* Salisb., *Eugenia oblongifolia* (O.Berg) Arechav., *Eugenia strigosa* (O.Berg) Arechav., *Eugenia zeylanica* Willd., *Luma arechavaletae* (Herter) Herter, *Luma costata* (Cambess.) Herter, *Luma dasyblasta* (O.Berg) Herter, *Luma strigosa* (O.Berg) Herter, *Myrtus brasiliana* L., *Myrtus willdenowii* Spreng, *Plinia pedunculata* L.f., *Plinia rubra* L., *Plinia tetrapetala* L., *Stenocalyx affinis* O.Berg, *Stenocalyx brunneus* O.Berg, *Stenocalyx costatus* (Cambess.) O.Berg, *Stenocalyx dasyblastus* O.Berg, *Stenocalyx glaber* O.Berg, *Stenocalyx impunctatus* O.Berg, *Stenocalyx lucidus* O.Berg, *Stenocalyx michelii* (Lam.) O.Berg, *Stenocalyx nhampiri* Barb. Rodr., *Stenocalyx oblongifolius* O.Berg, *Stenocalyx rhampiri* Barb.Rodr., *Stenocalyx ruber* (L.) Kausel, *Stenocalyx strigosus* O.Berg, *Stenocalyx uniflorus* (L.) Kausel, *Syzygium michelii* (Lam.) Duthie.

Common Names Pitanga, Suriname cherry, Brazilian cherry, Cayenne cherry.

Description Shrub or small (6–8 m) evergreen tree. Stem heavily branched, with dense foliage. Leaves opposite, simple, leathery, aromatic, ovate or ovate-lanceolate (2–5 × 1–3 cm), margin overall, bronze-pink color, then glossy deep green above, paler underneath; petiole 0.2–0.4 cm. Flowers axillary, bisexual, solitary or in pedunculate clusters (1.5–3 cm); calyx persistent, with 4 retroflexed lobes (0.4 cm); corolla with 4 obovate white petals (about 1 cm), ephemeral, numerous stamens (0.8 cm). Fruit globular berry flattened at both poles (1.5 × 2–4 cm), usually with 8 longitudinal ribs, color first green, then orange, finally red when ripe; skin thin, pulp red, juicy, sweet to acidulous, slightly resinous, containing 1 globular seed (Ø 0.5–1 cm), rarely 2–3.

Cultivar *Eugenia uniflora* exhibits considerable variability in fruit size, quality, productivity, and other characteristics. Some of the recognized cultivars are Lolita (medium-sized fruit, Ø 2.5 cm, weight 5.6 g, dark red color), Lorver (large fruit, Ø > 2.5 cm, weight 6.6 g, orange-red color), and Vermillion (large fruit, Ø > 2.5 cm, weight 7.3 g, orange-red color).

Related Species The genus *Eugenia* includes more than 1,000 taxa.

Geographic Distribution South America (N Argentina, Bolivia, E and S Brazil, Paraguay, Uruguay). **Habitat** Forest margins, dune systems, arid and semi-arid environments of NE Brazil, along riverbanks where it forms dense thickets in areas with medium to high levels of precipitation. **Range** Introduced in various countries with tropical, subtropical, and Mediterranean climates for ornamental and fruit cultivation; in some cases it has become invasive.

Beer-Making Parts (and Possible Uses) Fruits, leaves (?).

Toxicity The seeds have been reported as toxic to some people and therefore should be avoided.

Chemistry Fruit: energy 33 kcal, lipids 0.4 g, carbohydrates 7 g, protein 0.8 g, vitamin A 1,500 IU, vitamin C 26.3 mg, calcium 9 mg, iron 0.2 mg, magnesium 12 mg, sodium 3 mg, potassium 103 mg. Essential oil (leaves contain 32 compounds, representing 92.65% of the oil): sesquiterpenes 91.92% (curzerene 47.3%, γ-elemene 14.25%, trans-β-elemenone 10.4%).

Style American Pale Ale.

Beer Baca, Coruja (Porto Alegre, Brazil).

Source https://www.brewersfriend.com/other/pitanga/.

Eutrema japonicum (Miq.) Koidz.
BRASSICACEAE

Synonyms *Alliaria wasabi* (Maxim) Prantl, *Eutrema okinosimense* Taken., *Eutrema wasabi* Maxim., *Eutrema wasabii* Maxim., *Lunaria japonica* Miq., *Wasabia japonica* (Miq.) Matsum., *Wasabia pungens* Matsum., *Wasabia wasabi* (Siebold) Makino.

Common Names Japanese horseradish, wasabi.

Description Herb, perennial (20–75 cm), glabrous or sparsely hairy on upper parts. Rhizome fleshy (Ø 2 cm). Stem erect to decumbent, simple. Basal leaves rosulate; petiole (6–26 cm) dilated at base; lamina cordate to reniform (2.5–20 × 3–22 cm), base cordate, margin toothed, denticulate, slightly crenate, repand or subintermediate, with distinct apiculate callosities terminating the veins, apex rounded to obtuse; cauline leaves medium size with petiole 1–8 cm, lamina ovate to ovate-cordate (1.5–6 × 2–6 cm), with palmate veins, base and margins like basals, apex acute. Inflorescence raceme bracted throughout or only basally, slender, elongate at fruiting; fruit peduncles ascending to spreading, slender (1–5 cm). Flowers with oblong sepals (3–4 × 2–2.5 mm), deciduous; petals white, oblong-spatulate (6–9 × 2–3 mm); filaments white (3.5–5 mm); anthers oblong (0.6–0.8 mm); ovules 6–8 per ovary. Fruit siliqua, linear (1–2 cm × 1.5–2 mm), torulose. Seed oblong (2–3 × 1–1.5 mm).

Cultivar The two most commercially important are Daruma and Mazuma, but there are many others.

Related Species *Cardamine pseudowasabi* H.Shin & Y.D.Kim; the genus *Eutrema* includes 32 taxa.

Geographic Distribution Japan. **Habitat** Along mountain streams and in other cold, moist environments (0–2,500 m ASL). **Range** N Japan, China (part), Taiwan, Korea, New Zealand; in North America, the rainforests of the Oregon Coast and part of the Blue Ridge Mountains (North Carolina, Tennessee) provide the mix of conditions needed to grow this species, but at present only limited success in greenhouses or with hydroponics has been recorded.

Beer-Making Parts (and Possible Uses) Roots, leaves (?).

Chemistry Energy 292 kcal, water 31.7 g, lipids 0.6 g, carbohydrates 24 g, fiber 8 g, protein 4.8 g, vitamin A 35 IU, vitamin C 41.9 mg, vitamin B6 0.3 mg, magnesium 69 mg, sodium 17 mg, potassium 568 mg, calcium 128 mg, iron 1 mg. Aromatic compounds: 6-methylthiohexyl isothiocyanate, 7-methylthioheptyl isothiocyanate, 8-methylthiooctyl isothiocyanate.

Style Saison, Specialty IPA (White IPA).

Beer Wasabi Ale, Miyamori (Iwate, Japan).

Source https://www.brewersfriend.com/other/wasabi/.

Fagopyrum esculentum Moench
POLYGONACEAE

Synonyms *Fagopyrum cereale* Raf., *Fagopyrum dryandrii* Fenzl, *Fagopyrum emarginatum* (Roth) Meisn., *Fagopyrum emarginatum* Moench, *Fagopyrum sarracenicum* Dumort., *Fagopyrum vulgare* T.Nees, *Polygonum emarginatum* Roth, *Polygonum fagopyrum* L.

Common Names Buckwheat.

Description Herb, annual (2–6 dm). Stem erect, eventually reddened, branched above. Leaves with 1–3 cm long petiole and hastate triangular lamina (3–6 × 4–7 cm), acute, ochreous as long as stem diameter. Inflorescence in short raceme forming a leafy panicle to corymb. Flower with 3–4 mm perianth. Fruit achene, 5–6 mm, brown, eventually opaque.

Cultivar In North America there are many varieties of this species, including recent ones: Common (smaller seeds used for flour), Devyatka, Horizon (large seeds with high yields in North Dakota and Canada), Keukett (new variety with great commercial success), Koban, Koma (large seeds and higher yields than Mancan), Koto (available on the U.S. market since 2002, developed by Cornell University and Kade Research, more productive and stress resistant than Manisoba), Manisoba (more productive than Manor, a major variety of the 1990s, has large seeds required by the international market), Manor (dominant variety in the 1990s in the NE United States, has large seeds commercially in demand), Pennquad, Springfield (large seeds, tested in North Dakota), Tempest, Tokyo, and Winsor Royal. Of course, in other localities where the species is cultivated, there are cultivars adapted to local conditions (climate, resistance, use).

Related Species The genus *Fagopyrum* includes 27 taxa, among which *Fagopyrum tataricum* (L.) Gaertn. can be used in beer making.

Geographic Distribution C Asia. **Habitat** Cultivated in mountainous areas. **Range** Russia, China, Ukraine, United States, Kazakhstan, Poland, Japan, Brazil, Lithuania, France, Tanzania, Belgium, Nepal, Latvia, Bhutan, South Korea, Slovenia, Czech Republic, Estonia, Bosnia and Herzegovina, South Africa, Hungary, Croatia, Slovakia, Georgia, Moldova, Kyrgyzstan, Italy.

Beer-Making Parts (and Possible Uses) Seeds.

Toxicity A mild sensitization, called fagopyrinism, may result from consumption of fagopyrins; a symptom is a rash from exposure to sunlight. The leaves contain more fagopyrins than do the seeds, so the problem may occur more readily in animals eating the plant. However, it has also been reported in people who make heavy use of the young shoots (or their juice) as food. Cases of severe allergic reactions to this species and cases of cross-reactions to products containing this plant and other allergens have been reported (e.g., poppy seeds, rice, cashews, walnuts, sesame seeds).

Chemistry Energy 1,435 kJ (343 kcal), carbohydrates 71.5 g, fiber 10 g, lipids 3.4 g (saturated 0.741 g, monounsaturated 1.04 g, polyunsaturated 1.039 g, omega 3 0.078 g, omega 6 0.961 g), protein 13.25 g, thiamin (B1) 0.101 mg, riboflavin (B2) 0.425 mg, niacin (B3) 7.02 mg, pantothenic acid (B5) 1.233 mg, vitamin B6 0.21 mg, folate (B9) 30 μg, calcium 18 mg, iron 2.2 mg, magnesium 231 mg, manganese 1.3 mg, phosphorus 347 mg, potassium 460 mg, sodium 1 mg, zinc 2.4 mg, copper 1.1 mg, selenium 8.3 μg. According to other sources, seeds are composed as follows: starch 71–78%, protein 18%, iron 60–100 ppm, zinc 20–30 ppm, selenium 20–50 ppb., rutin 10–200 ppm, tannins 0.1–2%, catechin-7-O-glucoside, salicylaldehyde (2-hydroxybenzaldehyde), 2,5-dimethyl-4-ihydroxy-3(2H)-furanone, (E,E)-2,4-decadienal, phenylacetaldehyde, 2-methoxy-4-vinylphenol, (E)-2-nonenal, decanal, hexanal. Airborne parts: fagopyrins 0.4–0.6 mg/g (at least 6 different, but similar, substances).

Style American Amber Ale, Saison.

Beer Saracena, Birra del Borgo (Borgorose, Lazio, Italy).

Source Dabove 2015; Giaccone and Signoroni 2017; Heilshorn 2017; Hieronymus 2016b; Markowski 2015; https://www.bjcp.org (31A. Alternative Grain Beer); https://www.brewersfriend.com/other/buckwheat/; https://www.crea.gov.it/it/comunicati-stampa/il-grano-saraceno-per-la-vita-e-per-lo-sport; https://gazzagolosa.gazzetta.it/2016/07/09/birra-saracena/.

A Beer for Athletes
Of the many reasons brewers experiment with unusual ingredients, medicine and health are among the least frequent. However, in the creation of Saracena, an ale of little more than 3% ABV produced by Birra del Borgo, scientific research and experimentation played a central role. The impulse to produce a beer with *Fagopyrum tataricum* came from the Council for Research in Agriculture and Analysis of Agricultural Economics, which wanted to verify the claim made by some studies that the nutraceutical properties of this plant were more effective in the form of malt. This explains the decision to produce a beer with a low alcohol content—in order not to promote hypoglycemia—with large quantities of buckwheat malt and to test its effectiveness on a group of athletes from Trentino. The first results were positive. The consumption of this beer led to an increase in the production of antioxidant and anti-inflammatory substances, and the experimentation has continued. Beyond its medical aspects, buckwheat is interesting because it brings a pleasant spicy and vegetal note to the beer, which gives it great freshness.

Fagus grandifolia Ehrh.
FAGACEAE

Synonyms *Fagus alba* Raf., *Fagus americana* (Pers.) Sweet, *Fagus ferruginea* Aiton, *Fagus ferruginea* Dryand., *Fagus heterophylla* Raf., *Fagus nigra* Raf., *Fagus purpurea* Desf., *Fagus rotundifolia* Raf., *Fagus sylvestris* F.Michx., *Fagus virginiana* Wesm.

Common Names American beech, North American beech.

Description Tree, deciduous (20–35 m). Bark smooth, silver-gray. Winter branches easily distinguishable among North American trees, long and slender (15–20 × 2–3 mm), with 2 rows of bracts overlapping in buds; buds distinctly slender and long, cigar shaped, making them easily identifiable. Leaves dark green, simple, and sparsely toothed with small teeth in correspondence of which ends ribbing, 6–15 cm long, with short petiole. Monoecious species, with flowers of both sexes on the same tree. Fruit small and angled nut, carried in pairs in a 4-lobed shell covered with soft spines. The species reproduces by seed but also vegetatively through the emission of root suckers.

Subspecies *Fagus grandifolia* Ehrh. has two subspecies: the typical *Fagus grandifolia* Ehrh. subsp. *grandifolia* and *Fagus grandifolia* Ehrh. subsp. *mexicana* (Martínez) A.E.Murray.

Cultivar Several are known; for example, White Lightning.

Related Species The genus *Fagus* includes about 14 taxa, some of which are of hybridogenic origin.

Geographic Distribution E North America. **Habitat** Well-drained but humid slopes and valley floors. **Range** Original range expanded by cultivation for forestry purposes.

Beer-Making Parts (and Possible Uses) Wood (for smoking malt).

Wood With whitish sapwood, brown-to-reddish heartwood, medium-to-fine texture, straight grain, sometimes difficult to work by hand; takes on excellent polish. Suitable for bending after steaming, resistance to alteration varies. It has strong shrinkage; therefore, the seasoning must be accurate to avoid deformations and splits; WSG 0.65–0.90. Used for packaging, furniture, interior decoration, railway sleepers, flooring (to be treated with antiseptics), staves, various objects and tools.

Style Classic Smoked Beer.

Source Heilshorn 2017; https://www.bjcp.org (32A. Classic Style Smoked Beer).

Fagus sylvatica L.
FAGACEAE

Synonyms *Castanea fagus* Scop., *Fagus aenea* Dum. Cours., *Fagus asplenifolia* Dum.Cours., *Fagus cochleata* (Dippel) Domin, *Fagus comptoniifolia* Desf., *Fagus crispa* Dippel, *Fagus cristata* Dum.Cours., *Fagus cucullata* Dippel, *Fagus cuprea* Hurter ex A.DC., *Fagus incisa* Dippel, *Fagus pendula* (Lodd.) Dum. Cours., *Fagus purpurea* Dum.Cours., *Fagus quercoides* (Pers.) Dippel, *Fagus salicifolia* A.DC., *Fagus sylvestris* Gaertn., *Fagus variegata* A.DC.

Common Names European beech, common beech.

Description Deciduous tree (10–50 m). Bark dark gray, compact, in 1- to 3-year-old branches brown, shiny, glabrous, with gray punctiform lenticels. Buds acute (2.5 × 10 mm), brownish-black; brachyblasts short (1–2 cm) with ringed bark, pubescent. Leaves with 10–15 mm petiole and elliptic lamina (40–47 × 60–70 mm), base rounded, slightly asymmetrical, tip obtuse, bright green, margin nearly entire or obtusely crenate; upper leaf glabrous and glossy, under leaf with tufts of reddish hairs at rib angles. Male inflorescence short catkin (1.5–2 cm). Fruit reddish nucule (8 × 20 mm), completely enclosed in a 4- to 5-valved woody cupula with weak incurved or patent herbaceous spines (3–8 mm).

Subspecies Two subspecies have been described: *Fagus sylvatica* L. subsp. *sylvatica* (branches of the year rapidly glabrous, leaves broader toward the middle or in the lower half, 4–9 cm long, 5–9 pairs of secondary veins, margin entire and wavy, sometimes crenulated or more rarely toothed, petiole 1–1.5 cm, perianth of the male flower subdivided almost to the base, fruit 25 mm, scales of the cupula free and subulate) and *Fagus sylvatica* L. subsp. *orientalis* (Lipski) Greuter & Burdet (branches of the year pubescent, leaves broader in upper half, 6–12 cm, 7–12 pairs of secondary veins, perianth of male flower subdivided at most by one-third, fruit 20 mm, scales of basal part of cupula spatulate).

Cultivar Within the nominal subspecies there is great morphological variability (appearance, leaf color, shape) of interest for the ornamental use of the species. Among the more than 40 that have been described are the following. Variants in appearance: Cocleata (dwarf form, 4–5 m, crown conical), Dawyck (up to 25 m, crown columnar, 3 m wide maximum; Dawyck Gold variant with golden yellow leaves in spring and fall, Dawyck Purple with purple leaves), Pendula (main branches horizontal, secondary branches pendulous), Tortuosa (prostrate form with twisted and polycormic stem, widespread in Verzy in France, Suntel in Germany, Dalby-Soderskogs in Sweden, Calabria in Italy on the Pollino, Orsomarso, and Sila; slow-growing prostrate specimens, sinuous horizontal branches called "snake beeches"). Variants in leaf color: Purpurea (taxonomically defined form *Fagus sylvatica* purpurea, including erect or prostrate cultivars with more or less red leaves), Zlatia (golden yellow leaves that are yellowish-green at maturity). Variants in leaf shape and/or margin: Asplenifolia (leaves with incised lamina, sometimes pinnatifid, others with nontoothed margin), Cristata (leaves in clusters and curled with apex reflected downward to resemble cockscomb), Grandidentata (leaves with coarsely toothed margins to trigon teeth), Laciniata (leaves deeply and regularly toothed or lobate-pinnate, margin sometimes entire, considered an independent variety: var. *laciniata* Vignet = var. *heterophylla* Loud.), Latifolia (leaves up to 15 × 9–12 cm), Quercifolia (leaves of intermediate form Grandidentata × Laciniata). Variant in bark shape: *Fagus sylvatica* L. var. *quercoides* Pers. (bark fissured for raised plates and ridges; reported in Göttingen, Germany).

Related Species The genus *Fagus* includes about 14 taxa, some of which are of hybrid origin.

Geographic Distribution C Europe. **Habitat** Mesophilic forests (oceanic and suboceanic phytoclimate, absence of dry periods; mesophilic, sciaphilous species, adapted to fertile, fresh, deep, drained soils but also to less stony and poor soils if abundant atmospheric and edaphic moisture is present). **Range** Widely cultivated for forestry purposes throughout the range, widespread elsewhere (e.g., United States, Canada) for both ornamental and forestry purposes.

Beer-Making Parts (and Possible Uses) Wood (to dry and smoke malt and for casks), chips (to refine beer in lagering tanks).

Wood Light pinkish-brown in color, reddish-brown when steamed, normally appears undifferentiated; medullary rays very evident, particularly in radial sections where they appear as shiny speckles; WSG 0.65–0.85. The texture is fine, and the grain is mostly straight, so the wood is particularly fissile. Although compact, it is easy to work but is quickly altered if left in a damp environment. Widely used for furniture (especially if steamed) and furnishings, veneers, railway sleepers (after treatment with appropriate antiseptics), oars, and other splitting work. Valued as a fuel, both as firewood and charcoal.

Style Classic Smoked Beer, Lichtenhainer.

Beer Märzen, Schlenkerla (Bamberg, Germany).

Source Calagione et al. 2018; Cantwell and Bouckaert 2016; Cantwell and Bouckaert 2018; Daniels and Larson 2000; Hieronymus 2010; Jackson 1991; McGovern 2017; https://www.bjcp.org (32A. Classic Style Smoked Beer); https://www.schlenkerla.de/rauchbier/beschreibungi.html.

🍺 Bamberg Smoked Beers
Legend says that the intense smoky note that distinguishes Bamberg's beers is the result of a fire in Kaiserdom that smoked the malt from neighboring breweries. The truth is that for centuries malt has been dark and characterized by hints of an open fire, and only in modern malt producers has it been possible to produce grains of different colors and without the pyric note. But in the beer-making capital of Franconia, the tradition of using *rauch* malt, smoked with *Fagus sylvatica* wood, has been preserved. In the case of the Schlenkerla Brewery, the wood, coming from the French-Swiss Jura Mountains, is left to age for three years before being burned.

Ficus carica L.
MORACEAE

Synonyms *Caprificus insectifera* Gasp., *Caprificus leucocarpa* Gasp., *Caprificus oblongata* Gasp., *Caprificus pedunculata* (Miq.) Gasp., *Caprificus rugosa* (Miq.) Gasp., *Caprificus sphaerocarpa* Gasp., *Ficus albescens* Miq., *Ficus burdigalensis* Poit. & Turpin, *Ficus caprificus* Risso, *Ficus colchica* Grossh., *Ficus colombra* Gasp., *Ficus communis* Lam., *Ficus deliciosa* Gasp., *Ficus dottata* Gasp., *Ficus hypoleuca* Gasp., *Ficus hyrcana* Grossh., *Ficus kopetdagensis* Pachom., *Ficus latifolia* Salisb., *Ficus leucocarpa* Gasp., *Ficus macrocarpa* Gasp., *Ficus neapolitana* Miq., *Ficus pachycarpa* Gasp., *Ficus pedunculata* Miq., *Ficus polymorpha* Gasp., *Ficus praecox* Gasp., *Ficus regina* Miq., *Ficus rugosa* Miq., *Ficus silvestris* Risso.

Common Names Fig, common fig.

Description Small (3–10 m), short-lived, deciduous tree or shrub. Roots very expanded. Stem short, twisted; branches numerous, fragile, forming a flattened crown; bark ash-gray, thin, smooth on young branches, then slightly wrinkled; buds of 3 types, all present on 1-year-old branches: foliate, floriferous, and mixed; the former, small and often dormant, are in a lateral position; the floriferous are large and rounded; the latter, apical, are conical in shape. Leaves alternate, palmate-lobate (rarely simple), petiole 3–6 cm, lamina in 3–7 oblong lobes, unequal, expanded at the top, indented at the margin, with cordate or truncate base and lamina (5–10 × 8–15 cm), dark green, rough on top, pubescent and paler below, with strongly raised veins. Inflorescence syconium (2–5 cm), at the leaf axil, bearing monoecious pedicellate flowers enclosed within a fleshy pyriform receptacle of variable size and color (mostly green), provided with a narrow apical orifice (ostiole); male flowers, 3–5 stamens and orange-colored pollen, close to the ostiole; female flowers, ovary and lateral stylus ± long, below; pollination of brevistyle flowers by a specialized pronube insect (*Blastophaga psenes* L. - Hymenoptera), which penetrates the syconium through the ostiole and lays an egg in each ovary, turning it into a gall. Edible infructescence (fig) (at least in *Ficus carica* L. var. *domestica*) consisting of numerous small achenes (true fruit) located inside the enlarged syconium (stringy and inedible flesh in *Ficus carica* L. var. *caprificus*), which at maturity becomes green or violet in color. In *Ficus carica* var. *domestica*, there are 3 productions of syconium per year, with different ripening times, characters, and denominations: "fioroni" (flowering or productive figs) contain male and female gallicolous brevistyle flowers, develop from buds of the previous year, and ripen in June and July; "forniti" (mammon or true figs) contain male flowers (few) and female brevistyle, and longistyle flowers develop in the current year and mature in August and September; and "cratiri" (mamme or late figs) bear only gallicolous female flowers and form in the fall and over winter, maturing the following spring.

Subspecies Although there is no unequivocal agreement among botanists on the taxonomic rank to be assigned to the variability of *Ficus carica*, it is fairly clear that there are at least two distinct entities (even on ecological grounds) within the species. *Ficus carica* L. var. *domestica* Czern. & Rav. (*Ficus sativa* Fiori) would include all cultivars of cultivated figs, whereas *Ficus carica* L. var. *caprificus* Risso (*Ficus caprificus* Risso) represents the wild entity of this taxon, of great importance since it is the natural host of *Blastophaga* *psenes*, on which the production of figs and the sexual reproduction of the species depend.

Cultivar There are thousands of varieties of figs that, in the most climatically suitable areas (e.g., southern Italy), provide two to three annual productions (fiorini, forniti, cratiri). The first two are the most significant economically. In other areas (e.g., California), the fig provides two productions: the first ("breba" crop) develops in spring on the buds of the previous year, and the second (main fig crop) develops on the buds of the current year and matures in late summer into autumn. The second production is generally quantitatively and qualitatively superior, but some cultivars (e.g., Black Mission, Croisic, Ventura) produce a good first production. There are three varietal types of edible fig: the varieties with persistent (or common) figs have all-female flowers that do not need pollination to bear fruit because the fruit develops via a parthenocarpic mechanism (e.g., Black Mission, Brown Turkey, Brunswick, Celeste, Dottato or Kadota); the varieties with deciduous figs (or Smyrna) require cross-pollination by *Blastophaga psenes* with the pollen of *Ficus carica* var. *caprificus* to ripen the fruits, which fall if not fertilized (e.g., Calimyrna, Inchàrio, Lob Incir, Marabout, Smyrne, Zidi); intermediate varieties (San Pedro) produce a "breba" production without pollination but need pollination to produce a "main" production (e.g., King, Lampeira, San Pedro). Some important fig cultivars are Abicou, Adriatic, Alma, Australia, Black Genoa, Black Israel, Black Jack, Black Mission, Blue Giant, Bordissot Negra, Bornholms, Brogiotto bianco, Brogiotto nero, Brooke Japan, Brooke Red, Brown Turkey, Brunswick, Byadi, Calimyrna, Celeste, Col de dame blanc, Col de dame gris, Conadria, Dauphine, Desert King, Di Redo, Early Violet, Flanders, Genao, Györöki lapos, Hardy Chicago, Inca Gold, Italian Black, Italian Brooklyn White, Italian Honey, Italian White, Japan (BTM6), Jurupa, Kadota, King, Kunming, Larme de Jaune, Marseilles vs Black, Marseilles (Blanch), Mission, Negronne, Orourke, Osborn Prolific, Pop's Purple, Quantico, Sequoia, Sierra, Smith, Sunee2, Taiwan (TWA5), Tena, Texas Everbearing, Timla in Kumaon, Violette de Bordeaux, Wuhan, Yede Vern, Zidi, and Zöld óriás.

Related Species The genus *Ficus* includes more than 900 taxa.

Geographic Distribution Mediterranean-Turanian. **Habitat** Shady cliffs, walls. **Range** Turkey, Egypt, Algeria, Morocco, Tunisia, Iran, Syria, Albania, Greece, Italy, United States, Brazil.

Beer-Making Parts (and Possible Uses) Fruits (fresh or dried), leaves.

Chemistry Fresh fruits: energy 310 kJ (74 kcal), carbohydrates 19.18 g, sugars 16.26 g, fiber 2.9 g, lipids 0.30 g, protein 0.75 g, thiamine (B1) 0.06 mg, riboflavin (B2) 0.05 mg, niacin (B3) 0.4 mg, pantothenic acid (B5) 0.3 mg, vitamin B6 0.113 mg, folate (B9) 6 µg, choline 4.7 mg, vitamin C 2 mg, vitamin K 4.7 µg, calcium 35 mg, iron 0.37 mg, magnesium 17 mg, manganese 0.128 mg, phosphorus 14 mg, potassium 242 mg, sodium 1 mg, zinc 0.15 mg.

Style American Amber Ale, American Barleywine, American Brown Ale, American Pale Ale, American Stout, American Strong Ale, American Wheat or Rye Beer, Belgian Blond Ale, Belgian Dark Strong Ale, Belgian Dubbel, Belgian Golden Strong Ale, Belgian Specialty Ale, Bière de Garde, British Strong Ale, Brown Porter, English Barleywine, English IPA, Fruit Beer, Old Ale, Robust Porter, Russian Imperial Stout, Saison, Southern English Brown.

Beer Fich, Valscura (Caneva, Friuli-Venezia Giulia, Italy).

Source Giaccone and Signoroni 2017; Heilshorn 2017; Hieronymus 2016b; Jackson 1991; Josephson et al. 2016; Markowski 2015; McGovern 2017; https://www.bjcp.org (29. Fruit Beer); https://www.brewersfriend.com/other/vijgenblad/; https://www.sangabriel.it/it/birre/le-stagionali/birra-al-fico.

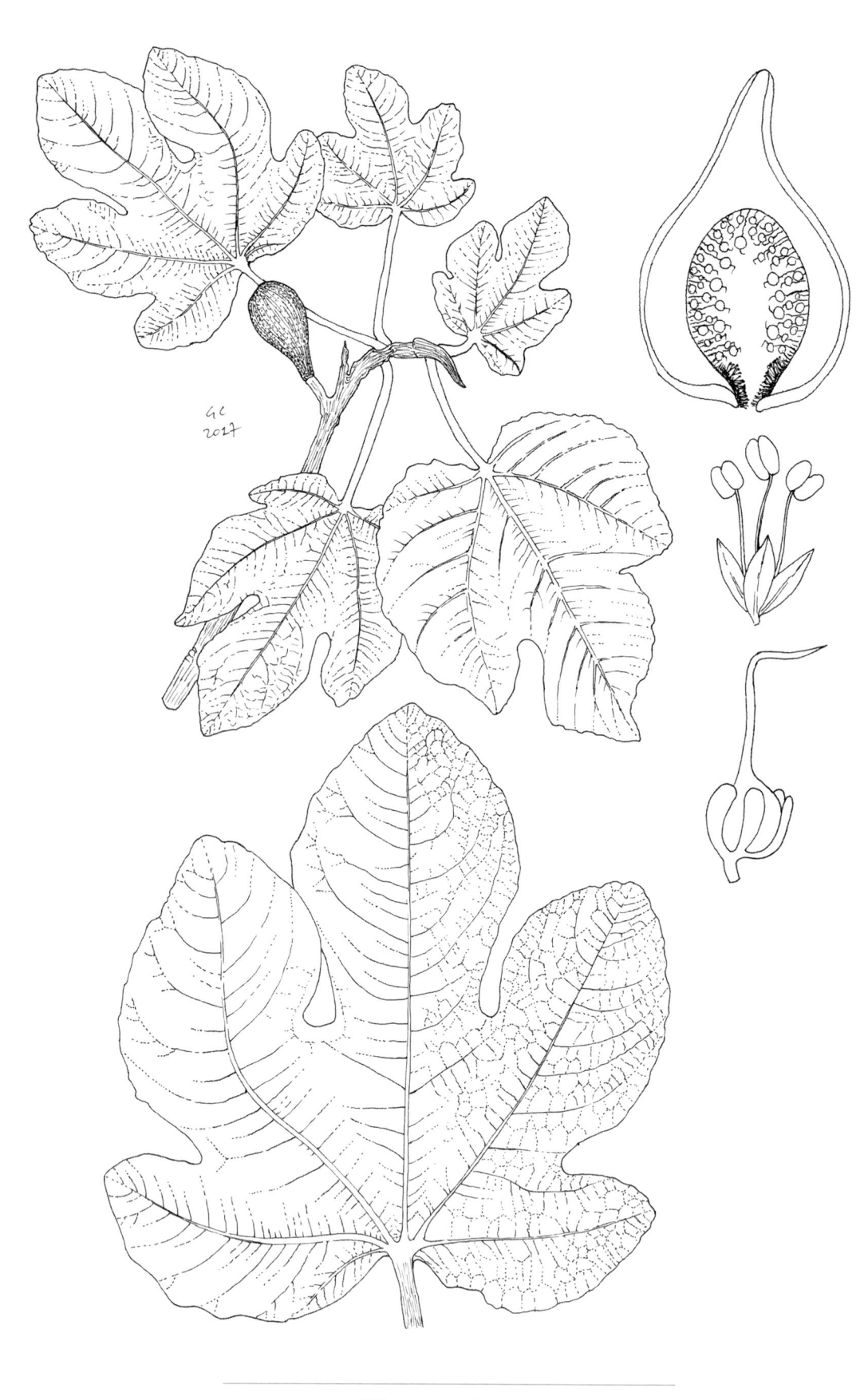

Ficus carica L.
MORACEAE

Filipendula ulmaria (L.) Maxim.
ROSACEAE

Synonyms *Filipendula denudata* (J.Presl & C.Presl) Fritsch, *Filipendula glauca* (Schultz) Asch. & Graebn. ex Dalla Torre & Sarnth., *Filipendula megalocarpa* Juz., *Filipendula subdenudata* Fritsch, *Spiraea contorta* Stokes, *Spiraea denudata* J.Presl & C.Presl, *Spiraea glauca* Schultz, *Spiraea odorata* Gray, *Spiraea palustris* (Moench) Salisb., *Spiraea quinqueloba* (Baumg.) Spreng, *Spiraea ulmaria* L., *Spiraea unguiculata* Dulac, *Thecanisia discolor* (W.D.J.Koch) Raf., *Thecanisia ulmaria* (L.) Raf., *Thecanisia ulmaria* (L.) Raf. ex B.D.Jacks., *Ulmaria denudata* (J.Presl & C.Presl) Opiz, *Ulmaria glauca* (Schultz) Fourr., *Ulmaria obtusiloba* Opiz, *Ulmaria palustris* Moench, *Ulmaria pentapetala* Gilib., *Ulmaria quinqueloba* Baumg., *Ulmaria spiraea-ulmaria* Hill, *Ulmaria ulmaria* (L.) Barnhart, *Ulmaria vulgaris* Hill.

Common Names Meadowsweet, mead wort, queen of the meadow, pride of the meadow, meadow-wort, meadow queen, lady of the meadow, dollof, meadsweet, bridewort.

Description Herb, perennial (5–10 dm). Stem erect, striate, glabrous. Leaves 1–4 dm long, imparipinnate, with 7–11 lanceolate segments (2–3 × 3–7 cm), toothed, between which are smaller intercalated segments (2–5 mm). Inflorescences dense corymbose cymes. Flowers with 2–5 mm white petals. Fruits follicles at maturity twisted into spirals.

Subspecies Two subspecies are recognized within the species: *Filipendula ulmaria* (L.) Maxim. subsp. *picbaueri* (Podp.) Smejkal and *Filipendula ulmaria* (L.) Maxim. subsp. *ulmaria*.

Related Species The genus *Filipendula* includes 17 taxa.

Geographic Distribution Euro-Siberian.
Habitat Wet meadows, swamps, riparian woods.
Range Introduced and naturalized species in North America.

Beer-Making Parts (and Possible Uses)
Flowers.

Toxicity No secondary or toxic effects are known with therapeutic doses except in particularly sensitive individuals; therefore, as a precautionary measure, its use is not recommended in individuals allergic to acetylsalicylic acid.

Chemistry Flower: salicylates (salicylaldehyde up to 70%, ethylsalicylate, methylsalicylate, methoxybenzaldehyde), flavonoids (36% spiraeoside, rutin, hyperoside, quercetin, kaempferol-4´-glucoside, avicularin), tannins (1–12% rugosin D, vanillin), phenolic glycosides (spiraein, monotropin, gaultherin), trace coumarin, mucilage, carbohydrates, vitamin C.

Style American Pale Ale, Belgian Specialty Ale, Blonde Ale, Braggot, California Common Beer, Holiday/Winter Special Spiced Beer, Lambic, Mild, Specialty Beer, Spice/Herb/Vegetable Beer.

Beer Reine des Près, Cantillon (Brussels, Belgium).

Source Heilshorn 2017; Hieronymus 2016b; McGovern 2017; Sparrow 2015; https://www.brewersfriend.com/other/meadowsweet/; http://honest-food.net/2016/05/05/gruit-herbal-beer/; https://www.nationalgeographic.it/food/2015/01/09/news/breve_storia_delle_bevute-2862950/?refresh_ce.

A Medicine That Tastes Like Almonds
As pointed out by Hieronymus, *Filipendula ulmaria* has been used for centuries by brewers throughout Europe as a flavoring plant. Its flowers give sweet notes similar to those of almonds. This species, long considered sacred, contains high quantities of salicylates; therefore, it is hypothesized that its use was also linked to archaic forms of medicine.

Foeniculum vulgare Mill.
APIACEAE

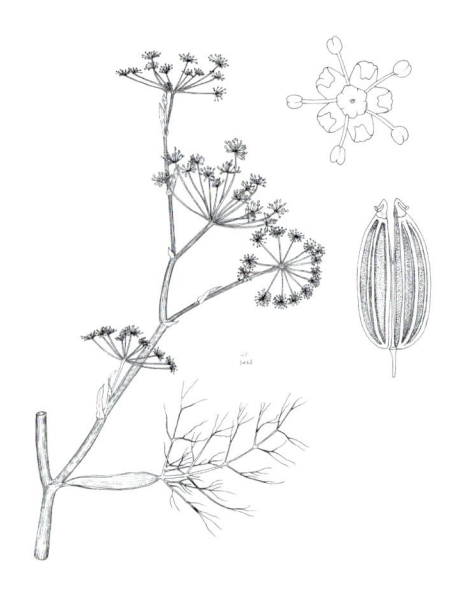

Synonyms *Anethum dulce* DC., *Anethum foeniculum* L., *Anethum minus* Gouan, *Anethum panmori* Roxb., *Anethum panmorium* Roxb. ex Fleming, *Anethum pannorium* Roxb., *Anethum rupestre* Salisb., *Foeniculum azoricum* Mill., *Foeniculum divaricatum* Griseb., *Foeniculum dulce* Mill., *Foeniculum giganteum* Lojac., *Foeniculum officinale* All., *Foeniculum panmorium* (Roxb.) DC., *Foeniculum piperitum* C.Presl, *Foeniculum rigidum* Brot. ex Steud., *Ligusticum foeniculum* (L.) Crantz, *Ligusticum foeniculum* (L.) Roth, *Meum foeniculum* (L.) Spreng., *Selinum foeniculum* E.H.L.Krause, *Seseli foeniculum* Koso-Pol.

Common Names Fennel.

Description Herb, biennial or perennial herb (4–15 dm), with an intense sweet aroma. Rhizome gnarled, annulated, whitish, horizontal. Stem erect, dark green, cylindrical, branched. Leaves 3–4 pinnate, divided into capillary laciniae, more than 10 mm long, soft, mostly yellowish. Inflorescence umbel (12–30 rays) compound, without involucre. Flowers with yellow petals. Fruit 4–7 mm.

Subspecies Within *Foeniculum vulgare* Mill. subsp. *vulgare*, taxonomists identify *Foeniculum vulgare* Mill. var. *azoricum* (Miller) Thell., cultivated as a vegetable, and *Foeniculum vulgare* Mill. var. *dulce* (Miller) Thell., cultivated for its aromatic seed. In Southern Italy, *Foeniculum vulgare* Mill. subsp. *piperitum* (Ucria) Coutinho, with its interesting aromatic characteristics, is widespread. It is traditionally harvested in uncultivated lands for its leafy shoots and seeds and used in traditional cooking and liquor making.

Cultivar Cultivars are distinguished according to morphology and time of harvesting. Examples include Autumn Carmo, Autumn Teseo F1, Early Prelude F1, Summer Rondo, Winter Serpico, and Winter Ulysses F1.

Related Species The *Foeniculum* genus includes three species: *Foeniculum scoparium* Quézel, *Foeniculum subinodorum* Maire, Weiller & Wilczek, and *Foeniculum vulgare* Mill.

Geographic Distribution S Mediterranean. **Habitat** There are cultivated varieties and wild plants of the species; the latter are found in arid wastelands and cultivated fields. **Range** India, Mexico, China, Iran, Bulgaria, Syria, Morocco, Egypt, Canada, Afghanistan (aggregated production of anise, star anise, fennel, and coriander), Italy.

Beer-Making Parts (and Possible Uses) Seeds, flowers, bulbs, stems, leaves.

Chemistry Seeds: energy 1,443 kJ (345 kcal), carbohydrates 52 g, fiber 40 g, lipids 14.9 g (saturated 0.5 g, monounsaturated 9.9 g, polyunsaturated 1.7 g), protein 15.8 g, thiamine (B1) 0.41 mg, riboflavin (B2) 0.35 mg, niacin (B3) 6.1 mg, vitamin B6 0.47 mg, vitamin C 21 mg, calcium 1196 mg, iron 18.5 mg, magnesium 385 mg, manganese 6.5 mg, phosphorus 487 mg, potassium 1694 mg, sodium 88 mg, zinc 4 mg.

Style American Pale Ale, Belgian Pale Ale, Bière de Garde, California Common Beer, Cream Ale, English Barleywine, Extra Special/Strong Bitter (ESB), Gruit, Holiday/Winter Special Spiced Beer, Irish Red Ale, Mild, Russian Imperial Stout, Specialty Beer, Spice/Herb/Vegetable Beer, Weizen/Weissbier, Winter Seasonal Beer, Witbier.

Beer Big Shrug, Fonta Flora (North Carolina, United States).

Source Calagione et al. 2018; Heilshorn 2017; Hieronymus 2016b; McGovern 2017; https://www.brewersfriend.com/other/fenel/; https://www.brewersfriend.com/other/fennel/.

Fragaria × *ananassa* (Duchesne ex Weston) Duchesne ex Rozier
ROSACEAE

Synonyms *Fragaria* × *ananassa* Duchesne, *Fragaria* × *cultorum* Thorsrud & Reisaeter, *Fragaria* × *grandiflora* Ehrh., *Fragaria* × *magna* auct., *Fragaria bathonica* Poit. & Turpin, *Fragaria bonariensis* Juss. ex Pers., *Fragaria calyculata* (Duchesne) Duchesne ex Steud., *Fragaria caroliniana* Poit. & Turpin, *Fragaria caroliniensis* Duchesne, *Fragaria chiloensis* auct., *Fragaria cuneifolia* Nutt. ex Howell, *Fragaria hybrida* Duchesne, *Fragaria latiuscula* Greene, *Fragaria suchiana* Poit. & Turpin, *Fragaria tincta* Duchesne, *Potentilla* × *ananassa* (Duchesne ex Weston) Mabb.

Common Names Strawberry, garden strawberry, pineapple strawberry.

Description Herb, perennial (1–5 dm). Stems short (± as long as basal leaves), woody, eventually prostrate, pauciflorous, with patent hairs; stoloniferous root at the ends forming new seedlings, allowing the plant to reproduce vegetatively (a feature widely used in propagation for commercial purposes). Leaves compound, with 3 leaflets coarsely toothed, forming a basal rosette. White flowers, consisting of 5 parts, generally hermaphrodite. False infructescence (strawberry) constituted by the fleshy receptacle (3–4 cm) that hosts numerous true fruits on the surface. Fruit achene containing a single seed.

Subspecies The wild progenitors of this hybrid are native to the Americas, *Fragaria virginiana* Mill. (wild strawberry, E North America) and *Fragaria chiloensis* (L.) Mill. (beach strawberry, Pacific coast from Alaska to California to South America), but the hybrid seems to have appeared spontaneously in Europe in 1750 as an accidental crossing; based on genetic analysis, some authors believe that the hybrid has a more complex structure since it contains genes from other species.

Cultivar Numerous: Alba, Albion, Alice, Alinta, Allstar, Altess, Amelia, Annapolis, Apollo, Archer, Aromas, Aromel, Asia, Atlas, Benton, Bogota, Bolero, Bountiful, Brunswick, Cabot, Calypso, Camarosa, Cambridge Favourite, Camino Real, Canoga, Cassandra, Cavendish, Chambly, Chandler, Christine, Clancy, Darselect, Delia, Delite, Delmarvel, Diamante, Earlibelle, Earliglow, Elegance, Elsanta, Elvira, Emily, Eros, Evangeline, Everest, Evie 2, Faith, Favors, Fenella, Firecracker, Flair, Flamenco, Fleurette, Florence, Florentina, Florina, Fort Laramie, Frel, Fruitful Summer, Furore, Gaviota, Glooscap, Governor Simcoe, Guardian, Hapil, Hecker, Hokowase, Honeoye, Hood, Itasca, Jewel, Judibell, Kent, L'Amour, Loran, Lucy, Mae, Malling Opal, Malling Pearl, Marshall, Matis, Mesabi, Midway, Mira, Mohawk, Monterey, Northeaster, Northeastern, Ogallala, Orléans, Oso Grande, Ozark Beauty, Palomar, Pandora, Pegasus, Pelican, Pink Panda, Pinnacle, Portola, Primetime, Puget Reliance, Puget Summer (Schwartze), Quinault, Rabunda, Rainier, Redchief, Redcrest, Redgauntlet, Redgem, Red Ruby (Samba), Rhapsody, Rosie, Roxana, Royal Sovereign, Sable, Saint Pierre, Sallybright, Samba (Red Ruby), San Andreas, Sapphire, Sasha, Scott, Seascape, Seneca, Senga Sengana, Sequoia, Shuksan, Sonata, Sophie, Stellarossa, Strasberry, Strawberry Festival, Sunrise, Surecrop, Sussette, Symphony, Tillamook, Titan, Totem, Tribute, Tristar, Valley Red, Variegata, Veestar, Ventana, Viktoriana, Wendy, Winona, Yamaska.

Related Species The genus *Fragaria* includes 23 taxa.

Geographic Distribution Hybrid fixed between *Fragaria virginiana* Mill. and *Fragaria chiloensis* (L.) Mill. **Habitat** Cultivated and sometimes subspontaneous. **Range** United States, Russia, Turkey, Spain, Egypt, Korea, Japan, Poland, Mexico, Germany, Italy.

Beer-Making Parts (and Possible Uses) False infructescence (fruit), juice.

Chemistry Water 89.9–92.4 g, total sugars 3.3–9.1 g (sucrose 0.2–2.5 g, fructose 1.7–3.5 g, glucose 1.4–3.1 g), protein 0.23 g, citric acid 420–1240 mg, malic acid 90–680 mg, ascorbic acid 26–120 mg, succinic acid 100 mg, tartaric acid 17 mg, pyruvic acid 5 mg, shikimic acid traces, phenols 58–210 mg/l, anthocyanins 55–145 mg/l, potassium 130 mg, sodium 6 mg, calcium 13 mg, magnesium 8 mg, iron 0.6 mg, zinc 0.2 mg.

Style American Amber Ale, American IPA, American Light Lager, American Pale Ale, American Stout, American Wheat Beer, American Wheat or Rye Beer, Baltic Porter, Belgian Blond Ale, Belgian Dubbel, Belgian Golden Strong Ale, Belgian Pale Ale, Belgian Specialty Ale, Berliner Weisse, Bière de Garde, Blonde ale, Braggot, California Common Beer, Cream Ale, Dortmunder Export, Dry Stout, Dunkelweizen, Fruit Beer, Fruit Lambic, German Pilsner (Pils), Kölsch, Robust Porter, Saison, Schwarzbier, Scottish Heavy 70/-, Specialty Beer, Specialty Fruit Beer, Sweet Stout, Weissbier, Weizenbock, Witbier.

Beer Magiuster, Montegioco (Montegioco, Piedmont, Italy).

Source Dabove 2015; Giaccone and Signoroni 2017; Heilshorn 2017; Hieronymus 2010; Hieronymus 2016b; Josephson et al. 2016; McGovern 2017; Sparrow 2015; https://www.bjcp.org (29A. Fruit Beer); https://www.brewersfriend.com/other/strawberries/; https://www.brewersfriend.com/other/strawberry/.

Fragaria vesca L.
ROSACEAE

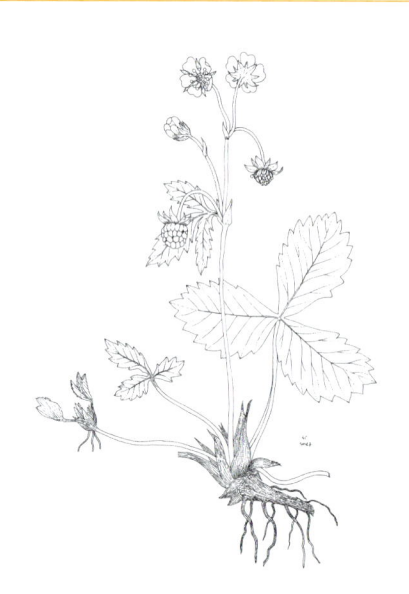

Synonyms *Dactylophyllum fragaria* Spenn., *Fragaria abnormis* Tratt., *Fragaria aliena* Weihe, *Fragaria alpina* (Weston) Steud., *Fragaria botryformis* Duchesne, *Fragaria chinensis* Losinsk., *Fragaria concolor* Kitag., *Fragaria eflagellis* Duchesne, *Fragaria eflagellis* (Weston) Duchesne ex Rozier, *Fragaria florentina* Poit. & Turpin, *Fragaria gillmanii* Clinton, *Fragaria hortensis* Duchesne, *Fragaria hortensis* (Weston) Duchesne ex Rozier, *Fragaria insularis* Rydb., *Fragaria mexicana* Schltdl., *Fragaria minor* Duchesne, *Fragaria monophylla* L., *Fragaria multiplex* Duchesne, *Fragaria multiplex* Duchesne ex Poit. & Turpin, *Fragaria muricata* L., *Fragaria nemoralis* Salisb., *Fragaria nuda* Pers., *Fragaria portentosa* Poit. & Turpin, *Fragaria retrorsa* Greene, *Fragaria roseiflora* Boulay, *Fragaria semperflorens* Duchesne, *Fragaria semperflorens* Duchesne ex Rozier, *Fragaria succulenta* Gilib., *Fragaria sylvestris* (L.) Duchesne, *Fragaria sylvestris* (L.) Weston, *Fragaria unifolia* Steud., *Fragaria vulgaris* Ehrh., *Potentilla vesca* (L.) Scop.

Common Names Wild strawberry, woodland strawberry, Alpine strawberry, European strawberry.

Description Herb, perennial, reptant (5–15 cm). Stem slightly lignified, short (2–4 cm), bearing the leaves in a rosette; scape floriferous with patent hairs; stolons aerial (1–many dm), rarely rooting, bearing a scale on each internode. Leaves erect, about as long as the scape; petiole 5–10 cm with dense erect-patent hairs; 3 segments, elliptic, the central sessile (15–30 × 35–75 mm) with 8–11 teeth on each side, with acute base and rounded apex, the lateral somewhat more connate, less toothed on inner side; bottom lamina whitish and sparsely pubescent, especially on veins. Inflorescence 1- to 3-flowered, usually with 1 sessile bract-like leaf; flowers erect at fruiting folded; laciniae of epicalyx linear; sepals triangular (2 × 3 mm); petals white (5 × 6 mm), erect-patent (cup-shaped corolla). Fruit red, ovate or spherical (5–20 mm), edible, easily detached from the calyx.

Subspecies Within the taxon *Fragaria vesca*, four subspecies are recognized: *Fragaria vesca* L. subsp. *americana* (Porter) Staudt, *Fragaria vesca* L. subsp. *bracteata* (A.Heller) Staudt, *Fragaria vesca* L. subsp. *californica* (Cham. & Schltdl.) Staudt, and *Fragaria vesca* L. subsp. *vesca*.

Related Species The genus *Fragaria* includes 23 taxa.

Geographic Distribution Euro-Siberian, later cosmopolitan. **Habitat** Beech forests, pine forests, fir forests, especially in clearings, hedges. **Range** Cosmopolitan (species mostly collected in natural or seminatural environments, rarely cultivated).

Beer-Making Parts (and Possible Uses) False infructescence (fruit).

Chemistry Fruit composition similar to that of *Fragaria × ananassa* with higher contents (% dry weight) of flavonoids 3%, phenolic acids 1.8%, tannins 4.3%, anthocyanins 11.2%.

Style IPA.

Source Calagione et al. 2018; Hieronymus 2016b; McGovern 2017 (the author generically uses the term "strawberry" with reference to North America, where there are 3 species belonging to the genus *Fragaria*: *Fragaria chiloensis* (L.) Mill., *Fragaria vesca* L., and *Fragaria virginiana* Mill.); https://www.omnipollo.com/beer/magic-411/.

Frangula purshiana Cooper
RHAMNACEAE

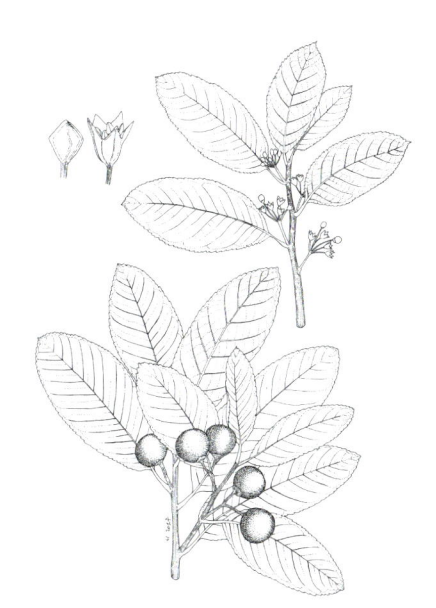

Synonyms *Cardiolepis obtusa* Raf., *Frangula anonifolia* (Greene) Grubov, *Rhamnus annonifolia* Greene, *Rhamnus purshiana* DC.

Common Names Cascara sagrada, cascara buckthorn, chittem bark.

Description Tree or shrub (up to 12 m). Stem with gray bark, young branches red to brown; terminal buds covered with bracts, dark hairiness. Leaves deciduous, petiole 6–23 mm; lamina 50–150 mm, broadly elliptic to obovate, thin, base obtuse or tapering, apex obtuse to truncate, margin irregularly toothed to entire, surfaces sparsely hairy to glabrous, rib prominent. Inflorescence (± 25 flowers); peduncles about 25 mm. Flowers hermaphrodite, hypanthium 3 mm; sepals 5; petals 5; stamens 5; stylus included. Fruit (10 mm) with 3 seeds, black in color.

Subspecies In addition to the nominal subspecies, *Frangula purshiana* Cooper ssp. *purshiana* (cascara sagrada), two other entities of subspecific rank are recognized: *Frangula purshiana* Cooper subsp. *anonifolia* (Greene) J.O.Sawyer & S.W.Edwards and *Frangula purshiana* Cooper subsp. *ultramafica* Sawyer & S.W.Edwards (caribou coffeeberry). According to some sources, the latter is synonymous with the nominal subspecies.

Related Species The genus *Frangula* includes 47 taxa of specific and subspecific rank.

Geographic Distribution NW North America (N California, Oregon, Washington, Idaho, Montana, British Columbia). **Habitat** Coniferous forests, chaparrals. **Range** Similar to geographic distribution but species also widespread elsewhere for ornamental purposes.

Beer-Making Parts (and Possible Uses) Bark.

Toxicity Bark is relatively easily harvested in spring or early summer. It must then be seasoned for at least one year before use; otherwise, it causes vomiting and severe diarrhea. It is toxic in excessive doses, especially in children.

Chemistry Numerous quinoid substances are contained in cascara bark; the substances responsible for its laxative action are the hydroxyanthracene glycosides, which include cascarosides A, B, C, and D; the bark of this species contains about 8% anthranoids, about 2/3 of which are cascarosides.

Style Dark Mild, Spice/Herb/Vegetable Beer, Strong Scotch Ale.

Source https://www.brewersfriend.com/other/cascara/.

0.1%, benzaldehyde 0.1%, (E)-2-nonenal 0.1%, octanol 0.1%, hexadecane 0.1%, nonanol 0.1%, heptadecane 0.1%, curcumene 0.1%, p-methylacetophenone 0.1%, methyl salicylate 0.1%, octadecane 0.1%, heptanoic acid 0.1%, eicosane 0.1%, undecanoic acid 0.1%, pentacosane 0.1%, hexanal < 0.1%, isoamyl acetate < 0.1%, amyl acetate < 0.1%, 2-heptanone < 0.1%, limonene < 0.1%, isoamyl alcohol < 0.1%, 1,8-cineole < 0.1%, ethyl hexanoate < 0.1%, terpinene < 0.1%, 5-methyl-3-heptanone < 0.1%, styrene < 0.1%, 1,2,4-trimethylbenzene < 0.1%, methyl heptanoate < 0.1%, octanal < 0.1%, tridecane < 0.1%, cis-rose oxide < 0.1%, A,P-dimethylstyrene < 0.1%, 2-ethylhexanol < 0.1%, 2-acetylfuran < 0.1%, (E,E)-3,5-octadien-2-one < 0.1%.

Style Herb Beer.

Source Heilshorn 2017; Hieronymus 2016b; Josephson et al. 2016.

Galium odoratum (L.) Scop.
RUBIACEAE

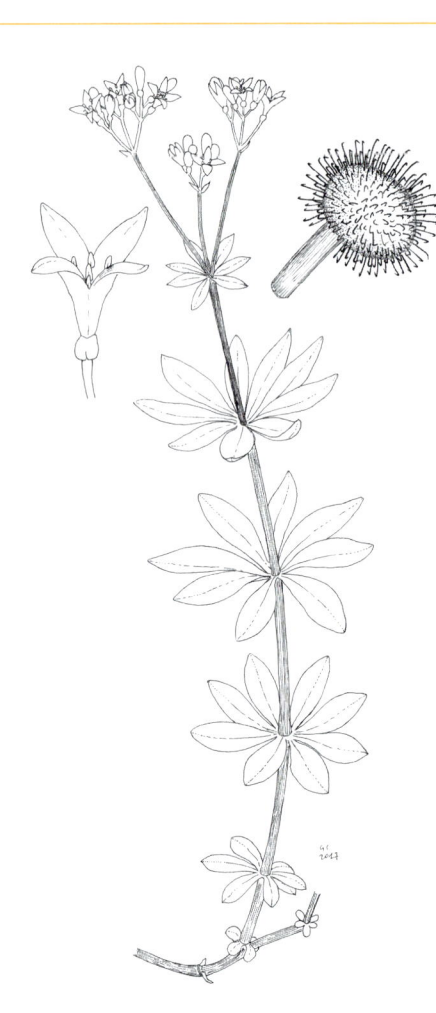

Synonyms *Asperula eugeniae* K.Richt., *Asperula odorata* L., *Asterophyllum asperula* Schimp. & Spenn., *Asterophyllum sylvaticum* Schimp. & Spenn., *Chlorostemma odoratum* (L.) Fourr.

Common Names Woodruff, sweet woodruff, wild baby's breath, master of the woods.

Description Herb, perennial (10–30 cm), fragrant. Rhizome horizontal underground, stems erect, quadrangular, hirsute at nodes, otherwise glabrous. Leaves in whorls of 6–9, linear-spatulate or narrowly oblanceolate (512 × 20–50 mm), acute. Flowers usually forming 3 corymbose inflorescences, long pedunculate; peduncles 1–3 mm, at fruiting 3–10 mm; corolla greenish-white, with 1.5 mm tube and 2–3 mm acute lobes. Fruit formed by 2 mericarps 2–3 mm, bristly.

Related Species The genus *Galium* includes more than 730 taxa.

Geographic Distribution Eurasian. **Habitat** Beech forests and other mesophilous broadleaf forests. **Range** Eurasia.

Beer-Making Parts (and Possible Uses) Leaves, inflorescences.

Toxicity Contains substances that may be poisonous in high doses.

Chemistry Essential oil (99 compounds identified): thymol 30.6%, iso-thymol 22.8%, iso-thymyl butyrate 5.5%, linalool 2.4%, α-humulene 2%, (Z)-3-hexenal 1.5%, epoxide humulene 1.5%, caryophyllene oxide 1.2%, β-ionone 1.1%, (Z)-3-hexenol 1%, isopinocanophon 1%, β-caryophyllene 1%, thymyl-valerate 0.9%, borneol 0.8%, β-selinene 0.8%, phytol 0.7%, hexanol 0.7%, 3,5-octadien-2-one 0.6%, trans-verbenol 0.6%, hexanal 0.5%, (E,E)-2,4-heptadienal 0.5%, intermedeol 0.5%, other (77) compounds 0.1–0.4%.

Style Belgian Dark Strong Ale, Berliner Weisse Asperula, Specialty Beer, Spice/Herb/Vegetable Beer.

Beer Berliner Weisse, Berliner Kindl (Berlin, Germany).

Source Fisher and Fisher 2017; Hieronymus 2010; Hieronymus 2016b; Jackson 1991; https://www.bjcp.org (17A. Berliner Weisse Asperula); https://www.brewersfriend.com/other/woodruf/; https://www.brewersfriend.com/other/woodruff/.

> 🍺 **Green Berliners**
> The most common use of woodruff in brewing is connected to the production of Berliner weisse, traditionally served with the addition of raspberry syrup or *Galium odoratum*. It is the latter that gives the beer its characteristic green color.

Garcinia indica (Thouars) Choisy
CLUSIACEAE

Synonyms *Brindonia oxycarpa* Thouars, *Garcinia purpurea* Roxb.

Common Names Kokum, Brindonia tallow tree, kokum butter tree.

Description Small tree (up to 10 m). Trunk with pale brown bark, smooth, fissured if old, flamed reddish; branches horizontal, opposite, subangular, glabrous; if incised they release yellowish latex. Leaves simple, opposite, decussate; petiole (0.7–1.5 cm) glabrous, slightly sheathed at base; lamina narrowly elliptic to narrowly obovate (6–12 × 1.5–3 cm), apex obtuse or acute to slightly acuminate, base acute to attenuate, margin slightly repand, papery to subcoriaceous, glabrous; secondary veins (4–7 pairs) with many intersecting veins, tertiary veins obscure. Flowers polygamodioecious. Male inflorescence in axillary and terminal racemes; female flowers solitary, terminal, or axillary. Fruit globular berry, smooth, with many seeds.

Cultivar They are numerous; examples include Borim 2, Gola 17, Hedode 1, Karsapal 5, Kharekhazan 1, Kokam Amruta, Kokam Hatis, Mashem 4, Parashate 3, Parashte 3, Pednem Kei 1, and Savoi Kamini 1.

Related Species According to different classifications, the genus *Garcinia* includes 50–300 species; taxa with similar applications to *Garcinia indica* include *Garcinia forbesii* King, *Garcinia gummi-gutta* (L.) Roxb., *Garcinia intermedia* (Pittier) Hammel, *Garcinia* × *mangostana* L., *Garcinia multiflora* Champ. ex Benth., and *Garcinia prainiana* King.

Geographic Distribution Endemic species of W India (Western Ghats). **Habitat** Undergrowth of disturbed evergreen and semi-evergreen forests (up to 700 m ASL). **Range** Original geographic distribution.

Beer-Making Parts (and Possible Uses) Fruits (fresh inner pulp) (?), fruits (dried outer coating called kokam) (?), seeds (from which oil is extracted).

Chemistry Seed oil (about 50% of dry seed weight): neutral lipids 88% (triglycerides 85%, free fatty acids 8% [stearic acid 38–56%, oleic acid 30–53%, palmitic acid, linoleic acid], diglycerides 3%), glycolipids 4% (digalactosyl diglyceride 40%, monogalactosyl diglyceride 20%), phospholipids 3% (phosphatidyl ethanolamine 75%). An interesting feature of this oil is the melting point (39.5–40°C), which makes it solid at room temperature; with the extraction of the oil from the seeds, a mush remains with 70% digestible nutrients, of which protein 9–17%, fiber 4%. Fruit (sweet and sour skin): hydroxycitric acid (HCA) 10–30%, garcinol, cyanidin-3-sambudioside, cyanidin-3-glucoside, other components.

Style Saison.

Beer Salted Kokum Saison, Great State Aleworks (Pune, India).

Source https://www.brewersfriend.com/other/kokum/.

Gaultheria procumbens L.
ERICACEAE

Synonyms *Brossaea procumbens* (L.) Kuntze, *Gaultheria humilis* Salisb., *Gautiera procumbens* (L.) Torr.

Common Names Wintergreen, checkerberry, eastern spicy-wintergreen, eastern teaberry, boxberry, American wintergreen, Canada tea, canterberry.

Description Shrub (5–20 cm), evergreen. Stem woody, rhizome creeping underground from which stolons and herbaceous branches emerge (5–15 cm). Leaves simple, briefly petiolate (2–5 mm), lamina coriaceous, oval, obovate or orbicular (2–5 × 1–3 cm), base cuneate or rounded, margin entire to toothed, flat or slightly repand, color bright dark green, reddish in winter, apex acute, obtuse or rounded. Flowers hermaphrodite, small (8–10 mm), urceolate, pendulous from a short stalk (1–3 mm) at the leaf axil; sepals 5, green in color, petals 5, pinkish-white, carpels 4–5, ovary superior, style 1, stamens 8–10. Fruit berry (6–9 × 7–10 mm), red in color, with many seeds (at least 20).

Related Species The genus *Gaultheria* includes 165 taxa. *Gaultheria humifusa* (Graham) Rydb. and *Gaultheria ovatifolia* A.Gray can be used in making beer.

Geographic Distribution E North America. **Habitat** Evergreen forests (conifers or oaks), clearings, marshes in acidic, nutrient-poor soils. **Range** Grown and sometimes naturalized in temperate climate areas.

Beer-Making Parts (and Possible Uses) Oil (extracted from leaves), fruits, leaves.

Toxicity Sensitivity to the main active component (aspirin-like) of the oil is manifested by allergic symptoms or asthma; 30 ml of oil are equivalent to 55.7 g of aspirin, so the risk of toxicity and overdose is possible even in small quantities and in nonallergic subjects.

Chemistry Essential oil (obtained by steam distillation of the leaves, previously macerated in hot water to activate the enzymatic reaction that produces methyl salicylate from a glycoside present in the leaves): methyl salicylates 98%, α-pinene, myrcene, δ-3-carene, limonene, 3,7-guaiadiene, δ-cadinene. Fruits: water 79.9%, dry matter 20.1%, lipids 0.23%, protein 0.8%, sugars 7.48%.

Style American Stout, Blonde Ale, Braggot, California Common Beer, Clone Beer, Holiday/Winter Special Spiced Beer, Mild, Specialty Beer, Spice/ Herb/ Vegetable Beer.

Beer Wintergreen IPA, Generations Brewing (Illinois, United States).

Source Fisher and Fisher 2017; Heilshorn 2017; Hieronymus 2016b; https://www.brewersfriend.com/other/wintergreen/.

Gaylussacia baccata (Wangenh.) K.Koch
ERICACEAE

Synonyms *Adnaria resinosa* (Torr. & A.Gray) Kuntze, *Andromeda baccata* Wangenh., *Decachaena baccata* (Wangenh.) Small, *Decamerium resinosum* Nutt., *Gaylussacia resinosa* (Aiton) Torr. & A.Gray, *Vaccinium resinosum* Aiton.

Common Names Black huckleberry, crackleberry.

Description Shrub (0.5–1 m), deciduous. Bark smooth, dark in adult branches, reddish in young branches. Leaves simple (20–50 × 10–25 mm), caducous, petiolate (1–4 mm), herbaceous, lamina elliptic, oblanceolate, ovate, base cuneate, apex round to obtuse, underside ± hairy, upper side with few hairs, main veins pinnate, margin entire. Flowers hermaphrodite, red, yellow, or green petals; green sepals; perianth parts fused to form a cupped or campanulate tube; 1 pistil; 10 stamens. Inflorescence raceme borne by older branches at leaf axil. Fruit fleshy drupe (Ø 6–7 mm), glabrous, black in color. Seeds about 2 mm.

Related Species The genus *Gaylussacia* includes about 60 species. In North America there are many with characteristics or possible uses similar to those of *Gaylussacia baccata*, including *Gaylussacia bigeloviana* (Fernald) Sorrie & Weakley, *Gaylussacia brachycera* (Michx.) Torr. & A.Gray, *Gaylussacia dumosa* (Andrews) Torr. & A.Gray, *Gaylussacia frondosa* (L.) Torr. & A.Gray, *Gaylussacia mosieri* Small, *Gaylussacia nana* (A.Gray) Small, *Gaylussacia orocola* (Small) Camp, *Gaylussacia tomentosa* (A.Gray) Pursh ex Small, and *Gaylussacia ursina* (M.A.Curtis) Torr. & A.Gray.

Geographic Distribution E North America. **Habitat** Forests, grasslands, pastures, scrub in sandy or rocky soils, roads, wetlands, bogs, acidic swamps. **Range** Plant cultivated in the N of the Northern Hemisphere.

Beer-Making Parts (and Possible Uses) Fruits.

Chemistry Fruits of *Gaylussacia brachycera* (Michx.) Torr. & A.Gray (no data available for *G. baccata*; % values in reference to daily intake): energy 37 kcal, lipids 0.1 g, sodium 10 mg, carbohydrates 9 g, protein 0.4 g, vitamin A 1%, vitamin C 4%, calcium 1%, iron 1%.

Style American IPA, American Pale Ale, American Stout, Blonde Ale, Foreign Extra Stout, Fruit Beer, Fruit Lambic, Russian Imperial Stout.

Beer Cantina Mirtilli, Crak (Campodarsego, Veneto, Italy).

Source Hieronymus 2016b; https://www.brewersfriend.com/other/huckleberrys/.

Gentiana andrewsii Griseb.
GENTIANACEAE

Synonyms *Cuttera catesbei* Raf., *Dasystephana andrewsii* (Griseb.) Small, *Pneumonanthe andrewsii* (Griseb.) W.A.Weber.

Common Names Closed bottle gentian, closed gentian, bottle gentian, Andrew's bottle gentian.

Description Herb (3–8 dm), perennial. Taproot robust. Stems numerous, straight, unbranched, smooth, circular in cross-section, green to violet in color. Leaves opposite, entire, glossy green (3–15 × 2–4 cm), ovoid-oblong, sessile, pointed; basal leaves often reduced to small, fused bracts, while median and upper leaves are progressively larger; upper 4–6 leaves verticillate at the base of a flower cluster. Inflorescence consisting of flowers in whorls, the terminal whorl with 2–10 flowers; flowers also at the axil of the upper leaves. Flower color blue-violet, 3–4.5 cm, sessile, closed, bottle shaped; calyx with lobes with slightly hairy margin. Fruit elliptical capsule, bivalve. Several seeds per fruit, oblong, shiny, whitish-brown in color.

Subspecies *Gentiana andrewsii* Griseb. var. *dakotica* A.Nelson is a recognized variety besides the typical *Gentiana andrewsii* Griseb. var. *andrewsii*.

Related Species The genus *Gentiana* includes approximately 470 taxa.

Geographic Distribution NE NC North America. **Habitat** Anthropogenic environments, meadows, usually wet fields. **Range** NE NC North America.

Beer-Making Parts (and Possible Uses)
Roots (substitute plant for *Gentiana lutea*).

Chemistry Species in the genus *Gentiana* often contain biologically active constituents, especially secoiridoid glycosides and xanthones. Examples include *Gentiana acaulis* L. (gentiakochianin, gentiacaulein), *Gentiana cruciata* L. (swertiamarin, gentiopicrin), *Gentiana kurroo* Royle (sweroside, swertiamarin, gentiopicroside), and *Gentiana lutea* L. (gentiopicroside, amarogentin, swertiamarin, amaroswerin, isogentisin, mangiferin, gentiopicrin, gentisin). These substances have known and demonstrated gastroprotective, hepatoprotective, anti-inflammatory, antibacterial, and cytotoxic actions. Some of these substances (or maybe others) could also be present in *Gentiana andrewsii*, but no specific information was found in the literature.

Style Spice Beer, Herb Beer.

Source Fisher and Fisher 2017; Heilshorn 2017.

Gentiana angustifolia Vill.
GENTIANACEAE

Synonyms *Ciminalis angustifolia* (Vill.) Holub, *Diploma angustifolia* Raf., *Ericala angustifolia* G.Don, *Ericoila angustifolia* (Vill.) Laínz, *Eyrythalia angustifolia* D.Don, *Gentiana caulescens* Lam., *Gentiana sabauda* Boiss. & Reut. ex Rchb.

Common Names Stemless gentian, short-stemmed gentian, trumpet gentian.

Description Herb (10 cm), perennial, glabrous. Stem reduced. Leaves linear-lanceolate, opposite, entire, without stipules. Flowers hermaphrodite, actinomorphic, pentamerous, solitary, terminal; calyx with lanceolate lobes, usually less than half the tube; corolla (4–7 cm) obconic, dark blue, with ± acute lobes, with small appendages located within the lobules of the corolla, throat and lobes not ciliated; androecium consisting of 5 stamens, inserted on the corolla tube, with connate anthers; gynoecium superior, unilocular, stigma bilobed. Fruit septicidal capsule.

Subspecies In addition to the typical subspecies, two taxa have been described that should be included in the variability of this species: *Gentiana angustifolia* Vill. subsp. *corbariensis* (Braun-Blanq.) Renob. and *Gentiana angustifolia* Vill. subsp. *occidentalis* (Jakow.) M.Laínz; their taxonomic position, however, is not completely clear.

Related Species The genus *Gentiana* includes about 470 taxa.

Geographic Distribution Alps, Pyrenees. **Habitat** Alpine meadows. **Range** Alps, Pyrenees.

Beer-Making Parts (and Possible Uses) Roots (substitute plant for *Gentiana lutea*).

Chemistry Species of the genus *Gentiana* often contain biologically active constituents, especially secoiridoid glycosides and xanthones. Examples include *Gentiana acaulis* L. (gentiakochianin, gentiacaulein), *Gentiana cruciata* L. (swertiamarin, gentiopicrin), *Gentiana kurroo* Royle (sweroside, swertiamarin, gentiopicroside), and *Gentiana lutea* L. (gentiopicroside, amarogentin, swertiamarin, amaroswerin, isogentisin, mangiferin, gentiopicrin, gentisin). These substances have known and demonstrated gastroprotective, hepatoprotective, anti-inflammatory, antibacterial, and cytotoxic actions. Some of these substances (or maybe others) could also be present in *Gentiana angustifolia*, but no specific information was found in the literature.

Style Spice Beer, Herb Beer.

Beer Quadro, Barbaforte (Folgaria, Trentino-Alto Adige, Italy).

Source Fisher and Fisher 2017; Giaccone and Signoroni 2018.

Gentiana lutea L.
GENTIANACEAE

Synonyms *Asterias hybrida* G.Don, *Asterias lutea* (L.) Borkh., *Coilantha biloba* Bercht. & J.Presl, *Gentiana major* Bubani, *Gentianusa lutea* (L.) Pohl.

Common Names Great yellow gentian, yellow gentian, bitter root, bitterwort, root, pale gentian.

Description Herb (4–15 dm), perennial, glabrous, ± glaucous. Root enlarged (Ø 1–3 cm), pale fleshed. Stem erect, simple. Basal leaves with 1 dm petiole and lanceolate lamina (1 × 2.5 dm) with 5 parallel main veins; cauline leaves (4–8 pairs) progressively sessile and reduced, the upper ones bract shaped, heart-shaped-acuminate (± 6 × 6 mm). Flowers in whorls: peduncle 1 cm, calyx reduced to a membranous bract (10 × 13 mm); corolla yellow with 2–3 lines of brown dots above, almost completely divided into 5–9 acute laciniae of 4 × 20–26 mm. Fruit 3–6 cm clavate capsule.

Subspecies *Gentiana lutea* L. includes four subspecific taxa: *Gentiana lutea* L. subsp. *lutea*, *Gentiana lutea* L. var. *aurantiaca* (M.Laínz) M.Laínz, *Gentiana lutea* L. subsp. *montserratii* (Vivant ex Greuter) Romo, and *Gentiana lutea* L. subsp. *symphyandra* (Murb.) Hayek.

Related Species The genus *Gentiana* includes more than 400 species, many of which have properties similar to *G. lutea*; in addition to those mentioned in this book, it is certain that *Gentiana saponaria* L. can also be used in making beer; the same is true for some species belonging to other genera in the Gentianaceae family (e.g., *Centaurium erythraea* Rafn); since *G. lutea* enjoys strict protection in much of its natural range, less rare species are also cultivated or used as a replacement.

Geographic Distribution Orophyte S Europe. **Habitat** Mountain meadows and pastures (calcareous). **Range** The species is widely cultivated within its natural range, but, owing to its commercial importance, several attempts (some successful) have been made to cultivate it in other areas.

BEER-MAKING PARTS (AND POSSIBLE USES) Roots (used before the introduction of hops), other parts of the plant (?).

CHEMISTRY Root: bittering principles are the secoiridoid glycosides amarogentin and gentiopicrin, gentiamarin, and gentiin; other components are dextrose, gentianic acid (gentisin), gentianose, levulose, and sucrose; amarogentin is one of the most bitter natural compounds known and is used as a scientific basis for measuring bitterness.

STYLE Ancient Ale, Gruit.

BEER Nature Genziana, Opperbacco (Notaresco, Abruzzo, Italy).

SOURCE Calagione et al. 2018; Dabove 2015; Fisher and Fisher 2017; Heilshorn 2017; Hieronymus 2016b; McGovern 2017; https://www.birraaltaquota.it/prodotti/birre-estrose/66-eva.html.

Geum urbanum L.
ROSACEAE

SYNONYMS *Caryophyllata officinalis* Moench, *Caryophyllata urbana* (L.) Scop., *Caryophyllata vulgaris* Lam., *Geum caryophyllata* Gilib., *Geum hederifolium* C.C.Gmel., *Geum hirtum* Wahlb., *Geum hyrcanum* C.A.Mey., *Geum ibericum* Besser ex Boiss., *Geum klettianum* Peterm., *Geum robustum* Schur, *Geum roylei* Wall. ex F.Bolle, *Geum* × *rubifolium* Lej., *Geum sordidum* Salisb., *Geum vidalii* Sennen, *Streptilon odoratum* Raf.

COMMON NAMES Wood avens, herb bennet, colewort, St. Benedict's herb.

DESCRIPTION Herb, perennial (5–10 dm). Stem erect, lower part with patent hairs, dichotomoramose. Lower leaves with lamina deeply tripartite or completely divided into 3 ovate or lanceolate segments (3–5 cm); apical leaves cuneate at base. Flowers 2–5 (Ø 1–1.5 cm); petals 4–7 mm yellow; carpophore null. Fruit achene (10 × 2 mm), hirsute with patent hairs, with beak (10 mm) curved into a hook.

RELATED SPECIES The genus *Geum* includes about 40 taxa.

GEOGRAPHIC DISTRIBUTION Circumboreal.
HABITAT Abandoned lands, deciduous woods.
RANGE Circumboreal.

BEER-MAKING PARTS (AND POSSIBLE USES) Apices (?), roots (?).

CHEMISTRY Aerial parts (% or mg/100 g): tannins 1.16–7.44%, flavan-3-ols ([+]-catechin 14.02–47.02 mg, [-]-epicatechin gallate 5.83–73.65 mg), phenolic acids (chlorogenic acid 15.63–289.51 mg, ellagic acid 12.77–125.84 mg, gallic acid 9.87–89.18 mg). Underground organs (% or mg/100 g): tannins 3.35.16%, flavani-3-ols ([+]-catechin 106.56–225.97 mg, [-]-epicatechin 29.36–55.99 mg, [-]-epicatechin gallate 4.35–33.58 mg, [-]-epigallocatechin 33, 69–129.93 mg, [-]-epigallocatechin gallate 50.20–111.08 mg), phenolic acids (ellagic acid 8.33–80.84 mg, gallic acid 39.12–121.83 mg), eugenol 5.61–14.54 mg.

STYLE Gruit.

SOURCE Hamilton 2014.

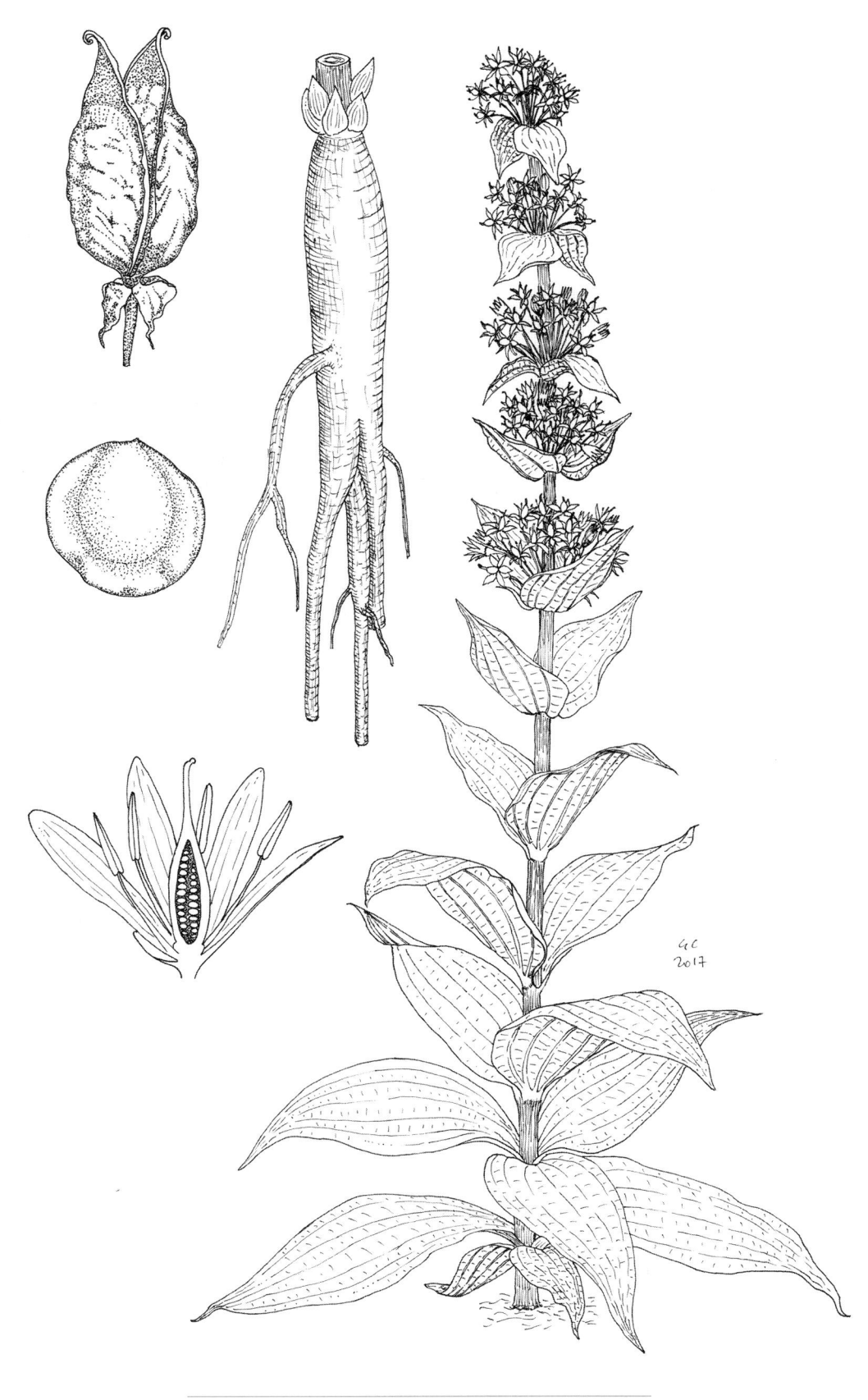

Gentiana lutea L.
GENTIANACEAE

Ginkgo biloba L.
GINKGOACEAE

Synonyms *Ginkgo macrophylla* K.Koch, *Salisburia biloba* (L.) Hoffmanns., *Salisburia macrophylla* Reyn.

Common Names Ginkgo, gingko, ginkgo tree, maiden-hair tree.

Description Tree, deciduous (5–30 m), habit monopodial, bark pale. Leaves deciduous, with 3–7 cm petiole and lamina (3–6 cm) bilobate fan shaped; veins regularly flabellate. Male flowers in catkins, female flowers clustered at the apex of an elongated peduncle. Seed (Ø 1.5–3 cm) containing embryo, edible if roasted, covered by a woody involucre (sclerotesta) in turn contained in a fleshy involucre (sarcotesta) of tegumental origin, yellowish or ± brown in color, fetid at maturity.

Cultivar There are numerous cultivated varieties of this species, including Autumn Gold, Barabits Nana, Beijing Gold, Elmwood, Fairmount, Fastigiata, Globosa, Golden Colonnade®, Golden Girl, Golden Globe, Gresham, Halka Ginkgo, Horizontalis, Jade Butterfly, Laciniata, Lakeview, Liberty Splendor, Magyar Ginkgo, Mariken, Mayfield, Presidential Gold®, Pendula, Princeton Sentry®, Santa Cruz, Saratoga, Tremonia, Troll, Tubifolia, Umbrella, ®Variegata, and Witches Broom.

Related Species The genus *Ginkgo* is monotypic.

Geographic Distribution E Asia. **Habitat** Cultivated for ornamental and practical uses. **Range** Cultivated everywhere as an ornamental and/or medicinal plant.

Beer-Making Parts (and Possible Uses) Leaves, seeds.

Toxicity If consumed in large quantities, the gametophyte contained in the seeds can cause poisoning by 4'-O-methylpyridoxine, which, being nonthermolabile, is resistant to cooking. Some subjects are sensitive to the substances contained in the sarcotesta, the fleshy outer shell; therefore, they must handle the seeds with gloves to avoid contact dermatitis and blisters; seeds without the sarcotesta are mostly safe. According to some authors, *Ginkgo biloba* extracts, often administered with anticoagulant substances, increase the risk of bleeding, but these claims are not supported by scientific evidence. Amentoflavone contained in *Ginkgo biloba* leaves is attributed to potential negative interactions with other drugs. Again, the supporting evidence appears insufficient. *Ginkgo biloba* leaves and the sarcotesta also contain ginkgolic acid, a highly allergenic long-chain alkylphenol (e.g., bilobol, adipostatin A); individuals with a history of allergy to ivy, mango, or cashew nuts are more likely to experience allergic reactions to ginkgo extracts. In addition, male individuals are highly allergenic owing to the pollen released, while females are not appreciated as street trees as their fruits litter the street.

Chemistry Leaves: flavonoids (ginkgolides A, B, C, J, M, quercitin, kaempferol), terpene lactones, amino acids (6-hydroxykynurenic acid), dimeric flavones (bilobetin, ginkgetin, isoginkgetin, sciadopitysin), proanthocyanidins, ginkgolic acid, ascorbic acid, carotenoids, bilobalide. Seeds: essential oil, fatty acids, tannin, resins.

Style American Pale Ale, Herbal Ale.

Source https://www.ratebeer.com/beer/blucreek-herbal-ale/21914/; https://www.brewersfriend.com/other/ginkgo/.

Glebionis coroonaria (L.) Cass. Ex Spach
ASTERACEAE

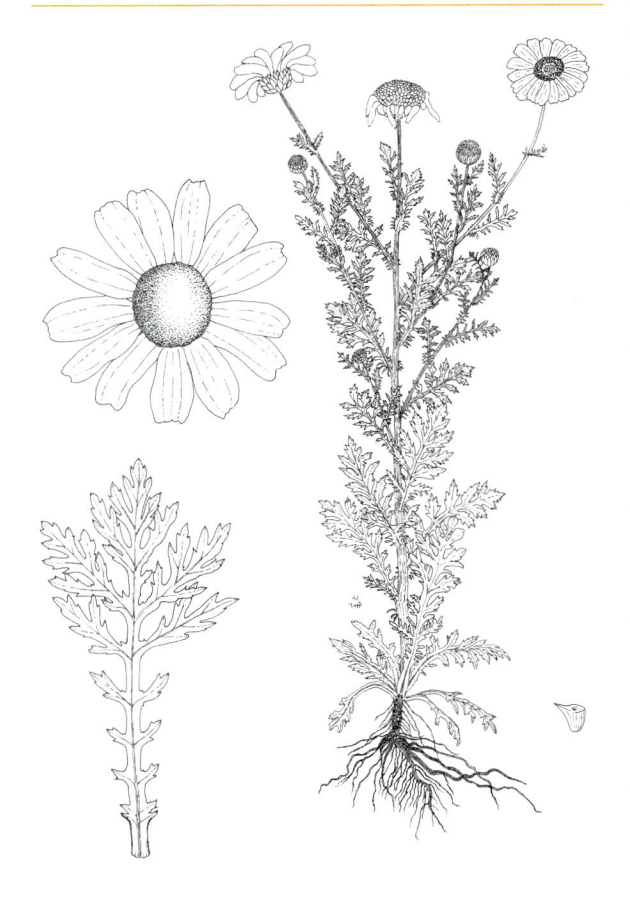

Synonyms *Buphthalmum oleraceum* Lour., *Chamaemelum coronarium* (L.) E.H.L.Krause, *Chrysanthemum breviradiatum* Hort. ex DC., *Chrysanthemum coronarium* L., *Chrysanthemum merinoanum* Pau, *Chrysanthemum roxburghii* Desf. ex Cass., *Chrysanthemum senecioides* Dunal ex DC., *Chrysanthemum spatiosum* (L.H.Bailey) L.H.Bailey, *Chrysanthemum speciosum* Brouss. ex Pers., *Dendranthema coronarium* (L.) M.R.Almeida, *Glebionis coronaria* (L.) Tzvelev, *Matricaria coronaria* (L.) Desr., *Pinardia coronaria* (L.) Less., *Pinardia roxburghii* (Desf. ex Cass.) Less., *Pyrethrum indicum* Roxb., *Pyrethrum roxburghii* Desf., *Xanthophthalmum coronarium* (L.) P.D.Sell, *Xanthophthalmum coronarium* (L.) Trehane ex Cullen.

Common Names Garland chrysanthemum, chrysanthemum greens, edible chrysanthemum, crown-daisy chrysanthemum, chop suey green, crown daisy, Japanese-green.

Description Herb, annual (20–60 cm), with strong aromatic odor. Stems erect, glabrous, very branched. Leaves sessile, light green, bipinnate-partite, divided into lanceolate lobes, those of second order often toothed. Inflorescence capitulum (30–50 mm) on peduncle enlarged at apex, formed by a central disc of tubular orange-yellow floscule, surrounded by ligulate yellow, white-yellow, or partially orange flowers; calyx scales oval, often with blackish edges. Fruits cypselae (2–3 mm), outer ones distinctly trigone, with winged angles, disc ones with 4 angles, compressed with posterior costa more pronounced than others.

Related Species The genus *Glebionis* includes only two species: *Glebionis coronaria* (L.) Cass. ex Spach, described here, and *Glebionis segetum* (L.) Fourr. Species belonging to related genera such as *Chrysanthemum indicum* L. and *Chrysanthemum morifolium* Ramat. may be used in making beer.

Geographic Distribution Steno-Mediterranean. **Habitat** Cultivated fields, fallow land, roadsides, ruins, sunny areas (0–600 m ASL). **Range** Infestant of cultivated fields.

Beer-Making Parts (and Possible Uses) Leaves, young shoots.

Chemistry Essential oil, flowers and leaves (161 components representing 87.2–96.5% of the total): myrcene 3.2–35.7%, (Z)-β-ocimene 0.6–23%, camphor 0.6–17.2%, cis-chrysanthenol 0–6.9%, cis-chrysanthenyl acetate 1.1–17.9%, isobornyl acetate 1.6–3.5%, (E)-β-farnesene 0–6%, germacrene D 0–8.7%, (E,E)-α-farnesene 0.7–12.4%.

Style Saison, Belgian Blond Ale, Belgian Tripel.

Source https://www.brewersfriend.com/other/chrysanthemum/; https://www.todayshomeowner.com/growing-edible-chrysanthemums/.

Glycyrrhiza echinata L.
FABACEAE

Synonyms *Glycyrrhiza inermis* Boros, *Glycyrrhiza macedonica* Boiss. and Orph.

Common Names Chinese licorice, German liquorice, hedgehog liquorice, Eastern European licorice, Hungarian licorice, Roman licorice.

Description Herb, perennial (4–10 dm), slimy. Stem erect, fluted, often bristly. Leaves 8–12 cm with a few oval or elliptical segments (1–2 × 3–5 cm) indented at the edge, rounded at the apex, often mucronate. Racemes ovoid (1–2 cm) at the leaf axil. Flower papilionaceous, blue-white corolla; vexillum 4–6 mm. Fruit legume (5–7 × 12–16 mm) covered with glandular bristles. Seeds 1–3.

Related Species The genus *Glycyrrhiza* includes 22 taxa.

Geographic Distribution SE Europe, Turkey, Syria, Palestine, Caucasus, Iran, E Asia. **Habitat** Arid grasslands. **Range** Likely Pontic. In other areas (SE Europe, Turkey, Syria, Palestine, Caucasus, Iran, E Asia), the plant is only naturalized.

Beer-Making Parts (and Possible Uses) Roots.

Chemistry Root extract, volatile fraction: (E)-2-heptenal 1.5%, 5-methyl-furfural 2.4%, (2E, 4E)-hepta-dienol 4.2%, (E)-2-octen-1-oral 7.4%, lavandulol 1.9%, decanal 1.6%, (2E, 4E)-nonadienal 2.1%, (2E, 4Z)-decadienal 6.4%, (2E, 4E)-decadienal 20.8%, β-caryophyllene 1.5%, spathulenol 4.3%, β-caryophyllene oxide 24.5%, humulene epoxide II 11.7%.

Style Porter, Stout.

Source Fisher and Fisher 2017.

Glycyrrhiza glabra L.
FABACEAE

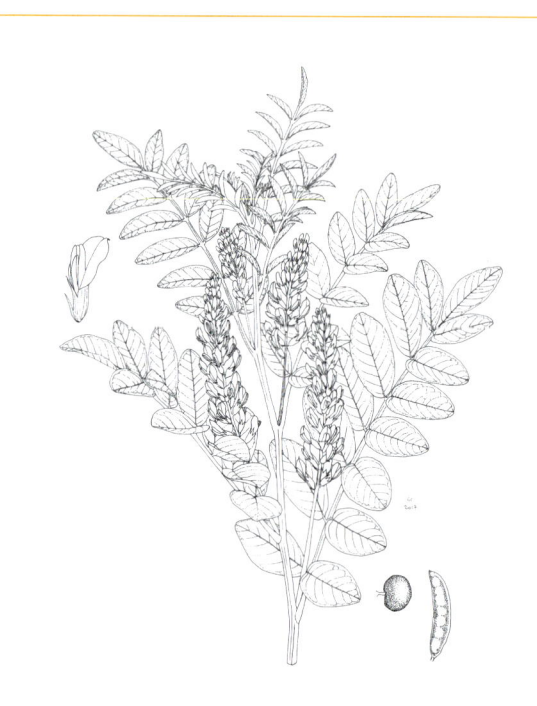

Synonyms *Glycyrrhiza brachycarpa* Boiss., *Glycyrrhiza glandulifera* Waldst. & Kit., *Glycyrrhiza hirsuta* Pall., *Glycyrrhiza pallida* Boiss., *Glycyrrhiza pallida* Boiss. & Noe, *Glycyrrhiza violacea* Boiss. & Noe.

Common Names Liquorice, licorice, sweet wood.

Description Herb, perennial (4–10 dm), slimy. Stem erect, fluted, often bristly. Leaves 8–12 cm with 9–15 oval or elliptic segments (1–2 × 3–5 cm), at apex rounded and often mucronate. Racemes smaller than the axillary leaf, elongate and ± linear. Flower papilionaceous, blue-white corolla; vexillum 8–12 mm.

Fruit legume (5–7 × 15–30 mm) glabrous or with few bristles. Seeds 2–6.

Related Species The genus *Glycyrrhiza* includes 22 taxa, some of which have traditional medicinal uses or even uses entirely analogous to *Glycyrrhiza glabra*, as in the case of *Glycyrrhiza echinata* L.

Geographic Distribution W Asia, Steno-Mediterranean. **Habitat** Arid grasslands. **Range** India, Iran, Afghanistan, China, Pakistan, Iraq, Azerbaijan, Uzbekistan, Turkmenistan, Turkey, Great Britain.

Beer-Making Parts (and Possible Uses) Roots.

Toxicity Only occasional intake is recommended (without exceeding 0.5 g/day of glycyrrhizin) because the active ingredient acts on the balance of mineral salts, causing water retention, increased blood pressure (through sodium reabsorption and potassium excretion), swelling, headache, and asthenia. People with hypertension, diabetes, or cirrhosis or who are pregnant or have recently given birth should avoid prolonged use of extracts of this plant. It has also been shown that the increase in blood pressure occurs only by administering the extract, not the root itself. Some healthy people (e.g., children, adults over 55 years of age) show a certain sensitivity to the active compound, especially if used at higher doses than those recommended and for long periods of time.

Chemistry Root: glycyrrhizin (glycyrrhizic acid, glycyrrhizinate; saponin glycoside; 30–60 times sweeter than sucrose) 10–25% of extract, flavonoids (liquirtin, isoliquertin, liquiritigenin, rhamnoliquirilin, glucoliquiritin apioside, prenyllicoflavone A, shinflavanone, shinpterocarpine, 1-methoxyphaseolin), lycopyranocoumarin, licoarylcoumarin, glysoflavone, coumarin-GU-12, phenolic constituents (semilicoisoflavone B, 1-methoxyfolinol, isoangustone A, lycorifenone), kanzonol R, volatile component (anethole 3%, pentanol, hexanol, linalool oxide A and B, tetramethyl pyrazine, terpinen-4-ol, a-terpineol, geraniol, and others), propionic acid, benzoic acid, ethyl linoleate, methyl ethyl ketin, 2,3-butanediol, furfuraldehyde, furfuryl formate, 1-methyl-2-formylpyrrole, trimethylpyrazie, maltol, and other compounds from the essential oil; phytoestrogens (isoflavene glabrene, isoflavene glabridin).

Style American IPA, American Light Lager, American Porter, American Stout, Baltic Porter, Belgian Dark Strong Ale, Belgian Dubbel, Belgian Pale Ale, British Strong Ale, Classic Rauchbier, Cream Ale, Foreign Extra Stout, German Pilsner (Pils), Imperial Stout, Oatmeal Stout, Old Ale, Other Smoked Beer, Russian Imperial Stout, Sweet Stout, Winter Seasonal Beer.

Beer Rusty Nail, Fremont (Washington, United States).

Source Dabove 2015; Fisher and Fisher 2017; Giaccone and Signoroni 2017; Heilshorn 2017; Hieronymus 2016b; Jackson 1991; Sparrow 2015; https://www.brewersfriend.com/other/liquorice/; https://www.brewersfriend.com/other/regaliz/; https://www.gruitale.com/bot_licorice.htm; https://www.sangabriel.it/it/birre/le-stagionali/nera-opitergium; https://unabirralgiorno.blogspot.it/2013/10/traquair-jacobite-ale.html.

Gossypium hirsutum L.
MALVACEAE

Synonyms *Gossypium birkinshawii* G.Watt, *Gossypium caespitosum* Tod., *Gossypium cavanillesianum* Tod., *Gossypium harrisii* G.Watt, *Gossypium jamaicense* Macfad., *Gossypium janiphifolium* Bello, *Gossypium lanceolatum* Tod., *Gossypium latifolium* Murray, *Gossypium marie-galante* G.Watt, *Gossypium mexicanum* Tod., *Gossypium nervosum* G.Watt, *Gossypium nicaraguense* Ram.Goyena, *Gossypium*

oligospermum Macfad., *Gossypium palmeri* G.Watt, *Gossypium parvifolium* Nutt. ex Seem., *Gossypium prostratum* Schumach. & Thonn., *Gossypium punctatum* Schumach. & Thonn., *Gossypium religiosum* L., *Gossypium rhorii* Tod., *Gossypium rufum* Scop., *Gossypium sandvicense* Parl., *Gossypium schottii* G.Watt, *Gossypium sericatum* Prokh., *Gossypium siamense* Tussac, *Gossypium tomentosum* Nutt. ex Seem., *Gossypium volubile* Ram. Goyena, *Hibiscus religiosus* (L.) Kuntze, *Xylon religiosum* (L.) Moench.

Common Names Cotton, upland cotton, Mexican cotton.

Description Herb, annual (cultivated), also shrub (1–2 m). Stem generally broadly branched, terete, hairy from stellate hairs. Leaves with terete petiole, 1/2 the length of the lamina (10–14 × 8–11 cm), this 3- to 5-lobed, lobes broadly ovate (4–10 cm), membranous, base cordate, apex acute to acuminate, glabrous or hairy surface; stipules subulate to falcate (5–20 mm). Flowers solitary in leaf axil; pedicels 2–4 cm, epicalyx segments ovate (to 4 cm); sepals almost completely joined into a cupuliform calyx; petals creamy yellow in color, 2–5 cm; staminal column 15 mm, glabrous; stylus somewhat longer than androecium; 3–5 stigmas. Fruit capsule 3- to 5-locular, broadly ovoid or subglobose, 2 4 cm, smooth, glabrous. Seeds 8–10 mm, comose, hairs (cotton) usually white in color.

Subspecies The two recognized varieties of this taxon are *Gossypium hirsutum* L. var. *hirsutum* and *Gossypium hirsutum* L. var. *taitense* (Parl.) Roberty.

Cultivar There are many cultivars of cotton. Some examples recently tested in producing countries: Pakistan (CIM-446, CIM-448, CIM-1100, Karishma, MNH-93, NIAB-78), South Africa (Candia BGRF, Carla, Delta 12 BRF, DP 1240 B2RF, DP 1531 B2RF, DP 1541 B2RF, PM 3225 B2RF).

Related Species The genus *Gossypium* includes about 60 taxa.

Geographic Distribution N America. **Habitat** Coastal, cultivated fields (0–20 m ASL). **Range** Plant cultivated in many countries (China, India, United States, Pakistan, Brazil, Uzbekistan, Turkey, Australia, Turkmenistan, Mexico), frequently wild or naturalized.

Beer-Making Parts (and Possible Uses) Fiber (used to stop small leaks from the spaces between barrel staves).

Chemistry Cotton fiber, almost entirely made of cellulose, contains, mainly on the outer surface, modest amounts of other chemical compounds: pectins, waxes, soluble salts, and sugars.

Style None known.

Source Cantwell and Bouckaert 2016; Cantwell and Bouckaert 2018.

Haematoxylum campechianum L.
FABACEAE

Synonyms *Cymbosepalum baronii* Baker.

Common Names Logwood, campeachy wood, bloodwood tree.

Description Small tree or tree (up to 15 m), often spiny and gnarled. Stem irregularly fluted and twisted (60 × 200–300 cm), extended into broad, long, straight branches, bark gray to brown, rather smooth, from which scales detach. Leaves alternate, paripinnate, distichous or fasciculate, on very short branches; stipules partially spine shaped; leaflets in 2–4 pairs, obcordate or obovate (10–35 × 5–25 mm), acute at base, notched at apex, with dense, glabrous veins. Flowers in racemes (5–20 cm) at the leaf axil (present or fallen), pentamerous, with sweet odor; deeply lobed calyx (4–5 mm); bright yellow petals (5–7 mm); 10 free stamens; ovary superior, slightly pedunculate, glabrous; stylus filiform. Fruit legume lanceolate, extremely flattened, 3–5 cm long, pointed at both ends, dehiscent not along sutures but along median of both sides, usually with 2 seeds.

RELATED SPECIES Congeners are *Haematoxylum brasiletto* H.Karst., *Haematoxylum dinteri* (Harms) Harms, and *Haematoxylum sousanum* R.Cruz D. & J.Jiménez Ram.

GEOGRAPHIC DISTRIBUTION Campeche (Mexico). **HABITAT** Lowland marshy areas often flooded by rivers but also calcareous hills as secondary arid forests. **RANGE** C America, West Indies, Fiji, Indonesia, Malaysia, Philippines, Singapore, India, SE Asia.

BEER-MAKING PARTS (AND POSSIBLE USES) Wood (from which the active ingredient haemoxylin is extracted).

TOXICITY Poisonous (used to adulterate porter beer with substances that simulate the effect of alcohol but are less expensive than raw materials such as malt, sugar, and molasses).

CHEMISTRY Hematoxylin is a compound extracted from the heartwood of this plant commonly used as a permanent histological colorant.

WOOD Wood (from which hematoxylin is extracted) very hard and heavy, WSG 0.950–1.085. Sapwood ring is thin, white or yellowish, and does not contain hemoxylin; heartwood turns bright red on exposure; is compact, with concatenated grains and a coarse but still smooth texture; has a pleasant violet odor and a sweet, astringent aroma; use as a work wood is severely limited by the irregularity of the trunk. The wood is strong but brittle, durable for outdoor use and when in contact with the ground. It is sometimes used for furniture and decoration because it lends itself to smoothness. A variety of pigments (gray, brown, violet, blue, black) are extracted from the heartwood and used to permanently dye textiles (silk, wool, cotton, nylon, rayon), leather, fur, feathers, paper, and bone and to make inks. Hematoxylin, the staining agent in logwood, is a histological dye for dyeing the cell nucleus; alcoholic solutions serve as an indicator for the titration of alkaloids.

STYLE Porter.

SOURCE Daniels 2016.

Helianthus annuus L.
ASTERACEAE

SYNONYMS *Helianthus aridus* Rydb., *Helianthus indicus* L., *Helianthus jaegeri* Heiser, *Helianthus lenticularis* Douglas, *Helianthus macrocarpus* DC., *Helianthus macrocarpus* DC. & A.DC., *Helianthus multiflorus* Hook., *Helianthus ovatus* Lehm., *Helianthus platycephalus* Cass., *Helianthus tubaeformis* Nutt.

COMMON NAMES Sunflower, common sunflower.

DESCRIPTION Herb, annual (10–30 dm). Stems erect, usually bristly. Leaves mostly cauline, alternate, petioles 2–20 cm; lamina lanceolate-ovate to ovate (10–40 × 5–40 cm), base cuneate to subcordate or cordate, margin serrate; underside usually ± bristly, sometimes punctate with glands. Inflorescences capitula 1–9; peduncles 2–20 cm; involucres hemispherical or broader, Ø 15–200 (or +) mm; phyllaries 20–100 (or +), ovate to lanceolate-ovate (13–25 × 3–8 mm), margins usually ciliate, apices abruptly narrowed, long acuminate, abaxial face usually hirsute to bristly, rarely glabrate or glabrous, usually punctate with glands; paleae 9–11 mm, 3-dentate, median tooth long acuminate, glabrous or bristly. Ray flowers (ligulate) 13–100 (or +); lamina 25–50 mm;

disc flowers (tubular) 150–1,000 (or +); corolla 5–8 mm (tube ± bulbous at base), lobes usually reddish, sometimes yellow; anthers brown to black, yellow to dark appendages (yellow stylar branches). Fruit cypsela (3–15 mm), glabrate; pappus consisting of 2 lanceolate scales (23.5 mm) and 0–4 obtuse scales (0.5–1 mm).

SUBSPECIES Despite considerable morphological variability, attempts to include *Helianthus annuus* in a single, widely accepted system of infraspecific classification so far have failed. Even the forms with red ligulate flowers, cultivated and sometimes spontaneous, would all seem to be attributable, together with the others, to a single mutant ancestor.

CULTIVAR They are numerous, including Autumn Beauty, Chianti, Earthwalker, Indian Blanket, Italian White, Lemon Eclair, Mammoth Russian, Moulin Rouge, Pro Cut Bicolor, Pro Cut Orange, Pro Cut Red and Lemon Bicolor, Stella Gold, Strawberry Blonde, Terra Cotta, The Joker, Tiffany, Tohoku Yae, Vanilla Ice, and Yellow Lite.

RELATED SPECIES The genus *Helianthus* includes 85 taxa.

GEOGRAPHIC DISTRIBUTION Canada, United States, Mexico. **HABITAT** Open environments (0–3,000 m ASL). **RANGE** Species cultivated worldwide (the only North American plant to have such wide agronomic success and global diffusion).

BEER-MAKING PARTS (AND POSSIBLE USES) Seeds (malted or not), heads.

CHEMISTRY Seeds: water 6–14%, protein 8–19%, lipids 22–36%, carbohydrates 13–21%, cellulose and ash 25–35%. Roasted seeds: energy 582 kcal, water 1.20 g, protein 19.33 g, lipids 49.8 g, carbohydrates 24.07 g, fiber 11.1 g, sugars 2.73 g, calcium 70 mg, iron 3.8 mg, magnesium 129 mg, phosphorus 1,155 mg, potassium 850 mg, sodium 3 mg, zinc 5.29 mg, vitamin C 1.4 mg, thiamine 0.106 mg, riboflavin 0.246 mg, niacin 7.042 mg, vitamin B-6 0.804 mg, folate 237 µg, vitamin A 9 IU, vitamin E 26.1 mg, vitamin K 2.7 µg.

STYLE Gluten-Free Beer.

SOURCE Calagione et al. 2018; Hieronymus 2016b; Josephson et al. 2016.

Heracleum maximum W. Bartram
APIACEAE

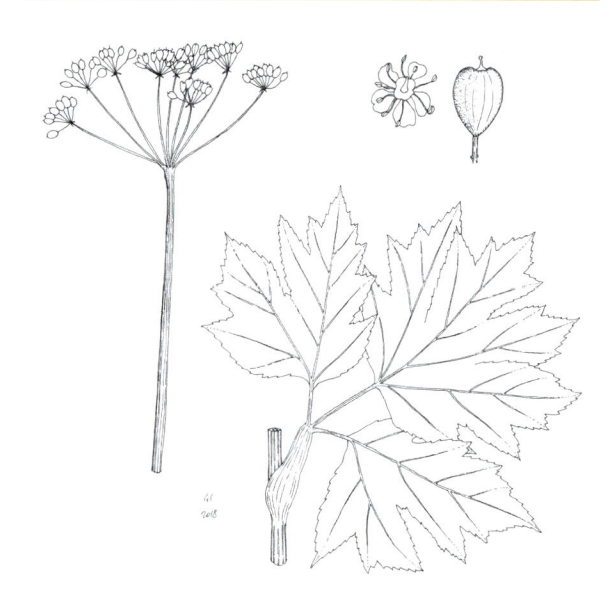

SYNONYMS None
COMMON NAMES Cow parsnip, Indian celery, Indian rhubarb.

DESCRIPTION Grass (1–3 m), stiff, strongly aromatic, tomentose. Stem rigid and succulent. Leaves green, lamina round to reniform, 2–5 dm, ternate, single leaf unit 1–4 dm, ovate to round, cordate, coarsely serrate and lobed; petiole 1–4 dm, broadly sheathing, the upper sheaths enlarged, without lamina. Inflorescence compound umbel, tomentose or long hairy, flattened or rounded apex; peduncle 5–20 cm; rays 15–30, 5–10 cm, unequal; pedicels 8–20 mm. Flowers with obovate petals, white in color. Fruit 8–12 mm, obovate to obcordate, ± hairy.

RELATED SPECIES The genus *Heracleum* includes about 74 taxa.

GEOGRAPHIC DISTRIBUTION North America. **HABITAT** Wet, wooded, or unwooded locations (0–2,700 m ASL). **RANGE** North America.

BEER-MAKING PARTS (AND POSSIBLE USES) Flowers, stems, seeds, roots.

TOXICITY The stem and leaves contain furocoumarins, chemicals responsible for the characteristic erythematous rash with blistering and subsequent hyperpigmentation caused by skin contact with the sap. These substances are photosensitive, that is active after exposure to ultraviolet light; therefore, maximum attention should be paid after cutting or mowing the plant on sunny days.

CHEMISTRY Essential oil, fruits (identified compounds 97.8% of total): octyl acetate 65.6%, octyl butyrate 7.9%, octanol 6.2%, (Z)-3-octenyl acetate 3.3%, octyl hexanoate 2.2%, hexyl butyrate 1.8%, myrcene 1.6%, octanal 0.9%, limonene 0.8%, decanal 0.7%, (Z)-phalcarinol 0.6%, hexyl acetate 0.4%, unknown compound 0.4%, α-pinene 0.3%, (Z)-4-octenyl butyrate 0.3%, decen-1-yl acetate 0.3%, (Z)-4-decen-1-yl acetate 0.3%, germacrene D 0.3%, prenol 0.2%, sabinene 0.2%, (Z)-5-octenol 0.2%, (Z)-5decen-1-yl acetate 0.2%, (E)-4-decen-1-yl acetate 0.2%, decyl acetate 0.2%, α-himachalene 0.2%, amorpha-4,11-diene 0.2%, α-zingiberene 0.2%, hexadecanoic acid 0.2%, (E)-2-hexenal 0.1%, hexanol 0.1%, α-thujene 0.1%, camphene 0.1%, nonanal 0.1%, (Z)-4-decenal 0.1%, (Z)-4-octenyl acetate 0.1%, decanol 0.1%, bornyl acetate 0.1%, nonyl acetate 0.1%, geranyl acetate 0.1%, β-cubebene 0.1%, hexyl hexanoate 0.1%, (Z)-6-decen-1-yl acetate 0.1%, 9-decen-1-yl acetate 0.1%, octyl 2-methylbutyrate 0.1%, amyl α-pyrone 0.1%, β-sesquiphellandrene 0.1%, (Z)-7-tetradecenal 0.1%, traces of the following compounds: tiglic aldehyde, prenal, hexanal, isopropyl butyrate, (E)-3-hexenol, (E)-2-hexenol, 2-methylbutyl acetate, isopropyl 2-methylbutyrate, isopropyl isovalerate, (Z)-4-nonene, nonane, heptanal, (E)-2-heptenal, isopropyl senecioate, β-pinene, octen-3-one, para-cymene, +β-phellandrene, cis-β-ocimene, trans-β-ocimene, isopropyl hexanoate, (E)-2-octenal, isoamyl butyrate, γ-terpinene, cis-sabinene hydrate, 4-nonanone, heptyl acetate, cis-limonene oxide, terpinen-4-ol, (Z)-7-decenal, (Z)-4-decenol, heptyl butyrate, octyl isobutyrate, citronellyl acetate, neryl acetate, α-copaene, β-caryophyllene, α-humulene, ar-curcumene, isodaucene, epi-cubebole, β-bisabolene, cis-α-bisabolene, cubebole, trans-γ-bisabolene, elemin, geranyl butyrate, (Z)-4-octenyl hexanoate, decyl butyrate, tetradecanal, (Z)-8-hydroxylinalyl butyrate, α-cadinol, germacra-4(15)-5,10(14)-trien-1-α-ol, geranyl hexanoate, tetradecanoic acid, octyl octanoate, decyl hexanoate, hexadecanal, hexadecanol, nonadecane, (Z)-9-octadecenal, methoxalene, iso-bbergaptene, oleyl alcohol, pimpinellin, tricosane, imperatorin, benzaldehyde. Essential oil, stems (identified compounds 98.4% of total): limonene 45.2%, sabinene 9.5%, α-thujene 7.7%, α-pinene 6%, myrcene 5.8%, (Z)-phalcarinol 3.3%, β-pinene 2.5%, γ-terpinene 1.5%, camphene 1.4%, bornyl acetate 1.4%, germacrene D 1.2%, amorphous-4,11-diene 1.1%, octyl acetate 0.8%, β-phellandrene 0.7%, α-himacalene 0.6%, geranyl acetate 0.5%, isodaucene 0.5%, para-cymene 0.4%, trans-β-ocimene 0.4%, terpinen-4-ol 0.4%, citronellyl acetate 0.4%, β-sesquiphellandrene 0.4%, isobutyl 2-methylbutyrate 0.3%, α-terpinene 0.3%, α-zingiberene 0.3%, isopropyl isovalerate 0.2%, isobutyl isovalerate 0.2%, isobutyl isovalerate 0.2%, cis-β-ocimene 0.2%, octanol 0.2%, terpinolene 0.2%, 2-methylbutyrate 0.2%, β-bourbonene 0.2%, isopropyl 2-methylbutyrate 0.1%, nonane 0.1%, octanal 0.1%, isoamyl isobutyrate 0.1%, cis-sabinene hydrate 0.1%, linalool 0.1%, 2-methylbutyl isovalerate 0.1%, trans-limonene oxide 0.1%, estragole 0.1%, decanal 0.1%, thymol methyl ether 0.1%, piperitone 0.1%, geraniol 0.1%, (E)-2-decenal 0.1%, myrtenyl acetate 0.1%, trans-carvil acetate 0.1%, α-terpinyl acetate 0.1%, α-copaene 0.1%, β-cubebene 0.1%, (Z)-4-decen-1yl acetate 0.1%, octyl butyrate 0.1%, phenylethyl isobutyrate 0.1%, β-caryophyllene 0.1%, α-humulene 0.1%, trans-β-farnesene 0.1%, arcurcumene 0.1%, phenylethyl isovalerate 0.1%, β-bisabolene 0.1%, cis-α-bisabolene 0.1%, germacra-4(15)-5,10(14)-trien-1-α-ol 0.1%, hexadecanol 0.1%, oleyl alcohol 0.1%, octadecanol 0.1%, oleyl acetate 0.1%, unknown compound 2 0.5%, unknown compound 3 0.3%; traces of the following compounds: unknown compound 1, isopropyl isobutyrate, (Z)-4-nonene, tricyclicene, isopropyl senecioate, 4-nonanone, isobutyl senecioate, 3-nonanone, nonanal, isoamyl isovalerate, β-thujone, cis-limonene oxide, hexyl isobutyrate, citronellal, borneol, α-terpineol, (Z)-4-decenal, (Z)-2-octyl acetate, trans-piperitol, trans-sabinyl acetate, benzyl butyrate, octyl isobutyrate, cis-carvil acetate, neryl acetate, (Z)-2-undecenal, daucene, benzyl 2-methyl butyrate, (Z)-5-dodecenal, benzyl isovalerate, (Z)-6-decen-1yl acetate, methyl eugenol, 2,5-dimethoxy-para-cymene, octyl 2-methylbutyrate, octyl isovalerate, γ-muurolene, epi-cubebole, geranyl isobutyrate, γ-cadinene, cubebole, 5-cadinene, trans-γ-bisabolene, elemine, spatulenol, octyl hexanoate, (Z)-8-hydroxylinalyl butyrate, 10-epi-γ-eudesmol, 1-epi-cubenol, τ-muurolol, cubenol, α-cadinol, isobicyclogermacrenal, hexahydropharnesylacetone, hexadecyl acetate, (E)-phalcarinol, (Z)-10-octadecenyl acetate, (Z)-vaccenyl acetate, (Z)-12-octadecenyl acetate, (Z)-13-octadecenyl acetate, octadecyl acetate, tricosane.

STYLE None verified.

SOURCE Hieronymus 2016b.

Hibiscus sabdariffa L.
MALVACEAE

Synonyms *Abelmoschus cruentus* (Bertol.) Walp., *Furcaria sabdariffa* Ulbr., *Hibiscus cruentus* Bertol., *Hibiscus fraternus* L., *Hibiscus palmatilobus* Baill., *Sabdariffa rubra* Kostel.

Common Names Hibiscus tea, roselle, rosella, sorrel, red sorrel, agua/rosa de Jamaica, Jamaica.

Description Herb, annual (up to 2 m). Taproot conical, thin, deep, glabrous. Stems erect, glabrous. Leaves with petiole (2–8 cm), covered with sparse, long pubescence; stipules linear (about 1 cm), sparsely villous; ovate, undivided lower leaves, upper leaves divided into 3 deep, lanceolate, palmate lobes (2–8 cm × 5–15 mm), margin serrate, apex obtuse or acuminate, base rounded to broadly cuneate; 3–5 main veins. Flowers solitary, axillary, nearly sessile; bracts 8–12, red, fleshy, lanceolate (5–10 × 2–3 mm), sparsely hirsute, with thorn-like appendages at the tip, base and calyx connate; calyx cup shaped (1 cm diameter), base 1/3 connate, lobes 5, triangular, gradually tapering (1–2 cm); flowers yellow, inner surface of base Ø 6–7 cm. Fruit capsule ovoid (Ø about 1.5 cm), densely hirsute, mericarps 5. Seeds reniform, glabrous.

Cultivar There are two main groups, according to some authors, identifiable as *Hibiscus sabdariffa* L. var. *sabdariffa* (calyxes red-yellow in color, swollen, edible, poor-quality fiber) and *Hibiscus sabdariffa* L. var. *altissima* Wester (calyxes inedible, high-quality fiber). Examples include Archer (White Sorrel), Rico, Temprano, and Victor.

Related Species The genus *Hibiscus* includes about 258 species.

Geographic Distribution W Africa (?), India (?), Malaysia (?). **Habitat** Plant widely cultivated in tropical and subtropical areas (0–600 m ASL) in well drained, sandy, fertile soils, in full sun. **Range** China, Thailand, Sudan, Mexico, Egypt, Senegal, Tanzania, Mali, Jamaica, Malaysia (cultivation is also intended for the production of fiber from the stem).

Beer-Making Parts (and Possible Uses) Calyxes.

Chemistry Dehydrated calyxes: energy 205 kJ (49 kcal), carbohydrates 11.31 g, lipids 0.64 g, protein 0.96 g, vitamin A 14 µg, thiamine (B1) 0.011 mg, riboflavin (B2) 0.028 mg, niacin (B3) 0.31 mg, vitamin C 12 mg, calcium 215 mg, iron 1.48 mg, magnesium 51 mg, phosphorus 37 mg, potassium 208 mg, sodium 6 mg, flavonoids (gossypetin, hibiscetin, sabdaretin), pigments: hibiscetin (daphniphylline), myrtiline (delphinidin 3-monoglucoside), chrysantenin (cyanidin 3-monoglucoside), delphinidin. Leaves: phenolic compounds (neochlorogenic acid, chlorogenic acid, cryptochlorogenic acid, caffeoylshikimic acid) and flavonoids (quercetin, kaempferol, and their derivatives). Flowers: anthocyanins and protocatechuic acid. Seeds: gamma-tocopherol.

Style American Pale Ale, American Wheat or Rye Beer, Blanche, Irish Red Ale, Saison.

Beer Fleur Sofronia, MC77 (Serrapetrona, Marche, Italy).

Source Heilshorn 2017; Hieronymus 2016b; https://www.brewersfriend.com/other/jamaica/.

Hibiscus syriacus L.
MALVACEAE

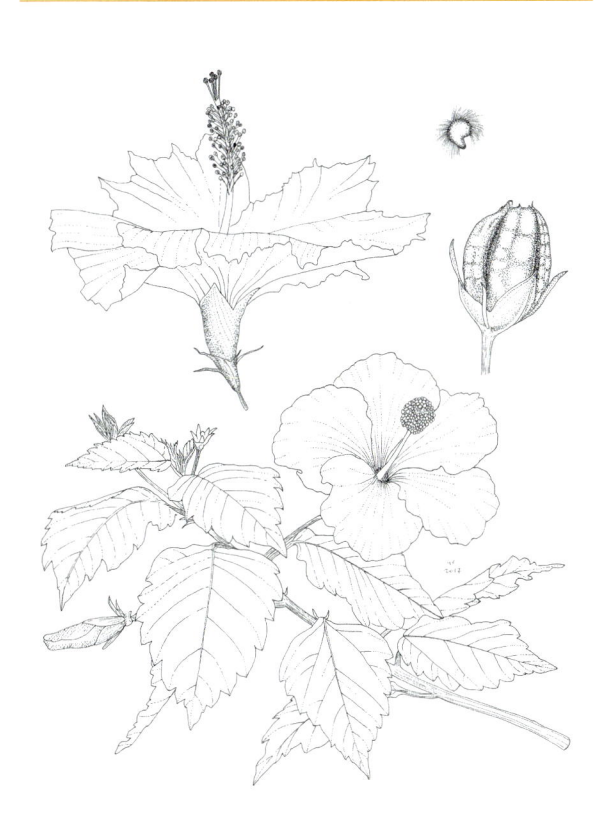

Synonyms *Althaea frutex* Mill., *Hibiscus rhombifolius* Cav., *Ketmia syriaca* (L.) Scop.

Common Names Hibiscus, rose of Sharon, Syrian ketmia, rose mallow, St. Joseph's rod.

Description Shrub (1–2 m), very branchy. Leaves deciduous, with lamina ± rhombic (4–7 cm) usually divided into 3 deep lobes, with ± parallel edges. Flowers axillary, solitary; epicalyx segments lanceolate; sepals triangular, fused in basal half; petals (4–5 cm) violet, pinkish, or whitish, with a dark spot at the base. Fruit oval capsule (10–15 × 15–25 mm), pubescent with dense star-shaped hairs.

Cultivar They are numerous, including Ardens, Blushing Bride, Diana, Hamabo, Lady Stanley, Lucy, Meehanii, Notwoodone, Oiseau Bleu (Blue Bird), Red Heart, William R. Smith, and Woodbridge.

Related Species *Hibiscus sabdariffa* L. belongs to the genus containing about 260 taxa, from whose dried calyxes the drink known as carcade is made; *Hibiscus rosa-sinensis* L. has also been used in making beer.

Geographic Distribution E Asia. **Habitat** Cultivated for hedges. **Range** Cosmopolitan.

Beer-Making Parts (and Possible Uses) Petals, leaves, bark.

Chemistry Flowers (mg/kg; g/kg): energy 360–3,530 kcal/kg, water 898 g, saponarin, isoxitexin, fumaric acid, ascorbic acid 40–405 mg, calcium 40–395 mg, cyanidin-3-sulfuroside, HHB, cyanidin-diglucoside, lipids 3.6–35.29 g, fiber 15.6–152.96 g, iron 17–165 mg, niacin 6–60 mg, nitrogen 640–6,275 mg, phosphorus 265–2,615 mg, riboflavin 0.4–4.7 mg, thiamine 0.3–3 mg, anthocyanins (delphinidin 3-sambubioside, delphinidin, cyanidin 3-sambubioside) 1.5%, mucilage 15%, galactose, arabinose, glucose, rhamnose, pectins 2%, galacturonic acid, xylose, mannose. Leaves: β-carotene 75–750 mg Bark: cyanidin-chloride, quercetin, azelaic acid, suberic acid, β-sitosterol, betulin. Seeds: malvalic acid, sterculic acid, dihydrosterculic acid.

Style American Amber Ale, American Barleywine, American Brown Ale, American IPA, American Light Lager, American Pale Ale, American Rye or Wheat Beer, Belgian Blond Ale, Belgian Dark Strong Ale, Belgian Dubbel, Belgian Pale Ale, Belgian Specialty Ale, Berliner Weisse, Blonde Ale, Brett Beer, British Strong Ale, California Common Beer, Düsseldorf Altbier, Extra Special/Strong Bitter (ESB), Fruit and Spice Beer, Fruit Beer, German Pilsner (Pils), Gueuze, Imperial Ipa, Irish Red Ale, Oktoberfest/Märzen, Saison, Special/Best/Premium Bitter, Specialty Beer, Spice/Herb/Vegetable Beer, Standard/Ordinary Bitter, Straight (Unblended) Lambic, Weizen/Weissbier, Witbier.

Source Hieronymus 2010; https://www.bjcp.org (30. Spiced Beer).

Hordeum vulgare L.
POACEAE

Synonyms *Frumentum sativum* E.H.L.Krause, *Hordeum bifarium* Roth, *Hordeum coeleste* (L.) P.Beauv., *Hordeum gymnodistichum* Duthie, *Hordeum hexastichon* L., *Hordeum himalayense* Schult., *Hordeum leptostachys* Griff., *Hordeum macrolepis* A.Braun, *Hordeum nigrum* Willd., *Hordeum polystichon* Haller, *Hordeum revelatum* (Körn.) A.Schulz, *Hordeum strobelense* Chiov., *Hordeum tetrastichum* Stokes.

Common Names Barley.

Description Herb, annual (5–15 dm). Root fasciculate, consisting of seminal primary root that develops at seed germination and adventitious roots arising from tillering culms that form from the base of the stem. Stem culm cylindrical, divided into 5–8 hollow internodes, separated by septa transverse to nodes; basal internodes generally shorter; tillering leads to the formation of 2–3 secondary culms for each primary root (distichous varieties tuft more than polychous varieties). Alternate leaves originate from the nodes, made of sheath (wrapping the culm), lamina, inconspicuous ligule, and longer auricles compared to those of other microtherm cereals; terminal leaf called "flag" is the smallest and wraps around the forming spikelets; underleaf lamina smooth, upper one with grooves where hygroscopic epidermal cells are present. Inflorescence composed spike characterized by short, zig-zag rachis, at the nodes of which (10–30 in number) are 3 uniform spikelets; spikelets arranged in 4 series (2 formed by central spikelets, the other 2 by lateral spikelets superimposed); spikelets formed by sterile glumes, reduced to simple hairy formations, which enclose within them the flowers protected by small fertile bracts: lemma (lower glumette) and palea (upper glumette); lemma envelops palea and at maturity both adhere to the caryopsis (hulled fruit), except in naked caryopsis varieties; in hexastic forms the 3 flowers are all fertile, so the spikelets have 3 rows of caryopses, whereas in distich only the central flower is fertile, so the spikelets will have only 2 rows of caryopses. Flower hermaphrodite, formed by 3 stamens and 2 hairy stigmas; 2 lodicules are also present. Pollination anemophilous. Fruit caryopsis with pericarp adherent to the seed, has ventral groove that can be more or less marked depending on the variety; color generally yellowish although some cultivars have whitish or even reddish or black caryopses; caryopsis size of this tetrastic taxon 3×10 mm (in other taxa hexastic and distichous on average $8-12 \times 3-4$ mm); weight of 1,000 hulled seeds 25–55 g (on average 45–50 g in the distichous, 35–45 g in the polychous); in the distichous cultivars the caryopses are larger in the central part of the spike, whereas in the polystichous the 2 central rows are symmetrical and uniform and the caryopses of the lateral rows are up to 20% smaller. The seed fruit of barley (and of other Poaceae), called the *caryopsis*, contains the project of the new plant (embryo) and a reserve of energy and materials

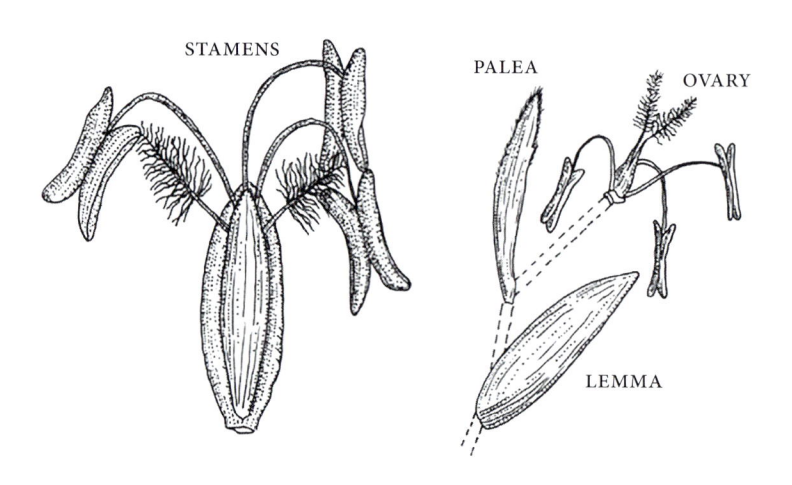

Structure of the single barley spikelet

(endosperm + aleurone) thanks to which growth and development of the new plant begins. The 2 distinct seminal units are enclosed by a complex system of integuments that has protective functions. Each of the 3 mentioned elements (embryo, reserves, involucre) has further attributes that allow each part of the caryopsis to carry out crucial functions in malting and brewing. The end with which the caryopsis connects to the rachis of the spike (proximal end) corresponds to the point at which the embryo is located. The outer involucre consists of 2 distinct structures that partially overlap: palea and lemma. The palea covers the ventral side of the caryopsis, whereas the lemma is on the dorsal side. The resta (or arista) is a ± long (10–14 mm), rough bristle attached to the distal end of the lemma, taking its origin from the midrib of the lemma. Although it contributes to photosynthesis during the growth of the plant, the aristas mostly detach during harvest. The caryopsis consists of distinct layers. The *pericarp*, the outermost layer, consists of a few layers of cells. The involucre just inside, called the *head*, is the seminal integument, which has the task of protecting the embryo and energy reserves with substances such as polyphenols (tannins), which are particularly concentrated in this layer. The head and pericarp are fused, forming a unique covering that prevents humidity and other environmental factors from affecting the underlying tissues. The integument is made of lignin, pentosans, hemicellulose, and silica, making it resistant to the enzymatic degradation of mashing and perfect to form the porous filtration bed for lautering (separation of mash from grain residues). Ferulic acid, a chemical compound present in the cell walls of the tegument and the aleuronic layer, where it binds with arabinoxylan, a hemicellulose, helping to keep the cell walls together, is important in barley and cereal crops. Ferulic acid (3-methoxy-4-hydroxycinnamic acid) is the precursor used by yeast to produce 4-vinyl-guaiacol (4VG), responsible for the clove flavor found in wheat beers. Below the head is the aleurone, another layer 2–3 cells thick, active in the mature caryopsis in the production of enzymes that mobilize the energy reserves of the endosperm to allow the development of the embryo and, more generally, germination. The starchy endosperm can represent up to 80% of the total weight of the caryopsis. Its internal structure consists of a mixture of starch granules, large (Ø 25 μ), about

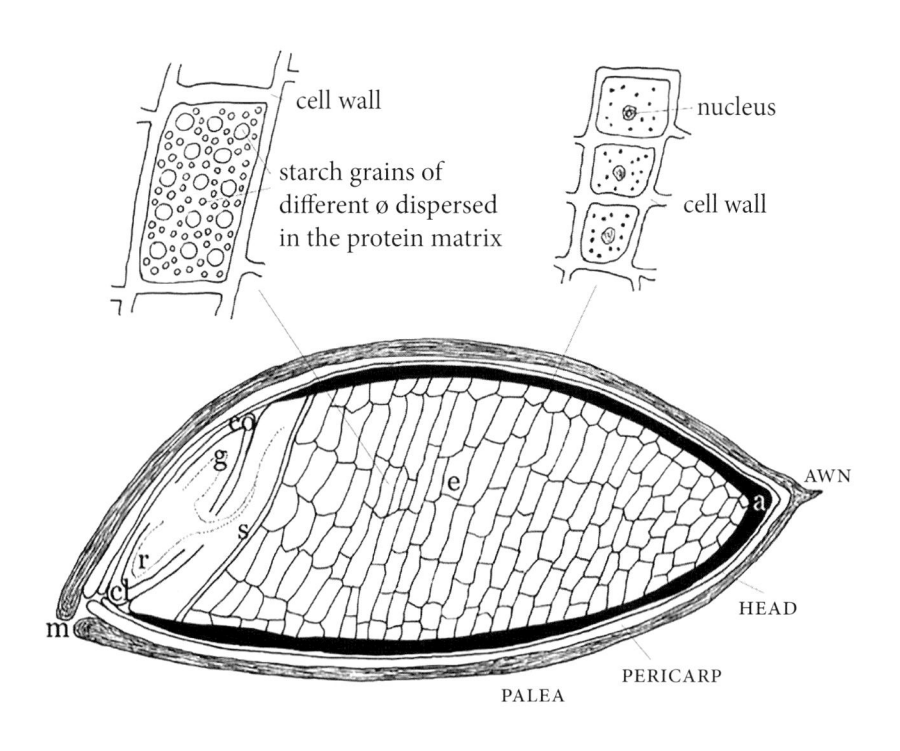

Longitudinal section of a barley caryopsis: a - aleurone; cl - coleorhiza; co - coleoptile; e - endosperm; g - shoot; m - micropyle; r - radicle; s - scutellum

90% of the total seed starch, and small (Ø 5 μ), about 10% of the starch, embedded in a protein matrix. The embryo is located at the proximal end of the caryopsis while in its quiescent state is very small (about 4% of the seed by weight). During hydration and germination, the embryo enzymatically breaks down the reserves contained in the endosperm, using them to generate new plant tissues. Although it is a mostly self-pollinated species, reproduction in barley requires 2 separate fertilizations (double fertilization). One of the 2 spermatic nuclei (haploid; 1n = 7) present in the pollen fertilizes the oosphere (haploid female cell; 1n = 7) giving rise to the zygote (diploid; 2n = 14), from which the embryo will form; the other instead fuses with 2 nuclei (polar nuclei) located at the center of the female gametophyte (embryo sac), forming the endosperm and aleurone (triploid; 3n = 21).

SUBSPECIES *Hordeum hexastichum* L.: spikelets arranged in 6 series (2 formed by the central spikelets, the other 4 by the lateral spikelets that remain next to each other, not overlapping); fruit caryopsis 3 × 10 mm. *Hordeum distichum* L.: spikelets arranged in 2 rows with erect, parallel awns up to 12 cm long; fruit caryopsis 4 × 10 mm. Barley can, however, be classified according to a number of other characteristics, including growing period: there are winter and spring varieties. The first ones overcome winter thanks to a form of dormancy, which requires a certain period of cold to be interrupted. On the contrary, spring varieties do not have dormancy and are sown in spring. To put it simply, winter varieties are more successful because of their higher productivity and lower water needs, as well as the higher protection they offer against soil erosion.

CULTIVAR Barley has been cultivated for thousands of years; therefore, it should not be surprising that hundreds of cultivars are known today. Moreover, in recent decades this plant has been subjected to intense programs of crossbreeding and selection in many countries, with the aim of creating new cultivars. The characteristics sought after, at least in relation to barley for beer making, are resistance to diseases, tolerance to aridity, higher production, and less susceptibility to early seed drop. When a new line includes all these agronomic requirements, it is tested for malting (modification, enzyme levels, ability to rapidly interrupt dormancy). After passing this barrier, it is time for evaluation of taste, aroma, and brewing performance in general and then, in case of a positive outcome, cultivation for commercial purposes. Among the historical varieties of most interest for brewing are Chevallier (variety selected from plants cultivated in Suffolk, dominated English production in the nineteenth century, and, after becoming the selective base for modern British varieties, it has been recently revaluated in Norwich for its aroma and disease resistance), Golden Promise (short, distichous variety with excellent agronomic and brewing performance, selected from the Maythorpe variety in 1956, it was widely used in brewing and distillation in the 1970s and 1980s; although new lines of this variety are now available, the original is still mainly valued in distilling), and Maris Otter (appreciated by brewers for the production of superior aromas and the easy release of the extract; less appreciated for its lower productivity; moreover, in the 2000s the rights were bought by a private company). Modern varieties of distichous barley bred in North America include AC

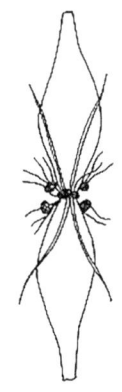

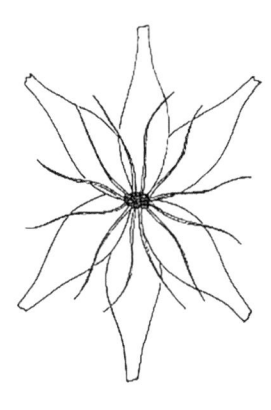

DIAGRAM OF THE GROWTH MODEL OF THE DISTICHOUS AND HEXASTICHOUS BARLEY

Metcalfe (developed in Manitoba, Canada, in 1986, a variety particularly suited to W North America, has high yields and remarkable disease resistance), CDC Meredith (developed in Sasktchewan, Canada, in 2008, a high-performing variety for yield, disease resistance, and enzyme power; suitable for the American light lager style), Charles (winter barley released in 2005 in Idaho, United States, gathering some success in western states but with winter survival problems in N United States), Conlon (created in North Dakota, United States, in 1988 and widely cultivated in the same state and in Montana, highly valued for its agronomic and brewing performances; variety suitable for the production of lager malt, in particular with regard to diastatic power and the presence of free amino nitrogen [FAN]), Conrad (released in 2005 in Fort Collins, Colorado, United States, offers very good production performance in irrigated conditions in the W United States), Expedition (spring variety created for the W United States, high production level and very good brewing qualities, in particular its low protein content and suitability for all-malt beers), Full Pint (made in Oregon, United States, conceived for the American craft brewing market; after some initial uncertainty, this spring variety is attracting more interest, despite its high protein level, for the good taste and body it gives to beer), Harrington (created in 1972 in Saskatchewan, Canada, it had remarkable success in the past and it is still cultivated, although it is giving way to more productive varieties; its current importance is in representing a standard for malting), Moravian 37 and Moravian 69 (developed first in Europe and then in the United States, they are mainly cultivated in the Intermountain West with vertical integration contracts for a multinational brewing company), Morex (hexastic variety released in Minnesota, United States, in 1978, it has been dominant for a long time, just like Laker and Robust, which preceded it; today it seems to be surpassed by other modern hexastic varieties such as Lacy and Tradition, better-performing agronomically), Pinnacle (released in North Dakota, United States, in 2007, tends to have a lower protein level than the varieties harvested in the same period), and Thoroughbred (winter hexastic variety produced in Virginia, United States, mainly cultivated in the E United States and much appreciated by craft brewers). Although in Europe there is great interest in new and more productive varieties, including those promoted by private companies that conduct the selection and testing (whereas in North America new varieties are mostly made by public research organizations), widespread attention, especially from the craft brewing world, is being paid to the historical ones whose consolidated brewing performances are considered superior to modern ones. In Great Britain are successful modern varieties such as Concerto (2008), NFC Tipple (2004), Optic (1995), Propino (2009), and Quench (2006); in Germany are historical varieties such as Barker (1996), Scarlet (1995), and Steffi (1989), as well as modern varieties such as Grace (2008), Marthe (2005), Propino (2009), Quench (2006), and Wintmalt (2007). Not all barley is made of pure varieties. So-called landraces are varietal mixtures that have changed over time, adapting to local conditions in a certain geographic area. Over the centuries, the selective action of a specific environment and of farmers who saved seeds from year to year for the next sowing determined this adaptation. To some extent this process was also favored by limited exchanges with other communities and territories. Seed exchanges certainly contributed to an increase in the variability of genetic heritages, but they fatally introduced elements of discontinuity in the landrace selection process. For this reason, few examples of landraces of great economic and technological importance remain today. Among these are Bere (Scotland), Hanna (Czech Republic), Manchurian (E Asia), and Oderbrucker (Germany). Moreover, the 1960s represented a watershed in the choice of varieties for brewing. If, in fact, all barley in the past was used indiscriminately in human and animal nutrition and in brewing, beginning in the 1960s, with the increase of scientific knowledge about the brewing process, a pool of morphological (distichous barley preferred to hexastichous barley), agronomic (disease resistance, high productivity) and technological characteristics started to be highlighted, and these characteristics directed the varietal choices of producers. On the other hand, the explosion of the homebrewing/craftbrewing movement has led to a rediscovery and enhancement of ecotypes, landraces, and local cultivars for this and many other plant species used in the production of beer, capable of giving craft beer a boost of personality and terroir. For example, Carambri, Nevada, Plaisant, and Prisma are the main distichous varieties, whereas Escourgeon is hexastic, used in the north of France and in Wallonia, Belgium, in the production of the famous *bière de garde*. Detailed knowledge of the variety and territory (and maybe even the grower) where a certain barley comes from (and for malt, maybe even the "vintage") should have increasing importance in the future for the brewer (and for the final consumer), as much so as knowledge of the varieties of hops used in the brewing of a certain beer, their origin, and the timing with which they were included in the production process. And this, in spite of the ironic words of John Mallett: "Barley isn't as sexy as hops."

RELATED SPECIES The genus *Hordeum* includes 46 taxa.

GEOGRAPHIC DISTRIBUTION *H. vulgare* and *H. hexastichum* from E Africa; *H. distichum* originally from Iran and Pamir. **HABITAT** Cultivated. **RANGE** Cultivated in Europe, Africa, Asia, India, Indochina, Malaysia, Australia, New Zealand, SW NC Pacific, North America, South America (main producers in decreasing order: Russia, France, Germany, Ukraine, Canada, Spain, United Kingdom, Turkey).

BEER-MAKING PARTS (AND POSSIBLE USES) Straw (used in wort filtration in N European brewing) and especially caryopses (as such or malted). Hundreds of types of malt and other products used in brewing can be obtained from the caryopsis of barley. The classification, at least in a rough way, of malts is possible according to the processes the caryopsis undergoes (stewed, caramelized, roasted), according to its enzymatic activity, or even according to the color of the wort it produces. However, similar products from different producers can show both overlapping qualities and unexpected differences depending on many factors (e.g., variation among batches, characteristic fluctuations of raw materials).

The following provides a brief description of the main types of malt.

Malts from standard processes. Malts of this group are produced through standard steeping, germination, and roasting techniques. They are all quite pale and have sufficient enzyme potential to convert their starches. Very pale standard malt is sometimes referred to as white malt and used as a base to produce roasted malts such as black or chocolate.

Pilsner Malt (1.2–2 SRM). Base malt designed for very pale all-malt beers. Traditional production involves the use of distichous barley with low protein content, low modification during germination, low temperatures, and high airflow stewing. This malt is very clear and very light, with moderate enzymatic potential, a characteristic herbaceous smell, and a taste of fresh wort, particularly evident in European pilsner-style beers. Dimethyl sulfide (DMS), with a characteristic taste of cooked corn or cabbage, is present in all malts, as its precursors (S-methylmethionine [SMM] and dimethyl sulfoxide [DMSO]) are created during malting but subsequently removed by heat treatments at higher temperatures. In malts subjected to low-temperature treatments, such as pilsner, this unpleasant taste can persist, universally considered a fault with the exception of some beer styles (e.g., German pilsner), in which a very low concentration can be accepted.

Pale Malt (1.6–2.8 SRM). Generic term used to indicate a wide range of light-colored base malts. North American maltsters produce pale malts with high enzyme potential and FAN, suitable for rapid carbohydrate conversion and proper yeast nutrition. These characteristics make the control of fermentability in all-malt beers quite difficult, as wort conversion occurs very rapidly. Maltsters in other areas of the world produce pale malt with moderate enzymatic potential and modification. Compared to pilsner, pale has a deeper malt flavor.

Pale Ale Malt (2.7–3.8 SRM). Base malt often used in the production of pale ale. Medium-high modification, darker than standard pale, optimized for use in single-temperature infusion mash, malt flavor evident but not excessive, with notes of biscuit or toast. The higher stewing temperatures used result in low DMS/DMSO potential. The agro-ecological conditions of the coasts of the United Kingdom seem to be particularly suitable for the cultivation of barleys destined to produce pale ale malts, perfect for the production of traditional British ales, characterized by pleasant biscuit notes and low alcohol content.

Vienna Malt (2.5–4 SRM). Responsible for the rich orange color of the beer and the flavor of traditional Märzen beer, it has enough enzyme power to convert up to 100% of the grist. In contrast to most crystal malts, the use of Vienna contributes a crisp, dry finish and a slightly toasty, nutty flavor to beers that pairs well with the spicy notes of the noble hops.

Munich Malt (3–20 SRM). This category covers a wide range of colors, with lighter versions having a sober and elegant character. Low enzymatic potential but sufficient to convert a wort when used at 100%. Many brewers add a small amount of Munich to the grains to complete the malty profile of their beers. The choice of color of Munich can also have significant consequences for the final product: refined and delicate in light versions, more character in darker versions.

Melanoidin Malt (17–25 SRM). Sweetish flavor similar to honey. Although some variants have sufficient enzymes to convert up to 100% of the starch, it is most commonly used at lower concentrations to prevent the flavor from becoming too pronounced when used as a primary malt. Also known as honey or brumalt, it is produced by reducing airflow during the final stages of germination and then drying at low temperatures to obtain significant amounts of products of Maillard reactions. Many brewers consider this malt a Super Munich, as it shares some flavor notes with Munich. Additions of up to 10% produce a distinct honey aroma in the beer.

Caramel Malts. Malts in this group are produced by loading green malt (wet and germinated but not dried) into a drum roaster. They are typically "mashed in the husk" and then roasted.

Vitreous Specialty Malts (1–12 SRM). Made using low temperature and high humidity to produce pale malts with vitreous endosperm. They have no enzyme potential and are used to improve foam hold, add body, or impart sweetness to beer. They are sold under a variety of trade names such as Cara-pils or Carapils (trademarked in the United States and elsewhere in the world) or as "dextrin malts."

Caramel/Crystal Malts (10–200 SRM). Caramel (also called crystal) malts are made by raising the temperature of green malt to convert starches and proteins into sugars and amino acids, respectively (both necessary reagents in Maillard reactions), in a standard dryer or drum roaster. Caramel malts contribute as much to the flavor as to the color of beer. Light- to medium-colored ones (20–60 SRM) are used when theirs is the desired main flavor of the beer, as they tend to taste cleaner; darker ones are generally used to provide supporting flavors but, if used in excess, can overpower a beer.

Roasted Malts. These are produced from pale malts roasted in a drum roaster, destroying their enzymatic potential. They are heated to produce colors ranging from light brown to dark brown. They are different from caramel malts and tend to have drier and astringent flavors; therefore, they should be used with moderation (rarely more than 10%).

Special Hybrid Malts (50–150 SRM). Produced by subjecting green malt first to caramelization and then to roasting, releasing flavors that combine characteristics of caramel and roasted malts. They can express deep, dark aromas of dried fruit (raisin, plum) and are often used in rich Belgian dubbel–style dark beers.

Biscuit Malts (20–30 SRM). Produced in a kiln at a high temperature (up to 227°C) producing flavors of bread crust and toast (typical of nut brown ale), they give a dry finish with characteristic aromas of hazelnut and toasted cookies (some similar to Vienna malt but more intense and without enzymatic potential). Biscuit and amber are similar, with biscuit being drier.

Amber Malt (20–36 SRM). Malt lightly roasted in the English traditional drum roaster, with flavors of caramel, baked bread, and hazelnut. Dry roasting forms pyrazine and pyrrole compounds, typically bitter. Amber works well in kegged mild ale, as the dry finish provides a counterpoint to the estery aromas created by the yeast.

Brown Malt (40–150 SRM). Similar to amber malt in terms of flavors of caramel, baked goods, and hazelnut but more intense in flavor and color, as a consequence of longer heat treatments. Production technique determines reactions present in very dark malts; therefore, brown is sometimes used to add depth and complexity to dark beers. In case it is used in excessive quantities, it can bring a not-so-pleasing astringency. Blown (or snap) malt, an ancient version of brown produced with the intense heat released by adding wood bundles to the fire, had a smoky flavor with caramel nuances, appreciated in porter beers before the introduction of drum roasters in the mid-1800s.

Chocolate Malt (350–500 SRM). Malt roasted in a drum roaster, provides a dark color (less than black) to the beer and a delicate burnt flavor (with notes of coffee and chocolate as products of the Maillard reaction) with a slight astringency.

Black Malt (435–550 SRM). Provides a huge addition of color to beer. High production temperatures develop acrid products that characterize stouts. Bitterness, dryness, and burntness are gradually attenuated in the roasting process, since as the temperature rises, the compounds are carbonized, imparting flavor and color. Therefore, dark malts and barley debittered by heat treatment have characteristics that allow the creation of dark but crisp beer. Astringency is pronounced when the barley husk is roasted; therefore, by using huskless barley it is possible to produce dark malt with insignificant bitterness.

Roasted Barley (300–650 SRM). Roasted barley is produced without previous malting, subjecting the caryopsis to a roasting process that corresponds chromatically to the chocolate–black range. Although usually milder than roasted malts, roasted barley is still characterized by flavors such as sour, dry, and roasted. Black barley is a key component of the flavor of dry Irish stouts.

Special-Process Malts. These are produced using processes that lead to a range of unique flavors and characteristics. The category includes acidulated, smoked, and peated malts. They are relatively expensive products intended for a fairly limited market.

Acidulated Malt (2.2–4 SRM). Acidulated malt (sauermalz) is produced by promoting the growth of lactic acid bacteria during germination by spraying acid wort on the malt before drying. It has a pungent taste, capable of reducing the pH value of wort while respecting the *Reinheitsgebot*. Used in small quantities as a spice, it brings a bright acidity that can be appreciated where sour and tart notes are required without subjecting the product to long aging in a cask with the traditional acidifying microflora. There are many variants.

Smoked Malt (2.5–5 SRM). Produced by completely or partially drying the malt by directly exposing it to the combustion gases of a wood fire, which gives an intense smoked flavor. Traditional versions (associated with the city of Bamberg, Franconia) use beech wood (*Fagus sylvatica*) as a fuel source. Other woods such as cherry (*Prunus* sp.) and alder (*Alnus* sp.) are used elsewhere in the world. Lightly smoked versions can be used for 100% of the grist. In general, smoked malt produced in the United States tends to be stronger than European malt.

Peated Malt (1.7–2.5 SRM/5–10 EBC). Malt smoked by direct exposure to peat combustion gases, primarily intended for Scotch whisky production.

TOXICITY Barley contains at least two dozen allergens, so it is possible for some individuals to experience adverse physical reactions after drinking beer. There is a distinction between a general allergy to barley and a specific sensitivity (intolerance) to gluten. Sensitivity to gluten can cause lacrimation, respiratory problems, and, in severe cases, gluten rash, a rash caused by an autoimmune response. Celiac disease, which triggers an immune response to gluten, is a serious autoimmune disorder that can damage the intestines to the point of preventing the absorption of nutrients and can sometimes cause tumors. What triggers the immune response is gliadin, a prolamin of wheat that, together with other proteins, forms gluten. Gliadin is quite similar to hordein (barley) and secalin (rye), prolamins (proteins containing the amino acid proline) of the Poaceae. Therefore, even though barley technically does not contain gliadin, beer may pose a problem for individuals with gluten intolerance or celiac disease. It is hoped that the use of a proline-specific endoprotease enzyme in mashing may solve this problem in the future.

CHEMISTRY Carioxides: energy 1,473 kJ (352 kcal), carbohydrates 77.7 g (sugars 0.8 g, fiber 15.6 g, lipids 1.2 g, protein 9.9 g, β-carotene 13 μg, lutein zeaxanthin 160 μg, thiamine (B1) 0.191 mg, riboflavin (B2) 0.114 mg, niacin (B3) 4.604 mg, pantothenic acid (B5) 0.282 mg, vitamin B6 0.26 mg, folate (B9) 23 μg, choline 37.8 mg, vitamin K 2.2 μg, calcium 29 mg, iron 2.5 mg, magnesium 79 mg, manganese 1.322 mg, phosphorus 221 mg, potassium 280 mg, sodium 9 mg, zinc 2.13 mg.

MALT The barley plant produces caryopses to perpetuate its progeny in time and space. In nature, the seed germinates to give life to a new plant once it has detached from its mother plant, assuming it finds suitable environmental conditions. The purpose of agriculture is to use the seed's capacity of diffusion and dispersion to obtain new plants and caryopses in large quantities. The purpose of malting is to exploit the biochemical transformations undergone by the content of the caryopsis during the germination process to obtain fermentable (and other) substances with which to produce beer. The raw materials of these processes are carbohydrates, proteins, and lipids, whereas enzymes are the tools capable of making these transformations. In the caryopsis, among others, two fundamental elements coexist: the embryo and the endosperm. The embryo is the miniature project of the new plant, whose development in the early stages of life (heterotrophic phase) depends on the reserves accumulated by the mother plant in the endosperm, a tightly packed mass of starch. The cells of the aleurone, the thin layer surrounding the endosperm, generate the enzymes that "unpack" the contents of the endosperm to release the sugars and FAN that the embryo needs for growth. Under normal conditions, the heterotrophic phase of germination is followed by the autotrophic phase, during which the seedling begins to perform photosynthesis, freeing itself from external sources for its supply of nutrients and energy. Malting abruptly interrupts this process, leaving sugars and FAN at the disposal of yeasts that will be used for brewing. In other words, the purpose of malting is to promote the optimal production of hydrolytic enzymes in the aleurone cells and to guide the controlled action of some of these enzymes to reduce impediments to the subsequent extraction of sugars and other useful substances during mashing.

ENZYMES AND MODIFICATION From the structural point of view, an enzyme is a protein molecule with a precise three-dimensional shape that allows it to interact with the reagents (or substrates) of a biochemical reaction, determining structural changes that result in new molecules (reaction products). While remaining unmodified in the reactions, enzymes act as biological catalysts, substances capable of accelerating biochemical reactions millions of times over. Because of this extraordinary catalytic ability and their extreme substrate specificity (and reaction type), enzymes are the keys that lock or unlock reactions and entire biochemical processes in living organisms. Despite their biological importance, enzymes are vulnerable to two factors: excessive heat and pH (too high or low). Each factor can cause the loss of molecular shape (denaturation) that underlies catalytic ability. The nomenclature of enzymes, characterized by their "-ase" suffix, depends on the substrates on which they act: amylases break down starch, beta gluconases break down beta glucans, proteases demolish proteins, etc. The specificity of enzymes can go as far as affecting a single type of bond or substrate: alpha amylases are effective only on alpha 1–4 bonds of glucose chains (but can do so at any position along the chain), whereas beta amylase can cleave only 1–4 bonds that bind two glucose units away from the chain termination; among beta gluconases, there is one enzyme for breaking 1–3 bonds and another for 1–4 bonding. Malting (similar to the first stage of germination) begins with hydration of the caryopsis, starting at the proximal end of the caryopsis, where the embryo is located. The hydrated embryo releases hormones (gibberellins), which stimulate scutellum and aleurone to produce enzymes that demolish the endosperm. The aleurone produces beta glucanase, protease, alpha amylase, and glucoamylase. The modification of the endosperm proceeds in stages: first the cell walls of the endosperm are demolished, then the protein matrix that surrounds the starch granules is demolished, and finally the partial hydrolysis of the starch occurs. The modification is complete when the endosperm is totally broken down, a fact detectable by the change of consistency of the content of the caryopsis from hard to floury-pasty. The only amylolytic enzyme present in barley before malting is beta (β) amylase. It exists in free or bound form. In the latter, it is bound to other compounds (e.g., Z protein) to be released only during germination or mashing, thanks to the proteolytic action. Alpha (α) amylase is produced by aleurone during germination, with glucoamylase (or alpha-glucosidase) and dextrinase. The diastatic power, that is the quantity of enzymes that remain at the end of the malting process (which includes many thermal processes and appears to be potentially capable of denaturing enzymes), is so important that it is noted in the packaging of malt with a certificate of analysis. In a nutshell, it is a measure of the capacity of a malt sample to produce sugar from a known quantity of a standardized starch solution. During malting, the endosperm is modified, and the starch granules contained within are exposed. In the preparation of wort, grain milling causes an exponential increase of the surface available to the enzymatic attack by the four amylases involved in the liberation of sugars composing the wort: maltose 41%, maltotriose 14%, maltrotetraose 6%, sucrose 6%, glucose and fructose 9%, dextrins 22%, and nonstarch polysaccharides (hemicellulose) 2%. Alpha and beta amylases are well known enzymes in brewing. Both attack the 1-4 bond of starch by breaking it down into shorter chains, but with different modes and products. Alpha amylase is an endo-amylase that hydrolyzes the α1–4 bond anywhere in the chain (except near an α1–6 bond), breaking the bonds in a randomized fashion and producing sugar chains of varying lengths. Beta amylase is an exo-amylase and can act only at the end of the chain, within three glucose units of the α1–6 bond, that is, by detaching maltose molecules from the non-reducing ends of the larger carbohydrate units. Yeast can ferment only glucose, maltose, and, in the case of lager strains, maltotriose, from malt starch. Any larger sugars will be found unmodified in the beer, giving it sweetness, flavor, and body. Amylases work synergistically during mashing to maximize saccharification. In fact, if it acted alone, α-amylase (optimum 65°C, pH 5.3) would produce about 20% fermentable sugars; the joint action of β-amylase (55°C, pH 5.7; denaturation over 68°C) brings fermentability to 70%, and the debranching enzyme to 80%. Given the low concentration of glucose, glucoamylase plays a marginal role in mashing. The control of mashing conditions is a crucial factor in favoring a particular amylase, therefore controlling wort production. Base malts usually have sufficient diastatic power to convert other starchy materials into fermentable sugars. However, in some cases it may be necessary to add external enzymes similar to those of malt.

Carbohydrates Photosynthesis is the process by which plants produce carbohydrates from water, carbon dioxide, and light. Carbohydrates serve as a material for building cellular structures and for storing energy. They consist of carbon, hydrogen, and oxygen. The simplest form is that of monosaccharides (single-molecule sugars), with the general chemical formula $C_x(H_2O)_x$, where x is generally greater than 3 but can range from 2 to 7. Monosaccharides can bind together to form larger, more complex molecules called disaccharides, trisaccharides, oligosaccharides, and polysaccharides. These complex molecules are referred to by several names: starches, cellulose, hemicellulose, and gums. These compounds have chemical, physical, and biological properties very different from one another. The benefit of carbohydrates in beer essentially depends on the ability of yeast to metabolize some of them, either in the process of cellular respiration or during fermentation. Yeast can metabolize only monosaccharides with six carbon atoms (hexoses) such as glucose, fructose, and galactose. It cannot ferment sugars with three, four, five, or seven carbon atoms. Yeast can also ferment the disaccharides sucrose and maltose and, in the case of lager yeast, the trisaccharide maltotriose. Any larger carbohydrate (e.g., maltotetraose) is considered a dextrin (oligosaccharide), a molecule that cannot be fermented.

Sugars Hexoses (monosaccharides with six carbon atoms), with the general chemical formula $C_6H_{12}O_6$, are the key molecules in brewing because they are preferred by yeast for their metabolism. They are also the building blocks of larger carbohydrates, also of relevant beer-making interest. Glucose is the basic nutrient for life on Earth. It is absorbed and metabolized by living cells, including yeast. Then there are fructose and galactose, both structural isomers of glucose. Fructose is commonly found in fruit and can also be produced from corn. Galactose is one of the two monosaccharide (the other is glucose) components of the disaccharide found in milk: lactose. Lactose is not fermentable by yeast. Galactose is fermentable if the lactose has previously been hydrolyzed through special enzymes. Aside from milk, lactose is found in small concentrations in the caryopsis during germination. The common cooking sugar, sucrose, a disaccharide composed of glucose and fructose, is also fermentable. Monosaccharides can be found in linear or cyclic configurations, owing to different arrangements of atomic bonds within the molecule. A single atomic (or chemical) bond is formed between two atoms sharing two electrons. If four electrons are shared between two atoms, a double bond is formed. Each type of atom can form a specific number of bonds under normal conditions: carbon forms four, oxygen two, hydrogen one. In the water molecule (H_2O), the two hydrogen atoms share their single electron with the two provided by the oxygen atom. Vice versa, in the carbon dioxide molecule (CO_2), the two electrons of each oxygen atom form a double bond with two of the four electrons provided by the carbon atom.

Sugar crystals are ordered conformations of single sugar molecules in their cyclic form. When placed in solution, these molecules are converted into their corresponding linear form. The transition from one molecular form to another involves a rearrangement of bonds without a change in the fundamental number or arrangement of atoms. Molecular bond formation between monosaccharides involves the loss of a water molecule. Each molecular bond between monosaccharides is identified by a numbered pair (e.g., 1–4). The two numbers indicate the number of the two carbon atoms (one for each monosaccharide) involved in the formation of these bonds. The numbering of the carbon atoms in a molecule is conventionally assigned in linear molecular form. For example, carbon #1 of a molecule is located at the end closest to the oxygen with the double bond. If you bond carbon #1 of a glucose molecule to carbon #4 of a second glucose molecule, you will form maltose. The bond will be of type 1–4. Binding a third glucose to maltose with an additional 1–4 bond will form maltotriose. The 1–4 bonding of glucose and fructose creates sucrose. Glucose and galactose bound 1–4 creates lactose. Glucose forms not only 1–4 bonds but also 1–6 bonds. While 1–4 bonds give linear oligo- or polysaccharide molecules, the presence of the 1–6 bond implies chain branching. Raffinose is a trisaccharide composed of galactose, fructose, and glucose that can constitute up to a quarter of the seminal reserves and is largely stored in the embryo (of beans or other foods such as cabbage, brussels sprouts, broccoli), acting as a starter for germination. The human digestive system cannot properly digest this sugar, which is metabolized by bacteria in the intestines causing flatulence. Sugars have a number of properties that are more or less known. For example, the association between sugars and sweetness is common. In reality, sugars have different levels of sweetness, according to the standard represented by sucrose (glucose is 80% of the sweetness of sucrose, fructose 170%, maltose 45%, and lactose 16%). Much higher sweetness values can be found among noncarbohydrate sweeteners (glycyrrhizin is 50 times sweeter than sucrose, stevia 200, aspartame 200, saccharin 450, and sucralose 600 times). Another intriguing property of sugars is the induced rotation of polarized light. A sucrose solution is dextrorotatory, meaning that polarized light passing through it deflects to the right. If sucrose is split into fructose and glucose, the solution becomes levogyrous; that is, it rotates to the left, since fructose is more levogyrous than glucose is dextrorotatory. The reversal of optical polarity reliably signals the splitting of sucrose during the production of inverted sugar, typical of some English brewing styles. A further interesting property of sugars is isomerization. The shape of organic compounds is crucial in determining their physical and chemical properties. Compounds having the same brute formula (e.g., $C_6H_{12}O_6$) can have different physical-chemical properties, as well as biological properties (such as fermentability). Isomers, similar but different molecules, are essentially of two main types: structural and stereo. Structural isomers have atoms bonded in a different order (e.g., ABCD and BCAD), whereas stereoisomers have the same general bonding order but are arranged differently (e.g., $_ABC^D$ and $_ABC^D$).

STARCHES Starches are rather long, branched polymers of glucose. As mentioned, the exclusive presence of 1–4 bonds implies the formation of linear chains (amylose), which, in barley malt, are about 2,000 glucose units long. Amylopectins, for a total molecular size of more than 100,000 glucose units, are composed of linear chains interrupted every 30 glucose units by 1–6 bonds, which branch the chains. The endosperm constitutes about 80% of the dry weight of the seed, about 65% (amylose 25%, amylopectin 75%) of the starch. The internal structure of the endosperm assimilates small starch granules (Ø 5 μ; 80–90% of the number of granules, 10% of starch by weight) and large starch granules (Ø 25 μ; 10–20% granules, 90% by weight) into a protein matrix. During malting, the small granules, exposing a large external surface, are broken down by alpha amylase and glucoamylase, producing glucose that feeds the growth of the acrospire (the epigeal portion of the seedling). The large granules, which contain most of the starch, are activated only by the enzymes, but their contents remain largely available for making beer.

NONSTARCH POLYSACCHARIDES The endosperm contains several nonamylaceous polysaccharides, mostly components of the cell walls enclosing the protein matrix of starch granules. Pentoses and hexoses contribute to the construction of the cell walls of the endosperm, which are mainly composed of beta glucans (75%), hemicellulose (20%), and cellulose (2%). Beta glucans are composed of beta-glucose molecules (not alpha-glucose) bound together by β1–3 and β1–4 bonds. Breaking down beta glucans by glucanases into oligosaccharides (and ultimately into glucose) exposes the endosperm to further protein and starch alteration. Two endo-beta-glucanases (endo-enzymes cut within the chain with exo-enzymes at the ends), thermolabile enzymes that denature at 65°C within minutes, cut chains at β1–4 bonds adjacent to β1–3 bonds at the reducing end of the chain, producing oligosaccharides of 3–4 glucose units, which are subsequently broken down into individual glucose molecules by other enzymes.

The major hemicellulose in barley cell walls is arabinoxylan. Hemicelluloses (or pentosans) are long-chain branched polymers composed mostly of pentose monosaccharides such as arabinose and xylose. Cellulose provides structural rigidity to the cell wall of the endosperm. It is a linear polymer of glucose with β1–4 bonds. The -OH groups present on the glucoses of one chain form hydrogen bonds with the oxygen atoms of the neighboring chain, holding the chains tightly together and forming microfibers (microfibrils) that are particularly resistant to tensile stress. Such fibers reinforce cell walls by forming a structure consistent with the matrix of beta glucan and hemicellulose. Phenolic acids (e.g., ferulic acid [3-methoxy-4-hydroxycinnamic acid] is found in barley and wheat. It is located in the cell walls of the tegument and aleurone, bound to arabinoxylan [a hemicellulose], contributing to cell wall stability. It is also used by yeast to produce 4-vinyl-guaiacol [4VG], which gives the typical clove aroma of wheat beers) hold this heterogeneous complex of substances together through the formation of specific bonds. Cellulose does not undergo enzymatic changes during malting or brewing but remains intact in spent grains.

PHENOLS Phenols are hydrocarbons with a ring-shaped molecular structure. Unlike carbohydrates, phenols lack oxygen atoms in their structure. The basic type of phenol is a hydrocarbon ring (C_6H_6) to which an oxygen atom is added to form a hydroxyl group (-OH), forming the molecule C_6H_5OH. Members of this category of substances often emit odors, which is why they are referred to as aromatic hydrocarbons. Single phenols can join to make bigger and more complex structures, polyphenols. These substances are present in the tegument and cell walls of barley and do not undergo the enzymatic action. Tannins are a group of polyphenols containing many hydroxyl groups (-OH), carboxyl groups (-COOH), and others, which have the property of forming bonds with proteins. In brewing they are very important because they are responsible for cloudiness, contribute to beer astringency, and have antioxidant properties. At low temperatures, cloudiness can be caused by the reversible bond of polyphenols and proteins; as soon as the temperature rises, this bond dissolves, restoring the beer's clarity. The presence of oxygen can favor the polymerization of these molecules, therefore causing an irreversible cloudiness.

Proteins As opposed to carbohydrates, made of carbon, hydrogen, and oxygen, proteins contain an additional chemical element: nitrogen (N). This element can form three bonds, compared to carbon's four. When ammonia (NH_3), loses one hydrogen atom, an amine group ($-NH_2$) is formed. This group, if bound to particular carbon compounds, creates amino acids, so called because in the same molecule are two functional groups, an amino group ($-NH_2$) and a carboxylic group (-COOH), with the general formula $H_2NRCHCOOH$, where R is a different organic substituent (for size, structure, and composition) for each amino acid. Amino acids bind to one other through functional groups forming a *peptide bond*. This bond joins the nitrogen of the amino group of an amino acid with the carbon of the carboxyl of another amino acid; this process involves the loss of a molecule of water. Because of this bond, amino acids can form chains with an increasing number of amino acid units (peptides, polypeptides, proteins). Living organisms use 20 different amino acids to build peptides and polypeptides. One or more polypeptide chains form a protein. The arrangement of amino acids along the polypeptide chain is not random but follows a strict sequence, specific for each protein. This sequence defines the physical structure and biochemical functions of the specific protein. Proteins can be classified according to several parameters, including solubility: albumins are soluble in water, globulins in saline solutions, and prolamins (e.g., hordein) in alcoholic solutions, whereas glutelins are insoluble. Barley proteins, according to their position and function within the seed, can be divided into two main groups: nonreserve proteins and reserve proteins. Nonreserve proteins (structural proteins and enzymes) include albumins, glutelins, and globulins. These include enzymes present in barley before malting (beta amylase, protein Z [an albumin responsible for foam formation in beer]). The degradation of endosperm proteins during germination is performed by enzymes such as endoproteases and endopeptidases (altogether there are about 40 enzymes that break down proteins from the inside) and exoenzymes (e.g., carboxypeptidases) that detach individual amino acids from the carboxyl end of the peptide chain. Reserve proteins (hordeins and globulins) provide peptides and amino acids (FAN) to the embryo. During the malting process, the protein matrix of the endosperm is hydrolyzed into polypeptides, oligopeptides, and free amino acids, providing the vast majority (70–90%) of FAN to the wort. The protein matrix also contains glutelins, rather large soluble peptides present in the wort that can release FAN during mashing. Barley produces caryopses with variable protein content according to a number of factors, including cultivar, growing environment, climatic conditions of the year, and dryness. In general, barley (and therefore malts) for brewing with low protein content are preferred because the high protein content produces more turbidity, provides substrates to unwanted microorganisms, and means less extract per weight unit.

Lipids Waxes, fats, fatty acids, vitamins, and sterols (e.g., cholesterol) belong to the heterogeneous lipid family of chemical compounds. Lipids have hydrophobic (repelled by water) and hydrophilic (attracted to water) molecular portions. Fatty acids, for example, are long-chain hydrocarbons with a carboxylic acid (-COOH) end. The acidic terminal is polar (hydrophilic), whereas the hydrocarbon chain is apolar (hydrophobic). The result is a molecule that attracts water at one end and repels it at the other. Thanks to this property, lipids can connect polar molecules to nonpolar molecules and participate in many cellular biochemical reactions. Fatty acids participate in the formation of cell walls and energy storage. Fatty acids in barley are represented as follows: linoleic acid 58%, palmitic acid 20%, oleic acid 13%, linolenic acid 8%, stearic acid 1%. About two-thirds of the lipids in barley are located between the endosperm and the aleurone, one-third in the embryo. About 75% of barley lipids are nonpolar; polar lipids include glycolipids, phospholipids, and fatty acids. Lipids are essential for yeast nutrition, but in minimal amounts. Excess lipid, because of the effect of specific enzymes (e.g., lipoxygenase), determines the oxidation of fatty acids into peroxides, contributing compounds (e.g., aldehydes) that cause the unpleasant sensation of staleness in the final product.

Flavor Substances Stewing or roasting is the last stage of malting. Most of the aromas associated with malt are the result of thermal reactions. Caramelization is the thermochemical degradation (pyrolysis) of sugars (120°C) with the formation of volatile products (caramel aroma) and colorants (caramel color). This process is similar to the Maillard reaction, of browning by chemical reaction, which can produce flavors and aromas similar to caramelization but at much lower temperatures. The reaction occurs in three steps. In the first, an amino acid and a sugar (the carbonyl group of the sugar with the amino group of the amino acid) combine (with the loss of a water molecule) to give an unstable compound. In the second step, the compound is isomerized (Amadori rearrangement), giving rise to a ketosamine (a combination of a ketone and an amine). The final step involves the ketosamine undergoing further transformations following three pathways and producing three different products. One pathway results in the further dehydration of ketosamine with the production of compounds similar to those of caramelization; the second combines dehydration with additional reactions with other amino acids, creating colored polymers called melanoidins; and in the third pathway, heterocyclic compounds active in flavor expression (e.g., pyranes such as maltol and isomaltol), as well as furans and furfurals, are formed by steps including Strecker degradation. Highly aromatic heterocyclic nitrogen compounds (e.g., nitrosamines) are most abundant in malt roasted at high temperatures (> 180°C). However, many more (about 10,000) compounds are formed during fermentation.

Chemical Composition of Malt Extract The type and distribution of sugars or other carbohydrates contained in malt extract differentiate this product from other fermentables (e.g., glucose). Extracts with a different sugar profile from the one obtained from mashing grains have different fermentation and taste characteristics. Among possible differences is the color of the wort. The one produced with malt extract is usually darker than the one produced with grains. The speed of fermentation is also notably lower than the one produced with grains, probably because of the scarcity of FAN needed for the metabolism of yeast. Another significant element is the high amount of dextrins (11–42% in the extract). Moreover, it seems that in many extracts, cheaper syrups are added (corn syrup, glucose syrup, etc.) for a content of sugars typical of wort (maltose, maltotriose, dextrins), which is overall modest. Three aspects characterize an extract: color, fermentability, and taste. The production of malt extracts involves the concentration of wort by removing a considerable quantity of water through a process of evaporation. Even if the process takes place in a vacuum to reduce the temperature, the wort undergoes a long thermal treatment that leads to the formation of melanoidins. Extracts with a high concentration of simple sugars tend to darken, as sugars are more easily converted into colored pigments, even after packaging. The undesired darkening reactions take place in an aqueous environment. On the contrary, dry extracts, properly kept in a cool and dry place, tend not to darken during storage. Another cause of darkening is the boiling of wort, more marked in the presence of extracts. The taste and fermentability of extracts are influenced, for better or worse, by their chemical composition. Some unpleasant aromas (off-flavor) are not uncommon in old extracts (e.g., oxidized sherry). It's a good reason to avoid aged extracts. The fermentability of an extract is a function of the so-called *apparent attenuation*, which is the difference between the initial density (OG) and the final density (FG) divided by the initial density. The result (0–1) in percentage—A% = (OG − FG)/(OG − 1) × 100—for extracts takes values of 50–80% but more frequently 55–65%. This parameter is important because it influences the alcoholic strength, body, and taste of the final product.

Style Most beer styles are made with barley malt.

Source Calagione et al. 2018; Cantwell and Bouckaert 2016; Cantwell and Bouckaert 2018; Dabove 2015; Fisher and Fisher 2017; Giaccone and Signoroni 2017; Heilshorn 2017; Hieronymus 2010; Hieronymus 2016a; Hieronymus 2016b; Jackson 1991; Josephson et al. 2016; Mallett 2014; Mallett 2016; McGovern 2017; Markowski 2015; Sparrow 2015; Steele 2012; https://www.giornaledellabirra.it/approfondimenti/orzi-da-birra/; https://beerlegends.com/grains-and-malt-extracts.

From Barley to Malt

To be used in brewing, barley needs to be transformed into malt. The main objective of this process is to activate the enzymes present in the seed and start the breakdown of starch into simpler sugars. Malting begins with maceration. Barley caryopses are immersed in water and left to soak until their humidity reaches a value between 42% and 45%. The endosperm is rehydrated, and the germination process begins. The second phase consists of the actual germination: seeds are moved to a ventilated place and kept at a constant temperature. Here the kernel begins to develop: the radicles sprout from the lower end, and the kernel is generated inside the sprout. During this phase, part of the starch is transformed into simpler sugars. The quantity of free simple sugars inside the seed determines the extent of its "modification." The more a malt is modified, the higher the percentage of disintegrated starch it contains. Today, most malts are highly modified to simplify the brewer's job during the mashing process and to make brewing techniques such as decoction superfluous. Once the desired modification is reached—which can be estimated according to the length of the sprout—seeds are dried: the barley, still wet, is subjected to jets of hot air that stop germination and dry the caryopsis. In relation to the temperature used, it is possible to obtain malts of different colors, a characteristic that will influence the color and aromatic apparatus of the beer. It is important to avoid "burning" the malt in this phase in order not to denature enzymes that will be needed in the following phases of mashing. During the Middle Ages, malt was produced at home, usually in modest quantities. Every operation was managed in an empirical way, without the help of technology, either to control the modification or to measure the drying temperature. (The thermometer was invented at the beginning of the eighteenth century, and its use in malt houses was not widespread until the 1820s). Drying and toasting were done in ovens heated with ferns, wood, and straw, fuels difficult to manage and used to obtain a malt that was basically dark, when not burnt, with a strong hint of smoke.

We do not see a major leap forward in malting technology until the mid-seventeenth century. In those years, a new fuel was introduced: coke coal. This allowed maltsters to control the temperature of ovens more easily and, consequently, the color of the malt. Thanks to this innovation, pale malt was created. Besides having a light color, it also has a higher quantity of sugars available for fermentation. What brewers still did not know (and was discovered only at the beginning of the 1900s by the French chemist Louis Camille Maillard) was that, in the presence of humidity, sugars exposed to heat combine with nitrogenous substances, giving rise to new compounds that give toasted and dark notes to foods. It is these compounds that influence the color, flavor, and many aromas of malt.

In Europe, the color of malt is usually measured in European Brewery Convention (EBC) degrees on a scale ranging from 4 to over 1,000. In the United States, the Standard Research Method (SRM) color scale is used.

By changing temperature, humidity, and period of exposure of caryopses to heat, different products are obtained, usually defined as base, caramel, roasted and specialty malts.

Hordeum vulgare L.
Poaceae

Humulus lupulus L.
CANNABACEAE

Synonyms *Humulus cordifolius* Miq., *Humulus vulgaris* Gilib., *Lupulus amarus* Gilib., *Lupulus communis* Gaertn., *Lupulus humulus* Mill.

Common Names Hop, common hop.

Description Liana, perennial, dioecious (3–7 m). Underground rhizome coarse, branched, perennial, from which scandent stems develop in spring, twisted to other plants, with 6 dark stripes in which are short, obtuse spines. Leaves opposite with oval stipules (7–10 mm), petiole and main veins spinulose, lamina with circular outline (Ø 5–20 cm), palmate-lobate with acute, aristulate teeth, divided into 3–5 deep lobes, the basal ones sometimes tending to split. Male plant flowers (staminiferous) have 5 yellowish-white petals (4 mm) and 5 stamens. Male inflorescence rich panicle, pendulous at the apex of the branches; female plant flowers (pistilliferous) placed 2 × 2 at the axil of ovate-acuminate leafy bracts (9 × 12 mm) united in catkins to form characteristic ovoid cones that surround the ovary surmounted by 2 elongated and hairy stigmas. Compound fruit (achenocone) formed by small fruits of about 3 mm, ashy, subrounded, surrounded by bracts heightened and then papery, with the surface covered by numerous glands secreting a yellow resinous substance (lupulin).

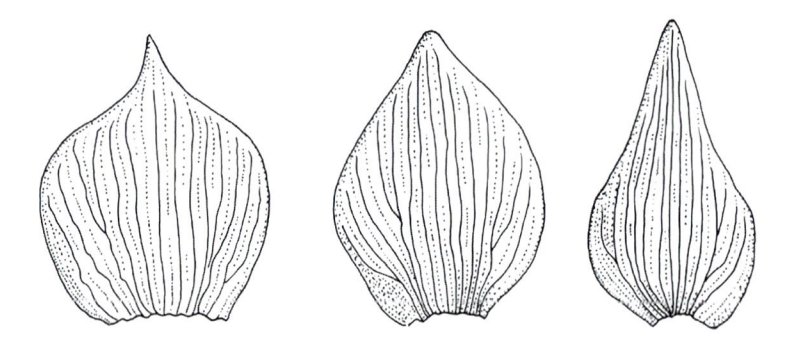

Morphological variability of the bracts of the achenocone of *Humulus lupulus*

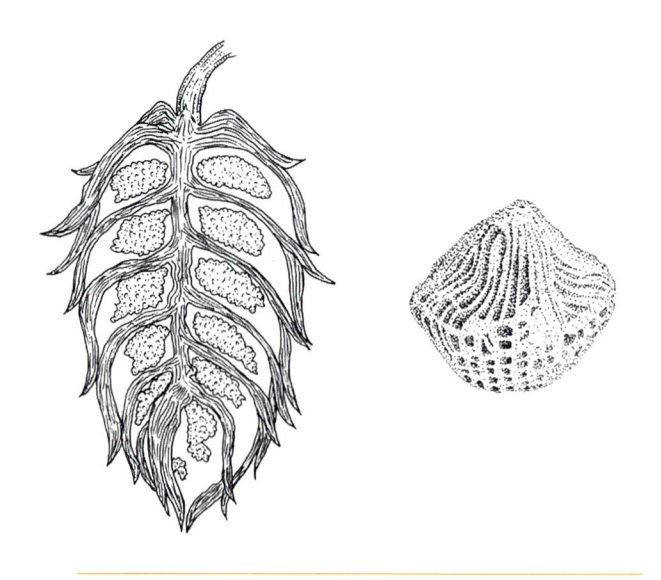

Longitudinal section of the achenocone of *Humulus lupulus* and shape of a single lupulin gland

Subspecies The genus *Humulus* is indigenous to the northern hemisphere in Europe, Asia, and North America and is thought to have originated in China. Within the genus *Humulus*, three entities of specific rank are recognized: *Humulus lupulus* L., *Humulus scandens* (Lour.) Merr. (= *Humulus japonicus* Siebold & Zucc.), with W Asian distribution, and *Humulus yunnanensis* Hu, endemic to the forests of Yunnan, China, at altitudes of 1,200–2,800 m ASL. The latter two species do not produce resinous achenocones, so they have no application in beer making. In contrast, *Humulus lupulus* produces female cones used in brewing. Wild *H. lupulus* plants have been classified based on leaf morphology and geographic distribution into five taxonomic varieties: *Humulus lupulus* L. var. *lupulus*, which includes European hops; *Humulus lupulus* L. var. *cordifolius* (Miq.) Maxim., the Japanese ones; *Humulus lupulus* L. var. *neomexicanus* A.Nelson & Cockerell; *Humulus lupulus* L. var. *pubescens* E.Small; and *Humulus lupulus* L. var. *lupuloides* E.Small, the North American ones. In addition to the aforementioned varieties, whose morphological and phytogeographic distinction is currently accepted, it should be noted that the *H. lupulus* populations of S Gansu and N Sichuan (China) may represent a distinct variety whose description has not yet been formalized. Further studies are therefore necessary to resolve the varietal complexity present in China, complicated by the introduction of var. *lupulus* for commercial purposes.

Hops have been cultivated for thousands of years in all of Europe. Genetic improvement has been based on European landraces such as Fuggle and Goldings (Great Britain), Saazer (Czech Republic), Tettnanger, Spalter, and Hallertauer Mfr. (Germany) because they are considered capable of providing the organoleptic qualities sought after by brewers. These genotypes are now known as "aroma hops," and Saaz continues to play an important role in the worldwide brewing industry today. The genetic improvement of hops, started by simple clonal selection from wild populations, has finally reached modern cultivars by means of guided hybridization processes. However, the repeated use of limited basic genetic resources, from cultivars considered of great value such as Northern Brewer or Brewer's Gold (Great Britain), has reduced the genetic variability in hop cultivars. Therefore, wild germplasm is of great importance in genetic improvement programs, as it can provide new resources to overcome the limited variability of modern hop cultivars.

Humulus lupulus L. var. *lupulus*. Stems with scattered pubescence at the nodes, usually < 15 hairs per 0.1 mm^2 in the most pubescent portion (excluding the angle of the petiole with the stem). Leaf blade usually < 20 hairs/cm on midrib, and < 25 glands/10 mm^2 between veins. *Habitat* Roadsides, moist groves, ruderal environments, forest mantles (0–2,000 m ASL). *Distribution* Native to Europe, naturalized following introduction for ornamental or beer-making purposes in many areas, including N Africa, N and NE Asia, China (Gansu, N Sichuan, Xinjiang), North America (Manitoba, New Brunswick, Nova Scotia, Ontario, Prince Edward Island, Quebec, California, Connecticut, Delaware, Illinois, Indiana, Maine, Maryland, Massachusetts, Michigan, Missouri, New Hampshire, New Jersey, New York, Ohio, Pennsylvania, Rhode Island, Vermont, Virginia, West Virginia, Wisconsin, Oregon). It is possible, if not highly probable, that spontaneous hybridization with some native varieties has occurred in some of these territories, particularly in North America.

Humulus lupulus L. var. *cordifolius* (Miq.) Maxim. According to some authors, this taxon deserves the taxonomic status of an autonomous species, such as *Humulus cordifolius* Miq. Perennial climber. Leaves 4–12 × 4–11 cm. Female flowers green. E Asian distribution, especially native to Hokkaido and NC Honshu (Japan); also cultivated in China.

Humulus lupulus L. var. *neomexicanus* A.Nelson & Cockerell. Stems relatively pubescent at nodes, usually > 15 hairs per 0.1 mm^2 in most pubescent portion (excluding corner of petiole with stem). Leaf laminae 10 cm or more, generally having at least 5 lobes; smaller laminae (about 5 cm) generally with > 3 easily visible veins branching from midrib (excluding proximal branching); surface generally with > 20 hairs/cm on midrib, > 25 glands/10 mm^2 between veins, abaxial glands exceptionally dense. *Habitat* On shrubs and trees on slopes, riverbanks, alluvial forests (300–3,000 m ASL). *Distribution* British Columbia, Manitoba, Saskatchewan, Arizona, California, Colorado, Kansas, Montana, Nebraska, Nevada, New Mexico, North Dakota, South Dakota, Texas, Utah, Wyoming, Mexico.

Humulus lupulus L. var. *pubescens* E.Small. Stems relatively pubescent at nodes, usually > 15 hairs/0.1 mm^2 in the most pubescent portion (excluding corner of small ones with stem). Leaf laminae 10 cm or more, usually having fewer than 5 lobes; minor laminae (about 5 cm) often with more than 3 easily visible veins branching from midrib (excluding proximal branching); abaxial surfaces very pubescent, > 100 hairs/cm along midrib, > 25 glands/mm^2 between veins, hairs present between the veins. *Habitat* Wetland scrub, woodlands (0–1,000 m ASL). *Distribution* Arkansas, Illinois, Indiana, Iowa, Kansas, Maryland, Minnesota, Missouri, Nebraska, New York, North Carolina, Ohio, Oklahoma, Pennsylvania, Virginia.

Humulus lupulus L. var. *lupuloides* E.Small (= *Humulus americanus* Nuttall). Stems relatively pubescent at nodes, usually > 15 hairs/0.1 mm^2 in most pubescent portion (excluding corner of petiole with stem). Leaf laminae 10 cm or more, generally with fewer than 5 lobes; minor laminae (about 5 cm) often with no more than 3 easily visible veins, branching from midrib (excluding proximal branching); abaxial surfaces not very pubescent, usually < 100 hairs/cm along midrib, > 25 glands/10 mm^2 between veins, hairs generally absent between veins. *Habitat* Wet scrublands, woodlands, riverbanks (0–2000 m ASL) *Distribution* Alberta, Manitoba, New Brunswick, Newfoundland and Labrador, Nova Scotia, Ontario, Prince Edward Island, Quebec, Saskatchewan, Connecticut, Delaware, Illinois, Indiana, Iowa, Kansas, Kentucky, Maine, Maryland, Massachusetts, Michigan, Minnesota, Montana, Nebraska, New Hampshire, New Jersey, New York, North Dakota, Ohio, Pennsylvania, Rhode Island, South Dakota, Vermont, Virginia, West Virginia, Wisconsin.

CULTIVAR It is estimated that hundreds have been obtained during the long cultivation history of this plant through more or less voluntary mechanisms of selection and back-crossing to wild varieties. Not all proposed cultivars have passed the test of time (or the market), but the more than 100 included here are among the most significant ones existing today, both for concrete beer-making applications and as basic material for new selection programs. In addition to the term "cultivar," these biological entities are indicated with terms such as "variety," "type," and in some cases, "landrace." Technically, perhaps the most suitable definition is *chemotype*, since distinctions are substantially made based on chemical profile, at least in relation to certain substances important in brewing.

Admiral English chemotype, used for bittering in traditional ale but also for fruity and citrus aroma if used at the end of boiling.

Agnus Czech chemotype, used for bittering because of its rather high alpha/beta acid ratio, herbaceous aromas, also suitable for dry hopping.

Ahtanum™ American chemotype with aromatic profile (floral, spicy, aromatic herbs, pine) compatible with "American C hops," used in late hopping and dry hopping.

Amarillo® Very successful chemotype, patented, apparently found in nature and therefore considered by some a landrace; its fruity aroma profile (citrus, melon, peach, etc.) makes it particularly sought after in American styles.

Apollo Patented chemotype, mainly for bittering, but with a complex of aromas suitable for dry hopping.

Aramis Chemotype from Strisselspalt × Whitbread Golding hybridization, spicy with hints of herbs, floral, citrus.

Aurora Slovenian chemotype, high alpha acids, floral, spicy.

Bramling Cross English chemotype, Bramling × Brewer's Gold hybrid, blackcurrant aroma.

Bravo Super alpha, bittering chemotype, growing success.

Brewer's Gold (U.S.) Chemotype with a strong blackberry aroma.

Calypso Chemotype with bitterness and aroma thanks to a fruity (pear, peach, cherry) and citrus aroma profile.

Cascade (NZ) Well-known American chemotype, less bitter than the U.S. version but with a complex aromatic profile apparently enhanced by the NZ environment.

Cascade (U.S.) American chemotype of exceptional floral aroma and taste, citrus, notes of grape; among the most popular American hops.

Celeia Slovenian chemotype derived from Styrian Golding, with fruity, citrus aromas and flavors; it is suitable for the lagering process and for the final phases of the ale process.

Centennial Chemotype with an original floral character that earned it the name Super Cascade and a growing popularity due to the explosion of IPAs.

Challenger English chemotype by antonomasia with a fruity and spicy character.

Chelan Chemotype for bittering derived from Galena, with a citrus aromatic profile.

Chinook Chemotype with dual bitter/aroma purpose, today particularly appreciated in dry hopping of

American hopped styles, in which it releases a complex fruity and pine resin aroma.

Citra® Chemotype with a flavor enriched by complex fruity notes (passion fruit, lychee, peach, gooseberry, etc.).

Cluster American chemotype considered to be among the most landrace-like; popular in the NW United States in the past.

Columbus One of three chemotypes (Columbus, Tomahawk, Zeus) so similar genetically to also be known by the collective name CTZ; this affinity is also found in their prevalent use for bitterness, although with distinct flavor profiles: fruity and spicy.

Crystal Chemotype derived from Hallertau Mittelfrüh × Cascade × Brewer's Gold, it is extremely flexible in use; from delicate spicy and floral notes to a strong pungent character.

Delta Chemotype from Fuggle × Cascade hybridization, with particular notes of wood, herbs, and citrus.

El Dorado® Chemotype from NW United States (Yakima Valley), dual purpose, with strong fruity aroma (pear, cherry) and notes of candied fruit.

Ella Chemotype potentially bitter because of its alpha acid content but recommended for its tropical, sweet, floral, fruity aroma (lemon, apricot, melon).

First Gold English chemotype derived from Whitbread Golding, interesting for its floral and fruity profile that is well suited to late boiling and dry hopping; interesting also for the small size of the plant (dwarf hops).

Fuggle English chemotype by antonomasia, with a fruity, spicy, and woody aromatic profile.

Galaxy Australian chemotype with a high content of alpha acids; however, mainly used for late hopping or dry hopping thanks to a fruity aromatic profile (passion fruit, citrus, apricot, melon, blackcurrant) of variable intensity.

Galena Bitter chemotype, today partially substituted by Super Galena but still used for its floral and citrus aroma in the production of pale lagers.

Glacier American chemotype with a characteristic low content of cohumulone and a delicate floral, citrus, fruity (peach), aroma.

Green Bullet Chemotype with a high content of alpha acids, today used also for its floral and fruity flavor (grape).

Hallertau Merkur Chemotype for bittering with earthy, floral, spicy aroma, used in the production of pale lagers.

Hallertau Mittelfrüh German chemotype considered a landrace, with herbal, spicy, woody aromas and flavors; widely used in the production of lagers.

Hallertau Taurus German chemotype with a delicate aroma, used in lager technology.

Hallertau Tradition German chemotype derived from Mittelfrüh, with a floral aroma (aromatic herbs) and superior production and agronomic performances.

Harmonie Czech chemotype, with balanced bitterness and higher persistence than Saaz; similar to the recent Bohemie chemotype (2010).

Helga Australian chemotype derived from Hallertau Mittelfrüh, has a spicy and herbal profiles similar to its parent.

Herkules German chemotype with a high content of alpha acids and high productivity.

Hersbrucker German chemotype with a spicy and aromatic herbal character, with citrus and fruity scents.

Horizon Chemotype genetically similar to Nugget, with a dual attitude for its floral and spicy aromatic profile.

Kazbek Czech chemotype obtained with Russian wild germplasm; characteristic strong and spicy aroma.

Kent Golding Traditional English chemotype, with an extraordinary aroma profile of lavender, spices, honey, thyme; earthy taste, delicate bitterness, sweet, silky, honey character; the quintessential English hops for ales and pale ales.

Liberty American triploid chemotype derived from Hallertau Mittelfrüh, with organoleptic characteristics similar to its prestigious parent.

Lublin Polish chemotype very similar to Saazer.

Magnum German chemotype rich in alpha acids, but less so than Herkules, by which it could be soon replaced.

Mandarina Bavaria German chemotype derived from Cascade and similar to it in aromatic profile although more intense, fruity with notes of aromatic herbs.

Marynka Polish chemotype used for bittering, also suitable for providing aroma thanks to its strong floral notes (roses).

Meridian American chemotype of unknown lineage but with original aroma of lemon cake and fruit punch.

Millennium Chemotype derived from Nugget, with a high content of alpha acids and a spicy aroma of crème caramel, yogurt, and toffee.

Mosaic™ Chemotype derived from Simcoe × Nugget hybridization with a complex fruity (mango, lemon, citrus, blueberry) and pine aroma.

Motueka New Zealand triploid chemotype derived from a Saaz × New Zealand unknown strain cross, with a citrus (lemon, lime) and tropical fruit profile.

Mt. Hood Triploid chemotype derived from Hallertau Mittelfrüh, today of decreasing importance.

Nelson Sauvin New Zealand chemotype derived from Smoothcone, has aromatic and gustatory characteristics incredibly similar to sauvignon blanc wine; considered too characterizing by some brewers, it should be used with care; gaining increasing popularity among craft brewers and home brewers for its uncommon characteristics (fruity, tropical, white wine notes) especially in American Pale Ale brewing.

Newport American chemotype rich in alpha acids, with a delicate aroma, which in dry hopping is sometimes pungent and resinous.

Northdown English chemotype obtained by successive crossings of Northern Brewer × Challenger × Target; relatively high content of alpha acids, in the past used for bittering but now replaced by other chemotypes, suitable for dry hopping.

Northern Brewer English chemotype derived from Canterbury Golding × Brewer's Gold crossbreed, with dual purpose, nowadays mostly cultivated in Germany and the United States, used in the brewing of many styles (European beers and ales, Lambic, Porter, California Common) for its wood, pine, and mint aroma.

Nugget American chemotype with a high content of alpha acids, with a pleasant aroma of aromatic herbs; typical in IPA and Imperial IPA.

Opal German chemotype with a spicy, woody, and slightly floral aroma profile.

Pacifica New Zealand chemotype derived from the crossbreed Hallertau Mittelfrüh × unknown New Zealand plant, has a spicy and floral aromatic profile, as well as citrus notes if used late in boiling.

Palisade® American chemotype derived from the free pollination of Swiss Tettnanger, selected for its floral, fruity, tropical aroma.

Perle German chemotype derived from Northern Brewer, presents a mint aroma, spicy but also used for bittering; also cultivated in NW United States.

Pilgrim English chemotype with single or dual purpose, unique for its citrus aroma (lemon, grapefruit).

Pioneer English chemotype with dual purpose, with intense citrus notes (lemon).

Polaris German chemotype with high alpha acids and essential oils, aroma of eucalyptus, peppermint, citrus.

Premiant Czech chemotype originated from Saaz, has a high content of alpha acids and a floral and citrus aromatic profile.

Pride of Ringwood Australian chemotype for bittering; lost importance as a hop for bittering, also because of the availability of more talented chemotypes, today it is revaluated for its interesting aroma profile (berries, citrus fruits).

Progress English chemotype for aroma, has some characteristics of a landrace that make it suitable for the production of classic English ales.

Rakau Chemotype from New Zealand with dual purpose thanks to its fruity aroma (tropical fruit, passion fruit, peach); particularly appreciated in the dry hopping of American Pale Ale.

Riwaka New Zealand chemotype derived by crossing Saazer × New Zealand germplasm, with pine and tropical aromas.

Rubin Czech chemotype derived from bitter Saaz but with overall inferior characteristics.

Saaz Traditional and main chemotype of the Czech Republic, appreciated for its pleasing and delicate aromatic and taste profile.

Saaz Late Czech chemotype derived from Saaz to try to reduce the annual alternation of alpha acid production.

Santium American chemotype developed to simulate the characteristics of Tettnang Tettnanger, with high levels of alpha acids, spicy and herbal aroma.

Saphir German chemotype, spicy and with notes of berries, citrus fruits.

Simcoe® American chemotype, fruity (blackcurrant, berries), citrusy, with hints of pine; aroma with a pungent hint of cat, in the past unwelcome but today sought after in American dry hopped styles.

Slàdek Czech chemotype derived from Saaz, with an alpha acid content similar to its progenitor and an appreciated floral and spicy aroma.

Smarago German chemotype with aroma and taste of classic hops, as well as notes of spices, aromatic herbs, wood.

Sorachi Ace Japanese chemotype derived from Saaz × Brewer's Gold crossbreed, with deep citrus (lemon, orange), coriander, dill, and oak aroma, considered excellent in several styles, especially those requiring a distinct lemon flavor.

Southern Cross New Zealand chemotype produced by crossing Smoothcone × Fuggle, with interesting citrus (lemon), spicy, and pine aromas that make it a dual-purpose hop often used in New Zealand lagers.

Sovereign English chemotype derived from Whitbread Golding, a dwarf morphotype, with an intense fruity aroma (peach).

Spalt Spalter German chemotype genetically similar to Saaz and Tettnanger, appreciated for its refined, spicy, and delicate aroma with notes of aromatic, woody, and floral herbs.

Spalter Select German chemotype with a spicy, floral, woody aroma.

Sterling American chemotype obtained by hybridization of Saaz × Cascade × other European chemotypes, with a spicy and citrus aroma, high content of alpha acids.

Strisselspalt French chemotype with an elegant, floral, spicy, citrusy (lemon) aroma, in the past much appreciated in the production of Michelob Pale Lager, nowadays in disuse.

Styrian Golding Chemotype from Slovenia derived from Fuggle, with a resinous, earthy, and white pepper aroma and taste.

Summer Australian chemotype similar to Sylva and derived from Saaz crossbreeds, with organoleptic characteristics similar to its parent but a lighter and fruity aroma, spicy and floral, with hints of tea.

Summit™ American chemotype with Zeus and Nugget among its parents, with citrus (grapefruit) aromas and flavors that make it a hop with dual purpose; however, it can sometimes have unwelcome notes of onion and garlic.

Super Galena American chemotype derived from Galena, with a high content of alpha acids and remarkable productivity.

Super Pride Australian chemotype derived from Pride of Ringwood, with a higher alpha acid content than the parent.

Sylva Australian chemotype similar to Summer and derived from Saaz crossings, it has a strong hopped character (with notes of cedar wood) and the typical spiciness of the prestigious parental.

Target English chemotype characterizing the taste of British beer productions, rich in alpha acids and suitable for dry hopping.

Tettnang Tettnanger German chemotype belonging to the Saazer group, with floral, citrus (bergamot), lily of the valley, Cognac, and chocolate aromas.

Tettnanger (U.S.) American chemotype apparently derived from Fuggle freely pollinated by Tettnang Tettnanger, has a floral, spicy, and wood aroma which lent it success in the past; today it is not very common.

Tillicum® American chemotype derived from Galena, with a pleasant and versatile aroma but used mainly for bitterness.

Tomahawk® One of the three chemotypes (Columbus, Tomahawk, Zeus) genetically similar enough to be known by the collective name CTZ; this affinity is also found in their prevalent use in bittering, although with distinct aromatic profiles: this one is citrusy, spicy, herbaceous, pungent, of black pepper, licorice, and curry.

Topaz Australian chemotype rich in alpha acids, therefore mainly used for bittering but, thanks to an aromatic and taste profile rich in fruity tones (berries, passion fruit), also revalued in late additions and dry hopping.

Triskel French chemotype derived from the crossbreed Strisselspalt × Yeoman, with a floral and citrus aromatic profile, particularly suited for styles such as American ale and Belgian ale; brewed with late additions and dry hopping in any case.

Tsingtao Flower Chinese chemotype derived from Cluster, with a floral and spicy aroma profile; the main hop cultivated in China.

Ultra American chemotype derived from Hallertau Mittelfrüh × Saazer crossbreed, with the delicate and pleasant characteristics of European hops.

Vanguard American chemotype derived from Mittelfrüh, with an aromatic profile of herbs and spices, similar to its parent.

Wakatu New Zealand triploid chemotype obtained by crossing Hallertau Mittelfrüh × New Zealand male, with dual purpose and a balanced floral and citrus (lime) aroma profile.

Warrior® American chemotype with a history similar to Simcoe, also obtained by spontaneous pollination; mainly used for bittering but with an interesting aromatic and taste profile (floral, spicy, woody, citrus [citron]) that led to an increase in the popularity of craft breweries and styles such as IPA.

Whitbread Golding English chemotype with a good level of alpha acids and a pronounced fruity aroma, characteristics that make it a typical hop with dual purpose.

Willamette American chemotype with delicate aromatic and spicy taste profiles, which make it versatile and useful for the production of different styles.

Zeus One of three chemotypes (Columbus, Tomahawk, Zeus) so genetically similar to be known by the collective name CTZ; this affinity is also found in their prevalent use in bittering, although with distinct aromatic profiles; this one, similar to the other CTZs, is very aromatic, even pungent, with strong citrus, spicy, and herbal notes.

Related Species Besides *Humulus lupulus* L., two other species are recognized within the genus *Humulus*: *Humulus scandens* (Lour.) Merr. and *Humulus yunnanensis* Hu.

Geographic Distribution Circumboreal, European-Caucasian. **Habitat** Wet woods, hedges. **Range** Germany, United States, China, Czech

Republic, Slovenia, Poland, United Kingdom, Australia, Spain, South Africa, France, Ukraine, New Zealand.

BEER-MAKING PARTS (AND POSSIBLE USES)
Female cones.

CHEMISTRY A dried hop achenocone contains: water 8–10%, cellulose 40–50%, protein up to 15%, ash 8%, pectins 2%, lipids and waxes up to 5%, polyphenols 2–5%, beta acids 0–10%, alpha acids 0–22%, essential oils 0.5–4%. The brewing process with regard to hops revolves primarily around alpha acids and essential oils, although not exclusively. Hops are believed to be involved in the composition of eight important beer characteristics: bitterness, aroma, flavor (combination of aroma and taste), mouthfeel, foam, Brussels lace, flavor stability, and antiseptic properties (inhibiting the development of potentially harmful microorganisms). Of most importance to the brewer is not the hop plant but the achenocone. The part of hops useful for brewing is an infructescence that develops from a female inflorescence. Along a central axis, two bracts (outer leaves) and four bracts (inner petals) are attached to each node. Lupulin glands—yellow, sticky, and aromatic—are concentrated at the base of the bracts and contain resins (hard and soft), essential oils, and polyphenols. The resins host alpha and beta acids. Alpha acids in particular, once isomerized during mashing, determine bitterness. Oils, on the contrary, are responsible for aroma and taste. The first compounds to be discovered were myrcene, humulene, linalool, linalyl-isonate, geraniol, and diterpenes. Subsequently, many other compounds were discovered (more than 400), some in concentrations so low as to be individually imperceptible by humans. The composition of the aroma mix of a hop changes significantly 15 days before harvesting, at the time of harvesting, during drying and storage, as well as during mashing, dry hopping, and, of course, beer aging. The aromas most frequently imparted by hops to beer are flowery and floral (rose, geranium, clove), citrusy (lemon, lime, bergamot, pomelo, cedar), fruity (tangerine, melon, mango, lychee, passion fruit, apple, banana, gooseberry, red and black currant, pear, peach, cherry), wine bouquet (e.g., sauvignon blanc), and woody (pine, woody, resinous). Essential oils: up to 4% of the hop achenocone contains hydrocarbons (50–80%), oxygenated hydrocarbons (20–50%), and sulfur compounds (about 1%). Hydrocarbons are volatile and not very water soluble, present in beer thanks only to late hopping or dry hopping. Oxygenated hydrocarbons, more soluble and aromatic, are therefore more present and stable in beer. Sulfur compounds, despite their low content, can greatly influence the aroma of beer. Known characteristics of the different chemotypes of hops are the contents of the four most important oils for brewing purposes: myrcene, caryophyllene, humulene, and farnesene. Myrcene is a monoterpene (10 carbon units); farnesene is a sesquiterpene (15 carbon units). Myrcene has an acerbic, herbaceous, resinous, fresh hop aroma, formerly considered undesirable, now sought after among American chemotypes (myrcene > 50% of essential oils in these selections) used for dry hopping. Caryophyllene and humulene are sesquiterpenes often evaluated together (humulene/caryophyllene 3:1 is the precursor of herbaceous and spicy aromas) that give the so-called noble aroma. When mature, hops release other monoterpenes in modest quantities (linalool, geraniol, nerol, citronellol, isobutyl isopyrate, limonene) that give important aromatic characteristics to beer (citrus, fruity, floral, woody) as a consequence of complex and little-known interactions with yeast (biotransformation). Although there is still much to discover about biotransformation, synergies between aromas, and the physical–chemical transformations that likely characterize the different phases of the brewing process and the interactions of these volatile substances with the human perception apparatus, some aromatic compounds present in beers coming from hops capable of producing specific aromas have been identified. These include the cis-rose oxide (Centennial) and caryophyll-3,8-dien-(13)dien-5-beta-ol, the basis of the cedarwood aroma. Our understanding of the role of key compounds such as myrcene, its high volatility in boiling, and the need for dry hopping (Cascade, Chinook) to create a strong American beer, very hoppy with hints of resinous pine, has deepened. The 4-mercapto-4-methylpentane-2-one (4MMP) responsible for the fruity aroma (grape, currant) was discovered, along with seven compounds that determine the aroma of blackcurrant, and two other thiols (in addition to 4MMP), 3-mercaptohexan-1-ol and 3-mercaptohexyl acetate (3MH and 3MHA), determinants in the aroma of passion fruit and sauvignon blanc wine. Two further thiols responsible for some wine aromas were also discovered. The New Zealand chemotype Nelson Sauvin owes its distinct exotic fruit and white wine aroma to the presence of 3-mercapto-4-methylpentan-1-ol (3M4MP) and 3-mercapto-4-methylpentan-acetate (3M4MPA). Table 5 lists other chemical compounds and their related flavorings.

TABLE 5. FLAVOR COMPOUNDS PREVALENT IN HOPS AND HOPPED BEERS

COMPOUND	AROMA
2-methylbutyric acid	cheese
3-methylbutyric acid (isovaleric acid)	cheese
3-mercaptohexan-1-ol (3MH)	blackcurrant, grapefruit
3-mercaptohexyl acetate (3MHA)	blackcurrant, grapefruit
3-mercapto-4-methylpentan-1-ol (3M4MP)	grapefruit, rhubarb
4-mercapto-4-methylpentan-2-one (4MMP)	blackcurrant
Alpha-pinene	pine, grass
Beta-ionone	floral, berries
Beta-pinene	pine, spices
Caryophylla-3,8(13)-dien-5-beta-ol	cedarwood
Caryophyllene	woody
Cis-3-hexenal	unripe fruit, leaves
Cis-rose oxide	fruity, grassy
Citral	sweet citrus, lemon
Citronellol	citrus, fruity
Ethyl-2-methylbutyrate	fruity
Ethyl-2-methylpropanoate	pineapple
Ethyl-3-methylbutanoate	fruity
Ethyl-4-methylpentanoate	fruity
Eudesmol	spicy
Farnesene	floral
Geraniol	floral, sweet, rose
Humulene	woody/pine
Isobutyl isobutyrate	fruity
Limonene	citric, orange
Linalool	floral, orange
Myrcene	sour, resinous
Nerol	rose, citrus
Terpineol	woody

Source: Hieronymus 2012.

TABLE 6. CHEMICAL COMPOSITION (KEY BEER-MAKING COMPONENTS) OF MAJOR HOP CHEMOTYPES

	ALPHA	BETA	MYRCENE	HUMULENE	TOTAL OILS	COHUMULONE	FARNESENE	CARYOPHYLLENE	FLORAL–ESTER	CITRUS FRUITS–PINE
Admiral	13-16	4.8-6.1	39-48	23-36	1-1.7	37-45	1.8-2.2	6.8-7.2		
Agnus	9-12	4-6.5	40-55	15-22	2-3	29-38	<1	6.8-7.2		
Ahtanum	5.2-6.3	5-6.5	50-55	16-20	0.8-1.2	30-35	<1	9-12		
Amarillo	8-11	6-7	68-70	9-11	1.5-1.9	21-24	2-4	2-4		
Apollo	15-19	5.5-8	30-50	20-35	1.5-2.5	24-28	<1	14-20		
Aramis	7.9-8.3	3.8-4.5	40	21	1.2-1.6	21.5-21.7	<1	8		
Aurora	7-9	3-5	35-53	20-27	0.9-1.4	23-28	6-9	4.8		
Bramling Cross	6-7.8	2.2-2.8	35-40	28-33	0.8-1.2	26-31	<1	14-18		
Bravo	14-17	3-4	25-50	18-20	1.6-2.4	29-34	<1	10-12		
Brewer's Gold (USA)	8-10	3.5-4.5	37-40	29-31	2.2-2.4	40-48	<1	7-7.5		
Calypso	12-14	5-6	30-45	20-35	1.6-2.5	40-42	<1	9-15		
Cascade (NZ)	6-8	5-5.5	53-60	14.5	1.1	37	6	5.4	2.8	5.9
Cascade (US)	4.5-7	4.8-7	45-60	8-13	0.7-1.4	33-40	3.7	3.6		
Celeia	4.5-7	2.5-3.5	27-33	20-35	0.6-1	27-31	3-7	8-9		
Centennial	9.5-11.5	3.5-4.5	45-55	10-18	1.5-2.3	29-30	<1	5-8		
Challenger	6.5-8.5	4-4.5	30-42	25-32	1-1.7	20-25	1-3	8-10		
Chelan	12-14.5	8.5-9.8	45-55	12-15	1.5-1.9	33-35	<1	9-12		
Chinook	12-14	3-4	35-40	18-23	1.7-2.7	29-35	<1	9-11		
Citra	11-13	3.5-4.5	60-65	11-13	2.2-2.8	22-24	0	6-8		
Cluster	5.5-8.5	4.5-5.5	45-55	15-18	0.4-0.8	37-43	<1	6-7		
Columbus	14-16.5	4-5	40-50	12-18	2-3	28-32	<1	9-11		
Crystal	3.5-5.5	4.5-6.5	40-65	18-24	1-1.5	20-26	<1	4-8		
Delta	5-5.7	5.5-7	25-40	30-40	0.5-1.1	22-24	<1	9-15		
El Dorado	14-16	7-8	55-60	10-15	2.5-2.8	28-33	<1	6-8		
Ella	14-16	4-4.5	33	1.2	2.9	36	13	15		
First Gold	5.6-9.3	2.3-4.1	24-27	20-24	0.7-1.5	32-34	2-4	6-7		
Fuggle	3-5.6	2-3	24-28	33-38	0.7-1.4	25-30	5-7	9-13		
Galaxy	13.5-15	5.8-6	33-42	1-2	2.4-2.7	35	3-4	9-12		
Galena	11-13.5	7.2-8.7	55-60	10-13	0.9-1.3	36-40	<1	4.5-6.5		
Glacier	5-6	7.6	33-62	24-36	0.7-1.6	11-13	<1	6.5-10		
Green Bullet	11-14	6.5-7	38	28	1.1	41-43	<1	9	2.3	7.9
Hallertau Merkur	10-14	3.5-7	25-35	35-50	1.4-1.9	17-22	<1	9-15		

	ALPHA	BETA	MYRCENE	HUMULENE	TOTAL OILS	COHUMULONE	FARNESENE	CARYOPHYLLENE	FLORAL–ESTER	CITRUS FRUITS–PINE
Hallertau Mittelfrüh	3-5.5	3-5	20-28	45-55	0.7-1.3	18-28	<1	10-15		
Hallertau Taurus	12-17	4-6	30-50	23-33	0.9-1.4	20-25	<1	6-11		
Hallertau Tradition	4-7	3-6	17-32	35-50	0.5-1	24-30	<1	10-15		
Harmonie	5-8	5-8	30-40	10-20	1-2	17-21	<1	6-11		
Helga	5-6	3.8-5.4	2-12	35-47	0.6-0.7	22-26	<1	10-14		
Herkules	12-17	4-5.5	30-50	30-45	1.6-2.4	32-38	<1	7-12		
Hersbrucker	1.5-4	2.5-6	15-30	20-30	0.5-1	17-25	<1	8-13		
Horizon	11-13	6.5-8.5	55-65	11-13	1.5-2	16-19	2.5-3.5	2.5-3.5		
Kazbek	5-8	4-6	40-50	20-35	0.9-1.8	35-40	<1	10-15		
Kent Golding	4-6.5	1.9-2.8	20-26	38-44	0.4-0.8	28-32	<1	12-16		
Liberty	3-5	3-4	20-40	35-40	0.6-1.2	24-30	<1	9-12		
Lublin	3-4.5	3-4	22-29	30-40	0.5-1.1	25-28	10-14	6-11		
Magnum	11-16	5-7	30-45	30-45	1.6-2.6	21-29	<1	8-12		
Mandarina Bavaria	7-10	5-7	71	5	2.1	33	1	1.7		
Marynka	6-12	10-13	28-31	26-33	1.8-2.2	26-33	1.8-2.2	11-12		
Meridian	6.5	9.5	30	8	1.1	45	<1	3.8		
Millenium	14-16.5	4.3-5.3	30-40	23-27	1.8-2.2	28-32	<1	9-12		
Mosaic	11-13.5	3.2-3.9	54	13	1.5	24-26	<1	6.4		
Motueka	6.5-7.5	5-5.5	48	3.6	0.8	29	12	2	4	18.3
Mt. Hood	4-7	5-8	30-40	30-38	1.2-1.7	21-23	<1	13-16		
Nelson Sauvin	12-13	6-8	21-23	35-37	1-1.2	22-26	<1	10-12	2.8	7.8
Newport	13.5-17	7.2-9.1	47-54	9-14	1.6-3.4	36-38	<1	4.5-7		
Northdown	7.5-9.5	5-5.5	23-29	40-45	1.2-2.5	24-30	<1	13-17		
Northern Brewer	6-10	3-5	50-65	35-50	1-1.6	27-32	<1	10-20		
Nugget	11-14	3-5.8	48-55	16-19	0.9-2.2	22-30	<1	7-10		
Opal	5-8	3.5-5.5	20-45	30-50	0.8-1.3	13-17	<1	8-15		
Pacifica	5-6	5.5-6	10-14	48-52	1	25	<1	17	1.6	6.9
Palisade	5.5-9.5	6-8	9-10	19-22	1.4-1.6	24-29	<1	16-18		
Perle	4-9	2.5-4.5	20-35	35-55	0.5-1.5	29-35	<1	10-20		
Pilgrim	9-13	4.3-5	30	17	1.2-2.4	36-38	0.3	7.3		
Pioneer	7-11	3.5-4	31-36	22-24	1-1.8	36	<1	7-8		
Polaris	19-23	5-7	50	22	4.4	27	<0.1	9		
Premiant	7-9	3.5-5.5	35-45	25-35	1-2	18-23	1-3	7-13		
Pride of Ringwood	7-11	4-6	25-50	2-8	0.9-2	32-39	<1	5-8		

	ALPHA	BETA	MYRCENE	HUMULENE	TOTAL OILS	COHUMULONE	FARNESENE	CARYOPHYLLENE	FLORAL–ESTER	CITRUS FRUITS–PINE
Progress	5-7	2-2.5	30-35	40-47	0.6-1.2	25-30	<1	12-15		
Rakau	10.8	4.6	56	16.3	2.1	25	4.5	5.2	1.2	5.7
Riwaka	4.5-6.5	4-5	68	9	0.8	29-36	1	4	2.8	5.9
Rubin	9-12	3.5-5	35-45	13-20	1-2	25-33	<1	7-10		
Saaz	3-6	4.5-8	25-40	15-25	0.4-1	23-26	14-20	10-12		
Saaz Late	3-7	3.8-6.8	25-35	15-20	0.5-1	20-24	15-20	6-9		
Santium	5.5-7	7-8.5	30-45	20-25	1.3-1.7	20-22	13-16	5-8		
Saphir	2-4.5	4-7	25-40	20-30	0.8-1.4	12-17	<1	9-14		
Simcoe	12-14	4-5	60-65	10-15	2-2.5	15-20	<1	5-8		
Sládek	4.5-6.5	4-6	40-50	20-30	1-2	25-30	<1	8-13		
Smaragd	4-6	3.5-5.5	20-40	30-50	0.4-0.8	13-18	<1	9-14		
Sorachi Ace	10-16	6-7	35	21-27	2-2.8	23	6	8-9		
Southern Cross	11-14	5-6	32	21	1.2	25-28	7.3	6.7	2.7	6.9
Sovereign	4.5-6.5	2.1-3.1	/	23	0.8	26-30	3.6	8.3		
Spalt Spalter	2.5-5.5	3-5	20-35	20-30	0.5-0.9	22-29	12-18	8-13		
Spalter Select	3-6.5	2.5-5	20-40	10-22	0.6-0.9	21-27	15-22	4-10		
Sterling	6-9	4-6	44-48	19-23	1.3-1.9	22-28	11-17	5-7		
Strisselspalt	1.8-2.5	3-6	35-52	12-21	0.6-0.8	20-25	<1	6-10		
Styrian Golding	4.5-6	2-3.5	27-33	20-35	0.5-1	25-30	3-5	7-10		
Summer	4-7	4.8-6.1	5-13	42-46	0.9-1.3	22-25	<1	14-15		
Summit	13-15.5	4-6	30-50	15-25	1.5-2.5	26-33	<1	10-15		
Super Galena	13-16	8-10	45-60	19-24	1.5-2.5	35-40	<1	6-14		
Super Pride	14-15	7-8	30-45	1-2	1.7-1.9	30-34	<1	7-9		
Sylva	4-7	3-5	17-23	19-26	0.5-1.1	23-28	23-25	6-9		
Target	9.5-12.5	4.3-5.7	45-55	17-22	1.2-1.4	35-40	<1	8-10		
Tettnang Tettnanger	2.5-5.5	3-5	20-35	22-32	0.5-0.9	22-28	16-24	6-11		
Tettnanger (US)	4-5	3.5-4.5	25-40	18-25	0.4-0.8	20-25	10-15	6-8		
Tilliticum	12-14.5	9.3-10.5	45-55	13-16	1.5-1.9	31-38	<1	7-8		
Tomahawk	14.5-17	4.5-5.5	50-60	9-15	2.5-3.5	28-35	<1	4-10		
Topaz	15-18	6-7	25-43	11-13	0.8-1.7	47-50	<1	10-11		
Triskel	8-9	4-4.7	60	13.5	1.5-2	20-23	<1	5.4		
Tsingtao Flower	6-8	3-4.2	45-55	15-18	0.4-0.8	35	<1	6-7		
Ultra	2-3.5	3-4.5	15-25	35-50	0.5-1	23-38	<1	10-15		
Vanguard	5.5-6	6-7	20-25	45-50	0.9-1.2	14-16	<1	12-14		

	ALPHA	BETA	MYRCENE	HUMULENE	TOTAL OILS	COHUMULONE	FARNESENE	CARYOPHYLLENE	FLORAL–ESTER	CITRUS FRUITS–PINE
Wakatu	6.5-8.5	8-9	35-36	16-17	0.9-1.1	28-30	6-7	7-9	3.2	9.5
Warrior	14-16.5	4.3-5.3	40-50	15-19	1.3-1.7	22-26	<1	9-11		
Whitbread Golding	5.4-7.7	2-2.5	43	29-44	0.9-1.4	25-36	1-3	12-14		
Willamette	4-6	3-4.5	30-40	20-27	1-1.5	30-35	5-6	7-8		
Zeus	12-16.5	4-6	25-65	10-25	1-2	27-35	<1	5-15		

Style Most styles are made with the addition of hops.

Source Calagione et al. 2018; Daniels and Larson 2000; Fisher and Fisher 2017; Giaccone and Signoroni 2017; Heilshorn 2017; Hieronymus 2016a; Hieronymus 2016b; Jackson 1991; Josephson et al. 2016; Markowski 2015; Sparrow 2015; Steele 2012; https://beerlegends.com/hops-varieties; https://www.erbemedicinali.netsons.org/index.php?option=com_content&view=article&id=44%3Aluppolo&catid=29%3Apiante&Itemid=29&lang=it; https://en.wikipedia.org/wiki/Hops; https://www.usahops.org/index.cfm?fuseaction=hop_farming&pageID=6; https://www.yakimavalleyhops.com/default.asp.

🍺 The Star of the Craft Revolution

If today we have such a wide range of beers available to choose from, it is surely thanks to the so-called craft revolution. This impressive movement began in the 1970s and has seen the birth and growth of an increasing number of small breweries that, in contrast with trends in previous decades, bet on and invested in beers full of character and flavor. Among these, the starring role is played by IPAs, a style of British origin characterized by a high alcohol content and a significant use of hops. The style began around the 1770s, when a small London brewery started to sell its October ale, an alcoholic and intensely hopped beer, to the East India Company, the holder of the exclusive right to export goods to India. Success was immediate, and the market grew rapidly. In the 1820s, the production moved to the city of Burton-on-Trent, where it was consecrated and where the name India Pale Ale was invented. After a period of remarkable fame, the style disappeared during the Second World War. It fell to American brewers between the 1970s and 1990s to rediscover and reinvent it, no longer using the austere English hops but the extroverted and aromatic American hops, which, with their notes of citrus and other fruits, fascinated consumers accustomed to flat and tasteless industrial lagers. Starting in the United States, IPAs have conquered the world and become the symbol of craft beer, and of a new way of drinking. After decades dominated by top-fermented blondes with no identity, these fragrant and bitter beers aroused the curiosity of consumers and stimulated, once again, the imagination of producers who, like their English predecessors two hundred years earlier, had new lands and new markets to conquer.

Humulus lupulus L.
CANNABACEAE

Hylocereus undatus (Haw.) Britton & Rose
CACTACEAE

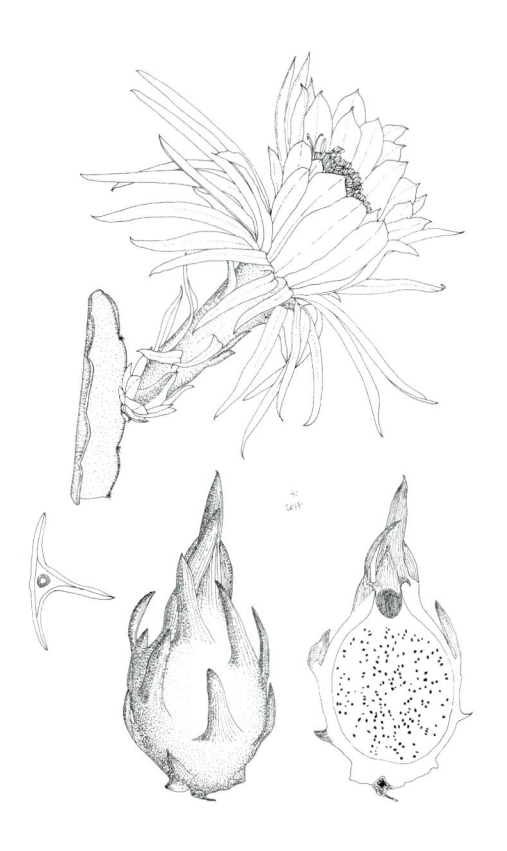

Synonyms *Cereus tricostatus* Rol.-Goss., *Cereus undatus* Pfeiff., *Cereus undulatus* D.Dietr., *Hylocereus guatemalensis* (Eichlam) Britton & Rose, *Hylocereus tricostatus* (Gosselin) Britton & Rose.

Common Names Dragonfruit.

Description Cactus, climbing (4–10 m), often epiphytic on other plants. Stems photosynthetic, green in color, succulent, stout, climbing, markedly angular, trialate, with wavy, scrawny margins, producing aerial roots. Leaves transformed into small, gray-brown, nonphotosynthetic spines in groups of 1–3 on stem crests. Flowers very large (25–30 cm), nocturnal, white in color, with many long, narrow petals. Fruit oblong (5–12.5 × 3.8–9 cm), red, smooth surface, covered with large carnose scales; whitish flesh containing numerous seeds. Seeds small, black in color.

Subspecies In addition to the nominal subspecies (*Hylocereus undatus* [Haw.] Britt. & Rose subsp. *undatus*), the subspecies *Hylocereus undatus* (Haw.) Britt. & Rose subsp. *luteocarpus* Cálix de Dios has been described.

Related Species Of the 15 taxa belonging to the genus *Hylocereus*, some species have possible applications in brewing: *Hylocereus costaricensis* (F.A.C. Weber) Britton & Rose, *Hylocereus trigonus* (Haw.) Saff., *Hylocereus ocamponis* (Salm-Dyck) Britton & Rose, and *Hylocereus megalanthus* (K. Schum. ex Vaupel) Ralf Bauer.

Geographic Distribution Mexico, Colombia. **Habitat** Tropical and subtropical forests. **Range** Australia, Cambodia, China, Colombia, Ecuador, Guatemala, Hawaii, Indonesia, Israel, Japan, Laos, Malaysia, Mexico, New Zealand, Nicaragua, Peru, Philippines, Spain, Sri Lanka, Taiwan, Thailand, SW United States, Vietnam.

Beer-Making Parts (and Possible Uses) Fruits, seeds (?).

Chemistry Fruits: water 89.4 g, protein 0.5 g, lipids 0.1 g, fiber 0.3 g, ash 0.5 g, calcium 6 mg, phosphorus 19 mg, niacin 0.2 mg, vitamin c 25 mg. Seeds (fatty acids): myristic acid 0.3%, palmitic acid 17.1%, stearic acid 4.37%, palmitoleic acid 0.61%, oleic acid 23.8%, cis-vaccenic acid 2.81%, linoleic acid 50.1%, linolenic acid 0.98%.

Style Berliner Weisse, Weizen/Weissbier.

Beer DFPF Berliner Weisse, J. Wakefield (Florida, United States).

Source https://www.brewersfriend.com/other/dragonfruit/.

Hyoscyamus albus L.
SOLANACEAE

Synonyms None
Common Names White henbane.

Description Herb, annual, biennial, or perennial (3–5 dm), fetid, densely villous with short hairs and 3 mm patent hairs, some glandular. Stem ascending, flaccid, simple, sometimes woody at base. Leaves all with 2–10 cm petiole; lamina ovate (6 × 8 cm in lower leaves, 1.5 × 2 cm in upper leaves), lobed. Flowers in leafy spikes ± unilateral; calyx campanulate with

7 mm tube and 3 mm teeth (at fruiting 2–2.5 cm with triangular teeth); corolla ± zygomorphic, 15–30 mm, color externally yellow, internally pale yellow and dark purple at bottom; filament and stylus purple; anthers and stigma yellow.

Related Species Of the 10 species belonging to the genus *Hyoscyamus*, one with similar uses in brewing is *Hyoscyamus niger* L.

Geographic Distribution Euro-Mediterranean. **Habitat** Walls, ruins, rubble. **Range** Euro-Mediterranean.

Beer-Making Parts (and Possible Uses) Leaves, seeds (?), roots (?).

Toxicity The plant, in therapeutic doses, exerts antispasmodic, calming, analgesic, narcotic, and mydriatic actions. It is recommended for spasmodic coughs, whooping cough, chronic bronchitis, neuralgia (especially trigeminal neuralgia), convulsions, chorea, hysteria, epilepsy, palpitations, and mental illness accompanied by agitation. Like many other Solanaceae species, all parts are highly toxic. Ingestion can lead to death. The Babylonians, Egyptians, Persians, Arabs, Greeks, and Romans used it as both a sedative and a poison; the active principles are identical to those of black henbane, with effects comparable to those of belladonna; however, whereas this one produces violent delirium, henbane induces drowsiness and then deep sleep with frightening dreams.

Chemistry The leaves, seeds, and roots contain mainly two alkaloids, hyoscyamine and scopolamine; during drying, hyoscyamine is in part transformed into atropine.

Style Gruit, findings in the Neolithic settlement of Skara Brae (Orkney Islands, Scotland, United Kingdom).

Source https://www.hacker-pschorr.com/it/il-nostro-marchio/requisito-di-purezza-bavarese.

Hyoscyamus niger L.
SOLANACEAE

Synonyms *Hyoscarpus niger* (L.) Dulac, *Hyoscyamus agrestis* Kit., *Hyoscyamus bohemicus* F.W.Schmidt, *Hyoscyamus pallidus* Waldst. & Kit. ex Willdenow.

Common Names Henbane, black henbane, stinking nightshade.

Description Herb, annual or biennial (3–8 dm), penetrating fetid odor, slimy. Stem erect, simple or branched. Leaves 5–20 cm, ovate or ovate-oblong, entire to angulate, sparsely toothed or deeply incised, basal leaves forming a rosette, petiolate, caulinar leaves amplexicaul sessile. Flowers subsessile, in dense unilateral spikes; bracts similar to leaves; calyx 1–1.5 cm, teeth triangular, acute; calyx at fruiting ventricose at base, with stinging teeth; corolla Ø 2–3 cm, slightly zygomorphic, pale yellow, usually with purple veins.

Related Species Of the 10 species belonging to the genus *Hyoscyamus*, one with similar uses in beer-making is *Hyoscyamus albus* L.

Geographic Distribution Eurasian. **Habitat** Rubble, under walls, garbage cans, sheepfolds. **Range** Eurasia.

Beer-Making Parts (and Possible Uses) Leaves, seeds (?).

Toxicity Although widely used in the past for medicinal purposes, for incense, and in the production of beer, tea, and wine, the leaves, like the seeds, are highly toxic or lethal, owing to the presence of several alkaloids; the ingestion of parts of the plant causes hallucinations, dilated pupils, restlessness, reddened skin, tachycardia, convulsions, vomiting, hypertension, fever, and coordination problems; side effects are dry mouth, confusion, locomotor and memory troubles; massive doses cause delirium, coma, respiratory paralysis, and death; moderate and medium doses are reputed to have intoxicating and aphrodisiac effects.

Chemistry Leaves, seeds (alkaloids 0.03–0.28%): hyoscyamine, scopolamine, other alkaloids.

Style Gruit, findings in the Neolithic settlement of Skara Brae (Orkney Islands, Scotland, United Kingdom).

Hyoscyamus albus L.—SOLANACEAE
Hyoscyamus niger L.—SOLANACEAE

Hyssopus officinalis L.
LAMIACEAE

Synonyms *Hyssopus alopecuroides* Fisch. ex Benth., *Hyssopus altissimus* Mill., *Hyssopus angustifolius* M. Bieb., *Hyssopus beugesiacus* Jord. & Fourr., *Hyssopus caucasicus* Spreng. ex Steud., *Hyssopus decumbens* Jord. & Fourr., *Hyssopus fischeri* Steud., *Hyssopus hirsutus* Hill, *Hyssopus judaeorum* Sennen, *Hyssopus myrtifolius* Desf., *Hyssopus orientalis* Adam ex Willd., *Hyssopus passionis* Sennen & Elias, *Hyssopus polycladus* Jord. & Fourr., *Hyssopus pubescens* Jord. & Fourr., *Hyssopus recticaulis* Jord. & Fourr., *Hyssopus ruber* Mill., *Hyssopus schleicheri* G.Don ex Loudon, *Hyssopus torresii* Sennen, *Hyssopus vulgaris* Bubani, *Thymus hyssopus* E.H.L.Krause.

Common Names Hyssop.

Description Shrub (3–5 dm) with fragrant odor. Stem woody below, ascending and branching, with very minute (0.1 mm) frizzy hairs. Leaves opposite, briefly petiolate, lanceolate (2–3 × 20–30 mm), usually revolute on the edges, then narrowly linear in appearance, dark green in color; floral leaves acute, not aristate. Flowers hermaphrodite; calyx generally blushing in color, conical, with 4 mm long tube and aristiform teeth (1–1.5 mm); corolla blue-violet in color (7–9 mm), upper labium bilobous, lower trilobous, strong aromatic smell similar to camphor, bitterish but pleasant taste; 4 long protruding stamina. Inflorescences axillary whorls (4–6 flowers), close together in unilateral terminal spike. Fruit tetrachenous ovoid or oblong, with three obtuse edges, glabrous and with minute dimples; each portion encloses 1 wrinkled black seed.

Subspecies The species includes four taxa of subspecific rank: *Hyssopus officinalis* L. subsp. *aristatus* (Godr.) Nyman, *Hyssopus officinalis* L. subsp. *canescens* (DC.) Nyman, *Hyssopus officinalis* L. subsp. *montanus* (Jord. & Fourr.) Briq., and *Hyssopus officinalis* L. subsp. *officinalis*.

Cultivar Perlay and many others.

Related Species The genus *Hyssopus* includes 10 taxa.

Geographic Distribution Orophyte Eurasian.
Habitat Rocky (calcareous) cliffs and pastures.
Range Mediterranean, Continental Europe, England.

Beer-Making Parts (and Possible Uses) Aerial parts.

Toxicity Intake of the essential oil may cause convulsions, diarrhea, and indigestion.

Chemistry Aerial parts (flavonoids): quercetin7-O-b-D-apiofuranosyl-(1→2)-b-d xylopyranoside, quercetin 7-O-b-D-apiofuranosyl-(1→2)-b-D-xylopyranoside 30-Ob-D-glucopyranoside, apigenin, apigenin 7-O-b-D-glucopyranoside (4), apigenin 7-O-b-D-glucuronopyranoside methyl ester, luteolin, apigenin 7-O-b-D-glucuronide, apigenin 7-O-b-D glucuronopyranoside butyl ester, luteolin 7-O-b-D glucopyranoside, diosmin, acacetin 7-O-a-L-rhamnopyranosyl-(1→6)-b-D-glucopyranoside. Essential oil (21 compounds; 95.6% of oil): pinocamphone 49.1%, β-pinene 18.4%, isopinocamphone 9.7%; myrtenol methyl ether, myrtenic acid, methyl myrtenate, pinic acid, cis-pinic acid, (+)-2-hydroxyisopinocamphone, pinonic acid, cis-pinonic acid.

Style Spice/Herb/Vegetable Beer, Extra Special/Strong Bitter (ESB), Old Ale.

Source Fisher and Fisher 2017; Heilshorn 2017; Hieronymus 2016b; https://www.brewersfriend.com/other/hyssop/.

Ilex paraguariensis A.St.-Hil.
AQUIFOLIACEAE

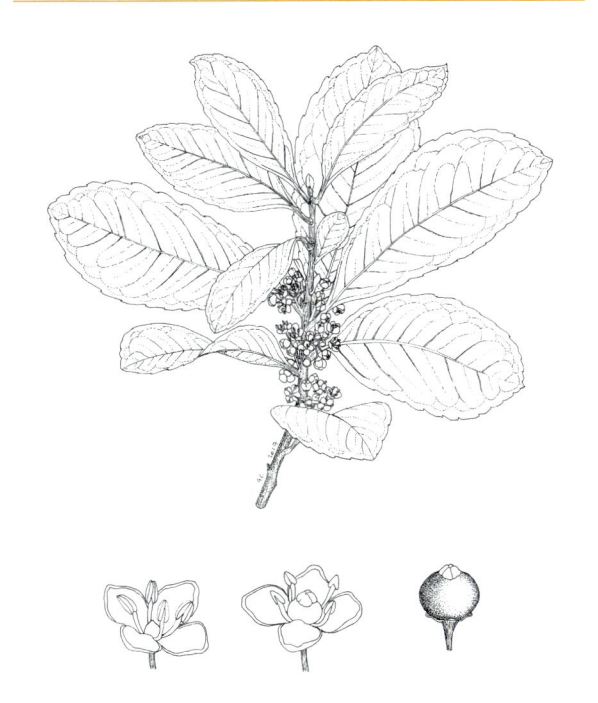

Synonyms *Ilex curitibensis* Miers, *Ilex domestica* Reissek, *Ilex mate* A.St.-Hil., *Ilex sorbilis* Reissek, *Ilex theaezans* Bonpl. ex Miers.

Common Names Mate, yerba mate, Paraguay tea.

Description Tree (up to 15 m), evergreen. Leaves perennial, alternate, leathery, olive green, lamina obovate (7–11 × 3–5.5 cm), base cuneate, margin serrate or slightly crenate-dentate, petiole about 15 mm. Flowers unisexual or hermaphrodite, small, in whorls of 4 elements, 4 petals grayish-white color. Inflorescences 1–15 flowers at the axil of the leaves. Fruit roundish drupe (Ø 4–6 mm), red or purple color. Seeds 4–5 per fruit.

Subspecies When considering a species with such a wide range, a remarkable morphological and ecological variability is to be expected; in fact, a series of varieties are described in the literature: *Ilex paraguariensis* A.St.-Hil. var. *acutifolia* Mart., *Ilex paraguariensis* A.St.-Hil. var. *angustifolia* Reissek, *Ilex paraguariensis* A.St.-Hil. var. *euneura* Loes., *Ilex paraguariensis* A.St.-Hil. var. *guaranina* Loes., *Ilex paraguariensis* A.St.-Hil. var. *latifolia* Reissek, *Ilex paraguariensis* A.St.-Hil. var. *longifolia* Reissek, and *Ilex paraguariensis* A.St.-Hil. var. *ulei* Loes.; a number of forms are also described: *Ilex paraguariensis* A.St.-Hil. f. *confusa* Loes., *Ilex paraguariensis* A.St.-Hil. f. *dasyprionata* Loes., *Ilex paraguariensis* A.St.-Hil. f. *domestica* (Reissek) Loes., *Ilex paraguariensis* A.St.-Hil. f. *latifolia* (Reissek) Chodat, *Ilex paraguariensis* A.St.-Hil. f. *microphylla* Reissek, *Ilex paraguariensis* A.St.- Hil. f. *parvifolia* Chodat, *Ilex paraguariensis* A.St.-Hil. f. *pubescens* Loes., and *Ilex paraguariensis* A.St.-Hil. f. *sorbilis* (Reissek) Loes. All this variability has at present been synonymized under the binomial *Ilex paraguariensis*.

Related Species The genus *Ilex* includes 512 taxa.

Geographic Distribution Argentina, Bolivia, Brazil (Rio Grande do Sul, Santa Catarina), Colombia, Ecuador, Paraguay, Uruguay. **Habitat** Stream banks, mountain forests at 500–700 m ASL in subtropical and temperate regions of South America. **Range** S Brazil (Rio Grande do Sul, Santa Catarina, Paraná, Mato Grosso do Sul), Argentina (Corrientes, Misiones), Paraguay, Uruguay.

Beer-Making Parts (and Possible Uses) Leaves.

Toxicity Consumption of hot mate infusion is associated with several types of oral cancer. Some studies relate the probability of these pathological forms to the temperature of administration of the infusion, making a possible role of the plant as a carcinogen unclear.

Chemistry Leaves: polyphenols (10–16% chlorogenic acid, neochlorogenic acid, isochlorogenic acid, 3,4-dicaffeolquinic acid, 3,5-dicaffeolquinic acid, 4,5-caffeolquinic acid, caffeic acid, catechin tannins 7–14%, flavonoids (quercetin, quercetin-3-O-glucoside, kaempferol, rutin), xanthine (caffeine 0.7–1.7%, theobromine 0.3–0.9%, trace theophylline), saponins (about 10 pentacyclic triterpenes, non-hemolytic matesaponins), other active ingredients (ursolic acid, trigonellins, α-amyrine, choline, 1,2-benzopyrene, sterols, 15 amino acids, sugars, β-carotene), vitamins (vitamin C, B1, B2, nicotinic acid), essential oil (32 identified compounds: aliphatic alcohols, aldehydes, ketones, terpenes, aliphatic hydrocarbons, etc.), minerals (potassium, magnesium, manganese).

Style American Light Lager, California Common Beer, Classic American Pilsner, English IPA, Spice/Herb/Vegetable Beer.

Source Dabove 2015; McGovern 2017; https://www.brewersfriend.com/other/mate/.

Illicium verum Hook.f.
SCHISANDRACEAE

Synonyms *Illicium san-ki* Perr.

Common Names Star anise.

Description Tree, evergreen (up to 15 m). Trunk Ø about 25 cm, white bark. Leaves shiny, leathery, in clusters of 3–6. Flowers solitary, yellowish-green, sometimes flushed pink to dark red; 7–12 tepals, up to 20 stamens and usually 7–9 carpels. Fruit star-shaped aggregate consisting of a ring of monoseminate carpels, dark reddish-brown in color, attached to a central column; fruit fleshy but becomes woody and wrinkled with drying.

Related Species The genus *Illicium* includes 34 taxa.

Geographic Distribution China, Vietnam. **Habitat** Cultivated in Asia. **Range** China, Vietnam, Jamaica, Laos, Philippines.

Beer-Making Parts (and Possible Uses) Fruits, seeds, leaves.

Chemistry Essential oil, fruits (5–8%), seeds, leaves: anethole 85–90%, phellandrene, safrole, terpineol.

Style American Amber Ale, American Light Lager, Baltic Porter, Belgian Dark Strong Ale, Belgian Specialty Ale, Belgian Tripel, Brown Porter, California Common Beer, Clone Beer, Doppelbock, Dunkelweizen, Dunkles Bock, Extra Special/Strong Bitter (ESB), Holiday/Winter Special Spiced Beer, Munich Dunkel, North German Altbier, Oatmeal Stout, Old Ale, Saison, Specialty Beer, Spice/Herb/Vegetable Beer, Strong Scotch Ale, Winter Seasonal Beer, Witbier.

Beer Piazza delle Erbe, Ofelia (Sovizzo, Veneto, Italy).

Source Calagione et al. 2018; Heilshorn 2017; Hieronymus 2010; Hieronymus 2016b; Markowski 2015; https://www.birramia.it/anice-stellato-250-gr.html#.V6w_osYcR9A; https://www.brewersfriend.com/other/stjerneanis/; https://www.fermentobirra.com/homebrewing/ricetta-fare-la-birra-in-casa/saison-estratti-grani/; https://www.mondobirra.org/homebrew2.htm.

Inula ensifolia L.
ASTERACEAE

Synonyms *Aster ensifolius* Scop.

Common Names Swordleaf inula, horseheal, slender-leaved elecampagne.

Description Herb, perennial (1–6 dm). Rhizome woody, creeping. Stems erect, pubescent. Leaves linear-ensiform (3–6 × 40–100 mm), entire, sessile, usually with 3–5 parallel veins. Inflorescence heads (Ø 3–5 cm) solitary, involucre hemispherical, scales lanceolate (2–3 × 8–10 mm), obtuse, herbaceous above; flowers yellow, orange in dried specimens, those ligulate 20–25 mm. Fruit achene, 1.5 mm.

Cultivar Two are Compacta and Gold Star.

Related Species The genus *Inula* includes 112 taxa.

Geographic Distribution SE Europe-Pontic. **Habitat** Arid steppe meadows, shores, consolidated stony grounds (80–1,000 m ASL). **Range** SE Europe-Pontic.

Beer-Making Parts (and Possible Uses) Roots (plant substitute for *Inula helenium*).

Chemistry No data available.

Style Spice Beer, Herb Beer.

Source Fisher and Fisher 2017.

Inula helenium L.
ASTERACEAE

Synonyms *Aster helenium* (L.) Scop., *Aster officinalis* All., *Corvisartia helenium* (L.) Mérat, *Helenium grandiflorum* Gilib.

Common Names Elecampane, horseheal, elfdock.

Description Herb, perennial (10–20 dm). Rhizome enlarged, fleshy, branched. Stem stout, basally covered with patent bristles, apically pubescent. Basal leaves ovate-spatulate, large (up to 70 cm), long petiolate, apex acute; cauline leaves alternate, lanceolate, thick, upper ones sessile, cordate-amplexicaul, all acute, serrate on edges, gray-tomentose on underleaf. Inflorescence corymb consisting of large (Ø 6–7 cm) pedunculated flower heads, hemispherical involucre composed of 3–4 series of imbricated, tomentose scales, the outer ones resembling small leaves, spatulate, curved outward; disc florets golden yellow, ligulate ray florets (3 cm × 1–2 mm). Fruit achene cylindrical, glabrous, with pappus (about 10 mm).

Subspecies In addition to the typical subspecies, *Inula helenium* L. subsp. *orgyalis* (Boiss.) Grierson is known.

Related Species In the genus *Inula*, 112 taxa are recognized.

Geographic Distribution Orophyte SE Europe. **Habitat** Moors, hedges, wet meadows (500–1,200 m ASL). **Range** Widely cultivated species for ornamental and medicinal purposes, naturalized almost everywhere.

Beer-Making Parts (and Possible Uses) Roots.

Toxicity Medium toxicity on internal organs owing to the presence of lactones; contact dermatitis and irritation of mucous membranes owing to antolactone, which binds to skin proteins.

Chemistry Root: sesquiterpene lactones, helenin (alant camphor), essential oil, alanthol, inula essence (crystalline substance), phytoncides, alantopicrin, sterol, inulin, mucilage, pectin, ascorbic acid.

Style Mum.

Beer Elecampane, Scratch Brewing (Illinois, United States).

Source Fisher and Fisher 2017; Hieronymus 2010; Hieronymus 2016b.

Inula magnifica Lipsky
ASTERACEAE

Common Names Giant inula, giant fleabane.

Description Herb, perennial (up to 1.8 m). Stems (Ø 7 mm) erect, pubescent to white hairs, ribbed, forming a shrub Ø up to 1 m. Leaves large and pubescent; lower leaves oblong-elliptic or ovate (up to 50 × 25 cm), with petiole up to 36 cm; upper leaves smaller, sessile, broadly toothed, upper leaf sparsely hairy, underleaf covered with dense, long, white hairs and small yellow glands. Inflorescence heads (Ø 9–15 cm) yellow, clustered in cymes; involucre 45–65 mm, rays lanceolate, densely hairy, upper ones reddish-violet; disc florets yellow, tubular, 10 mm; ray florets yellow, 2.5–3 times the length of involucral rays, 38–53 × 1.5 mm, with 4–8 veins, 2–3 small grooves, and scattered hairs. Seeds linear (3 × 0.5 mm), ribbed, crown (about 3.5 times seed length) yellowish color.

Cultivar A rather well-known cultivar of this species is Sonnenstrahl.

Related Species The genus *Inula* includes 112 taxa.

Geographic Distribution E Caucasus. **Habitat** Associations with tall grasses, sparse forests, open, mixed forests, shrublands, meadows, from montane to subalpine zones (1,250–2,100 m ASL). **Range** Cultivated for ornamental purposes.

Beer-Making Parts (and Possible Uses) Roots (substitute plant for *Inula helenium*).

Chemistry Little information available. Root: alantholactone, isoalantholactone.

Style Spice Beer, Herb Beer.

Source Fisher and Fisher 2017.

Ipomoea batatas (L.) Lam.
CONVOLVULACEAE

Synonyms *Batatas edulis* (Thunb.) Choisy, *Batatas wallii* C. Morren, *Convolvulus apiculata* M.Martens & Galeotti, *Convolvulus attenuatus* M.Martens & Galeotti, *Convolvulus batatas* L., *Convolvulus candicans* Sol. ex Sims, *Convolvulus edulis* Thunb., *Convolvulus esculentus* Salisb., *Convolvulus varius* Vell., *Ipomoea davidsoniae* Standl., *Ipomoea edulis* (Thunb.) Makino, *Ipomoea mucronata* Schery, *Ipomoea purpusii* House, *Ipomoea vulsa* House, *Ipomoea wallii* (C. Morren) Hemsl.

Common Names Sweet potato, yam (the same common name is also used to indicate species belonging to the genus *Dioscorea*).

Description Herb, perennial (cultivated as an annual), tuberous. Stem herbaceous, desiccating annually, forming a shoot (up to 4 m), usually slender, prostrate, containing milky juice, lateral branches usually unbranched. Leaves lamina ovate-cordate, carried by long petioles, veins palmate, angular or lobed, depending on variety, green or purplish color. Flowers rare, white or pale violet in color, axillary, funnel shaped, single or in cymes on short peduncles. Fruit round pod. Seeds 1–4 per fruit, flattened, coated, angular.

Cultivar Of the approximately 7,000 of this species, some are cultivated for the tuber, others for ornamental purposes. Cultivars cultivated for the tuber: Acadian, Allgold/Okla. 240, Americana, Apache, Australian Canner, Ayamurasaki, Baker/V 2158, Beauregard, Bonara, Canbake/G-52–15–1, Caro-Gold, Carolina Bunch, Carolina Nugget, Carolina Ruby, Caromex, Carver, Centennial/L-3–77, Chipper, Covington NC98–608, Cliett Bunch Puerto Rico/Georgia Bunch Puerto Rico, Coastal Red, Coppergold, Cordner, Creole, Darby, Don Juan, Earlyport, Earlysweet/T-3, Eureka, Evangeline, Excel, GA90–16, Garnet, Georgia Jet, Georgia Red/T-6, Gold Rush, Golden Belle, Goldmar, Hannah Sweet, Hayman White, Heartogold, Hernandez, HiDry, Hoolehua Gold, Hoolehua Red, Hopi/HM-122, Iliua, Japanese/Oriental, Jersey Orange/Orange Little Stern, Jersey Red, Jersey Yellow, Jewel, Kandee/K1716, Kona B, Kote Buki, Lakan/L-0–123, Mameyita, Maryland Golden, Miguela, Murasaki, Murff Bush Puerto Rico, Nancy Gold, Nancy Hall, Nemagold/Okla. 46, Northern Star, Nugget/NC-171, O'Henry, Okla. 46, Oklamar/Okla. 52, Oklamex Red, Onokeo, Onolena/HES number 14, Orlis, Owairaka Red, Papota, Pelican Processor/L-5/L-4–5, Pope, Puerto Rico 198/Porto Rican/Puerto Rican, Purple Heart/Okinawa, Queen Mary/L-126, Ranger, Rapoza, Red Diane, Red Garnet, Red Jewel, Red Nancy, Redglow, Redgold/Okla. 26, Redmar/Md 2416, Regal, Resisto, Rojo Blanco, Rose Centennial, Ruddy, Scarlet, Shore Gold, Southern Delite, Stokes Purple, Sumor, Sunnyside, Sweet Red, Tango, Tanhoma, Toka Toka Gold, Topaz, Travis, UPLSP-1, UPLSP-2, U.P.R. number 3, U.P.R. number 7, Vardaman, Virginian/V-53, VSP-5, VSP-6, Waimanalo Red, White Delite, White Triumph, Whitestar, Yellow Yam. Cultivated for ornamental purposes: Black Heart/Ace of Spades/Purple Heart, Blackie, Bronze Beauty, Copper, Freckles, Gold Finger, Ivory Jewel, Lady Fingers, Marguerite/Chartreuse/Sulfur, Mini Blackie, NCORNSP011MNLC/Illusion® Midnight Lace, NCORNSP012EMLC/Illusion® Emerald Lace, Purple Tuber, Seki Blakhrt/Chillin™/Blackberry Heart, Sidekick Black, Sidekick Lime, Sweet Caroline Bewitched Purple/PP18574, Sweet Caroline Bronze/PP15437, Sweet Caroline Green, Sweet Caroline Green Yellow, Sweet Caroline Light Green, Sweet Caroline Purple, Sweet Caroline Red, Sweet Caroline Sweetheart Light Green, Sweet Caroline Sweetheart Red, Sweet Georgia Heart Purple, Terrace Lime, Tricolor, Vardaman.

Related Species Of the more than 460 species described for the genus *Ipomoea*, some have recognized food uses; no information about beer-making uses is available.

Geographic Distribution Tropical America. **Habitat** Tropical and subtropical climates (relatively constant rainfall and high temperatures), deep, fertile, well-drained soils. **Range** Introduced and cultivated in many tropical and subtropical countries, including India, China, Philippines, South Sea Islands, New Zealand, Australia, Japan, Hawaii, North America.

Beer-Making Parts (and Possible Uses) Tubers (raw or roasted), leaves (?); these have remarkable bioactive properties.

Toxicity Roots may have a laxative effect owing to the presence of hypomein.

Chemistry Tubers: energy 360 kJ (86 kcal), carbohydrates 20.1 g (starch 12.7 g, lipids 0.1 g, protein 1.6 g, vitamin A 709 µg, beta-carotene 8509 µg, thiamine (B1) 0.1 mg, riboflavin (B2) 0.1 mg, niacin (B3) 0.61 mg, pantothenic acid (B5) 0.8 mg, vitamin B6 0.2 mg, folate (B9) 11 µg, vitamin C 2.4 mg, vitamin E 0.26 mg, calcium 30.0 mg, iron 0.6 mg, magnesium 25.0 mg, phosphorus 47.0 mg, potassium 337 mg, sodium 55 mg, zinc 0.3 mg. Leaves (major bioactive components): triterpenes/steroids, alkaloids, anthraquinones, coumarins, flavonoids, saponins, tannins, phenolic acids.

Style Robust Porter, Baltic Porter, Bière de Garde, Imperial Stout.

Beer Sweet Potato Stout, Lazy Magnolia (Mississippi, United States).

Source Hieronymus 2016b; Jackson 1991; Josephson et al. 2016; https://www.brewersfriend.com/other/yam/.

Iris × germanica L.
IRIDACEAE

Synonyms *Iris × alba* Savi, *Iris × amoena* DC., *Iris × atroviolacea* Lange, *Iris × australis* Tod., *Iris × belouinii* Bois & Cornuault, *Iris × biliottii* Foster, *Iris × buiana* Prodán, *Iris × croatica* Prodán, *Iris × cypriana* Foster & Baker, *Iris × deflexa* Knowles & West., *Iris × florentina* L., *Iris × florentinoides* Prodán ex Nyar., *Iris × humei* G.Don, *Iris × laciniata* Berg, *Iris × lurida* Aiton, *Iris × macrantha* Simonet, *Iris × mesopotamica* Dykes, *Iris × murorum* Gaterau, *Iris × neglecta* Hornem., *Iris × nepalensis* Wall. ex Lindl., *Iris × nyaradyana* Prodán, *Iris × officinalis* Salisb., *Iris × piatrae* Prodán, *Iris × redouteana* Spach, *Iris × repanda* Berg, *Iris × rothschildii* Degen, *Iris × sambucina* L., *Iris × spectabilis* Salisb., *Iris × squalens* L., *Iris × superba* Berg, *Iris × tardiflora* Berg, *Iris × trojana* A.Kern. ex Stapf, *Iris × varbossania* K.Malý, *Iris × venusta* J.Booth ex Berg, *Iris × violacea* Savi, *Iris × vulgaris* Pohl.

Common Names Iris.

Description Herb, perennial (5–10 dm), rhizomatous. Root fasciculate. Rhizome fleshy, creeping, horizontal, fragrant; stem scape cylindrical (Ø 1.5 cm) or slightly compressed, erect, bearing 2–4 flowers (about 10 cm) accompanied by broad bracts. Leaves linear (1.5–4 × 20–40 cm), parallel, innervate, abruptly narrowed at acute tip, almost all basal; cauline shorter, scarious and clasping the stem. Flowers subsessile, deep blue-violet, odorless or lightly fragrant, spathe white-membranous in upper half; perigonium (10–12 cm) actinomorphic, consisting of 6 tepals: 3 outer (4 × 7.5 cm) curved and wavy with laciniae bearing in the center a barb consisting of a line of yellowish hairs; 3 inner (3–4 × 6–7 cm) are curved with lighter shades, wavy and crepe-like at the margin, laciniae spatulate, narrowed at the base, bearing styluses with tube longer than the ovary and about 1/3 the size of the laciniae, almost completely enveloped in the spathe; laciniae violet; hair lines yellowish; stamens white, with anthers about as long as the filament; stylus 3.5 cm with divergent lobes; ovary inferior. Fruit fusiform trigonal capsule (usually abortive) containing black, discoidal, shiny seeds.

Related Species The genus *Iris* comprises about 390 taxa.

Geographic Distribution Origin unknown.
Habitat Synanthropic environments. **Range** Cultivated for ornamental purposes.

Beer-Making Parts (and Possible Uses) Rhizomes.

Chemistry Rhizome: starch, mucilage, essential oil with strong violet odor (irone), soft resin (camphor of Ireos), carotenoids, aldehydes, esters of various acids, ketone, glycoside (iridine).

Style English IPA.

Source https://www.brewersfriend.com/other/iris/ (generic reference to *Iris*; this particular species is described here).

Jasminum officinale L.
OLEACEAE

Synonyms *Jasminum affine* Royle ex Lindl.

Common Names Jasmine, common jasmine, summer jasmine, poet's jasmine, white jasmine, true jasmine, jessamine.

Description Shrub (to 5 m), ± deciduous, with arching branches. Leaves opposite, imparipinnate, with lanceolate-acuminate segments 1–6 cm long, the apical one much larger than the lateral ones and often somewhat curved to a sickle. Flowers fragrant in terminal pauciflorus cymes; calyx with teeth larger than the tube; corolla white, sometimes suffused with pink, with 2 cm tube and 4–5 patent lobes.

Related Species The genus *Jasminum* includes 220 taxa, including *Jasminum grandiflorum* L., with probable beer-making applications.

Geographic Distribution SW Asiatic (Caucasus, N Iran, Afghanistan, Pakistan, Himalayas, Tajikistan, India, Nepal, W China [Guizhou, Sichuan, Xizang, Tibet, Yunnan]). **Habitat** Cultivated and often wild species. **Range** Widely cultivated for ornamental and medicinal purposes, naturalized in France, Italy, Portugal, Romania, Ex-Yugoslavia, Algeria, Florida, West Indies.

Beer-Making Parts (and Possible Uses) Flowers.

Toxicity Jasmine oil is nontoxic, nonirritating, and generally nonsensitizing, although some subjects may have an allergic reaction; because of its emmenagogic properties, it is not recommended during pregnancy; used in high doses, it prevents concentration because of its relaxing action.

Chemistry Essential oil (more than 100 constituents): benzyl acetate, linalool, benzyl alcohol, indole, benzyl benzoate, cis-jasmone, geraniol, methyl anthranilate; trace components: cresol, farnesol, cis-3-hexenyl benzoate, eugenol, nerol, ceosol, benzoic acid, benzaldehyde, γ-terpineol, nerolidol, isoitol, phytol, etc.

Style Amber Ale, Braggot, Russian Imperial Stout.

Beer La 16, Math (Barberino Tavarnelle, Tuscany, Italy).

Source Hieronymus 2016b; McGovern 2017; https://www.brewersfriend.com/other/jasmin/.

Juglans nigra L.
JUGLANDACEAE

Synonyms *Wallia nigra* (L.) Alef.

Common Names Black walnut.

Description Tree, deciduous (30–50 m), with a globular, light green crown. Trunk cylindrical and straight, with strong apical dominance and thin branches, bark grayish-brown with blackish shades, with age it becomes deeply incised and furrowed by lines that form a longitudinal network. Leaves alternate, fragrant but not aromatic, with petiole expanded at the base, imparipinnate, 20–60 cm long, with 9–23 ovate-acuminate or ovate-lanceolate segments with rounded or subcuneate base with serrated margin; terminal segment reduced or often missing (paripinnate); upper surface glabrous, below slightly tomentose. Plant monoecious. Male flowers greenish, without perianth and with numerous stamens, collected in axillary pendulous catkins, 5–10 cm long, bracteate, solitary or paired on previous year's branches; female flowers less visible, arranged in 3- to 5-flowered racemes, on current year's branches; ovary inferior and stigma bifid. Fruits globose false drupes, Ø 3–4 cm, solitary or paired; epicarp wrinkled, pleasantly aromatic, 1 cm thick, endocarp woody, rough, warty, deeply longitudinally furrowed, blackish; seed closely concamerate.

Cultivar Approximately 100 cultivars of *Juglans nigra* have been selected and named on the basis of a series of morphological, ecological, ecophysiological, and phytosanitary characteristics. Some are Cochrane, Cornell, Elmer Myers, Huber, Myers, Ohio, Snyder, Sparrow, Stambaugh, Thomas, and Wiard. There is also a wide range of natural hybrids (e.g., × *Juglans intermedia* Carr = *J. nigra* Iris × *J. regia*) and artificial ones (Royal = *J. nigra* Iris × *Juglans hindsii* Jeps. ex R.E. Sm).

Related Species There are 22 species recognized as belonging to the genus *Juglans*.

Geographic Distribution North America. **Habitat** Rich woods. **Range** Tree cultivated for fruit and wood, sometimes naturalized.

Beer-Making Parts (and Possible Uses) Wood; fruits edible but not used in beer making.

Chemistry Fruits: energy 2,586 kJ (618 kcal), water 4.56 g, carbohydrates 9.91 g, starch 0.24 g, sugars 1.10 g, fiber 6.8 g, lipids 59 g (saturated 3.37 g, monounsaturated 15 g, polyunsaturated 35.08 g), omega-3 2 g, omega-6 33.1 g, protein 24.1 g, vitamin A 2 µg, thiamine (B1) 0.057 mg, riboflavin (B2) 0, 13 mg, niacin (B3) 0.47 mg, pantothenic acid (B5) 1.66 mg, vitamin B6 0.583 mg, folate (B9) 31 µg, vitamin C 1.7 mg, vitamin E 1.8 mg, vitamin K 2.7 µg, calcium 61 mg, iron 3.12 mg, magnesium 201 mg, manganese 3.9 mg, phosphorus 513 mg, potassium 523 mg, sodium 2 mg, zinc 3.37 mg.

Wood Whitish sapwood clearly differentiated from the heartwood, which is more or less dark brown and variegated. Wood easy to season and easy to work; WSG 0.60–0.70. Uses similar to those of walnut but of lesser value.

Style Smoked Beer.

Source Hieronymus 2016b; Josephson et al. 2016; https://www.bjcp.org(32A.ClassicStyleSmokedBeer); https://byo.com/hops/item/306-brewing-smoked-beers-tips-from-the-pros.

Juglans regia L.
JUGLANDACEAE

Synonyms *Juglans duclouxiana* Dode, *Juglans fallax* Dode, *Juglans kamaonia* Dode, *Juglans orientis* Dode, *Juglans sinensis* (C.DC.) Dode.

Common Names Persian walnut, English walnut, common walnut.

Description Tree, deciduous (10–15 m). Leaves with fragrant odor, imparipinnate, with 5–9 elliptic or oblanceolate segments, the 3 apical 2–5 × 5–10 cm, the basal ones progressively smaller, entire, briefly acuminate. Inflorescence male catkins, pendulous (1 × 5–8 cm), the female short (1- to 5-flowered), apical. Fruit oval or globular (4–6 cm) with green fleshy epicarp (hull), brown when ripe, hardened endocarp (shell), and edible seed (kernel).

Cultivar There are numerous cultivars of this species in virtually every country where cultivation is practiced. Examples include Franquette, Lara, Marbot, Mayette, Mellanaise, and Parisienne.

Related Species There are 22 species recognized as belonging to the genus *Juglans*.

Geographic Distribution SW Asia (?). **Habitat** Scrubland, degraded woodland edges, cool ruderal environments, riverbanks, avenues, gardens. **Range** Tree cultivated for its fruit and wood, sometimes naturalized.

Beer-Making Parts (and Possible Uses) Wood (for barrels and malt smoking), edible fruits used in beer making.

Chemistry Fruits: lipids 62–71% (of the nut), saturated (% of total fatty acids): palmitic acid 5.2–7.3%, stearic acid 2.6–3.7%; unsaturated: oleic acid 21.2–40.2%, linoleic acid 43.9%–60.1%, linolenic acid 6.9–11.5%.

Wood Whitish sapwood clearly differentiated from the brown heartwood, sometimes with dark variegations, which increase its value. Wood easy to work both with a saw or a cutter, assumes perfect smoothness and has good duration, however attacked by various insects; WSG 0.63–0.75. Sought after for fine work of all kinds: furniture, decorative panels and veneers, rifle stocks, floor strips, small luxury items for which the marbled roots are much appreciated.

Style American Amber Ale, American Barleywine, American Brown Ale, American Stout, Baltic Porter, British Brown Ale, Brown Porter, Dunkles Weissbier, Northern English Brown Ale, Robust Porter, Saison, Spice/Herb/Vegetable Beer.

Beer Parthenope, Birrificio Sorrento (Massa Lubrense, Campania, Italy).

Source Calagione et al. 2018; Cantwell and Bouckaert 2016; Cantwell and Bouckaert 2018; Daniels and Larson 2000; Hieronymus 2016b; Josephson et al. 2016; https://www.bjcp.org (32A. Classic Style Smoked Beer); https://www.brewersfriend.com/other/walnuts/; https://byo.com/hops/item/306-brewing-smoked-beers-tips-from-the-pros; nut/.

Juniperus californica Carrière
CUPRESSACEAE

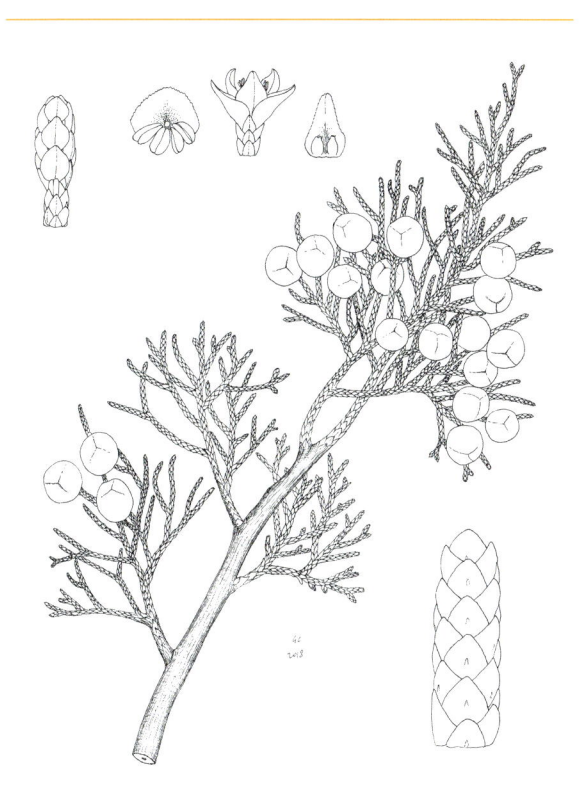

Synonyms *Juniperus cedrosiana* Kellogg, *Juniperus cerrosianus* Kellogg, *Juniperus pyriformis* A.Murray bis, *Sabina californica* (Carrière) Antoine.

Common Names California juniper.

Description Shrub or tree (up to 8 m), dioecious (rarely monoecious). Trunk multicaule (rarely monocaulescent); crown rounded. Bark gray in color, exfoliating into thin strips, those on branches smooth. Branches scattered to ascending, erect, terete, as wide as the length of the scaly leaves. Leaves light green, abaxial glands elliptic to ovate, conspicuous, exudate absent, margins denticulate; leaves needle-like 3–5 mm, not adaxially glaucous; leaves squamiform 1–2 mm, not overlapping or rarely overlapping for about 1/5 of their length, usually flattened, apex acute to obtuse, closely appressed. Female cones mature in 1 year, on straight peduncles, globose (7–13 mm), bluish-brown, glaucous, fibrous, with 1–2 seeds. Seeds 5–7 mm.

Cultivar This is the only *Juniperus* species in the Western Hemisphere to have two chemotypes that differ in essential oil composition, although in the absence of significant morphological differences and

Beer Corcolocia, Plotegher Brewery (Besenello, Trentino-Alto Adige, Italy).

Source Cantwell and Bouckaert 2016; Cantwell and Bouckaert 2018; Dabove 2015; https://www.brewersfriend.com/other/larch/.

Laurus nobilis L.
LAURACEAE

Synonyms None
Common Names Laurel, sweet bay laurel, bay laurel.

Description Shrub or tree (1–10 m), evergreen, dioecious. Young green branches with lenticels (0.5–1.5 mm) elongated longitudinally. Leaves evergreen, aromatic, leathery, dark green and glossy above, alternate, with 6–10 mm arcuate petiole and lamina mostly appressed to stem, narrowly elliptic or oblanceolate (6–12 × 2–4 cm), acuminate at apex, irregularly eroded and mostly wavy at margin. Axillary umbels on 1 cm peduncles; flowers yellow, the male with 8–12 stamens, the female with 4 staminoids and 1 ovary superior. Fruit ovoid berry, black in color when ripe.

Related Species Only four species belong to the genus *Laurus*: *Laurus azorica* (Seub.) Franco, *Laurus chinensis* Blume, *Laurus melissifolia* Walter, and, of course, *Laurus nobilis* L.

Geographic Distribution Steno-Mediterranean.
Habitat Sunny locations in the olive belt. **Range** Widely cultivated for ornamental and medicinal purposes.

Beer-Making Parts (and Possible Uses) Leaves.

Toxicity The leaves contain substances with a pleasant odor (essential oils). These are low-molecular-weight compounds, volatile at room temperature, and lipophilic. These physical and chemical characteristics allow these substances to easily overcome the blood–brain barrier, being able to generate in adults states of confusion and neurological disorders and in children much more serious compromises. The plant (leaves or fruits as the case may be) should be used with caution in the preparation of herbal teas, soups, roasted meats, and berry liqueurs.

Chemistry Essential oil, leaves (81 compounds representing 98.74% of total oil): monocyclic monoterpenes 1,8-cineole 58.59%, α-terpenyl acetate 8.82%, terpinene-4-ol 4.25%; bicyclic monoterpenes: α-pinene 3.39%, β-pinene 3.25%, sabinene 3.32%; acyclic monoterpenes: linalool 0.19%, myrcenol 0.10%; sesquiterpenes o-cymene 1.30%, p-cymene 1.83%, cumin aldehyde 0.24%, dimethylstyrene 0.08%, eugenol 0.16%, methyl eugenol 0.05%, carvacrol 0.05%.

Style Herb Beer.

Source Hamilton 2014; Heilshorn 2017; Hieronymus 2012; https://www.birradegliamici.com/news-birraie/birre-strane/.

Lavandula angustifolia Mill.
LAMIACEAE

Synonyms *Lavandula delphinensis* Jord. ex Billot, *Lavandula fragrans* Salisb., *Lavandula officinalis* Chaix, *Lavandula spica* L., *Lavandula vulgaris* Lam.

Common Names English lavender, true lavander.

Description Shrub (3–18 dm), gray-tomentose color, with pleasant scent. Stems woody, erect; young branches herbaceous, pubescent. Leaves linear (15–40 × 1.5–2 mm), usually revolute at the edge. Spikes 3–8 cm, ± long stalked, with 6- to 12-flowered whorls eventually spaced; bracts membranous, rhombic, narrowed to an elongate tip (3–4 × 6–8 mm) with 5–7 divergent fan-shaped nerves; bracts null or reduced; calyx 4–5 mm; corolla violet-purple 9–12 mm.

Subspecies In addition to the typical subspecies, *Lavandula angustifolia* Mill. subsp. *pyrenaica* (DC.) Guinea is recognized.

Cultivar In addition to morphological characteristics, the distinction between cultivars is based on the composition of the essential oil produced (chemotypes). Examples include Munstead, Raya, Silver, and Vera.

Related Species The genus *Lavandula* includes 57 taxa, some of hybridized origin such as the well-known lavandin (*Lavandula* × *intermedia* Emeric ex Loisel.), known for its abundant production of oil, however of inferior quality compared to that of *L. angustifolia* and of which no use is known in making beer.

Geographic Distribution W Steno-Mediterranean. **Habitat** Low scrub, garrigue. **Range** Widely cultivated for ornamental and medicinal purposes.

Beer-Making Parts (and Possible Uses) Flowering tops, flowers, leaves, stems.

Chemistry Essential oil: linalyl acetate 47.56%, linalool 28.06%, lavandulyl acetate 4.34%, α-terpineol 3.75%, geranyl acetate 1.94%, caryophyllene oxide 1.38%, 1,8-cineole 1.14%, β-caryophyllene 0.93%, borneol 0.85%, epi-α-cadinol 0.70%, nerol 0.59%, terpinen-4-ol 0.56%, β-myrcene 0.55%, limonene 0.55%, 1-otten-3-ol 0.53%.

Style American Barleywine, American IPA, American Light Lager, American Pale Ale, American Porter, American Stout, American Wheat Beer, American Wheat or Rye Beer, Belgian Blond Ale, Belgian Pale Ale, Belgian Specialty Ale, Belgian Tripel, Bière de Garde, Blonde Ale, Brown Porter, California Common Beer, Cream Ale, Dunkelweizen, English IPA, Experimental Beer, Extra Special/Strong Bitter (ESB), Foreign Extra Stout, Fruit Beer, Gruit, Holiday/Winter Special Spiced Beer, Imperial IPA, Kölsch, Metheglin, North German Altbier, Old Ale, Saison, Special/Best/Premium Bitter, Specialty Beer, Specialty IPA (Black IPA), Spice/Herb/Vegetable Beer, Sweet Stout, Weizen/Weissbier, Witbier.

Beer Novi Luna, Birrificio Maiella (Pretoro, Abruzzo, Italy).

Source Dabove 2015; Fisher and Fisher 2017; Giaccone and Signoroni 2017; Heilshorn 2017; Hieronymus 2016b; Josephson et al. 2016; https://www.birradegliamici.com/news-birraie/birre-strane/; https://www.brewersfriend.com/other/lavanda/; https://www.brewersfriend.com/other/lavender/.

Lavandula latifolia Medik.
LAMIACEAE

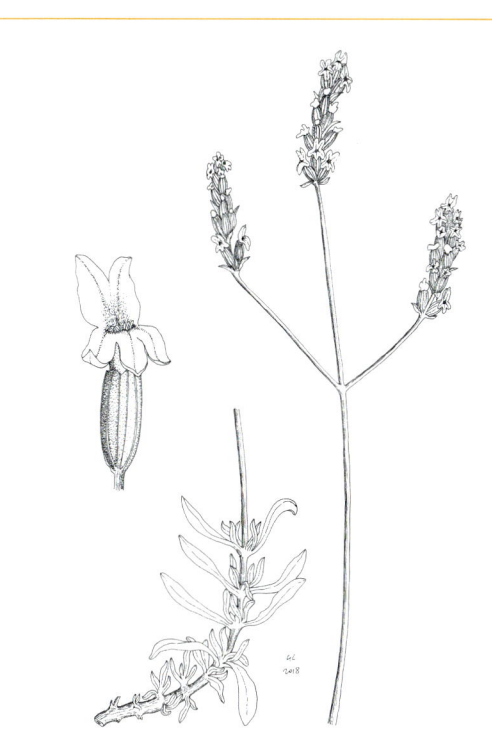

Synonyms *Lavandula cladophora* Gand., *Lavandula decipiens* Gand., *Lavandula erigens* Jord. & Fourr., *Lavandula guinardii* Gand., *Lavandula inclinans* Jord. & Fourr., *Lavandula interrupta* Jord. & Fourr., *Lavandula ovata* Steud.

Common Names Broadleaved lavender, spike lavender, Portuguese lavender.

Description Shrub (30–80 cm). Stem with branches densely tomentose with stellate hairs. Leaves verticillate at base of branches, widely spaced apically, narrowly lanceolate to linear (20–40 × 2–5 mm), densely tomentose with stellate hairs, base attenuate to petiolate, margin entire and revolute, apex obtuse to acute. Inflorescences 4- to 6-flowered whorls, lax, spikelet 7–8 cm, pedunculate (15–25 cm), terminally interrupted. Flowers with linear bracts, ± as long as the corolla; bracts linear, shorter than the calyx; calyx tubular, straight, 5–6 mm, densely tomentose for star hairs, 13-veined, 5-dentate; posterior tooth larger than the other 4; corolla 1–1.1 cm, densely tomentose; upper labellum straight, with lobes divaricated almost at right angles, ovate, apically obtuse; lower petal lobes subcircular.

Cultivar Some exist but are usually hybrids with related species.

Related Species The genus *Lavandula* includes 57 taxa.

Geographic Distribution W Steno-Mediterranean. **Habitat** Arid, calcareous (less frequently siliceous), bushy slopes (0–1,000 m ASL). **Range** W Steno-Mediterranean.

Beer-Making Parts (and Possible Uses) Flowering tops (plant substitute for *Lavandula angustifolia*).

Chemistry Essential oil (99.01% of total oil): 1,8-cineole 41.96%, linalool 30.34%, camphora 9.27%, p-pinene 2.49%, α-pinene 2.01%, borneol 1.61%, α-terpineol 1.34%, limonene 1.32%, sabinene 0.75%, terpinen-4-ol 0.63%, myrcene 0.63%, cis-sabinene hydrate 0, 50%, camphene 0.49%, caryophyllene oxide 0.46%, β-caryophyllene 0.44%, α-muurulol 0.38%, myrtenal 0.33%, trans-sabinol 0.31%, pinocarvone 0.31%, β-cis-ocimene 0.28%, γ-terpinene 0.27%, α-bisabolol 0.23%, terpinolene 0.20%, β-cymene 0.19%, β-cimen-8-ol 0.17%, α-terpinene 0.16%, germacrene D 0.15%, verbenone 0.13%, nopinone 0.12%, hexyl-2-methylbutanoate 0.12%, α-campholenal 0.12%, trans-β-farnesene 0.11%, carvone 0.11%, 7-cadinene 0.11%, hexyl-iso-valerate 0.09%, cis-linalool oxide 0.08%, α-gurjunene 0.07%, 1-cadinene 0.07%, trans-linalool oxide 0.06%, thuja-2,4(10)-diene 0.06%, p-trans-ocimene 0.06%, isobornyl formate 0.06%, tricyclene 0.05%, sabina ketone 0.05%, ledol 0.05%, hexyl butanoate 0.05%, cis-a-bisabolene 0.05%, cis-p-menth-2-en-1-ol 0.04%, α-phellandrene 0.04%, viridiflorol 0.03%, trans-carveol 0.03%, α-thujene 0.03%, germacrene-D-4-ol 0.02%, linalool acetate 0.01%, elemol 0.01%.

Style Herb Beer.

Source Fisher and Fisher 2017.

Ledum palustre L. subsp. *groenlandicum* (Oeder) Hultén
ERICACEAE

Synonyms *Rhododendron groenlandicum* (Oeder) Kron & Judd.

Common Names Labrador tea.

Description Differences from nominal subspecies, described next: leaves elliptic-oblong, 2.5–5 times longer than wide; midrib usually hidden by tomentum. Stamens 5–14.

Subspecies *Ledum palustre* L. subsp. *decumbens* (Aiton) Hultén, *Ledum palustre* L. subsp. *groenlandicum* (Oeder) Hultén, *Ledum palustre* L. subsp. *palustre*.

Related Species The genus *Ledum* has eight taxa.

Geographic Distribution N America, Greenland. **Habitat** Swamps, marshes, muskeg, tundra, dwarf shrub communities. **Range** Beyond natural geographic distribution, locally naturalized in Great Britain and Germany.

Beer-Making Parts (and Possible Uses)
Leaves, flowering tops.

Toxicity Plant containing alkaloids (e.g., andromedotoxin) whose ingestion in large doses can cause problems such as cardiovascular effects (hypotension, sinus bradycardia, bradyarrhythmia, atrioventricular block), nausea and vomiting, and a change in consciousness; the long boiling practiced by Native North Americans seems to remove the toxin.

Chemistry Essential oil, flowering apex (more than 160 compounds identified). Monoterpenes: sabinene 0.05–35%, β-pinene 0.05–8.4%, p-cymene 0.2–3.4%, limonene 0.3–67%, camphene 1.3%, α-terpinene 2.3%, terpinolene 1.5%; oxygenated monoterpenes: terpinen-4-ol 0.5–5.1%, myrtenal 0.3–3.8%, bornyl acetate 0.3–8.4%, trans-pinocarveol 2.3%, isopiperitenol A and B, mentadienehydroperoxides, limonene, β-selinene; sesquiterpenes: β-selinene 2.3–35.4%, α-selinene 0.3–9.9%, β-caryophyllene traces-3.1%. β-elemene 1.6%, δ-cadinene 2.2%, β-bisabolene 0–10.8%, germacrene B 0–9.4%; oxygenated sesquiterpenes: eudesma-3,11-dien-2-one 0.2–8.9%, germacrene 29.3%.

Style Gruit, Imperial Gruit.

Beer Big Land Lager, YellowBelly (Newfoundland, Canada).

Source Calagione et al. 2018; Heilshorn 2017; https://www.gruitale.com/bot_labradortea.htm.

Ledum palustre L. subsp. *palustre*
ERICACEAE

Synonyms *Rhododendron tomentosum* Harmaja.

Common Names Marsh rosemary, wild rosemary.

Description Shrub, evergreen (up to 120 cm), decumbent to erect; young twigs ferruginous-tomentose. Leaves alternate, briefly petiolate, linear to elliptic-oblong (12–50 × 1.5–12 mm), ferruginous-tomentose below, deflexed in winter, margin revolute. Flowers pentamerous, in terminal umbrella-like inflorescences (corymbose racemes); flowers numerous, peduncles 5–25 mm, glandular-verrucose, often also ferruginous-tomentose at first, erect at flowering, deflected at fruiting; petals 4–8 mm, obovate, patent, white; ovary and capsule warty-glandular.

Subspecies In addition to the typical one, two subspecies are recognized as belonging to this taxon: *Ledum palustre* L. subsp. *decumbens* (Aiton) Hultén and *Ledum palustre* L. subsp. *groenlandicum* (Oeder) Hultén.

Related Species The genus *Ledum* has eight taxa.

Geographic Distribution Circumboreal, N Europe. **Habitat** Swamps, heaths, coniferous forests. **Range** Beyond the geographic distribution, species cultivated for ornamental and medicinal purposes and sometimes naturalized in N Eurasia.

Beer-Making Parts (and Possible Uses)
Leaves, flowering tops.

Toxicity Plant containing ledel, a narcotic toxin whose ingestion can cause health problems with widely varying symptoms.

Chemistry Essential oil, flowering apix: palustrole 15.9–53.5%, ledol 11.8–18.3%, γ-terpineol 0–31.2%, p-cymene 0.1–13.9%, lepalone 0.7–6.5%, lepalol 1.0–6.5%, cyclocolorenone 1.0–6.4%.

Style American Pale Ale, Düsseldorf Altbier, Gruit.

Beer Rosemary, Tomos and Lilford (Llandow, Wales).

Source Heilshorn 2017; Hieronymus 2012; Hieronymus 2016b; McGovern 2017; https://www.brewersfriend.com/other/finnmarkspors/; https://www.brewersfriend.com/other/suopursu/; https://www.gruitale.com/bot_wildrosemary.htm.

Lens culinaris Medik
FABACEAE

Ledum palustre L. subsp. *palustre*
ERICACEAE

Synonyms *Ervum lens* L., *Ervum lens* Wall., *Lens esculenta* Moench, *Lens lens* Huth, *Vicia lens* (L.) Coss. & Germ.

Common Names Lentil.

Description Herb, annual (30–75 cm), slender, semi-erect. Stem single or branched, bushy. Leaves compound, pinnate, up to 14 sessile, ovate to lanceolate leaflets (15–40 mm); each leaf has 2 small basal stipules and ends in a tendril. Flowers small (< 12 mm), white to light purple to dark purple in color, usually 2 or in short racemes of 1–4. Fruit legume (12–20 mm) flat, smooth, containing 1–2 seeds. Seeds lentiform, seminal integument highly variable color (green, brown, black), even with purple or black spots or mottling.

Subspecies In addition to the nominal one (*Lens culinaris* Medik. subsp. *culinaris*), at least the following four subspecies are recognized: *Lens culinaris* Medik. subsp. *microsperma* (Baumg) N.F.Mattos, *Lens culinaris* Medik. subsp. *odemensis* (Ladiz.) M.E.Ferguson & al., *Lens culinaris* Medik. subsp. *orientalis* (Boiss.) Ponert, and *Lens culinaris* Medik. subsp. *tomentosus* (Ladiz.) M.E.Ferguson & al.

Cultivar Since this plant has been cultivated for thousands of years, in quite varied environmental and climatic conditions and with a wide genetic base guaranteed by different botanical subspecies, it is not surprising that there are hundreds of lentil cultivars. For reasons of space, only the Italian ones that have obtained recognition and the North American ones have been included. Italian cultivars: Altamura Lentil Traditional Food Product (PAT), Black Lentil of Enna Hills Slow Food Presidium (PSF), Black Lentil of Leonforte or Monti Erei PAT, Castelluccio di Norcia Lentil PGI and PDO, Colfiorito Lentil PAT, Onano Lentil PAT, Rascino Lentil PAT and PSF, Santo Stefano di Sessanio Lentil PAT and PSF, Ustica Lentil PAT and PSF, Valle Agricola Lentil PAT, Ventotene Lentil PAT, Villalba Lentil PAT and PSF. North American Cultivars: Black Beluga, Brewer, Canary, CDC Glamis, CDC Grandora, CDC Greenland, CDC Impact, CDC Impala, CDC Imperial, CDC Impress, CDC Improve, CDC LeMay, CDC Maxim, CDC Milestone, CDC Plato, CDC Redberry, CDC Redwing, CDC Robin, CDC Rosetown, CDC Rouleau, CDC Sedley, CDC Sovereign, CDC Vantage, CDC Viceroy, Crimson, Emerald, Eston, French Green (Puy), Ivory, Laird, Mason, Merrit, Moitree, Morton, Palouse, Pardina (Spanish brown), Pennell, Petite Golden, Red Chief, Richlea, Riveland.

Related Species The genus *Lens* has 11 taxa.

Geographic Distribution Mediterranean, SW Asia. **Habitat** Cultivated plant. **Range** Cultivated in more than 50 countries; main producers: Canada, India, Turkey, Australia, United States, China, Bangladesh, Iran, Nepal.

Beer-Making Parts (and Possible Uses) Seeds.

Chemistry Energy 1,477 kJ (353 kcal), water 8.3 g, carbohydrates 63 g, sugars 2 g, fiber 10.7 g, lipids 1 g, protein 25 g, thiamine (B1) 0.87 mg, riboflavin (B2) 0.211 mg, niacin (B3) 2.605 mg, pantothenic acid (B5) 2.14 mg, vitamin B6 0.54 mg, folate (B9) 479 µg, vitamin C 4.5 mg, calcium 56 mg, iron 6.5 mg, magnesium 47 mg, phosphorus 281 mg, potassium 677 mg, sodium 6 mg, zinc 3.3 mg.

Style Robust Porter.

Source https://www.brewersfriend.com/other/lentils/.

Leonurus cardiaca L.
LAMIACEAE

Synonyms *Cardiaca crispa* (Murray) Moench, *Cardiaca stachys* Medik., *Cardiaca trilobata* Lam., *Cardiaca vulgaris* Moench, *Lamium cardiaca* (L.) Baill., *Leonurus aconitifolius* Schltdl. ex Ledeb., *Leonurus campestris* Andrz. ex Benth., *Leonurus canescens* Dumort., *Leonurus crispus* Murray, *Leonurus discolor* W.D.J.Koch, *Leonurus illyricus* Benth., *Leonurus lacerus* Lindl., *Leonurus neglectus* Schrank, *Leonurus trilobatus* (Lam) Dulac, *Stachys triloba* Stokes.

Common Names Motherwort.

Description Herb, perennial (30–150 cm). Rhizome oblique, short. Stems quadrangular, erect, stout, usually branched, glabrescent on the faces, with short hairs (0.1–0.5 mm) appressed to the corners. Leaves opposite, long petiolate, palmate-nerved, lamina subglabrous and dark upper leaf, woolly-gray underleaf; lower leaves palmate-incised lamina (4–12 × 1–4 cm) with 5–7 acute, irregularly toothed lobes; upper leaves minor, elliptic-lanceolate, cuneate at base, with 3–5 irregular lobes. Inflorescence spiciform in dense, spaced whorls. Flowers hermaphrodite, numerous at the axil of the upper leaves, long outgrowing the spike; floral bracts 2–3 mm, linear, spinescent and setaceous; calyx 6–8 mm, campanulate tube with 5 marked veins and 5 teeth, the lower ones 2.5–4 mm deflected, the upper ones erect, all with spiny tips; corolla bilabiate, white or pinkish in color, 8–12 mm, with tube longer than calyx, inside with ring of hairs; upper lip entire, erect, densely villous on back; lower lip trilobed, purple stained, lateral lobes small; stamens 4, didynamous, with hairy filaments, anthers glabrous; stigma bifid. Fruit schizocarp with 4 mericarps (achenes or nucules), 2.5 × 1.6 mm, light brown, trigonous, truncate and hairy at apex.

Related Species The genus *Leonurus* includes 25 taxa.

Geographic Distribution Asia. **Habitat** Damp soils, manure, ruins, rubble, riverbanks, roadsides, nitrophilous environments (0–1,400 m ASL). **Range** Subcosmopolitan (beyond the natural geographic distribution, also Europe and Americas).

Beer-Making Parts (and Possible Uses) Flowers, leaves (used as a bittering agent).

Toxicity Not recommended in pregnancy.

Chemistry Essential oil (98.2% of the total): germacrene D 12%, epi-cedrol 9.7%, α-humulene 9.2%, dehydro-1,8-cineole 8.9%, spatulenol 8.8%, caryophyllene oxide 4.6%, β-pinene 4.4%, α-pinene 4.3%, δ-cadinene 4.2%, β-caryophyllene 3.2%, limonene 2.4%, (E)-β-ionone 2.4%, p-cymene 2.2%, terpinolene 2.2%, (E)-α-ionone 2.2%, geranyl acetone 2.2%, trans-sesquisabinene hydrate 2.2%, α-bisabolol 2.1%, 6,10,14-tirpmethyl-2-pentadecanone 1.8%, allo-aromadrene epoxide 1.7%, dihydroedulane 1.3%, eugenyl acetate 1.2%, (2z, 6e)-farnesyl acetate 1.1%, methyl linolea to 0.6%, 3-methylbutanal 0.3%, (E,E)-2,4-hexadienal 0.3%, β-humulene 0.3%, β-selinene 0.3%, presilfiperforlan-8-ol 0.3%, 8-cedren-13-ol acetate 0.3%, heneicosane 0.3%, p-menta-1,5-dien-8-ol 0.2%, dodecane 0.2%, (E)-nerolidol 0.2%, (2E,6E)-farnesol 0.2%, benzaldehyde 0.1%, phenylacetaldehyde 0.1%, β-atlantol 0.1%. Aerial parts: terpenes (monoterpenosides [iridoids]: leonuride, ajugoside, galiridoside, reptoside; bitter clerodane and type-labdane diterpenoids: leocardine, furanolabdane derivatives [marrubiin-like]: leosibiricin 2.6–3.2 mg/g, 19-acetoxypregaleopsin; triterpenes: ursolic acid 0.26%, oleanolic acid, corosolic acid, euscaphic acid, ilelatiphol D); pyrrolidine-like alkaloids (stachydrin 0.5–1.5%, betonicin, turicin, guanidino-derived imines [leonurine 0.0068%], amino choline); flavonoids (quercetin, kaempferol O-glycosides [rutin, hyperoside, quercitrin, isoquercitrin, astragalin, apigenin], quinqueloside); flavonoid C-glycosides (vitexin, isovitexin); free aglycones (quercetin, kaempferol, genkwanin); phenylpropanoid glycosides (lavandulifolioside: arabinoside of verbascosidephenolic acids: chlorogenic, rosmarinic, caffeic, p-coumaric, p-hydroxybenzoic, vanillic, ferulic acid); phenolic glycosides (caffeic acid 4-rutinoside 0.1%); volatile oils 0.02% sesquiterpenes (germacrene D 28.3–32.0%, epicedrole, β-caryophyllene 7.0–8.9%, α-caryophyllene 6.5–8.8%, spatulenol); monoterpenes (α-pinene 0.5–1.0%, dehydro-1,8-cineole); sterols 0.02% (β-sitosterol, stigmasterol); tannins 5–9%.

Style Herb Beer.

Source Heilshorn 2017; Hieronymus 2016b.

Leptospermum scoparium J.R.Forst & G.Forst
MYRTACEAE

Synonyms *Leptospermum bullatum* Fitzh., *Leptospermum humifusum* A.Cunn. ex Schauer, *Leptospermum multiflorum* Cav., *Leptospermum nichollsii* Dorr.Sm., *Leptospermum obliquum* Colla, *Leptospermum oxycedrus* Schauer, *Melaleuca scoparia* (J.R.Forst. & G.Forst.) L.f., *Melaleuca tenuifolia* J.C.Wendl.

Common Names Manuka, mānuka, manuka myrtle, New Zealand tea, broom teatree.

Description Shrub or small tree (to about 4 m). Trunk with bark detached in long strips; branches and young leaves ± covered with silky hairs. Leaves subsessile, about 4–20 × 1–4 mm, of 2 forms on different plants, narrowly lanceolate or ovate, leathery, stiff, acute, prickly, erect to patent. Flowers axillary or occasionally terminal on branches, ± sessile, usually solitary (Ø 8–12 mm), usually glabrous; hypanthium (2–3 mm) broadly turbinate; calyx lobes ± triangular, caducous; petals about 6 mm (4–7 mm), ± suborbicular, usually white, rarely pink or red, patent; stamens (2.5–3.5 mm) about 20, filaments slenderer than the stylus, the latter thin; ovary glabrous. Fruit capsule 5-locular (3–7 × 4–10 mm, Ø 6–9 mm), woody, long persistent; valves distinctly exserted beyond edge of receptacle, first dome shaped, then enlarged.

Cultivar Many have been selected for ornamental use, for example: Kiwi, Nichollsii Nanum, Red Damask, and Silver Sheen. Many others are available in New Zealand but show better performance elsewhere owing to the absence of specific fungal pests. Hybrids with resistant Australian *Leptospermum* species have been made to overcome this adversity.

Related Species The genus *Leptospermum* includes 88 taxa.

Geographic Distribution Australia (New South Wales, Victoria, Tasmania), New Zealand. **Habitat** Shrublands in rocky and sandy soils, often in riparian environments along fast-flowing streams. **Range** Beyond the natural geographic distribution, present and invasive in Europe (Great Britain), United States (Hawaii), Africa (Uganda), India.

Beer-Making Parts (and Possible Uses) Wood (for brewing and to smoke malt), leaves (to flavor beer according to an ancient recipe tried by James Cook's sailors during their first voyage around the world after being introduced to the properties of the plant by the Maori people).

Chemistry Essential oil, leaves: α-pinene 0.7%, β-pinene 0.3%, myrcene 0.3%, p-cymene 0.3%, 1,8-cineole 0.7%, β-ocimene 0.2%, linalool 0.3%, terpinene-4-ol 0.11%, α-terpineol 0.1%, citronellyl formate 0.2%, trans-methyl cinnamate/α-cubebene 2.8%, γ-ylangene + α-copaene 5.7%, β-elemene 0.4%, α-gurjunene 0.8%, b-caryophyllene 1.5%, aromadendrene 1.7%, α-humulene 3.6%, α-amorphene 2.6%, β-selinene 1.8%, α-selinene/viridiflorene 3.0%, α-muurolene 0.9%, γ-cadinene 0.2%, trans-calamenene 15.6%, δ-cadinene 4.5%, flavesone 8.2%, caryophyllene epoxide 0.7%, isoleptospermone 6.2%, leptospermone 16.6%, β-eud-hexmol 0.8%.

Wood Light brown, heavy, firm, fine texture and straight grain, hard but elastic, durable (WSG 0.75). Used for carriage work, handles, small fine objects and cabinetmaking, for smoking meat and fish.

Style Herb Beer (Manuka Beer).

Source Cantwell and Bouckaert 2016; Cantwell and Bouckaert 2018.

Leucanthemum vulgare (Vaill.) Lam.
ASTERACEAE

Synonyms *Chamaemelum leucanthemum* (L.) E.H.L.Krause, *Chrysanthemum ircutianum* Turcz., *Chrysanthemum lanceolatum* Pers., *Chrysanthemum lanceolatum* Vest, *Chrysanthemum leucanthemum* L., *Chrysanthemum praecox* (M.Bieb.) DC., *Chrysanthemum pratense* Salisb., *Chrysanthemum sylvestre* Willd., *Chrysanthemum vulgare* (Lam) Gaterau, *Leucanthemum lanceolatum* DC., *Leucanthemum praecox* (Horvatić) Villard, *Matricaria leucanthemum* (L.) Desr., *Matricaria leucanthemum* (L.) Scop., *Pontia heterophylla* (Willd.) Bubani, *Pontia vulgaris* Bubani, *Pyrethrum leucanthemum* (L.) Franch., *Tanacetum leucanthemum* (L.) Sch.Bip.

Common Names Ox-eye daisy, oxeye daisy.

Description Herb, perennial (20–100 cm). Stems woody at the base, glabrous or ± hairy, grooved, simple or sparsely branched at the base. Rhizome creeping, from which green shoots depart with reddish stripes along the grooves. Basal leaves 10–35 × 20–80 mm, clustered in a rosette; lamina obovate-cuneate to spatulate, incised-lobed (3–7), entire or rounded, gradually narrowing to form a petiole (10–30 mm), often winged; leaves cauline, middle and upper leaves semi-amplexicaul, the upper ones sessile, lamina oblong-ovate, pinnatifid with irregularly spaced lobes. Inflorescence flower head (Ø 4–7 cm) enclosed by 3–4 series of ovate-oblong-lanceolate imbricate scales (5–8 mm) with membranaceous purple margin (rarely concolorous). Flowers ligulate white (10–25 mm), sympetal, zygomorphic, 2–3 denticles at apex; inner flowers tetracyclic flosculus forming disc Ø 4–7 cm, yellow, apex divided into 5 lobes; peripheral flowers female, the central ones hermaphrodite; stylus single, stigma bifid, ovary inferior bicarpellary concrescent containing 1 ovule. Fruit cypsela (1.5–2.5 mm) oval-oblong, with 10 ribs forming resiniferous canals; fruits produced in outer ray bear rudimentary crowned pappi.

Subspecies There are four subspecific rank taxa of the species *Leucanthemum vulgare*: *Leucanthemum vulgare* (Vaill.) Lam. subsp. *eliasii* (Sennen & Pau) Sennen and Pau, *Leucanthemum vulgare* (Vaill.) Lam. subsp. *parviceps* (Briq. & Cavill.) Vogt & Greuter, *Leucanthemum vulgare* (Vaill.) Lam. subsp. *pujiulae* Sennen, and *Leucanthemum vulgare* (Vaill.) Lam. subsp. *vulgare*.

Cultivar There are some ornamental cultivars of hybridogenic origin with related species.

Related Species The genus *Leucanthemum* includes 53 taxa.

Geographic Distribution Eurasian, Eurasian-Mediterranean, Eurosiberian. **Habitat** Grasslands, wet meadows, roadsides and paths, banks of rivers and streams (0–1,500 m ASL). **Range** Similar to geographic distribution but with extensions to other continents.

Beer-Making Parts (and Possible Uses) Flowers, leaves.

Chemistry Essential oil, aerial parts: 1,8-cineole 1.20%, verbenyl acetate 6.08%, lavandulyl acetate 20.63%, M-isopropoxyaniline 5.14%, α-terpineol 3.33%, α-amorphene 1.47%, neryl acetate 1.42%, caryophyllene oxide 3.34%, α-cadinol 10.52%, torreyol 3.03%, β-guaiene 2.17%, β-eudesmol 10.13%, caryophyllene-II 1.22%, β-spatulenol 1.56%.

Style Herb Beer (?).

Source Hieronymus 2016b.

Levisticum officinale W.D.J.Koch
APIACEAE

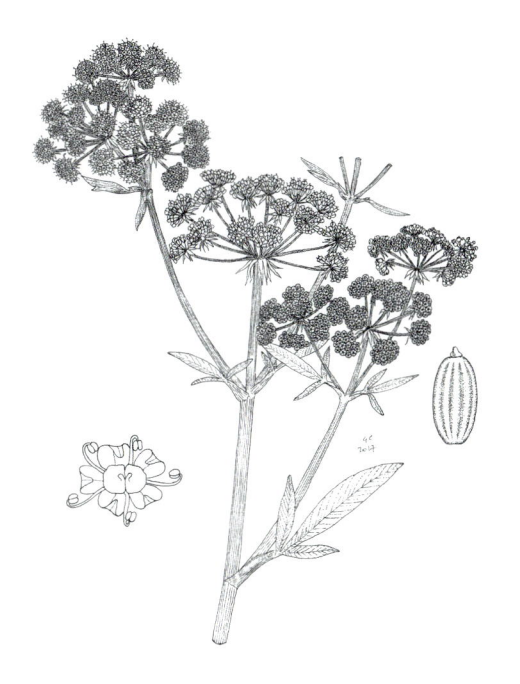

Synonyms *Angelica levisticum* (L.) All., *Angelica paludapifolia* Lam., *Hipposelinum levisticum* (L.) Britton & Rose, *Ligusticum levisticum* L., *Selinum levisticum* (L.) E.H.L.Krause.

Common Names Lovage, garden lovage.

Description Herb, perennial (up to 2 m), rhizomatous. Stem erect, hollow, striate, glabrous, celery-smelling, large rhizome containing a yellowish juice. Basal leaves (30–70 cm) long petiolate, 2–3 slightly pinnate, obovate-rhomboidal segments deeply toothed in apical half; cauline leaves sessile with long amplexicaul sheath. Flowers hermaphroditic, 5 petals small (1–2 mm), greenish-yellow. Inflorescence umbels, 12–20 rays, bracts and bracteoles lanceolate folded downward. Fruits 2 elliptical achenes (5–7 mm) provided with lateral ribs, aromatic when ripe.

Related Species The genus *Levisticum* is monotypic.

Geographic Distribution SW Asia. **Habitat** Cultivated areas and ruderal environments in damp soils, alpine pastures, hedges near watercourses. **Range** Well beyond its original geographic distribution, the plant is wild in Europe and widely cultivated for a variety of agro-food and industrial applications.

Beer-Making Parts (and Possible Uses) Edible parts: flowers, leaves, rhizome, seeds, stem. It has not been possible to determine which parts of the plant have been used in making beer. In general, they have the following edible uses: raw or cooked leaves and stems as a flavorful condiment in salads, soups, stews; fresh or dried leaves; blanched or raw young stems, like celery; raw or cooked seeds, whole or ground, with strong yeasty flavor, used as a flavoring in cakes, soups, salads; cooked rhizome (2–3 years old), strong flavor, used as a flavoring or cooked as a vegetable, also grated; an infusion can be made from the dried leaves or from the grated roots.

Chemistry Essential oil, miscellaneous parts: β-phellandrene 42.5%, α-terpineol 27.9%, cis-β-ocimene 7.5%, dehydro-1,8-cineole 6.8%, γ-terpinene 4.9%, ortho-cymene 1.6%, α-terpenyl acetate 1.6%, unknown 1.3%, α-thujene 1.2%, α-pinene 1.2%, α-terpinene 1%, trans-verbenol 0.8%, camphene 0.6%, β-pinene 0.6%, neryl acetate 0.5%, ethyl hydroquinone 0.4%.

Style Saison.

Source https://www.brewersfriend.com/other/lubczyk/.

Lindera benzoin (L.) Blume
LAURACEAE

Synonyms *Laurus benzoin* L.

Common Names Spicebush, Benjamin bush.

Description Shrub or small tree (up to 5 m). Young branches glabrous or sparsely pubescent. Leaves horizontal to ascending, strongly aromatic throughout the growing season; petiole about 10 mm, glabrous or pubescent; lamina obovate, smaller leaves usually elliptic (4–15 × 2–6 cm), membranous, base cuneate, margins ciliate, apices rounded to sharp in larger leaves; underleaf glabrous to densely pubescent, upper leaf glabrous except for a few hairs along the main rib. Fruit oblong drupe, about 10 mm; fruit peduncles from previous seasons do not persist on stem, thin, 3–5 mm, apices not conspicuously enlarged.

Subspecies In addition to the typical form, *Lindera benzoin* (L.) Blume f. *xanthocarpa* (G.S.Torr.) Rehder is recognized.

Cultivar Green Gold, Rubra, and Xanthocarpa are some of the cultivars of this species.

Related Species The genus *Lindera* includes about 60 taxa of different taxonomic rank (species, varieties, forms).

Geographic Distribution Canada (Ontario), United States (Alaska, Arkansas, Connecticut, Deleware, District of Columbia, Florida, Georgia, Illinois, Indiana, Kansas, Kentucky, Louisiana, Maine, Maryland, Massachusetts, Michigan, Mississippi, Missouri, New Hampshire, New Jersey, New York, North Carolina, Ohio, Oklahoma, Pennsylvania, Rhode Island, South Carolina, Tennessee, Texas, Vermont, Virginia, West Virginia). **Habitat** Riverbanks, moist woodlands, marsh margins, uplands, especially with exposed limestone (0–1,200 m ASL). **Range** Similar to natural geographic distribution.

Beer-Making Parts (and Possible Uses) Leaves, branches, fruits.

Chemistry Fruits: water 62.3%, protein 4.50%, fiber 1.97%, ash 2.17%, nonnitrogenous extract 9.92%, lignin 1.52%, cellulose 2.26%, tannins 0.51%, magnesium 0.05%, phosphorus 0.11%. Essential oil, leaves (total identified components 94.57%): 6-methyl-5-epten-2-one 42.94%, β-caryophyllene 7.74%, bicyclogermacrene 5.12%, δ-cadinene 4.92%, (E)-nerolidol 4.57%, (E)-β-farnesene 3.48%, β-bisabolene 2.84%, β-bisabolenol 2.33%, (E,E)-farnesol 1.87%, α-phellandrene 1.81%, γ-cadinene 1.64%, α-bisabolol 1.32%, α-cadinol 1.22%, 1,8-cineole 1.18%, τ-cadinol 1.17%, α-trans-bergamotene 1.14%, β-elemene 0.98%, caryophyllene oxide 0.98%, cis-3-hexanol 0.9%, (Z,E)-farnesyl acetate 0.85%, α-humulene 0.7%, α-muurolene 0.62%, citronellal 0.53%, citronellol 0.48%, trans-α-bisabolene 0.48%, sabinene 0.47%, 6-methyl-3,5-heptadien-2-one 0.29%, δ-elemene 0.28%, p-cymene 0.25%, γ-muurolene 0.25%, citronellyl acetate 0.22%, α-muurolol 0.21%, cis-α-bisabolene 0.2%, α-gurjunene 0.17%, β-bourbonene 0.16%, α-cadinene 0.15%, germacrene D 0.13%.

Style Herb Beer, Bière de Mars.

Source Heilshorn 2017; Hieronymus 2016b; Josephson et al. 2016.

Linum usitatissimum L.
LINACEAE

Synonyms *Linum crepitans* (Boenn.) Dumort., *Linum humile* Mill., *Linum indehiscens* (Neilr.) Vavilov & Elladi.

Common Names Flax, common flax, linseed.

Description Herb, annual (30–100 cm). Taproot. Stems erect, glaucous, tenacious, simple, sometimes branched. Lower leaves lanceolate-blunt (3–4 × 20–30 mm), glaucous green, sessile or briefly petiolate, alternate and rarely opposite, 3-nerved, upper leaves smaller and linear-acuminate. Flowers in terminal raceme corymbiform, lax, and on terminal or axillary peduncles; sepals 5 (6–9 mm) ciliate, glandular, ovate-acute, margin membranous; petals pentamerous, blue, veined with darker lines, caducous, subcrenate, obovate, 3 times as long as calyx (12–15 mm); stamens fertile 5, plus 5 abortive without anther; styli 5 free, stigma narrow and elongate club shaped. Fruit capsule subspherical (Ø 6–9 mm) enveloped by persistent calyx, containing 5–10 lodges. Seeds brown, oval-compressed.

Subspecies Besides the typical variety, *Linum usitatissimum* L. var. *stenophyllum* (Boiss.) Rech.f. is known.

Cultivar Numerous all over the world including, as an example, these from New Zealand: Aohanga (Awanga), Arawa, Atarau, Ate, Atewheke, Awahou, Hūhiroa, Kauhangaroa, Kōhunga, Māeneene, Makaweroa, Matawai Taniwha, Mawaru, Motu-o-nui, Ngaro, Ngutunui, Opiki, Oue, Pango, Paoa, Parekoritawa, Paretaniwha, Potaka, Rangiwaho, Raumoa, Ruahine, Ruapani, Ruawai, Taeore, Taiore, Takaiapu, Tākirikau, Tāne-ā-wai, Taniwha, Tapamangu, Tāpoto, Tārere, Taumataua, Te Aue Davis, Te Mata, Te Tatua, Tukura, Tupurupuru, Turingawari, Tūtaewheke, Waihirere, Whakaari, Wharanui, Whararīki, and Whareongaonga.

Related Species The genus *Linum* includes about 170 species.

Geographic Distribution Europe, Caucasus, subcosmopolitan. **Habitat** Cultivated or subspontaneous in fallow land (0–1,800 m ASL). **Range** Russia, Kazakhstan, India, China, Canada, United States.

Beer-Making Parts (and Possible Uses) Seeds.

Chemistry Seeds (g or mg/100 g; amino acids g/100 g protein): α-linolenic acid 22.8 g, linoleic acid 5.9 g, oleic acid 7.3 g, stearic acid 1.3 g, palmitic acid 2.1 g, calcium 236 mg, magnesium 431 mg, phosphorus 622 mg, potassium 831 mg, sodium 27 mg, zinc 4 mg, copper 1 mg, iron 5 mg, manganese 3 mg, glutamic acid 19.6 g, aspartic acid 9.3 g, arginine 9.2 g, glycine 5.8 g, cysteine 1.1 g, histidine 2.2 g, isoleucine 4.0 g, leucine 5.8 g, lysine 4.0 g, methionine 1.5 g, proline 3.5 g, serine 4.5 g, threonine 3.6 g, tryptophan 1.8 g, tyrosine 2.3 g, valine 4.6 g, γ-tocopherol 522 mg, α-tocopherol 7 mg, δ-tocopherol 10 mg, ascorbic acid 0.5 g, thiamine (B1) 0.5 mg, riboflavin (B2) 0.2 mg, niacin (nicotinic acid) 3.2 mg, pyridoxine (B6) 0.6 mg, pantothenic acid 0.6 mg, arabinoxylan 1.2 mg, rhamnogalacturonan 0.4 mg, fiber 4.3–8.6 g, ferulic acid 10.9 mg, chlorogenic acid 7.5 mg, gallic acid 2.8 mg, secoisolariciresinol 165 mg, laricinesol 1.7 mg, pinoresinol 0.8 mg.

Style English IPA.

Source https://www.brewersfriend.com/other/linaza/.

Litchi chinensis Sonn.
SAPINDACEAE

Synonyms *Corvinia litschi* Stadtm. ex Willemet, *Euphoria didyma* Blanco, *Euphoria sinensis* J.F. Gmel., *Nephelium chinense* (Sonn.) Druce, *Nephelium litchi* Cambess., *Nephelium litchi* Steud., *Scytalia chinensis* Gaertn.

Common Names Lychee, litchi, liechee, liche, lizhi, li zhi, lichee.

Description Tree (up to 30 m), evergreen, long-lived. Trunk short, stocky; crown broad or rounded depending on cultivar. Leaves alternate, compound, with 2–5 oblong leaflets (5–15 cm); young shoots characteristically red-brown in color, light green to dark green when mature. Inflorescence panicle pluribranched, each with 1-plus leaves and up to 3,000 flowers and 5–80 fruits at maturity. Flowers small, apetalous, yellowish-white, functionally male or female; male 6–10 staminoids; female 6–10 staminoids and 1 functional bicarpellate pistil. Fruit extremely variable depending on cultivar, round, ovoid, heart shaped (Ø 2–3.5 cm); skin smooth to rough with characteristic protuberances, thick or thin, color pink-red to bright pink to purple-red; flesh (aril) represents an outgrowth of the outer cells of the seed tegument, and in the best cultivars can constitute up to 80% of the fruit weight; aril translucent white in color, juicy or compact, sweet and aromatic. Seed 1 per fruit, dark brown in color (6–12 × 10–23 mm); some cultivars are seedless.

Subspecies In addition to the nominal subspecies (*Litchi chinensis* Sonn. subsp. *chinensis*), some botanists recognize the subspecies *Litchi chinensis* Sonn. subsp. *javensis* Leenh. (Java, Malaysia).

Cultivar Getting a sense of the many cultivars of litchi is not easy because the original name in Chinese is transliterated in different ways and because of the characteristic production by the same cultivar of morphologically different fruits in different environmental conditions. Among the most common ones are Mauritius (South Africa), Groff (Hawaii), Sanyuehong, Baitangying, Baila, Shuidong, Feizixiao, Dazou, Heiye, Nuomici, Guiwei, Huaizhi, Lanzhu, Chenzi (China), Shahi, Dehra Dun, Early Large Red, Kalkattia, Rose Scented (India), Kway Mai Pink (or Bosworth 3), Brewster (or Chen-Tze or Chenzi), Sweetheart, and Tai-so (or Tai-tsao) (Italy).

Related Species Besides *L. chinensis* Sonn., the genus *Litchi* includes two species: *Litchi philippinensis* Radlk. ex Whitford and *Litchi sinensis* Sonner.

Geographic Distribution China, Vietnam, Malaysia. **Habitat** Tropical and subtropical forests, humid lowlands. **Range** China, SE Asia, Pakistan, Bangladesh, India, Japan, United States (California, Hawaii, Texas, Florida), Mexico, Brazil, Australia, Madagascar, South Africa, Israel.

Beer-Making Parts (and Possible Uses) Fruits.

Chemistry Peeled fruits: energy 276 kJ (66 kcal), carbohydrates 16.53 g, sugars 15.23 g, fiber 1.3 g, lipids 0.44 g, protein 0.83 g, thiamine (B1) 0.011 mg, riboflavin (B2) 0.065 mg, niacin (B3) 0.603 mg, vitamin B6 0.1 mg, folate (B9) 14 µg, vitamin C 71.5 mg, calcium 5 mg, iron 0.13 mg, magnesium 10 mg, manganese 0.055 mg, phosphorus 31 mg, potassium 171 mg, sodium 1 mg, zinc 0.07 mg.

Style American Brown Ale, American Wheat or Rye Beer, Berliner Weisse, California Common Beer, Fruit Beer, Saison, Standard American Lager, Standard/Ordinary Bitter.

Source https://www.brewersfriend.com/other/lychee/; https://www.brewersfriend.com/other/lichy/; https://www.brewersfriend.com/other/licky/.

Lolium temulentum L.
POACEAE

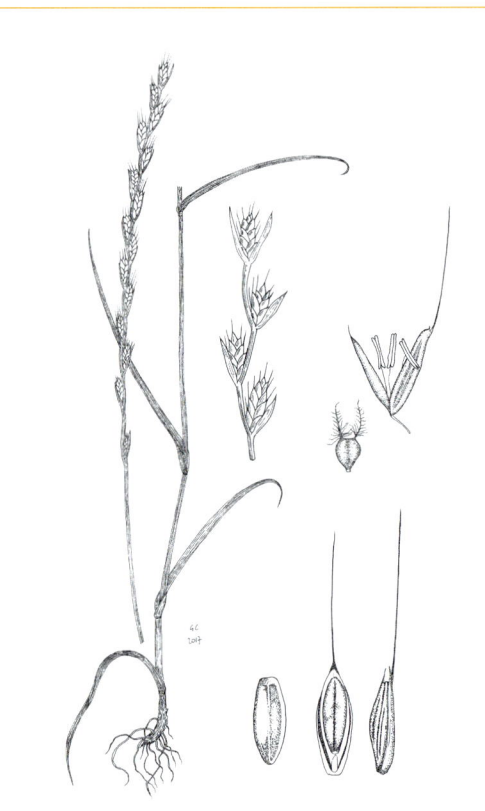

Synonyms *Bromus temulentus* Bernh., *Craepalia temulenta* (L.) Schrank, *Lolium annuum* Lam., *Lolium arvense* With., *Lolium asperum* Roth, *Lolium berteronianum* Steud., *Lolium cuneatum* Nevski, *Lolium decipiens* Dumort., *Lolium gracile* Dumort., *Lolium infelix* Rouville, *Lolium lucidum* Dumort., *Lolium maximum* Willd., *Lolium pseudolinicola*

Gennari, *Lolium robustum* Rchb., *Lolium speciosum* Steven ex M.Bieb., *Lolium triticoides* Janka.

COMMON NAMES Darnel, poison darnel, darnel ryegrass, cockle.

DESCRIPTION Herb, annual (2–7 dm). Culms generally kneeling at the base, then erect and somewhat incurved. Leaves glabrous, with lamina 4–10 mm wide, flat; ligule 1 mm. Inflorescence linear spike; spikelets 3- to 8-flowered; single glume (except apical spikelet) 12–15 mm long (more than the remainder of the spikelet); lemma 6 mm with apical awns 8–10 mm long.

SUBSPECIES There are two subspecies of this species (in Italy): the typical *Lolium temulentum* L. subsp. *temulentum* (culms 5–7 dm, straight, robust spike, oval-lanceolate lemma, usually aristate) and *Lolium temulentum* L. subsp. *gussonei* (Parl.) Pign. (culms 2–6 dm, gracile and curved, spikelet gracile, lemma narrowly lanceolate, muticulate).

RELATED SPECIES The genus *Lolium* includes about 17 taxa.

GEOGRAPHIC DISTRIBUTION Subcosmopolitan. **HABITAT** Infesting species of cereal crops. **RANGE** Subcosmopolitan.

BEER-MAKING PARTS (AND POSSIBLE USES)
Caryopses infested by fungi of the genus *Claviceps* sp. (used to adulterate porter beer with substances simulating the effect of alcohol but less expensive than raw materials such as malt, sugar, molasses).

TOXICITY Poisonous plant (seeds and ear) owing to the presence of toxic alkaloids; symptoms of poisoning are ataxia, dizziness, apathy, mydriasis, nausea, vomiting, gastroenteritis, and diarrhea; rarely fatal.

CHEMISTRY The alkaloids temulin and loliin are probably responsible for the toxicity of this plant; it has also been suggested that *Claviceps*, an ascomycete that lives in the seed, is responsible for the toxicity.

STYLE Porter.

SOURCE Daniels 2016.

Lonicera caerulea L.
CAPRIFOLIACEAE

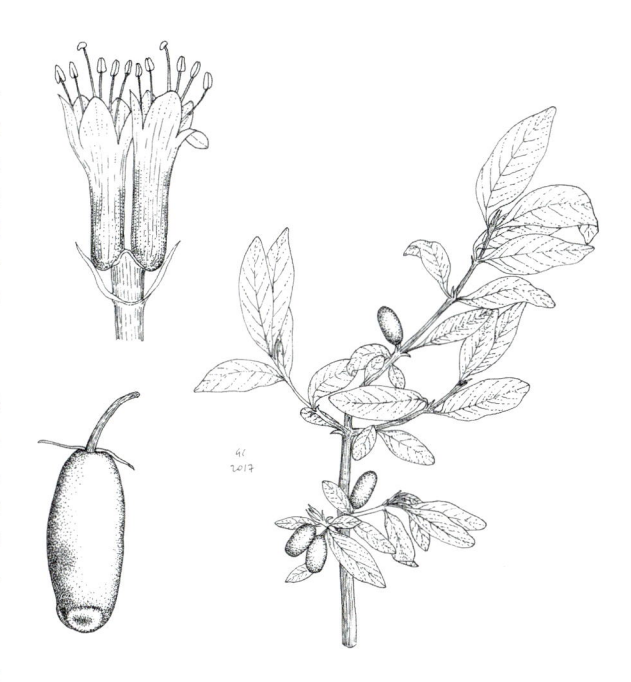

SYNONYMS *Caprifolium caeruleum* (L.) Lam., *Euchylia caerulea* (L.) Dulac, *Isika coerulea* (L.) Medik., *Xylosteon caeruleum* (L.) Dum.Cours.

COMMON NAMES Camerise, honeyberry, blue-berried honeysuckle, sweetberry honeysuckle.

DESCRIPTION Shrub (5–15 dm), deciduous, with shallow, spreading roots, branched crown. Stems slender, erect, woody, opposite, inserted almost at right angles; twigs flexible, reddish-brown; bark of old branches reddish-brown, peeling into thin leaves; buds opposite, sessile, small, ogival-sharp; wood hard white, with ivory pith and vaguely alcoholic odor. Leaves opposite, briefly petiolate (2 mm), lamina elliptic (3–5 cm), apex rounded, margin entire, ± hairy when young (underleaf), then glabrous, glaucous and ciliate on the edge. Flowers sparse, patent or pendulous, paired on single hairy peduncle (about 1 cm), linear green bracts ciliate at base of ovary; calyx with short teeth; corolla actinomorphic, whitish-yellow (12–15 mm), 5-lobed, with hairy tube; 5 stamens, equal to or larger than corolla, anthers yellow; ovaries of flower pair fully concrescent. Fruits ellipsoidal, blue, pruinose berries (5–8 × 6–10 mm), retaining the remnants of 2 appressed stigmas at apex. Seeds minute, numerous.

(*Lycium barbarum* Big Lifeberry®); it is likely that other cultivars are actually interspecific hybrids of *Lycium barbarum* × *chinense*.

Related Species *Lycium chinense* Mill.; the genus *Lycium* includes about 94 taxa.

Geographic Distribution SE Europe, SW Asia. **Habitat** Cultivated. **Range** Cultivated and naturalized in various European countries.

Beer-Making Parts (and Possible Uses) Fruits, leaves (?).

Chemistry Dried fruits: polysaccharide complexes (5–8%, 6 saccharides: arabinose, rhamnose, xylose, mannose, galactose, glucose, galacturonic acid; 18 amino acids), carotenoids, phenylpropanoids, phenolic compounds (caffeic acid 110.84 mg, chlorogenic acid 113.18 mg, coumaric acid 111.32 mg, ferulic acid 125.8 mg, flavonol hyperoxide 116.27 mg, quercetin-diglycoside 66 mg, kaempferol-3-O-rutinoside 11.30 mg, gallic acid 15.31 mg, catechin 118.76, epicatechin 229.18 mg, vitamin C 48.94 mg.

Style Weizen/Weissbier.

Source https://www.brewersfriend.com/other/goji/; https://www.ratebeer.com/beer/goji-beer/155579/.

Lycopersicon esculentum Mill.
SOLANACEAE

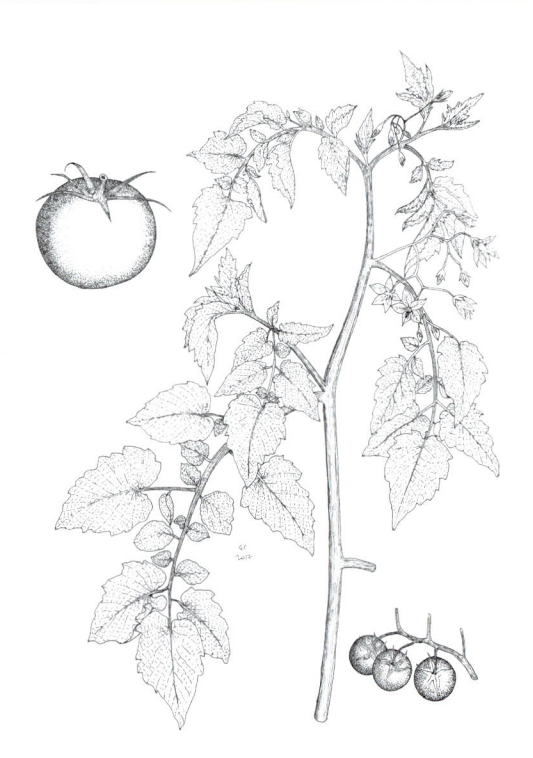

Synonyms *Lycopersicon humboldtii* (Willd.) Dunal, *Scubulon humboldtii* (Willd.) Raf., *Solanum humboldtii* Willd.

Common Names Tomato.

Description Herb, annual (3–10 dm). Stem glandular, carrot scented, weak, incapable of standing erect, ± tubular or compressed, with appressed pubescence and rare bristles. Leaves usually 2-pinnate, perimeter oval (8–15 × 13–25 cm); oval segments very unequal in size, some 3–6 cm long interspersed with others 1 cm long or less. Inflorescence cymes 3- to 20-flowered with incurved peduncles. Flower with calyx almost completely divided into linear laciniae; corolla (Ø 1–2 cm) yellowish. Fruit edible berry (tomato) of highly variable shape and size, red in color at maturity. Seeds numerous.

Subspecies Besides the nominal taxon, *Lycopersicon esculentum* Mill. subsp. *esculentum*, within the same species, the taxa *Lycopersicon esculentum* Mill. subsp. *galeni* (Mill.) Luckwill and *Lycopersicon esculentum* Mill. var. *leptophyllum* (Dunal) D'Arcy are recognized.

Cultivar There are about 7,500 cultivars, bred for different purposes and to grow in different environmental conditions. Since a complete list of cultivars of this species is beyond the needs and purposes of this book, I mention only the characteristics of the main "technical" categories, based essentially on the shape and size of the fruit: beefsteak tomatoes (fruit Ø 10 cm or more, kidney shaped, thin skin, not very durable and therefore not suitable for large-scale trade), plum tomatoes (fruit 7–9 × 4–5 cm, low water content, suitable for making sauces and preserves), cherry tomatoes (Ø 1–2 cm, roundish), cluster tomatoes (small, oblong fruits, a variant of plum tomatoes), Campari tomatoes (known for sweetness and juiciness, low acidity, absence of flouriness, larger than cherry tomatoes and smaller than plum tomatoes), tomberry tomatoes (tiny fruits, Ø about 5 mm), ox heart tomatoes (large, shape similar to a big strawberry), pear tomatoes (based on San Marzano type with elongated fruit, particularly suitable for Italian cooking), and slicing or globe tomatoes (fruit Ø 5–6 cm, most common varieties for fresh consumption but also for industrial processing).

Tomato varieties can also be classified into determinate and indeterminate categories, according to the model of development and growth and to the ripening of fruits; determinate varieties have a defined bush-like growth and ripen fruits more or less at the same time; understandably they are suitable for mechanized harvesting and industrial processing; indeterminate varieties develop as creepers and produce fruits nonstop until winter; they are preferred by small-scale growers because of the scalarity of production; there are also varieties with intermediate characteristics. The poor taste and lack of sugar of modern commercial varieties are the result of crossings aimed at producing tomatoes with a uniformly ripening red color; this change occurred after the discovery, in the mid-twentieth century, of a mutant phenotype, "u," that ripened uniformly; crossings containing this gene do not have the typical green ring around the peduncle (responsible for the unpleasant splits in the upper part of the fruit), but the old varieties produce more sugar (10–20%) during the ripening of the fruits, which are sweeter and tastier; the "u" mutation encodes for a factor that produces defective low-density chloroplasts, light green in color, and represses the accumulation of sugars. In addition, the mutant chloroplasts of the fruit are remodeled early during ripening into chromoplasts, known to lack chlorophyll but active in the synthesis and accumulation of lycopene, β-carotene, and other metabolites that constitute the sensory and nutritional characteristics of the ripe fruit.

Related Species The most significant taxa belonging to the genus *Lycopersicon* are *Lycopersicon cheesmanii* Riley, *Lycopersicon chilense* Dunal, *Lycopersicon chmielewskii* Rick, *Lycopersicon hirsutum* Dunal, *Lycopersicon parviflorum* Rick, Kesickii, Forbes & Holle, *Lycopersicon pennellii* (Correll) D'Arcy, *Lycopersicon peruvianum* (L.) Miller, *Lycopersicon peruvianum* (L.) Miller var. *humifusum*, and *Lycopersicon pimpinellifolium* (B. Juss.) Miller; this classification is not universally accepted, as shown by the fact that, according to other scientists, the genus *Lycopersicon* would include, at a specific level, only five species: *Lycopersicon chmielewskii* C.M.Rick, Kesicki, Fobes & M.Holle, *Lycopersicon esculentum* Mill., *Lycopersicon hirsutum* Dunal, *Lycopersicon parviflorum* C.M.Rick, Kesicki, Fobes & M.Holle, and *Lycopersicon pennellii* (Correll) D'Arcy; whether and to what extent these taxa may have applications in making beer remains to be verified.

Geographic Distribution CS America.
Habitat Widely cultivated everywhere and often subspontaneous. **Range** China, India, United States, Turkey, Egypt, Iran, Italy, Spain.

Beer-Making Parts (and Possible Uses) Fruits (unripe, ripe, dried).

Toxicity Leaves, stems, and unripe green fruits contain small amounts of the toxic alkaloids tomatine and solanine, alkaloids also present in potato leaves; ingesting these parts can be dangerous and even fatal in large quantities, but the toxic elements are usually present only in fairly low amounts.

Chemistry Energy 74 kJ (18 kcal), water 94.5 g, carbohydrates 3.9 g, sugars 2.6 g, fiber 1.2 g, lipids 0.2 g, protein 0.9 g, vitamin A 42 µg, beta-carotene 449 µg, lutein zeaxanthin 123 µg, thiamine (B1) 0.037 mg, niacin (B3) 0.594 mg, vitamin B6 0.08 mg, vitamin C 14 mg, vitamin E 0.54 mg, vitamin K 7.9 µg, magnesium 11 mg, manganese 0.114 mg, phosphorus 24 mg, potassium 237 mg, lycopene 2,573 µg.

Style Dark Strong Ale, Golden Ale, IPA, Saison.

Beer Italian Tomato Ale, Carrobiolo (Monza, Lombardy, Italy).

Source Calagione et al. 2018; Giaccone and Signoroni 2017; Josephson et al. 2016; https://www.brewersfriend.com/other/tomato/; https://www.brewersfriend.com/other/tomatoes/.

Almost a Pizza Beer
Pietro Fontana is a talented brewer, capable of playing on subtle balances like few others in Italy. In his hands, even the most daring ideas acquire elegance and harmony. Among his wild ideas are putting dried porcini mushrooms in the recipe of an imperial stout and, perhaps most wild, brewing a beer with tomatoes, oregano, and basil. Italian tomato ale is a golden ale marked by these ingredients but pleasing to drink. Oregano and basil come into play during the boiling process, whereas tomatoes are added during the second day of fermentation to extract the most aromatic parts.

Malus angustifolia (Aiton) Michx.
ROSACEAE

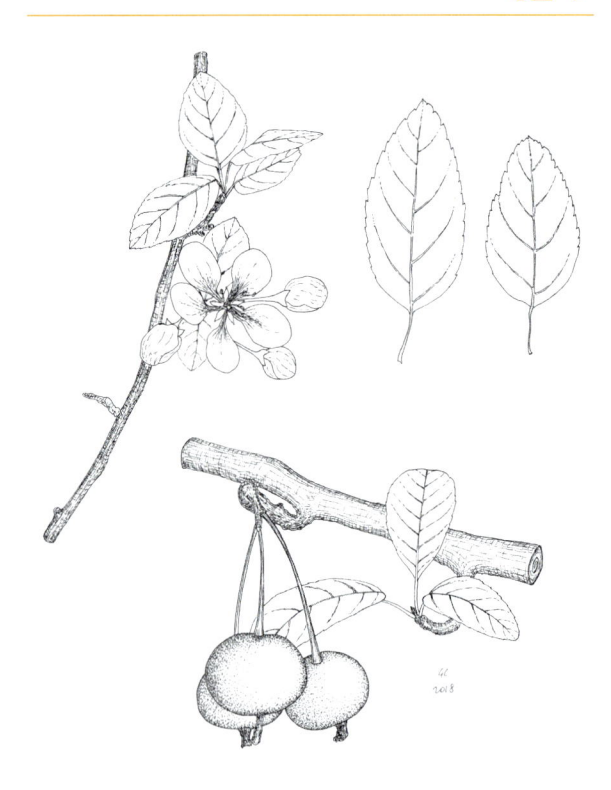

Synonyms *Pyrus angustifolia* Aiton.

Common Names Crabapple, southern crabapple, narrow-leaved crabapple.

Description Tree or shrub (1–10 m). Trunk Ø 5–25 cm; bark reddish-brown to gray, longitudinally fissured in plates; young branches reddish-brown, sparsely puberulent, then gray or grayish-brown, glabrescent; flowering branches develop as short spurs or spines (5–60 mm). Buds reddish-brown, ovoid, 1.6 mm, margins of scales tomentose. Leaves conduplicate in buds, heteromorphic; stipules deciduous, linear-lanceolate (2–5 mm), apex acuminate; leaves of sterile branches vigorous: petiole 10–25 mm, villous, sometimes tomentose; lamina elliptic, oval or ovate, sometimes triangular-ovate (3.5–8 × 1.5–5 cm), base cuneate, sometimes rounded, margins sometimes slightly lobed, crenate, crenate-serrate, serrate, or entire, sometimes doubly serrate, apex rounded, broadly acute, sometimes apiculate, underleaf glabrous (villous only on veins), upper leaf glabrous; leaves of fertile branches: petiole 3–25 mm, villous, sometimes glabrous; lamina elliptic or oblong, sometimes ovate, obovate, or lanceolate (9–65 × 7–30 mm), base cuneate, sometimes rounded, margins nonlobed, crenate, crenate-serrate, or entire, apex rounded (or acute), underleaf glabrous (veins villous), upper leaf glabrous. Inflorescence panicle corymbiform; peduncles absent; bracts sometimes persistent, filiform (1–6 mm). Pedicels 10–40 mm, glabrous, sometimes slightly villous. Flower (Ø 20–30 mm) hypanthium glabrous, rarely slightly villous; sepals triangular (2–5 mm), equal to tube, apex acuminate, underleaf glabrous, upper leaf canescent-tomentose; petals pink, sometimes toward white, oblong to narrowly obovate (10–22 mm), claws 2–3 mm, margins entire, sinuate or fimbriate, apex rounded; 20 stamens (7–14 mm), anthers pink, emerging before dehiscence; 5 styles, basally connate (6–15 mm), usually slightly longer than stamens, tomentose in the proximal 1/3. Fruit pome green or yellow-green, subglobose (Ø 10–30 mm), sepals persistent, erect. Seeds dark brown.

Subspecies This taxon has two varieties: *Malus angustifolia* (Aiton) Michx. var. *angustifolia* and *Malus angustifolia* (Aiton) Michx. var. *puberula* (Rehder) Rehder.

Related Species The genus *Malus* includes 71 taxa, most of which produce edible fruits.

Geographic Distribution United States (Alabama, Arkansas, Delaware, District of Columbia, Florida, Georgia, Illinois, Kentucky, Louisiana, Michigan, Missouri, New Jersey, North Carolina, Ohio, Pennsylvania, South Carolina, Tennessee, West Virginia). **Habitat** Open woodlands, scrublands, hillsides, stream and lake banks, floodplains, terraces, roadsides, open fields, sandy soils, loamy soils (10–700 m ASL). **Range** Similar to geographic distribution.

Beer-Making Parts (and Possible Uses)
Fruits (hard and acidic, often used for preserves, cider, jellies).

Chemistry Fruits: water 87.0%, protein 0.75%, fiber 2.2%, ash 0.41%, nonnitrogenous extract 8.84%, lignin 1.55%, cellulose 2.01%, tannins 0.61%, calcium 0.01%, magnesium 0.01%, phosphorus 0.02%.

Style Fruit Beer.

Source Heilshorn 2017; Hieronymus 2016b.

Malus coronaria (L.) Mill.
ROSACEAE

Synonyms *Pyrus bracteata* (Rehder) L.H.Bailey, *Pyrus coronaria* L.

Common Names Crabapple, American crabapple, sweet crabapple, fragrant crabapple, garland tree.

Description Tree or shrub (2.5–10 m). Stem Ø 1–45 cm; bark reddish-brown to gray, longitudinally fissured in plates; young branches reddish-brown to dark brown, with orange lenticels, pubescent, glabrescent; flower sprouts become spurred or thorny (10–100 mm). Buds reddish-brown, ovoid, 1–6 mm, perulae with tomentose margins. Leaves conduplicate in buds, heteromorphic; stipules deciduous, linear-lanceolate (3–7 mm), apex acuminate; leaves of sterile shoots vigorous: petiole (7–30 mm) glabrous or villous, sometimes puberulent; lamina ovate or triangular-ovate, sometimes lanceolate (2.5–10 × 1.5–8 cm), base rounded or cordate-rounded, sometimes cuneate, margins ± lobed, sometimes nonlobed, serrate, sometimes doubly serrate or crenate-serrate, apices broadly acute or acute, sometimes rounded, apiculate, lower surface glabrous (villous only on nerves), upper surface glabrous; leaves of fertile shoots: petiole 5–25 mm, glabrous or villous; lamina ovate, triangular-ovate, or lanceolate, sometimes ovate or elliptic (15–85 × 10–60 mm), base rounded, cordate, or even cuneate-rounded or cuneate, margins lobed or nonlobed, serrate, sometimes crenate-serrate, doubly serrate or entire, apices acute or broadly acute; underleaf glabrous (veins villous), upper leaf glabrous. Inflorescence panicle corymbiform; peduncles absent; bracts sometimes persistent, filiform (3–10 mm). Pedicels 10–40 mm, glabrous, sometimes villous. Flower (Ø 25–40 mm) hypanthium glabrous, sometimes puberulent; sepals triangular (3–7 mm), equal to or slightly longer than tube, apex acute or acuminate, lower face glabrous or glabrescent, upper face canescent-tomentose; petals pink, toward white, oblong-obovate, ovate, or oblong (11–20 mm), claws 2–4 mm, margins entire, sinuate or fimbriate, apex rounded; 20 stamens (8–14 mm), anthers pink, salmon, or purple before dehiscence; 5 styli, basally connate (8–16 mm), equal to or slightly longer than stamens, villous in the proximal ½. Fruit pome green or yellow-green, depressed-globose (Ø 15–55 mm); sepals persistent, erect. Seeds dark brown in color.

Subspecies There are two varieties of this taxon: *Malus coronaria* (L.) Mill. var. *coronaria* and *Malus coronaria* (L.) Mill. var. *dasycalyx* Rehder.

Cultivar There are some with double flowers, often used for ornamental purposes.

Related Species The genus *Malus* includes 71 taxa, most of which produce edible fruits.

Geographic Distribution Canada (Ontario), United States (Alaska, Arkansas, Delaware, District of Columbia, Georgia, Illinois, Indiana, Kansas, Kentucky, Louisiana, Maryland, Michigan, Missouri, New Jersey, New York, North Carolina, Ohio, Pennsylvania, South Carolina, Tennessee, Virginia, Wisconsin). **Habitat** Open woodlands, forest canopies, thickets, stream banks, fields, fences, roadsides (50–1,000 m ASL). **Range** Similar to geographic distribution.

Beer-Making Parts (and Possible Uses)
Fruits (raw—edible only when fully ripe—or cooked, fresh or preserved, hard and acid, mostly used for jams; if buried in the ground in winter, they lose most of their acidity by spring; rich in pectin, they can be added to fruits lacking pectin to produce jams).

Chemistry No data available.

Style Fruit Beer.

Source Heilshorn 2017; Hieronymus 2016b.

Malus domestica Borkh.
ROSACEAE

Synonyms *Pyrus malus* L.

Common Names Apple tree.

Description Tree (3–15 m), deciduous. Leaves with ovate lamina (25–40 × 35–80 mm), serrate or crenulate all around, acute, rarely acuminate, upper leaf glabrous, lower especially at base and along veins densely tomentose even in summer; petiole 1/3–1/2 the size of lamina. Flowers 3–7 in umbel-like cymes; peduncles and calyx tube densely tomentose; petals obovate (10–20 mm); styluses glabrous. Fruit subrounded edible pome (apple).

Cultivar Of the more than 7,500 known cultivars, only a small number are noted here, divided by main type of use. Cultivars for fresh consumption: Adams Pearmain, Aia Ilu, Airlie Red Flesh, Akane, Åkerö, Alkmene AGM, Allington Pippin, Ambrosia, Anna, Annurca, Apollo, Ariane, Arkansas Black, Ashmead's Kernel AGM, Aurora Golden Gala, Autumn Glory, Bailey, Bardsey Island Apple, Beauty of Bath, Ben Davis, Beverly Hills, Bloody Ploughman, Bottle Greening, Braeburn, Bravo de Esmolfe, Breedon Pippin, Brina, Cameo, Carroll, Carter's Blue, Champion, Charles Ross, Chelmsford Wonder, Chiver's Delight, Clivia, Cornish Gilliflower, Cortland, Court Pendu Plat, Cox's Orange Pippin, Crimson Delight, Crimson Gold, Crispin, D'Arcy Spice, Delblush, Delcorf AGM, Delfloga, Delflopion, Delrouval, Deltana, Devonshire Quarreden, Discovery AGM, Dorsett Golden, Dougherty/Red Dougherty, Early Victoria, Egle, Egremont Russet AGM, Ein Shemer, Ellison's Orange AGM, Empire, Enterprise, Envy, Epicure, Fiesta AGM, Fireside, Flamenco, Florina, Flower of Kent, Fortune AGM/Laxton's Fortune, Fuji, Gala/Royal Gala AGM, Garden Royal, Gascoyne's Scarlet, Geheimrat Dr. Oldenburg, George Cave, Gloster, Golden Delicious AGM, Golden Noble AGM, Golden Orange, Golden Supreme, Goldspur, Gordon, Gradirose, Green Cheese, Greensleeves AGM, Hawaii, Herefordshire Russet, Heyer 12, Honeycrisp, Honeygold, Jazz/Scifresh, Jonagold AGM, Jupiter AGM, Kalmar Glasäpple, Kanzi/Nicoter, Katy, Kerry Pippin, Kidd's Orange Red AGM, King, King of the Pippins AGM, King Russet AGM, Lady Alice, Laxton's Superb, Liberty, Limelight, Liveland Raspberry Apple, Lodi, Lord Lambourne AGM, Macoun, Mantet, Margil, May Queen, Melba, Melon, Melrose, Merton Worcester, Miller's Seedling, Mollie's Delicious, Mother/American Mother, Mutsu, My Jewel, Newell-Kimzey/Airlie Red Flesh, Norfolk Royal, Opal, Orin, Orleans Reinette, Ozark Gold, Pacific Rose, Pam's Delight, Paula Red, Pink Pearl, Pixie AGM, Porter's, Prima, Pristine, Rajka, Red Delicious, Red Prince, Ribston Pippin AGM, Rubens/Civni, Santana, Saturn, Scrumptious, Smokehouse, Son ya, Splendour/Splendor, St. Edmund's Pippin AGM, Star of Devon, Stark Earliest, Streifling Herbst, Sturmer Pippin, Summerfree, Sunset AGM, Sweet Sixteen, SweeTango, Teser, Tydeman's Early Worcester, Tydeman's Late Orange, Wealthy, Westfield Seek-No-Further, Wijcik McIntosh, Worcester Pearmain AGM, Wyken Pippin.

Cultivars for cooking: Antonovka, Arthur Turner AGM, Baldwin, Ballyfatten, Beacon, Belle de Boskoop AGM, Birgit Bonnier, Bismarck, Blenheim Orange AGM, Bramley/Bramley's Seedling AGM, Calville Blanc d'Hiver, Carolina Red June, Catshead, Claygate Pearmain AGM, Cripps Pink/Pink Lady, Criterion, Duchess of Oldenburg, Dudley Winter, Dummellor's Seedling AGM, Edward VII AGM, Elstar AGM, Emneth Early AGM, Esopus Spitzenburg, Falstaff AGM, George Neal AGM, Ginger Gold, Glockenapfel,

Gragg/Aka Red Gragg, Granny Smith, Gravenstein, Grenadier AGM, Haralson, Harrison Cider Apple, Howgate Wonder, Idared AGM, Irish Peach, James Grieve AGM, Jonathan, Junaluska, Junami, Karmijn de Sonnaville, Lane's Prince Albert AGM, Lord Derby, Maiden's Blush, Malinda, Manks Codlin, McIntosh, Newton Wonder AGM, Newtown Pippin/Albemarle Pippin, Nickajack, Peasgood's Nonsuch AGM, Pinova, Pott's Seedling, Pound Sweet, Red Astrachan, Rev. W. Wilks, Rome Beauty, Roxbury Russet, Spartan, Warner's King AGM, White Transparent, Winter Queen, Wolf River, Zestar.

Cultivars for cider making: Baldwin, Brown Snout, Dabinett, Dymock Red, Ellis Bitter, Foxwhelp, Golden Russet, Golden Spire, Grimes Golden, Hagloe Crab, Hangdown, Kingston Black, Knobbed Russet, Muscadet de Dieppe, Newtown Pippin, Northern Spy, Redstreak, Rhode Island Greening, Ros Picant, Roxbury Russet, Slack-ma-Girdle, Snow apple (Fameuse), Stayman, Stoke Red, Styre, Tolman Sweet, Tom Putt, Topaz, Tremlett's Bitter, Twenty Ounce, Vista Bella, Wagener, Winesap, Woodcock, Yeovil Sour, York Imperial AGM.

RELATED SPECIES The genus Malus includes 71 taxa, most of which produce edible fruit: *Malus angustifolia* (Aiton) Michx., *Malus angustifolia* (Aiton) Michx. var. *puberula* (Rehder) Rehder, *Malus* × *arnoldiana* (Rehder) Rehder, *Malus asiatica* Nakai, *Malus baccata* (L.) Borkh., *Malus baccata* (L.) Borkh. var. *gracilis* (Rehder) T.C.Ku, *Malus bracteata* Rehder, *Malus chitralensis* Vassilcz., *Malus coronaria* (L.) Mill., *Malus coronaria* (L.) Mill. var. *dasycalyx* Rehder, *Malus daochengensis* C.L.Li, *Malus dasyphylla* Borkh., *Malus* × *dawsoniana* Rehder, *Malus domestica* Borkh., *Malus doumeri* (Bois) A.Chev., *Malus florentina* (Zuccagni) C.K.Schneid., *Malus floribunda* Siebold ex Van Houtte, *Malus fusca* (Raf.) C.K. Schneid., *Malus glabrata* Rehder, *Malus glaucescens* Rehder, *Malus halliana* Koehne, *Malus honanensis* Rehder, *Malus hupehensis* (Pamp.) Rehder, *Malus ioensis* (Alph.Wood) Britton, *Malus ioensis* (Alph. Wood) Britton var. *texana* Rehder, *Malus jinxianensis* J.Q.Deng & J.Y.Hong, *Malus kansuensis* (Batalin) C.K.Schneid., *Malus kansuensis* (Batalin) C.K. Schneid. var. *calva* (Rehder) T.C.Ku & Spongberg, *Malus kirghisorum* Al.Fed. & Fed., *Malus komarovii* (Sarg.) Rehder, *Malus lancifolia* Rehder, *Malus leiocalyca* S.Z.Huang, *Malus mandshurica* (Maxim) Kom. ex Juz., *Malus manshurica* (Maxim) Kom. ex Skvortsov, *Malus melliana* (Hand.-Mazz.) Rehder, *Malus micromalus* Makino, *Malus montana* Uglitzk., *Malus muliensis* T.C.Ku, *Malus niedzwetzkyana* Dieck ex Koehne, *Malus ombrophila* Hand.Mazz., *Malus orientalis* Uglitzk. ex Juz., *Malus pallasiana* Juz., *Malus platycarpa* Rehder, *Malus praecox* (Pall.) Borkh., *Malus prattii* (Hemsl.) C.K.Schneid., *Malus* × *prunifolia* (Willd.) Borkh., *Malus pumi la* Mill., *Malus pumila* Mill. f. *apetala* (Münchh.) C.K. Schneid., *Malus* × *purpurea* (E.Barbier) Rehder, *Malus* × *robusta* (Carrière) Rehder, *Malus rockii* Rehder, *Malus sachalinensis* Kom. ex Juz., *Malus sargentii* Rehder, *Malus sieboldii* (Regel) Rehder, *Malus sieboldii* (Regel) Rehder var. *zumi* (Matsum) Asami, *Malus sieversii* (Ledeb.) M.Roem., *Malus sikkimensis* (Wenz.) Koehne ex C.K.Schneid., *Malus soulardii* (L.H.Bailey) Britton, *Malus* × *spectabilis* (Sol.) Borkh., *Malus spontanea* (Makino) Makino, *Malus sylvestris* (L.) Mill., *Malus toringo* (Siebold) Siebold ex de Vriese, *Malus toringoides* (Rehder) Hughes, *Malus transitoria* (Batalin) C.K.Schneid., *Malus transitoria* (Batalin) C.K.Schneid. var. *centralasiatica* (Vassilcz.) T.T.Yu, *Malus trilobata* (Labill. ex Poir.) C.K.Schneid., *Malus tschonoskii* (Maxim) C.K. Schneid., *Malus turkmenorum* Juz. & Popov, *Malus yunnanensis* (Franch.) C.K.Schneid., *Malus yunnanensis* (Franch.) C.K.Schneid. var. *veitchii* Rehder, *Malus zumi* (Matsum) Rehder.

GEOGRAPHIC DISTRIBUTION Complex of hybridized forms between *Malus sylvestris* and Near Eastern species stabilized by cultivation. **HABITAT** Cultivated. **RANGE** China, United States, Turkey, Poland, Italy.

BEER-MAKING PARTS (AND POSSIBLE USES) Fruits (pulp, juice), fermented juice (cider), wood (for smoking malt).

CHEMISTRY Fruits (including peel): energy 218 kJ (52 kcal), water 85.56 g, carbohydrates 13.81 g, sugars 10.39 g, fiber 2.4 g, lipids 0.17 g, protein 0.26 g, vitamin A 3 μg, beta-carotene 27 μg, zeaxanthin lutein 29 μg, thiamin (B1) 0.017 mg, riboflavin (B2) 0.026 mg, niacin (B3) 0.091 mg, pantothenic acid (B5) 0.061 mg, vitamin B6 0.041 mg, folate (B9) 3 μg, vitamin C 4.6 mg, vitamin E 0.18 mg, vitamin K 2.2 μg, calcium 6 mg, iron 0.12 mg, magnesium 5 mg, manganese 0.035 mg, phosphorus 11 mg, potassium 107 mg, sodium 1 mg, zinc 0.04 mg, fluoride 3.3 μg.

WOOD Reddish-colored wood, fine and homogeneous texture, with mostly straight grain, WSG 0.75. Suitable for fine work, scoring, turnery, veneering, woodcutting.

Style Fruit Beer, Smoked Ale, Sour Ale.

Beer Dike, Birrificio Pinerolese (Pinerolo, Piedmont, Italy).

Source Calagione et al. 2018; Cantwell and Bouckaert 2016; Cantwell and Bouckaert 2018; Daniels and Larson 2000; Giaccone and Signoroni 2017; Heilshorn 2017; Hieronymus 2010; Hieronymus 2016b; Josephson et al. 2016; McGovern 2017; https://www.birramoretti.it/le-birre-di-casa/birra-moretti-alla-friulana/; https://www.bjcp.org (29. Fruit Beer; 32A. Classic Style Smoked Beer); https://www.brewersfriend.com/other/applewood/.

Malus fusca (Raf.) C.K.Schneid.
ROSACEAE

Synonyms *Malus diversifolia* (Bong.) M.Roem., *Pyrus fusca* Raf., *Pyrus rivularis* Douglas ex Hook.

Common Names Crabapple, Oregon crabapple, Pacific crabapple.

Description Tree or shrub (5–20 m). Stem Ø 12–40 cm; bark gray to reddish-brown, smooth when young, scaly and deeply fissured when mature; young branches reddish, puberulent when young, then reddish-brown or gray and glabrous; lateral flowering shoots become spurs (10–50 mm). Buds red-brown, ovoid, 1.5–4 mm, perulae margins ciliate. Leaves conduplicate in buds, isomorphic; stipules deciduous, narrowly lanceolate, 1–5 mm, apex acuminate; petiole 10–30 mm, tomentose, glabrescent; lamina ovate, sometimes ovate, elliptic, or lanceolate (3–11 × 1–4 cm), base rounded to cuneate, margins entire or 3-lobed, serrate to doubly serrate, sometimes serrate, apex acute or acuminate, underleaf glabrous, upper leaf puberulent. Inflorescence panicle corymbiform; peduncle absent; bracts absent; bracteoles sometimes persistent, filiform, 1–4 mm; pedicels 15–40 mm, villous or glabrous. Flowers Ø 15–20 mm; hypanthium glabrous or tomentose; sepals triangular, 3–6 mm, shorter than tube, apex apiculate, lower surface glabrous, upper surface canescent-tomentose; petals white, sometimes pink, orbiculate to obovate (6–15 mm), claw 1.5–2 mm, margins eroded or undulate, apex rounded; about 20 stamens (4–6 mm), anthers white before dehiscence; styles 3–4, connate in the proximal 1/3, 6–7 mm, longer than stamens, glabrous. Fruit pome color yellow to violet-red, oblong, sometimes ovoid or obovoid (Ø 6–13 mm); sepals deciduous, sometimes late. Seeds reddish-brown in color.

Related Species The genus *Malus* includes 71 taxa, most of which produce edible fruits.

Geographic Distribution Canada (British Columbia), United States (Alaska, California, Oregon, Washington). **Habitat** Open, moist places, open forests of *Picea sitkensis*, forest mantles, beach edges, sea cliffs, swamps, salt marshes, glades (0–600 m ASL). **Range** Corresponding to geographic distribution.

Beer-Making Parts (and Possible Uses) Fruits.

Chemistry No information available.

Style Fruit Beer.

Source Heilshorn 2017; Hieronymus 2016b.

Malus ioensis (Alph.Wood) Britton
ROSACEAE

Synonyms *Pyrus ioensis* (Alph.Wood) Carruth, *Pyrus ioensis* (Alph.Wood) L.H.Bailey.

Common Names Crabapple, prairie crabapple, Iowa crabapple, western crabapple.

Description Tree or shrub (2–10 m). Stem Ø 6–45 cm; bark reddish-brown to gray, with narrow reddish-brown scales, sometimes initially glabrous; young branches reddish-gray, tomentose, then reddish-brown or dark brown and glabrous or slightly puberulent; flowering branches spurred or spiny (10–60 mm). Buds reddish-brown, ovoid, 3–4 mm, margins of scales tomentose. Leaves conduplicate in buds, heteromorphic; stipules deciduous, narrowly lanceolate, 3–8 mm, apex acute; sterile shoot leaves vigorous: petiole 15–30 mm, tomentose; lamina ovate, sometimes triangular-ovate or ovate, 3–12 × 1.5–6 cm, base cuneate to rounded, margins usually lobed, serrate, crenate-serrate, or doubly serrate, apex acute, broadly acute or rounded, underleaf tomentose (villous only in veins), upper leaf glabrous or slightly villous; leaves of fertile shoots: petiole 5–25 mm, tomentose, sometimes puberulent; lamina elliptic, ovate or ovoid (2–5 × 1–2 cm), base cuneate or rounded, sometimes truncate-rounded, margins lobed or nonlobed, serrate, doubly serrate, crenate-serrate, apex acute, broadly acute or rounded, underleaf usually tomentose, upper glabrous. Inflorescence panicle corymbiform; peduncles absent; bracts sometimes persistent, filiform, 3–10 mm. Pedicels 15–30 mm, tomentose, sometimes puberulent or sparsely strigose. Flowers fragrant, Ø 35–50 mm, hypanthium tomentose; sepals triangular, 3–7 mm, equal to tube, apex acuminate, surfaces canescent-tomentose; petals pink, sometimes turning white, oblong to narrowly obovate (11–22 mm), claw 2–4 mm, margins entire, sinuate or fimbriate, apex rounded; 20 stamens (8–16 mm), anthers emerge before dehiscence; 5 stigmas, basally connate, 9–15 mm, ± equal to stamens, tomentose in proximal 1/3. Fruit pome color green, sometimes yellow, depressed-globose or globose, Ø 20–30 mm, waxy, sepals persistent, erect. Seeds dark brown in color.

Subspecies This taxon has two varieties: *Malus ioensis* (Alph.Wood) Britton var. *ioensis* and *Malus ioensis* (Alph.Wood) Britton var. *texana* Rehder.

Cultivar Plena (Bechtel's crab) is an example of the few known belonging to this species.

Related Species The genus *Malus* includes 71 taxa, most of which produce edible fruit.

Geographic Distribution United States (Arkansas, Illinois, Indiana, Iowa, Kansas, Kentucky, Louisiana, Michigan, Minnesota, Missouri, Nebraska, Ohio, Oklahoma, South Dakota, Texas, Wisconsin). **Habitat** Woodlands, thickets, salt marshes, hills, banks and levees, floodplains, roadsides, open fields, fence lines (100–300 m ASL). **Range** North America, sporadically grown elsewhere as an ornamental plant for its beautiful late spring blooms; cultivars with double flowers have also been selected for this purpose (Bechtel's crab, f. *fimbriata* A.D. Slavin).

Beer-Making Parts (and Possible Uses) Fruits (extremely acrid taste and high pectin content that allows the production of yellow jelly).

Chemistry No information available.

Style Fruit Beer.

Source Heilshorn 2017; Hieronymus 2016b.

Malus sylvestris (L.) Mill.
ROSACEAE

Synonyms *Malus acerba* Mrat, *Malus communis* L. subsp. *sylvestris* (L.) Gams, *Malus malus* (L.) Britton, *Pyrus acerba* (Mrat) DC., *Pyrus malus* L. var. *sylvestris* L.

Common Names Crab apple.

Description Tree, small (3–10 m). Trunk slender, more or less straight; main branches strong and patent; twigs initially greenish, then reddish-brown, hardened at the apex, usually prickly; crown globular and densely leafy; bark gray-brown, rather smooth in youth, flaking into plates at maturity; buds spirally arranged, ovate-rounded and blunt, with many red-brown perulae. Leaves alternate, simple, lamina coriaceous ovate-rounded (25–40 × 35–80 mm), dark opaque green, glabrous on both sides (initially slightly pubescent on the underleaf), apex attenuated acute, margin densely and minutely serrated (mostly doubly so) or crenate; petiole usually shorter than the lamina, at most as long as the same, pubescent only when young; secondary veins curved and converging toward the apex, very evident below. Flowers hermaphrodite in umbrella-shaped cymes or pauciflor corymbs (3–7 flowers) borne by brachyblasts, with peduncles glabrous or slightly tomentose, sepals (up to 7 mm) acuminate, patent, or, more often, folded back, internally hairy-felted; petals 5, subrounded (10–15 mm), usually not overlapping, white (only in buds, sometimes pinkish externally); stamens with yellow anthers; ovary with 5 stigmas joined together at base. Fruit pome globose or globose-ovoid (Ø 2–3 cm), greenish-yellow in color, sometimes with reddish streaks, fragrant when ripe but sour-tasting and inedible, doubly umbilicate, with residual erect-patent calyxes; pulp lacking stony cells (sclereids). Seeds about 10, ovoid-compressed, white internally, covered at maturity with a brown integument.

Related Species The genus *Malus* includes about 71 taxa, of which some have widespread food use and others also beer-making applications.

Geographic Distribution C Europe, European-Caucasian. **Habitat** Mesophilous, heliophilous species, solitary individuals in pure hardwood forests or mixed with conifers, at the margins or in clearings, in loamy or sandy soils rich in humus and well drained, in harsh winter climates without late frosts at flowering (0–1,400 m ASL). **Range** Cultivated in Europe and on other continents (where it sometimes tends to grow wild), especially as rootstock of *Malus domestica*.

Beer-Making Parts (and Possible Uses) Fruits.

Toxicity Seeds contain modest quantities of cyanide and can cause poisoning, potentially serious, if consumed in high doses.

Chemistry Fruits: water 80%, ash 1.0%, lipids 0.18%, pectin 1.3%, fiber 3.2%, protein 0.2%, carbohydrates 8.4%, calcium 19.8 mg, sodium 2.8 mg, potassium 213 mg, phosphorus 16.1 mg, magnesium 7.8 mg, iron 4.0 mg, vitamin C 8.1 mg, vitamin A 0.92 mg.

Style American Wheat or Rye Beer.

Source https://www.brewersfriend.com/other/crabapple/.

Malus toringo (Siebold) Siebold ex de Vriese
ROSACEAE

Synonyms *Malus hupehensis* Koidz., *Sorbus toringo* Siebold.

Common Names Crabapple, Japanese crabapple, Toringo crabapple, Siebold crabapple.

Description Tree, sometimes shrub (3–10 m). Trunk up to 30 cm in diameter; bark purplish-brown, smooth, irregularly fissured with time; young branches orange-brown, puberulent, becoming dark purple or purplish-brown and glabrous; flowering shoots (5–35 mm) form spurs, rarely spines. Buds purplish-brown, ovoid, 20–30 mm, glabrous or only the terminal perulae with puberulent margins. Leaves conduplicate in buds, isomorphic; stipules deciduous, lanceolate, 4–6 mm, apex acute or acuminate; petiole 10–25 mm, puberulent; lamina narrowly elliptic, elliptic, or ovate (3–7 × 2–4 cm), base broadly cuneate or rounded, margins usually 3-lobed, serrate, apex acute, underleaf puberulent when young, veins puberulent when mature, upper leaf glabrous. Inflorescences umbelliform panicles; peduncles absent; bracts deciduous, filiform, 4–5 mm. Pedicels 20–25 mm, pubescent or subglabrous. Flowers Ø 20–30 mm; hypanthium subglabrous or pubescent; sepals lanceolate, 6–9 mm, up to 2 times longer than tube, apex caudate-acuminate, underleaf glabrous, upper tomentose; petals white, sometimes pink-white, elliptic-obovate, 15–18 mm, claw 1–2 mm, margins entire, sometimes crenulated, apices rounded; 20 stamens (7–9 mm), anthers yellow before dehiscence; stigma 3 (or 4–5), basally connate to 1/2 the length, 8–10 mm, slightly longer than stamens, proximally villous or woolly. Fruit pome red or brownish-yellow, subglobose, Ø 6–10 mm; sepals deciduous. Seeds reddish-brown in color.

Related Species The genus *Malus* includes 71 taxa, most of which produce edible fruits.

Geographic Distribution Temperate Asia (China, Japan, Korea). **Habitat** Slopes, mixed forest, shrubland (100–2,000 m ASL). **Range** Asia, North America (naturalized), Europe (naturalized).

Beer-Making Parts (and Possible Uses) Fruits.

Chemistry No information available.

Style Fruit Beer.

Source Heilshorn 2017; Hieronymus 2016b.

Mandragora officinalis Mill.
SOLANACEAE

Synonyms *Atropa mandragora* Sm., *Mandragora autumnalis* Bertol.

Common Names Mandrake.

Description Herb, perennial (10–30 cm), fetid odor. Root coarse, blackish, taproot, often branched, with vaguely anthropomorphic appearance. Stem null or very short. Leaves all in basal rosette (2–3 × 5–7 cm) at flowering, later elongated, briefly petiolate, subglabrous, wrinkled-reticulate; lamina oblanceolate-spatulate, entire or notched-rounded at margin, apex acuminate, midrib thickened. Hermaphrodite flowers actinomorphic, solitary, in groups of 12–60 in the center of the rosette, on pubescent peduncles (1.5–2 cm), increasing at fruiting; calyx gamosepalous, turbinate, hirsute, increasing and persistent at fruiting, tube 5 mm divided into 5 triangular or linear lobes of 6–8 mm; corolla gamopetalous, campanulate (1.8–6 cm), pale blue-purple, with reticulate veins, 5 triangular broad lobes, tube funnel shaped (2–2.5 cm); androecium 5 stamens soldered to corolla tube, hairy filaments at base; stigma 1, longer than stamens; stigma bilobate or capitate, yellow; ovary superior biloculate. Fruit berry ellipsoid (16–25 × 13–21 mm), yellow-orange, fetid, blackish color when dry, containing numerous small ± reniform, alveolate seeds.

Related Species In addition to the taxon reported here, the genus *Mandragora* includes two recognized species: *Mandragora caulescens* C.B.Clarke and *Mandragora officinarum* L.

Geographic Distribution Steno-Mediterranean.
Habitat Fields, arid and sunny wastelands, preferably on calcareous substratum (0–600 m ASL).
Range Usually not cultivated.

Beer-Making Parts (and Possible Uses) Leaves (?), fruits (?).

Toxicity Extremely toxic plant owing to the presence of a complex system of alkaloids.

Chemistry The main active chemical components of mandrake are the following alkaloids: scopolamine (most prevalent component), atropine, apoatropine, hyoscyamine, mandragorine. The total alkaloid content varies among plant organs: stem low levels, fresh root 0.3–0.4%, dried root 0.2–0.6%, fruit low levels, dried leaves medium levels.

Style Old Ale.

Source Heilshorn 2017; https://www.brewersfriend.com/other/mandrake/.

Mangifera indica L.
ANACARDIACEAE

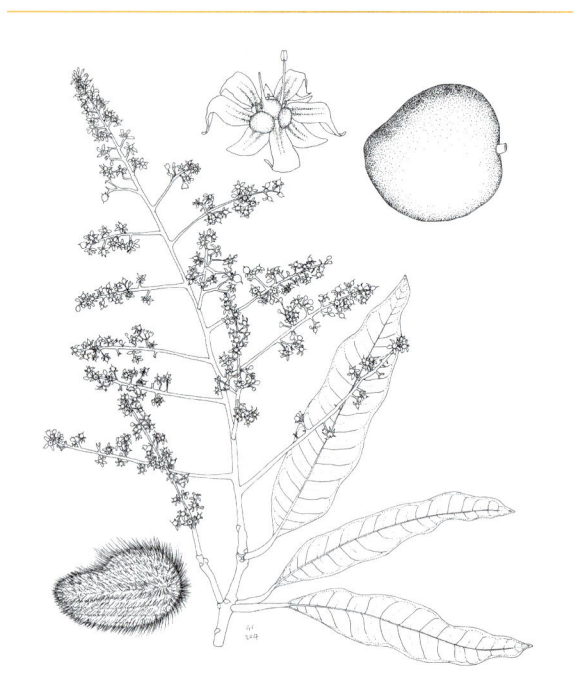

Synonyms *Mangifera austroyunnanensis* Hu.

Common Name Mango.

Description Tree, evergreen (15–30 m), long-lived. Leaves alternate, simple, leathery, oblong-lanceolate (29–30 × 3–5 cm) on flowering branches, up to 50 cm on sterile branches. Young leaves red, aging bright green above and paler below, with white or yellow veins. Inflorescence branched panicle with very small (4 mm) greenish-white or greenish-pink flowers; male and hermaphrodite flowers borne by the same tree; flowers actinomorphic, usually with 5 petals, mottled with red; usually only 1 fertile stamen per flower, the other 4 sterile; flower has a 5-lobed disc between petals and stamens. Fruit drupe irregularly ovate and slightly compressed, 8–30 cm long, attached to the thickest end of the peduncle; epicarp (skin) smooth, greenish-yellow, sometimes colored red, the underlying yellow-orange flesh varies in texture, when ripe sweet, soft, juicy, and fibreless (in cultivated varieties). Single seed, compressed-ovoid, embedded in the white fibrous inner layer of the fruit.

Cultivar There are hundreds of cultivars, among which Tommy Atkins excels, more for its resistance to damage during transport than for its organoleptic qualities. Other varieties, divided by country of selection or cultivation, are found in Australia (B74, Brooks, Haden, Irwin, Keitt, Kensington Pride, Kent, Nam Doc Mai, Palmer, R2E2), Bangladesh (Ashini, Fazli, Himsagar, Khirshapat, Langra, Lokhon-bhog, Raj-bhog), Burma (Aug Din, Ma Chit Su, Sein Ta Lone, Shwe Hin Tha), Brazil (Coquinho, Haden, Manga Espada, Manga Rosa, Palmer, Tommy Atkins), Cambodia (Cambodian), Cameroon (Améliorée du Cameroun), Caribbean (Amélie, Black [Blackie], Bombay, Dou-douce, East Indian, Graham, Haden, Julie [St. Julian], Long, Madame Francis, Mango blanco, Rose, Spice-Box, Starch), China (Baiyu, Guixiang, Huangpi, Huangyu, Macheco, Sannian, Yuexi), Costa Rica (Haden, Irwin, Keitt, Mora, Tommy Atkins), Ecuador (Ambassador, Alfonso, Ataulfo, Criollos, Haden, Julie, Keitt, Kent, Reina, Tommy Atkins), Egypt (Alfonso, Beid El Agl, Fuss Oweis, Hindi, Hindi Besennara, Oweisi, Taymoor, Zebdiah), Philippines (Apple, Carabaoo Kinalabaw, Indian, Paho, Pahohutan, Piko), Guatemala (Haden, Kent, Tommy Atkins), Haiti (Baptiste, Corne, Fil, Francine [Madame Francis], Labiche, Muscas, Poirier, Rosalie), India (Alampur Baneshan, Alphonso, Amrapali, Badami, Badshahpasand, Bangalora, Banganapalli, Bombay, Bombay Green, Chausa, Dusehri, Cheruku Rasalu, Chinna Rasalu, Fajli, Fajri Kalan, Fernandian, Gaddamar, Gulabkhas, Himayath, Himsagar, Husan-Nara, Imam Pasand, Jehangir, Kalami, Kesar, Kishen Bhog, Komanga, Lalbaug, Langra, Langra Benarsi, Maldah, Malgis, Mallika, Mankur, Mankurad, Moovandan, Mulgoba, Nattuma, Neelum, Ottu Mangai, Pairi, Pedda Rasalu, Priyor, Rajapuri, Raspuri, Ratna, Roomani, Safeda, Sammar Bahisht, Surkha, Suvarnarekha, Totapuri, Vanraj, Zardalu), Indonesia (Arumanis, Gadung, Golek, Manalagi), Israel (Haden, Keitt, Kent, Maya, Nimrod, Palmer, Tommy Atkins), Italy (Glenn, Keitt, Kensington Pride, Kent, Maya, Osteen, Tommy Atkins, Van Dyke), Kenya (Batwi, Boubo, Ngowe), Malaysia (Apple Mango, Apple Rumani, Arumanis, Golek, Kuala Selangor, Maha-65, Malgoa, Tok Boon), Mali (Amelie, Kent), Mexico (Ataulfo, Criollo, Haden, Irwin, Kent, Manila, Niño, Oro, Palmer, Petakon, Sensation, Tommy Atkins, Van Dyke), Pakistan (Alphonso, Anwar Rataul, BaganPali, Chausa, Desi, Dusehri, Fajri, Langra, Muhammad Wole, Neelum, Saroli, Sindhri), Peru (Criollos, Haden, Keitt, Kent, Tommy Atkins), Reunion (Auguste, Carotte, Jose, Lucie), Singapore (Apple Mango, Arumanis, Golek, Kaem Yao, Mangga Dadol), United States (Florida: Alampur Baneshan, Alice, Alphonso, Anderson, Angie, Bailey's Marvel, Bennet Alphonso, Beverly, Bombay, Brahm Kai Meu, Brooks, Carabao, Carrie, Chok Anon, Cogshall, Cushman, Dot, Duncan, Earlygold, East Indian, Edward, Eldon, Emerald, Fairchild, Fascell, Florigon, Ford, Gary, Gaylour, Glenn, Gold Nugget, Golden Lippens, Graham, Haden, Hatcher, Ice Cream, Irwin, Ivory, Jakarta, Jean Ellen, Julie, Keitt, Kensington Pride, Kent, Lancetilla, Langra Benarsi, Lippens, Mallika, Manilita, Mendoza, Mulgoba, Nam Doc Mai, Nam Tam Teen, Neelum, Nu Wun Chan, Okrung, Osteen, Palmer, Parvin, Pascual, Philippine, Pickering, Po Pyu Kalay, Rosigold, Ruby, Rutledge, Saigon, Sensation, Sophie Fry, Southern Blush, Spirit of '76, Springfels, Sunset, Suwon Tip, Tebow, Toledo, Tom Dang, Tommy Atkins, Torbert, Turpentine, Valencia Pride, Van Dyke, Zill; Hawaii: Gouveia, Hawaiian Common, Hawaiian Dwarf, Kurahige, Mapulehu, Momi K, Pope, Rapoza, Sugai, Turpentine), South Africa (Fascell, Haden, Keitt, Kent, Sensation, Tommy Atkins, Zill), Sudan (Alfonso, Bez el-Anza, Oweisi, Taymoor), Sri Lanka (Dampara, Hingurakgoda, Karutha Colomban, Malwana amba, Parrot Mango and Peterpasand, Petti amba, Rata amba, Vellai Colomban, Willard), Tanzania (Boribo Muyini, Dodo, Mawazo, Sindano), Taiwan (JinHwang, Red JinHwang, Tainong N°1), Thailand (Brahm Kai Meu, Khaew Sawei, Nam Dok Mai, Okrong, Rad), Venezuela (Haden, Keitt, Kent, Tommy Atkins), Vietnam (Bình Định Elephant mango, Cao Lãnh Cát Chu mango, Hoà Lộc Sand mango).

Related Species Almost all species of the genus *Mangifera* have edible fruits, including *Mangifera caesia* Jack, *Mangifera casturi* Kosterm., *Mangifera laurina* Blume, *Mangifera lineariflia* (Mukh.) Kosterm., *Mangifera odorata* Griff., *Mangifera persiciforma* C.Y.Wu & T.L.Ming, *Mangifera rubropetala* Kosterm., *Mangifera siamensis* Warb. ex W. G. Craib, and *Mangifera sylvatica* Roxb.

Geographic Distribution S Asia (Bangladesh, India, Pakistan). **Habitat** Cultivated in tropical areas. **Range** India, China, Thailand, Indonesia, Mexico, Pakistan, Brazil.

Beer-Making Parts (and Possible Uses) Fruits.

Chemistry Fruits (pulp): energy 250 kJ (60 kcal), carbohydrates 15 g, sugars 13.7 g, fiber 1.6 g, lipids 0.38 g, protein 0.82 g, vitamin A 54 µg, beta-carotene 640 µg, lutein zeaxanthin 23 µg, thiamine (B1) 0.028 mg, riboflavin (B2) 0.038 mg, niacin (B3) 0.669 mg, pantothenic acid (B5) 0.197 mg, vitamin B6 0.119 mg, folate (B9) 43 µg, choline 7.6 mg, vitamin C 36.4 mg, vitamin E 0.9 mg, vitamin K 4.2 µg, calcium 11 mg, iron 0.16 mg, magnesium 10 mg, manganese 0.063 mg, phosphorus 14 mg, potassium 168 mg, sodium 1 mg, zinc 0.09 mg.

STYLE American Amber Ale, American Barleywine, American IPA, American Light Lager, American Pale Ale, American Wheat Beer, American Wheat or Rye Beer, Belgian Specialty Ale, Berliner Weisse, Blonde Ale, California Common Beer, Classic American Pilsner, Cream Ale, Double IPA, Dry Stout, Fruit Beer, Fruit IPA, Fruit Lambic, German Pilsner (Pils), Gueuze, Imperial IPA, Kölsch, Mixed-fermentation Sour Beer, Saison, Specialty Beer, Specialty IPA (Belgian IPA), Spice/Herb/Vegetable Beer, Weizen/Weissbier, Witbier.

BEER Felix, Kashmir (Filignano, Molise, Italy).

SOURCE Calagione et al. 2018; Hieronymus 2016b; https://www.bjcp.org (29. Fruit Beer); https://www.brewersfriend.com/other/mango/; https://www.craftbeer.com/news/beer-release/oregon-fruit-products-collaborates-with-breakside-brewery; https://www.omnipollo.com/beer/white/.

Manihot esculenta Crantz
EUPHORBIACEAE

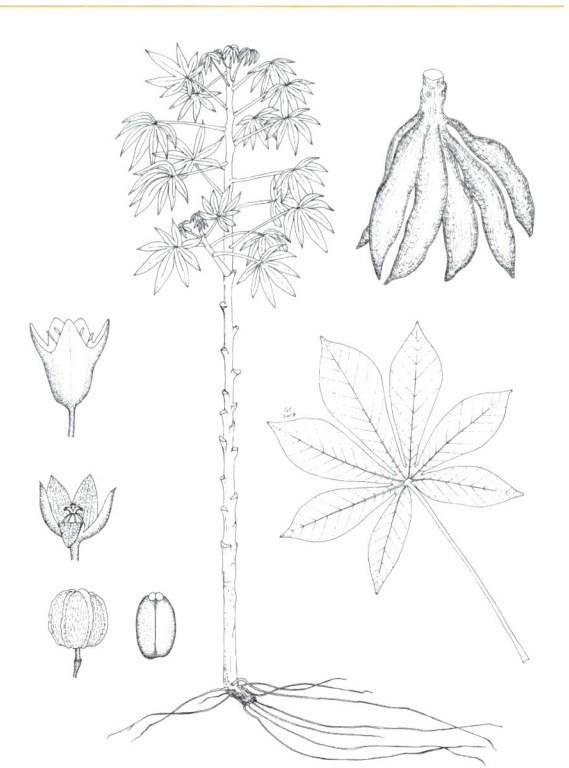

SYNONYMS *Janipha aipi* (Pohl) J.Presl, *Janipha manihot* (L.) Kunth, *Jatropha aipi* (Pohl) A.Moller, *Jatropha diffusa* (Pohl) Steud., *Jatropha digitiformis* (Pohl) Steud., *Jatropha dulcis* J.F.Gmel., *Jatropha flabellifolia* (Pohl) Steud., *Jatropha loureiroi* (Pohl) Steud., *Jatropha manihot* L., *Jatropha mitis* Rottb., *Jatropha paniculata* Ruiz & Pav. ex Pax, *Jatropha silvestris* Vell., *Jatropha stipulata* Vell., *Mandioca aipi* (Pohl) Link, *Mandioca dulcis* (J.F.Gmel.) D.Parodi, *Mandioca utilissima* (Pohl) Link, *Manihot aipi* Pohl, *Manihot aypi* Spruce, *Manihot cannabina* Sweet, *Manihot diffusa* Pohl, *Manihot digitiformis* Pohl, *Manihot dulcis* (J.F.Gmel.) Pax, *Manihot dulcis* (J.F.Gmel.) Baill., *Manihot edule* A.Rich., *Manihot edulis* A.Rich., *Manihot flabellifolia* Pohl, *Manihot flexuosa* Pax & K.Hoffm., *Manihot loureiroi* Pohl, *Manihot melanobasis* Müll.Arg., *Manihot sprucei* Pax, *Manihot utilissima* Pohl.

COMMON NAMES Cassava, yuca, manioc, mandioca, Brazilian arrowroot.

DESCRIPTION Shrub or tree (up to 7 m), semi-woody, often grown as an annual and propagated vegetatively once tubers are harvested. Stem single or few, sparsely branched; branches light green to reddish, nodes reddish. Bark smooth, yellowish-gray to creamy-green. Leaves with long petiole (up to 60 cm), light green to red; lamina basally attached or slightly peltate (2 mm), dark green above, grayish green below; lamina deeply divided into 3–7 narrow lobes (2.9–12.5 times as long as wide), central part not lobed, usually short. Inflorescence lax, with 3–5 flowers in fascicles; peduncles green to red. Flowers inconspicuous, male flowers with calyx divided about halfway, green with white to reddish lobes, white filaments, yellow anthers; female flowers with green calyx with red margins and midrib; ovary with 6 longitudinal connections, green with pink to orange stripes; pistil and stigma white. Fruit capsule subglobose, small (Ø about 1 cm), light yellow, white, dark brown, 6-angled, with 6 longitudinal wings. Seed up to 12 mm long. Root tubers (in cultivated varieties take 9–18 months to reach harvest size), in groups of 4–8 at the base of the stem, elongated, 15–30 × 5–10 cm.

CULTIVAR Cultivated varieties of this species have been selected in each of the traditional growing areas.

Related Species Approximately 128 taxa belonging to the genus *Manihot* are recognized today.

Geographic Distribution South America.
Habitat Species cultivated in equatorial areas.
Range Nigeria, Thailand, Indonesia, Brazil, Democratic Republic of Congo.

Beer-Making Parts (and Possible Uses) Tubers.

Toxicity Fresh roots, skin, and leaves contain cyanide compounds (linamarin, lotaustralin, hydrocyanic acid) at potentially toxic levels. Consumption must therefore be preceded by a rather laborious treatment, which can include roasting, soaking, fermentation (peeling tubers and boiling however seem sufficient to remove toxic compounds). Toxic elements are decomposed by linamarase, an enzyme naturally present in the plant, releasing hydrogen cyanide (HCN). According to the quantity of cyanogenic glycosides, cassava varieties are classified as sweet (20 mg HCN per kg) or bitter (50 times as much; i.e., 1 g/kg). Cultivation conditions (e.g., drought) can influence the presence of toxins; 25 mg of cyanogenic glycoside (about 2.5 mg of cyanide) is enough to kill a mouse; 50–60 mg is the lethal dose for an adult human being. Various illnesses are linked to cyanide excess, including acute intoxication (dizziness, vomiting, collapse, death; treatable by injection of thiosulfate, which converts cyanide to thiocyanate), goiter, ataxia (neurological disorder affecting the ability to walk), and chronic pancreatitis. However, societies that traditionally consume cassava have learned that certain treatments (soaking, cooking, fermentation, etc.) make the root edible. There are several techniques, though not all equally effective, to reduce the danger of this food. The most used in South America consists of flouring the roots, mixing the flour with water to form a sort of dough, then putting it to rest in a thin layer in a container for some hours, during which the linamarase converts cyanogenic glycosides into hydrocyanic acid, which is released into the atmosphere, making the flour safe for human consumption. In W Africa, people allow the roots to ferment for a few days before cooking. Projects are underway to select varieties completely free of toxins and rich in vitamins and minerals.

Chemistry Energy 1,507 kJ (360 kcal), water 12.6 g, protein 0.6 g, lipids 0.2 g, carbohydrates 86.3 g, trace soluble sugars, fiber 0.4 g, sodium 4 mg, potassium 20 mg, iron 1 mg, calcium 12 mg, magnesium 21 mg, phosphorus 12 mg, riboflavin 0.1 mg, vitamin A 13 IU, vitamin C 20.6 mg, vitamin B6 0.1 mg.

Style Cauim, Cicha, Kasiri (also Kaschiri or Cassava Beer), Nihamanci (also Nijimanche or Nijiamanchi), and Sakurá are traditional beers of the indigenous people of South America whose fermentation is propitiated by the enzyme amylase contained in human saliva.

Source Jackson 1991; Josephson et al. 2016; McGovern 2017; https://en.wikipedia.org/wiki/Nihamanchi; https://www.encyclopedia.com/topic/cassava.aspx.

Marrubium incanum Desr.
LAMIACEAE

Synonyms None

Common Names Horehound, common horehound, white horehound, common white horehound, green pompon horehound.

Description Herb, perennial (up to 50 cm) covered with dense woolly-candid tomentum, faint odor. Stems ascending, usually simple. Leaves petiolate (2 cm), lamina elliptic, toothed at margin, with sunken nerves above and protruding below. Inflorescences dense whorls, 20- to 25-flowered, at the axil of normal leaves. Flowers with linear bracts (8–9 mm); calyx with 7–8 mm tube and 5 teeth at the end diverging (2–6 mm); corolla (13–14 mm), projecting from calyx teeth, milky-white in color.

Related Species The genus *Marrubium* includes more than 50 species.

Geographic Distribution NE Mediterranean **Habitat** Arid and uncultivated pastures (0–1,200 m ASL). **Range** NE Mediterranean.

Beer-Making Parts (and Possible Uses) Flowering tops.

Chemistry Essential oil (99.8% of total): limonene 0.48%, geijerene 0.53%, carvacrol 7.47%, α-copaene 1.02%, β-bourbonene 0.52%, β-elemene 0.05%, E-caryophyllene 29.57%, β-copaene 0.55%, α-humulene 11.08%, γ-muurolene 0.05%, germacrene D 28.75%, bicyclogermacrene 0.15%, germacrene A 3.47%, γ-cadinene 0.05%, δ-amorphene 1.32%, caryophyllene oxide 1.15%, α-cadinol 13.59%.

Style Herb Beer.

Source Fisher and Fisher 2017; Heilshorn 2017.

Marrubium vulgare L.
LAMIACEAE

Synonyms *Marrubium apulum* Ten., *Marrubium ballotoides* Boiss. & Balansa, *Marrubium germanicum* Schrank ex Steud., *Marrubium hamatum* Kunth, *Marrubium uncinatum* Stokes, *Marrubium vaillantii* Coss. & Germ.

Common Names White horehound, common horehound.

Description Herb, perennial (2–4 dm), gray or woolly-white in color, fragrant. Stems erect, simple. Leaves with 1–2 cm petiole progressively enlarged into an oval, rounded or kidney-shaped lamina (the largest 2–3 × 2 cm), irregularly crenate at the edge, with sinking veins (emerging below). Inflorescences 20- to 30-flowered whorls, forming spaced subspherical glomerules at the axil of normal leaves; bracts 2–3 mm, linear; calyx with 3–5 mm tube and 10 2 mm subspinous teeth; corolla whitish, 6–7 mm.

Related Species The genus *Marrubium* includes more than 50 species.

Geographic Distribution Euro-Mediterranean-Siberian, later subcosmopolitan. **Habitat** Grasslands, ruins, arid pastures. **Range** Subcosmopolitan.

Beer-Making Parts (and Possible Uses) Aerial flowering parts.

Chemistry Essential oil (50 compounds, 82.46% of oil): 4,8,12,16-tetramethyl heptadecan-4-olide 16.97%, germacrene D-4-ol 9.61%, α-pinéne 9.37%, phytol 4.87%, dehydro-sabine ketone 4.12%, piperitone 3.27%, δ-cadinene 3.13%, 1-octen-3-ol 2.35%, benzaldehyde 2.31%.

Style American Brown Ale, Extra Special/Strong Bitter (ESB), Gruit, Irish Red Ale, Specialty Beer, Spice/Herb/Vegetable Beer.

Source Fisher and Fisher 2017; Hieronymus 2016b; Steele,2012; https://birrapertutti.wordpress.com/2016/03/15/blanche/; https://www.brewersfriend.com/other/horehound/; https://it.wikipedia.org/wiki/Marrubium_vulgare; https://www.guidabirreartigianali.it/birre-blanche-o-bianche.html.

Matricaria chamomilla L.
ASTERACEAE

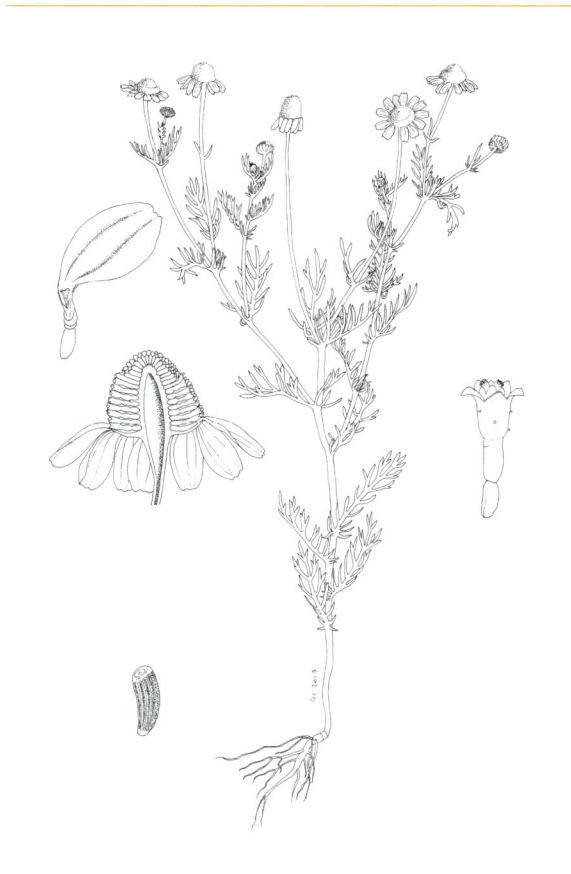

Synonyms *Chamaemelum suaveolens* E.H.L.Krause, *Chamaemelum vulgare* Bubani, *Chamomilla courrantiana* (DC.) K.Koch, *Chamomilla officinalis* K.Koch, *Chamomilla recutita* (L.) Rausch., *Chamomilla vulgaris* Gray, *Chrysanthemum chamomilla* (L.) Bernh., *Chrysanthemum suaveolens* (L.) Cav., *Courrantia chamomilloides* Sch. Bip., *Matricaria bayeri* Kanitz, *Matricaria capitellata* Batt. & Pit., *Matricaria courrantiana* DC., *Matricaria exigua* Tuntas, *Matricaria kochiana* Sch.Bip., *Matricaria pusilla* Willd., *Matricaria recutita* L., *Matricaria salina* (Schur) Schur, *Matricaria suaveolens* L.

Common Names Chamomile, Italian camomilla, German chamomile, Hungarian chamomile, wild chamomile, scented mayweed, common chamomile.

Description Herb, annual (1–5 dm), fragrant. Stems erect or ascending, branched at least above, glabrous. Leaves 2- to 3-pinnate with lanceolate outline (1.5 × 3–5 cm), reduced to linear laciniae (less than 0.5 mm wide). Inflorescences flower heads (Ø 1.5–2 cm) on 2–6 cm peduncles, bearing 1–2 linear bracts; receptacle conical, about twice as long as wide, hollow; involucre very flared, cup shaped (Ø 6 mm). Flowers ligulate white (2 × 6 mm), the tubular ones yellow. Fruit achene (1–2 mm), smooth on back, slightly curved, on concave side 4–5 ± distinct ribs; dimples absent; pappus absent or forming a short crown above the fruit.

Subspecies In addition to the typical variety, *Matricaria chamomilla* L. var. *chamomilla*, a second variety is recognized within this taxon: *Matricaria chamomilla* L. var. *coronata* J.Gay ex Boiss.

Cultivar Four chemotypes of chamomile are distinguished based on the content of (-)-α-bisabolol and its oxides: type A (dominant bisabolol oxide A), type B (dominant bisabolol oxide B), type C (dominant [-]-α-bisabolol), and type D ([-]-α-bisabolol and bisabolol oxides A and B in a 1:1 ratio).

Related Species Approximately 26 taxa are considered to belong to the genus *Matricaria*.

Geographic Distribution SE Asia (?), later subcosmopolitan. **Habitat** Species infesting cereal crops. **Range** Poland, Hungary, Germany, Argentina, Czech Republic, Slovakia.

Beer-Making Parts (and Possible Uses) Aerial flowering parts.

Chemistry *Matricaria chamomilla* contains a large group of therapeutically active and interesting compound classes: sesquiterpenes, flavonoids, coumarins (herniarin, umbelliferone, other minor ones), and polyacetylene are considered the most important constituents. Eleven bioactive phenolic compounds (erniarin, umbelliferone, chlorogenic acid, caffeic acid, apigenin, apigenin-7-O-glucoside, luteolin, luteolin-7-O-glucoside, quercetin, rutin, naringenin) have been found in chamomile extract. More than 120 chemical compounds have been identified in the flowers as secondary metabolites, including 28 terpenoids, 36 flavonoids, and 52 additional compounds with potential pharmacological activity. Components such as α-bisabolol and cyclic ethers are antimicrobial, umbelliferone is fungistatic, and chamazulene and α-bisabolol are antiseptic. Chamomile also has effective antileishmaniasis activity. Essential oil, flowers and flower heads: (E)-β-farnesene 4.9–8.1%, terpene alcohol (farnesol), chamazulene 2.3–10.9%, α-bisabolol 4.8–11.3%, α-bisabolol oxide A 25.5–28.7%, α-bisabolol oxide B 12.2–30.9%.

Style American IPA, Belgian Tripel, English IPA, Specialty IPA (White IPA), Spice/Herb/Vegetable Beer, Sweet Stout, Weissbier, Witbier.

Beer Camilla, Trunasse (Centallo, Piedmont, Italy).

Source Dabove 2015; Hamilton 2013; Hamilton 2014; Heilshorn 2017; Hieronymus 2010; Fisher and Fisher 2017; McGovern 2017; https://www.brewers-friend.com/other/chamomille/.

Matricaria discoidea DC.
ASTERACEAE

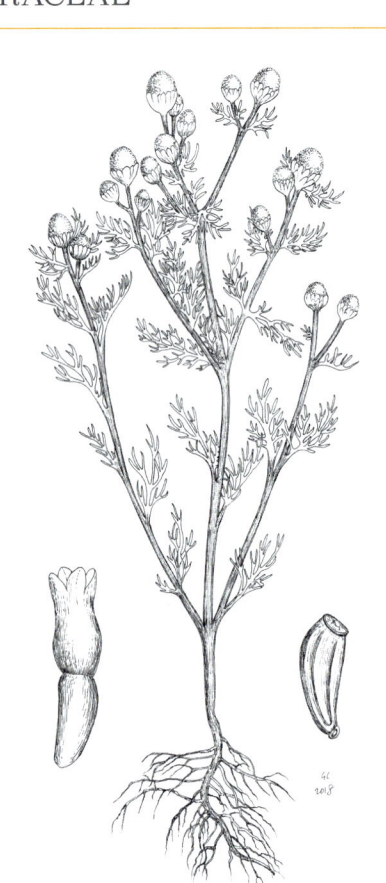

Synonyms *Akylopsis suaveolens* (Pursh) Lehm., *Anthemis inconspicua* Fisch. ex Herder, *Cenocline pauciflora* K.Koch, *Chamomilla discoidea* (DC.) J.Gay. ex A.Braun, *Chamomilla suaveolens* (Pursh) Rydb., *Matricaria graveolens* (Pursh) Asch.

Common Names Pineapple weed.

Description Herb, annual (up to 30 cm). Root fibrous. Stems erect, much branched, densely leafy, with odor similar to chamomile but more persistent. Lower leaves in a rosette, the cauline ones 2- to 3-pinnate, alternate, sessile, divided into thin laciniae, glabrous, dark green, strongly aromatic. Flowers all tubular, corolla 4-lobed, color greenish. Inflorescence head conical, solitary at the end of twigs, lacking ligulate flowers; peduncle swollen at apex, involucre hemispherical of lanceolate or oval bracts, with scarious, translucent margins; receptacle hollow. Fruits achenes, glabrous, with 2 marginal nerves and 1 on inner face.

Related Species About 26 taxa are considered to belong to the genus *Matricaria*.

Geographic Distribution Subcosmopolitan. **Habitat** Grasslands, paths, boulders, rubble, trampled environments, nitrophilous, near houses, stables (0–2,200 m ASL). **Range** Subcosmopolitan.

Beer-Making Parts (and Possible Uses) Flowers.

Chemistry Essential oil, aerial parts (36 compounds, 86.7% of total oil): myrcene 28.5%, (E)-β-farnesene 23.4%, germacrene D 6.8%, geranyl isovalerate 6.4%, (Z)-enyn-dicycloether 8.1%.

Style None verified (?).

Source Heilshorn 2017; Hieronymus 2016b.

Matteuccia struthiopteris (L.) Tod.
ONOCLEACEAE

Synonyms *Matteuccia nodulosa* Fernald, *Matteuccia pensylvanica* (Willd.) Raymond, *Onoclea pensylvanica* (Willd.) Sm., *Pteretis pensylvanica* (Willd.) Fernald, *Struthiopteris nodulosa* Desv., *Struthiopteris pensylvanica* Willd.

Common Names Fiddlehead, ostrich fern.

Description Herb, perennial, pteridophytic (30–140 cm). Rhizome stout, short, erect, with long hypogeal stolons, through which it expands rapidly and extensively. Fronds fasciculate, dimorphic, bipinnate; sterile fronds numerous, outer, soft, deciduous, erect or shallowly sloping (30–140 cm), arranged in a rosette to form a regular funnel. Stipe short (1–2 cm), stout, with furrows continuing to upper half of rachis without scales; lamina oblong-lanceolate, light green, with maximum width at top, gradually decreasing at base, provided with numerous long, narrowly lanceolate, bearing obtuse, entire or crenulate pinnules on edges. Fertile fronds (1–6) inner, late, wintering, shorter than sterile ones, erect and rigid, greenish, brown at spore maturity, with relatively long stipe and contracted lamina (15–35 cm), with crumpled pinnules. Sori orbicular, confluent, with squamiform, laciniate-fimbriate, caducous pseudoindusium. Sporangia thin walled with annulus. Spores monolete.

Subspecies In addition to the typical variety (var. *struthiopteris*), *Matteuccia struthiopteris* (L.) Tod. has two additional subspecific units: *Matteuccia struthiopteris* (L.) Tod. var. *pensylvanica* (Willd.) C.V.Morton and *Matteuccia struthiopteris* (L.) Tod. subsp. *sinuata* (Thunb.) Á.Löve & D.Löve.

Related Species Besides the one described here, the genus *Matteuccia* includes only one species: *Matteuccia intermedia* C. Chr.

Geographic Distribution Circumboreal.
Habitat Wet woods, shady streams, in acidified soil (100–1,550 m ASL). **Range** Circumboreal.

Beer-Making Parts (and Possible Uses) Sprouts (6–10 cm).

Toxicity Although some species of fern are toxic or carcinogenic, *Matteuccia struthiopteris* is considered safe for humans; it is often eaten raw or cooked without problems or with a slight laxative effect. Gastrointestinal disorders have been reported connected to the consumption of this plant. Some researchers believe there is a thermolabile toxin (not yet identified), whereas according to others, such effects would derive from a pathogen (unknown at the moment) present on the consumed leaves; in both cases, a 10- to 15-minute cooking would seem sufficient to solve the problem.

Chemistry Vitamin A 568 µg, vitamin C 36.8 mg, niacin 4 mg, riboflavin 0.182 mg, thiamine 0.022 mg, phosphorus 91 mg, iron 1.82 mg, magnesium 37.5 mg, zinc 0.55 mg, calcium 32, sodium chloride traces, high fiber content; content of phenolic compounds similar to that of blueberry, omega-6/omega-3 fatty acids ratio about 4.

Style American IPA, Irish Red Ale.

Source https://www.brewersfriend.com/other/fiddleheads/.

Melissa officinalis L.
LAMIACEAE

Synonyms *Faucibarba officinalis* (L.) Dulac, *Melissa altissima* Sm., *Melissa bicornis* Klokov, *Melissa cordifolia* Pers., *Melissa corsica* Benth., *Melissa foliosa* Opiz ex Rchb., *Melissa graveolens* Host, *Melissa hirsuta* Hornem., *Melissa occidentalis* Raf. ex Benth., *Melissa romana* Mill., *Melissa taurica* Benth., *Mutelia officinalis* (L.) Gren. ex Mutel, *Thymus melissa* E.H.L.Krause.

Common Names Balm, lemon balm, honey plant.

Description Herb, perennial (30–80 cm), with a pleasant lemon odor. Short rhizome. Stems erect, quadrangular, branched, with patently bristly hairs, arranged on the edges, especially dense at the nodes. Leaves petiolate, opposite, yellowish-green, with ovate lamina; lower ones often heart shaped, margin crenate-dentate (6–14 teeth per side). Flowers on short peduncles, collected in sparse glomerules, at the axil of leaves and lanceolate bracts; calyx tubular, hairy, campanulate, bilabiate; corolla tubular, sympetalous, bilabiate, upper lip notched, lower trilobate with median lobe more developed than the 2 lateral ones, color yellowish, white-pink after fertilization. Fruit tetrachene containing 4 small nucules (achenes), dark brown in color (1.6–2 × 0.8–1 mm).

Subspecies In addition to *Melissa officinalis* L. subsp. *officinalis*, there is a second subspecies: *Melissa officinalis* L. subsp. *inodora* Bornm.

Cultivar *Melissa officinalis* L. var. *aurea*, *Melissa officinalis* L. var. *citronella*, *Melissa officinalis* L. var. *lemonella*, *Melissa officinalis* L. var. *lime*, *Melissa officinalis* L. var. *quedlinburger*, *Melissa officinalis* L. var. *quedlinburger* niederliegende, *Melissa officinalis* L. var. *variegate*.

Related Species The genus *Melissa* includes other species, some of which have similar uses to *M. officinalis*: *Melissa axillaris* (Benth.) Bakh.f., *Melissa flava* Benth. and *Melissa yunnanensis* C.Y.Wu & Y.C.Huang.

Geographic Distribution Eurasian-Mediterranean, W Asian. **Habitat** Hedges, dry, sunny places, near houses, in ruderal places (0–1,000 m ASL). **Range** Hungary, Egypt, Italy, Ireland.

Beer-Making Parts (and Possible Uses) Leaves.

Toxicity At high concentrations, the essential oil can cause problems; it is not recommended for subjects with hypothyroidism (because of the presumed antagonism with thyroid stimulating hormone [TSH]), people with hypertension, or people with glaucoma (because it is considered capable of raising intraocular pressure.

Chemistry Essential oil (33 components; 89.3% of oil): citronellal 14.40%, caryophyllene oxide 11%, geraniol acetate 10.2%, β-caryophyllene 8.2%, isogeraniol 6.4%, nerol acetate 5.1%, cis-p-met-2 en-7-ol 3.8%, nerol 3, 5%, carane 2.32%, menthol 2.2%, camphene 2.1%, aromadendrene oxide 1.8%, pimara-7,15-dien-3-one 1.8%, 2-pinen-4-one 1.75%, cycloisolengifolene 1.7%, patchoulene 1.6%, aromadendrene oxide 1.6%, germanicol 1.2%, geraniol 1%, verbenol 0.9%, cis-mirtanol 0.9%, colest-5-en-7-ol 0.9%, 1R-α-pinene 0.7%, cinerone 0.7%, longifolene 0.7%, himachalene 0.7%, α-pinene 0.6%, cis-z-bisabolene oxide 0.6%, verbenone 0.6%, andropholide 0.6%, himachala-2,4-diene 0.6%, cubenol 0.6%, lupan-3-ol acetate 0.5%.

Style Belgian Specialty Ale, Spice/Herb/Vegetable Beer.

Beer Meilé, Herba Monstrum (Galbiate, Lombardy, Italy).

Source Fisher and Fisher 2017; Heilshorn 2017; Hieronymus 2016a; Hieronymus 2016b; Josephson et al. 2016; https://www.brewersfriend.com/other/citronmelisse/; https://www.brewersfriend.com/other/sitronmelisse/.

Mentha × *piperita* L.
LAMIACEAE

Synonyms *Mentha* × *adspersa* Moench, *Mentha* × *balsamea* Willd., *Mentha* × *banatica* Heinr.Braun,

Mentha × *braousiana* Pérard, *Mentha* × *citrata* Ehrh., *Mentha* × *concinna* Pérard, *Mentha* × *crispula* Wender., *Mentha* × *durandoana* Malinv. ex Batt., *Mentha* × *exaltata* Heinr.Braun, *Mentha* × *fraseri* Druce, *Mentha* × *glabrata* Vahl, *Mentha* × *hercynica* Röhl., *Mentha* × *heuffelii* Heinr.Braun, *Mentha* × *hircina* Hull, *Mentha* × *hircina* J.Fraser, *Mentha* × *hirtescens* Heinr. Braun & Topitz, *Mentha* × *hudsoniana* Heinr.Braun, *Mentha* × *kahirina* Forssk., *Mentha* × *langii* Geiger ex T.Nees, *Mentha* × *napolitana* Ten., *Mentha* × *nigricans* Mill., *Mentha* × *odora* Salisb., *Mentha* × *odorata* Sole, *Mentha* × *officinalis* Hull, *Mentha* × *pimentum* Nees ex Bluff and Fingerh., *Mentha* × *piperoides* Malinv., *Mentha* × *schultzii* Boutigny ex F.W.Schultz.

Common Names Mint.

Description Herb, perennial (3–10 dm), with strongly aromatic odor. Ascending stems often reddened, nearly glabrous, hairy only at stem angles and on nerves of lower leaf. Leaves ovate to lanceolate-elliptic, 1.8–3 times longer than wide (2–4 cm), serrate, rounded to wedge shaped at base; petiole 3–12 mm. Flowers in a broad spike formed by clustered whorls, sometimes with 1–3 whorls ± distant; calyx cylindrical (tubular) with glabrous base, 3–4 mm; corolla pinkish or violet.

Cultivar Many are cultivated for ornamental purposes (Candymint, Chocolate Mint, Citrata [Eau de Cologne Mint, Grapefruit Mint, Lemon Mint, Orange Mint], Crispa, Lavender Mint, Lime Mint, Variegata) or commercial purposes (Bulgarian population #2, Clone 11-6-22, Clone 80-121-33, Dulgo Pole, Mitcham Digne 38, Mitcham Ribecourt 19, Todd's#x2019, Todd's Mitcham, Zefir).

Related Species Of the many species recognized by taxonomists in the genus *Mentha*, some are of hybridogenic origin: *Mentha alaica* Boriss., *Mentha alopecuroides* Hull, *Mentha aquatica* L., *Mentha arvensis* L., *Mentha atrolilacina* B.J.Conn & D.J.Duval, *Mentha australis* R.Br., *Mentha canadensis* L., *Mentha cardiaca* J. Gerard ex Baker, *Mentha* × *carinthiaca* Host, *Mentha cervina* L., *Mentha cunninghamii* (Benth.) Benth., *Mentha dahurica* Fisch. ex Benth., *Mentha* × *dalmatica* Tausch, *Mentha darvasica* Boriss., *Mentha diemenica* Spreng, *Mentha* × *dumetorum* Schult., *Mentha gattefossei* Maire, *Mentha* × *gayeri* Trautm., *Mentha* × *gentilis* L., *Mentha grandiflora* Benth., *Mentha japonica* (Miq.) Makino, *Mentha* × *kuemmerlei* Trautm., *Mentha laxiflora* Benth, *Mentha* × *locyana* Borbás, *Mentha longifolia* (L.) L. (10 between subsp. and var.), *Mentha* × *maximilianea* F.W.Schultz, *Mentha micrantha* (Fisch. ex Benth.) Heinr.Braun, *Mentha nemorosa* Willd., *Mentha pamiroalaica* Boriss, *Mentha pulegium* L., *Mentha* × *pyramidalis* Ten., *Mentha requienii* Benth., *Mentha* × *rotundifolia* (L.) Huds., *Mentha royleana* Wall. ex Benth., *Mentha royleana* Wall. ex Benth. var. *afghanica* (Murata) Rech.f., *Mentha royleana* Wall. ex Benth. var. *detonsa* (Briq.) Rech.f., *Mentha saturejoides* R.Br., *Mentha* × *smithiana* R.A.Graham, *Mentha suaveolens* Ehrh., *Mentha suaveolens* Ehrh. subsp. *timija* (Coss. ex Briq.) Harley, *Mentha* × *verticillata* L., *Mentha* × *villosa* Huds., *Mentha* × *villosa-nervata* Opiz). Several have potential medicinal uses and produce distinctive essential oils.

Geographic Distribution Complex of hybridized forms between *Mentha aquatica* L. and *Mentha spicata* L. stabilized by cultivation. **Habitat** Plant cultivated. **Range** Cultivated everywhere.

Beer-Making Parts (and Possible Uses) Leaves, flowers.

Toxicity The presumed toxicity of this plant to the liver and other organs of the human body is controversial and appears to be largely dose dependent.

Chemistry Essential oil (99.37% oil): menthol 53.28%, menthyl acetate 15.1%, menthofuran 11.18% 1.8 cineole 6.69%, neomenthol 2.79%, menthone 2.45%, (z)-caryophyllene 2.06%, germacrene D 2.01%, neomentyl acetate 0.65%, isomentyl acetate 0.61%, β-pinene 0.58%, cis-sabinene hydrate 0.5%, β-bourbonene 0.37%, α-pinene 0.32%, E-β-farnesene 0.3%, sabinene 0.26%, bicyclogermacrene 0.22%.

Style American Amber Ale, American IPA, American Pale Ale, American Porter, American Stout, American Wheat or Rye Beer, Apple Wine, Baltic Porter, Belgian Blond Ale, Belgian Dark Strong Ale, Bière de Garde, Blonde Ale, Bohemian Pilsner, Brett Beer, Brown Porter, Cream Ale, Dry Stout, Experimental Beer, Extra Special/Strong Bitter (ESB), Foreign Extra Stout, German Pilsner (Pils), Holiday/Winter Special Spiced Beer, Kölsch, Oatmeal Stout, Robust Porter, Russian Imperial Stout, Saison, Specialty Beer, Spice/Herb/Vegetable Beer, Strong Scotch Ale, Sweet Stout, Weizen/Weissbier.

Beer Bianca Piperita, Opperbacco (Notaresco, Abruzzo, Italy).

Source Dabove 2015; Fisher and Fisher 2017; Hieronymus 2010; Hieronymus 2012; Hieronymus 2016b; Josephson et al. 2016; McGovern 2017; https://www.brewersfriend.com/other/mint/; https://www.omnipollo.com/beer/7/; https://unabirralgiorno.blogspot.it/2014/02/alesmith-speedway-stout.html.

Mentha pulegium L.
LAMIACEAE

Synonyms *Calamintha fenzlii* Vis., *Melissa pulegium* (L.) Griseb., *Mentha albarracinensis* Pau, *Mentha aromatica* Salisb., *Mentha aucheri* Pérard, *Mentha daghestanica* Boriss., *Mentha erinoides* Heldr., *Mentha exigua* L., *Mentha gibraltarica* Willd., *Mentha hirtiflora* Opiz ex Heinr.Braun, *Mentha montana* Lowe ex Benth., *Mentha pulegioides* Dumort., *Mentha subtomentella* Heinr.Braun, *Mentha tomentella* Hoffmanns. & Link, *Mentha tomentosa* Sm., *Micromeria fenzlii* Regel, *Micromeria maritima* Yild. Sadikoglu & M.Keskin, *Minthe pulegia* (L.) St.-Lag, *Pulegium aromaticum* Gray, *Pulegium daghestanicum* (Boriss.) Holub, *Pulegium erectum* Mill., *Pulegium heterophyllum* Opiz ex Boenn., *Pulegium micranthum* Claus, *Pulegium pubescens* Opiz ex Boenn., *Pulegium tomentellum* C.Presl, *Pulegium vulgare* Mill., *Satureja fenzlii* (Vis.) K.Malý, *Thymus bidentatus* Stokes.

Common Names Pennyroyal.

Description Herb, perennial (15–60 cm), rhizomatous, very aromatic, velvety-pelose or glabrescent. Stems ascending, quadrangular in section, sometimes lignified at the base, rooting at the nodes, with internodes generally longer than the leaves. Leaves small (5–10 × 10–30 mm), velvety, briefly petiolate (2–4 mm), ovate or oblong, sparsely serrate or entire on edges, usually ± arcuate and shower folded, apex obtuse or subacute; bract leaves sessile, resembling cauline leaves, shorter than floral whorls, often forming whorls at the axil of lower leaves. Inflorescences globose whorls (Ø 1–1.5 cm) of 10–30 hermaphrodite flowers, spaced, at axil of bract leaves. Flower with calyx (up to 3 mm) tubular, often purple, hirsute-glandular, 10 ribs, 5 unequal lanceolate-lesiniform teeth, then ± bilabiate, fitted with an inner ring of hairs; corolla (4.5–6 mm) lilac pink, rarely white, pubescent on the outside, 4 lobes ± equal; stamens 4, very prominent; ovary superior, 4-locular; stigma bifid. Fruit composed of 4 nucules (tetrachene), ellipsoid, wrinkled, light brown, enclosed in persistent calyx.

Cultivar There are different chemotypes of this species.

Related Species The genus *Mentha* includes 56 taxa, some of which are hybridogenic; several of these, besides those described here, have a significant production of aromatic substances with potential beer-making applications.

Geographic Distribution Euro-Mediterranean, subcosmopolitan. **Habitat** Edges of ditches and roads, riverbeds, ephemeral ponds, wet or dry grassy places periodically flooded (0–1,200 m ASL). **Range** Subcosmopolitan.

Beer-Making Parts (and Possible Uses) Leaves (?), inflorescences (?).

Chemistry Essential oil (53 components): pulegone 43.5%, piperitone 12.2%, p-menthane-1,2,3-triol 6.5%, γ-elemenene 3.6%, guaiene (cis-β) 3%, carvacrol acetate 2.6%, phenyl ethyl alcohol 2.4%.

Style Mum.

Source Heilshorn 2017; Hieronymus 2010.

Mentha spicata L.
LAMIACEAE

Synonyms *Mentha atrata* Schur, *Mentha brevispicata* Lehm., *Mentha cordato-ovata* Opiz, *Mentha crispa* L., *Mentha crispata* Schrad. ex Willd., *Mentha glabra* Mill., *Mentha hortensis* Opiz ex Fresen., *Mentha inarimensis* Guss., *Mentha integerrima* Mattei & Lojac., *Mentha lacerata* Opiz, *Mentha laciniosa* Schur, *Mentha laevigata* Willd., *Mentha lejeuneana* Opiz, *Mentha lejeunii* Opiz ex Rchb., *Mentha michelii* Ten. ex Rchb., *Mentha ocymiodora* Opiz, *Mentha pectinata* Raf., *Mentha piperella* (Lej. & Courtois) Opiz ex Lej. & Courtois, *Mentha pudina* Buch.-Ham. ex Benth., *Mentha romana* Bubani, *Mentha rosanii* Ten., *Mentha rubicunda* Heinr.Braun & Topitz, *Mentha sepincola* Holuby, *Mentha tauschii* Heinr.Braun, *Mentha tenuiflora* Opiz, *Mentha tenuifolia* Opiz ex Rchb., *Mentha tenuis* Michx., *Mentha undulata* Willd., *Mentha viridifolia* Pérard, *Mentha viridis* (L.) L., *Mentha walteriana* Opiz.

Common Names Spearmint.

Description Herb, perennial (3–10 dm), with aromatic odor. Stem erect or ascending, often stoloniferous; stem, leaves, bracts, peduncles, and calyxes hairy. Leaves wrinkled, ovate to subrounded (1.5–3 × 5–9 cm) with maximum width toward base, toothed or crenate-toothed, acute at apex; on underleaf simple hairs mixed with branched hairs with basal cells Ø 30–47 μm. Inflorescences contiguous whorls forming apical spikes (0.7 × 3–9 cm), fusiform, pointed at apex; calyx conical 1–2 mm, gray-tomentose, with triangular-elongated teeth; corolla whitish or pinkish. Fruit schizocarp composed of 4 nucules (achenes).

Subspecies Besides the typical one, the subspecies *Mentha spicata* L. subsp. *condensata* (Briq.) Greuter & Burdet is also recognized.

Cultivar *Mentha spicata* L. cv. *nana* is used in the mint tea typical of Morocco.

Related Species The genus *Mentha* includes 56 taxa, some of which are hybridogenic in nature; several of these, beyond those described here, have a significant production of aromatic substances with potential beer-making applications.

Geographic Distribution Euro-Mediterranean (probably allotetraploid hybrid of *Mentha longifolia* [L.] Hudson × *Mentha suaveolens* Ehrh. stabilized with cultivation). **Habitat** Crops, roadside verges, meadows, etc. **Range** Cultivated and often wild species in temperate climates.

Beer-Making Parts (and Possible Uses) Leaves.

Toxicity Despite the many positive effects attributable to this species, there is evidence of adverse effects and toxicity (notable histopathological changes in kidney, liver, uterus tissues), allergic skin reactions, and reduced male libido apparently related to hormonal interactions.

Chemistry Essential oil (34 compounds, 99.9% of total): carvone 40.8%, limonene 20.8%, 1,8-cineole 17%, spathulenol 10.1%, β-pinene 2.2%, cis-dihydro-carvone 1.9%, dihydrocarveol 1.7%, cis-sabinene hydrate 1.6%, α-pinene 1.4%, sabinene 1.4%, 4-terpineol 1.3%, myrcene 1.1%, 3-octanol 1%, β-bourbonene 0.9%, p-cymene 0.8%, cis-carveol 0.6%, α-terpineol 0.5%, linalool 0.4%, δ-terpineol 0.4%, trans-carveol 0.4%, pulegone 0.3%, β-elemene 0.3%, β-caryophyllene 1.2%, germacrene D 0.2%, germacrene A 0.2%, camphene 0.2%, (Z)-β-ocimene 0.2%, cis-p-menth-2-en-1-ol 0.1%, cis-limonene oxide 0.1%, trans-limonene oxide 0.1%, borneol 0.1%, isobornyl acetate

0.1%, iso-dihydrocarveol acetate 0.2%, caryophyllene oxide 0.3%.

Style American Barleywine, American IPA, American Pale Ale, American Stout, American Wheat or Rye Beer, Blonde Ale, Brown Porter, Dry Stout, English Barleywine, Foreign Extra Stout, Imperial IPA, Old Ale, Saison, Specialty Beer, Sweet Stout.

Source Calagione et al. 2018; Fisher and Fisher 2017; Heilshorn 2017; Hieronymus 2016b; https://www.brewersfriend.com/other/spearamint/; https://www.brewersfriend.com/other/spearmint/.

Mentha suaveolens Ehrh.
LAMIACEAE

Synonyms *Mentha barcinonensis* Sennen, *Mentha bellojocensis* Gillot, *Mentha bofillii* Sennen, *Mentha germanica* Déségl. & T.Durand, *Mentha insularis* Req., *Mentha krockeri* Strail, *Mentha linnaei* Déségl. & T.Durand, *Mentha macrostachya* Ten., *Mentha meduanensis* Déségl. & T.Durand, *Mentha mexicana* M.Martens & Galeotti, *Mentha mucronulata* Opiz, *Mentha neglecta* Ten., *Mentha oblongifolia* (Lej. ex Malinv.) Strail, *Mentha pachystachya* Timb.-Lagr. & Marcais, *Mentha ripartii* Déségl. & T.Durand, *Mentha roseiflora* Sennen, *Mentha rugosa* Lam., *Mentha sepium* (Déségl. & T.Durand) Heinr.Braun, *Mentha willdenowii* Déségl. & T.Durand.

Common Names Apple mint, round-leaved mint, pineapple mint.

Description Herb, perennial (up to 90 cm), with strong acrid odor. Stem erect or ascending, woody at base, branched, stoloniferous, quadrangular, with variable hairiness. Leaves oblong, obovate (2–4 × 6–7 cm), sometimes orbicular, rounded at apex, margin toothed or crenate-dentate, gray-green color more intense on upper leaf, velvety and wrinkled, sessile, opposite, veins evident, corrugated and tomentose on underleaf due to white, simple, branched cottony hairs. Inflorescences dichasial, composed of verticils (0.7 × 3–9 cm), close together and dense, fusiform, pointed at apex. Flowers fragrant, numerous, hermaphrodite, briefly pedunculate; bracts (3–5 mm) ovate to lanceolate; calyx persistent (1–2 mm), gamosepalous, conical, gray-tomentose, with 5 ciliated triangular teeth; corolla 3–4 mm with 4 lobes, whitish or pinkish in color; stamens protruding, stigma bifid. Fruit schizocarp, microbasarium consisting of 4 monosperm achenes.

Subspecies *Mentha suaveolens* Ehrh. subsp. *suaveolens* and *Mentha suaveolens* Ehrh. subsp. *timija* (Coss. ex Briq.) Harley are the two subspecies of this taxon.

Related Species The genus *Mentha* includes 56 taxa, some of which are hybridogenic in nature; several of these, beyond those described here, have a significant production of aromatic substances with potential beer-making applications; beer-making use has been ascertained for *Mentha aquatica* L. and *Mentha canadensis* L.

Geographic Distribution Euro-Mediterranean. **Habitat** Crops, edges of fields, ditches, and wetlands. **Range** Euro-Mediterranean.

Beer-Making Parts (and Possible Uses) Flowering tops.

Chemistry Essential oil (components identified 99.77%): limonene 31.25%, γ-terpinene 0.01%, linalool 0.32%, borneol 0.68%, cis-dihydrocarvone 0.15%, trans-dihydrocarvone 0.85%, cis-carveol 2.31%, carvone 50.59%, dihydrocarveol acetate 0.36%, trans-carvil acetate 0.35%, piperitenone 0.12%, cis-carvil acetate 0.44%, β-bourbonene 1.27%, cis-jasmone 1.27%, trans-β-caryophyllene 2.56%, β-cedrene 0.32%, β-copaene 0.28%, α-humulene 1.24%, γ-muurolene 0.61%, germacrene D 2.04%, bicyclogemacrene 0.35%, γ-cadinene 0.13%, trans-calamenene 0.45%, α-cadinene 0.09%, caryophyllene oxide 0.30%, cubenol-1,10-di-epi 0.31%, epi-α-cadinol 0.50%, epi-α-muurolol 0.34%, manoyl oxide 0.28%.

STYLE Herb Beer (?).

SOURCE Fisher and Fisher 2017; Josephson et al. 2016.

Metrosideros excelsa Sol. ex Gaertn.
MYRTACEAE

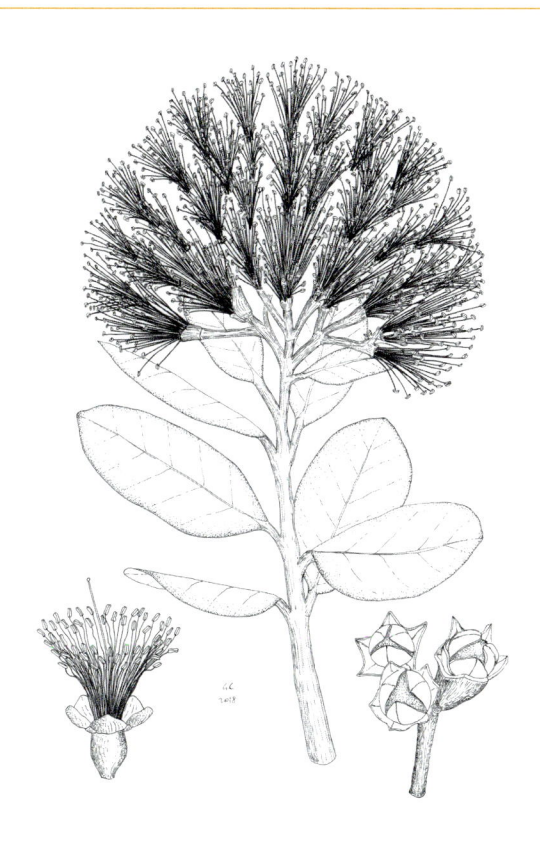

SYNONYMS *Metrosideros tomentosa* A.Rich., *Nania tomentosa* (A.Rich.) Kuntze.

COMMON NAMES Pohutukawa, pōhutukawa, New Zealand pohutukawa, New Zealand Christmas tree, New Zealand Christmas bush, iron tree.

DESCRIPTION Tree (up to 20 m). Trunk (Ø up to 2 m) with spreading branches; branches stout, tomentose. Leaves on short, stout petioles; lamina 2.5–10 × 2.5–5 cm, elliptic to oblong, acute or obtuse, leathery, thick, with white tomentum beneath (young plants occasionally glabrous beneath). Inflorescence compound top with ∞ flowers; peduncles stout, tomentose. Flowers with obconic receptacle; sepals deltoid; petals red, oblong; stamens ∞, color red, 3–4 cm; ovary adnate to receptacle. Fruit capsule (7–9 mm), tomentose, distinctly exserted, 3-valve, loculicidal.

CULTIVAR At least 39 are known, including Aurea, Blockhouse Bay, Butterscotch, Centennial, Christmas Cheer, Dalese, Fire Mountain, Firestone, Flame Crest, Gold Finger, Gold Nugget, Golden Dawn, Hauraki, Kopere, Lighthouse, Manukau, Maori Princess, Midas, Mini Christmas, Moon Maiden, Mt. Maunganui, Octopussy, Ohope, Parnell, Pink Lady, Plus Four, Pouawa, Rangitoto, Royal Flame, Scarlet Pimpernel, Sunglow, Tamaki, Te Kaha, Titirangi, Upper Hutt, Variegata, Vibrance, Whakarewarewa, and White Caps.

RELATED SPECIES The genus *Metrosideros* includes 67 accepted taxa.

GEOGRAPHIC DISTRIBUTION Endemic species of New Zealand (especially North Island). **HABITAT** Wide variety of natural habitats (shrublands, river and lake banks, cliffs, coasts, arid environments, plains, hills, ridges, etc.), synanthropic (roadsides, trails, eroded environments, etc.) on different substrates (clays, gravel, rocky outcrops, sands, silt, stony soils, etc.). **RANGE** Beyond its natural range, the species is widespread for ornamental purposes and frequently naturalized elsewhere, such as in Norfolk Island, SE Australia (Sydney area), United States (California), South Africa, and Europe (Spain), to mention just a few examples.

BEER-MAKING PARTS (AND POSSIBLE USES) Wood (tested in beer making).

WOOD Color rose-brown, fine texture, hard, compact (WSG 0.88).

STYLE Saison/Farmhouse Ale (known examples produced with the addition of *M. excelsa* honey and not with plant parts).

SOURCE Cantwell and Bouckaert 2016; Cantwell and Bouckaert 2018.

Monarda citriodora Cerv. ex Lag.
LAMIACEAE

Synonyms *Monarda aristata* Nutt., *Monarda dispersa* Small, *Monarda tenuiaristata* Small.

Common Names Lemon mint, lemon beebalm.

Description Herb (30–75 cm), annual or biennial. Stem unbranched or sparsely branched; central stem light green, rather stiff, terete or angular, densely pubescent with short hairs. Leaves opposite along stem; lamina 2.5–7.5 × 0.6–2 cm, ± elliptic or elliptic-oblong, smooth along margins, sessile or degrading into a short petiole; upper leaf light to medium green and sparsely canescent, the lower pale green and pubescent with short hairs. Inflorescences 2-plus verticillate ± terminal along central stem; below each inflorescence ther is a verticil of floral bracts, color light pink to violet-pink, veined with green, looking and sized like true leaves (± oblong) with apical barbs. Flowers about 2 cm, corolla light pink or violet-pink, deeply bilabiate; calyx tubular, green to violet-green, with 5 deltate-linear teeth ending in long barbs; reproductive organs (2 stamens, 1 pistil) enclosed by corolla; upper lip narrow, hairy at apex, acts as protective shield, lower one has 3 terminal lobes suitable to receive pollinators; corolla with stripes inside. Fruits achenes.

Subspecies The taxon includes two entities of sub-specific rank: *Monarda citriodora* Cerv. ex Lag. var. *austromontana* (Epling) B.L.Turner and *Monarda citriodora* Cerv. ex Lag. var. *citriodora* (both are reported to be of beer-making interest).

Related Species The genus *Monarda* includes 31 taxa.

Geographic Distribution S United States, Mexico. **Habitat** Grasslands, roadsides, sunny places, preferably in clay soils. **Range** S United States, Mexico.

Beer-Making Parts (and Possible Uses) Flowering tops (plant substitute for *Monarda didyma*).

Chemistry Essential oil, flowers: thymol 44.6%, 1,8-cineole 23.61%, α-phellandrene 4.815%, β-cymene 4.019%, 1R-α-pinene 3.478%, carvacrol 3.21%, γ-terpinene 2.574%, β-pinene 1.645%, unknown compound 1.222%, α-terpineol 0.955%, α-terpinene 0.95%, citral 0.919%, unknown compound 0.877%, 4-terpineol 0.872%, (E)-2-decenal 0.688%, β-myrcene 0.664%, linalool 0.504%, caryophyllene 0.481%, geraniol 0.417%, borneol 0.381%, 3-carene 0.364%, unknown compound 0.358%, cis-ocimene 0.321%, unknown compound 0.313%, sabinene 0.242%, camphene 0.217%, terpinolene 0.131%, δ-cadinene 0.117%, heptane 0.115%, β-cubebene 0.109%.

Style Herb Beer.

Source Fisher and Fisher 2017.

Monarda clinopodia L.
LAMIACEAE

Synonyms *Monarda allophylla* Michx., *Monarda altissima* Willd., *Monarda clinopodifolia* L., *Monarda glabra* Lam., *Monarda rugosa* Aiton, *Monardella caroliniana* Benth.

Common Names Bee balm, basil beebalm, wild bergamot, white bergamot.

Description Herb, perennial (to about 1 m), aromatic. Leaves simple, opposite, petiolate (2–3.5 cm), petiole covered only with scattered, diffused hairs (up to 1 mm), lamina ± ovate, often wider than 4 cm and 2 times as long as wide, margin serrate, apex acuminate. Inflorescence top terminal solitary, flower head-like. Flowers fragrant, corolla white or yellow-brown in color, up to 3 cm, upper lip slender, nearly straight, glabrous to puberulent; stamens past upper lip; bract leaves mostly green but whitish at base.

Related Species The genus *Monarda* includes 31 taxa.

Geographic Distribution E United States. **Habitat** Woods, thickets. **Range** Similar to geographic distribution.

Beer-Making Parts (and Possible Uses) Leaves, flowers.

Chemistry No information available.

Style Herb Beer.

Source Hieronymus 2016b; Josephson et al. 2016.

Monarda didyma L.
LAMIACEAE

Synonyms *Monarda contorta* C.Morren, *Monarda kalmiana* Pursh, *Monarda oswegoensis* W.Barton, *Monarda purpurascens* Wender., *Monarda purpurea* Lam.

Common Names Scarlet beebalm, horsemint, Oswego tea, bee balm.

Description Herb, annual. Stems subglabrous, villous at the nodes and at the apex along the corners, glabrescent. Leaves petiolate (up to 2.5 cm), nearly missing apically, base slightly dilated; lamina ovate-lanceolate (up to 10 × 4.5 cm), papery, upper leaf villous or glabrescent, underleaf sparsely glandular, villous on veins, base rounded, margin unequally serrate, apex acuminate. Inflorescence verticillate in terminal flower heads (Ø up to 6 cm); bracts short, petiolate, leaf-like, margin entire, red, shorter than flower heads; bracts linear-subulate (about 10 × 1.5 mm), long caudate, puberulent, red; pedicel about 1 mm, puberulent; calyx slightly curved (about 10 × 2.5 mm), purple-red when dry, veins pubescent, interior sparsely hirsute; teeth equal, subulate-triangular, about 1 mm, apex spinescent; corolla purple-red, about 2.5 cm, puberulent; upper lip straight, slightly curved outward, margin entire; lower lip enlarged, with middle lobe narrower, emarginate.

Cultivar Examples include Adam, Cambridge Scarlet, and Jacob Cline.

Related Species The genus *Monarda* includes 31 taxa of varying taxonomic rank.

Geographic Distribution E United States. **Habitat** Woodlands, gardens, sunny borders, cultivated fields. **Range** North America, China, Europe.

Beer-Making Parts (and Possible Uses) Flowers, leaves.

Chemistry Essential oil (components identified 99.8%): α-pinene 0.8%, camphene 3.4%, myrcene 3.7%, δ-3-carene 4.4%, α-phellandrene 1.1%, p-cymene 10.3%, limonene 1.4%, β-phellandrene 0.2%, terpinolene 9.2%, 1-octen-3-ol 2.5%, linalool 0.7%, α-terpineol 0.1%, thymol methyl ether 0.2%, thymol 59.3%, α-copaene 0.3%, β-bourbonene 0.1%, α-humulene 0.2%, epi-bicyclic-sesqui-phellandrene 0.2%, germacrene D 1.1%.

Style Herb Beer (?).

Source Fisher and Fisher 2017; Josephson et al. 2016.

Monarda fistulosa L.
LAMIACEAE

Synonyms *Monarda affinis* Link, *Monarda albiflora* C.Morren, *Monarda barbata* Wender., *Monarda coerulea* Benth., *Monarda commutata* Wender., *Monarda cristata* Benth., *Monarda dubia* Benth., *Monarda hybrida* Wender., *Monarda involucrata* Wender., *Monarda lilacina* Wender., *Monarda longifolia* Lam., *Monarda oblongata* Aiton, *Monarda purpurascens* Dum.Cours., *Monarda purpurea* Pursh, *Monarda scabra* Beck, *Monarda undulata* Tausch ex Rchb., *Monarda urticifolia* Tausch, *Monarda varians* W.Barton, *Monarda violacea* Desf., *Monarda virginalis* Vilm., *Pycnanthemum monardella* Michx.

Common Names Wild bee-balm, wild bergamot, horsemint.

Description Herb, perennial (5–12 dm). Stem reddish color ± punctuated with purple-red, branched apically, densely pubescent with white hairs, nodes glabrous or villous. Leaves with 2–15 mm petiole; lamina lanceolate-ovate to ovate, about 80 × 3 mm, base rounded to subtruncate, margin unequally serrate, apex acuminate. Inflorescences whorls in terminal heads (Ø up to 5 cm); floral leaves leaf-like, reduced, densely pubescent, glandular, briefly petiolate or subsessile, margin entire; bracts linear, about 1 cm, curved upward, hairy, glandular. Flowers with peduncle about 1 mm, puberulent; calyx tubular, narrow, 7–9 mm, pubescent, brown, glandular, with white barbs at opening; teeth subulate, equal, about 1 mm, apex spinescent; corolla purple-red, 3–4 × 7–9 mm, densely pubescent, glandular; upper lip curved slightly inward, entire; lower lip ± patent. Fruit achene obovoid, truncate.

Subspecies *Monarda fistulosa* L. var. *fistulosa* is one of the three subspecific entities of this taxon, the other two being *Monarda fistulosa* L. var. *menthifolia* (Graham) Fernald and *Monarda fistulosa* L. var. *mollis* (L.) L.

Cultivar There are many, but it is advisable to use local varieties, at least in natural areas of geographic distribution.

Related Species The genus *Monarda* includes 31 taxa.

Geographic Distribution United States, Canada. **Habitat** Upland forests, thickets, prairies. **Range** United States, Canada.

Beer-Making Parts (and Possible Uses) Flowering tops (plant substitute for *Monarda didyma*).

Chemistry Essential oil, aerial parts (identified components 98.11%): α-thujene 2.32%, α-pinene 2.28%, β-pinene 3.07%, myrcene 8.7%, δ-carene 3.57%, α-phellandrene 13.7%, p-cymene 13.27%, β-phellandrene 16.87%, methyl carvacrol 3.76%, thymol 28.38%, germacrene D 2.23%.

Style Herb Beer.

Source Fisher and Fisher 2017.

Monarda punctata L.
LAMIACEAE

Synonyms *Monarda lutea* Michx.

Common Names Dotted beebalm, horsemint, spotted beebalm.

Description Herb, perennial (3–10 dm). Stem thinly canescent, square sectioned. Leaves lanceolate or narrowly oblong, 2–8 cm, ± hairy. Inflorescences head-like, small, compact, 2- to 5-flowered. Flowers terminal and axillary, tubular, 13- to 15-nerved; corolla light yellow dotted with purple, strongly bilabiate; upper lip entire, narrow, strongly arched, lower wider; stamens included in upper lip.

Subspecies *Monarda punctata* L. comprises seven varieties: *Monarda punctata* L. var. *punctata* (typical variety), *Monarda punctata* L. var. *arkansana* (E.M.McClint. & Epling) Shinners, *Monarda punctata* L. var. *coryi* (E.M.McClint. & Epling) Shinners, *Monarda punctata* L. var. *intermedia* (E.M.McClint. & Epling) Waterf, *Monarda punctata* L. var. *lasiodonta* A.Gray, *Monarda punctata* L. var. *occidentalis* (Epling) E.J.Palmer & Steyerm., and *Monarda punctata* L. var. *villicaulis* (Pennell) Shinners.

Cultivar There are several, but it is advisable to use local varieties, at least in natural areas.

Related Species The genus *Monarda* includes 31 taxa.

Geographic Distribution United States, NE Canada. **Habitat** Arid, sandy soils. **Range** United States, NE Canada.

Beer-Making Parts (and Possible Uses) Flowering tops (plant substitute for *Monarda didyma*).

Chemistry No information available.

Style Herb Beer.

Source Fisher and Fisher 2017.

Morus alba L.
MORACEAE

Synonyms *Morus atropurpurea* Roxb., *Morus chinensis* Lodd. ex Loudon, *Morus intermedia* Perr., *Morus latifolia* Poir., *Morus multicaulis* (Perr.) Perr.

Common Names White mulberry, mulberry.

Description Tree, deciduous (4–8 m). Trunk with large irregular branches forming an enlarged globular crown; often pruned originating branches of equal size open like a fan; bark in youth grayish-yellowish color, almost smooth, then brownish color and longitudinal grooves; buds ovoid, small, pointed. Leaves alternate, nearly distichous, petiole fluted (20–30 mm), lamina entire, soft, ovate-acute (5–8 × 7–10 cm), slightly cordate, glabrous and shiny on both sides, except for short tufts of whitish hairs on underleaf at insertion of secondary and tertiary veins, edge irregularly toothed; sucker leaves deeply tripartite (3–5 lobes). Flowers monoecious (to a lesser extent hermaphrodite), united in catkin inflorescences: male cylindrical (2–4 cm), short peduncle, 4-parted perianth with 4 stamens;

female globular (1–2 cm), equally pedunculated, 4- to 5-parted perianth, 1 ovary and 2 stigmas. Fruits small (Ø 1.5–2 mm), fleshy pseudodrupae, each with 1 seed, whitish, more rarely pinkish, reddish, or even black, sweet when ripe. Infructescence syncarp, sorose (mulberry), oval-rounded (1–2 cm), pedunculate.

Subspecies In addition to the typical one, *Morus alba* L. includes a second variety: *Morus alba* L. var. *tatarica* (L.) Loudon.

Cultivar A small sample of the numerous cultivars of this species existing in the world includes Cheongil, Gaeryang, Guksang, K2, NG1, S14, S41, S30, S34, S36, S54, Shimheungppong, Somok, Whicaso, and Yongcheon.

Related Species There are 19 taxa belonging to the genus *Morus*.

Geographic Distribution E Asia (China). **Habitat** Widely cultivated in the past for silkworm breeding, sometimes subspontaneous on abandoned and uncultivated lands; in fresh, deep, and permeable soils, not clayey, and without humidity stagnation (0–700 m ASL). **Range** Cultivated worldwide in warm temperate climates.

Beer-Making Parts (and Possible Uses) Infructescence.

Chemistry Fruits: water 87.2%, sugars 2310 mg, lipids 1.16%, protein 1.62%, phosphorus 178.2 mg, potassium 954.2 mg, calcium 177.3 mg, sodium 41.7 mg, magnesium 168.9 mg, zinc 3.5 mg, iron 3.25 mg, manganes 0.61 mg, selenium 0.0067 mg, anthocyanins 116.67 mg, polyphenols 189.67 mg, flavonoids 90 mg.

Style American Stout, American Wheat Beer, American Wheat or Rye Beer, Belgian Specialty Ale, California Common Beer, Dunkelweizen, Fruit Beer, Scottish Export, Sweet Stout.

Beer Gelsobirra, Bi-Du (Rodero, Lombardy, Italy).

Source Calagione et al. 2018; Dabove 2015; Josephson et al. 2016; https://www.bjcp.org (29. Fruit Beer); https://www.brewersfriend.com/other/amora/; https://www.brewersfriend.com/other/mulberries/.

Morus nigra L.
MORACEAE

Synonyms None

Common Names Black mulberry.

Description Tree (4–8 m), deciduous, similar to the previous species. Differs as follows: pubescent twigs, more rounded, stiff leaves, 5–15 mm petiole, lamina deeply cordate at base, rough above, more densely pubescent below, subsessile sorosis, larger (2–2.5 cm), almost black in color at maturity, with somewhat less sweet flavor.

Cultivar Genotypes 64USA06, 64USA08, 64USA10, Emiralem, Salihli, and Tire are only a modest example of the wide variety of cultivars of this species.

Related Species There are 19 taxa belonging to the genus *Morus*.

Geographic Distribution SW Asia (Iran). **Habitat** Widely cultivated in the past for fruit and rarely subspontaneous. **Range** Cultivated worldwide in warm temperate climates.

Beer-Making Parts (and Possible Uses) Infructescence.

Chemistry Fruits: energy 180 kJ (43 kcal), carbohydrates 9.8 g (sugars 8.1 g, fiber 1.7 g), lipids 0.39 g, protein 1.44 g, thiamin (B1) 0.029 mg, riboflavin (B2) 0.101 mg, niacin (B3) 0.62 mg, vitamin B6 0.05 mg, folate (B9) 6 µg, choline 12.3 mg, vitamin C 36.4 mg, calcium 39 mg, iron 1.85 mg, magnesium 18 mg, phosphorus 38 mg, potassium 194 mg, zinc 0.12 mg.

Style American Stout, American Wheat Beer, American Wheat or Rye Beer, Belgian Specialty Ale, California Common Beer, Dunkelweizen, Fruit Beer, Scottish Export, Sweet Stout.

Source Calagione et al. 2018; Dabove 2015; https://www.bjcp.org (29. Fruit Beer); https://www.brewersfriend.com/other/amora/; https://www.brewersfriend.com/other/mulberries/.

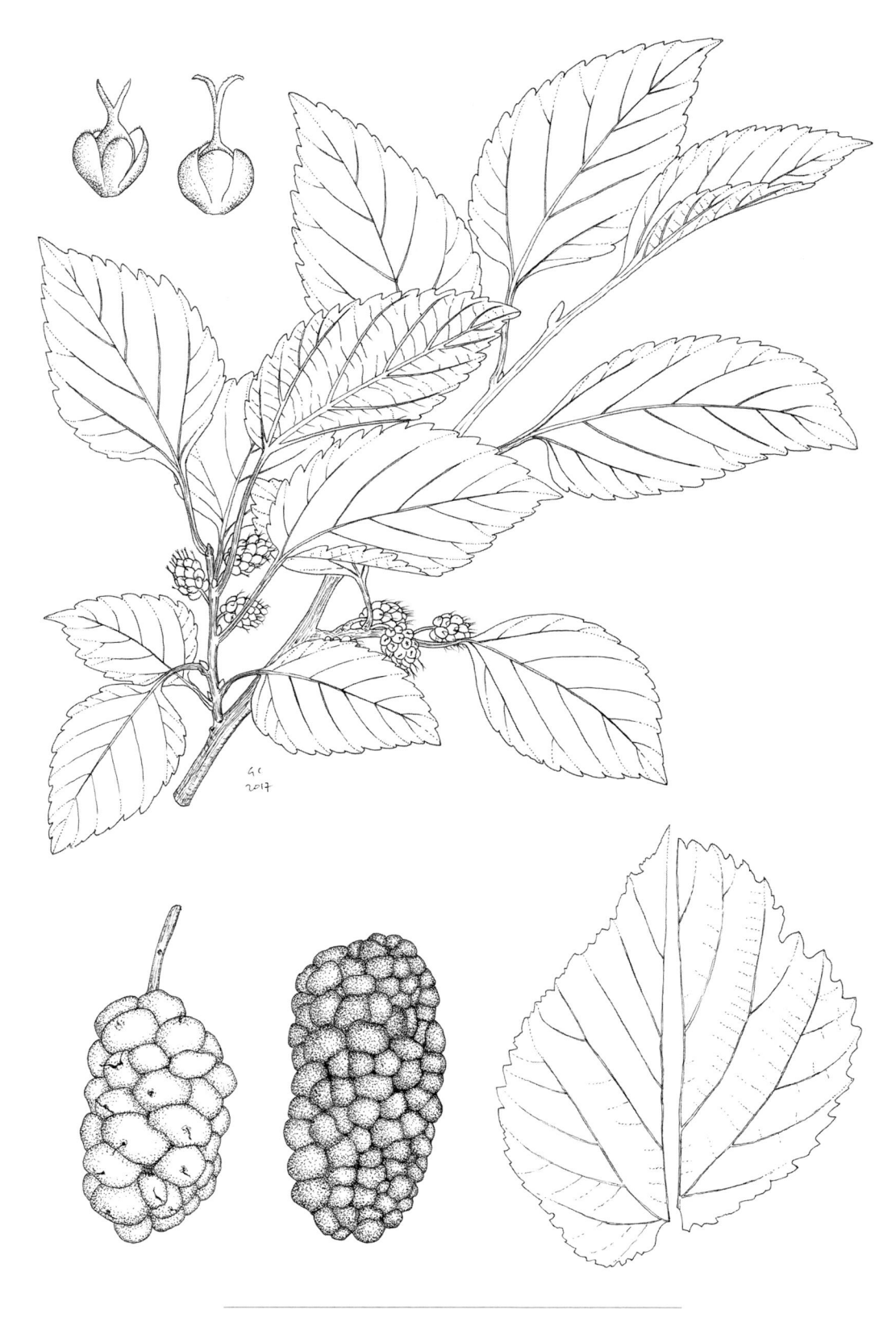

Morus alba L.—MORACEAE
Morus nigra L.—MORACEAE

Murraya koenigii (L.) Spreng.
RUTACEAE

Synonyms *Bergera koenigii* (L.) Roxb.

Common Names Curry tree, curry leaf-tree.

Description Tree, medium (4–10 m). Trunk Ø 30–80 cm. Leaves aromatic, compound, imparipinnate, with 11–21 oblong-lanceolate leaflets (1–2 × 2–4 cm), margin entire, apex emarginate, base oblique. Flowers fragrant, small (Ø 1.12 cm), hermaphrodite, white, collected in terminal panicles (60–90 flowers). Fruit subglobose berry (Ø 2.5 cm), purple-black color shimmering when ripe, edible pulp containing 1–2 seeds.

Cultivar There are many, each adapted to the particular environmental conditions of parts of the geographic distribution.

Related Species In addition to *Murraya koenigii* (L.) Spreng., 7 other species belong to the genus *Murraya*: *Murraya alata* Drake, *Murraya crenulata* (Turcz.) Oliv., *Murraya euchrestifolia* Hayata, *Murraya kwangsiensis* (C.C.Huang) C.C.Huang, *Murraya microphylla* (Merr. & Chun) Swingle, *Murraya paniculata* (L.) Jack, and *Murraya tetramera* C.C.Huang.

Geographic Distribution India. **Habitat** Evergreen and deciduous forests as an understory element. **Range** India, Sri Lanka, Bangladesh.

Beer-Making Parts (and Possible Uses) Leaves.

Chemistry Essential oil, fresh leaves (34 compounds): α-pinene 51.7%, sabinene 10.5%, β-pinene 9.8%, β-caryophyllene 5.5%, limonene 5.4%, bornyl acetate 1.8%, terpinen-4-ol 1.3%, γ-terpinene 1.2%, α-humulene 1.2%. Mature leaves contain water 63.2%, total nitrogen 1.15%, lipids 6.15%, total sugars 18.92%, starch 14.6%, fiber 6.8%, ash 13.06%; also a large number of secondary metabolites such as alkaloids (murrayastine, murrayalin, pypayafolinecarbazole), triterpenoids (cyclomahanimbine, tetrahydroromahanmbine), coumarins (murrayone, imperatoxin), and other compounds (mahanimbicin, bicyclomahanimbicin, phebalosin). Fruits (pulp): water 64.9%, sugars 9.76%, vitamin C 13.35% C, phosphorus 1.97%, potassium 0.082%, calcium 0.811%, magnesium 0.166%, iron 0.007%, acids trace tannins trace.

Style American IPA, Smoked Beer, Witbier.

Source Calagione et al. 2018; Heilshorn 2017; https://www.brewersfriend.com/other/curry/.

Musa acuminata × balbisiana Colla
MUSACEAE

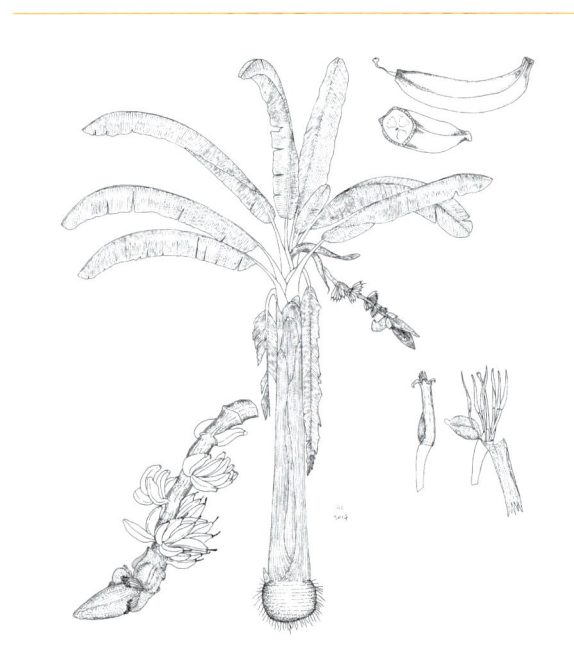

Synonyms None

Common Names Banana tree.

Description Herb, evergreen, perennial, with an arborescent habit (1.5–6 m). Stem (pseudostem) erect, fleshy, watery, consisting of broad leaf sheaths spirally arranged and overlapping (Ø up to 30 cm at the base), emerging from a more or less underground corm; laminae large (1–2 × 0.6 m). Inflorescence horizontal, then pendulous, with male flowers located distally to the floral axis and protected by leathery bracts; female flowers proximal. Fruit berry elongated (10–25 × 2–4 cm), cylindrical section or angled, green, yellow, or brown, with negative gravitropism (grows in the opposite direction to that of the force of gravity), with or without seeds (most of the cultivated varieties are seedless as they are triploid).

CULTIVAR The classification of the 300–1,000 known cultivars of banana is rather complex. Almost all modern cultivars of banana and plantain (cooking banana), mostly seedless, are polyploid hybrids of two species with seeds: *Musa acuminata* Colla and *Musa balbisiana* Colla. The classification is based on a system that indicates the degree of genetic inheritance of each of the two parent species and the number of chromosomes (ploidy). Bananas derive mainly from *Musa acuminata*, whereas plantains derive from *Musa balbisiana*. Linnaeus originally classified banana plants into two species (not hybrids as the currently accepted nomenclature indicates): *Musa × paradisiaca* L. (plantain) and *Musa × sapientum* L. (banana), now considered synonyms. Further research in SE Asia led to the description and naming of numerous other species, subspecies, and varieties. Later it was discovered that many cultivars were actually hybrids of two wild species, *Musa acuminata* and *Musa balbisiana* (both described in 1820 by the Italian botanist Luigi Colla), including Linnaeus's "species" now synonymized into *Musa × paradisiaca* L. In 1955, Norman Simmonds and Ken Shepherd proposed abandoning the traditional Latin names for cultivated bananas anticipating the *International Code of Nomenclature for Cultivated Plants*, which, in addition to the Latin names of the *International Code of Nomenclature for Algae, Fungi, and Plants*, gives cultivars names in a commonly spoken language and organizes them into groups. Today, cultivars derived from *M. acuminata* and *M. balbisiana* are classified using two criteria. The first is the number of chromosomes: diploid, triploid, or tetraploid. The second is the relationship to ancestral species, which can be determined by genetic analysis or by a point system (0–5), relative to several morphological characters, developed by Simmonds and Shepherd (pseudostem color, petiolar canal, stem, pedicels, ovule, bract elbow, bract bend, bract shape, bract apex, bract color, bract scarification, free tepal in male flower, male flower color, color of markings). The total score for a cultivar is around 15 if all characters belong to *M. acuminata* and around 75 if to *M. balbisiana*; intermediate scores suggest mixed origin. To the already intricate nomenclature (with numerous synonyms for the most common cultivars and even the same name for different cultivars) has recently been added the development of the so-called somaclones, clones obtained by causing mutations to cells of existing cultivars obtained by micropropagation and selecting the most promising individuals for one or two characteristics. Groups are named using a combination of the letters A and B, the number of which indicates ploidy, the proportion of A and B the contribution of ancestral species: group AA (*Musa acuminata* diploid, wild plants and cultivars: Chingan banana, Lacatan banana, Lady Finger banana [Sugar banana], Pisang jari buaya [Crocodile fingers banana], Señorita banana [Arnibal banana, Cariñosa, Cuarenta dias, Monkoy, Pisang Empat Puluh Hari, Pisang Lampung], Sinwobogi banana); group AAA (triploid *Musa acuminata*, wild plants and cultivars: Cavendish subgroup Dwarf Cavendish, Dwarf Red banana, East African Highland bananas, Giant Cavendish [Williams], Grand Nain [Chiquita], Gros Michel banana, Masak Hijau, Red Dacca, Robusta); group AAAA (*Musa acuminata* tetraploide, wild plants and cultivars: Bodles Altafort banana, Golden Beauty banana); group AAAB (*Musa × paradisiaca*, tetraploid cultivars: Atan banana, Goldfinger banana); group AAB (*Musa × paradisiaca*, triploid cultivars: French plantain, Green French banana, Horn plantain and Rhino Horn banana, Iholena, Latundan banana [Apple banana, Silk banana], Maoli-Popo'ulu, Maqueño banana, Mysore banana, Nendran banana, Pink French banana, Pisang Raja banana, Pisang Seribu banana, Pome banana, Popoulu banana, Prata-anã banana, Tiger banana]; AABB group (*Musa × paradisiaca*, tetraploid cultivars: Kalamagol banana, Pisang Awak [Ducasse banana]); group AB (*Musa × paradisiaca*, diploid cultivars: Ney Poovan banana); ABB group (*Musa × paradisiaca*, triploid cultivars: Benedetta banana, Blue Java banana [Ash plantain, Dukuru, Ice Cream banana, Ney mannan, Pata hina, Vata], Bluggoe banana, Cardaba banana, Pelipita banana [Pelipia, Pilipia], Saba banana [Cardaba, Dippig], Silver Bluggoe banana); group ABBB (*Musa × paradisiaca*, tetraploid cultivars: Tiparot banana); group BB (*Musa balbisiana*, diploid wild bananas); group BBB (*Musa balbisiana*, wild bananas and triploid cultivars: Kluai Lep Chang Kut).

Related Species The genus *Musa* includes approximately 90 taxa including the parental species of *Musa acuminata* × *balbisiana* Colla: *Musa acuminata* Colla and *Musa balbisiana* Colla.

Geographic Distribution Complex of hybrids stabilized by cultivation. **Habitat** Cultivated. **Range** India, China, Uganda, Philippines, Ecuador, Brazil, Indonesia, Colombia, Cameroon, Ghana.

Beer-Making Parts (and Possible Uses) Fruits.

Chemistry Energy 89 kcal, lipids 0.3 g, carbohydrates 23 g, fiber 2.6 g, sugars 12 g, protein 1.1 g, sodium 1 mg, potassium 358 mg, calcium 5 mg, magnesium 27 mg, iron 0.3 mg, vitamin A 64 IU, vitamin C 8.7 mg, vitamin B6 0.4 mg.

Style Berliner Weisse, Fruit Beer, Fruit Lambic.

Source Dabove 2015; Giaccone and Signoroni 2017; Hieronymus 2010; Jackson 1991; McGovern 2017; https://aob.oxfordjournals.org/content/88/6/1017.full.pdf; https://www.birradegliamici.com/news-birraie/birre-strane/; https://www.bjcp.org (29. Fruit Beer).

Myrica gale L.
MYRICACEAE

Synonyms *Gale belgica* Dumort., *Gale commune* J.Presl, *Gale palustris* Chev., *Gale portugalensis* Chev., *Gale uliginosa* Spach, *Myrica brabantica* Gray, *Myrica palustris* Lam.

Common Names Bog-myrtle, bog myrtle, sweetgale, sweet gale.

Description Shrub (up to 1 m), with resinous odor. Branches reddish-brown in color. Leaves deciduous, cuneate-oblanceolate (2–5 cm long), sometimes serrated toward apex, spirally arranged, grayish-green in color. Inflorescences male and female catkins on different plants; catkins develop before leaves; male flowers with 4 anthers, female with 1 style.

Related Species The genus *Myrica* includes 21 taxa.

Geographic Distribution W Europe, N America. **Habitat** Swamps, shallow waters. **Range** Similar to geographic distribution.

Beer-Making Parts (and Possible Uses) Twigs, leaves (containing various terpenes also found in hops), flowers, flower cones.

Toxicity The plant contains toxic flavoglycosides and considerable quantities of bitter tannins. It is considered a natural abortifacient; therefore, it should not be ingested by pregnant women. In sensitive subjects it can cause skin irritations. The essential oil extracted from the seeds is toxic and cannot be used in foods.

Chemistry Essential oil, leaves (53 compounds; 95.24% of total; 30- and 60-minute extractions): α-thujene 0.53–0.19%, α-pinene 3.89–1.51%, β-pinene 0.34–0.15%, myrcene 23.18–12.14%, α-phellandrene 9.90–6.49%, α-terpinene 0.14–0.11%, p-cymene 4.47–2.36%, limonene 11.20–6.75%, β-phellandrene 1.52–0.98%, (Z)-α-ocimene 4.31–2.89%, (E)-β-ocimene 4.53–3.16%, γ-terpinene 0.34–0.22%, terpinolene 0.15–0.12%, isopentyl isovalerate 0.27–0.22%, δ-elemene 0.45–0.43%, α-copaene 0.38–0.37%, geranyl acetate 0.83–0.73%, (Z)-α-bergamotene 0.85–0.95%, β-caryophyllene 9.31–10.97%, (E)-α-bergamotene 0.33–0.37%, β-gurjunene 0.58–0.53%, α-humulene 1.54–1.32%, (E)-cadin-1(6),4-diene 0.25–0.48%, γ-gurjunene 2.57–2.70%, γ-curcumene 0.65–1.51%, β-selinene 0–0.38%, ar-curcumene 0.75–0.81%, δ-selinene 0.40–0.40%, α-selinene 0.25–0.38%, (E,E)-α-farnesene 1.78–1.75%, β-bisabolene 0.41–0.51%, β-curcumene 0.96–1.57%, δ-cadinene 1.64–2.25%, (E)-cadin-1(2),4-diene 0.39–0.33%, (E)-γ-bisabolene 0–0.28%, (E)-nerolidol 0.25–0.73%, caryophyllene oxide 3.47–9.94%, humulene epoxide 0.38–1.34%, non-identified A 0.22–1.20%, eremoligenol 1.15–6.66%, tau-muurolol 0–0.33%, β-eudesmol 0–0.22%, α-eudesmol 0–0.31%, unidentified D 0.22–1.43%, γ-eudesmol acetate 0–0.43%, (2E,6E)-farnesol 0–0.19%, unidentified E 0.25–1.22%, pentadecanal 0.21–1.15%, α-eudesmol acetate 0–0.24%, rimuene 0–0.17%.

Style Gruit, Imperial Gruit, Strong Scotch Ale.

Beer Season of Epiphany, Jopen (Haarlem, The Netherlands).

Source Calagione et al. 2018; Giaccone and Signoroni 2017; Heilshorn 2017; Hieronymus 2012; Hieronymus 2016b; Jackson 1991; Josephson et al. 2016; McGovern 2017; Steele 2012; https://www.brewers-friend.com/other/bogmyrtle/; https://www.gruitale.com/bot_bogmyrtle.htm.

🍺 Gruit

Gruit was a mixture of aromatic herbs, spices, flowers, roots, resins, leaves, and barks used for centuries to bitter and flavor beer in continental Europe. Some of the herbs contained in gruit also had antiseptic properties, therefore helping to prevent the dreaded bacterial contamination of wort and beer. According to reliable sources, it was not uncommon for gruit to contain plant species known to have euphoric or even hallucinogenic effects, qualities particularly appreciated, together with the intoxicating power of ethanol, during religious rites. The composition of gruit was anything but constant, as it was usually an expression of the spontaneous flora of the territory where it was collected, composed, and used. Some species, however, almost always had a prominent place in gruit, and brewers rarely renounced using them. Species such as *Myrica gale*, *Ledum palustre*, and *Achillea millefolium* gave beer a dry and slightly bitter taste, as well as moderate narcotic properties. By adding other herbs in controlled doses, each supplier obtained a specific composition of their own gruit, whose exact formula, just like those of some of the most popular modern beverages, was kept secret. The economic importance of gruit was such that in the tenth century, strict state and fiscal control on its production and sale was institutionalized (*gruitrecht*), meaning these activities could be done only by authorized citizens. Gruit ales were narcotic, aphrodisiac, psychotropic, and euphoric, the exact opposite of hopped ale, which tended to be sedative and anaphrodisiac. Therefore, the complex implications should not be surprising, including the religious ones connected to the Protestant Reformation, which would have fed the long conflict between two opposite ways of understanding brewing. The advent of hops, capable of including many properties useful for brewing (aroma, bitterness, antiseptic) in a single plant, decreed the gradual but constant marginalization of gruit, until the middle of the fifteenth century, when the decline of this brewing practice led to its almost total disappearance. The craft revolution, contributing to the recovery from oblivion of many forgotten brewing techniques and styles, has played a role in the recovery of culture and technology, giving a new perspective to the term "gruit," which today also indicates a family of beers. Far-sighted artisanal producers, whose activities are not necessarily located in the traditional territories of gruit, have found a way to reintroduce the ancient mix of aromatic herbs (or, more frequently, single herbs) to the practice of brewing, and even to imagine new uses for those plants that history seemed to have relegated to the yellowed pages of old manuals.

Myristica fragrans Houtt.
MYRISTICACEAE

Synonyms *Aruana silvestris* Burm.f., *Myristica aromatica* Lam., *Myristica aromatica* Sw., *Myristica moschata* Thunb., *Myristica officinalis* L.f., *Palala fragrans* (Houtt.) Kuntze.

Common Names Nutmeg.

Description Tree, evergreen (5–20 m), aromatic. Bark contains pink or red sap. Leaves (5–15 × 2–7 cm) alternate, pointed, dark green, arranged along branches and borne by 1 cm long petioles, glossy on upper leaf. Flowers dioecious, pale yellow, waxy, fleshy, bell shaped. Male flowers 5–7 mm long in clusters of 1–10; female flowers up to 1 cm long in clusters of 1–3; occasionally both sexes found on the same tree. Fruit oval or pyriform, drooping, yellow, smooth, 6–9 cm long, with longitudinal ridge and fleshy skin; when ripe, skin splits into 2 halves revealing a shiny seed, ovoid (20–30 × 15–18 mm), violet-brown in color, surrounded by a net of red or crimson leathery tissue; the brown seed represents nutmeg, whereas

Myrica gale L.
MYRICACEAE

another spice called mace is obtained from the red involucre.

Cultivar The variability of cultivars of this species is expressed in factors such as the speed of growth and the size and shape of leaves, flowers, and fruits; for production, the number of fruits, their weight (factors usually negatively correlated), and the weight ratio among the parts composing them are very important. Some cultivars developed in India include Konkan Shrimathi, Konkan Sugandha, and Konkan Swad.

Related Species In addition to *Myristica fragrans* Houtt., the genus *Myristica* includes 10 species, including *Myristica argentea* Warb., *Myristica fatua* Houtt. (= *Virola surinamensis* [Rol. ex Rottb.] Warb.), *Myristica malabarica* Lam., *Myristica otoba* Humb. & Bonpl. (= *Otoba novogranatensis* Moldenke), *Myristica speciosa* Warb., and *Myristica succedanea* Blume.

Geographic Distribution Banda Islands (Moluccas Archipelago, Indonesia). **Habitat** Species grown in tropical climates. **Range** Indonesia, India, Malaysia, Papua New Guinea, Sri Lanka, Caribbean Islands (Grenada, St. Vincent and the Grenadines).

Beer-Making Parts (and Possible Uses) Seeds, reddish arils covering the seeds (give two different spices).

Toxicity In low doses, *Myristica fragrans* does not cause physiological or neurological problems; however, in high doses it produces effects (convulsions, palpitations, nausea, dehydration, migraine, widespread pains, delirium, hallucinations, rarely death) that can last up to several days, because of the presence of myristicin, a psychoactive substance. Once erroneously considered an abortifacient spice, nutmeg appears to be safe for culinary use, even during pregnancy. However, it inhibits the production of prostaglandins and contains hallucinogens that can cause problems to the fetus when consumed in large quantities. Highly neurotoxic to dogs.

Chemistry Essential oil, seeds (37 compounds; 91% of the oil): α-pinene 8.5%, β-pinene 3.5%, myrcene 2.5%, γ-terpinene 2%, α-terpinene 9.8%, limonene 8.8%, trans-β-ocimene 0.7%, γ-terpinene 9.9%, terpinolene 5.1%, linalool 0.7%, terpineol-4 21.3%, β-phenyl alcohol 8.1%, D-citronellol 0.2%, isobornyl acetate 0.2%, safrole 0.4%, 2,6-dimethyl 2,6-octadiene 0.4%, eugenol 0.2%, geranyl acetate 0.4%, methyl eugenol 0.1%, trans-β-caryophyllene 0.3%, trans-α-bergamotene 0.1%, trans-β-farnesene 0.1%, α-humelene 0.1%, δ-cadinene naphthalene 0.1%, saychellene 0.1%, D-germacrene 0.1%, trans-methyl isoeugenol 0.1%, α-muurolene 0.4%, β-bisabolene 0.1%, myristicin 6.8%, elemicin 0.6%, guaiol 0.1%, octacosane 0.1%, hexacosane 0.1%, heptacosane 0.1%, elcosane 0.1%, octadecane 1.3%.

Style Apple Pie Spice, Alternative Sugar Beer, American Amber Ale, American Barleywine, American Brown Ale, American IPA, American Pale Ale, American Porter, American Stout, American Strong Ale, American Wheat or Rye Beer, Autumn Seasonal Beer, Baltic Porter, Belgian Blond Ale, Belgian Dark Strong Ale, Belgian Dubbel, Belgian Golden Strong Ale, Belgian Pale Ale, Belgian Specialty Ale, Belgian Tripel, Berliner Weisse, Bière de Garde, Blonde Ale, Braggot, British Brown Ale, Brown Porter, California Common Beer, Classic American Pilsner, Classic Rauchbier, Clone Beer, Cream Ale, Dry Stout, Dunkelweizen, Düsseldorf Altbier, English Barleywine, Experimental Beer, Extra Special/Strong Bitter (ESB), Flanders Brown Ale/Oud Bruin, Flanders Red Ale, Foreign Extra Stout, Fruit and Spice Beer, Fruit Beer, Fruit Lambic, Gruit, Holiday/Winter Special Spiced Beer, Imperial Stout, Irish Red Ale, Kölsch, Metheglin, Mild, Mixed-style Beer, Northern English brown Ale, Oatmeal Stout, Oktoberfest/Märzen, Old Ale, Other Smoked Beer, Robust Porter, Russian Imperial Stout, Saison, Schwarzbier, Scottish Export, Scottish Export 80/-, Southern English Brown, Special/Best/Premium Bitter, Specialty Beer, Specialty Fruit Beer, Specialty IPA (Black IPA), Spice/Herb/Vegetable Beer, Standard American Lager, Straight (unblended) Lambic, Strong Bitter, Strong Scotch Ale, Sweet Stout, Traditional Bock, Trappist Single, Wee Heavy, Weizen/Weissbier, Weizenbock, Winter Seasonal Beer, Witbier, Wood-aged Beer.

Beer 10-Lords-A-Leaping, the Bruery (California, United States).

Source Heilshorn 2017; Hieronymus 2010; https://www.beeradvocate.com/beer/style/72/; https://www.bjcp.org (29B. Fruit and Spice Beer; 30B. Autumn Seasonal Beer); https://www.brewersfriend.com/other/nutmeg/.

Myrocarpus fastigiatus Allemao
FABACEAE

Synonyms None

Common Names Cabreúva.

Description Tree (2.5–25 m). Stem with cylindrical, lenticulate branches. Leaves compound 7- to 12-foliolate; petiole (6–14 mm), rachis (1.9–6 cm) and petiole (1.2–3.7 mm) sericeous to glabrous; rachis sparsely tomentose; leaflets 1.9–3.3 × 1–1.9 cm, oblong, elliptic, apex usually obtuse and retuse, rarely emarginate, base attenuated, margin entire, leaf variously colored, sparsely sericeous or glabrous. Inflorescences twinned racemes, fasciculate, axillary, on aphyllous branches, axis ferruginous-tomentose in color, shorter than leaves; bracts linear 0.5–0.7 mm; peduncle sparsely ferruginous-tomentose, 1.0–1.9 mm. Flowers 2.0–2.5 mm, pale yellowish; calyx campanulate (1.6–2.8 × 1–1.6 mm), laciniae obtuse, externally densely ferruginous-tomentose; petals 1.4–3.8 × 0.4 mm, claw about 0.8 mm, spatulate, apex truncate, base attenuate, glabrous; stigma two different lengths; gynoecium 2–4.5 mm, sparsely sericeous, stigma punctiform. Fruit samara (3.5–7.7 × 0.7–1.5 cm), yellowish or brownish in color; seminal area 4–6.4 mm, brownish, scaly, base asymmetrical, calyx ± persistent, apex obtuse, sometimes apiculate.

Related Species There are five species belonging to the genus *Myrocarpus*: *Myrocarpus emarginatus* A.L.B.Sartori & A.M.G.Azevedo, *Myrocarpus fastigiatus* Allemao, *Myrocarpus frondosus* Allemao, *Myrocarpus leprosus* Pickel, and *Myrocarpus venezuelensis* Rudd.

Geographic Distribution Brazil (from Pernambuco to Rio de Janeiro). **Habitat** In NE Brazil it is one of the inland mountain forest species, being tolerant to lower temperatures. **Range** Brazil (from Pernambuco to Rio de Janeiro).

Beer-Making Parts (and Possible Uses) Wood for barrels (first holding cachaça, then for aging beer).

Wood Differentiated with white-yellowish sapwood and brown-yellowish or chocolate heartwood, oily appearance, with darker variegations and golden reflections, slightly scented. Fine texture and varied fibers, often giving a striped effect. Difficult to work but amenable to good cleaning (WSG 0.90–1.10). Used for permanent constructions, flooring, handles, turnery, plywood, veneers.

Style Saison.

Source Cantwell and Bouckaert 2016; Cantwell and Bouckaert 2018 (although these two sources include the scientific name of the species *Myroxylon peruiferum* L.f. as corresponding to the common name cabreùva, it is more likely that it identifies the species *Myrocarpus frondosus* Allemao and *Myrocarpus fastigiatus* Allemao, cited here).

Myrocarpus frondosus Allemao
FABACEAE

Synonyms *Leptolobium punctatum* Benth., *Myrocarpus paraguariensis* Hallier f.

Common Names Cabreùva.

Description Tree (6–27.5 m). Trunk (Ø 60–90 cm) with cylindrical, rarely quadrangular, lenticulate branches. Leaves compound with 5–8 leaflets; petiole (9–20 mm), rachis (1.7–6.5 cm) and petiole (2–4 mm) glabrous, sometimes sparsely tomentose; leaflets (3.7–5.6 × 1.3–3 cm) elliptic to ovate, apex acuminate, mucronate, retuse, base attenuate, asymmetrical, rounded, rarely subcordate, margin entire or crenate, sinuous or entire, leaves of different green colors, both glabrescent, veins conspicuous in the upper one. Inflorescence raceme simple, axillary, or terminal, sometimes on aphyllous branches, ± as long as leaves, axis red-tomentose, rarely red-sericeous; bracts deltoid (0.5–0.8 mm), concave, externally red-tomentose; pedicel (1.5–2 mm) and calyx ferruginous-sericeous. Flowers 2.6–4 mm, white or greenish-white; calyx campanulate (2.5–4 × 1.9–3.4 mm), laciniae obtuse, rarely acute; petals 4–6.3 × 0.8–1.5 mm, with claw 1.5–2.6 mm, elliptic, apex obtuse, base attenuate, glabrous, rarely sericeous; stamens two different lengths; gynoecium 3.5–7 mm, glabrous, stigma punctiform. Fruit samara (4.2–8 × 0.8–1.8 cm), yellowish color, seminal area 3–7 mm, brownish, scaly, base asymmetrical, calyx persistent, rarely also stamens, apex acute, apiculate.

Related Species The genus *Myrocarpus* includes five species: *Myrocarpus emarginatus* A.L.B.Sartori & A.M.G.Azevedo, *Myrocarpus fastigiatus* Allemao, *Myrocarpus frondosus* Allemao, *Myrocarpus leprosus* Pickel, and *Myrocarpus venezuelensis* Rudd.

Geographic Distribution Bolivia, Brazil (Paraná). **Habitat** Remarkable ecological variability but absent in the Cerrado; forest vault species, in both primary and secondary forests; in Brazil and in Atlantic rainforests and semi-deciduous forests of the Pará. **Range** Bolivia, Brazil (Paraná).

Beer-Making Parts (and Possible Uses) Wood for casks (first holding cachaça, then for aging beer).

Wood Differentiated, with white-yellowish sapwood and yellowish-brown or chocolate heartwood, oily appearance, with darker variegations and golden reflections, slightly scented. Fine texture and varied fibers, often giving a striped effect. Difficult to work but amenable to good cleaning (WSG 0.90–1.10). Used for permanent constructions, flooring, handles, turnery, plywood, veneers.

Style Saison.

Source Cantwell and Bouckaert 2016; Cantwell and Bouckaert 2018 (although these two sources include the scientific name of the species *Myroxylon peruiferum* L.f. as corresponding to the common name cabreúva, it is more likely that it identifies the species *Myrocarpus frondosus* Allemao and *Myrocarpus fastigiatus* Allemao, cited here).

Myroxylon peruiferum L.f.
FABACEAE

Synonyms *Myrospermum pedicellatum* Lam.

Common Names Cabreuva (probable common name of *Myrocarpus fastigiatus* and/or *Myrocarpus frondosus)*, quina or quina-quina (common name of *Myroxylon peruiferum*).

Description Tree (6–30 m). Leaves 5- to 15-foliolate, densely to sparsely ferruginous-tomentose, petiole 0.5–3.4 cm, rachis 3.2–14 cm; leaflets petioles

1–6 mm; leaflets 2.9–10.5 × 1–3.3 cm, oblong, broadly elliptic to ovate, apex obtuse, retuse to acuminate, base oblique, rounded or attenuated to subcordate, terminal leaflet narrowly elliptic, obovate, or rhombic, base ovate, margin sinuous, entire, rarely crenate, upper surface glabrous or sparsely tomentose on midrib, lower surface sparsely tomentose on lamina, tomentose on midrib, shiny, prominent veins, conspicuous pellucid spots and stripes throughout lamina. Inflorescence raceme axillary or terminal, longer than leaves, axis ferruginous-tomentose, 11–20 cm; bracts deltate, apex acute, base truncate, concave, ascending, about 1 mm; pedicel 9–16 mm, grayish or ferruginous-tomentose. Flowers 7–9 mm; calyx 5–7 × 9.6–11.4 mm, grayish-tomentose externally; vexillum 3.3–4.6 × 6.5–9.7 mm, claw 4–6.6 mm, broadly ovate, apex rounded or emarginate, base attenuated to rounded, margin entire, externally glabrous (rarely with trichomes at apex); other petals 5.9–10.3 × 0.7–1.5 mm, lanceolate, apex acute, base attenuated, margin sinuous, sometimes externally sericeous at apex and margin; stamens about 9 mm, filaments about 4 mm, anthers about 5.2 mm; ovary about 6.5 mm, briefly stipitate. Fruit samara 5.7–9.8 cm; seminal chamber 1–2 cm, elliptic.

Related Species In addition to *Myroxylon peruiferum* L.f., the species *Myroxylon balsamiferum* Harms, *Myroxylon balsamum* (L.) Harms (two varieties known for this taxon) and *Myroxylon nitidum* (Hell.) Kuntze are known for the genus *Myroxylon*.

Geographic Distribution CS America (NW Argentina, Bolivia, Brazil, Colombia, Costa Rica, Ecuador, El Salvador, Guyana, Honduras, SC Mexico, Nicaragua, Peru). **Habitat** Mesophilic forests and arid habitats (540–2,000 m ASL). In Bolivia, species known for dry forests in stony and/or rocky soils and for residual forests in low valleys. **Range** America (from Mexico to N Argentina to S Brazil).

Beer-Making Parts (and Possible Uses) Wood for barrels (first holding cachaça, then for aging beer).

Toxicity Although severe allergic reactions are uncommon, wood from this species has been reported to cause skin and respiratory irritation.

Wood Clearly differentiated, with white sapwood and orange, reddish-brown, or purple heartwood, sometimes variegated, pleasantly scented, fine to medium texture and varied grain. Hard, difficult to work but amenable to good cleaning, long lasting (WSG 0.85–1). Used for civil and naval constructions, carriage works, floorboards, carpentry, veneers, ship's masts, and for the extraction of oleoresin, known as tolu balsam.

Style Saison.

Source Cantwell and Bouckaert 2016; Cantwell and Bouckaert 2018 (although these two sources state the scientific name of the species described here, it is possible that the common name cabreùva more properly corresponds to *Myrocarpus frondosus* Allemao and/or *Myrocarpus fastigiatus* Allemao, also of the *Fabaceae* family).

Myrrhis odorata (L.) Scop.
APIACEAE

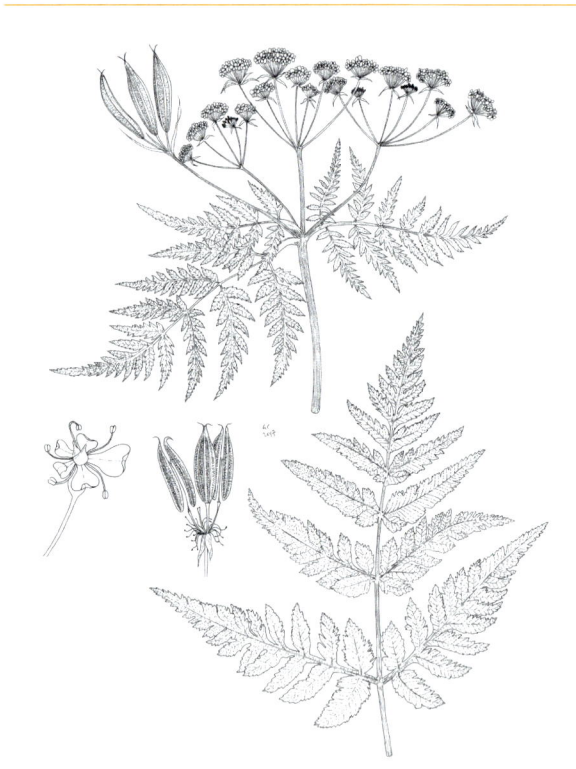

Synonyms *Chaerophyllum odoratum* (L.) Crantz, *Lindera odorata* (L.) Asch., *Scandix odorata* L., *Selinum myrrhis* E.H.L.Krause.

Common Names Sweet cicely.

Description Herb, perennial (5–12 dm, up to 20 dm), aromatic. Stems erect, hollow, pubescent, coppiced above. Leaves slightly pubescent, similar in shape to fern fronds, outline triangular, 4-pinnate and pinnate-partite, with serrated oval laciniae. Flowers with wedge-shaped-obovate petals, white in color; large hermaphrodite flowers borne by stout petioles, smaller male flowers borne by slender petioles. Inflorescence apical umbel with 4–15 pubescent rays with dense frizzy hairs, involucre absent; secondary umbel with involucre (4–5 reflexed bracts). Fruit achene fusiform, compressed and ribbed, covered with rough hairiness, shiny black color at maturity, exudes strong anise odor.

Related Species Besides the species cited here, only *Myrrhis nuda* (Torr.) Greene belongs to the genus *Myrrhis*.

Geographic Distribution Orophyte S Europe. **Habitat** Mountain and subalpine meadows, in damp places (1,000–2,100 m ASL). **Range** Widely cultivated species.

Beer-Making Parts (and Possible Uses) Leaves, seeds, roots.

Chemistry Essential oil: p-cymene 62%, α-terpinene 8.9%, δ-cadinene 5.3%, patchoulene 4%, pyrazine 3.1%, α-selinene 3.1%, limonene 2.1%, champhene 1.9%, seychellene 1.8%, α-patchoulene 1.7%, β-pinene 1.1%, calamenene 1%, α-gurjunene 0.9%, β-caryophyllene 0.7%, bi-cyclo-sesqui-phellandrene 0.4%, epi-zonarene 0.2%, α-phellandrene 0.2%.

Style Spice/Herb/Vegetable Beer.

Source Hieronymus 2016b; https://www.brewersfriend.com/other/saksankirveli/.

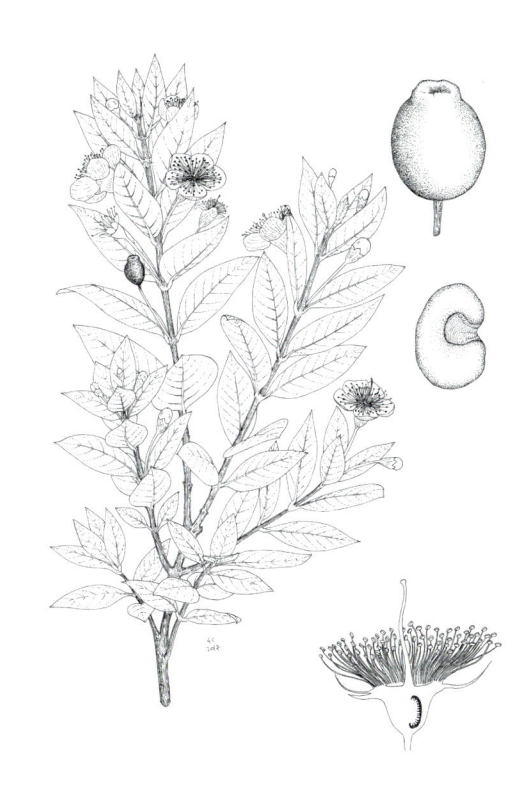

Myrtus communis L.
MYRTACEAE

Synonyms *Myrtus acuta* Mill., *Myrtus acutifolia* (L.) Sennen & Teodoro, *Myrtus augustini* Sennen & Teodoro, *Myrtus baetica* (L.) Mill., *Myrtus baui* Sennen & Teodoro, *Myrtus belgica* (L.) Mill., *Myrtus borbonis* Sennen, *Myrtus briquetii* (Sennen & Teodoro) Sennen & Teodoro, *Myrtus christinae* (Sennen & Teodoro) Sennen & Teodoro, *Myrtus eusebii* (Sennen & Teodoro) Sennen & Teodoro, *Myrtus gervasii* (Sennen & Teodoro) Sennen & Teodoro, *Myrtus italica* Mill., *Myrtus josephi* Sennen & Teodoro, *Myrtus littoralis* Salisb., *Myrtus macrophylla* J.St.-Hil., *Myrtus media* Hoffmanns., *Myrtus microphylla* J.St.-Hil., *Myrtus minima* Mill., *Myrtus mirifolia* Sennen & Teodoro, *Myrtus oerstedeana* O.Berg, *Myrtus petriludovici* (Sennen & Teodoro) Sennen & Teodoro, *Myrtus rodesi* Sennen & Teodoro, *Myrtus romana* (L.) Hoffmanns., *Myrtus romanifolia* J.St.-Hil., *Myrtus sparsifolia* O.Berg, *Myrtus theodori* Sennen, *Myrtus veneris* Bubani, *Myrtus vidalii* (Sennen & Teodoro) Sennen & Teodoro.

Common Names Myrtle, true myrtle.

Description Shrub (5–25 dm), evergreen, with resinous aromatic scent. Bark pinkish with longitudinal fracture, desquamating in fibrous bundles; branches opposite. Leaves opposite, leathery, sessile;

lamina lanceolate or elliptic (8–12 × 20–32 mm). Flowers solitary or paired at leaf axils; peduncles 12–18 mm; petals white, subrounded (7 mm); stamens numerous. Fruit berry ellipsoid or subspherical (6–9 mm), surmounted by the rudiments of the calyx.

Subspecies Besides the nominal subspecies, there is also *Myrtus communis* L. subsp. *tarentina* (L.) Nyman.

Cultivar There are many cultivars of *Myrtus communis*: Acutifolia, Italica, Microphylla, Nana, Tarentina (considered as a subspecies), Variegata; some with intermediate characters of Microphylla × Communis, Microphylla × Communis × Tarentina, Tarentina × Communis, Variegata × Communis, Variegata × Microphylla, Variegata × Tarentina.

Related Species Besides *Myrtus communis* L., the genus *Myrtus* includes 2 species: *Myrtus nivelii* Batt. & Trab. and *Myrtus phyllireaefolia* (A.Rich.) Kuntze.

Geographic Distribution Steno-Mediterranean. **Habitat** Mediterranean shrubland. **Range** Plant grown in all areas with a Mediterranean climate.

Beer-Making Parts (and Possible Uses) Leaves, fruits.

Chemistry Essential oil, leaves: α-pinene 37.8%, 1,8-cineole 23.1%, limonene 17.1%, linalool 10.1%. Other compounds in leaves: flavonoids (quercetin, catechin, myricetin), myrtucommulone A, myrtucommulone B, semi-myrtucommulone, glycosides (gallomyrtucommulone A, gallomyrtucommulone B, gallomyrtucommulone C, gallomyrtucommulone D). Fruit: volatile oils, oleic acid 67.07%, palmitic acid 10.24%, stearic acid 8.19%, tannins, sugars, flavonoids, organic acids (citric acid, malic acid). Essential oil, flowers (1.75% of the extract): α-pinene 48.54%, 1.2 cineol 14.75%, myrtenal 5.01%, myrtenol 4.01%, myrtenil acetate 3.45%, myrcene 2.09%, linalool 2.01%, geraniol 1.67%.

Style American Pale Ale, Gruit, Spice/Herb/Vegetable Beer.

Beer Mutta Affumiada, Birrificio di Cagliari (Cagliari, Sardinia, Italy).

Source Heilshorn 2017; https://www.brewersfriend.com/other/mirto/; https://www.inbirrerya.com/2007/10/13/i-trappisti-e-la-birra-parte-3/.

Nasturtium officinale R.Br.
BRASSICACEAE

Synonyms *Arabis nasturtium* Clairv., *Baeumerta nasturtium* P.Gaertn., B.Mey. & Schreb., *Baeumerta nasturtium-aquaticum* (L.) Hayek, *Cardamine aquatica* (Garsault) Nieuwl., *Cardamine fontana* Lam., *Cardamine nasturtium* (Moench) Kuntze, *Cardamine nasturtium-aquaticum* (L.) Borbás, *Cardaminum nasturtium* Moench, *Crucifera fontana* E.H.L.Krause, *Nasturtium fontanum* Asch., *Nasturtium nasturtium-aquaticum* (L.) H.Karst., *Nasturtium siifolium* Rchb., *Radicula nasturtium* (Moench) Druce, *Radicula nasturtium-aquaticum* (L.) Britten & Rendle, *Rorippa nasturtium* (Moench) Beck, *Rorippa nasturtium-aquaticum* (L.) Hayek, *Rorippa officinalis* (W.T. Aiton) P.Royen, *Sisymbrium nasturtium* (Moench) Willd., *Sisymbrium nasturtium-aquaticum* L.

Common Names Watercress.

Description Plant, perennial, herbaceous (3–4 dm). Stem ascending, glabrous, branched at the top. Basal leaves with 3–5 cm petiole, 2–3 pairs of progressively enlarged lateral segments, the larger ones patent or ± connate, ovate, obscurely serrate at margin, larger reniform terminal segment; cauline leaves with shorter petiole and subrounded or oval terminal segment. Flowers with brownish sepals, the 2 inner ones saccate at the base; petals milky white, more or less patent; stigmas with yellow anthers. Inflorescence shortened raceme. Fruit siliqua (2 × 13–18 mm). Seeds in 2 sets.

Related Species The genus *Nasturtium* includes 15 species.

Geographic Distribution Cosmopolitan. **Habitat** Still and running waters, banks of streams (0–2,500 m ASL). **Range** Cosmopolitan.

Beer-Making Parts (and Possible Uses) Leaves, flowering tops.

Chemistry Energy 32.29 kcal/100 g, water 87.5%, ash 0.36%, lipids 0.81%, protein 2.85%, fiber 1.06%, carbohydrates 7.4%. Essential oil, leaves (9 compounds; 97% of total oil): myristicin 57.6%, α-terpinolene 8.9%, limonene 6.7%. Stems (8 compounds; 100% of the oil): caryophyllene oxide 37.2%, p-cymene-8-ol 17.6%, α-terpinolene 15.2%, limonene 11.8%. Essential oil, flowers (15 compounds; 94.7% of oil): limonene 43.6%, α-terpinolene 19.7%, p-cymene-8-ol 7.6%, caryophyllene oxide 6.7%.

Style Saison.

Source https://www.brewersfriend.com/other/nasturtium/.

Nepeta cataria L.
LAMIACEAE

Synonyms *Calamintha albiflora* Vaniot, *Cataria vulgaris* Gaterau, *Glechoma cataria* (L.) Kuntze, *Glechoma macrura* (Ledeb. ex Spreng.) Kuntze, *Nepeta bodinieri* Vaniot, *Nepeta ceretana* Sennen, *Nepeta citriodora* Dumort., *Nepeta laurentii* Sennen, *Nepeta macrura* Ledeb. ex Spreng, *Nepeta minor* Mill., *Nepeta mollis* Salisb., *Nepeta tomentosa* Vitman.

Common Names Catnip, catswort, catmint.

Description Plant, perennial, herbaceous (3.5–20 dm), with characteristic minty odor. Tuberous roots from which rhizomes branch. Stem vigorous, erect, tetragonal, covered with more or less scattered and patent hairs, sometimes hooked (0.2–0.45 mm), branched, more or less reddened in the abaxial part. Leaves opposite, petiolate, in the lower petioles 1–3.5 cm long, gradually shorter in the middle and upper ones; upper leaf shiny, light green, subglabrous, underleaf grayish owing to felty pubescence; lamina triangular-ovate with cordate base (2.7–12 × 1.1–6.3 cm), margin dentate-crenate. Leaf bracts petiolate (0.3–2.8 mm), leaf-like, the lower 7–36 × 4.8–22 mm, the upper lanceolate with ± dentate-crenate margin. Inflorescence verticillate (9–23.5 mm). Flowers hermaphrodite on peduncles 3.1–10.6 mm, bracts linear, shorter than calyx (2.5–4.4 × 0.2–0.5 mm); calyx zygomorphic (5–8 mm), violet-green, covered with gray hairs, 5 subequal teeth shorter than tube and furrowed by 15 raised nerves; coralline tube curved, covered with ± patent hairs, just shorter than calyx (4.7–7.2 mm); corolla whitish or pale lilac in color, 5.7–9 mm; upper lip bilobed, lower trilobed (2 lateral and 1 larger central), with lilac macules and dense patent hairs; stamens filamentous, white, with purple-reddish anthers; stigma 4.9–8.4 mm. Fruit nucule brown in color (1.1–1.7 × 0.6–1 mm), smooth, covered with short hairs at apex.

Related Species The genus *Nepeta* has about 280 taxa.

Geographic Distribution Mediterranean. **Habitat** Grasslands, edges of hillside paths, rubble, old walls and canopies of deciduous forests, preferably in well-drained soils (0–1,200 m ASL). **Range** Mediterranean.

Beer-Making Parts (and Possible Uses) Leaves.

Chemistry Essential oil: 4a-α,7-α,7a-β-nepetalactone 55-58%, 4a-α,7a-β,7a-α-nepetalactone 30–31.2%.

Style American Amber Ale, California Common Beer, Cream Ale, Sweet Stout.

Beer Pussycat, Birrificio Lambrate (Milan, Lombardy, Italy).

Source Heilshorn 2017; https://birrificiolambrate.com/pussycat/; https://www.brewersfriend.com/other/catnip/.

Nicotiana tabacum L.
SOLANACEAE

Synonyms *Nicotiana chinensis* Fisch. ex Lehm., *Nicotiana latissima* Mill., *Nicotiana mexicana* Schltdl., *Nicotiana pilosa* Dunal.

Common Names Tobacco.

Description Herb, annual or perennial (8–30 dm), pubescent-glandulous and viscous. Stem erect, branched only in the inflorescence. Leaves elliptic or ± ovate-acuminate, the largest 2–5 dm. Inflorescence panicle apical corymbiform. Flower with 1–2 cm tubular calyx; corolla 3–5 cm, ± red, pinkish, or yellowish. Fruit ovoid capsule (1.5–2 cm).

Subspecies *Nicotiana tabacum* L. includes four varieties: the typical, *Nicotiana tabacum* L. var. *tabacum*, *Nicotiana tabacum* L. var. *fruticosa* (L.) Hook.f., *Nicotiana tabacum* L. var. *loxensis* (Kunth) Kuntze, and *Nicotiana tabacum* L. var. *virginica* (C.Agardh) Comes.

Cultivar There are hundreds, including Aromatic Fire-Cured, Brightleaf Tobacco, Burley, Cavendish, Corojo, Criollo, Dokha, Ecuadorian Sumatra, Habano, Habano 2000, Latakia, Maduro, Oriental Tobacco, Perique, Shade Tobacco, Thuoc lao, Type 22, White Burley, Wild Tobacco, CC 13, CC 27, CC 33, CC 37, CC 67, CC 143, CC 700, GF 318, GL 395, K 149, K 326, K 346, K 394, K 399, K 730, McNAIR 944, NC 55, NC 72 1994 PVPA, NC 102, NC 196, NC 291, NC 297, NC 299, NC 606, NC 810, NC 925, PVH 1118, PVH 1452, PVH 2110, PVH2275, PVH 2310, RG 17, SPEIGHT 168, SPEIGHT 210, SPEIGHT 220, SPEIGHT 225, SPEIGHT 227, SPEIGHT 234, SPEIGHT 236, SPEIGHT G-28, SPEIGHT G-70, SPEIGHT H-20, SPEIGHT NF 3, VA 119, and Y1.

Related Species Of the more than 60 species in the genus *Nicotiana*, in addition to *N. tabacum*, the most commercially important is *Nicotiana rustica* L.

Geographic Distribution N America. **Habitat** Cultivated species. **Range** China, Brazil, India, United States, Indonesia, Pakistan, Malawi, Argentina, Zambia, Mozambique.

Beer-Making Parts (and Possible Uses) Leaves (used to adulterate porter beer with substances that simulate the effect of alcohol but are less expensive than raw materials such as malt, sugar, and molasses).

Toxicity The leaves of this plant, as is well known, contain 1–3% nicotine, a very toxic alkaloid.

Chemistry Nicotine 1–3%, ammonium 0.019–0.159%, glutamine 0.020–0.041%, asparagine 0.016–0.111%, acetic acid 0.022–0.194%, malic acid 2.43–6.75%, citric acid 0.78–8.22%, oxalic acid 0.81–3.16%, essential oils 0.140–0.248%, resins 8.94–11.28%, sugars 6.77–22.09%, pectin 6.77–12.41%, fiber 6.63–21.79%, ash 10.81–24.53%, calcium 2.22–8.01%, potassium 2.33–5.22%, magnesium 0.36–1.29%, chlorine 0.26–0.84%, phosphorus 0.47–0.57%, sulfur 1.23–3.34%.

Style Porter.

Beer Keto Reporter, Birra del Borgo (Borgorose, Lazio, Italy).

Source Calagione et al. 2018; Dabove 2015; Daniels 2016; Serna-Saldivar 2010; https://en.wikipedia.org/wiki/Nicotiana_rustica.

🍺 Birra del Borgo's Experimentation
Despite being toxic, tobacco is used as a flavoring by some breweries. The first to try it in Italy (and probably in the world) was Birra del Borgo, which in 2008 used Kentucky tobacco leaves produced in Tuscany for its Keto Reporter, a dark beer of English inspiration. The leaves of this plant, left to infuse for a few hours in the beer at the end of fermentation, give the final product a light, spicy, and astringent sensation.

Ochroma pyramidale (Cav. ex Lam.) Urb.
MALVACEAE

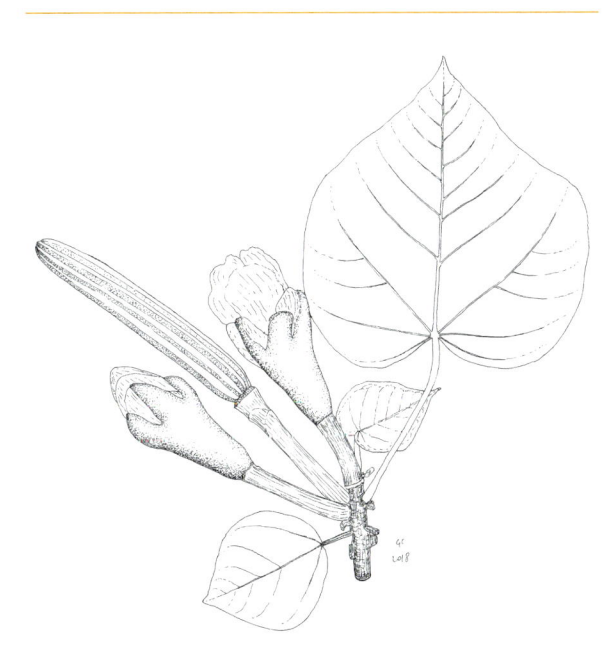

Synonyms *Bombax angulata* Sessé and Moc., *Bombax angulatum* Sessé and Moc., *Bombax pyramidale* Cav. ex Lam., *Ochroma bicolor* Rowlee, *Ochroma bolivianum* Rowlee, *Ochroma concolor* Rowlee, *Ochroma grandiflora* Rowlee, *Ochroma lagopus* Sw., *Ochroma limonense* Rowlee, *Ochroma limonensis* Rowlee, *Ochroma obtusa* Rowlee, *Ochroma obtusum* Rowlee, *Ochroma peruvianum* I.M.Johnst., *Ochroma tomentosum* Humb. & Bonpl. ex Willd., *Ochroma velutina* Rowlee.

Common Names Balsam, balsa wood tree, bobwood, corkwood, cork tree, down tree.

Description Tree, deciduous or evergreen (up to 30–50 m). Trunk straight, usually short, cylindrical (Ø up to 100–180 cm), with short buttresses in older specimens; smooth bark, mottled grayish-white in color; broad, expanded crown. Leaves spirally arranged, simple; stipules broadly lanceolate (about 1.5×1 cm); petiole 3–40 cm; lamina ovate, slightly 3- to 5-lobed ($10–40 \times 11–35$ cm), base cordate, apex acute or acuminate, margin wavy, glabrescent above, hairy below, veins palmate or pinnate with 7–9 pairs of lateral veins. Flowers solitary, axillary, bisexual, regular, pentamerous; pedicel 4–11 cm; calyx tubular (8–12 cm), with unequal lobes (2.5–4 cm), hairy inside and out; petals ($11–15 \times 5$ cm) whitish; stamens numerous, joined to the petals at their base, united in a staminal column (10–12.5 cm) briefly 5-lobed, bearing sessile anthers; ovary superficial, 5-carpellar, club-shaped stylus (9–10 cm), stigma spiraled. Fruit capsule oblong ($12–25 \times 2.5$ cm), ribbed, 5-valve, dehiscent, densely woolly inside, containing many seeds. Seeds pyriform ($4–5 \times 1.5$ mm), covered with abundant light brown fluff.

Related Species The genus *Ochroma* is monotypic (owing to the high morphological variability, in the past, up to 11 species have been attributed to it).

Geographic Distribution CS tropical America, from S Mexico to Bolivia up to Ecuador. **Habitat** Typically pioneer species, capable of colonizing deforested areas; in natural conditions it is found up to 1,000 m ASL, in areas with average annual rainfall of 1,250–3,000 mm and temperatures of 22–28°C; it tolerates a dry season up to five months with relative humidity not less than 75%; it prefers flat alluvial areas, with volcanic, rich, deep, well-drained soils; in less suitable areas, it produces higher-density wood (more than 160 kg/m^3), which has little commercial interest. **Range** Beyond its natural range, the species has been introduced in many countries with tropical climates, in Asia (Indonesia, Malaysia), Africa (Cameroon, Zimbabwe), and South Africa, sometimes becoming naturalized (Cameroon) in thickets and secondary forests.

Beer-Making Parts (and Possible Uses) Wood for barrels (first containing cachaça, then for aging beer).

Wood Known for its lightness (WSG 0.08–0.25), widely used for floaters, modeling material; white or brownish color with pinkish hues, glossy, silky appearance, straight grain and coarse texture, easy to work, not durable.

Style South American Beer (?).

Source Cantwell and Bouckaert 2016; Cantwell and Bouckaert 2018.

Ocimum basilicum L.
LAMIACEAE

Synonyms *Ocimum album* L., *Ocimum anisatum* Benth., *Ocimum barrelieri* Roth, *Ocimum bullatum* Lam., *Ocimum caryophyllatum* Roxb., *Ocimum chevalieri* Briq., *Ocimum ciliare* B.Heyne ex Hook.f., *Ocimum ciliatum* Hornem., *Ocimum citrodorum* Blanco, *Ocimum cochleatum* Desf., *Ocimum dentatum* Moench, *Ocimum hispidum* Lam., *Ocimum integerrimum* Willd., *Ocimum lanceolatum* Schumach. & Thonn., *Ocimum laxum* Vahl ex Benth., *Ocimum medium* Mill., *Ocimum nigrum* Thouars ex Benth., *Ocimum scabrum* Wight ex Hook.f., *Ocimum simile* N.E.Br., *Ocimum thyrsiflorum* L., *Plectranthus barrelieri* (Roth) Spreng.

Common Names Basil, great basil, Saint-Joseph's-wort.

Description Plant, annual, herbaceous (20–40 cm), very aromatic. Taproot. Stems erect, quadrangular section, much branched, tending to become woody at the base. Leaves opposite, petiolate, glabrous, glossy, often bullous, 3–5 cm long; lamina oval-lanceolate, entire or sparsely toothed, attenuated at base, apex subobtuse or acuminate. Inflorescence raceme much elongated at the end of anthesis, flowers clustered in 4- to 6-flowered whorls at the axil of small bracts. Flower with bell-shaped calyx (5–6 mm), persistent, pubescent on the outside, upper lip broad, reticulate-leaved, suborbicular, concave with mucronate apex, lower lip divided into 4 small acute teeth; corolla white ± pinkish in color, twice as long as the calyx (10–12 mm), bilabiate, with upper lip 4-lobed, ± flat, the lower one entire or irregularly fringed; stamens 4, didynamous protruding, lying on the lower lip of the corolla. Fruit consisting of 4 small achenes (tetrachenium) oval, foveolate, dark brown in color.

Cultivar Some examples: *Ocimum americanum* Lemon, *Ocimum americanum* Lime, *Ocimum basilicum*, *Ocimum basilicum* Boxwood, *Ocimum basilicum* Cinnamon, *Ocimum basilicum* Genovese Gigante, *Ocimum basilicum* Greek Yevani, *Ocimum basilicum* Lettuce Leaf, *Ocimum basilicum* Licorice, *Ocimum basilicum* Magical Michael, *Ocimum basilicum* Mammoth, *Ocimum basilicum* Nufar F1, *Ocimum basilicum* Osmin Purple, *Ocimum basilicum* Piccolo, *Ocimum basilicum* Purple Ruffles, *Ocimum basilicum* Purpurascens, *Ocimum basilicum* Red Rubin, *Ocimum basilicum* Spicy Globe, *Ocimum basilicum* var. citriodora Mrs. Burns, *Ocimum basilicum* var. minimum, *Ocimum basilicum* var. thyrsiflorum, *Ocimum basilicum* var. thyrsiflorum Siam Queen; hybrids: *Ocimum × citriodorum* Lesbos, *Ocimum × citriodorum* Thai, *Ocimum basilicum × americanum* Spice, *Ocimum basilicum × americanum* Sweet Dani, *Ocimum kilimandscharicum × basilicum* African Blue.

Related Species Of the 76 taxa belonging to the genus *Ocimum*, some have food applications similar to *Ocimum basilicum*: *Ocimum × citriodorum* Vis. (= *Ocimum × africanum* Lour.), *Ocimum americanum* L., *Ocimum basilicum* L. var. *thyrsiflorum* (L.) Benth., *Ocimum gratissimum* L., *Ocimum kilimandscharicum* Gürke, *Ocimum tenuiflorum* L.; beer-making applications are known for *Ocimum × africanum* Lour.

Geographic Distribution Asia. **Habitat** Fresh, well-drained soils in sunny or semi-shady locations (0–600 m ASL). **Range** Cultivated throughout the world.

Beer-Making Parts (and Possible Uses) Leaves, whole plant.

Chemistry Energy 94 kJ (22 kcal), water 92.06 g, carbohydrates 2.65 g, fiber 1.6 g, lipids 0.64 g, protein 3.15 g, vitamin A 264 µg, beta-carotene 3142 µg, thiamine (B1) 0.034 mg, riboflavin (B2) 0.076 mg, niacin (B3) 0.902 mg, pantothenic acid (B5) 0.209 mg, vitamin B6 0.155 mg, folate (B9) 68 µg, choline 11.4 mg, vitamin C 18 mg, vitamin E 0.8 mg, vitamin K 414.8 µg, calcium 177 mg, iron 3.17 mg, magnesium 64 mg, manganese 1.148 mg, phosphorus 56 mg, potassium 295 mg, sodium 4 mg, zinc 0.81 mg Essential oil: citronellol, linalool, myrcene, pinene, ocimene, terpineol, linalyl acetate, phenkyl acetate, trans-ocimene, 1,8-cineole, camphor octane, methyl eugenol, methyl chavicol, eugenol, beta-caryophyllene, bergamotene, eugenol, methyl cinnamate, phenylpropanoids, trans-β-ocimene.

Style Herb Beer, Basil Ale, Liquorice Basil Schwarzbier.

Beer Genova, Birrificio La Superba (Busalla, Liguria, Italy).

Source Calagione et al. 2018; Fisher and Fisher 2017; Giaccone and Signoroni 2017; Heilshorn 2017; Hieronymus 2016b; Josephson et al. 2016; https://www.brewersfriend.com/other/basil/.

Ocimum tenuiflorum L.
LAMIACEAE

Synonyms *Geniosporum tenuiflorum* (L.) Merr., *Lumnitzera tenuiflora* (L.) Spreng, *Moschosma tenuiflorum* (L.) Heynh., *Ocimum anisodorum* F.Muell., *Ocimum caryophyllinum* F.Muell., *Ocimum hirsutum* Benth., *Ocimum inodorum* Burm.f., *Ocimum monachorum* L., *Ocimum sanctum* L., *Ocimum scutellarioides* Willd. ex Benth., *Ocimum subserratum* B.Heyne ex Hook.f., *Ocimum tomentosum* Lam., *Plectranthus monachorum* (L.) Spreng.

Common Names Holy basil, tulasi, tulsi, thulasi.

Description Shrub, small (up to 1 m), much branched. Stems erect, woody at base, sparsely hairy. Leaf petiolate (1–2.5 cm), lamina oblong (2.5–5.5 × 1–3 cm), puberulent, glandular, hairy on veins, base cuneate to rounded, margin superficially wavy-serrate, apex obtuse. Inflorescence whorl 6-flowered, in thyrsus or panicle (6–8 cm) pedunculate, terminal; bracts sessile, cordate (1.5 × 1.5 mm), apex acute; peduncle 1–1.5 cm. Flower pedunculate (2.5 mm); calyx campanulate (2.5 mm), villous, tube 1.5 mm; median tooth of upper lip broadly oblate, acute; lateral teeth broadly triangular, shorter than lower lip teeth, spinescent; lower lip teeth lanceolate, apex spinescent; calyx at fruiting 6 × 4 mm, conspicuously veined; corolla white to reddish in color, 3 mm, slightly exserted, sparsely puberulent; tube about 2 mm, upper lip less than 1 × 2.5 mm, lobes ovate; lower lip oblong, about 1 × 0.6 mm, flat; stamens slightly exert, free; posterior filaments puberulent at base. Fruit nucule brown, ovoid (1 × 0.7 mm), glandular-foveolate.

Cultivar Kapoor Tulsi, Rama Tulsi, Vana Tulsi (*Ocimum gratissimum* L.).

Related Species Of the 76 taxa belonging to the genus *Ocimum*, some have similar food applications to *Ocimum basilicum*: *Ocimum americanum* L., *Ocimum basilicum* L. var. *thyrsiflorum* (L.) Benth., *Ocimum gratissimum* L., *Ocimum kilimandscharicum* Gürke, *Ocimum tenuiflorum* L., and *Ocimum* × *citriodorum* Vis. (= *Ocimum* × *africanum* Lour.).

Geographic Distribution India. **Habitat** Dry, sandy environments. **Range** China (Hainan, Sichuan, Taiwan), Cambodia, India, Indonesia, Laos, Malaysia, Myanmar, Philippines, Thailand, Vietnam, Africa, SW Asia, Australia.

Beer-Making Parts (and Possible Uses) Leaves, whole plant.

Chemistry Essential oil, leaves (24 components): eugenol 25.3–51.5%, b-caryophyllene 1.2–25.4%, trans-β-guaiene 9.4–19.2%, (E)-α-bergamotene 1.1–2.8%, caryophyllene oxide 1.7–3.9%, 1,8-cineole 2.2–9.2%, E-methyl cinnamate 0.1–8.7%, 1.10 di-epi-cubenol 1.3–2.6%, trans-β-farnesene 1.6–4.1%.

Style Porter, Triple, Wild Ale.

Source Josephson et al. 2016; https://www.brewersfriend.com/other/tulsi/.

Olea europaea L.
OLEACEAE

Synonyms *Olea alba* Lam. ex Steud., *Olea amygdalina* Gouan, *Olea angulosa* Gouan, *Olea argentata* Clemente ex Steud., *Olea atrorubens* Gouan, *Olea bifera* Raf., *Olea brevifolia* Raf., *Olea cajetana* Petagna, *Olea cayana* Raf., *Olea craniomorpha* Gouan, *Olea ferruginea* (Aiton) Steud., *Olea gallica* Mill., *Olea hispanica* Mill., *Olea lancifolia* Moench, *Olea longifolia* (Aiton) Steud., *Olea lorentii* Hochst., *Olea obliqua* (Aiton) Steud., *Olea oblonga* Gouan, *Olea odorata* Rozier ex Roem. and Schult., *Olea officinarum* Crantz, *Olea oleaster* Hoffmanns. & Link, *Olea polymorpha* Risso ex Schult., *Olea praecox* Gouan, *Olea racemosa* Gouan, *Olea regia* Rozier ex Roem. & Schult., *Olea sativa* Weston, *Olea sphaerica* Gouan, *Olea sylvestris* Mill., *Olea variegata* Gouan, *Olea viridula* Gouan, *Phillyrea lorentii* Walp.

Common Names Olive tree.

Description Tree (8–15 m), evergreen, very long-lived. Taproots in the first 3 years, then adventitious, extensive, and superficial. Stem initially cylindrical, erect, then expanded at the base, irregular, sinuous, knotty, often hollow; branches assurgent, angular twigs, sometimes spinescent in wild forms, with dense, expanded crown, silvery-gray in color; bark gray-green, smooth until about 10 years old, then gnarled, rough with deep furrows, and cracked into quadrangular plates; stump forms globular structures, from which numerous basal suckers are emitted each year; buds mostly axillary. Leaves simple, opposite, leathery, lanceolate, attenuated at the base in a short petiole, acuminate at the apex, margin entire, often revolute; upper leaf opaque, color green-glaucous, glabrous, underleaf paler, midrib prominent, color silky-silvery with star hairs. Flowers hermaphrodite collected in short, sparse axillary panicles; calyx persistent with 4 teeth, corolla funnel shaped with short tube consisting of 4 whitish petals, joined together at the base; 2 protruding stamina with large yellow anthers; ovary superior, stylus bilobate. Fruits ovoid drupes (olives), green to yellow, purple to purplish-black in color, with oily mesocarp and tapered woody, wrinkled kernel.

Subspecies *Olea europaea* L. includes six subspecies: *Olea europaea* L. subsp. *cerasiformis* G.Kunkel & Sunding, *Olea europaea* L. subsp. *cuspidata* (Wall. & G.Don) Cif., *Olea europaea* L. subsp. *europaea*, *Olea europaea* L. subsp. *guanchica* P.Vargas & al., *Olea europaea* L. subsp. *laperrinei* (Batt. & Trab.) Cif., and *Olea europaea* L. subsp. *maroccana* (Greuter & Burdet) P. Vargas & al.

Cultivar With *Olea europaea* L. having been cultivated for millennia, there are hundreds of cultivars throughout the Mediterranean basin. Greece and Italy have the primacy of varietal richness. In Italy, there are about 500 cultivars. Among the most renowned ones are Bianchera, Biancolilla, Bosana, Canino, Carboncella, Carolea, Carpellese, Casaliva, Cellina di Nardò, Cerasuola, Coratina, Dolce Agogia, Dolce di Rossano, Dritta, Frantoio, Gentile di Chieti, Gentile di Larino, Grossa di Cassano, Leccino, Maiatica, Moraiolo, Nocellara, Nostrana, Ogliarola Barese, Ottobratica, Pendolino, Pisciottana, Provenzale, Raja, Razzola, Rosciola, Santagatese, Sinopolese, Taggiasca, and Tondina.

Related Species The genus *Olea* includes 42 taxa.

Geographic Distribution Steno-Mediterranean. **Habitat** Frugal, thermophilic, heliophilous, tolerant of salinity, sensitive to low temperatures, prefers dry environments and climates, loose, coarse, or shallow soils, with rocky outcrops (0–900 m ASL). **Range** Cultivated throughout the world in areas with Mediterranean climates.

Beer-Making Parts (and Possible Uses) Fruits, leaves.

Chemistry Extra virgin olive oil: energy 3,699 kJ (884 kcal), lipids 100 g, calcium 1 mg, sodium 2 mg, potassium 1 mg, iron 0.56 mg, betaine 0.1 mg, alpha-tocopherol (Vit. E) 14.35 mg, phylloquinone (Vit. K) 60.2 µg, choline (Vit. J) 0.3 mg, beta-tocopherol 0.11 mg, gamma-tocopherol 0.83 mg, monounsaturated fatty acids 72.9 g, polyunsaturated fatty acids 10.5 g, saturated fatty acids 13.8 g, phytosterols 221 mg.

Style Fruit Beer, Blonde Ale.

Beer Montefollia, Nadir (San Remo, Liguria, Italy).

Source Calagione et al. 2018; https://www.brewersfriend.com/other/olives/.

Opuntia ficus-indica (L.) Mill.
CACTACEAE

Synonyms *Cactus ficus-indica* L., *Cactus opuntia* L., *Opuntia arcei* Cárdenas, *Opuntia castillae* Griffiths, *Opuntia chinensis* (Roxb.) K.Koch, *Opuntia cordobensis* Speg, *Opuntia ficus-barbarica* A.Berger, *Opuntia incarnadilla* Griffiths, *Opuntia megacantha* Salm None Dyck, *Opuntia vulgaris* Mill.

Common Names Fig opuntia, Barbary fig, cactus pear, spineless cactus, prickly pear.

Description Tree (1–5 m), succulent. Stem woody, coppery; branches obovate or elliptic (10–20 × 20–50 cm), glaucous green, with woolly areoles bearing rare and usually isolated whitish spines, up to 1 cm long. Flowers hermaphrodite, actinomorphic, yellow, Ø 6–7 cm; ovary inferior; hypanthium roughly cylindrical, tepals and stamens very numerous. Fruit ovoid, 5–9 cm long, edible. Seed about 5 mm long.

Subspecies Besides the typical variety, for this species *Opuntia ficus-indica* L. var. *gymnocarpa* (F.A.C.Weber) Speg. is also known.

Cultivar In Italy, three cultivars are well known: Gialla, Bianca, and Rossa (or Sanguigna).

Related Species Of the more than 200 species recognized in the genus *Opuntia*, many produce edible fruits.

Geographic Distribution Neotropical (Mexico?). **Habitat** Arid places. **Range** Cultivated for hedges and fruit production and widely naturalized.

Beer-Making Parts (and Possible Uses) Fruits.

Chemistry Fruits (pulp): water 94.40±2.61%, protein 1.45±0.08%, lipids 0.7±0.08%, ash 1±0.01%, sucrose 0.19%, glucose 29%, fructose 24%, calcium 12.4 mg/100 g, potassium 199 mg, sodium 1.09 mg, magnesium 18.8 mg.

Style Fruit Beer.

Beer Indica, Irias (Torrenova, Sicily, Italy).

Source Calagione et al. 2018; Hieronymus 2016b; McGovern 2017; https://www.birramoretti.it/le-birre-di-casa/birra-moretti-alla-pugliese/; https://www.bjcp.org (29. Fruit Beer).

Opuntia joconostle F.A.C.Weber
CACTACEAE

Synonyms None

Common Names Xoconostle. Under the trade name of Xoconostle (sour-fruited cactus) in Mexico, fruits of plants belonging to at least 10 species are included, nine of which are members of the genus *Opuntia* (*Opuntia durangensis* Britton & Rose, *Opuntia elizondoana* E.Sanchez & Villaseñor, *Opuntia heliabravoana* Scheinvar, *Opuntia joconostle* F.A.C. Weber, *Opuntia leucotricha* DC, *Opuntia matudae* Scheinvar, *Opuntia oligacantha* C.F. Förster, *Opuntia spinulifera* Salm-Dyck, *Opuntia zamudioi* Scheinvar) and one belonging to the genus *Cylindropuntia* (*Cylindropuntia imbricata* [Haw.] F.M.Knuth). Other *Opuntia* species also belong to this group (*Opuntia galleguiana* Scheinvar & Olalde, *Opuntia leiascheinveriana* Martínez & Gallegos, *Opuntia sainaltense* Scheinvar, *Opuntia tezontepecana* Scheinvar & Gallegos), as well as some hybrids between the already mentioned taxa (e.g., *Opuntia leucotricha* Salm-Dyck × *Opuntia joconostle* F.A.C.Weber) and species belonging to related genera (e.g., *Corynopuntia reflexispina* [Wiggins & Rollins] Backeb.). Among these species, the most frequently cultivated and commercialized are *Opuntia joconostle* F.A.C.Weber ex Diguet cv. Cuaresmeño and *Opuntia matudae* Scheinvar cv. Rosa.

Description Tree, small (up to 2.5 m). Trunk with usually compact, bushy branching, often forming a spreading canopy on top of a straight, cylindrical trunk (Ø about 20 cm), grayish color; branches small, oval, ± regular, epidermis shiny, light green to yellow. Spines white, unequal. Flower yellow. Fruits subglobular (Ø about 2 cm), with fragrant, slightly acidulous flesh, pinkish color.

Cultivar The most common is *Opuntia joconostle* Weber cv. Cuaresmeño.

Related Species Of the more than 200 species recognized in the genus *Opuntia*, many produce edible fruits.

Geographic Distribution Mexico (Jalisco, Querétaro, Michoacán). **Habitat** Highlands (about 1,500 m ASL) with a temperate climate and regular rainfall. **Range** Widely cultivated in Mexico for its fruit.

Beer-Making Parts (and Possible Uses) Fruits.

Chemistry Fresh fruits: energy 29.48–34.76 kcal, water 87.3–89.05 g, carbohydrates 5.81–7.98 g, protein 0.71–1.56 g, lipids 0.10 g, fiber 2.35–4.28 g, ash 0.49–0.65 g.

Style Clone Beer, Experimental Beer, Mixed-Style Beer.

Source https://www.brewersfriend.com/other/xoconostle/.

Opuntia matudae Scheinvar
CACTACEAE

Synonyms None

Common Names Xoconostle.

Description Shrub or tree (1.5–4.5 m). Stem defined, broad; branches narrowly obovate to obovate (20–25 × 10–15 cm), grayish blue-green in color, epidermis glabrous, areoles arranged in 13–14 spiral series, usually with a purple spot inside, more than 2 cm apart, with grayish felt, glochids (2–3 mm) pinkish-chestnut in color. Spines (1–8) usually in all areoles, unequal (0.7–3.5 cm), thin, flexible, some crooked, the lower ones reflexed, some with bent base, the median and upper ones greater, usually interlaced with each other, divergent, grayish-white or yellowish color with translucent apex. Flower (5–7 × 8 cm at anthesis) bright yellow with red spots, then pink and red. Fruit ellipsoid, pyriform (2.5–4 × 1.5–2.5 cm), with deep umbilical scar, epicarp purple, mesocarp reddish-pink; areoles spineless, with grayish wool and pinkish-brown glochids; acidic, edible. Seeds discoid, with a thin, well-marked lateral aril (4 × 3 × 2 mm), whitish with pink tinge.

Cultivar The most common is *Opuntia matudae* Scheinvar cv. Rosa.

Related Species Of the more than 200 species recognized in the genus *Opuntia*, many produce edible fruits.

Geographic Distribution Mexico (Hidalgo). **Habitat** Arid hills. **Range** Widely cultivated in Mexico for the fruit.

Beer-Making Parts (and Possible Uses) Fruits.

Chemistry Fresh fruit (pulp): water 94.11 g, carbohydrates 3.93 g, protein 0.56 g, lipids 0.04 g, fiber 1.74 g, ash 0.11 g.

Style Clone Beer, Experimental Beer, Mixed-Style Beer.

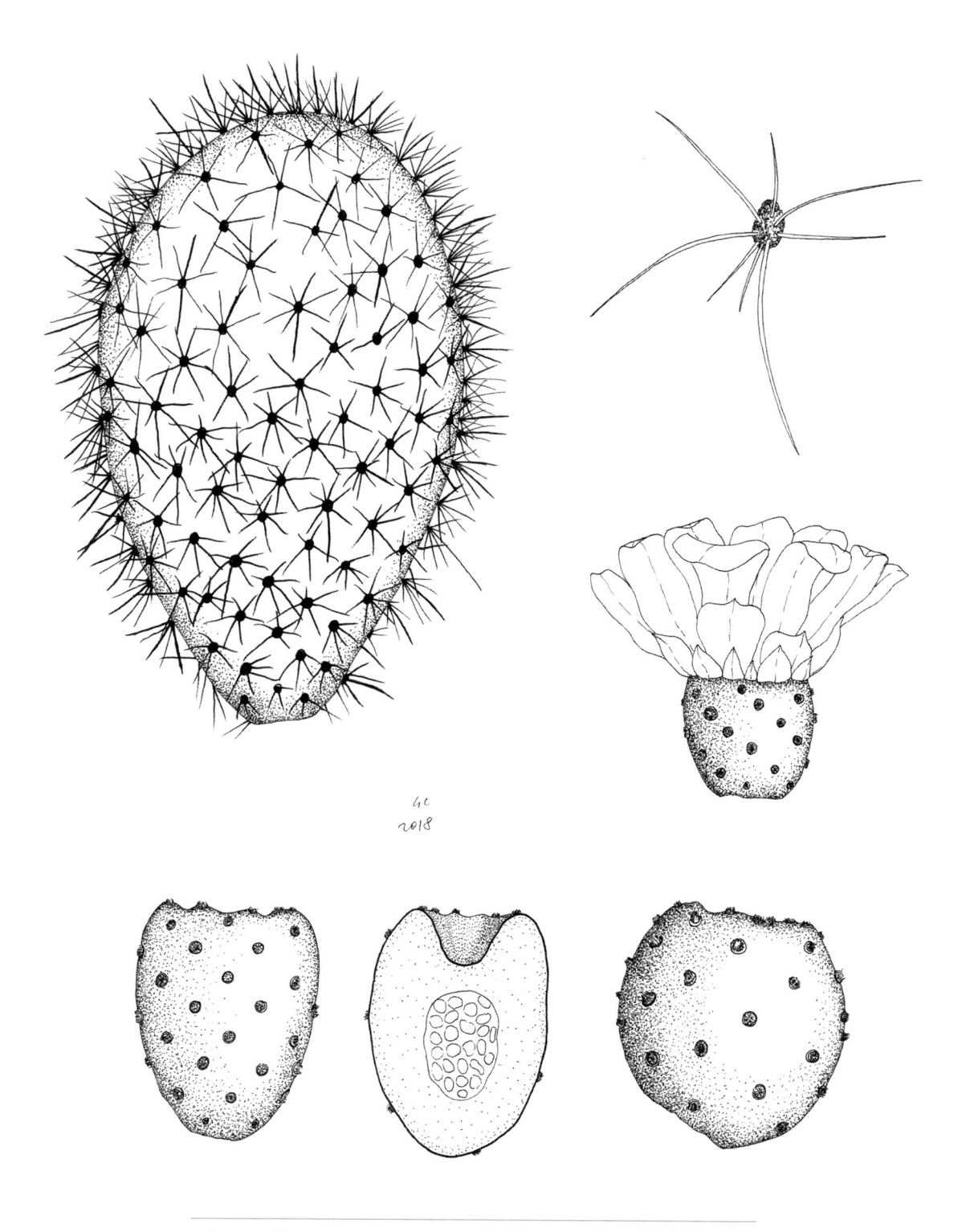

Opuntia joconostle F.A.C.Weber—CACTACEAE
Opuntia matudae Scheinvar—CACTACEAE

Source https://www.brewersfriend.com/other/xoconostle/.

Origanum majorana L.
LAMIACEAE

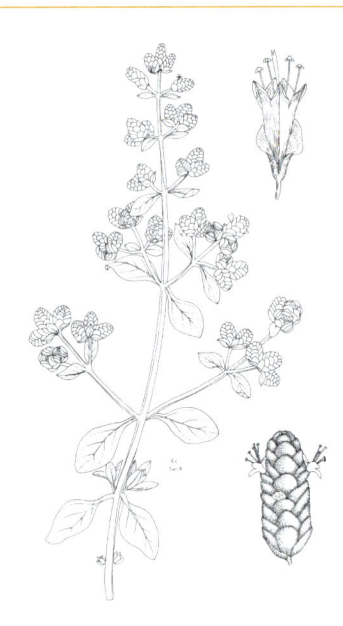

Synonyms *Amaracus majorana* (L.) Schinz & Thell., *Majorana dubia* (Boiss.) Briq., *Majorana fragrans* Raf., *Majorana hortensis* Moench, *Majorana mexicana* M. Martens & Galeotti, *Majorana ovalifolia* Stokes, *Majorana ovatifolia* Stokes, *Majorana suffruticosa* Raf., *Majorana tenuifolia* Gray, *Majorana tenuifolia* Raf., *Majorana uncinata* Stokes, *Origanum confertum* Savi, *Origanum dubium* Boiss., *Origanum majoranoides* Willd., *Origanum salvifolium* Roth, *Thymus majorana* (L.) Kuntze.

Common Names Marjoram.

Description Herb (2–6 dm), perennial, growing annually. Stem erect, pubescent, woody at the base, branched-corymbose in the inflorescence with dense 0.1–0.2 mm papillose hairs and scattered 1–1.5 mm patent bristles. Leaves ovate or ovate-lanceolate (8–20 × 5–10 mm), petiolate, obtuse or rounded at the base. Inflorescences dense, ovate spikes (5–6 × 7–9 mm); bracts ovate-rhomboid, the lower ones 3 × 4.5 mm, ciliate on the edge; calyx 2–3 mm; corolla white or pinkish (4.5–5.5 mm).

Cultivar Some names: *Origanum majorana* cv. Aurea, *Origanum majorana* cv. Golden, *Origanum majorana* cv. Herb or Undershrub, *Origanum majorana* cv. Hortensis, *Origanum majorana* cv. Sweet, *Origanum majorana* cv. Sweet Asti, *Origanum majorana* cv. Vulgare Variegata.

Related Species Of the more than 60 taxa (species and subspecies) recognized in the genus *Origanum*, *Origanum laevigatum* Boiss., *Origanum microphyllum* (Benth.) Vogel, *Origanum rotundifolium* Boiss., *Origanum scabrum* Boiss. & Heldr., and *Origanum vulgare* L. are used for food.

Geographic Distribution Saharo-Syndonesian (Cyprus, S Turkey). **Habitat** Crops, roadsides. **Range** Widely cultivated.

Beer-Making Parts (and Possible Uses) Leaves.

Chemistry Essential oil, leaves (sabinyl chemotype): cis-sabinene hydrate/linalool 60.6%, sabinene 9.4%, α-thujene 0.5%, trans-sabinene hydrate 4.1%, α-pinene 1.1%, camphene 0.5%, β-pinene 0.6%, myrcene 2%, limonene/β-phellandrene 3.3%, 1,8-cineole 0.6%, (Z)-β-ocimene 0.1%, γ-terpinene 0.2%, borneol 1.8%, terpinen-4-ol 0.3%, α-terpineol 3.7%, cis-sabinene acetate hydrate 1%, linalyl acetate 3.4%, bornyl acetate 0.2%, β-caryophyllene 2.3%, α-humulene 0.1%, bicyclo-germacrene 2.6%, germacrene-D-4-ol 0.3%.

Style Belgian Specialty Ale, Belgian Tripel, Bière de Garde, Spice/Herb/Vegetable Beer.

Source Fisher and Fisher 2017; Heilshorn 2017; Hieronymus 2012; Hieronymus 2016b; https://www.brewersfriend.com/other/majoram/; https://www.brewersfriend.com/other/marjoram/.

Origanum syriacum L.
LAMIACEAE

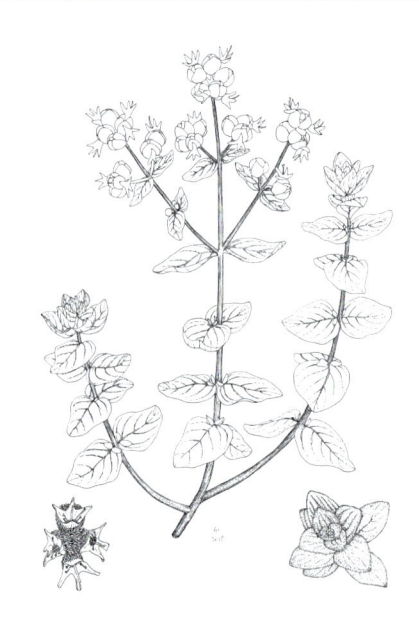

Synonyms *Majorana crassa* Moench, *Majorana crassifolia* Benth., *Majorana scutellifolia* Stokes, *Majorana syriaca* (L.) Raf., *Origanum maru* L., *Origanum vestitum* E.D.Clarke, *Schizocalyx syriacus* (L.) Scheele, *Zatarendia egyptiaca* Raf.

Common Names Za'atar, zatar (this common name can indicate a series of species belonging to different genera of the Lamiaceae family, such as *Origanum*, *Satureja*, *Thymbra*, and *Thymus*, or even mixes of different spices), Egyptian oregano, bible hyssop, biblicalhyssop, Lebanese oregano, Syrian oregano, wild marjoram.

Description Shrub, perennial (60–80 cm to 1 m). Stems woody, multiple, square sectioned, hairy covered with hairy leaves that give the entire plant a grayish color and help it cope with dry conditions. Leaves opposite, simple, small (up to 2.5 cm), ovate, entire, gray-green color, covered with short hairs, with pleasant and intense scent of thyme. Inflorescences dense spikes. Flowers bilabiate, tubular, small, not particularly showy, corolla white or light pink; floral bracts hairy. Black fruits.

Subspecies Besides the typical subspecies, *Origanum syriacum* L. subsp. *syriacum*, two subspecies are recognized: *Origanum syriacum* L. subsp. *bevanii* (Holmes) Greuter & Burdet and *Origanum syriacum* L. subsp. *sinaicum* (Boiss.) Greuter & Burdet.

Related Species Of the more than 60 taxa (species and subspecies) recognized in the genus *Origanum*, several entities have established food and beer-making uses.

Geographic Distribution Middle East (Lebanon, Syria, Turkey, Egypt [Sinai Peninsula]), Mediterranean coastal areas. **Habitat** Slopes in rocky (calcareous) and arid soils, often in partial shade (200–2,700 m ASL). **Range** Occasionally cultivated in the area of origin owing to increasing demand.

Beer-Making Parts (and Possible Uses) Flowering tops, leaves.

Chemistry Essential oil: carvacrol 53.9 ± 21.6%, p-cymene 17.7 ± 20.1%, α-thujene 0.8 ± 0.3%, α-pinene 0.6 ± 0.2%, sabinene + 1.0 ± 0.4%, myrcene 1.4 ± 0.6%, α-phellandrene 0.1 ± 0.1%, α-terpinene 1.1 ± 0.5%, limonene 0.3 ± 0.2%, γ-terpinene 8.0 ± 3.4%, trans-sabinene hydrate 0.9 ± 0.2%, cis-sabinene hydrate 0.4 ± 0.1%, terpinen-4-ol 0.2 ± 0.1%, thymoquinone 4.5 ± 3.1%, thymol 0.4 ± 0.2%, β-caryophyllene 3.5 ± 1.3%, α-humulene 0.4 ± 0.3%, β-bisabolene 1.9 ± 1.8%, germacrene B 0.4 ± 0.2%, caryophyllene oxide 0.5 ± 0.4%.

Style Ancient Ale.

Source McGovern 2017; https://en.wikipedia.org/wiki/Dogfish_Head_Brewery.

Origanum vulgare L.
LAMIACEAE

Synonyms *Mentha formosana* (C.Marquand) S.S.Ying, *Micromeria formosana* C.Marquand, *Origanum albiflorum* K.Koch, *Origanum americanum* Raf., *Origanum anglicum* Hill, *Origanum barcense* Simonk., *Origanum capitatum* Willd. ex Benth., *Origanum creticum* L., *Origanum decipiens* Wallr. ex Benth., *Origanum dilatatum* Klokov, *Origanum elegans* Sennen, *Origanum latifolium* Mill., *Origanum laxiflorum* Royle ex Benth., *Origanum loureiroi* Kostel., *Origanum micranthum* Colla, *Origanum nutans* Willd. ex Benth., *Origanum officinale* Gueldenst., *Origanum orientale* Mill., *Origanum puberulum* (Beck) Klokov, *Origanum serpylliforme* Fisch. & C.A.Mey., *Origanum stoloniferum* Besser ex Rchb., *Origanum thymiflorum* Rchb., *Origanum venosum* Willd. ex Benth., *Origanum watsonii* A.Schlag. ex J.A.Schmidt, *Oroga heracleotica* Raf., *Thymus origanum* Kuntze.

COMMON NAMES Oregano.

DESCRIPTION Shrub, perennial (20–60 cm), at maturity semi-arborescent, variable aspect, pleasant scent. Rhizome branched, woody, creeping horizontally, from which the stems develop, often lying down in the basal part, then erect, with quadrangular section, branched at the top, reddish color, covered with thick patent hair. Leaves opposite, petiole short, lamina oval-elongated, maximum width toward base rounded, major nerves hairy, margin entire or slightly toothed, hairy. Flowers at axil of sessile leaf-like bracts, on short peduncles, united in terminal heads purple-pink, rarely white; calyx tubular campanulate, symmetrical, 5 teeth 1/3 length of tube; corolla with upper lip emarginate, the lower one longer, trilobed. Fruits tetramerous with subspherical mericarps, apex acute, with 2 whitish triangles in the insertion surface, dark brown in color.

SUBSPECIES In addition to the nominal subspecies, five subspecies are recognized: *Origanum vulgare* L. subsp. *glandulosum* (Desf.) Ietsw., *Origanum vulgare* L. subsp. *gracile* (K.Koch) Ietsw., *Origanum vulgare* L. subsp. *hirtum* (Link) Ietsw., *Origanum vulgare* L. subsp. *virens* (Hoffmanns. and Link) Ietsw., and *Origanum vulgare* L. subsp. *viridulum* (Martrin-Donos) Nyman (= *Origanum heracleoticum* L.); the last subspecies has the most use in brewing.

CULTIVAR They are numerous, including *Origanum vulgare* cv. Aureum, *Origanum vulgare* cv. Aureum Crispum, *Origanum vulgare* cv. Compactum, *Origanum vulgare* cv. Variegata, *Origanum vulgare* L. subsp. *hirtum* (Link) Ietsw. (Greek oregano), *Origanum vulgaris* cv. Kaliteri, and *Origanum* × *majoricum* (Italian oregano).

RELATED SPECIES Of the more than 60 taxa (species and subspecies) recognized in the genus *Origanum*, besides *Origanum vulgare* L., the following are of alimentary use: *Origanum laevigatum* Boiss., *Origanum majorana* L., *Origanum microphyllum* (Benth.) Vogel, *Origanum rotundifolium* Boiss., *Origanum scabrum* Boiss., and *Origanum scabrum* Boiss. & Heldr.

GEOGRAPHIC DISTRIBUTION Eurasian. **HABITAT** Arid and sunny places, sparse woods and stony places, cliffs, especially in chalky soils. **RANGE** Species cultivated in temperate regions.

BEER-MAKING PARTS (AND POSSIBLE USES) Leaves, inflorescences.

CHEMISTRY Essential oil, dried leaves (167 compounds; 93.9% of oil): α-thujene 0.2%, α-pinene 0.3%, sabinene 4.8%, myrcene 0.7%, δ2-carene 0.5%, o-cymene 0.2%, p-cymene 3.8%, (Z)-β-ocimene 0.8%, (E)-β-ocimene 0.8%, γ-terpinene 0.6%, terpinolene 0.2%, allo-ocimene 1.1%, 1,8-cineole 3.9%, (Z)-sabinyl acetate 0.2%, (Z)-linalool oxide 0.1%, trans-sabinene hydrate 0.2%, linalool 0.9%, trans-p-menth-2-en-1-ol 0.3%, cis-p-menth-2-en-1-ol 0.4%, trans-verbenol 0.1%, sabina ketone 0.3%, borneol 0.2%, cis-isopulegone trace, terpinen-4-ol 5.6%, cis-piperitol 0.1%, α-terpineol 2.3%, myrtenol 0.2%, trans-piperitol 0.1%, cuminaldehyde 0.2%, pulegone 0.1%, phellandral 0.1%, cumin alcohol 0.1%, p-mentha-1,4-dien-7-ol 0.1%, bicycloelemene 0.5%, α-cubebene 0.1%, α-copaene 0.2%, β-bourbonene 1.5%, β-elemene 0.3%, β-caryophyllene 8.8%, γ-elemene 1.3%, α-humulene 1.1%, allo-aromadendrene 0.8%, germacrene D 3.3%, α-amorphene trace, γ-muurolene 0.4%, epi-bicyclosesquiphellandrene 0.8%, cis β-guajene 0.1%, α-muurolene 0.3%, (E,E)-α-farnesene 0.3%, β-bisabolene 0.7%, 5-cadinene 1.7%, α-cadinene 0.2%, germacrene B 0.7%, cis-α-copaen-8-ol 0.4%, ledol 0.4%, spatulenol 18.6%, caryophyllene oxide 1.4%, viridiflorol 0.7%, caryophylladienol I 0.6%, t-cadinol 0.7%, torreyol 0.4%, α-cadinol 1.7%, carvacrol methyl ether 0.2%, thymol 0.5%, carvacrol 11.7%, eugenol 0.1%, cryptone trace, cis-jasmone 0.1%, (E)-β-ionone 0.2%, hexahydrofarnesyl acetone 0.1%, trace octane, tetracosane 0.1%, pentacosane 0.2%, hexacosane 0.1%, heptacosane 0.2%, octacosane 0.1%, nonacosane 0.2%, triacontane trace, entriacontane 0.2%, dotriacontane trace, tritriacontane 0.1%, 1-octent-3-ol 1.3%, octan-3-ol 0.3%, dihydroedulan II 0.9%.

STYLE American Amber Ale, American IPA, American Pale Ale, American Stout, Bière de Garde, Imperial IPA, Old Ale, Russian Imperial Stout, Saison, Spice/Herb/Vegetable Beer.

SOURCE Fisher and Fisher 2017; Hieronymus 2016b; https://www.brewersfriend.com/other/oregano/.

Oryza sativa L.
POACEAE

Synonyms *Oryza communissima* Lour., *Oryza formosana* Masam. & Suzuki, *Oryza glutinosa* Lour., *Oryza montana* Lour., *Oryza perennis* Moench, *Oryza plena* (Prain) N.P.Chowdhury, *Oryza praecox* Lour., *Oryza rubribarbis* (Desv.) Steud.

Common Names Rice.

Description Herb, annual (7–15 dm). Culms erect, curved in inflorescence, often quite branched. Leaves with 10–15 mm wide lamina and enlarged sheath; ligule membranous, white, elongate (to 20 mm). Inflorescence panicle slender, drooping on one side, 2–3 dm long; spikelets 1-flowered; glumes very short (1–2 mm); lemma hardened, ovate-elliptic (7–9 mm), bearing a remnant of highly variable length.

Subspecies In addition to the nominal variety, *Oryza sativa* L., there are four varieties: *Oryza sativa* L. var. *atrobrunnea* (Gustchin) Portères, *Oryza sativa* L. var. *melanoglumella* (Gustchin) Portères, *Oryza sativa* L. var. *mulayana* (Gustchin) Portères, and *Oryza sativa* L. var. *rubriglumella* (Gustchin) Portères.

Cultivar In all the main areas where rice is cultivated, many cultivars have been selected. In some rice-growing areas, the varietal richness is surprising, such as in India, where it is estimated that there are more than 82,700 landraces (the so-called folk rice). The main cultivars are as follows: Africa (Abakaliki, African rice [*Oryza glaberrima*], Ekpoma, New Rice for Africa, Ofada), Australia (Doongara, Illabong, Koshihikari, Kyeema, Langi, Opus, Reiziq, Sherpa, Topaz), Bangladesh (Akia Beruin Red, Akia Beruin White, Balam Dhan, Balam-small Red, Balam-small White, Bashmoti Rice, BINA Dhan, Binni, BRRI Dhan, Chinigura, Digha Dhan, Hail Girvi, Hamim, Hori Dhan, Kalijira, Kalo Beruin, Katari Bhog, Kathali Beruin Red, Kathali Beruin White, Khara Beruin, Khato Dosh, Lapha, Lathial-7 Red, Lathial-7 White, Loha Sura, Matichak, Miniket, Modhu Beruin Red, Modhu Beruin White, Mou Beruin, Moulata, Najir Shail, Pajam, Pakh Beruin, Raujan-1, Raujan-2, Sakhorkhana, Shail Dhan, Tepu Dhan, Thakur Bhog), Burma (Black glutinous rice, Byat, Emahta, Letyezin, Midon, Ngasein, Paw hsan hmwe, White glutinous rice), Cambodia (Bonla Pdao, Cammalis, Long Rice, Neang Khon, Neang Minh, Phka Khnei, Phka Malis, Phka Romdul, Red Rice, Sen Kro Ob, Sen Pi Dao), Canada (Wild Rice), China (Black Asian, Manchurian Rice [*Zizania latifolia* (Griseb.) Turcz. ex Stapf], *Oryza rufipogon* Griff., Ponlai, Wild Rice), Dominican Republic (Cristal 100, Idiaf 1, Inglés Corto, Inglés Largo, Juma 57, Juma 58, Juma 66, Juma 67, Prosequisa 4, Prosequisa 5, Prosequisa 10, Toño Brea, Yocahú CFX-18), France (Camargue red rice), Greece (Bella, Fancy, Fino/Nychaki, Lais, Soupé), India (74 count, Aizon, Ambemohar, Annapoornna, Atop, Basmati, Bhut Muri [Kelas], Champaa Rice, Clearfield, Dubraj, Gandhasala, Gobindobhog, Hansraj, Hasan Serai, HMT, Idly, Jay Shrirama, Joha, Jyothi, Kamini, Katta Sambar, Laxmi Bhog, Minicate, Super Minicate, Molakolukulu, Navara, Patna, Ponni, Poreiton Chakhau, Pusa Basmati 1121, Pusa, Raja Hansa, Ranjit, Rosematta, Sona Masuri, Thimmasamundaram Mollakolukulu, Tulaipanji, Chitti Mutyalu, BR-2655, IR-30864, Jaya, Kagga of Manikatta, Thanu, Tunga, Nalihati, Padmakeshari, Kalamoti, Balami, Kalajiri, Tulasibasa, Pimpudibasa, Swarna, Nalabainsi, Pateni, Arikirai, Aryan, Boli Ari, Cheera Thouvan, Chembaavu, Chenjeera, Chennellu, Cherumadan, Chitteni, Chuvanna Choman, D1 [Uma], Dhebi, Gandhakasala, Jeerakasala, Kaattaazhi, Kariyadukkan, Karutharikannan, Keeripallan, Kochathikkalaari, Kodukayama, Kozhivalan, Kumaro-Athikkalaari, Kunjinellu, Kunjuvithu, Kuppakayama, Kuttadan, Malakkannan, Malakkaran, Matta Palakkad, Modan, Moorkhan, Mundakan, Mundodan, Mundot, Munnayan, Nallachennellu, Njavara, Odachan, Oryssa, Palliyaran, Paravalappan, Ponnariyan, Punjakayama, Rajadhani, Rajakayama, Ratha Choodi, Sughikayama, Thavalakkannan, Thonnuran, Thottam, Thovvan, Undakayama, Vachan, Vattan, Vayalthoova, Vellachoman, Vellariyan, Vellathovvan, Velutharikannan, Wayanad Kayamma, Bahadur, Bora, IR-8, Jaha, Jahingia, Kushal, Laodubi, Malbhog, Maniram, Manuhar, Molakolukulu, Naldubi, Sali, Suwagmoni, Aduthurai [ADT-1 to 49], Akshayadhan, Ambasamudharam [ASD-1 to 19], Amsipiti Dhan, Aravan Kuruva, Ariyan Nel, Arubatham Samba [CO-21], Aruvadhan Kodai, Arwa, Basmati Tukda, Trichy 3, Bhatta Dhan, Biagunda Dhan, Bod Dhan, Chengalpattu Sirumani, Chennel, Chithiraikar [Pondy], Chithiraikkar, Chomala, Chot Dhan, Cochin Samba, Coimbatore, Eravapandi, Gandakesala, Improved Samba Mahsuri, Improved White Ponni, IR-20, Molakolukulu, IR-50, Molakolukulu, Jil Jil Vaigunda, Jirkudai, Jogarnath Dhan, Kadaikazhuthan, Kaividhai Samba, Kalarpaalai,

Kalarpalai, Kalinga III, Kaliyan Samba, Kallimadaiyan, Kallundai, Kallurundaiyan, Kamban Samba, Kandasel/Kandasali, Kappakkar, Kappa Samba, Kar Samba, Karthigai Samba, Karunguruvai, Karuppu Nel, Karuthakkar, Katarni, Kattanur Nel, Katta Samba, Kattukuthalam, Kattu Samba, Kattu Vaniyam, Kitchili Samba, Kollan Samba, Kollikkar, Konakkuruvai, Koomvalai, Kouni Nel, Kudaivazhai, Kudavazhai, Kudhiraival Samba, Kullakaar, Kundri Manisamba, Kunthali, Kurangu Samba, Kuruvai, Kuruvai Kalanchiyam, Kuruvaikalayan [ASD-4], Kuzhiyadichan, Lakshmi Kajal, Lendhi Dhan, Mahate, Mal-bhog, Manakathai, Mansoori, Mappillai Samba, Maranel, Mathimuni, Mattai 110, Mattaikkar, Mattaikkuruvai [ADT-26], Mottakur, Murugangar Nel, Muttakkar, Nalla Manisamba, Navara, Neelasamba, Ninni Dhan, Norungan, Njavara, Oazhava, Katrazhai, Ondrarai Samba, Oldisaur Dhan, Ottadai, Paddy, Parwmal, Pathrakali, Pattaraikkar, Pattar Pisin, Periyavari, Perungar, Pisini, Pitchavari, Ponni Rice, Poongar, Poovan Samba, Puzhuthiikar, Puzhuthikal, Puzhuthi Samba, Rasagadam, Rongalachi Dhan, Sadakar, Siraga samba, Samba, Samba Mosanam, Sandikar, Sanna Samba [ADT-13], Seela Rice, Seeraga Samba, Selam Samba, Sembilipanni, Sempalai, Sempalai [D.K.M.], Sigappu Jermany, Sigappu Kuzhiyadichan, Sivappu Chithiraikar, Sivappu Kuruvikar, Soolaikuruvai, Sooran Kuruvai, Sornavali, Sornavari, Sureka, Surti Kolam/Kolam, Thangam Samba, Thidakkal, Thinni, Thooyamallee, Trichy 3, Tulsi-manjari Black, Vadan Samba, Vaigunda, Valla Arakkan, Vangu Vellai, Varadhan, Varalan, Varappu Kudainchan, Vasaramundan, Veer Adangan, Veethivadangan [ASD-3], Velchi, Veliyan, Vellai Chithiraikkar, Vellaikkariyan, Vellaikkuruvai, Vellaikkuruvai [Sobanapuram—Thuraiyur], Vellai Nel, Vellai Poonkar, Alai, Amkel, Antarbhet, Are emo, Adansilpa, Agarali, Agniban, Aguripak, Anshphal, Arabaihar, Arabaihar, Are hat Punko, Aroar, Asanliya, Bahurupi, Begun bichi, Bhaaluki, Bhaludubraj, Bhasa Kalmi, Bhasa manik, Birpana, Badamphul, Badshabhog, Bahdurbhog, Baigan monjia, Bakui, Balam, Balaramsal, Banglaptanai, Baranali, Barshalakshmi, Baskathi, Basmoti-370, Baspata, Bhimsal, Bhogdhan, Bhutia, Birahi, Birai, Bodimani, Chamarmani, Chamatkar, Chinakamini, Chandrakanata, Deko 2, Denta, Dheku, Dhundhuni, Dudheswar, Dudheswar Sundarban, Danaguri, Dandkhani, Dandsal, Danti, Dayalmadina, Debsundari, Dehradun Gandheswari, Desi Jarhan Baihar, Desidhan, Dhapa, Dharaial, Dhusari, Dolle Kartick, Dope, Dubraj, Dudheswar, Dudhkalma, Dwarkasal, Fulmugri, Gangabaru, Garikhajara, Geligeti, Ghios, Ghurghupaijam, Gitanjali, Gobindabhog, Gopalbhog, Goradhan, Janglijata, Haldichuri, Haludgati, Hamai, Hamilton, Harinkajli, Hatidhan, Hendebaihar, Heruajoha, HJP 110, HJP 72, HJP 73, HJP 77, HJP sahebbhog, HMT, Hogla, HP 203, Hamiltan, Hangara, Hormanona, Itanagar, Jabaful, Jalkamini, Jamainaru, Jasmine, Jata, Jhili, Jhinge sal, Jhingesal L, Jhuli, Jhulur, Jhumpuri, Jigiresamba, Jugal, Jabra, Jaldhapa, Jaldubi, JP 57, Jubri dhan, Kaleti, Kaloboro, Kumragore, Kabirajsal, Kabirajsal Odisha, Kaggavat, Khara, Kahndagiri, Kaika, Kalachipta, Kalalahi, Kalamdani, Kalodhan, Kalodhapa, Kalogoda, Kalojira, Kalonunia, Kaltura, Kalavat, Kaminibhog, Kamolsankari, Kanakchur, Kanchafulo, Kankri, Karigavole, Karikagga, Karni, Kasiphul, Kasuabinni, Katarangi, Kataribhog, Katki, Kendu Manjia, Kerala Sundar, Kesab sal, Khadwak, Kolajoha, Kolkiala pateni, Komol, Krishanabhog, Kuji Pateni, Kute Patnai, Laghu, Lakhsman sal, Lalbadshabhog, Lalbasmati, Laldhan Patela, Laldhapa, Lalgetu, Lalkalam, Lalsita, Lathisal, Lusri, Lalbahal, Laldhan, Lalsaru, Lalu, Langlamuthi, Lohoindi, Maianguri, Malabati, Maliagiri, Marikhas, Meghi, Meghnad Dumru, Mohini 2, Motabaihar, Moulo, Machakanata, Madhumala, Malsiara, Marichsal, Markali, Mastaer patnai, Matidhan, Matla, Medi, Mehadi, Mohonbhog, Mugojai, Nagaland kalo, Nalipankhia, Nikkodan, Nivar, Nonakshitish, Nagaland, Nagalanmd Sada, Nageswari, Nagra patnai, Narayan 21, Narayan kamini, Narkelchari, Naryan Patnai, Neli emo, Orasal, P 64 From Meghalya, P 66 From Meghalaya, Pakistani Basmati, Pakri, Pankha Gura, Pari, Paru, Pateni, Patikalam, Patnai 23, Payti emo, Putkhalai, Parbal, Parimolsana, Phakirmoni, Pokali, Rabansal, Radhatilak, Radhuni pagal, Raghu sal, Ramchandrabhog, Ramigelli, Rangi Dhan, Rani akanda, Radha emo, Rahspanjar, Ramjiara, Ranikajal, Rupali, Rupsal, Sadagetu, Sabita, Sabraj, Sada Chenga, Sada Kalam, Safari, Salkele, Samba sole, Sapuri, Sararaj, Sarukala, Shatia, Shu Kalma, Silkote, Sita sal, Sitabhog, Siulee, Sole, Soler Pona, Sonalu, Sonasari, Srikamal, Sundari, Super sita, Talmugur, Tangar sal, Teranga, Thubi, Tilakkachari, Tulsa, Tulsibas, Tulsimanjari, Tulsmukul, Tilakasturi, Toragoda, Tulaipanji, Vogalaya, White Harisnakar, White ponny), Indonesia (Anak Daro, Andel Jaran, Angkong, Batang Lembang, Batang Ombilin, Batang Piaman, Bengawan, Cempo, Ceredek, Ceredek asal Gaduang Surian, Ceredek asal Talang Babungo, Ceredek Merah, Ceredek Putih asal Tanjung Balik, Cianjur Pandanwangi, Cisokan, Dharma Ayu, Ekor Kuda, Engseng, Gropak,

Gundelan, Indra mayu, Induk Ayam, Jambur Urai, Kebo, Ketan Lusi, Ketan Tawon, Longong, Markoti, Melati, Merong, Ondel, Padi Boy, Padi Cere Kuning, Padi Cere Unggul, Padi Gadu, Padi Gandamana, Padi Hitam, Padi Kidangsari, Padi Konyal, Padi Kuning, Padi Kutu, Padi Mendali, Padi Mentik, Padi Mentik Wangi, Padi Pandan Wangi, Padi Parak, Padi Putih, Padi Sari Wangi, Padi Sri Wulan, Padi Suntiang, Padi Wangi Lokal, Pemuda Idaman, Peta, Rajalele, Rangka Madu, Rejung Kuning, Rijal, Rojolele, Sawah Kelai, Siherang, Sikadedek, Simenep, Sri Kuning, Srimulih, Tambun Data, Temanggung black rice, Tembaga, Tjina, Tumpang Karyo, Tunjung, Umbul-umbul, Untup, Wulu), Iran (Ambaroo, Binam, Domsiah, Gerdeh, Gharib, Hasan Sarai, Hasani, Hashemi, Kamfiruzi, Lenjan, Salari, Sang Tarom, Tarom), Italy (Arborio, Ariete, Baldo, Balilla, Carnaroli, Lido, Maratelli, Originario, Padano, Ribe, Roma, Sant'Andrea, Selenio, Thaibonnet, Venere, Vialone Nano), Japan (Akebono, Akitakomachi, Asahi, Domannaka, Haenuki, Hanaechizen, Hinohikari, Hitomebore, Hoshinoyume, Kamachi, Kinuhikari, Kirara397, Koshihikari, Nihonbare, Phoosphoos takuba, Sasanishiki, Tsuyahime), Laos (Sticky rice), Pakistan (Basmati 198, Basmati 370, Basmati 385, Basmati 515, Basmati 2000, Basmati Pak [Kernal Basmati], D-98, DR 82, DR 83, Himsha Basmati 67, Irri-6 Non Basmati Long Grain Rice, Irri-9 Non Basmati Long Grain Rice, C-9 Non basmati Long Grain Rice, Kainaat [1121], Kasha Basmati 167, Khushbu, KS 282 Non Basmati Long Grain Rice, KSK 133, KSK 434, Moomal, PK 386 Long Grain Rice, PS 2 Non Basmati Extra Long Grain Rice, PSP 2001, SD-D34 Basmati, Shadab [R-1, A-1, B-1, B-2], Shaheen Basmati, Shahi Basmati 451, Super Basmati, Unique Basmati 15, Zaiqa HT-1215), Portugal (Ariete, Arroz da terra, Ponta rubra, Valtejo), Philippines (7 Tonner, Angelika, Azucena, Baysilanon, Chong-ak, Dinorado, Ifugao Rice, Imbuucan, IR-64, IR-841, Kalinayan, Maharlika, Malagkit, Mestiso, Milagrosa pino, Ominio, R-238, Sampaguita, Segadis Milagrosa, Sinandomeng, Wagwag, V-10, V-160, White Rose, another 80 cultivars including Red, Brown, Black, White, Glutinous rice), Spain (Albufera, Bahia, Balilla × Sollana, Bomba, Fonsa, Gleva, Guadiamar, J.Sendra, Marisma, Puntal, Senia), Sri Lanka (Badhabath, Kaluheenati, Keeri Samba, Madathawalu, Nadu Rice, Samba, Sri Lankan Red Rice, Sri Lankan white rice, Supiri Nadu, Supiri Samba, Suwandhel, Kurulu Thuda), Taiwan (Taichung 65), Thailand (Black glutinous rice, Brown rice, Jasmine rice, Red Cargo rice, Riceberry, White glutinous rice, White rice), United States (Akitakomachi rice, California New Variety rice, Calmochi rice, Calrose rice, Carolina Gold, Charleston Gold, Long grain rice, Nishiki rice, Pecan rice, Popcorn rice, Texas rice, Texas Wild Rice, Wehani rice), Vietnam (Dự Hương, Nàng Thơm Chợ Đào, Nếp cái hoa vàng, Nếp cẩm, Nếp Tú Lệ, Tài Nguyên, Tám Xoan).

RELATED SPECIES In addition to *Oryza sativa* L., the genus *Oryza* has 17 species: *Oryza australiensis* Domin, *Oryza barthii* A.Chev., *Oryza brachyantha* A.Chev. & Roehr., *Oryza eichingeri* Peter, *Oryza glaberrima* Steud., *Oryza grandiglumis* (Döll) Prodoehl, *Oryza latifolia* Desv., *Oryza longiglumis* Jansen, *Oryza longistaminata* A.Chev. & Roehr., *Oryza meyeriana* (Zoll. & Moritzi) Baill, *Oryza minuta* J.Presl, *Oryza neocaledonica* Morat, *Oryza officinalis* Wall. ex Watt, *Oryza punctata* Kotschy ex Steud., *Oryza ridleyi* Hook.f., *Oryza rufipogon* Griff., and *Oryza schlechteri* Pilg.

GEOGRAPHIC DISTRIBUTION SE Asia. **HABITAT** Cereal cultivated. **RANGE** China, India, Indonesia, Vietnam, Thailand, Bangladesh, Myanmar, Philippines, Brazil, Japan, Pakistan, Cambodia, United States, South Korea, Egypt, Nepal, Nigeria, Madagascar, Sri Lanka, Laos, Australia, Italy.

BEER-MAKING PARTS (AND POSSIBLE USES) Caryopsis, husk.

CHEMISTRY White rice, long grain, cooked without salt: energy 540 kJ (130 kcal), water 68.44 g, carbohydrates 28.1 g, sugars 0.05 g, fiber 0.4 g, lipids 0.28 g, protein 2.69 g, thiamin (B1) 0.02 mg, riboflavin (B2) 0.013 mg, niacin (B3) 0.4 mg, vitamin B6 0.093 mg, calcium 10 mg, iron 0.2 mg, magnesium 12 mg, phosphorus 43 mg, potassium 35 mg, sodium 1 mg, zinc 0.049 mg. White rice, long grain, uncooked: energy 1,527 kJ (365 kcal), water 11.61 g, carbohydrates 80 g, sugars 0.12 g, fiber 1.3 g, lipids 0.66 g, protein 7.13 g, thiamin (B1) 0.0701 mg, riboflavin (B2) 0.0149 mg, niacin (B3) 1.62 mg, pantothenic acid (B5) 1.014 mg, vitamin B6 0.164 mg, calcium 28 mg, iron 0.80 mg, magnesium 25 mg, manganese 1.088 mg, phosphorus 115 mg, potassium 115 mg, zinc 1.09 mg.

STYLE Roggenbier (German Rye Beer), Specialty Beer, Weizen/Weissbier.

BEER Hell Rice, BSA (Vercelli, Piedmont, Italy).

Source Dabove 2015; Heilshorn 2017; Hieronymus 2010; Hieronymus 2016b; Jackson 1991; Josephson et al. 2016; Markowski 2015; McGovern 2017; https://www.bjcp.org (31A. Alternative Grain Beer); https://www.brewersfriend.com/other/risskall/; https://www.omnipollo.com/beer/nathalius/; https://www.omnipollo.com/beer/premium-remix/.

Rice

In industrial breweries, rice is often preferred over barley malt because it is a significantly cheaper source of carbohydrates. However, it is also widely used by craft breweries that operate in areas where it is traditionally cultivated. An example of this are the breweries of Vercelli, which made *Oryza sativa* one of the favored ingredients in their production. The addition of rice gives beer an appealingly sweet and delicate note. Once threshing is completed, the caryopsis, unlike other cereals, is still "dressed" (paddy); that is, covered by the glumes. To separate the grain from its outer husk, it is necessary to subject it to an operation called dehusking, and, before marketing, to other treatments (not necessarily all) such as bleaching, polishing, shining, and oiling. The main by-product of these operations, the husk, is used in craft breweries in the "kettles," where there is a high percentage of starchy substances other than barley. In these cases, husks are extremely useful in the constitution of a filtering bed that allows an efficient separation of mash from grains. However, grain obtained from the paddy has lost its germinability and therefore cannot be subjected to malting. In Asian countries, where rice is often the main cereal of the brewing process and therefore for the conversion of starch into fermentable sugars, it is not possible to use the amylolytic enzymes provided by other cereals, and an alternative system to malting is used. In short, it involves the inoculation of rice with certain molds, selected strains of ascomycete fungi, the most common of which is *Aspergillus oryzae* (Ahlburg) E.Cohn (Trichocomaceae), which have the capacity to saccharify rice. Once the molds have accomplished their task, all that remains to be done is to activate fermentation.

Panax ginseng C.A.Mey.
ARALIACEAE

Synonyms *Aralia ginseng* (C.A.Mey.) Baill., *Panax chin-seng* Nees, *Panax verus* Oken.

Common Names Ginseng, Chinese ginseng.

Description Herb, perennial (30–60 cm). Roots 1- to 2-fasciculate, fusiform or cylindrical. Leaves 3–6, verticillate at stem apex, palmate-compound; base of petiole without stipules or stipular appendages; leaflets 3–5, membranous, abaxially glabrous, adaxially sparsely bristly (trichomes about 1 mm), base broadly cuneate, margin densely serrulate, apex long acuminate; central leaflet elliptic to oblong-elliptic (8–12 × 3–5 cm), lateral leaflets ovate to rhombic-ovate (2–4 × 1.5–3 cm). Inflorescence terminal solitary umbel, with 30–50 flowers; peduncle 15–30 cm, usually longer than petiole; pedicels 0.8–1.5 cm; ovaries 2-carpelled; stigma 2, distinct. Fruit red, compressed-globose (4–5 × 6–7 mm). Seeds reniform.

Cultivar They are numerous, including Chungsun, Chunpoong, Gopoong, Gumpoong, Hwangsookjong, Jakyungjong, Kumpoong, Sunhyang, Sunpoong, Sunwon, Sunwoon, and Yunpoong.

Related Species Of the 13 species in the genus *Panax*, others have similar properties to *Panax ginseng*:

***Oryza sativa* L.**
POACEAE

Panax bipinnatifidus Seem., *Panax bipinnatifidus* Seem. var. *angustifolius* (Burkill) J.Wen, *Panax japonicus* (T.Nees) C.A.Mey., *Panax notoginseng* (Burkill) F.H.Chen, *Panax pseudoginseng* Wall., *Panax quinquefolius* L., *Panax sokpayensis* Shiva K.Sharma & Pandit, *Panax stipuleanatus* H.T.Tsai & K.M.Feng, *Panax trifolius* L., *Panax vietnamensis* Ha & Grushv., *Panax wangianus* S.C.Sun, and *Panax zingiberensis* C.Y.Wu & Feng.

GEOGRAPHIC DISTRIBUTION E Asia (China, Korea). **HABITAT** Mixed forests, deciduous forests. **RANGE** China, Korea, Russia.

BEER-MAKING PARTS (AND POSSIBLE USES) Roots (every part of the plant has a pharmacological action; however, the root is the most commonly used because it has a higher ginsenoside content).

CHEMISTRY The most active components of ginseng are a group of steroidal saponins, called ginsenosides, whose mechanism of action is still largely unknown. It is believed that they are transformed into pharmacologically active substances (compounds K and M4) by intestinal microorganisms. Compound K is derived from protopanaxadiol ginsenoside, whereas M4 is derived from protopanaxatriol ginsenoside. More recent research has shown that ginsenosides act as a pro-drug for these metabolites and that the ginsenoside Rg2 regulates the activity of the human 5-hydroxytryptamine3A channel receptor. Other components include methylxanthines (caffeine, theophylline, theobromine, and others), 20(S)-ginsenoside Rg3, acetylene alcohol, aglycones, alpha-maltosyl-beta-D-fructofuranoside, antioxidants, chikusetsaponin-L8, citral, tetracyclic triterpenoid saponins, essential amino acids (especially arginine), fatty acids, ginsenan polysaccharide, ginsenosides (e.g., F1, F2, F3, R0, Ra1, Ra2, Rb1, Rb2, Rb3, Rc, Rd, Rd2, Re, Rf, Rg1, Rg2, Rg3, Rh1, Rh2, Rh3, Rs4), notoginsenoside-Fe, ginsenoside-Ia, ginsenoside-Ra1, ginsenoside-Re, ginsenoside-Rg2 (20R), ginsenoside-Rh1 (20R), ginsenoside-Rh1 (20S), ginsenoside-Rs3, notoginsenoside-R4, oleanolic acid, panaxadial, panaxans, panaxatriol, panaxosides, panaxydol, panaxytriol, peptidoglycans, polyacetylene, polyacetylenic compounds, poly-furanosyl-pyranosyl-saccharides, polysaccharides, protopanaxatriol, protopanaxadiol ginsenosides, quinqueginsin, saponins, terpineol, triterpenoids, vitamins (especially vitamin C and group B), volatile oils (β-elemene, panaxydol, panaxytriol, falcarinol, limonene), aluminum, calcium, cobalt, copper, iron, manganese, magnesium, molybdenum, phosphorus, potassium, sodium, vanadium, zinc.

STYLE American Pale Ale, Saison.

SOURCE Fisher and Fisher 2017; Hieronymus 2016b; https://www.brewersfriend.com/other/ginseng/.

Panax quinquefolius L.
ARALIACEAE

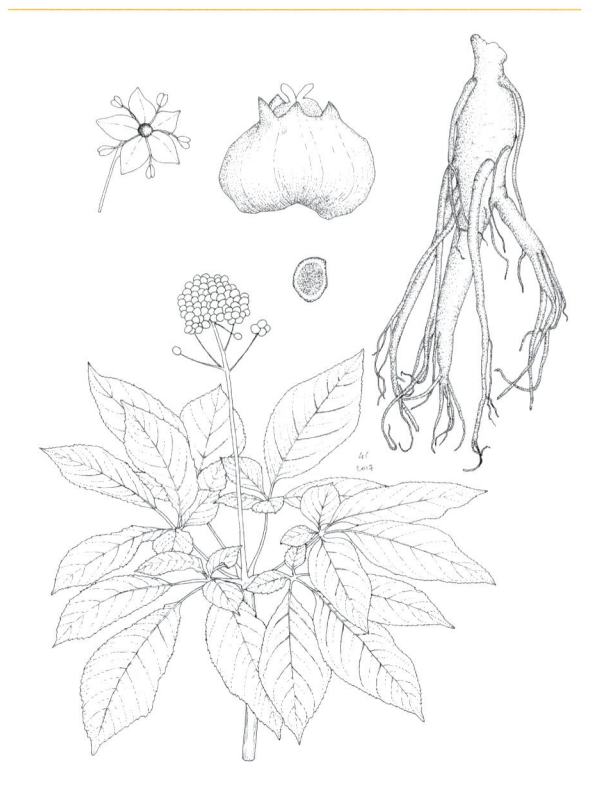

SYNONYMS *Aralia quinquefolia* (L.) Decne. & Planch., *Ginseng quinquefolium* (L.) Alph.Wood, *Panax americanus* (Raf.) Raf., *Panax cuneatus* Raf.

COMMON NAMES American ginseng.

DESCRIPTION Shrub, perennial (2–6 dm), aromatic, slow growing. Solitary stem with a single inflorescence. Root pale yellow to brown in color, fusiform (up to 10 × 2.5 cm), with transverse striations (fine grooves) and longitudinal wrinkles; older roots fork shaped or branched. Leaves in the seedling 3, collected in a single whorl, each compound and palmate; beginning in the third year, the plant produces 3–5 compound leaves, each consisting of 3–7 leaflets 5–10 cm long, the basal smaller than the upper ones; leaflets serrate, wedge shaped, and tapering to a point at the apex, with small, stiff hairs along the marginal veins; leaves dark green in summer, yellow in fall. Inflorescence greenish-white consisting of a group of 4–40 flowers on a single peduncle 5–13 cm long. Flower 2–3 mm long; calyx green with 5 teeth; 5 yellow-green petals; stamens 5, short, with oblong anthers; stigma divided into 2 at the end (looks like 2 stamens); ovary inferior. Fruit drupe, bright red in color (1 cm). Seeds 2–3 per fruit.

RELATED SPECIES Of the 13 species belonging to the genus *Panax*, others have properties similar to those of *Panax quinquefolius*: *Panax bipinnatifidus* Seem., *Panax bipinnatifidus* Seem. var. *angustifolius* (Burkill) J.Wen, *Panax ginseng* C.A.Mey, *Panax japonicus* (T.Nees) C.A.Mey., *Panax notoginseng* (Burkill) F.H.Chen, *Panax pseudoginseng* Wall., *Panax sokpayensis* Shiva K.Sharma & Pandit, *Panax stipuleanatus* H.T.Tsai & K.M.Feng, *Panax trifolius* L., *Panax vietnamensis* Ha & Grushv., *Panax wangianus* S.C.Sun, and *Panax zingiberensis* C.Y.Wu & Feng.

GEOGRAPHIC DISTRIBUTION North America (SE Canada [Ontario, Quebec], United States [Alabama, Arkansas, Connecticut, Delaware, Georgia, Illinois, Indiana, Iowa, Kentucky, Maine, Massachusetts, Maryland, Michigan, Minnesota, Mississippi, Missouri, Nebraska, New Hampshire, New Jersey, New York, North Carolina, Ohio, Oklahoma, Pennsylvania, South Carolina, South Dakota, Tennessee, Vermont, Virginia, West Virginia, Wisconsin]). **HABITAT** Temperate forests. **RANGE** North America (Canada [Ontario], United States [Appalachians, Ozarks, Pennsylvania, New York, Marathon County, Wisconsin]).

BEER-MAKING PARTS (AND POSSIBLE USES) Roots.

CHEMISTRY Similar to *Panax ginseng*, the species *Panax quinquefolius* contains dammaran-type ginsenosides (saponins) as the main biologically active agents (primarily 20[S]-protopanaxadium [PPD] and 20[S]-protopanaxatriol [PPT]). *Panax quinquefolius* contains high levels of ginsenosides Rb1 and Rd (PPD) and higher levels of Re (PPT) than in *Panax ginseng*.

STYLE Herbal Ale.

SOURCE Fisher and Fisher 2017; Hieronymus 2016b; https://www.ratebeer.com/beer/blucreek-herbal-ale/21914/.

Panax trifolius L.
ARALIACEAE

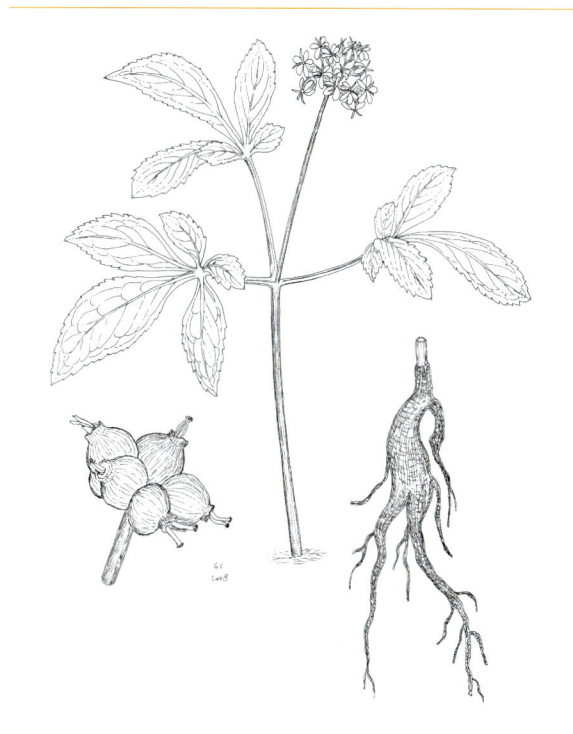

SYNONYMS *Aralia trifolia* (L.) Decne. & Planch., *Aralia triphylla* Poir., *Ginseng trifolium* (L.) Alph.Wood, *Panax lanceolatus* Raf., *Panax pusillus* Sims.

COMMON NAMES Dwarf ginseng.

DESCRIPTION Herb, perennial (10–20 cm), erect. Root rounded. Leaves in a single whorl, finely toothed, palmate, divided into 3–5 sessile leaflets. Flowers white to pink in color, pentamerous. Inflorescence solitary, rounded umbel. Fruit is a yellow berry.

RELATED SPECIES Of the 13 belonging to the genus *Panax*, others have properties similar to *Panax trifolius*: *Panax bipinnatifidus* Seem., *Panax bipinnatifidus* Seem. var. *angustifolius* (Burkill) J.Wen, *Panax ginseng* C.A.Mey, *Panax japonicus* (T.Nees) C.A. Mey., *Panax notoginseng* (Burkill) F.H.Chen, *Panax pseudoginseng* Wall., *Panax quinquefolius* L., *Panax sokpayensis* Shiva K.Sharma & Pandit, *Panax stipuleanatus* H.T.Tsai & K.M.Feng, *Panax vietnamensis* Ha & Grushv., *Panax wangianus* S.C.Sun, and *Panax zingiberensis* C.Y.Wu & Feng.

GEOGRAPHIC DISTRIBUTION E Canada, E United States. **HABITAT** Woodlands, depressions, partially shady environments in rich soils. **RANGE** E Canada, E United States.

BEER-MAKING PARTS (AND POSSIBLE USES) Roots.

CHEMISTRY Leaves: flavonoids (kaempferol-3,7-dirhamnoside, kaempferol-3-gluco-7-rhamnoside), ginsenosides (ginsenoside Rd, ginsenoside Rc, ginsenoside Rb3, notoginsenoside-Fe), aglycones (kaempferol, [20S]-protopanaxadiol).

STYLE None known.

SOURCE Fisher and Fisher 2017; Hieronymus 2016b.

Panicum miliaceum L.
POACEAE

SYNONYMS *Leptoloma miliacea* (L.) Smyth, *Panicum asperrimum* Fisch., *Panicum asperrimum* Fischer ex Jacq., *Panicum densepilosum* Steud., *Panicum ruderale* (Kitag.) D.M.Chang.

COMMON NAMES Millet, broomcorn millet, common millet, broomtail millet, hog millet, kashfi millet, red millet, white millet.

DESCRIPTION Herb, annual (6–12 dm). Culms erect or kneeling at base, stout, often branched above. Leaves usually with elongated hairs, lamina up to 1 mm wide and ligule of hairs. Inflorescence panicle broad, ± corymbose or fan shaped, usually drooping on one side; branches slender, rough; lower glume 1.5–2 mm; upper glume and lemma nearly equal, 3–3.5 mm.

CULTIVAR Numerous cultivars of *Panicum miliaceum* are used around the world, including the following: Australia (Red French Millet, White French Millet), United States (Deerbrook, Early Fourtune, Turghai, Yellow Manitoba), and Canada (Crown).

RELATED SPECIES Of the more than 440 species recognized as belonging to the genus *Panicum*, none, other than *Panicum miliaceum*, is used on a large scale for human consumption.

GEOGRAPHIC DISTRIBUTION C Asia. **HABITAT** Cereal cultivated. **RANGE** India, Nepal, Russia, Ukraine, Middle East, Turkey, Romania, United States.

BEER-MAKING PARTS (AND POSSIBLE USES) Caryopses.

CHEMISTRY Raw caryopsis: energy 378 kcal, water 8.67 g, protein 11.02 g, lipids 4.22 g, carbohydrates 72.85 g, fiber 8.5 g, calcium 8 mg, iron 3.01 mg, magnesium 114 mg, phosphorus 285 mg, potassium 195 mg, sodium 5 mg, zinc 1.68 mg, thiamine 0.421 mg, riboflavin 0.29 mg, niacin 4.720 mg, vitamin B6 0.384 mg, folate 85 µg, vitamin E 0.05 mg, vitamin K 0.9 µg, saturated fatty acids 0.723 g, monounsaturated fatty acids 0.773 g, polyunsaturated fatty acids 2.134 g.

STYLE Cereal used in the first known beer in China, Oatmeal Stout, Special/Best/Premium Bitter.

SOURCE Dabove 2015; Hieronymus 2010; Jackson 1991; McGovern 2017; Steele 2012; https://www.bjcp.org (31A. Alternative Grain Beer); https://www.brewersfriend.com/other/millet/; https://www.fermentobirra.com/la-birra-preistorica-cina/.

Papaver somniferum L.
PAPAVERACEAE

SYNONYMS None

COMMON NAMES Opium poppy.

DESCRIPTION Herb, annual (3–12 dm), subglabrous. Stems erect, stout, glaucous. Caulescent leaves ovate or lanceolate, 7–12 dm long, cordate-amplexicaule at base, deeply lobed at edge, wavy, waxy; obtuse lobes not terminating in a bristle. Flowers terminal on elongated peduncles; bud pendulous (8 × 15–20 mm); sepals early deciduous; petals 4, 3–5 cm long, wavy,

Panicum miliaceum L.
POACEAE

white, pinkish, or violet, with a dark blotch at base, often lobed at edge; stamens numerous with white filaments; stigmatic disc with 8–18 rays. Fruit capsule subspherical (3–5 × 4–8 cm).

Subspecies Besides the nominal subspecies, there is also *Papaver somniferum* L. subsp. *setigerum* (DC.) Arcang.

Cultivar *Papaver somniferum* has many cultivars that vary in color, number and shape of tepals, number of flowers and fruits, number and color of seeds, alkaloid content, etc. *Papaver somniferum* group *Paeoniflorum* is a double-flowered subtype, while *Papaver somniferum* group *Laciniatum* is a double-flowered subtype with strongly lobed petals. Norman and Przemko varieties have a low morphine content (< 1%) and higher concentrations of other alkaloids, whereas most other varieties, including those for ornamental use or seed production, have a morphine content of around 10%.

Related Species Of the 71 taxa of specific or subspecific rank accepted for the genus *Papaver*, several contain small amounts of alkaloids.

Geographic Distribution Euro-Mediterranean, later subcosmopolitan. **Habitat** Cultivated as a medicinal or ornamental plant and wild. **Range** Seed producers: Czech Republic, Spain, Hungary, Turkey, Germany, France, Palestine, Romania, Croatia, Austria, Serbia, Netherlands, Slovakia, Macedonia, Poland. Producers of legal drugs (e.g., morphine, codeine) from the opium poppy: Australia (Tasmania), Turkey, India. Producer of illegal drugs: Afghanistan. In many countries, cultivation is widespread for ornamental purposes but prohibited or subject to special permits.

Beer-Making Parts (and Possible Uses) Latex from immature capsule (illegal drug), seeds (food use; low opiate content), oil (food use; very low opiate content).

Toxicity Opiates are highly toxic and addictive.

Chemistry Dried latex (opium): this product contains a class of alkaloids (opiates) that include morphine (8–14%), thebaine, codeine, papaverine, noscapine, and oripavine. Seeds: energy 6,367 kcal, water 4.76%, oil 41.86%, protein 11.94%, fiber 24.73%, ash 4.92%, aluminum 12.6 ppm, boron 69.4 ppm, calcium 8756.8 ppm, cadmium 0.3 ppm, chromium 3.5 ppm, copper 13.2 ppm, iron 64, 1 ppm, potassium 10,535.7 ppm, lithium 6.7 ppm, magnesium 3,872.2 ppm, manganese 62.3 ppm, sodium 664.5 ppm, nickel 1.6 ppm, phosphorus 9,375.9 ppm, lead 1.4 ppm, strontium 86.0 ppm, vanadium 25.2 ppm, zinc 29.6 ppm. Oil: palmitic acid 15.5%, stearic acid 3.65%, oleic acid 23.45%, linoleic acid 57.18%, linolenic acid 0.39%, α-tocopherol 33.6 ppm, β-tocopherol 553.2 ppm, δ-tocopherol 10.4 ppm.

Style Porter.

Source Daniels 2016; Daniels and Larson 2000; https://www.hacker-pschorr.com/it/il-nostro-marchio/requisito-di-purezza-bavarese.

Passiflora edulis Sims
PASSIFLORACEAE

Synonyms *Passiflora gratissima* A. St.-Hil., *Passiflora incarnata* L., *Passiflora iodocarpa* Barb. Rodr., *Passiflora middletoniana* J.Paxton, *Passiflora pallidiflora* Bertol., *Passiflora picroderma* Barb. Rodr., *Passiflora pomifera* M.Roem., *Passiflora rigidula* J.Jacq., *Passiflora rubricaulis* Jacq., *Passiflora vernicosa* Barb. Rodr.

Common Names Passion fruit, maracuya, grenadille, maracujá.

Description Plant, perennial, woody, vigorous, climbing (up to 15 m). Stem herbaceous, striate, with simple tendrils up to 10 cm long. Leaves alternate (to 13 × 15 cm), ± deeply 3-lobed, slightly leathery, glossy green or yellow-green above, dull light green below; 2 glands at petiole apex; margin finely toothed; linear stipules present (about 1 cm). Flowers solitary, Ø up to 7 cm; petals white; corona with filaments up to 2.5 cm long, in 4–5 rows, white, purple at base; 5 stamens with broad anthers, ovary with 3-parted stigma forming a prominent structure. Fruit pepo, ovoid to spherical, Ø 4–5 cm, yellow, greenish-yellow, or purple; flesh juicy, containing about 250 small brown or black seeds.

Subspecies Within *Passiflora edulis* Sims, the following taxa of subspecific rank are recognized: *Passiflora edulis* Sims f. *flavicarpa* O.Deg. and *Passiflora edulis* Sims var. *kerii* (Spreng.) Mast.

Cultivar There are numerous cultivars and many hybrids based on *Passiflora edulis*. Although probably incomplete, the following list offers a glimpse of the varietal complexity of this species: *P. edulis* f. *edulis* × *P. edulis* f. *flavicarpa*, *P. edulis* f. *flavicarpa* × *P. incarnata*, *P. edulis* ♀ × *P. laurifolia* ♂, *P. edulis* ♂ × *P. maliformis* ♀, *P. edulis* ♀ × *P. quadrangularis* ♂, *P. edulis* ♂ × *P. subpeltata* ♀, *P. edulis* Alice, *P. edulis* Australian Purple, *P. edulis* Bali Hai, *P. edulis* Black Beauty, *P. edulis* Black Knight, *P. edulis* Boi, *P. edulis* Brasileira Amarilla, *P. edulis* Brasileira Rosada, *P. edulis* Brazilian Golden, *P. edulis* Common Purple, *P. edulis* Comun C, *P. edulis* Corrego Rico, *P. edulis* Crackerjack, *P. edulis* Dvoploda, *P. edulis* E-23, *P. edulis* Eche, *P. edulis* Ecke Select, *P. edulis* Ecuador Lace, *P. edulis* Ecuadorian Gold, *P. edulis* Edgehill, *P. edulis* Espino Strain, *P. edulis* Florida Clone, *P. edulis* Frederick (*P. edulis* Brazilian Gold × *P. edulis* Kahuna), *P. edulis* Fredrick → *P. edulis* Frederick, *P. edulis* Frosty, *P. edulis* Gema de Ovo, *P. edulis* Globe, *P. edulis* Gold Star, *P. edulis* Golden Giant, *P. edulis* Golden Nugget, *P. edulis* Granadilla, *P. edulis* Grande, *P. edulis* Guaratingueta, *P. edulis* Guassu, *P. edulis* Hawaiiana, *P. edulis* Hawaiian Yellow, *P. edulis* Jundiai Amarelo, *P. edulis* Jundiai Vermelho, *P. edulis* Kahuna, *P. edulis* Kapoho Selection, *P. edulis* Knight (= *P. edulis* Black Knight), *P. edulis* Lacey (*P. edulis* f. *edulis* × *P. edulis* f. *flavicarpa*), *P. edulis* Lanphiers Stermer, *P. edulis* M-21471 A (*P. edulis* f. *edulis* × *P. edulis* f. *flavicarpa*), *P. edulis* Maloya (*P. edulis* f. *edulis* × *P. edulis* f. *flavicarpa*), *P. edulis* Mammoth Purple Granadilla, *P. edulis* Marmelo, *P. edulis* McCain, *P. edulis* Mirim, *P. edulis* Miudo Parana, *P. edulis* Muico, *P. edulis* Nancy Garrison, *P. edulis* Ned Kelly, *P. edulis* Nelly Kelly → *P. edulis* Australian Purple, *P. edulis* Nichols, *P. edulis* Nina, *P. edulis* Noels Special, *P. edulis* Norfolk, *P. edulis* Ouropretano, *P. edulis* Panama Gold, *P. edulis* Panama Red, *P. edulis* Patrick, *P. edulis* Paul Ecke, *P. edulis* Perfecta, *P. edulis* Peroba, *P. edulis* Peroba Roxo, *P. edulis* Pintado, *P. edulis* Possum Purple (*P. edulis* f. *edulis* × *P. edulis* f. *flavicarpa*), *P. edulis* Pratt Hybrid, *P. edulis* Purple Champion, *P. edulis* Purple Giant, *P. edulis* Purple Gold, *P. edulis* Rainbow Sweet, *P. edulis* Red Giant, *P. edulis* Red-Possum Trot, *P. edulis* Red Riveira (*P. edulis* f. *edulis* × *P. edulis* f. *flavicarpa*), *P. edulis* Red Rover (*P. edulis* Brazilian Gold × *P. edulis* Kahuna), *P. edulis* Red Sunset, *P. edulis* Redlands Pink, *P. edulis* Redlands Triangular, *P. edulis* Redondo (= *P. edulis* Mirim), *P. edulis* Roxo Pequeno, *P. edulis* Roxo Silvestre, *P. edulis* Sao Sebastiao, *P. edulis* Selection E-23 (= *P. edulis* E-23), *P. edulis* Sevcik Selection, *P. edulis* Sevick Selection → *P. edulis* Sevcik Selection, *P. edulis* Silv. Jundiai, *P. edulis* Sunnypash, *P. edulis* Supersweet (*P. edulis* f. *edulis* × *P. edulis* f. *flavicarpa*), *P. edulis* Supreme, *P. edulis* Sweepur, *P. edulis* Taiwan Yellow, *P. edulis* Ubatuba, *P. edulis* University Round Selection (*P. edulis* Waimanolo Selection × *P. edulis* Yee Selection), *P. edulis* University Selection B-74 (= *P. edulis* B-74), *P. edulis* Vista, *P. edulis* Waimanolo Selection, *P. edulis* Yee Selection.

Related Species The genus *Passiflora* has more than 500 recognized species.

Geographic Distribution South America.
Habitat Tropical and subtropical warm areas.
Range Cultivated in tropical and subtropical warm areas.

Beer-Making Parts (and Possible Uses) Fruits.

Chemistry Fruits: energy 406 kJ (97 kcal), water 72.9%, carbohydrates 22.4 g, sugars 11.2 g, fiber 10.4 g, lipids 0.7 g, protein 2.2 g, vitamin A 64 µg, beta-carotene 743 µg, riboflavin (B2) 0.13 mg, niacin (B3) 1.5 mg, vitamin B6 0.1 mg, folate (B9) 14 µg, choline 7.6 mg, vitamin C 30 mg, vitamin K 0.7 µg, calcium 12 mg, iron 1.6 mg, magnesium 29 mg, phosphorus 68 mg, potassium 348 mg, sodium 28 mg, zinc 0.1 mg.

Style American IPA, American Pale Ale, American Wheat or Rye Beer, Berliner Weisse, California Common Beer, Flanders Red Ale, Fruit Beer, Gueuze, Imperial IPA, Saison.

Beer Hillbilly Wine, Siren (Finchampstead, England, United Kingdom).

Source Heilshorn 2017; Hieronymus 2016b; https://www.bjcp.org (29A. Fruit Beer: Belville Passion Fruit); https://www.brewersfriend.com/other/pajonsfrukt/; https://www.brewersfriend.com/other/passionfruit/; https://www.omnipollo.com/beer/42/; https://www.ratebeer.com/beer/belville-passion-fruit/47035/.

Paullinia cupana Kunth
SAPINDACEAE

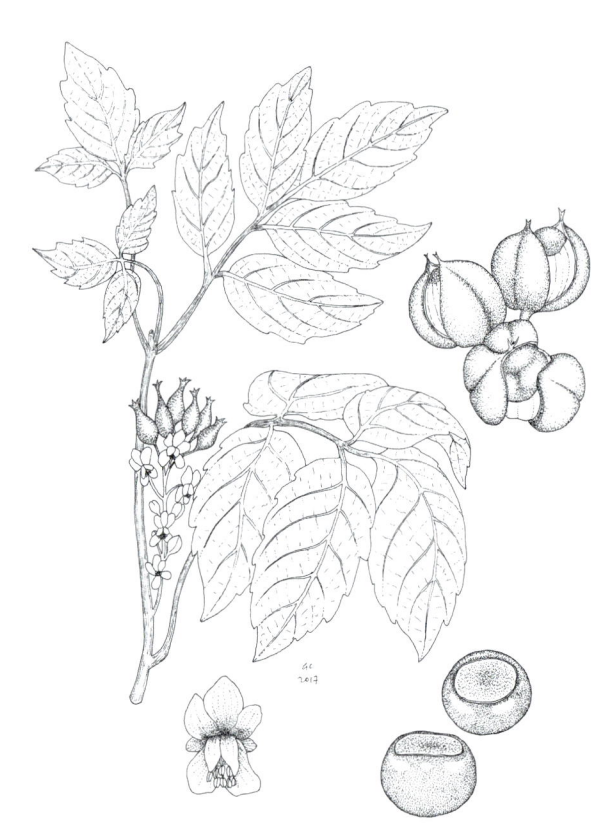

Synonyms *Paullinia sorbilis* Mart.

Common Names Guarana.

Description Shrub, tree, or climber (up to 13 m); in cultivation the plant is kept as a sapling or shrub (2–3 m) to facilitate seed collection. Bark light green on young branches, brown on trunk and larger branches. Leaves compound, alternate, with short petiole, leathery, glossy green on the upper leaf, dull green underneath. Flowers white, short stalked, single or, more often, collected in panicles of up to 15–20. Fruit trilocular drupe, very firm, with fiery red parchment epicarp, whitish parenchymatous mesocarp, woody endocarp that becomes pea-sized after drying.

Subspecies Some researchers recognize two varieties of this species: *Paullinia cupana* Kunth var. *cupana* and *Paullinia cupana* Kunth var. *sorbilis* Mart. However, the taxonomic and genetic relationships between the two varietal entities are unclear, as is the role of domestication in their formation and spread; according to some authors, they should be synonymous with *Paullinia cupana*.

Related Species Of the 167 known species in the genus *Paullinia*, only *Paullinia yoco* R.E. Schult. & Killip has similar properties and traditional uses to those of *Paullinia cupana*.

Geographic Distribution S America. **Habitat** Rainforests. **Range** Brazil.

Beer-Making Parts (and Possible Uses) Seeds.

Chemistry Seeds: adenine, ash 14,200 ppm, caffeine 9,100–76,000 ppm, catechutannic acid, choline, D-catechin, lipids 30,000 ppm, guanine, hypoxanthine, mucilage, protein 98,600 ppm, resins 70,000 ppm, saponins, starch 50,000–60,000 ppm, tannin 50,000–120,000 ppm, theobromine 200–400 ppm, theophylline 0–2,500 ppm, thymbonine, xanthine.

Style Herbal Ale.

Source https://www.ratebeer.com/beer/blucreek-herbal-ale/21914/.

Pelargonium tomentosum Jacq.
GERANIACEAE

Synonyms None

Common Names Geranium.

Description Shrub (up to 1 m), aromatic. Stem with hair-covered branches scattered in all directions. Leaves simple, gray-green color, trilobed lamina, covered with soft hairs and numerous glandular hairs giving velvety appearance, minty smell. Flowers small; petals white in color with purple markings. Inflorescence very branched and persistent.

Cultivar There are several cultivars of this taxon.

Related Species More than 400 species belong to the genus *Pelargonium*. Among these related to *Pelargonium tomentosum* Jacq. is *Pelargonium graveolens* L'Hér. (= *Pelargonium asperum* Ehrh. ex Spreng.), both species represented in the iconography and from which essential oil with numerous therapeutic properties is extracted.

Geographic Distribution South Africa (on the Hottentots Holland Mountains near Somerset West, on the Riviersonderend Mountains near Grayton, on the Langeberg Range from Swellendam to Riversdale). **Habitat** Mountain environments, in shady, humid areas, at the edge of forests, in ravines near watercourses, in sandy soils derived from sandstone. **Range** Cultivated as an ornamental and medicinal plant.

Beer-Making Parts (and Possible Uses)
Flowers, stems, leaves.

Chemistry Essential oil: tricyclene 0.4%, α-pinene 0.5%, camphene 0.1%, myrcene 0.8%, limonene 1.5%, β-phellandrene 0.6%, p-cymene 1.1%, menthone 41.1%, isomentone 49.3%, neomenthol 0.3%, β-caryophyllene 0.4%, neoisomenthol 0.6%, α-humulene 0.1%, cryptone 0.2%, α-terpineol 0.1%, ledene 0.1%, germacrene D 0.1%, β-bisabolene 0.1%, piperitone 0.5%, piperitenone 0.1%, caryophyllene oxide 0.2%.

Style American Amber Ale, Belgian Pale Ale, Strong Bitter.

Source Hieronymus 2016b (reference is to the genus *Pelargonium* without specific rank indications); https://www.brewersfriend.com/other/geranium/ (although it appears to be established that *Geranium* as a brewing ingredient is used not only for the aroma imparted to some beers by Cascade hops; among the more than 400 species belonging to this genus, it has not been possible to identify those possibly used for beer making. Moreover, the common term "geranium" can also refer to members of the genus *Pelargonium* [Geraniaceae], which further complicates species identification; the one described here must therefore be considered an example of the genus *Pelargonium*).

Pennisetum glaucum (L.) R.Br.
POACEAE

Synonyms *Alopecurus typhoides* Burm.f., *Cenchrus americanus* (L.) Morrone, *Cenchrus spicatus* (L.) Cav., *Chaetochloa glauca* (L.) Scribn., *Chamaeraphis glauca* (L.) Kuntze, *Holcus racemosus* Forssk., *Holcus spicatus* L., *Ixophorus glaucus* (L.) Nash, *Panicum americanum* L., *Panicum coeruleum* Mill., *Panicum glaucum* L., *Panicum indicum* Mill., *Panicum involucratum* Roxb., *Panicum sericeum* Aiton, *Panicum spicatum* (L.) Roxb., *Penicillaria arabica* A.Braun, *Penicillaria ciliata* Willd., *Penicillaria deflexa* Andersson, *Penicillaria deflexa* Andersson ex A.Braun, *Penicillaria involucrata* (Roxb.) Schult., *Penicillaria mossambicensis* Müll.Berol, *Penicillaria nigritarum* Schltdl., *Penicillaria plukenetii* Link, *Penicillaria solitaria* Stokes, *Penicillaria spicata* (L.) Willd., *Penicillaria typhoidea* (Burm) Schltdl., *Penicillaria willldenowii* Klotzsch ex.A.Braun and C.D.Bouché, *Pennisetum albicauda* Stapf & C.E.Hubb., *Pennisetum americanum* (L.) Leeke, *Pennisetum ancylochaete* Stapf & C.E.Hubb., *Pennisetum aureum* Link, *Pennisetum cereale* Trin., *Pennisetum cinereum* Stapf & C.E.Hubb., *Pennisetum echinurus* (K.Schum) Stapf & C.E.Hubb., *Pennisetum gambiense* Stapf & C.E.Hubb., *Pennisetum gibbosum* Stapf & C.E.Hubb., *Pennisetum leonis* Stapf & C.E.Hubb., *Pennisetum linnaei* Kunth, *Pennisetum maiwa* Stapf & C.E.Hubb., *Pennisetum malacochaete* Stapf & C.E.Hubb., *Pennisetum megastachyum* Steud., *Pennisetum nigritarum* (Schltdl.) T.Durand & Schinz, *Pennisetum nigritarum* (Schltdl.) T.Durand & Schinz var. *deflexum* (A.Braun) T.Durand & Schinz, *Pennisetum nigritarum* (Schltdl.) T.Durand & Schinz var. *macrostachyum* (A.Braun) T.Durand & Schinz, *Pennisetum plukenetii* (Link) T.Durand & Schinz, *Pennisetum pycnostachyum* Stapf & C.E.Hubb., *Pennisetum solitarium* Stokes, *Pennisetum spicatum* (L.) Körn., *Pennisetum typhoides* (Burm.f.) Stapf & C.E.Hubb., *Pennisetum typhoideum* Rich., *Phleum africanum* Lour., *Setaria glauca* (L.) P.Beauv., *Setaria sericea* (Sol.) P.Beauv., *Setariopsis glauca* (L.) Samp.

Common Names Pearl millet.

Description Herb, annual (up to 3 m). Culms stout, densely pubescent at nodes and below inflorescence. Leaf with a lax, smooth sheath; lamina 20–100 × 2–5 cm, both surfaces and margins scabrous; base subcordate; ligules 2–3 mm. Inflorescence linear to broadly elliptic, dense (40–50 × 1.5–2.5 cm); axis densely pubescent; involucre persistent, enclosing 1–9 spikelets, basal axis pubescent, 1–25 mm; awn usually shorter than spikelets, nearly glabrous to densely plumose. Spikelets obovate (3.5–4.5 mm); lower glume minute (about 1 mm), upper glume 1.5–2 mm, 3-veined; lower flower staminate, lemma about 2.5 mm, 5-veined, margins membranous and ciliate, palea subtly papery, puberulum; upper lemma 5- to 7-veined, subtly papery, puberulum, margins ciliate, tip obtuse; anthers with tuft of short hairs at end. Caryopsis ovoid (3–4 mm).

Cultivar There are numerous cultivars for the caryopsis; for example: ACC-1022-12SPT, BONKOK-SHORT, D2P29, DMR22, DMR43, EX-BORNO, IMV11-3-3SPT, LCIC9702, LCIC9703-27. Selected cultivars are also cultivated for ornamental purposes, including Jade Princess, Purple Barron, Purple Jester, and Purple Majesty.

Related Species The genus *Pennisetum* includes about 86 taxa.

Geographic Distribution Africa.
Habitat Cultivated cereal. **Range** Introduced in Asia, Europe, and elsewhere.

Beer-Making Parts (and Possible Uses) Caryopses malted or not.

Chemistry Caryopses: energy 4,498 kcal, starch 64.78%, protein 16.66%, fiber 2.30%, ash 1.95%, calcium 0.06%, phosphorus 0.32%.

Style None verified.

Source Fisher and Fisher 2017.

Perilla frutescens (L.) Britton
LAMIACEAE

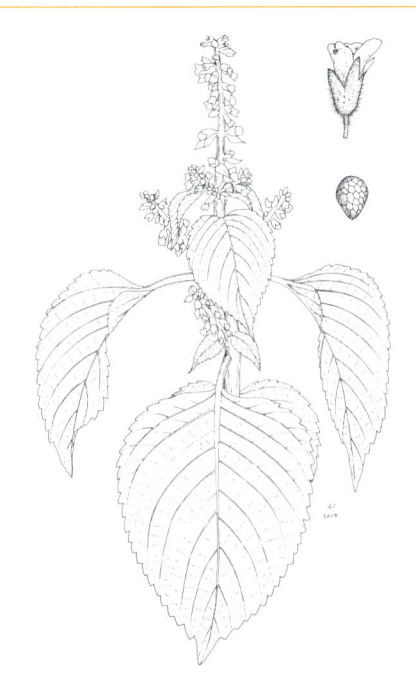

Synonyms *Melissa cretica* Lour., *Melissa maxima* Ard., *Mentha perilloides* Lam., *Ocimum frutescens* L., *Perilla albiflora* Odash., *Perilla avium* Dunn, *Perilla ocymoides* L., *Perilla shimadae* Kudô, *Perilla urticifolia* Salisb.

Common Names Perilla-mint, perilla, wild sesame, shiso.

Description Herb, perennial (up to 1 m), pleasantly aromatic. Stem erect, branched, sparsely hairy. Leaves opposite, petiolate (25–80 mm), lamina simple, oblong-ovate (45–130 × 28–100 mm), green or purple variegated (depending on variety), with glandular hairs, margin with rounded teeth, base cuneate, truncate or rounded, apex acute-acuminate. Inflorescences elongated terminal or axillary racemes. Flower zygomorphic, hermaphrodite, calyx tubiform, corolla consisting of 5 petals (3–4 mm) joined to form a tube, color white to blue-purple, upper petal trilabiate, ovary superior, 4 stamens. Fruit schizocarp tetramerous, dry indehiscent, mericarp 1.2–1.8 mm.

Subspecies Within the species *Perilla frutescens*, the following varieties and forms are recognized: *Perilla frutescens* (L.) Britton var. *frutescens* (deulkkae, kkaennip), *Perilla frutescens* (L.) Britton var. *crispa* (Thunb.) H.Deane (shiso or tía tô), *Perilla frutescens* (L.) Britton var. *crispa* (Thunb.) H.Deane f. *purpurea* (akajiso, red shiso), *Perilla frutescens* (L.) Britton var. *crispa* (Thunb.) H.Deane f. *viridis* (aojiso, green shiso), and *Perilla frutescens* (L.) Britton var. *hirtella* (Nakai) Makino (lemon perilla).

Cultivar While waiting for an organic classification and varietal delimitation, it is possible to refer to the Subspecies section. Some cultivars are Anyu, Areum, Baegsang, Baekkwang, Daesil, Daeyeop, Daeyu, Dasil, Hyangim, Kwangim, Okdong, Saeyeupsil, Yangsan, Yeupsil, and Yujin.

Related Species The genus *Perilla* is monotypic.

Geographic Distribution Himalayas, China. **Habitat** Mountainous areas. **Range** China, Japan, Korea, India, Vietnam.

Beer-Making Parts (and Possible Uses) Leaves, seeds, oil (from seeds).

Chemistry Essential oil, aerial parts (15 compounds; 88.1% of oil): perillaketone 35.6%, isoegomachetone 35.1%, β-caryophyllene 4.3%, (Z,E)-α-farnesene 2.7%, benzaldehyde 2.3%, linalool 1.7%, germacrene D 1.2%, 1-otten-3-ol 1%, caryophyllene oxide 0.9%, perillene 0.8%, α-copaene 0.8%, humulene epoxide 0.6%, hexahydrofarnesyl acetone 0.6%, spatulenol 0.5%, elsholtzia ketone traces.

Style American Wheat Beer.

Beer Shiso Ale, Atsugi-shi (Kanagawa, Japan).

Source Josephson et al. 2016; https://www.brewersfriend.com/other/shiso/.

Persea indica (L.) Spreng.
LAURACEAE

Synonyms *Borbonia indica* J.Presl, *Laurus indica* L.

Common Names Vinhatico, viñátigo.

Description Tree, evergreen (over 20 m). Trunk robust, straight, often surrounded by suckers at the base, branching into a broad, dense crown; bark dark gray, cracked; in young branches greenish color and characteristic lenticels. Leaves aromatic, simple, alternate, oblong-lanceolate, wide (up to 20 × 7 cm), slightly leathery; smooth apex, entire margin sometimes folded; lamina dark green on the upper leaf, a little bit lighter on the underside (as it ages, the color of the leaves turns yellow, orange, or reddish; this characteristic gives the species the appearance of an autumn tree all year round); leaf stalk (2–3 cm) usually yellowish. Flowers are simple and singularly inconspicuous (1 cm), formed by 6 small greenish-white petals; they are grouped in small terminal inflorescences, a little more showy than single flowers. Fruit elliptical fleshy drupe (about 2.5 × 2 cm), green, violet, or bluish when ripe, containing a small nucule with 1 seed.

RELATED SPECIES The genus *Persea* includes about 120 taxa.

GEOGRAPHIC DISTRIBUTION Endemic species of Macaronesia (Canary Islands, Madeira). **HABITAT** Characteristic element of the laurel forests of Macaronesia, typical forest of the humid areas of the Atlantic islands, together with species such as *Laurus novocanariensis* Rivas Mart., Lousã, Fern. Prieto, E.Días, J.C.Costa & C.Aguiar (Lauraceae), *Ocotea foetens* (Aiton) Baill. (Lauraceae), *Apollonias barbujana* (Cav.) Bornm. (Lauraceae), etc. It prefers shady areas but can also be present in sunny environments. Intolerant to wind, it requires constantly humid soils such as those found at the bottom of ravines or in river valleys (600–1,000 m ASL). **RANGE** Canary Islands, Madeira, Azores.

BEER-MAKING PARTS (AND POSSIBLE USES) Wood for casks (first they hold cachaça, then beer for aging).

TOXICITY Sap toxic to small mammals.

WOOD Dark raw wood (WSG 0.51–0.57).

STYLE None known.

SOURCE Cantwell and Bouckaert 2016; Cantwell and Bouckaert 2018 (It is possible that, once again, a common name has produced some confusion with respect to the correct scientific identification of the species to which it belongs. Indeed, vinhatico is a common name that some sources associate with *Persea indica*, the species cited here. According to these sources, the tree should be part of a contingent of South American species, but its geographical origin is different. Moreover, very little information is available about the characteristics and uses (mainly food) of the wood of this species, maybe because of the toxicity of the sap. It seems plausible that the term "vinhatico" is therefore more properly applied to South American woody legumes belonging to the genera *Plathymenia* and *Pithecolobium*).

Petroselinum crispum (Mill.) Fuss
APIACEAE

SYNONYMS *Apium crispum* Mill., *Apium petroselinum* L., *Carum petroselinum* (L.) Benth. & Hook.f., *Cnidium petroselinum* DC., *Petroselinum hortense* Hoffm., *Petroselinum sativum* Hoffm., *Petroselinum vulgare* Lag, *Peucedanum petroselinum* (L.) Desf., *Selinum petroselinum* (L.) E.H.L.Krause, *Wydleria portoricensis* DC.

COMMON NAMES Parsley, garden parsley.

DESCRIPTION Herb (up to 75 cm), annual (tropical or subtropical climate) or biennial (temperate climate). Taproot robust, yellowish-white in color. Stem flowering, appears in second year, erect, tubular, weakly ribbed. Leaves glabrous, lamina outline triangularly indented, 2- to 3-pinnate (10–25 cm), with numerous leaflets (1–3 cm) in rosette form the first year; cauline leaves simply pinnate. Inflorescence umbel (Ø 3–10 cm), involucre null or 1–3 linear bracts, often deciduous, formed by 8–21 rays bearing numerous small flowers (2 mm) with 5 petals (0.7 mm), color yellowish-white. Fruit ovoid (2–3 mm), with prominent remnants of the stigma at the apex.

Subspecies Two known subspecies: *Petroselinum crispum* (Mill.) Fuss subsp. *crispum* and *Petroselinum crispum* (Mill.) Fuss subsp. *giganteum* (Pau) Dobignard.

Cultivar
There are two groups of cultivated varieties: leaf parsley and root parsley. Leaf parsley can be flat-leafed (Italian: *Petroselinum crispum* var. *neapolitanum*), similar to wild individuals and with a stronger aroma, or curly-leafed (*Petroselinum crispum* var. *crispum*), usually used as a garnish; a third type, cultivated in Southern Italy, has thick leaf stalks similar to celery. Root parsley (*Petroselinum crispum* var. *tuberosum*) produces a big taproot, mainly used in cooking in EC Europe; also consumed raw.

Related Species Two species belong to this genus: *Petroselinum crispum* (Mill.) Fuss and *Petroselinum segetum* (L.) W.D.J.Koch.

Geographic Distribution C Mediterranean.
Habitat Crops, near gardens. **Range** Widely cultivated all over the world.

Beer-Making Parts (and Possible Uses)
Leaves, stems.

Chemistry Fresh leaves: energy 151 kJ (36 kcal), carbohydrates 6.33 g, sugars 0.85 g, fiber 3.3 g, lipids 0.79 g, protein 2.97 g, vitamin A 421 µg, beta-carotene 5054 µg, lutein zeaxanthin 5561 µg, thiamin (B1) 0.086 mg, riboflavin (B2) 0.09 mg, niacin (B3) 1.313 mg, pantothenic acid (B5) 0.4 mg, vitamin B6 0.09 mg, folate (B9) 152 µg, vitamin C 133 mg, vitamin E 0.75 mg, vitamin K 1,640 µg, calcium 138 mg, iron 6.2 mg, magnesium 50 mg, manganese 0.16 mg, phosphorus 58 mg, potassium 554 mg, sodium 56 mg, zinc 1.07 mg Essential oil, leaves (36 compounds; 96% of total oil): p-1,3,8-menthatriene 40.0–44.6%, β-phellandrene 15.1–16.9%, myristicin 13–13.1%, myrcene 6.5–7%; terpinolene, limonene, 1-methyl-4-isopropylbenzene; β- and α-pinene, each 0.6%–4.2%; other constituents under 0.8%. Essential oil, roots: apiol 34.5%, myristicin 28.8%, terpinolene 13.2%, β-phellandrene 4.6%. Essential oil, seeds: apiol, myristicin, safrole, 2,3,4,5-tetramethoxy-1-allylbenzene, santene, α-thujene, camphene, β-pinene, α-phellandrene, β-phellandrene, limonene, γ-caryophyllene, α-pinene, terpinolene.

Style American IPA, American Pale Ale, Bière de Garde, Saison, Specialty Beer, Strong Bitter.

Source Hieronymus 2016b; https://www.brewersfriend.com/other/parsley/; https://www.brewersfriend.com/other/parsley/.

Phaseolus vulgaris L.
FABACEAE

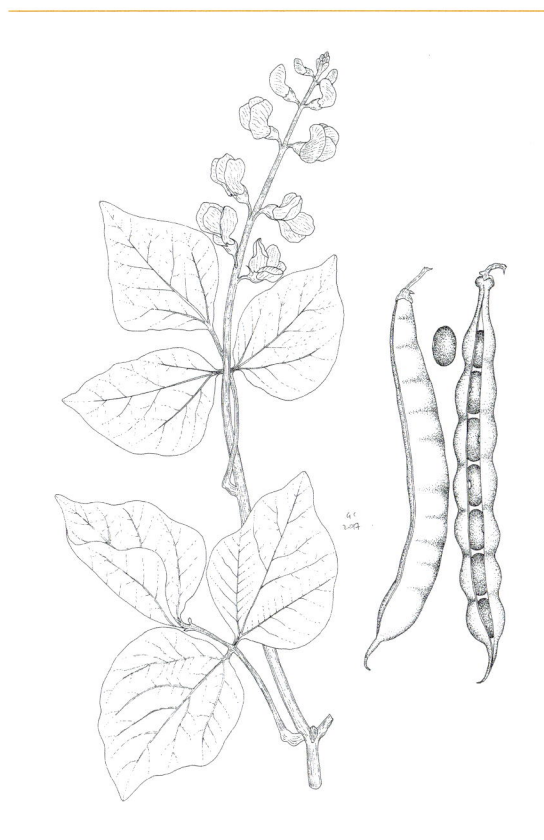

Synonyms *Phaseolus aborigineus* Burkart, *Phaseolus communis* Pritz., *Phaseolus communis* Pritzel, *Phaseolus compessus* DC., *Phaseolus esculentus* Salisb., *Phaseolus nanus* L. & Jusl.

Common Names Common bean, string bean, field bean, flageolet bean, French bean, garden bean, green bean, haricot bean, pop bean, snap bean, snap.

Description Herb, annual (5–30 dm). Stems herbaceous climbing, voluble. Leaves with 3 ovate-acuminate segments (3–8 × 5–12 cm) with minor stipules at base. Inflorescences short racemes. Flowers (1–1.5 cm) usually paired on erect-patent peduncles; corolla yellowish-green, mottled pinkish or purple, vexillum folded back. Fruit flattened legume (1–2 × 5–15 cm). Seeds reniform.

Subspecies Besides the typical variety (var. *vulgaris*), *Phaseolus vulgaris* L. var. *aborigineus* (Burkart) Baudet is recognized.

Cultivar Main varieties cultivated in Italy. Seed varieties: Bianco di Spagna (fagiolana), Bingo, Blason de Biella, Blu della Valsassina, Borlotto di Vigevano Nano, Borlotto Lingua di Fuoco, Borlotto Lingua di Fuoco Nano, Borlotto Suprema dwarf, Cannellin Scaramanzin negrè, Cannellino or Lingot, Cantare, Castagnaio fejuolo marron's, Elegante fagiolo, Fagiolo maggiolino, Fagiolo patrone, Fejuolo pacificus el drammoso cotenna, Fesciela lamon negrucc fagiolos de Biella, Garfagnana, Giallorino della Lamon (Lucian Fejuol), Meraviglia di Venezia black, Romano Pole, Sossai Extra Large (protected variety), Stregonta, Stregonta Nano, Superbo Migliorato. Green bean varieties: Beurre de Rocquencourt, Bobis a Grano Bianco and Bobis a Grano Nero, Bobis Bianco, Cornetto Largo Giallo and Cornetto Largo Verde, Nano Burro mangiatutto, Nerina mangiatutto, Paguro fagiolato mangiatutto, Prelude dwarf mangiatutto, Slenderette mangiatutto, Superpresto mangiatutto, Trionfo Violetto mangiatutto, Wade mangiatutto, Yellow and Green Anellino. American seed cultivars: Appaloosa, Black Turtle, Calypso, Cranberry, Dragon Tongue, Flageolet, Kidney, Pea, Pink, Pinto, Rattlesnake, Tongue of Fire, White, Yellow.

Related Species Of the more than 100 taxa belonging to the genus *Phaseolus*, some have food applications: *Phaseolus acutifolius* A.Gray, *Phaseolus coccineus* L., *Phaseolus lunatus* L., and *Phaseolus polyanthus* Greenman (= *Phaseolus coccineus* L. subsp. *polyanthus* [Greenm] Marechal et al.).

Geographic Distribution America. **Habitat** Commonly cultivated and sometimes subspontaneous in gardens but never wild. **Range** Green beans: China, Indonesia, Turkey, India, Thailand, Egypt, Morocco, Italy, Spain, Mexico. Dried beans: India, Brazil, Myanmar, China, United States, Mexico, Tanzania, Uganda, Kenya, Argentina.

Beer-Making Parts (and Possible Uses) Immature fruits (green beans), seeds.

Chemistry Green beans: energy 75 kJ (18 kcal), water 90.5 g, protein 2.1 g, lipids 0.1 g, sugars 2.4 g, fiber 2.9 g, sodium 2 mg, potassium 280 mg, iron 0.9 mg, calcium 35 mg, phosphorus 48 mg, thiamin 0.07 mg, riboflavin 0.15 mg, niacin 0.8 mg, vitamin A 41 µg, vitamin C 6 mg. Beans, dried seeds: energy 1,216 kJ (291 kcal), water 10.5 g, protein 23.6 g, lipids 2 g, carbohydrates 40 g, sugars 3.5 g, fiber 17.5 g, sodium 4 mg, potassium 1445 mg, iron 8 mg, calcium 135 mg, phosphorus 450 mg, magnesium 170 mg, zinc 3.4 mg, copper 0.7 mg, selenium 16 µg, thiamine 0.4 mg, riboflavin 0.1 mg, niacin 2.3 mg, vitamin A 3 µg, vitamin C 3 mg.

Style Golden Ale, Mum.

Beer Jack-a, Nadir (San Remo, Liguria, Italy).

Source Dabove 2015; Daniels 2016; Heilshorn 2017; Hieronymus 2010; Steele 2012; https://www.microbirrifici.org/Arte_Birraia_Borlotta_birra.aspx.

Phoenix dactylifera L.
ARECACEAE

Synonyms *Palma dactylifera* (L.) Mill., *Phoenix chevalieri* D.Rivera, S.Ríos & Obón, *Phoenix iberica* D.Rivera, S.Ríos & Obón.

Common Names Date, date palm.

DESCRIPTION Tree (over 30 m). Stem and new leaves originate from a single terminal bud at the apex of the stem. Root grows from base of trunk, sometimes 50 cm above ground level, largest about 1.5 cm thick. Leaves huge (up to 7 m) with relatively short petiole (50 cm), pinnate, leaflets (50–60 pairs) long and narrow, attached to a stiff rachis (midrib); leaves with average life span of 3–7 years. Inflorescence (produced at the axil of a 1-year-old leaf) branched spadix enclosed in a hard spathe that opens when flowers are mature. Male flowers cream-colored, waxy, with 6 stamens and no carpels; female flowers whitish, with 6 rudimentary stamens and 3 carpels. Fruits yellow to reddish-brown in color, each with a single seed. Seed up to 2.5 cm long, deeply grooved, with very hard endosperm.

CULTIVAR They are numerous (more than 100 only in Iraq). The most common are Aabel (Libya), Abid Rahim (Sudan), Ajwah (Saudi Arabia), Al-Khunaizi (Saudi Arabia), Amir Hajj (or Amer Hajj; Iraq), Barakawi (Sudan), Barhee (or Barhi), Bireir (Sudan), Dabbas (United Arab Emirates), Datça (Turkey), Deglet Noor (Libya, Algeria, United States, Tunisia), Derrie (or Dayri; S Iraq), Empress (California, United States), Fardh (or Fard; Oman), Ftimi (or Alligue; Tunisia), Haleema (Libya), Hayany (or Hayani; Egypt), Holwah (or Halawi), Honey, Iteema (Algeria), Kenta (Tunisia), Khadrawi (or Khadrawy; Saudi Arabia), Khalasah (Saudi Arabia), Khastawi (or Khusatawi or Kustawy; Iraq), Khenaizi (United Arab Emirates), Lulu (United Arab Emirates), Maktoom, Manakbir, Mazafati (or Mozafati; Iran), Medjool (or Majdool; Morocco, United States, Israel, Saudi Arabia, Jordan, Palestinian Territories), Mgmaget Ayuob (Libya), Migraf (or Mejraf; S Yemen), Mishriq (Sudan, Saudi Arabia), Nabtat-seyf (Saudi Arabia), Piarom, Rotab (Saudi Arabia), Sag'ai (Saudi Arabia), Saidy (Libya), Sayer (or Sayir), Sukkary (Saudi Arabia), Sellaj (Saudi Arabia), Sinhala (Sri Lanka), Tagyat (Libya), Tamej (Libya), Thoory (or Thuri; Algeria), Umeljwary (Libya), Umelkhashab (Saudi Arabia), Zaghloul, Zahidi (Egypt).

RELATED SPECIES Of the 15 species recognized in the genus *Phoenix*, in addition to *Phoenix dactylifera* L., at least nine produce edible fruits: *Phoenix acaulis* Roxb., *Phoenix atlantica* A.Chev., *Phoenix caespitosa* Chiov., *Phoenix loureiroi* Kunth, *Phoenix paludosa* Roxb., *Phoenix pusilla* Gaertn., *Phoenix reclinata* Jacq., *Phoenix rupicola* T.Anderson, and *Phoenix sylvestris* (L.) Roxb.

GEOGRAPHIC DISTRIBUTION Morocco, Palestine. **HABITAT** Desert and pre-desert areas. **RANGE** Egypt, Iran, Saudi Arabia, Iraq, Pakistan, United Arab Emirates, Algeria, Sudan, Oman.

BEER-MAKING PARTS (AND POSSIBLE USES) Dates, date sugar.

CHEMISTRY Fruits (cultivar Daglet Noor): energy 1,178 kJ (282 kcal), water 20.53 g, carbohydrates 75.03 g, sugars 63.35 g, fiber 8 g, lipids 0.39 g, protein 2.45 g, beta-carotene 6 μg, zeaxanthin lutein 75 μg, vitamin A 10 IU, thiamine (B1) 0.052 mg, riboflavin (B2) 0.066 mg, niacin (B3) 1.274 mg, pantothenic acid (B5) 0.589 mg, vitamin B6 0.165 mg, folate (B9) 19 μg, vitamin C 0.4 mg, vitamin E 0.05 mg, vitamin K 2.7 μg, calcium 39 mg, iron 1.02 mg, magnesium 43 mg, manganese 0.262 mg, phosphorus 62 mg, potassium 656 mg, sodium 2 mg, zinc 0.29 mg.

STYLE American Barleywine, American Brown Ale, American Porter, American Strong Ale, American Wheat or Rye Beer, Belgian Blond Ale, Belgian Dark Strong Ale, Belgian Dubbel, Belgian Specialty Ale, Bière de Garde, British Brown Ale, British Strong Ale, English Barleywine, Experimental Beer, Flanders Red Ale, Fruit Beer, Holiday/Winter Special Spiced Beer, Northern English Brown Ale, Old Ale, Russian Imperial Stout, Saison, Southern English Brown Ale, Specialty Beer, Sweet Stout.

BEER Brother Dewey's Date Night, College Street Brewhouse (Arizona, United States).

SOURCE Calagione et al. 2018; Cantwell and Bouckaert 2016; Cantwell and Bouckaert 2018; Heilshorn 2017; Hieronymus 2016a; Hieronymus 2016b; McGovern 2017; https://www.bjcp.org (29. Fruit Beer); https://www.brewersfriend.com/other/dates/.

Phyllocladus aspleniifolius (Labill.) Hook.f.
PODOCARPACEAE

Synonyms *Brownetera aspleniifolia* (Labill.) Tratt., *Phyllocladus glaucus* Carrière, *Phyllocladus rhomboidalis* Rich., *Phyllocladus serratifolius* Nois. ex Henkel & Hochst., *Podocarpus aspleniifolius* Labill., *Thalamia asplenifolia* (Labill.) Spreng.

Common Names Celery top pine.

Description Tree (up to 30 m, usually < 20 m). Trunk (Ø up to 1 m), single or double, bark brown (often appears black in moist conditions) to dark gray or reddish brown, with many lenticels, in older specimens the bark is usually darker and appears in rectangular scales; crown dense, dark green, young leaves light green. Leaves usually grouped at the ends of branches, reduced to minute scales along the edges of flattened twigs (cladodes), wedge or diamond shaped with rounded lobes and small teeth; surface of cladodes glossy dark green with numerous conspicuous veins from the central rib to the distal margin. Male and female strobili inconspicuous, produced separately, sometimes on separate trees; pollen cones cylindrical (to 0.5 cm), single or a few together on lateral branches; female strobili usually in groups of 3–4 in a short spike or on the margin of a cladode; each strobilus consists of an ovule at the axil of a bract that becomes fleshy and pink to red in color. Seed at maturity hard, greenish-black in color, surrounded by a white fleshy aril.

Related Species The genus *Phyllocladus* includes four species: *Phyllocladus aspleniifolius* (Labill.) Hook.f., *Phyllocladus hypophyllus* Hook.f., *Phyllocladus toatoa* Molloy, and *Phyllocladus trichomanoides* D.Don.

Geographic Distribution W SW Tasmania, with small areas on Maria Island, Bruny Island, Tasman Peninsula, Blue Tier (North East Highlands); the species was present on King Island until World War II, then cleared to establish military settlements. **Habitat** Common floristic component of the Tasmanian cold rainforest, where it associates with *Nothofagus cunninghamii* (Hook.) Oerst. (Nothofagaceae), *Atherosperma moschatum* Labill. (Atherospermataceae), *Eucryphia lucida* (Labill.) Baill. (Cunoniaceae), and *Acacia melanoxylon* R.Br. (Fabaceae); also found in high open mixed forests (wet sclerophyll forests), where *Eucalyptus* species (Myrtaceae) such as *Eucalyptus ambigua* A.Cunn. ex DC., *Eucalyptus ovata* Labill, *Eucalyptus delegatensis* F.Muell. ex R.T.Baker, *Eucalyptus regnans* F.Muell., and *Eucalyptus obliqua* L'Hér. have a typical rainforest understory; the species occurs also in open forests, where it is associated with *Leptospermum lanigerum* (Aiton) Sm. (Myrtaceae), *Pittosporum bicolor* Hook. (Pittosporaceae), and *Anodopetalum biglandulosum* A.Cunn. ex Hook.f. (Cunoniaceae). Optimal range is 0–1,000 m ASL, with a temperate climate (warmest/coldest month 18–20°C/0–3°C), moderate to high frost incidence (light snowfall is typical) and rainfall of 1,000–3,000 mm/year; topography and soil type appear relatively unimportant. **Range** Tasmania, grown in botanical gardens worldwide.

Beer-Making Parts (and Possible Uses) Apical sprigs (in the travel report of his search for the disappeared La Pérouse, the French botanist Labillardier mentions this species because it was used by sailors who landed in Tasmania to produce a sort of spruce beer, already considered an effective remedy against scurvy at the time of James Cook.

Chemistry Essential oil, leaves: α-pinene 44.5%, phyllocladene 23.9%, 8-p-hydroxyisopimarene 5%, β-caryophyllene 2.9%, myrcene 2.3%, β-phellandrene 1.9%, limonene 1.2%, spathulenol 1.2%, p-pinene 1.1%, 15-isopimaradiene 1.1%, bicyclogermacrene 0.9%, terpinolene 0.7%, α-humulene 0.6%, (E)-nerolidol 0.5%, camphene 0.3%, δ-cadinene 0.3%, caryophyllene oxide 0.3%, β-bisabolol 0.2%, tricyclicene 0.1%, γ-terpinene 0.1%, terpinen-4-ol 0.1%, ar-curcumene 0.1%, globulol 0.1%, viridiflorol 0.1%, rimuene 0.1%, isophyllocladene 0.1%, α-phenchene trace, sabinene trace, 6-3-carene trace, α-terpinene trace, p-cymene trace, α-terpineol trace.

STYLE Herb Beer, Spruce Beer.

SOURCE Duyker 2003.

Phyllostachys edulis (Carrière) J.Houz.
POACEAE

SYNONYMS *Bambusa heterocycla* Carrière, *Phyllostachys heterocycla* (Carrière) Matsum., *Phyllostachys pubescens* J.Houz., *Phyllostachys tubaeformiis* (S.Y.Wang) Ohrnberger.

COMMON NAMES Moso bamboo, tortoise-shell bamboo.

DESCRIPTION Herb, perennial (10–35 m). Rhizomes elongate, leptomorphic. Stems culms, erect, woody, internodes semicylindrical, hollow, yellow or light green, distally pubescent; lateral branches dendroid, sheaths deciduous, leathery, bristly, with red, silky hairiness; sheath lamina linear, reflexed, pubescent. Leaves caulescent deciduous, sheath glabrous on the surface or puberulent, hairs ciliate; ligule membrane ciliolate, pubescent on the abaxial surface, base with short petiole-like connection to the sheath, lanceolate; leaf blade 8–10 cm × 8–10 mm; vein with 8 secondary veins with distinct transverse veins, margins scabrous, apex acuminate. Inflorescence fasciculate, without axillary buds at base of spikelet, last bract covering a compact bundle of spikelets; fertile spikelets sessile with 1 fertile flower, spikelet lanceolate, laterally compressed (25–27 mm), opening at maturity; lower glume absent or obscure, persistent, shorter than spikelet, upper glume oblong, papery, without hull, main venation ciliolate, apex acute. Flower fertile with ovate lemma (20–25 mm), papery, without hull (10–11 veins); puberulent, apex acuminate; palea papery, surface rough, apex dentate, lodicules 3, veined, ciliate; anthers 3, stigma 3; ovary umbonate. Fruit caryopsis with adherent pericarp.

CULTIVAR *Phyllostachys edulis* Bicolor, *Phyllostachys edulis* Heterocycla, *Phyllostachys edulis* Kikko (or Kikko-Chiku), *Phyllostachys edulis* Nabeshimana, *Phyllostachys edulis* Okina, *Phyllostachys edulis* Subconvexa, *Phyllostachys edulis* Tao Kiang.

RELATED SPECIES The genus *Phyllostachys* has 54 taxa.

GEOGRAPHIC DISTRIBUTION Caucasus, China, Indochina. **HABITAT** Woodlands on mountain slopes (over 1,600 m ASL). **RANGE** E Asia.

BEER-MAKING PARTS (AND POSSIBLE USES) Young sprouts.

CHEMISTRY Raw sprouts: energy 27 kcal, carbohydrates 5.2 g, protein 2.6 g, lipids 0.3 g, fiber 2.2 g, folate 7 μg, niacin 0.6 mg, pantothenic acid 0.161 mg, pyridoxine 0.24 mg, riboflavin 0.07 mg, thiamine 0.15 mg, vitamin C 4 mg, vitamin A 20 IU, vitamin E 1 mg, sodium 4 mg, potassium 533 mg, calcium 13 mg, copper 0.19 mg, iron 0.5 mg, magnesium 3 mg, manganese 0.262 mg, phosphorus 59 mg, selenium 0.8 μg, zinc 1.1 mg, β-carotene 12 μg. Other compounds isolated from *Phyllostachys edulis*: p-coumaric acid, caffeic acid, ferulic acid, chlorogenic acid (3-[3,4-dihydroxycinnamoyl] quinic acid), 3-O-(3'-methylcaffeoyl) quinic acid, 5-O-caffeoyl-4-methylquinic acid, 3-O-caffeoyl-1-methylquinic acid, tricine, 7-O-methyltricine, glycosylated flavones, orientin, isoorientin, vitexin, isovitexin, 5,7,3´-trihydroxy-6-C-β-D-digitoxopyranosyl-4´-O-β-D-glucopyranosyl flavonoside, 5,3´,4´-trihydroxy-7-O-β-D-glucopyranosyl flavonoside, 5,4´-dihydroxy-3´,5´,-dimethoxy-7-O-β-D-glucopyranosyl flavonoside, 5,7,3´,4´-trihydroxy-6-C-(α-L-rhamnopyranosyl-[1→6])-β-D-glucopyranosyl flavonoside.

Style Premium Lager.

Source https://bamboobeer.ca/index.html.

Physalis peruviana L.
SOLANACEAE

Synonyms *Alkekengi pubescens* Moench, *Boberella peruviana* (L.) E.H.L.Krause, *Physalis esculenta* Salisb., *Physalis latifolia* Lam., *Physalis tomentosa* Medik.

Common Names Cape gooseberry, goldenberry, physalis.

Description Herb, annual (temperate climates) or perennial shrub (tropical climates), 1–1.6 m. Stem diffusely branched. Leaves heart-shaped lamina, velvety, margin with a few teeth. Flowers hermaphrodite, bell shaped (15–20 mm), downward facing, yellow with dark purple inner spots; when the flower has fallen, the calyx expands to form a beige involucre that includes the entire fruit. Fruit diclesium (berry included in the growing perianth) Ø 12–20 mm, yellow to bright orange in color, sweet when ripe, with characteristic sour tomato flavor. Seeds discoidal (1.7–2 mm), compressed, reticulate, whitish to brownish color.

Cultivar There are few varieties and genotypes of this species, selected in various countries to adapt to different climates of specific regions (ecotypes). Generally, in South America (Brazil, Chile, Ecuador, Peru), as well as in Mexico and in the Caribbean (Costa Rica, Guatemala, Guadeloupe), the traditional wild varieties prevail, characterized by high genetic variability and superior organoleptic qualities of the fruit. In Colombia, cultivars such as South Africa and Kenya, with bigger fruits but of lower quality, have recently been introduced. In countries with advanced agriculture, specific cultivars are used: Australia (Golden Nugget, New Sugar Giant), United States (Giallo Grosso, Giant, Giant Groundcherry, Giant Poha Berry, Golden Berry, Golden Berry-Long Ashton, Goldenberry, Peace).

Related Species The genus *Physalis* includes about 130 species, some of which produce edible fruits; *Physalis philadelphica* Lam. has established beer-making applications.

Geographic Distribution Chile, Peru. **Habitat** Tropical forests, forest edges, riparian environments, fallow land at altitudes of 500–3,000 m ASL. **Range** Widely cultivated species in South America, other tropical and subtropical climate districts (Australia, China, India, Malaysia, Philippines), and in some temperate climate areas (Great Britain, New Zealand). Shows a tendency to invasiveness.

Beer-Making Parts (and Possible Uses) Leaves, twigs, fruits (?).

Chemistry Fruits: energy 222 kJ (53 kcal), lipids 0.7 g, carbohydrates 11 g, protein 1.9 g, vitamin A 720 IU, vitamin C 11 mg, calcium 9 mg, iron 1 mg, phosphorus 40 mg.

Style Only one example with parts of this plant is known, from corn, flavored by John Boston (first recognized Australian brewer, 1794).

Source Jackson 1991.

Picea abies (L.) H.Karst.
PINACEAE

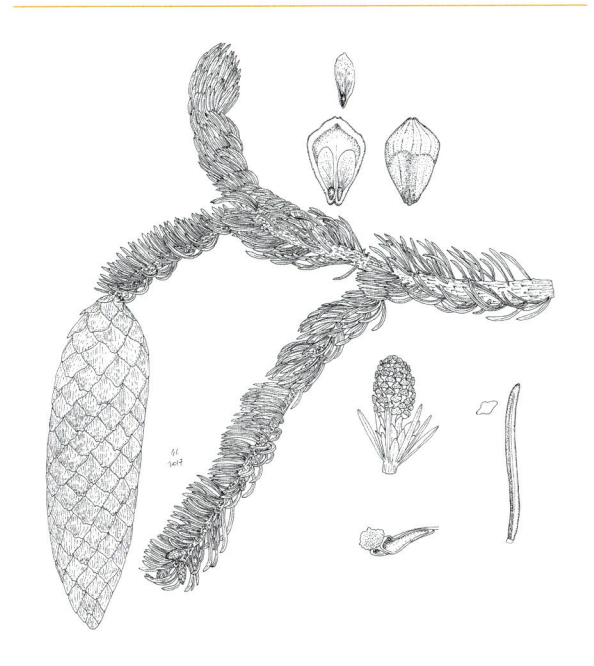

Synonyms *Abies alpestris* Brügger, *Abies carpatica* (Loudon) Ravenscr., *Abies cinerea* Borkh., *Abies clambrasiliana* Lavallée, *Abies clanbrassiliana* P.Lawson, *Abies coerulescens* K.Koch, *Abies conica* Lavallée, *Abies elegans* Sm. ex J.Knight, *Abies eremita* K.Koch, *Abies erythrocarpa* (Purk.) Nyman, *Abies excelsa* (Lam) Poir., *Abies extrema* Th.Fr., *Abies finedonensis* Gordon, *Abies gigantea* Sm. ex Carrière, *Abies gregoryana* H.Low. ex Gordon, *Abies inverta* R.Sm. ex Gordon, *Abies lemoniana* Booth ex Gordon, *Abies medioxima* C.Lawson, *Abies minuta* Poir., *Abies montana* Nyman, *Abies parvula* Knight, *Abies picea* Mill., *Abies subarctica* (Schur) Nyman, *Abies viminalis* Wahlenb., *Picea alpestris* (Brügger) Stein, *Picea cranstonii* Beissn., *Picea elegantissima* Beissn., *Picea excelsa* (Lam) Link, *Picea finedonensis* Beissn., *Picea gregoryana* Beissn., *Picea integrisquamis* (Carrière) Chiov., *Picea maxwellii* Beissn., *Picea montana* Schur, *Picea remontii* Beissn., *Picea rubra* A.Dietr., *Picea subarctica* Schur, *Picea velebitica* Simonk. ex Kümmerle, *Picea viminalis* (Alstr.) Beissn., *Picea vulgaris* Link, *Pinus abies* L., *Pinus excelsa* Lam., *Pinus sativa* Lam., *Pinus viminalis* Alstr.

Common Names Spruce, Norway spruce.

Description Tree, evergreen (30–60 m), long-lived (more than 500 years). Trunk straight, slender, circumference up to 2 m; crown dark green in color; thin bark, reddish in young plants, with membranous scales; rhytidome in adult specimens not very thick, brownish-gray in color, with irregular or round plates; first-order branches short, never large, in the apical third ascending, in the middle third horizontally typically arched, in the lower third ± descending; second-order branches and young twigs slender, subglabrous or hairy, brownish or yellowish-orange in color, bearing conical (6 mm) nonresinous buds, covered with reflexed reddish-brown perulae. Leaves needle-like (15–25 mm), persistent (up to 10 years), tetragonal-rhomboidal section, dark green, shiny or dull, stomatal lines (2–3 per side) not very evident; needles inserted on raised pads enveloping the twig, straight or curved upward, with slightly prickly mucronate tip (not always), arranged all around the twig or even rare or absent in the lower part. Male cones reddish in color, then yellow-pink at anthesis, usually at apex of previous year's twigs and in upper third of crown, horizontal or ascending, bending downward at end of anthesis; female cones at apex of lateral twigs in highest part of crown, sessile, cylindrical, dark red in color, erect until fertilization, then pendulous; strobili cylindrical, tapering at apex, straight or curved (5.5–20 × 3–4 cm), green when immature, brown when mature, made by thin, coriaceous scales, persistent, rhombic shape with rounded or toothed apex; bracts invisible, lanceolate, toothed at apex; each scale bears 2 brown seeds (3–5 mm) surrounded by thin, shiny wing.

Subspecies In addition to the nominal variety, *Picea abies* (L.) H.Karst. var. *acuminata* (Beck) Dallim. & A.B.Jacks is recognized.

Related Species The genus *Picea* has about 60 recognized species, some of which have potential beer-making applications.

Geographic Distribution Eurosiberian. **Habitat** Alpine species (400–2,200 m ASL), cold hardy, heliophilous, needs atmospheric humidity during the vegetative period and well-distributed rainfall, prefers fresh soils originating from sandstones and shales but indifferent to pH, suffers from water stagnation. **Range** Europe, in three main chorological areas: Sarmantic-Baltic (Fennoscandia, N Russia up to the Urals, lowland forests), Hercynian-Carpathian sector (Sudeten and Bohemian Forest, N Poland, W Thuringia, Bavarian Forest, S Carpathians, Transylvanian forests), Alpine-Balkan sector (Alps, except Maritimes, two small relict stations of the N Apennines, Foce di Campolino in the Abetone Forest, Cerreto Pass).

BEER-MAKING PARTS (AND POSSIBLE USES) Young apices, twigs (for flavoring; to filter the wort of sahti, the traditional Finnish beer), buds, sap.

CHEMISTRY Essential oil, wood: 4-hydroxy-4-methyl-2-pentanone 29.42%, α-cedrol 26.98%, Δ3-carene 6.08%, terpinen-4-ol 5.42%, α-humulene 3.79%, isopulegol acetate 3.76%, thujopsene 2.85%. Essential oil, bark: di(2-ethylhexyl)phthalate 30.91%, cyclohexane 12.89%, caryophyllene oxide 8.90%, α-pinene 4.59%, geranyl-linalool 3.66%, thunbergol 3.52%.

STYLE American Amber Ale, American Brown Ale, American Lager, American Light Lager, California Common Beer, Holiday/Winter Special Spiced Beer, Mild, Sahti, Spice/Herb/Vegetable Beer, Spruce Beer, Weissbier, Witbier, Wood-aged Beer.

SOURCE Cantwell and Bouckaert 2016; Cantwell and Bouckaert 2018; Fisher and Fisher 2017; Heilshorn 2017; Josephson et al. 2016; https://www.brewersfriend.com/other/spruce/; https://www.brewersfriend.com/other/sprucetip/.

🍺 The Tradition of Spruce Beer

The type of beer known as spruce beer in the United States and the United Kingdom includes all beers with fir shoots among their ingredients. Its origin can be traced back to Germany in the fifteenth to sixteenth centuries, when beers with high alcohol by volume, dark and spiced, with the resinous needles of the plant were produced. Known in the British isles thanks to trade with what was then Prussia, spruce beers went by three names: Danzig bier, with reference to the city (Gdansk) from which ships departed; black bier, so called because of the dark color of the beverage; and spruce bier, from the German *Sprossenbier*, which refers to the leaf buds (or sprouts) of the Norway spruce. In ancient times, the must was left to boil for many hours (up to 20 according to some documents), developing a brown color because of the caramelization of sugars. Fermentation took place in open vats and was mainly done by lactic bacteria and molds. The resinous notes of the spruce tree were added at the end of fermentation by adding a decoction of sprouts. The success of this type of beer in the United Kingdom was so great that between the eighteenth and nineteenth centuries, many brewers began to produce it on their own. The style also arrived with settlers on the American continent, where new forms were developed, with some inspiration coming from traditional Native American ways of using spruce plants.

Picea glauca (Moench) Voss
PINACEAE

SYNONYMS *Abies arctica* A.Murraybis, *Abies canadensis* Mill., *Abies coerulea* Lodd. ex J.Forbes, *Abies laxa* (Münchh.) K.Koch, *Abies virescens* R.Hinterh. & J.Hinterh., *Picea acutissima* Beissn., *Picea alba* (Münchh.) Link, *Picea coerulea* (Lodd. ex J.Forbes) Link, *Picea laxa* Sarg, *Pinus alba* (Münchh.) Aiton, *Pinus glauca* Moench, *Pinus laxa* (Münchh.) Ehrh., *Pinus tetragona* Moench.

COMMON NAMES Alaska White spruce, white spruce, western white spruce, Porsild spruce, Black Hills spruce.

DESCRIPTION Tree (up to 30 m). Trunk Ø up to 1 m; crown broadly conical to spire-like; bark gray-brown; branches slightly drooping; twigs not drooping, rather elongated, pink-brown, glabrous; buds dark orange, 3–6 mm, rounded apex. Leaves (0.8–2.5 cm) 4-angular in cross-section, stiff, blue-green in color, bearing stomata over entire surface, apex pointed. Female cones 2.5–8 cm; scales fan shaped, widest near the rounded apex, 9–13 mm, flexuous, apex margin ± entire, apex extending 0.5–3 mm beyond the shape of the seed wing.

SUBSPECIES Despite a taxonomy complicated by the tendency of this species to hybridize with *Picea engelmannii* Parry ex Engelm. where the two entities co-exist, in addition to *Picea glauca* (Moench) Voss var. *glauca*, two varieties of probable hybridogenic origin have been identified: *Picea glauca* (Moench) Voss var. *albertiana* (S.Br.) Beissn. (= *Picea* × *albertiana* S.Br. *pro parte*) and *Picea glauca* (Moench) Voss var. *porsildii* Raup (= *Picea* × *albertiana* S.Br. *pro parte*). According to some authors, the characters used are not significant or constant enough to justify this taxonomic–nomenclatural arrangement.

RELATED SPECIES The genus *Picea* has about 60 recognized species, some of which have potential beer-making applications.

GEOGRAPHIC DISTRIBUTION Canada, N United States (including Alaska). **HABITAT** Muskeg, swamps, river margins to mountain slopes (0–1,000 m ASL). **RANGE** Canada, N United States (including Alaska).

BEER-MAKING PARTS (AND POSSIBLE USES) Wood (for fermenters), buds.

CHEMISTRY Essential oil, needles and twigs: α-pinene, β-pinene, myrcene, limonene, terpinolene, camphene, β-phellandrene, camphora, bornyl acetate.

WOOD Undifferentiated resinous wood, whitish color, lightly grained (WSG 0.32–0.53); valued for regularity of structure and behavior in construction and joinery while having many secondary uses (pulp, wood pulp, furniture, paneling, wood wool, etc.).

STYLE Spruce Beer.

SOURCE Cantwell and Bouckaert 2016; Cantwell and Bouckaert 2018; Fisher and Fisher 2017; Heilshorn 2017.

Picea mariana (Mill.) Britton, Sterns & Poggenb.
PINACEAE

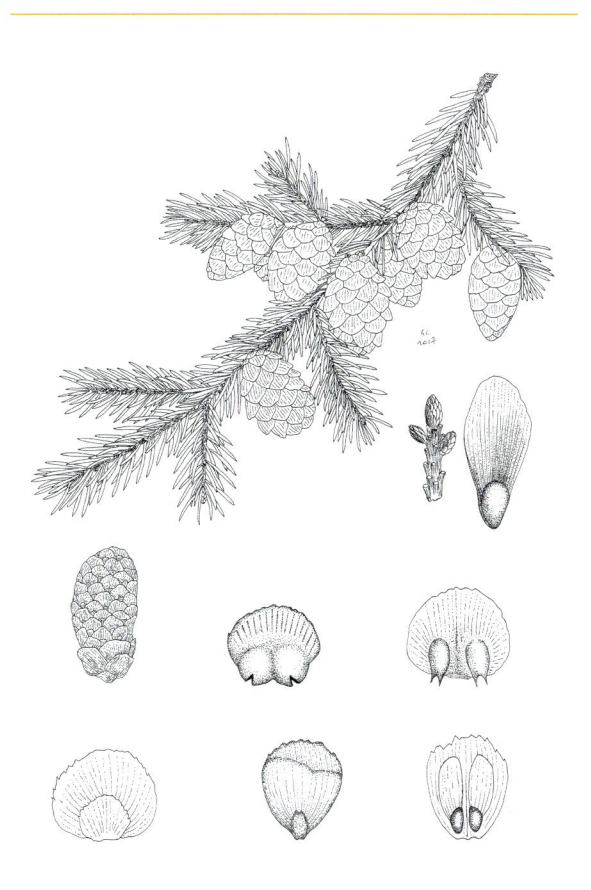

SYNONYMS *Abies denticulata* Michx, *Abies mariana* Mill., *Abies nigra* (Castigl.) Du Roi, *Peuce rubra* Rich., *Picea brevifolia* Peck, *Picea ericoides* Bean, *Picea nigra* (Du Roi) Link, *Pinus denticulata* (Michx) Muhl., *Pinus mariana* (Mill.) Du Roi.

COMMON NAMES Black spruce.

DESCRIPTION Shrub (near tree line) or tree (up to 25 m), evergreen. Canopy narrowly conical to spiciform or irregularly subcylindrical; branches short and drooping, frequently layered; twigs not drooping, slender, yellow-brown, hairy. Bark gray-brown. Leaves needle-like, evergreen, 0.6–2 cm long, 4-angular, stiff and blunt tipped, waxy, pale blue-green. Female cones 1.5–3.5 cm long, fusiform, purple-brown at maturity; cone scales fan shaped, wider near apex, 8–12 mm long, rigid, margin at apex irregularly toothed.

RELATED SPECIES The genus *Picea* has about 60 recognized species, some of which have potential beer-making applications.

Geographic Distribution N America. **Habitat** Wet or flooded boreal forests, muskeg. **Range** North America (Canada, NE United States).

Beer-Making Parts (and Possible Uses) Young shoots, buds.

Chemistry Essential oil, bark: α-pinene 40.6%, β-pinene 33.9%, β-phellandrene 4.8%, 3-carene 4.1%, limonene 4%. Bark hydrosol: α-terpineol 29.3%, trans-pinocarveol 5.2%, terpinen-4-ol 5%, verbenone 4.9%, borneol 4.9%, pinocarvone 4.6%. Essential oil, needles: bornil acetate 37%, α-pinene 16%, camphene 10%, β-pinene 6.5%, limonene 6.5%. Hydrosol needles: α-terpineol 14.8%, borneol 13.5%, bornil acetate 7.8%, terpinen-4-ol 6.5%.

Style Spruce Beer.

Source Heilshorn 2017; Jackson 1991; https://www.bjcp.org (30. Spiced Beer).

Picea pungens Engelm.
PINACEAE

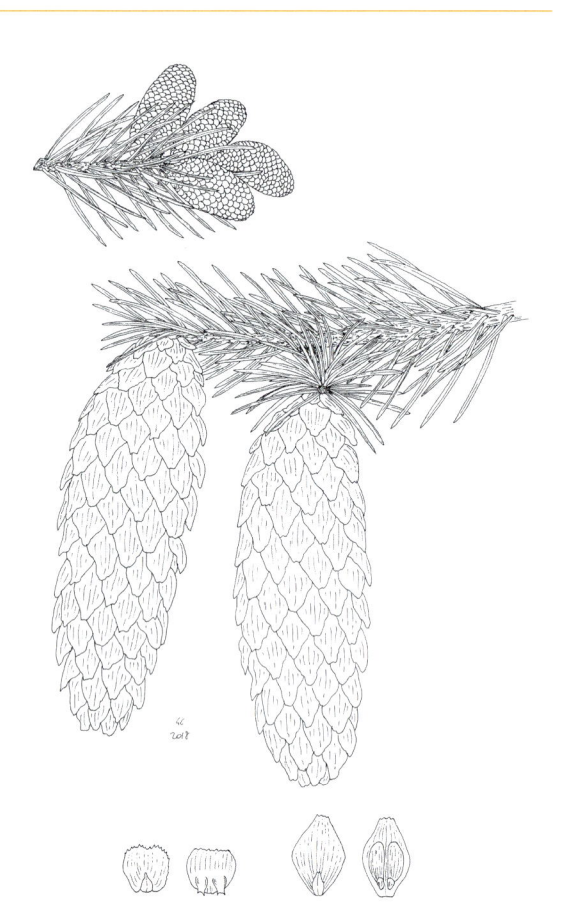

Synonyms *Abies parlatorei* Dallim. & A.B.Jacks., *Picea commutata* Beissn., *Picea parryana* (André) Sarg, *Pinus armata* Voss.

Common Names Blue spruce, Colorado blue spruce.

Description Tree, evergreen (up to 50 m). Trunk Ø up to 1.5 m; crown broadly conical; bark gray-brown; branches slightly to strongly drooping; twigs not drooping, stiff, yellow-brown, usually glabrous; buds orange–dark brown, 6–12 mm, apex rounded to acute. Leaves 1.6–3 cm, 4-angular in cross-section, stiff, blue-green in color, with stomata on all surfaces, apex surmounted by a thorn. Female cones (5–12 cm), scales elliptical to diamond shaped, wider in lower half, 10–15 mm, rather stiff, margin eroded at apex, apex extended 8–10 mm beyond seed wing.

Cultivar Aurea, Bakeri, Caerulea, Compacta, Glauca, Glauca Pendula, Hoopsli, Hunnewelliana, Koster, Moerheimi' Ruys, Thomsen, Viridis.

Related Species The genus *Picea* has about 60 recognized species, some of which have potential beer-making applications.

Geographic Distribution WC United States (Arizona, Colorado, Idaho, New Mexico, Utah, Wyoming). **Habitat** Mountain forests (1,800–3,000 m ASL). **Range** WC United States (Arizona, Colorado, Idaho, New Mexico, Utah, Wyoming).

Beer-Making Parts (and Possible Uses) Vegetative apices, buds.

Chemistry Essential oil (91.77% of total components): bornilacetate 29.40%, camphora 26.43%, β-myrcene 7.47%, camphene 7.01%, naphthalene 5.47%, α-pinene 4.06%, borneol 3.57%, 2-cyclohexene-1-one 1.52%, 1,3-benzenediamine 0.95%, 3-cyclohexene-1-methanol 0.71%, hexo-methylcamophenyll 0.57%, 1,8-cineole 0.48%, β-elemene 0.42%, bicyclic (3,1,0) hexan-2-ol 0.40%, ethanol 0.40%, caryophyllene 0.36%, γ-cadinene 0.33%, germacrene D 0.33%, α-terpinolene 0.30%, linalool 0.24%, 6-otten-1-ol 0.23%, δ-cadinene 0.21%, sabinene hydrate 0.20%, copaene 0.14%, γ-muurolene 0.14%, 2-cyclohexen-1-ol 0.12%, bicyclo (4,4,0) dec-1-ene 0.10%, verbenone 0.08%, benzene 1-methyl-4 0.06%, 1,6-cyclodecadiene 0.04%, calamenene 0.03%.

Style Spruce Beer.

Beer Poor Richard's Spruce Ale, Yards Brewing Company (Pennsylvania, United States).

Source Fisher and Fisher 2017; Heilshorn 2017.

Picea rubens Sarg.
PINACEAE

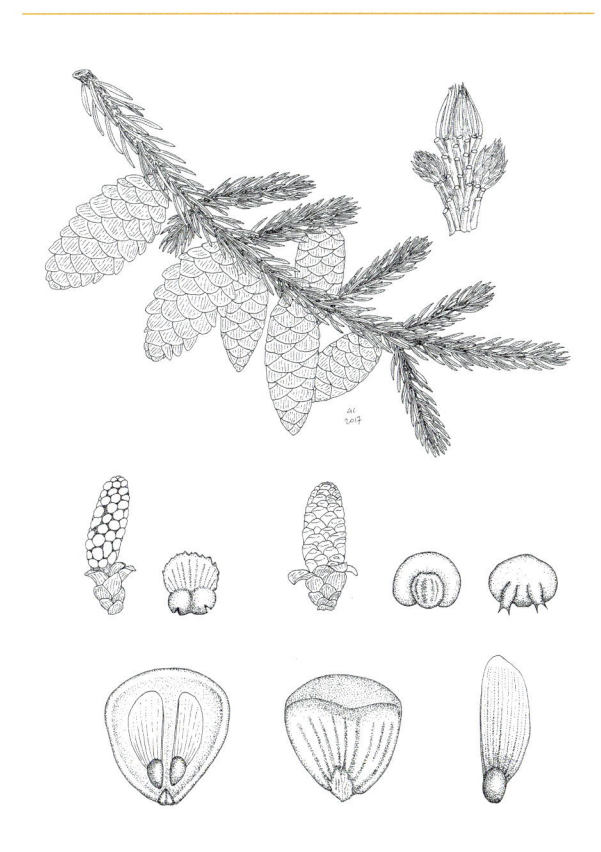

Synonyms *Abies rubra* (Du Roi) Poir., *Picea americana* Suringar, *Picea australis* Small.

Common Names Red spruce.

Description Tree (up to 25 m), evergreen. Trunk Ø up to 60 cm (larger only in Southern Appalachians). Leaves needle-like, 4-sided, dark, glossy, yellow-green, about 1.3 cm long, single on all sides of twigs and branches; slender young needles covered with a reddish fuzz for the first year; this characteristic with the short, curved needles helps distinguish the species from other spruces. Seeds small (about 31,000/100 g).

Related Species The genus *Picea* has about 60 recognized species, some of which have potential beer-making applications.

Geographic Distribution NE North America. **Habitat** Mountain forests, in surface soils (pioneer species). **Range** North America.

Beer-Making Parts (and Possible Uses) Young shoots, buds.

Chemistry Essential oil, leaves (98.6% of total components): β-phellandrene 29%, α-pinene 18.6%, β-pinene 14.9%, myrcene 10.2%, δ-3-carene 5.9%, limonene 3.7%, terpinolene 2.5%, sabinene 1.8%, δ-cadinene 1.3%, terpinen-4-ol 1.1%, γ-terpinene 1%, p-cymene 0.8%, α-thujone 0.7%, α-thujene 0.6%, α-terpinene 0.6%, borneol 0.5%, piperitone 0.5%, γ-cadinene 0.5%, camphene 0.4%, camphora 0.4%, styrene 0.3%, α-terpineol 0.3%, α-phellandrene 0.2%, linalool 0.2%, isoamyl isovalerate 0.2%, α-copaene 0.2%, α-muurolene 0.2%, cembrene 0.2%, bornilacetate 0.15%, p-cymenene 0.1%, β-thujone 0.1%, trans-p-mentha-2,8-dien-1-ol 0.1%, trans-pinocarveol 0.1%, cryptone 0.1%, myrtenal 0.1%, methyl thymol 0.1%, α-cubebene 0.1%, α-longipinene 0.1%, cis-muurola-3,5-diene 0.1%, cis-muurola-4(14),5-diene 0.1%, γ-muurolene 0.1%, santene 0.05%, tricyclicene 0.05%, camphene hydrate 0.05%, β-caryophyllene 0.05%, α-humulene 0.05%, trans-cadin-1-(6),4-diene 0.05%, α-selinene 0.05%, α-cadinol 0.05%, manool 0.05%, traces of the following compounds: hexanal, octane, heptanal, α-phenchene, thuja-2,4(10)-diene, isoterpinolene, germacrene D, (E)-calamenene, α-cadinene, cadinol, α-muurolol, 8,13-abietadiene.

Style Spruce Beer.

Source Heilshorn 2017; https://www.bjcp.org (30. Spiced Beer).

Picea sitchensis (Bong.) Carrière
PINACEAE

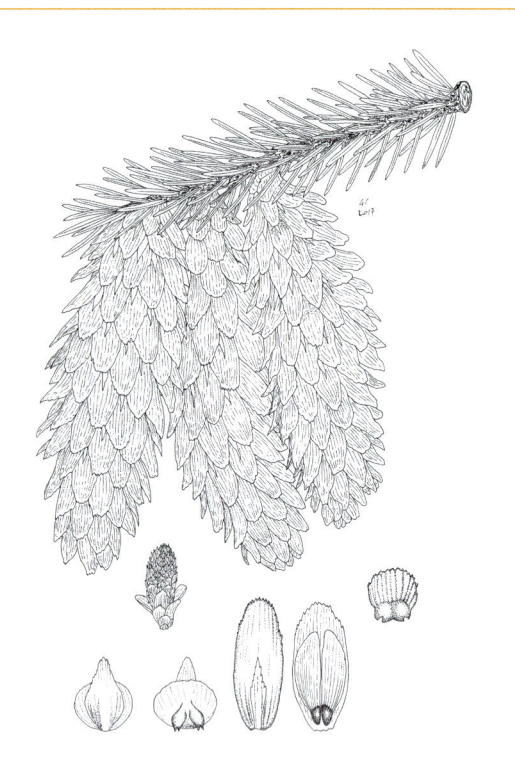

Synonyms *Abies falcata* Raf., *Abies merkiana* Fisch. ex Parl., *Abies sitchensis* (Bong.) Lindl. & Gordon, *Abies trigona* Raf., *Picea falcata* (Raf.) J.V.Suringar, *Picea menziesii* (Douglas ex D.Don) Carrière, *Pinus menziesii* Douglas ex D.Don, *Pinus sitchensis* Bong, *Sequoia rafinesquei* Carrière.

Common Names Sitka spruce.

Description Tree (up to 60 m), evergreen. Bark gray and smooth on small trunks, dark violet-brown on older trunks. Leaves needle-like, yellowish-green to bluish-green, stiff, very pointed, 2.5–4 cm long, with white lines of stomata on upper surface. Female cones 2.5–10 cm long, pendulous, with very thin, rounded, irregularly toothed scales.

Related Species The genus *Picea* has about 60 recognized species, some of which have potential beer-making applications.

Geographic Distribution W North America. **Habitat** Forests on floodplains, marine terraces, headlands. **Range** North America.

Beer-Making Parts (and Possible Uses) Young shoots, buds.

Chemistry Essential oil (98.6% of total components): β-phellandrene 29%, α-pinene 18.6%, β-pinene 14.9%, myrcene 10.2%, δ-3-carene 5.9%, limonene 3.7%, terpinolene 2.5%, sabinene 1.8%, δ-cadinene 1.3%, terpinen-4-ol 1.1%, γ-terpinene 1%, p-cymene 0.8%, α-thujone 0.7%, α-thujene 0.6%, α-terpinene 0.6%, borneol 0.5%, piperitone 0.5%, γ-cadinene 0.5%, camphene 0.4%, camphora 0.4%, styrene 0.3%, α-terpineol 0.3%, α-phellandrene 0.2%, linalool 0.2%, isoamyl isovalerate 0.2%, α-copaene 0.2%, α-muurolene 0.2%, cembrene 0.2%, bornilacetate 0.15%, p-cymenene 0.1%, β-thujone 0.1%, trans-p-mentha-2,8-dien-1-ol 0.1%, trans-pinocarveol 0.1%, cryptone 0.1%, myrtenal 0.1%, methyl thymol 0.1%, α-cubebene 0.1%, α-longipinene 0.1%, cis-muurola-3,5-diene 0.1%, cis-muurola-4(14),5-diene 0.1%, γ-muurolene 0.1%, santene 0.05%, tricyclicene 0.05%, camphene hydrate 0.05%, β-caryophyllene 0.05%, α-humulene 0.05%, trans-cadin-1-(6),4-diene 0.05%, α-selinene 0.05%, α-cadinol 0.05%, manool 0.05%, traces of the following compounds: hexanal, octane, heptanal, α-phenchene, thuja-2,4(10)-diene, isoterpinolene, germacrene D, (E)-calamenene, α-cadinene, cadinol, α-muurolol, 8,13-abietadiene.

Style Spruce Beer.

Source Heilshorn 2017; Hieronymus 2016b; https://www.bjcp.org (30. Spiced Beer).

Pimenta dioica (L.) Merr.
MYRTACEAE

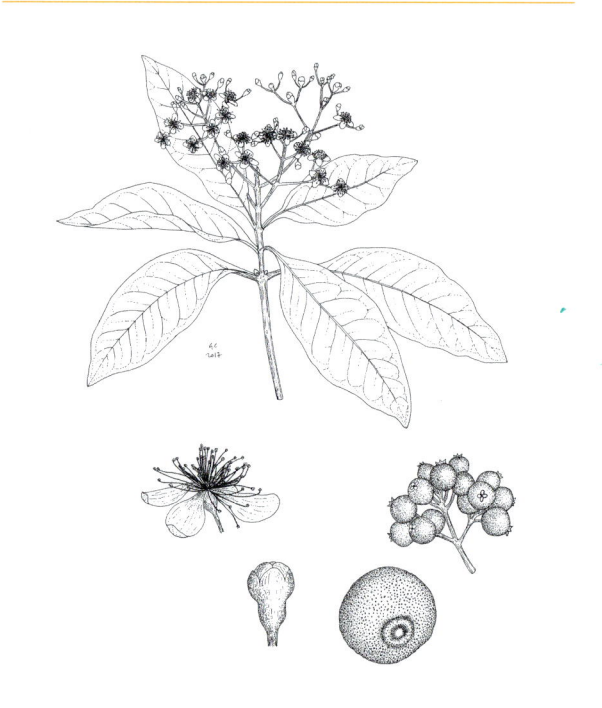

Synonyms *Caryophyllus pimenta* (L.) Mill., *Eugenia micrantha* Bertol., *Eugenia pimenta* (L.) DC., *Evanesca micrantha* Bertol., *Myrtus dioica* L., *Myrtus pimenta* L., *Myrtus piperita* Sessé & Moc., *Pimenta communis* Benth. & Hook.f., *Pimenta officinalis* Lindl., *Pimenta vulgaris* Bello, *Pimenta vulgaris* Lindl., *Pimentus geminata* Raf.

Common Names Allspice, Jamaica pepper, pepper, myrtle pepper, pimenta, Turkish yenibahar, newspice (the plant has an aroma and flavor that resemble a mixture of cinnamon, cloves, and nutmeg from which the English name allspice is derived).

Description Tree, evergreen (up to 15 m). Bark light brown. Leaves simple, opposite, entire, oblong-elliptical, 6–20 cm long, dotted with pellucid glands that release the aroma of the spice when crushed. Flowers small, whitish, with distinctive aroma, grouped in cymes. Flower structurally hermaphroditic but functionally dioecious; trees that do not bear fruit are male; their flowers have more than 100 stamens whereas female flowers have about 50; receptacle with 4 cream-colored calyx lobes, which expand at anthesis and are persistent in the fruit; petals 4, whitish, soon deciduous; stigma white with yellow stigma; in female individuals, stigma slightly shorter, yellow; ovary inferior bicarpellate, usually with 1 ovule per cell. Fruit berry, ripens 3–4 months after flowering; for use as a spice it is harvested when fully developed but still green. Seeds reniform, 2 per fruit.

Related Species Of the 16 species recognized in the genus *Pimenta*, besides *Pimenta dioica*, *Pimenta guatemalensis* (Lundell) Lundell, *Pimenta haitiensis* (Urb.) Landrum, and *Pimenta jamaicensis* (Britton & Harris) Proctor are also used locally as spice; others may be as well. However, *Pimenta racemosa* (Mill.) J.W.Moore (= *Pimenta tabasco* [Willd. ex Schltdl. and Cham] Lundell) is used most.

Geographic Distribution Jamaica, Mexico, Honduras, Guatemala, Costa Rica, Cuba, Caribbean Islands. **Habitat** Rainforests. **Range** Jamaica, C America.

Beer-Making Parts (and Possible Uses) Fruits, seeds, leaves.

Chemistry Seeds: energy 263 kcal, carbohydrates 72.12 g, protein 6.09 g, lipids 8.69 g, fiber 21.6 g, folate 36 μg, niacin 2.86 mg, pantothenic acid 0.210 mg, pyridoxine 0.21 mg, riboflavin 0.063 mg, thiamine 0.101 mg, vitamin A 540 IU, vitamin C 39.2 mg, sodium 77 mg, potassium 1,044 mg, calcium 661 mg, copper 0.553 mg, iron 7.06 mg, magnesium 135 mg, manganese 2.943 mg, phosphorus 113 mg, zinc 1.01 mg. Essential oil, fruits: eugenol 87%, β-selinene 0.2%, 1,8-cineole 3.3%, γ-terpinene 0.2%, β-caryophyllene 2.5%, α-terpineol 0.2%, α-humulene 1.6%, calamenene 0.1%, p-cymene 0.7%, caryophyllene oxide 0.1%, terpinen-4-ol 0.5%, α-copaene 0.1%, terpinolene 0.5%, γ-muurolene 0.1%, δ-cadinene 0.4%, β-phellandrene 0.1%, guaiene 0.4%, β-pinene 0.1%, limonene 0.4%, α-terpinene 0.1%, α-phellandrene 0.4%, γ-cardinene 0.1%, camphene 0.2%, p-dimethylstyrene 0.1%, β-elemene 0.2%, humulene oxide 0.1%, myrcene 0.2%, α-pinene 0.2%. Essential oil leaves (supercritical CO_2 extraction): methyl chavicol 0.31%, α-muurolene 0.05%, thymol 1.82%, calamenene + γ-cadinene 0.05%, carvacrol 1.08%, caryophyllene oxide 0.07%, eugenol 93.87%, cadinol 0.17%, P-caryophyllene 1.79%, α-cadinol 0.17%, α-humulene 0.35%, α-amorphene 0.37%. Essential oil leaves (steam distillation): α-pinene 0.56%, γ-terpinene 0.56%, myrcene 0.19%, terpinolene 1.38%, α-phellandrene 1.12%, menthol 0.56%, p-cymene 1.87%, methyl caviculus 0.09%, 1,8-cineole 14.5%, carvone 0.1%, limonene 0.1%, thymol 1%, carvacrol 1%, δ-cadinene 5.49%, eugenol 28.04%, cadine-1,4-diene 0.49%, β-caryophyllene 1%, α-calacorene 1.23%, α-humulene 10.12%, caryophyllene oxide 2.69%, allo-aromadendrene 2.13%, α-eudesmol 0.52%, α-amorphene 2.77%, β-eudesmol 0.82%, α-muurolene 1.76%, cadinol 6.64%.

Style American Pale Ale, Barleywine, Belgian Dark Strong Ale, Holiday/Winter Special Spiced Beer, Pumpkin Ale.

Beer Paranormal Imperial Pumpkin Ale, Flying Monkeys (Ontario, Canada).

Source Calagione et al. 2018; Josephson et al. 2016; McGovern 2017; https://www.beeradvocate.com/beer/style/72/; https://www.brewersfriend.com/other/piment/; https://www.brewersfriend.com/other/pimenta/.

Pimpinella anisum L.
APIACEAE

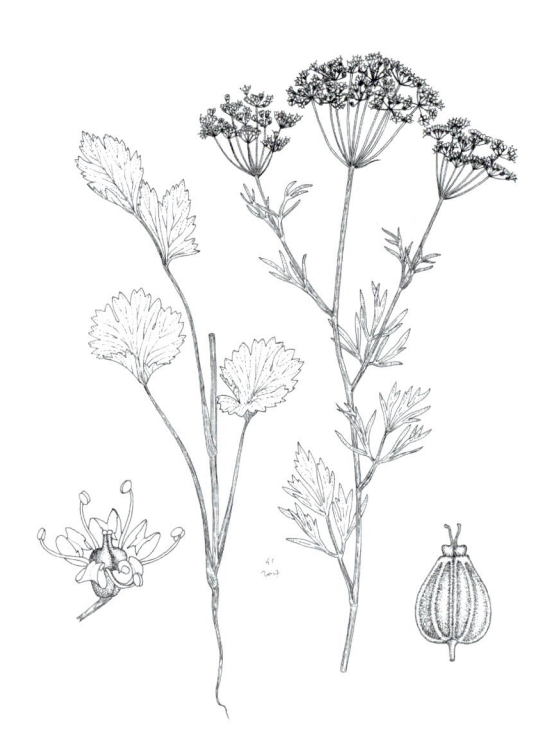

Synonyms *Anisum odoratum* Raf., *Anisum officinale* DC., *Anisum officinarum* Moench, *Anisum vulgare* Gaertn., *Apium anisum* (L.) Crantz, *Carum anisum* (L.) Baill., *Ptychotis vargasiana* DC., *Selinum anisum* (L.) E.H.L.Krause, *Seseli gilliesii* Hook. & Arn., *Sison anisum* (L.) Spreng, *Tragium anisum* (L.) Link.

Common Names Anice, anise.

Description Herb, annual (1–5 dm). Stem striate, pubescent. Leaves primordial with reniform lamina, basal pinnate with 3–5 ovate segments, cauline 2- to 3-pinnate with linear-lanceolate segments. Inflorescence umbel with 7–15 rays; bracts absent, rarely 1; bracts filiform; petals white. Fruit pyriform (3–5 mm), strongly scented, with dense appressed hairiness.

Related Species The genus *Pimpinella* includes more than 110 recognized species, but none besides *Pimpinella anisum* and *Pimpinella anisoides* V.Brig. (endemic to CS Italy) seems to have significant food applications.

Geographic Distribution Asia (?). **Habitat** Cultivated plant, rarely wild in uncultivated land. **Range** Europe, Egypt, Syria, Iran, Mexico, Turkey, South America, Russia.

Beer-Making Parts (and Possible Uses) Fruits, seeds.

Chemistry Essential oil, fruits (21 compounds): trans-anethole 76.9–93.7%, γ-himacalene 0.4–8.2%, trans-pseudoisoeugenol 2-methylbutyrate 0.4–6.4%, p-anisaldehyde trace-5.4%, methylcavicol 0.5–2.3%.

Style Gruit.

Beer Pimpinella, Le Fate (Comunanza, Marche, Italy).

Source Calagione et al. 2018; Heilshorn 2017; Hieronymus 2012; Hieronymus 2016b; Jackson 1991; Josephson et al. 2016; McGovern 2017.

Pinus caribaea Morelet
PINACEAE

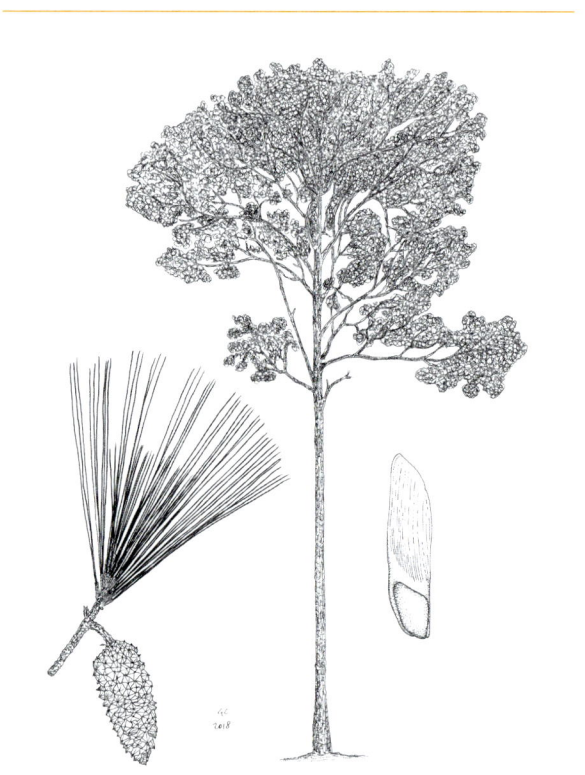

Synonyms *Pinus recurvata* Rowlee.

Common Names Honduran yellow pine, pitch pine, Caribbean pitch pine, Cuban pine, Honduras pine, Nicaragua pine, Caribbean pine, slash pine.

Description Tree (20–30 m, up to 35 m). Trunk (Ø 50–80 cm, up to 1 m occasionally) usually straight and well formed; lower branches broad, horizontal, and drooping; upper branches often ascending to form a rounded to pyramidal crown. Leaves needle-like (15–25 cm × about 1.5 mm), crowded and spreading at the end of twigs, remain attached for 2 years in bundles of 3–5, stiff, serrulate, dark or yellowish-green, slightly shiny, with stomata in whitish-colored lines on all surfaces. Male strobili many, sessile, in short clusters toward ends of twigs, mostly in lower part of foliage; mature female cones usually reflexed, symmetrical; cone scales reflexed, thin, flat, chocolate-brown color on inner surface. Seeds narrowly ovoid, about twice as long as wide, pointed at both ends, 3-angular, about 6 × 3 mm, color black speckled with gray or light brown.

Subspecies This taxon includes three varieties: the typical *Pinus caribaea* Morelet var. *caribaea*, *Pinus caribaea* Morelet var. *bahamensis* (Griseb.) W.H.Barrett & Golfari, and *Pinus caribaea* Morelet var. *hondurensis* (Sénécl.) W.H.Barrett & Golfari.

Related Species The genus *Pinus* consists of more than 175 taxa (of different rank), many of which, being resinous plants, produce resins and volatile substances with potential applications in beer making.

Geographic Distribution Bahamas, Colombia, Cuba, Guatemala, Honduras, Mexico, Nicaragua, Panama. **Habitat** Tropical- and subtropical-climate environments, in well-drained soils (0–1,500 m ASL). **Range** Beyond its natural range, the tree has been introduced and sometimes naturalized in Australia, Brazil, Canada, Costa Rica, Gambia, Ghana, Guyana, India, Indonesia, Jamaica, Kenya, Madagascar, Malawi, Malaysia, Mozambique, Nigeria, Philippines, Puerto Rico, Sierra Leone, South Africa, Sri Lanka, Sudan, Suriname, Tanzania, Trinidad and Tobago, Uganda, United States, Venezuela, Zambia, and Zimbabwe.

Beer-Making Parts (and Possible Uses) Resin obtained from this and other species (Greek pitch, rosin, or colomea) is used to coat some containers (fermenters, barrels) used in beer making.

Chemistry The incision of pine trees produces turpentine, which, when subjected to distillation, provides the essence of turpentine (volatile fraction) and rosin (distillation residue). The latter is a transparent solid, yellowish in color, with the following chemical composition: abietic acid 90%, dihydroabietic acid, dehydroabietic acid (the last two together about 10%).

Insoluble in water, it was used to caulk (waterproof) the planking of marine hulls and to treat the cordage.

Style None known.

Source Cantwell and Bouckaert 2016; Cantwell and Bouckaert 2018.

Pinus nigra J.F.Arnold subsp. *laricio* Maire
PINACEAE

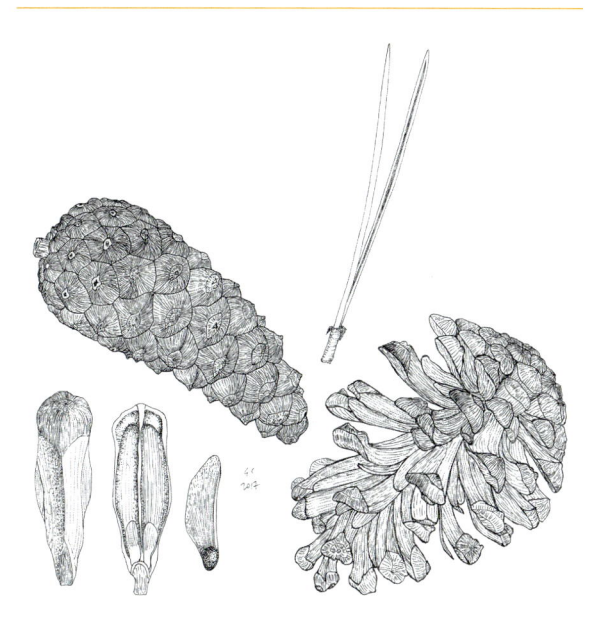

Synonyms *Pinus calabrica* Gordon, *Pinus caramanica* Bosc. ex Loudon, *Pinus italica* Herter, *Pinus karamana* Mast., *Pinus laricio* Poir. (note that some of the synonyms listed here may be formally invalid or illegitimate under the *International Code of Nomenclature for Algae, Fungi, and Plants*).

Common Names Calabrian pine.

Description Tree (10–20 m). Bark color gray-violet, desquamating in tangential layers. Horizontal branches. Leaves needle shaped (1.8 × 85–130 mm), stiff, dense, subspinose. Fruit ovoid pinecone (4 × 6 cm) with weakly mucronate scales; catkins in terminal clusters, reddish-yellow (3 × 16 mm). Seed winged.

Subspecies The recognized entities within the *Pinus nigra* cycle are *Pinus nigra* J.F.Arnold subsp. *dalmatica* (Vis.) Franco, *Pinus nigra* J.F.Arnold subsp. *laricio* Maire, *Pinus nigra* J.F.Arnold subsp. *nigra*, *Pinus nigra* J.F.Arnold subsp. *pallasiana* (Lamb.) Holmboe, and *Pinus nigra* J.F.Arnold subsp. *salzmannii* (Dunal) Franco.

RELATED SPECIES The genus *Pinus* comprises more than 175 taxa (of different taxonomic rank), many of which, being resinous plants, produce resins and volatile substances with potential applications in beer making.

GEOGRAPHIC DISTRIBUTION Calabria, Sicily (Etna), Corsica. **HABITAT** Mountainous forests, generally on siliceous rocks and lava. **RANGE** Calabria, Sicily (Etna), Corsica.

BEER-MAKING PARTS (AND POSSIBLE USES) Young apices, wood for barrels (uncertain use in beer making).

CHEMISTRY Essential oil, leaves (27 compounds, 97.9% of the oil): manool oxide 38%, germacrene D 16.7%, δ-cadinene 9%, (E)-caryophyllene 8.9%.

WOOD Strongly resinous, with yellowish sapwood and brownish or pinkish-brown heartwood (WSG 0.40–0.80). Suitable for construction, ship masts and marine pieces, planks, poles, and railway sleepers that must be treated with antiseptics because of their easy alteration by fungi and insects.

STYLE American IPA, American Pale Ale, Baltic Porter, Blonde Ale, British Brown Ale, Fruit Beer.

SOURCE Cantwell and Bouckaert 2016; Cantwell and Bouckaert 2018; https://www.brewersfriend.com/other/pine/; https://www.brewersfriend.com/other/tallskott/ (the species described here was chosen as an example of *Pinus* sp., a small tribute to an endemism of my home region of Calabria, although it is likely that the bibliographic references given here, or at least some of them, are more properly attributable to other species such as *Pinus sylvestris* L. or *Pinus mugo* Turra); https://honest-food.net/2016/05/05/gruit-herbal-beer/.

Pinus palustris Mill.
PINACEAE

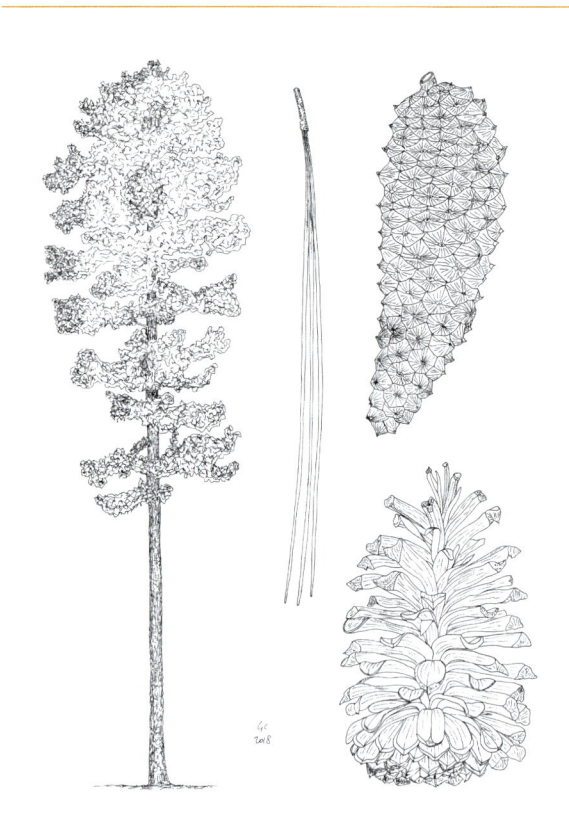

SYNONYMS *Pinus australis* Michx.f.

COMMON NAMES Longleaf pine.

DESCRIPTION Tree, evergreen (up to 47 m). Trunk Ø up to 1.2 m, straight; crown rounded; bark orange-brown, with large rectangular plates; branches diffuse-descending, curved upward at the ends; twigs stiff (Ø up to 2 cm), orange-brown, then darker brown, rough; buds ovoid, silvery-white, 3–4 cm; scales narrow, margins frayed. Leaves (20–45 cm × about 1.5 mm) 2–3 per fascicle, diffuse-recurved, persistent for 2 years, slightly twisted, shiny yellow-green, all surfaces with fine stomatal lines, margins finely serrated, apex suddenly acute to acuminate; sheath 2–3 cm, persistent at base. Pollen cones cylindrical (30–80 mm), purplish. Seminal cones mature in 2 years, then rapidly shed seeds and fall off, solitary or paired toward twig end, symmetrical (15–25 cm), dull brown, sessile (rarely briefly pedunculate); apophysis dull, slightly thickened and raised, almost rhombic; central umbo, broadly triangular, with a short, rigid, reflexed spine. Seeds truncate-obovoid; body about 10 mm, light brown with dark blotches; wing 30–40 mm.

Related Species The genus *Pinus* consists of more than 175 taxa (of varying rank), many of which, being resinous plants, produce resins and volatiles with potential applications in beer making.

Geographic Distribution SE United States. **Habitat** Arid sandy plateaus, sandy hills, lowland forests (0–700 m ASL). **Range** SE United States.

Beer-Making Parts (and Possible Uses) Resin obtained from this and other species (Greek pitch, rosin, or colomea) is used to coat some vessels (fermenters, barrels) used in beer making.

Chemistry The incision of pine trees produces turpentine which, when subjected to distillation, provides the essence of turpentine (volatile fraction) and rosin (distillation residue). The latter is a transparent solid, yellowish in color, with the following chemical composition: abietic acid 90%, dihydroabietic acid, dehydroabietic acid (the last 2 together about 10%). Insoluble in water, it was used to caulk (waterproof) the planking of marine hulls and to treat the cordage.

Style None known.

Source Cantwell and Bouckaert 2016; Cantwell and Bouckaert 2018.

Pinus ponderosa Douglas ex C.Lawson
PINACEAE

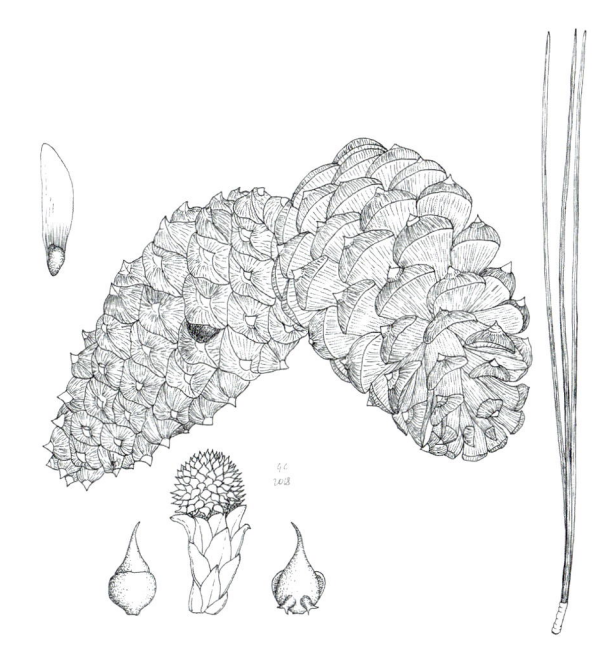

Synonyms *Pinus beardsleyi* A.Murray bis, *Pinus benthamiana* Hartw., *Pinus craigiana* A.Murray bis, *Pinus parryana* Gordon, *Pinus washoensis* H.Mason and Stockw.

Common Names Ponderosa pine, western yellow pine.

Description Tree, up to 72 m. Trunk Ø up to 2.5 m, straight. Crown broadly conical to rounded. Bark yellow to red-brown, deeply and irregularly grooved, covered with broadly rectangular plates. Branches descending to spreading-ascending; twigs stiff (up to 2 cm thick), orange-brown in color, darker with age, rough. Buds ovoid, 1 × 2 cm, red-brown, very resinous; scale margins edged in white. Leaves 2–5 per fascicle, scattered to erect, persistent (2–7 years old), 7–30 cm × 1–2 mm, slightly twisted, flexible, deep yellow-green, all surfaces with conspicuous stomatal lines, margins serrate, apex abruptly to narrowly acute or acuminate; sheath 1.5–3 cm, base persistent. Male cones ellipsoid-cylindrical (1.5–3.5 cm), color yellow or red. Female cones mature in 2 years, shedding seeds soon after, leaving a rosette of scales on twigs, solitary or rarely in pairs, diffuse to reflexed, symmetrical to slightly asymmetrical, conical-ovoid before opening, broadly ovoid when open, 5–15 cm, mostly reddish brown, sessile to nearly sessile, scales in steep spirals of 5–7 per line viewed from the side, those of cones just before and after dropping diffuse and reflexed, therefore well separated from adjacent scales; apophyses dull to shiny, thickened, variously raised and transversely keeled; central umbo, usually pyramidal to truncate, rarely depressed, barely acute, either with a very short tip or with a stiff spike or sting. Seeds ellipsoid-obovoid, body 3–9 mm, brown to yellow-brown, often mottled darker, wing 15–25 mm.

Subspecies In addition to the typical variety, *Pinus ponderosa* Douglas ex C.Lawson var. *ponderosa*, the variety *Pinus ponderosa* Douglas ex C.Lawson var. *scopulorum* Engelmann is recognized. According to some authors, a distinct variety, *Pinus ponderosa* Douglas ex C.Lawson var. *arizonica* (Engelmann) Shaw, should also be recognized. According to other sources, the entities of subspecific rank of this taxon would be even more. In addition to those mentioned, the following should be added: *Pinus ponderosa* Douglas ex C.Lawson subsp. *brachyptera* Engelm. (Southwestern ponderosa pine), *Pinus ponderosa* Douglas ex C.Lawson subsp. *critchfieldiana* Robert

Z. Callaham (Pacific ponderosa pine), *Pinus ponderosa* Douglas ex C.Lawson var. *pacifica* J.R.Haller & Vivrette (Pacific ponderosa pine), *Pinus ponderosa* Douglas ex C.Lawson subsp. *readiana* Robert Z. Callaham (Central High Plains ponderosa pine), *Pinus ponderosa* Douglas ex C.Lawson var. *washoensis* (H.Mason & Stockw.) J.R.Haller & Vivrette (Washoe pine). Therefore, the entire group seems to need clear taxonomic resolution.

Related Species The genus *Pinus* includes more than 175 taxa of different taxonomic rank, many of which, being resinous plants, produce resins and volatile substances with potential applications in beer making.

Geographic Distribution Canada (British Columbia), United States (Arizona, California, Colorado, Idaho, Montana, Nebraska, Nevada, New Mexico, North Dakota, Oklahoma, Oregon, South Dakota, Texas, Utah, Washington, Wyoming), Mexico. **Habitat** Slopes, canyons, canyon edges, plateaus, western Great Plains, Rocky Mountains, open dry montane forests (0–3,000 m ASL). **Range** Similar to geographic distribution (Canada, United States, Mexico).

Beer-Making Parts (and Possible Uses) Twigs, seeds, inner bark (roasted cambium takes on an interesting vanilla aroma).

Chemistry Turpentine from *Pinus ponderosa* (different origins). SW Idaho: l-α-pinene 6%, l-β-pinene 30%, β-myrcene 3%, d-Δ3-carene 40%, dl-limonene 10%, terpinolene 2%, cadienene 5%. C Montana: l-α-pinene 3% l-β-pinene 25%, β-myrcene 5%, d-Δ3-carene 47% dl-limonene 6%, terpinolene 4%, longifolene 2%, cadinene 5%. SE-Wyoming: l-α-pinene 2%, l-β-pinene 31%, β-myrcene 4%, d-Δ3-carene 40%, dl-limonene 4%, terpinolene 1%, longifolene 10%, cadinene 5%. SW Nebraska: dl-α-pinene 2%, l-β-pinene 32%, β-myrcene 3%, d-Δ3-carene 38%, dl-limonene 7%, longifolene 12%, cadinene 4%. SE Colorado: DL-α-pinene 5%, L-β-pinene 17%, β-myrcene 10%, D-Δ3-carene 40%, DL-limonene 15%, longifolene 8%.

Style Herb Beer.

Source Hieronymus 2016b; Josephson et al. 2016.

Pinus strobus L.
PINACEAE

Synonyms *Leucopitys strobus* (L.) Nieuwl., *Pinus nivea* Booth ex Carrière, *Strobus weymouthiana* Opiz.

Common Names Eastern white pine, northern white pine.

Description Tree (up to 67 m). Trunk Ø up to 1.8 m, straight; crown conical, then rounded or flattened. Bark gray-brown, deeply furrowed, with long, irregularly rectangular plates. Branches spiral, diffuse-curved; twigs slender, light red-brown, glabrous or puberulent, aging gray and ± smooth. Buds ovoid-cylindrical, light red-brown, 0.4–0.5 cm, slightly resinous. Leaves 5 per fascicles, spreading to ascending, persistent 2–3 years, 6–10 cm × 0.7–1 mm, straight, slightly twisted, flexible, dark green to blue-green in color, stomatal lines evident only on upper leaf, margins finely serrulate, apex suddenly acute to briefly acuminate; sheath 1–1.5 cm, early deciduous. Pollen cones ellipsoid, 10–15 mm, yellow. Female cones mature in 2 years, shedding seeds and then dropping, clustered, pendulous, symmetrical, cylindrical to lanceolate-cylindrical or ellipsoid-cylindrical before opening, ellipsoid-cylindrical to cylindrical or lanceolate-cylindrical when open,

7–20 cm, grayish-brown to light brown in color with purple or gray highlights, peduncle 2–3 cm; apophysis slightly raised, resinous at end; terminal umbo, low. Seeds compressed broadly obliquely obovoid; body 5–6 mm, red-brown mottled with black; wing 1.8–2.5 cm, light brown.

SUBSPECIES The taxon *Pinus strobus* L. includes the typical variety (var. *strobus*) and the variety *Pinus strobus* L. var. *chiapensis* Martínez.

CULTIVAR They have mainly ornamental applications: Blue Shag, Compacta, Contorta, Fastigiata, Glauca, Hillside Winter Gold, Nana, Pendula, Sea Urchin, UConn.

RELATED SPECIES The genus *Pinus* includes more than 175 taxa of different taxonomic rank, many of which, being resinous plants, produce resins and volatile substances with potential applications in brewing; the species *Pinus mugo* Turra and *Pinus sylvestris* L. have proven applications in beer making.

GEOGRAPHIC DISTRIBUTION Canada (Manitoba, New Brunswick, Newfoundland, Nova Scotia, Ontario, Prince Edward Island, Quebec), St. Pierre and Miquelon, United States (Connecticut, Delaware, Georgia, Illinois, Indiana, Iowa, Kentucky, Maine, Missouri, Massachusetts, Michigan, Minnesota, New Hampshire, New Jersey, New York, North Carolina, Pennsylvania, Ohio, Rhode Island, South Carolina, Tennessee, Vermont, Virginia, West Virginia, Wisconsin), Mexico, Guatemala. **HABITAT** Mesic to arid sites (0–1,500 m ASL). **RANGE** Overlapping with natural geographic distribution but species also cultivated elsewhere.

BEER-MAKING PARTS (AND POSSIBLE USES) Twigs, seeds, inner bark (roasted cambium takes on an interesting vanilla aroma).

CHEMISTRY Leaves have vitamin C content five times that of *Citrus limon*, so they are used in herbal teas. Composition, wood: sitosterol, lignan (lariciresinol), 5 flavonoid (5-hydroxy-7-methoxyflavanone [pinostrobin], 6-methylpinostrobin, 5-hydroxy-7-methoxyflavanone [tectocrisin], 6-methyltectocrisin, 5,7-dihydroxy-6-methylflavone [strobopinin]), 4 stilbenes (3-methoxy-5-hydroxystilbene [pinosylvin-monomethyl ether], 3-methoxy-5-hydroxy-7,8-dihydrostilbene [dihydropinosylvin-monomethyl ether], 3,5-dimethoxy-7,8-dihydrostilbene [dihydropinosylvin-dimethyl ether], 3,5-dimethoxystilbene [pinosylvin-dimethyl ether]), 3 diterpenes (isopimaric acid, dehydroabietic acid, copalic acid).

STYLE Herb Beer.

SOURCE Hieronymus 2016b; Josephson et al. 2016.

Piper cubeba L.f.
PIPERACEAE

SYNONYMS None

COMMON NAMES Cubeb, tailed pepper, java pepper.

DESCRIPTION Shrub, climbing (up to 6 m). Stem ascending, round sectioned, ash colored, rooting at nodes. Leaves ovate-oblong (10–16 × 3.5–5 cm), long pointed, leathery, smooth, dark green. Flowers subsessile to short-pedunculate, clustered in narrow spike inflorescences terminal to branches. Fruit drupe (similar to black pepper drupe but with persistent peduncle even when dried); pericarp in dried fruit wrinkled, grayish-brown to black in color. Single seed, white, oily (contains numerous essential oils).

RELATED SPECIES The genus *Piper* has about 1,500 species. Among them, besides the better-known *Piper nigrum*, used as a spice, *Piper auritum* Kunth and *Piper aduncum* L. are frequently used as adulterants of *Piper methysticum*. Other interesting taxa are *Piper methysticum* G.Forst. var. *wichmannii* (C.DC.) Lebot (wild ancestor of the cultivated variety), *Piper* *excelsum* G.Forst. (endemic to New Zealand and with traditional uses different from those of *P. methysticum*), and *Piper cubeba* L.f., cited here.

Geographic Distribution Java, Sumatra. **Habitat** Tropical forests. **Range** Indonesia (Java, Sumatra, other Indonesian islands), Africa (Sierra Leone, Democratic Republic of Congo), India.

Beer-Making Parts (and Possible Uses) Dried fruits.

Chemistry Essential oil, ripe fruits (11.8% v/w): sabinene 9.1%, β-elemene 9.4%, β-caryophyllene 3.1%, epi-cubebole 4.3%, cubebole 5.6%. Leaves (0.9% v/w): trans-sabinene hydrate 8.2%, β-caryophyllene 5.0%, epi-cubebole 4.2%, γ-cadinene 16.6%, cubebole 4.8%.

Style Bière de Garde, Blanche, Metheglin.

Beer Cubeb, Stranger Fellows (British Columbia, Canada).

Source https://www.brewersfriend.com/other/cubeb/.

Piper methysticum G.Forst.
PIPERACEAE

Synonyms *Methysticum methysticum* (G.Forst.) A.Lyons.

Common Names Kava, kawa, kava-kava.

Description Shrub (2–4 m), perennial, dioecious, woody. Root enlarged at or just below ground level, from which numerous shoots form. Stems erect (Ø 1–3 cm), green, dark red or dark purple in color. Leaves alternate, deciduous, petiolate (2–7 cm), lamina heart shaped (10–30 × 8–23 cm), margin entire, apex acute, glabrous to finely pubescent, veins palmate; stipules broad, persistent. Inflorescence spike, axillary or opposite leaf, but smaller; peduncle 1.5 cm; spike 3–9 cm long, with small unisexual flowers lacking sepals or petals; male spike bears numerous flowers with 2 short stamens; female spike bears flowers with a single basal ovule in a unilocular ovary surmounted by a stigma. Fruit (rarely produced) berry containing 1 seed.

Subspecies The taxon includes a second variety, *Piper methysticum* G.Forst. var. *wichmannii* (C.DC.) Lebot, besides the nominal one, *Piper methysticum* G.Forst. var. *methysticum*.

Cultivar They are numerous and frequently distinguished in technical types (chemotypes?): noble kava and non-noble kava. The second category includes the so-called kava tudei (or two-day), medicinal kava, and wild kava (*Piper wichmanii*). Traditionally, only noble kava was used for regular consumption (and spread in Pacific archipelagos) because of the favorable mix of kavalactones and other compounds, which produces the desired effects with a low potential for negative side effects (nausea, after effects).

Related Species The genus *Piper* has about 1,500 species. Among them, in addition to the better-known *Piper nigrum*, used as a spice, *Piper auritum* Kunth and *Piper aduncum* L. are frequently used as adulterants of *Piper methysticum*. Other interesting taxa are *Piper methysticum* G.Forst. var. *wichmannii* (C.DC.) Lebot (wild ancestor of the cultivated variety), *Piper excelsum* G.Forst. (endemic to New Zealand, with traditional uses different from those of *P. methysticum*), and *Piper cubeba* L.f.

Geographic Distribution Vanuatu. **Habitat** Well-drained soils in mountainous areas with little wind; species sciaphilous when young, heliophilous when adult. **Range** Vanuatu, Fiji Islands, Germany, Papua New Guinea, Samoa, Solomon Islands, Tonga.

Beer-Making Parts (and Possible Uses) Roots.

Toxicity The aerial parts of the plant are known to be toxic. The root can cause liver toxicity and skin effects.

Chemistry Dried root: starch 43%, fiber 20%, kavalactone 15% (collar 10%, basal part of stem 5%), water 12%, sugars 3.2%, protein 3.6%, minerals 3.2%. Root (rhizome): kavalactones (11-hydroxy-12-methodehydrokavain, 7,8-dihydro-5-hydroxykavain, 11,12-dimethoxyhydrokavain, methysticin, dihydromethysticin, kavain, 7,8-dihydrokavain, 5,6-dihydromethysticin, 5,6-dihydrokavain, yangonin, 5,6,7,8-tetrahydroyangonin, 5,6-dihydroyangonin, 7,8-dihydroyangonin, 10-methoxyangonin, 11-methoxyangonin, 11-hydroxyyangonin, 5-hydrokavaine, 11-methoxy-12-hydroxide hydrokavaine), flavokavin A, flavokavin B, flavokavin C, dihydrokavain-5-ol, cuproic acid, cinnamalketone, methylenedioxy-3,4-cinnamalketone, 4-oxononanoic acid, benzoic acid, phenyl acetic acid, dihydrocinnamic acid, cinnamic acid, pipermethvstine, 1-(meta-methoxycinnamoyl)pyrrolidine, 1-cinnamoylpyrro-lidine.

Style Foreign Extra Stout, Russian Imperial Stout.

Source Heilshorn 2017; https://www.brewersfriend.com/other/kawa/.

Piper nigrum L.
PIPERACEAE

Synonyms None

Common Names Black pepper, black peppercorn.

Description Plant, climbing (over 10 m). Once the main stem has grown, many side branches form to create a bushy column. The plant develops short adventitious roots to connect to available supports. Leaves alternate, lanceolate, tapering toward the tip, dark green and glossy above, light green below. Flowers whitish to yellow-green. Inflorescence spike consisting of 50–150 flowers gathered in groups along a pendulous floral axis. Fruit spherical, berry-like, Ø up to 6 mm, color first green, then red when ripe. Seed 1 per fruit. Infructescence carrying 50–60 fruits.

Cultivar All types of pepper are obtained from the same plant: white pepper consists only of the seed by removal of the outer shells after a period of soaking; green pepper is obtained from unripe seeds treated to remain green; orange (or red) pepper is obtained from ripe fruit placed in brine and vinegar; black pepper from unripe drupes is treated to turn the color dark brown. Pink pepper is obtained from two completely different species: *Schinus molle* L. and *Schinus terebinthifolia* Raddi (Anacardiaceae).

Related Species *Piper baccatum* C.DC., *Piper longum* L., *Piper retrofractum* Vahl; the genus *Piper* has about 1,500 species in total.

Geographic Distribution W Ghats (Kerala, India). **Habitat** Tropical forests. **Range** Cultivated in all tropical areas of the world, in particular Vietnam, Indonesia, Brazil.

Beer-Making Parts (and Possible Uses) Fruits, seeds.

Chemistry Essential oil, fruits: β-caryophyllene, limonene, β-pinene, δ-3-carene, sabinene, α-pinene, eugenol, terpinen-4-ol, edicariol, β-eudesmol, caryophyllene oxide.

Style Gruit, White IPA.

Beer Wayan, Baladin (Piozzo, Piedmont, Italy).

Source Calagione et al. 2018; Dabove 2015; Giaccone and Signoroni 2017; Josephson et al. 2016; Markowski 2015; McGovern 2017; Steele 2012; https://www.fermentobirra.com/homebrewing/birre-speziali/.

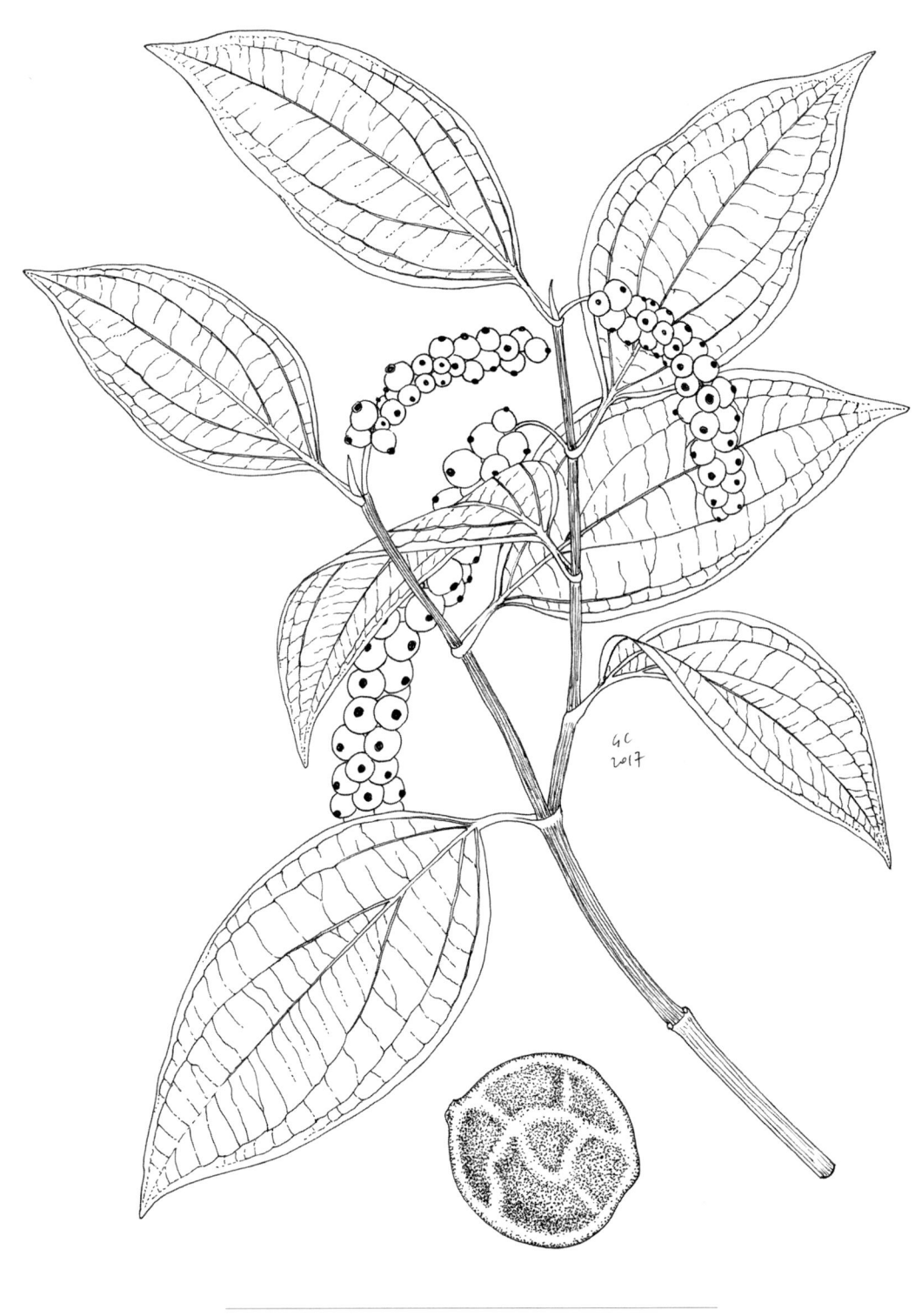

***Piper nigrum* L.**
PIPERACEAE

Pistacia lentiscus L.
ANACARDIACEAE

Synonyms *Lentiscus vulgaris* Fourr., *Terebinthus lentiscus* (L.) Moench.

Common Names Mastic.

Description Shrub (1–3 m), evergreen, rarely tree-like (6–8 m), with strong resinous odor. Canopy globular, dense, densely branched. Bark scaly, ashen in young branches, trunk reddish-brown; wood pinkish in color. Leaves alternate, compound, paripinnate, glabrous, dark green, 6–10 elliptic-lanceolate obtuse segments (up to 30 mm), margin entire, apex obtuse, coriaceous, glabrous, small apical mucro, rachis slightly winged. Flowers unisexual, actinomorphic, pentamerous, tetracyclic. Inflorescences short, dense, cylindrical panicles in the axils of the leaves of the previous year's branches. Male flowers with 4–5 stamens, 1 rudimentary pistil, conspicuous for the presence of bright red stamens; female flowers green with ovary superior, petals absent. Fruit globular or lenticular drupe (Ø 4–5 mm), fleshy, reddish color tending to black at maturity. Seed 1 per fruit.

Subspecies Within the intraspecific variability of *Pistacia lentiscus*, *Pistacia lentiscus* L. var. *chia* (Desf. Ex Poiret) DC. (= *Pistacia chia* Desf.) was described with reference to the plants (only male individuals) cultivated to extract the famous mastic on the Greek island of Chios.

Cultivar *Pistacia lentiscus* L. var. *chia* (Desf. ex Poiret) DC. (?).

Related Species Of the 13 species recognized in the genus *Pistacia*, at least *Pistacia terebinthus* L. and *Pistacia raportae* Burnat have some affinities with *P. lentiscus*. However, it is possible that other congeners have similar properties.

Geographic Distribution S Mediterranean, Steno-Mediterranean, Macaronesian. **Habitat** Species heliophilous, thermophilic, xerophilous, prefers acid substrates; typical component of the evergreen Mediterranean scrub (0–600 m ASL), dominant in the phases of scrub degradation, especially because of the effect of fire. **Range** Similar to geographic distribution (Mediterranean coasts).

Beer-Making Parts (and Possible Uses) Dried resin (mastic), fruits (?).

Chemistry Mastic (oleoresin containing about 2% volatile oil): α-masticoresins, β-masticoresins, masticin, mastic acid, masticoresene, tannins. Mastic essential oil (more than 70 compounds): α-pinene, myrcene, caryophyllene, β-pinene, linalool, germacrene D. Essential oil, leaves (43 components; 97.4% of oil): germanicol 12.8%, thunbergol 8.8%, himachalene 7.4%, trans-squalene 6.7%, terpinyl propionate 6.7%, 3,3-dimenthol 6.2%, cadine-1,4-diene 5.1%. Essential oil, fruits: 3-carene 54.1%, α-pinene 7.64%. Essential oil, seeds (production 35.37% of raw material): α-tocopherol 7.59 mg/g, β-tocopherol 0.47 mg/g, γ-tocopherol 0.48 mg/g, oleic acid 51.06% TFA, linoleic acid 20.71% TFA, palmitic acid 23.52% TFA.

Style No specific indication.

Source https://www.brewersfriend.com/other/mastic/.

Pistacia vera L.
ANACARDIACEAE

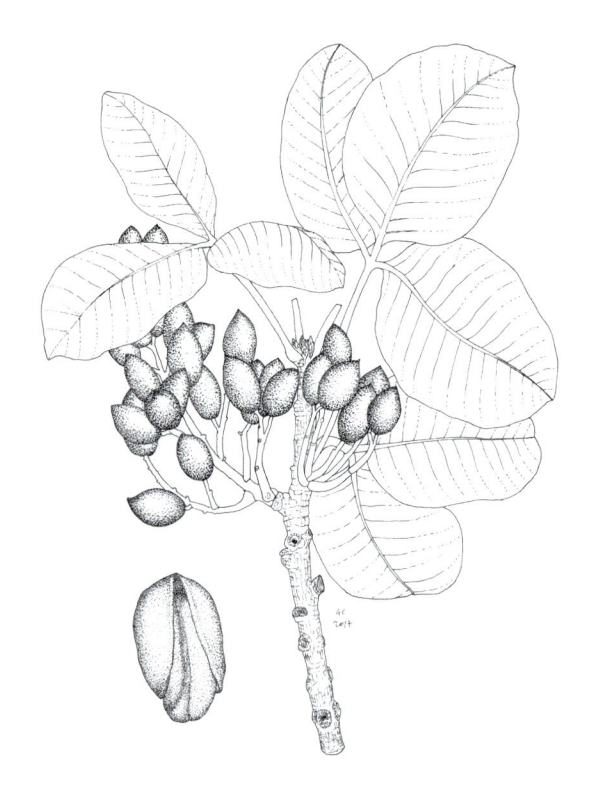

Synonyms None

Common Names Pistachio, pistachio nut.

Description Tree (1–5 m), resinous odor. Bark reddish-brown, young branches pubescent and with linear longitudinal lenticels (1 mm). Leaves leathery, dark green above, grayish-green below, usually with 1–5 elliptical or lanceolate segments (greater than 16–22 × 30–45 mm), rounded at apex. Inflorescence panicle pyramidal. Flowers brown. Fruits ovoid drupes (8–9 × 20 mm), greenish, then dark red, on 4–7 mm peduncles. Edible seed.

Cultivar They are numerous, including Achoury, Aegina, Ajamy, Alemi, Aria, Avidon, Ayimi, Batoury, Bayazi, Bianca, Bianca Regina, Boundoky, Cappuccia, Cerasola, Damghan, El Bataury, Ghermeza, Ghiandolara, Gialla, Golden Hills, Insolia, Iraq, Kalehghouchi, Kerman, Larnaka, Latwhardy, Lost Hills, Kastel, Marawhy, Mateur, Momtaz, Muntaz, Obiad, Ogah, Ouleimy, Peters, Pignatone, Pontikas, Randy, Rashti, Red Jalap, Shasti, Silvana, White Ouleimy.

Related Species Of the 13 species recognized in the genus *Pistacia*, at least *Pistacia terebinthus* L. and *Pistacia raportae* Burnat show some affinities with *P. lentiscus*. It is possible that other congeners have similar properties.

Geographic Distribution Mediterranean. **Habitat** Cultivated plant. **Range** Iran, United States, Turkey, China.

Beer-Making Parts (and Possible Uses) Seeds.

Toxicity This species, like other Anacardiaceae, contains urushiol, an irritant that may cause allergic reactions. Toxicity risks can also derive from aflatoxins (carcinogenic) produced by fungi of the genus *Aspergillus* (Ascomycota) such as *Aspergillus flavus* Link and *Aspergillus parasiticus* Speare, which can be caused by poor management of this species' fruits and of food products in general.

Chemistry Seeds: energy 2,351 kJ (562 kcal), carbohydrates 27.51 g, sugars 7.66 g, fiber 10.3 g, lipids 45.39 g (saturated 5.556 g, monounsaturated 23.82 g, polyunsaturated 13.744 g), protein 20.27 g, vitamin A lutein zeaxanthin 1205 µg, thiamine (B1) 0.87 mg, riboflavin (B2) 0.16 mg, niacin (B3) 1.3 mg, pantothenic acid (B5) 0.52 mg, vitamin B6 1.7 mg, folate (B9) 51 µg, vitamin C 5.6 mg, vitamin E 2.3 mg, vitamin K 13.2 µg, calcium 105 mg, iron 3.92 mg, magnesium 121 mg, manganese 1.2 mg, phosphorus 490 mg, potassium 1025 mg, zinc 2.2 mg.

Style American Brown Ale, American Pale Ale, California Common Beer, Fruit Beer, Imperial Stout, Specialty Beer, Sweet Stout.

Beer Supermassive Imperial Stout, Crak + Põhjala (Campodarsego, Veneto, Italy).

Source https://www.brewersfriend.com/other/pistachio/; https://www.brewersfriend.com/other/pistachios/.

Pisum sativum L.
FABACEAE

Synonyms *Lathyrus oleraceus* Lam., *Pisum arvense* L., *Pisum vulgare* Judz., *Pisum vulgare* Jundz.

Common Names Pea.

Description Herb, annual (2–20 dm), glabrous, glaucous. Stems cylindrical, prostrate, or voluble. Leaves major (medium) with toothed semiamplexicaule stipules (up to 15 × 35 mm) and 2–4 oval segments (16–18 × 22–24 mm); apical 3-branched cirrus; lower leaves minor; upper leaves 4–6 lanceolate segments and 3- to 7-branched cirrus. Flowers isolated (or in 2–3) on 3–4 cm peduncles; calyx 13 mm; corolla with purple-blackish wings (18 mm) and purple vexillum (corolla niveous white or whitish in cultivated forms), heart shaped (up to 36 × 25 mm), bilobate with 3–4 mm incision. Fruit legume (10–12 × 50–80 mm), seeds subspherical (Ø 3–10 mm), green (brown or black in wild forms).

Subspecies Within *Pisum sativum*, L. three subspecies are recognized: *Pisum sativum* L. subsp. *brevipedunculatum* (P.Davis & Meikle) Ponert, *Pisum sativum* L. subsp. *elatius* (M.Bieb.) Asch. & Graebn, and *Pisum sativum* L. subsp. *sativum*.

Cultivar Thousands of pea cultivars are known worldwide. The European catalog of species and varieties authorized for cultivation lists 1,390 (514 forage, 776 horticultural). The intervarietal distinction is based on morphological characters (e.g., shape and color of seeds, pods, leaves, stems; plant height; presence of anthocyanins; shape of starch granules) and disease resistance. Among the most popular horticultural pea cultivars, examples include Blauwschokker, Caracatus, Carouby de Maussane, Corne de bélier, Douce Provence, Early Snap, Express Alaska, Merveille d'Amerique, Merveille de Kelvedon, Nain très hâtif d'Annonay, Petit Provençal, Prince Albert, Roi des conserves, Serpette amélioré, Serpette Cent pour Un, Sugar Snap, Téléphone climbing Thomas Laxton.

Related Species Besides *Pisum sativum*, five other taxa are recognized in the genus *Pisum*: *Pisum abyssinicum* A.Braun, *Pisum ensifolium* (Lapeyr.) E.H.L.Krause, *Pisum fulvum* Sibth. & Sm., *Pisum heterophyllum* (L.) E.H.L.Krause, *Pisum hirsutum* (L.) E.H.L.Krause, and *Pisum pumilio* (Meikle) Greuter. According to some authors, these taxa would be subspecies of *Pisum sativum*.

Geographic Distribution Subcosmopolitan. **Habitat** Crops, fallow land. **Range** Canada, China, Russia, India, United States, France, Ukraine, Iran, Australia, Germany, Ethiopia, Spain, United Kingdom, Egypt, Morocco, Turkey, Hungary, Italy, Algeria, Peru.

Beer-Making Parts (and Possible Uses) Seeds.

Chemistry Seeds: energy 81 kcal, carbohydrates 14 g, fiber 5 g, sugars 6 g, protein 5 g, lipids 0.4 g, sodium 5 mg, potassium 244 mg, calcium 25 mg, iron 1.5 mg, magnesium 33 mg, vitamin A 765 IU, vitamin C 40 mg, vitamin B6 0.2 mg.

Style None known.

Source Steele 2012.

Plantago lanceolata L.
PLANTAGINACEAE

Synonyms *Arnoglossum lanceolatum* (L.) Gray, *Lagopus lanceolatus* (L.) Fourr., *Plantago sinuata* Lam.

Common Names English plantain, narrowleaf plantain, ribwort plantain, ribleaf, lamb's tongue.

Description Herb, perennial (20–50 cm), extremely polymorphic. Rhizome fibrous, short, thick, roots fasciculate. Leaves in basal rosette, long, straight, lanceolate, margin entire or toothed, short petiole, usually glabrous but sometimes very hairy; leaf blade traversed by 5 parallel, well-marked main veins. Flower scape emerging from leaf rosette, covered with bristly hairs, aphyllous, 5 longitudinal striations. Inflorescence terminal spike, oval or conical, consisting of numerous flowers closely appressed to one other. Flower develops in correspondence with brown membranous bract; calyx composed of 2 free sepals and 2 joined, straight, with green central rib; corolla tubular and funnel shaped, whitish, divided into brownish lanceolate lobes; 4 stamina equipped with long filaments greater than corolla, surmounted by anthers yellow, then orange. Fruit capsule with transverse dehiscence (pyxis), oval, tiny, brown in color, containing 1–2 shiny seeds with concave inner face.

Related Species There are 185 taxa of specific and subspecific rank recognized in the genus *Plantago*.

Geographic Distribution Eurasian-cosmopolitan. **Habitat** Rustic, ubiquitous, generally synanthropic species, adapted to almost all climates and soils; meadows, pastures, wastelands, rubble, roadsides, gardens (0–2,000 m ASL). **Range** Cosmopolitan (mostly collected in natural or seminatural environments, rarely cultivated).

Beer-Making Parts (and Possible Uses) Leaves (used for bittering), flowers, seeds.

Chemistry Active substances: iridoid glycosides (2–3% total; aucubin, catalpol, asperuloside, globularin, desacetyl asperuloside-methylester acid), mucilagins (2–6.5% total; arabinogalactan, glucomannan, rhamnogalacturonan), flavonoids (apigenin, luteolin, and derivatives of the 2 as apigenin-6,8-di-C-glucoside, luteolin-7-O-glucuronide, luteolin-7-O-glucoside, 7-O-glucuronide-3´-glucoside, apigenin-7-O-glucuronyl-glycoside, luteolin-7-O-glucuronyl-glycoside, apigenin-7-O-glucoside, 7-O-glucuronide), tannins 6.5%, carboxylic phenolic acids (p-hydroxybenzoic acid, protocatechuic acid, gentisinic acid, chlorogenic acid, neochlorogenic acid, etc.), coumarins (aesculetin, etc.), saponins, volatile oils, inorganic compounds (salicylic acid 1%, mineral salts with zinc and potassium). Essential oil: fatty acids 28–52.1% (palmitic acid 15.3–32%), oxidized monoterpenes 4.3–13.2% (linalool 2.7–3.5%), aldehydes + ketones 6.9–10% (pentyl vinyl ketone 2–3.4%), alcohols 3.8–9.2% (1-octen-3-ol 2.4–8.2%), apocarotenoids 1.5–2.3%.

Style None verified.

Source Hieronymus 2016b; https://www.brewersfriend.com/other/ribwort/.

Plantago major L.
PLANTAGINACEAE

Synonyms *Plantago borysthenica* Wissjul., *Plantago dregeana* Decne., *Plantago gigas* H. Lév., *Plantago jehohlensis* Koidz., *Plantago latifolia* Salisb., *Plantago macronipponica* Yamam., *Plantago sawadai* (Yamam) Yamam., *Plantago villifera* Kitag.

Common Names Broadleaf plantain, Englishman's foot, white man's foot, common plantain, grand plantain.

Description Herb, perennial (10–30 cm). Rhizome short with numerous fine rootlets, with 1 or more flower scapes erect or ascending, or kneeling down and then ascending. Leaves exclusively basal, forming a rosette and mostly appressed to the ground; lamina oval or elliptical enlarged, narrowing at the base into a petiole that sheathes the rhizome at its insertion; apex obtuse (sometimes acute), surface glabrous or rarely just pubescent, with veins that remain parallel to the margin meeting at the petiole and apex. Inflorescence cylindrical spike, green-yellow-rust color, carried at the apex of the scape, composed of numerous small hermaphrodite flowers; calyx formed by 4 oval sepals, joined at the base; corolla membranaceous, tubular, at the fauces forming 4 patent lobes, scarious; 4 stamina with large anthers protruding from the corolla; floral bracts green in color on the back. Fruit capsule oval-oblong (pyxidium), with circumscissile dehiscence, enclosed in calyx and corolla, containing 4–25 dark, wrinkled seeds.

Subspecies The taxon *Plantago major* L. includes the typical subspecies (*Plantago major* L. subsp. *major*) and two other subspecies: *Plantago major* L. subsp. *intermedia* (Gilib.) Lange and *Plantago major* L. subsp. *winteri* (Wirtg.) W.Ludw.

Related Species There are 185 taxa of specific and subspecific rank recognized in the genus *Plantago*.

Geographic Distribution Eurasian, later sub-cosmopolitan. **Habitat** Edges of paths, roads, wastelands, rubble, fields, wet and nitrate-rich soils (0–1,500 m ASL). **Range** Subcosmopolitan.

Beer-Making Parts (and Possible Uses) Leaves (used for bittering), flowers, seeds.

Chemistry Leaves: mucilage, tannins, mineral salts, glucides, allantonin, apigenin, aucubin, baicalein, vitamin C, linoleic acid, oleanolic acid, silicic acid, sorbitol. The numerous therapeutic properties attributed to this plant are mostly associated with the following categories of bioactive compounds: flavonoids, alkaloids, terpenoids, phenolic compounds (caffeic acid derivatives), iridoid glycosides, fatty acids, polysaccharides, and vitamins. All parts of the plant (seeds, leaves, flowers, root) of *Plantago major* contain a variety of chemical compounds considered bioactive: flavonoids (flavones, luteolin, apigenin, baicalein, hispidulin, plantaginin, scutallarein, luteolin 7-glucoside, hispidulin 7-glucuronide, luteolin 7-diglucoside, apigenin 7-glucoside, nepetin 7-glucoside, luteolin 6-hydroxy 4-methoxy 7-galactoside, homoplantaginin, aucubin, baicalein, leuteolin, baicalin), alkaloids (indicain, plantagonin), terpenoids (loliolide, ursolic acid, oleanolic acid, sitosterol acid, β-glycyrrhetinic acid), caffeic acid derivatives (plantamajoside, acteoside), iridoid glycosides (aucubin, asperuloside, majoroside, 10-hydroxymajoroside, 10-acetoxymajoroside, catapole, gardoside, geniposidic acid, melittoside), fatty acids (lignoceric acid, palmitic acid, stearic acid, oleic acid, linoleic acid, linolenic acid, myristic acid, 9-hydroxy-cis-11-octadecenoic acid, arachidic acid, behenic acid), polysaccharides (xyloso, arabinose, galacturonic acid, galactose, glucuronic acid, rhamnose, glucose, highly esterified pectin, arabinogalactan), vitamins (vitamin C, carotenoids).

Style Herb Beer.

Source Heilshorn 2017; Hieronymus 2016b.

Platanus occidentalis L.
PLATANACEAE

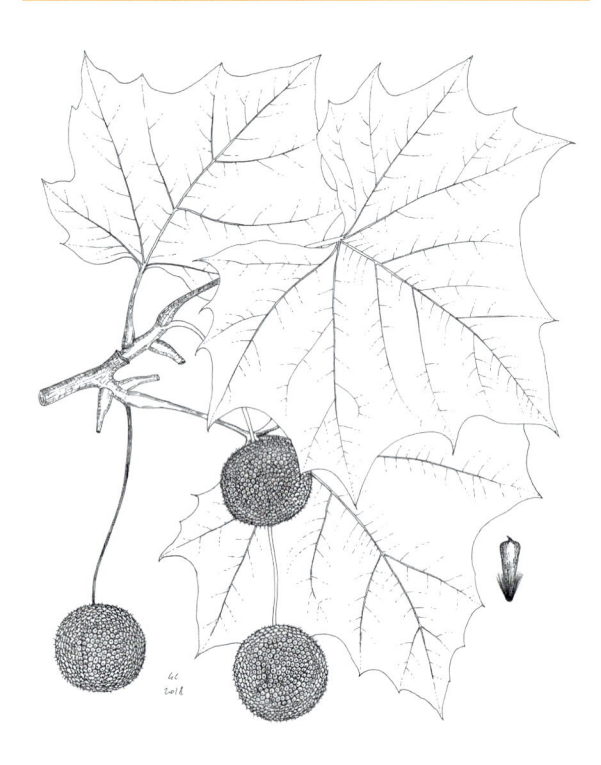

Synonyms *Platanus densicoma* Dode, *Platanus integrifolia* K.Koch, *Platanus lobata* Moench, *Platanus pyramidalis* Dippel.

Common Names Sycamore, American plane tree.

Description Tree, imposing (over 50 m). Trunk straight, unbranched to considerable height or branched below or multicaule (Ø up to more than 4 m). Leaf blade light green, usually not deeply 3- to 5(7)-lobed, occasionally nonlobed (6–20+ × 6–25+ cm; up to 30 × 40 cm on suckers), not especially thick; lobes usually wider than long, basal lobes usually smaller, often strongly reflexed, sinuses broadly and gently concave, depth of distal sinus mostly less than 1/2 the distance from sinus to base of lamina, terminal lobe 1/2–2/3 the length of the lamina; margins entire to coarsely serrate, apex usually acuminate; surface glabrous, the lower surface often persistently tomentose along the veins; stipules entire to coarsely serrate. Inflorescences pistillate flower heads (1–2); at fruiting Ø 25–30 mm; peduncle up to 15 cm. Fruit achene 7–10 mm, basal hairs nearly as long.

Subspecies Specimens with leaves smaller and wider than long, with mostly entire lobes, have been named *Platanus occidentalis* L. var. *glabrata* (Fernald) Sargent. Other specimens with more deeply lobed lamina, a long cuneate base, and decurrent on the petiole have been named *Platanus occidentalis* L. var. *attenuata* Sargent. Not all taxonomists agree on the existence of these taxa of subspecific rank.

Related Species The genus *Platanus* includes 10 accepted taxa: *Platanus acerifolia* (Aiton) Willd., *Platanus gentryi* Nixon & J.M.Poole, *Platanus kerrii* Gagnep., *Platanus mexicana* Moric., *Platanus mexicana* Moric. var. *interior* Nixon & J.M.Poole, *Platanus occidentalis* L., *Platanus orientalis* L., *Platanus racemosa* Nutt., *Platanus rzedowskii* Nixon & J.M.Poole, and *Platanus wrightii* S.Watson.

Geographic Distribution Canada, United States, Mexico. **Habitat** Alluvial soils near rivers and lakes, but also wet ravines, sometimes on plateaus, in calcareous soils (0–950 m ASL). **Range** Beyond the natural range, the species (or its hybrids) is cultivated in parks, gardens, tree-lined avenues.

Beer-Making Parts (and Possible Uses)
Wood (experimented with in beer making), bark (experimented with in beer making).

Wood With whitish sapwood and pinkish-brown heartwood, it has nice speckles on the radial sections (WSG about 0.55). It is used for furniture, staves, brushes, small objects; the marbled or figured pieces often have the commercial denomination of lacewood or buttonwood.

Style None known.

Source Cantwell and Bouckaert 2016; Cantwell and Bouckaert 2018; Hieronymus 2016b; Josephson et al. 2016.

Plathymenia foliolosa Benth.
FABACEAE

Synonyms *Pirottantha modesta* Speg.

Common Names Vinhatico.

Description Shrub or tree (3–40 m), with lenticels. Branches, petioles, and leaf rachis glabrous, tomentose, or woolly. Leaf compound with 3–12 pairs of leaflets, each with 6–20 pairs of linear to elliptic leaflets, 3–19 × 1–11 mm, apex retuse, rarely obtuse, base asymmetrical, leaves of different colors; lower and upper leaf woolly to glabrescent; margin sparsely ciliate; tertiary veins evident. Flowers pedunculate, 3–5 mm; calyx 1–2 mm, glabrous to sparsely woolly; corolla 3–5 mm, glabrous or woolly, at apex of petals; stamens with glandular anthers; ovary stipitate, densely woolly. Fruit cryptoloment oblong (7–18 × 1.5–4 cm). Seed obovoid to obovoid-oblong.

Related Species The genus includes two species: *Plathymenia foliolosa* Benth. and *Plathymenia reticulata* Benth. According to some recent revisions, however, they are synonyms and therefore a single species.

Geographic Distribution Brazil, Paraguay, Argentina. **Habitat** Species of the Brazilian Cerrado, the Atlantic forest (Mata Atlântica), and Amazonia; also present in Paraguay and Argentina. **Range** Brazil, Paraguay, Argentina.

Beer-Making Parts (and Possible Uses) Wood for casks (first they contain cachaça, then beer for aging).

Wood Differentiated, with whitish sapwood and bright yellow heartwood turning to brown after exposure to air, often variegated (WSG 0.60), easy to work and to clean, suitable for carpentry, window frames, floor strips, carpentry, boats.

Style None known.

Source Cantwell and Bouckaert 2016; Cantwell and Bouckaert 2018 (It is possible that, as in other cases in this volume, the use of the common name may cause some confusion with respect to the correct scientific identification of the species to which this plant belongs. Indeed, "vinhatico" is a common name that some sources associate with *Persea indica* [L.] Spreng. According to other sources, the term is more properly applied to South American woody legumes such as *Plathymenia foliolosa* Benth., *Plathymenia reticulata* Benth., and *Zygia bangii* Barneby & J.W.Grimes).

Podocarpus totara G.Benn. ex D.Don
PODOCARPACEAE

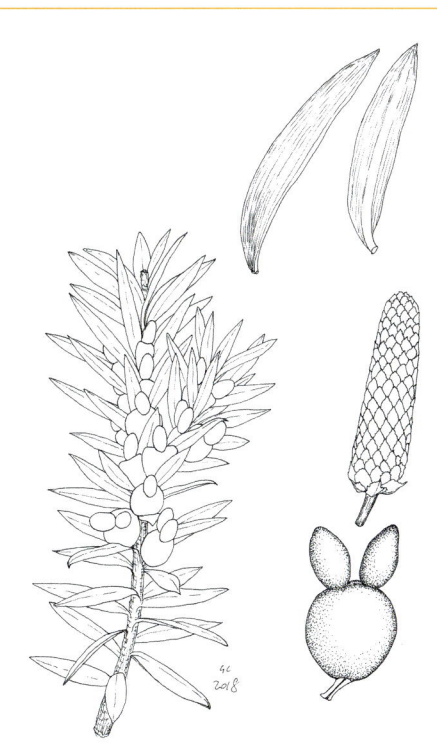

Synonyms *Nageia totara* (G.Benn. ex D.Don) F.Muell., *Podocarpus variegatus* auct.

Common Names Totara.

Description Tree (up to 30 m), evergreen. Trunk (Ø up to 2 m) with thick, fibrous bark, furrowed by long strips that wrap around it until they fall off. Leaves brown to dark green, young about 2 cm × 1–2 mm, adults 1.5–3 cm × 3–4 mm, linear-lanceolate, straight or slightly falcate, acute, prickly, leathery, patent, sessile, with distinct or obscure midrib. Male cones 1–1.5 cm, solitary to 4 together on a short peduncle, surrounded by broad, rigid scales; female branches axillary, flowers solitary or paired, peduncle 2–3 mm; receptacle formed by 2–4 scales, acute and free at apex, usually red in color, swollen and succulent, occasionally dry. Seeds subglobose when mature or ovoid-oblong, 3–5 mm.

Subspecies In addition to the typical variety (*Podocarpus totara* G. Benn. ex D. Don var. *totara*), the taxon *Podocarpus totara* G. Benn. ex D. Don var. *waihoensis* Wardle has been described but is not recognized by some authors, although it is a hybrid between the nominal variety and *Podocarpus acutifolius* Kirk.

Related Species The genus *Podocarpus* includes more than 110 taxa.

Geographic Distribution New Zealand. **Habitat** Lowland to subalpine forests (0–600 m ASL) on both main islands; the *waihoensis* variety is found on the west coast of the South Island, from the Waiho River to the Cascade Falls. **Range** Cultivated as an ornamental species in the native range and as a curiosity in botanical gardens and herbaria around the world.

Beer-Making Parts (and Possible Uses) Wood tested for use in beer making.

Wood Differentiated, with narrow sapwood white-yellowish or light brown in color, heartwood reddish-brown, fine and uniform texture, straight edge; fairly easy to work; slow and sometimes irregular drying (in the kiln often resinous exudations), low shrinkage; not high mechanical properties, heartwood resistant to alterations, sapwood less resistant (WSG 0.42–0.56). Used for glazed doors, carpentry, boats, marine works and general construction, fine figured pieces for furniture and fine works.

Style None verified.

Source Cantwell and Bouckaert 2016; Cantwell and Bouckaert 2018.

Populus nigra L.
SALICACEAE

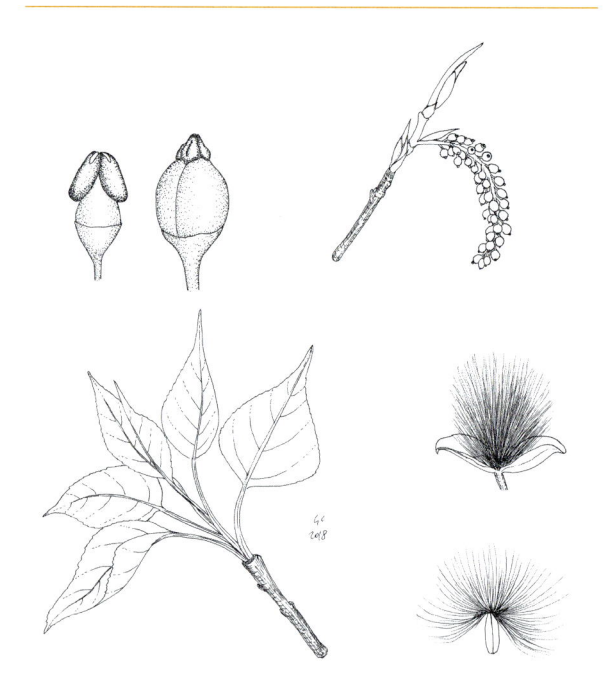

Synonyms *Aigiros nigra* (L.) Nieuwl., *Populus caudina* Ten., *Populus neapolitana* Ten., *Populus pyramidalis* Rozier, *Populus sosnowskyi* Grossh., *Populus thevestina* Dode.

Common Names Poplar.

Description Tree (up to 30 m), rarely shrub, not particularly long-lived (90–100 years). Trunk erect, pyramidal, or columnar (Ø up to 1 m), straight, often deformed by conspicuous bumps; bark smooth at first, then cracked or deeply fissured, grayish-white in young individuals, then brownish-gray; crown expanded (in the typical form), branched at the top; twigs subcylindrical or angular, green or reddish; buds brownish, viscous, small, glabrous. Leaf brachyblast (flowering twigs) with petiole (2–6 cm) lacking basal glands, lamina smooth, glabrous, dark green, upper leaf shiny, underleaf dull yellowish-green, with raised veins; lamina triangular-rhomboidal (5–7 × 4–6 cm), obtuse at base, margin notched (not at base), apex acute or acuminate; leaves turion (elongating branches), same characteristics, larger and usually triangular. Male and female flowers on separate individuals (dioecious species). Inflorescence male catkin (4–9 × 1 cm), preceding foliation, up to 30 stamens per flower, anthers initially reddish, then purple, finally black after pollen fall; female catkins longer and gracile, pendulous, greenish, without style, stigmas yellow; both have laciniate floral bracts. Fruits glabrous bivalve capsules. Seeds very small, provided with white cottony pappus for anemophilous dissemination.

SUBSPECIES In addition to the typical subspecies, *Populus nigra* L. subsp. *nigra*, the taxa *Populus nigra* L. subsp. *betulifolia* (Pursh) Wettst. and *Populus nigra* L. var. *italica* Münchh. are recognized.

RELATED SPECIES The genus *Populus* includes 116 taxa of specific, hybrid, and subspecific rank; among these, species such as *Populus balsamifera* L., *Populus* × *canadensis* Moench, *Populus* × *candicans* Aiton, *Populus deltoides* Marshall, *Populus tacamahacca* Mill., and *Populus tremula* L. have recognized uses in beer making.

GEOGRAPHIC DISTRIBUTION Paleotemperate.
HABITAT Near rivers, lakes, wet soils, fresh and deep, even periodically flooded, poor sandy and gravelly soils (if water table reachable by roots), 0–1,200 m ASL; tree heliophilous, moderately thermophilic.
RANGE Cultivated (also for ornamental purposes).

BEER-MAKING PARTS (AND POSSIBLE USES) Wood for barrels (in Roman times; uncertain use in beer making), buds.

CHEMISTRY Essential oil, buds: ar-curcume ne 27.18%, γ-curcumene 9.5%, δ-cadinene 7.74%, τ-cadinol 5.99%, γ-cadinene 5.11%, α-cadinol 3.82%, benzyl alcohol 3.22%, trans-α-bergamotene 2, 87%, γ-muurolene 2.31%, trans and cis-calamene 2.3%, α-muurolene 2.09%, δ-himachalene 1.97%, (cis,as)-α-farnesene 1.91%, allo-aromadendrene 1.57%, α-copaene 1.43%, cts-α-bergamotene 1.39%, δ-cadinol 1.29%, α-bisabolol 1.28%, epicubenol 1.24%, epizonarene 1.14%, di-e-pi-1,10-cubenol 0.86%, phenylethyl benzoate 0.86%, α-calacorene 0.78%, aliphatic ester 0.75%, 7ah,10bh-cadin-1(6),4-diene 0.71%, α-cadinene 0.69%, β-bisabolol 0.63%, cadin-1,4-diene 0.57%, tricosane 0.5%, aliphatic component 0.49%, benzyl acetate 0.48%, γ-amorphene 0.39%, 2-phenylethyl acetate 0.36%, cts-β-farnesene 0.36%, α-ylanglene 0.35%, 2-phenylethanol 0.34%, methyl 4,7,10,13-hexadecatetraenote 0.34%, phenethyl 2-methylbutyrate 0.33%, (-)-spatulenol 0.32%, ledol 0.28%, muurola-4,10(14)-dien-1-p-ol 0.27%, benzyl benzoate 0.26%, p-vinylguaiacol 0.24%, cadalene 0.24%, δ-eudesmene 0.23%, cubenol 0.23%, (-)-gleenol 0.22%, eugenol 0.2%, germacrene D 0.2%, α-cedrene 0.17%, trans-γ-β isabolene 0.16%, 3-hydroxy-2-methyl-cyclopent-2-enone 0.14%, α-humulene 0.13%, methyl geramate 0.12%, (E,E)-farnesol 0.11%, aliphatic component 0.11%, benzaldehyde 0.1%, cts-geraniol 0.1%, ar-turmerol 0.1%, acetophenone 0.09%, renyl acetate 0.07%, trans-2-methyl-2-butenoic acid 0.07%, s-cyclocitral 0.07%, viridiflorol 0.07%, benzyl salicylate 0.07%, trans-nerolidol 0.06%, caryophyllene oxide 0.06%, benzyl 3-methyl butanoate 0.05%, traces of the following compounds: 2-methylbutanoic acid, α-thujene, camphene, eucalyptol (1,8-cineole), cis-β-ocimene, trans-linalool oxide, cis-linalool oxide, p-cymene, cts-pinocarveol, benzoic acid, methyl salicylate, α-terpineol, 7-epi-sesquithujene, β-copaene, trans-ζ-α-bisabolene epoxide, (+)-spatulenol.

WOOD Whitish in sapwood and brown in heartwood, WSG 0.40–0.50, without particular merits, used for current purposes and as fuel.

STYLE None verified.

SOURCE Cantwell and Bouckaert 2016; Cantwell and Bouckaert 2018; Heilshorn 2017.

Prosopis glandulosa Torr.
FABACEAE

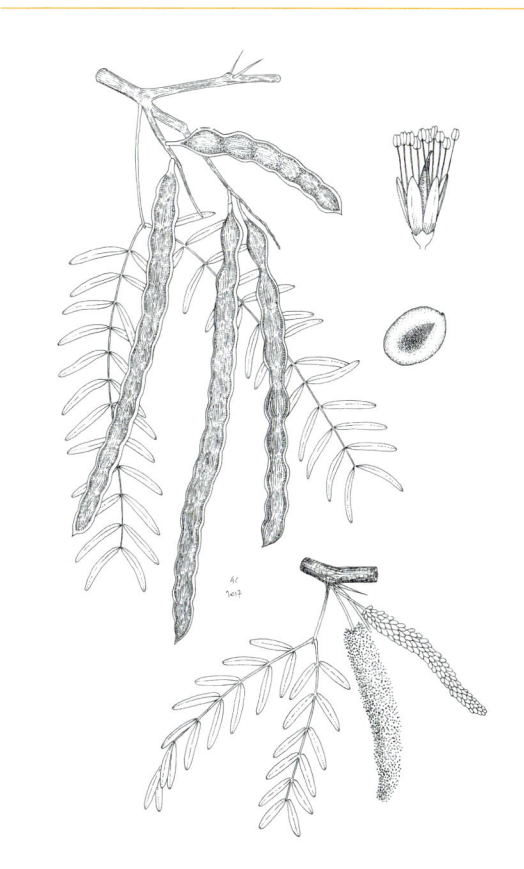

Synonyms *Algarobia glandulosa* (Torr.) Cooper, *Algarobia glandulosa* (Torr.) Torr. & A.Gray, *Neltuma constricta* (Sarg.) Britton & Rose, *Neltuma glandulosa* (Torr.) Britton & Rose, *Neltuma neomexicana* Britton, *Neltuma neomexicana* Britton & Rose, *Prosopis chilensis* sensu auct., *Prosopis juliflora* sensu auct., *Prosopis odorata* Torr. & Frém.

Common Names Mesquite.

Description Shrub or tree (3–9 m), deciduous, with axillary spines (1–4.5 cm long) mostly solitary. Root system deep (to 18.3 m). Leaves compound, glabrous; petiole (with rachis when extant) 2–15 cm long; pinnae 6–17 cm long; leaflets in 6–17 pairs, spaced on rachis about 7–18 mm, linear or oblong, obtuse, glabrous, subcoriaceous, with prominent veins on underleaf; midrib frequently lighter in color (1.5–6.3 cm × 1.5–4.5 mm, or 5–15 times longer than wide). Inflorescence raceme spiciform, 5–14 cm long, multiflorous; flowers with petals 2.5–3.5 mm long; ovary stipitate, villous. Fruit legume straight (8–30 cm × 5–13 mm), rarely subfalcate, compressed, glabrous, straw-yellow or tinged with violet, striate, with strong, short or elongate apex, with 5–18 oblique to longitudinal seeds.

Subspecies Two varieties are recognized within *Prosopis glandulosa*: *Prosopis glandulosa* Torr. var. *torreyana* (Benson) Johnston (mainly in the deserts and arid areas of the SW United States and N Mexico) and *Prosopis glandulosa* Torr. var. *glandulosa* (from Mexico northward to Kansas and eastward to Louisiana). Note that in North America, the common name "mesquite" is used to indicate several species of *Prosopis*.

Related Species There are 53 taxa of specific and subspecific rank recognized in the genus *Prosopis*. Some may have similar properties and applications to those of *P. glandulosa*; *Prosopis pallida* (Willd.) Kunth has established beer-making uses.

Geographic Distribution United States, Mexico. **Habitat** Many arid areas of the world. **Range** The species has been introduced and become invasive in several countries (e.g., Australia, South Africa, Botswana, Namibia), often together with the related *Prosopis velutina* Wooton.

Beer-Making Parts (and Possible Uses) Wood (to smoke malt); boiled and/or roasted seeds, edible, to flavor beer.

Chemistry Seeds contain 20–50% sugars but no starch and therefore do not require mashing.

Wood Differentiated, with thin, greenish-yellow sapwood and red-brown heartwood with purple undertones. Rather coarse texture and irregular grain. Easy to work and bring to excellent cleaning; very durable, WSG 0.65–0.90. Used for railroad ties, road paving, carriage-house work; excellent fuel.

Style American Light Lager, California Common Beer, Saison.

Source Cantwell and Bouckaert 2016; Cantwell and Bouckaert 2018; Daniels and Larson 2000; McGovern 2017; https://www.bjcp.org (32A. Classic Style Smoked Beer).

Prunus armeniaca L.
ROSACEAE

Synonyms *Armeniaca vulgaris* Lam.

Common Names Apricot.

Description Tree (3–10 m), deciduous. Branches shiny, glabrous, ± reddened, smelling like apple. Leaves with petiole 2–5 cm long (usually with glands ± developing into leafy appendages) and lamina slightly heart shaped, acuminate (7–8 × 8–9 cm), glabrous and glossy above, with ovate base and serrated margin. Flowers 1–2, preceding leaves; petals white or pinkish, obovate or ± round (10–15 mm), anthers yellow. Fruit drupe, yellow-orange in color, Ø 3–6 cm, edible (apricot).

Subspecies Within *Prunus armeniaca*, several entities of varietal rank have been described whose taxonomic interpretation is still controversial.

Cultivar They are numerous; here are some (excluding hybrids with congeners): Autumn Royal, Flore Pleno, Maximum, Pendula, and Variegata.

Related Species The genus *Prunus* counts about 280 species, many of which produce edible fruits.

Geographic Distribution Armenia. **Habitat** Cultivated tree. **Range** Uzbekistan, Turkey, Iran, Italy, Algeria, France, Pakistan, Spain, Greece, Japan.

Beer-Making Parts (and Possible Uses) Fruits.

Chemistry Fruits: energy 201 kJ (48 kcal), carbohydrates 11 g, sugars 9 g, fiber 2 g, lipids 0.4 g, protein 1.4 g, vitamin A 96 µg, beta-carotene 1094 µg, lutein zeaxanthin 89 µg, thiamine (B1) 0.03 mg, riboflavin (B2) 0.04 mg, niacin (B3) 0.6 mg, pantothenic acid (B5) 0.24 mg, vitamin B6 0.054 mg, folate (B9) 9 µg, vitamin C 10 mg, vitamin E 0.89 mg, vitamin K 3.3 µg, calcium 13 mg, iron 0.4 mg, magnesium 10 mg, manganese 0.077 mg, phosphorus 23 mg, potassium 259 mg, sodium 1 mg, zinc 0.2 mg.

Style Fruit Beer, Lambic, Saison.

Beer Bliss, Wild Beer (Shepton Mallet, England, United Kingdom).

Source Dabove 2015; Giaccone and Signoroni 2017; Hieronymus 2016b; Josephson et al. 2016; McGovern 2017; Sparrow 2015; Steele 2012; https://www.bjcp.org (29. Fruit Beer); https://www.oregonfruit.com/fruit-brewing/category/products.

Prunus avium (L.) L.
ROSACEAE

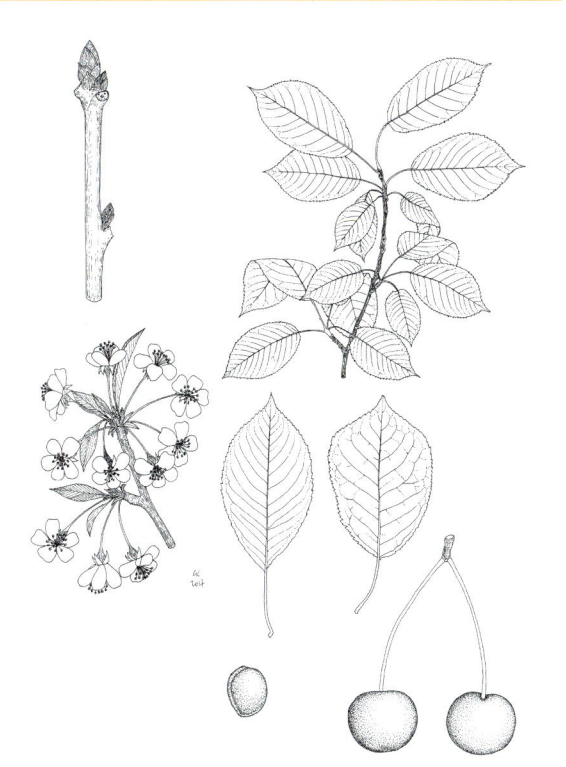

Synonyms *Cerasus avium* (L.) Moench, *Druparia avium* (L.) Clairv.

Common Names Cherry.

Description Tree (3–20 m), deciduous. Bark smooth and ± shiny in young branches (Ø 5–10 mm), glabrous, with 3 × 15 mm transverse lenticels; lacerations in ribbons that roll up; wood smelling of apple; gummy secretion exudes from wounds. Leaves usually pendulous, sparsely pubescent on branches, dark green above, pale below; lamina oblanceolate (6–8 × 12–15 cm), rarely ovate (7–12 × 10–15 cm), toothed; petiole 2–4 cm with 2–4 red glands upward. Inflorescence umbel pauciflor; flower with peduncles 3–5 cm, petals 10–15 mm white. Fruit subspherical drupe, Ø 1–3 cm, dark red, edible (cherry).

Subspecies Within *Prunus avium*, several forms and varieties have been described whose taxonomic interpretation remains unclear.

Cultivar Some of the following have been awarded by the Royal Horticultural Society: Amanogawa, Autumnalis (*Prunus × subhirtella*), Autumnalis Rosea (*Prunus × subhirtella*), Avium, Bella di Garbagna, Bigaraux, Colorata (*Prunus padus*), Durone di Vignola, Kanzan, Kiku-shidare-zakura, Kursar, Morello (*Prunus cerasus*), Okamé (*Prunus × incam*), Pandora, Pendula Rosea, Pendula Rubra, Pink Perfection, Plena (Grandiflora), Praecox (*Prunus incisa*), Prunus × cisterna, Prunus avium (wild cherry), *Prunus sargentii* (Sargent cherry), *Prunus serrula* (Tibetan cherry), Schaerbeek (cultivar from the surroundings of Brussels used in the past for spontaneous-fermentation beers and now practically extinct), Shirofugen, Shirotai, Shōgetsu, Spire, Stella, Ukon. Door County is used for some spontaneously fermented beers; Bing is an important North American cultivar.

Related Species The genus *Prunus* has about 280 species, many of which produce edible fruit.

Geographic Distribution Pontic (?). **Habitat** Cultivated tree. **Range** Turkey, United States, Iran, Spain, Italy.

Beer-Making Parts (and Possible Uses) Fruits, wood (for barrels and to smoke malt).

Chemistry Fruits: energy 263 kJ (63 kcal), carbohydrates 16 g, sugars 12.8 g, fiber 2.1 g, lipids 0.2 g, protein 1.1 g, vitamin A 3 µg, beta-carotene 38 µg, lutein zeaxanthin 85 µg, thiamine (B1) 0.027 mg, riboflavin (B2) 0.033 mg, niacin (B3) 0.154 mg, pantothenic acid (B5) 0.199 mg, vitamin B6 0.049 mg, folate (B9) 4 µg, choline 6.1 mg, vitamin C 7 mg, vitamin K, 2.1 µg,

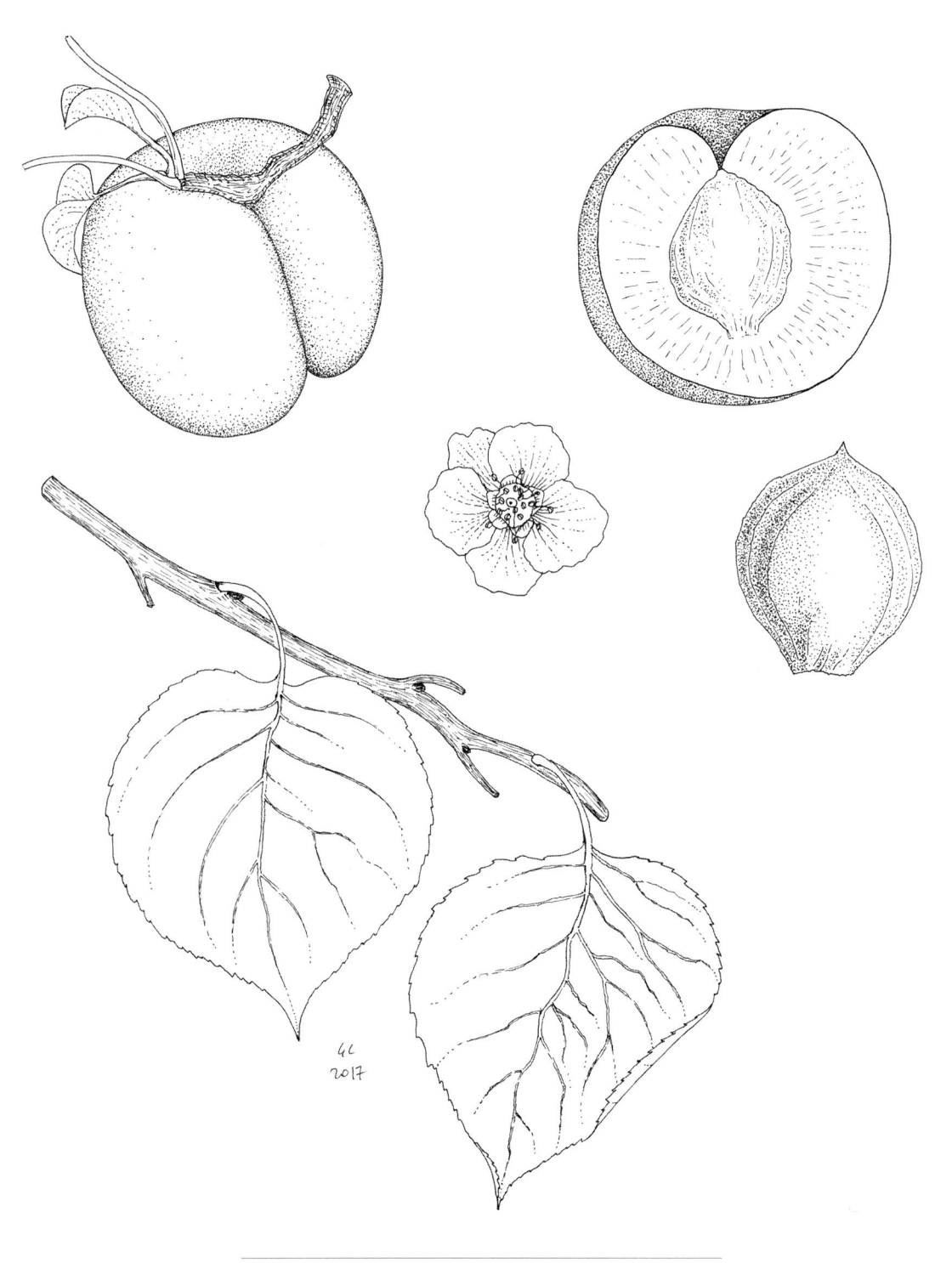

Prunus armeniaca L.
ROSACEAE

calcium 13 mg, iron 0.36 mg, magnesium 11 mg, manganese 0.07 mg, phosphorus 21 mg, potassium 222 mg, zinc 0.07 mg.

Wood Hardwood brown-pink, sapwood tending to yellowish, WSG 0.60, straight grain, fine texture; it is easy to work and can be cleaned well, but it twists and splits during seasoning and is not very durable. Valued for furniture, fine work, boxes, scientific instruments.

Style Baltic Porter, Berliner Weisse, Robust Porter, Specialty Beer.

Source Calagione et al. 2018; Cantwell and Bouckaert 2016; Cantwell and Bouckaert 2018; Dabove 2015; Giaccone and Signoroni 2017; Heilshorn 2017; Hieronymus 2010; Hieronymus 2016b; Jackson 1991; McGovern 2017; Sparrow 2015; Steele 2012; https://www.bjcp.org (29. Fruit Beer; 32A. Classic Style Smoked Beer).

Prunus cerasus L.
ROSACEAE

Synonyms *Cerasus acida* (Ehrh.) Borkh., *Cerasus austera* (L.) Borkh., *Cerasus austera* (L.) M.Roem., *Cerasus collina* Lej. & Courtois, *Cerasus fruticosa* Pall., *Cerasus vulgaris* Mill., *Druparia cerasus* (L.) Clairv., *Prunus acida* Ehrh., *Prunus aestiva* Salisb., *Prunus austera* (L.) Ehrh.

Common Names Sour cherry, tart cherry, dwarf cherry.

Description Tree, deciduous (5–10 m); if wild, often shrubby. Bark glabrous, possibly reddened; propagation by underground stolons. Leaves with petiole generally without glands, ovate or lanceolate lamina (3–7 × 6–12 cm), finely toothed. Flowers 2–4 on 2–4 cm peduncles; petals white, obovate or subrounded (8–12 mm). Fruit red (Ø 1 cm), edible (amarena, tart cherry).

Subspecies Within this taxon, several varieties have been described whose taxonomic interpretation remains unclear.

Cultivar Among the many cultivars famous in the brewing world is Schaarbeekse Kriek, cultivated in Belgium and an essential ingredient of lambic kriek.

Related Species The genus *Prunus* has about 280 species, many of which produce edible fruits.

Geographic Distribution Pontic (?). It is believed that *Prunus cerasus* originated as a natural hybrid, then stabilized, of *Prunus avium* and *Prunus fruticosa* Pall. on the Iranian Plateau or in Eastern Europe, where the two species come in contact; *Prunus fruticosa* would have contributed with the small size and acid taste of the fruit. **Habitat** Cultivated tree. **Range** Turkey, Russia, Ukraine, Poland, Iran, Serbia, Hungary, United States, Uzbekistan, Azerbaijan.

Beer-Making Parts (and Possible Uses) Fruits.

Chemistry Fruits: energy 209 kJ (50 kcal), carbohydrates 12.2 g, sugars 8.5 g, fiber 1.6 g, lipids 0.3 g, protein 1 g, vitamin A 64 µg, beta-carotene 770 µg, lutein zeaxanthin 85 µg, thiamin (B1) 0.03 mg, riboflavin (B2) 0.04 mg, niacin (B3) 0.4 mg, pantothenic acid (B5) 0.143 mg, vitamin B6 0.044 mg, folate (B9) 8 µg, choline 6.1 mg, vitamin C 10 mg, vitamin K 2.1 µg, calcium 16 mg, iron 0.32 mg, magnesium 9 mg, manganese 0.112 mg, phosphorus 15 mg, potassium 173 mg, sodium 3 mg, zinc 0.1 mg.

Style American Wild Ale, Baltic Porter, Fruit Beer, Kriek Lambic.

Beer Oude Kriek, 3 Fonteinen (Beersel, Belgium).

Source Calagione et al. 2018; Dabove 2015; Giaccone and Signoroni 2017; Hieronymus 2016a; Hieronymus 2016b; Jackson 1991; Sparrow 2015; https://www.brewersfriend.com/other/griottes/.

The Kriek Tradition
Kriek is a type of spontaneously fermented beer usually brewed in Belgium in Pajottenland, the area southwest of Brussels crossed by Senne River. It is started from a lambic—beer obtained by fermenting a wort of barley and wheat with yeasts (mainly belonging to the *Brettanomyces* genus) and bacteria naturally present in the brewery and in the wood of casks—to which *Prunus cerasus* fruits are added. The traditional variety used is Schaerbeek, a rather small black cherry, with a large stone and a deep red sourish pulp, increasingly rare in the vicinity of Brussels. Drupes are added directly into the cask, in a percentage ranging from 20% to 30% of the wort. The addition of the sugars contained in sour cherries triggers an intense fermentation, followed by a period of aging and maceration. It is during this phase that color, aroma, and tannins (from the skin and stone) are extracted from the fruit and benzaldehyde is formed, an aromatic complex that gives kriek its typical hint of almond. Once bottled, the addition of young lambic allows the beer to referment and be carbonated.

Prunus domestica L.
ROSACEAE

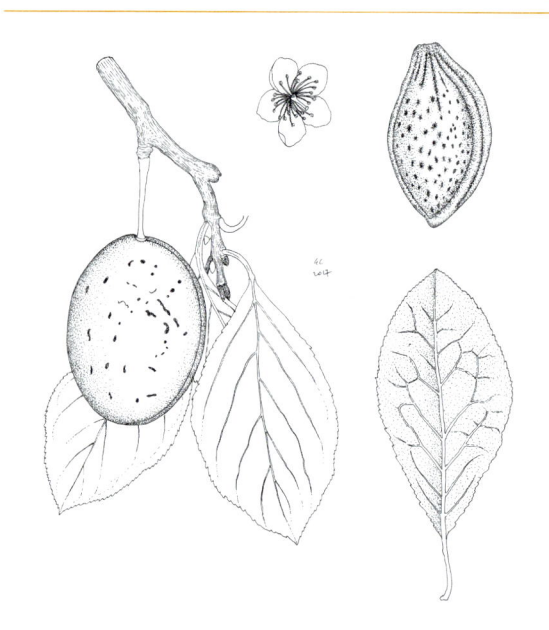

Synonyms *Druparia prunus* Clairv., *Prunus communis* Huds., *Prunus oeconomica* Borkh.

Common Names Plum (not all fruits listed here belong to the *Prunus domestica* species).

Description Tree, deciduous (2–10 m). Bark red-brown; young branches pubescent, then glabrous; wood with apple scent. Leaves with oblanceolate lamina (3–4 × 6–8 cm), notched all around, rounded or obtuse at apex, with broadly cuneate base, pubescent below; petiole 1/3–2/5 the size of the lamina. Flowers 2–3 in glomerules on 1–2 cm peduncles; petals white (7–12 mm). Fruit drupe (2–7 cm), edible (plum).

Subspecies In addition to the typical subspecies, *Prunus domestica* L. subsp. *domestica*, taxonomists recognize two taxa of subspecific rank: *Prunus domestica* L. subsp. *insititia* (L.) Bonnier & Layens and *Prunus domestica* L. subsp. *italica* (Borkh.) Gams. *The European Garden Flora* recognizes only three subspecies within *Prunus domestica*; however, other scholars advocate recognizing more subspecies: *Prunus domestica* L. ssp. *domestica* (including ssp. *oeconomica* [Borkh.] C.K.Schneid.), *Prunus domestica* L. ssp. *insititia* (L.) Bonnier & Layens, *Prunus domestica* L. ssp. *intermedia* Poir., *Prunus domestica* L. subsp. *italica* (Borkh.) Gams (including spp. *claudiana* Poir. and ssp. *rotunda* J.F.Macbr.), *Prunus domestica* L. ssp. *pomariorum* (Boutigny) H.L.Werneck (= *Prunus domestica* L. ssp. *insititia* [L.] Bonnier and Layens?),

Prunus domestica L. ssp. *prisca* Ettingsh. & J.S.Gardner, and *Prunus domestica* L. ssp. *syriaca* Borkh. Complicating the taxonomic arrangement is the ease of crossing subspecies, resulting in the formation of numerous intermediate forms for fruit characters, such as sweetness, acidity, and color (blue-purple, red, orange, yellow, green).

Cultivar A sample of cultivars of *Prunus domestica* (some awarded by the Royal Horticultural Society): Blue Rock, Blue Tit, Czar, Damson, Imperial Gage, Jefferson, Laxton's Delight, Lombard, Mallard, Marjory's Seedling, Maynard, Opal, Oullins Gage, Pershore, Victoria, Yellow Egg.

Related Species The genus *Prunus* has about 280 species, many of which produce edible fruits.

Geographic Distribution Origin uncertain. **Habitat** Cultivated plant. **Range** Cultivated.

Beer-Making Parts (and Possible Uses) Fresh and dried fruits, juice.

Chemistry Fruits: energy 46 kcal, water 87.23 g, protein 0.7 g, lipids 0.28 g, carbohydrates 11.42 g, fiber 1.4 g, sugars 9.92 g, calcium 6 mg, iron 0.17 mg, magnesium 7 mg, phosphorus 16 mg, potassium 157 mg, zinc 0.10 mg, vitamin C 9.5 mg, thiamine 0.028 mg, riboflavin 0.026 mg, niacin 0.417 mg, vitamin B6 0.029 mg, folate 5 μg, vitamin A 17 μg, vitamin E 0.26 mg, vitamin K 6.4 μg.

Style American Amber Ale, American Barleywine, American Stout, Belgian Specialty Ale, Blonde Ale, Brown Porter, California Common Beer, English IPA, Flanders Brown Ale/Oud Bruin, Fruit Beer, Fruit Lambic, Northern English Brown Ale, Russian Imperial Stout, Saison, Witbier.

Beer Beerbrugna, LoverBeer (Marentino, Piedmont, Italy) (a wild sour aged fruit ale produced with the addition of ramassin, an ecotype of damson plum typical of Bronda Valley in Piedmont, Italy).

Source Calagione et al. 2018; Giaccone and Signoroni 2017; Heilshorn 2017; Hieronymus 2016b; Josephson et al. 2016; Markowski 2015; McGovern 2017; https://www.bjcp.org (29. Fruit Beer); https://www.brewersfriend.com/other/plum/; https://www.brewersfriend.com/other/plums/; https://www.oregonfruit.com/fruit-brewing/category/products.

Prunus domestica × *armeniaca* (hybrids *Prunus* sp. × *Prunus* sp.)
ROSACEAE

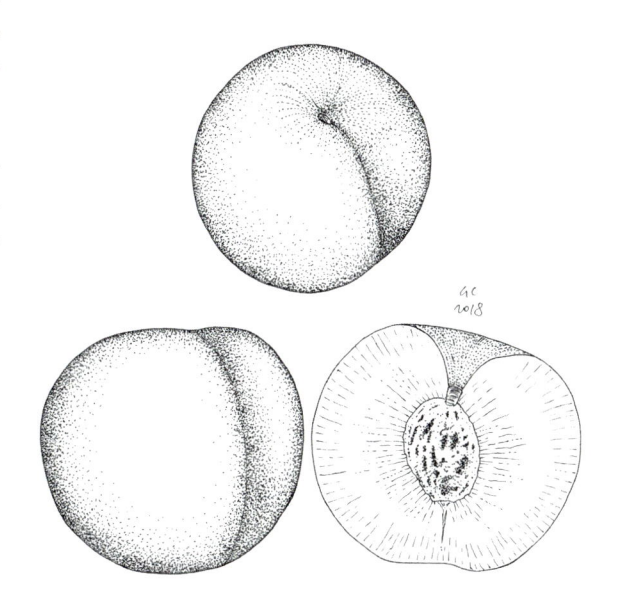

Synonyms None

Common Names Pluot, aprium, apriplum, plumcot.

Description Tree, medium-sized. Leaves vary in shape according to variety, elliptical, glabrous, subcoriaceous, oblong or obovate (5–15 cm); petiole 1–2.5 cm. Flowers usually white (Ø 1.5–2.5 cm). Fruit drupes (90–140 g) round to heart shaped, depending on variety; epicarp color violet to red, flesh yellow, pink, or red. Plumcot (or apriplum): natural plumcots/apriplums have been known for hundreds of years from regions of the world where both species were grown from seed. The name "plumcot" was coined by Luther Burbank (1849–1926), a botanist who selected more than 800 varieties of cultivated plants. Plumcot can be propagated by grafting. Pluot: the generations of hybrids following the first and are genetically 1/4 (25%) apricot and 3/4 (75%) plum. The outward appearance of the fruit, with smooth skin, is very similar to that of the plum. Plums were developed at the end of the twentieth century by Floyd Zaiger, a biologist who founded a company specializing in the production of new cultivars. Aprium: under this denomination are a complex of hybrids that show prevalent morphological traits of apricot. Genetically they are 1/4 (25%) plum and 3/4 (75%) apricot. Fruits have the external appearance of apricots. The flesh is usually dense, notable for its sweet taste owing to high levels of fructose and other sugars.

Pseudotsuga menziesii (Mirb.) Franco
PINACEAE

Synonyms *Abies douglasii* (Sabine ex D.Don) Lindl., *Abies drummondii* Gordon, *Abies menziesii* Mirb., *Abies mucronata* Raf., *Abies obliqua* Bong. ex Gordon, *Abies obliquata* Raf. ex Gordon, *Abies standishiana* K.Koch, *Abies taxifolia* C.Presl, *Abietia douglasii* (Sabine ex D.Don) A.H.Kent, *Picea douglasii* (Sabine ex D.Don) Link, *Pinus douglasii* Sabine ex D.Don, *Pseudotsuga douglasii* (Sabine ex D.Don) Carrière, *Pseudotsuga mucronata* (Raf.) Sudw. ex Holz., *Pseudotsuga taxifolia* (Lindl.) Britton, *Pseudotsuga vancouverensis* Flous, *Tsuga douglasii* (Sabine ex D.Don) Carrière.

Common Names Douglas fir, Oregon pine.

Description Tree (up to 100 m). Trunk straight (Ø up to 4.4 m); crown narrow to broadly conical, flattening with age; sprouts slender, pubescent, becoming glabrous. Leaves 15–40 × 1–1.5 mm, yellow-green to dark green or bluish, apex obtuse to acute. Male cones yellow-red. Female cones 4–10 × 3–3.5 cm. Seeds 5–6 mm, wing longer than seminal body.

Subspecies In addition to *Pseudotsuga menziesii* (Mirb.) Franco var. *menziesii*, the variety *Pseudotsuga menziesii* (Mirb.) Franco var. *glauca* (Beissn.) Franco is recognized.

Related Species The genus *Pseudotsuga* includes four entities of specific rank: *Pseudotsuga japonica* (Shiras.) Beissn., *Pseudotsuga macrocarpa* (Vasey) Mayr, *Pseudotsuga menziesii* (Mirb.) Franco (with the typical variety and var. *glauca* [Beissn.] Franco), and *Pseudotsuga sinensis* Dode (three varieties: *brevifolia* (W.C.Cheng & L.K.Fu) Farjon & Silba, *gaussenii* [Flous] Silba, and *sinensis*).

Geographic Distribution Distribution of both varieties: W North America (Canada [Alberta, British Columbia], United States [California, Oregon, Washington, Nevada, Arizona, Colorado, Idaho, Montana, New Mexico, Texas, Utah, Wyoming], Mexico). **Habitat** Coniferous or mixed forests. **Range** Tree grown in North America, Europe, Asia for forestry purposes.

Beer-Making Parts (and Possible Uses) Wood for fermentation vats and casks (uncertain use in beer making).

Wood Whitish sapwood and pinkish-brown heartwood, fine texture, straight and regular grain, with modest shrinkage, easy to work with the saw, planer, and sheeter; WSG 0.48–0.55. Durable and valuable for beams, sleepers, planks, poles, carpentry, furniture, fixtures, packaging, plywood.

Style None verified.

Source Cantwell and Bouckaert 2016; Cantwell and Bouckaert 2018.

Psidium guajava L.
MYRTACEAE

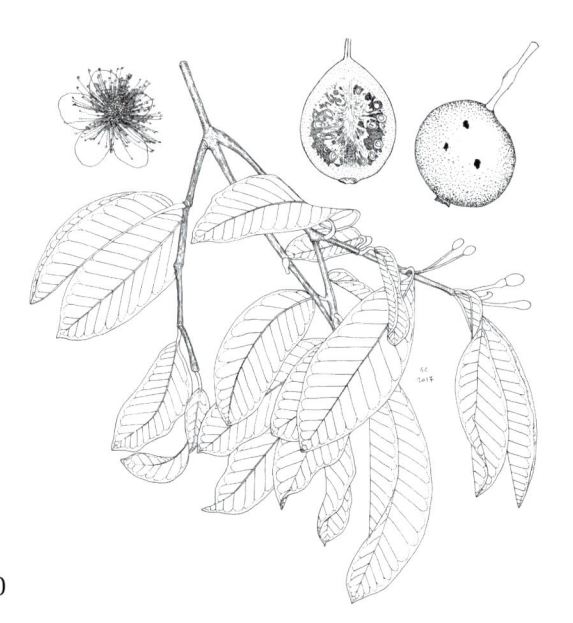

Synonyms *Guajava pumila* (Vahl) Kuntze, *Guajava pyrifera* (L.) Kuntze, *Myrtus guajava* (L.) Kuntze, *Psidium angustifolium* Lam., *Psidium cujavillus* Burm.f., *Psidium cujavus* L., *Psidium fragrans* Macfad., *Psidium guava* Griseb., *Psidium igatemyense* Barb. Rodr., *Psidium igatemyensis* Barb.Rodr., *Psidium intermedium* Zipp. ex Blume, *Psidium pomiferum* L., *Psidium prostratum* O.Berg, *Psidium pumilum* Vahl, *Psidium pyriferum* L., *Psidium sapidissimum* Jacq., *Psidium vulgare* Rich., *Syzygium ellipticum* K.Schum. & Lauterb.

Common Names Guava, common guava, yellow guava, lemon guava.

Description Shrub or tree (3–10 m), evergreen. Stem twisted, branches numerous, bark pale to reddish-brown, thin, smooth, continuously defoliating. Root system generally very extensive and shallow, frequently well beyond the canopy, with some deep roots but not a distinct taproot. Leaves opposite, simple, stipules absent, petiole short (3–10 mm); lamina thick and leathery, oblong to elliptic (5–15 × 4–6 cm), apex obtuse to distinctly acuminate, base rounded to subcuneate, margin entire, dull gray to yellow-green above, slightly paler below, veins prominent, glands punctate. Inflorescence axillary, 1–3 flowers, pedicels about 2 cm long, bracts 2, linear. Flower with calyx irregularly divided into 2–4 lobes, whitish and sparsely hairy inside; petals 4–5, white, linear-ovate, about 2 cm long, delicate; stamens numerous, filaments pale white, about 12 mm long, erect or spreading, anthers straw colored; ovary inferior, ovules numerous, stigma about 10 cm long, green, capitate. Fruit berry ovoid or pyriform, 4–12 cm long, weight up to 500 g; skin color yellow when ripe, sometimes tinged with red; flesh juicy, creamy white to creamy yellow to pink or red; mesocarp thick, edible; soft flesh contains numerous kidney-shaped or flattened seeds, cream to brown; exterior of fruit is carnose and center is seed-rich flesh.

Cultivar Varietal characteristics of guava are not as distinct as in other cultivated species because the frequent use of multiplication from seed tends to mix them. Some cultivars are Allahabad Safeda, Allahabadi Surkha, Apple Guava, Arka Mridula, Chittidar, Fruits of Sebia, Hafshi, Harijha, Lucknow 49 (or Sardar), and Seedless.

Related Species Among the 118 taxa recognized in the genus *Psidium* and the 26 of the related genus *Myrciaria*, some species have known food applications: *Myrciaria floribunda* (H.West ex Willd.) O.Berg (rumberry, guavaberry), *Psidium acutangulum* Mart. ex DC. (para guava), *Psidium cattleianum* Afzel. ex Sabine (cattley guava, strawberry guava, guayaba peruana, Peru guayabita), *Psidium friedrichsthalianum* (O.Berg) Nied. (Costa Rican guava), and *Psidium guineense* Sw. (Brazilian guava, guisaro).

Geographic Distribution Colombia, Mexico, Peru, United States. **Habitat** Plant grown in many tropical countries. **Range** Widely cultivated in tropical and subtropical areas of the planet. In some cases (e.g., Galapagos Islands), the species can become invasive.

Beer-Making Parts (and Possible Uses) Fruits.

Chemistry Fruits: energy 84 kJ, water 86.1 g, ash 600 mg, fiber 5.4 g, carbohydrates 9.54 g, sugars 11.88 g, protein 820 mg, lipids 600 mg, potassium 284 mg, phosphorus 25 mg, magnesium 10 mg, calcium 20 mg, iron 310 µg, copper 103 µg, zinc 230 µg, sodium 3 mg, manganese 144 µg, selenium 0.6 µg, vitamin C 183.5 mg, vitamin B1 50 µg, vitamin B2 50 µg, vitamin B3 1.2 mg, vitamin B5 150 µg, vitamin B6 143 µg, vitamin B9 14 µg, vitamin A 792 IU, vitamin E 1.12 mg, fatty acids (saturated 172 mg, monounsaturated 55 mg, polyunsaturated 253 mg).

Style American Amber Ale, American Wheat or Rye Beer, California Common Beer, Fruit Beer, Fruit Lambic, Saison, Specialty Beer, Standard American Lager, Witbier.

Beer Terreux Frucht Guava, the Breuery (California, United States).

Source Hieronymus 2016b; McGovern 2017; https://www.bjcp.org (29. Fruit Beer); https://www.brewersfriend.com/other/guava/.

Pterocarpus santalinus L.f.
FABACEAE

Synonyms *Lingoum santalinum* (L.f.) Kuntze.

Common Names Red sanders, red sandalwood, saunderswood, almug.

Description Tree, small to medium (10–20 m), deciduous. Crown rounded, rather dense. Stem generally erect (Ø 30–50 cm). Bark dark brown in color, with rectangular plates, deeply fissured when mature. Leaves pinnate-compound, leaflets 3–5, elliptic (4–8 cm), upper leaf green, underleaf pale green, covered with fine appressed hairs, margin entire, apex emarginate. Inflorescences broad pauciflor racemes, axillary or terminal. Flowers papilionate, corolla yellow, about 2 cm. Fruits winged, subcircular legumes (Ø about 3.5 cm), brown when ripe, on short petioles. Seeds 1–2 per legume.

Related Species There are 72 species included in the genus *Pterocarpus*.

Geographic Distribution S Eastern Ghats (India).
Habitat Arid deciduous forests. **Range** India, China, Pakistan, Sri Lanka, Taiwan.

Beer-Making Parts (and Possible Uses)
Wood (ground heartwood in powder form is used as a red dye in many foods, spice mixtures, and sauces).

Chemistry Wood (heartwood): santalin (santalic acid) 16% (santalin A and santalin B), isoflavonoids, terpenoids, phenolic compounds, beta-sitosterol, lupeol, (-)-epicatechin, 6-OH-1-methyl-3´,4´,5´-trimethoxy-adenoside-4-O-rhamnoside, 6,4´-dihydroxy adenoside-4-O-neohesperidoside, 4´,5-dihydroxy-7-methyl isoflavone, 3´-O-beta-D-glucoside, pterocarpol, pterocarptriol, ispterocarpolonem pterocarpo-diolones, beta-eudesmol, cryptomeridol, ethers, alkali, santal, pterocarpine, hompterocarpine, tannin, triterpenes.

Wood Color red-orange, then changing to dark red, WSG 0.80–0.85; heartwood extremely hard, dark purple color, used for furniture and fine objects, engraving, and turnery, as well as for dyeing purposes because it contains an alcohol-soluble principle to which it gives a red color (yellow in ether, violet in bases); being harmless, it is used as a food coloring.

Style Mum.

Source Hieronymus 2010.

Pterocarpus soyauxii Taub.
FABACEAE

Synonyms None

Common Names African padauk, vermillion.

Description Tree, evergreen (up to 55 m), sometimes deciduous. Trunk straight, cylindrical (Ø up to 140–200 cm), unbranched up to 20–30 m; bark gray-brown to brown, peeling off in irregular flakes, exuding abundant red gum from the cuts; crown dome shaped, open; young shoots brown, tomentose. Leaves alternate, compound, imparipinnate, with 7–17 leaflets; stipules linear, to 2 cm, hairy, early caducous; petiole 1–3.5 cm, rachis 3.5–16.5 cm, brown, densely hairy; leaflets petioles 3–5 mm, superficially furrowed; leaflets alternate to nearly opposite, obovate to elliptic (2.5–9 × 1.5–4 cm), base rounded to obtuse, apex usually acuminate and mucronate, leathery, glabrous. Inflorescence panicle axillary or terminal, much branched (10–35 cm), densely hairy, brown; bracts linear, caducous at anthesis. Flowers hermaphrodite, papilionaceous; peduncle 3–19 mm, hairy; calyx campanulate (about 7.5 mm), densely hairy, with 5 triangular 1–2.5 mm teeth, the upper 2 ± connate; corolla with clawed petals, bright yellow or orange-yellow, vexillum nearly circular to 13 mm × 10 mm, wings to 12 mm, keel to 9.5 mm; stamens 10, connate, to 8.5 mm, the upper sometimes free; ovary superficial; stigma to 4 mm, glabrous. Fruit legume circular, flattened, indehiscent (Ø 4.5–9 cm), peduncle up to 1 cm, with finely veined wing with wavy or braided margin, finely hairy, glossy brown, 1-seeded. Seeds reniform, flat or slightly thickened (12–16 × 5–7 mm), smooth, red when fresh, then dark brown or black.

Related Species The genus *Pterocarpus* includes 72 taxa.

Geographic Distribution C Africa, W tropical Africa. **Habitat** Isolated specimens or in small groups in deciduous or evergreen forests (0–500 m), in moist but well-drained and deep soils, with an average annual rainfall of 1,500–1,700 mm and an average annual temperature of 23°C. **Range** Gabon, Cameroon.

Beer-Making Parts (and Possible Uses) Barrel wood (in Roman times; uncertain use in beer making).

Wood Differentiated, with whitish sapwood (not used) and coral red heartwood, darker with time (WSG 0.65–0.90); medium texture, straight grain, fissile, easy to work, long lasting, suitable for outdoor work and construction and for fine work, cabinetmaking, handles, staves for barrels.

Style None verified.

Source Cantwell and Bouckaert 2016; Cantwell and Bouckaert 2018.

Punica granatum L.
LYTHRACEAE

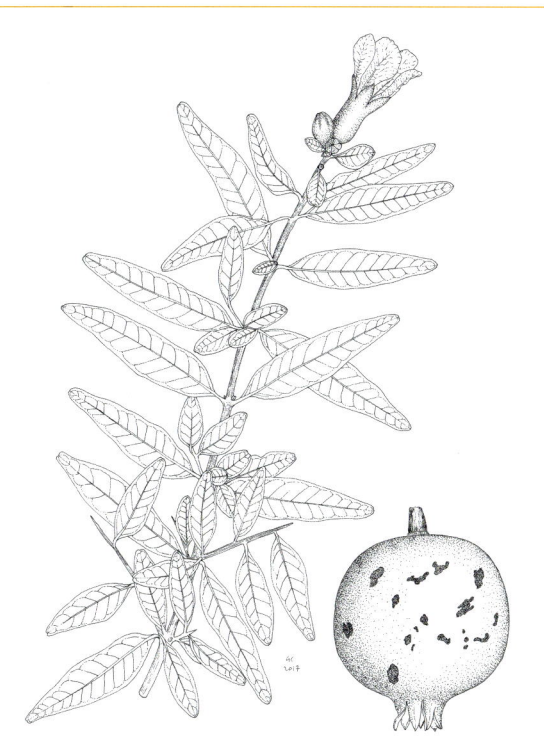

Synonyms *Punica nana* L.

Common Names Pomegranate.

Description Tree, deciduous, small (2–4 m); wild, often shrubby. Branches frequently spinescent. Leaves deciduous, oblanceolate or obovate (1 × 4–6 cm), leathery, usually rounded at apex. Flowers hermaphrodite, actinomorphic, isolated, axillary; sepals and petals (5–8) free; calyx coriaceous, reddened; petals red (rarely pinkish or white), 1–2 × 2–3 cm; stamens very numerous; ovary inferior with numerous overlapping locules. Fruit berry (Ø 6–12 cm). Seeds numerous with gelatinous outer pulp.

Cultivar *Punica granatum* has more than 500 nominal cultivars, of which many should be synonymized since different names often indicate the same genotype. The characteristics on which differentiation among genotypes is based are size of the fruit, color of the exocarp (from yellow to purple, from pink to red), color of the seed coating (from white to red), hardness of the seed, juice content, and juice acidity, sweetness, an astringency.

RELATED SPECIES The only other taxon belonging to the genus, *Punica protopunica* Balf. f., is a species endemic to the island of Socotra (Yemen), whose bitter fruits are not particularly appreciated either by animals or, even less, by humans.

GEOGRAPHIC DISTRIBUTION SW Asia. **HABITAT** Plant cultivated for ornamentation and for the fruit. **RANGE** Middle East, Caucasus, N Africa, Indian subcontinent, C Asia, SE Asia, Mediterranean, United States (Arizona, California).

BEER-MAKING PARTS (AND POSSIBLE USES) Fruits.

CHEMISTRY Fruits: energy 346 kJ (83 kcal), carbohydrates 18.7 g, sugars 13.67 g, fiber 4 g, lipids 1.17 g, protein 1.67 g, thiamine (B1) 0.067 mg, riboflavin (B2) 0.053 mg, niacin (B3) 0.293 mg, pantothenic acid (B5) 0.377 mg, vitamin B6 0.075 mg, folate (B9) 38 µg, choline 7.6 mg, vitamin C 10.2 mg, vitamin E 0.6 mg, vitamin K 16.4 µg, calcium 10 mg, iron 0.3 mg, magnesium 12 mg, manganese 0.119 mg, phosphorus 36 mg, potassium 236 mg, sodium 3 mg, zinc 0.35 mg.

STYLE American Pale Ale, American Wheat Beer, Ancient Ale, Blonde Ale, California Common Beer, Fruit Beer, Light American Lager, Mild, Saison, Spice/Herb/Vegetable Beer, Standard American Lager.

BEER Super Tramp 5, DecimoPrimo (Trinitapoli, Apulia, Italy).

SOURCE Hieronymus 2016b; McGovern 2017; https://www.bjcp.org (29. Fruit Beer); https://www.birradegliamici.com/news-birraie/birre-strane/; https://www.brewersfriend.com/other/pomegranate/.

Pyrus communis L.
ROSACEAE

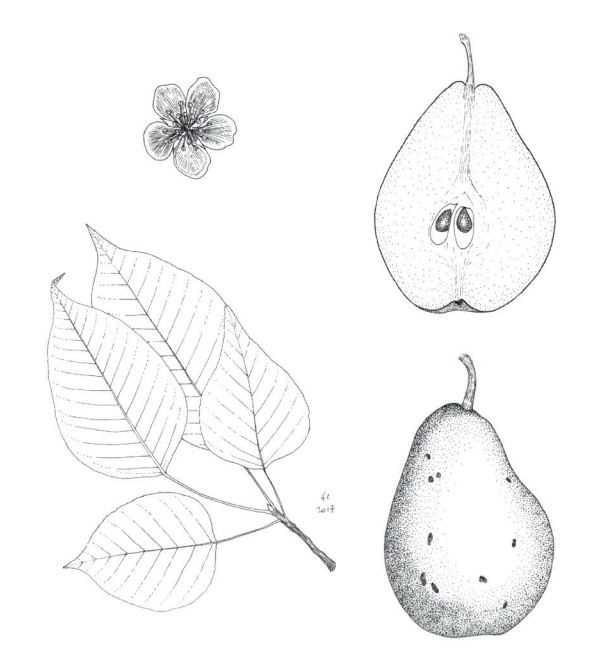

SYNONYMS *Pyrus sativa* DC.

COMMON NAMES Pear.

DESCRIPTION Tree (5–20 m), deciduous. Branches not spinescent, tomentose when young. Leaves with ovate or elliptic lamina (3–4 × 4–7 cm), briefly acuminate at apex, margin notched, young ones pubescent, then glabrous and shiny above; petiole usually smaller than lamina (3–7 cm). Flowers 2–8 in umbrella-like cymes; sepals lanceolate (3 × 7 mm), petals white, obovate-subround (10–13 mm), glabrous. Fruit pome (5–16 cm), edible (pear).

CULTIVAR Of the numerous cultivars of pear tree, the following are the most significant: Abate Fetel (Italy), Ayers (United States; *Pyrus communis* × *Pyrus pyrifolia* [Burm.f.] Nakai), Bambinella (Malta), Beth, Beurré Bosc (or Kaiser Alexander, Bosc, Kaiser; France), Beurré d'Anjou (or d'Anjou; France), Beurré Hardy/Gellerts Butterbirne, Blake's Pride (United States), Blanquilla (Spain), Butirra Precoce Morettini (Italy), Carmen, Clara Frijs (Denmark), Concorde (England), Conference (England), Corella (Australia), Coscia (Italy), Don Guindo (Spain), Doyenné du Comice (France), Dr. Jules Guyot, Forelle, Glou Morceau (Belgium), Gorham (United States), Harrow Delight (Canada), Harrow Sweet (Canada), Joséphine de Malines (Belgium), Kieffer (United States),

Laxton's Superb (England), Louise Bonne of Jersey, Luscious (United States), Merton Pride (England), Onward (England), Orient (United States; P. *communis* × *P. pyrifolia*), Packham's Triumph (Australia), Pineapple (United States; *P. communis* × *P. pyrifolia*), Red Bartlett (United States), Rocha (Portugal), Rosemarie (South Africa), Seckel (United States), Starkrimson (United States), Sudduth, Summer Beauty, Taylor's Gold (New Zealand), Triomphe de Vienne, Williams' Bon Chrétien (or Bartlett or Williams; England).

Related Species There are 76 taxa included in the genus *Pyrus*. Some, besides *Pyrus communis* of course, produce edible fruits and have potential commercial uses; *Pyrus pyraster* (L.) Burgsd. is a species for beer making.

Geographic Distribution Hybridized complex fixed by cultivation. **Habitat** Cultivated. **Range** China, United States, Italy, Argentina, Turkey.

Beer-Making Parts (and Possible Uses) Fruits, wood (for smoking malt).

Chemistry Fruits: energy 57 kcal, water 83.96 g, protein 0.36 g, lipids 0.14 g, carbohydrates 15.23 g, fiber 3.1 g, sugars 9.75 g, calcium 9 mg, iron 0.18 mg, magnesium 7 mg, phosphorus 12 mg, potassium 116 mg, sodium 1 mg, zinc 0.10 mg, vitamin C 4.3 mg, thiamin 0.012 mg, riboflavin 0.026 mg, niacin 0.161 mg, vitamin B6 0.029 mg, folate 7 µg, vitamin A 1 µg, vitamin A 25 IU, vitamin E 0.12 mg, vitamin K 4.4 µg.

Style American Pale Ale, Belgian Specialty Ale, Belgian Tripel, Brett Beer, California Common Beer, Fruit Beer, Fruit Lambic, Imperial IPA, Old Ale, Other Smoked Beer, Saison.

Source Cantwell and Bouckaert 2016; Cantwell and Bouckaert 2018; Giaccone and Signoroni 2017; Heilshorn 2017; Hieronymus 2016b; Josephson et al. 2016; https://www.bjcp.org (29. Fruit Beer); https://www.brewersfriend.com/other/pear/; https://www.brewersfriend.com/other/pears/.

Quassia amara L.
SIMAROUBACEAE

Synonyms None

Common Names Quassia, amargo, bitter ash, bitter wood.

Description Shrub or small tree (3–8 m). Stem, twigs, rachis, veins usually reddish color. Leaves alternate, compound, 5–16 cm long, imparipinnate, with 3–5 leaflets, each with pointed apex, narrow base, winged-leaf rachis. Flowers (2.5–3.5 cm) with 5 petals, bright red on outside, white on inside, which remain joined for a long time forming a spiraled rotate cylinder from which 10 stamens protrude. Inflorescence panicle (length 15–25 cm). Fruit aggregate of 5 elliptic or obovate drupes (8–15 mm), black in color, attached to a fleshy receptacle. Seed 1 per drupe.

Related Species The genus *Quassia* contains 36 species.

Geographic Distribution Costa Rica, Nicaragua, Panama, Brazil, Peru, Venezuela, Suriname, Colombia, Argentina, French Guyana, Guyana. **Habitat** Undergrowth species in rainforests and other moist habitats. **Range** Grown frequently in humid tropical areas.

Beer-Making Parts (and Possible Uses) Bark, wood (the bitter substance contained in the bark and wood is used as a base for angostura bitters and flavoring in gin drinks; it is also used as a substitute for hops in beer making).

Toxicity Although the European Commission Scientific Committee on Food has extended the authorization for the use of quassin as a food component, and although the U.S. Food and Drug Administration has granted quassia Generally Recognized as Safe (GRAS) status, several studies (conducted mostly on guinea pigs) question the safety of this product.

Chemistry *Quassia amara* contains many active substances, including alkaloids, triterpenes, and bitter principles. The latter are quassinoids (triterpenoid compounds), present in amounts of about 0.25%, of which 0.1–0.15% are quassin, neoquassin, 18-hydroxyquassin, and simalicalactone D. Other quassinoids present in the wood are isoquassin, paraine, quassimarin, quassinol, and quassol. Chemically, quassinoids are triterpene-δ-lactones found in the Simaroubaceae family and largely responsible for its biological and pharmacological properties. More than 170 quassinoids have been identified and characterized. Those of quassia are largely C-20, that is, with 20 carbon atoms (quassin, neoquassin, 18-hydroxy-quassin, isoquassin, paraine, quassinol, quassol), whereas simalicalactone D is a C-25 and quassimarin a C-27. In addition to these, other active compounds have been identified, namely indol-alkaloids of the beta-carboline family, specifically 1-vinyl-4,8-dimethoxy-beta-carbolene, 1-methoxycarbonyl-betacarboline, 3-methylcantin-2,6-dione.

Style Scotch Ale, Scottish Ale.

Source Daniels 2016; Steele 2012.

Quercus afares Pomel
FAGACEAE

Synonyms *Quercus castaneifolia* Coss. ex J.Gay.

Common Names Afares oak.

Description Tree (up to 25 m), deciduous. Trunk, twigs, and buds rough, somewhat suberose, deeply furrowed; twigs first tomentose, then glabrous; terminal buds tomentose, with persistent stipules. Leaves 7–14 × 4–6 cm, oblong-lanceolate; apex slightly attenuated, base rounded; margin with 8–12 pairs of triangular teeth with a short mucro, with rounded sinuses; upper leaf shiny green, with star-shaped hairs, lower leaf with short whitish tomentum, especially along veins; 8–15 pairs of veins; petiole hairy, 0.5–1 cm. Fruits acorns 4 cm, 1–3 (sometimes more) on a short peduncle (1 cm); seed included for 1/3 in an acorn cup that looks like that of *Quercus cerris*; matures in 2 years.

Related Species More than 600 taxa are believed to belong to the genus *Quercus*.

Geographic Distribution Algeria, Tunisia.
Habitat Coastal upland environments. **Range** Algeria, Tunisia.

Beer-Making Parts (and Possible Uses) Wood (for barrels).

Wood Heavy, with remarkable mechanical properties.

Style None known.

Source Cantwell and Bouckaert 2016; Cantwell and Bouckaert 2018.

Quercus alba L.
FAGACEAE

Synonyms *Quercus candida* Steud., *Quercus nigrescens* Raf., *Quercus ramosa* Dippel, *Quercus retusa* Raf.

Common Names White oak.

Description Tree, deciduous (up to 25 m). Trunk (Ø up to 1–1.3 m) with pale gray, flaky bark; twigs green or reddish, then gray (Ø 2–4 mm), initially pubescent, then glabrous; buds reddish-dark brown, ovoid, about 3 mm, apex obtuse, glabrous. Leaves with petiole 4–30 mm; lamina obovate to narrowly elliptic or narrowly obovate (79–230 × 40–165 mm), base narrowly cuneate to acute, margins moderately to deeply lobed, lobes often narrow, rounded distally, sinuses extending for 1/3–7/8 of the distance from midrib, secondary veins arcuate, divergent, 3–7 per side, apex broadly rounded or ovate; underleaf light green, with numerous whitish or reddish erect hairs, upper leaf light green, dull or shiny. Fruits acorns 1–3, subsessile or on peduncles to 25–50 mm; acorn cup hemispherical, covering 1/4 of nut, scales densely appressed, finely grayish-tomentose. Nut light brown in color, ovoid-ellipsoid or oblong (12–25 × 9–18 mm), glabrous; cotyledons distinct.

Subspecies The taxonomy of the genus *Quercus* and the recognition of individual species frequently present considerable complexity because entities described as different species on a morphological basis continue to interbreed freely, resulting in a number of individuals with intermediate characters. *Quercus alba*, for example, crosses with *Quercus macrocarpa*, *Quercus stellata*, and *Quercus montana*, three North American species whose ranges overlap with that of *Quercus alba*.

Related Species Of the more than 600 taxa recognized as belonging to the genus *Quercus*, it is conceivable that many more than those cited in this book have (at least potential) similar applications.

Geographic Distribution E North America (SE Canada, EC United States). **Habitat** Mesophilic forests. **Range** North America.

Beer-Making Parts (and Possible Uses) Bark, wood (for barrels and malt smoking), shavings, spirals; acorns detannized by fermentation (one year), leaves (green or dried).

Chemistry Seeds: water 47.3%, protein 3.31%, fiber 1.30%, ash 1.39%, nonnitrogenous extract 43.37%, lignin 1.39%, cellulose 1.71%, tannins 2.94%, calcium traces, magnesium 0.05%, phosphorus 0.08%. Phenolic compounds contained in *Quercus alba* wood include hail/roburin E, castalagin/vescalagin, gallic acid, monogalloyl glucose (glucogallin), valoneic acid dilactone, digalloyl glucose, trigalloyl glucose, rhamnose ellagic acid, quercitrin, ellagic acid.

Wood Differentiated, with whitish sapwood and gray-brown heartwood, hard, with strong shrinkage, good mechanical strength and very durable, easy to work. It is widely used in staves for spirit barrels, floors, and furniture.

Style American Amber Ale, American Barleywine, American Brown Ale, American IPA, American Pale Ale, American Stout, Baltic Porter, Belgian Blond Ale, Belgian Dark Strong Ale, Belgian Golden Strong Ale, Belgian Specialty Ale, Belgian Tripel, Blonde Ale, Braggot, Brown Porter, California Common Beer, Cream Ale, Extra Special/Strong Bitter (Esb), Flanders Brown Ale/Oud Bruin, Flanders Red Ale, Foreign Extra Stout, Imperial IPA, Imperial Stout, Irish Red Ale, Mixed-fermentation Sour Beer, Oatmeal Stout, Old Ale, Other Smoked Beer, Robust Porter, Russian Imperial Stout, Saison, Special/Best/Premium Bitter, Specialty Beer, Strong Scotch Ale, Wheatwine, Wood-aged Beer.

Source Calagione et al. 2018; Hieronymus 2010; Hieronymus 2016b; Josephson et al. 2016; Sparrow 2015; https://www.bjcp.org (33A. Wood-Aged Beer; 33B. Specialty Wood-Aged Beer); https://www.brewersfriend.com/other/oak/; https://www.quercusalbabeer.com/.

Wood and Beer

Before steel appeared in breweries, most containers used for brewing, particularly fermentation and maturation, were made of wood. Vats and casks—small, large, enormous—have been part of the brewer's daily toolkit for centuries. The advent of steel has radically changed the practice of brewing, yet wood has never completely disappeared, and the craft revolution has given new dignity to this material. Actually, it would be more appropriate to talk about *materials*, in the plural, considering the numerous species from which wood with very different properties can be obtained or used in various aspects in brewing. Wood is not inert at all and can contribute with many aromatic substances—for example, vanillin or molecules releasing toasted aromas—to the final characteristics of the product. If the wood is new, it will give beer intense, evident notes and therefore can certainly be considered a full-fledged beer ingredient, as it makes a contribution to flavor and aroma. This is why many artisans choose to ferment, mature, and age their beers in casks made of different woods, in particular *Quercus alba* and *Quercus petraea*, from which the term "oaked" is derived, referring to the passage of beer in oak casks. Some beers, when tasted, betray the complexity of the intense resinous notes released by coniferous wood, used for thousands of years in northern fermentations. Others are dominated by the exotic aromas of tropical scented woods, such as *Bulnesia sarmientoi*, chosen by Sam Calagione (Dogfish Head Brewery) for two specially commissioned 10,000-gallon tanks (almost 38,000 liters), the largest wooden containers built in the United States since the Prohibition era. Besides being a specific ingredient, each type of wood, thanks to its porosity, guarantees the micro-oxygenation needed for the most alcoholic and full-bodied beers, which require long periods of aging (and thus are considered wood aged) before being marketed. In lambic beers, wood plays a central role, both because spontaneous fermentation takes place in casks and because the porous walls of the casks host the microflora responsible for the aging process. The porosity of wood is incredibly functional in absorbing the aromas and flavors of what is being aged in the cask (e.g., whisky, bourbon, sherry, port, cachaça, wine, soy sauce), making it possible for such aromas and flavors to be passed on to beer when the cask takes on a second life in beer making.

Quercus bicolor Willd.
FAGACEAE

Synonyms *Quercus mollis* Raf., *Quercus paludosa* Petz. & G.Kirchn., *Quercus platanoides* (Lam.) Sudw.

Common Names Swamp white oak.

Description Tree, deciduous (up to 30 m). Trunk with dark gray bark, flaky or flat plates; twigs light to dark brown (Ø 2–4 mm), glabrous; buds light to dark brown, globose to ovoid, 2–3 mm, glabrous. Leaves with petiole 4–30 mm; lamina obovate to narrowly elliptic or narrowly obovate (79–215 × 40–160 mm), base narrowly cuneate to acute, margins regularly toothed or entire with teeth in distal 1/2 only, or moderately to deeply lobed, sometimes lobed proximally and toothed distally, secondary veins arcuate, divergent, 3–7 on each side, apex broadly rounded or ovate; underleaf light green or whitish, with minute appressed-stellate hairs, erect, 1- to 4-radiated, velvety to the touch; upper leaf dark green, shiny, glabrous. Fruits acorns 1–3 (rarely 5) mm, on soft axillary peduncles 20–70 mm; acorn cup hemispherical or turbinate (10–15 × 15–25 mm), covering 1/2–3/4 of the nut, scales tightly appressed, finely grayish-tomentose, those near the edge often with short, stiff, spiny barbs, irregularly recurved and sometimes branched, emerging from tubercles. Nut color light brown, ovoid-ellipsoid or oblong (12–25 × 9–18 mm), glabrous; cotyledons distinct.

***Quercus alba* L.**
FAGACEAE

Related Species More than 600 taxa are believed to belong to the genus *Quercus*.

Geographic Distribution SE Canada, NE United States. **Habitat** Swampy forests, moist slopes, poorly drained uplands (0–1,000 m ASL). **Range** SE Canada, NE United States.

Beer-Making Parts (and Possible Uses) Wood (for barrels).

Wood Differentiated, with white sapwood and gray-brown heartwood, hard, high shrinkage, good mechanical strength and very durable, easy to work. It is considered superior to American red oak (*Quercus rubra* L. and related taxa) and is widely used in staves for spirit barrels, floors, furniture, and furnishings.

Style None known.

Source Cantwell and Bouckaert 2016; Cantwell and Bouckaert 2018.

Quercus canariensis Willd.
FAGACEAE

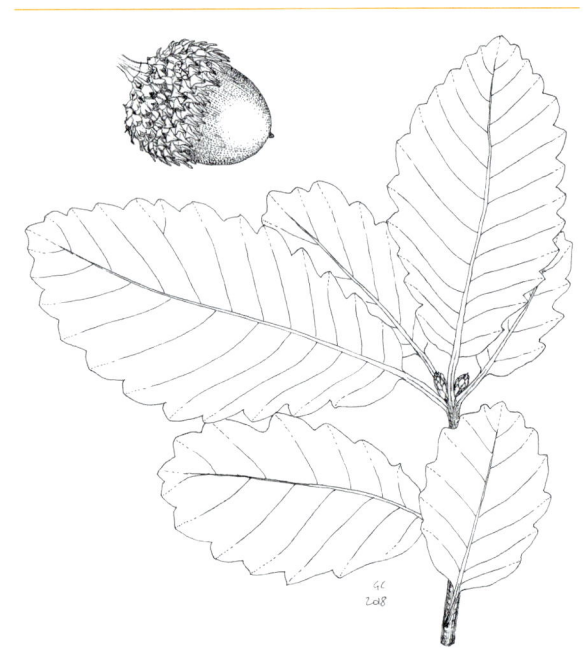

Synonyms *Quercus baetica* (Webb) Villar, *Quercus corymbifolia* Ehrenb. ex Boiss., *Quercus cypri* Kotschy ex A.DC., *Quercus esculenta* K.Koch, *Quercus gibraltarica* K.Koch, *Quercus mirbeckii* Durieu, *Quercus nordafricana* Villar, *Quercus salzmanniana* (Webb) Cout.

Common Names Mirbeck's oak, Kabylie's oak, Canarian oak, Algerian oak.

Description Tree (20–30 m). Trunk (Ø 1 m) with blackish, fissured, thick bark; twigs gray-green, first densely pubescent, then smooth and glabrous; buds narrowly conical, 7 mm, with light brown scales covered with white hairs; crown broad. Leaves (6–18 × 4–7 cm) semi-evergreen, obovate to oblong or ellipsoidal; young ones tomentose-lanate owing to long reddish hairs, free, never stellate, then glabrous except at the axil of the veins; adult leaves dull dark green in color, glabrous above, glabrous below, with hairs along the midrib; base subcordate; margins with 7–14 pairs of shallowly dentate lobes; 12–14 pairs of prominent lateral veins, diverging from the main vein less than 32°; tertiary veins conspicuous; petiole 1.5–2.5 cm, dark pink, pubescent at first, soon glabrous. Male flowers in 4–8 cm catkins; pistillate flowers on short pedicels; perianth pubescent, with 6 short lobes; stigmas 3–4. Fruits acorns, 2–3, ovoid-cylindrical (2.5–3.5 cm), briefly pedunculate (0.5–1 cm); 1/3 covered by hemispherical acorn cup with tomentose, lanceolate, protruding scales; mature in 1 year.

Subspecies Within the considerable variability of *Quercus canariensis* Willd., several varieties have been described, including *Quercus canariensis* Willd. var. *salzmanniana* (Webb) C.Vicioso (leaves 12 cm, elliptic, elongate, flat, very leathery, with slightly wavy margins), *Quercus canariensis* Willd. var. *mirbeckii* (Durieu) C.Vicioso (leaves over 20 cm, oblong, base attenuated, margins wavy with numerous lobes, petiole 4 cm), *Quercus canariensis* Willd. var. *carpinifolia* C. Vicioso (leaves 8 cm, elliptical-oboval or obtriangular, with pointed lobes), *Quercus canariensis* Willd. var. *elongate* (leaves 10 × 2–4 cm, narrow, with banded whitish hairs beneath; lobes 5 short (less than 1 cm), rounded on each side; 4–8 pairs of veins; petiole 1.2–1.5 cm, hairy; bud oboval, slightly projecting, 7 × 5 cm), *Quercus canariensis* Willd. var. *fissa* (few pointed lobes with sinuses up to halfway up the lamina), *Quercus canariensis* Willd. var. *ovata* (oval leaves 7 × 4–5 cm, petiole 4 cm), *Quercus canariensis* Willd. var. *pseudocastanea* (chestnut-like leaves), *Quercus canariensis* Willd. var. *suborbicularis* (leaves 15 × 10 cm, elliptical; few regular, shallow lobes). Currently, however, these entities have been synonymized into *Quercus canariensis* Willd.

RELATED SPECIES More than 600 taxa are believed to belong to the genus *Quercus*.

GEOGRAPHIC DISTRIBUTION ES Spain, Algeria, Morocco, Tunisia. **HABITAT** Calcareous but also heavy and clayey substrate environments (700–1,000 m ASL). **RANGE** ES Spain, Algeria, Morocco, Tunisia.

BEER-MAKING PARTS (AND POSSIBLE USES) Wood (for barrels).

WOOD No information available.

STYLE None known.

SOURCE Cantwell and Bouckaert 2016; Cantwell and Bouckaert 2018.

Quercus frainetto Ten.
FAGACEAE

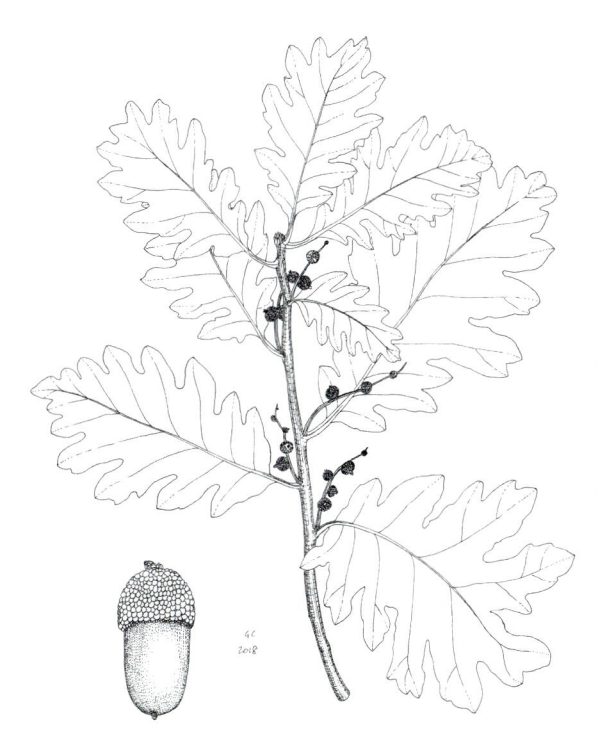

SYNONYMS
Quercus apennina Loisel., *Quercus byzantina* Borbás, *Quercus conferta* Kit., *Quercus conferta* (Kit.) Vuk., *Quercus esculiformis* O.Schwarz, *Quercus farnetto* Ten., *Quercus hungarica* Hubeny, *Quercus pannonica* Endl., *Quercus slavonica* Kit. ex Borbás, *Quercus spectabilis* Kit. ex Simonk., *Quercus strigosa* Wierzb. ex Rochel.

COMMON NAMES Hungarian oak.

DESCRIPTION Tree (up to 25–30 m). Trunk straight (Ø up to 2 m); bark furrowed, light gray; twigs light brown-green, slightly pubescent; buds ovoid, 1 cm, reddish at first, then light gray-brown; crown rounded. Leaves (10–25 × 8–15 cm) crowded at ends of twigs, obovate, ± oblong; base heart shaped to auriculate, apex rounded (often with 3 small lobes), lamina deeply lobed, with 7–10 lobes on each side, bright dark green, slightly rough underneath; color grayish tomentose, soft underneath; petiole hairy, 0.2–1.2 cm. Male flowers on catkins (5–8 cm), golden-brown color. Fruits acorns, 2 cm, covered for 1/3–1/2 by the acorn cup on a short stalk, covered with broad, pubescent, appressed scales; ripen in 1 year.

RELATED SPECIES More than 600 taxa are believed to belong to the genus *Quercus*.

GEOGRAPHIC DISTRIBUTION S Italy, Hungary, Balkans, S Black Sea. **HABITAT** Mesophilic woods. **RANGE** S Italy, Hungary, Balkans, S Black Sea.

BEER-MAKING PARTS (AND POSSIBLE USES) Wood (for barrels).

WOOD Differentiated, with large sapwood, it has no characteristics that distinguish it significantly from *Quercus petraea*; it is used for beams, staves, carpentry boards, and agricultural tools.

STYLE None known.

SOURCE Cantwell and Bouckaert 2016; Cantwell and Bouckaert 2018.

Quercus garryana Douglas ex Hook.
FAGACEAE

Synonyms *Quercus jacobi* R.Br.ter, *Quercus neaei* Liebm., *Quercus patula* Hansen.

Common Names Oregon white oak, Garry oak.

Description Tree (up to 15–20 m) or shrub (0.1–3 m), deciduous. Trunk solitary (tree) or multicaule (shrub); bark light gray or nearly white, scaly; twigs brown, red, or yellowish (Ø 2–4 mm), densely puberulent or glabrous; buds brown or yellowish, ovoid or fusiform, acute apex (2–12 mm), glandular-puberulent or densely pubescent. Leaves with petiole 4–10 mm; lamina obovate, elliptic, or subrounded, moderately to deeply lobed (25–140 × 15–85 mm), base rounded-attenuated or cuneate, rarely subcordate, often unequal, margins with sinuses generally reaching more than 1/2 the distance from the midrib, lobes oblong or spatulate, obtuse, rounded or blunt, major lobes usually with 2–3 sublobes or teeth, veins often ending in retuse teeth, secondary veins yellowish, 4–7 per side, most distal vein often branched within distal lobes, apex broadly rounded; underleaf light green or waxy yellowish, often felted to the touch, densely to sparsely covered with simple hairs (0.1–1 mm), fasciculate, semi-erect or erect, 2- to 8-radiated, secondary veins protruding; upper leaf bright green or dark, shiny or somewhat hairy owing to scattered stellate hairs. Fruits acorns, 1–3, subsessile, rarely on peduncles to 10–20 mm; cupule cup shaped or hemispherical (4–10 × 12–22 mm); scales yellowish or dark reddish, often long and acute near cupule edge, moderately or sparsely tuberculate, canescent or tomentulose. Nut color light brown, oblong to globular (12–40 × 10–22 mm), apex blunt or rounded, glabrous or often persistently puberulent; cotyledons distinct.

Subspecies This entity includes three varieties: *Quercus garryana* Douglas ex Hook. subsp. *garryana*, *Quercus garryana* Douglas ex Hook. var. *fruticosa* (Engelm) Govaerts, and *Quercus garryana* Douglas ex Hook. var. *semota* Jeps.

Related Species More than 600 taxa are believed to belong to the genus *Quercus*.

Geographic Distribution W Canada (British Columbia), W United States (California, Oregon, Washington). **Habitat** Oak woodlands, edges of *Sequoia sempervirens* forests, mixed evergreen forests (0–800 m ASL). **Range** W Canada (British Columbia), W United States (California, Oregon, Washington).

Beer-Making Parts (and Possible Uses) Wood (for barrels).

Wood Differentiated, with white sapwood and brown-gray heartwood, hard, high shrinkage, good mechanical strength and very durable, easy to work. It is judged superior to American red oak (*Quercus rubra* L. and related taxa) and is widely used in staves for spirit barrels, flooring, furnishings, and furniture.

Style None known.

Source Cantwell and Bouckaert 2016; Cantwell and Bouckaert 2018.

Quercus lyrata Walter
FAGACEAE

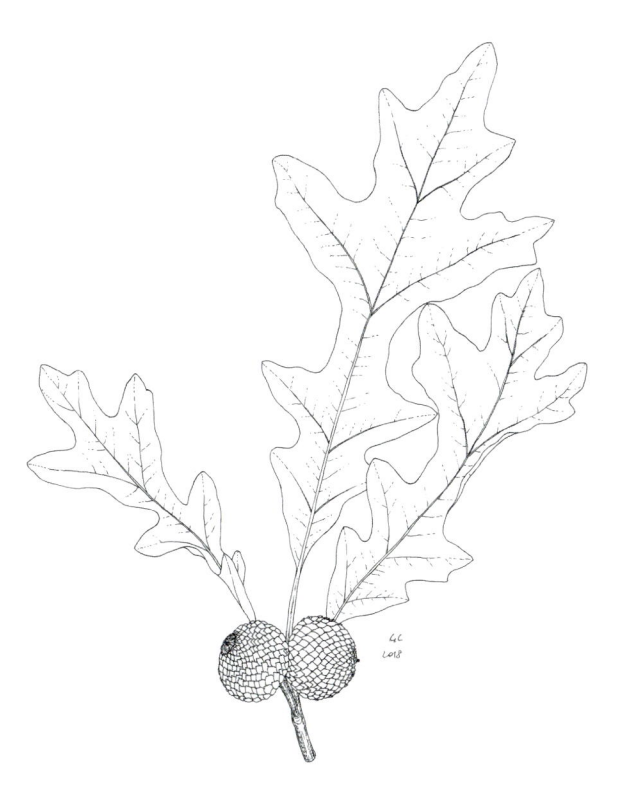

Synonyms *Scolodrys lyrata* (Walter) Raf.

Common Names Overcup oak.

Description Tree, deciduous (up to 20 m). Trunk with pale gray bark tinged with red, with thick plates below scales; twigs (Ø 2–4 mm) grayish or reddish, villous, soon glabrous; buds 3 mm, gray with pubescence. Leaves with villous petioles (8–25 mm); lamina obovate or broadly obovate (100–200 × 50–120 mm), base narrowly cuneate to acute, margins moderately to deeply lobed, lobes somewhat distinctly angular or spatulate, often with 2–3 teeth, sinuses close to midrib, secondary veins arcuate, divergent, 3–7 on each side, apex broadly rounded or ovate; underleaf light green or glaucous, tomentose, persistent or early caducous tomentum, upper leaf dark green or dull gray, sparsely puberulent to glabrous. Fruits acorns, 1–2, axillary peduncles to 40 mm; cupule cup shaped, curly, or spheroidal (15–20 × 20–30 m), usually completely covering the nut or only the visible apex, rarely covering only 1/2 the nut, orifice smaller than nut diameter, often irregularly opening at maturity, scales tightly appressed, especially at margin, laterally connate, broadly triangular, convex-tuberculate, finely grayish-tomentose. Nut light brown or grayish, ovoid-ellipsoid or oblong (15–50 × 10–40 mm), finely puberulent or woolly; cotyledons distinct.

Related Species More than 600 taxa are believed to belong to the genus *Quercus*.

Geographic Distribution SE United States. **Habitat** Lowlands, humid forests, riparian forests, swampy forests, periodically flooded areas (0–200 m ASL). **Range** SE United States.

Beer-Making Parts (and Possible Uses) Wood (for barrels).

Wood Differentiated, with white sapwood and brown-gray heartwood, hard, high shrinkage, good mechanical strength and very durable, easy to work. It is considered superior to American red oak (*Quercus rubra* L. and related taxa) and is widely used in staves for spirit barrels, floors, furniture, and furnishings.

Style None known.

Source Cantwell and Bouckaert 2016; Cantwell and Bouckaert 2018.

Quercus macranthera Fisch. & C.A.Mey. ex Hohen.
FAGACEAE

Synonyms None

Common Names Caucasus oak, Persian oak.

Description Tree, deciduous (up to 25–30 m). Trunk with purple-gray, plaque-like, thick bark; twigs orange-brown, tomentose, stiff, then glabrous; buds 1–1.5 cm, dark red-brown, hairy at end. Leaves (10–20 × 7–13 cm) obovate-lanceolate to ovate; apex pointed, base cuneate; tough, leathery; margin with 8–12 pairs of shallow lobes; minor lobes near apex; lamina gray-green above, gray-tomentose below; petiole pubescent, 1–2 cm. Male flowers in catkins (5–8 cm), pubescent. Fruits acorns, 2.5 cm, ovoid, ± sessile, 1/2 covered by the cupule, with small lanceolate, hairy scales; ripen in 1 year.

Subspecies In addition to the typical subspecies, *Quercus macranthera* Fisch. & C.A.Mey. ex Hohen. subsp. *macranthera*, *Quercus macranthera* Fisch. & C.A.Mey. ex Hohen. subsp. *syspirensis* (K.Koch) Menitsky is recognized.

Related Species More than 600 taxa are believed to belong to the genus *Quercus*.

Geographic Distribution Caucasus, Turkey, N Iran, S Caspian Sea, Georgia, Armenia. **Habitat** Arid areas; soil indifferent. **Range** Caucasus, N Iran, S Caspian Sea, Armenia.

Beer-Making Parts (and Possible Uses) Wood (for barrels).

Wood No information available.

Style None known.

Source Cantwell and Bouckaert 2016; Cantwell and Bouckaert 2018.

Quercus macrocarpa Michx.
FAGACEAE

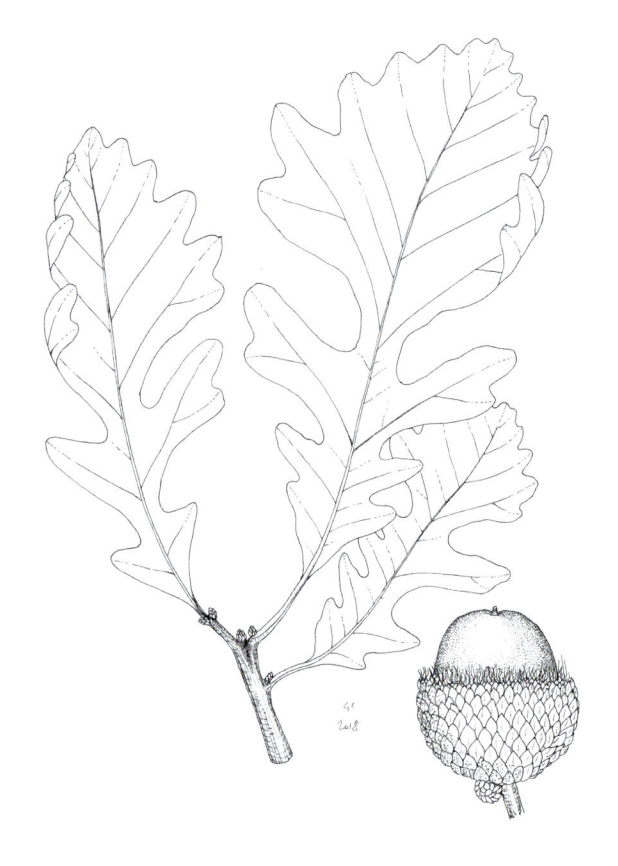

Synonyms *Cerris macrocarpa* (Michx.) Raf., *Cerris oliviformis* (F.Michx.) Raf., *Quercus oliviformis* F.Michx.

Common Names Bur oak, burr oak, mossy-cup oak.

Description Tree, deciduous (up to 50 m). Trunk with dark gray, scaly or flat bark; twigs (Ø 2–4 mm) grayish or reddish, often forming broad, flat, radiating, suberose, finely pubescent wings; buds 2–6 mm, glabrous. Leaves with petiole 6–30 mm; lamina obovate to narrowly elliptic or narrowly obovate, often fiddle shaped, 50–310 × 40–160 mm, base rounded to cuneate, margins moderately to deeply lobed, toothed, sinuses deeper at midleaf (at least in the proximal 2/3), sinuses nearly reaching midrib, longer lobes becoming less deep or simple teeth distally; proximally compound lobes, less deep, secondary veins arcuate, diverging, 4–10 on each side, apex broadly rounded or ovate; underside light green or whitish, with minute appressed stellate hairs forming a dense, rarely scattered, tomentum; upper leaf dark green or dull gray, sparsely puberulent to glabrous. Fruits acorns, 1–3 on a stiff peduncle (0–25 mm); acorn cup hemispherical or turbinate (8–50 × 10–60 mm), covering 1/2–7/8 of the nut or more, scales appressed, laterally connate, broadly triangular, keeled, tuberculate, color finely grayish, tomentose, those near the margin often with soft barbs 5–10 mm or more, forming a fringe around the nut; nut light brown or grayish color, ovoid-ellipsoid or oblong (15–50 × 10–40 mm), finely puberulent or woolly. Seeds with distinct cotyledons.

SUBSPECIES In addition to the typical variety, the variety *Quercus macrocarpa* Michx. var. *depressa* (Nutt.) Engelm is recognized.

RELATED SPECIES More than 600 taxa are believed to belong to the genus *Quercus*.

GEOGRAPHIC DISTRIBUTION C Canada, EC United States. **HABITAT** Depressions, riparian escarpments, poorly drained areas, grasslands, generally on limestone or calcareous clays, NW of range also on dry slopes, mountain ranges, dry grasslands (0–1,000 m ASL). **RANGE** C Canada, EC United States.

BEER-MAKING PARTS (AND POSSIBLE USES) Wood (for barrels); fruits (to activate fermentation).

CHEMISTRY Fruits: lipids 4.8%, protein 4.3%, carbohydrates 45.9%, phosphorus 0.1%, calcium 0.08%, magnesium 0.06%.

WOOD Differentiated, with white sapwood and brown-gray heartwood, hard, with strong shrinkage, good mechanical strength and very durable, easy to work. It is judged superior to American red oak (*Quercus rubra* L. and related taxa) and is widely used in staves for spirit barrels, flooring, furnishings, and furniture.

STYLE None known.

SOURCE Cantwell and Bouckaert 2016; Cantwell and Bouckaert 2018.

Quercus michauxii Nutt.
FAGACEAE

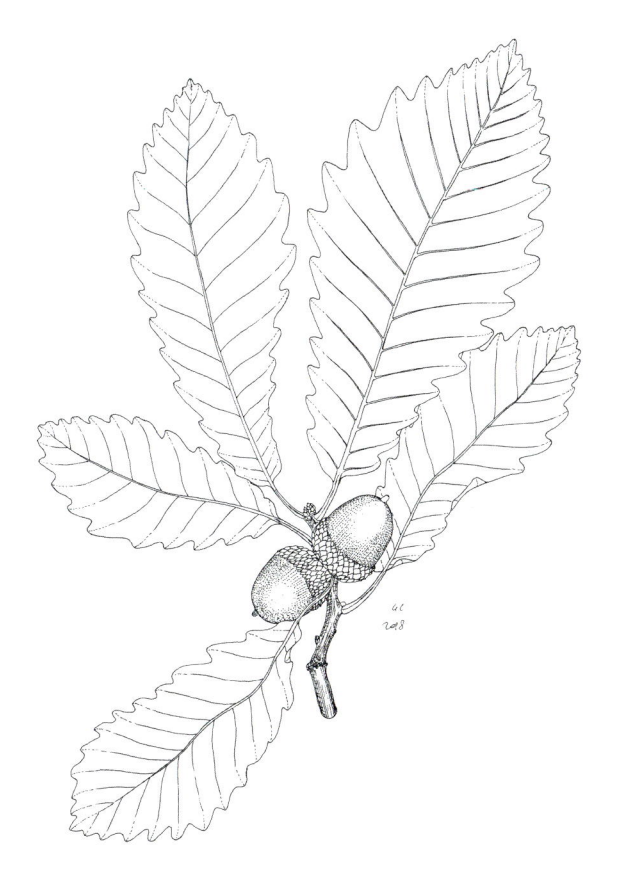

SYNONYMS *Quercus houstoniana* C.H.Mull., *Quercus prinus* L.

COMMON NAMES Basket oak, cow oak, swamp chestnut oak.

DESCRIPTION Tree, deciduous (up to 20 m). Trunk with light brown or gray, flaky bark; twigs brown or reddish-brown (Ø 2–3 mm), with scattered hairs or glabrous; buds reddish-brown, ovoid, apex rounded or acute, glabrous or minutely puberulent. Leaves with petiole 5–20 mm; lamina broadly obovate or elliptic (60–280 × 50–180 mm), base rounded-acuminate or broadly cuneate, margins regularly toothed, teeth rounded, toothed or acuminate, secondary veins 15–20 per side, parallel, straight or somewhat curved, apex broadly rounded or acuminate; underside light green or yellowish, felted to the touch owing to conspicuous or minute hairs, erect, 1- to 4-rayed, upper leaf glabrous or with minute, simple, fasciculated hairs. Fruit acorns, 1–3, subsessile or more often on axillary peduncles to 20–30 mm; acorn cup hemispherical or briefly cylindrical (15–25 × 25–40 mm), covering 1/2 of the nut or more, scales loosely appressed, distinct at base, gray or light brown, moderately to strongly tuberculate, tips silky-tomentose. Nut light brown in color, ovoid or cylindrical (25–35 × 20–25 mm), glabrous; cotyledons distinct.

Related Species More than 600 taxa are believed to belong to the genus *Quercus*.

Geographic Distribution SE United States. **Habitat** Depressions, sandy and swampy forests, in a variety of soils (0–600 m ASL). **Range** SE United States.

Beer-Making Parts (and Possible Uses) Wood (for barrels).

Wood Differentiated, with white sapwood and brown-gray heartwood, hard, high shrinkage, good mechanical strength and very durable, easy to work. It is considered superior to American red oak (*Quercus rubra* L. and related taxa) and is widely used in staves for spirit barrels, floors, furniture, and furnishings.

Style None known.

Source Cantwell and Bouckaert 2016; Cantwell and Bouckaert 2018.

Quercus mongolica Fisch. ex Ledeb.
FAGACEAE

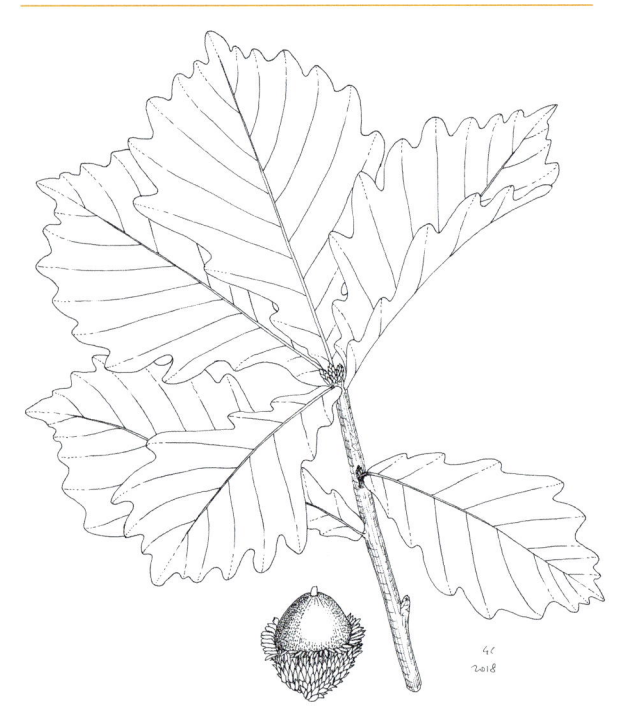

Synonyms *Quercus kirinensis* Nakai.

Common Names Mongolian oak.

Description Tree (20–35 m). Canopy ovoid, sometimes flattened; bark grayish, deeply furrowed; twigs glabrous, angled, brown, lenticellate. Leaves (10–30 × 4–11 cm) sometimes decaying, obovate, leathery, margin toothed (7–10 broad teeth per side); apex pointed, base rounded or auriculate; upper leaf glabrous, lower with simple hairs along veins; leaves crowded at distal end of twig; 10–18 pairs of hairy veins, tertiary veins evident; petiole short (2–8 mm). Flowers with perianth 6-lobed; rachis of male inflorescence glabrous. Fruits acorns (1.4–2.2 × 1.1–1.6 cm), ovoid, yellowish, hairless except apex, 1/3–1/2 covered by the cupule; cupule scales glabrous, gibbous; mature in 1 year.

Subspecies Along with the typical subspecies, *Quercus mongolica* Fisch. ex Ledeb. subsp. *mongolica*, there is *Quercus mongolica* Fisch. ex Ledeb. subsp. *crispula* (Blume) Menitsky.

Related Species More than 600 taxa are believed to belong to the genus *Quercus*.

Geographic Distribution China, Japan, Korea, Mongolia, E Russia, Sakhalin Islands. **Habitat** Acidophilous plant (200–2,500 m ASL). **Range** China, Japan, Korea, Mongolia, E-Russia, Sakhalin Islands, Europe.

Beer-Making Parts (and Possible Uses) Wood (for barrels).

Wood Has characteristics and uses similar to those of high-quality wood from European deciduous oaks (see *Quercus robur*).

Style None known.

Source Cantwell and Bouckaert 2016; Cantwell and Bouckaert 2018.

Quercus montana Willd.
FAGACEAE

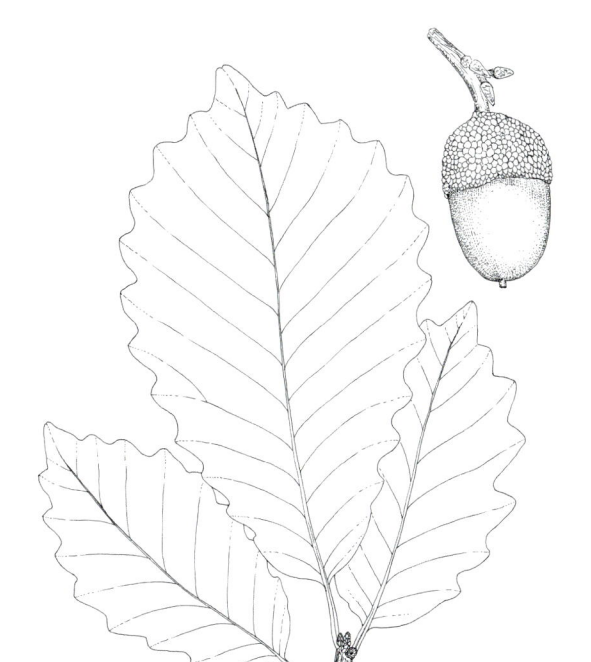

Synonyms *Quercus carolineana* Dippel, *Quercus carpinifolia* Raf., *Quercus granulata* Raf., *Quercus longifolia* Raf., *Quercus monticola* Petz. & G.Kirchn., *Quercus versicolor* Raf.

Common Names Chestnut oak, mountain chestnut oak, rock chestnut oak.

Description Tree, deciduous (up to 30 m). Trunk with dark gray or brown bark, hard, with deep V-shaped furrows; twigs (Ø 2–4 mm) light brown, glabrous; buds light brown, ovoid (3–6 mm), occasionally acute apex, glabrous. Leaves with petiole 3–30 mm; lamina obovate to narrowly elliptic or narrowly obovate (100–220 × 60–120 mm), base subacute or rounded-acuminate, often unequal, margins regularly toothed, teeth rounded or rarely rather acute, secondary veins ± parallel, straight or moderately curved, 10–16 on each side, apex broadly acuminate; underleaf light green, apparently glabrous but with scattered minute asymmetrical hairs and larger, erect, simple or fasciculate hairs, usually visible along the veins; upper leaf dark green, shiny, glabrous or with scattered minute simple hairs. Fruits acorns, 1–3, subsessile or on a peduncle (8–25 mm); cupule shallowly cupped to hemispherical or deeply calyciform, edge thin, often flared and wavy, helmet shaped (9–15 × 18–25 mm), scales often in concentric or transverse rows, laterally connate, gray, broadly ovate, tips reddish, glabrous. Nut color light brown, ovoid-ellipsoid (15–30 × 10–25 mm), glabrous; cotyledons distinct.

Related Species More than 600 taxa are believed to belong to the genus *Quercus*.

Geographic Distribution EC United States. **Habitat** Rocky plateau forests, dry ridges, mixed deciduous forests in shallow soils (0–1,400 m ASL). **Range** EC United States.

Beer-Making Parts (and Possible Uses) Wood (for barrels).

Wood Differentiated, with white sapwood and brown-gray heartwood, hard, with strong shrinkage, good mechanical strength and very durable, easy to work. It is judged superior to American red oak (*Quercus rubra* L. and related taxa) and is widely used in staves for spirit barrels, floors, and furniture.

Style None known.

Source Cantwell and Bouckaert 2016; Cantwell and Bouckaert 2018 (this taxon has mistakenly been named *Quercus prinus* by these authors).

Quercus muehlenbergii Engelm.
FAGACEAE

Synonyms *Quercus alexanderi* Britton, *Quercus brayi* Small, *Quercus sentenelensis* C.H.Mull.

Common Names Chinkapin oak, chinquapin oak, yellow chestnut oak.

Description Tree, deciduous (up to 30 m), occasionally large shrub (about 3 m) in arid areas. Trunk with gray, thin, papery bark; twigs brownish (Ø 1.5–4 mm), sparsely finely pubescent, soon glabrous, graying by second year; buds brown to red-brown in color, subrounded to broadly ovoid (20–40 × 10–25 mm), rounded apex, sparsely pubescent. Leaves with petiole 7–37 mm, lamina usually obovate, sometimes lanceolate to oblanceolate (32–210 × 10–106 mm), leathery, base truncate to cuneate, margins regularly wavy, toothed or slightly lobed, teeth or lobes rounded or acute-acuminate, secondary veins usually 9–16 on each side, ± parallel, apex briefly acute to acuminate or apiculate; underside glabrous or light green in color, glabrous in appearance but with minute 6- to 10-ray star hairs, scattered or dense, appressed, symmetrical; upper leaf dark green, shiny, glabrous. Fruits acorns, 1–2, subsessile or on an 8 mm axillary peduncle; cupule hemispherical or slightly cup shaped (4–12 × 8–22 mm), covering 1/4–1/2 of the nut, base rounded, margin usually thin, scales closely appressed, moderately to prominently tuberculate, uniform gray to short pubescence. Nut light brown in color, oblong to ovoid (13–28 × 10–16 mm), with distinct cotyledons.

Related Species More than 600 taxa are believed to belong to the genus *Quercus*.

Geographic Distribution Canada (Ontario), EC United States, Mexico (Coahuila, Nuevo León, Hidalgo, Tamaulipas). **Habitat** Mixed deciduous forests, scrub and woodland, sometimes restricted to N slopes and riparian habitats in the W part of the geographic distribution; in calcareous soils, rarely in other substrates (0–2,300 m ASL). **Range** Canada (Ontario), EC United States, Mexico (Coahuila, Nuevo León, Hidalgo, Tamaulipas).

Beer-Making Parts (and Possible Uses) Wood (for barrels).

Wood Differentiated, with white sapwood and gray-brown heartwood, hard, high shrinkage, good mechanical strength and very durable, easy to work. It is judged superior to American red oak (*Quercus rubra* L. and related taxa) and is widely used in staves for spirit barrels, floors, and furniture.

Style None known.

Source Cantwell and Bouckaert 2016; Cantwell and Bouckaert 2018.

Quercus petraea (Matt.) Liebl.
FAGACEAE

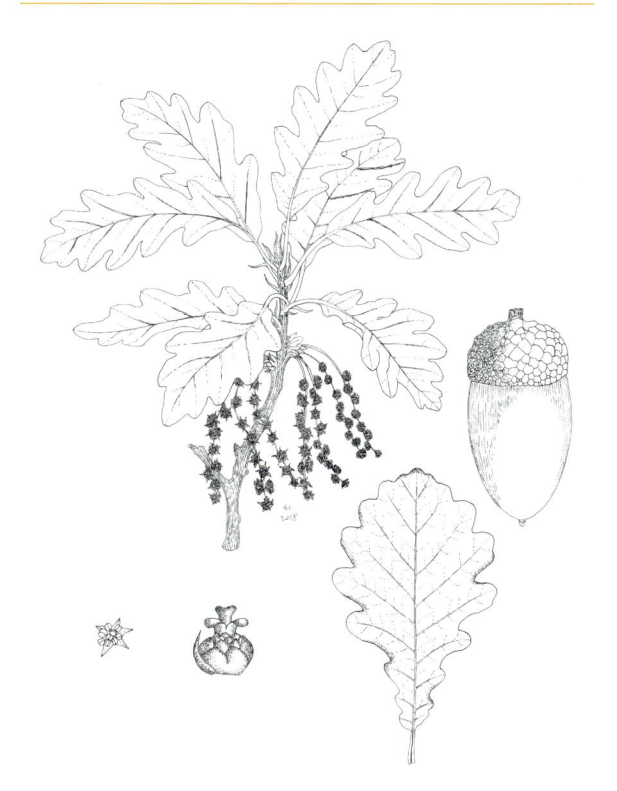

Synonyms *Quercus brevipedunculata* Cariot & St.Lag, *Quercus calcarea* Troitsky, *Quercus columbaria* Vuk., *Quercus coriacea* Bechst., *Quercus coronensis* Schur, *Quercus decipiens* Behlen, *Quercus dispar* Raf., *Quercus durinus* Raf., *Quercus erythroneura* Vuk., *Quercus esculus* L., *Quercus longipetiolata* Schur, *Quercus mas* Thore, *Quercus mespilifolia* Wallr., *Quercus peraffinis* Gand., *Quercus petiolata* Schur, *Quercus regalis* Burnett ex Endl., *Quercus sessiliflora* Salisb., *Quercus sessilis* Ehrh. ex Schur, *Quercus spathulifolia* Vuk., *Quercus sphaerocarpa* Vuk., *Quercus sublobata* Kit. Regarding *Quercus sessiliflora* Salisb., Sparrow (2015) distinguishes *Quercus petraea* taxonomically, biogeographically, and technologically from *Quercus sessiliflora*. They are actually the same taxonomic entity but can take on different wood technological characteristics under different growth conditions.

Common Names Sessile oak, durmast oak.

Description Tree (up to 30 m). Bark gray-brown, on 2- to 5-year-old branches reddish-brown, shiny, compact, with 1 mm transverse lenticels; year branches glabrous. Leaves glabrous, lamina obovate (4–7 × 8–11 cm), base acute, apex rounded, 5–7 shallow lateral lobes, usually rounded and wider than long; petiole 10–18 mm. Fruits sessile acorns, 1–3 at leaf axil, oval or subspherical (Ø 1–2 cm), with cupule about 1/2 as covering, formed by lanceolate imbricate scales.

Subspecies In addition to the nominal subspecies, *Quercus petraea* (Matt.) Liebl. subsp. *petraea*, numerous taxa of varying subspecific rank have been described because of the great variability in this group. Those currently accepted are *Quercus petraea* (Matt.) Liebl. subsp. *huguetiana* Franco & G.López, *Quercus petraea* (Matt.) Liebl. subsp. *iberica* (Steven ex M.Bieb.) Krassiln., and *Quercus petraea* (Matt.) Liebl. subsp. *pinnatiloba* (K.Koch) Menitsky. For southern Calabria, the entity *Quercus petraea* (Matt.) Liebl. subsp. *austrotyrrhenica* Brullo, Guarino, and Siracusa has been described, at the moment placed in synonymy with the nominal subspecies.

Related Species Of the more than 600 taxa recognized as belonging to the genus *Quercus*, it is conceivable that even more than those cited in this book have (at least potential) similar applications.

Geographic Distribution European-Sub-Atlantic. **Habitat** Acid soil forests. **Range** Europe.

Beer-Making Parts (and Possible Uses) Bark, wood (for barrels and malt smoking), shavings, chips, spirals, leaves (gruit).

Chemistry Wood: ellagitannins (roburin a 5.4%, roburin b 8.1%, roburin c 5.3%, grandinin 11.3%, roburin d 10.2%, vescalagin 17.4%, roburin e 9.3%, castalagin 33%), fatty acids (acetic acid, propanoic acid, 2-methylpropanoic acid, butanoic acid, 2- and 3-methylbutanoic acid, pentanoic acid, hexanoic acid, octanoic acid, nonanoic acid, decanoic acid, dodecanoic acid, tetradecanoic acid, pentadecanoic acid, hexadecanoic acid, linoleic acid, linolenic acid, stearic acid), furans (furfural, 2-furanmethanol, 5-methylfurfural, 2-furanoic acid, 5-hydroxymethylfurfural), simple phenols (benzaldehyde, benzophenone, phenol, O-methyl eugenol, eugenol, gaiacol, vanillin, vanilloyl methyl acetone, propiovanillone, 2-[4´-gaiacil]-ethanol, 3-[4´-gaiacil]-propanol, syringaldehyde, acetosyringone, α-hydroxypropiosyringone, coniferaldehyde), lactones (2[5H]-furanone, trans-β-methyl-γ-octalactone, cis-β-methyl-γ-octalactone, γ-decalactone, γ-butyrolactone, 3-methyl-2[5H]-furanone, δ-decalactone, mevalonic lactone [tetrahydro-4-hydroxy-4-methyl-2H-pyran-2one]), alcohols (phenylethanol, 2,3-butanediol, propanediol, decanol, glycerol, [E]-farnesol), norisoprenoids (3-oxo-α-ionol, 4-oxo-β-ionol, 4,5-dihydrovomipholiol), miscellaneous compounds (2,3-octanedione, benzoic acid, 2-nonenal, [E,E]-nona-2,4-dienal, 2,4-decadienal).

Wood Differentiated, heartwood WSG 0.70–0.80. Excellent for heavy constructions, maritime works, railway sleepers, staves, strips for floors, etc. It also provides excellent firewood and good coal.

Style American Amber Ale, American Barleywine, American Brown Ale, American IPA, American Pale Ale, American Stout, Baltic Porter, Belgian Blond Ale, Belgian Dark Strong Ale, Belgian Golden Strong Ale, Belgian Specialty Ale, Belgian Tripel, Blonde Ale, Braggot, Brown Porter, California Common Beer, Cream Ale, Extra Special/Strong Bitter (ESB), Flanders Brown Ale/Oud Bruin, Flanders Red Ale, Foreign Extra Stout, Gruit, Imperial IPA, Imperial stout, irish red ale, Mixed-fermentation Sour Beer, Oatmeal Stout, Old Ale, Other Smoked Beer, Robust Porter, Russian Imperial Stout, Saison, Special/Best/Premium Bitter, Specialty Beer, Strong Scotch Ale, Wheatwine, Wood-aged Beer.

Source Calagione et al. 2018; Cantwell and Bouckaert 2016; Cantwell and Bouckaert 2018; Dabove 2015; Jackson 1991; Sparrow 2015; Steele 2012; https://www.bjcp.org (33A. Wood-Aged Beer; 33B. Specialty Wood-Aged Beer); https://www.birradegliamici.com/news-birraie/birre-strane/; https://www.brewersfriend.com/other/oak/; https://www.ricettariomedievale.it/2014/02/da-httpricetta-riomedievale.html.

Quercus petraea L. subsp. *iberica* (Steven ex M.Bieb.) Krassiln.
FAGACEAE

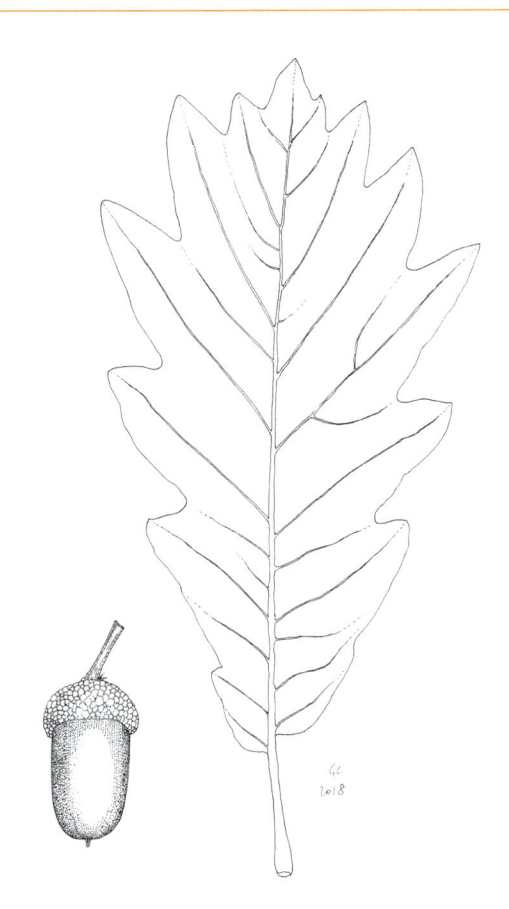

Synonyms *Quercus colchica* Czeczott, *Quercus dshorochensis* K.Koch, *Quercus hypochrysa* Steven, *Quercus iberica* Steven ex M.Bieb., *Quercus kochiana* O. Schwarz, *Quercus kozlowskyi* Woronow ex Grossh., *Quercus lamprophyllos* K.Koch, *Quercus polycarpa* Schur, *Quercus rhodopaea* Velen., *Quercus sorocarpa* Woronow ex Maleev, *Quercus szowitzii* Wenz.

Common Names Georgian oak.

Description Tree or sapling (quite variable size). Bark whitish-gray, grooved in long plates; twigs pubescent, then glabrous; buds large, pointed, brown in color. Leaves (8–10 × 4–4.5 cm) rather leathery, obovate, base rounded or slightly cordate; upper leaf glabrous, dark green, underleaf briefly and sparsely hairy, with long light brown hairs along midrib; 6–10 pairs of shallow, rounded lobes; some sinus veins near base or absent; few secondary veins, not parallel; petiole tomentose at beginning. Male flowers in long catkins; female flowers solitary on short inflorescences. Fruits acorns, 3–4, 1.5–2 cm, peduncle short (to 5 mm), 1/2 covered by the cupule, this one 1–1.5 cm in diameter, thick, with closely appressed scales. This taxon is often confused with *Quercus petraea*, from which it differs in having a pedunculated acorn and larger, more leathery, pubescent leaves below, with more small lobes (not reaching 1/3 of a half width of lamina).

Related Species More than 600 taxa are believed to belong to the genus *Quercus*.

Geographic Distribution Transcaucasia, Balkans, NW Turkey, Crimea, N Iran. **Habitat** Sunny places; edaphoindifferent species (up to 1,800 m ASL). **Range** Transcaucasia, Balkans, NW Turkey, Crimea, N Iran.

Beer-Making Parts (and Possible Uses) Wood (for barrels).

Wood It has characteristics very similar to those of *Quercus petraea*.

Style None known.

Source Cantwell and Bouckaert 2016; Cantwell and Bouckaert 2018.

Quercus pubescens Willd.
FAGACEAE

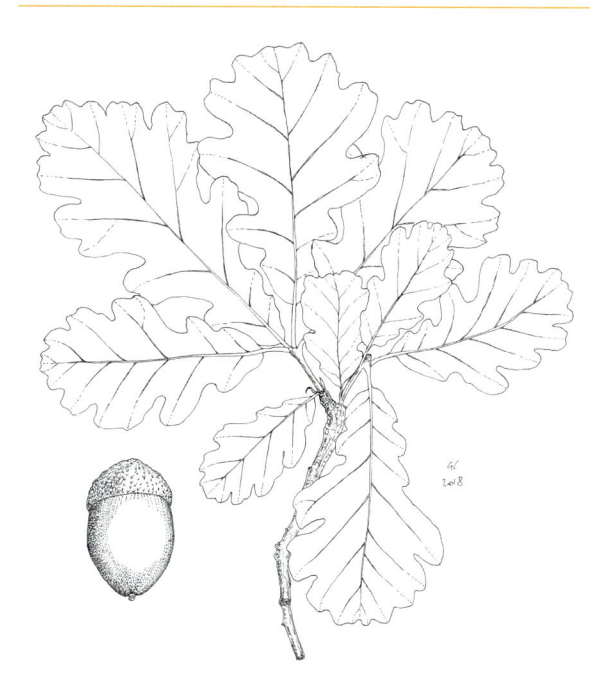

Synonyms *Eriodrys lanata* Raf., *Quercus aegilops* Mill., *Quercus amplifolia* Guss., *Quercus appenina* Lam., *Quercus aspera* Bosc, *Quercus asperata* Pers., *Quercus bacunensis* Vuk., *Quercus banja* Endl, *Quercus bellojocensis* Ghent., *Quercus brachyloba* Jord., *Quercus brachyphylla* Kotschy, *Quercus brachyphylloides* Vuk., Quercus *brandisii* (Vuk.) Vuk., *Quercus brevifolia* Kotschy ex A.DC., *Quercus buccarana* Vuk, *Quercus budayana* Haberle ex Heuff., *Quercus budensis* (Borbás) Borbás, *Quercus collina* Schleich. ex Endl., *Quercus conglomerata* Pers., *Quercus croatica* Vuk., *Quercus cupaniana* Guss, *Quercus dalmatica* Radic, *Quercus diversifrons* Borbás, *Quercus erythrolepis* (Vuk.) Vuk., *Quercus humilis* Mill., *Quercus ilicifolia* Koord. & Valeton ex Seemen, *Quercus lacinifolia* Vuk., *Quercus laciniosa* Boreau, *Quercus lanuginosa* (Lam) Thuill. (illegitimate), *Quercus macrostipulata* Guss. ex Parl., *Quercus menesiensis* Kit., *Quercus microbalanos* Boreau, *Quercus microlepis* Vuk., *Quercus oxycarpa* Raddi, *Quercus pinnatifida* C.C.Gmel., *Quercus pseudoaegilopsis* Petz. & G.Kirchn., *Quercus pusilla* Vuk., *Quercus rufa* Vuk., *Quercus schulzei* (Vuk.) Vuk., *Quercus sectifolia* Vuk., *Quercus stenolepis* Vuk., *Quercus* × *subspicata* (A.Camus) C.Vicioso, *Quercus sulcata* Vuk., *Quercus susedana* Vuk., *Quercus tenorei* (A.DC.) Borzí, *Quercus torulosa* (Vuk.) Raddi, *Quercus virgiliana* (Ten.) Ten., *Quercus vukotinocicii* Borbás.

Common Names Downy oak, pubescent oak.

Description Tree (15–20 m), deciduous. Trunk (Ø up to 1 m) short, twisted; crown open, rounded; bark dark gray, furrowed in rough square plates; twigs grayish with short hairs, then glossy brown; buds ovoid, pointed, pubescent, gray-brown at tip, 3 mm. Leaves (4–13 × 3.5–5 cm) with rounded or slightly crenate apex; base cordate or truncate, narrow; margin with 4–8 pairs of rounded lobes and deep sinuses; upper leaf light green, slightly hairy (trichomes uniseriate, solitary, fasciculate); underleaf gray-green, pubescent with fasciculate hairs; young leaves pubescent on both faces; secondary veins, 5–8 pairs, prominent below; tertiary veins evident; petiole (0.7–1.2 cm) hairy. Flowers 3–5, styluses with narrow stigmas. Fruits acorns, 1–3 cm, sessile or on short tomentose peduncle, 1/3 covered by the cupule; basal scar slightly convex, stylopodium pubescent; cupule with appressed scales, gray in color, pubescent; mature in 1 year.

SUBSPECIES *Quercus pubescens* Willd. constitutes a cycle of entities whose persistent tendency to hybridization with related taxa makes the classification particularly difficult. According to some authors, in this group the following typical subspecies are recognizable: *Quercus pubescens* Willd. subsp. *crispata* (Steven) Greuter & Burdet, *Quercus pubescens* Willd. subsp. *pubescens*, and *Quercus pubescens* Willd. subsp. *subpyrenaica* (Villar) Rivas Mart. & C.Saenz.

RELATED SPECIES More than 600 taxa are believed to belong to the genus *Quercus*.

GEOGRAPHIC DISTRIBUTION S Europe, W Asia, Caucasus. **HABITAT** Warm environments, sub-Mediterranean climate, prefers calcareous soils (0–1,500 m ASL). **RANGE** S Europe, W Asia, Caucasus.

BEER-MAKING PARTS (AND POSSIBLE USES) Wood (for barrels).

WOOD Without notable histological differences of common oak and oak wood; in general, heavier and with greater shrinkage, consequently not very suitable for fine work, although it is excellent for maritime and boat constructions, railway sleepers, and cartwrights' work and tools, as well as for firewood and coal.

STYLE None known.

SOURCE Cantwell and Bouckaert 2016; Cantwell and Bouckaert 2018 (These authors report the binomial *Quercus lanugnosa*, probably an erroneous reference to *Quercus lanuginosa*; other authors have used that binomial, illegitimately, to indicate different entities. Therefore, the most correct interpretation of this record is probably *Quercus pubescens* Willd.).

Quercus robur L. subsp. *robur*
FAGACEAE

SYNONYMS *Quercus abbreviata* Vuk., *Quercus acutiloba* Borbás, *Quercus aesculus* Boiss., *Quercus aestivalis* Steven, *Quercus afghanistanensis* K.Koch, *Quercus altissima* Petz. & G.Kirchn., *Quercus arenaria* Borbás, *Quercus argentea* Morogues, *Quercus asterotricha* Borbás & Csató, *Quercus atropurpurea* K.Koch, *Quercus atrosanguinea* K.Koch, *Quercus atrovirens* Sm., *Quercus auzin* Secondat ex Bosc., *Quercus avellanoides* Vuk., *Quercus axillaris* Schur, *Quercus bedoi* Borbás, *Quercus bellogradensis* Borbás, *Quercus brevipes* A.Kern., *Quercus brevipes* Borbás, *Quercus bruttia* Borbás, *Quercus castanoides* Vuk., *Quercus comptoniifolia* K.Koch, *Quercus concordia* K.Koch, *Quercus condensata* Schur, *Quercus coriifolia* Vuk., *Quercus crispa* Vuk., *Quercus cunisecta* Borbás, *Quercus cuprea* K.Koch, *Quercus cupressoides* K.Koch, *Quercus cylindracea* Guss. ex Parl., *Quercus dilatata* A.Kern., *Quercus dissecta* K.Koch, *Quercus ettingeri* Vuk., *Quercus extensa* (Schur) Schur, *Quercus extremadurensis* O.Schwarz, *Quercus falkenbergensis* Booth ex Loudon, *Quercus farinosa* Vuk., *Quercus fastigiata* Lam., *Quercus femina* Mill., *Quercus fennessii* A.DC., *Quercus filicifolia* A.DC., *Quercus filipendula* Schloss. & Vuk., *Quercus foemida* Mill., *Quercus fructipendula* Schrank, *Quercus geltowiensis* K.Koch, *Quercus germanica* Lasch, *Quercus haas* Kotschy, *Quercus hentzei* Petz. & G.Kirchn., *Quercus horizontalis* Dippel, *Quercus hyemalis* Steven, *Quercus laciniata* Lodd., *Quercus lasistan* Kotschy ex A.DC., *Quercus longiglans* Debeaux, *Quercus longipedunculata* Cariot & St.-Lag, *Quercus louettii* Dippel, *Quercus lucorum* Vuk., *Quercus malacophylla* (Schur) Schur, *Quercus microcarpa* Lapeyr., *Quercus monorensis* Simonk., *Quercus nigricans* K.Koch, *Quercus ochracea* Morogues, *Quercus paleacea* Desf., *Quercus palmata* Vuk, *Quercus parmenteria* Mutel, *Quercus pectinata* K.Koch, *Quercus pedemontana* Colla, *Quercus pedunculata* Hoffm., *Quercus pendula* (Neill) Lodd., *Quercus pendulina* Kit., *Quercus pilosa* (Schur) Simonk., *Quercus pseudopedunculata* Vuk., *Quercus pseudoschorochensis* Boiss., *Quercus pseudosessilis* Schur, *Quercus pseudotscharakensis* Kotschy ex A.DC., *Quercus pulverulenta* K.Koch, *Quercus purpurea* Lodd. ex Loudon, *Quercus pyramidalis* C.C.Gmel., *Quercus racemosa* Lam., *Quercus rubens* Petz. & G.Kirchn., *Quercus rubicunda* Dippel, *Quercus rumelica* Griseb. & Schenk, *Quercus scolopendrifolia* K.Koch, *Quercus sieboldii* Dippel, *Quercus speciensis* Dippel, *Quercus subvelutina* Schur,

Quercus svecica Borbás, *Quercus tardiflora* Czern. ex Stev., *Quercus tennesi* Wesm., *Quercus tetracarpa* Vuk., *Quercus tozzae* Dippel, *Quercus tricolor* Petz. & G.Kirchn., *Quercus turbinata* Kit., *Quercus viminalis* Bosc, *Quercus virgata* Martrin-Donos, *Quercus vulgaris* Bubani, *Quercus welandii* Simonk.

COMMON NAMES Pedunculate oak, English oak.

DESCRIPTION Tree (5–25 m). Bark gray-brown with longitudinal cracks; 2- to 5-year-old branches shiny, gray, or gray-brown, pruinose, with rare 1–2 mm transverse lenticels. Leaves subsessile (petiole 1–5 mm) with obcuneate blade, the largest 7–9 × 12–13 cm, base auriculate, apex rounded, with 4–6 rounded lobes on each side. Fruits acorns, subspherical oval or elongate (2–3 cm), in groups of 2–3, sessile but borne by a common peduncle 2–5 cm long; cupule covering 1/4–1/2 with rhombic scales, the central ones wider than marginal.

SUBSPECIES In addition to the nominal subspecies, *Quercus robur* L. subsp. *robur*, the variability of this group has been framed into other taxa, among which *Quercus robur* L. subsp. *brutia* (Ten.) O.Schwarz, *Quercus robur* L. subsp. *imeretina* (Steven ex Woronow) Menitsky, and *Quercus robur* L. subsp. *pedunculiflora* (K.Koch) Menitsky are currently accepted.

RELATED SPECIES Of the more than 600 taxa recognized as belonging to the genus *Quercus*, it is conceivable that many more than those cited in this book have similar (at least potential) applications.

GEOGRAPHIC DISTRIBUTION European-Caucasian. **HABITAT** Forests in rich, ± neutral soils with a high water table. **RANGE** Europe, Caucasus, Asia.

BEER-MAKING PARTS (AND POSSIBLE USES) Bark, wood (for barrels and malt smoking), chips, cubes, pieces of staves.

CHEMISTRY Wood: ellagitannins (roburin a 6.3%, roburin b 7.9%, roburin c 4.8%, grandinin 10.7%, roburin d 9.7%, vescalagin 19%, roburin e 8.6%, castalagin 33%), fatty acids (acetic acid, propanoic acid, 2-methylpropanoic acid, butanoic acid, 2and 3-methylbutanoic acid, pentanoic acid, hexanoic acid, octanoic acid, nonanoic acid, decanoic acid, dodecanoic acid, tetradecanoic acid, pentadecanoic acid, hexadecanoic acid, linoleic acid, linolenic acid, stearic acid), furans (furfural, 2-furanmethanol, 5-methylfurfural, 2-furanoic acid, 5-hydroxymethylfurfural), simple phenols (benzaldehyde, benzophenone, phenol, O-methyl eugenol, eugenol, gaiacol, vanillin, vanilloyl methyl acetone, propiovanillone, 2-[4'-gaiacil]-ethanol, 3-[4'-gaiacil]-propanol, syringaldehyde, acetosyringone, α-hydroxypropiosyringone, coniferaldehyde), lactones (2-[5H]-furanone, trans-β-methyl-γ-octalactone, cis-β-methyl-γ-octalactone, γ-decalactone, γ-butyrolactone, 3-methyl-2[5H]-furanone, δ-decalactone, mevalonic lactone [tetrahydro-4-hydroxy-4-methyl-2H-pyran-2-one]), alcohols (phenylethanol, 2,3-butanediol, propanediol, decanol, glycerol, [E]-farnesol), norisoprenoids (3-oxo-α-ionol, 4-oxo-β-ionol, 4,5-dihydrovomipholiol), miscellaneous compounds (2,3-octanedione, benzoic acid, 2-nonenal, [E,E]-nona-2,4-dienal, 2,4-decadienal).

WOOD Differentiated, with yellowish sapwood and brown heartwood (WSG 0.60–0.80) furrowed by conspicuous medullary rays that in the radial sections give rise to evident shiny speckles. Coarse texture and mostly straight grain. Resistant to alteration (provided the sapwood is removed, which is easily attacked by fungi and insects) and adaptable to various uses: beams for large constructions, squared and figured pieces for ships, railway sleepers, strips for floors, staves, splitting work, furniture, veneers for plywood, firewood, charcoal.

STYLE American Amber Ale, American Barleywine, American Brown Ale, American IPA, American Pale Ale, American Stout, Baltic Porter, Belgian Blond Ale, Belgian Dark Strong Ale, Belgian Golden Strong Ale, Belgian Specialty Ale, Belgian Tripel, Blonde Ale, Braggot, Brown Porter, California Common Beer, Cream Ale, Extra Special/Strong Bitter (ESB), Flanders Brown Ale/Oud Bruin, Flanders Red Ale, Foreign Extra Stout, Gruit, Imperial IPA, Imperial Stout, Irish Red Ale, Mixed-fermentation Sour Beer, Oatmeal Stout, Old Ale, Other Smoked Beer, Robust Porter, Russian Imperial Stout, Saison, Special/Best/Premium Bitter, Specialty Beer, Strong Scotch Ale, Wheatwine, Wood-Aged Beer.

SOURCE Cantwell and Bouckaert 2016; Cantwell and Bouckaert 2018; Daniels and Larson 2000; Hieronymus 2012; Sparrow 2015; https://www.bjcp.org (33A. Wood-Aged Beer; 33B. Specialty Wood-Aged Beer); https://www.birradegliamici.com

/news-birraie/birre-strane/; https://www.brewersfriend.com/other/oak/; https://www.ricettariome-dievale.it/2014/02/da-httpricettariomedievale.html.

Quercus robur L. subsp. *imeretina* (Steven ex Woronow) Menitsky
FAGACEAE

Synonyms *Quercus imeretina* Steven ex Woronow.

Common Names None

Description Tree (up to 15 m). Leaves (5–14 × 2.5–5 cm) oblong to obovate-oblong, narrow; base cordate or strongly auriculate; upper leaf glabrous, smooth, green, underleaf light green; margin with 6–8 pairs of entire lobes, with sinuses up to 1/3 from midrib; secondary veins at 40–50° to midrib; petiole 1–4 mm, nearly hidden by auricles. Fruits acorns, narrowly cylindrical (3 × 1.5 cm); cupule shallow and thin, with flat, appressed, gray-tomentose scales; peduncle to 10 cm and more.

Related Species More than 600 taxa are believed to belong to the genus *Quercus*.

Geographic Distribution W Caucasus, Transcaucasia (Azerbaijan, Armenia, Georgia, Russia). **Habitat** Forests. **Range** W Caucasus, Transcaucasia (Azerbaijan, Armenia, Georgia, Russia).

Beer-Making Parts (and Possible Uses) Wood (for barrels).

Wood It has characteristics similar to those of *Quercus robur* ssp. *robur*.

Style None known.

Source Cantwell and Bouckaert 2016; Cantwell and Bouckaert 2018.

Quercus robur L. subsp. *pedunculiflora* (K.Koch) Menitsky
FAGACEAE

Synonyms *Quercus erucifolia* Steven, *Quercus kurdica* Wenz., *Quercus longipes* Steven, *Quercus mestensis* Bondev & Gancev, Quercus *pedunculiflora* K.Koch, Quercus *pinnatipartita* (Boiss.) O.Schwarz.

Common Names None

Description Tree (up to 15 m), deciduous. Bark gray, furrowed. Leaves (8–17 × 6–9 cm) similar to those of *Quercus robur* ssp. *robur* but somewhat thicker, dark green, slightly glabrous, glabrous above, tomentose yellow-gray below; 4–5 pairs of lobes with deep sinuses; secondary veins at 60–80° to midrib, petiole up to 2 cm. Fruits acorns (2–3 x 1.5–2 cm) with tomentose stylopodium, long peduncle; 1/3 covered by cupule (Ø 1.8–2.5 cm), with warty, appressed scales, yellow-rust color.

Related Species More than 600 taxa are believed to belong to the genus *Quercus*.

Geographic Distribution Asia Minor, Caucasus, Balkans. **Habitat** Very drought-tolerant species (up to 1,800 m ASL). **Range** Asia Minor, Caucasus, Balkans.

Beer-Making Parts (and Possible Uses) Wood (for barrels).

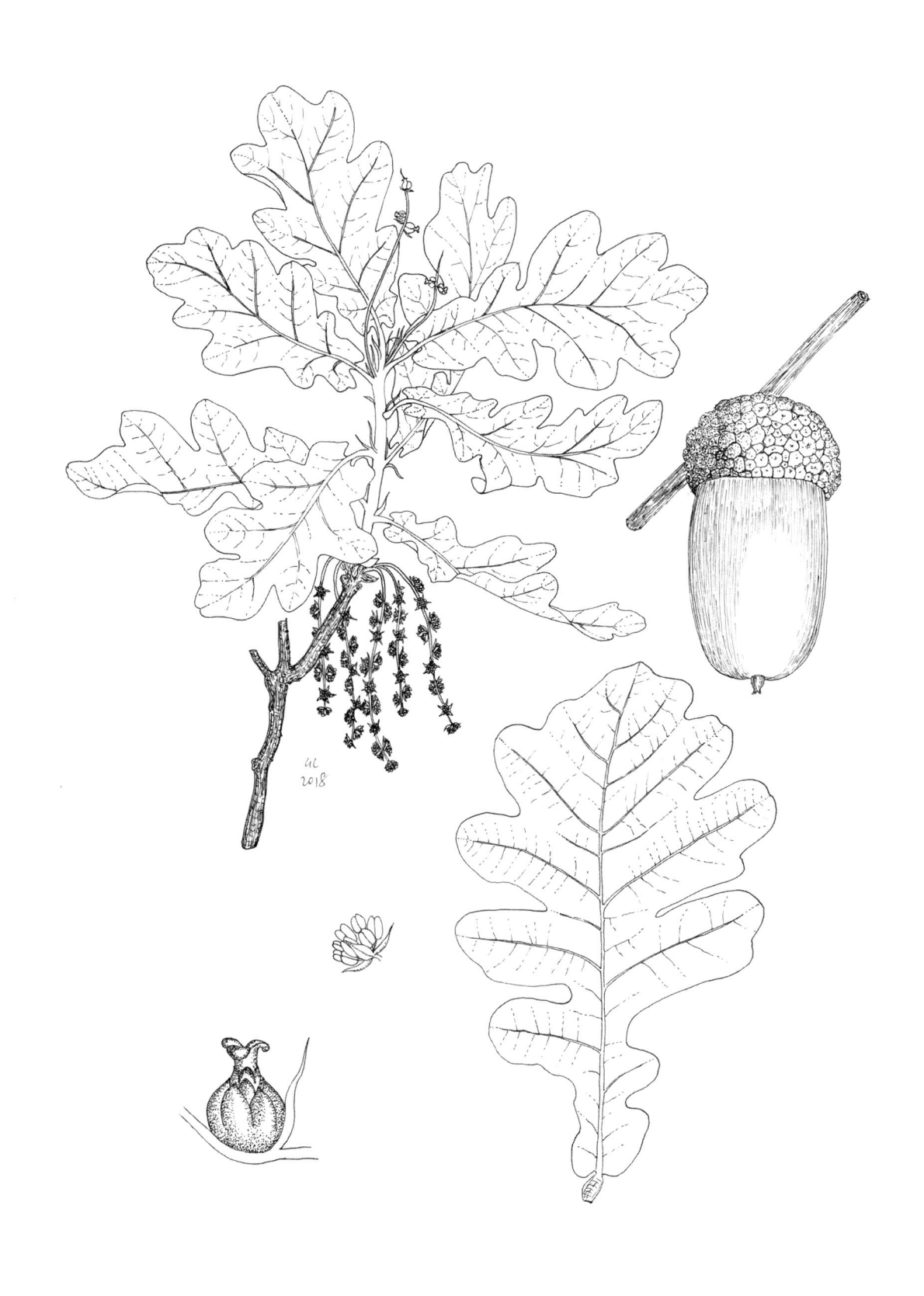

Quercus robur L. subsp. *robur* (incl. subsp. *imeretina* [Steven ex Woronow] Menitsky) -FAGACEAE

Rhus chinensis Mill.
ANACARDIACEAE

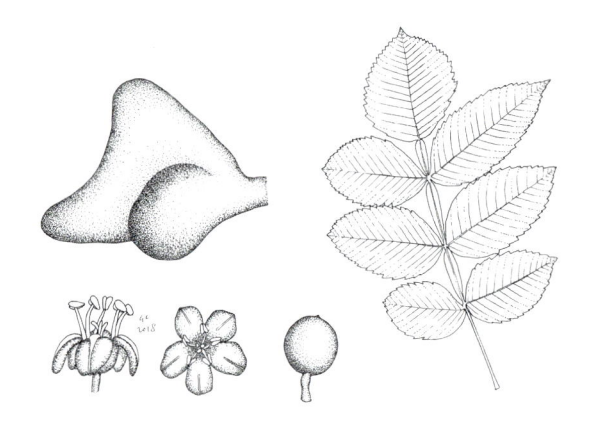

Synonyms *Rhus amela* D.Don, *Rhus semialata* Murray, *Schinus indicus* Burm.f.

Common Names Chinese sumac, Chinese gall, nutgall tree.

Description Tree or shrub (2–10 m). Main stem Ø 6–18 cm, ferruginous pubescent branches with lenticels. Leaves compound, imparipinnate, rachis broadly winged to nonwinged, ferruginous pubescent; leaflets (5–13), lamina ovate to oblong, size increasing toward apex (6–12 × 3–7 cm), above dark green, sparsely pubescent or glabrescent, below light green, glaucous, ferruginous pubescent, base rounded to cuneate in terminal leaflet, margin toothed, often crenate, apex acute. Flowers white; male ones with minutely pubescent calyx, long ovate lobes (about 1 mm), with ciliate margins; petals obovate-oblong (about 2 mm); staminal filaments 2 mm, anthers ovoid (0.7 mm); disc annular; ovary reduced to absent; female flowers with calyx lobes about 0.6 mm, petal elliptic-ovate, about 1.6 mm; staminodes much reduced; disc annular; ovary ovoid, about 1 mm, densely pubescent, styluses 3, stigma capitate. Inflorescences multibranched, densely ferruginous, pubescent, the male ones 30–40 cm, the female ones shorter; peduncle about 1 mm, minutely pubescent. Fruit drupe globular, slightly compressed (Ø 4–5 mm), glandular-pubescent hairiness, red at maturity.

Subspecies Besides the nominal variety, there is *Rhus chinensis* Mill. var. *roxburghii* (DC.) Rehder, easily distinguishable from the nominal variety by its nonwinged leaf rachis.

Related Species Of the 146 species of the genus *Rhus*, some are reported to have beer-making uses; other species with known food uses are *Rhus aromatica* Aiton, *Rhus glabra* L., *Rhus integrifolia* (Nutt.) Benth. & Hook. f. ex Rothr., *Rhus muelleri* Standl. & F.A.Barkley, *Rhus ovata* S.Watson, and *Rhus typhina* L.

Geographic Distribution E Asia (China, Japan, Korea, India, Bhutan, Malaysia, Thailand, Laos, Cambodia, Vietnam, Indonesia). **Habitat** Open sites, often along watercourses, from lowlands to montane forests (up to 2,800 m ASL). **Range** E Asia.

Beer-Making Parts (and Possible Uses) Many parts of the plant (e.g., fruits, leaves) are edible or have medicinal applications; however, for the extraction of tannin (and the consequent beer-making use), the leaf galls produced by the aphid *Melaphis chinensis* Bell (Hemiptera) are most often used.

Toxicity Sap seems to be irritating (causing a skin rash) in some subjects.

Chemistry Galls: gallic acid, galloylglucopyranose, di-galloyl glucopyranose, tri-galloyl glucopyranose, calcium, chromium, iron, manganese, phosphorus, tin, zinc.

Style Hefewiessbier.

Source Hieronymus 2010.

Rhus coriaria L.
ANACARDIACEAE

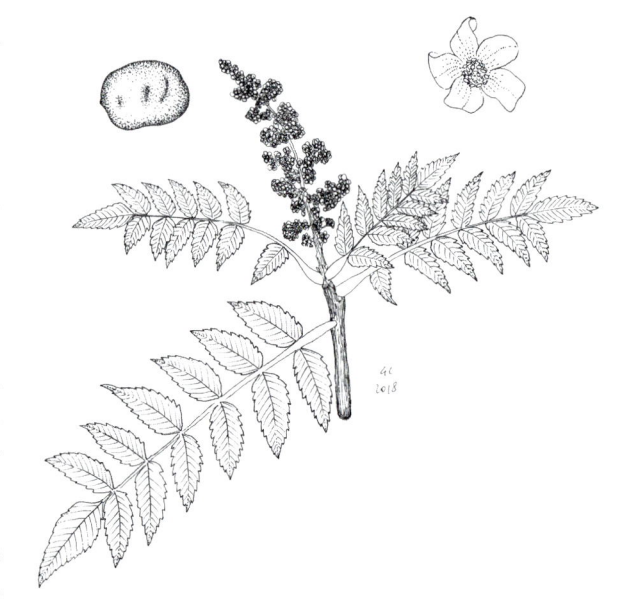

Synonyms *Toxicodendron coriaria* (L.) Kuntze.

Common Names Elm-leafed sumac, scarlet sumac, upland sumac, mountain sumac, Sicilian sumac, sumach, sumak, summak, Tanner's sumach.

Description Shrub (1–4 m), usually evergreen. Leaves (10–18 cm) compound, imparipinnate, 9–21 lanceolate leaflets, margin serrated, rachis winged. Flower with greenish, villous sepals; petals greenish-white. Inflorescence linear panicle about as long as the leaves. Fruit spherical drupe (Ø 4–6 mm), with short glandular hairs.

Related Species Of the 146 species of the genus *Rhus*, some are believed to have beer-making uses; other species with known alimentary uses are *Rhus aromatica* Aiton, *Rhus glabra* L., *Rhus integrifolia* (Nutt.) Benth. & Hook. f. ex Rothr., *Rhus ovata* S.Watson, *Rhus typhina* L., and *Rhus muelleri* Standl. & F.A.Barkley.

Geographic Distribution S Mediterranean.
Habitat Arid grasslands, often as a relic of ancient crops aimed at extracting tannin from leaves. **Range** S Mediterranean.

Beer-Making Parts (and Possible Uses)
Fruits (?), seed oil (?); despite the recognized toxicity of some parts of the plant, some Middle Eastern countries use it in traditional cooking; leaves (?) are used to extract tannin (a stabilizer of turbidity like the previous species?) and different pigments.

Toxicity The plant contains toxic substances that may cause severe irritation in some individuals. Both the sap and fruits are poisonous.

Chemistry Essential oil, leaves (60 compounds): β-caryophyllene 0.33–16.95%, patchoulane 3.08–23.87%. Bark and branch oil (63 compounds): β-caryophyllene 12.35–21.91%, cembrene 10.71–26.50%. Fruit oil (pericarp, 85 compounds): limonene 0.17–9.49%, nonanal 10.77–13.09%, (Z)-2-decanal 9.90–42.35%. Astringency usually provided by 2 components: tannins (gallotannins 4%), organic acids (malic, citric, tartaric, succinic, maleic, fumaric, ascorbic).

Style American Brown Ale, Belgian Specialty Ale, Gose, Saison, Witbier.

Source McGovern 2017; https://www.brewersfriend.com/other/sumac/.

Rhus trilobata Nutt.
ANACARDIACEAE

Synonyms None

Common Names Sumac, squawbush, skunkbush sumac, lemonberry sumac.

Description Shrub (0.5–2.5 m). Stems arching and ascending, forming a rounded or ascending crown, often with a diameter greater than the height of the plant. Roots deep and extensively branched, with widespread superficial woody rhizomes, with connections between plants up to 9 m apart; new shoots emerge from rhizomes and root. Leaves deciduous, alternate, composed of 3 leaflets variable in size, shape, lobing, margins; leaflets sessile, ovate to rhomboidal, more or less cuneate at base, coarsely toothed, usually glabrous and shiny above, terminal leaflet 3–6.5 cm; summer leaves green, in autumn orange or reddish. Flowers yellowish to whitish, in small dense inflorescences borne by short lateral shoots; flowers bisexual and unisexual on the same plant (polygamous species); male flowers (staminate) in yellowish catkins, female flowers (pistillate) in short, bright yellow inflorescences at the ends of branches. Anthesis precedes foliation. Fruits drupes, Ø 5–7 mm, red at maturity and sparsely hairy, each containing a single nucule.

Subspecies The great extension of the range and the variety of habitats to which the species has adapted have produced a large number of ecotypes with distinct morphological characteristics (e.g., growth form, height, shape and size of leaves, shape of fruit, pubescence), with a series of intermediate forms in which the ranges of these subspecific groups overlap. According to some authors, this variability can be traced to three varieties (*Rhus trilobata* Nutt. var. *malacophylla* [Greene] Munz, *Rhus trilobata* Nutt. var. *racemulosa* [Greene] F.A.Barkley, and *Rhus trilobata* Nutt. var. *trilobata*), whereas according to others, it would include six entities (*Rhus trilobata* Nutt. var. *anisophylla* [Greene] Jepson, *Rhus trilobata* Nutt. var. *pilosissima* Engelm., *Rhus trilobata* Nutt. var. *quinata* [Greene] Jepson, *Rhus trilobata* Nutt. var. *racemulosa* [Greene] Barkl., *Rhus trilobata* Nutt. var. *simplicifolia* [Greene] Barkl., and *Rhus trilobata* Nutt. var. *trilobata*).

Cultivar Few are known, used for ornamental purposes or environmental regeneration (e.g., the Bighorn cultivar, native to Wyoming, United States).

Related Species Of the 146 species of the genus *Rhus*, some are known to have beer-making uses; other species with known food uses are *Rhus aromatica* Aiton, *Rhus glabra* L., *Rhus integrifolia* (Nutt.) Benth. & Hook. f. ex Rothr., *Rhus muelleri* Standl. & F.A.Barkley, *Rhus ovata* S.Watson, and *Rhus typhina* L.; of the species *Rhus aromatica* Aiton, *Rhus copallinum* L., *Rhus glabra* L., and *Rhus typhina* L. are also known for beer-making uses.

Geographic Distribution United States (Arizona, Arkansas, California, Colorado, Idaho, Kansas, Montana, Nebraska, Nevada, New Mexico, North Dakota, Oklahoma, Oregon, South Dakota, Texas, Utah, Washington, Wyoming), Canada (Alberta, Saskatchewan), Mexico (Baja California). **Habitat** Hilly and rocky slopes, arid, sandy hills, along streams, canyons and wetlands, grassy uplands, open scrubland (1,000–2,700 m ASL). **Range** Similar to natural geographic distribution.

Beer-Making Parts (and Possible Uses) Fruits, leaves, tender tops (added at the end of boiling to impart color and lemon flavor).

Toxicity Recent research shows low or zero acute toxicity of *Rhus trilobata* extract, at least at the quantities tested. Although these results are reassuring for the use of the above-mentioned extract (also for medical purposes), further studies are needed to verify with certainty the absence of chronic toxicological damage.

Chemistry Fruits (active compounds): gallic acid, tannic acid.

Style Herb Beer.

Source Hieronymus 2016b.

Ribes nigrum L.
GROSSULARIACEAE

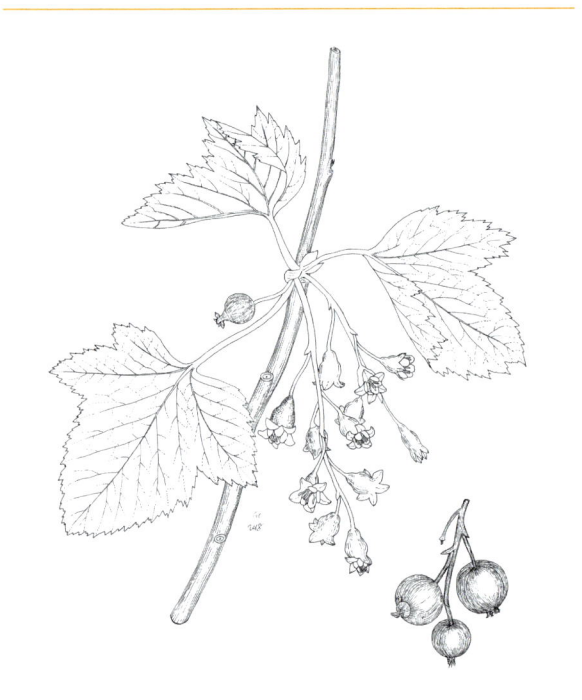

Synonyms *Botrycarpum nigrum* (L.) A.Rich, *Botrycarpum nigrum* (L.) Spach, *Grossularia nigra* (L.) Rupr., *Ribes cyathiforme* Pojark, *Ribes olidum* Moench, *Ribes pauciflorum* Turcz. ex Ledeb., *Ribesium nigrum* (L.) Medik.

Common Names Black currant.

Description Shrub, deciduous (1–2 m). Bark smooth, pale to reddish in young stems, dark in old stems. Leaves 3- to 5-lobed, toothed, glabrous above and with numerous sessile glands below, viscous-oily, aromatic. Inflorescence raceme patent with lax, distinctly pedicellate flowers. Flowers bell shaped, with reflexed calyx laciniae, tomentose; petals small, erect. Fruit berry, black in color, sweetish, aromatic.

Cultivar Examples include Ben Alder, Ben Gaire, Ben Hope, Ben Lomond, Ben Sarek, Bona, Boskoop Giant, CaCanska CRNA, Cotswold Cross, Tiben, and Wellington XXX. There are many others, but some are actually hybrids with other species in the genus *Ribes*.

Related Species Of the approximately 228 species belonging to the genus *Ribes*, only the three cited in this book seem to have beer-making applications.

Geographic Distribution Eurasian. **Habitat** Mountain woods. **Range** Eurasia.

Beer-Making Parts (and Possible Uses) Fruits.

Chemistry Fruits: two cases (I and II) are recorded, showing different chemical compounds and composition. Fruits I: sugars 9% (fructose 45%, glucose 40%, sucrose 15%), soluble solids 15.4%, acids 5.1% (citric acid 88%, malic acid 12%), ascorbic acid 67–204 mg/100 g. Fruits II: energy 264 kJ (63 kcal), water 82 g, carbohydrates 15.4 g, lipids 0.4 g, protein 1.4 g, thiamine (B1) 0.05 mg, riboflavin (B2) 0.05 mg, niacin (B3) 0.3 mg, pantothenic acid (B5) 0.398 mg, vitamin B6 0.066 mg, vitamin C 181 mg, vitamin E 1 mg, calcium 55 mg, iron 1.54 mg, magnesium 24 mg, manganese 0.256 mg, phosphorus 59 mg, potassium 322 mg, sodium 2 mg, zinc 0.27 mg.

Style Fruit Beer, Weissbier.

Beer Cassissona, Birrificio Italiano (Lurago Marinone, Lombardy, Italy).

Source Calagione et al. 2018; Dabove 2015; Giaccone and Signoroni 2017; Hieronymus 2010; Hieronymus 2016b; Jackson 1991; Josephson et al. 2016; Sparrow 2015; https://www.bjcp.org (29. Fruit Beer); https://www.oregonfruit.com/fruit-brewing/category/products.

Ribes rubrum L.
GROSSULARIACEAE

Synonyms *Grossularia rubra* (L.) Scop., *Ribes sylvestre* (Lam.) Mert. & Koch, *Ribes vulgare* Lam.

Common Names Redcurrant, red currant.

Description Shrub, deciduous (1–1.5 m). Stem unarmed, erect branches, those of the year herbaceous, green in color, the previous ones scaly, lignified, light brown in color. Leaves simple, petiolate (about 6 cm), lamina (6 cm) usually wider than long, palmate (3–5 lobes, each with ± rounded teeth or tiny apical mucro), ± cordate at base, glabrous above, subglabrous to pubescent below. Flowers pedicellate, basal bract smaller than pedicel, hermaphrodite, pentamerous, rotate, yellowish-green in color; sepals glabrous, reddish, twice as long as the small petals. Inflorescence sparse raceme patent-pendant, with basal bract. Fruit berry, red (yellowish-white in some cultivars), glabrous, acidic flavor.

Subspecies Of the many varieties described within this taxon, only *Ribes rubrum* L. var. *alaskanum* (Berger) B.Boivin is currently accepted in addition to the nominal variety.

Cultivar Some examples: Detvan, Heinemann's RS, Hron, Jonkheer van Tets, Red Dutch, Red Lake, Redpoll, Rolan, Rondom, Rovada.

Related Species Of the approximately 228 species belonging to the genus *Ribes*, only the three cited in this book appear to have beer-making applications.

Geographic Distribution C Europe. **Habitat** Wet woods, cool places, hedges, preferably in fertile basic soils; often wild from cultivated plants (0–2,000 m ASL). **Range** Widely cultivated in temperate climate areas.

Beer-Making Parts (and Possible Uses) Fruits.

Chemistry Fruits: energy 234 kJ (56 kcal), carbohydrates 13.8 g, sugars 7.37 g, fiber 4.3 g, lipids 0.2 g, protein 1.4 g, thiamine (B1) 0.04 mg, riboflavin (B2) 0.05 mg, niacin (B3) 0.1 mg, pantothenic acid (B5) 0.064 mg, vitamin B6 0.07 mg, folate (B9) 8 µg, choline 7.6 mg, vitamin C 41 mg, vitamin E 0.1 mg, vitamin K 11 µg, calcium 33 mg, iron 1 mg, magnesium 13 mg, manganese 0.186 mg, phosphorus 44 mg, potassium 275 mg, sodium 1 mg, zinc 0.23 mg.

Style Belgian Specialty Ale, Experimental Beer.

Beer Currant Affair, Allagash (Maine, United States).

Source Hieronymus 2016b; Josephson et al. 2016; https://www.brewersfriend.com/other/groselha/; https://www.brewersfriend.com/other/redcurrants/.

Ribes uva-crispa L.
GROSSULARIACEAE

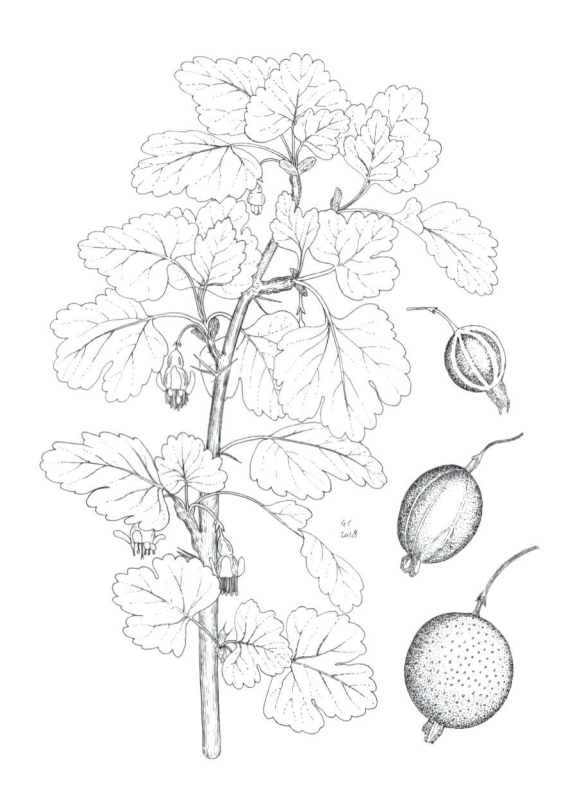

Synonyms *Grossularia glandulosetosa* Opiz, *Grossularia hirsuta* Mill., *Grossularia intermedia* Opiz, *Grossularia pubescens* Opiz, *Grossularia reclinata* (L.) Mill., *Grossularia spinosa* (Lam.) Rupr., *Grossularia uva-crispa* (L.) Mill., *Oxyacanthus uva-crispa* (L.) Chevall., *Ribes caucasicum* Adams ex Schult., *Ribes grossularia* L., *Ribes reclinatum* L., *Ribes spinosum* Lam.

Common Names Gooseberry.

Description Shrub (5–15 dm). Branches intricate, with 1–3 thorns at the nodes and silky spines on young branches. Leaves incised-lobed, glabrous or hairy. Flowers solitary or in groups of 2–3, briefly pedunculate; sepals 5–7 mm, yellow or yellow-green, sometimes purple; ovary inferior, stylus bifid. Fruit berry, variable size, glabrous, pubescent or silky-glandular.

Cultivar Some examples: Ajax, Bedford Yellow, Careless, Careless 30240, Careless VT 5120, Cousens Seedling, Goutrays Earliest, Guy seedling, Heart of Oak, Jolly Amylers, Jubilee Careless, Lord Audley, Lord Elco, May Dulle, Mitre, Nailer, Rubuste Nool, Scotch Red Rough, Victorià, White Eagle.

Related Species Of the approximately 228 species belonging to the genus *Ribes*, only the three cited in this book appear to have beer-making applications.

Geographic Distribution Eurasian. **Habitat** Mountain woods and pastures. **Range** Widely cultivated in temperate climate areas.

Beer-Making Parts (and Possible Uses) Fruits.

Chemistry Fruits: energy 184 kJ (44 kcal), water 87.87 g, carbohydrates 10.18 g, fiber 4.3 g, lipids 0.58 g, protein 0.88 g, vitamin A 15 μg, thiamine (B1) 0.04 mg, riboflavin (B2) 0.03 mg, niacin (B3) 0.3 mg, pantothenic acid (B5) 0.286 mg, vitamin B6 0.08 mg, folate (B9) 6 μg, vitamin C 27.7 mg, vitamin E 0.37 mg, calcium 25 mg, iron 0.31 mg, magnesium 10 mg, manganese 0.144 mg, phosphorus 27 mg, potassium 198 mg, sodium 1 mg, zinc 0.12 mg.

Style Fruit Beer, Fruit Lambic, Holiday/Winter Special Spiced Beer, Old Ale, Saison, Specialty Beer, Weissbier.

Beer Gooseberry Cove, Buxton (England, United Kingdom).

Source Heilshorn 2017; Hieronymus 2016b; Josephson et al. 2016; https://www.bjcp.org (29. Fruit Beer); https://www.brewersfriend.com/other/gooseberries/; https://www.craftbeer.com/news/beer-release/oregon-fruit-products-collaborates-with-breakside-brewery.

Robinia pseudoacacia L.
FABACEAE

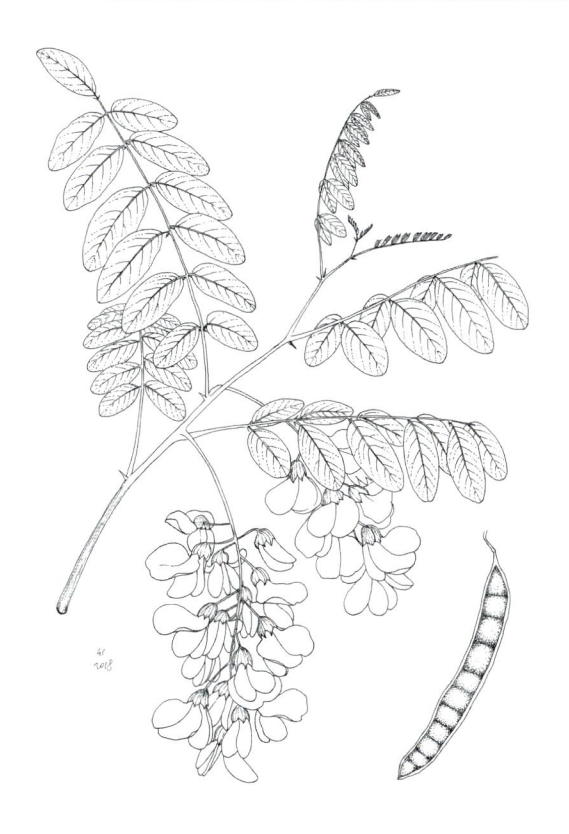

Synonyms *Robinia pringlei* Rose, *Robinia pseudacacia* L.

Common Names Black locust.

Description Tree or shrub (2–25 m), deciduous, spinescent. Roots emit root shoots, very numerous, which spread rapidly, colonizing all available space. Stems (Ø up to 50 cm) erect, often forked, branches smooth, crown branched, bark wrinkled, gray-brown, longitudinally fissured with age. Leaves alternate, imparipinnate, 6–7 pairs of briefly petiolate oval segments, margin entire, pale green, glabrous, with stipules transformed into strong sickle-shaped spines. Flowers intensely fragrant, joined in dense pendulous racemes, leafy at the base; calyx velvety, broadly bell shaped, pale green, pubescent; corolla papilionaceous, white, rarely pink. Fruits legumes (5–10 cm), smooth, leathery, compressed, dehiscent, red-brown at maturity, remain on the plant all winter; contain 3–10 seeds. Seeds reniform, very hard, brown in color.

Subspecies Within the taxon *Robinia pseudoacacia* L., the following variants are recognized: *Robinia pseudoacacia* L. f. *monophylla-pendula* (Dieck) Voss, *Robinia pseudoacacia* L. var. *pendula* (Ortega) Loudon, and *Robinia pseudoacacia* L. var. *pyramidalis* (Pepin) C.K.Schneid.

Related Species The genus *Robinia* includes eight taxa of specific rank: *Robinia elliottii* (Chapm.) Ashe, *Robinia hartwigii* Koehne, acc. to Ashe, *Robinia hispida* L. (in addition to the typical variety, also var. *fertilis* [Ashe] R.T.Clausen, var. *kelseyi* [Hutch.] Isely, var. *nana* [Elliott] DC., var. *rosea* Pursh), *Robinia luxurians* (Dieck) Rydb., *Robinia margarettae* Ashe, *Robinia neomexicana* A.Gray (var. *tipica* and var. *rusbyi* [Wooton & Standl.] Peabody), *Robinia pseudoacacia* L. (with its internal variability already illustrated), and *Robinia viscosa* Vent. (typical var. and var. *hartwigii* [Koehne] Ashe).

Geographic Distribution E North America (Appalachians). **Habitat** Frugal and extremely adaptable species, indifferent to the substrate, provided it is well drained, with a certain preference for acidic soils, heliophilous, tending to form dense thickets. It is an invasive species for its speed of growth and its imposing root system, which emits strong suckers, spreading at the expense of native species. It grows in pure coppice woods, along escarpments, in uncultivated places, and in hedges (0–1,500 m ASL). **Range** Widely spread beyond its original area for ornamental purposes, forestry, or for the consolidation of railways and/or roadsides, it has often become extremely invasive and even naturalized.

Beer-Making Parts (and Possible Uses) Wood for barrels (uncertain use in beer making), flowers.

Wood Differentiated, with yellow sapwood and brown heartwood (WSG 0.75–0.85), of great mechanical strength and long life; suitable for poles, carriage bodies, mine struts, barrel staves, gears, floorboards; it is also an excellent combustible wood.

Chemistry Essential oil, flower parts: lauric acid 11.35%, caproic acid 11.06%, isomitylitol 9.53%, butylated hydroxylanisole 7.12%, 10-methyl-2-(10h)-phenazinone 6.25%, palmitic acid 3.74%, 2-methoxy-2,9-dimethyl benzyl norbornene 3.11%, linoleic acid 2.95%, ambrettolide 2%, 1,2-dihydro-1-methyl-2,3,1-benzodiazaborine 1.78%, pentadecanoic acid 1.54%, 11,14-octadecadienoic acid methyl ester 1.46%, myristic acid 1.31%, 3-hydroxybenzaldehyde 1.28%, butyl myristate 1.22%,

2-hydroxyfluorene 1.21%, 5-hydroxy-1,4-dimethoxy-anthraquinone 1.17%, 2,4-dimethoxythiophenol 1.08%, sclerodione 0.98%, maltol 0.97% (2α,3β,5α)-aspidofractinine-3-methanol 0.91%, daphnetine 0.76%, 1,5,5-trimethyl-6-methylene-cyclohexene 0.73%, methyl benzoate 0.71%, ethylidene diacetate 0.71%, 2,3,5-trimethylphenylsulfone 0.69%, arthol 0.33%.

STYLE Herb Beer.

SOURCE Cantwell and Bouckaert 2016; Cantwell and Bouckaert 2018; Heilshorn 2017.

Rosa × damascena Herrm.
ROSACEAE

SYNONYMS None

COMMON NAMES Damask rose.

DESCRIPTION Shrub (up to 15 dm). Stems arcuate, spiny, bushy. Leaves compound, imparipinnate (5–7 leaflets), dark green in color. Flower with pink, medium, semi-double or double petals, damask fragrance; buds with long sepals. Inflorescences group small clusters of flowers; also, solitary flowers.

SUBSPECIES The taxon *Rosa × damascena* occurs in two recognized forms: *Rosa × damascena* Herrm. f. *trigintipetala* (Dieck) R.Keller and *Rosa × damascena* Herrm. f. *versicolor* (Weston) Rehder.

CULTIVAR Some of the many cultivars of this hybrid: Celsiana, Gloire de Guilan, Katzantik (or Bulgarian Rose, from which the famous essence is extracted), Marie Louise, Mme Hardy, Quatre Saisons Continue, Rose d'Hivers, Rose de Puteaux, York and Lancaster.

RELATED SPECIES Of the about 435 species recognized in the genus *Rosa*, many likely have a potential use in beer-making similar to the one cited here for some of these taxa.

GEOGRAPHIC DISTRIBUTION This is a tetraploid hybrid, *Rosa moschata* Herrm. × *Rosa gallica* L., obtained from an unknown plant before 1560.
HABITAT Cultivated plant. **RANGE** Cultivated.

BEER-MAKING PARTS (AND POSSIBLE USES)
Petals, flower buds.

CHEMISTRY Essential oil, dried petals (57 components; 75.3% of the oil): aliphatic hydrocarbons 56.4% (heneicosane 19.7%, nonadecane 13%, tricosane 11.3%, pentacosane 5.3%, eicosane 2.5%), oxygenated monoterpenes 14.7% (citronellol 7.1%, geraniol 4.1%, geranyl acetate 0.8%, linalool 0.5%, geranyl formate 0.5%, 2-phenyl ethyl alcohol 0.4%). Rose water from dried petals (48 components; 65.3% of rose water): aliphatic hydrocarbons 46.3% (heneicosane 15.7%, nonadecane 8.4%, tricosane 9.3%, pentacosane 5.1%, eicosane 2.4%), oxygenated monoterpenes 8.7% (2-phenyl ethyl alcohol 7.1%, geraniol 2.5%, citronellol 2.2%, geranyl formate 1.5%, linalool 0.7%). Essential oil, fresh petals: citronellol 15.9–35.3%, geraniol 8.3–32.2%, nerol 4–9.6%, nonadecane 4.5–16%, heneicosane 2.6–7.9%. Rose water from fresh petals: 2-phenyl-ethyl-alcohol 66.2–80.7%, citronellol 1.8–5.5%, geraniol 3.3%–7.9%.

STYLE American Brown Ale, Belgian Dubbel, Belgian Tripel, Spice/Herb/Vegetable Beer.

SOURCE https://www.brewersfriend.com/other/damasco/; https://www.ratebeer.com/beer/beerstorming-bs0064-rose-minh/431124/.

Rosa canina L.
ROSACEAE

Synonyms *Rosa* × *ancarensis* Pau & Merino, *Rosa* × *trichoneura* Ripart ex Crép., *Rosa acanthina* Déségl. & Ozanon, *Rosa actinodroma* Gand., *Rosa adenocalyx* Gren., *Rosa amansii* Déségl. & Ripart, *Rosa beatricis* Burnat & Gremli, *Rosa belgradensis* Pančić, *Rosa burnatii* (H.Christ) Burnat & Gremli, *Rosa caucasica* Pall., *Rosa cinerascens* Cariot, *Rosa cinerosa* Déségl., *Rosa cladoleia* Ripart ex Crép., *Rosa condensata* Puget, *Rosa curticola* Puget ex Déségl., *Rosa dilucida* Déségl. & Ozanon, *Rosa dumetorum* auct., *Rosa dumosa* Salisb., *Rosa edita* Déségl., *Rosa erythrantha* Boreau, *Rosa firma* Puget, *Rosa fissispina* Wierzb. ex Heuff., *Rosa flexibilis* Déségl., *Rosa flexuosa* Raf., *Rosa frivaldskyi* Heinr.Braun, *Rosa frondosa* Steven ex Spreng., *Rosa glaucescens* Desv. ex Mérat, *Rosa heterostyla* Chrshan., *Rosa hispidula* Ripart ex Déségl., *Rosa inconspicua* Déségl., *Rosa insignis* Déségl. & Ripart, *Rosa istrica* Degen, *Rosa kalmiussica* Chrshan. & Lasebna, *Rosa keissleriana* Sennen, *Rosa lioclada* Boullu, *Rosa litigiosa* Crép., *Rosa longituba* Debeaux, *Rosa lutetiana* Léman, *Rosa macroacantha* Ripart ex Déségl., *Rosa mandonii* Déségl., *Rosa marisensis* Simonk. & Heinr.Braun, *Rosa mollardiana* Moutin, *Rosa montivaga* Déségl., *Rosa mucronulata* Déségl. ex Godet, *Rosa nemophila* Déségl. & Ozanon, *Rosa nitens* Desv. ex Mérat, *Rosa nitens* Vuk., *Rosa oblonga* Déségl. & Ripart, *Rosa oblongata* Opiz, *Rosa oreades* Cottet & Castella, *Rosa platyphylla* A.Rau, *Rosa podolica* Tratt., *Rosa polyodon* Gand., *Rosa pratincola* Heinr.Braun, *Rosa prutensis* Chrshan., *Rosa psilophylla* A.Rau, *Rosa pubens* Déségl. & Ozanon, *Rosa ramosissima* Déségl., *Rosa rougeonensis* Ozanon, *Rosa rubescens* Ripart ex Déségl., *Rosa sarmentacea* Sw., *Rosa sarmentacea* Woods, *Rosa senticosa* Ach., *Rosa separabilis* Déségl., *Rosa sphaerica* Gren., *Rosa sphaeroidea* Ripart ex Déségl., *Rosa spinetorum* Déségl. & Ozanon, *Rosa spuria* Puget ex Déségl., *Rosa squarrosa* auct. angl., *Rosa stipularis* Mérat, *Rosa sylvularum* Ripart ex Déségl., *Rosa syntrichostyla* Ripart ex Déségl., *Rosa timbaliana* Debeaux, *Rosa touranginiana* Déségl. & Ripart, *Rosa venosa* Sw., *Rosa willibaldii* Chrshan.

Common Names Dog rose.

Description Shrub (3–20 dm). Stem woody, glabrous, with red spines, robust, arched, with an elongated base, compressed. Leaves with lanceolate stipules (3 × 15 mm), on flowering stems broader (4 × 12 mm); leaf segments 5–7, elliptic, ovate or obovate (9–25 × 13–40 mm); hairiness and denticulation variable; teeth 12–33 in number. Flowers in clusters of 1–3 (Ø 4–7 cm), usually overtaken by leaves; peduncles 20–25 mm; sepals (15–18 mm) laciniate, after flowering sloping and rapidly drooping; petals obcuneate-bilobed (19–25 × 20–25 mm), pinkish on lobes, ± white on the rest; disc ± conical with orifice larger than 1 mm; styluses mostly woolly and elongate, forming a cylindrical column. Fruit pyriform (1–2 cm), red.

Subspecies Within *Rosa canina*, a myriad of forms and varieties have been described, all of which are still waiting for a definitive taxonomic arrangement.

Cultivar They are so numerous and continuously increasing that even trying to draw up a simple list is unrealistic.

Related Species Of the approximately 435 species recognized in the genus *Rosa*, many probably have a potential for beer-making application similar to that cited here for some of these taxa.

Geographic Distribution Paleotemperate.
Habitat Degraded woodlands, shrublands, hedgerows. **Range** Europe, Africa, Asia.

Beer-Making Parts (and Possible Uses) Flower buds, fruits.

Chemistry Fruits: ash, oils, energy, fiber, protein, ascorbic acid, dimethyl sulfite (DMS), minerals (potassium, phosphorus, magnesium, calcium, iron). Essential oil, seeds: palmitic acid 3.17%, stearic acid 2.47%, oleic acid 16.73%, linoleic acid 54.41%, linolenic acid 17.14%, arachidic acid 2.11%, minerals, iron.

Style American Amber Ale, American IPA, Belgian Blond Ale, Belgian Dubbel, Belgian Pale Ale, Belgian Specialty Ale, Bière de Garde, Cream Ale, Dunkles Bock, Holiday/Winter Special Spiced Beer, Saison, Specialty Beer, Spice/Herb/Vegetable Beer, Witbier.

Source Calagione et al. 2018; Dabove 2015; Heilshorn 2017; Jackson 1991; Fisher and Fisher 2017; https://www.bjcp.org (30. Spiced Beer); https://www.brewersfriend.com/other/nyper/; https://www.brewersfriend.com/other/rose/; https://www.brewers-friend.com/other/rosehips/; https://www.fermento-birra.com/homebrewing/birre-speziali/.

Rosa rugosa Thunb.
ROSACEAE

Synonyms *Rosa andreae* Lange, *Rosa coruscans* Waitz ex Link, *Rosa ferox* Lawrance, *Rosa ferox* Sol., *Rosa pubescens* Baker, *Rosa regeliana* Linden & André.

Common Names Japanese rose, rugose rose, sitka rose.

Description Shrub (10–25 dm) forming dense brush. Stem erect, sometimes arcuate, densely branched; bark greenish-white when young, purplish-black with age, densely tomentose or puberulent; infrastipular spines paired, erect, rigid, subulate, 10 × 4 mm, internode spines similar, densely intermixed with acicular and stipitate glands. Leaves 7–11 cm; stipules 20–30 × 4–7 mm, auricles flared, 4–6 mm, margins entire or unevenly serrate, sessile-glandular, surface rugose, glabrous or pubescent, glandular; petiole and rachis with spines usually curved, subequal, pubescent, sparsely glandular; leaflets 5–9, terminal with petiole 8–18 mm, lamina elliptic to ovate, rarely obovate (20–55 × 10–35 mm), leathery, base cuneate to obtuse, margins usually 1-crenate-serrate, ± glandular, teeth 11–17 on each side, sometimes surmounted by a gland, apex obtuse to acute; underleaf gray-green, deeply innervate, pubescent, upper leaf dark green, glossy, glabrous. Inflorescence corymb, 1- to 4-flowered; peduncle erect, sometimes reflexed (with mature hips), rigid, 10–30 mm, pubescent, sometimes bristly, sparsely sessile-glandular or stipitate-glandular; bracts 2, ovate-lanceolate, 12–15 × 2–5 mm, margins entire with few hairs, ± glandular, surface pubescent, glandular. Flowers Ø 6–9 cm; hypanthium depressed-globose (6–8 × 5–6 mm), glabrous, sometimes ± bristly, rarely glandular, collar 1–2 × 4–5 mm; sepals erect, ovate-lanceolate (20–37 × 4–6 mm), ends 4–5 × 1–2 mm, margins entire, rarely pinnatifid, tomentose, glandular, underleaf pubescent, sessile-glandular and stipitate-glandular, often silky; petals single, sometimes double (in certain cultivars), purplish, pink, or white in color, 35–50 × 30–45 mm; carpels 48–60, stylus exsert 1–2 mm beyond the stylar orifice of the hypanthium disk (Ø 5–10 mm). Hip scarlet in color, depressed-globose, 18–20 × 20–25 mm, leathery, glabrous, ± bristly, ± glandular, collar 1–2 × 4–5 mm; sepals persistent, erect. Achenes 40, 4–6 × 2–4.5 mm.

Subspecies In addition to the typical form, *Rosa rugosa* Thunb. f. *alboplena* (Rehder) Rehder and *Rosa rugosa* Thunb. f. *rosea* (Rehder) Rehder are recognized.

Cultivar More than 50 cultivars and hybrids of this species are known. Some examples: Alba, Alboplena, Blanc Double de Coubert, F. J. Grootendorst, Frau Dagmar Hastrup, Grootendorst Pink, Hansa, Jens Munk, Therese Bugnet, Topaz Jewel.

Related Species The genus *Rosa* has 435 taxa; *Rosa arkansana* Porter, *Rosa palustris* Marshall, and *Rosa virginiana* Mill. have established uses in brewing.

Geographic Distribution Asia (N China, Japan, Siberia). **Habitat** Disturbed areas, roadsides, fields, maritime dunes and cliffs, shorelines, riparian stands (0–1,000 m ASL). **Range** Asia, North America, Europe (naturalized).

Beer-Making Parts (and Possible Uses) Mature cinorrhoids, flowers, roots.

Chemistry Composition (µg/mg) in terms of phenolic acids and flavonoid glycosides: caffeic acid 0.70, gentisic acid 0.17, protocatechuic acid 1.10, gallic acid 8.30, salicylic acid 0.20, sinapic acid 0.21, p-coumaric acid 2.04, 3-hydroxybenzoic acid 0.07, 4-hydroxybenzoic acid 1.37, quercetin-3-L-arabinofuranoside (avicularin) 3.08, guercetin-3-O-galactoside (hyperoside) 1.30, quercetin-3-O-glucoside (isoquercitrin) 1.95, quercetin-4´-O-glucoside (spiraeoside) 0.14, kaempferol-3-O-(6″-O-(E)-p-coumaroyl)-glucoside (tiliroside) 4.36, kaempferol-3-O-glucoside (astragalin) 0.38, quercetin-3-O-rhamnoside (quercitrin) 0.30.

Style Rose Beer (?).

Source Fisher and Fisher 2017; Hieronymus 2016b; Josephson et al. 2016.

Rosmarinus officinalis L.
LAMIACEAE

Synonyms *Rosmarinus angustifolius* Mill., *Rosmarinus flexuosus* Jord. & Fourr., *Rosmarinus latifolius* Mill., *Rosmarinus laxiflorus* Noë, *Rosmarinus laxiflorus* Noë ex Lange, *Rosmarinus palaui* (O.Bolòs & Molin.) Rivas Mart. & M.J.Costa, *Rosmarinus prostratus* Mazziari, *Rosmarinus rigidus* Jord. & Fourr., *Rosmarinus serotinus* Loscos, *Rosmarinus tenuifolius* Jord. & Fourr., *Salvia fasciculata* Fernald, *Salvia rosmarinus* Schleid.

Common Names Rosemary.

Description Shrub or bush (3–20 dm) with an intense aromatic scent. Stem with prostrate or ascending branches, sometimes erect. Bark light brown. Leaves linear, revolute at the edge, then apparently 2–3 × 15–28 mm, dark green and shiny above, white and tomentose below; usually clustered in axillary bundles. Inflorescence short raceme, 4- to 16-flowered. Flower calyx pubescent bilabiate, 5–6 mm, divided by 1/3; corolla 10–12 mm, light blue or lilac, rarely pinkish or white.

Subspecies Taxonomic classification of the considerable variability of this species should be considered provisional. According to some sources, the significant subspecies would be at least three: *Rosmarinus officinalis* L. subsp. *officinalis*, *Rosmarinus officinalis* L. subsp. *palaui* (O.Bolòs & Molin.) Rivas Mart. & M.J.Costa, and *Rosmarinus officinalis* L. subsp. *valentinus* P.P.Ferrer, A.Guillén & Gòmez Nav.

Cultivar According to recent studies, there are at least five chemotypes of this species. Among the numerous cultivars selected for ornamental purposes, the most common are Albus (white flowers), Arp (leaves light green color, lemon aroma), Aureus (leaves spotted with yellow), Benenden Blue (narrow leaves, dark green color), Blue Boy (dwarf, small leaves), Blue Rain (pink flowers), Gold Dust (dark green leaves with more golden streaks than Golden Rain), Golden Rain (green leaves with yellow streaks), Haifa (low and small form, white flowers), Irene (low, slender, creeping, deep blue flowers), Ken Taylor (shrub), Lockwood de Forest (prostrate, derived from Tuscan Blue), Majorica Pink (pink flowers), Miss Jessop's Upright (distinctive fastigiate tall form, wider leaves), Pinkie (pink leaves), Prostratus (appressed to the ground), Pyramidalis (or Erectus, fastigiate form, pale blue flowers), Remembrance (or Gallipoli, from the Turkish Peninsula), Roseus (pink flowers), Salem (light blue flowers, cold hardy, similar to Arp), Severn Sea (spreading, low growing, arching branches, dark violet flowers), Sudbury Blue (blue flowers), Tuscan Blue (traditional upright hardy form), and Wilma's Gold (yellow leaves).

Related Species In addition to *Rosmarinus officinalis* L., the genus includes two species, *Rosmarinus eriocalyx* Jord. & Fourr. and *Rosmarinus tomentosus* Hub.-Mor. & Maire, and two hybrids, *Rosmarinus* × *lavandulaceus* Noë (= R. eriocalyx × R. officinalis) and *Rosmarinus* × *mendizabalii* Sagredo ex Rosua (= R. officinalis × R. tomentosus). There are no known beer-making uses for these other taxa.

Geographic Distribution Steno-Mediterranean. **Habitat** Scrubs and garigues (limestone). **Range** Widely cultivated for ornamentation and seasoning.

Beer-Making Parts (and Possible Uses) Leaves, flowers.

Chemistry Essential oil: α-pinene 21.3%, camphene 8.7%, β-pinene 4.7%, β-myrcene 1.3%, p-cymene 1.4%, cineol 28.5%, β-terpinene 0.3%, terpinolene 0.3%, camphor 27.7%, borneol 2.5%, α-terpineol 0.7%, bornyl acetate 1.3%, caryophyllene 1.1%.

Style American Amber Ale, American Barleywine, American Brown Ale, American IPA, American Pale Ale, American Stout, American Wheat or Rye Beer, Baltic Porter, Belgian Golden Strong Ale, Belgian Specialty Ale, Belgian Tripel, Bière de Garde, British Golden Ale, Brown Porter, California Common Beer, English Barleywine, English IPA, Experimental Beer, Extra Special/Strong Bitter (ESB), Gruit, Holiday/Winter Special Spiced Beer, Imperial IPA, Irish Red Ale, Kölsch, Metheglin, Old Ale, Other Smoked Beer, Roggenbier (German Rye Beer), Saison, Scottish Export 80/-, Semi-sweet Mead, Specialty Beer, Specialty IPA (Belgian IPA, White IPA), Spice/Herb/Vegetable Beer, Strong Bitter, Winter Seasonal Beer, Witbier.

Beer XXI Quattro, ECB (Rome, Lazio, Italy).

Source Calagione et al. 2018; Fisher and Fisher 2017; Giaccone and Signoroni 2017; Hamilton 2014; Hieronymus 2016b; Josephson et al. 2016; McGovern 2017; Serna-Saldivar 2010; Steele 2012; https://www.brewersfriend.com/other/rosemary/.

Rubia tinctorum L.
RUBIACEAE

Synonyms *Galium rubia* E.H.L.Krause, *Rubia acaliculata* Cav., *Rubia iberica* (Fisch. ex DC.) K.Koch, *Rubia sativa* Guadagno, *Rubia sylvestris* Mill.

Common Names Common madder, dyer's madder.

Description Herb (3–10 dm), rough on stems, leaves on short downward-pointing spines. Stem tetragonal, with 4 cartilaginous wings 1–2 mm across, flexible but tough, scandent. Leaves deciduous, verticillate to 6, usually lanceolate (6–25 × 14–50 mm), with 3 reticulate veins, erect-patent and soft when young, then ± reflexed, leathery, shiny. Flower with subnullus calyx; corolla (5–7 mm) white or yellowish, with 4 rounded laciniae, joined only at the base. Inflorescence panicle axillary or terminal. Fruit black, shiny, spherical berry (Ø 4 mm).

Related Species For the 87 species belonging to the genus *Rubia*, there are no beer-making applications other than that for *R. tinctorum*.

Geographic Distribution WC Asia. **Habitat** Scrubland, hedgerows. **Range** Widely cultivated for dyeing purposes and often wild.

Beer-Making Parts (and Possible Uses) Roots (?).

Toxicity Owing to the presence of the alkaloid lucidine, a suspected carcinogen, *Rubia tinctorum* cannot be used in food or cosmetics.

Chemistry Root: purpurin, mollugin, 1-hydroxy-2-methylanthraquinone, 2-ethoxymethylanthraquinone, rubiadin, 1,3-dihydroxyanthraquinone, 7-hydroxy-2-methylanthraquinone, polishin, 1-methoxymethylanthraquinone, 2,6-dihydroxyanthraquinone, lucidin-3-O-primveroside, alizarin, lucidin-O-ethyl ether, 1-hydroxy-2-hydroxymethylanthraquinone 3-glucoside, 2-hydroxymethylanthraquinone 3-glucoside, 3,8-dihydroxy-2-hydroxymethylanthraquinone 3-glucoside, ruberitric acid, quinizarin, iridoid asperuloside, tectoquinone, nordamnacanthal, 1-hydroxy-2-methoxyanthraquinone, 1,3-dihydroxy-2-ethoxymethylanthraquinone, scopoletin. Essential oil, aerial parts (34 components; 96.4% of the oil): pentadecanal 20.2%, tridecanal 16.7%, globulol 7.8%.

Style Mum.

Source Hieronymus 2010.

Rosmarinus officinalis L.
LAMIACEAE

Rubus arcticus L.
ROSACEAE

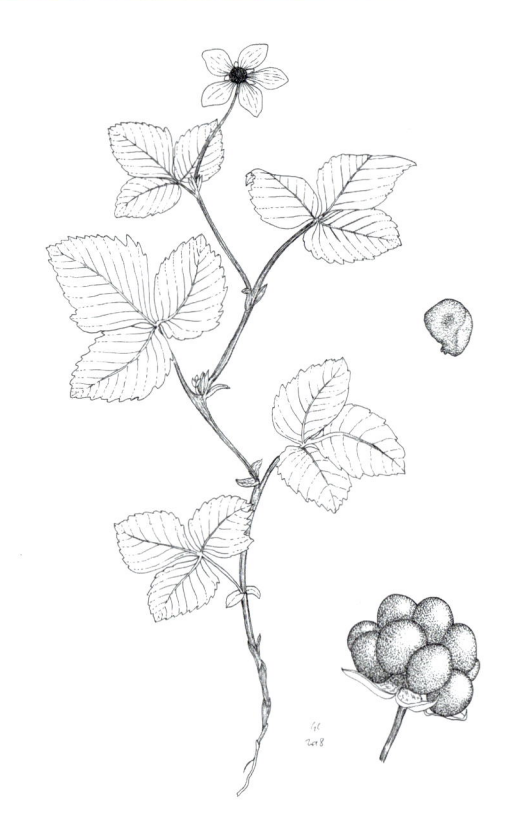

Synonyms *Cylastis arcticus* (L.) Raf. ex B.D.Jacks.

Common Names Arctic bramble, arctic raspberry, dwarf raspberry.

Description Shrub, small (10–30 cm). Roots associated with a small, creeping, semiwoody rhizome. Stems slender, ascending, green or brown in color, branched, covered with fine hairs. Leaves composed of 3 leaflets; longer petiole hairy, leaflets rhombic to obovate in shape, margin irregularly serrated, sometimes superficially incised. Flowers small (Ø about 15 mm), apical, solitary (1–3); petals bright pink; sepals 5–10, bending backward as fruit enlarges; axillary pedicel covered with dense hairs; stamens erect, shorter than petals; pistils 20, lower than stamens. Fruit (polydrupe?) small globular berry (Ø < 1 cm), bright red in color.

Subspecies Besides the typical subspecies, *Rubus arcticus* L. subsp. *arcticus*, *Rubus arcticus* L. subsp. *acaulis* (Michx.) Focke, and *Rubus arcticus* L. subsp. *stellatus* (Sm.) B.Boivin are recognized.

Cultivar Examples of varietal strains of *Rubus arcticus* include 39, 57, Anna, Aura, Astra, Beata, E1, Elpee, Marika, Mespi, Muuruska, Pima, and Susanna (in cultivation, combinations of several cultivars are often used because of self-incompatibility).

Related Species Of the more than 1,500 species belonging to the genus *Rubus*, many produce edible fruits and therefore are potentially useful to the beer-making process; however, just a few are known to be used for this purpose.

Geographic Distribution Alaska, Canada, N United States (Oregon, Colorado, Michigan, Maine), Scandinavia, Russia, Great Britain, Netherlands, Poland, Belarus, Mongolia, NE China, N Korea, Estonia, Lithuania. **Habitat** Acidic soils with a high content of organic matter, in woods, slopes, gullies (0–1,200 m ASL). **Range** Circumboreal.

Beer-Making Parts (and Possible Uses) Fruits.

Chemistry Fruits: soluble solids 6.6–7.5%, titratable acidity 0.27–0.48%, ascorbic acid 12–19 mg/100 g.

Style Fruit Beer.

Source Jackson 1991.

Rubus argutus Link
ROSACEAE

Synonyms *Rubus abundiflorus* L.H.Bailey, *Rubus betulifolius* Small, *Rubus floridensis* L.H.Bailey, *Rubus floridus* Tratt., *Rubus incisifrons* L.H.Bailey,

Rubus louisianus A.Berger, *Rubus penetrans* L.H.Bailey, *Rubus rhodophyllus* Rydb.

Common Names Blackberry, sawtooth blackberry, tall blackberry, highbush blackberry.

Description Shrub (up to 2.5 m). Branches often curved to touch the ground at the tip, slightly woody, light green to dark red in color, stiff, angular, furrowed or ridged, glabrous; sharp spines, less than 6 mm long, straight to slightly curved. Leaves alternate, palmate-compound; 1-year branches have palmate leaves with 5 elliptic to ovate leaflets (7.5–12.5 × 2.5–6 cm), toothed; 2-year branches have palmate leaves with 3 leaflets similar to younger ones but slightly smaller; color above green or yellow-green, jagged along veins; leaves glabrous or with scattered appressed hairs, underside light green with prominent veins; underside veins glabrous or with appressed hairs; leaflets petioles, light green or yellow-green, glabrous or covered with appressed hairs; leaf petiole 5–10 cm, with small spines; at base of petiole a pair of deciduous linear stipules (less than 1.5 cm long). Inflorescence elongated raceme (5–15 cm) with 5–20 flowers; raceme bracts small and deciduous. Flowers (about 2–2.5 cm) hermaphrodite, with 5 white, oblong to elliptic petals; 5 light green, lanceolate, recurved sepals, covered with dense appressed white hairs. Fruits compound drupes, red to black when ripe, globoid to ovoid, 0.8–1.7 cm long, juicy; each fruit contains a single yellow seed.

Related Species Of the more than 1,500 species belonging to the genus *Rubus*, many produce edible fruits and therefore are potentially useful to the beer-making process; however, just a few are known to be used for this purpose.

Geographic Distribution N America.
Habitat Preforest environments, disturbed habitats, sands. **Range** N America; invasive in Australia, New Zealand, and elsewhere.

Beer-Making Parts (and Possible Uses) Fruits.

Chemistry Fruits (composition similar to *Rubus laciniatus*): energy 180 kJ (43 kcal), carbohydrates 9.61 g, sugars 4.88 g, fiber 5.3 g, lipids 0.49 g, protein 1.39 g, vitamin A 214 IU, thiamin (B1) 0.02 mg, riboflavin (B2) 0.026 mg, niacin (B3) 0.646 mg, vitamin B6 0.03 mg, folate (B9) 25 µg, vitamin C 21 mg, vitamin E 1.17 mg, vitamin K 19.8 µg, calcium 29 mg, iron 0.62 mg, magnesium 20 mg, phosphorus 22 mg, potassium 162 mg, sodium 1 mg, zinc 0.53 mg.

Style Fruit Beer, Sour Ale.

Beer Thicket, Side Project (Missouri, United States).

Source Hieronymus 2016b (generic reference to *Rubus*, no specifics given); https://www.craftbeer.com/news/beer-release/oregon-fruit-products-collaborates-with-breakside-brewery.

Rubus chamaemorus L.
ROSACEAE

Synonyms *Chamaemorus anglica* Clus. ex Greene, *Chamaemorus anglicus* Greene, *Chamaemorus chamaemorus* (L.) House, *Chamaemorus norvegicus* Greene, *Chamaemorus norwegica* Clus. ex Greene, *Rubus nubis* Gray, *Rubus pseudochamaemorus* Tolm., *Rubus yessoicus* Kuntze.

Common Names Cloudberry, bakeapple, knotberry, knoutberry, low-bush salmonberry.

Description Herb, perennial (10–25 cm), rhizomatous, dioecious. Rhizomes (up to 10 m long) develop about 10–15 cm below the surface, forming large, dense formations. Stem single, slender. Leaves (1–4) alternate, long petiolate (1–7 cm), lamina (2–5 × 3–7 cm) reniform, pentalobate (5–7 lobes), deeply cordate at base, wrinkled, margin serrate; petiole and underleaf covered with glandular hairs. Flower solitary, white, 5 petals (8–12 mm) obovate, 5 sepals; peduncle and sepals covered with glandular hairs. Fruit polydrupe composed of 4–25 fleshy drupes, semitransparent orange color, soft, tasty and fragrant when ripe, sweet flavor with an astringent note. Seed 1 for each drupelet.

Cultivar A real diversification in cultivars is not known, but it seems certain that the chemical characteristics of the fruit change according to the environment of origin. Some Nordic countries are trying clonal selection to make the cultivation of this plant profitable in environments otherwise uncultivable.

Related Species Of the more than 1,500 species belonging to the genus *Rubus*, many produce edible fruits and therefore are potentially useful to the beer-making process; however, just a few are known to be used for this purpose.

Geographic Distribution Circumboreal, circumpolar. **Habitat** Wet sparse forests, alpine heaths, sphagnum bogs. **Range** N Eurasia, N America.

Beer-Making Parts (and Possible Uses) Fruits.

Chemistry Fruits: energy 51 kcal, water 87 g, protein 2.4 g, lipids 0.8 g, carbohydrates 8.6 g, calcium 18 mg, iron 0.7 mg, phosphorus 35 mg, vitamin C 158 mg, thiamin 0.05 mg, riboflavin 0.07 mg, niacin 0.9 mg, vitamin A 210 IU, benzoic acid, vanillin (and derivatives), ellagic acid, flavonols (quercetin, kaempferol, myricetin), hydroxycinnamic acids (p-coumaric, caffeic, ferulic), p-hydroxybenzoic acid, rubyxanthin (3-hydroxy-[-carotene]), phytoestrogens. Essential oil, seeds (9.1–12.4%): triacylglycerols (linoleic, linolenic, oleic, palmitic acid) 95%, carotenes, tocopherol.

Style American Amber Ale, American IPA, Fruit Beer.

Beer My Name Is Little Ingrid, Brewdog (Fraserburgh, Scotland, United Kingdom).

Source Dabove 2015; Fisher and Fisher 2017; Jackson 1991; McGovern 2017; https://www.brewersfriend.com/other/cloudberries/; https://www.brewersfriend.com/other/moroshka/.

Rubus Chehalem × Olallie
ROSACEAE

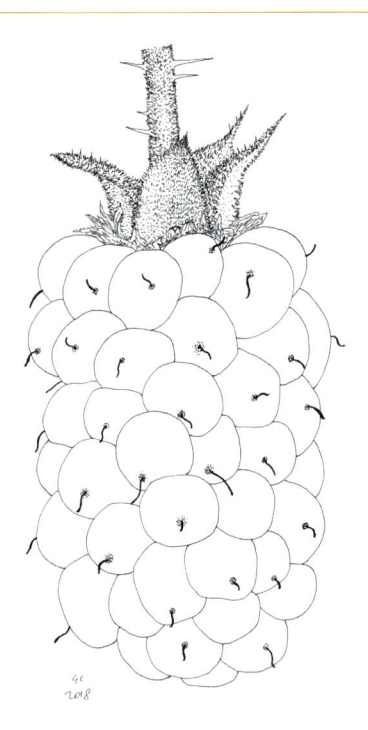

Synonyms None

Common Names Marionberry.

Description Shrub with a morphology very similar to *Rubus ursinus*. Branches up to 6 m long, very vigorous, with strong thorns; fruiting laterals long and strong, very productive. Fruit polydrupe, large (3–4 g), shiny, black, conical-elongated shape, longer (up to 2.5–3 cm) than wide; flavorful, very aromatic, slightly acid, earthy with traces of sweetness; much appreciated by consumers both for direct consumption and for processing.

Cultivar Cross between *Rubus* Chehalem (*Rubus armeniacus* × *Rubus ursinus*) and *Rubus* Olallie (Blackberry Black Logan × *Rubus caesius*) obtained in 1956 by the U.S. Department of Agriculture, Agricultural Research Service, crossbreeding program in cooperation with Oregon State University.

Related Species Of more than 1,500 species belonging to the genus *Rubus*, many produce edible fruits and therefore are potentially useful to the beer-making process; however, just a few are known to be used for this purpose.

Geographic Distribution United States (Oregon). **Habitat** Cultivated. **Range** United States (about 50% of Oregon blackberry production is represented by this hybrid).

Beer-Making Parts (and Possible Uses)
Fruits.

Chemistry Fruits (in syrup): energy 89 kcal, protein 0.74 g, carbohydrates 22.22 g, fiber 3 g, sugars 19.26 g, calcium 15 mg, iron 0.27 mg, vitamin C 2.7 mg, vitamin K, anthocyanins, gallic acid, rutin.

Style American Porter, Fruit Beer, Northern English Brown Ale.

Beer Marionberry Sour, Rogue (Oregon, United States).

Source https://www.brewersfriend.com/other/marionberries/.

Rubus idaeus L.
ROSACEAE

Synonyms *Batidaea idea* (L.) Nieuwl., *Batidaea vulgaris* Nieuwl., *Batidea peramoena* Greene, *Rubus acanthocladus* Borbàs, *Rubus buschii* (Rozanova) Grossh., *Rubus chrysocarpus* Čelak. ex Gayer, *Rubus* × *euroasiaticus* Sinkova, *Rubus fragrans* Salisb., *Rubus frambaesianus* Lam., *Rubus glaber* Mill. ex Simonk., *Rubus greeneanus* L.H.Bailey, *Rubus leesii* Bab., *Rubus obtusifolius* Willd., *Rubus sericeus* Gilib., *Rubus vulgatus* Rozanova.

Common Names Raspberry, red raspberry, European raspberry.

Description Shrub (1–2 m). Stem woody, erect, branched, often spiny; shoots of the year herbaceous, glabrous. Leaves compound with 3–5 pinnate segments; stipules linear; petiole 2–4 cm, spiny; segments lanceolate (2–4 × 4–6 cm), irregularly serrate on edges, acuminate, white-tomentose below; sucker leaves up to 25 cm long, with segments up to 10 × 15 cm, sometimes 3-lobed. Flowers in cymes, 2–5 florets; peduncles 1 cm; sepals triangular, 3 × 10 mm, folded down after flowering; petals narrowly obovate, white. Fruit subspherical polydrupe (Ø 1 cm), red-felted, with a deep hollow corresponding to the receptacle (pyramidal) that remains attached to the peduncle.

Subspecies The common name "raspberry" is a textbook case of the confusion that can be generated by the use of the common name rather than the scientific name. In fact, the beer-making raspberry referred to by various sources corresponds to several species belonging to the genus *Rubus* subgenus *Idaeobatus* (*Rubus crataegifolius* Bunge [Asian raspberry], *Rubus gunnianus* Hook. [Tasmanian alpine raspberry], *Rubus idaeus* L. [European red raspberry], *Rubus leucodermis* Douglas ex Torr. and A.Gray [Whitebark or Western raspberry, Blue raspberry, Black raspberry], *Rubus occidentalis* L. [Black raspberry], *Rubus parvifolius* L. [Australian native raspberry], *Rubus phoenicolasius* Maxim. [Wine raspberry or Wineberry], *Rubus rosifolius* Sm. [Mauritius raspberry], *Rubus strigosus* Michx. [American red raspberry], *Rubus ellipticus* Sm. [Yellow Himalayan raspberry]) and to species of the genus *Rubus* belonging to other subgenera (*Rubus arcticus* L. [Arctic raspberry, subgenus *Cyclactis*], *Rubus deliciosus* Torr. [Boulder raspberry, subgenus *Anoplobatus*], *Rubus nivalis* Douglas ex Hook. [Snow raspberry, subgenus *Chamaebatus*], *Rubus odoratus* L. [Flowering raspberry, subgenus *Anoplobatus*], *Rubus sieboldii* Blume [Molucca raspberry, subgenus *Malachobatus*]). In addition, there is considerable internal variability within the taxon *Rubus idaeus*, formalized by the description of forms and subspecies such as *Rubus idaeus* L. f. *albus* (Weston) Rehder, *Rubus idaeus* L. f. *biflorus* (Weston) Rehder, *Rubus idaeus* L. f. *laevis* (Weston) Rehder, *Rubus idaeus* L. f. *sterilis* (Kóhler) Focke, *Rubus idaeus* L. f. *succineus* Rehder, *Rubus idaeus* L. f. *tonsus* Fernald, *Rubus idaeus* L. subsp. *melanolasius* Dieck ex Focke, and *Rubus idaeus* L. subsp. *strigosus* (Michx.) Focke.

Cultivar The common name "raspberry" refers to a large group of plants grown for fruit throughout temperate regions of the world. Many modern commercial red raspberry cultivars are derived from hybrids between *Rubus idaeus* and *Rubus strigosus*, species so closely related that some taxonomists consider them subspecies of *Rubus idaeus* (*R. idaeus* subsp. *idaeus* and *R. idaeus* subsp. *strigosus*). Recent crosses have produced erect, thornless cultivars. The black raspberry (*Rubus occidentalis* L.) is grown for fresh and frozen fruit and for canning. Purple raspberries have been obtained by artificial hybridization (but also natural in some cases, such as in Vermont, where the two parental species coexist) between red and black raspberries. Although the commercial production of purple raspberries is rare, at least one cultivar, Columbian, is quite widespread in Canada. Both red and black raspberries have natural (or artificial) albino variants, with a pale yellow color owing to the presence of recessive genes that prevent the production of anthocyanin pigments. The fruits of these plants are called golden raspberries or yellow raspberries. Red raspberries have also been crossed with many species belonging to subgenera of the genus *Rubus*, producing a number of hybrids (loganberry, boysenberry, tayberry) with many Asian species as well. Several cultivars have been selected, divided into groups according to the time of production and the color of infructescence: red cultivars with early summer fruiting (Boyne, Cascade Dawn, Fertödi Venus, Glen Clova, Glen Moy, Killarney, Malahat, Malling Exploit, Malling Jewel, Rubin Bulgarski, Titan, Willamette), red cultivars with midsummer fruiting (Chemainus, Cowichan, Cuthbert, Glen Ample, Glen Prosen, Lloyd George, Meeker, Newburgh, Ripley, Saanich, Skeena), red cultivars with late-summer fruiting (Cascade Delight, Coho, Fertödi Rubina, Leo, Malling Admiral, Octavia, Schoenemann, Tulameen), red cultivars with fall fruiting (Amity, Augusta, Autumn Bliss, Caroline, Fertödi Kétszertermö, Heritage, Imara, Joan J. Thornless, Josephine, Kwanza, Kweli, Mapema, Rafiki, Ripley, Summit, Zeva Herbsternte), golden/yellow cultivars with fall fruiting (Anne, Fallgold, Fertödi Aranyfürt, Golden Queen, Goldenwest, Honey Queen, Jambo, Kiwi Gold), purple cultivars (Brandywine, Glencoe, Purple, Royalty), and black cultivars (Black, Black Hawk, Bristol, Cumberland, Jewel, Munger, Ohio Everbearer, Scepter).

Related Species Of the more than 1,500 species belonging to the genus *Rubus*, many produce edible fruits and therefore are potentially useful to the beer-making process; however, just a few are known to be used for this purpose.

Geographic Distribution Circumboreal.
Habitat Glades, mantles, forest clearings; nitrophilous species. **Range** Russia, Poland, Serbia, United States, Mexico.

Beer-Making Parts (and Possible Uses)
Fruits.

Chemistry Fruits: energy 220 kJ (53 kcal), water 85.8 g, carbohydrates 11.94 g, sugars 4.42 g, fiber 6.5 g, lipids 0.65 g, protein 1.2 g, thiamin (B1) 0.032 mg, riboflavin (B2) 0.038 mg, niacin (B3) 0.598 mg, pantothenic acid (B5) 0.329 mg, vitamin B6 0.055 mg, folate (B9) 21 µg, choline 12.3 mg, vitamin C 26.2 mg, vitamin E 0.87 mg, vitamin K 7.8 µg, calcium 25 mg, iron 0.69 mg, magnesium 22 mg, manganese 0.67 mg, phosphorus 29 mg, potassium 151 mg, zinc 0.42 mg.

Style American Amber Ale, American Brown Ale, American IPA, American Pale Ale, American Porter, American Stout, American Strong Ale, American Wheat or Rye Beer, Baltic Porter, Belgian Blond Ale, Belgian Dark Strong Ale, Belgian Pale Ale, Belgian Specialty Ale, Belgian Tripel, Berliner Weisse, Blonde Ale, Brown Porter, California Common Beer, Classic American Pilsner, Cream Ale, Czech Pale Lager, Dry Stout, English IPA, Flanders Brown Ale/Oud Bruin, Foreign Extra Stout, Fruit And Spice Beer, Fruit Beer, Fruit Lambic, German Helles Exportbier, Holiday/Winter Special Spiced Beer, Irish Red Ale, Mild, Mixed-fermentation Sour Beer, Oatmeal Stout, Oktoberfest/Märzen, Other Smoked Beer, Robust Porter, Roggenbier (German Rye Beer), Russian Imperial Stout, Saison, Scottish Light 60/-, Specialty Beer, Specialty Fruit Beer, Specialty IPA (Belgian IPA), Straight (Unblended) Lambic, Sweet Stout, Weizen/Weissbier, Weizenbock, Wild Specialty Beer, Witbier.

Beer Rosé de Gambrinus, Cantillon (Brussels, Belgium).

Source
Calagione et al. 2018; Cantwell and Bouckaert 2016; Cantwell and Bouckaert 2018; Dabove 2015; Fisher and Fisher 2017; Heilshorn 2017; Hieronymus 2010; Hieronymus 2016b; Jackson 1991; Josephson et al. 2016; Sparrow 2015; https://www.bjcp.org (29A. Fruit Beer); https://www.brewersfriend.com/other/rasberries; https://www.brewersfriend.com/other/rasberry/; https://www.omnipollo.com/beer/bianca-raspberry/.

***Rubus idaeus* L.**
ROSACEAE

Rubus laciniatus (Weston) Willd.
ROSACEAE

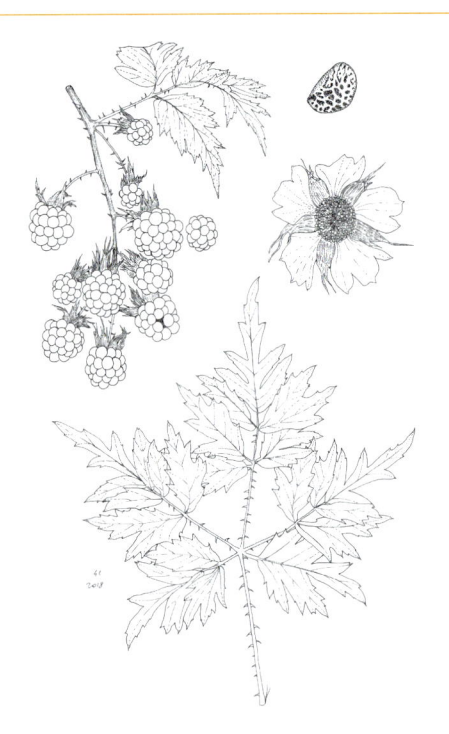

Synonyms None

Common Names Blackberry.

Description Shrub, semi-erect to erect (up to 3 m), with branches often arched downward, often forming dense formations. Stems covered at the end with numerous curved and rigid thorns; stems usually biennial, the sterile ones in the first year develop from buds at ground level or below, producing only leaves; lateral branches develop at the axils of the previous ones during the second year and carry both flowers and leaves. Leaves evergreen, composed of 5 palmate or pinnate-compound leaflets; leaves green on both sides, hairy underneath; leaflets laciniate to dissected. Flowers hermaphrodite, white to pinkish, clustered in paniculate compound cymes. Fruit round polydrupe (Ø 2 cm), glossy black, consisting of several sweet, succulent drupes.

Subspecies *Rubus laciniatus* (Weston) Willd. f. *laciniatus* represents the typical form of the taxon, but a second form, *Rubus laciniatus* (Weston) Willd. f. *elegans* (Bean) Rehder, is also known.

Related Species Of the more than 1,500 species belonging to the genus *Rubus*, many produce edible fruits and therefore are potentially useful to the beer-making process; however, just a few are known to be used for this purpose.

Geographic Distribution North America. **Habitat** Preforest, disturbed habitats. **Range** North America.

Beer-Making Parts (and Possible Uses) Fruits.

Chemistry Fruits: energy 180 kJ (43 kcal), carbohydrates 9.61 g, sugars 4.88 g, fiber 5.3 g, lipids 0.49 g, protein 1.39 g, vitamin A 214 IU, thiamin (B1) 0.02 mg, riboflavin (B2) 0.026 mg, niacin (B3) 0.646 mg, vitamin B6 0.03 mg, folate (B9) 25 µg, vitamin C 21 mg, vitamin E 1.17 mg, vitamin K 19.8 µg, calcium 29 mg, iron 0.62 mg, magnesium 20 mg, phosphorus 22 mg, potassium 162 mg, sodium 1 mg, zinc 0.53 mg.

Style Fruit Beer.

Source https://www.craftbeer.com/news/beer-release/oregon-fruit-products-collaborates-with-breakside-brewery.

Rubus loganobaccus L.H.Bailey
ROSACEAE

Synonyms *Rubus* × *loganobaccus* L.H.Bailey (= *Rubus ursinus* × *R. idaeus*), *Rubus ursinus* Cham. & Schltdl. var. *loganobaccus* (L.H.Bailey) L.H.Bailey.

Common Names Loganberry.

Description Shrub low, branches (up to 5 m) almost prostrate, entangled or climbing. Flower color white. Hexaploid hybrid produced by pollinating an octoploid plant of *Rubus ursinus* (cultivar Aughinbaugh) with the diploid red raspberry (*Rubus idaeus*). The plant and fruit have the general appearance of *Rubus ursinus* rather than *Rubus idaeus*. The color of the fruit, however, is dark red, not black as in *Rubus ursinus*. This plant is grown commercially and for gardening purposes.

Related Species Of the more than 1,500 species belonging to the genus *Rubus*, many produce edible fruits and therefore are potentially useful to the beer-making process; however, just a few are known to be used for this purpose.

Geographic Distribution Artificial hybrid. **Habitat** Temperate climates. **Range** North America.

Beer-Making Parts (and Possible Uses) Fruits.

Chemistry Frozen fruits: energy 55 kcal, water 84.61 g, protein 1.52 g, lipids 0.31 g, carbohydrates 13.02 g, fiber 5.3 g, sugars 7.7 g, calcium 26 mg, iron 0.64 mg, magnesium 21 mg, phosphorus 26 mg, potassium 145 mg, sodium 1 mg, zinc 0.34 mg, vitamin C 15.3 mg, thiamine 0.05 mg, riboflavin 0.034 mg, niacin 0.84 mg, vitamin B6 0.065 mg, folate 26 µg, vitamin A 2 µg, vitamin A 35 IU, vitamin E 0.87 mg, vitamin K 7.8 µg, saturated fatty acids 0.011 g, monounsaturated fatty acids 0.03 g, polyunsaturated fatty acids 0.176 g.

Style California Common Beer, Wheat Ale.

Beer Loganberry, Revolution Brewing (Illinois, United States).

Source https://www.brewersfriend.com/other/loganberries/.

Rubus parviflorus Nutt.
ROSACEAE

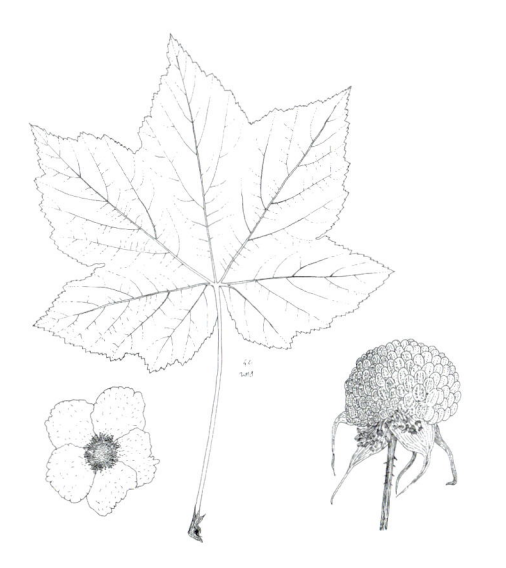

Synonyms *Bossekia nutkana* Greene, *Bossekia parviflora* (Nutt.) Greene, *Rubacer parviflorum* (Nutt.) Rydb., *Rubus nutkanus* Moc. ex Ser., *Rubus velutinus* Hook. & Arn.

Common Names Thimbleberry.

Description Shrub (5–30 dm) with no thorns. Stem biennial, erect, sparsely hairy, glabrescent, moderately glandular, glands yellowish to reddish in color, not pruinose. Leaves deciduous, simple, stipules lanceolate to ovate (5–15 mm); lamina orbiculate to reniform (5–20 × 5–25 cm), base cordate, palmate (3- to 7-lobed), margins irregularly to doubly serrate, apex briefly acuminate to obtuse, underside glabrous to densely hairy, sparsely to moderately stipitate-glandular, glands yellowish to reddish in color. Inflorescences terminal and axillary (1–15), cymiform to thyrsiform; peduncles sparsely to moderately hairy and stipitate-glandular. Flowers hermaphrodite; petals white, broadly obovate (10–28 mm); filaments filiform; ovaries distally densely hairy, stylus clavate, glabrous. Fruit hemispherical polydrupe (1–1.8 cm), red in color; drupes 50–60.

Subspecies There are two known varieties of this taxon: *Rubus parviflorus* Nutt. var. *parviflorus* and *Rubus parviflorus* Nutt. var. *fraserianus* J.K.Henry.

Related Species Of the more than 1,500 species belonging to the genus *Rubus*, many produce edible fruits and therefore are potentially useful to the beer-making process; however, just a few are known to be used for this purpose.

Geographic Distribution CE United States. **Habitat** Disturbed areas (0–400 m ASL). **Range** United States (Delaware, Illinois, Iowa, Massachusetts, Missouri, Nebraska, New Jersey, Ohio, Virginia), Asia, Australia.

Beer-Making Parts (and Possible Uses) Fruits.

Chemistry Fruits (active compounds): cyanidin-3-glucoside, cyanidin-3-rutinoside, pelargonidin-3-glucoside.

Style Fruit Beer.

Source Fisher and Fisher 2017; McGovern 2017; https://www.brewersfriend.com/other/thimbleberry/.

Rubus phoenicolasius Maxim.
ROSACEAE

Synonyms None

Common Names Japanese wineberry, wine raspberry, wineberry, dewberry.

Description Shrub (1–3 m). Stems erect at first, then bent, densely pubescent, covered with reddish-brown bristles, pedunculated glandules and scattered spines. Leaves imparipinnate, 3-foliate (rarely 5-foliolate); petiole 3–6 cm, that of the terminal leaflet 2–3 cm, lateral leaflets subsessile, rachis pubescent due to bristles, glands, and spines; stipules linear (5–8 mm), pubescent, with intertwined glandular hairs; leaflets laminae ovate, broadly ovate, or rhombic, rarely elliptic (4–10 × 2–7 cm), below densely gray-tomentose, setose, with stipitate glands, with scattered minute needle-like spines, above pubescent or pubescent only along veins, base rounded to subcordate, margin irregularly coarsely serrate, usually incised; terminal leaflet often lobed, apex acute to acuminate. Inflorescences terminal or axillary, racemes short, pauciflor, terminal racemes 6–10 cm, the lateral ones shorter; rachis, pedicels, and lower surface of calyx densely pubescent, bristly, with stipitate glands; bracts lanceolate (5–8 mm), pubescent, with stipitate glands, margin entire or apically 2-lobed; pedicel 0.5–1.5 cm. Flowers Ø 6–10 mm, sepals erect, lanceolate (1–1.5 cm × 4–7 mm), apex caudate; petals purple-red, obovate-spatulate or suborbicular, much longer than sepals, base with long claws and pubescent; stamens many, almost as long as petals; pistils barely longer than the stamens; ovary glabrous or puberulent. Fruit aggregate orange or red, subglobose (Ø about 1 cm), glabrous.

Cultivar Strains of this species have been used in crossing programs of the genus *Rubus* aimed at improving production performance in North America and elsewhere.

Related Species Of the more than 1,500 species belonging to the genus *Rubus*, many produce edible fruits and therefore are potentially useful to the beer-making process; however, just a few are known to be used for this purpose.

Geographic Distribution N China, Japan, Korea.
Habitat Mountain valleys, roadsides, low to medium altitudes. **Range** Asia, Europe, North America.

Beer-Making Parts (and Possible Uses)
Fruits.

Chemistry Fruits: energy 727.2 kcal, carbohydrates 34.5 g, fiber 0.6 g, sugars 20.3 g, protein 33.4 g lipids 40.3 g, cholesterol 129.2 mg, sodium 235 mg.

Style Weissbier.

Source https://www.brewersfriend.com/other/wineberries/.

Rubus spectabilis Pursh
ROSACEAE

Synonyms *Parmena spectabilis* (Pursh) Greene, *Rubus franciscanus* Rydb., *Rubus stenopetalus* Cham.

Common Names Salmonberry.

Description Shrub (10–40 dm), generally spiny. Stems erect to arching, glabrous or sparsely to densely hairy, glandless or sparsely to densely stipitate-glandular, not pruinose; bark papery with age; spines absent or scattered to dense, erect, slender (1–5 mm), broad to narrow base. Leaves deciduous, ternate; stipules filiform to linear (3–10 mm); terminal leaflet ovate (4–15 × 3.5–15 cm), base truncate, rounded to cordate, lobed, margins coarsely to doubly serrate, apex acute to acuminate; underside without spines or with spines erect on midrib, moderately to densely hairy, without glands or rarely stipitate-glandular along midrib. Inflorescences terminal and axillary, 1- to 2-flowered; peduncles without spines or with scattered, erect spines, moderately to densely hairy, without glands or rarely stipitate-glandular. Flowers bisexual, petals pink to magenta in color, broadly to narrowly obovate (10–30 mm); filaments laminar; ovaries glabrous. Fruits yellow, orange, or red, globose to ovoid (1–2 cm); drupes 20–80.

Related Species Of the more than 1,500 species belonging to the genus *Rubus*, many produce edible fruits and therefore are potentially useful for making beer; however, just a few are known to be used for this purpose.

Geographic Distribution W North America. **Habitat** Forests, forest canopies, swamps, shorelines, roadsides, disturbed areas, moist to wet soils (0–2,000 m ASL). **Range** Canada (British Columbia), United States (Alaska, California, Idaho, Oregon, Washington), Asia (Japan), Europe (introduced).

Beer-Making Parts (and Possible Uses) Fruits.

Chemistry Fruits: energy 47 kcal, water 88.21 g, protein 0.85 g, lipids 0.33 g, carbohydrates 10.05 g, fiber 1.9 g, sugars 3.66 g, calcium 13 mg, iron 0.40 mg, magnesium 15 mg, phosphorus 27 mg, potassium 110 mg, sodium 14 mg, zinc 0.28 mg, vitamin C 9.2 mg, thiamin 0.041 mg, riboflavin 0.062 mg, niacin 0.466 mg, vitamin B6 0.078 mg, folate 17 µg, vitamin A 50 µg, vitamin A 496 IU, vitamin E 1.61 mg, vitamin K 14.8 µg.

Style Oud Bruin.

Source Josephson et al. 2016; McGovern 2017; https://www.brewersfriend.com/other/salmonberrys/.

Rubus ulmifolius Schott
ROSACEAE

Synonyms *Rubus abruptus* Lindl., *Rubus aetneus* Tornab., *Rubus albescens* Boulay and Gillot, *Rubus appenninus* Evers, *Rubus bellidiflorus* hort. ex K.Koch, *Rubus bujedanus* Sennen & T.S.Elias, *Rubus castellanus* Sennen & T.S.Elias, *Rubus cocullotinus* Evers, *Rubus crispulus* Gand., *Rubus discolor* Syme, *Rubus discolor* Weihe & Nees, *Rubus edouardii* Sennen, *Rubus gerundensis* Sennen, *Rubus hispanicus* Willk., *Rubus inermis* A.Beek, *Rubus karstianus* Borbàs, *Rubus legionensis* Gand., *Rubus lejeunei* Weihe ex Lej., *Rubus longipetiolatus* Sennen, *Rubus minutiflorus* Lange, *Rubus oculus-junonis* Gand., *Rubus panormitanus* Tineo, *Rubus rusticanus* Mercier, *Rubus segobricensis* Pau, *Rubus siculus* C.Presl, *Rubus sinusifolius* Sennen, *Rubus × tridentinus* Evers, *Rubus valentinus* Pau.

Common Names Elm-leaf blackberry.

Description Shrub, evergreen (5–15 dm). Suckers glabrous or with appressed pubescence, pruinose, glaucous, pentagonal-grooved, on the ribs spiny and ± reddened; spines straight or ± falciform. Leaves with 3–5 segments palmate, the lesser elliptic (1.5–3 × 3–4.5 cm), the greater obovate (5 × 6 cm) or orbiculate-acuminate (5–6 cm); on flowering branches sometimes only leaves with 3 segments, the lateral deeply bilobed; upper leaf dark green, leathery, subglabrous, lower white-tomentose. Inflorescence panicle pyramidal; axis with straight or falcate spines, without glands; sepals 3-angular (3 × 7 mm), white-tomentose, at fruiting folded downward; petals ovate (9 × 13 mm), pinkish; stamens and styluses white or pinkish, about equal in length. Fruit subspherical polydrupe (Ø 1 cm), glossy black, consisting of several drupelets.

Subspecies *Rubus ulmifolius* Schott includes forms and varieties ± taxonomically accepted: *Rubus ulmifolius* Schott f. *bellidiflorus* (hort. ex Petz. & G.Kirchn.) Voss, *Rubus ulmifolius* Schott f. *inermis* (Willd.) Rehder, and *Rubus ulmifolius* Schott f. *variegatus* (G.Nicholson) Rehder are the accepted forms, whereas *Rubus ulmifolius* Schott var. *chlorocarpus* (Boreau ex Genev.) Focke, *Rubus ulmifolius* Schott var. *petiolulatus* (Timb.-Lagr. ex Malbr.) Sudre, *Rubus ulmifolius* Schott var. *polyanchus* Sudre, and *Rubus ulmifolius* Schott var. *villosispinus* Sudre are awaiting clarification of their taxonomic position.

Related Species Of the more than 1,500 species belonging to the genus *Rubus*, many produce edible fruits and therefore are potentially useful to the beer-making process; however, just a few are known to be used for this purpose.

Geographic Distribution Euro-Mediterranean. **Habitat** Hedges, fallow land, coppices. **Range** Europe, N Africa, United States (California, Nevada, New Jersey, Oregon, Washington), South America.

Beer-Making Parts (and Possible Uses) Fruits, leaves (?).

Chemistry Fruits: water 38.41%, vitamin C 16.33 mg, phenolic acids 942.38 mg, flavonols 85.61 mg, anthocyanins 298.18 mg, total phenolic compounds 1326.17 mg (gallic acid 268.72 mg, quercetin 3-galactoside 5.44 mg, quercetin 3-glucoside 18.18 mg, quercetin 3-rutinoside 6.45 mg, cyanidin 3-glucoside 86.73 mg, pelargonidin 3-rutinoside 4.23 mg, cyanidin 3-glycoside 19.49 mg). Leaves: quercetin, kaempferol, trifolin (kaempferol-3-0-galactoside), hyperin (quercetin-3-0-galactoside), chlorogenic acid, caffeic acid, kaempferol-3-0-caffeoyl ester, glucose-caffeoyl ester, free fatty acids.

Style Fruit Lambic.

Source https://www.craftbeer.com/news/beer-release/oregon-fruit-products-collaborates-with-breakside-brewery.

Rubus ursinus Cham. & Schltdl.
ROSACEAE

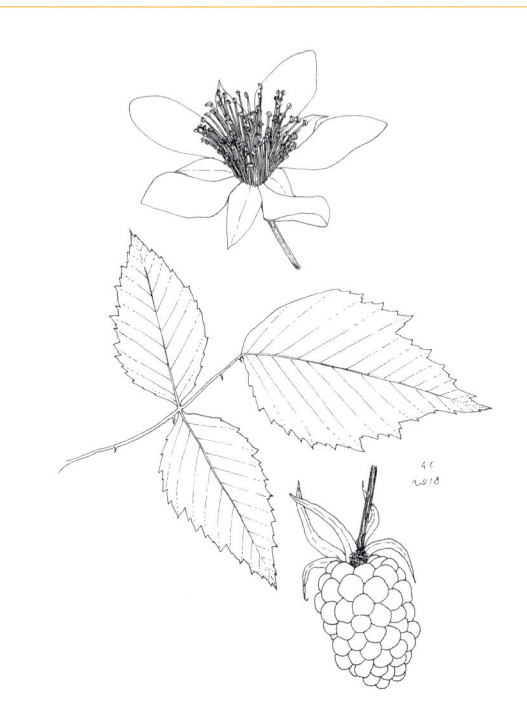

Synonyms *Parmena menziesii* (Hook.) Greene, *Rubus menziesii* Hook.

Common Names Blackberry.

Description Shrub, evergreen, also climbing (5–6 m). Stems densely spiny, greenish color when young, brown when mature. Leaves alternate, spiny, deeply lobed to palmate, pale green below; biennial stems, sterile the first year, develop from buds at ground level or below, producing only leaves; lateral branches develop at the axil of the former during the second year and bear both flowers and leaves. Flowers hermaphrodite in clusters of 2–15 toward the end of the flowering branch. Fruit polydrupe, red and hard when immature, black and shiny when mature, oblong to conical, somewhat shaggy, up to 2 cm long; drupes sweet and fragrant when mature.

Subspecies Three subspecific taxa are included in *Rubus ursinus*: *Rubus ursinus* Cham. & Schltdl.

subsp. *macropetalus* (Douglas ex Hook.) R.L.Taylor & MacBryde, *Rubus ursinus* Cham. & Schltdl. subsp. *ursinus*, and *Rubus ursinus* Cham. & Schltdl. var. *sirbenus* (L.H.Bailey) J.T.Howell.

Cultivar Rather than providing just a varietal picture within the taxon, it is worth highlighting the role of this species as a parent of commercially successful cultivated hybrids (e.g., loganberry, boysenberry).

Related Species Of the more than 1,500 species belonging to the genus *Rubus*, many produce edible fruits and therefore are potentially useful to the beer-making process; however, just a few are known to be used for this purpose.

Geographic Distribution SW North America. **Habitat** Fallow land, coppices, burned areas, scrub, shrubland, disturbed areas, dry to moist soils (0–1,600 m ASL). **Range** Canada (British Columbia), United States (California, Idaho, Montana, Oregon, Washington), Mexico (Baja California).

Beer-Making Parts (and Possible Uses) Fruits.

Chemistry Fruit juice: energy 38 kcal, water 90.9 g, protein 0.3 g, lipids 0.6 g, carbohydrates 7.8 g, fiber 0.1 g, sugars 7.7 g, calcium 12 mg, iron 0.48 mg, magnesium 21 mg, phosphorus 12 mg, potassium 135 mg, sodium 1 mg, zinc 0.41 mg, vitamin C 11.3 mg, thiamine 0.012 mg, riboflavin 0.018 mg, niacin 0.446 mg, vitamin B6 0.021 mg, folate 10 µg, vitamin A 6 µg, vitamin A 123 IU, vitamin E 0.9 mg, vitamin K 15.2 µg, saturated fatty acids 0.018 g, monounsaturated fatty acids 0.058 g, polyunsaturated fatty acids 0.344 g.

Style Fruit Beer.

Source Heilshorn 2017; Josephson et al. 2016 (this source does not offer specific rank details but only the common name "blackberry"); https://www.craftbeer.com/news/beer-release/oregon-fruit-products-collaborates-with-breakside-brewery.

Rubus ursinus × idaeus Boysen
ROSACEAE

Synonyms None

Common Names Boysenberry.

Description Shrub, deciduous, low, branches long (up to 3 m), usually drooping. Leaves compound, with 3–5 ovate or obovate lamina leaflets, margins serrate, apex acute. Flowers hermaphrodite, whitish or pinkish. Inflorescence raceme. Flower white. Fruit polydrupe of elongated conical shape, considerable size (3–4 × 2–3 cm) and weight (about 8 g), delicate, thin epicarp, sweet flavor, dark reddish-purple color when ripe, loses juice very easily and starts decaying a few days after harvest.

Related Species Of the more than 1,500 species belonging to the genus *Rubus*, many produce edible fruits and therefore are potentially useful to the beer-making process; however, just a few are known to be used for this purpose.

Geographic Distribution Artificial hybrid obtained by Rudolph Boysen on a farm in N California. In the 1920s, George M. Darrow of the USDA and Walter Knott, a farmer from S California and an expert in berries, found the hybrid on Boysen's abandoned farm. Knott was the first to resume commercial cultivation of the plant on his farm in Buena Park and to give the common name "boysenberry" to the fruit in honor of its inventor. After having been successfully cultivated, this plant has suffered a strong decline because of susceptibility to diseases, poor shelf life of the fruit, and a short productive season. These characteristics have relegated it to farmers' markets and to its use in jellies and preserves. New cultivars, however, free from these problems, have given new hope for the production and commercialization of this fruit. **Habitat** Cultivated. **Range** New Zealand, United States (California).

Beer-Making Parts (and Possible Uses) Fruits.

Chemistry Fruits: energy 28.62 kcal, carbohydrates 22 g, lipids 0.11 g, protein 0.68 g, fiber 1.49 g, sugars 3.52 g, vitamin A 40 IU, vitamin C 3.1 mg, calcium 15.63 mg, iron 0.62 g, sodium 3.36 g, potassium 90 mg, magnesium 11 mg, anthocyanins 120–160 mg, ellagic acid 5.98 mg/g on a dry weight basis.

Style American Wheat, Berliner Weisse, Fruit Beer, Rye Beer.

Beer Berry Bu, De Garde Brewing (Oregon, United States).

Source https://articles.latimes.com/2010/may/27/food/la-fo-boysenberry-20100527; https://www.brewersfriend.com/other/boysenberry/.

Rumex acetosa L.
POLYGONACEAE

Synonyms *Acetosa agrestis* Raf., *Acetosa amplexicaulis* Raf., *Acetosa angustata* Raf., *Acetosa bidentula* Raf., *Acetosa fontanopaludosa* (Kalela) Holub, *Acetosa hastifolia* Schur, *Acetosa hastulata* Raf., *Acetosa magna* Gilib., *Acetosa officinalis* Gueldenst. ex Ledeb., *Acetosa olitoria* Raf., *Acetosa pratensis* Mill., *Acetosa subalpina* Schur, *Rumex biformis* Lange, *Rumex fontanopaludosus* Kalela.

Common Names Sorrel, garden sorrel.

Description Herb, perennial (40–110 cm), with slender horizontal rhizome. Stem erect, fluted, reddish below, leafy to below inflorescence, branched at inflorescence. Leaves basal petiolate, 1.5 times the size of the lamina, this 2–3 times longer than wide (2–2.5 × 5–7 cm), hastate-lanceolate; cauline leaves 3–6 times as long as wide, the upper leaves sessile, with divergent lobes, acute or rounded, directed downward but not amplexicaul. Inflorescence compact, broadly branched, consisting of numerous small flowers. Flowers with green perianth consisting of 6 elements, 4 outer round-ovoid, with hard, round or tetragonal outgrowths, inner ones rolled at fruiting. Fruits shiny black achenes protected by persistent rust-red involucre.

Subspecies The taxon *Rumex acetosa* L. includes the typical subspecies, *Rumex acetosa* L. subsp. *acetosa*, and two other taxa: *Rumex acetosa* L. subsp. *hibernicus* (Rech.f.) Akeroyd and *Rumex acetosa* L. subsp. *vinealis* (Timb.-Lagr. & Jeanb.) O.Bolòs & Vigo.

Related Species There are 183 known species of the genus *Rumex*; *Rumex obtusifolius* L. has known beer-making uses.

Geographic Distribution Circumboreal-Eurosiberian. **Habitat** Fertile and fertilized meadows, along ditches and in pastures, prefers rich clayey soils (0–2,000 m ASL). **Range** Circumboreal-Eurosiberian; news of cultivation in France.

Beer-Making Parts (and Possible Uses) Aerial parts, leaves, roots.

Toxicity Aerial parts of the plant contain large quantities of oxalates, which in high doses can be toxic.

Chemistry Leaves (g/100 g dry matter): water 76% (fresh weight), ash 10.0%, lipids 2.5%, protein 25%, fiber 12.9%, nitrogen-free extract 50%, sodium 5 mg, potassium 440 mg, calcium 1,071 mg, magnesium 104.2 mg, iron 15 mg, oxalic acid (total) 1.34 mg, oxalic acid (soluble) 0.45 mg, tannic acid 3 mg, phytic phosphate 63.2 mg, anthraquinones, flavonoids, polysaccharides, vanillic acid, beta-carotene.

Style Saison.

Source Heilshorn 2017; Hieronymus 2016b; Josephson et al. 2016; https://www.brewersfriend.com/other/sorrel/.

Ruta graveolens L.
RUTACEAE

Synonyms *Ruta hortensis* Mill.

Common Names Rue, Common rue, Herb-of-grace.

Description Shrub (40–100 cm), suffruticose perennial, glabrous, glandular above. Stem lignified only at the base, silvery color, branches erect. Leaves alternate, green-glaucous, petiole 2–4 cm, lamina reniform and 2- to 3-pinnate-compound, segments spatulate, apex obtuse or mucronate, texture slightly fleshy, dotted with intensely fragrant glands. Inflorescence raceme with lanceolate leaf-like bracts. Flowers small (1–2 times the size of the coccarium), borne by short peduncles; sepals acute persistent, 4 petals (sometimes 5 in central flowers), yellow or greenish-yellow, concave, slightly toothed and wavy at the edge; ovary superficial. Fruit capsule, glabrous, subspherical, 4–5 carpels rugose, with obtuse-rounded apical teeth.

Related Species The genus *Ruta* has 13 recognized species, but only this one is used in brewing.

Geographic Distribution S Europe, S Siberia. **Habitat** Stony, grassy, shrubby areas in full sun, calcareous, dry soils (0–1,100 m ASL); often cultivated and wild. **Range** Discontinuous; in strong decline throughout distribution area.

Beer-Making Parts (and Possible Uses) Aerial parts.

Toxicity Toxic plant owing to the content of furocoumarins and rutarins and to the quinolone alkaloids present in the essential oil, with an unpleasant smell, which is collected in the leaf bladders. Taken in excessive doses, it has serious effects with potentially lethal consequences. Handling it is discouraged to avoid the risk of skin reddening, swelling, and blisters.

Chemistry Aerial parts: furanoacridones (several), 2 acridone alkaloids (arborinine and evoxanthine), coumarins (umbelliferone, scopoletin, psoralen, xanthotoxin, isopimpinellin, rutamarin, rutacultin), alkaloids (skimmianine, kokusaginin, 6-methoxydictamnin, edulinin), limonoids, 2 furanocoumarins, 1 quinoline alkaloid, 4 quinolones alkaloids, chalepensin (also present in root and stem). Essential oil: undecan-2-one 46.8%, nonan-2-one 18.8%.

Style Spice/Herb/Vegetable Beer.

Source Dabove 2015; https://www.brewersfriend.com/other/rue/.

Saccharum officinarum L.
POACEAE

Synonyms *Arundo saccharifera* Garsault, *Saccharum atrorubens* Cuzent & Pancher ex Drake, *Saccharum fragile* Cuzent & Pancher ex Drake, *Saccharum glabrum* Cuzent & Pancher ex Drake, *Saccharum luzonicum* Cuzent & Pancher ex Drake, *Saccharum monandrum* Rottb., *Saccharum obscurum* Cuzent & Pancher ex Drake, *Saccharum occidentale* Sw., *Saccharum rubicundum* Cuzent & Pancher ex Drake, *Saccharum violaceum* Tussac.

Common Names Sugarcane.

Description Herb, perennial (3–6 m). Stems (culms) erect or ascending, swollen at nodes (20–40), caespitose, in cross-section round or polygonal (Ø 2–5 cm), nodes glabrous. Leaves mostly cauline, distichous; leaf sheath smooth, glabrous, open, hairy above; leaf blade linear (70–150 × 4–6 cm), flat, glabrous, midrib broad, white, margin finely serrate; ligule 2–3 mm, ciliate. Inflorescence solitary, terminal (50–100 cm), panicle with branches (10 and up) open; axis glabrous, hairy at nodes. Flowers hermaphrodite, spikelets pedicellate, conspicuously hairy (2–3 times spikelet length), 3.5–4 mm long, with 1 fertile flower, disarticulate at rachis nodes; lower glume oblong, uniformly rigid, dorsum glabrous, margins membranous and ciliate above, apex acuminate; lower lemma oblong-lanceolate, subequal to glumes; upper lemma linear; lodicules glabrous; stamens 3, stylus bifid. Fruit caryopsis.

Cultivar There are many worldwide. As an example, the following are some used only in Benin (W Africa): Aleke Kpikpa, Alekebogoun, Alekedoudou, Alekeolomihessou, Alekeolomiwe, Archibi, Areke Tourawa, Arekebaki, Arekefari, Atinwlinwlin, Azeleke, Azeleke Assi, Azeleke Assou, Azeleke Gokpitikpiti, Azeleke Huinihuini, Azeleke Veve, Azeleke Vovo, Azeleke Wiwi, Azelekedaho, Azelekeyibo, Daweleke, Founfoun Djin, Founfoun You, Founfounyibo, Gartin Dombourou, Gartin fonharoun, Gartin Wonka, Gbaglo, Gnakani Mori, Gnankani Mori, Gnankanigue, Gounleke, Ipeokanga Ignonye, Ipeokanga Ipegni, Ireke Doudou, Ireke Oniandoudou, Kantooma, Karai Itchire, Karai Kpare, Karaikoukpeti, Karaipieri, Karaisoori, Karakouhouloumou, Karakoukpeto, Konakri, Leke Akparon, Leke Doudou, Leke Fefe, Leke Founfoun, Leke Kpikpa, Leke Mamoui, Leke Vee, Leke Vovo, Leke Wewe, Lekekoklodj Onon, Lekewewe, Ogniguin, Semeleke, Semeleke Vovo, Semelekewiwi, Soucletchi Mamoui, Soucletin Djin, Soucletin He, Soucletin Hi, Wandanwandan, Yeke, Yeke Fonton, Yeke Foufou.

Related Species Among the 38 species belonging to the genus *Saccharum*, only this one has known beer-making uses.

Geographic Distribution SE Asia. **Habitat** Cultivated for sugar in subtropical areas. **Range** Brazil, India, China, Thailand, Pakistan, Mexico, Colombia, Philippines, Australia, Argentina.

Beer-Making Parts (and Possible Uses) Sap, molasses, dark sugar.

Toxicity Some parts of the plant may be toxic to domestic animals owing to the presence of hydrocyanic acid.

Chemistry Raw sugar: energy 1,590 kJ (380 kcal), carbohydrates 98.09 g (dextrose 1.35 g, fructose 1.11 g, sucrose 94.56 g), protein 0.12 g, water 1.34 g, ash 0.45 g, calcium 83 mg, sodium 28 mg, phosphorus 4 mg, potassium 133 mg, iron 0.71 mg, magnesium 9 mg, zinc 0.03 mg, copper 0.047 mg, manganese 0.064 mg, selenium 1.2 μg, betaine 0.1 mg, niacin (vitamin B3) 0.11 mg, pantothenic acid (vitamin B5) 0.132 mg, pyridoxine (vitamin B6) 0.041 mg, folate 1 μg, choline (vitamin J) 2.3 mg. Inflorescence: energy 25 kcal, water 91, protein 4.6 g, lipids 0.4 g, carbohydrates 3 g, ash 1 g, calcium 40 mg, phosphorus 80 mg, iron 2 mg, thiamine 0.08 mg, vitamin C 50 mg. Leaf: energy 75 kcal, water 77.5 g, protein 1.8 g, lipids 0.8 g, carbohydrates 17.7 g, fiber 3 g, ash 2 g. Stem (culm): energy 62 kcal, water 82.5 g, protein 0.6 g, lipids 0.1 g, carbohydrates 16.5 g (mostly sucrose), fiber 3.1 g, ash 0.3 g, calcium 8 mg, phosphorus 6 mg, iron 1.4 mg, thiamin 0.02 mg, riboflavin 0.01 mg, niacin 0.10 mg, vitamin C 3 mg.

Style Baltic Porter, Belgian Dubbel, Imperial IPA, Northern English Brown Ale, Oatmeal Stout.

Beer Tripel, Westmalle (Westmalle, Belgium).

Source Calagione et al. 2018; Dabove 2015; Heilshorn 2017; Hieronymus 2010; Hieronymus 2016a; Jackson 1991; Markowski 2015; McGovern 2017; Steele 2012; https://www.bjcp.org (30. Spiced Beer; 30B. Autumn Seasonal Beer).

Sugar in the Beers of Belgium
In his seminal book, *Les Trappistes*, Jef van den Steen relates the tradition of Trappist breweries in Belgium to add a small amount of sugar to their more alcoholic products with the goal of increasing the sugar density of the wort without weighing it down with dextrins, proteins, and nonfermentable minerals. Sugar is added at the end of boiling, and during fermentation it is almost completely transformed into alcohol and carbon dioxide. The first Trappist brewery to use sugar was Westmalle, which since 1922 has been using it in a candied form in some of its productions. Besides Westmalle, sugar is used by Orval (candied), Westvleteren (caramelized and crystallized), and Rochefort (raw).

Saccharum officinarum L.
POACEAE

Salicornia europaea L.
AMARANTHACEAE

Synonyms *Salicornia annua* Sm., *Salicornia biennis* Afzel. ex Sm., *Salicornia brachystachya* (G.Mey.) D.König, *Salicornia herbacea* L., *Salicornia stricta* Dumort.

Common Names Glasswort, marsh samphire.

Description Herb, annual (10–35 cm). Stems erect, very branched; branches green, succulent. Leaves undeveloped, scaly (up to 1.5 mm), base united in a shield, margin membranous, apex acute. Inflorescence briefly pedunculate, spicate (1–5 cm). Flowers axillary, 3 per bract, median flower larger, situated slightly above lateral ones; perianth fleshy, obconic; stamens exserted, anthers oblong; ovary ovoid; stigma papillate. Fruit with membranous pericarp. Seed cylindrical-ovoid (Ø about 1.5 mm).

Subspecies In addition to the typical one (*Salicornia europaea* L. var. *europaea*), three varieties are known in this taxon: *Salicornia europaea* L. var. *polystachya* (W.D.J.Koch) Fernald, *Salicornia europaea* L. var. *prona* (Lunell) B.Boivin, and *Salicornia europaea* L. var. *simplex* (Pursh) Fernald.

Related Species The genus *Salicornia* includes about 30 taxa, some of which have alimurgical applications.

Geographic Distribution Europe, Russia, North America, SW Asia, India, Japan, Korea. **Habitat** Alkaline and saline soils, salt-lake shores, beaches. **Range** Europe, Russia, North America, India, Japan, SW Asia, Korea.

Beer-Making Parts (and Possible Uses) Young stems.

Chemistry Stem: high levels of ascorbic and dehydroascorbic acids (100 mg), good source of carotenoids (5 mg). Seeds: lipids 26–33%, protein 31%, fiber + ash 5–7%. Seed oil: linoleic acid 75.6%, oleic acid 13.0%, palmitic acid 7.0%, linolenic acid 2.6%, stearic acid 2.4%.

Style Gose, Gueuze.

Source Heilshorn 2017; https://www.brewersfriend.com/other/samphire/.

Salix alba L.
SALICACEAE

Synonyms *Argorips alba* Raf., *Argorips cerulea* Raf., *Salix pameachiana* Barratt.

Common Names Willow.

Description Tree (up to 20–30 m). Trunk straight (Ø up to 60 cm), bark more or less pale gray, soon cracked, cordate lengthwise when mature; branches erect, twigs slender and flexible, bark reddish-green to reddish-brown, golden-yellow in var. *vitellina*. Leaves petiolate (about 1 cm), lamina lanceolate-acuminate (5–10 × 1–2 cm), upper leaf glabrescent, slightly shiny, underleaf silky-silvery with dense appressed hairiness, arranged parallel to the conspicuous midrib as much as the secondary veins; margin finely toothed, base cuneate, apex slightly asymmetrical; stipules only on spears, narrow and deciduous. Inflorescences catkins contemporary with leaves (dioecious species); male 6–7 × 1 cm, pluriflorus, with 2 stamens and yellow anthers with hairy filaments; female shorter, ovary glabrous, elongate, pyriform. Fruit capsule glabrous, subsessile, conical (up to 6 mm).

Subspecies *Salix alba* L. subsp. *micans* (Andersson) Rech.f., *Salix alba* L. var. *caerulea* (Sm.) Sm., *Salix alba* L. var. *vitellina* (L.) Stokes.

Related Species The genus *Salix* includes approximately 630 accepted taxa, including species, subspecies, varieties, and hybrids.

Geographic Distribution Paleotemperate.
Habitat Riparian; tolerates periodic flooding, loose, silty, or sandy, moist soils (0–1,500 m ASL).
Range Eurasia.

Beer-Making Parts (and Possible Uses) Wood for barrel rims, twigs for filtering wort in traditional Scandinavian beer making, bark.

Toxicity Toxic to those allergic to aspirin.

Wood White or pinkish color, fine and regular texture, WSG 0.40–0.50, used for agricultural tools, clogs, small house utensils, matches; provides charcoal for gunpowder.

Style Ancient Ale, Wood-Aged Beer.

Source Cantwell and Bouckaert 2016; Cantwell and Bouckaert 2018; Hieronymus 2016b; McGovern 2017.

Salix laevigata Bebb
SALICACEAE

Synonyms None

Common Names Red willow, polished willow.

Description Tree (2–15 m). Trunk with branches flexible to highly brittle at base, gray-brown to yellow-brown, glabrous or hairy; branches yellow-brown or red-brown, glabrous, densely hairy, velvety or hairy, nodes hairy. Leaves with rudimentary stipules, ± leaf-like (apex acute, acuminate, rounded or convex) or absent; petiole (slightly or deeply grooved on upper leaf, sometimes with glands at base of lamina), 3.5–18 mm, pubescent to glabrescent adaxially; lamina narrowly oblong, narrowly elliptic, lanceolate or obovate (53–190 × 11–35 mm), 2.8–9 times longer than wide, base convex, subcordate, rounded or cuneate, margins crenate, entire or finely serrate, apex acuminate, acute, or caudate; underside glabrous or pubescent, hairs diffuse, white and/or ferrugineous in color, upper side slightly or very shiny, glabrous or pubescent, midrib sometimes villous; proximal part of lamina with entire margin; juvenile lamina glabrous or moderately covered with long silky hairs or hairy, hairs white and/or ferrugineous in color. Male catkin inflorescence (31–83 × 7–13 mm), on flowering twigs 2–26 mm; female catkin inflorescence (28–79 × 6–11 mm), on flowering twigs 3–14 mm; floral bract 1.6–3.4 mm, apex rounded or acute, irregularly toothed or entire, on underleaf sparsely to densely hairy proximally or throughout, hairs wavy; pistillate bracts deciduous after flowering. Flowers staminate abaxially with nectaries 0.4–0.6 mm, adaxially with nectaries oblong, square, or ovate, 0.3–0.6 mm, nectaries separate; stamens 3–7; filaments (sometimes basally connate), hairy in the proximal 1/2 or basally; anthers 0.4–0.6 mm. Flowers adaxially pistillate with square nectaries, 0.2–0.6 mm; ovary pyriform, obturbinate, or ellipsoid, beak slightly inflated below stylus; ovules 12–24 per ovary; stylus 0.1–0.2 mm; stigma 0.2–0.28 mm. Fruit capsule 3–5.5 mm.

RELATED SPECIES The genus *Salix* includes about 630 accepted taxa, including species, subspecies, varieties, and hybrids.

GEOGRAPHIC DISTRIBUTION United States (Arizona, California, Nevada, Oregon, Utah), Mexico (Baja California, Baja California Sur). **HABITAT** Riparian forests, seepage areas, springs, brackish lakeshores, canyons, ditches (0–2,200 m ASL). **RANGE** Species usually not cultivated.

BEER-MAKING PARTS (AND POSSIBLE USES) Twigs, bark.

TOXICITY Toxic to those allergic to acetylsalicylic acid.

CHEMISTRY Bark: salicin 0.08–12.6% (metabolic precursor of salicylic acid, easily acetylated to acetylsalicylic acid).

STYLE None known.

SOURCE Hieronymus 2016b.

Salvia hispanica L.
LAMIACEAE

SYNONYMS *Kiosmina hispanica* (L.) Raf., *Salvia chia* Colla, *Salvia prysmatica* Cav., *Salvia schiedeana* Stapf, *Salvia tetragona* Moench.

COMMON NAMES Chia, Spanish sage.

DESCRIPTION Herb, annual (up to 1.75 m), erect or ascending. Stem and branches quadrangular, villous, and shaggy. Leaves opposite, slender, petiole 1–6 cm, thin; lamina oblong-lanceolate to ovate (3–8 × 1–4.5 cm), obtuse or abruptly attenuated at base, margin entire at base, serrate or serrulate elsewhere, apex acute or acuminate; both leaf sides pubescent. Inflorescence whorl of 6–10 flowers forming a dense (internodes 2–5 mm) terminal false spike, 5–25 × 1.5 cm; bracts ovate-acuminate, 6–8 mm, persistent. Flowers hermaphrodite, calyx tubular, sepals 2-dentate, 6–8 mm (8–11 in fruiting), densely hairy; corolla tubular (4.5–5.5 mm), petals 2-labiate, color blue or blue-purple; 4 stamina didynamous; disc prominent; ovary superficial with dichotomous stylus. Fruits 4 schizocarpic nucules, each ellipsoid, 1.8 mm, spotted with black and gray.

Cultivar There are numerous cultivars and natural varieties of this species. Natural variability appears to be concentrated in the mountains of W Mexico.

Related Species Of the 1,037 taxa (including species, subspecies, and hybrids) belonging to the genus *Salvia*, only the three cited here (*Salvia hispanica*, *Salvia officinalis*, and *Salvia sclarea*) have beer-making use; given the wealth of medicinal and aromatic plants contained in this genus, it is conceivable that other taxa may have beer-making uses.

Geographic Distribution S Mexico, N Guatemala. **Habitat** Dry or moist meadows, often in open, rocky places, sometimes along sandy banks of streams. **Range** C Mexico, Guatemala, United States (S California, SE Texas), NW Argentina, W Java, Malaysia, Singapore.

Beer-Making Parts (and Possible Uses) Seeds.

Toxicity Based on available data, despite some dissenting opinions, chia seeds used up to 5% in bread are not expected to produce adverse effects to human health; acute toxicity and mutagenicity tests have not been positive; further data are needed to ascertain the nonhazardousness of this species.

Chemistry Seeds: energy 486 kcal, lipids 31 g (saturated fatty acids 3.3 g, polyunsaturated fatty acids 24 g, monounsaturated fatty acids 2.3 g), sodium 16 mg, potassium 407 mg, carbohydrates 42 g, fiber 34 g, protein 17 g, vitamin A 54 IU, vitamin C 1.6 mg, calcium 631 mg, iron 7.7 mg, magnesium 335 mg.

Style American Wheat or Rye Beer, English IPA, Witbier.

Source https://www.brewersfriend.com/other/chia/; https://www.globinmed.com/index.php?option=com_content&view=article&id=62991:salvia-hispanica-.

Salvia officinalis L.
LAMIACEAE

Synonyms *Oboskon cretica* (L.) Raf., *Salvia chromatica* Hoffmanns., *Salvia clusii* Vilm., *Salvia cretica* L., *Salvia crispa* Ten., *Salvia digyna* Stokes, *Salvia papillosa* Hoffmanns., *Salvia tricolor* Vilm.

Common Names Sage, garden sage, common sage, culinary sage.

Description Shrub (2–4 dm), gray tomentose, with aromatic odor. Stem woody at the base, branched, with patent hairs. Leaves with a 10–15 mm petiole, lanceolate (1 × 2–3 cm), obtuse, crenate lamina at the edge. Inflorescences verticillaster, 5- to 10-flowered, ± unilateral, the lower one wrapped in a pair of bract leaves. Flower calyx ferrugineous with 5–7 mm tube and 4–6 mm teeth; corolla purplish (rarely pinkish or white) with 10–15 mm tube and 7–10 mm upper lip.

Subspecies *Salvia officinalis* includes four taxa: *Salvia officinalis* L. subsp. *gallica* (W.Lippert) Reales, D.Rivera & Obón, *Salvia officinalis* L. subsp. *lavandulifolia* (Vahl) Gams, *Salvia officinalis* L. subsp. *officinalis*, and *Salvia officinalis* L. subsp. *oxyodon* (Webb & Heldr.) Reales, D.Rivera & Obón.

Cultivar There are many, among which the following should be mentioned: Alba, Aurea, Berggarten, Extrakta, Icterina, Lavandulaefolia, Purpurascens (or Purpurea), and Tricolor (the last two have α-humulene and β-pinene contents similar to certain chemotypes of hops).

Related Species Of the 1,037 taxa (including species, subspecies, and hybrids) belonging to the genus *Salvia*, only the three cited here (*Salvia hispanica*, *Salvia officinalis*, and *Salvia sclarea*) have beer-making use; given the wealth of medicinal and aromatic plants contained in this genus, it is conceivable that other taxa may have beer-making uses.

Geographic Distribution Steno-Mediterranean (E). **Habitat** Arid cliffs, stony grounds (calcareous). **Range** Widely cultivated for alimentary and aromatic use.

Beer-Making Parts (and Possible Uses) Leaves.

Toxicity Some of the components of the essential oil may be toxic at high doses.

Chemistry Leaves: energy 483 kJ (116 kcal), water 66.4 g, protein 3.9 g, lipids 4.6 g, sugars 15.6 g, sodium 4 mg, potassium 390 mg, calcium 600 mg, phosphorus 33 mg, thiamine 0.11 mg, vitamin A 215 μg, vitamin C traces. Essential oil (40 compounds; 97.1% of the oil): borneol 2.6%, bornyl acetate 2.6%, (E)-β-caryophyllene 3.4%, α-humulene 3.3%, viridiflorol 5.6%, α-pinene 3.5%, camphene 5.3%, 1,8-cineole 11.9%, α-thujone 21%, camphor 23.9%.

Style American IPA, American Pale Ale, American Wheat or Rye Beer, Belgian Blond Ale, Belgian Dark Strong Ale, Belgian Golden Strong Ale, Belgian Pale Ale, Belgian Specialty Ale, Belgian Tripel, Berliner Weisse, Bière de Garde, California Common Beer, Dark Mild, Fruit Beer, Holiday/Winter Special Spiced Beer, Imperial IPA, Irish Red Ale, Metheglin, North German Altbier, Oatmeal Stout, Robust Porter, Saison, Special/Best/Premium Bitter, Specialty Beer, Spice/Herb/Vegetable Beer, Strong Bitter, Strong Scotch Ale, Weizen/Weissbier, White IPA, Witbier.

Beer Lola, Renton (Fano, Marche, Italy).

Source Fisher and Fisher 2017; Heilshorn 2017; Hieronymus 2012; Hieronymus 2016b; Josephson et al. 2016; Markowski 2015; Steele 2012; https://www.brewersfriend.com/other/sage/; https://www.gruitale.com/bot_sage.htm.

Salvia sclarea L.
LAMIACEAE

Synonyms *Aethiopis sclarea* (L.) Fourr., *Salvia altilabrosa* Pan, *Salvia calostachya* Gand., *Salvia coarctata* Vahl, *Salvia foetida* Lam., *Salvia lucana* Cavara & Grande, *Salvia pamirica* Gand., *Salvia simsiana* Schult., *Salvia turkestanica* Noter, *Sclarea vulgaris* Mill.

Common Names Clary, clary sage.

Description Herb, biennial (over 1 m). Stem erect, enlarged, quadrangular, ribbed, hairy with frizzy hairs. Leaves velvety, the lower ones with lamina 4–12 × 7–18 cm, in rosette in the first year, the cauline ones smaller, irregularly toothed. Inflorescence broad with erect-patent branches; bracts membranous, purplish, 2–3 cm, larger than corolla. Flower calyx with bristly tube and spinulose teeth; corolla pinkish or lilaceus (15–20 mm).

Related Species Of the 1,037 taxa (including species, subspecies, and hybrids) belonging to the genus

Salvia, only the three cited here (*Salvia hispanica*, *Salvia officinalis*, and *Salvia sclarea*) have been found to have beer-making use; given the wealth of medicinal and aromatic plants contained in this genus, it is conceivable that other taxa may have beer-making uses.

GEOGRAPHIC DISTRIBUTION Euro-Mediterranean. **HABITAT** Arid slopes and thickets (0–1,000 m ASL). **RANGE** Euro-Mediterranean.

BEER-MAKING PARTS (AND POSSIBLE USES) Twigs (give muscatel aroma), leaves, flowers, aerial parts (bittering).

CHEMISTRY Essential oil (about 70 compounds; 94.2% of the oil): linalyl acetate 39.2%, linalool 12.5%, germacrene D 11.4%, a-terpineol 5.5%, geranyl acetate 3.5%, (E)-caryophyllene 2.4%, bicyclogermacrene 1.2%, myrcene 0.7%, nerol 1.1%, 1-phenyl-2,4-pentadiins 1.2%, thymol 1.5%, carvacrol 1.3%, neryl acetate 1.9%, a-copaene 1%, sclareol 1.2%.

STYLE Gruit, Sage Ale.

BEER Rex Grue, Montegioco (Montegioco, Piedmont, Italy).

SOURCE Dabove 2015; Fisher and Fisher 2017; Heilshorn 2017; Hieronymus 2016b; Josephson et al. 2016; https://honest-food.net/2016/05/05/gruit-herbal-beer/.

Sambucus canadensis L.
ADOXACEAE

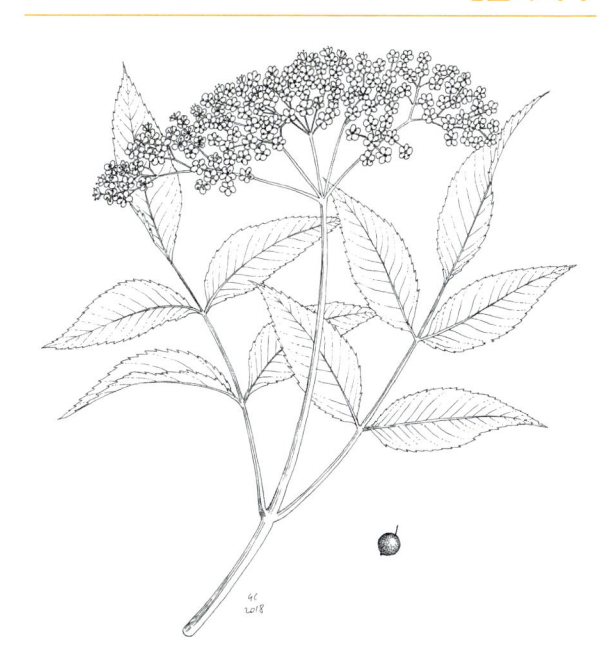

SYNONYMS *Aralia sololensis* Donn.Sm., *Sambucus mexicana* C.Presl ex DC., *Sambucus oreopola* Donn. Sm., *Sambucus simpsonii* Rehder.

COMMON NAMES Common elderberry, Arizona elderberry, American elder, sweet elder, wild elder, tree of music, danewort, walewort, New Mexican elderberry, velvet-leaf elder, hairy blue elderberry, dwarf elder.

DESCRIPTION Shrub (2–8 m) or rarely small single-stem tree. Young stems soft and rich in pith but wood hard; bark thin, grayish to dark brown, irregularly furrowed and ridged. Leaves pinnate-compound deciduous, opposite, about 15–35 cm, imparipinnate, with 3–9 serrate leaflets 2–15 cm long, often with a long petiole, frequently asymmetrical at base. Inflorescence flat at apex, Ø 4–30 cm, wider than long. Flowers bisexual, corollas small, white to cream, rotate, 5-lobed, with a pleasant though slightly rancid odor. Fruit pod shaped, Ø 5–6 mm, blue-black to purple-black in color at maturity with a waxy-white patina. Seeds 3–5 per fruit.

RELATED SPECIES The genus *Sambucus* consists of about 28 species (or subspecies) accepted by science. Only *Sambucus nigra* and *Sambucus canadensis*, however, seem to have established beer-making use.

GEOGRAPHIC DISTRIBUTION Canada, United States, Mexico. **HABITAT** Wet, well-drained, sunny habitats: forest habitats (slopes, canyons, cliff bases, riparian banks), open arid habitats (shrublands, montane scrub, *Pinus-Juniperus* forests, *Pinus ponderosa* woodlands), ruderal environments (fences, roads), 0–3,000 m ASL. **RANGE** Canada, United States, Mexico.

BEER-MAKING PARTS (AND POSSIBLE USES) Flowers, ripe fruits (*Sambucus nigra* substitute plant).

TOXICITY Leaves, bark, roots, and unripe fruits are toxic.

CHEMISTRY Fruits: water 76.4%, protein 2.63%, fiber 4.24%, ash 1.30%, nonnitrogenous extract 12.37%, lignin 3.63%, cellulose 2.49%, tannins 0.64%, calcium 0.03%, magnesium 0.05%, phosphorus 0.08%.

STYLE Herb Beer, Witbier.

SOURCE Fisher and Fisher 2017; Josephson et al. 2016; McGovern 2017.

Sambucus nigra L.
ADOXACEAE

Synonyms *Sambucus graveolens* Willd.

Common Names Elder, elderberry, black elder, European elder, European elderberry, European black elderberry.

Description Shrub or small tree (1–8 m), with fetid odor. Young branches green with longitudinal lenticels (1.5–3 mm); bark brown with longitudinal fractures and deep furrows (5–8 mm). Leaves opposite imparipinnate with 5–7 elliptic or lanceolate, acuminate segments, the largest 10–13 × 4–6 cm, serrate. Inflorescence umbelliform (Ø 1–2 dm) with numerous milky-white flowers; calyx subnullus; corolla (Ø 5 mm) with subnullus tube and 5 rounded lobes; stamens 5 with yellow anthers. Fruit drupe subspherical (Ø 5–6 mm), black-purple, shiny.

Subspecies *Sambucus nigra* is a complex of closely related European, Asian, and North American species that some taxonomists believe can be ascribed to subspecific or even varietal rank: *Sambucus nigra* L. subsp. *nigra*, *Sambucus canadensis* L. (= *Sambucus nigra* L. var. *canadensis* [L.] B.L.Turner, *Sambucus nigra* L. subsp. *canadensis* [L.] Bolli, *Sambucus mexicana* C.Presl ex DC.), *Sambucus lanceolata* R.Br. (= *Sambucus nigra* L. subsp. *maderensis* [Lowe] Bolli), *Sambucus palmensis* Link (= *Sambucus nigra* L. subsp. *palmensis* [Link] Bolli), and *Sambucus cerulea* Raf.

Cultivar Some examples: Aurea, Gerda (or Black Beauty), Laciniata.

Related Species The genus *Sambucus* consists of about 28 species (or subspecies) accepted by science. Only *Sambucus nigra* and *Sambucus canadensis*, however, seem to have an established beer-making use.

Geographic Distribution European-Caucasian.
Habitat Wet woods, clearings, coppices, hedges.
Range Eurasia, North America.

Beer-Making Parts (and Possible Uses) Flowers, fruits.

Toxicity Fruits poisonous when unripe, edible when ripe.

Chemistry Fruits: sugars 8.88%, pectin 0.16%, fiber 1.65%, citric acid 1.3%, ash 0.92%, anthocyanins 863.9 mg/L, vitamin C 34.1 mg/100 g. Volatile components of the fruit: phenylacetaldehyde 32.3%, benzaldehyde 7.9%, ethyl linoleate 5.4%, 4-vinyl guaiacol 4.9%, linalool 4.5%, phenyl ethyl alcohol 4.1%, methyl hexadecanoate 2%, ethyl hexadecanoate 2%, hexahydro farnesyl acetone 1.1%, decane 2.5%, (E)-β-damascenone 2.5%, 2-phenyl ethyl acetate 1.6%, hexanal 1.4%, isoamyl acetate 1.4%, limonene 1%, 2-nonanone 3%, nonanal 1.2%, calarene 0.9%, bornyl acetate 1%, heptadecane 1.1%, methyl linoleate 1.1%.

Style Belgian Blond Ale, Bitter, California Common Beer, Experimental Beer, Flanders Brown Ale/Oud Bruin, Oatmeal Stout, Weizen/Weissbier, Witbier. Flowers: American Amber Ale, American IPA, American Lager, American Pale Ale, American Pilsner, American Wheat or Rye Beer, Belgian Specialty Sle, Bitter, English IPA, Fruit Beer, Gueuze, Kölsch, Mild, Robust Porter, Spice/Herb/Vegetable Beer, Stout, Witbier.

Beer Mamouche, Cantillon (Brussels, Belgium).

Source Calagione et al. 2018; Dabove 2015; Fisher and Fisher 2017; Heilshorn 2017; Hieronymus 2010; Hieronymus 2016b; Jackson 1991; Markowski 2015; https://www.brewersfriend.com/other/elderberries/; https://www.brewersfriend.com/other/elderflowers/; https://www.williamsbrosbrew.com/beer/ebulum (Ebulum Elderberry Black Ale).

Sanguisorba minor Scop.
ROSACEAE

Synonyms *Pimpinella minor* (Scop.) Lam., *Pimpinella sanguisorba* (L.) Gaertn., *Poterion sanguisorbens* St.Lag., *Poterium anceps* Ball, *Poterium collinum* Salisb., *Poterium dictyocarpum* Spach, *Poterium duriaei* Spach, *Poterium fontanesii* Spach, *Poterium glaucescens* Rchb., *Poterium magnolii* Willk., *Poterium maroccanum* Coss., *Poterium minus* (Scop.) Gray, *Poterium sanguisorba* L., *Poterium vulgare* Hill ex Cuatrec., *Sanguisorba dictyocarpa* (Spach) Franch., *Sanguisorba dictyocarpa* Spach ex T.Durand & Pittier, *Sanguisorba duriaei* (Spach) A.Braun & C.D.Bouch, *Sanguisorba fontanesii* (Spach) A.Braun & C.D.Bouch, *Sanguisorba glaucescens* (Rchb.) Ces., *Sanguisorba guestphalica* (Boenn. ex Rchb.) T.Durand & Pittier, *Sanguisorba maroccana* (Coss.) Maire, *Sanguisorba pimpinella* Spenn., *Sanguisorba poterium* F.H.Wigg., *Sanguisorba sanguisorba* (L.) Britton, *Sanguisorba vulgaris* Hill, *Sanguisorba vulgaris* Hill ex Pau.

Common Names Salad burnet, small burnet.

Description Herb, perennial (20–60 cm), evergreen. Rhizome woody, enlarged. Stems erect, striate, simple or branched above, sometimes hairy below, often reddish in color. Leaves basal, clustered in a rosette (10–20 cm), pinnate, with 5–17 petiolate leaflets; leaf segments 1 cm, elliptic, with scattered hairs on both sides, rarely glabrous, margin with 4–6 acute teeth per side; upper leaf green, underleaf glabrous; cauline leaves gradually smaller and with decreasing number of leaflets. Inflorescence spherical-oval spike (about 2 cm) capituliform, apical. Flowers apetalous with 4 ovate sepaloid laciniae, green or reddish in color with white margin; upper female flowers with pinkish or red plumose stigma, lower male flowers with numerous yellow, long, protruding stamens; in the central part there are also a number of hermaphrodite flowers whose stamens do not protrude from the calyx. Infructescence pometum, obovoid (2.8–6 × 2–3.8 mm), consisting of the receptacle (hypanthium), urceolate, growing, dry, with 4 ribs, and with reticulate sides or ± acute ridges, containing 1–3 achenes inside.

Subspecies In addition to the typical one, *Sanguisorba minor* Scop. subsp. *minor*, there are four recognized subspecies: *Sanguisorba minor* Scop. subsp. *balearica* (Bourg. ex Nyman) Munoz Garm. & C.Navarro, *Sanguisorba minor* Scop. subsp. *lasiocarpa* (Boiss. & Hausskn.) Nordborg, *Sanguisorba minor* Scop. subsp. *magnolii* (Spach) Briq., and *Sanguisorba minor* Scop. subsp. *verrucosa* (Ehrenb. ex Decne.) Holmboe.

Related Species The genus *Sanguisorba* includes 33 taxa.

Geographic Distribution Paleotemperate subcosmopolitan. **Habitat** Arid environments, fallow land, forage fields, roadsides, ruderal areas (0–1,300 m ASL, rarely up to 2,000 m ASL). **Range** Paleotemperate subcosmopolitan.

Beer-Making Parts (and Possible Uses) Flowering tops (can substitute for *Borago officinalis*).

Chemistry Essential oil, leaves (93.2% of total oil): linalool 73%, (E,E)-farnesyl acetate 13.4%, nonadecane 11.2%, docosanc 11%, β-caryophyllene 9.7%, nonanal 8.5%, germacrene D 5.4%, caryophyllene oxide 5.1%, tetradecane 3.8%, dodecane 3.6%, (E)-α-damascenone 3.2%, eicosane 2.6%, tridecane 2.4%, octadecane 2.4%, enicosane 2.3%, heptadecane 1.2%, hexadecane 0.1%.

Style Herb Beer, Spice Beer.

Source Fisher and Fisher 2017.

Sassafras albidum (Nutt.) Nees
LAURACEAE

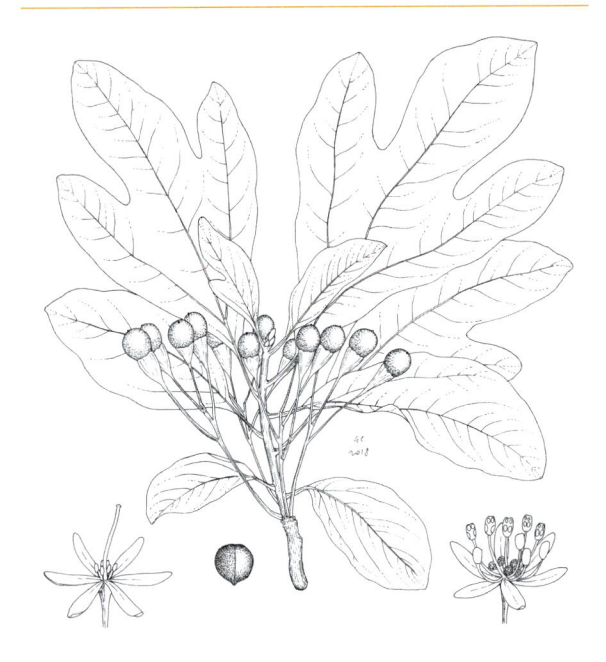

Synonyms *Laurus sassafras* L., *Sassafras officinalis* T.Nees & C.H.Eberm., *Sassafras variifolium* Kuntze.

Common Names Sassafras, white sassafras.

Description Tree (up to 35 m). Young trunks light green with olive-colored spotting. Leaf blade ovate to elliptic, unlobed or 2- to 3-lobed (rarely more), 10–16 × 5–10 cm, apex obtuse to acute. Inflorescence up to 5 cm, silky-pubescent; floral bract up to 1 cm. Flowers fragrant, glabrous; tepals greenish-yellow; flowers staminate with 3 inner stamens with 2 conspicuous glands at base of slender filament, pistillodes usually absent; flowers pistillate with 6 staminoids, stylus slender, 2–3 mm, stigma capitate. Fruit drupe about 1 cm, peduncle reddish color, club shaped, ± fleshy.

Subspecies Some taxa of subspecific rank have been described based on leaf pubescence and color of young shoots, but most taxonomists do not recognize them as valid.

Related Species Of the five species recognized in the genus *Sassafras*, only *S. albidum* is known to have a beer-making use.

Geographic Distribution Canada (Ontario), United States (Alabama, Arkansas, Connecticut, Delaware, District of Columbia, Florida, Georgia, Illinois, Indiana, Iowa, Kansas, Kentucky, Louisiana, Maine, Maryland, Massachusetts, Michigan, Missouri, Mississipi, New Hampshire, New Jersey, New York, North Carolina, Ohio, Oklahoma, Pennsylvania, Rhode Island, South Carolina, Tenneesse, Texas, Vermont, Virginia, West Virginia). **Habitat** Forests, thickets, hedgerows, abandoned fields, disturbed areas (0–1,500 m ASL). **Range** North America.

Beer-Making Parts (and Possible Uses)
Roots, bark, leaves, wood (for smoking malt).

Toxicity Safrole (p-allylin ethylene dioxybenzene), the main component (80%) of sassafras oil, is suspected of causing contact dermatitis and hallucinogenic effects, especially in large doses; it is also thought to be carcinogenic and hepatotoxic.

Chemistry Essential oil (30 compounds): safrole 85%, camphor 3.25%, methyleugenol 1.1%.

Style American Stout, Belgian Blond Ale, Belgian Specialty Ale, Brown Porter, California Common Beer, Experimental Beer, Extra Special/Strong Bitter (ESB), Holiday/Winter Special Spiced Beer, Mum, Northern English Brown, Robust Porter, Specialty Beer, Spice/Herb/Vegetable Beer, Sweet Stout.

Beer Ten, Surly (Minnesota, United States).

Source Daniels and Larson 2000; Heilshorn 2017; Hieronymus 2010; Hieronymus 2016b; https://www.brewersfriend.com/other/rootbeer/; https://www.brewersfriend.com/other/rootbeerextract/; https://www.brewersfriend.com/other/sassafras/.

Satureja hortensis L.
LAMIACEAE

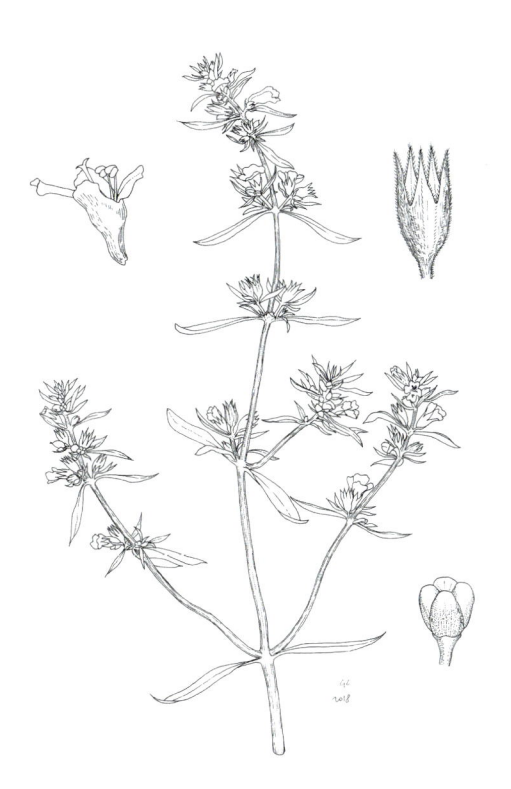

Synonyms *Clinopodium hortense* (L.) Kuntze, *Clinopodium pachyphyllum* (K.Koch) Kuntze, *Satureja altaica* Boriss., *Satureja brachiata* Stokes, *Satureja filicaulis* Schott ex Boiss., *Satureja litwinowii* Schmalh. ex Lipsky, *Satureja officinarum* Crantz, *Satureja pachyphylla* K.Koch, *Satureja zuvandica* D.A.Kapan., *Thymus cunila* E.H.L.Krause.

Common Names Summer savory.

Description Herb, annual (10–30 cm), with pleasant aromatic scent, usually ± colored violet. Stem erect, branched-corymbose. Leaves linear-lanceolate (2–4 × 10–25 mm), the upper ones linear, entire. Inflorescences 2- to 5-flowered whorls, at the axil of ± normal leaves, longer than these. Flower with campanulate calyx (3–4 mm) with teeth as long as the tube; corolla (4–7 mm), lilac color ± faded.

Cultivar One is Saturn; presumably others exist.

Related Species The genus *Satureja* includes 60 taxa.

Geographic Distribution W Asia, Euro-Mediterranean. **Habitat** Arid grasslands and calcareous walls (0–1,300 m ASL). **Range** W Asia, Euro-Mediterranean.

Beer-Making Parts (and Possible Uses) Flowering tops.

Chemistry Energy: 272 kcal, lipids 6 g, carbohydrates 69 g, fiber 46 g, protein 7 g, magnesium 377 mg, sodium 24 mg, calcium 2132 mg, iron 37.9 mg, potassium 1,051 mg, vitamin A 5,130 IU, vitamin C 50 mg, vitamin B6 1.8 mg. Essential oil: γ-terpinene 40.9%, carvacrol 39.3%.

Style Herb Beer (?).

Source Fisher and Fisher 2017; Hieronymus 2016b; McGovern 2017.

Satureja montana L.
LAMIACEAE

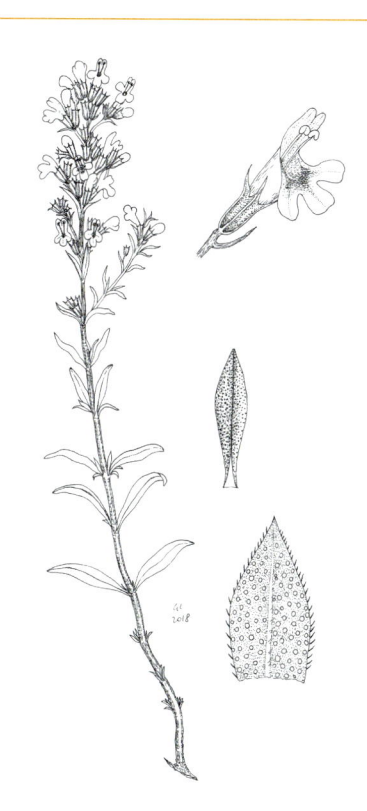

Synonyms *Clinopodium montanum* (L.) Kuntze, *Micromeria montana* (L.) Rchb., *Micromeria pygmaea* Rchb., *Micromeria variegata* Rchb., *Satureja brevis* Jord. & Fourr., *Satureja ciliata* Avé-Lall., *Satureja flexuosa* Jord. & Fourr., *Satureja hyssopifolia* Bertol., *Satureja karstiana* Justin, *Satureja mucronifolia* Stokes, *Satureja ovalifolia* Huter, Porta & Rigo, *Satureja petraea* Jord. & Fourr., *Satureja pollinonis* Huter,

Porta & Rigo, *Satureja provincialis* Jord. & Fourr., *Satureja pyrenaica* Jord. & Fourr., *Satureja rigidula* Jord. & Fourr., *Satureja trifida* Moench, *Saturiastrum montanum* (L.) Fourr., *Saturiastrum petraeum* Fourr.

Common Names Winter savory, mountain savory.

Description Herb, perennial (2–5 dm), with aromatic odor. Stems tetragonal, pubescent all around, woody at base. Leaves linear-lanceolate, the lower ones 2–3 × 10–25 mm, bristly at the margin (especially at the base) and sparsely glandular, usually at the axil bearing a bundle of 2–8 reduced leaves (rarely a short branch). Inflorescences whorls with 2–3 (rarely more) flowers. Flowers with pale pinkish corolla, tube 3–4 mm, upper lip 4–5 mm, dark red, lower lip 3 mm with violet spots.

Subspecies The taxon includes four subspecies: *Satureja montana* L. subsp. *macedonica* (Formánek) Baden, *Satureja montana* L. subsp. *montana*, *Satureja montana* L. subsp. *pisidia* (Wettst.) Šilić, and *Satureja montana* L. subsp. *variegata* (Host) P.W.Ball.

Related Species The genus *Satureja* includes 60 taxa.

Geographic Distribution Orophyte S Mediterranean. **Habitat** Arid steppe meadows, in serpentine and limestone soils (0–1,300 m ASL). **Range** Orophyte S Mediterranean.

Beer-Making Parts (and Possible Uses) Flowering tops.

Chemistry Essential oil: carvacrol 83.40%, γ-terpinene 9.62%, α-terpinene 1.70%, thymol methyl ether 1.12%, cymene(p) 0.75%, myrcene 0.61%, α-pinene 0.58%, linalool 0.44%, camphene 0.43%, (Z)-β-ocimene 0.39%, α-phellandrene 0.38%, α-thujene 0.35%, carvacrol acetate 0.28%, β-pinene 0.27%, limonene 0.23%, octanol 0.17%, bisabolene 0.14%, δ-3-carene 0.13%, hexadecane 0.11%, (E)-β-ocimene 0.08%.

Style Saison.

Beer Ardiva, Bionoć (Primiero San Martino di Castrozza, Trentino-Alto Adige, Italy).

Source Fisher and Fisher 2017.

Schinus molle L.
ANACARDIACEAE

Synonyms *Schinus angustifolia* Sessé & Moc., *Schinus bituminosus* Salisb., *Schinus huigan* Molina, *Schinus occidentalis* Sessé & Moc.

Common Names Peruvian pepper, American pepper, Peruvian peppertree, false pepper, pepper tree, peppercorn tree, Californian pepper tree, Peruvian mastic.

Description Tree (up to 15 m), evergreen. Bark rough, twisted, grayish, drips sap. Branches leaning. Leaves compound, imparipinnate, 8–25 × 4–9 cm, consisting of 19–41 alternate, linear-lanceolate leaflets, 15–27 mm long. Male and female flowers on separate plants (dioecious); flowers small, greenish-white, numerous clustered in conical panicles at the ends of pendulous branches. Fruits subspherical drupes (Ø 5–7 mm), green, then pink, reddish, or violet in color, clustered in dense infructescences of hundreds of units that can remain on the plant for long periods. Seed woody. Bark, leaves, and fruit aromatic when crushed.

Related Species The genus *Schinus* includes 38 taxa, including *Schinus terebinthifolia* Raddi (also toxic).

Geographic Distribution Peru. **Habitat** Cultivated for ornamental purposes in warm climates. **Range** Cultivated for ornamental purposes in warm climates and frequently invasive.

Beer-Making Parts (and Possible Uses)
Fruits, leaves (?).

Toxicity The toxicity and repellency of some parts of the plant for various insects have led to the hypothesis of a potential toxicity to humans; however, this has yet to be demonstrated by scientific data.

Chemistry Essential oil, fruits: p-cymene 32.8%, β-pinene 19%, α-terpinene 18.3%, limonene 7.19%, borneol 2.74%, methyl caprylate 2.44%, α-phellandrene 1.32%, α-pinene 1.04%, terpineol 0.86%, carvotanacetone 0.85%, calamenene 0.85%, eucalyptol 0.83%, (E)-2,3-epoxycarane 0.77%, 3-penten-1-ol 0.76%, cis-p-ment-2,8-dienol 0.7%, 2-cyclohexen-1-one, 6-methyl-3-(1-methylethyl) 0.7%, terpinyl acetate 0.69%, (E)-3-caren-2-ol 0.67%, 1,5,7-octatrien-3-o-lo-2,6-dimethyl 0.64%, dehydroxy-isocalamendiol 0.45%, (2E,8E)-2,8-decadiene 0.27%, α-thujenal 0.26%, α-calacorene 0.23%, terpinolene 0.21%, 1-epten-4-ol 0.2%, camphene 0.19%, fellandral 0.18%, limonene-1,2-epoxide 0.16%, carveol 0.15%, 3-carene 0.14%, tricyclicene 0.06%, β-eudesmene 0.06%, (+)-δ-cadinene 0.06%, β-cis-ocimene 0.05%, pinocanfone 0.05%, copaene 0.04%, ethyl caproate 0.03%, α-canfolenal 0.03%, p-cumicaldehyde 0.03%, T-gurjunene 0.03%. Essential oil, leaves: β-pinene 31.1%, α-pinene 22.7%, γ-cadinene 6%, epi-α-cadinol 5.6%, β-caryophyllene 4.7%, allo-aromadendrene 4%, germacrene D 3.9%, δ-cadinene 3.9%, bicyclogermacrene 3%, limonene + phellandrene 2.6%, 1,10-di-epi-cubenol 1.7%, globulol 0.8%, α-cadinol 0.8%, α-muurolene 0.7%, (E)-β-ocimene 0.6%, α-humulene 0.6%, spathulenol 0.6%, camphene 0.5%, α-cadinene 0.5%, viridiflorol 0.4%, myrcene 0.3%, α-gurjunene 0.3%, α-thujene 0.2%, sabinene 0.2%, terpinen-4-ol 0.2%, β-elemene 0.2%, 1-epicubenol 0.2%, α-muurolene 0.2%, tricycline 0.1%, α-phellandrene 0.1%, p-cymene 0.1%, γ-terpinene 0.1%, α-terpineol 0.1%, α-copaene 0.1%, viridiflorene 0.1%, trans-cadin-1(2),4-diene 0.1%, α-terpinene traces, 1,8-cineol traces, (Z)-β-ocimene traces, terpinolene traces, cis-muurola-3,5-diene traces.

Style IPA, Spice Beer.

Beer Pink IPA, Almond '22 (Loreto Aprutino, Abruzzo, Italy).

Source Giaccone and Signoroni 2017; McGovern 2017; https://www.fermentobirra.com/homebrewing/birrespeziali/.

Secale cereale L.
POACEAE

Synonyms *Secale ancestrale* (Zhuk.) Zhuk., *Secale arundinaceum* Trautv., *Secale strictum* C.Presl, *Secale triflorum* P.Beauv., *Secale turkestanicum* Bensin, *Triticum cereale* (L.) Salisb., *Triticum ramosum* Weigel, *Triticum secale* Link, *Triticum strictum* C. Presl.

Common Names Rye.

Description Herb (10–18 dm), annual or biennial. Densely bushy, culms erect, with blackened nodes. Leaves pubescent with flat lamina 4–7 mm wide and truncate ligule (1–1.5 mm). Inflorescence spiky, regularly distichous, with stout rachis, hairy beneath spikelets; spikelets 2- to 3-flowered; glumes 10–11 mm; lemma 15–20 mm, lengthened into a 3 cm awn; anthers yellow, 5 mm. Fruit caryopsis (2×7 mm).

Cultivar Many are available: AC Remington, Ardmore, Aroostook, Cougar, Dakold, Dankowski Nowe (Danko), Elbon, Frederick, Hancock, Maton, Metzi, Musketeer, Noble Foundation, Oklahoma, Prima, Puma, Rosen, Rymin, and Von Lochow are some U.S., Canadian, and German cultivars.

Related Species In addition to *Secale cereale* L., eight other taxa belong to the genus *Secale*: *Secale africanum* Stapf, *Secale anatolicum* Boiss., *Secale ciliatiglume* (Boiss.) Grossh., *Secale × derzhavinii* Tzvelev, *Secale montanum* Guss., *Secale segetale* (Zhuk.) Roshev., *Secale sylvestre* Host, and *Secale vavilovii* Grossh. However, none is known to have any beer-making application.

Geographic Distribution C Asia. **Habitat** Fields, wastelands, ruins. **Range** Germany, Poland, Russia, Belarus, China, Ukraine, Denmark, Turkey, Canada, Spain.

Beer-Making Parts (and Possible Uses)
Caryopsis as is or malted (rye malt color range 2.8–3.7 SRM); straw (burned to smoke the malt). Rye malt has some similarities to wheat malt but has a distinctive spicy flavor that it transfers to the beer; it combines well with the flavor of American hops (judging by the number of beers made this way). Like wheat

and oats, rye lacks integuments that contribute to the chewy, dense, viscous properties useful in lautering; as with wheat and barley, it can be malted using a variety of techniques (roasted rye malts are available in colors up to 250 SRM and presumably higher if desired).

TOXICITY As with wheat and barley, rye contains gluten, making it unsuitable for consumption by individuals with celiac disease. Moreover, the plant is very sensitive to the fungus *Claviceps purpurea* (Fr.) Tul. (Clavicipitaceae, Ascomycota), commonly called ergot or ergot fungi. The consumption of rye or other infected cereals causes ergotism, a serious disease caused by the LSD-like toxins produced by the fungus. Today it is no longer a common disease thanks to efforts made to ensure food safety and the selection of cultivars resistant to the fungus.

CHEMISTRY Caryopsis: energy 338 kcal, lipids 1.6 g (saturated fatty acids 0.2 g, polyunsaturated fatty acids 0.8 g, monounsaturated fatty acids 0.2 g), sodium 2 mg, potassium 510 mg, carbohydrates 76 g, fiber 15 g, sugars 1 g, protein 10 g, vitamin A 11 IU, vitamin B6 0.3 mg, calcium 24 mg, iron 2.6 mg, magnesium 110 mg.

STYLE Roggenbier, Rye Ale.

BEER Blou, Les Bières du Grand St. Bernard (Gignod, Valle d'Aosta, Italy).

SOURCE Calagione et al. 2018; Dabove 2015; Daniels and Larson 2000; Fisher and Fisher 2017; Giaccone and Signoroni 2017; Heilshorn 2017; Hieronymus 2010; Hieronymus 2016b; Jackson 1991; Josephson et al. 2016; Markowski 2015; McGovern 2017; https://www.bjcp.org (31A. Alternative Grain Beer).

An Ancient Grain
Rye is one of the grains most closely associated with the history of humans, agriculture, and beer. Its use in the production of the "beverage of Ceres" is surely very ancient, but it became more widespread in the Middle Ages. In Germany, a type called roggenbier that makes distinctive use of *Secale cereale* has been produced for a long time. The introduction of the *Reinheitsgebot* in 1516 limited the production of this style, which practically disappeared until, in 1988, a group of breweries decided to start producing roggenbier again, obtaining a sort of dunkelweizen.

Sedum roseum (L.) Scop.
CRASSULACEAE

SYNONYMS *Rhodiola arctica* Boriss., *Rhodiola borealis* Boriss., *Rhodiola elongata* (Ledeb.) Fisch. & C.A.Mey., *Rhodiola iremelica* Boriss., *Rhodiola krivochizhinii* Sipliv., *Rhodiola maxima* Nakai, *Rhodiola minor* Mill., *Rhodiola odora* Salisb., *Rhodiola odorata* Lam., *Rhodiola roanensis* (Britton) Britton, *Rhodiola rosea* L., *Rhodiola sachalinensis* Boriss., *Rhodiola scopolii* A.Kern. ex Simonk., *Sedum altaicum* G.Don, *Sedum arcticum* (Boriss.) Rønning, *Sedum elongatum* Ledeb., *Sedum roanense* Britton, *Sedum sachalinense* (Boriss.) Vorosch., *Tetradium odoratum* (Lam.) Dulac, *Tolmachevia krivochizhinii* (Sipliv.) Á.Löve & D.Löve.

COMMON NAMES Rhodiola rosea, golden root, roseroot.

DESCRIPTION Herb, perennial (15–30 cm). Stem erect, simple, stout, glabrous, glaucous, tuberous, with violet scent, bearing closely spaced leaves. Leaves scattered, very close together, flat, ovate-acuminate, sessile and rounded at the base, toothed in the distal half, erect or erect-patent. Flowers dioecious, greenish or reddish in color, pedicellate; calyx consisting of 4 sepals, small, lanceolate; corolla with 4 elliptical petals, overlying the calyx; androecium with 8 stamens, projecting; gynoecium with 4 carpels, linear-acuminate, with tip curved outward. Inflorescence narrow corymb with twigs arranged in a spiral.

Secale cereale L.
POACEAE

Cultivar At the moment, not many are known (e.g., Mattmark, grown in Switzerland); however, the growing interest in the properties of this plant and the considerable demand for plant material may mean that others will soon be selected as part of the process of domestication necessary to undertake cultivation on a large scale.

Related Species The genus *Sedum* includes about 420 taxa.

Geographic Distribution Vosges (Hohneck), Alps, French and Spanish Pyrenees, NC Europe, W Siberia, North America. **Habitat** Rocky mountain environments. **Range** Cultivation of this species is still in an early stage, but it is assumed that the range will overlap with the natural range and perhaps beyond.

Beer-Making Parts (and Possible Uses) Roots.

Chemistry Essential oil: linalool 2.7%, 6,6-dimethyl-cyclo [3,1,1] hept-2-ene-2-carboxyaldehyde 1.0%, myrtenol 36.9%, trans-pinocarveol 16.1%, octanol 13.6%, geraniol 12.7%, cumin alcohol 12.1%, myrtanol 1.0%, perilla alcohol 1.7%, dihydrocumin alcohol 2.1%.

Style Barleywine.

Source https://en.wikipedia.org/wiki/Dogfish_Head_Brewery.

Sequoia sempervirens (D.Don) Endl.
CUPRESSACEAE

Synonyms *Condylocarpus sempervirens* Salisb. ex Carrière, *Gigantabies taxifolia* J.Nelson, *Schubertia sempervirens* (Lamb.) Spach, *Sequoia pyramidata* Carrière, *Sequoia religiosa* C.Presl, *Sequoia taxifolia* K.Koch, *Steinhauera sempervirens* (D.Don) Voss, *Taxodium nutkaense* Lamb. ex Endl., *Taxodium sempervirens* D.Don.

Common Names Redwood.

Description Tree, large (up to 110 m). Trunk Ø up to 9 m; crown conical and monopodial when young, narrowed conical when adult; bark reddish-brown, up to about 35 cm thick, fibrous, furrowed; branches downward facing to slightly ascending. Leaves 1–30 mm, usually with stomata on both sides, those on dominant ascending twigs and fertile shoots strongly appressed, short, lanceolate to deltate, those on horizontal or downward-facing twigs mostly linear to linear-lanceolate, divergent and arranged in 2 ranks, with 2 abaxial strips of prominent white stomata. Male cones globose to ovoid, 2–5 mm, borne singly on short terminal or axillary stems; female cones 1.3–3.5 cm. Seeds flattened, 3–6 mm, leathery.

Related Species The genus *Sequoia* is monospecific.

Geographic Distribution United States (California, Oregon). **Habitat** Coastal redwood forests (0–300 m, occasionally up to 1,000 m ASL). **Range** Cultivated as a forest plant.

Beer-Making Parts (and Possible Uses) Young apices, wood (tried for barrels and wine vessels with little success; uncertain use in beer making).

Chemistry Essential oil: α-phellandrene 29.6%, dl-limonene 15.6%, α-pinene 8.65%, β-germacrene 4.87%, terpinene-4-ol 4.35%, α-terpenyl acetate 3.65%, germacrene D 2, 54%, γ-terpinene 1.82%, phytol 1.53%, m-cymene 1.45%, thymol acetate 1.45%, cis-β-ocimene 1.35%, terpinolene 1.32%, carvacrol 1.21%, p-cymene 1.2%, β-citronellol 1.12%, α-humulene 1.08%, citronellyl pentanoate 1.06%, trans-piperitol 0.92%, E,E-α-farnesene 0.89%, phenchone 0.85%, sabinene 0.75%, trans-β-ocimene 0.75%, spathulenol 0.75%, borneol 0.73%, p-cimen-8-ol 0.73%, trans-β-farnesene 0.73%, thymol 0.69%, cis-piperitol 0.65%, α-cadinol 0.65%, b-myrcene 0.64%, β-caryophyllene 0.53%, eugenol 0.43%, geraniol 0.34%, α-terpineol 0.26%, trans-α-bergamotene 0.26%.

Wood Differentiated, with narrow whitish sapwood and red heartwood (WSG 0.45–0.48), extremely durable and resistant to alteration, not attacked by insects; valuable for building and road construction, fixtures, carpentry, crossbars, cigar boxes.

Style Apices: Herb Beer. Wood: no known style.

Source Cantwell and Bouckaert 2016; Cantwell and Bouckaert 2018; Heilshorn 2017; https://broom02.revolvy.com/main/index.php?s=Moonlight%20Brewing%20Company; https://honest-food.net/2016/05/05/gruit-herbal-beer/.

Silybum marianum (L.) Gaertn.
ASTERACEAE

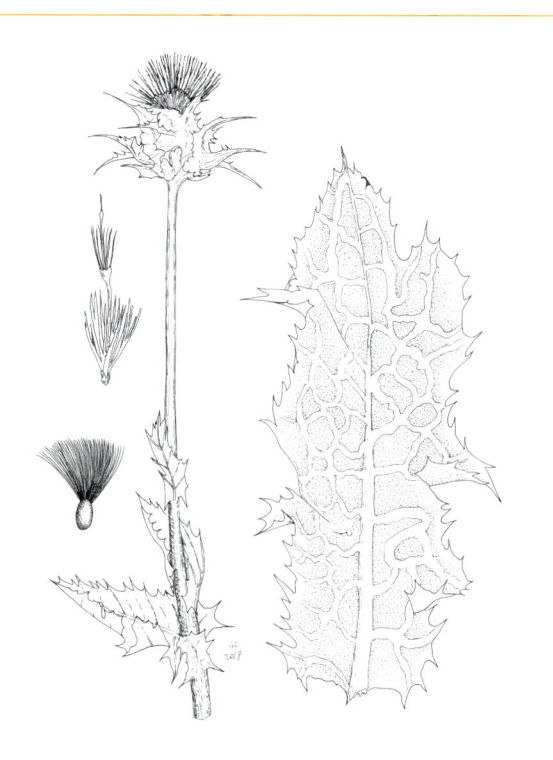

Synonyms *Carduus lactifolius* Stokes, *Carduus mariae* Crantz, *Carduus marianus* L., *Carduus versicolor* Salisb., *Carthamus maculatus* (Scop.) Lam., *Centaurea dalmatica* Fraas, *Cirsium maculatum* Scop., *Mariana lactea* Hill, *Silybum maculatum* (Scop.) Moench, *Silybum mariae* (Crantz) Gray, *Silybum pygmaeum* Cass.

Common Names Cardus marianus, milk thistle, blessed milkthistle, Marian thistle, Mary thistle, Saint Mary's thistle, Mediterranean milk thistle, variegated thistle, Scotch thistle.

Description Herb, biennial (3–15 dm). Stem erect, simple or with a few branches, bare and spidery above. Lower leaves large (2–4 dm), glossy, and leathery, white variegated, ± lanceolate-lobed, with strong spines; upper leaves smaller, amplexicaul, toothed. Inflorescences flower heads (Ø 4–7 cm) isolated on long peduncles; involucre ovate; scales with basal portion lanceolate, above which is an enlarged appendage with a strong ± patent apical spine and minor lateral spinules; corolla purple in color. Fruits achenes (6–7 mm) with white pappus.

Related Species Besides the species cited here, the genus *Silybum* includes only one other taxon, *Silybum eburneum* Coss. & Durieu.

Geographic Distribution Mediterranean-Turanic. **Habitat** Ruins, hedges, along roads. **Range** Mediterranean-Turanic; this plant has become a pest in various areas of the world in synanthropic environments.

Beer-Making Parts (and Possible Uses) Seeds, leaves, young shoots (give bitterness), dried plant (for smoking malt).

Toxicity Higher concentrations of mycotoxins (up to 37 mg/kg) have been found in dietary supplements derived from this plant than in similar products derived from other species.

Chemistry Silymarin, a flavonolignan complex, is present at a reduced extent (about 4–6%) in the seeds and in much greater amounts (65–80%) in the extract, which also contains 20–35% fatty acids (including linoleic acid). Silymarin is a complex mixture of polyphenolic molecules, including seven similar flavonolignans (silybine A, silybine B, isosilybine A, isosilybine B, silychristine, isosilychristine, silydianine) and a flavonoid (taxifolin).

Style None known.

Source Calagione et al. 2018; Daniels and Larson 2000; Fisher and Fisher 2017; Hieronymus 2016b; Josephson et al. 2016.

Sinapis alba L.
BRASSICACEAE

Synonyms *Bonannia officinalis* C.Presl, *Brassica alba* (L.) Rabenh., *Brassica foliosa* (Willd.) Samp., *Brassica hirta* Moench, *Crucifera lampsana* E.H.L.Krause, *Eruca alba* (L.) Noulet, *Leucosinapis alba* (L.) Spach, *Napus leucosinapsis* K.F.Schimp. & Spenn., *Raphanus albus* (L.) Crantz, *Rhamphospermum album* (L.) Andrz. ex Rchb., *Sinapis foliosa* Willd.

Common Names White mustard, yellow mustard.

Description Herb, annual (20–80 cm). Taproot. Stems erect, striate, branched, ± covered with reflexed hairs. Leaves petiolate, basal ones lyrate, lamina with serrated outline, cauline ones smaller but similar, with terminal segment larger than lateral ones. Inflorescence elongated multifloral raceme. Flowers pedunculate, corolla 4 spatulate saffron-yellow petals (6 × 9 mm), base abruptly narrowed into a slender appendage; calyx of 4 linear yellowish sepals (about 4 mm). Fruit siliqua (2–3 cm × 3–4 mm), pedunculate, densely setose, with long ensiform rostrum (10–25 mm), slightly arcuate. Seeds (Ø 2–3.5 mm) yellowish-white to brownish color.

Subspecies Besides the nominal one, *Sinapis alba* L. subsp. *alba*, there are two subspecies: *Sinapis alba* L. subsp. *dissecta* (Lag.) Simonk. and *Sinapis alba* L. subsp. *mairei* (H.Lindb.) Maire.

Cultivar They are numerous, particularly in countries where its cultivation is traditionally practiced. In general, traditional cultivars have high contents of glucosinolates and erucic acid (e.g., Polish Nakielska) compared to "modern" cultivars of recent selection and release (e.g., Polish Bamberka).

Related Species In addition to *Sinapis alba* L., five other species are recognized by taxonomists in the genus *Sinapis* (*Sinapis allionii* Jacq., *Sinapis arvensis* L., *Sinapis circinata* Desf., *Sinapis flexuosa* Poir., *Sinapis pubescens* L.), but none has known beer-making use.

Geographic Distribution Mediterranean. **Habitat** Cultivated fields, wastelands, roadsides, ruins, often cultivated and subspontaneous (0–800 m ASL). **Range** Mediterranean.

Beer-Making Parts (and Possible Uses) Seeds.

Chemistry Essential oil (30 compounds; 98.75% of total oil): benzyl isothiocyanate 64.89%, benzyl nitrile 12.05%, thymol 7.2%, 2-phenyl isothiocyanate to 6.5%, limonene 4.73%, 1-butenyl isothiocyanate 0.59%, p-eudesmol 0.4%, p-pinene 0.3%, ethylte-tradecanoate 0.28%, 5-methylhexanonitrile 0.22%, furfural 0.21%, indole 0.18%, 2-methylpropyl isothiocyanate 0.11%, benzaldehyde 0.1%, flamenol 0.1%, isophitol 0.1%, benzenepropanonitrile 0.09%, 5-methylthiopentyl isothiocyanate 0.09%, 4-methyl pentyl isothiocyanate 0.08%, heptadecane 0.08%.

Style Spice/Herb/Vegetable Beer.

Source Calagione et al. 2018; Hieronymus 2010; https://www.brewersfriend.com/other/sinapinsiemen/.

Smilax ornata Lem.
SMILACEAE

Synonyms *Smilax regelii* Killip & C.V.Morton.

Common Names Sarsaparilla, Honduran sarsaparilla, Jamaican sarsaparilla.

Description Shrub, perennial (up to 15 m), climbing. Rhizomatous root. Stem semiwoody, basally acutely quadrangular, with stout spines (up to 1.2 cm), compressed, straight or curved; upper branches acutely quadrangular, sometimes narrowly winged at the corners. Leaves basal with petiole up to 7 cm, aculeate; leaves basal, often very large (up to 30 × 20 cm), broad, ovate to narrowly oblong, rounded or briefly acuminate at apex, ± slightly cordate or hastate at base; upper leaves smaller, oblong-lanceolate or ovate, acute at base, papery, glabrous, often armed below with short, recurved spines, the larger leaves 7-nerved, the smaller 5-nerved, bright green or light green when dry. Flowers staminate on peduncles up to 6.5 cm, pedicels 7–12 mm; perianth segments lanceolate, 3.5–5 mm; filaments up to 1.2 mm; anthers 2–2.5 mm; pistillate flowers with peduncles up to 10 cm, compressed; pedicels 9–19 mm. Fruits globose (Ø up to 1.3 cm), black.

Subspecies In addition to the black-fruited form, a white-fruited form is known (*Smilax ornata* Lem. f. *albida* [Killip & Morton] Standl. & Steyerm. = *Smilax ornata* Lem. var. *albida* Killip & Standl.), found near Tela, on the Atlantic coast of Honduras.

Related Species Of the 257 species belonging to the genus *Smilax*, many have ethnobotanical applications (food, medicine, etc.), but only the one described here seems to have beer-making uses, although the common name (the term "sarsaparilla" usually refers to several species of the genus *Smilax*) may lead to some confusion about the exact species used.

Geographic Distribution Mexico, Central America (Nicaragua, Honduras, Belize, Guatemala).
Habitat Forests and thickets (0–1,500 m ASL).
Range Central America, South America.

Beer-Making Parts (and Possible Uses) Roots.

Chemistry Substances contained in the root (considering the not clearly univocal relationship between botanical species and commercial product, it seems more correct to indicate the main components of the latter): sarsaponin, sarsaparilloside, arsasapogenin, starch, flavonoids, sarsapaco acid.

Style American Brown Ale, American IPA, American Porter, American Stout, Belgian Specialty Ale, Braggot, Brown Porter, California Common Beer, Clone Beer, Cream Ale, Dark Mild, Dry Stout, English Barleywine, Extra Special/Strong Bitter (ESB), Holiday/Winter Special Spiced Beer, Mild, Northern English brown, Oatmeal Stout, Specialty Beer, Spice/Herb/Vegetable Beer, Sweet Stout, Wee Heavy.

Beer Sarsaparilla Six, Dark Horse (Michigan, United States).

Source Heilshorn 2017; https://www.brewersfriend.com/other/sarsaparilla/.

Solanum tuberosum L.
SOLANACEAE

Synonyms *Solanum andigenum* Juz. & Bukasov, *Solanum aquinas* Bukasov, *Solanum chiloense* Berthault, *Solanum chilotanum* Hawkes, *Solanum cultum* Berthault, *Solanum diemii* Brücher, *Solanum fonckii* Phil., *Solanum kesselbrenneri* Juz. & Bukasov, *Solanum leptostigma* Juz. & Buk., *Solanum molinae* Juz., *Solanum oceanicum* Brücher, *Solanum ochoanum* Lechn., *Solanum sanmartiniense* Brucher, *Solanum subandigena* Hawkes, *Solanum tascalense* Brucher, *Solanum zykinii* Lechn.

Common Names Potato.

Description Herb, annual (3–9 dm). Ascending, pubescent stems with underground stolons that produce starchy tubers (potatoes) at the end of the season. Leaves imparipinnate with 7–13 ovate-lanceolate segments (1–3 × 5–8 cm), acuminate, interspersed with irregular lobes. Inflorescence few-flowered cyme; peduncles 2–3 cm, usually articulate. Flower calyx 5–10 mm; corolla (Ø 2–3 cm) bell shaped, usually white, rarely mottled pinkish, purplish, or purple; anthers yellow or orange (6–7 mm). Fruit subspherical berry (Ø 2–4 cm), green to purple, poisonous.

Subspecies In addition to the typical subspecies, *Solanum tuberosum* L. subsp. *tuberosum*, science recognizes *Solanum tuberosum* L. subsp. *andigena* (Juz. & Bukasov) Hawkes; a botanical variety, *Solanum tuberosum* L. var. *longibaccatum* (Bukasov) Ochoa is also recognized, as well as different botanical forms such as *Solanum tuberosum* L. f. *ccompis* (Bukasov) Ochoa, *Solanum tuberosum* L. f. *cevallosii* (Bukasov) Ochoa, *Solanum tuberosum* L. f. *huacalajra* (Bukasov) Ochoa, and *Solanum tuberosum* L. f. *pallidum* (Bukasov) Ochoa.

Cultivar There are more than 4,000 varieties of potato, grouped according to the external color of the tuber: rust, red, white, yellow, or purple. For food purposes, they are frequently differentiated according to their waxyness: floury (20–22% starch) or waxy (16–18% starch). Another distinction can be made according to the ratio of the two biochemical components of starch: amylose and amylopectin. Varieties with a higher content of amylopectin, a strongly branched molecule, hold their shape better when boiled. The most common are Adirondack Blue, Adirondack Red, Agata, Almond, Alpine Russet, Alturas, Amandine, Annabelle, Anya, Arran Victory, Atlantic, Austrian Crescent, Avalanche, Bamberg, Bannock Russet, Belle de Fontenay, BF-15, Bildtstar, Bintje, Blazer Russet, Blue Congo, Bonnotte, British Queen, Cabritas, Camota, Canela Russet, Cara, Carola, Chelina, Chiloé, Cielo, Clavela Blanca, Désirée, Estima, Fianna, Fingerling, Flava, French Fingerling, German Butterball, Golden Wonder, Goldrush, Home Guard, Innovator, Irish Cobbler, Irish Lumper, Jersey Royal, Kennebec, Kerr's Pink, Kestrel, Keuka Gold, King Edward, Kipfler, Lady Balfour, Langlade, Linda Potato, Marcy, Marfona, Maris Piper, Marquis, Megachip, Melody, Monalisa, Nicola, Norgold Russet, Pachacoña, Pike, Pink Eye, Pink Fir Apple, Primura, Ranger Russet, Ratte, Record, Red La Soda, Red Norland, Red Pontiac, Rooster, Russet Burbank, Russet Norkotah, Selma, Shepody, Sieglinde, Silverton Russet, Sirco, Snowden, Stobrawa, Superior, Tick, Umatilla Russet, Up to Date, Villetta Rose, Vitelotte, Vivaldi, Yellow Finn, and Yukon Gold.

Related Species The genus *Solanum* includes about 1,230 taxa, some of which have food uses; *Solanum melongena* L. also has beer-making use.

Geographic Distribution South America.
Habitat Cultivated plant. **Range** China, India, Russia, Ukraine, United States, Germany, Bangladesh, France, Netherlands, Poland.

Beer-Making Parts (and Possible Uses) Tubers.

Toxicity The toxicity of the plant (not the tuber) of *Solanum tuberosum* is extreme and is related to the presence of glycoalkaloids such as solanine and chaconine. Ingestion produces gastrointestinal, hepatic, and cardiac damage and can cause death by cardiorespiratory arrest when consumed in large quantities. Externally, the juice of the plant or even simple contact can cause skin irritation and blisters. Similar symptoms can appear through the ingestion of green or sprouted potatoes, or even young buds.

Chemistry Tubers: energy 77 kcal, lipids 0.1 g, carbohydrates 17 g, fiber 2.2 g, sugars 0.8 g, protein 2 g, sodium 6 mg, potassium 421 mg, calcium 12 mg, iron 0.8 mg, magnesium 23 mg, vitamin A 2 IU, vitamin C 19.7 mg, vitamin B6 0.3 mg.

Style American Amber Ale, Belgian Specialty Ale, Blonde Ale, Robust Porter, Russian Imperial Stout, Saison, Spice/Herb/Vegetable Beer, Witbier.

Beer La Prima, Bodibeer (Roccavione, Piedmont, Italy).

Source Hieronymus 2010; Hieronymus 2016b; Josephson et al. 2016; McGovern 2017; https://www.brewersfriend.com/other/potatoes/; https://www.brewersfriend.com/other/potato/; https://www.eticamente.net/24977/birra-quali-sostanze-sono-contenute-al-suo-interno.html.

Solidago virgaurea L.
ASTERACEAE

Synonyms *Aster virgaurea* (L.) Kuntze, *Dectis decurrens* Raf., *Doria virgaurea* Scop., *Solidago cantoniensis* Lour., *Solidago corsica* (Rouy) A.W.Hill, *Solidago minor* Mill., *Solidago nudiflora* Viv., *Solidago vulgaris* Lam.

Common Names European goldenrod, woundwort.

Description Herb, perennial (1–8 dm). Rhizome oblique, reddish-brown in color. Stem erect, subglabrous, sparsely pubescent, and striated above. Lower leaves with winged petiole 5–8 cm long and lanceolate lamina (2–3 × 7–9 cm to 8 × 15 cm in sterile rosettes), acute, with connivent denticles; cauline leaves facing ± toward the same side, progressively reduced and sessile. Inflorescence flower head on 1–3 mm pubescent peduncle; involucre cylindrical; scales 6–8 mm. Flowers yellow, those ligulate 10–16 mm (ligule 5–10 mm), those tubular 7–9 mm. Fruit achene (3 mm) ribbed, pubescent.

Subspecies In addition to the typical subspecies, *Solidago virgaurea* L. subsp. *virgaurea*, 15 subspecies (and varieties) are recognized: *Solidago virgaurea* L. subsp. *alpestris* (Waldst. and Kit.) Gremli, *Solidago virgaurea* L. subsp. *armena* (Grossh.) Greuter, *Solidago virgaurea* L. subsp. *asiatica* Kitam. ex Hara, *Solidago virgaurea* L. subsp. *caucasica* (Kem.-Nath.) Greuter, *Solidago virgaurea* L. subsp. *dahurica* (Kitag.) Kitag., *Solidago virgaurea* L. subsp. *gigantea* (Nakai) Kitam., *Solidago virgaurea* L. var. *insularis* (Kitam.) Hara, *Solidago virgaurea* L. subsp. *jailarum* (Juz.) Tzvelev, *Solidago virgaurea* L. subsp. *lapponica* (With.) Tzvelev, *Solidago virgaurea* L. subsp. *macrorrhiza* (Lange) Nyman, *Solidago virgaurea* L. subsp. *minuta* (L.) Arcang, *Solidago virgaurea* L. subsp. *stenophylla* (G.E.Schultz) Tzvelev, *Solidago virgaurea* L. subsp. *talyschensis* (Tzvelev) Sennikov, *Solidago virgaurea* L. subsp. *taurica* (Juz.) Tzvelev, and *Solidago virgaurea* L. subsp. *turfosa* (Woronow ex Grossh.) Greuter.

Related Species The genus *Solidago* includes about 160 taxa, but only *Solidago virgaurea* L. has known beer-making applications.

Geographic Distribution Circumboreal.
Habitat Woods, thickets, pastures. **Range** Circumboreal.

Beer-Making Parts (and Possible Uses) Flowering tops.

Chemistry Flowering tops: flavonoids (quercetin, kaempferol, astragalin, rutoside) 1.5%, anthocyanidins, triterpene saponins of the oleanane type up to 2%, phenols glycosides bisdesmosides leiocarposide 0.08–0.48% and virgaureoside A, diterpenoid lactones of the type cis-clerodane, phenolic acids (caffeic acid, chlorogenic acid 0.2–0.4%, ferulic acid, sinapic acid, vanillinic acid), essential oils (cadinene, α-pinene, β-pinene, myrcene, limonene, sabinene, germacrene D).

Style American Brown Ale, Belgian Specialty Ale.

Source https://www.brewersfriend.com/other/goldenrod/.

Sorbus aucuparia L.
ROSACEAE

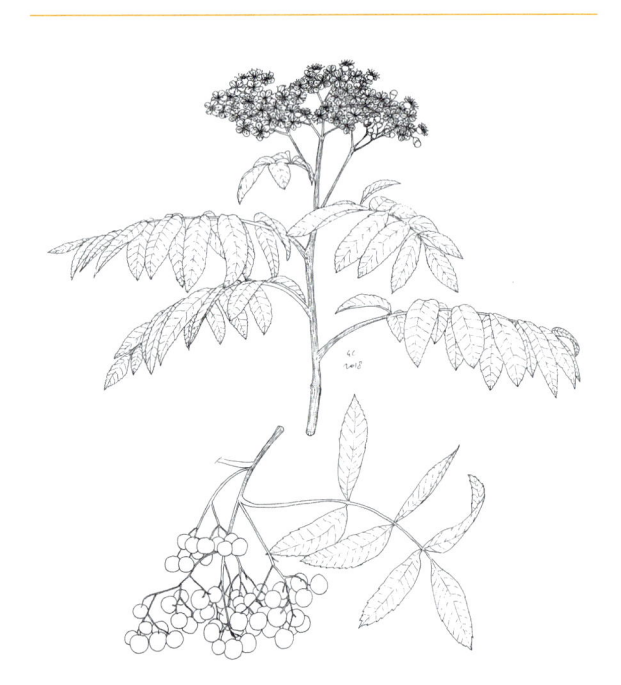

Synonyms *Aucuparia sylvestris* Medik., *Mespilus aucuparia* (L.) Scop., *Pyrenia aucuparia* (L.) Clairv., *Pyrus aucuparia* (L.) Gaertn., *Pyrus rossica* A.D. Danilov, *Sorbus altaica* Koehne, *Sorbus amurensis* Koehne, *Sorbus anadyrensis* Kom., *Sorbus boissieri* C.K.Schneid., *Sorbus camschatcensis* Kom., *Sorbus glabrata* (Wimm. & Grab.) Hedl., *Sorbus pohuashanensis* (Hance) Hedl., *Sorbus polaris* Koehne.

Common Names Mountain ash, European mountain ash, rowanberry tree.

Description Shrub or tree (1–15 m) with young pubescent branches; bark brown, smooth, with linear lenticels (1–2 mm). Leaves composed of 13–15 lanceolate segments (11–14 × 27–55 mm), sparsely pubescent below, especially on nerves. Inflorescence corymb. Flowers numerous, floral peduncles with appressed or subglabrous hairs; sepals connivent; petals white (5 mm). Fruit ovoid (5–6 × 6–8 mm), scarlet red at maturity.

Subspecies In addition to the nominal subspecies, *Sorbus aucuparia* L. subsp. *aucuparia*, four subspecies are recognized: *Sorbus aucuparia* L. subsp. *maderensis* (Lowe) McAll., *Sorbus aucuparia* L. subsp. *pohuashanensis* (Hance) McAll., *Sorbus aucuparia* L. subsp. *praemorsa* (Guss.) Nyman, *Sorbus aucuparia* L. subsp. *sibirica* (Hedl.) Krylov. One botanical form is also recognized: *Sorbus aucuparia* L. f. *xanthocarpa* (Hartwig & Rümpler) Rehder.

Related Species The genus *Sorbus* includes about 260 taxa of specific, subspecific, or hybrid rank, of which some have edible fruits and thus potential beer-making use.

Geographic Distribution Europe. **Habitat** Mountain woods (beech forests, fir forests) and subalpine woods, rhododendron bushes (600–2,100 m ASL). **Range** Eurasia.

Beer-Making Parts (and Possible Uses) Fruits, twigs (used to filter wort in traditional brewing in Scandinavian countries).

Toxicity Large quantities of raw fruits can cause intoxication symptoms such as vomiting, probably owing to cyanide compounds that can cause respiratory problems and even death.

Chemistry Whole fruits (% of dry matter): ash 5.1%, protein 9.3%, lipids 6.2%, fiber 8.9%, sugars 6.6%, calcium 2 mg/g, phosphorus 2.1 mg/g, potassium 15.8 mg/g, magnesium 1.6 mg/g, vitamin C 131.2 mg, β-carotene 99 μg/g. Pulp (% of dry matter): ash 10.2%, protein 6.4%, lipids 2.3%, fiber 5.3%, sugars 17.7%, calcium 1.8 mg/g, phosphorus 1.1 mg/g, potassium 15.8 mg/g, magnesium 0.9 mg/g.

Style Sahti, Saison.

Source Cantwell and Bouckaert 2016; Cantwell and Bouckaert 2018; Heilshorn 2017; https://www.brewersfriend.com/other/rowanberries/.

Sorbus domestica L.
ROSACEAE

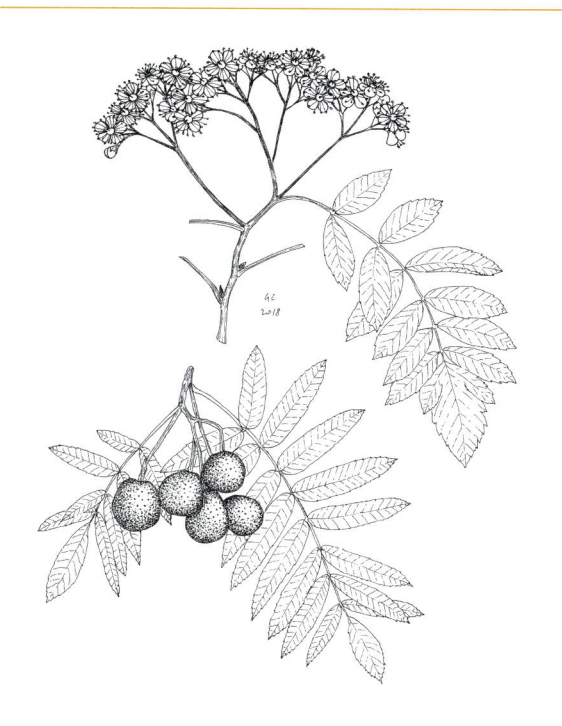

Synonyms *Cormus domestica* (L.) Spach, *Pyrus domestica* (L.) Ehrh., *Pyrus domestica* (L.) Sm., *Pyrus serrulata* hort., *Pyrus sorbus* Gaertn.

Common Names Service tree.

Description Tree (5–20 m) with young pubescent branches; bark brown, incised, with linear lenticels (1–2 mm). Leaves composed of 13–15 lanceolate segments (11–14 × 27–55 mm), sparsely pubescent below, especially on nerves. Inflorescence corymb. Flowers numerous, flower peduncles first tomentose, then glabrescent; sepals patent after flowering; petals (5 mm) white. Fruit pyriform (2–3 cm), sweet when ripe (sorb fruit).

Cultivar The cultivars of *Sorbus domestica*, in many cases genotypes propagated by vegetative reproduction, are numerous in all areas where the species is spread as a spontaneous plant and/or cultivated. As an example, some genotypes from Sicily include 104SRB003A, 61SR-B001P, 66SRB002S, 66SRB017S, 66SRB022S, 66SRB023S, 67SRB001A, 67SRB002A, 67SRB003A, 67SRB004A, 67SRB005A, 67SRB008A, 67SRB009A, 67SRB010A, 69SRB001S, 69SRB002S Sorbo Natalino, 69SRB003S (SN2), 69SRB004P, 69SRB004P (SN3), 69SRB005S, 69SRB006P (SN5), 69SRB008P (SN7), 69SRB011S, 69SRB012S, 71SRB003CD, 71SRB005A Austarica, 71SRB006CD, 71SRB009A Rowan Tardivo, 90SRB050G, DCASRB001, and Zorba d'Inverno Kolymbetra.

Related Species The genus *Sorbus* includes about 260 taxa of specific, subspecific, or hybrid rank, of which some have edible fruits and thus potential beer-making use.

Geographic Distribution Euro-Mediterranean. **Habitat** Sub-Mediterranean woods, also cultivated (0–800 m ASL). **Range** CS Europe, N Turkey.

Beer-Making Parts (and Possible Uses) Fruits.

Chemistry Fruits (pulp): protein 0.44–0.65 g/kg, lipids (not analyzed), sugars 14–16%, vitamin C 0.89–0.98 mg/kg, vitamin E 1–2.35 mg/kg, phenolic compounds (62), potassium, calcium, sodium, zinc, copper, magnesium, chromium, iron.

Style Saison.

Source Giaccone and Signoroni 2017; https://www.brewersfriend.com/other/rowanberries/.

Sorbus torminalis (L.) Crantz
ROSACEAE

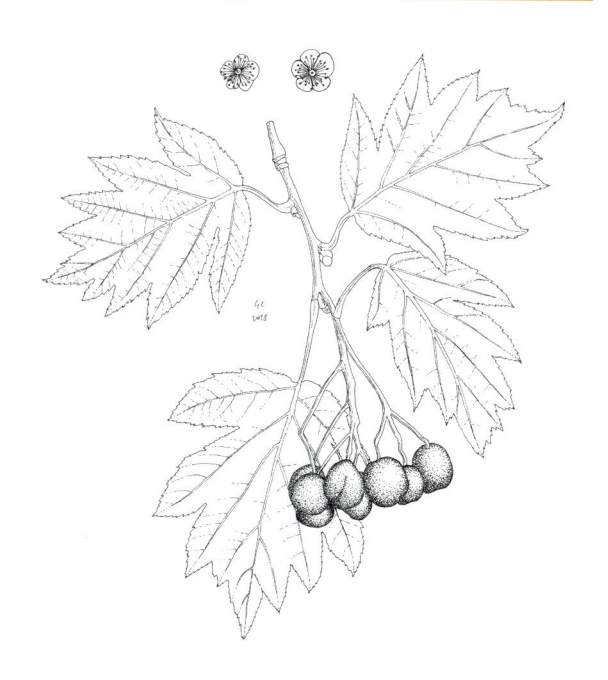

Synonyms *Aria torminalis* (L.) Beck, *Crataegus torminalis* L., *Hahnia torminalis* (L.) Medik., *Pyrus*

torminalis (L.) Ehrh., *Sorbus orientalis* Schenb.-Tem., *Torminalis clusii* (M.Roem.) K.R.Robertson & J.B.Phipps, *Torminaria clusii* M.Roem.

Common Names Wild service tree, chequers, checker tree.

Description Tree, small (up to 7 m) or medium height (up to 15–20 m), deciduous. Trunk with grayish-brown, shiny, somewhat angular branches; crown globular, irregular, dense, flattened; trunk erect, sometimes polycormic, branches ascending; bark brownish-gray, smooth, and sprinkled with pale, elliptical lenticels in young specimens; wrinkled, irregularly cracked, and fissured in older ones. Leaves alternate, long petiolate and ornamented with linear stipules, hairy-glandular (10–12 cm), lamina ovate-lobate, subcordate at base, 3–4 pairs of lobes deep, acute, irregularly toothed at margins, central lobe acuminate hairy when young; upper leaf bright green, underleaf paler, both glabrous when mature, with 3–5 pairs of prominent secondary veins, tending to blood red in autumn. Flowers on tomentose peduncles, hermaphrodite, fragrant, clustered in sparse, branched, erect corymbs; calyx hairy with 5 triangular, caducous laciniae; corolla of 5 creamy white, obovate, clawless petals; numerous stamina as long as petals, anthers yellow, 2 styluses connate almost to the middle. Fruits small ovoid knobs with a sour taste, reddish, then rust-brown in color, dotted with lenticels; endocarp membranous with 4 red-brown trigone seeds.

Related Species The genus *Sorbus* includes about 260 taxa of specific, subspecific, or hybrid rank, some of which have edible fruits and thus potential beer-making use.

Geographic Distribution Paleotemperate.
Habitat Deciduous woodlands, transitional woodlands with Mediterranean evergreen forests, shrublands. Sunny exposures, acidic or subacidic soils, clayey, deep, soils, but also adapts well to calcareous and stony substrates, intolerant to shade (0–1,000 m ASL). **Range** Eurasia, N Africa, Middle East.

Beer-Making Parts (and Possible Uses) Fruits.

Chemistry Fruits (active compounds µg/g dry weight): protocatechuic acid 13.7 µg, ferulic acid 27.8 µg, amentoflavone 15.8 µg, quercitrin-3-glucoside 13.6 µg, hyperoside 10.4 µg; below the quantifiable limit are the following compounds: gallic acid, kaempferol-3-O-glucoside, quercitrin, rutin, catechin, p-hydroxybenzoic acid, 2,5-dihydroxybenzoic acid, vanillic acid, cinnamic acid, caffeic acid, syringic acid, o-coumaric acid, p-coumaric acid, 3,4-dimethoxycinnamic acid, sinapic acid, apigenin, apigenin-7-O-glucoside, baicalin, baicalein, apiin, daidzein, naringenin, vitexin, genistein, isorhamnetin, luteolin, luteolin-7-O-glucoside, myricetin, kaempferol, epigallocatechin gallate, chrysoeriol, quercetin, epicatechin, umbelliferone, scopoletin, matairesinol, secoisolariciresinol.

Style Gruit (?).

Source https://en.wikipedia.org/wiki/Sorbus_torminalis; https://www.floraitaliae.actaplantarum.org/viewtopic.php?t=25799.

Sorghum bicolor (L.) Moench
POACEAE

Synonyms *Agrostis nigricans* (Ruiz & Pav.) Poir., *Andropogon besseri* Kunth, *Andropogon bicolor* (L.) Roxb., *Andropogon caffrorum* (Thunb.) Kunth, *Andropogon compactus* Brot., *Andropogon dulcis* Burm.f., *Andropogon niger* (Ard.) Kunth, *Andropogon saccharatrus* Kunth, *Andropogon sorghum* (L.) Brot., *Andropogon subglabrescens* Steud., *Andropogon truchmenorum* Walp., *Andropogon usorum* Steud., *Andropogon vulgare* (Pers.) Balansa, *Andropogon vulgaris* Raspail, *Holcus arduinii* J.F.Gmel., *Holcus bicolor* L., *Holcus cafer* Ard., *Holcus caffrorum* (Retz.) Thunb., *Holcus cernuus* Ard., *Holcus compactus* Lam., *Holcus dochna* Forssk., *Holcus dora* Mieg, *Holcus durra* Forssk., *Holcus niger* Ard., *Holcus rubens* Gaertn., *Holcus saccharatus* L., *Holcus sorghum* L., *Milium bicolor* (L.) Cav., *Milium compactum* (Lam.) Cav., *Milium maximum* Cav., *Milium nigricans* Ruiz & Pav., *Milium sorghum* (L.) Cav., *Panicum caffrorum* Retz., *Rhaphis sorghum* (L.) Roberty, *Sorghum ankolib* (Hack.) Stapf, *Sorghum anomalum* Desv., *Sorghum arduinii* (Gmel.) J.Jacq., *Sorghum basiplicatum* Chiov., *Sorghum basutorum* Snowden, *Sorghum caffrorum* (Retz.) P.Beauv., *Sorghum campanum* Ten. & Guss., *Sorghum caudatum* (Hack.) Stapf, *Sorghum centroplicatum* Chiov., *Sorghum cernuum* (Ard.) Host, *Sorghum compactum* Lag., *Sorghum conspicuum* Snowden, *Sorghum coriaceum* Snowden, *Sorghum dochna* (Forssk.) Snowden, *Sorghum dora* (Mieg) Cuoco, *Sorghum dulcicaule* Snowden, *Sorghum dura* Griseb., *Sorghum durra* (Forssk.) Trab., *Sorghum elegans* (Körn.) Snowden, *Sorghum eplicatum* Chiov., *Sorghum exsertum* Snowden, *Sorghum gambicum* Snowden, *Sorghum giganteum*

Edgew., *Sorghum glabrescens* (Steud.) Schweinf. & Asch., *Sorghum glycychylum* Pass., *Sorghum guineense* Stapf, *Sorghum japonicum* (Hack.) Roshev., *Sorghum margaritiferum* Stapf, *Sorghum medioplicatum* Chiov., *Sorghum melaleucum* Stapf, *Sorghum melanocarpum* Huber, *Sorghum mellitum* Snowden, *Sorghum membranaceum* Chiov., *Sorghum miliiforme* (Hack.) Snowden, *Sorghum nankinense* Huber, *Sorghum nervosum* Besser ex Schult. & Schult.f., *Sorghum nigricans* (Ruiz & Pav.) Snowden, *Sorghum nigrum* (Ard.) Roem. & Schult., *Sorghum notabile* Snowden, *Sorghum papyrascens* Stapf, *Sorghum rigidum* Snowden, *Sorghum rollii* Chiov., *Sorghum roxburghii* Stapf, *Sorghum saccharatum* (L.) Moench, *Sorghum sativum* (Hack.) Trab., *Sorghum simulans* Snowden, *Sorghum splendidum* (Hack.) Snowden, *Sorghum subglabrescens* (Steud.) Schweinf. & Asch., *Sorghum tataricum* Huber, *Sorghum technicum* (Körn.) Roshev., *Sorghum technicum* (Körn.) Trab., *Sorghum truchmenorum* K.Koch, *Sorghum usorum* Nees.

Common Names Sorghum, great millet.

Description Herb, annual (1–3 m). Culm robust, erect, full, Ø up to 10 mm. Leaves lanceolate, 2–7 cm wide. Inflorescence panicle highly developed (2–4 dm). Spikelets not detached at maturity. Fruit caryopsis spherical to obovoid (3–6 mm).

Subspecies *Sorghum bicolor* (L.) Moench var. *charisianum* (Busse & Pilg.) Snowden appears to be one of the two varieties, together with the nominal *Sorghum bicolor* (L.) Moench subsp. *bicolor*, accepted for this taxon.

Cultivar Hundreds are known.

Related Species There are 32 species (some of hybrid origin) belonging to the genus *Sorghum*.

Geographic Distribution Paleotropical. **Habitat** Cultivated and sometimes wild. **Range** Africa (Nigeria), Mexico, United States.

Beer-Making Parts (and Possible Uses) Caryopses.

Chemistry Caryopses: energy 1,377 kJ (329 kcal), carbohydrates 72.1 g, fiber 6.7 g, lipids 3.5 g, protein 10.6 g, thiamine (B1) 0.33 mg, riboflavin (B2) 0.1 mg, niacin (B3) 3.7 mg, pantothenic acid (B5) 0.4 mg, vitamin B6 0.44 mg, folate (B9) 20 μg, calcium 13 mg, iron 3.4 mg, magnesium 165 mg, manganese 1.6 mg, phosphorus 289 mg, potassium 363 mg, sodium 2 mg, zinc 1.7 mg.

Style Doppelbock, Robust Porter, Russian Imperial Stout.

Beer Zulu, Herba Monstrum (Galbiate, Lombardy, Italy).

Source Calagione et al. 2018; Fisher and Fisher 2017; Hieronymus 2016b; Jackson 1991; McGovern 2017; https://www.bjcp.org (30. Spiced Beer; 31A. Alternative Grain Beer); https://www.brewersfriend.com/other/sorghum/; https://www.brewersfriend.com/other/sorgum/.

Spinacia oleracea L.
AMARANTHACEAE

Synonyms *Spinacia glabra* Mill., *Spinacia inermis* Moench, *Spinacia spinosa* Moench.

Common Names Spinach.

Description Herb, annual (3–8 dm), dark green in color. Stems erect, branched, leafy to apex. Leaves petiolate (1–2 cm), lamina triangular (2–4 × 3–6 cm), underleaf with enlarged hairs at apex. Flowers dioecious, 4- to 5-merous in leafy spikes (10 mm) with 2–4 spines or unarmed.

Cultivar The many varieties of spinach can be grouped into old and modern. The older ones have

Sorghum bicolor (L.) Moench
POACEAE

narrower leaves and tend to have a stronger, more bitter taste; modern varieties have wider leaves and a milder taste. There are three basic types of spinach: Savoy (dark green, curled leaves; Bloomsdale [the best-selling fresh spinach in the United States], Merlo Nero [sweet Italian variety], Viroflay [very large and productive spinach]); Flat or Smooth-Leaf (large, smooth leaves that are easier to clean than those of Savoy; this type is preferred in industrial processes [cans, frozen foods, soups, etc.]; Giant Noble); and Semi-Savoy (hybrid variety with slightly rippled leaves, for fresh consumption and industrial processing; Tyee Hybrid).

Related Species The only known species belonging to the genus *Spinacia*, besides the one described here, is *Spinacia turkestanica* Iljin.

Geographic Distribution SW Asia (?). **Habitat** Cultivated plant. **Range** China, United States, Japan, Turkey, Indonesia, France, Pakistan.

Beer-Making Parts (and Possible Uses) Leaves.

Chemistry Leaves: energy 23 kcal, lipids 0.4 g, carbohydrates 3.6 g, fiber 2.2 g, sugars 0.4 g, protein 2.9 g, vitamin A 9,377 IU, vitamin C 28.1 mg, vitamin B6 0.2 mg, sodium 79 mg, potassium 558 mg, iron 2.7 mg, calcium 99 mg, magnesium 79 mg.

Style Dry Stout.

Source https://www.brewersfriend.com/other/spinach/.

Spondias mombin L.
ANACARDIACEAE

Synonyms *Spondias aurantiaca* Schumach. & Thonn., *Spondias dubia* A.Rich., *Spondias graveolens* Macfad., *Spondias lutea* L., *Spondias oghigee* G.Don, *Spondias pseudomyrobalanus* Tussac.

Common Names Cajà, hog plum.

Description Tree (up to 30 m). Plantule with taproot, probably persistent in the mature tree, which also has a shallower root system. Trunk with grayish-brown, thick, wrinkled, often deeply grooved bark with blunt, spine-like extroversions; trunk with branches 2–10 m above ground level forming a broad canopy (Ø up to 15 m). Leaves alternate, compound, imparipinnate; stipules absent; rachis 30–70 cm; leaflets 5–10 pairs, elliptic (5–11 × 2–5 cm); apex long acuminate, asymmetrical, truncate or cuneate; margin entire, glabrous or slightly puberulent. Inflorescence panicle branched, terminal, bearing male, female, and hermaphrodite flowers. Flowers with 5 sepals, briefly deltoid (0.5–1 cm); petals 5, white or yellow, oblong (3 mm); stamens 10, inserted under a fleshy disc; ovary superficial (1–2 mm); styluses 4, short, erect. Fruit drupe ovoid or ellipsoid (3–4 × 2–2.5 cm), dull orange or yellow-brown; in groups of 1–20; epicarp thin, mesocarp juicy, orange or yellow, 3–6 mm thick; endocarp broad, with soft, fibrous, intricate covering, around 4–5 seeds. Seeds small.

Related Species The genus *Spondias* includes 12 species; of these, only for *Spondias mombin* has known beer-making applications.

Geographic Distribution Central and South America (Argentina, Bolivia, Chile, Colombia, Ecuador, French Guiana, Guatemala, Guyana, Honduras, Mexico, Nicaragua, Panama, Paraguay, Peru, Puerto Rico, Suriname, Uruguay, Venezuela). **Habitat** Wet climates of tropical forests, often in secondary vegetation derived from lowland evergreen forests or semideciduous forests. **Range** Central and South America (Argentina, Bolivia, Chile, Colombia, Ecuador, French Guyana, Guatemala, Guyana, Honduras, Mexico, Nicaragua, Panama, Paraguay, Peru, Puerto Rico, Suriname, Uruguay, Venezuela), Africa (Central African Republic, Democratic Republic of the Congo, Gabon, Gambia), India.

Beer-Making Parts (and Possible Uses) Fruits.

Chemistry Fruits: energy 201 kJ (47 kcal), protein 0.6 g, carbohydrates 9 g, lipids 0.6 g, vitamin C 15 mg, citric acid 1 g.

Style Saison.

Source McGovern 2017; https://www.brewersfriend.com/other/caja/.

Stachys officinalis (L.) Trevis.
LAMIACEAE

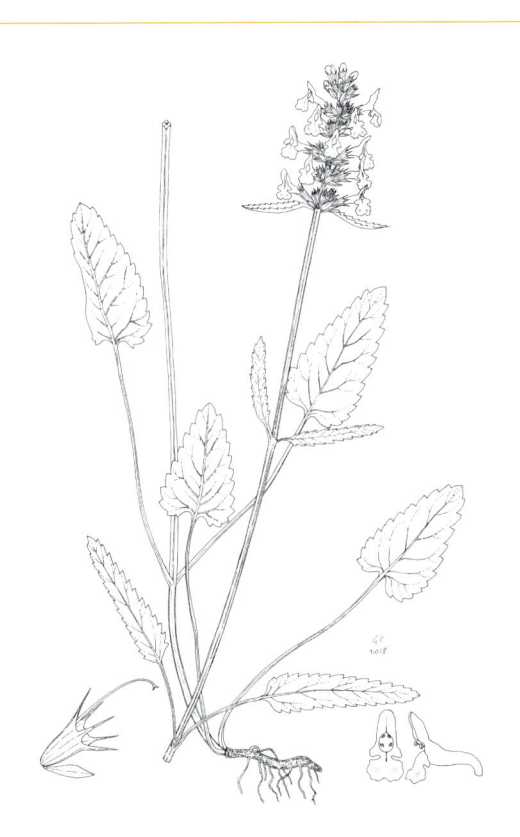

Synonyms *Betonica affinis* Wender., *Betonica alpigena* Schur, *Betonica angustifolia* Jord. & Fourr., *Betonica bjelorussica* Kossko ex Klok., *Betonica bjelorussica* Kossko ex Klokov, *Betonica brachydonta* Klokov, *Betonica brachystachya* Jord. & Fourr., *Betonica bulgarica* Degen & Nejceff, *Betonica clementei* Pérez Lara, *Betonica danica* Mill., *Betonica densiflora* Schur, *Betonica drymophila* Jord. & Fourr., *Betonica foliosa* C.Presl, *Betonica fusca* Klokov, *Betonica glabrata* K.Koch, *Betonica glabriflora* Borbás, *Betonica grandifolia* Jord. & Fourr., *Betonica hylebium* Jord. & Fourr., *Betonica incana* Mill., *Betonica laxata* Jord. & Fourr., *Betonica legitima* Link, *Betonica leiocalyx* Jord. & Fourr., *Betonica monieri* Gouan, *Betonica montana* Lej., *Betonica monticola* Jord. & Fourr., *Betonica nemorosa* Jord. & Fourr., *Betonica nutans* Kit. ex Schult., *Betonica occitana* Jord. & Fourr., *Betonica officinalis* L., *Betonica parvula* Jord. & Fourr., *Betonica pe raucta* Klokov, *Betonica polyclada* Jord. & Fourr., *Betonica pratensis* Jord. & Fourr., *Betonica psilostachys* Jord. & Fourr., *Betonica purpurea* Bubani, *Betonica pyrenaica* Jord. & Fourr., *Betonica recurva* Jord. & Fourr., *Betonica recurvidens* Peterm., *Betonica rigida* Jord. & Fourr., *Betonica rusticana* Jord. & Fourr., *Betonica sabauda* Jord. & Fourr., *Betonica stricticaulis* Jord. & Fourr., *Betonica subcarnea* Jord. & Fourr., *Betonica valdepubens* Jord. & Fourr., *Betonica validula* Jord. & Fourr., *Betonica virescens* Jord. & Fourr., *Betonica virgultorum* Jord. & Fourr., *Betonica vulgaris* Rota, *Stachys bulgarica* (Degen & Nejceff) Hayek, *Stachys danica* (Mill.) Schinz & Thell., *Stachys glabriflora* (Borbás) Rossi, *Stachys monieri* (Gouan) P.W.Ball, *Stachys officinalis* (L.) Trevis. ex Briq.

Common Names Common hedgenettle, betony, purple betony, wood betony, bishopwort, bishop's wort.

Description Herb, perennial (2–4 dm), rhizomatous, pleasantly fragrant. Stems hairy, erect, quadrangular in cross-section. Leaves basal with long, hairy petiole, lamina ovate-cordate, crenate with deep ± rounded teeth, wrinkled on upper leaf with pronounced veins; leaves cauline subsessile, or barely petiolate, with triangular teeth and variable hairiness; in upper part of caulis, they gradually turn to bracts. Inflorescence verticillaster in the form of a dense, interrupted spike. Flowers with bell-shaped calyx with 5 sepals; corolla pink or purple in color, white tube slightly curved; upper lip straight or curved inward, hairy on the outside; lower lip trilobed with central major lobe toothed on the edge and 2 lateral rounded; stamens 4, internal to the coralline tube, carrying brown-violet anthers; pistil with pinkish stylus and divided stigma. Fruits polyachene trigones, brown in color.

Subspecies Within *Stachys officinalis* (L.) Trevis., five taxa (of different taxonomic rank) are recognized: *Stachys officinalis* (L.) Trevis. var. *algeriensis* (Noë) Cout, *Stachys officinalis* (L.) Trevis. subsp. *haussknechtii* (Nyman) Greuter & Burdet, *Stachys officinalis* (L.) Trevis. subsp. *officinalis*, *Stachys officinalis* (L.) Trevis. subsp. *serotina* (Host) Hayek, and *Stachys officinalis* (L.) Trevis. subsp. *velebitica* (A.Kern.) Hayek.

Related Species The genus *Stachys* includes more than 420 taxa. Of these, many have medicinal and culinary applications, but none appears to have a specific beer-making use apart from the species mentioned here and *Stachys recta* L.

Geographic Distribution European-Caucasian. **Habitat** Meadows, arid meadows, pastures, alpine meadows, hilly and mountainous deciduous forests, wooded glades, especially in siliceous soils (0–1,800 m ASL). **Range** Europe-Caucasus.

Beer-Making Parts (and Possible Uses) Flowering tops, leaves.

Chemistry Essential oil, aerial parts (190 components; 97.9% of total oil): germacrene D 19.9%, β-caryophyllene 14.1%, α-humulene 7.5%.

Style Herb Beer, Mum.

Source Fisher and Fisher 2017; Heilshorn 2017; Hieronymus 2010; Hieronymus 2016b (this author also cites another species, *Stachys floridana* Shuttlew. ex Benth., whose leaves, unlike those of other taxa in the genus *Stachys*, would not be sufficiently bitter).

Stevia rebaudiana (Bertoni) Bertoni
ASTERACEAE

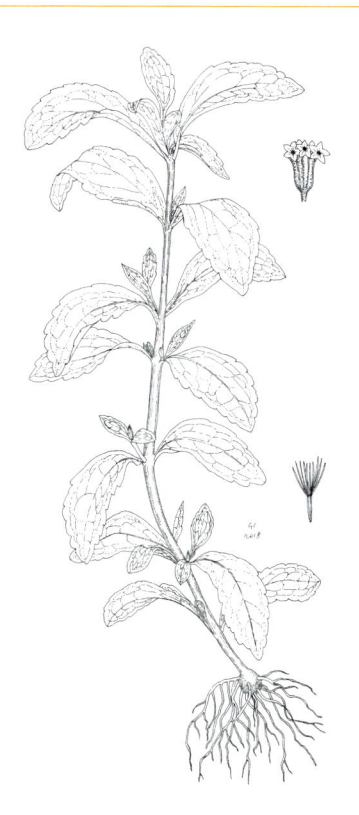

Synonyms *Eupatorium rebaudianum* Bertoni.

Common Names Candyleaf, sweetleaf, sweet leaf, sugarleaf.

Description Herb (7–10 dm), annual or perennial, stoloniferous, with tapering fibrous roots. Stem woody at the bottom, cylindrical, leafy from the base, covered in dense fuzz with simple, starry hairs. Basal leaves larger (5 × 2 cm), opposite, alternate, subsessile, lamina oval-lanceolate, slightly lobed at the apex, color dull green, upper leaf sprinkled with simple hairs; central vein well marked compared to the secondary ones, which branch about 1/3 of the way up from the junction, margin crenate-dentate; petiole very short, semiamplexicaule, with presence of hairs along the margin; median and upper leaves progressively smaller; at the leaf axil 2 secondary leaflets, lamina oval-lanceolate, margin sometimes entire, alternate, and also covered with simple hairs; leaf bracts just mentioned, perpendicular to secondary leaflets. Inflorescence corymb or corymbous panicle, bearing very small flowers. Flowers white with pinkish fauces, pentamerous and hermaphrodite; floral bracts shorter than corolla tube. Seeds achenes with pappus (about 3 mm).

Related Species Of the 47 known species belonging to the genus *Stevia*, only the one cited here seems to have beer-making applications.

Geographic Distribution Paraguay, Brazil. **Habitat** Mountain environments. **Range** E Asia (China, Korea, Taiwan, Thailand, Malaysia), Brazil, Colombia, Peru, Paraguay, Uruguay, Israel.

Beer-Making Parts (and Possible Uses) Leaves.

Chemistry Leaves: carbohydrates 63.1%, protein 10.73%, lipids 6.13%, ash 12.06%, fiber 5.03%, sugars 4.5%, vitamin C 14.98 mg, vitamin B2 0.43 mg, folic acid 52.18 mg, calcium 464.4 mg, phosphorus 11.4 mg, sodium 190 mg, potassium 1,800 mg, iron 55.3 mg, magnesium 349 mg, zinc 1.5 mg. Leaves also contain 4–20% diterpene glycosides, responsible for the great sweetening power of this plant; stevioside represents 4–13% stevia glycosides, and, despite its great sweetening power (150–250 times that of sucrose), it has a bitter aftertaste, whereas rebaudioside A has none, representing 2–4% of fresh leaf, with an even greater sweetening power (250–450 times that of sucrose).

Style Brown Porter, Cream Ale, Sweet Stout.

Source https://www.brewersfriend.com/other/stevia/.

Strychnos nux-vomica L.
LOGANIACEAE

Synonyms *Strychnos spireana* Dop.

Common Names Strychnine tree, nux vomica, poison nut, semen strychnos, Quaker buttons.

Description Tree, medium sized (up to 20–25 m). Trunk short, curved, thick; branches irregular, covered with smooth, thin bark; young shoots dark green, glossy. Leaves simple, shortly petiolate (5–13 mm), petiole glabrous, opposite, stipulate, lamina elliptic or elliptic-ovate (5–14 × 2–9 cm), base acute to attenuate, apex acute or obtuse, margin entire, glabrous, shiny, leathery; 3–5 veins from base, prominent, glabrous, intercurrent veins reticulate. Flowers hermaphrodite, small, sessile, tomentose, bracts 5 mm; bracteoles 1.5 mm; calyx lobes 4 (1 mm), triangular, acute, pubescent; corolla greenish-yellow in color, tube 10 mm, longer than the 4 corolla lobes (4 × 1.5 mm), oblong, acute; stamens 4, anthers sessile, inserted at the fauces of the corolla tube; ovary 2-carpellar, ovules many per carpel. Inflorescences small sessile apical cymes. Fruit globular berry (Ø 3–5 cm), yellowish-red or orange in color, glabrous, has a rigid involucre filled with a white gelatinous substance, containing 3–5 seeds. Seeds orbicular (Ø 2–2.5 cm), coin shaped, densely covered with appressed, shiny, satiny, greenish-white hairs.

Related Species Of the approximately 170 species belonging to the genus *Strychnos*, none is currently known to have any beer-making application.

Geographic Distribution Sri Lanka, India, Burma. **Habitat** Edges of dense forests, riverbanks, sea coasts, especially in organic or organic-sandy soils, also in lateritic and organic-clay soils (1,200–3,500 mm rainfall/year). **Range** Beyond the range in India and SE Asia, the species has been introduced and locally naturalized in tropical Africa (Ghana, Sudan) and cultivated in the Philippines.

Beer-Making Parts (and Possible Uses) Seeds, roots, bark (plant used in the past to adulterate porter beer with substances that could simulate the effect of alcohol and bitterness but were less expensive than raw materials such as malt, sugar, and molasses).

Toxicity Extremely poisonous because of the presence of toxic alkaloids. Strychnine, the main active compound, is one of the most powerful poisons (30–120 mg lethal dose for humans); it acts on the nervous and muscular systems, causing a sudden tetanic contraction.

Chemistry Dried seeds (alkaloids 2.6–3%): strychnine 1.25–1.5%, brucin 1.7%, vomicin, igasurin, α-colubrin, β-colubrin, 3-methoxyicajine, protostrychnine, novacin, n-oxystrychnine, pseudostrychnine, isostrychnine, chlorogenic acid, loganin.

Style Porter.

Source Daniels 2016; Daniels and Larson 2000; Steele 2012.

Syringa vulgaris L.
OLEACEAE

Synonyms *Lilac caerulea* (Jonst.) Lunell, *Lilac vulgaris* (L.) Lam., *Liliacum album* (Weston) Renault, *Liliacum vulgare* (L.) Renault, *Syringa alba* (Weston) A.Die tr. ex Dippel, *Syringa albiflora* Opiz, *Syringa amoena* K.Koch, *Syringa bicolor* K.Koch, *Syringa caerulea* Jonst., *Syringa carlsruhensis* K.Koch, *Syringa cordifolia* Stokes, *Syringa marliensis* K.Koch, *Syringa nigricans* K.Koch, *Syringa notgeri* K.Koch, *Syringa philemon* K.Koch, *Syringa rhodopea* Velen., *Syringa versaliensis* K.Koch, *Syringa virginalis* K.Koch.

Common Names Lilac, common lilac.

Description Shrub or small tree (2–7 m), deciduous. Stem (Ø up to 20 cm) multistemmed, sometimes forming small clonal patches from root suckers; bark gray to gray-brown, smooth in young shoots, longitudinally furrowed and desquamating in older stems. Leaves simple (4–12 × 3–8 cm), opposite or in whorls of 3, petiolate (1–3 cm), light green to glaucous, ovate to cordate, pinnate vein, apex mucronate, margin entire. Flower tubular, corolla (6–10 mm), with 4-lobed fauces (5–8 mm), color lilac to mauve, occasionally white. Inflorescences dense terminal panicles (8–18 cm). Fruit dry, smooth, brown capsule (1–2 cm), dehiscent in 2 to release biwinged seeds.

Cultivar They are numerous. Some that have won Royal Horticultural Society prizes are Andenken an Ludwig Späth, Firmament, Katherine Havemeyer, Madame Lemoine, and Vestale.

Related Species The genus *Syringa* consists of about 20 species, none of which, other than the one described here, has known beer-making applications.

Geographic Distribution E Europe (Balkans). **Habitat** Thickets, hedges. **Range** Cultivated for ornamental purposes.

Beer-Making Parts (and Possible Uses) Flowers.

Chemistry Essential oil, flowers: (E)-ocimene, lilac aldehyde, benzyl methyl ether, 1,4-dimethoxybenzene, indole, anisaldehyde, 8-oxolinalool, cinnamic alcohol, elemine.

Style Berliner Weisse, Saison.

Source Heilshorn 2017; https://www.brewersfriend.com/other/lilac/.

Syzygium aromaticum (L.) Merr. & L.M.Perry
MYRTACEAE

Synonyms *Caryophyllus aromaticus* L., *Caryophyllus hortensis* Noronha, *Caryophyllus silvestris* Teijsm. ex Hassk., *Eugenia caryophyllata* Thunb., *Eugenia caryophyllus* (Spreng.) Bullock & S.G.Harrison, *Jambosa caryophyllus* (Thunb.) Nied., *Myrtus caryophyllus* Spreng.

Common Names Clove, cloves, clovetree.

Description Tree, evergreen (8–30 m). Branches numerous, semi-erect. Leaves pink when young, shiny, elliptic to oblong-elliptic (10 × 5 cm), closely attenuated at base, glabrous, with numerous oil glands on underside, emit clove aroma when crushed. Inflorescences cymose terminal cymes, each peduncle bearing 3–4 stalked flowers. Flowers small, hermaphrodite, self-fertile, sepals minute, triangular, petals (about 6 mm) purplish, early deciduous; stamens grayish-yellow. Fruit dark purple, oblong-ellipsoid (2–2.5 × 1.3 cm) or olive shaped, juicy, usually 1 seed. Many of the plant parts are aromatic (leaves, flowers, bark). Flower buds unopened green, pinkish, or bright crimson color; dried, brown color.

Related Species The genus *Syzygium* has 1,157 accepted taxa, but none, other than *S. aromaticum*, has documented beer-making applications.

Geographic Distribution Moluccas Islands (Indonesia). **Habitat** Tropical climates and humus-rich soils. **Range** Moluccas, Borneo, Madagascar, tropical E Africa (Zanzibar).

Beer-Making Parts (and Possible Uses) Flower buds (singly or as a component of allspice).

Chemistry Dried flower buds: energy 274 kcal, lipids 13 g, carbohydrates 66 g, fiber 34 g, sugars 2.4 g, protein 6 g, magnesium 259 mg, potassium 1.02 g, sodium 277 mg, calcium 632 mg, iron 11.8 mg, vitamin C 0.2 mg, vitamin B6 0.4 mg, vitamin A 160 IU. Essential oil, dried flower buds: eugenol 75-88%, acetyl eugenol 4-15%, β-caryophyllene 5-14%, caviculus, (Z)-isoeugenol, (E)-isoeugenol, benzyl acetate, α-pinene, β-pinene, limonene, α-ylangene, γ-caryophyllene, α-caryophyllene (humulene), caryophyllene epoxide, caryophyllene oxide, caryophyll-3(12),7(13)-dien-6α-ol, caryophyll-3(12), 6-dien-4-ol, 4,4-dimethyltricyclo-trideca-8-ene-1-ol, caryophylla-4(12),8(13)-dien-5β-ol, caryophylla-3,8(13)-dien-5α-ol, caryophylla-3,8(13)-dien5β-ol, α-copaene, α-cubebene, farnesol, benzaldehyde, m-methoxybenzaldehyde, benzyl alcohol, heptan-2-one, octan-2-one, naphthalene, 2,6-dihydroxy-4-methoxyacetophenone, methylxanthoxylin, quercetin, kaempferol, kaempferide, rhamnetin, kaempferol-3-O-β-D-glucoside, quercetin-3-O-β-D-glucoside, quercetin-3-O-β-D-galactoside, quercetin-3,4'-O-β-D-diglucoside, ellagitannins, eugenin, gallic acid, ellagic acid, 3-caffeol, 4-caffeol, 3-p-coumarol, 3-feruloquinic acid, ferulic acid, p-hydroxybenzoic acid, caffeic acid, salicylic acid, syringic acid, vanillic acid, gentisic acid, protocatechuic acid, p-coumaric acid, oleanolic acid, crataegolic acid, β-sitosterol, stigmasterol, campesterol, glucose, xylose, arabinose. Essential oil, leaves (16 compounds): eugenol 94.41%, β-caryophyllene 2.91%.

Style Belgian Golden Strong Ale, Belgian Specialty Ale, Christmas Ale, Experimental Beer, Fruit Beer, Gruit, Old Ale, Russian Imperial Stout, Specialty Beer, Spice/Herb/Vegetable Beer.

Beer Merry Maker, Samuel Adams (Massachusetts, United States).

Source Calagione et al. 2018; McGovern 2017; https://www.bjcp.org (29B. Fruit and Spice Beer; 30B. Autumn Seasonal Beer); https://www.beeradvocate.com/beer/style/72/; https://www.brewersfriend.com/other/clove/; https://www.sangabriel.it/it/birre/le-stagionali/del-tempio.

Tagetes erecta L.
ASTERACEAE

Synonyms *Tagetes corymbosa* Sweet, *Tagetes ernstii* H.Rob. & Nicolson, *Tagetes excelsa* Soule, *Tagetes heterocarpha* Rydb., *Tagetes patula* L., *Tagetes remotiflora* Kunze, *Tagetes tenuifolia* Millsp.

Common Names African marigold, Aztec marigold, big marigold, American marigold.

Description Herb, annual (2–4 dm). Stems short, often branched at the base. Leaves imparipinnate with ± deeply toothed segments. Inflorescence large flower head (Ø 2–8 cm). Flowers ligulate, numerous, protruding, velvety, and mostly brightly colored.

Related Species The genus *Tagetes* includes 53 taxa of which only the species mentioned here have beer-making applications.

Geographic Distribution NC America (Mexico, Guatemala). **Habitat** Species cultivated for ornamental purposes and sometimes subspontaneous. **Range** Cultivated.

Beer-Making Parts (and Possible Uses) Inflorescences.

Chemistry Essential oil, leaves and stems (33 compounds): β-caryophyllene 8.5%, terpinolene 18.4%, (E)-ocimenone 12.6%, (Z)-β-ocimene 10.4%, piperitenone 10.4%, (Z)-ocimenone 5.5%, limonene 6.2%. Essential oil, flowers (34 compounds): β-caryophyllene 35.2%, terpinolene 6.3%, (E)-ocimenone 9.8%, (Z)-β-ocimene 13.7%, piperitenone 2.6%, (Z)-ocimenone 7.7%, limonene 2.5%.

Style Saison.

Source https://www.brewersfriend.com/other/marigold/ (*Tagetes erecta* L. was chosen as representative of *Tagetes* ssp.).

Tagetes lucida Cav.
ASTERACEAE

Synonyms *Tagetes anethina* Sessé & Moc., *Tagetes florida* Sweet, *Tagetes gilletii* De Wild., *Tagetes lucida* (Sweet) Voss, *Tagetes pineda* La Llave, *Tagetes schiedeana* Less., *Tagetes seleri* Rydb.

Common Names Marigold, Mexican marigold, Mexican mint marigold, Mexican tarragon, sweet mace, sweetscented marigold, Texas tarragon.

Description Shrub, perennial (up to 80 cm). Stem semiwoody at base, with ± erect or bushy habit. Leaves sessile, lamina linear to oblong or lanceolate (4–9 × 1 cm), glabrous, bright green, margin in the distal 1/2. Inflorescences flower heads (Ø about 1.5 cm), apical, each bearing in the outer part 3–5 flowers widely ligulate, orange or golden yellow. Hermaphrodite flowers. Fruit cypsela (containing 1 seed), linear, thin, bearing a pappus (pappus + cypsela about 1 cm).

Cultivar Numerous.

Related Species The genus *Tagetes* includes 53 taxa of which only the species mentioned here have beer-making applications.

Geographic Distribution NC America (Mexico, Guatemala). **Habitat** Woods, hillsides, stony slopes. **Range** Widely cultivated in warm and temperate climates for ornamental and medicinal purposes.

Beer-Making Parts (and Possible Uses) Flowers (spicy character), leaves (bittering).

Toxicity The species is known for its cytotoxicity owing to the presence of coumarins.

Chemistry Aerial parts (pharmacologically active compounds): coumarin constituents: 6,7,8-trimethoxycoumarin (dimethyl-fraxetine), herniarin (7-methoxycoumarin), scoparone, 2,3,7-8-dihydroxy coumarin, umbelliferone (7-hydroxycoumarin), 6-hydroxy-7-methoxycoumarin, esculetin (6,7-dihydroxy-coumarin), scopoletin (6-methoxy-7-hydroxycoumarin); flavonoids (patuletin, isorhamnetin, quercetagetin 3-O-arabinosyl-galactoside, isorhamnetin 7-O-glucoside, quercetin, rutin, quercetagenin, naringenin).

Style None verified.

Source Hieronymus 2016b; https://en.wikipedia.org/wiki/Tagetes_lucida.

Tagetes minuta L.
ASTERACEAE

SYNONYMS *Tagetes bonariensis* Pers., *Tagetes glandulifera* Schrank, *Tagetes glandulosa* Schrank ex Link, *Tagetes porophyllum* Vell., *Tagetes tinctoria* Hornsch.

COMMON NAMES Huacatay, Muster John Henry, southern marigold, Stinking Roger, wild marigold, black mint.

DESCRIPTION Herb, annual (5–25 dm), with intense, unpleasant odor. Stem ascending, cylindrical, glabrous, corymbose above. Leaves opposite (5–12 cm), pinnate, with 11–13 lanceolate-linear segments (5–10 × 30–70 mm), serrate, acute, lamina with dark glands. Inflorescence corymb of flower heads. Flower involucre cylindrical (2 × 10 mm), yellow mottled with brown; flowers pale yellow, those ligulate (3–4) protruding 1–2 mm, all others tubular, pubescent.

RELATED SPECIES The genus *Tagetes* includes 53 taxa, of which only the species mentioned here have beer-making applications.

GEOGRAPHIC DISTRIBUTION South America.
HABITAT Rubble, grassy wastelands, roadsides.
RANGE Species naturalized in many countries.

BEER-MAKING PARTS (AND POSSIBLE USES)
Flowers (?), leaves (?).

CHEMISTRY Essential oil, leaves: β-phellandrene 2.6%, limonene 13.1%, β-ocimene 11.5%, dihydrotagetone 42.9%, tagetone 16.8%, tagetenone 11.7%. Essential oil, flowers: β-phellandrene 1.5%, limonene 7%, β-ocimene 45.4%, dihydrotagetone 7.7%, tagetone 5.6%, tagetenone 25%.

STYLE Wood-Aged Beer.

SOURCE https://www.brewersfriend.com/other/huacatay/.

Tamarindus indica L.
FABACEAE

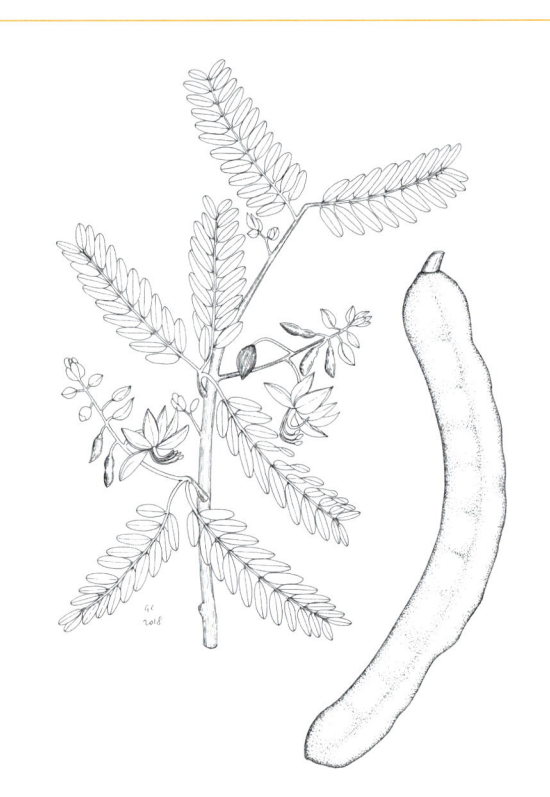

SYNONYMS *Tamarindus occidentalis* Gaertn., *Tamarindus officinalis* Hook., *Tamarindus umbrosa* Salisb.

COMMON NAMES Tamarind.

DESCRIPTION Tree (up to 30 m). Broad crown (Ø up to 12 m). Leaves (up to 15 cm) compound, numerous small segments that close at night, paired along a central axis. Flowers Ø 2.5 cm, 3 gold-colored petals with red veins, 2 small filiform petals barely visible. Inflorescence up to about 20 cm. Fruits brown in color, briefly hairy, sausage shaped; edible acidulous flesh.

Cultivar There are different ones, each adapted to the particular environmental conditions of the cultivation area.

Related Species The genus *Tamarindus* is monotypical; that is, it includes only this single species.

Geographic Distribution Tropical Africa, Madagascar. **Habitat** Tropical forests with an arid season, thickets, arboreal grasslands, often along rivers. **Range** Species grown wild in tropical countries of Africa, South America, Asia, Australia.

Beer-Making Parts (and Possible Uses) Fruit pulp.

Chemistry Fruits (pulp): energy 1,000 kJ (239 kcal), carbohydrates 62.5 g (sugars 57.4 g, fiber 5.1 g), lipids 0.6 g, protein 2.8 g, thiamine (B1) 0.428 mg, riboflavin (B2) 0.152 mg, niacin (B3) 1.938 mg, pantothenic acid (B5) 0.143 mg, vitamin B6 0.066 mg, folate (B9) 14 µg, choline 8.6 mg, vitamin C 3.5 mg, vitamin E 0.1 mg, vitamin K 2.8 µg, calcium 74 mg, iron 2.8 mg, magnesium 92 mg, phosphorus 113 mg, potassium 628 mg, sodium 28 mg, zinc 0.1 mg.

Style American IPA, Belgian Specialty Ale, English IPA, Fruit Beer, Gueuze, Kölsch, Robust Porter, Saison, Specialty IPA (Belgian IPA), Witbier.

Beer Mayan Apocalypse Judgment Day, Lost Abbey (California, United States).

Source Hieronymus 2016b; https://www.brewersfriend.com/other/tamarind/.

Tanacetum balsamita L.
ASTERACEAE

Synonyms *Balsamita major* Desf., *Balsamita suaveolens* Pers., *Balsamita vulgaris* Willd., *Chamaemelum balsamita* (L.) E.H.L.Krause, *Chrysanthemum balsamita* (L.) Baill., *Chrysanthemum grande* (L.) Hook.f., *Chrysanthemum grandiflorum* (Desf.) Dum.Cours., *Chrysanthemum majus* (Desf.) Asch., *Chrysanthemum tanacetifolium* (Desr.) Dum.Cours., *Chrysanthemum tanacetum* Vis., *Leucanthemum balsamita* (L.) Over, *Matricaria balsamita* (L.) Desr., *Pyrethrum majus* (Desf.) Tzvelev.

Common Names Costmary.

Description Herb, perennial (5–10 dm), aromatic. Stoloniferous rhizome. Stems erect, branched-corymbose above. Leaves leathery, the basal and lower cauline entire, slightly ovate, petiolate, up to 20 cm long, margin crenate or serrate; upper cauline leaves similar but progressively smaller upward where they become sessile and often ± amplexicaul; margin coarsely toothed, sometimes pinnate leaves with 1–2 pairs of oblong segments at the base. Inflorescences discoid heads (Ø 5–8 mm), numerous (30–100), in broad terminal corymbs. Flowers all tubular, greenish-yellow in color.

Related Species The genus *Tanacetum* has about 180 taxa.

Geographic Distribution W Asia. **Habitat** Synanthropic habitats. **Range** Cultivated for seasoning and sometimes wild in gardens.

Beer-Making Parts (and Possible Uses) Flowering tops, leaves.

Chemistry Essential oil, leaves: bornil acetate 47.7%, pinocarvone 27.1%, camphor 9.3%, terpinolene 5.4%. Essential oil, flowers: bornil acetate 55.2%, pinocarvone 34.2%, camphor 2.8%, terpinolene 2.0%. Essential oil, stems: bornil acetate 49.2%, pinocarvone 28%, camphor 9.5%, terpinolene 6%.

Style None known.

Source Fisher and Fisher 2017; Hieronymus 2016b.

Tanacetum vulgare L.
ASTERACEAE

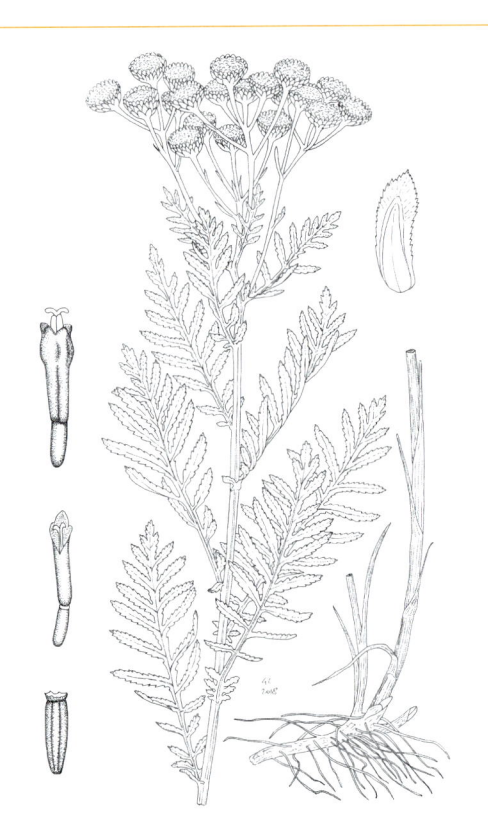

Synonyms *Chamaemelum tanacetum* (Vis.) E.H.L.Krause, *Chrysanthemum asiaticum* Vorosch., *Pyrethrum vulgare* (L.) Boiss., *Tanacetum boreale* Fisch. ex DC., *Tanacetum crispum* Steud.

Common Names Tansy.

Description Herb, perennial (6–12 dm), strongly aromatic. Rhizome creeping and branching, woody. Stems erect, leafy, striate, branched at the top. Leaves petiolate (5–15 cm), alternate, glabrous, with 15–23 pinnately parted segments, serrate on edge; sheathing cauline leaves have small glands on underleaf. Inflorescences compact terminal corymbs dense with discoidal flower heads (Ø about 1 cm), long stalked. Flowers tubular, yellow-gold; peripheral flosculi, around the margin of the flower head, female with tridentate corolla, the central ones hermaphrodite with 5-toothed corolla; involucral bracts lanceolate-obtuse and scarious at the margin, very appressed. Fruits achenes (about 2 mm) with 4–5 longitudinal ribs, dotted with glands; pappus with a small, irregularly furrowed crown.

Cultivar Of this species, chemotypes are known according to the dominant component of their essential oil: camphor chemotype (camphor 22.3–41.4%, 1,8-cineole 10.6–26.4%); α-thujone chemotype (α-thujone 25.7–71.5%, 1,8-cineole 11.3–22.3%); myrtenol chemotype (1,8-cineole 24.5–32.7%, camphor 8.3–23.8%); artemisia ketone chemotype (artemisia ketone 30.5%, camphor 23%).

Related Species The genus *Tanacetum* includes about 180 taxa. Among them, only the two species mentioned here have a known beer-making application; among the others, that of *T. vulgare* is now obsolete (or to be avoided because of the toxicity of the plant).

Geographic Distribution Eurasian. **Habitat** Along ditches, uncultivated lands, riverbanks, prairies, especially in acidic soils of mountains (up to 1,600 m ASL). **Range** Eurasian.

Beer-Making Parts (and Possible Uses) Inflorescences, leaves.

Toxicity Owing to the high toxicity of thujone, the essential oil contained in high quantities in this species, the use of this species has been abandoned.

Chemistry Essential oil (83 compounds, 98.4% of the oil), chemotype 1: trans-chrysanthenone 41.4%, 1,8-cineole 9.6%, β-pinene 6.5%, α-pinene 5%, 6-camphenone 4.6%; chemotype 2: β-thujone 47.2%, trans-chrysanthenyl acetate 30.7%.

Style Spice/Herb/Vegetable Beer.

Source https://www.brewersfriend.com/other/tansy/.

Taraxacum campylodes G.E.Haglund
ASTERACEAE

Synonyms *Crepis taraxacum* (L.) Stokes, *Leontodon taraxacum* L., *Taraxacum officinale* (L.) Weber ex F.H.Wigg., *Taraxacum subspathulatum* A.J.Richards, *Taraxacum vulgare* (Lam.) Schrank.

Common Names Common dandelion, dandelion.

Description Herb, perennial (15–60 cm). Taproot first undivided, then branched, without tunica, at collar enveloped in bluish or blackish scales. Leaves ± dark green, with reticulate veins usually coarsely toothed or lobed, rarely fully entire or incised to the midrib; petiole conspicuous, sometimes broadly winged. Inflorescence flower head (Ø 2.5–4 cm); outer involucral scales linear, usually folded downward; corolla bright yellow. Fruit achene, whitish-gray to purplish-brown, from the middle upward covered with spines or at least tuberculate; beak at least twice as long as the achene.

Related Species The genus *Taraxacum* has about 2,336 taxa, but only *T. campylodes* appears to be used in beer making.

Geographic Distribution Circumboreal. **Habitat** Forest clearings, fertilized meadows, ruderal environments (synanthropic species). **Range** Cosmopolitan.

Beer-Making Parts (and Possible Uses) Whole plant, dried flowers, roots, leaves.

Chemistry Plant: inosite, asparagine, saponin, palmitic acid, oleic acid, linoleic acid, resinic acids, choline. Flowers: inulin, taraxacin. Latex: taraxin, inosite, taraxacinar. Root: taraxacin, taraxacerin, beta-amyrin, taraxasterol, free sterols, luteolin, rutin, hiperoside, quercetin, caffeic acid, chlorogenic acid, catechic tannins, sterols, triterpenes, carotenoids, coumarins, mucilage. Leaves: glycosides, alkaloids, phenolic compounds, tannins, flavonoids, proteins, potassium, calcium, sodium, iron, zinc, cadmium, copper.

Style Blonde Ale, California Common Beer, Experimental Beer.

Beer KM0, Curtense (Passirano, Lombardy, Italy).

Source Heilshorn 2017; Hieronymus 2016a; Hieronymus 2016b; Josephson et al. 2016; Markowski 2015; https://www.botanical.com/botanical/mgmh/d/dandel08.html; https://brooklynbrewshop.com/directions/Brooklyn_Brew_Shop_Dandelion_Gruit_Instructions.pdf; https://www.newbelgium.com/beer/dandelion-ale; https://shebrewsgoodale.wordpress.com/2011/05/13/brewing-with-dandelions/.

Taxodium distichum (L.) Rich.
CUPRESSACEAE

Synonyms *Cuprespinnata disticha* (L.) J.Nelson, *Cupressepinnata disticha* (L.) J.Nelson, *Cupressus americana* Catesby ex Endl., *Cupressus disticha* L., *Cupressus montezumae* Humb. & Bonpl. ex Parl., *Glyptostrobus columnaris* Carrière, *Schubertia disticha* (L.) Mirb., *Taxodium denudatum* Carrière, *Taxodium knightii* K.Koch, *Taxodium pyramidatum* Beissn.

Common Names Baldcypress, bald-cypress, cypress, southern-cypress, white-cypress, tidewater red-cypress, gulf-cypress, red-cypress, swamp cypress.

Description Tree, deciduous (up to 40 m). Roots carrying root tubercles emerging from the soil

Taraxacum campylodes G.E.Haglund
ASTERACEAE

(pneumatophores) that allow the oxygenation of the submerged parts, even for long periods. Trunk enlarged at the base and conspicuously reinforced; crown monopodial and conical when young, then irregularly flattened (branched and divided to the point that the main axis cannot be identified); dimorphic shoots, some long with linear to lanceolate leaves, others short bearing at one end 2 rows of strictly linear leaves, decurrent, laterally divergent, at the other end lanceolate to deltate leaves, closely appressed. Male cones in pendulous panicles (25 cm × 2–3 mm); female cones Ø 1.5–4 cm.

Subspecies In addition to the typical *Taxodium distichum* (L.) Rich. subsp. *distichum*, a second variety, *Taxodium distichum* (L.) Rich. var. *imbricatum* (Nutt.) Croom, is also known.

Related Species Besides the one cited here, the genus *Taxodium* includes only one other taxon, *Taxodium huegelii* C.Lawson.

Geographic Distribution SE United States (Alabama, Arkansas, Delaware, Florida, Georgia, Illinois, Indiana, Kentucky, Louisiana, Maryland, Mississipi, Missouri, North Carolina, Oklahoma, South Carolina, Tennesse, Texas, Virginia). **Habitat** The species grows in flat, alluvial soils, usually at elevations below 50 m ASL, sometimes up to 500 m, mostly in inland or subcoastal riparian environments subject to periodic freshwater flooding (poor tolerance of salinity) and in warm climates (good resistance to cold) with an average annual rainfall between 760 and 1,630 mm. **Range** Cultivated everywhere (Europe, Asia, North America, subtropical climate areas) for ornamental purposes.

Beer-Making Parts (and Possible Uses) Wood (for barrels and fermentation vats).

Wood Differentiated, with thin whitish sapwood and yellowish-brown to chocolate-colored heartwood, light (WSG 0.35–0.55), soft, easy to work, resistant to alteration and therefore used in the production of poles, railway sleepers, construction, staves, shingles.

Style None verified.

Source Cantwell and Bouckaert 2016; Cantwell and Bouckaert 2018.

Tectona grandis L.f.
LAMIACEAE

Synonyms *Jatus grandis* (L.f.) Kuntze, *Tectona theca* Lour., *Theka grandis* (L.f.) Lam.

Common Names Teak.

Description Tree, deciduous (up to 30 m). Root shallow (often no more than 50 cm deep) but extending laterally up to 15 m from trunk. Trunk frequently spurred and grooved to about 15 m, below first branches; bark dark, distinctly fibrous, with shallow longitudinal grooves. Leaves very broad, falling for 3–4 months during the second half of the dry season, leaving branches bare; upper leaf shiny, hairy, underleaf conspicuously ribbed (30 × 20 cm to 1 m). Flowers small (about 8 mm), white to mauve in color, organized in large inflorescences (about 45 cm) on higher or unshaded branches. Fruit drupe with 4 chambers, round, hard, and woody, enclosed in a swollen, blister-like involucre; light green at first, brown at maturity. Seeds 0–4 per fruit.

Related Species Two other species belong to the genus *Tectona*: *Tectona hamiltoniana* Wall. and *Tectona philippinensis* Benth. & Hook.f.

Geographic Distribution India, Indonesia, Laos, Myanmar, Thailand. **Habitat** Pioneer species in warm and humid environments with tropical climate (with significant differences between wet and dry seasons), in deep, well-drained, fertile, alluvial soils. **RANGE** Antigua, Barbuda, Bangladesh, Barbados, Brazil, Brunei, Cambodia, China, Côte d'Ivoire, Cuba, Dominica, Dominican Republic, Guadeloupe, Jamaica, Kenya, Mauritius, Nepal, Nigeria, Pakistan, Panama, Philippines, Puerto Rico, South Africa, Sri Lanka, St. Lucia, St. Vincent, Granada, Granadine, Tanzania, Togo, Trinidad, Tobago, Uganda, United States, Vietnam, Virgin Islands.

BEER-MAKING PARTS (AND POSSIBLE USES)
Wood (for barrels).

WOOD Differentiated, with thin whitish sapwood and golden-brown or greenish-brown heartwood, with exposure to air it turns toward darker shades and variegated blackish, oily appearance, coarse texture, straight grain, WSG 0.55–0.80. Fairly easy to work but quickly makes tools blunt, slow drying but without difficulty, holds firm well. This characteristic and the excellent natural resistance to alterations make it valuable for all-weather naval and hydraulic constructions, as well as for tanks for chemical industries, floors, fixtures, fine furniture, and cabinet panels.

STYLE Various styles (not defined).

SOURCE Steele 2012.

Theobroma cacao L.
MALVACEAE

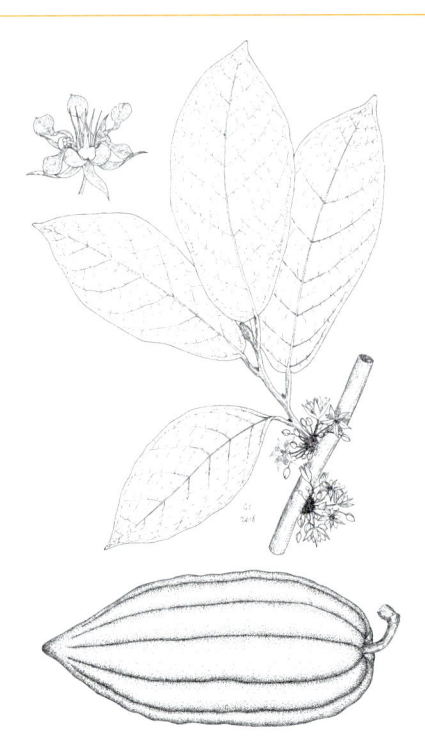

SYNONYMS *Cacao minar* Gaertn., *Cacao minus* Gaertn., *Cacao sativa* Aubl., *Cacao theobroma* Tussac, *Theobroma integerrima* Stokes, *Theobroma kalagua* De Wild., *Theobroma leiocarpum* Bernoulli, *Theobroma pentagonum* Bernoulli, *Theobroma saltzmanniana* Bernoulli, *Theobroma sapidum* Pittier, *Theobroma sativa* (Aubl.) Lign. & Le Bey, *Theobroma sativum* (Aubl.) Lign. & Bey.

COMMON NAMES Cacao tree, cocoa tree.

DESCRIPTION Tree (5–10 m), semideciduous, cauliflorus. Main trunk short, branches in whorls of 5, dimorphic; vertical ones have leaves arranged with 5/8 phyllotaxis, lateral ones with 1/2 phyllotaxis. Leaves with petiole equipped with 2 pulvinus, one at base and the other at leaf insertion point; stipules 2, deciduous; lamina elliptic-oblong or obovate-oblong, simple, 10–45 cm long, usually smooth, sometimes hairy, rounded and obtuse at base, pointed at apex. Inflorescence dichasium; primary peduncle very short, thick, and lignified. Flowers with peduncle 1–4 cm long; sepals 5, triangular, whitish or reddish; petals 5, joined at base in a cup-like structure, whitish-yellow with adaxial violet bands; ligules spatulate, yellowish; stamens 5, fertile, alternating with 5 staminoids, the 2 whorls united to form a tube; anthers 2, stamens fused; ovary superior with single stylus terminating in 5 glutinous stigmas. Fruit variable in shape, ovoid to oblong, sometimes pointed and constricted at base or nearly spherical, with 10 grooves of which 5 are prominent. Placentation axial, seeds lodged in a mucilage, flattened or round, with white or purple cotyledons.

SUBSPECIES Besides the nominal one, the subspecies *Theobroma cacao* L. subsp. *sphaerocarpum* (A.Chev.) Cuatrec has been described.

CULTIVAR There are three varietal groups of this species: Criollo cocoa or Noble cocoa (*Theobroma cacao* L. ssp. *cacao*), with white, very aromatic seeds, reduced bitterness (Mayan seed), native to Mexico, spread throughout Central America and N South America (Ecuador, Venezuela), shows sensitivity to adversities, scarce production (less than 10% of world production) but of high quality, used for the production of fine chocolate; Forastero cocoa or consumer cocoa (*Theobroma cacao* L. ssp. *sphaerocarpum* [A.Chev.] Cuatrec.), with violet seeds having a strong and bitter taste, resistant to adversities, very productive (about 80% of world production), cultivated in W Africa, Brazil, SE Asia; Trinitario cocoa (*Theobroma* *cacao* × *sphaerocarpum*), native of lower Amazonia (Trinidad), presents intermediate characteristics between the two parental taxa, cultivated in Mexico, Trinidad, Caribbean, Colombia, Venezuela, SE Asia (10% of world production).

Related Species The genus *Theobroma* has 25 taxa; only *Theobroma cacao* has known beer-making applications.

Geographic Distribution NC South America. **Habitat** Tropical climate and rich, well-drained soil. **Range** Ivory Coast, Ghana, Indonesia, Nigeria, Brazil, Cameroon, Ecuador, Colombia, Mexico, Papua New Guinea, Malaysia, Dominican Republic, Peru, Venezuela, Sierra Leone, Togo, Argentina, Philippines, Republic of Congo, Solomon Islands.

Beer-Making Parts (and Possible Uses) Dried fermented seeds.

Chemistry Cocoa butter: palmitic acid 24.4–26.2%, stearic acid 34.4–35.4%, oleic acid 37.7–38.1%, linoleic acid 2.1%. Fresh seeds: water 7.78%, protein 13.78%, theobromine 1.39%, lipids 44.52%, starch 23.36%, cellulose 4.67%, minerals 4.5%. Dried and shelled seeds: water 5.5%, protein 13.93%, theobromine 1.52%, lipids 49.38%, starch 22.36%, cellulose 3.86%, mineral salts 3.45%.

Style Foreign Extra Stout, Robust Porter.

Beer Theobroma, Dogfish Head (Delaware, United States).

Source Calagione et al. 2018; Heilshorn 2017; McGovern 2017; https://www.bjcp.org (30. Spiced Beer); https://www.beer-shop.it/young-s/49-young-s-double-chocolate-stout-50cl.html; https://www.omnipollo.com/beer/symzonia/.

Beer and Cocoa

Chocolate stouts are a popular and successful style in various parts of the world. They are dark beers, of high fermentation, with an alcohol by volume of 4-6%, produced by adding cocoa in many forms. The use of cocoa in brewing seems to be ancient: according to some studies, it goes as far back as 600 BCE. According to research conducted by John Henderson of Cornell University, it seems cocoa was used by the people of Mesoamerica to produce a beverage similar to beer, with the addition of fermented cacao beans giving a chocolate aroma. This was likely the origin of today's famous drink.

Thymus × *citriodorus* (Pers.) Schreb.
LAMIACEAE

Synonyms None

Common Names Lemon thyme, citrus thyme, creeping lemon thyme, lemon-scented thyme.

Description Shrub, low (15–20 cm). Stem woody at the base, erect. Leaves broader and more rounded than parent species (*Thymus pulegioides* L., *Thymus vulgaris* L.), lemon flavored, golden yellow variegated color (cultivar Bertram Anderson) or golden green (cultivar Golden Dwarf). Small flowers, purple or lilac color.

Cultivar The most common ones are Bertram Anderson and Golden Dwarf.

Related Species Among species, subspecies, and hybrids, there are 354 taxa in the genus *Thymus*. Of these, only those cited here have documented beer-making uses.

Geographic Distribution According to some sources, a hybrid between *Thymus pulegioides* L. and *Thymus vulgaris* L. grows wild on the Mediterranean coasts; according to others, it is not known in nature. **Habitat** Cultivated in well-drained sandy and loamy soils, neutral to basic pH, bright, even windy environments (but not exposed to marine aerosol).

Range Cultivated for ornamental purposes and for fragrance.

Beer-Making Parts (and Possible Uses) Flowering tops, leaves.

Chemistry Essential oil (%): terpinolene 71.0±1.76, α-terpineol 20.03±0.08, linalool 1.61±0.02, bornyl acetate 0.99±0.09, sabinene 0.49±0.63, limonene 0.57±0.19, γ None terpinene 0.14±0.02, m-mentha-1,8-diene, (+) 0.08, 1,8-cineole 0.38, p-cymene 0.11, 1-octen-3-ol acetate 0.23±0.01, 1-octen-3-ol 0.15±0.01, β-bourbonene 0.21±0.002, caryophyllene-e 0.68±0.03, borneol 0.62±0.06, α-terpinyl acetate 0.14±0.08, trans-carveol 0.21±0.04, geraniol 0.05, thymol 0.04±0.01, caryophyllene oxide 0.42±0.02, carvacrol 0.67±0.04.

Style Saison.

Source Josephson et al. 2016; https://www.brewers-friend.com/other/citrontimjan/.

Thymus serpyllum L.
LAMIACEAE

Synonyms *Cunila thymoides* L., *Hedeoma thymoides* (L.) Pers., *Origanum serpyllum* (L.) Kuntze, *Serpyllum angustifolium* (Pers.) Fourr., *Serpyllum citriodora* Pall., *Serpyllum vulgare* Fourr., *Thymus adscendens* Bernh. ex Link, *Thymus affinis* Vis., *Thymus albidus* Opiz, *Thymus angulosus* Dulac, *Thymus angustus* Opiz ex Déségl., *Thymus apricus* Opiz, *Thymus aureus* auct., *Thymus barbatus* Opiz, *Thymus beneschianus* Opiz, *Thymus borbasii* Borbás, *Thymus calcicolus* Schur, *Thymus carstiensis* (Velen.) Ronniger, *Thymus caucasicus* Willd. ex Benth., *Thymus ciliatus* Lam., *Thymus citratus* Dumort., *Thymus communis* Kitt., *Thymus concolor* Opiz, *Thymus decumbens* Bernh. ex Rchb., *Thymus deflexus* Benth., *Thymus elatus* Schrad. ex Rchb., *Thymus ellipticus* Opiz, *Thymus elongatus* Opiz, *Thymus erioclados* Borbás, *Thymus exserens* Ehrh. ex Link, *Thymus flogellicaulis* A.Kern., *Thymus gizellae* Borbás, *Thymus gratissimus* Dufour ex Willk. & Lange, *Thymus hackelianus* Opiz, *Thymus hausmannii* Heinr.Braun, *Thymus hornungianus* Opiz, *Thymus includens* Ehrh. ex Rchb., *Thymus interruptus* Opiz, *Thymus kollmunzerianus* Opiz ex Benth., *Thymus kratzmannianus* Opiz, *Thymus laevigatus* Vahl, *Thymus linearifolius* Heinr.Braun, *Thymus longistylus* Opiz, *Thymus lucidus* Willd., *Thymus macrophyllus* Heinr.Braun, *Thymus majoranifolius* Desf., *Thymus minutus* Opiz, *Thymus muscosus* Zaver., *Thymus procerus* Opiz ex Benth., *Thymus procumbens* Benth. ex Opiz, *Thymus pseudoserpyllum* Rchb. ex Benth., *Thymus pumilus* Gueldenst. ex Ledeb., *Thymus pusillus* Gueldenst. ex Ledeb., *Thymus pusio* Dichtl, *Thymus pycnotrichus* (R.Uechtr.) Ronniger, *Thymus radoi* Borbás, *Thymus raripilus* Dichtl, *Thymus reflexus* Lej., *Thymus reichelianus* Opiz, *Thymus rigidulus* Kerguélen, *Thymus sanioi* Borbás, *Thymus serbicus* Petrovic, *Thymus serratus* Opiz, *Thymus simplex* Kitt., *Thymus subcitratus* Schreb., *Thymus subhirsutus* Borbás & Heinr. Braun, *Thymus variabilis* Hoffmanns. & Link, *Thymus wierzbickianus* Opiz, *Thymus wondracekianus* Opiz, *Ziziphora thymoides* (L.) Roem. & Schult.

Common Names Breckland thyme, wild thyme, creeping thyme.

Description Shrub (3–6 cm), prostrate. Stem woody at base, creeping, rooting at nodes, with erect floriferous apices; year stems long repent, holotrichous with deflected hairs (0.3–0.4 mm). Leaves leathery, sessile, small (but much > than the internode above, so that the entire floriferous branch appears densely banded with pairs of close leaves), laminae narrowly lanceolate, progressively elongated toward the apex, glabrous, with strong veins. Inflorescences dense, spherical, ovate or ± elongate, and interrupted at base. Flower calyx 3–5 mm; corolla purple or pinkish (5–6 mm). Fruit included in persistent calyx with fauces occupied by a tuft of white, cottony hairs.

Subspecies In addition to the nominal subspecies, *Thymus serpyllum* L. subsp. *tanaensis* (Hyl.) Jalas has been described.

Related Species Among species, subspecies, and hybrids, there are 354 taxa in the genus *Thymus*. Of these, only those cited here have documented beer-making uses.

Geographic Distribution C Europe. **Habitat** Arid steppe meadows, scree slopes, and sunny cliffs. **Range** C Europe.

Beer-Making Parts (and Possible Uses) Aerial parts.

Chemistry Essential oil (34 compounds): γ-terpinene 21.9–22.7%, p-cymene 20.7–21.1%, thymol 18.7%, germacrene D 5.1–6%.

Style Ancient Ale, Mum, Sour Ale.

Beer Serpilla, LoverBeer (Marentino, Piedmont, Italy).

Source Fisher and Fisher 2017; Hieronymus 2010; McGovern 2017.

Thymus vulgaris L.
LAMIACEAE

Synonyms *Origanum thymus* Kuntze, *Origanum webbianum* (Rouy) Kuntze, *Thymus chinensis* K.Koch, *Thymus ilerdensis* González ex Costa, *Thymus sublaxus* Rouy, *Thymus webbianus* Rouy.

Common Names Common thyme, German thyme, garden thyme, thyme.

Description Shrub or bush (10–60 cm), perennial, fragrant. Stems quadrangular erect, coppiced, tending to lignify after 4–5 years, forming dense bushes, grayish or whitish green in color; branches lignified with brown bark. Leaves first revolute only on the edge, lanceolate (7–9 × 3 mm), then revolute tube shaped and apparently linear, opposite, sessile or briefly petiolate, gray-green color, lighter underleaf owing to the presence of hairs. Inflorescence subspherical or elongate, with lanceolate bracts similar to leaves but smaller. Flower calyx (3–4 mm) with 10–13 nerves and convex tube on the back, velvety, with 2 ciliate lips (the upper trifid with joined teeth over more than 1/2 the height, the lower bifid with lanceolate-lesiniform teeth, separated by a deep sinus); corolla pinkish-white in color (5–6 mm), with protruding, straight tube, bilabiate, with straight, notched upper lip, the lower trilobed; stamens 4, protruding and diverging, nearly equal, anthers bilocular, ellipsoidal; stylus bifid, with short, diverging laciniae. Fruit tetrachene, consisting of smooth ovoid nucules.

Subspecies Besides the nominal subspecies, *Thymus vulgaris* L. subsp. *aestivus* (Reut. ex Willk.) A.Bolòs & O.Bolòs has been described.

Cultivar Different chemotypes of *Thymus vulgaris* are known (e.g., chemotype linalool, chemotype 1,8 cineol). Although the differentiation has a genetic basis, in many cases a strong environmental influence (e.g., altitude, rainfall) has been demonstrated on the composition of the essential oil.

Related Species Among species, subspecies, and hybrids, there are 354 taxa in the genus *Thymus*. Of these, only those cited here have documented uses in beer making.

Geographic Distribution Steno-Mediterranean. **Habitat** Garigues, arid slopes, frequently along dry coastal hills, rarer inland (0–800 m ASL). **Range** Widely cultivated as an aromatic plant.

Beer-Making Parts (and Possible Uses) Aerial parts.

Chemistry Chemotype linalool: α-pinene 0.97%, camphene 1.9%, sabinene 0.79%, β-pinene 0.37%, α-myrcene 1.43%, α-terpinene 0.97%, p-cymene 9.28%, limonene 3.46%, 1,8-cineole 1.58%, γ-terpinene 2.29%, E-sabinene hydrate 2.21%, γ-terpinolene 0.49%, linalool 42.57%, camphor 2.62%, borneol 0.65%, terpin-4-ol 5.47%, α-terpineol 2.02%, trans-dihydrocarvone 0.52%, bornyl/isobornyl acetate 0.4%, thymol 2.46%, δ-elemene 0.62%, α-terpenyl acetate 11.35%, β-caryophyllene 1.29%, β-elemol 0.81%, caryophyllene oxide 1.62%. Chemotype 1,8-cineole: α-pinene 1.78%, β-pinene 3.27%, α-myrcene 7.3%, α-terpinene 1.86%, p-cymene 2.89%, limonene 3.13%, 1,8-cineole 22.08%, γ-terpinene 4.24%, E-sabinene hydrate 13.1%, γ-terpinolene 0.71%, linalool 2.86%, terpin-4-ol 9.32%, α-terpineol 4.73%, thymol 2.62%, δ-elemene 0.72%, perillol acetate 5.67%, α-terpenyl acetate 1.09%, β-caryophyllene 6.59%, β-elemol 0.5%, caryophyllene oxide 1.96%.

Style American Amber Ale, American Barleywine, American IPA, American Pale Ale, Belgian Blond Ale, Belgian Dark Strong Ale, Belgian Pale Ale, Belgian Specialty Ale, Belgian Tripel, Bière de Garde, California common Beer, Doppelbock, English Barleywine, Experimental Beer, Extra Special/Strong Bitter (ESB), Holiday/Winter Special Spiced Beer, Imperial IPA, Russian Imperial Stout, Saison, Specialty Beer, Spice/Herb/Vegetable Beer, Standard/Ordinary Bitter, Strong Bitter, Weizen/Weissbier, Witbier.

Source Calagione et al. 2018; Hieronymus 2016b; Fisher and Fisher 2017; McGovern 2017; https://www.brewersfriend.com/other/thym/; https://www.brewersfriend.com/other/thyme/.

Tilia × europaea L.
MALVACEAE

Synonyms *Tilia × intermedia* DC., *Tilia × vulgaris* B.Heyne.

Common Names Common lime, common linden.

Description Tree (30–40 m). Young trunks slightly hairy to glabrous. Leaves 5-costate, broadly ovate (5–7.5 × 3.5–5.5 cm), sparsely hairy, with clusters of hairs at the axils of the veins, obliquely cordate or rather truncate at the base, serrate, suddenly acuminate-acute; petiole 1.5–3.5 cm, densely hairy. Buds umbellate, with 5–10 flowers, peduncle as long as or slightly longer than bract, glabrous; bracts oblong-oblanceolate (6.5–7.5 × 2–2.5 cm), obtuse, glabrous. Flowers Ø about 1.5 cm; peduncle 1.2–1.5 cm; sepals narrowly ovate, about 6 × 2.5 mm, hairy on the outside, acute; petals oblanceolate, about 7 mm, obtuse; stamens 30–35, filaments as long as petals; ovary globose, about 2 mm, densely hairy; stylus 3 mm, glabrous. Fruit broadly ovoid-subglobose, tomentose, apiculate.

Related Species The genus *Tilia* includes about 64 taxa.

Geographic Distribution Europe, Russia (natural hybrid between *Tilia cordata* Mill. and *Tilia* *platyphyllos* Scop. present in a sporadic manner where the ranges of the two parental species overlap). **Habitat** Mesophyllous forests. **Range** Cultivated for ornamental purposes.

Beer-Making Parts (and Possible Uses)
Flowers (?).

Chemistry Flowers: carbohydrates (mucilaginous polysaccharides, arabinose, galactose, rhamnose, glucose, mannose, xylose, galacturonic acid, glucuronic acid), amino acids (alanine, cysteine, cystine, isoleucine, leucine, phenylalanine, serine), flavonoids (kaempferol, quercetin, myricetin, tiliroside), caffeic acid, chlorogenic acid, p-coumaric acid, volatile oils (alkanes, phenolic alcohols, esters, citral terpenes, citronellal, citronellol, eugenol, limonene, nerol, α-pinene, terpineol, farnesol), other constituents (saponins, tannins, phytosterol).

Style Belgian Dark Strong Ale, Belgian Specialty Ale.

Source Heilshorn 2017; McGovern 2017; https://www.brewersfriend.com/other/tilleul/.

Trifolium pratense L.
FABACEAE

Synonyms *Trifolium borysthenicum* Gruner, *Trifolium bracteatum* Schousb., *Trifolium lenkoranicum* (Grossh.) Roskov, *Trifolium ukrainicum* Opperman.

Common Names Red clover, cowgrass, purple clover.

Description Herb, perennial (10–60 cm), of limited longevity (2–3 years). Woody rhizome wrapped in dark sheaths, tubercles present in lateral roots for nitrogen fixation. Stems erect, briefly creeping, simple. Leaves long petiolate, trifoliate, ovate or elliptic, stipules elongate, with terminal rests, upper leaf characterized by whitish V-shaped pattern. Inflorescences globular heads at axils of upper leaves. Flowers stalked or subsessile, corolla color light red, carmine, or milky. Fruits indehiscent legumes (camara) included in the calyx, with membranous pericarp and a single ovoid seed, smooth, yellowish or brown in color.

Subspecies *Trifolium pratense* constitutes an extremely polymorphic group. Some authors recognize four subspecies: *Trifolium pratense* L. subsp. *pratense* (corolla color pinkish-violet, reduced or no hairiness, stipules lanceolate, awns 1/3–1/4 as long as the lamina, solitary flower heads, common), *Trifolium pratense* L. subsp. *nivale* (W.D.J.Koch) Ces. (corolla milky color, suffused with pinkish toward apex, plant robust, densely hairy, stipules ovate with awns 1/4–1/6 as long of lamina, apical heads paired), *Trifolium pratense* L. subsp. *semipurpureum* (Strobl) Pignatti (small, single rose-violet-colored flower heads, modest plant size [5–10 cm], with dense appressed hairiness, ovate stipules 1/4–1/6 the length of lamina), and *Trifolium pratense* L. subsp. *sativum* (Schreb.) Schübl. & G.Martens. According to other authors, the variability of *Trifolium pratense* should be framed in taxa of varietal rank: *Trifolium pratense* L. var. *pratense*, *Trifolium pratense* L. var. *americanum* Harz, *Trifolium pratense* L. var. *maritimum* Zabel, and *Trifolium pratense* L. var. *sativum* Schreb.

Related Species The genus *Trifolium* includes about 340 taxa of which it is unclear how many and which ones have some beer-making application; clear information of possible beer-making uses is available for *Trifolium repens* L.

Geographic Distribution Subcosmopolitan.
Habitat Meadows, pastures, uncultivated land, resistant to cold, prefers clayey soils (0–2,600 m ASL).
Range Cultivated for fodder and frequently spontaneous in temperate climate areas.

BEER-MAKING PARTS (AND POSSIBLE USES)
Inflorescences.

CHEMISTRY Phytoestrogens 0.17% (formononetin, genistein, daidzein, biochanin A, sitosterol), essential oil (methyl salicylate, etc.), phenolic glycosides (salicylic acid), coumarins, cyanogenic glycosides, silica, choline, lecithin, vitamin A, vitamin C, group B vitamins, calcium, chromium, iron, magnesium.

STYLE American Pale Ale, American Wheat or Rye Beer, Holiday/Winter Special Spiced Beer, Robust Porter, Spice/Herb/Vegetable Beer.

SOURCE Hieronymus 2016b; https://www.brewersfriend.com/other/clover/ (*Trifolium pratense* L. was chosen as representative of *Trifolium* sp.).

Trigonella foenum-graecum L.
FABACEAE

SYNONYMS *Trigonella tibetana* (Alef.) Vassilcz.

COMMON NAMES Fenugreek.

DESCRIPTION Herb, annual (up to 50 cm). Branched stems. Leaves trifoliate, oblong-lanceolate leaflets (up to 5 cm). Flowers yellowish (12–18 mm), 1–2 at the leaf axil. Fruit legume, sickle shaped to nearly straight, flattened, with pronounced beak (legume 50–110 mm, beak 10–35 mm); each legume containing 10–20 seeds. Seeds brownish color, oblong-rhomboidal (about 3 mm), with deep groove dividing them into 2 unequal lobes.

RELATED SPECIES There are about 100 taxa in the genus *Trigonella*, but only *Trigonella foenum-graecum* seems to have some beer-making application.

GEOGRAPHIC DISTRIBUTION Middle East, N Africa. **HABITAT** Fallow land. **RANGE** India, Afghanistan, Pakistan, Iran, Nepal, Bangladesh, Argentina, Egypt, France, Spain, Turkey, Morocco, N Italy.

BEER-MAKING PARTS (AND POSSIBLE USES)
Leaves, seeds.

CHEMISTRY Seeds: energy 1,540 kJ, water 7.8 g, protein 28.2 g, lipids 5.9 g, carbohydrates 46.5 g, fiber 8 g, ash 3.6 g (calcium 220 mg, phosphorus 358 mg, iron 24.2 mg), β-carotene 55 µg, thiamine 0.32 mg, riboflavin 0.30 mg, niacin 1.5 mg, tryptophan 274 mg, alkaloids (trigonelline), oleoresins 0.02%, sotolone (3-hydroxy-4,5-dimethyl-2[5H]-furanone, responsible for the characteristic fenugreek aroma), diosgenin 0.5–2% (steroid used for the synthesis of oral contraceptive hormones and corticosteroids). Fresh leaves: water 87.6 g, protein 4.6 g, lipids 0.2 g, carbohydrates 4.8 g, fiber 1.4 g, ash 1.4 g (calcium 150 mg, phosphorus 48 mg). Essential oil, seeds: dihydroactinidiolide 7.5%, 2,3-dihydrobenzofuran 7.5%, 1-hexanol 7.5%, aniline 2.5%, calamenene 2.5%, calarene 2.5%, camphor 2.5%, diphenylamine 2,5%, dodecanoic acid 2.5%, β-elemene 2.5%, δ-elemene 2.5%, eugenol 2.5%, heptanal 2.5%, heptanoic acid 2.5%, 2-heptanone 2.5%, hexadecane 2.5%, 1-hexadecene 2.5%, 2-hexylfuran 2.5%, methylcyclohexyl acetate 2.5%, ε-muurolene 2.5%, γ-muurolene 2.5%, γ-nonalactone 2.5%, 3-octen-2-one 2.5%, pentadecane 2.5%, phenol 2.5%, tetradecane 2.5%, thymol 2.5%, decanoic acid 0.5%, 1-dodecene 0.5%, β-ionone 0.5%, 3-hydroxy-4,5-dimethyl-2(5H)-furanone 0.1%.

STYLE American Amber Ale, American Brown Ale, American IPA, American Pale Ale, Belgian Dark Strong Ale, Cream Ale, Experimental Beer, Holiday/Winter Special Spiced Beer, Metheglin, Other Smoked Beer, Robust Porter, Roggenbier (German Rye Beer), Saison, Specialty Beer, Spice/Herb/Vegetable Beer, Strong Bitter.

SOURCE https://www.brewersfriend.com/other/fenugreek/.

Triticum aestivum L.
POACEAE

Synonyms *Frumentum triticum* E.H.L.Krause, *Triticum antiquorum* (Heer) Udachin, *Triticum asiaticum* Kudr., *Triticum clavatum* Seidl ex Opiz, *Triticum erinaceum* Hornem., *Triticum horstianum* Clemente, *Triticum hybernum* L., *Triticum imberbe* Desv., *Triticum inflatum* Kudr., *Triticum koeleri* Clemente, *Triticum linnaeanum* Lag., *Triticum pulverulentum* Hornem., *Triticum quadratum* Mill., *Triticum sativum* Lam., *Triticum vavilovii* Jakubz., *Triticum velutinum* Schübl., *Triticum vulgare* Vill.

Common Names Wheat, common wheat, bread wheat.

Description Herb, annual (60–100 cm), caespitose. Internodes of culms empty. Leaves with glabrous or pubescent sheath, with hairs ciliate at mouth; auricles falcate, ligule (1 mm) membrane ciliate; lamina (10–60 cm × 10–15 mm) pubescent. Inflorescence composed of a raceme, single, linear or oblong, bilateral, 5–18 cm long; rachis stout, internodes 2–3 mm, flattened, glabrous on the surface, ciliate at the margins; spikelets appressed to the rachis, ascending, solitary, the fertile ones sessile; fertile spikelets include 2–4 fertile flowers, diminishing toward the apex; spikelet ovate, laterally compressed, 10–15 × 9–18 mm, persistent on the plant; internodes of rachilla 1–1.2 mm long; glumes similar, shorter than spikelet; glume low (6–11 mm), leathery, with double fairing above and winged on the hull, with 5–9 veins; surface of glume low, glabrous, puberulent or villous; apex with unilateral, truncate, muticate, or aristate tooth; remnant 0–40 mm long; upper glume ovate, 6–11 mm long (50–70% length of adjacent fertile lemma), coriaceous, double keel, 5–9 veins, lateral ones diverging at apex, surface glabrous, pubescent, or villous; apex with unilateral truncate, muticate, or aristate teeth; awns 0–40 mm long; fertile lemma ovate, 12–15 mm long, papery, keeled, with 5–9 veins, apex acute, muticate, or aristate (0–150 mm long); palea with 2 veins, keel winged, ciliate; apical sterile flowers similar to fertile ones although poorly developed; anthers 3, ovary with a fleshy appendage below the insertion of the stylus, pubescent at the apex. Fruit caryopsis with adherent pericarp, oblong, furrowed on hilar face, 5–7 mm long, hairy at apex.

Subspecies The only taxon of subspecific rank recognized in the species *Triticum aestivum* L., besides the typical var. *aestivum*, is *Triticum aestivum* L. var. *vavilovii* (Tumanian) Sears.

Cultivar Little attention is usually paid to wheat varieties used to make malt, but an increasing number of brewers want to know the variety of grain their malt is made from. Wheat types are defined by protein level (hard or soft), color (red or white), and planting time (winter or summer). The red color of the caryopsis (bran color actually) in red wheat comes from the higher phenol and tannin contents than are present in white wheat. Hard wheat contains a higher level of protein than does common wheat. The size of the grains can vary; red grains tend to be physically smaller than white grains and therefore more difficult to mill. Considering this species has been cultivated for thousands of years, spreading throughout a range of climatic and edaphic conditions, it should come as no surprise that there are hundreds (maybe thousands) of cultivars. In every country where the plant is cultivated, cultivars adapted to local conditions have been traditionally selected, to which have been added others obtained according to modern scientific criteria and, more recently, by means of recombinant DNA technology. Since it is impossible to give a complete list of common wheat cultivars here, I will limit myself to two examples, from Canada and Italy. Traditional cultivars of Canada: Bishop (registration year 1904), Canthatch (1959), Canuck (1974), Canus (1935), Chinook (1952), Cypress (1962), Early Red Fife (1932), Garnet (1925), Glenlea (1972), Hard Red Calcutta (1890), Kitchener (1911), Ladoga (1888), Lemhi 62 (1968), Manitou (1965), Marquis (1910), Napayo (1972), Neepawa (1969), Norquay, Park (1963), Pitic 62 (1969), Prelude (1913), Preston (1895), Red Bobs 222 (1926), Red Fife (1885), Rescue (1946), Reward (1928), Ruby (1920), Saunders (1947), Selkirk (1953), Sinton (1975), Springfield (1972), Stanley (1895), Thatcher (1935). Cultivars mostly cultivated in Italy: Aubusson, Aztec, Bolero, Bologna, Guadalupe, Salmone, Tibet, Zena.

Related Species Of the 32 taxa currently recognized within the genus *Triticum*, only the six species cited here appear to have documented beer-making applications: *Triticum aestivum* L., *Triticum dicoccon* (Schrank) Schübl., *Triticum durum* Desf., *Triticum monococcum* L., *Triticum spelta* L., and *Triticum turanicum* Jakubz.

Geographic Distribution Europe, Africa, Asia, Australia and New Zealand, NS America. **Habitat** Cereal widely grown in all temperate and tropical climates. **Range** China, India, Russia, United States, France, Canada, Germany, Pakistan, Australia, Ukraine, Turkey, Kazakhstan, Great Britain, Iran, Poland, Egypt, Argentina, Uzbekistan, Morocco, Italy, Denmark, Romania, Brazil, Afghanistan, Spain, Hungary, Czech Republic, Mexico, Bulgaria, Syria, Turkmenistan, Algeria, Ethiopia, Sweden, Serbia, Lithuania, Azerbaijan, South Africa, Belarus, Belgium.

Beer-Making Parts (and Possible Uses)
Caryopses as they are or malted (wheat malt color range: 1.5–3.5 SRM), straw (for aromatization; to filter wort for Sahti, the traditional Finnish beer; used in ancient times as fuel for smoking malt but with unappreciated results in terms of quality), bran. By applying the same techniques used for malting barley, it is possible to obtain a variety of products from wheat malt, including toasted and roasted forms; wheat also has functional properties for beer, such as a high protein content, which promotes the formation and persistence of foam. Many wheat beers are unfiltered, and some of the properties that characterize wheat beer styles (e.g., cloudiness, bready aroma) may be attributable to both wheat and yeast. Although wheat is used in a range of styles (e.g., Berliner weiss, weizen), it can be difficult to process in both the malt house and the brewery. High levels of protein and gum may require more effort from an enzymatic perspective. In this regard, it appears that the use of giberellic acid in malting seems to be more common with wheat than with barley. More intensive mashing regimes in the brewery may be required to use wheat malts (especially those that are poorly modified). In wort containing very high percentages of wheat malt, the use of rice husks can help make the wort bed more porous and aid the filtration process.

Chemistry Caryopsis (whole wheat flour): energy 340 kcal, water 10.74 g, protein 13.21 g, lipids 2.5 g, carbohydrates 71.97 g, fiber 10.7 g, sugars 0.41 g, calcium 34 mg, iron 3.6 mg, magnesium 137 mg, phosphorus 357 mg, potassium 363 mg, sodium 2 mg, zinc 2.6 mg, thiamine 0.502 mg, riboflavin 0.165 mg, niacin 4.957 mg, vitamin B6 0.407 mg, folate 44 µg, vitamin A 9 IU, vitamin E 0.71 mg, vitamin K 1.9 µg.

Style Berliner Weiss, Sahti, Weizen, Wit.

Beer Schneider Weisse Tap 6, Schneider (Munich, Germany).

Source Calagione et al. 2018; Cantwell and Bouckaert 2016; Cantwell and Bouckaert 2018; Dabove 2015; Daniels and Larson 2000; Fisher and Fisher 2017; Giaccone and Signoroni 2017; Hieronymus 2010; Hieronymus 2016a; Hieronymus 2016b; Jackson 1991; Josephson et al. 2016; Markowski 2015; McGovern 2017; Sparrow 2015; Steele 2012; https://beer-legends.com/grains-and-maltextracts; beerlegends.com/wheat-grain.

Weizen and Other Wheat Beers
Together with barley, wheat is the most used cereal in the production of beer. Its use is linked to a relatively small number of styles developed in Belgium and Germany. In Germany, the presence of *Triticum aestivum* among beer ingredients has been limited since 1516 with the introduction of the *Reinheitsgebot*, which forbade producers to use it and allowed only clergy and nobles to produce weizenbier. In this style, wheat is malted and can make up to 60% of the total composition of the grist. In blanche styles, the wheat is instead raw and usually present in a lesser quantity (maximum 30%). Other styles that use wheat are Berliner weisse, gose, gratzer (in which the cereal is smoked), and lambic.

Triticum dicoccon (Schrank) Schübl.
POACEAE

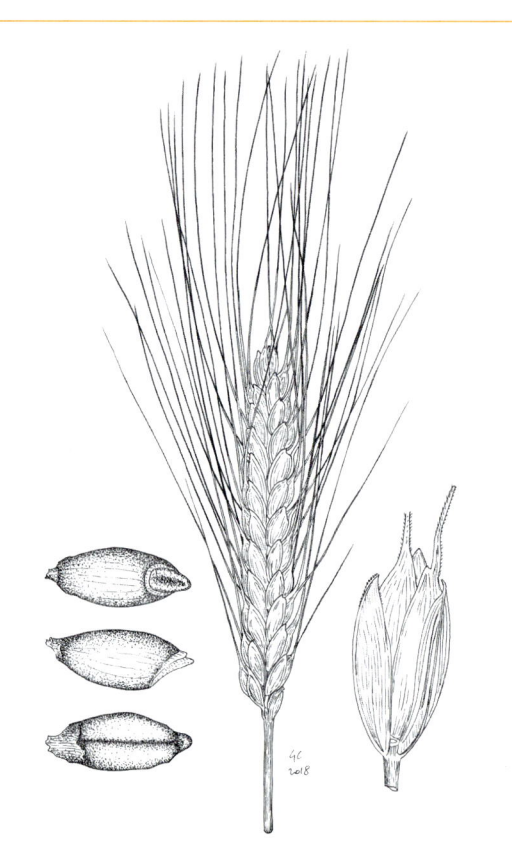

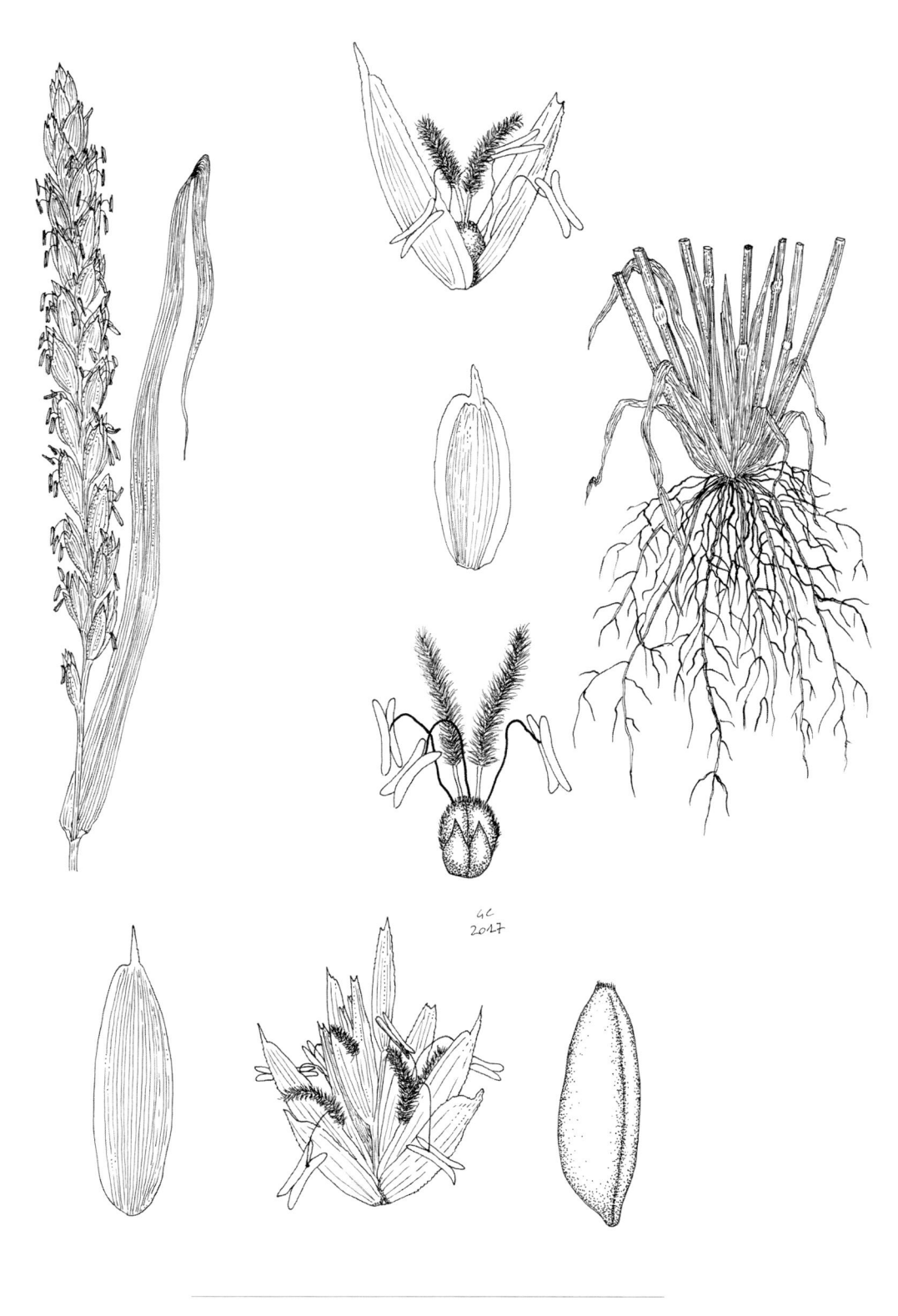

Triticum aestivum L.
POACEAE

Synonyms *Spelta amylea* (Ser.) Ser., *Triticum amyleum* Ser., *Triticum armeniacum* (Stolet.) Nevski, *Triticum arras* Hochst., *Triticum atratum* Host, *Triticum cienfuegos* Lag., *Triticum farrum* Bayle-Bar., *Triticum gaertnerianum* Lag., *Triticum ispahanicum* Heslot, *Triticum karamyschevii* Nevski, *Triticum palaeocolchicum* Menabde, *Triticum palaecocolchicum* (Menabde) L.B.Cai, *Triticum subspontaneum* (Tzvelev) Czerep., *Triticum tricoccum* Schübl., *Triticum volgense* (Flaksb.) Nevski.

Common Names Emmer, emmer wheat, hulled wheat.

Description Herb, annual (80–100 cm). Culm with glabrous or pubescent nodes. Leaf sheath glabrous or hairy, outer margin hairy, auricles falcate, ligule membrane ciliate; lamina (30–60 cm × 10–20 mm) with scaberulous surface. Inflorescence single raceme, bilateral, 3–10 cm × 10–13 mm, rachis brittle at nodes, flattened, glabrous or pubescent on surface and margins, hairs (1–2 mm) especially at internodes; spikelets appressed to rachis, internodes of rachis 3 mm long, falling off with the spikelet above; spikelets solitary, fertile ones sessile, comprising 2 fertile flowers, declining toward apex; spikelet oblong, 9–12 mm long, laterally compressed, dropping entire, deciduous with accessory structures of branches; glumes similar, shorter than spikelet; lower glume oblong, 7–10 mm long (same as upper glume), leathery, single keel running through entire length, 5–9 veins, apex cropped; upper glume oblong, 7–10 mm long (90–100% the length of adjacent fertile lemma), leathery, single keel running through entire length, 5–9 veins, lateral ones diverging at apex, this one cropped; fertile lemma elliptic, 9–12 mm long, leathery, without keel, 9–11 veins, apex aristate (awns 100–150 mm long); palea with 2 veins, keel ciliate, apical sterile flowers (1–2) resembling fertile ones although poorly developed; anthers 3, ovary with fleshy appendage below insertion of stylus, pubescent at apex. Fruit caryopsis with adherent pericarp, 7–9 mm long, hairy at apex.

Related Species Of the 32 taxa currently recognized within the genus *Triticum*, only the six species cited here appear to have documented beer-making applications: *Triticum aestivum* L., *Triticum dicoccon* (Schrank) Schübl., *Triticum durum* Desf., *Triticum monococcum* L., *Triticum spelta* L., and *Triticum turanicum* Jakubz.

Geographic Distribution SE Europe, temperate Asia (Caucasus, W Asia), Arabia, India. **Habitat** Cultivated in warm climates. **Range** Armenia, Morocco, Spain (Asturias), Czech Republic, Slovakia, Albania, Turkey, Switzerland, Germany, Greece, Italy, United States, Ethiopia.

Beer-Making Parts (and Possible Uses) Caryoxides as such or malted.

Chemistry Caryopsis: energy 1,400 kJ (335 kcal), water 10.4 g, protein 15.1 g, lipids 2.5 g, carbohydrates 67.1 g (starch 58.5 g, sugars 2.7 g, fiber 6.8 g), sodium 18 mg, potassium 440 mg, iron 0.7 mg, calcium 43 mg, phosphorus 420 mg.

Style Saison.

Source Dabove 2015; Giaccone and Signoroni 2017; Hieronymus 2010; Markowski 2015; McGovern 2017; Hieronymus 2016b; https://www.bjcp.org (31A. Alternative Grain Beer).

Triticum durum Desf.
POACEAE

Synonyms *Triticum alatum* Peterm., *Triticum bauhinii* Lag., *Triticum candissimum* Bayle-Bar., *Triticum cevallos* Lag., *Triticum cochleare* Lag., *Triticum*

fastuosum Lag., *Triticum hordeiforme* Host, *Triticum platystachyum* Lag., *Triticum pruinosum* Hornem., *Triticum pyramidale* Percival, *Triticum rimpaui* Mackey, *Triticum siculum* Roem. & Schult., *Triticum tomentosum* Bayle-Bar., *Triticum venulosum* Ser., *Triticum villosum* Host.

Common Names Durum wheat, pasta wheat, macaroni wheat, hard wheat.

Description Herb, annual (60–150 cm), caespitose. Internodes of culms empty or full. Leaf sheath at mouth bearing ciliated hairs, auricles falcate; ligule with ciliate membrane; lamina (30–60 cm × 10–20 mm) with glabrous or hairy surface, sparsely villous, margin smooth. Inflorescence single raceme, bilateral, 4–11 cm long, rachis stout, flattened, glabrous on surface, apex hairy, ciliate on margins; spikelets leaning against rachis, this with oblong internodes, 3–4 mm long; spikelets ascending, solitary, fertile ones sessile, comprising 2–3 fertile flowers, diminishing toward apex; spikelets oblong, laterally compressed, 10–15 × 8–15 mm, persisting on the plant; glumes similar, shorter than the spikelet; lower glume ovate, 8–12 mm long (as the upper), leathery, double keel running through it entirely, 5–9 veins, surface glabrous or pubescent, at the apex with a unilateral tooth, truncate; upper glume ovate, 8–12 mm long (90–100% the length of adjacent fertile lemma), coriaceous, double keel running through entire length, winged on keel, 5–9 veins, lateral ones divergent at apex; upper surface of glume glabrous or pubescent, apex with unilateral, truncate tooth; fertile lemma elliptic, 10–12 mm long, leathery, without keel, 9–15 veins, apex acute, aristate, awns 80–150 mm long; palea with 2 veins, keel ciliate; apical sterile flowers resembling fertile ones though undeveloped; anthers 3, 4 mm long, ovary with fleshy appendage below insertion of stylus, pubescent at apex. Fruit caryopsis with adherent pericarp, 8 mm long, villous at apex.

Cultivar There is an unspecified number of cultivars of durum wheat around the world. Also, similarly to common wheat, the long tradition of cultivation and geographical spreading have led to the selection of a large number of cultivars more or less adapted to specific pedoclimatic conditions and resistant to adversities. Besides the traditional Italian cultivars, such as Senatore Cappelli and Leonessa, which are making a strong comeback, new varieties of foreign origin are gaining the interest of farmers in Italy, including Acadur, Alemanno, Anco Marzio, Antalis, Aureus, Claudius, Core, Duilio, Dylan, Furio Camillo, Gibraltar, Hector, Iride, Kanakis, Karalis, Kenobi, Marakas, Marcus Aurelius, Monastir, Odysseus, Opera, Pigreco, Prospero, Ramirez, Saragolla, Simeto, Svevo, Teodorico, Tirex, Tito Flavio, and Vespucci.

Related Species Of the 32 taxa currently recognized within the genus *Triticum*, only the six species cited here appear to have documented beer-making applications: *Triticum aestivum* L., *Triticum dicoccon* (Schrank) Schübl., *Triticum durum* Desf., *Triticum monococcum* L., *Triticum spelta* L., and *Triticum turanicum* Jakubz.

Geographic Distribution SE Europe, NES Africa, Asia, India, South America. **Habitat** Cereal widely cultivated. **Range** Middle East, W Europe, North America, N Africa, South America.

Beer-Making Parts (and Possible Uses) Caryopses.

Chemistry Carioxides: energy 339 kcal, lipids 2.5 g, carbohydrates 71 g, protein 14 g, sodium 2 mg, potassium 431 mg, magnesium 144 mg, calcium 34 mg, iron 3.5 mg, vitamin B6 0.4 mg.

Style Ale, Ancient Ale, Blonde Ale.

Beer Arsa, Birranova (Triggianello, Apulia, Italy).

Source Dabove 2015; McGovern 2017; https://beerlegends.com/grains-and-malt-extracts; https://beerlegends.com/wheat-grain; https://www.birraaltaquota.it/prodotti/birre-estrose/70-leonessa.html; https://www.birraaltaquota.it/prodotti/birre-estrose/91-anastasia.html.

Agricultural Tradition and Beer-Making Innovation

Italy has a multimillennial tradition of cultivating the land. The acumen of its farmers and the varied environmental conditions of the *bel paese* have led, over time, to the selection of an enormous heritage of plant biodiversity, a quantity of cultivars adapted to specific environmental conditions that today, fortunately, are being rediscovered, surveyed, collected, and exchanged with renewed enthusiasm. This precious heritage concerns not only vegetables and fruits but also several cereals. Conservation, if not an end in itself, always implies some form of valorization. And what better use could cereals have than beer production? Spelt, present above all in Central Italy, and durum wheat, widely cultivated in the center-south, are excellent examples of how local cultivars or ecotypes are gradually finding their place in craft brewing. Only local, small-scale agriculture can give beer the cultural and cultivation coordinates of a territory, giving the finished product character and personality that cannot be reproduced elsewhere. Thus, thanks to craft beer, the short agricultural chain can hope to live a second youth . . . with all due respect for globalization.

Triticum monococcum L.
POACEAE

Synonyms *Aegilops hordeiformis* Steud., *Agropyron pubescens* (M.Bieb.) Schischk., *Crithodium monococcum* (L.) Á.Löve, *Nivieria monococca* (L.) Ser., *Triticum hornemannii* Clemente, *Triticum pubescens* M.Bieb., *Triticum sinskajae* Filat. & Kurkiev, *Triticum spontaneum* Flaksb., *Triticum tenax* Hausskn.

Common Names Einkorn wheat, Enkir.

Description Herb, annual (45–80 cm). Culms solitary or caespitose, nodes villous. Leaf sheath at mouth bearing ciliate hairs, auricles falcate, ligule membrane ciliate; lamina 2–5 mm wide, surface puberulent, margins smooth or scaberulous. Inflorescence single raceme, bilateral, 2.5–4 cm long; rachis brittle on nodes late, flattened, glabrous, or pubescent at margins; spikelets leaning against rachis, internodes oblong, 2–2.5 mm long, falling off with overlying spikelet; spikelets solitary, fertile ones sessile, comprising 1–2 fertile flowers, diminishing at apex; spikelets oblong, laterally compressed, 10 mm long, falling off whole, deciduous with accessory structures; glumes similar, shorter than spikelets; lower glume oblong, 6–8 mm long (as long as upper glume), leathery, double keel throughout its length, 5–9 veins, apex toothed; upper glume oblong, 6–8 mm long (90–100% the length of adjacent fertile lemma), leathery, double keel throughout its length, 5–9 veins, lateral ones diverging at apex, apex dentate; fertile lemma elliptic, 8–10 mm long, leathery, keeled, 9–11 veins, apex dentate, aristate, awns 30–80 mm long; palea 2 veins, keel ciliate; apical sterile flower (1) resembling fertile ones though undeveloped; anthers 3, ovary with fleshy appendage below insertion of stylus, pubescent at apex. Fruit caryopsis with adherent pericarp, villous at apex.

Subspecies Besides the nominal variety, *Triticum monococcum* L. var. *monococcum*, the variety *Triticum monococcum* L. var. *reuteri* (Flaksb.) K.Hammer & A.Szabó is recognized.

Cultivar Long-cultivated species are usually characterized by a remarkable taxonomic and nomenclatural complexity. *Triticum monococcum* L. is not an exception; while according to some authors it would include only two varieties, according to others, there would be many more varieties or cultivars. Cultivars include *Triticum monococcum* L. var. *eredvianum* Zhuk, *Triticum monococcum* L. var. *flavescens* Körn., *Triticum monococcum* L. var. *hornemanii* Clem., *Triticum monococcum* L. var. *leatissimum* Körn., *Triticum monococcum* L. var. *macedonicum* Papag., *Triticum monococcum* L. var. *monococcum*, *Triticum monococcum* L. var. *pseudohornemanii* Dekapr. & Menabde, *Triticum monococcum* L. var. *ratschinicum* Dekapr. & Menabde, and *Triticum monococcum* L. var. *vulgare* Körn.

Related Species Of the 32 taxa currently recognized within the genus *Triticum*, only the six species cited here appear to have documented beer-making applications: *Triticum aestivum* L., *Triticum dicoccon* (Schrank) Schübl., *Triticum durum* Desf., *Triticum monococcum* L., *Triticum spelta* L., and *Triticum turanicum* Jakubz.

Geographic Distribution Europe, Africa, Asia. **Habitat** Cultivated. **Range** France (N Provence), India, Italy, Morocco, former Yugoslavia, Turkey.

Beer-Making Parts (and Possible Uses) Carioxides.

Chemistry Caryopsis (flour): starch 62 g, fiber 8.7 g, lipids 2.8 g, protein 18 g, phosphorus 415 mg, potassium 390 mg, sulfur 190 mg, manganese 4.4 mg, thiamine 0.5 mg, riboflavin 0.45 mg, niacin 3.1 mg, pyridoxine 3.1 mg, vitamins group B 0.49 mg.

Style Weissbier.

Beer Alba, BABB (Manerbio, Lombardy, Italy).

Source Calagione et al. 2018; Dabove 2015; https://www.bjcp.org (31A. Alternative Grain Beer).

Triticum spelta L.
POACEAE

Synonyms *Spelta vulgaris* Ser., *Triticum arias* Clemente, *Triticum elymoides* Hornem., *Triticum forskalei* Clemente, *Triticum palmovae* G.I.Ivanov, *Triticum speltiforme* Seidl ex Opiz, *Triticum zea* Host, *Zeia spelta* (L.) Lunell.

Common Names Spelt wheat.

Description Herb, annual (60–120 cm). Stem culm. Leaf sheath ciliate at mouth, ligule membranaceous, ciliate; lamina (30–60 cm × 10–15 mm) glabrous or hairy, sparsely villous. Inflorescence single raceme, bilateral, 10–15 cm long, rachis brittle at nodes, flattened, ciliate at margins; spikelets appressed to rachis, internodes of rachis clavate, 5–6 mm long, falling with spikelet above; spikelets solitary, fertile sessile, comprising 2 fertile flowers, diminishing toward apex; spikelet oblong, laterally compressed, 12–15 mm long, deciduous entire with accessory structures; glumes similar, shorter than the spikelet, the lower oblong, 8–12 mm long (as the upper glume), leathery, double keel running through it entirely, 5–9 veins, glabrous or hairy, apex toothed, truncate; upper glume oblong, 8–12 mm long (90–100% the length of adjacent fertile lemma), coriaceous, double keel running entirely through it, 5–9 veins, lateral ones diverging at apex, surface glabrous or pubescent, apex dentate, truncate; fertile lemma elliptic, 7 mm long, leathery, keeled, 9–11 veins, surface pubescent, above villous, apex aristate, awns 6–60 mm long; palea 2 veins, keel ciliate; apical sterile flowers resembling fertile ones though undeveloped, 2–3; anthers 3, ovary with fleshy appendage below insertion of stylus, pubescent at apex. Fruit caryopsis with adherent pericarp, ellipsoid, 7–10 mm long, hairy at apex.

Cultivar Some varietal names are Altgold, H 9227, H 9228, Hercule, KR 489-11-15, Lueg, Oberkulmer, Ostro, Ostroschwarzer, and Rouquin.

Related Species Of the 32 taxa currently recognized within the genus *Triticum*, only the six species cited here appear to have documented beer-making applications: *Triticum aestivum* L., *Triticum dicoccon* (Schrank) Schübl., *Triticum durum* Desf., *Triticum monococcum* L., *Triticum spelta* L., and *Triticum turanicum* Jakubz.

Geographic Distribution CSE Europe, NS Africa, EW Asia. **Habitat** Cultivated. **Range** France, Germany, Switzerland.

Beer-Making Parts (and Possible Uses) Caryopsis.

Chemistry Caryopsis (macronutrients): starch 63.8 g, sugars 2.1 g, fiber 10.9 g, protein 15.6 g, lipids 2.5 g.

Style Saison, Weissbier.

Beer Cento per Cento, Petrognola (Piazza al Serchio, Tuscany, Italy).

Source Calagione et al. 2018; Fisher and Fisher 2017; Giaccone and Signoroni 2017; Hieronymus 2016b; Markowski 2015; https://www.birraaltaquota.it/prodotti/birre-estrose/70-leonessa.html; https://www.bjcp.org(31A. Alternative GrainBeer); https://www.brewersfriend.com/other/spelt/.

Triticum turanicum Jakubz.
POACEAE

Synonyms *Triticum percivalianum* Parodi, *Triticum percivalii* E.Schiem.

Common Names Khorasan wheat, oriental wheat, Kamut®.

Description Herb, annual (66–110 cm), bushy. Internodes of culms with a small or full lumen. Leaf sheath with falcate auricles; ligule membranaceous, ciliate; lamina 10–15 mm wide, surface pubescent. Inflorescence raceme, single, oblong, bilateral (9–15 cm × 10–11 mm), rachis strong, flattened, 1.3–3 mm wide, ciliate at margins; spikelets appressed to rachis, slender, rachis internodes wedge shaped, 5–6.5 mm long; spikelets ascending, solitary, fertile ones sessile, comprising 2–3 flowers, decreasing toward apex; spikelets oblong, laterally compressed, 12–15 mm long, persistent on the plant, flower callus pubescent; similar glumes as long as the flowers; lower glume ovate, 12–15 mm long (as long as the upper glume), leathery, pale, double keel running through it entirely, 5–9 veins, lateral ones prominent, surface pubescent, apex with a unilateral, truncate tooth; upper glume ovate, 12–15 mm long (as long as adjacent fertile lemma), coriaceous, surface pubescent, pale, double keel running through it entirely, 5–9 veins, lateral ones prominent and divergent at apex, apex with unilateral, truncate tooth; fertile lemma elliptic, 12–15 mm long, coriaceous, hulless, 5–9 veins, apex acute, aristate, awns 140–160 mm long; palea narrow to lemma, coriaceous, 2 veins, with winged keel, ciliate; apical sterile flowers similar to fertile ones although undeveloped; anthers 3, ovary with fleshy appendage below insertion of stylus, pubescent at apex. Fruit caryopsis with adherent pericarp, ellipsoid, 10.5–12 mm long, hairy at apex.

Cultivar Kamut (cultivar with registered trademark), Khorasan.

Related Species Of the 32 taxa currently recognized within the genus *Triticum*, only the six species cited here appear to have documented beer-making applications: *Triticum aestivum* L., *Triticum dicoccon* (Schrank) Schübl., *Triticum durum* Desf., *Triticum monococcum* L., *Triticum spelta* L., and *Triticum turanicum* Jakubz.

Geographic Distribution W Temperate Asia, Caucasus. **Habitat** Cultivated. **Range** Europe, Iran, Middle East (other regions), United States (Montana), Canada (S Saskatchewan, SE Alberta), Australia.

Beer-Making Parts (and Possible Uses) Caryopsis.

Chemistry Caryopses: energy 1,411 kJ (337 kcal), water 10.95 g, carbohydrates 70.38 g (starch 52.41 g, fiber 9.1 g), lipids 2.2 g (saturated 0.192 g, monounsaturated 0.214 g, polyunsaturated 0.616 g), protein 14.7 g, thiamine (B1) 0.591 mg, riboflavin (B2) 0.178 mg, niacin (B3) 6.35 mg, pantothenic acid (B5) 0.9 mg, vitamin B6 0.255 mg, vitamin E 0.6 mg, iron 4.41 mg, magnesium 134 mg, manganese 2.86 mg, phosphorus 386 mg, potassium 446 mg, zinc 3.68 mg.

Style Saison, Weissbier.

Beer Nora, Baladin (Piozzo, Piedmont, Italy).

Source Calagione et al. 2018; Dabove 2015; https://www.baladin.it/it/productdisplay/nora; https://it.wikipedia.org/wiki/Kamut.

Tropaeolum majus L.
TROPAEOLACEAE

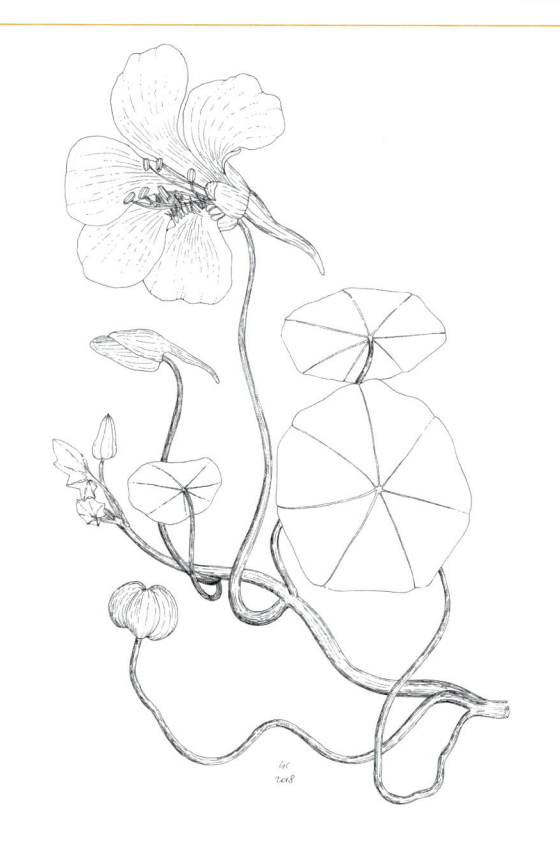

Synonyms *Cardamindum majus* Moench, *Tropaeolum pinnatum* Andrews, *Tropaeolum quinquelobum* Bergius, *Trophaeum majus* (L.) Kuntze.

Common Names Garden nasturtium, Indian cress, monks cress.

Description Herb, annual (15–250 cm). Stem creeping, climbing, glabrous or nearly so. Leaf lamina orbicular to reniform (Ø 3–12 cm), peltate, 9 main nerves radiating from petiole, margin variously angled, sinuate or entire, underleaf usually papillose. Flowers axillary, solitary, yellow-orange, purple, burgundy, or creamy-white (Ø 2.5–6 cm), peduncle 6–18 cm, receptacle cup shaped; sepals 5, oblong-lanceolate (1.5–2 × 0.5–0.7 cm), spur 2.5–3.5 cm, straight or curved; petals 5, apex usually rounded, sometimes pointed or notched; 2 apical petals (2.5–5 × 1–1.8 cm), margin often entire; 3 basal petals with margin deeply fringed on claw; stamens 8, distinct, unequal; ovary 3-lobed; stylus 1; stigma linear, 3-lobed. Fruit oblate (± spheroidal but flattened at poles), Ø 1.5–2 cm, dehiscent into 3 monosperm mericarps at maturity. Seed Ø 5–8 mm.

Cultivar There are many, including African Queen, Alaska Cherry Rose, Alaska Mix, Alaska Red Shades, Alaska Salmon, Alaska Scarlet, Apricot Gleam, Apricot Trifle, Apricot Twist, Banana Split, Black, Black Velvet, Burning Embers, Burpee S Tall Mix, Buttercream, Campari N Soda, Canary, Canary Creeper, Capuchina Mezcla de Noche y Dia, Caribbean Cocktail, Cherries Jubilee, Cherry Rose, Cherry Rose Jewel, Climbing Amazon Jewel, Climbing Mixed, Cobra, Cook's Custom Mix, Copper Sunset Cream, Creamsicle, Crimson Beauty, Curly Mix, Darjeeling, Dayglow Mix, Diamant Des Abendlandes, Double, Double Delight Apricot, Double Delight Cream, Double Dwarf Jewel Mix, Double Gleam Hybrid, Double Gleam Mix, Double Kroshka, Dwarf Cherry Rose, Dwarf Compact Mixed, Dwarf Jewel Blend, Empress Of India, Feuervogel, Fiesta Blend, Fordhook Favorites Mix, Forest Flame, Gem Dwarf, Gleam Mix, Gleaming Mahogany, Glorious Gleam, Golden Emperor, Golden Gleam, Golden Jewel, Golden King, Hermine Grashoff, Indian Chief, Jewel of Africa, Kaleidoscope Mix, Kappertjes, King Theodore, Lipstick, Little Firebirds, Mahogany, Mahogany Jewel, Mahogany Velvet, Margaret Long, Mastuerzo, Milkmaid, Moon Gleam, Moonlight, Mounding Vanilla Berry, Night and Day, Orchid Flame, Out of Africa, Out of Africa Yellow, Papaya Cream, Park S Fragrant Giants, Peach Melba, Peach Melba Superior, Peach Schnapps, Peaches and Cream, Primrose Jewel, Princess of India, Queen Victoria, Red Wonder, Ruffled Apricot, Salmon Mousse, Saucy Rascal, Scarlet, Scarlet Climber, Scarlet Emperor, Scarlet Gleam, Spitfire, St. Clements, Strawberries, Strawberry Cream, Strawberry Ice, Summer Charm, Sunset Pink, Tall Mixed Colors, Tall Single Mixed Colors, Tall Trailing Mix, Tangerine Dream, Tip Top, Tip Top Alaska, Tip Top Apricot, Tip Top Mahogany, Tip Top Mix, Tip Top Velvet, Tom Thumb Black Velvet, Tom Thumb Mixed, Top Flowering Lemon Yellow, Trailing Mixed Colours, Tutti Frutti Mix, Vanilla, Vanilla Berry, Variegated Alaska, Vesuvius, Whirlybird Cherry, Whirlybird Cherry Rose, Whirlybird Cream, Whirlybird Gold, Whirlybird Mahogany Red, Whirlybird Mix, Whirlybird Tangerine, and Wina.

Related Species About 95 species (or subspecies) belonging to the genus *Tropaeolum* are known, but only the one cited here seems to have beer-making uses.

Geographic Distribution Andes (Bolivia to Colombia). **Habitat** Roadside, landfills, beaches, coastal cliffs, shady ravines, intermittent streams (0–300 m ASL). **Range** Species grown for ornamental purposes or food and naturalized (also invasive) in South America, United States, Europe, Asia, Africa, Australia.

BEER-MAKING PARTS (AND POSSIBLE USES)
Flowers (?); all parts of the plant are edible (taste vaguely similar to that of cress, *Lepidium sativum* L., Brassicaceae, family similar to that of Tropaeolaceae); fruits used as substitute for capers (*Capparis* sp., Capparaceae), leaves (slightly spicy) used to flavor pasta dishes and salads.

CHEMISTRY Essential oil (92.0% of total oil): benzyl isothiocyanate 82.5%, benzene acetonitrile 3.9%, 2-phenylethylisovalerate 2.9%. Flowers: lutein 300–510 µg/g, violaxanthin, anteraxanthin, zeaxanthin, zeinoxanthin, β-cryptoxanthin, α-carotene, β-carotene. Leaves: lutein 136 ± 18 µg/g, β-carotene 69 ± 7 µg/g, violaxanthin 74 ± 23 µg/g, neoxanthin 48 ± 13 µg/g.

STYLE Saison.

SOURCE Fisher and Fisher 2017; Hieronymus 2016b; https://www.brewersfriend.com/other/nasturtium/.

Turnera diffusa Willd. ex Schult.
PASSIFLORACEAE

SYNONYMS *Turnera aphrodisiaca* Ward, *Turnera humifusa* Endl. ex Walp., *Turnera pringlei* Rose.

COMMON NAMES Damiana, damiane.

DESCRIPTION Shrub (1–2 m), perennial. Young branches reddish, buds grayish with appressed or floccose pubescence, white. Leaves aromatic, obovate to lanceolate (10–25 × 4–10 mm), shortly petiolate, apex obtuse or acute, base short, cuneate; margins serrate or deeply 2- to 10-dentate on each side; veins ascending, usually strong, straight, and simple; upper leaf smooth, pale green, the lower glabrous, with a few hairs on the veins, or densely tomentose over the entire surface. Flowers yellowish. Fruit small, globose.

RELATED SPECIES The genus *Turnera* includes 128 taxa of which only *T. diffusa* has known beer-making applications.

GEOGRAPHIC DISTRIBUTION United States (S Texas), Mexico, Caribbean, South America. **HABITAT** Dry sandy and rocky places. **RANGE** United States (S Texas), Mexico, Caribbean, South America.

BEER-MAKING PARTS (AND POSSIBLE USES)
Leaves.

CHEMISTRY Leaves: albuminoids 15 g, chlorophyll 8 g, resins 6.5 g, tannins 3.5 g, calcium 1.81 g, potassium 0.776 g, essential oil 0.75 g, arbutin 0.7 g, magnesium 204 mg, phosphorus 115 mg, vitamin C 98.3 mg, iron 88 mg, cineole 80 mg, cobalt 12.9 mg, p-cymol 10 mg, β-carotene 7.1 mg, sodium 5.8 mg, manganese 5.4 mg, chromium 3.1 mg, tin 1.1 mg, thiamine 0.18 mg, barterine, γ-cadinene, calamenene, kaoutchouc, α-copaene, p-cymene, damianine, gonzalitosin-I, hexacosanol-1, luteolin-8-C-α-L-rhamopyranosyl(1->2)-quinovopyranoside, methoxyflavone, niacin, α-pinene, β-pinene, riboflavin, selenium, β-sitosterol, tetraphylline-B, thymol, triacontane, zinc.

STYLE Spice/Herb/Vegetable Beer.

SOURCE Heilshorn 2017; https://www.brewersfriend.com/other/damiana/.

Typha latifolia L.
TYPHACEAE

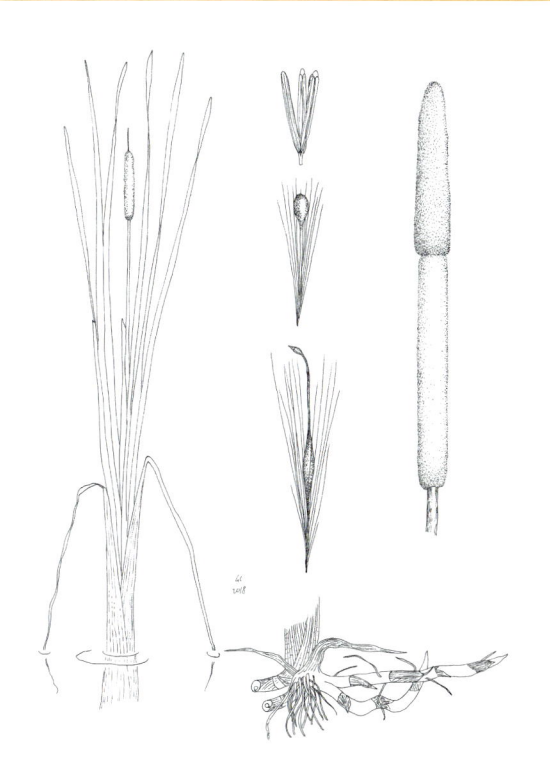

SYNONYMS *Massula latifolia* (L.) Dulac, *Typha ambigua* Schur ex Rohrb., *Typha crassa* Raf., *Typha elongata* Dudley, *Typha engelmannii* A.Br. ex Rohrb., *Typha intermedia* Schur, *Typha major* Curtis, *Typha palustris* Bubani, *Typha pendula* Fisch. ex Sond., *Typha remotiuscula* Schur, *Typha spathulifolia* Kronf.

COMMON NAMES Common cattail.

DESCRIPTION Herb (up to 2.5 m), perennial. Horizontal rhizome elongated; stem erect, simple. Leaves bluish-green, glabrous, linear (8–25 mm wide), sheathing, parallelinervate, the upper ones up to the top of the inflorescence. Inflorescence formed by 2 overlapping monoecious spikes (spadices), the lower female (15–25 cm), cylindrical, contiguous to the upper male inflorescence, narrower, whitish, conical, about as long as the female spike or shorter. Male flowers with numerous clusters of stamens surrounded by short bristles, with 1–3 basifixed anthers 2–3 mm long; female flowers very small, closely appressed around stem, lacking bracts, with perianth formed of fine bristles; ovary long stalked, stigma spatulate, brown color when mature, exceeding bristles. Infructescence spike, reddish-brownish color, cylindrical, formed by thousands of fusiform cypselae (about 1–1.5 mm), longitudinally furrowed; pericarp not adherent to seed, dehiscent at maturity. Seeds dispersed by transparent bristles of the perianth, forming a kind of parachute.

RELATED SPECIES The genus *Typha* has about 40 taxa.

GEOGRAPHIC DISTRIBUTION Cosmopolitan. **HABITAT** Swamps, ponds, ditches. **RANGE** Cosmopolitan.

BEER-MAKING PARTS (AND POSSIBLE USES) Leaves (spongy interior used to stop small leaks from barrels); leathery leaves are usually used to make mats and stuff flasks, carboys, chairs, and boats.

STYLE None known.

SOURCE Cantwell and Bouckaert 2016; Cantwell and Bouckaert 2018; McGovern 2017.

Ulmus minor Mill.
ULMACEAE

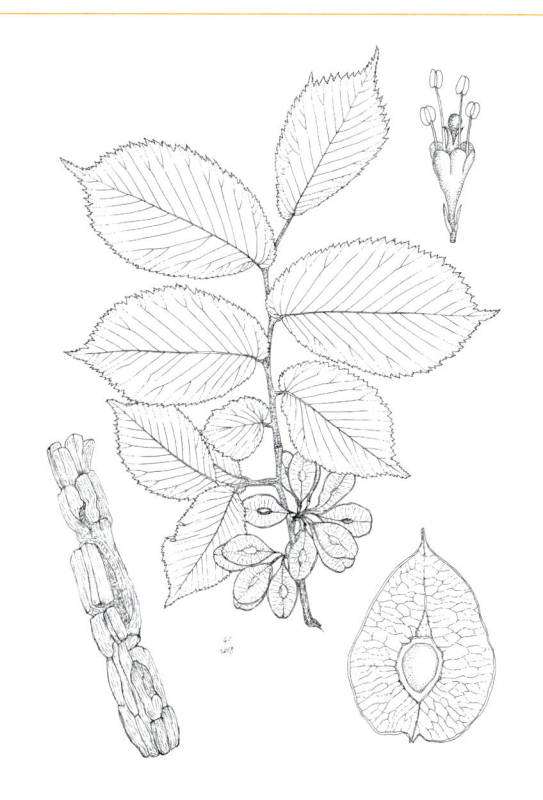

SYNONYMS *Ulmus angustifolia* (Weston) Weston, *Ulmus araxina* Takht., *Ulmus diversifolia* Melville, *Ulmus foliacea* Gilib., *Ulmus fungosa* (Aiton) Dum. Cours., *Ulmus georgica* Schchian, *Ulmus grossheimii*

Takht., *Ulmus plotii* Druce, *Ulmus procera* Salisb., *Ulmus reticulata* Dumort., *Ulmus rotundifolia* Carrière, *Ulmus sarniensis* (Loudon) H.H.Bancr., *Ulmus sativa* Mill., *Ulmus suberosa* Moench, *Ulmus uzbekistanica* Drobow, *Ulmus wheatleyi* (Simon-Louis) Druce, *Ulmus wyssotzkyi* Kotov.

Common Names Elm.

Description Tree, deciduous (up to 30 m). Erect trunk, branched at the top, wide dark green crown; bark smooth and gray at first, then longitudinally fissured until it becomes suberose and brownish. Leaves alternate (up to 10 cm), briefly petiolate, elliptic-acuminate, margin toothed, lamina typically asymmetrical to the midrib, 7–12 pairs of secondary veins, upper leaf glabrous, shiny, underleaf paler, sparsely hairy at the bifurcation of the veins. Flowers hermaphrodite, formed by an involucre enclosing 4–6 stamens with red-brown anthers and 1 pistil; united in axillary glomerules, preceding foliation. Fruit elliptical samara with seed in the apical half, gathered in clusters.

Related Species The genus *Ulmus* includes 43 taxa.

Geographic Distribution European-Caucasian. **Habitat** Woods, hedges, fallow land. **Range** Europe, Caucasus.

Beer-Making Parts (and Possible Uses) Wood (tried for wine containers with little success; uncertain use in beer making), wood (for smoking malt).

Wood Differentiated, with white-rose sapwood clearly distinguished from heartwood, brown or chocolate color (WSG 0.46–0.70). Used for underwater construction, wheel spokes, wheelwright work, floorboards, furniture, turned objects, carving, veneer, decorative veneers.

Style None verified.

Source Cantwell and Bouckaert 2016; Cantwell and Bouckaert 2018; Daniels and Larson 2000 (widely distributed species that well represents the genus *Ulmus* in Europe).

Umbellularia californica (Hook. & Arn.) Nutt.
LAURACEAE

Synonyms *Oreodaphne californica* (Hook. & Arn.) Nees, *Tetranthera californica* Hook. & Arn.

Common Names California bay laurel, California bay, California laurel, Oregon myrtle, pepperwood, spicebush, cinnamon bush, peppernut tree, headache tree, mountain laurel, balm of heaven.

Description Tree (up to 30–35 m). Stem Ø up to 80 cm, bark green to red-brown. Leaves narrowly ovate to oblong (3–10 × 1.5–3 cm), glossy, yellow-green, dotted with minute glands; underside glabrous, sparsely hairy, appressed, or minutely tomentose from gray hairs, upper side glabrous; petiole smaller than lamina. Inflorescences at terminal axils, umbelliform, pedunculate, 5- to 10-flowered, bracts about 7 mm. Flower with 6 perianth elements, oblong-ovate (3–4.5 mm); stamens 9, staminodes 3, anthers 4-celled. Fruit usually 1, round-ovoid to oliviform (Ø 2–2.5 cm), color green, dark purple when dry.

Related Species *Umbellularia* is a monotypic genus; therefore, *U. californica* is the only species belonging to this genus.

Geographic Distribution Endemic species of California and S Oregon. **Habitat** Canyons, valleys, chaparrals (< 1,600 m ASL). **Range** California, Oregon.

Beer-Making Parts (and Possible Uses)
Fruits, seeds, leaves. Flesh and inner kernel of the fruit were once used as food; fruit dried in the sun until the outer fleshy part was detached from the seed; lower third of the dried flesh edible, while the upper part, thinner, contains a higher concentration of acrylic oil; seeds usually consumed after roasting since roasting removes most of the pungency, leaving only a hint of acidity; leaves used as a substitute for laurel leaves.

Toxicity Leaves and the oil extracted from them are irritating to mucous membranes owing to the presence of umbellulone and can cause dermatitis; because of the presence of safrole, the leaves are suspected to have carcinogenic activity.

Chemistry On the composition of fruits and seeds, no information was found. Essential oil, leaves (97.1% of total oil): α-pinene 0.1%, β-pinene 0.1%, sabinene 0.1%, myrcene 0.2%, α-terpinene 0.2%, limonene 0.1%, 1,8-cineole 19.5%, γ-terpinene 0.3%, p-cymene 2.1%, terpinolene 0.1%, α,p-dimethylstyrene 0.1%, trans-sabinene hydrate 0.1%, camphor 0.2%, linalool 0.4%, cis-sabinene hydrate 0.1%, trans-p-menth-2-en-1-ol 0.1%, terpinen-4-ol 6.6%, cis-p-menth-2-en-1-ol 0.1%, umbellulone 36.7%, δ-terpineol 0.6%, α-terpineol 6.5%, borneol 0.1%, β-bisabolene 2.2%, phellandral 0.1%, (E)-α-bisabolene 0.5, ar-curcumene 0.1%, nerol 0.1%, 2-tridecanone 0.1%, p-cymen-8-ol 0.2%, cuminyl acetate 0.1%, caryophyllene oxide trace, methyleugenol 8.4%, (E)-nerolidol 0.3%, p-mentha-1,4-dien-7-ol 0.3%, elemol 0.4%, cuminol 0.5%, 51 cis-p-menth-3-en-1,2-diol trace, eugenol 0.4%, γ-eudesmol 0.2%, thymol 7.8%, carvacrol trace, eleminacin 0.1%, α-eudesmol 0.1%, β-eudesmol 0.2%, chavicol 0.2%, dodecanoic acid 0.3%, hexadecanoic acid 0.1%.

Style Gruit.

Source Hieronymus 2016b; https://honest-food.net/2016/05/05/gruit-herbal-beer/.

Urtica dioica L.
URTICACEAE

Synonyms *Urtica galeopsifolia* Wierzb. ex Opiz, *Urtica tibetica* W.T.Wang.

Common Names Nettle, stinging nettle.

Description Herb, perennial (3–12 dm), rough in appearance and deep green in color. Rhizomes long, stoloniferous, branched, creeping just below the surface, from which rise numerous stout, erect, striated stems, obtusely square in section, reddish or yellowish, usually unbranched, covered with urticating hairs. Leaves opposite, petiole smaller than lamina, longer than wide (> 5 cm), lamina ovate-lanceolate to lanceolate-linear, base heart shaped, apex narrowed into an acute tooth, margin coarsely toothed, apical tooth longer than adjacent ones, surface roughened, sprinkled with short simple hairs mixed with long, stiff, stinging hairs; 4 stipules free (> 3 mm), pubescent on both faces. Dioecious plant. Inflorescences simple or branched racemes, in whorls at the axils of the upper leaves, pendulous or curved in female plants, usually patent in male plants, shorter but always longer than the relative petiole (2–5 cm). Small flowers grouped in glomerules, greenish-yellow or reddish color; male ones with 4 hirsute-pubescent tepals, 4 stamens; female ones unequal, 2 internal tepals entirely pubescent, larger than the external ones, stigmas apically reddened. Fruit ovoid-elliptic diclesium, olive-brown in color, with apical tuft of hairs.

SUBSPECIES Besides *Urtica dioica* L. subsp. *dioica*, the species includes several other subspecies (*Urtica dioica* L. subsp. *afghanica* Chrtek, *Urtica dioica* L. subsp. *gracilis* [Aiton] Selander, *Urtica dioica* L. subsp. *holosericea* [Nutt.] Thorne) and a variety (*Urtica dioica* L. var. *sicula* [Gasp. ex Guss.] Wedd.).

RELATED SPECIES The genus *Urtica* includes about 58 taxa, some of which are known to be used in beer making.

GEOGRAPHIC DISTRIBUTION Subcosmopolitan. **HABITAT** Grasslands, woods, nitrophilous anthropized areas, roadsides (0–2,300 m ASL). **RANGE** Subcosmopolitan.

BEER-MAKING PARTS (AND POSSIBLE USES) Leaves.

CHEMISTRY Leaf extract: 2,4-di-t-butylphenol 5.28%, tributyl ester phosphoric acid 4.12%, 8-methyl-heptadecane 1.2%, 1-heptadecene 2.15%, eicosane 2.83%, neophytadiene 25.21%, 3,7,11,15-tetramethyl-2-hexadecyl ester 1.63%, phthaleic acid 8.15%, 2,6,10,15-tetramethylheptadecane 1.17%, olean-18-ene 2.25%, 3,5-di-tert-butyl-ortho-benzoquinone 1.28%, 2,6,10,14-tetramethylpentadecane 1.45%, dibutylphthaleate 7.37%, heneicosane 2.26%, hexacosane 2.04%, bis(2-ethyl hexyl)maleate 6.32%, nonacosane 2.72%, pentacosane 1.51%, 1,2,-benzenedicarboxylic acid 7.69%, 2-tert-butyl-4,6-bis(3,5-di-tert-butyl-4-hydroxy-benzyl) phenol 3.42%.

STYLE American Amber Ale, American IPA, American Pale Ale, California Common Beer, Other Smoked Beer, Spice/Herb/Vegetable Beer.

SOURCE Fisher and Fisher 2017; Heilshorn 2017; Hieronymus 2016b; Josephson et al. 2016; https://www.brewersfriend.com/other/nettels/.

Urtica urens L.
URTICACEAE

SYNONYMS *Urtica trianae* Rusby.

COMMON NAMES Small nettle.

DESCRIPTION Herb, annual (10–80 cm), monoecious. Root white, fusiform, fleshy. Stems erect, tetragonal, ± branched, green or reddish, with a few short urticating hairs. Leaves opposite, with petioles (12.5 cm), puberulent; lamina ovate or elliptic (1.2–6 × 0.6–3 cm), base cordate or rounded, margin deeply 6- to 11-dentate with sharp teeth and apical tooth equal to 2 adjacent ones; upper leaf with only urticating hairs, underleaf with sparse urticating hairs on major veins (5 veins); numerous roundish, whitish cystoliths; at each node are 4 stipules (1–2.5 mm), free, lanceolate, toothed and ciliate on edges. Inflorescences spiciform racemes (0.5–2.5 cm), axillary, subsessile, subequal to or shorter than leaf petioles, on which numerous female flowers are placed proximally and a few male flowers distally. Flowers briefly pedicellate, in bud about 1.2 mm; perianth monochlamydeous with a few scattered hairs, the male ones in 4 equal laciniae, ovate, greenish color on the back, yellowish-white and membranous at the margin and apex, connate at the base, at anthesis abruptly open, allowing the 4 longest and most patent stamens to throw far the pollen contained in the small, round, yellowish, bilocular anthers; 4 laciniae of female flowers unequal, inner ones oval, concave, much larger and accrescent, with urticating hair on greenish back and ovary inside. Fruit diclesium ovoid, compressed, greenish (about 0.8–1 mm), enclosed in the laciniae of the accrescent perianth.

Related Species The genus *Urtica* includes about 58 taxa, some of which are known to be used in beer making.

Geographic Distribution Subcosmopolitan. **Habitat** Forest margins, roadsides, synanthropic environments (0–3,000 m ASL). **Range** Subcosmopolitan.

Beer-Making Parts (and Possible Uses) Leaves (substitutive plant for *Urtica dioica*).

Chemistry Leaves (% of dry weight): dry matter 90.03 ± 7.97%, ash 17.81 ±0.32%, carbohydrates 15.87 ± 1.97%, protein 22.5 ±0.14%, lipids 2.79%, total sugars 1.97 ±0.10 mg/mL, carotenoids 0.68 ± 0.03 mg/g, anthocyanins 4.21 ±0.13 mg/g, fiber 63.80 ±4.89 g /100 g fresh weight, calcium 19.7 g/100 g dry matter, phosphorus 143.35 mg/100 g DM, iron 403 mg/100 g DM, manganese 72 mg/100 g DM, magnesium 143 mg/100 g DM, zinc 84 mg/100 g DM, copper 19 mg/100 g DM; antioxidants (5 relative to alcoholic extract): chlorogenic acid 21.34%, myricetin 19.72%, japonic acid 19.22%, p-cumaroyl malate 17.36%, synapic acid 13.75%, p-hydroxybenzoic acid 8.58%.

Style Herb Beer.

Source Fisher and Fisher 2017.

Vaccinium caespitosum Michx.
ERICACEAE

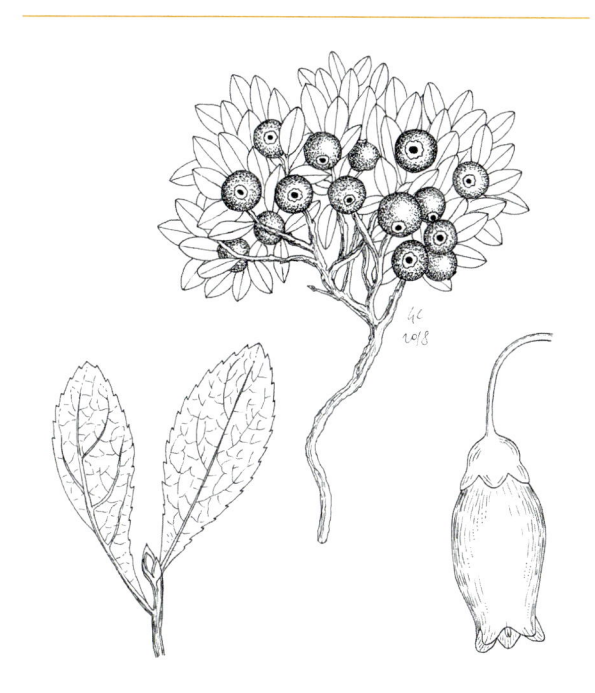

Synonyms *Vaccinium arbuscula* (A.Gray) Merriam, *Vaccinium nivictum* Camp.

Common Names Bilberry, dwarf bilberry, dwarf huckleberry.

Description Shrub (0.3–6 dm) forming dense colonies, superficially rhizomatous. Shoots yellow-green, reddish-green, or reddish-brown, circular or somewhat angular section, finely puberulent or rarely glabrous. Leaf laminae green, usually oblanceolate, sometimes obovate or narrowly elliptic (10–30 × 3–12 mm), margins serrate from apex to at least half of lamina, underleaf usually glandular, upper leaf glabrous. Flower with calyx pale green, vestigial lobes glabrous; corolla white, white with pink stripes, or pink, cylindrical-urceolate to globular (4–7 × 3–5 mm), glabrous; filaments glabrous. Fruits berries, glaucous, blue or rarely dull black (Ø 5–9 mm). Seeds about 1 mm.

Subspecies In addition to the typical variety, *Vaccinium caespitosum* Michx. var. *caespitosum*, *Vaccinium caespitosum* Michx. var. *paludicola* (Camp) Hultén has been recognized.

Related Species Of the 265 taxa known for the genus *Vaccinium*, several species have recognized beer-making applications: *Vaccinium caespitosum* Michx., *Vaccinium corymbosum* L., *Vaccinium* *deliciosum* Piper, *Vaccinium macrocarpon* Aiton, *Vaccinium membranaceum* Douglas ex Torr., *Vaccinium myrtillus* L., *Vaccinium ovalifolium* Sm., *Vaccinium parvifolium* Sm., *Vaccinium uliginosum* L., and *Vaccinium vitis-idaea* L. Given the synonymous ambiguity resulting from the common names, there may be others, including *Vaccinium erythrocarpum* Michx., *Vaccinium microcarpum* (Turcz. ex Rupr.) Schmalh., and *Vaccinium oxycoccos* L.

Geographic Distribution North America (Canada, United States, Mexico), Central America (Guatemala). **Habitat** Open habitats, usually arid, from plains to subalpine plains (0–4,500 m ASL). **Range** North America (Canada, United States, Mexico), Central America (Guatemala).

Beer-Making Parts (and Possible Uses) Fruits.

Chemistry Fairly sweet fruit owing to high concentrations of monosaccharides and disaccharides; lipid content about 3.8%.

Style Fruit Beer.

Beer Draco (Montegioco, Piedmont, Italy).

Source Hieronymus 2016b; https://www.brewersfriend.com/other/bilberries/; https://www.omnipollo.com/beer/90000/.

Vaccinium corymbosum L.
ERICACEAE

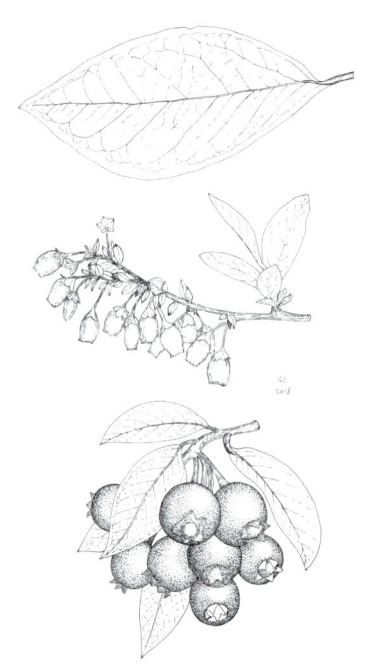

Synonyms *Cyanococcus corymbosus* (L.) Rydb.

Common Names Blueberry, high-bush blueberry.

Description Shrub, deciduous (2–4 m), forms dense colonies. Branches warty, glabrous, yellow-green in color. Leaves alternate, simple, narrow to elliptic or ovate, 3.8–8.2 cm long, pubescent at least on underleaf veins, slightly waxy above, margin ciliate, entire to toothed. Flowers in clusters of 8–10, 6–12 mm long, urn shaped, white, with 5 petals. Fruits berries (Ø 5–12 mm), blue to blue-black in color. Seeds numerous.

Subspecies In addition to the nominal variety, *Vaccinium corymbosum* L. var. *corymbosum*, there is *Vaccinium corymbosum* L. var. *albiflorum* (Hook.) Fernald.

Cultivar Numerous cultivars of this species are grown widely outside North America, including Bluecrop, Bluejay, Blueray, Bonifacy, Bonus, Brigitta Blue, Chandler, Chanticleer, Coville, Croatan, Darrow, Duke, Earliblue, Herbert, Jersey, Nelson, Northland, Patriot, Spartan, and Toro.

Related Species Of the 265 taxa known for the genus *Vaccinium*, several species have recognized beer-making applications: *Vaccinium caespitosum* Michx., *Vaccinium corymbosum* L., *Vaccinium deliciosum* Piper, *Vaccinium macrocarpon* Aiton, *Vaccinium membranaceum* Douglas ex Torr., *Vaccinium myrtillus* L., *Vaccinium ovalifolium* Sm., *Vaccinium parvifolium* Sm., *Vaccinium uliginosum* L., and *Vaccinium vitis-idaea* L. Given the synonymous ambiguity resulting from the common names, there may be others, including *Vaccinium erythrocarpum* Michx., *Vaccinium microcarpum* (Turcz. ex Rupr.) Schmalh., and *Vaccinium oxycoccos* L.

Geographic Distribution E North America.
Habitat Wet environments (marshy areas, sandy banks of ponds and streams, wet plains, birch thickets, barren pine forests, bayheads, heather meadows, upland forests, ravines, mountain peaks). **Range** North America, Europe (United Kingdom, Netherlands), Asia (Japan), New Zealand.

Beer-Making Parts (and Possible Uses) Fruits.

Chemistry Fresh fruits: water 83.4 g, protein 0.6 g, lipids 0.6 g, sugars 15 g, ash 0.3 g, vitamin B1 0.02 mg, vitamin B2 0.02 mg, niacin (vitamin PP) 0.3 mg, vitamin C 16 mg, vitamin A 289 IU, calcium 16 mg, phosphorus 13 mg, iron 0.8 mg.

STYLE Fruit Beer.

BEER Flora, Hill Farmstead (Vermont, United States).

SOURCE Heilshorn 2017; Hieronymus 2016b; Josephson et al. 2016; Steele 2012; https://www.oregonfruit.com/fruit-brewing/category/products.

Vaccinium deliciosum Piper
ERICACEAE

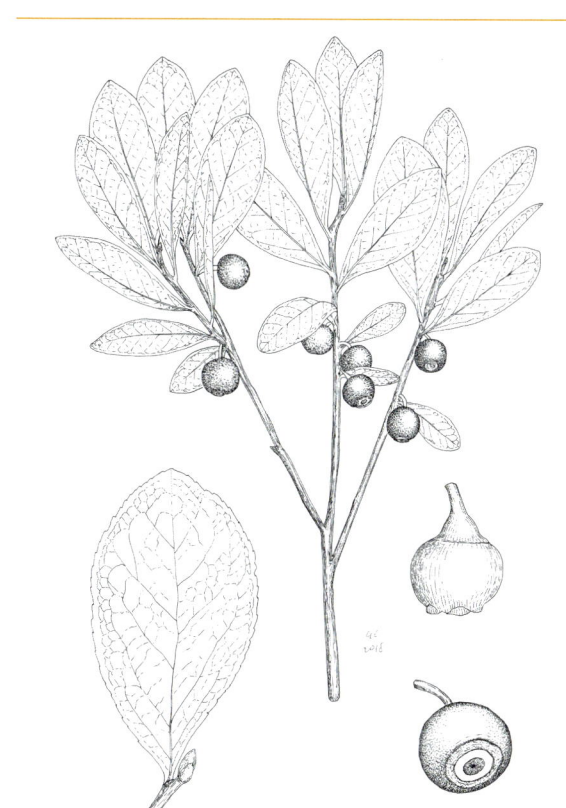

SYNONYMS None

COMMON NAMES Bilberry, cascade bilberry, cascade blueberry, blueleaf huckleberry.

DESCRIPTION Shrub (0.5–15 dm), deciduous, rhizomatous, forming small nuclei to extensive colonies. Young stems glaucous green in color, ± circular in section, rarely angled, usually glabrous, rarely hairy along veins or puberulent. Leaf laminae usually glabrous, obovate, oblanceolate, or rarely elliptic (17–35 × 9–17 mm), margins serrate for at least 2/3 distal, surfaces glabrous, needle-like or rarely glandular throughout, often hairy-glandular along the midrib. Flower calyx, glaucous, lobes indistinct or shallow, glabrous; corolla pink, creamy pink, or red, globular to globular-urceolate (4–6 × 5–7 mm); filaments glabrous. Fruits berries, blue, glaucous, sometimes dull black, brown, or red (Ø 9–13 mm). Seeds about 1 mm.

RELATED SPECIES Of the 265 taxa known for the genus *Vaccinium*, several species have recognized beer-making applications: *Vaccinium caespitosum* Michx., *Vaccinium corymbosum* L., *Vaccinium deliciosum* Piper, *Vaccinium macrocarpon* Aiton, *Vaccinium membranaceum* Douglas ex Torr., *Vaccinium myrtillus* L., *Vaccinium ovalifolium* Sm., *Vaccinium parvifolium* Sm., *Vaccinium uliginosum* L., and *Vaccinium vitis-idaea* L. Given the synonymous ambiguity resulting from the common names, there may be others, including *Vaccinium erythrocarpum* Michx., *Vaccinium microcarpum* (Turcz. ex Rupr.) Schmalh., and *Vaccinium oxycoccos* L.

GEOGRAPHIC DISTRIBUTION Canada (British Columbia), United States (California, Idaho, Oregon, Washington). **HABITAT** Alpine meadows, subalpine coniferous forests, breccias (600–2,000 m ASL). **RANGE** Canada (British Columbia), United States (California, Idaho, Oregon, Washington).

BEER-MAKING PARTS (AND POSSIBLE USES) Fruits.

CHEMISTRY Fruits (bilberry juice): energy 50 kcal, protein 0.20 g, carbohydrates 12.4 g, fiber 1 g, sugar 8.28 g, sodium 3 mg, vitamin C 1.2 mg, vitamin A 120 IU.

STYLE American IPA, American Pale Ale, American Stout, Blonde Ale, Foreign Extra Stout, Fruit Beer, Fruit Lambic, Porter, Russian Imperial Stout.

BEER Mathias Dahlgren Matölen, Situna Brygghus (Arlandastad, Sweden).

SOURCE Hieronymus 2016b; https://www.brewersfriend.com/other/bilberries/; https://www.brewersfriend.com/other/huckleberrys/; https://www.omnipollo.com/beer/90000/.

Vaccinium macrocarpon Aiton
ERICACEAE

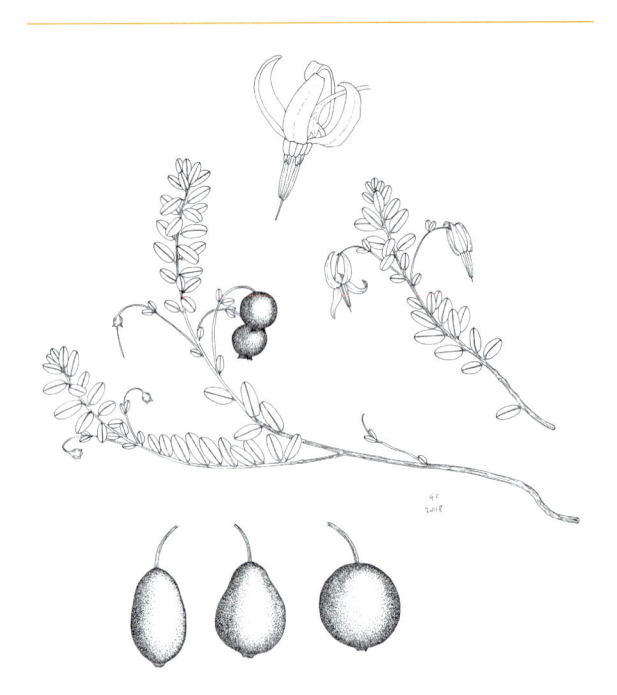

Synonyms *Oxycoca macrocarpa* (Aiton) Raf., *Oxycoccus macrocarpus* (Aiton) Pers., *Schollera macrocarpa* (Aiton) Steud., *Schollera macrocarpos* (Aiton) Britton.

Common Names Cranberry, american cranberry, canneberge.

Description Shrub (0.4–1.5 dm), often ascending. Leaves with laminae glaucous below, green above, usually narrowly elliptic to elliptic, rarely oblong, 2–55 mm, margin entire, slightly revolute. Inflorescence at the axils of leaf-like bracts at the base of the year's shoots. Peduncle with nodes, thin, 2–3 cm, bearing bracts; bracts 2, greenish-white, scale-like, 1–2 mm wide. Flower with relatively small calyx lobes; corolla strongly reflexed at anthesis, white to pink in color; filaments villous; anther tubule 1–2 mm. Fruits subspherical to pyriform berries (Ø 9–14 mm), smooth, color red to pink.

Cultivar There are many, including 6, 20, 35, 41, AA4, AJ, AR2, AW2, Bain 1, Bain 2, Bain 3, Bain 4, Bain 5, Bain 6, Bain 7, Bain 8, Bain 9, Bain 10, Bain 11, Bain Favorite #1, Bain Favorite #2, Bain McFarlin, BD, BE4, Beckwith, Ben Lear, Bergman, Biron Selection, Centennial, Centerville, CN, Cropper, Crowley, DF5, Drever, Early Black, Early Richard, Eastern Variety, FN Searles, Franklin, Foxboro Howes, Gebhardt Beauty, HA, Habelman, Habelman #2, Hollistar Red, Holliston, Howes, Matthews, McFarlin, Middleboro, New Jersey 10, Norman Le Munyon, Paradise Meadow, Pilgrim, Prolific, Rezin, Rezin McFarlin, Round Howes, Searles, Stankovich, Stanley, Stevens, Thunder Lake 3, Thunder Lake 4, Wales Henry, Wilcox, WSU 61, WSU 77, and WSU 108.

Related Species Of the 265 taxa known for the genus *Vaccinium*, several species have recognized beer-making applications: *Vaccinium caespitosum* Michx., *Vaccinium corymbosum* L., *Vaccinium deliciosum* Piper, *Vaccinium macrocarpon* Aiton, *Vaccinium membranaceum* Douglas ex Torr., *Vaccinium myrtillus* L., *Vaccinium ovalifolium* Sm., *Vaccinium parvifolium* Sm., *Vaccinium uliginosum* L., and *Vaccinium vitis-idaea* L. Given the synonymous ambiguity resulting from the common names, there may be others, including *Vaccinium erythrocarpum* Michx., *Vaccinium microcarpum* (Turcz. ex Rupr.) Schmalh., and *Vaccinium oxycoccos* L.

Geographic Distribution E North America. **Habitat** Wet environments (swamps, marshes, wet shorelines, headlands), 0–1,400 m ASL. **Range** E North America, W North America (British Columbia, Oregon, Washington), Europe.

Beer-Making Parts (and Possible Uses) Fresh or dried fruits (craisins).

Chemistry Fruits: energy 46 kcal, water 87.32 g, lipids 0.1 g (polyunsaturated fatty acids 0.1 g), carbohydrates 12 g (fiber 4.6 g, sugars 4 g), protein 0.4 g, sodium 2 mg, potassium 85 mg, calcium 8 mg, magnesium 6 mg, iron 0.3 mg, vitamin A 60 IU, vitamin C 13.3 mg, vitamin B6 0.1 mg, thiamine 0.012 mg, riboflavin 0.020 mg, vitamin E 1.2 mg, catechin traces, myrcetin 4.33 mg, quercetin 14.02 mg.

Style American Amber Ale, American Barleywine, American Pale Ale, American Stout, American Wheat or Rye Beer, Ancient Ale, Autumn Seasonal Beer, Baltic Porter, Belgian Dark Strong Ale, Belgian Pale Ale, Berliner Weisse, Braggot, Brown Porter, California Common Beer, Cream Ale, English Barleywine, Extra Special/Strong Bitter (ESB), Fruit Beer, Fruit Lambic, Imperial Stout, Irish Red Ale, Kölsch, Metheglin, Northern English Brown, Old Ale, Russian Imperial Stout, Saison, Scottish Export 80/-, Southern English Brown, Specialty Beer, Specialty

Fruit Beer, Spice/Herb/Vegetable Beer, Strong Scotch Ale, Weissbier, Weizen/Weissbier, Witbier.

Beer Serendipity, New Glarus (Wisconsin, United States).

Source Cantwell and Bouckaert 2016; Cantwell and Bouckaert 2018; Heilshorn 2017; Hieronymus 2016b; Jackson 1991; McGovern 2017; https://www.brewersfriend.com/other/canneberge/; https://www.brewersfriend.com/other/cranberries/; https://fr.wikipedia.org/wiki/Canneberge; https://www.omni-pollo.com/beer/en-el-bosque/; https://www.oregon-fruit.com/fruit-brewing/category/products.

Vaccinium membranaceum Douglas ex Torr.
ERICACEAE

Synonyms None

Common Names Thinleaf huckleberry, tall huckleberry, big huckleberry, mountain huckleberry, squaretwig blueberry, black huckleberry.

Description Shrub, perennial (2–30 dm), nonrhizomatous in modest to extensive formations. Year stems yellow-green or reddish-green in color, circular in cross-section or lightly angled, glabrous or hairy in line. Leaves green, lamina broadly elliptic to ovate (25–50 × 11–23 mm), margins serrate, underside glandular. Flower calyx green, obscurely lobed, glabrous; corolla white, cream, yellowish-pink, or bronze, globose to urceolate (3–5 × 5–7 mm), glaucous; filaments glabrous. Fruit berries shiny, dull black, or dark purple, rarely red or white (Ø 9–13 mm). Seeds about 1 mm.

Subspecies *Vaccinium membranaceum* Douglas ex Torr. var. *membranaceum* and *Vaccinium membranaceum* Douglas ex Torr. var. *rigidum* (Hook.) Fernald are two varieties of this taxon.

Related Species Of the 265 taxa known for the genus *Vaccinium*, several species have recognized beer-making applications: *Vaccinium caespitosum* Michx., *Vaccinium corymbosum* L., *Vaccinium deliciosum* Piper, *Vaccinium macrocarpon* Aiton, *Vaccinium membranaceum* Douglas ex Torr., *Vaccinium myrtillus* L., *Vaccinium ovalifolium* Sm., *Vaccinium parvifolium* Sm., *Vaccinium uliginosum* L., and *Vaccinium vitis-idaea* L. Given the synonymous ambiguity resulting from the common names, there may be others, including *Vaccinium erythrocarpum* Michx., *Vaccinium microcarpum* (Turcz. ex Rupr.) Schmalh., and *Vaccinium oxycoccos* L.

Geographic Distribution E North America (United States, Canada). **Habitat** Coniferous forests, especially those periodically cut, rubble, subalpine fir forests, alpine heathlands (900–3,500 m ASL). **Range** E North America (United States, Canada).

Beer-Making Parts (and Possible Uses) Fruits.

Chemistry Generic composition of fruits of species of the genus *Vaccinium*: energy 622 kcal, water 83.2 g, carbohydrates 15.3 g, lipids 0.5 g, protein 0.7 g, anthocyanins 110–153 mg.

Style Fruit Beer.

Source Hieronymus 2016b; https://www.brewersfriend.com/other/bilberries/; https://www.omnipollo.com/beer/90000/.

Vaccinium myrtillus L.
ERICACEAE

Synonyms *Vaccinium oreophilum* Rydb., *Vitis-idaea myrtillus* (L.) Moench.

Common Names Bilberry, whortleberry.

Description Shrub, deciduous (1–4 dm). Elongated underground stem with reddish bark; erect branches ± twisted, green, angular or narrowly winged. Leaves deciduous with 1 mm petioles, lamina ovate or elliptic (10–15 × 20–26 mm), acute, serrated at edge, green on both sides, without glands. Flowers isolated at leaf axils, pendulous; petioles 4–7 mm, usually reddened; corolla urceolate (4 × 5 mm), greenish and tinged with pinkish-wine color. Fruit berry, subspherical (Ø 4–6 mm), bluish, pruinose.

Related Species Of the 265 taxa known for the genus *Vaccinium*, several species have recognized beer-making applications: *Vaccinium caespitosum* Michx., *Vaccinium corymbosum* L., *Vaccinium deliciosum* Piper, *Vaccinium macrocarpon* Aiton, *Vaccinium membranaceum* Douglas ex Torr., *Vaccinium myrtillus* L., *Vaccinium ovalifolium* Sm., *Vaccinium parvifolium* Sm., *Vaccinium uliginosum* L., and *Vaccinium vitis-idaea* L. Given the synonymous ambiguity resulting from the common names, there may be others, including *Vaccinium erythrocarpum* Michx., *Vaccinium microcarpum* (Turcz. ex Rupr.) Schmalh., and *Vaccinium oxycoccos* L.

Geographic Distribution Circumboreal. **Habitat** Woods, heaths, shrublands, pastures always in humid acidic soil. **Range** Circumboreal.

Beer-Making Parts (and Possible Uses) Fruits.

Chemistry Fruit juice (100 ml): energy 50 kcal, protein 0.2 g, carbohydrates 12.4 g, fiber 1 g, sugars 8.28 g, sodium 3 mg, vitamin C 1.2 mg, vitamin A 120 IU, anthocyanins 358 mg/100 g, phenols 603 mg/100 g.

Style Fruit Beer.

Source Dabove 2015; https://www.birramoretti.it/le-birre-di-casa/birra-moretti-alla-piemontese/; https://www.bjcp.org (29A. Fruit Beer); https://www.omnipollo.com/beer/90000/.

Vaccinium ovalifolium Sm.
ERICACEAE

Synonyms None

Common Names Early blueberry, oval-leaf huckleberry, bilberry.

Description Shrub (3–40 dm), nonrhizomatous, compact, suckering when disturbed, rarely forming extensive colonies. Young stems yellow-green or golden-brown, glabrous, usually circular or slightly angled in section, glabrous, sometimes hairy on lines. Leaf laminae light green or glabrous below, slightly darker above, ovate to elliptic, rarely obovate (25–39 × 16–20 mm), margins entire to obscurely serrate, lower surface glabrous, not glandular (sometimes hairy or glandular along midrib), upper surface usually glabrous (sometimes hairy and/or glandular). Flower calyx light green or glaucous, lobes vestigial or absent, glabrous; corolla pink, bronze-pink, or greenish-white, globose, sometimes urceolate (5–7 × 4–5 mm), glabrous; filaments glabrous or basally hairy. Fruits berries, blue, dull purple-black, or black, sometimes glaucous (Ø 8–10 mm). Seeds about 1 mm.

Related Species Of the 265 taxa known for the genus *Vaccinium*, several species have recognized beer-making applications: *Vaccinium caespitosum* Michx., *Vaccinium corymbosum* L., *Vaccinium deliciosum* Piper, *Vaccinium macrocarpon* Aiton, *Vaccinium membranaceum* Douglas ex Torr., *Vaccinium myrtillus* L., *Vaccinium ovalifolium* Sm., *Vaccinium parvifolium* Sm., *Vaccinium uliginosum* L., and

Vaccinium myrtillus L.
ERICACEAE

Vaccinium vitis-idaea L. Given the synonymous ambiguity resulting from the common names, there may be others, including *Vaccinium erythrocarpum* Michx., *Vaccinium microcarpum* (Turcz. ex Rupr.) Schmalh., and *Vaccinium oxycoccos* L.

GEOGRAPHIC DISTRIBUTION North America (N United States, Canada), Asia (Japan, Kamchatka). **HABITAT** Wet, mesic coniferous forests, transitional habitats adjacent to these forests, repeatedly cut coniferous forests, roadsides, edges of coniferous forests, peaty slopes (0–2,100 m ASL). **RANGE** North America (N United States, Canada), Asia (Japan, Kamchatka).

BEER-MAKING PARTS (AND POSSIBLE USES) Fruits.

CHEMISTRY Fresh fruit composition (referring generically to "huckleberries"): energy 37 kcal, water 90.7 g, protein 0.4 g, lipids 0.1 g, carbohydrates 8.7 g, calcium 15 mg, iron 0.3 mg, sodium 10 mg, vitamin C 2.8 mg, thiamine 0.01 mg, riboflavin 0.03 mg, niacin 0.3 mg, vitamin A 79 IU.

STYLE Fruit Beer.

SOURCE Hieronymus 2016b; https://www.brewersfriend.com/other/bilberries/; https://www.omnipollo.com/beer/90000/.

Vaccinium parvifolium Sm.
ERICACEAE

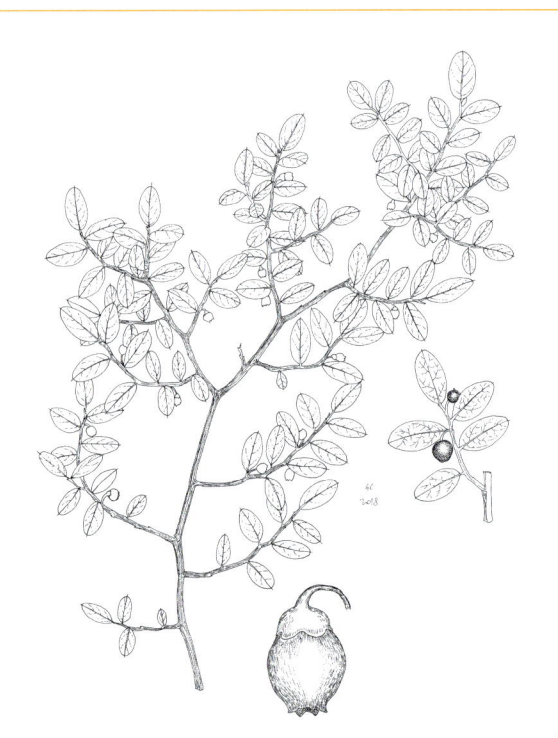

SYNONYMS None

COMMON NAMES Red huckleberry.

DESCRIPTION Shrub (10–70 dm), nonrhizomatous, canopy forming, sometimes suckering if damaged. Season stems green, strongly angled section, glabrous or minutely puberulent in lines; lateral branches often up to 75° apart. Leaf laminae dark green, ovate to oblong-elliptic (13–25 × 8–14 mm), margins entire, surface puberulent or glabrous below, glabrous above. Flower calyx light green, lobes open, distinct, broadly ovate (0.4–0.6 mm), glabrous; corolla pink, bronze, or yellowish-green, globose to urceolate (4–6 × 3–5 mm), glabrous; filaments glabrous. Fruit berry red, sometimes faintly glaucous, translucent (Ø 7–10 mm). Seeds about 1 mm.

RELATED SPECIES Of the 265 taxa known for the genus *Vaccinium*, several species have recognized beer-making applications: *Vaccinium caespitosum* Michx., *Vaccinium corymbosum* L., *Vaccinium deliciosum* Piper, *Vaccinium macrocarpon* Aiton, *Vaccinium membranaceum* Douglas ex Torr., *Vaccinium myrtillus* L., *Vaccinium ovalifolium* Sm., *Vaccinium parvifolium* Sm., *Vaccinium uliginosum* L., and *Vaccinium vitis-idaea* L. Given the synonymous ambiguity resulting from the common names, there may be others, including *Vaccinium erythrocarpum* Michx., *Vaccinium microcarpum* (Turcz. ex Rupr.) Schmalh., and *Vaccinium oxycoccos* L.

GEOGRAPHIC DISTRIBUTION W North America (Canada [British Columbia], United States [Alaska, California, Oregon, Washington]). **HABITAT** Coniferous forests, often on logs and trunks, disturbed areas (0–1,100 m ASL). **RANGE** W North America (Canada [British Columbia], United States [Alaska, California, Oregon, Washington]).

BEER-MAKING PARTS (AND POSSIBLE USES) Fruits.

CHEMISTRY Fresh fruit composition (referring generically to "huckleberries"): energy 37 kcal, water 90.7 g, protein 0.4 g, lipids 0.1 g, carbohydrates 8.7 g, calcium 15 mg, iron 0.3 mg, sodium 10 mg, vitamin C 2.8 mg, thiamine 0.01 mg, riboflavin 0.03 mg, niacin 0.3 mg, vitamin A 79 IU.

STYLE American IPA, American Pale Ale, American Stout, Blonde Ale, Foreign Extra Stout, Fruit Beer, Fruit Lambic, Russian Imperial Stout.

SOURCE Hieronymus 2016b; https://www.brewersfriend.com/other/huckleberrys/.

Vaccinium uliginosum L.
ERICACEAE

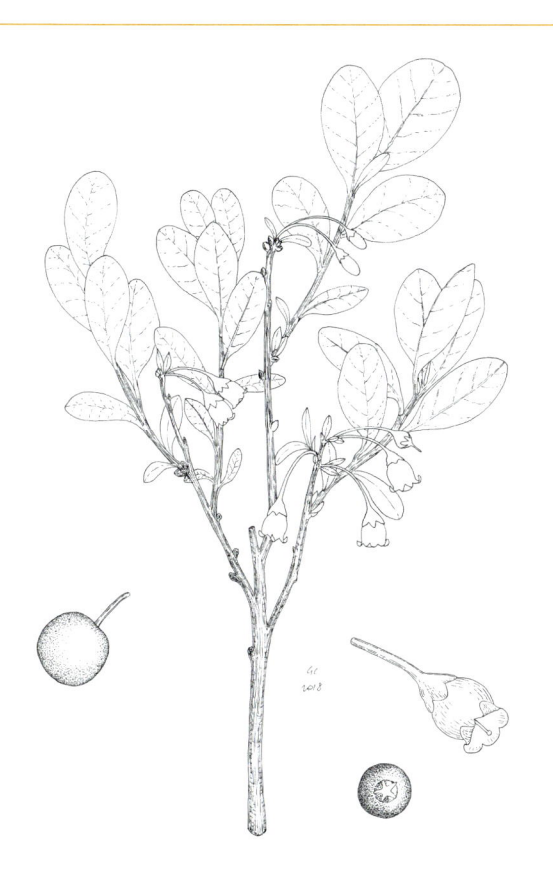

Synonyms *Myrtillus uliginosus* (L.) Drejer, *Vaccinium gaultherioides* Bigelow, *Vaccinium occidentale* A.Gray, *Vaccinium pedris* Holub, *Vaccinium pubescens* Wormsk. ex Hornem.

Common Names Bilberry, bog bilberry, bog blueberry, northern bilberry, western blueberry, bog whortleberry.

Description Shrub, deciduous, rhizomatous, heavily branched (1–10 dm). Young stems circular in section, puberulent to glabrous. Leaves scattered; petiole about 2 mm, puberulent; lamina obovate to elliptic to oblong (1–3 × 0.6–1.5 cm), papery, underleaf glabrous, puberulent, upper subglabrous, secondary veins 3–5 pairs, fine veins evident especially below, base cuneate or broadly cuneate, margin flat, entire, with 1 basal gland per side, apex rounded, sometimes retuse. Inflorescence fasciculate, terminal, 1- to 3-flowered; bracts caducous (1.5–2.5 mm). Flowers 4- to 5-merous; peduncle about 5 mm, glabrous; hypanthium about 0.8 mm, glabrous; calyx lobes 4–5, triangular-ovate, about 1 mm; corolla greenish-white, broadly urceolate (about 5 mm), glabrous; lobes triangular (about 1 mm); filaments about 1 mm, glabrous; anthers 1.5 mm. Fruits berries 4- to 5-loculate, bluish-purple, subglobose or ellipsoidal (Ø about 1 cm).

Subspecies In addition to the typical subspecies, *Vaccinium uliginosum* L. subsp. *uliginosum*, some authors frame intraspecific variability in entities of different rank; for example, *Vaccinium uliginosum* L. f. *langeanum* (Malte) Polunin, *Vaccinium uliginosum* L. subsp. *microphyllum* (Lange) Hultén, and *Vaccinium uliginosum* L. var. *salicinum* (Cham. & Schltdl.) Hultén.

Related Species Of the 265 taxa known for the genus *Vaccinium*, several species have recognized beer-making applications: *Vaccinium caespitosum* Michx., *Vaccinium corymbosum* L., *Vaccinium deliciosum* Piper, *Vaccinium macrocarpon* Aiton, *Vaccinium membranaceum* Douglas ex Torr., *Vaccinium myrtillus* L., *Vaccinium ovalifolium* Sm., *Vaccinium parvifolium* Sm., *Vaccinium uliginosum* L., and *Vaccinium vitis-idaea* L. Given the synonymous ambiguity descending from the common names, there may be others, including *Vaccinium erythrocarpum* Michx., *Vaccinium microcarpum* (Turcz. ex Rupr.) Schmalh., and *Vaccinium oxycoccos* L.

Geographic Distribution Circumboreal. **Habitat** Pastures, subalpine heaths, shrublands (acidophilous). **Range** Circumboreal.

Beer-Making Parts (and Possible Uses) Fruits.

Chemistry Fruit composition (referred generically to "blueberries"): energy 57 kcal, water 84.21 g, protein 0.74 g, lipids 0.33 g, carbohydrates 14.49 g, fiber 2.4 g, sugars 9.96 g, calcium 6 mg, iron 0.28 mg, magnesium 6 mg, phosphorus 12 mg, potassium 77 mg, sodium 1 mg, zinc 0.16 mg, vitamin C 9.7 mg, thiamine 0.037 mg, riboflavin 0.041 mg, niacin 0.418 mg, vitamin B6 0.052 mg, folate 6 µg, vitamin A 3 µg, vitamin A 54 IU, vitamin E 0.57 mg, vitamin K 19.3 µg.

Style Fruit Beer.

Source https://www.brewersfriend.com/other/bilberries/; https://www.omnipollo.com/beer/90000/.

Vaccinium vitis-idaea L.
ERICACEAE

Synonyms *Rhodococcum vitis-idaea* Avrorin, *Vaccinium jesoense* Miq., *Vitis-idaea punctata* Moench.

Common Names Lingonberry, cowberry, partridgeberry.

Description Shrub (8–10 cm), perennial, densely colonial. Semiwoody suckering stems (perennial stems) and creeping stems (stolons) developing horizontally (up to 2 m), in marshy areas forming a dense network of shoots and roots; previous year's branches circular in section, puberulent, not warty. Leaf laminae pale green, glandular below, bright green above, elliptic to obovate (5–18 × 3–9 mm), glabrous, leathery, margins entire, slightly revolute. Flower corolla pink-white (3–5 mm); filaments puberulent; peduncle 4–6 mm. Fruit berry red in color (Ø 8–10 mm).

Subspecies Besides the nominal subspecies, *Vaccinium vitis-idaea* L. subsp. *vitis-idaea*, a second subspecies is recognized: *Vaccinium vitis-idaea* L. subsp. *minus* (Lodd., G.Lodd. & W.Lodd.) Hultén.

Cultivar Erntedank, Regal, and Splendor are just a few of the cultivars of this species.

Related Species Of the 265 taxa known for the genus *Vaccinium*, several species have recognized beer-making applications: *Vaccinium caespitosum* Michx., *Vaccinium corymbosum* L., *Vaccinium deliciosum* Piper, *Vaccinium macrocarpon* Aiton, *Vaccinium membranaceum* Douglas ex Torr., *Vaccinium myrtillus* L., *Vaccinium ovalifolium* Sm., *Vaccinium parvifolium* Sm., *Vaccinium uliginosum* L., and *Vaccinium vitis-idaea* L. Given the synonymous ambiguity resulting from the common names, there may be others, including *Vaccinium erythrocarpum* Michx., *Vaccinium microcarpum* (Turcz. ex Rupr.) Schmalh., and *Vaccinium oxycoccos* L.

Geographic Distribution Circumboreal. **Habitat** Taiga (boreal forests) in stands of jack pine (*Pinus banksiana* Lamb.), muskegs, elevated marshes, arid and rocky environments, lichen-covered forests, exposed habitats, heaths, mountain heaths, headlands, tundra, cliffs, mountain peaks (0–1,800 m ASL). **Range** Circumboreal.

Beer-Making Parts (and Possible Uses) Fruits.

Chemistry Fruits (phenolic compounds): proanthocyanidin B, catechin, cyanidin-3-galactoside, epicatechin, cyanidin-3-glucoside, cyanidin-3-arabinoside, proanthocyanidin A, ferulic acid-exoxide, 2″-caffeoyl-arbutin, coumaroyl-hexose-hydroxyphenol, caffeoyl-hexose-hydroxyphenol, quercetin-3-O-β-galactoside, coumaroyl-hexose-hydroxyphenol, quercetin-3-0-β-glucoside, quercetin-0-(hexose-deoxyhexoside), quercetin-3-0-β-xyloside, quercetin-3-O-α-arabinoside, quercetin-3-O-α-arabinofuranoside, quercetin-3-O-α-rhamnoside (quercitrin), kaempferol-pentoside, kaempferol-deoxyhexoside, quercetin-3-0-(4″-HMG)-α-rhamnoside, kaempferol-(HMG) rhamnoside.

Style American Brown Ale, Ancient Ale, Belgian Dubbel, California Common Beer, Fruit Beer, Mixed-Style Beer, Weissbier.

Beer Spontanlingonberry, Mikkeller (Copenhagen, Denmark).

Source McGovern 2017; https://www.brewersfriend.com/other/cowberries/.

Valeriana officinalis L.
CAPRIFOLIACEAE

Synonyms *Valeriana alternifolia* Bunge, *Valeriana baltica* Pleijel, *Valeriana chinensis* Kreyer ex Kom., *Valeriana dubia* Bunge, *Valeriana exaltata* J.C. Mikan, *Valeriana leiocarpa* Kitag., *Valeriana nipponica* Nakai ex Kitag., *Valeriana stubendorfii* Kreyer ex Kom., *Valeriana subbipinnatifolia* A.I.Baranov, *Valeriana tianschanica* Kreyer ex Hand.-Mazz.

Common Names All-heal.

Description Herb, perennial (30–150 cm). Rhizome short, ovoid, brownish color. Stems stout, erect, hollow, glabrous, cylindrical, fluted, sparsely branched, leafy. Leaves (6–13 opposite pairs) compound, imparipinnate, leaflets (6–7 pairs similar, with marked teeth) lanceolate, terminal leaflet larger than lateral leaflets; lower leaves petiolate, upper leaves sessile and less divided, hairiness scattered by ± appressed hairs along underleaf veins. Inflorescence apical corymb, dense, divided into 3 branches. Flowers slightly fragrant, hermaphrodite, asymmetrical; calyx divided into 5 teeth; corolla tubular, pinkish, rarely white, divided into 5 unequal lobes; stamens 3, pistil with inferior 3-carpellar ovary. Fruits ovoid achenes, striate, with feathery bristles resulting from modification of calyx teeth at maturity, enveloped in desiccated calyx.

Subspecies Two subspecific taxa: *Valeriana officinalis* L. subsp. *collina* (Wallr.) Nyman and *Valeriana officinalis* L. subsp. *officinalis*.

Cultivar Available cultivars include Anthose, Anton, Mehrfahrig, and Select.

Related Species The genus *Valeriana* has more than 300 taxa.

Geographic Distribution E Europe. **Habitat** Wet meadows, bushes, riverbanks, from plains to the mountains up to 1,400 m ASL. **Range** E Europe.

Beer-Making Parts (and Possible Uses) Leaves, flowers, roots.

Chemistry Essential oil: bornyl acetate 24.59%, valerenal 11.85%, camphene 10.10%, α-pinene 4.52%, cedrandiol 4.37%, spathulenol 4.32%, phenchene 3.84%, (-)-spatulenol 3.49%, myrtenyl acetate 2.49%, α-gurjunene 2.15%, β-pinene 1.97%, bicyclogermacrene 1.60%, α-eudesmol 1.19%, isovalerenic acid 0.70%, Z-valerenyl acetate 0.56%, zingiberene 0.07%, zingiberene 0.09%, γ-terpinene 0.07%, γ-elemene 0.24%, veridifloral 0.73%, veridifloral 0.36%, valeric acid, 3-methyl, ethyl ester 0.11%, valerenyl hexanoate 0.75%, valencene 0.12%, tricyclicene 0.14%, thymyl methyl ether 0.11%, thymohydroguinone dimethyl ether 0.39%, terpinen-4-ol 0.18%, sabinene 0.11%, β-ionone 0.55%, β-eudesmol 0.21%, β-elemol 0.15%, β-elemene 0.08%, β-caryophyllene 1.06%, patchouli alcohol 1.13%, o-cymene 0.10%, nerolidol-epoxycetate 2.38%, nepetalactol 0.32%, naphthalene 0.17%, myrtenol 0.23%, methyl linolenate 0.35%, linalyl isobutyrate 0.20%, limonene 1.04%, ledol 0.43%, isovaleric acid, isobutyl ester 0.03%, isopathulenol 0.62%, isoamylvalerate 1.29%, hexyl isovalerate 0.12%, globulol 1.01%, germacrene-D 0.60%, farnesene epoxide 0.07%, E-valerenyl isovalerate 0.19%, eudesma-4(14), 11-diene 0.29%, epicubenol 0.27%, endobornyl acetate 0.82%, cubenol 0.18%, cis-limonene oxide 0.11%, caryophyllene oxide 0.71%, caryophyllene 0.13%, carvacrol methyl ether 0.02%, bornyl isovalerate 0.16%, α-terpinolene 0.04%, α-terpinenyl acetate 0.22%, α-terpinene 0.02%, aromadendrene 0.86%, aristolene 0.06%, α-phellandrene 0.10%, α-humulene 0.11%, α-caryophyllene 0.02%, α-bisabolol 0.54%, δ-verbenone 0.20%, δ-nerolidole 0.01%, (+)-cycloisosativene

0.08%, (-)-myrtenyl acetate 0.59%, (-)-borneol 0.45%, (-)-aristolene 0.34%.

Style None known.

Source Fisher and Fisher 2017; Hieronymus 2016b.

Vanilla planifolia Jacks. ex Andrews
ORCHIDACEAE

Synonyms *Epidendrum rubrum* Lam., *Notylia planifolia* (Jacks. ex Andrews) Conz., *Notylia sativa* (Schiede) Conz., *Vanilla bampsiana* Geerinck, *Vanilla duckei* Huber, *Vanilla rubra* (Lam.) Urb., *Vanilla sativa* Schiede, *Vanilla sylvestris* Schiede, *Vanilla viridiflora* Blume.

Common Names Vanilla, flat-leaved vanilla, Tahitian vanilla, West Indian vanilla.

Description Epiphytic climbing plant (over 30 m). Stems branched, thick, fleshy, circular in section. Leaves fleshy, elliptic-oblong or ovate-elliptic, acute to subacuminate, gradually narrowed at the base, arranged at regular intervals along the entire stem. Inflorescence 5–7 cm long, axillary, subsessile, racemose, containing up to 20 flowers, each with a sharp, ovate-triangular bract at the base. Flowers (Ø 5 cm) greenish-yellow in color, fleshy, fragrant, opening in succession (present all year if grown for the fruit that produces vanilla essence), short-lived (they last only 1 day and are pollinated by insects in the wild but must be pollinated manually during the morning if the fruit is to be obtained; plants are self-fertile, with pollination requiring simply the transfer of pollen from the anther to the stigma; if the flower is not pollinated it falls off the next day; in nature there is less than a 1% chance that the flower will be pollinated). The fruits, produced only by mature plants, usually greater than 3 m, are long capsules (15–23 cm) from which vanilla essence is obtained; they are treated (fermented and dried) and contain thousands of small black seeds.

Cultivar In its area of origin, *Vanilla planifolia* has a great variety of aromatic expressions based on the variation of four phenolic compounds (vanillin, vanillic acid, p-hydroxybenzaldehyde, p-hydroxybenzoic acid).

Related Species The genus *Vanilla* includes about 108 taxa, but beer-making applications are known only for *Vanilla planifolia*.

Geographic Distribution Mexico, C America. **Habitat** Tropical forests. **Range** C America, Madagascar.

Beer-Making Parts (and Possible Uses) Fruits, seeds.

Chemistry Fruits (active compounds): vanillin, vanillic acid, 4-hydroxybenzaldehyde, 4-hydroxybenzoic acid.

Style American Amber Ale, American Barleywine, American Brown Ale, American IPA, American Light Lager, American Pale Ale, American Porter, American Stout, American Wheat Beer, American Wheat or Rye Beer, Autumn Seasonal Beer, Baltic Porter, Belgian Dark Strong Ale, Belgian Dubbel, Belgian Golden Strong Ale, Belgian Pale Ale, Belgian Specialty Ale, Belgian Tripel, Blonde Ale, Braggot, Brown Porter, California Common Beer, Cream Ale, Dark American Lager, Dark Mild, Doppelbock, Dry Stout, Dunkelweizen, English Barleywine, English Porter, Experimental Beer, Extra Special/Strong Bitter (ESB), Foreign Extra Stout, Fruit and Spice Beer, Fruit Beer, Fruit Lambic, Holiday/Winter Special Spiced Beer, Imperial IPA, Imperial Stout, Irish Red Ale,

Irish Stout, Kölsch, Metheglin, Mild, Munich Dunkel, North German Altbier, Northern English Brown, Oatmeal Stout, Oktoberfest/Märzen, Old Ale, Other Smoked Beer, Premium American Lager, Robust Porter, Russian Imperial Stout, Saison, Scottish Export, Scottish Light 60/-, Southern English Brown, Specialty Beer, Specialty Wood-aged Beer, Spice/Herb/Vegetable Beer, Strong Scotch Ale, Sweet Stout, Traditional Bock, Vienna Lager, Weissbier, Weizenbock, Winter Seasonal Beer, Witbier, Wood-aged Beer.

Beer AphroRhum, Dieu Du Ciel (Quebec, Canada).

Source Calagione et al. 2018; Giaccone and Signoroni 2017; https://www.birraaltaquota.it/prodotti/birre-estrose/62-croco.html; https://www.brewersfriend.com/other/vainilla/; https://www.brewersfriend.com/other/vaniljestang/; https://www.brewersfriend.com/other/vanilla/; https://www.fermento-birra.com/homebrewing/birre-speziali/; https://www.omnipollo.com/beer/90000/; https://www.omnipollo.com/beer/symzonia/.

Verbena officinalis L.
VERBENACEAE

Synonyms *Verbena adulterina* Hausskn., *Verbena domingensis* Urb., *Verbena macrostachya* F.Muell., *Verbena riparia* Raf. ex Small & A.Heller, *Verbena rumelica* Velen., *Verbena spuria* L., *Verbena vulgaris* Bubani, *Vitex × adulterina* Hausskn.

Common Names Common vervain, common verbena.

Description Herb, perennial (15–60 cm). Stems woody only at the base, ascending, 4-angular-sided, pubescent on the edges. Leaves spatulate (1–2 × 3–5 cm), lobed to pinnate, the upper ones reduced and ± entire; veins protruding below. Inflorescence spike variable in length (3–6 cm at flowering, 10–25 cm at fruiting) with 1–2 pairs of arcuate basal branches. Flowers with lanceolate-acuminate bracts; calyx 1.5 mm; corolla 4 mm, slightly bilabiate, rose-violet color on edge. Fruit tetrachene (each 1.5–2 mm), with 4–5 longitudinal stripes on the back.

Subspecies In addition to the nominal variety, *Verbena officinalis* L. var. *officinalis*, a second variety, *Verbena officinalis* L. var. *africana* (R.Fern. & Verdc.) Munir, is recognized by taxonomists.

Related Species The genus *Verbena* includes about 107 taxa, 14 of which are of hybridogenic origin; *Verbena hastata* L. has beer-making applications.

Geographic Distribution Cosmopolitan. **Habitat** Roadside, trampled environments; synanthropic species. **Range** Cosmopolitan.

Beer-Making Parts (and Possible Uses) Inflorescences.

Chemistry Active compounds (some still unknown): verbenalin, verbenin, aucubin, bastatoside, bitter principles, tannins, artemetin, volatile oil (citral, geraniol, limonene, verbenone), mucilage, unidentified alkaloids, saponins, adenosine, β-carotene, ursolic acid, 3-α-24-dihydroxy-oleandic acid, 7-α-22S-dihydroxysitosterol.

Style Saison.

Source https://www.brewersfriend.com/other/verbina/.

Vigna angularis (Willd.) Ohwi & H.Ohashi
FABACEAE

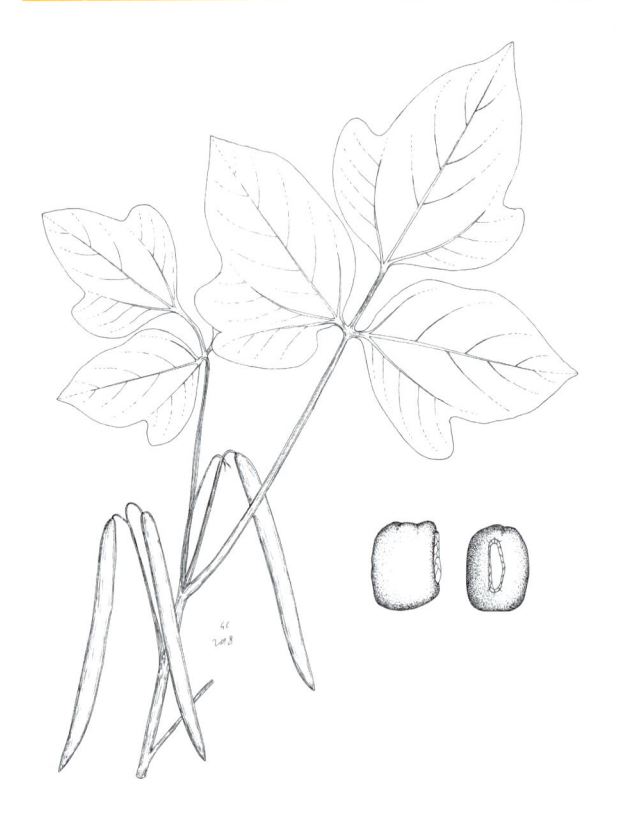

Synonyms *Azukia angularis* (Willd.) Ohwi, *Dolichos angularis* Willd., *Phaseolus angularis* (Willd.) W.Wight, *Phaseolus chrysanthos* Savi.

Common Names Adzuki bean, azuki bean.

Description Herb, annual, upright (30–90 cm), climbing or bushy. Root formed by a taproot and lateral roots with many nodules. Stem color green or purplish, hairy, angular. Leaves trifoliate, leaflets usually ovate or rhomboid-ovate (5–10 × 5–8 cm), sparsely hairy on both surfaces, apex broadly triangular or subrounded, lateral leaflets oblique, entire or slightly 3-lobed; the central leaflet long petiolate; stipules peltate, lanceolate, acuminate, with basal appendages. Inflorescence racemose axillary, short, with 6–20 stalked flowers. Flower with bell-shaped calyx (3–4 mm), 5 pale yellow petals (15 mm), oblate or kidney-shaped banner with beveled apex, wings greater than keel, slightly clawed and auriculate, keel asymmetrical with spirally curved apex and clawed base. Fruit legume pendulous, linear, cylindrical (5–8 × 0.3–0.5 cm), glabrous, blackish when ripe, bearing 5–10 seeds. Seed smooth, oblong, grayish-brown, variegated black at maturity (3.7 × 2.9 × 2.6 mm).

Subspecies Within the taxon *Vigna angularis* (Willd.) Ohwi & H.Ohashi, a variety differentiated from the nominal one, *Vigna angularis* (Willd.) Ohwi & H.Ohashi var. *nipponensis* (Ohwi) Ohwi & H.Ohashi, has been identified.

Cultivar Numerous cultivars of *Vigna angularis* have been recorded, varying in maturity time, seed color, and plant habitus; intermediate types between cultivated, and wild plants have also been found.

Related Species The genus *Vigna* includes more than 150 taxa, some of which are used in human nutrition (*Vigna mungo* [L.] Hepper, *Vigna radiata* [L.] R.Wilczek, *Vigna unguiculata* [L.] Walp.), but no beer-making uses are known.

Geographic Distribution Japan.
Habitat Cultivated in subtropical and tropical high-altitude environments. **Range** Japan, China, Manchuria, Korea, Sarawak, S Asia, Angola, Democratic Republic of the Congo, Kenya, Thailand, New Zealand, S United States, South America.

Beer-Making Parts (and Possible Uses) Seeds.

Chemistry Dried seeds: energy 1,411 kJ, water 10.8 g, protein 19.9 g, lipids 0.6 g, carbohydrates 64.4 g, fiber 7.8 g, ash 4.3 g.

Style Oatmeal Stout.

Source https://www.bjcp.org (16B. Oatmeal Stout); https://untappd.com/b/infusion-brewing-company-adzuki-bean/1543217/photos.

Viola odorata L.
VIOLACEAE

Synonyms *Viola wiedemannii* Boiss.

Common Names Wood violet, sweet violet, English violet, common violet, florist's violet, garden violet.

Description Herb (5–15 cm), perennial, acaulescent, pubescent, rhizomatous, with frail underground stolons and aerial stolons, rooting at the nodes, elongate, creeping, flowering in the second year. Leaves basal only, in rosette, dark-bluish green, broadly ovate-cordate or reniform, obtuse and crenulate, with maximum width at mid- and deep basal indentation, stipules lanceolate-oval, 4–5 mm wide, with short glandular bangs. Flowers zygomorphic pentamerous, typically fragrant, large (15–20 mm), with calyx with obtuse ovate sepals, prolonged posteriorly in appendages; corolla deep purple, rarely pink or white, with straight spur of equal color (about 6 mm); lateral petals folded downward and close to the larger lower one. Fruits subglobose loculicidal 3-valve capsules.

Related Species The genus *Viola* includes 550 taxa, many of which are fragrant and capable of producing an interesting essential oil; however, it is unclear if any are used in beer making.

Geographic Distribution Euro-Mediterranean.
Habitat Grassy and uncultivated places, woodland margins, hedges, meadows; often cultivated in gardens and wild (0–1,200 m ASL). **Range** Europe, America, Asia.

Beer-Making Parts (and Possible Uses)
Essence, flowers (?).

Chemistry Essential oil, leaves (25 compounds, 92.77% of the oil): butyl-2-ethylhexylphthalate 30.1%, 5,6,7,7α-tetrahydro-4,4,7 α-trimethyl-2(4 H)-benzofuranone 12.03%, veridiflorol 6.31%, 1-hexadecene 4.53%, hexadecane 4.25%, geraniol 3.21%, linalool 3.06%, citronellal 2.96%, 9-19-cycloanost-6-ene-3,7-diol 2.88%, spatulenol 2.54%, methyl salicylate 2.32%, tridecane 2.23%, 1,8-cineole 1.92%, 1-hexicosene 1.81%, cis-3-hexenol 1.42%, heneicosane 1.41%, α-pinene 1.31%, 1-dodecanol 1.29%, cis-2-hexenal 1.12%, undecanal 1.11%, hexadecanoic acid 1.01%, camphor 0.92%, 1-octadecene 0.92%, β-pinene 0.62%, salicylaldehyde 0.46%.

Style Imperial IPA.

Source Dabove 2015; Hieronymus 2016b; https://www.brewersfriend.com/other/violessens/; https://www.birrificio.it/wp-content/uploads/2017/02/fleurette.pdf (*Viola odorata* L. was chosen as representative of *Viola* sp.).

Vitis apyrena Schult.
VITACEAE

Synonyms *Vitis vinifera* L. var. *apyrena* L.

Common Names Zante currants, corinth raisins, corinthian raisins, currants, Thompson seedless.

Description Morphological characteristics similar to those of *Vitis vinifera*. Leaves are medium sized, heart shaped, oblong, pentalobed, with deep sinuses. Wild vines are usually dioecious (male plants and female plants). Flowers of this variety (Black Corinth; Italian: *Corinto Nera*) have well-developed anthers and small, undeveloped ovaries. Clusters small (average weight 180 g, 91–269 g), cylindrical, winged. Fruit berry (grape) very small (0.34–0.60 g), round, reddish-black color; epicarp very thin, mesocarp juicy and delicate. Fruits dry (0.091–0.139 g) dark brown or black color. Seeds practically absent (seedless grapes), except in occasionally large berries.

Related Species There are about 100 taxa in the genus *Vitis*, but only a few, besides *Vitis apyrena* and *Vitis vinifera*, produce fruits with organoleptic qualities suitable for human consumption and therefore have potential beer-making applications.

Geographic Distribution Corinth (Greece). **Habitat** Cultivated species. **Range** Greece, United States (California), South Africa, Australia.

Beer-Making Parts (and Possible Uses) Dried berries.

Chemistry Fruits: energy 1,184 kJ (283 kcal), carbohydrates 74.08 g (sugars 67.28 g; fiber 6.8 g), lipids 0.27 g, protein 4.08 g, thiamin (B1) 0.16 mg, riboflavin (B2) 0.142 mg, niacin (B3) 1.615 mg, pantothenic acid (B5) 0.045 mg, vitamin B6 0.296 mg, folate (B9) 10 µg, choline 10.6 mg, vitamin C 4.7 mg, vitamin E 0.11 mg, vitamin K 3.3 µg, calcium 86 mg, iron 3.26 mg, magnesium 41 mg, manganese 0.469 mg, phosphorus 125 mg, potassium 892 mg, sodium 8 mg, zinc 0.66 mg.

Style Belgian Dark Strong Ale, Belgian Specialty Ale, Flanders Brown Ale/Oud Bruin, Mixed-fermentation Sour Beer, Saison, Sweet Mead.

Source Calagione et al. 2018; https://www.brewersfriend.com/other/currants/.

Vitis vinifera L.
VITACEAE

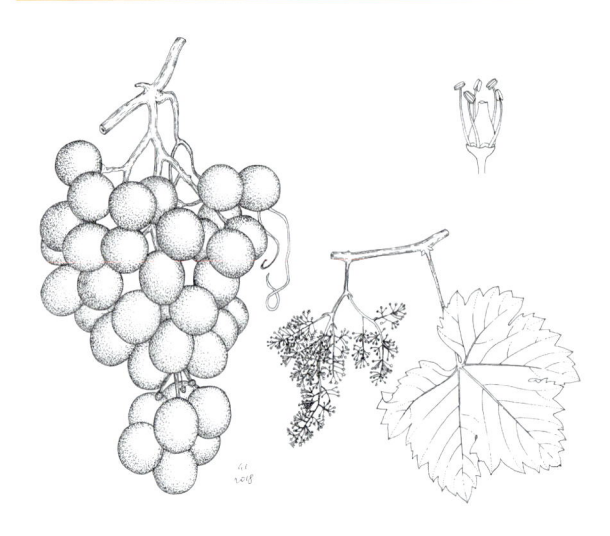

Synonyms *Cissus vinifera* (L.) Kuntze, *Vitis sylvestris* C.C.Gmel.

Common Names Grapevine.

Description Lianose plant (up to 40 m). Stem woody, climbing; branches brown or reddish, striate, usually glabrous in the basal portion with prehensile cirri. Leaves alternate, mostly opposite either a cirrus (lower ones) or an inflorescence (upper ones); lamina heart or kidney shaped in outline (5–15 cm), deeply divided into 3–5 palmate lobes and irregularly toothed all around; hairiness scarce or none above, more developed below and often persistent but not forming a continuous cobweb-like layer. Inflorescence panicle dense, fragrant; calyx reduced to 5 denticles, petals (5 mm) greenish, forming a cap that falls off as the flower opens. Fruit berry (grape) ellipsoid, spherical, or other shape (6–22 mm). Seeds 1–4, pyriform, with very tough shell, oil-rich kernel (8–11%).

Subspecies *Vitis vinifera* L. var. *multiloba* (Raf.) Kuntze, *Vitis vinifera* L. var. *palmata* (Vahl) Kuntze, and *Vitis vinifera* L. var. *vinifera* are three varieties of this taxon.

Cultivar Hundreds of vines are studied morphologically in detail by a specific science (ampelography); some have been used in beer making (Cabernet, Cabernet Franc, Dolcetto, Merlot, Moscato, Pinot Grigio, Riesling, Sangiovese; other grapes from Aglianico di Taurasi, Biancolella d'Ischia, Cannonau, Falanghina, Langhe, Moscato di Trani, Pallagrello);

others are not yet but could be soon considering the growing success of Italian Grape Ale (IGA), which blends wort from cereals with grape must.

Related Species There are about 100 taxa in the genus *Vitis*, but only few, besides *Vitis vinifera* and *Vitis apyrena*, produce fruits with organoleptic qualities suitable for human consumption and therefore have some potential beer-making applications; among these, *Vitis arizonica* Engelm., *Vitis californica* Benth., and *Vitis girdiana* Munson are used in beer making.

Geographic Distribution Origin uncertain. **Habitat** Cultivated. **Range** Cultivated from temperate to subtropical areas.

Beer-Making Parts (and Possible Uses) Grape must, raisins, pomace; cooked must, sapa (Sardinian cooked must of Cannonau, Malvasia, Nasco, Vermentino); casks that have contained wine (Amarone di Valpolicella, Barolo, Champagne, Franciacorta, Sassicaia, Taurasi); shoots and pruning residues (to smoke malt).

Chemistry Grapes (pulp + skin): water 70–85%, sugars (L-rhamnose, L-arabinose, L-xylose, L-ribose, D-fructose, sucrose) 15–20%, organic acids (tartaric acid, malic acid, citric acid), anthocyanins, tannins (flavanols, flavandiols, quercetin, isoquercitin), L-arabano, D-galactan, pectins, nitrogenous substances (ammonium salts, amino acids, proteins), minerals (potassium, calcium, sodium, phosphates, sulfates, chlorides), pruine (oleanolic acid, alcohols, esters, fatty acids, long chain aldehydes), odoriferous substances (linalool, α-terpineol), vitamins (vitamin C 90 mg/kg, vitamins group B), enzymes (invertase, oxidoreductase, pectolytic enzymes).

Style American Amber Ale, American Barleywine, American Brown Ale, American IPA, American Pale Ale, American Stout, American Strong Ale, American Wheat or Rye Beer, Baltic Porter, Belgian Blond Ale, Belgian Dark Strong Ale, Belgian Dubbel, Belgian Golden Strong Ale, Belgian Pale Ale, Belgian Specialty Ale, Belgian Tripel, British Strong Ale, Brown Porter, California Common Beer, Dark Mild, English IPA, Flanders Brown Ale/Oud Bruin, Flanders Red Ale, Foreign Extra Stout, Fruit Beer, Holiday/Winter Special Spiced Beer, Imperial Stout, Irish Red Ale, Italian Grape Ale (IGA), Metheglin, Northern English Brown Ale, Oatmeal Stout, Old Ale, Other Smoked Beer, Robust Porter, Russian Imperial Stout, Saison, Southern English Brown, Specialty Beer, Specialty IPA (Brown IPA, Red IPA), Specialty Wood-Aged Beer, Spice/Herb/Vegetable Beer, Strong Scotch Ale, Sweet Stout, Trappist Single, Wood-Aged Beer.

Beer BB10, Barley (Maracalagonis, Sardegna, Italy).

Source Calagione et al. 2018; Cantwell and Bouckaert 2016; Cantwell and Bouckaert 2018; Giaccone and Signoroni 2017; Heilshorn 2017; Hieronymus 2010; Hieronymus 2016a; Hieronymus 2016b; McGovern 2017; Sparrow 2015; https://www.bjcp.org (29. Fruit Beer; 32A. Classic Style Smoked Beer); https://www.brewersfriend.com/other/cabernet/; https://www.brewersfriend.com/other/mazsola/; https://www.brewersfriend.com/other/muskat/; https://www.brewersfriend.com/other/raisin/; https://www.brewersfriend.com/other/raisins/; https://www.omnipollo.com/beer/1-618/; https://www.omnipollo.com/beer/stolen-fruit/.

Italian Grape Ales

After experimenting with the addition of chestnuts, other fruits, and wild herbs, Italian brewers turned to grapes, using many varieties of *Vitis vinifera* in their creations. The first to attempt this encounter between beer and grapes was Nicola Perra, a Sardinian from Maracalagonis (Cagliari), who introduced the *sapa* (cooked must) of Cannonau into his productions: the result was an aromatic profile hitherto unknown, which adds vinous, intensely fruity or spicy sensations, and often a soft and juicy freshness otherwise difficult to find. After Nicola, many followed this path, trying a variety of methods. There are those who use the grapes (fresh or dried), those who use the must—concentrated, cryo-concentrated, stabilized, or in fermentation, and those who use the skins after pressing. A new universe has opened up, so much so that in 2016, the Beer Judge Certification Program, the main body for categorizing beer styles, accepted Italy's proposal to include among the recognized styles the first all-Italian one: Italian Grape Ale (IGA). Today dozens of breweries have at least one IGA in their range, and some dedicate a part of their production exclusively to this style. Among the main characteristics of the Italian way of using grapes is the choice to opt for local grape varieties, when not autochthonous, and to choose them by looking for the same peculiarities that winemakers look for.

Xanthosoma sagittifolium (L.) Schott
ARACEAE

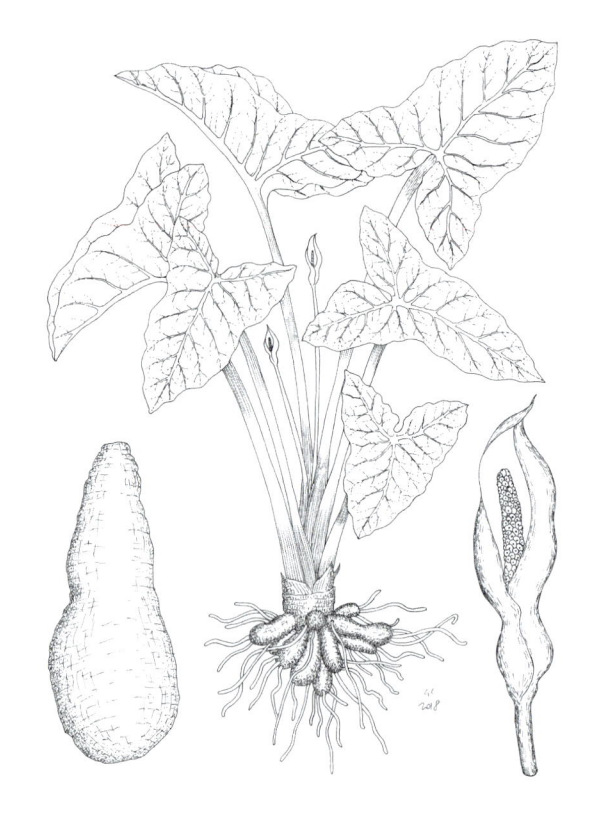

Synonyms *Alocasia talihan* Elmer ex Merr., *Arum sagittifolium* L., *Arum xanthorrhizon* Jacq., *Caladium edule* G.Mey., *Caladium mafaffa* Engl., *Caladium sagittifolium* (L.) Vent., *Caladium utile* Engl., *Caladium xanthorrhizon* (Jacq.) Willd., *Xanthosoma appendiculatum* Schott, *Xanthosoma atrovirens* K.Koch & C.D.Bouché, *Xanthosoma aureum* E.G.Gonç., *Xanthosoma blandum* Schott, *Xanthosoma edule* (G.Mey.) Schott, *Xanthosoma ianthinum* K.Koch & C.D.Bouché, *Xanthosoma jacquinii* Schott, *Xanthosoma maculatum* G.Nicholson, *Xanthosoma mafaffa* Schott, *Xanthosoma monstruosum* E.G.Gonç., *Xanthosoma nigrum* Stellfeld, *Xanthosoma panduriforme* E.G.Gonç., *Xanthosoma peregrinum* Griseb., *Xanthosoma roseum* Schott, *Xanthosoma utile* K.Koch & C.D.Bouché, *Xanthosoma violaceum* Schott, *Xanthosoma wallisii* Linden, *Xanthosoma xantharrhizon* (Jacq.) K.Koch.

Common Names Elephant ear, arrowleaf elephant ear, cocoyam, malanga, taioba.

Description Herb (up to 2 m), perennial, robust. Thickened underground tuberous stem (corm) from which numerous minor secondary tubers depart from slender rhizomes; these organs exude a milky, watery sap when cut. Leaves emerge from the central bud of the main corm with overlapping enveloping bases; petiole (up to 1.5 m) succulent, rounded near the leaf blade to which it attaches between the 2 lobes of the leaf margin, grooved in the basal portion; lamina (up to 1 × 1 m) sagittate to heart shaped, glabrous, light green with waxy, powdery covering, broadly angled to a wide point at the tip, deeply 2-lobed (sagittate) at the base; veins evident. Inflorescence on fleshy stem shorter than leaf petiole; upper part of stem (spadix) bearing tiny cream-colored flowers densely packed, male above, female below; spadix surrounded by broad bract (spathe), greenish-white, boat shaped with rolled margins. Fruit small yellow berry.

Cultivar There are several cultivars of this species.

Related Species Of the 76 taxa belonging to the genus *Xanthosoma*, none other than *Xanthosoma sagittifolium* appears to have beer-making applications.

Geographic Distribution N South America. **Habitat** Disturbed wetlands, mesic pine forests, wet ditches, swamps, and freshwater springs. **Range** South America, United States (Alabama, Texas, Florida, Puerto Rico, Virgin Islands), Caribbean, Asia, Africa, New Zealand, Pacific Islands.

Beer-Making Parts (and Possible Uses) Tubers.

Chemistry Tubers: energy 556 kJ, water 70–77%, protein 1.3–3.7%, lipids 0.2–0.4%, carbohydrates 17–26%, fiber 0.6–1.9%, starch 17–34.5%, ash 0.6–1.3%, calcium 20 mg, iron, 1 mg, thiamine 1.1 mg, riboflavin 0.03 mg, niacin 0.0005 mg, vitamin C 6–10 mg.

Style American Pale Ale.

Source McGovern 2017; https://www.brewersfriend.com/other/taioba/.

Zanthoxylum bungeanum Maxim.
RUTACEAE

Synonyms *Fagara bungei* (Planch.) M.Hiroe, *Zanthoxylum acanthophyllum* Hayata, *Zanthoxylum argyi* H. Lév., *Zanthoxylum bungei* Planch., *Zanthoxylum bungei* Planch. & Linden ex Hance, *Zanthoxylum fraxinoides* Hemsl., *Zanthoxylum piperitum* Benn., *Zanthoxylum podocarpum* Hemsl., *Zanthoxylum setosum* Hemsl., *Zanthoxylum simulans* Hance.

Common Names Japanese pepper, chopi, Korean pepper.

Description Tree, deciduous, small (3–7 m). Stems spiny; cauline spines with flat base; young shoots pubescent. Leaves compound with 5–13 leaflets; rachis margined; leaflets sessile, opposite, ovate, elliptic, or rarely lanceolate, sometimes suborbicular near base of leaf rachis (2–7 × 1–4.5 cm), both surfaces pubescent or with underleaf flocculent along midrib; midrib impressed on upper leaf, margin crenate. Inflorescence axillary, terminal on lateral branches; rachis and pedicels pubescent or glabrous. Flower with perianth in 2 irregular series or 1 series, with 6–8 ± undifferentiated yellowish-green tepals; male flowers with 5–8 stamens and a rudimentary 2-lobed gynoecium; female flowers with 2–5 carpels. Fruits purplish-red follicles (Ø 4–5 mm), pustular glandular, apex with short or absent beak. Seeds 3.5–4.5 mm.

Subspecies *Zanthoxylum bungeanum* Maxim. var. *zimmermannii* (Rehder & E.H.Wilson) C.C.Huang is another variety in addition to the nominal one.

Cultivar They are mostly recognized by the color of the pericarp. Examples include Da Hongpao, Hanyuan HuaJiao, and Yuexigong Jiao.

Related Species Of the approximately 195 taxa belonging to the genus *Zanthoxylum*, none other than the *Z. bungeanum* is reported to have beer-making uses.

Geographic Distribution W Sichuan (China). **Habitat** Semidesert temperate areas, plains, arid hills, montane forests. **Range** China, former Soviet Union.

Beer-Making Parts (and Possible Uses) Fruits.

Chemistry Essential oil, pericarp, hydrodistillation (HD) extraction: linalool 25.99%, limonene 19.34%, linalyl anthranilate 12.22%, 4-terpinenol 10.49%, eucalyptol 6.53%, α-terpineol 5.02%. Essential oil pericarp, supercritical fluid CO_2 extraction (SFE): nonanoic acid 21.43%, γ-terpinene 14.51%, eucalyptol 13.45%, α-terpineol 5.83%, caryophyllene oxide 5.48%.

Style Session Pepper IPA, Spiced Pale Ale.

Beer Timpa, Yblon (Ragusa, Sicily, Italy); Movembeer, La Buttiga Craft Brewery (Piacenza, Emilia-Romagna, Italy).

Source Calagione et al. 2018.

Zea mays L.
POACEAE

Synonyms *Mays vulgaris* Ser., *Mayzea vestita* Raf., *Zea alba* Mill., *Zea americana* Mill., *Zea amylacea* Sturtev., *Zea amyleosaccharata* Sturtev. ex L.H.Bailey, *Zea canina* S.Watson, *Zea curagua* Molina, *Zea erythrolepis* Bonaf., *Zea everta* Sturtev., *Zea glumacea* Larrañaga, *Zea gracillima* Voss, *Zea hirta* Bonaf., *Zea*

Zingiber officinale Roscoe
ZINGIBERACEAE

Synonyms *Amomum zingiber* L., *Curcuma longifolia* Wall., *Zingiber cholmondeleyi* (F.M.Bailey) K.Schum., *Zingiber majus* Rumph., *Zingiber missionis* Wall., *Zingiber sichuanense* Z.Y.Zhu, S.L.Zhang & S.X.Chen.

Common Names Ginger.

Description Herb, perennial (up to 1.2 m). Rhizome thickened, branched, brown in color, pale yellow interior, lemon flavor, spicy. Leaves in the form of shoots (pseudostems) grow annually from buds on the rhizome; pseudostems formed by a series of leaf sheaths tightly wrapped around each other, with narrow laminae (up to 7 × 1.9 cm), green, with alternate arrangement. Inflorescence, borne by separate shorter stems, cone-shaped spike composed of a series of leaf-like bracts, greenish to yellowish in color. Flowers come out from the edge of the bract, pale yellow in color with a violet lip that has yellowish dots and streaks.

Cultivar There are many in each area where the species is cultivated.

Related Species The genus *Zingiber* includes 146 taxa, but only *Z. officinale* is used in brewing.

Geographic Distribution S Asia, India. **Habitat** Tropical forests. **Range** India, China, Nepal, Indonesia, Thailand.

Beer-Making Parts (and Possible Uses) Rhizomes.

Chemistry Roots: energy 80 kcal, lipids 0.8 g (saturated fatty acids 0.2 g, polyunsaturated fatty acids 0.2 g, monounsaturated fatty acids 0.2 g), carbohydrates 18 g, fiber 2 g, sugars 1.7 g, protein 1.8 g, sodium 13 mg, potassium 415 mg, magnesium 43 mg, calcium 16 mg, iron 0.6 mg, vitamin B6 0.2 mg, vitamin C 5 mg.

Style Apple Pie Spice, Foreign Extra Stout, Gruit, Holiday/Winter Special Spiced Beer, Russian Imperial Stout, Specialty Beer, Spice/Herb/Vegetable Beer.

Beer Zingibeer, Doppio Malto (Erba, Lombardy, Italy).

Source Calagione et al. 2018; Dabove 2015; Fisher and Fisher 2017; Hieronymus 2010; Hieronymus 2012; Hieronymus 2016a; Hieronymus 2016b; Josephson et al. 2016; Markowski 2015; McGovern 2017; https://www.baladin.it/it/productdisplay/nora; https://www.beeradvocate.com/beer/style/72/; https://www.bjcp.org (29B. Fruit and Spice Beer; 30B. Autumn Seasonal Beer); https://lefthandbrewing.com/beers/great-juju/; https://www.ratebeer.com/beer/hitachino-nest-real-ginger-ale/48838/; https://www.williamsbrosbrew.com/beerboard/bottles/fraoch-heather-ale.

Zizania palustris L.
POACEAE

Synonyms *Melinum palustre* (L.) Link.

Common Names Wild rice.

Description Herb, annual, monoecious, aquatic. Stem (culm) erect (< 3 m), Ø 1 cm, puberulent at nodes. Leaf with membranous, narrow ligule; lamina 14–45 cm × 6–40 mm. Inflorescence panicle (2–5 dm), lower branches spreading to ascending, bearing staminate spikelets, upper branches ascending to erect, bearing pistillate flowers; staminate spikelets (6–8 mm), usually drooping; vestigial glumes, forming a kind of collar; flower 1; lemma linear, 5-venate, acuminate or aristate; palea 3-venate, stamens 6; pistillate flower (10–20 mm) usually ascending to erect, cylindrical, angular, firm; glumes usually absent; flower 1, lemma 3-venate, long tapering, aristate (awn 5–40 mm). Fruit caryopsis 1–2 cm, cylindrical, violet-black in color.

Related Species The genus *Zizania* includes four species: *Zizania aquatica* L. (also *Zizania aquatica* L. var. *interior* Fassett), *Zizania latifolia* (Griseb.) Turcz. ex Stapf, *Zizania palustris* L., and *Zizania texana* Hitchc.

Geographic Distribution Canada, United States. **Habitat** Wet meadows, shallow ponds, lakeshores (< 1,200 m ASL). **Range** Corresponds roughly to the geographic distribution. Attempts at cultivation also in some swampy areas of Hungary, Australia, and France.

Beer-Making Parts (and Possible Uses) Seeds.

Chemistry Caryopsis: energy 356 kcal, protein 15.56 g, carbohydrates 75.56 g, fiber 6.7 g, sugars 2.22 g, iron 2.22 mg, phosphorus 433 mg, zinc 6.67 mg, niacin 8.889 mg, vitamin B6 0.444 mg.

Style (?).

Source Hieronymus 2016b (the most common species of this genus in North America; *Zizania aquatica* L. and *Zizania texana* Hitchc. are rarer).

Zygia bangii Barneby & J.W.Grimes
FABACEAE

Synonyms *Pithecellobium laxiflorum* Rusby.

Common Names Vinhatico.

Description Tree (height unknown). One-year-old branches gray in color, glabrous except for spiciform racemes of long, narrow, red flowers, axillary on same-aged leaves. Leaves stiffly papery, pale olivaceous in color; upper leaf slightly darker and smooth; leaflets 14 per leaf; peduncle ± 12–20 mm; rachis ± 9–13 cm; leaflets with ovate-elliptic laminae, base broadly cuneate, asymmetrically obtusely acuminate, distal ones ± 7–9 × 2.5–3.5 cm, about 2.6–2.8 times longer than wide; venation asymmetrically pinnate, the main one subcentrally straight, the secondary ones 5–7 per side, the proximal one only slightly longer and stronger than the others, the tertiary venation weak, prominent only on the underleaf. Inflorescence spike (with ± 17 cm peduncle), bracts lanceolate ± 0.4 mm, soon dry and deciduous. Flowers briefly pedicellate, pedicels drum shaped (0.3–0.5 × 0.7–0.8 mm); perianth minutely and finely puberulent, calyx nerveless, corolla striate; calyx membranous, crateriform (about 0.5 mm), first shallowly dentate, then broken; corolla subtubular (9–9.5 × 2 mm), lobes ovate ± 1.5 mm; androecium 54-merous, 28 mm, tube ± 7 mm, intrastaminal disc 0.7 mm; ovary glabrous, symmetrically conical at apex. Fruit legume (?).

Related Species The genus *Zygia* includes about 70 taxa.

Geographic Distribution Bolivia (Yungas).
Habitat Tropical forests. **Range** Bolivia (Yungas).

Beer-Making Parts (and Possible Uses)
Wood for barrels (first containing cachaça, then used for aging beer).

Wood No information available.

Style South American Beers (?).

Source Cantwell and Bouckaert 2016; Cantwell and Bouckaert 2018 (It is possible that, once again, a common name has produced some confusion with respect to the correct scientific identification of the species to which it belongs. Indeed, "vinhatico" is the common name that some sources associate with *Persea indica* [L.] Spreng. According to others, the term is more properly applied to South American woody legumes such as *Plathymenia foliolosa* Benth., *Plathymenia reticulata* Benth., and *Zygia bangii* Barneby & J.W.Grimes).

Other Beer-Making Plant Species

In addition to the plants described in this book, others have certainly been used in beer making, in the distant past or recently, and it is easy to imagine that others could be the object of beer-making experimentation in the near future. At the time of writing, the species excluded from this first attempt to include all plant species used in beer making worldwide belong to the following genera:

Abies—PINACEAE
Acer—SAPINDACEAE
Actinidia—ACTINIDIACEAE
Aframomum—ZINGIBERACEAE
Alliaria—BRASSICACEAE
Alnus—BETULACEAE
Amelanchier—ROSACEAE
Aralia—ARALIACEAE
Aronia—ROSACEAE
Artemisia—ASTERACEAE
Asperula—RUBIACEAE
Bacopa—PLANTAGINACEAE
Bellis—ASTERACEAE
Berberis—BERBERIDACEAE
Betula—BETULACEAE
Boswellia—BURSERACEAE
Brassica—BRASSICACEAE
Cakile—BRASSICACEAE
Capparis—CAPPARACEAE
Capsella—BRASSICACEAE
Capsicum—SOLANACEAE
Carya—JUGLANDACEAE
Celtis—CANNABACEAE
Cercocarpus—ROSACEAE
Citrus—RUTACEAE
Cola—MALVACEAE
Comptonia—MYRICACEAE
Crataegus—ROSACEAE
Diospyros—EBENACEAE
Elaeagnus—ELAEAGNACEAE
Enterolobium—FABACEAE
Eragrostis—POACEAE
Eriodictyon—BORAGINACEAE
Eruca—BRASSICACEAE
Eschscholzia—PAPAVERACEAE
Fragaria—ROSACEAE
Fraxinus—OLEACEAE
Gentiana—GENTIANACEAE
Handroanthus—BIGNONIACEAE
Hibiscus—MALVACEAE
Hydrastis—RANUNCULACEAE
Ilex—AQUIFOLIACEAE
Impatiens—BALSAMINACEAE
Lactuca—ASTERACEAE
Lamium—LAMIACEAE
Ligustrum—OLEACEAE
Lilium—LILIACEAE
Lonicera—CAPRIFOLIACEAE
Melilotus—FABACEAE
Mentha—LAMIACEAE
Nardostachys—CAPRIFOLIACEAE
Nelumbo—NELUMBONACEAE
Nymphaea—NYMPHAEACEAE
Ocimum—LAMIACEAE
Pastinaca—APIACEAE
Pausinystalia—RUBIACEAE
Pedicularis—OROBANCHACEAE
Peumus—MONIMIACEAE
Phragmites—POACEAE
Physalis—SOLANACEAE
Phytolacca—PHYTOLACCACEAE
Pinus—PINACEAE
Pistacia—ANACARDIACEAE
Podophyllum—BERBERIDACEAE
Populus—SALICACEAE
Pouteria—SAPOTACEAE
Prosopis—FABACEAE
Prunus—ROSACEAE
Ptychopetalum—OLACACEAE
Pulmonaria—BORAGINACEAE

Pycnanthemum—LAMIACEAE
Pyrus—ROSACEAE
Reynoutria—POLYGONACEAE
Rhamnus—RHAMNACEAE
Rhus—ANACARDIACEAE
Rosa—ROSACEAE
Rubus—ROSACEAE
Rumex—POLYGONACEAE
Sapindus—SAPINDACEAE
Schisandra—SCHISANDRACEAE
Scirpus—CYPERACEAE
Scutellaria—LAMIACEAE
Sesamum—PEDALIACEAE
Setaria—POACEAE
Solanum—SOLANACEAE
Stachys—LAMIACEAE
Stellaria—CARYOPHYLLACEAE
Suaeda—AMARANTHACEAE
Swertia—GENTIANACEAE
Tragopogon—ASTERACEAE
Trichosanthes—CUCURBITACEAE
Trifolium—FABACEAE
Tsuga—PINACEAE
Tunilla—CACTACEAE
Tussilago—ASTERACEAE
Ulex—FABACEAE
Vaccinium—ERICACEAE
Verbena—VERBENACEAE
Veronica—PLANTAGINACEAE
Viburnum—ADOXACEAE
Viscum—SANTALACEAE
Vitex—LAMIACEAE
Vitis—VITACEAE
Ziziphus—RHAMNACEAE

It is estimated that the contingent of beer-making plants still to be identified amounts to 200–300 taxa. It is conceivable that further research could determine a significant increase in the size of this list of "new" beer-making plants, maybe even enough to justify another edition of *The Botany of Beer* or, who knows, maybe even a second volume.

Glossary

ABAXIAL: on the lower side of the leaf.

ABV: alcohol by volume.

ACCRESCENT: growing larger after flowering.

ACHENE: an indehiscent dry fruit with a thin, leathery pericarp, adherent but not attached to the seed.

ACHENOCONE: an indehiscent compound fruit, derived from more than one flower, with a cone-like structure formed by small fruits enveloped by spiraled or imbricate bracts, sometimes deciduous (*Alnus glutinosa*, *Humulus lupulus*).

ACICULAR: elongated, narrow, and pointed, similar to a sewing needle.

ACTINOMORPHIC: describes a flower with an axis of symmetry on which the various parts of the flower are arranged in a regular manner.

ACUMINATE: describes an organ, usually a leaf or fruit, that ends in a point.

ACUTE: (apex-) describes an organ thinned and pointed at the end.

ADAXIAL: on the upper side of the leaf.

ADNATE: describes organs belonging to different whorls or to organs morphologically different but still more or less fused together.

ALBEDO: the innermost, white, spongy part of the rind of a citrus fruit.

ALLOGAMOUS: cross-fertilized.

ALTERNATE: describes leaves arranged singly (one per node) on the axis of the stem.

AMPLEXICAULE: a leaf or bract that fits directly on the stem, enveloping it.

ANDROECIUM: the complex of stamens of a flower, constituting the male sexual apparatus.

ANEMOCHOROUS: disseminated by wind.

ANEMOPHILOUS: pollinated by wind.

ANGULAR: describes an organ forming angles.

ANNUAL: a plant that completes its life cycle in less than a year.

ANNULAR: with a ring shape or appearance.

ANNULUS: the annular structure of angiosperms on which the proximal parts of the cyclic organs (e.g., petals, stamens) are joined; also, the ring-shaped line of thickened cells that develops at the margin of the sporangia of numerous pteridophytes, acting as a hinge for the opening of the sporangia and the dispersal of the spores.

ANTHER: the top portion of the stamen where pollen is formed, usually divided into microsporangia.

ANTHESIS: a plant's flowering period.

APETALOUS: describes a flower lacking a corolla.

APEX: of the crown of a plant. Also, the terminal portion of an organ.

APHYLLOUS: destitute of foliage leaves.

APICAL: placed at the apex of another organ.

AREOLES: points on the blades of cacti from which spines usually develop.

ARIL: the total or partial involucre of the seed of some plants formed by a hyperplasia of the funiculus.

ARISTATE: describes an organ with bristles or awns.

ARISTIFORM: in the shape of an awn.

ARISTULATE: describes an organ with a small awn or beard.

ARMED: describes an organ (or plant) with spines.

AROMATIC: producing volatile odorous substances.

ARTICLE: a cactus stem unit.

ASCENDING: describes a stem or branch that, initially arranged horizontally, curves upward.

ASEXUAL: reproduction without exchange of genetic material.

ASSURGENT: vegetative behavior of a tree whose branches develop with a tendency to be vertical.

ATTENUATED: describes a leaf base that gradually narrows toward the petiole.

AURICLE: a semicircular expansion at the leaf base.

AURICULATE: describes leaves or stipules with auricles at the base.

AUTOGAMOUS: describes an organism capable of self-pollination.

AWN: a bristle-shaped extension typical of the bracts of the spikelet of certain Poaceae.

AXIL: the upper corner formed between the axis and any organ that departs from it (e.g., the leaf axil usually houses a bud).

AXILLARY: of an organ in the axilla.

AXIS: the main centerline of plant or organ development (e.g., main stem).

BASAL: placed at the base of an organ.

BASE: the proximal part of a plant or organ.

BASIFIXED: describes an anther whose filament is inserted at its basal part.

BEAK: the elongated portion at the apex of some fruits (e.g., cypselae, follicles, achenes, siliques), normally without seeds.

BERRY: a fleshy fruit, soft in every part, containing numerous seeds.

BIENNIAL: a term usually referring to a plant that lives two years; in the first it completes only its vegetative development; in the second it flowers and bears fruit.

BIFID: divided in two, forming a fork of two flaps.

BILABIATE: describes a calyx or corolla with joined tube-like elements in the lower part and divided into two lips in the upper.

BILOBED: formed by two lobes.

BIPINNATE: describes a leaf with pinnate leaflets.

BIVALVE: formed by two valves.

BRACHYBLAST: a shortened branch with close internodes.

BRACT: a modified leaf that usually surrounds flowers or inflorescences.

BRACTEOLE: a small bract.

BRANCH: a growth off the main axis of the stem.

BRANCHED: with several branches.

BRISTLE: a short, pointed hair, sometimes stiff.

BUD: a formation to protect the meristematic tissues during the period of vegetative rest.

BULB: an underground, shortened, ovoid stem with fascicled roots and a bud surrounded by distinctive fleshy leaves (cataphylls) and protected on the outside by dry leaves.

BULBIL: a small bulb that forms, in addition to underground, in the axil of the leaves or in the floral zone and that, in addition to its reserve function, fulfills the function of vegetative reproduction.

BUSH: an aggregate of branches from the base of the stem.

BUSHY: with a bush-like shape or structure.

BUTTRESS: a tabular expansion widening the base of trees, frequent in tropical environments.

CALYX: the outermost involucre of the floral perianth formed by the sepals, usually green in color.

CAMARA: an indehiscent legume of the genus *Trifolium.*

CAMPANULATE: bell shaped.

CANESCENT: describes a plant organ (stem, leaf, or flower) that turns whitish.

CAPILLARY: a long, very slender organ.

CAPITULIFORM: describes inflorescences that have the shape of or resemble that of the capitulum.

CAPITULUM: an inflorescence typical of the Asteraceae, characterized by a wide receptacle on which numerous sessile flowers are inserted, closely appressed to each other, externally wrapped by bracts, which constitute the involucre.

CAPSULE: a dry dehiscent fruit, containing numerous seeds, that derives from a pluricarpellar ovary; different types are distinguished according to their modalities of opening.

CARNEOUS: of a color similar to meat or flesh.

CARNOSE: fleshy with a soft consistency.

CARPEL: the female organ of the flower, one of the fertile metamorphosed leaves of the gynoecium that contains the ovules.

CARTILAGINOUS: describes a plant with solid tissue with structural functions.

CARYOPSIS: an indehiscent dry fruit, containing a single seed, typical of the Graminaceae.

CATKIN: an inflorescence consisting of unisexual, sessile flowers arranged on an elongated, flexible axis.

CAULIFLORY: describes a woody plant that produces flowers on the trunk or on old branches.

CAULINE: relative to the stem.

CAULIS: a stem.

CESPITOSE: describes a plant branching from below, dense and bushy owing to the presence of numerous stems converging at the base.

CHAMAEPHYTE: a small shrub that keeps its buds on lignified branches at a height of no more than 20 centimeters above the ground.

CILIA: filiform appendages of small size.

CILIATE: equipped with cilia.

CINEREOUS: grayish or ashy in color.

CIRRUS (or **TENDRIL**): a long, thin, prehensile organ derived from the transformation of the caulis, leaves, or roots, that allows nonfleeting plants to cling to supports in order to climb.

CLADODE: a modified stem or branch, without leaves or with very few leaves, that flattens out (sometimes enlarging as a water reservoir) when carrying out photosynthesis.

CLAVATE: club shaped.

CLAW: the slenderest portion of the petal closest to the point of insertion of the corolla with the thalamus or receptacle.

CLIMBER: a slender, non-self-supporting stem that rests on other supports, plants or otherwise.

CLUSTER (or **RACEME**): an undefined inflorescence with an elongated axis bearing numerous stalked flowers.

COCCARIUM: a schizocarpic fruit made up of small fruits derived from carpels, which initially grow together and then separate, opening along the margins or along the midrib; if it is made up of three carpels, it is called tricoccus.

COLLAR: the point of union (and functional link) between root and stem.

COLUMNAR: cylindrical, column-like in shape.

COMPOUND: made up of several parts; used to describe leaves that, depending on the subdivisions of the lamina, can be simple and compound (e.g., pinnate-compound, when formed by several units arranged on a pinnate base structure).

COMPRESSED: describes an organ compacted in a given direction.

CONCRESCENCE: an organ grown together with other previously isolated parts.

CONE: a flower or inflorescence similar to a spike, with persistent woody bracts and axes.

CONFLUENT: describes leaves that join at their bases, uniting to form a single leaf.

CONICAL: cone shaped.

CONNATE (or **CONJOINED**): born together; describes plant organs (e.g., anthers, stipules, leaves) born and grown together, united at the base.

CONNIVENT: describes an organ in which one part fits into another without allowing the boundary between them to be seen.

CONSTRICTED: with a sudden reduction in cross-section.

CONTORTED: folded, twisted.

CORDATE: heart shaped. Also, a leaf base divided into two heart-shaped lobes at the petiole junction.

CORIACEOUS: leathery, with a leather-like consistency.

CORM: the whole of the root system, leaves, and stem of higher plants; also, the short, enlarged underground organ of some plants, resembling a bulb but without buds, similar to a tuber but with the reserve substances located in the enlarged caulis, not in the scales (bulbotuber).

COROLLA: the whorl of the perianth formed by elements called petals, usually colored; may be reduced or absent.

CORTEX: the involucre part of the stem and its branches.

CORYMB: an undefined inflorescence in which the flowers, inserted with their peduncles at various heights on the common axis, all terminate at the same level.

CORYMBOSE: corymb shaped.

COSTA: a longitudinal prominence on the surface of an organ.

COSTAPALMATE: describes a leaf with a well-defined rib but with radially organized leaflets (palmate).

COSTATE: with protrusions and ribs.

COTTONY: covered with soft, long, frizzy hair.

COTYLEDONS (or **EMBRYOPHYLLS**): embryonic leaves, the first leaves formed inside the seed, which will sprout at germination and often contain a reserve of nutrients.

CRACKED: with more or less deep, irregular fissures.

CREEPER: a woody plant with a long flexible stem that grows thicker with age.

CREEPING: an organ (e.g., stem) that grows horizontally on the ground.

CRENATE: a leaf margin or other organ with small teeth and a rounded apex.

CRENULATE: describes an organ with crenulations.

CROWN: the complex of the top leafy branches of woody plants.

CRYPTO-: a prefix meaning "hidden."

CULM: the generally herbaceous stem of the Gramineae, characteristically articulated, fistulous, hollow in the internodes, and solid only at the nodes.

CULTIGEN: a plant cultivated voluntarily, modified or selected by humans; represents the result of plant cultivation in horticulture, agriculture, and forestry.

CUNEATE: wedge shaped.

CUPULIFORM: dome or cup shaped.

CURVED: of an organ bent in a curved form.

CUSP: an organ that, narrowing from the base toward the apex, ends in a stiff, thin tip that is fairly short.

CYME: a type of inflorescence, also called cymose or defined, with branches of varying lengths terminating in a flower that blocks the elongation; different types of cymes are recognized: uniparous, biparous, multiparous; also indicates the apical part of a plant whose axis terminates in a flower.

CYMIFORM: similar in appearance and structure to a cyme.

CYMOSE (or **SYMPODIAL**): a simple definite inflorescence in which the secondary branches are as long as or longer than the primary branches.

CYPSELA: a simple achene-like fruit with a pericarp surmounted at the apex by awns, hairs, scales, or bristles.

CYSTOLITH: a calcareous incrustation forming on the cell wall within some cells.

DECIDUOUS: describes a tree or shrub that loses its leaves during the unfavorable season. Also, an organ that tends to fall off (e.g., leaf).

DECUMBENT: tending to a horizontal position.

DECUSSATE: describes opposite organs in which each pair is inserted in an alternate position to the previous one.

DEFLECTED: describes a stem that, growing to a certain height, curves toward the ground in an arc.

DEHISCENCE: the opening of an organ (e.g., fruit) when ripe to release its contents (e.g., seeds).

DEHISCENT: describes an organ (e.g., fruit) that opens when ripe to release its contents (e.g., seeds).

DENSE: describes an organ (e.g., inflorescence) whose units (e.g., flowers) are inserted on an axis with a high density.

DENTATE: describes an organ with highly incised margins.

DENTICULATE: describes an organ with slightly incised margins.

DIACHENIUM: a double achene.

DIADELPHY: describes filaments (stamens) united into two groups.

DIALYPETAL: describes a flower whose corolla consists of separate petals.

DIALYSEPAL: describes a calyx with sepals that are all free and completely separated from each other.

DICHASIUM: a definite inflorescence in which, below the terminal flower, two symmetrical flowering branches develop, branching that may be repeated several times.

DICHOTOMOUS: describes an apical bifurcation into two divergent axes, two equal, opposite branches.

DICLESIUM: a berry included in the ascending perianth.

DIDYNAMOUS: describes the androecium (or stamens) of a flower that has four free stamen filaments, arranged in two pairs, one of which is longer than the other.

DIFFERENTIATED: describes wood with a visible distinction between sapwood and heartwood.

DIGITATE: describes an organ whose parts, which all originate from the same point, diverge like the fingers of a hand.

DIMORPHIC: describes an organ with two different forms, even on the same plant.

DIOECIOUS: describes a plant with unisexual flowers on different individuals (male and female plants are present).

DISSEMINATION: the natural process that allows seeds to be dispersed.

DISTICHOUS: of organs inserted on two opposite sides of a common axis, alternately on the same plane but at different heights.

DIVERGENT: describes a position of development of two organs that are increasingly distant from each other.

DORSIFIXED: describes an anther whose dorsal part of the connective is inserted on the filament.

DOME (or **CUPULA**): a scaly or spiny involucre that surrounds and protects the fruits of the Fagaceae (beech, oak, chestnut).

DOUBLE WINGED: double samara (e.g., *Acer*).

DRUPE: a fruit with a membranous epicarp, fleshy mesocarp, and woody endocarp (e.g., stone) containing one to two seeds.

DRUPELET: a small drupe.

ELLIPSOID: with a shape similar to that of a tridimensional ellipse.

ELLIPSOIDAL: ellipsoid in shape.

ELLIPTICAL: shaped like an ellipse.

ELONGATED: longer than wide.

EMARGINATED: describes the apical margin of an organ with a notch forming a shallow sinus.

EMBRYO: the set of elements in a seed that constitute a future plant.

ENDOCARP: the tissue forming the innermost layer of the fruit; derived from the inner epidermis of the ovary, it is not always distinct and may have different consistencies (fleshy, hard).

ENDOSPERM: the parenchymal tissue of the seed containing reserve materials that the embryo uses during germination.

ENSIFORM: a saber-shaped organ (flattened surface, sharp margin, sharp apex).

ENTIRE: with a linear and continuous margin.

EPHEMERAL: not very durable, short-lived.

EPICALYX (or **CALYX**): a verticil of small sepaloid bracts or stipules outside and immediately below the calyx, enveloping it in an independent or partially attached involucre.

EPICARP (or **EXOCARP**): originates from the outer epidermis of the ovary and forms the outer part of the fruit.

EPIPHYTE: a plant that lives on another plant without parasitizing it.

EPISPERM: the tissue covering a seed.

ERECT: in an upright position.

EROSE: of a perianth or leaf with a jagged margin, irregularly indented, or with small unequal sinuosity, as though nibbled away.

EVERGREEN: with persistent leaves, i.e., lasting all year round.

EXCRESCENCE: an epidermic outgrowth, especially one caused by disease or abnormality.

EXFOLIATION: of bark, the flaking of which is characterized by the detachment, often followed by the falling off, of one or more subparallel superficial layers.

EXSERTED: of a stamen or style protruding from the corolla tube.

FALCIFORM: sickle shaped.

FAN: free amino nitrogen, a measure of the concentration of amino acids and small peptides (consisting of 1–3 amino acids) that can be used by yeasts for growth and reproduction.

FASCES: a set of grouped organs (e.g., bundled leaves).

FASCICULATED: describes organs aggregated to form tight groups (e.g., fasciculated root).

FAUCES: the opening of the tubular part of a calyx or corolla, situated at the boundary between the flap and the tube.

FEATHERY: with a shape resembling a feather.

FELT: a thickening of hair.

FELTED: felt covered.

FEMALE: the part of the flower (or individual) that produces the ovary and ovules.

FILAMENT: the pedicel of the stamen, usually elongated and thin, supporting the anther.

FILIFORM: thread shaped.

FIMBRIATE (or **FRINGED**): describes an organ with a laciniate, fringed margin owing to the division of the marginal zone into fine, parallel, filamentous segments.

FLATTENED: wider than long.

FLOSCULE: a tubular, pentamerous, actinomorphic flower, centrally located in the flower head of the Asteraceae.

FLOWER: the apparatus of reproductive organs of higher plants.

FLUTED: describes an organ with a shallow longitudinal groove.

FOLIACEUS: a flattened organ with the appearance of a leaf (e.g., cladode).

FOLIOSE: of an axis with numerous and conspicuous leaves.

FOLLICLE: a dry, dehiscent fruit, monocarpellar, more or less elongated, opening at maturity by a longitudinal (usually ventral) slit.

FOVEOLATE: an organ with many small grooves or pits (foveolae).

FRACTURE: an irregular breakage of woody plant organs.

FRAYED: forming fronds.

FRUCTIFEROUS: having a fruit-like nature or appearance, with a lignified stem, even at the top.

FRUIT: an ovary developed after fertilization.

FUSIFORM: roughly ellipsoidal in shape.

GALBULUS: the pseudofruit of some Cupressaceae, consisting of woody scales and bracts, normally peltate and closely appressed, enclosing seeds.

GALL: an outgrowth developing on various organs (leaves, roots, fruits, flowers) that, as a result of the reaction produced by attacks from a foreign organism (e.g., insect), appears as a histological structure profoundly modified in comparison to the normal organ.

GAMOPETALOUS: a corolla with fused petals.

GAMOSEPALOUS: a calyx with fused sepals.

GLABRESCENT: almost completely hairless.

GLABROUS: hairless.

GLAND: an organ secreting substances processed by the plant.

GLANDULAR: having glands.

GLAUCESCENT: green-gray color.

GLAUCOUS: describes a plant or organ that is blue-green in color because of the pruinose layer.

GLOBOSE: globular in shape.

GLOCHIDS: small, thin, thorn-like hairs of some Cactaceae, particularly of the genus *Opuntia*, in which, in groups of 6–12, they cover every part of the plant to protect it from predation by herbivores; being easily detachable, they penetrate the skin, causing irritation.

GLOMERULUS: a grouping of elements that together form a spheroidal whole (e.g., sessile flowers of an inflorescence).

GLUME: one of the sterile bracts at the base of the spikelets of Poaceae or one of the sterile or fertile bracts at the base of the flowers of the spikelets of Cyperaceae.

GOITROUS: of a plant organ with conspicuous enlargement.

GRANULAR: containing granules.

GRAVITROPISM (or **GEOTROPISM**): the growth of an organ that orients its axis according to the direction of gravity, positive toward the center of the earth, negative in the opposite direction.

GREENISH: more or less green in color.

HAIR: a filamentous extroversion of the epidermis (many types: unicellular, multicellular, live, dead, simple, branched, glandular, needle-like, root hairs, etc.).

HAIRY: an organ (leaf, stem) covered with hair.

HASTATE: a leaf, the lamina of which is hollow at the base, forming two lobes facing outward.

HEARTWOOD: the innermost part of the wood of trees and shrubs whose tissues consist of dead cells.

HELIOPHILOUS: a species that prefers bright, sunny environments for its growth.

HEMIPARASITE: a plant that is parasitic for water and mineral salts, not for the products of photosynthesis.

HEMISPHERICAL: with the shape of a half-sphere.

HERB: a short plant with green, nonwoody stem.

HERBACEOUS: with the consistency of herbs.

HERMAPHRODITE: an organism or organ containing both sexual organs.

HEXASTICHOUS: a spike with six rows of grains (*Hordeum*).

HIP: multiple fruit (= aggregate) formed by apocarps immersed in the hypanthium or receptacle not divided into more than one cavity (e.g., *Rosa*).

HIRSUTE: thick with bristly hairs.

HISPID: covered with hard, rough, bristly hair.

HULL: the lower portion of the corolla of the Fabaceae flower, consisting of the two lower petals enveloping the sexual organs; the conformation of a plant organ analogous to the hull of a boat (e.g., seed with a keeled back); the outer green, fleshy part of some fruits (e.g., walnut, almond).

HYALINE: thin, translucent.

HYDROCHOROUS: the dispersion of seeds and fruits by water.

HYPANTHIUM: a formation of receptacular and partly perianthal origin, tubular or cup shaped, totally or partially enveloping an inferior or semi-inferior ovary (*Rosa*).

HYPOCARP: a fleshy structure that forms below the fruit.

HYPOGEUM: underground.

IMBRICATE: describes leaves, petals, scales, or bracts that cover each other like the scales of a fish or the tiles of a roof (*Juniperus*).

IMPARIPINNATE: describes a pinnate leaf bearing an isolated leaflet at the apex of the rachis.

INCISED: of the margin of leaves, or other organs, provided with deep teeth. Also, a leaf divided into several lobes, separated by incisions that reach up to the middle of the leaf without reaching the midrib.

INCLUDED: describes one organ contained within another (e.g., fruit included in the calyx, small fruit not exceeding the size of the calyx).

INCONSPICUOUS: of an organ, structure, appearance, etc. not or little evident.

INDEHISCENT: an organ (e.g., fruit) that does not open when ripe to release its contents (e.g., seeds).

INDUSIUM: the membrane, often deciduous, that covers and protects the spores in many species of ferns.

INFERIOR: said of the ovary when it is immersed in the receptacle.

INFLORESCENCE: a set of flowers arranged in various ways on a primary axis and secondary axes.

INFRUCTESCENCE: a set of fruits arranged in various ways, arising from the flowers of an inflorescence and sometimes taking on the appearance of a single fruit, as, for example, in the blackberry.

INFUNDIBULIFORM: funnel shaped.

INTERNODE: the portion of the stem between two successive nodes.

INVOLUCRAL BRACTS: bracts arranged in an involucre at the base of a small umbel in a compound umbel.

INVOLUCRE: a group of bracts at the base of a flower head, a small umbel or an umbel, or under one or more flowers.

ISOLATED: an organ separated from the complex to which it belongs.

IUCN: International Union for Conservation of Nature.

IVORY: whitish or cream-colored.

JUNCIFORM: reed-like in shape.

KEELED: hull shaped.

KNEE-SHAPED: bent like a knee.

LABELLUM: the larger and more developed tepal of an orchid, with vexillary functions.

LACERATED: describes an organ with notches and jagged edges.

LACINIA: the narrow and elongated division of a laminar organ (e.g., leaves, petals).

LACINIATE: describes an organ with reduced laciniae.

LACTICIFEROUS DUCT: an internal structure that produces and carries the latex produced by the plant.

LAMINA: the broad part of the leaf.

LANCEOLATE: of a laminar plant organ with a lance-shaped outline.

LATERAL: describes an organ placed at the side of an axis of the plant.

LATEX: the milky juice of certain plants contained in special anastomosed vessels.

LAX: of an organ (e.g., inflorescence) whose units (e.g., flowers) are inserted on an axis with a low density.

LEAF: a more or less laminar appendage of the stem with photosynthetic and other functions.

LEAFLET: each of the parts into which a compound leaf is divided.

LEATHERY: with the texture of leather.

LEGUME: a dry dehiscent fruit originating from a monocarpellar ovary and containing numerous seeds; when ripe it opens along two dehiscence lines; it differs from the siliqua by the absence of the intermediate septum.

LEMMA: the lower glume typical of the Poaceae, i.e., a sterile bract similar to the glumes, but more internal to them, that surrounds the palea and its flower.

LENTICEL: a formation of secondary tegument tissue, permeable to gases, present on the branches of many woody plants.

LEPT-: a prefix used in compound words meaning narrow, thin, or puny.

LEPTOMORPHIC: generally indicates a gracile or tenuous (typically thin) character, shape, or structure.

LESINIFORM: describes an organ with a strictly linear and acute shape.

LIANOUS: forming climbing stems.

LIGNIFIED: describes an organ that has acquired the consistency of wood.

LIGULA: a small emergence at the base of the leaves; in the Poaceae, it is the membranous lamina between the sheath and the leaf blade; in the Asteraceae, the corolla may be ligulate.

LIGULATE: strap shaped.

LIMB: a main branch of a stem.

LINEAR: describes an organ about 10 times longer than wide.

LIP: one of the two parts of a bilabiate corolla.

LOBATE: describes a flattened organ (e.g., leaf) divided into several parts by shallow incisions (e.g., bi-, tri-, tetra-, or quadrilobate).

LOBE: a part separated from the indentation.

LOCULE: a cavity of the pericarp (or anthers) divided by septa.

LODICULE: the remains of the perianth reduced to 1–3 small, inconspicuous scales that, inserted at the base of the ovary of the Poaceae, cause the glumes (lemma and palea) to open at the anthesis and swell to allow pollination.

LOMENT: a pericarp or dry fruit with a single carpel and transverse constrictions, which at maturity splits completely into articles containing seeds.

LONGITUDINAL: lengthwise.

LYRATE: describes a lyre-shaped organ, usually a leaf, divided transversely into lobes in a manner reminiscent of a lyre.

MACRO-: a prefix meaning large or coarse.

MACROBLAST: a branch with spaced leaves and normal or elongated internodes.

MACROSPOROPHYLL: a sporophyll (modified leaf) bearing ovules.

MACULA: a differently colored mark on the surface of plant organs.

MALE: the part of the flower (or individual) that produces pollen.

MARGIN: the boundary of the leaf blade.

MEIOSIS: a process of cell division involving only reproductive cells, whereby a mother cell forms four daughter cells, each different from the others, containing half of the parent's genetic heritage.

MERICARP: one of the portions, containing a single seed, into which a schizocarpic fruit is divided when ripe.

-MEROUS: a suffix preceded by a number form (e.g., tetra-, penta-) that describes the number of units making up a certain organ (e.g., tetramerous, pentamerous).

MESOCARP: the part of the fruit between the epicarp and endocarp.

MICROBASARIUM: a schizocarpic fruit consisting of mericarps derived from two bilocular carpels, concrescent with a single style, that separate at maturity.

MICROSPOROPHYLL: a sporophyll (modified fertile leaf) bearing one or more microsporangia, pollen sacs (stamen in Angiospermae, scales of male cones in Gymnospermae).

MONOCLINE (or **HERMAPHRODITE** or **BISEXUAL**): describes a plant or flower with stamens and pistils. Also, a hermaphrodite flower.

MONOECIOUS: describes a plant with male and female flowers on the same individual.

MONOLETE: describes the spores of pteridophytes, characterized by a single line indicating the vertical axis along which the mother spore divided into four.

MONOPODIAL: describes the branching of a tree whose main axis has indefinite growth and branches of different orders with limited development and subordinate to the trunk.

MONOSPERMOUS: single-seeded.

MONOTYPIC: describes a genus consisting of only one species.

MONOVULAR: with a single egg.

MUCRO: the hard, prickly tip of any organ.

MUCRONATE: ending in a sharp point or mucro.

MULTICAULE: with many stems.

MULTIFLORA: with many flowers.

MUTICATE: without an awn.

NECTAR: a sugary secretion produced by the nectary.

NECTARY: a glandular organ for secreting nectar.

NITROPHILIC: describes an organism that prefers nitrogen-rich soil.

NODE: a transverse plane at the point of insertion of the bud.

NUCULE: a nutlet, small nut.

NULLUS: describes an organ or part of it that is missing or reduced until it disappears.

NUT: a single indehiscent, dry fruit with a distinct pericarp separated from the seed; improperly used to describe the fruit of *Juglans regia* (walnut), which is a drupe.

OB-: a prefix indicating an inversion of the normal form or position.

OBCONIC: in the shape of an inverted cone.

OBCORDATE: of a leaf in the shape of an inverted heart, wider in the upper half.

OBCUNEATE: in the shape of an inverted wedge.

OBLANCEOLATE: of an organ (e.g., leaf) with the flap narrowed at its point of attachment (base).

OBLATE: of an organ slightly flattened at the poles.

OBLONG: of an organ, laminar with an elliptical outline and parallel lateral margins.

OBOVATE (or **OBOVOID**): describes an organ shaped like the outline of an inverted egg.

OBSCURE: indistinct, not macroscopically visible.

OBTUSE: describes a nonacute form. Also, referring to the end of an organ (e.g., leaf) that is sloped and beveled, with an angle greater than 90 degrees.

OGIVAL: ogive shaped (e.g., the tapered anterior portion of a projectile).

OPERCULUM: a kind of movable cover with which some dry fruits (e.g., *Eucalyptus*) are equipped.

OPPOSITE: describes one of two organs (e.g., leaves) arranged at the same internode at 180 degrees.

ORBICULAR (or **CIRCULAR** or **DISCIFORM**): describes an organ (leaf, seed, stigma) that is more or less circular in shape.

ORDER: a hierarchical level of plant organs or parts of them (e.g., first-order ribs, second-order ribs); the term also has other meanings.

OROPHYTE: a mountain plant.

OSTIOLE: the narrow apical orifice of the syconium.

OVAL: describes a laminar organ with a contour comparable to that of the longitudinal section of an egg.

OVARY: the part of the flower of the Angiospermae that contains the ovules and after fertilization turns into fruit.

OVOID (or **OVATE**): egg shaped.

PALEA: the fertile bract, innermost, typical of the Poaceae, that encloses the flower.

PALMATE: describes an organ (leaf) divided into lobes like the fingers of the palm of the hand.

PALMINERVED: describes a leaf with main veins that have the same point of origin and are arranged like the spread fingers of a hand.

PANICLE (or **THYRSUS, CLUSTER,** or **COMPOUND RACEME**): a compound inflorescence, which is a raceme in which along the main axis, instead of single flower pedicels, branched lateral axes are inserted (raceme, spike, flower head, corymb, or other inflorescences) and each branching is always shorter than the one from which it derives.

PAPILIONACEOUS: describes an irregular flower or corolla consisting of five petals: one upper petal (vexillum or standard), two lateral petals (wings) in a median position, and two lower petals (keels) forming the hull that houses the reproductive organs.

PAPILLA: a unicellular epidermal protuberance that is shaped like a short, conical hair, rounded at the apex, giving a velvety appearance to some petals and leaves.

PAPPUS: a light feathery or scaly appendage that crowns the top of some fruits and seeds to aid their dispersal.

PARIPINNATE: describes a pinnate leaf with an even number of leaflets.

PARTITE: of a leaf with deep incisions up to half the distance between the margin and the midrib (e.g., tripartite: three incisions).

PATENT: describes an organ arranged to form a 90-degree angle with the axis from which it departs.

PAUCIFLORATE: describes an inflorescence (or plant) with few flowers.

PEDUNCLE: an axillary organ supporting the flower.

PELLUCID: of a semitransparent, translucent, hyaline organ (wing, leaf).

PELTATE: a shield-shaped organ; a peltate leaf blade has the petiole inserted in the middle of the underleaf.

PENDULOUS: hanging.

PENNINERVED: describes a leaf with a single main vein from which lateral veins branch off, like bird feathers.

PENTA-: a prefix used to indicate the number 5 (e.g., pentapetalous).

PENTAMEROUS: describes an organ consisting of five units.

PEPO: a simple, indehiscent fruit, berry-like with a leathery or sometimes woody epicarp, a fleshy or juicy mesocarp, and a deliquescent endocarp when ripe.

PERENNIAL: with a multiyear duration.

PERIANTH: the set of floral whorls surrounding or enveloping the flowers, usually consisting of a calyx and corolla.

PERICARP: the part of the fruit derived from the wall of the pistil forming the stratified involucre of the fruit; comprises, from the outside, the epicarp (exocarp), mesocarp, and endocarp; the pericarp may be dry or fleshy.

PERIGONIUM: a perianth not differentiated into calyx and corolla.

PERSISTENT: describes an organ that lasts beyond the fulfilment of its function.

PERULA: one of the scaly leaves that protect the embryonic buds and usually fall off when the buds hatch.

PETAL: an element of the corolla originating from a metamorphosed leaf.

PETALOID: petal-like in texture and color.

PETIOLATE: describes a leaf with a petiole.

PETIOLE: an axillary organ used to support the leaf.

PHYLLARY: of the outer involucral bracts of some flower heads of the Asteraceae.

PHYLLODY: a petiole flattened and transformed as a leaf blade, of which it also assumes the function (*Acacia*, *Citrus*).

PHYLLOTAXIS: an arrangement of leaves on a branch taking into account the number per node and their angle of divergence (AD).

PINNA: in ferns, the first division of the rachis of the frond, corresponding to the first-order leaflets of the compound leaf.

PINNATE: describes a compound leaf with the leaflets inserted on both sides of the midrib, like the barbs of a feather.

PINNATE-PARTITE: describes a leaf with incisions from the margin to half the lamina.

PINNATE-PINNATISECT: describes a leaf with deep incisions from the margin of the lamina almost reaching the midrib.

PINNULA: the last subdivision of fern leaves; also, the second- and third-order leaflets of pinnate leaves.

PISTIL: the female organ of a flower (gynoecium), formed by one or more carpels; consists of the ovary, style, and stigma.

PLACENTATION: the position of the placenta (tissue of the carpel leaves in which the ovules originate) and thus the arrangement of the ovules in the ovary.

PLANTULE: a seedling originating from a seed that has not yet emitted leaves the same as those of the adult plant.

PLEONANTIC (or **POLYCARPIC**): describes the behavior of plants that flower and fruit several times during the vegetative cycle.

POD: a popular term for a legume, or an organ shaped similarly to a legume pod.

POLLEN: the tiny granules, called pollen grains, that fill the anthers of the stamens and are used for pollination.

POLY-: a prefix indicating a plurality of organs (e.g., polyachene: a set of aggregated achenes).

POLYACHENE: a schizocarpic fruit, from an inferior ovary, with achene-like monocarps, which at maturity separate longitudinally from each other and remain attached to a central longitudinal axis (columella or carpophore), dehiscent or indehiscent (Apiaceae, Geraniaceae).

POLYCYSTIC: with organs arranged in several rows.

POLYDRUPE: an infructescence consisting of several drupes.

POLYGAMODIOECIOUS: describes populations made up of individuals of the same species, hermaphrodite and male on some plants, hermaphrodite and female on others.

POLYGAMOUS: describes a species that bears hermaphroditic and unisexual flowers of a single type on the same individual (gynomonoecious, andromonoecious, and trimonoecious plants) or on different individuals (gynodioecious, androdioecious, and trioecious plants).

POLYMORPH: of variable shape depending on various factors (environment, light, climate).

POME: a false fruit in which the fleshy and juicy part is formed by the receptacle, which is highly developed after fertilization; the true fruit is formed by the inner core; it is typical of the Rosaceae (apple, pear).

PROCUMBENT: extending prostrate, bent toward the ground but tending to rise in the apical terminations.

PROSTRATE: developed, lying at ground level.

PRUINESCENCE: a waxy coating covering aerial parts of plants; impermeable to water, serves to protect against moisture and prevent excessive transpiration.

PRUINOSE: covered with white powdery granules.

PSEUDO-: a prefix used in the formation of compound words to indicate resemblance or falsity.

PSEUDOCARP: a false fruit, formed from parts not having a carpel origin.

PSEUDODRUPE: a simple indehiscent fruit with a hardened, often woody pericarp, wrapped in a fleshy or leathery exocarp, and missing an endocarp; a drupe from an inferior ovary.

PSEUDOSTEM: a false stem consisting of leaf sheaths closely joined together.

PTERIDOPHYTE: a vascular plant (fern, equisetum, lycopodium, selaginella) with spores rather than seeds.

PUBERULENT: with very short, fine, erect, and sparsely dense hairs, hardly visible without a lens.

PUBESCENCE: dense hair.

PUBESCENT: describes an organ covered with dense hairs.

PULVINUS: a cell structure capable of determining the movement of certain plant organs (e.g., leaf fins); also, the hemispherical cushion shape assumed by some shrubs.

PURPUREUS: bright red in color.

PYRAMIDAL: pyramid shaped.

PYRIFORM: pear shaped.

PYXIDIUM (or **PYXIS**): a capsule that opens at maturity through a transverse slit, causing the detachment of an apical cover called the operculum.

QUADRANGULAR: an organ that forms a section with four corners.

QUADRI-: a prefix indicating the number 4 in compound words (e.g., quadrilobed: four lobes).

RACEME (or **CLUSTER**): an undefined inflorescence with an elongated axis bearing stalked flowers.

RACEMOID: raceme-like.

RACHILLA: a secondary rachis, such as the axis of the spikelets of Poaceae and Cyperaceae where the flowers fit.

RACHIS: a common axillary organ in compound organs (e.g., leaves, inflorescences).

RAY: a unit bearing flowers in an umbel inflorescence.

RAYED: describes a flower head with a crown of ligulate outer flowers.

RECEPTACLE: the apical part of the floral peduncle on which the various elements of the flower are inserted.

REFLEX: of an organ (flap, branch, anther, petiole, hair, etc.) that, as it develops, folds longitudinally outward and downward, at an angle of about 180 degrees from its normal position.

REGULAR: describes a flower radially symmetrical and capable of being bisected into two or more similar planes.

RENIFORM: kidney shaped.

REPAND: describes a leaf margin that is slightly wavy, less than sinuate.

RETUSE: describes a leaf margin with a shallow incision on a rounded apex.

REVOLUTE: of an organ, or part of it, rolled down or back.

RHIZOME: a metamorphosed underground stem acting as a reserve organ, capable of producing adventitious roots and stems; usually enlarged and horizontal in development.

RHOMBIC: in the shape of a rhombus.

ROSEHIP: the false fruit of *Rosa* consisting of a fleshy, colored receptacle containing the true fruits (achenes).

ROSETTE: an arrangement of leaves radiating from the base of the stem outward, close to or grazing the ground level.

ROSTELLUM: a small beak-shaped tip, an extension of one of the stigmatic lobes (the sterile one) of Orchidaceae on which the retinaculum is located.

ROSTRUM: an elongated beak-shaped part.

ROTATE: describes a gamopetalous corolla with petals fused into a short tube and patent lobes.

SACCATE: describes a sack-shaped organ, deeply concave.

SAGITTATE: arrow shaped; some triangular leaves pointed at the apex and with elongated base lobes.

SAMARA: a type of achene characterized by a wing-shaped extension of the pericarp.

SAP: the watery solution circulating in the conductive tissues of plants.

SAPWOOD: soft wood generated by cambium in trees and shrubs during the annual period of resumption of growth.

SCALE: a special leaf with a protective function, of variable shape and consistency (e.g., hard, leathery, fleshy); in some cases, the scales play an important role in reproduction, as in Gymnospermae (conifers), which have strobili consisting of male (polleniferous) and female (ovuliferous) scales. Also, a flattened structure.

SCANDENT: describes a climbing stem equipped with specialized organs (tendrils, cirri, suckers) to attach itself to for support.

SCAPE: an elongated floral axis, often leafless but with bracts and scales, ending in a flower or inflorescence, surrounded at the base by leaves of the same origin, a bulb, or rhizome.

SCAPOSE: forming a well-defined stem (tree).

SCAR (LEAF): the mark left at the point where a leaf detaches.

SCARIOUS: describes an organ having the consistency of a membranous scale.

SCHIZOCARP: a category of syncarpous dehiscent pluricarpellar dry fruits that, when ripe, split into several indehiscent unicarpellar monosperm units (mericarps).

SCLEROPHYLLOUS: describes plants with stiff evergreen leaves adapted for living in hot, arid environments.

SCULPTURE: a more or less characteristic pattern of relief and grooves in the bark of trees.

SEED: an organ consisting of the fertilized and matured ovum containing the embryo, intended to propagate Spermatophytes in time and space.

SEEDLESS: describes a fruit lacking seeds.

SEGMENT: a part or subdivision of an organ (e.g., leaf segment); in a compound leaf it indicates the leaflets.

SELF-POLLINATION (or **SELF-FERTILIZATION**): occurs between two gametes, male and female, from the same hermaphrodite individual as a result of self-pollination.

SEMI-: a prefix meaning half (e.g., semi-amplexicaule, semi-prostrate).

SEMI-AMPLEXICAULE: describes a sessile leaf that only half embraces the stem by means of its basal flaps.

SEMI-INFERIOR: describes an ovary inserted in an intermediate position between superior and inferior.

SENESCENCE: aging.

SEPAL: one of the elements forming the calyx of a flower, usually green in color.

SEPALOID: sepal shaped.

SEPTATE: of fruit (e.g., capsule) that at maturity divides into longitudinal units.

SEPTUM: the dividing layer of some organs (e.g., the siliqua of *Sinapis pubescens*).

SERICEOUS: covered with fine, shiny hairs.

SERRATE: with a serrated leaf margin.

SERRATED: with a toothed margin, with teeth pointing toward the apex.

SERRULATE: with a finely serrated leaf margin.

SESSILE: describes an organ attached directly to the base, without a stalk.

SEXUAL: describes reproduction through an exchange of genetic material.

SHEATH: the widened basal part of a leaf that wraps around the stem.

SHOOT: a young scaly sprout developing from an underground stem (*Asparagus*).

SHRUB: a woody plant branching from the base and rising 1–5 meters. Also, a woody plant consisting of several stems.

SIDE: a surface of the leaf blade; a single leaf contains an upper leaf and an underleaf, which sometimes have very different characteristics.

SILICLE: a bilocular dehiscent dry fruit, as long as wide, otherwise similar to the siliqua.

SILIQUA: a bilocular dehiscent dry fruit, longer than wide, with a dehiscence from the bottom upward releasing the intermediate septum to which the seeds are attached; distinguished from the legume by the presence of the intermediate septum.

SIMPLE: a generic term excluding multiplicity (e.g., simple, unbranched stem); for leaves it is the opposite of "compound."

SINUS: the hollow between two flat organs, as between the lobes of leaves or bifid petals.

SOIL INDIFFERENT: describes a plant species with no particular predilection for soil characteristics.

SOROSIS: a compound fruit consisting of more than two small, fleshy, drupe-like fruits, developed on a stalk or wrapped in a fleshy, growing perianth and joined to resemble a single fruit (*Morus, Ananas*).

SORUS: in ferns, the group of sporangia produced by mature sporophytes, located on the underside of the fronds.

SPADIX: a racemose or indefinite inflorescence like a spike, but with an enlarged and fleshy axis, characterized by a large bract called a spathe that wraps around it.

SPATHE: a large, conspicuous bract enveloping an inflorescence or single flower from the base.

SPATULATE: describes an organ (e.g., leaf) narrow at the base, broadened and rounded at the apex.

SPICIFORM: describes a spike-shaped inflorescence.

SPIKE: a thorn of cutaneous origin or formed by the epidermis (e.g., *Rosa*). Also, a type of racemose or indefinite inflorescence with an elongated main axis on which sessile flowers are inserted.

SPIKELET: an inflorescence typical of Cyperaceae and Poaceae that can be uni- or plurifloral, consisting at the base of 1–3 opposite bracts (glumes) and an axis (rachilla) with a variable number of nodes on each of which a flower is distally inserted; the perianth is reduced to 2–3 membranous lodicules or is absent of lodicules; the flowers are usually surrounded by 2 bracts (lower glumella or lemma, upper glumella or palea).

SPINESCENT: tending to form thorns.

SPINULE: a small spine.

SPINY (or **SPINULOSE**): describes an organ with spines or thorns.

SPIRAL: of organs (e.g., leaves) arranged in a spiral along an axis of the plant.

SPONGY: describes a porous organ or tissue.

SPORANGIUM: an organ made of tiny sacs in which spores originate and are contained.

SPORE: a haploid, single-celled, reproductive structure that originates through meiosis and, as it develops, produces the gametophyte.

SPOROPHYLL: a fertile leaf bearing the sporangia in which spores are formed; in Angiosperms, microsporophylls (stamens) and macrosporophylls (carpels).

SPUR: a hollow, conical, or cylindrical extension of the calyx or corolla; in orchids, it is the tubular termination of the labellum.

SQUAMIFORM: of a laminar organ in the shape of scales.

STAMEN: the male reproductive organ of the flower, consisting of an axile part (filament) and the anther in which the pollen is formed.

STAMINOID (or **STAMINODE**): a stamen that has lost its pollen-forming function or a stamen-like structure that sometimes performs vexillary functions.

STEM: the aerial part of the plant axis.

STIGMA: the organ of the female reproductive apparatus of a flower that surmounts the stylus and whose purpose is to capture pollen (with appropriate structures or sticky materials) and promote its germination so that it can fertilize the ovules contained in the ovary.

STIPITATE: describes an organ with a basal pedicel (stipe).

STIPULE: a leaf expansion at the base of the leaf stalk.

STOLON: a particular branch characteristic of some species that, when resting on the ground, produces shoots that can in turn emit roots and thus generate new seedlings (*Fragaria*, *Viola*).

STONE: woody endocarp of the drupe.

STRIAE: linear marks in relief or of a different color from that of the background, of varying length, usually oriented in the same direction.

STRIATE: describes organ-bearing striae.

STROBILUS: a reproductive structure of conifers consisting of more or less woody scales at the axil of which the male gametophyte (pollen) and the female gametophyte (ovum) are formed.

STROPHIOLE: the fleshy or spongy part of the seed that forms above the raphe.

STYLUS: usually an axile extension that connects the ovary with the stigma in the female reproductive organ of the flower.

SUB-: a prefix indicating an attenuated quality of resemblance (e.g., subcordate, subequal, suberect, subglabrous, subglobose, subnullus, subopposite, suborbicular, subobtuse, subrounded, subsessile, subspherical, subspatulate, subsessile, subspinose, subterete).

SUBEROSE: describes a stem covered with cork (secondary epidermal tissue).

SUBULATE: cylindrical, gradually narrowing to a pointed terminal region (lesiniform-like).

SUCCULENT: a plant or parts thereof characterized by water-rich tissues, particularly suitable for arid environments (Cactaceae).

SUCKER: a branch originating from an adventitious bud, often at ground level.

SUCKERING: that emits suckers.

SUFFRUTESCENT: describes a perennial plant, woody at the base only, with annual herbaceous branches that dry up after fruiting.

SUFFRUTICOSE: with a woody stem only at the base and herbaceous in the rest.

SUPERIOR: an organ, usually floral, placed above others; used specifically to denote the ovary placed above the receptacle.

SWOLLEN: bulging, distended.

SYCONIUM (or **SYCONUS**): the inflorescence and infructescence of the fig tree.

SYNANDRODE: a concrescence of sterile staminodes.

SYNANDROUS: describes a complex of stamens united and fused by filaments and anthers.

SYNANTHROPIC: describes a species (or habitat) capable of living in anthropized ecosystems by adapting to environmental conditions created or modified by humans.

SYNCARP: a multiple fruit derived from the syncarp gynoecium, formed by joining the carpels together on a fleshy axis, with concrescent small fruits (apocarp).

TAPERED: describes a stem whose diameter progressively reduces toward the apex.

TAPERING: with a taproot shape.

TAPROOT: a long, tapering main root that develops more than secondary and lateral roots.

TAXA: plural of taxon.

TAXON: a systematic group belonging to any rank (e.g., family, genus, species) and characterized by a name (which is a monomial above the species rank, a binomial in the case of species, and a polynomial below the species rank).

TENACIOUS: strong, resistant.

TENDRIL (or **CIRRUS**): a prehensile organ, slender and long, derived from the transformation of the caulis, leaves, or roots, which allows nonvoluble plants to wrap themselves around supports for climbing.

TEPAL: a constituent element of the perigonium of the flower (perianth not differentiated into calyx and corolla).

TERETE: describes a circular organ in cross-section.

TERMINAL: placed at the top, at the apex (of leaf, stem, etc.).

TETRACHENIUM: an indehiscent dry fruit, schizocarp, also called a microbasarium, which separates into four monocarpellar achene-like portions at maturity.

TETRAGON: with four corners.

TETRAMER: a flower with a calyx and corolla made up of four elements.

TETRASTICHOUS: arranged in four vertical rows.

THORN: an organ modified to form an acuminate structure; there are thorns of different origins: twigs (e.g., cauline thorn in *Prunus spinosa*), leaves (e.g., leaf thorn in *Opuntia ficus-indica*), stipules (e.g., *Robinia pseudoacacia*), and prickles of epidermal origin (e.g., *Rosa*).

THORNLESS: without thorns.

THYRSIFORM: shaped like a thyrsus.

THYRSUS: a type of compound inflorescence, heterogeneous, usually ovoid in shape, in which the main axis of undefined type (racemose) bears numerous lateral and opposite dichasial branches, simple or compound.

TILLERING: the development of new stems at the base (or collar) of an herbaceous plant.

TOMENTOSE: covered with dense, intertwined hairs.

TOMENTUM: a set of dense, intertwined hairs.

TOOTH: a notch or dentiform protrusion.

TRANSVERSE: describes a plane orthogonal (or horizontal) to the main axis of the plant or part of it.

TREE: a woody plant that produces a main trunk and a more or less distinct crown.

TREELINE: the altitudinal limit beyond which conditions are inhospitable to tree life.

TRI-: a prefix indicating the number 3 in compound words (e.g., trilobate: three-lobed).

TRICHOMES (or **HAIRS**): epidermal appendages formed by one or more cells.

TRICOCCOUS: describes a fruit consisting of three carpels.

TRIFID: divided into three parts (stem, corolla, compound fruit).

TRIFOLIATE: describes a leaf composed of three leaflets.

TRIGONE: an elongated organ with a triangular cross-section.

TRILOBATE: divided into three lobes.

TRIMEROUS: describes a floral or leaf verticil composed of three pieces or a multiple of three.

TRIPLOID: describes a cell or organism with three haploid chromosome sets.

TRUNCATED: describes an organ that ends abruptly, as if by a clean cut.

TUBE: the lower portion of the calyx or corolla.

TUBER: a plant organ, more or less globular, with few or no leaves, enlarged by the accumulation of reserve material, resulting from the transformation of an underground stem.

TUBULOSE: tubular shaped.

UMBEL: an indefinite inflorescence in which the flowers are carried by peduncles (rays) of equal length, all departing from the same point (simple umbel).

UMBELLATE: describes a plant or its inflorescence whose flowers are arranged in such a way as to resemble an umbrella. Also, describes an umbel making up a compound umbel.

UMBILICUS: the central part, generally depressed, of a structure.

UMBO: the mamillary protuberance of some organs (e.g., the scales of the strobili).

UMBONATE: having a protuberance.

UNDULATED: describes a wavy leaf margin.

UNI-: a prefix indicating the number 1 in compound words (e.g., unilocular: with one cavity).

UNILATERAL: with organs carried on one side only.

UNILOCULAR: describes an organ formed by a single cavity.

UNISEXUAL (**DICLINOUS**): of a flower having only stamens or only pistils.

UNISEXUAL FLOWERS: flowers containing only a gynoecium or androecium (pistilliferous or staminiferous).

URCEOLATE: describes the gamopetalous corolla of a flower, with almost no tube, a bulging flap in the middle, and narrowed at the fauces in the form of a jar (*Erica*, *Arbutus*, *Vaccinum*); also, a gamosepalous calyx with a relatively large tube, bulging, and a little-developed flap (*Hyoscyamus*).

VALVE: each of the elements by which a dehiscent dry fruit opens.

VEGETATIVE REPRODUCTION: the generation of new individuals without an exchange of genetic material (e.g., root suckers).

VEINS: leaf scaffold made up of cribro-vascular bundles; there are sometimes main and secondary ones; depending on the arrangement, the leaf can be defined as parallelinervate, penninervate, or palminervate.

VENTRICOSE: dilated, swollen, and generally pointing downward.

VERTICIL (or **WHORL**): a set of flowers or leaves inserted in groups of three or more on the same node of the main axis.

VERTICILLASTER: a false verticil formed by several cymes surrounding a branch; the flowers appear to be arranged in a verticillate whorl as found in Lamiaceae.

VERTICILLATE: arranged in a whorl.

VESTIGIAL: describes an organ or structure that was once functional and is currently present in an approximate, rudimentary form.

VEXILLUM: the upper petal of the corolla in Fabaceae.

VISCOUS: covered with a sticky substance.

VOLUBLE: describes a plant with a thin, weak, herbaceous stem, endowed with a winding motion to attach itself to a support; the direction of the spiral is usually constant, so there are right-handed species (clockwise movement) and left-handed species (counterclockwise movement).

WING: a laminar expansion of various types of organ; also, two of the five petals of the papilionaceous flower.

WINGED: describes an organ with thin, wing-like expansions.

WOOD: the complex plant tissue that forms the vascular system for transporting raw sap; construction material made from the trunk of some trees.

WOODY: with the texture of wood.

WOOLLY: covered with soft, wiry hair.

WRINKLED: forming wrinkles, folds.

ZYGOMORPHIC: describes a flower with a plane of symmetry such that the right half is specular to the left.

Bibliography

Abdel-Sattar E., Zaitoun A. A., Farag M. A., El Gayed S. H., Harraz F. M. H. 2010. "Chemical composition, insecticidal and insect repellent activity of *Schinus molle* L. leaf and fruit essential oils against Trogoderma granarium and Tribolium castaneum." *Natural Product Research* 24 (3): 226–235.

Abegaz B. M., Kebede T. 1995. "Geshoidin: a bitter principle of *Rhamnus prinoides* and other constituents of the leaves." *Bulletin of the Chemical Society of Ethiopia* 9 (2): 107–114.

Abram V., Čeh B., Vidmar M., Hercezi M., Lazić N., Bucik V., Možina S. S., Košir I. J., Kač M., Demšar L., Poklar Ulrih N. 2014. "A comparison of antioxidant and antimicrobial activity between hop leaves and hop cones." *Industrial Crops and Products* 64: 124–134.

Adekunle O. K., Acharya R., Singh B. 2007. "Toxicity of pure compounds isolated from *Tagetes minuta* oil to *Meloidogyne incognita*." *Australasian Plant Disease Notes* 2: 101–104.

Adom M. B., Taher M., Mutalabisin M. F., Amri M. S., Kudos M. B. A., Sulaiman M. W. A. W., Sengupta P., Susanti D. 2017. "Chemical constituents and medical benefits of *Plantago major*." *Biomedicine & Pharmacotherapy* 96: 348–360.

Ağalar H. G., Demirci B., Başer K. H. C. 2014. "The volatile compounds of elderberries (*Sambucus nigra* L.)." *Natural Volatiles & Essential Oils* 1 (1): 51–54.

Ahmedt D. 1990. "Isolation and structural studies of the chemical constituents of *Buxus* species." Ph.D. dissertation. University of Karachi, H. E. J. Research Institute of Chemistry, Karachi City, Sindh, Pakistan.

Akhbari M., Batooli H., Kashi F. J. 2012. "Composition of essential oil and biological activity of extracts of *Viola odorata* L. from Central Iran." *Natural Product Research* 26 (9): 802–809.

Al-Andal A., Moustafa M., Alruman S. 2017. "Taxonomic variation among *Schinus molle* L. plants associated with a slight change in elevation." *Bangladesh Journal of Plant Taxonomy* 24 (2): 205–214.

Albulushi S. M. A., Al Saidi H., Amaresh N., Mullaicharam A. R. 2014. "Study of physicochemical properties, antibacterial and GC-MS analysis of essential oil of the aniseed (*Pimpinella anisum* Linn.) in Oman." *Research & Reviews: Journal of Pharmacognosy and Phytochemistry* 2 (4): 24–33.

Aliu S., Rusinovci I., Fetahu S., Salihu S., Zogaj R. 2012. "Nutritive and mineral composition in a collection of *Cucurbita pepo* L. grown in Kosova." *Food and Nutrition Sciences* 3: 634–638.

Alizadehl M., Aghaeil M., Saadatian M., Sharifian I. 2012. "Chemical composition of essential oil of *Artemisia vulgaris* from West Azerbaijan, Iran." *Electronic Journal of Environmental, Agricultural and Food Chemistry* 11 (5): 493–496.

Al-Qudah M. A., Al-Jaber H. I., Muhaidat R., Hussein E., Al Abdel Hamid A., Al-Smadi M. L., Abaza I. F., Afifi F. U., Abu-Orabi S. T. 2011. "Chemical composition and antimicrobial activity of the essential oil from *Sinapis alba* L. and *Sinapis arvensis* L. (Brassicaceae) growing wild in Jordan." *Research Journal of Pharmaceutical, Biological and Chemical Sciences* 2 (4): 1135–1144.

Al-Snafi A. E. 2015. "The chemical constituents and pharmacological effects of C*henopodium album*: an overview." *International Journal of Pharmacological Screening Methods* 5 (1): 10–17.

———. 2016. "The constituents and pharmacology effects of *Cnicus benedictus*: a review." *Pharmaceutical and Chemical Journal* 3 (2): 129–135.

Amiri H. 2011. "Volatile constituents and antioxidant activity of flowers, stems and leaves of *Nasturtium officinale* R. Br." *Natural Product Research* 26 (2): 109–115.

Amri I., Hanana M., Jamoussi B., Hamrouni L. 2017. "Essential oils of *Pinus nigra* J.F. Arnold subsp. *laricio* Maire: chemical composition and study of their herbicidal potential." *Arabian Journal of Chemistry* 10 (suppl 2): S3877–S3882.

Aouinti F., Zidane H., Tahri M., Wathelet J.-P., El Bachiri A. 2014. "Chemical composition, mineral contents and antioxidant activity of fruits of *Pistacia lentiscus* L. from Eastern Morocco." *Journal of Materials and Environmental Science* 5 (1): 199–206.

Archak S., Gaikwad A. B., Gautam D., Rao E. V. V. B., Swamy K. R. M., Karihaloo J. L. 2003. "DNA fingerprinting of Indian cashew (*Anacardium occidentale* L.) varieties using RAPD and ISSR techniques." *Euphytica* 230: 397–404.

Argus G. W. 2007. "*Salix* (Salicaceae) distribution maps and a synopsis of their classification in North America, north of Mexico." *Harvard Papers in Botany* 12 (2): 335–368.

Arunkumar A. N., Joshi G. 2014. "*Pterocarpus santalinus* (Red Sanders) an endemic, endangered tree of India: current status, improvement and the future." *Journal of Tropical Forestry and Environment* 4 (2): 1–10.

Asgarpanah J., Ariamanesh A. 2015. "Phytochemistry and pharmacological properties of *Myrtus communis* L." *Indian Journal of Traditional Knowledge* 14 (1): 82–87.

Ata A., Naz S., Choudhary M. I., Ur-Rahman A., Senerc B., Turkozc S. 2002. "New triterpenoidal alkaloids from *Buxus sempervirens*." *Zeitschrift für Naturforschung* 57c: 21–28.

Avigad B., Danin A. 1972. *Flowers of Jerusalem*. E. Jerusalem: Lewin-Epstein.

Azamthulla M., Balasubramanian R., Kavimani S. 2015. "A review on *Pterocarpus santalinus* Linn." *World Journal of Pharmaceutical Research* 4 (2): 282–292.

Azimova S. S., Glushenkova A. I. 2012. *Lipids, Lipophilic Components and Essential Oils from Plant Sources*. London: Springer.

Aziz M., Anwar M., Uddin Z., Amanat H., Ayub H., Jadoon S. 2013. "Nutrition comparison between genus of Apple (*Malus sylvestris* and *Malus domestica*) to show which cultivar is best for the province of Balochistan." *Journal of Asian Scientific Research* 3 (4): 417–424.

Azizov U. M., Khadzhieva U. A., Rakhimov D. A., Mezhlumyan L. G., Salikhov S. A. 2012. "Chemical composition of dry extract of *Arctium lappa* roots." *Chemistry of Natural Compounds* 47 (6): 1038–1039.

Bachrouch O., Mediouni-Ben Jemâa J., Waness Wissem A., Talou T., Marzouk B., Abderraba M. 2010. "Composition and insecticidal activity of essential oil from *Pistacia lentiscus* L. against *Ectomyelois ceratoniae* Zeller and *Ephestia kuehniella* Zeller (Lepidoptera: Pyralidae)." *Journal of Stored Products Research* 46: 242–247.

Badgujar S. B., Patel V. V., Bandivdekar A. H. 2014. "*Foeniculum vulgare* Mill: a review of its botany, phytochemistry, pharmacology, contemporary application, and toxicology." *BioMed Research International* 2014: 842674.

Bais S., Gill N. S., Rana N., Shandil S. 2014. "A phytopharmacological review on a medicinal plant: *Juniperus communis*." *International Scholarly Research Notices* 2014: 634723.

Bajer T., Janda V., Bajerová P., Kremr D., Eisner A., Ventura K. 2016. "Chemical composition of essential oils from *Plantago lanceolata* L. leaves extracted by hydrodistillation." *Journal of Food Science and Technology* 53 (3): 1576–1584.

Barragan Ferrer D., Rimantas Venskutonis P., Talou T., Zebibe B., Barragan Ferrer J. M., Merah O. 2016. "Bioactive compounds and antioxidant properties of *Myrrhis odorata* deodorized residue leaves extracts from Lithuania and France origins." *Pharmaceutical and Chemical Journal* 3 (3): 43–48.

Barroso J. G., Pedro L. G., Figueiredo C., Pais M. S. S., Scheffer J. J. C. 1992. "Seasonal variation in the composition of the essential oil of *Crithmum maritimum* L." *Flavour and Fragrance Journal* 7 (3): 147–150.

Basar S. 2005. "Comparative studies on the essential oils, pyrolysates and boswellic acids of *Boswellia carterii* Birdw., *Boswellia serrata* Roxb., *Boswellia frereana* Birdw., *Boswellia neglecta* S. Moore and *Boswellia rivae* Engl." Dissertation for the fulfillment of the requirements for the degree of Doctor Rerum Naturalium. Universität Hamburg, Germany.

Basas J., López J. F., De Las Heras F. X. C. 2015. "Labdane-type diterpenoids from *Juniperus communis* needles." *Industrial Crops and Products* 76: 333–345.

Basch E., Bent S., Foppa I., Haskmi S., Kroll D., Mele M., Szapary P., Ulbricht C., Vora M., Yong S. 2006. "Marigold (*Calendula officinalis* L.): an evidence-based systematic review by the natural standard research collaboration." *Journal of Herbal Pharmacotherapy* 6 (3/4): 135–159.

Baser K. H. C., Demirci B., Dönmez A. A. 2003. "Composition of the essential oil of *Perilla frutescens* (L.) Britton from Turkey." *Flavour and Fragrance Journal* 18: 122–123.

Baser K. H. C., Ozek. T., Kırımer N. A., Orhan D., Ergun F. 2004. "Composition of the essential oils of *Galium aparine* L. and *Galium odoratum* (L.) Scop. from Turkey." *Journal of Essential Oil Research* 16: 305–307.

Bastos A. O., Moreira I., Furlan A. C., Fraga A. L., De Oliveira R. P., De Oliveira E. 2005. "Composição química, digestibilidade dos nutrientes e da energia de diferentes milhetos (*Pennisetum glaucum* [L.] R. Brown) em suínos." *Revista Brasileira de Zootecnia* 34 (2): 520–528.

Baudar P. 2018. *The Wildcrafting Brewer: Creating Unique Drinks and Boozy Concoctions from Nature's Ingredients. Primitive Beers, Country Wines, Herbal Meads, Natural Sodas, and More*. White River Junction, Vt.: Chelsea Green.

Bayramoglu E. E. E. 2006. "Antibacterial activity of *Myrtus communis* essential oil used in soaking." *Journal of the Society of Leather Technologists and Chemists* 90 (5): 217–219.

Behxhet M., Dashnor N., Avni H. 2016. "Chemical composition of the essential oils of *Juniperus communis* subsp. *alpina* (Suter) Čelak (Cupressaceae)." *Macedonian Pharmaceutical Bulletin* 62: 479–480.

Belleau F., Collin G. 1993. "Composition of the essential oil of *Ledum groenlandicum*." *Phytochemistry* 33 (1): 117–121.

Belocchi A., Cecchini C., Fornara M., Mazzon V., Mortaro R., Quaranta F. 2016. "Frumento duro, le varietà consigliate." *Agricoltura* 24. 16 September.

Benyelles B., Allali H., Fekih N., Touaibia M., Muselli A., Djabou N., El Amine Dib M., Tabti B., Costa J. 2015. "Chemical composition of the volatile components of *Tropaeolum majus* L. (Garden nasturtium) from North Western Algeria." *PhytoChem & BioSub Journal* 9 (3): 92–97.

Bernacchia R., Preti R., Vinci G. 2014. "Chemical composition and health benefits of flaxseed." *Austin Journal of Nutrition and Food* 2 (8): 1045.

Bertão M. R., Moraes M. C., Palmieri D. A., Pereira Silva L., Da Silva R. M. G. 2016. "Cytotoxicity, genotoxicity and antioxidant activity of extracts from *Capsicum* spp. Res." *Journal of Medicinal Plants Research* 10 (4): 265–275.

Bertucci B., Lo Conte N. 2018. *La tua birra fatta in casa con il PLC*. Albino (Bergamo), Italy: Sandit Libri.

Bhalla P., Bajpai V. K. 2017. "Chemical composition and antibacterial action of *Robinia pseudoacacia* L. flower essential oil on membrane permeability of foodborne pathogens." *Journal of Essential Oil Bearing Plants* 20 (3): 632–645.

Bicchi C., Nano G. M., Frattini C. 1982. "On the composition of the essential oils of *Artemisia genipi* Weber and *Artemisia umbelliformis* Lam." *European Food Research and Technology* 175 (3): 182–185.

Bijelic S. M., Golosin B. R., Todorovic J. I. N., Cerovic S. B., Popovic B. M. 2011. "Physicochemical fruit characteristics of cornelian cherry (*Cornus mas* L.) genotypes from Serbia." *HortScience* 46 (6) : 849–853.

Bilalis D., Karkanis A., Patsiali S., Agriogianni M., Konstantas A., Triantafyllidis V. 2011. "Performance of wheat varieties (*Triticum aestivum* L.) under conservation tillage practices in organic agriculture." *Notulae Botanicae Horti Agrobotanici Cluj-Napoca* 39 (2): 28–33.

Blomme E., Van Calenbergh S., Maggi F., Risseeuw M., De Vos F. 2015. "Chemical analysis and biological activity of *Schizogyne sericea* and *Lonicera caerulea*." Master's thesis. Ghent University, Flanders, Belgium, University of Camerino, Marche, Italy.

Bonner F. T. 1974. *Chemical Components of Some Southern Fruits and Seeds*. U.S. Forest Service Research Note SO-193. New Orleans, La.: U.S. Department of Agriculture.

Booth D. 1826. *The Art of Brewing*. London: F. G. Mason.

Bos R., Woerdenbag H. J., Kayser O., Quax W. J. 2007. "Essential oil constituents of *Piper cubeba* L. fils. from Indonesia." *Journal of Essential Oil Research* 19: 14–17.

Bouaziz M. A., Rassaoui R., Besbes S. 2014. "Chemical composition, functional properties, and effect of inulin from Tunisian *Agave americana* L. leaves on textural qualities of pectin gel." *Journal of Chemistry* 2014: 758697.

Bourgou S., Rahali F. Z., Ourghemmi I., Tounsi M. S. 2012. "Changes of peel essential oil composition of four Tunisian Citrus during fruit maturation." *Scientific World Journal* 2012: 528593.

Boutkhil S., El Idrissi M., Amechrouq A., Chbicheb A., Chakir S., El Badaoui K. 2009. "Chemical composition and antimicrobial activity of crude, aqueous, ethanol extracts and essential oils of *Dysphania ambrosioides* (L.) Mosyakin & Clemants." *Acta Botanica Gallica* 156 (2): 201–209.

Brandis D. 1909. *Indian Trees*. London: Archibald Constable.

Brandolini A., Hidalgo A., Moscaritolo S. 2008. "Chemical composition and pasting properties of einkorn (*Triticum monococcum* L. subsp. *monococcum*) whole meal flour." *Journal of Cereal Science* 47 (3): 599–609.

Brenneisen R. 2007. "Chemistry and analysis of phytocannabinoids and other cannabis constituents." In *Marijuana and the Cannabinoids*, ed. M. A. ElSohly. Totowa, N.J.: Humana.

Bresson J.-L., Flynn A., Heinonen M., Hulshof K., Korhonen H., Lagiou P., Løvik M., Marchelli R., Martin A., Moseley B., Przyrembel H., Salminen S., Strain J. J., Strobel S., Tetens I., Van den Berg H., Van Loveren H., Verhagen H. 2009. "Opinion on the safety of 'Chia seeds (*Salvia hispanica* L.) and ground whole Chia seeds' as a food ingredient. Scientific opinion of the Panel on Dietetic Products, Nutrition and Allergies (Question No. EFSA-Q-2008-008) adopted on 13 March 2009." *EFSA Journal* 996: 1–26.

Brickell C. D., Alexander C., Cubey J. J., David J. C., Hoffman M. H. A., Leslie A. C., Valéry Malécot Jin X. 2016. *International Code of Nomenclature for Cultivated Plants (ICNCP or Cultivated Plant Code) Incorporating the Rules and Recommendations for Naming Plants in Cultivation*, 9th ed. Adopted by the International Union of Biological Sciences International Commission for the Nomenclature of Cultivated Plants, International Society for Horticultural Science.

Britton N. L., Rose J. N. 1919. *The Cactaceae: Descriptions and Illustrations of Plants of the Cactus Family*, vol. 1. Washington, D.C.: Carnegie Institution of Washington.

Brophy J. J., Goldsack R. J. 2003. "Chemistry of the Australian gymnosperms. Part 5: Leaf essential oils of some endemic Tasmanian gymnosperms: *Diselma archeri*, *Lagarostrobos franklinii*, *Microcachrys tetragona* and *Phyllocladus aspleniifolius*." *Journal of Essential Oil Research* 15: 217–220.

Buhner S. H. 1998. *Sacred and Herbal Healing Beer: The Secret of Ancient Fermentation*. Boulder, Colo.: Brewers.

Bumblauskiene L., Jakstas V., Janulis V., Mazdzieriene R., Ragazinskiene O. 2009. "Preliminary analysis on essential oil composition of *Perilla* L. cultivated in Lithuania." *Acta Poloniae Pharmaceutica & Drug Research* 66 (4): 409–413.

Calagione S., Alstrom J., Alstrom T. 2018. *Project Extreme Brewing: An Enthusiast's Guide to Extreme Brewing at Home.* Beverly, Mass.: Quarry.

Calegari L., Gatto D. A., Stangerlin D. M., Martins S. V., Agnes C. C., Durlo M. A. 2009. "Caracterização de povoamentos de *Myrocarpus frondosus* M. Allemão na região central do Rio Grande do Sul." *Revista Científica Eletrônica de Engenharia Florestal* 8 (14): 18–28.

Cantwell D., Bouckaert P. 2016. *Wood and Beer: A Brewer's Guide.* Boulder, Colo.: Brewers.

——. 2018. *Legno e birra. Guida per il birraio.* Milan: Edizioni LSWR.

Caruso G. 2011. *Guida al riconoscimento di alberi, arbusti, cespugli e liane del Parco Nazionale della Sila con iconografia botanica ed appendici didattiche.* Lorica di San Giovanni in Fiore (Cosenza), Italy: Parco Nazionale della Sila-Calabria Trekking. PN Sila.

——. 2015. *Andar per piante tra terra e mare. Escursioni botaniche sulle coste della Calabria.* Königstein, Germany: Koeltz Scientific.

Cattani D. J. 2016. "Selection of a perennial grain for seed productivity across years: intermediate wheatgrass as a test species." *Canadian Journal of Plant Science* 97 (3).

Cautela D., Pirrello A. G., Esposito C., Minasi P. 2004. "Caratteristiche compositive del chinotto (*Citrus myrtifolia*): Parte I." *Essenze Derivati Agrumari* 74: 49–55.

Čerenak A., Pavlovič M., Oset Luskar M., Košir I. Z. 2011. "Characterisation of Slovenian hop *(Humulus lupulus* L.) varieties by analysis of essential oil." *Hop Bulletin* 18: 27–32.

Čerenak A., Satovic Z., Jakse J., Luthar Z., Čarovic-Stanko K., Javornik B. 2009. "Identification of QTLs for alpha acid content and yield in hop (*Humulus lupulus* L.)." *Euphytica* 170: 141–154.

Čerenak A., Satovic Z., Javornik B. 2006. "Genetic mapping of hop (*Humulus lupulus* L.) applied to the detection of QTLs for alpha acid content." *Genome* 49 (5) 485–494.

Cettolin G., Baini G., Governa P., Borgonetti V., Barbato C., Cocetta V., Montopoli M., Miraldi E., Giachetti D., Calamai L., Biagi M. 2017. "*Calluna vulgaris* Hull: chemical composition and biological properties." S. I. Fit. Young Researchers Project, November 20, Republic of San Marino.

Chadwick L. R., Nikolic D., Burdette J. E., Overk C. R., Bolton J. L., Van Breemen R. B., Fröhlich R., Fong H. H. S., Farnsworth N. R., Pauli G. F. 2004. "Estrogens and congeners from spent hops (*Humulus lupulus*)." *Journal of Natural Products* 67 (12): 2024–2032.

Chadwick L. R., Pauli G. F., Farnsworth N. R. 2006. "The pharmacognosy of *Humulus lupulus* L. (hops) with an emphasis on estrogenic properties." *Phytomedicine* 13: 119–131.

Chamorro E. R., Ballerini G., Sequeira A. F., Velasco G.A, Zalazara M. F. 2008. "Chemical composition of essential oil from *Tagetes minuta* L. leaves and flowers." *Journal of the Argentine Chemical Society* 96 (1–2): 80–86.

Chávez-Servín J. L., Cabrera-Baeza H. F., Jiménez Ugalde E. A., Mercado-Luna A., De La Torre-Carbot K., Escobar-García K., Barreyro A. A., Serrano-Arellano J., García-Gasca T. 2017. "Comparison of chemical composition and growth of amaranth (*Amaranthus hypochondriacus*) between greenhouse and open field systems." *International Journal of Agriculture and Biology* 19: 577–583.

Chen Z.-Y., Lin Y.-S., Liu X.-M., Cheng J.-R., Yang C.-Y. 2017. "Chemical composition and antioxidant activities of five samples of *Prunus mume* Umezu from different factories in South and East China." *Journal of Food Quality* 2017: 4878926.

Cheong M. W., Chong Z. S., Liu S. Q., Zhou W., Curran P., Yu B. 2012. "Characterisation of calamansi (*Citrus microcarpa*). Part I: Volatiles, aromatic profiles and phenolic acids in the peel." *Food Chemistry* 134: 686–695.

Cheong M. W., Zhu D., Sng J., Liu S. Q., Zhou W., Curran P., Yu B. 2012. "Characterisation of calamansi (*Citrus microcarpa*). Part II: Volatiles, physicochemical properties and non-volatiles in the juice." *Food Chemistry* 134: 696–703.

Chizzola R. 2010. "Composition of essential oil from *Daucus carota* ssp. *carota* growing wild in Vienna." *Journal of Essential Oil-Bearing Plants* 13 (1): 12–19.

Choudhrury D., Sahu J. K., Sharm G. D. 2012. "Bamboo shoot: microbiology, biochemistry and technology of fermentation: a review." *Indian Journal of Traditional Knowledge* 11 (2): 242–249.

Chrenková M., Gálová Z., Čerešňáková Z., Sommer A. 2001. "Nutritional and biological value of spelt wheat." Special number (*Proceedings of the International Scientific Conference on the Occasion of the 55th Anniversary of the Slovak Agricultural University in Nitra*), Acta Fytotechnica et Zootechnica 4.

Chryssavgi G., Vassiliki P., Athanasios M., Kibouris T., Michael K. 2008. "Essential oil composition of *Pistacia lentiscus* L. and *Myrtus communis* L.: evaluation of antioxidant capacity of methanolic extracts." *Food Chemistry* 107: 1120–1130.

Chudiwal A. K., Jain D. P., Somani R. S. 2010. "*Alpinia galanga* Willd. An overview on phyto-phamacological properties." *Indian Journal of Natural Products and Resources* 1 (2): 143–149.

Chudnoff M. 1979. *Tropical Timbers of the World*. Agricultural Handbook Number 607. Washington, D.C.: U.S. Department of Agriculture, Forest Service.

Clayton W. D., Vorontsova M. S., Harman K. T., Williamson H. 2006–2018. *GrassBase – The Online World Grass Flora*. http://www.kew.org/data/grasses-db.html.

Coenraads P. J., Dusinska M., Lilienblum W., Nielsen E., Platzek T., Rastogi S. C., Rousselle C., Van Benthem J., Bernard A., Giménez-Arnau A., Vanhaecke T. 2015. *Opinion on the fragrance ingredients* Tagetes minuta *and* T. patula *extracts and essential oils (phototoxicity only)*. Luxembourg: European Commission, Scientific Committee on Consumer Safety.

Coiffard L., Piron-Frenet M., Amicel L. 1993. "Geographical variations of the constituents of the essential oil of *Crithmum maritimum* L., Apiaceae." *International Journal of Cosmetic Science* 15: 15–21.

Collin G. 2015. "Aromas from Quebec. IV. Chemical composition of the essential oil of *Ledum groenlandicum*: a review." *American Journal of Essential Oils and Natural Products* 2 (3): 6–11.

Cömert Önder F., Ay M., Sarker S. D. 2013. "Comparative study of antioxidant properties and total phenolic content of the extracts of *Humulus lupulus* L. and quantification of bioactive components by LC-MS/MS and GC-MS." *Journal of Agricultural and Food Chemistry* 61: 10498–10506.

Contreras L. E., Jaimez O. J., Castañeda O. A., Añorve M. J., Villanueva R. S. 2011. "Sensory profile and chemical composition of *Opuntia joconostle* from Hidalgo, Mexico." *Journal of Stored Products and Postharvest Research* 2 (2): 37–39.

Cornell M. 2010. *Amber, Gold and Black: The History of Britain's Great Beers*. Stroud, UK: History.

Couladis M., Tzakou O., Mimica-Dukic N., Jancic R., Stojanovic D. 2002. "Essential oil *of Salvia officinalis* L. from Serbia and Montenegro." *Flavour and Fragrance Journal* 17: 119–126.

Cubero J. 2012. "Hops (*Humulus lupulus* L.) and beer: benefits on the sleep." *Journal of Sleep Disorders and Therapy* 1 (1): 1–3.

Čurná V., Lacko-Bartošová M. 2017. "Chemical composition and nutritional value of emmer wheat (*Triticum dicoccon* Schrank): a review." *Journal of Central European Agriculture* 18 (1) : 117–134.

Dabove L. 2015. *La birra non esiste*. Milan: Altreconomia Edizioni.

Daniels R. 2016. *Progettare grandi birre. La guida definitiva per produrre gli stili classici della birra*. Milan: Edizioni LSWR.

Daniels R., Larson G. 2000. *Smoked Beers*. Boulder, Colo.: Brewers.

Danilova T. V., Danilov S. S., Karlov G. I. 2003. "Assessment of genetic polymorphism in hop (*Humulus lupulus* L.) cultivars by ISSR-PCR analysis." *Russian Journal of Genetics* 39 (11): 1252–1257.

Das A., Santhy K. S. 2015. "Chemical characterisation of *Alpinia galanga* (L.) Willd by GC–MS, XRD, FTIR and UV-VIS spectroscopic methods." *International Journal of Pharmaceutical Sciences and Research* 7 (9): 499–501.

De Aguiar A. C., Pereira Coutinho J., Barbero G. F., Teixeira Godoy H., Martínez J. 2016. "Comparative study of capsaicinoid composition in *Capsicum* peppers grown in Brazil." *International Journal of Food Properties* 19 (6): 1292–1302.

De Carvalho M. G., Carvalho Cranchi D., De Carvalho A. G. 1996. "Chemical constituents from *Pinus strobus* var. chiapensis." *Journal of the Brazilian Chemical Society* 7 (3): 187–191.

De Falco E., Mancini E., Roscigno G., Mignola E., Taglialatela-Scafati O., Senatore F. 2013. "Chemical composition and biological activity of essential oils of *Origanum vulgare* L. subsp. *vulgare* L. under different growth conditions." *Molecules* 18: 14948–14960.

De Keukeleire J., Janssens I., Heyerick A., Ghekiere G., Cambie J., Roldan-Ruiz I., Van Bockstaele E., De Keukeleire D. 2007. "Relevance of organic farming and effect of climatological conditions on the formation of α-acids, β-acids, desmethylxanthohumol, and xanthohumol in hop (*Humulus lupulus* L.)." *Journal of Agricultural and Food Chemistry* 55: 61–66.

De Oliveira G., Brunsell N. A., Sutherlin C. E., Crews T. E., DeHaan L. R. 2018. "Energy, water and carbon exchange over a perennial kernza wheatgrass crop." *Agricultural and Forest Meteorology* 249: 120–137.

Debnath S. C., McRae K. B. 2001. "In vitro culture of lingonberry (*Vaccinium vitis-idaea* L.)." *Small Fruits Review* 1 (3): 3–19.

Deeds S. 2013. *Brewing Engineering: Great Beer Through Applied Science*, 2nd ed. Scotts Valley, Calif.: CreateSpace.

DeLong J. M., Prange R. K. 2008. "Fiddlehead fronds: nutrient rich delicacy." *Chronica Horticulturae* 48 (1): 12–15.

Derbal W., Zerizer A., Gerard J. 2013. "Caractérisation physico-mécanique de trois espècese bois algériens en vue de la fabrication de carrelets 3-plis pour des menuiseries intérieures." Poster presentation. Deuxièmes journées scientifiques du Groupement de recherches Sciences du Bois. Champs sur Marne, 19–21 November.

Dessi M. A., Deiana M., Rosa A., Piredda M., Cottiglia F., Bonsignore L., Deidda D., Pompei R., Corongiu F. P. 2001. "Antioxidant activity of extracts from plants growing in Sardinia." *Phytotherapy Research* 15: 511–518.

Dhifi W., Jelali N., Chaabani E., Beji M., Fatnassi S., Omri S., Mnif W. 2013. "Chemical composition of lentisk (*Pistacia lentiscus* L.) seed oil." *African Journal of Agricultural Research* 8 (16): 1395–1400.

Dias M. I., Barros L., Dueñas M., Pereira E., Carvalho A. M., Alves R. C., Oliveira M. B. P. P., Santos-Buelga C., Ferreira I. C. 2013. "Chemical composition of wild and commercial *Achillea millefolium* L. and bioactivity of the methanolic extract, infusion and decoction." *Food Chemistry* 141: 4152–4160.

Díaz C., Quesada S., Brenes O., Aguilar G., Cicció J. F. 2008. "Chemical composition of *Schinus molle* essential oil and its cytotoxic activity on tumour cell lines." *Natural Product Research* 22 (17) : 1521–1534.

Diguet L. 1928. *Les Cactacées Utiles du Mexique. Archives d'Histoire Naturelle IV*. Paris: Société Nationale d'Acclimatation de France.

Ditchkoff S. S., Lewis J. S., Lin J. C., Muntifering R. B., Chappelka A. H. 2009. "Nutritive quality of highbush blackberry (*Rubus argutus*) exposed to tropospheric ozone." *Rangeland Ecology & Management* 62: 364–370.

Đorđević A. S. 2015. "Chemical composition of *Hypericum perforatum* L. essential oil." *Advanced Technologies* 4 (1): 64–68.

Dorji K., Yapwattanaphun C. 2011. "Morphological identification of mandarin (*Citrus reticulata* Blanco) in Bhutan." *Kasetsart Journal (Natural Science)* 45: 793–802.

Douglas M. H., Van Klink J. W., Smallfield B. M., Perry N. B., Anderson R. E., Johnstone P., Weavers R. T. 2004. "Essential oils from New Zealand manuka: triketone and other chemotypes of *Leptospermum scoparium*." *Phytochemistry* 65: 1255–1264.

Duman E., Özcan M. M. 2017. "The chemical composition of *Achillea wilhelmsii*, *Leucanthemum vulgare* and *Thymus citriodorus* essential oils." *Journal of Essential Oil Bearing Plants* 20 (5): 1310–1319.

Dung N. X., Cu L. D., Khiên P. V., Muselli A., Casanova J., Barthel A., Leclercq P. A. 2011. "Volatile constituents of the stem and leaf oils of *Eupatorium coelestinum* L. from Vietnam." *Journal of Essential Oil Research* 10 (5): 478–482.

Ďurechová D., Kačániová M., Terentjeva M., Petrová J., Hleba L., Kata I. 2016. "Antibacterial activity of *Drosera rotundifolia* L. against Gram-positive and Gram-negative bacteria." *Journal of Microbiology, Biotechnology and Food Sciences* 5 (special 1): 20–22.

Duyker E. 2003. *Citizen Labillardière. A Naturalist's Life in Revolution and Exploration*. Melbourne, Australia: Miegunyah.

Ebadollahi A. 2011. "Chemical constituents and toxicity of *Agastache foeniculum* (Pursh) Kuntze essential oil against two stored-product insect pests." *Chilean Journal of Agricultural Research* 71 (2): 212–217.

Eggink P. M. 2013. "A taste of pepper: genetics, biochemistry and prediction of sweet pepper flavor." PhD dissertation. Wageningen University, Wageningen, Netherlands.

Ek S., Kartimo H., Mattila S., Tolonen A. 2006. "Characterization of phenolic compounds from lingonberry (*Vaccinium vitis-idaea*)." *Journal of Agricultural and Food Chemistry* 54: 9834–9842.

Ekpélikpézé O. S., Agre P., Dossou-Aminon I., Adjatin A., Dassou A., Dansi A. 2016. "Characterization of sugarcane (*Saccharum officinarum* L.) cultivars of Republic of Benin." *International Journal of Current Research in Biosciences and Plant Biology* 5 (3): 147–156.

Elfekih S., Abderrabba M. A. 2008. "Effect of *Myrtus communis* essential oil on the Mediterranean fruit fly mating behaviour." *Biosciences Biotechnology Research Asia* 5 (2): 527–540.

El-Ghorab A. H. 2006. "The chemical composition of the *Mentha pulegium* L. essential oil from Egypt and its antioxidant activity." *Journal of Essential Oil Bearing Plants* 9 (2): 183–195.

El-Kalamouni C., Venskutonis P. R., Zebib B., Merah O., Raynaud C., Talou T. 2017. "Antioxidant and antimicrobial activities of the essential oil of *Achillea millefolium* L. grown in France." *Medicines* 4: 30.

El-Kashoury El-Sayeda A., El-Askary H. I., Kandil Z. A., Salem M. A. 2014. "Chemical composition of the essential oil and botanical study of the flowers of *Mentha suaveolens*." *Pharmaceutical Biology* 52 (6): 688–697.

Elmaci Y., Altuğ T. 2002. "Flavour evaluation of three black mulberry (*Morus nigra*) cultivars using GC/MS, chemical and sensory data." *Journal of the Science of Food and Agriculture* 82: 632–635.

Ertürk Ü., Mert C., Soylu A. 2006. "Chemical composition of fruits of some important chestnut cultivars." *Brazilian Archives of Biology and Technology* 49 (2): 183–188.

Eryigit T., Okut N., Ekici K., Yildirim B. 2014. "Chemical composition and antibacterial activities of *Juniperus horizontalis* essential oil." *Canadian Journal of Plant Science* 94: 323–327.

Escarnot E., Jacquemin J.-M., Agneessens R., Paquot M. 2012. "Comparative study of the content and profiles of macronutrients in spelt and wheat: a review." *Biotechnology, Agronomy Society and Environment* 16 (2): 243–256.

Escudero N. L., De Arellano M. L., Luco J. M., Gimenez M. S., Mucciarelli S. I. 2004. "Comparison of the chemical composition and nutritional value of *Amaranthus cruentus* flour and its protein concentrate." *Plant Foods for Human Nutrition* 59: 15–21.

Esmaeili A., Masoudi S., Masnabadi N., Rustaiyan A. H. 2010. "Chemical constituents of the essential oil of *Sanguisorba minor* Scop. leaves, from Iran." *Journal of Medicinal Plants* 9 (35): 67–70.

Eun-Sun H., Gun-Hee K. 2013. "Safety evaluation of *Chrysanthemum indicum* L. flower oil by assessing acute oral toxicity, micronucleus abnormalities, and mutagenicity." *Preventive Nutrition and Food Science* 18 (2): 111–116.

Eyres G., Dufour J. P. 2009. "Hop essential oil: analysis, chemical composition and odor characteristics." *Beer Composition and Properties* 22: 239–254.

Eyres G. T., Marriott P. J., Dufour J. P. 2007. "Comparison of odor-active compounds in the spicy fraction of hop (*Humulus lupulus* L.) essential oil from four different varieties." *Journal of Agricultural and Food Chemistry* 55: 6252–6261.

Falasca A., Caprari C., De Felice V., Fortini P., Saviano G., Zollo F., Iorizzi M. 2016. "GC-MS analysis of the essential oil of *Juniperus communis* L. berries growing wild in the Molise region: seasonal variability and in vitro antifugal activity." *Biochemical Systematics and Ecology* 69: 166–175.

Farag M. A., Wessjohann L. A. 2012. "Volatiles profiling in medicinal licorice roots using steam distillation and solid-phase microextraction (SPME) coupled to chemometrics." *Journal of Food Science* 77 (11): 1179–1184.

Farhoudi R. 2013. "Chemical constituents and antioxidant properties of *Matricaria recutita* and *Chamaemelum nobile* essential oil growing wild in the south west of Iran." *Journal of Essential Oil Bearing Plants* 16 (4): 531–537.

Fathiazad F., Hamedeyazdan S. 2011. "A review on *Hyssopus officinalis* L.: composition and biological activities." *African Journal of Pharmacy and Pharmacology* 5 (17): 1959–1966.

Fernandez De Simon B., Sanz M., Cadahia E., Poveda P., Broto M. 2006. "Chemical characterization of oak heartwood from Spanish forests of *Quercus pyrenaica* (Wild.). Ellagitannins, low molecular weight phenolic, and volatile compounds." *Journal of Agricultural and Food Chemistry* 54: 8314–8321.

Ferrer-Gallego P., Ferrer-Gallego R., Rosello R., Peris J. B., Guillen A., Gomez J., Laguna E. 2014. "A new subspecies of *Rosmarinus officinalis* (Lamiaceae) from the eastern sector of the Iberian Peninsula. *Phytotaxa* 172 (2): 61–70.

Fischer G., Almanza-Merchán P. J., Miranda D. 2014. "Importancia y cultivo de la uchuva (*Physalis peruviana* L.)." *Revista Brasileira de Fruticultura* 36 (1): 1–15.

Fisher J., Fisher D. 2017. *Coltiva i tuoi ingredienti per la birra. Come coltivare, preparare e utilizzare i tuoi luppoli, i tuoi malti e le tue erbe per la birra*. Milan: Edizioni LSWR.

Fisk J. R., Hoover E. 2015. *Wild Fruits of Minnesota: A Field Guide*. Minneapolis: College of Food, Agriculture and Natural Resource Science, University of Minnesota.

Fortesa A. M., Testillano P. S., Risueño M. C., Paisa M. S. 2002. "Studies on callose and cutin during the expression of competence and determination for organogenic nodule formation from internodes of *Humulus lupulus* var. Nugget." *Physiologia Plantarum* 116: 113–120.

Fox G. P. 2009. "Chemical Composition in Barley Grains and Malt Quality." In *Genetics and Improvement of Barley Malt Quality*, ed. Zhang G. and Li C. Hangzhou, China: Zejiang University Press.

Francenzon N., Stevanovic T. 2017. "Chemical composition of essential oil and hydrosol from *Picea mariana* bark residue." *BioResources* 12 (2): 2635–2645.

Frank A., Kugler E. 2001. "Hybrids and cultivars of passion flowers: a checklist for the genus *Passiflora*." *Journal of the Association of Passion Flower Enthusiasts*. https://www.passionflow.co.uk/wp-content/uploads/2017/09/passiflora-hybrid-list-2001.pdf.

Fraternale D., Giampieri L., Bucchini A., Ricci D. 2009. "Antioxidant activity of *Prunus spinosa* L. fruit juice." *Italian Journal of Food Science* 21 (3): 337–346.

Gabriele B., Fazio A., Dugo P., Costa R., Mondello L. 2009. "Essential oil composition of *Citrus medica* L. cv. Diamante (Diamante citron) determined after using different extraction methods." *Journal of Separation Science* 32: 99–108.

Gahlawat D.K, Jakhar S., Dahiya P. 2014. "*Murraya koenigii* (L.) Spreng: an ethnobotanical, phytochemical and pharmacological review." *Journal of Pharmacognosy and Phytochemistry* 3 (3): 109–119.

Gamble G. R. 2003. "Variation in surface chemical constituents of cotton (*Gossypium hirsutum*) fiber as a function of maturity." *Journal of Agricultural and Food Chemistry* 51 (27): 7995–7998.

Garcia G., Tissandié L., Filippi J.-J., Tomi F. 2017. "New pinane derivatives found in essential oils of *Calocedrus decurrens*." *Molecules* 22: 921.

García-Herrera P., Cortes Sánchez-Mata M. C., Cámara M., Tardío J., Olmedilla-Alonso B. 2014. "Carotenoid content of wild edible young shoots traditionally consumed in Spain (*Asparagus acutifolius* L., *Humulus lupulus* L., *Bryonia dioica* Jacq. and *Tamus communis* L.)." *Journal of the Science of Food and Agriculture* 93 (7): 1692–1698.

Giaccone L., Signoroni E. 2017. *Il piacere della birra*. Bra (Cuneo), Italy: Slow Food Editore.

Giampieri F., Tulipani S., Alvarez-Suarez J. M., Quiles J. L., Mezzetti B., Battino M. 2012. "The strawberry: composition, nutritional quality, and impact on human health." *Nutrition* 28: 9–19.

Giordano G. 1964. *I legnami del mondo*. Milan: Ceschina.

Giuffrè A. M. 2007. "Chemical composition of purple passion fruit (*Passiflora edulis* Sims var. edulis) seed oil." *Rivista Italiana Sostanze Grasse* 84 (2): 87–93.

Gloser V., Baláž M., Svoboda P. 2011. "Analysis of anatomical and functional traits of xylem in *Humulus lupulus* L. stems." *Plant, Soil and Environment* 57 (7): 338–343.

Gogorcena Y., Ortiz J. M. 1989. "Characterisation of sour orange (*Citrus aurantium*) cultivars." *Journal of the Science of Food and Agriculture* 48: 275–284.

Góra A., Skórska C., Sitwonska J., Prażmo Z., Krysińska-Traczyk E., Urbanowicz B., Dutkiewicz J. 2004. "Exposure of hop growers to bioaerosols." *Annals of Agricultural and Environmental Medicine* 11: 129–138.

Gudipati V. 2003. "Leaf essential oils of coriander (*Coriandrum sativum* L.) cultivars." *Indian Perfumer* 47 (1): 35–37.

Güney M., Tulin Oz A., Kafkas E. 2015. "Comparison of lipids, fatty acids and volatile compounds of various kumquat species using HS/GC/MS/FID techniques." *Journal of the Science of Food and Agriculture* 95: 1268–1273.

Gupta M., Abu-Ghannam N., Gallaghar E. 2010. "Barley for brewing: characteristic changes during malting, brewing and applications of its by-products." *Comprehensive Reviews in Food Science and Food Safety* 9: 318–328.

Habibi Lahigi S., Amini K., Moradi P., Asaadi K. 2001. "Investigating the chemical composition of different parts extracts of bipod nettle *Urtica dioica* L. in Tonekabon region." *Iranian Journal of Plant Physiology* 2 (1): 339–342.

Hajdari A., Mustafa B., Nebija D., Miftari E., Quave C.L, Novak J. 2015. "Chemical composition of *Juniperus communis* L. cone essential oil and its variability among wild populations in Kosovo." *Chemistry & Biodiversity* 12: 1706–1717.

Hakki Alma M., Eyyüp K., Murat E., Altuntas E., Karaman S., Emel Diraz E. 2012. "Chemical composition of seed oil from Turkish *Prunus mahaleb* L." *Analytical Chemistry Letters* 2 (3): 182–185.

Hakki Kalyoncu I. H., Ersoy N., Yilmaz M. 2009. "Physico-chemical and nutritional properties of cornelian cherry fruits (*Cornus mas* L.) grown in Turkey." *Asian Journal of Chemistry* 21 (8): 6555–6561.

Hamerski L., Vieira Somner G., Tamaio N. 2013. "*Paullinia cupana* Kunth (Sapindaceae): a review of its ethnopharmacology, phytochemistry and pharmacology." *Journal of Medicinal Plants Research* 7 (30): 2221–2229.

Hamidi M., Ziaee M., Delashoub M., Marjani M., Karimitabar F., Khorami A., Ahmadi N. A. 2015. "The effects of essential oil of *Lavandula angustifolia* on sperm parameters quality and reproductive hormones in rats exposed to cadmium." *Journal of Reports in Pharmaceutical Sciences* 4 (2): 121–128.

Hamilton A. 2014. *Brewing Britain: The Quest for the Perfect Pint*. London: Bantam.

Harborne, J. B., Hall, E. 1964. "Plant polyphenolis-XIII. The systematic distribution and origin of anthocyanins containing branched trisaccharides." *Phytochemistry* 3: 453–463.

Hasbal G., Yilmaz-Ozden T., Can A. 2015. "Antioxidant and antiacetylcholinesterase activities of *Sorbus torminalis* (L.) Crantz (wild service tree) fruits." *Journal of Food and Drug Analysis* 23: 57–62.

Hashim F.J., Shawkat M. S., Al-Rikabi A. A. N. 2013. "The role of *Curcuma longa* rhizomes ethanolic extract on human lymphocytes treated by bickel by using G banding technique and karyotyping." *Malaysian Journal of Fundamental and Applied Sciences* 9 (2): 105–109.

Hegerhorst D. F., Weber D. J., McArthur E. D., Khan A. J. 1987. "Chemical analysis and comparison of subspecies of *Chrysothamnus nauseosus* and other related species." *Biochemical Systematics and Ecology* 15 (2): 201–208.

Heiberg N., Måge F., Haffner K. 1992. "Chemical composition of ten blackcurrant (*Ribes nigrum* L.) cultivars." *Acta Agriculturae Scandinavica, Section B: Soil & Plant Science* 42 (4): 251–254.

Heilshorn B. 2017. *Against All Hops: Techniques and Phylosophy for Creating Extraordinary Botanical Beers*. Salem, Mass.: Page Street.

Henning J. A., Steiner J.J, Hummer K. E. 2004. "Genetic diversity among world hop accessions grown in the USA." *Crop Science* 44: 411–417.

Hethelyi E., Korany K., Galambosi B., Domokos J., Palinkas J. 2005. "Chemical composition of the essential oil from rhizomes of *Rhodiola rosea* L. growing in Finland." *Journal of Essential Oil Research* 17: 628–629.

Hieronymus S. 2010. *Brewing with Wheat: The Wit and Weizen of World Wheat Beer Styles*. Boulder, Colo.: Brewers.

——. 2012. *For the Love of Hops: The Practical Guide to Aroma, Bitterness and the Culture of Hops*. Boulder, Colo.: Brewers.

——. 2016a. *Le birre del Belgio I. Degustare e produrre birre trappiste, d'abbazia e strong Belgian ale*. Milan: Edizioni LSWR.

——. 2016b. *Brewing Local*. Boulder, Colo.: Brewers.

Hniličková H., Hnilička F., Krofta K. 2007. "Determining the saturation irradiance and photosynthetic capacity for new perspective varieties of hop (*Humulus lupulus* L.)." *Cereal Research Communications* 35 (2): 461–464.

Holubec V., Krivka P. 2006. *The Caucasus and Its Flowers*. Czech Republic: Loxia.

Hsu T.-W., Su M.-H. 2013. "A taxonomic revision of *Rhus chinensis* Mill. (Anacardiaceae) in Taiwan." *Taiwan Journal of Forest Science* 28 (3): 145–151.

Huang X.-L., Liu M.-D., Li J.-Y., Zhou X.-D., M ten Cate J. 2012. "Chemical composition of *Galla chinensis* extract and the effect of its main component(s) on the prevention of enamel demineralization in vitro." *International Journal of Oral Science* 4: 146–151.

Hulten E. 1968. *Flora of Alaska and Neighboring Territories: A Manual of the Vascular Plants*. Stanford, Calif.: Stanford University Press.

Hummer K. E., Pomper K. W., Postman J., Graham C. J., Stover E., Mercure E. W., Aradhya M., Crisosto C. H., Ferguson L., Thompson M. M., Byers P., Zee F. 2012. "Emerging Fruit Crops." In *Fruit Breeding*, ed. M. L. Badenes and D. H. Byrne. New York: Springer.

Iamonico D., Galasso G., Banfi E., Ardenghi N. M. G. 2015. "Typification of Linnaean names in the genus *Vitis* (Vitaceae)." *International Association for Plant Taxonomy* 64 (5): 1048–1050.

Ibrahim H., Khalid N., Hussin K. 2007. "Cultivated gingers of peninsular Malaysia: utilization, profiles and micropropagation." *Garden's Bullettin Singapore* 1–2: 71–88.

Ibrahim L., Bassal A., El Ezzi A., El Ajouz N., Ismail A., Karaky L., Kfoury L., Sassine Y., Zeineddine A., Ibrahim S. K. 2012. "Characterization and identification of *Origanum* spp. from Lebanon using morphological descriptors." *World Research Journal of Agricultural Biotechnology* 1 (1): 4–9.

Ietswaart J. H. 1980. *A Taxonomic Revision of the Genus Origanum (Labiatae)*. Leiden, Netherlands: Leiden University Press.

Ilieș D.-C., Rădulescu V., Duțu L. 2014. "Volatile constituents from the flowers of two species of honeysuckle (*Lonicera japonica* and *Lonicera caprifolium*)." *Farmacia* 62 (1): 194–201.

Ismail A., Hanana M., Jamoussi B., Hamrouni L. 2014. "Essential oils of *Pinus nigra* J. F. Arnold subsp. *laricio* Maire: chemical composition and study of their herbicidal potential." *Arabian Journal of Chemistry* 10 (2): S3877–S3882.

Jackson M. 1991. *The New World Guide to Beer*. London: Apple.

Jaimand K., Rezaee M. B. 2005. "Chemical constituents of essential oils from *Tanacetum balsamita* L. ssp. *balsamitoides* (Schultz-Bip.) Grierson from Iran." *Journal of Essential Oil Research* 17 (5): 565–566.

Jaksc J., Bandelj D., Javornik B. 2002. "Eleven new microsatellites for hop (*Humulus lupulus* L.)." *Molecular Ecology Notes* 2: 544–546.

Jakse J., Štajner N., Luthar Z., Jeltsch J. M., Javornik B. 2011. "Development of transcript-associated microsatellite markers for diversity and linkage mapping studies in hop (*Humulus lupulus* L.)." *Molecular Breeding* 28: 227–239.

Janati S. S. F., Beheshti H. R., Feizy J., Fahim N. K. 2012. "Chemical composition of lemon (*Citrus limon*) and peels its considerations as animal food." *GIDA* 37 (5): 267–271.

Javornik B., Jake J., Štajner N., Kozjak P. 2005. "Molecular genetic hop (*Humulus lupulus* L.) research in Slovenia." *Acta Horticulturae* 668: 31–34.

Jensen E. C., Anderson D. J., Zasada J. C., Tappeiner J. C. 1995. *The Reproductive Ecology of Broadleaved Trees and Shrubs: Salmonberry, Rubus Spectabilis Pursh*. Corvallis: Forest Research Laboratory of Oregon State University.

Jepson Flora Project. 2018. *Jepson eFlora*. Berkely: University of California Press.

Jurková M., Kellner V., Hašková D., Čulík J., Čejka P., Horák P., Dvořák J. 2011. "Hops: an abundant source of antioxidants. Methods to assessment of antioxidant activity of hop matrix." *Kvasný průmysl* 57 (10): 366–370.

Josephson M., Kleidon A., Tockstein R. 2016. *The Homebrewer's Almanac: A Seasonal Guide to Making Your Own Beer from Scratch*. New York: Countryman.

Jurikova T., Mlcek J., Skrovankova S., Balla S., Sochor J., Baron M., Sumczynski D. 2016. "Black crowberry (*Empetrum nigrum* L.) flavonoids and their health promoting activity." *Molecules* 21 (12): 1685.

Jurikova T., Rop O., Micek J., Sochor J., Balla S., Szekeres L., Hegedusova A., Hubalek J., Vojtech A., Kizek R. 2012. "Phenolic profile of edible honeysuckle berries (genus *Lonicera*) and their biological effects." *Molecules* 17: 61–79.

Kac J., Vovk T. 2007. "Sensitive electrochemical detection method for α-acid, β-acids and xanthohumol in hops (*Humulus lupulus* L.)." *Journal of Chromatography* 850: 531–537.

Kac J., Zakrajšek J., Mlinarič A., Kreft S., Filipic M. 2007. "Determination of xanthohumol in hops (*Humulus lupulus* L.) by nonaqueous CE." *Electrophoresis* 28: 965–969.

Kamal G. M., Anwar F., Hussain A. I., Sarri N., Ashraf M. Y. 2011. "Yield and chemical composition of *Citrus* essential oils as affected by drying pretreatment of peels." *International Food Research Journal* 18 (4): 1275–1282.

Karabín M., Hudcová T., Jelínek L., Dostálek P. 2012. "The importance of hop prenylflavonids for human health." *Chemicke Listy* 106: 1095–1103.

Karlsson Strese E. M., Lundström M., Hagenblad J., W Leino M. 2014. "Genetic diversity in remnant Swedish hop

(*Humulus lupulus* L.) yards from the 15th to 18th century." *Economic Botany* 68 (3): 231–245.

Karol M. K. 2013. "Chemistry and Biology of Cinchona Alkaloids." In *Natural Products*, ed. K. G. Ramawat and J. M. Mérillon, 605–641. Berlin: Springer-Verlag.

Karp K., Starast M., Varnik R. 1997. "The Arctic bramble (*Rubus arcticus* L.). The most profitable wild berry in Estonia." *Baltic Forestry* 3 (2): 47–52.

Katiyar R. S., Nainwal R. C., Singh D., Tewari S. K. 2015. "Genetic characterization and performance of cultivars of Damask rose (*Rosa damascena* Mill L.) in sodic soil." *Society for Scientific Development in Agriculture and Technology. Meerut (U. P.) India* 10 (special 5) 2643–2645.

Kaur H., Bose S. K., Vadekeetil A., Geeta, Harjai K., Richa. 2017. "Essential oil composition and antibacterial activity of flowers of *Achillea filipendulina*." *International Journal of Pharmaceutical Sciences and Drug Research* 9 (4): 182–186.

Kaya D. A., Kirici S., Giray S., Inan M. 2010. "Essential oil composition of cultivated valerian (*Valeriana officinalis* L.) in Cukurova region." *Proceedings of the Third International Conference on Advanced Materials and Systems, Bucharest, Romania, September 16–18*, 393–398.

Knaflewski M., Kałużewicz A., Chen W., Zaworska A., Krzesiński W. 2014. "Suitability of sixteen asparagus cultivars for growing in polish environmental conditions." *Journal of Horticultural Research* 22 (2): 151–157.

Koçak A., Kiliç Ö. 2014. "Identification of essential oil composition of four *Picea* Mill. (Pinaceae) species from Canada." *Journal of Agricultural Science and Technology B* 4: 209–214.

Koelling J., Coles M. C., Matthews P. D., Schwekendiek A. 2011. "Development of new microsatellite markers (SSRs) for *Humulus lupulus*." *Molecular Breeding* 30 (1): 479–484.

Kolenc Z., Vodnik D., Mandelc S., Javornik B., Kastelec D., Cerenak A. 2016. "Hop (*Humulus lupulus* L.) response mechanisms in drought stress: proteomic analysis with physiology." *Plant Physiology and Biochemistry* 105: 67–78.

Kornyšova O., Stanius Ž., Obelevičius K., Ragažinskienė O., Skrzydlewska E., Maruška A. 2009. "Capillary zone electrophoresis method for determination of bitter (α- and β-) acids in hop (*Humulus lupulus* L.) cone extracts." *Advances in Medical Sciences* 54: 41–46.

Kossouoh C., Moudachirou M. 2007. "Essential oil chemical composition of *Annona muricata* L. leaves from Benin." *Journal of Essential Oil Research* 19: 307–309.

Kraft T. F. B., Dey M., Rogers R. B., Ribnicky D. M., Gipp D. M., Cefalu W. T., Raskin I., Lila M. A. 2008. "Phytochemical composition and metabolic performance enhancing activity of dietary berries traditionally used by Native North Americans." *Journal of Agricultural and Food Chemistry* 56 (3): 654–660.

Król B. 2012. "Yield and chemical composition of flower heads of selected cultivars of pot marigold (*Calendula officinalis* L.)." *Acta Scientiarum Polonorum Hortorum Cultus* 11 (1): 215–225.

Kuczerenko A., Krawczyk M., Przybył J. L., Geszprych A., Angielczyk M., Bączek K., Węglarz Z. 2011. "Morphological and chemical variability within the population of common avens (*Geum urbanum* L.)." *Kerva Polonica* 57 (2): 17–21.

Kuhl J. C., DeBoer V. L. 2008. "Genetic diversity of rhubarb cultivars." *Journal of the American Society for Horticultural Science* 133 (4): 587–592.

Kulczyński B., Gramza-Michałowska A. 2016. "Goji berry (*Lycium barbarum*) composition and health effects: a review." *Polish Journal of Food and Nutrition Science* 66 (2): 67–75.

Kumar D., Arya V., Bhat Z., Khan N. A., Prasad D. N. 2012. "The genus *Crataegus*: chemical and pharmacological perspectives." *Revista Brasileira de Farmacognosia* 22 (5): 1187–1200.

Kumaran K. 2014. "Production potential of annatto (*Bixa orellana* L.) as a source of natural edible dye." *International Workshop on Natural Dyes, Hyderabad, India, March 5–7*.

Laferriere J. E., Weber C. W., Kohlhepp E. A. 1991. "Use and nutritional composition of some traditional Mountain Pima plant foods." *Journal of Ethnobiology and Ethnomedicine* 11 (1): 93–114.

Lahigi S. H., Amini K., Moradi P., Asaadi K. 2011. "Investigating the chemical composition of different parts extracts of bipod nettle *Urtica dioica* L. in Tonekabon region." *Iranian Journal of Plant Physiology* 2 (1): 339–342.

Lalli J. Y. Y., Viljoen A. M., Basner H. C. K., Demirci B., Özek T. 2006. "The essential oil composition and chemotaxonomical appraisal of South African pelargoniums (Geraniaceae)." *Journal of Essential Oil Research* 18: 89–105.

Langeland K. A., Cherry H. M., McCormick C. M., Craddock Burks. 2008. *Xanthosoma sagittifolium (L.) Schott: Identification and Biology of Non-native Plants in Florida's Natural Areas*, 2nd ed. Gainesville, Fla.: IFAS, University of Florida.

Lan-Phi N. T., Shimamura T., Ukeda U., Sawamura M., 2009 "Chemical and aroma profiles of yuzu (*Citrus junos*) peel oils of different cultivars." *Food Chemistry* 115 (3): 1042–1047.

Lawal O. A., Ogunwande I. A., Olorunloba O. F., Opoku A. R. 2014. "The essential oils of *Chrysanthemum morifolium* Ramat. from Nigeria." *American Journal of Essential Oils and Natural Products* 2 (1): 63–66.

Lazarević J. S., Đorđević A. S., Kitić D. V., Zlatković B. K., Stojanović G. S. 2013. "Chemical composition and antimicrobial activity of the essential oil of *Stachys officinalis* (L.) Trevis. (Lamiaceae)." *Chemistry & Biodiversity* 10 (7): 1335–49.

Lee J., Dossett M., Finn C. E. 2012. "*Rubus* fruit phenolic research: the good, the bad, and the confusing." *Food Chemistry* 130: 785–796.

Lee J. H., Park K. H., Lee M.-H., Kim H.-T., Seo W. D., Young Kim J., Baek I.-Y., Jang D. S., Ha T. J. 2013. "Identification, characterisation, and quantification of phenolic compounds in the antioxidant activity-containing fraction from the seeds of Korean perilla (*Perilla frutescens*) cultivars." *Food Chemistry* 136: 843–852.

Lee T. M., Der Marderosian A. H. 1988. "Studies on the constituents of dwarf ginseng." *Phytotherapy Research* 2 (4): 165–169.

Leffingwell J. C. 1999. "Leaf Chemistry: Basic Chemical Constituents of Tobacco Leaf and Differences Among Tobacco Types." In *Tobacco: Production, Chemistry, and Technology*, ed. D. Layten Davis and Mark T. Nielsen, 265–284. Hoboken, N.J.: Blackwell.

Leizer C., Ribnicky D., Poulev A., Dushenkov S., Raskin I. 2000. "The composition of hemp seed oil and its potential as an important source of nutrition." *Journal of Nutraceuticals, Functional & Medical Foods* 2 (4): 35–53.

Liang L., Wu X., Zhu M., Zhao W., Li F., Zou Y., Yang L. 2012. "Chemical composition, nutritional value, and antioxidant activities of eight mulberry cultivars from China." *Pharmacognosy Review* 8: 215–224.

Liu P. 2012. "Composition of hawthorn (*Crataegus* spp.) fruits and leaves and emblic leafflower (*Phyllanthus emblica*) fruits." PhD dissertation. University of Turku, Department of Biochemistry and Food Chemistry and Functional Foods Forum, Turku, Finland.

Liu Q., Li D., Wang W., Wang D., Meng X., Wang Y. 2016. "Chemical composition and antioxidant activity of essential oils and methanol extracts of different parts from *Juniperus rigida* Siebold & Zucc." *Chemistry & Biodiversity* 13 (9): 1240–1250.

Lombard K., McCarver K., Franklin J. T., Acharya R., Bates T. 2014. "What's hop'pening in northwest New Mexico? Hops (*Humulus lupulus*) trials summary 2009 to 2014." Annual Conference of the American Society for Horticultural Science.

Lombardo G., Schicchi R., Marino P., Palla F. 2011. "Genetic analysis of *Citrus aurantium* L. (Rutaceae) cultivars by ISSR molecular markers." *Plant Biosystems* 146 (1): 19–26.

Lopes D., Kolodziejczyk P. P. 2005. "Essential oil composition of pineapple-weed (*Matricaria discoidea* DC.) grown in Canada." *Journal of Essential Oil Bearing Plants* 8 (2): 178–182.

Lourdes F., Sanchez C., Bravo R., Rodríguez A. B., Barriga C., Romero E., Cubero J. 2012. "The sedative effect of non-alcoholic beer in healthy female nurses." *Acta Physiologica Hungarica* 7: 1–6.

Lu Z. G., Li X. H., Li W. 2011. "Chemical composition of antibacterial activity of essential oil from *Monarda citriodora* flowers." *Advanced Materials Research Online* 183–185: 920–923.

Lubsandorzhieva P. B., Taraskin V. V., Nagaslaeva O. V., Radnaeva L. D. 2015. "Chemical composition of essential oil of the herbal remedy." *World Journal of Pharmacy and Pharmaceutical Sciences* 4 (5): 246–252.

Lukas B. 2010. "Molecular and phytochemical analyses of the genus *Origanum* L. (Lamiaceae)." PhD dissertation. Universität Wien, Austria.

Lukas B., Schmiderer C., Franz C., Novak J. 2009. "Composition of essential oil compounds from different Syrian populations of *Origanum syriacum* L. (Lamiaceae*)*." *Journal of Agricultural and Food Chemistry* 57: 1362–1365.

Ma B., Chen J., Zheng H., Fang T., Ogutu C., Li S., Han Y., Wu B. 2015. "Comparative assessment of sugar and malic acid composition in cultivated and wild apples." *Food Chemistry* 172: 86–91.

Ma B., Yuan Y., Gao M., Li C., Ogutu C., Li M., Ma F. 2018. "Determination of predominant organic acid components in *Malus* species: correlation with apple domestication." *Metabolites* 8 (4): 74.

Määttä-Riihinen, K. R., Kamal-Eldin, A., Törrönen, A. R. 2004. "Identification and quantification of phenolic compounds in berries of *Fragaria* and *Rubus* species (family Rosaceae)." *Journal of Agricultural and Food Chemistry* 52: 6178–6187.

Majić B., Šola I., Likić S., Juranović Cindrić I., Rusak G. 2015. "Characterisation of *Sorbus domestica* L. bark, fruits and seeds: nutrient composition and antioxidant activity." *Food Technology and Biotechnology* 53 (4): 463–471.

Malakyan M. H., Bajinyan S. A., Vardevanyan L. A., Mairapetyan S. A., Tadevosyan A. H., Yeghiazaryan D. E. 2009. "Anti-radiation and anti-radical activity of hydroponically produced hop extract." *Acta Horticulturae* 848: 253–258.

Mallett J. 2014. Malt. *A Practical Guide from Field to Brewhouse*. Boulder, Colo.: Brewers.

———. 2016. *Gli ingredienti della birra. Il malto. La guida pratica dal campo al birrificio*. Milan: Edizioni LSWR.

Marakoglu T., Arslan D., Ozcan M., Haydar Hacıseferogullari H. 2005. "Proximate composition and technological properties of fresh blackthorn (*Prunus spinosa* L. subsp *dasyphylla* Schur.) fruits." *Journal of Food Engineering* 68 (2): 137–142.

Marcinek K., Krejpcio Z. 2015. "*Stevia rebaudiana* Bertoni. Chemical composition and functional properties." *Acta Scientiarum Polonorum Technologia Alimentaria* 14 (2) : 145–152.

Markowski P. 2015. *La birre del Belgio II. Degustare e produrre bière de garde e saison*. Milan: Edizioni LSWR.

Marshall J. A., Knapp S., Davey M. R., Power J. B., Cocking E. C., Bennett M. D., Cox A. V. 2001. "Molecular systematics of Solanum section Lycopersicum (lycopersicon) using the nuclear ITS rDNA region." *Theoretical and Applied Genetics* 103: 1216–1222.

Martínez-González C. R., Luna-Vega I., Gallegos-Vázquez C., García-Sandoval R. 2015. "*Opuntia delafuentiana* (Cactaceae: Opuntioideae), a new xoconostle from Central Mexico." *Phytotaxa* 231 (3): 230–244.

Massara M., Sameh B. K., Bardaa S., Sahnoun Z., Rebai T. 2016. "Chemical composition, phytochemical constituents, antioxidant and anti-inflammatory activities of *Urtica urens* L. leaves." *Archives of Physiology and Biochemistry* 123 (2): 1–12.

Mattarelli P., Epifano F., Minardi P., Di Vito M., Modesto M., Barbanti L., Bellardi M. G. 2017. "Chemical composition and antimicrobial activity of essential oils from aerial parts of *Monarda didyma* and *Monarda fistulosa* cultivated in Italy." *Journal of Essential Oil Bearing Plants* 20 (1): 76–86.

McAdam E. L., Freeman J. S., Whittock S. P., Buck E. J., Jakše J., Cerenak A., Javornik B., Kilian A., Wang C. H., Andersen D., Vaillancourt R. E., Carling J., Beatson R., Graham L., Graham D., Darby., Koutoulis A. 2013. "Quantitative trait loci in hop (*Humulus lupulus* L.) reveal complex genetic architecture underlying variation in sex, yield and cone chemistry." *BMC Genomics* 14 (1): 1–66.

McGovern P. E. 2017. Ancient Brews: Rediscovered and Re-created. New York: Norton.

McMillan T., Cornett G. 2018. "Il menù del futuro? Alghe, insetti e hamburger vegetali … al sangue." *National Geographic Italia* 47–57.

McNeill J., Barrie F. R., Buck W. R., Demoulin V., Greuter W., Hawksworth D. L., Herendeen P. S., Knapp S., Marhold K., Prado J., Prud'Homme Van Reine W. F., Smith G. F., Wiersema J. H., Turland N. J. 2012. *International Code of Nomenclature for Algae, Fungi, and Plants (Melbourne Code)*. Köeningstein, Germany: Koeltz Scientific.

Medeiros De Oliveira H. 2014. "O gênero *Paullinia* L. (Sapindaceae) no Acre, Brasil." Dissertação de Mestrado. Programa de Pós-graduação stricto sensu, Instituto De Pesquisas Jardim Botânico do Rio de Janeiro, Escola Nacional de Botânica Tropical, Rio de Janeiro, Brasil.

Meeker E. 1883. *Hop Culture in the United States*. Puyallup, Wash.: E. Meeker.

Mejia-Barajas J. A., Del Rio R. E. N., Martinez-Muñoz R. E., Flores-Garcia A., Martinez-Pacheco M. M. 2012. "Cytotoxic activity in *Tagetes lucida* Cav. Emir." *Journal of the Science of Food and Agriculture* 24 (2): 142–147.

Mendes I. C.A, Paviani T. I. 1997. "Morfo-anatomia comparada das folhas do par vicariante *Plathymenia foliolosa* Benth. e *Plathymenia reticulata* Benth. (Leguminosae-Mimosoideae)." *Revista Brasileira de Botânica* 20 (2): 185–195.

Méndez-Tovar I., Novak J., Sponza S., Herrero B., Asensio-S-Manzanera M. C. 2016. "Variability in essential oil composition of wild populations of Labiatae species collected in Spain." *Industrial Crops and Products* 79: 18–28.

Meng L., Lozano Y., Bombarda I., Gaydou E. M., Li B. 2009. "Polyphenol extraction from eight *Perilla frutescens* cultivars." *Comptes Rendus Chimie* 12: 602–611.

Menichini F., Tundis R., Bonesi M., De Cindio B., Loizzo M. R., Conforti F., Statti G. A., Menabeni R., Bettini R., Menichini F. 2011. "Chemical composition and bioactivity of *Citrus medica* L. cv. Diamante essential oil obtained by hydrodistillation, cold-pressing and supercritical carbon dioxide extraction." *Natural Product Research* 25 (8): 789–799.

Menz G., Aldred P., Vriesekoop F. 2011. "Growth and survival of foodborne pathogens in beer." *Journal of Food Protection* 74 (10): 1670–1675.

Mercadante A. Z., Steck A., Pfander H. 1997. "Isolation and identification of new apocarotenoids from annatto (*Bixa orellana*) seeds." *Journal of Agricultural and Food Chemistry* 45: 1050–1054.

Mharti F. Z., Lyoussi B., Abdellaoui A. 2011. "Antibacterial activity of the essential oils of *Pistacia lentiscus* used in Moroccan folkloric medicine." *Natural Product Communications* 6 (10): 1505–1506.

Miles P. D. 2009. *Specific Gravity and Other Properties of Wood and Bark for 156 Tree Species Found in North America*. Research Note NRS-38. Newtown Square, Pa.: U.S. Department of Agriculture, Forest Service, Northern Research Station.

Milivojević J., Rakonjac V., Fotirić Akšić M., Bogdanović Pristov J., Maksimović V. 2013. "Classification and fingerprinting of different berries based on biochemical profiling and antioxidant capacity." *Pesquisa Agropecuária Brasileira* 48 (9) : 1285–1294.

Milligan S. R., Kalita J. C., Heyerick A., Rong H., De Cooman L., De Keukeleire D. 1999. "Identification of a potent phytoestrogen in hops (*Humulus lupulus* L.) and beer." *Journal of Clinical Endocrinology & Metabolism* 83 (6): 2249–2252.

Mirov N. T., Iloff P. M. 1958. "Chemical composition of gum turpentines of pines XXIX. A report on *Pinus ponderosa* from five localities: Central Idaho, Central Montana, Southeastern Wyoming, Northwestern Nebraska, and Central Eastern Colorado." *Journal of the American Pharmaceutical Association* 47 (6): 404–409.

Mockutë D., Bernotienë G., Judþentienë A. 2005. "Chemical composition of essential oils of *Glechoma hederacea* L. growing wild in Vilnius district." *Chemija (Vilnius)* 16 (3–4): 47–50.

Mockute D., Judzentiene A. 2003. "The myrtenol chemotype of essential oil of *Tanacetum vulgare* L. var. vulgare (tansy) growing wild in the Vilnius region." *Chemija (Vilnius)* 14 (2): 103–107.

——. 2004. "Composition of the essential oils of *Tanacetum vulgare* L. growing wild in Vilnius district (Lithuania)." *Journal of Essential Oil Research* 16 (6): 550–553.

Mohanraj R., Sivasankar S. 2014. "Sweet potato (*Ipomoea batatas* [L.] Lam): a valuable medicinal food: a review." *Journal of Medicinal Food* 17 (7): 733–741.

Mondello F., Marella A. M., Di Vito M. 2014. Il Congresso Nazionale per la ricerca sugli olii essenziali. Riassunti. Terni, Umbria, Italy, November 14–16.

Mongelli A., Rodolfi M., Ganino T., Marieschi M., Dall'Asta C., Bruni R. 2015. "Italian hop germoplasm: characterization of wild *Humulus lupulus* L. genotypes from Northern Italy by means of phytochemical, morphological traits and multivariate data analysis." *Industrial Crops and Products* 70: 16–27.

Monschein M., Iglesias Neira J., Kunert O., Bucar F. 2010. "Phytochemistry of heather (*Calluna vulgaris* [L.] Hull) and its altitudinal alteration." *Phytochemistry Reviews* 9 (2): 205–215.

Montemurro V. n.d. *L'Estratto di Bergamotto: Una nuova risorsa per il trattamento dell'ipercolesterolemia.* http://www.fitelab.it/doc/relazioni_scilla/bergamotto_scilla.pdf.

Moradalizadeh M., Akhgar M. R., Rajaei P., Faghihi-Zarandi A. 2012. "Chemical composition of the essential oils of *Levisticum officinale* growing wild in Iran." *Chemistry of Natural Compounds* 47 (6): 1007–1009.

Morales P., Barros L., Ramírez-Moreno E., Santos-Buelga C., Ferreira I. C. F. R. 2015. "Xoconostle fruit (*Opuntia matudae* Scheinvar cv. Rosa) by-products as potential functional ingredients." *Food Chemistry* 185 (9): 289–322.

Morales P., Ramírez-Moreno E., De Cortes Sanchez-Mata M., Carvalho A. M., Ferreira I. C. F. R. 2012. "Nutritional and antioxidant properties of pulp and seeds of two xoconostle cultivars (*Opuntia joconostle* F. A. C. Weber ex Diguet and Opuntia matudae Scheinvar) of high consumption in Mexico." *Food Research International* 46: 279–285.

Mosedale J. R., Feuillat F., Baumes R., Dupouey J. L., Puech J. L. 1998. "Variability of wood extractives among *Quercus robur* and *Quercus petraea* trees from mixed stands and their relation to wood anatomy and leaf morphology." *Canadian Journal of Forest Research* 28: 994–1006.

Moyer R. A., Hummer K. M., Finn C. E., Frei B., Wrolstad R. E. 2002. "Anthocyanins, phenolics, and antioxidant capacity in diverse small fruits: *Vaccinium*, *Rubus*, and *Ribes*." *Journal of Agricultural and Food Chemistry* 50: 519–525.

Mrkonjić Z. O., Nađpal J. D., Beara I. N., Sabo V. S. A., Četojević-Simin D. D., Mimica-Dukić N. M., Lesjak M. M. 2017. "Phenolic profiling and bioactivities of fresh fruits and jam of *Sorbus* species." *Journal of the Serbian Chemical Society* 82 (6): 651–664.

Mudura E., Tofană M., Muste S., Păucean A., Socaci S. 2009. "The evaluation of hop utilisation in brewing process." *Journal of Agroalimentary Processes and Technologies* 15 (2): 249–252.

Muntean L. S., Cernea S., Salontai A., Morar G., Vârban D., Muntean L., Duda M. M., Muntean S., Vârban R., Tofană M. 2011. "The hop cultivar 'Ardeal' (new name of 'Cluj Superalfa' cultivar)." *Hop and Medicinal Plants* 19 (1–2): 7–14.

Murakami A., Darby P., Javornik B., Pais M. S. S., Seigner E., Lutz A., Svoboda P. 2006. "Microsatellite DNA analysis of wild hops, *Humulus lupulus* L." *Genetic Resources and Crop Evolution* 53: 1553–1562.

Musso T., Drago M. 2013. Baladin. *La birra artigianale è tutta colpa di Teo*. Milan: Feltrinelli.

Nagel J., Culley L. K., Lu Y., Liu E., Matthews P. D., Stevens J. F., Page J. E. 2008. "EST analysis of hop glandular trichomes identifies an O-methyltransferase that catalyzes the biosynthesis of xanthohumol." *Plant Cell* 20: 186–200.

Najda A., Dyduch M. 2009. "Chemical diversity within strawberry (*Fragaria vesca* L.) species." *Herba Polonica* 55 (3): 140–146.

Nakatani N., Inatani R., Ohta H., Nishioka A. 1986. "Chemical constituents of peppers (*Piper* spp.) and application to food preservation: naturally occurring antioxidative compounds." *Environmental Health Perspectives* 67: 135–142.

Nakov G., Stamatovska V., Vasileva N., Necinova L. 2016. "Nutritional properties of einkorn wheat (*Triticum monococcum* L.): review." Conference paper, 55th Science Conference of Ruse University, Bulgaria, 381–384.

Narain N., Bora P. S., Holschuh H. J., Da S. Vasconcelos M. A. 2001. "Physical and chemical composition of carambola fruit (*Averrhoa carambola* L.) at three stages of maturity." *CyTA Journal of Food* 3 (3): 144–148.

Natarajana P., Kattab S., Andreic I., Babu Rao Ambatia V., Leonida M., Haasa G. J. 2008. "Positive antibacterial co-action between hop (*Humulus lupulus*) constituents and selected antibiotics." *Phytomedicine: International Journal of Phytotherapy and Phytopharmacology* 15: 194–201.

Natsume S., Takagi H., Shiraishi A., Murata J., Toyonaga H., Patzak J., Takagi M., Yaegashi H., Uemura A., Mitsuoka C., Yoshida K., Krofta K., Satake H., Terauchi R., Ono E. 2014. "The draft genome of hop (*Humulus lupulus*), an essence for brewing." *Plant and Cell Physiology* 56 (3): 1–14.

Naz S., Jabeen S., Ilyas S., Manzoor F., Aslam F., Ali A. 2010. "Antibacterial activity of *Curcuma longa* varieties against different strains of bacteria." *Pakistan Journal of Botany* 42 (1): 455–462.

Negbi M. 1999. *Saffron: Crocus sativus L.* Amsterdam: Harwood Academic.

Nemati Z., Talebi E., Khosravinezhad M., Golkari H. 2018. "Chemical composition and antioxidant activity of Iranian *Satureja montana*." *Science International* 6 (1): 39–43.

Nesom G. 2000. *Skunkbush*: Rhus trilobata *Nutt*. Plant guide. Washington, D.C.: U.S. Department of Agriculture, Natural Resources Conservation Service.

Nghiem C. T., Jiang G. L., Shen K. F., Wang Z. 2016. "Effect of dose fertilizer and cultivars to the active compound glyceryl trioleate of coix lacryma-jobi L." *Journal of Agricultural Science* 38 (3): 261–268.

Niknejad F., Mohammadi M., Khomeiri M., Razavi S. H., Alami M. 2015. "Antifungal and antioxidant effects of hops (*Humulus lupulus* L.) flower extracts." *Advances in Environmental Biology* 8 (24): 395–401.

Nikolić M., Marković T., Mojović M., Pejin B., Savić A., Perić T., Marković D., Stević T., Soković M. 2013. "Chemical composition and biological activity of *Gaultheria procumbens* L. essential oil." *Industrial Crops and Products* 49: 561–567.

Nikonova L. P. 1973. "Lactones of *Inula magnifica*." *Chemistry of Natural Compounds* 9 (4): 528–528.

Nohrstedt A., Butterweck V. 1997. "Biologically active and other chemical constituents of the herb of *Hypericum perforatum* L." *Pharmacopsychiatry* 30 (2): 129–134.

Novak J., Lucas B., Chlodwig F. 2008. "The essential oil composition of wild growing sweet marjoram (*Origanum majorana* L., Lamiaceae) from Cyprus-three chemotypes." *Journal of Essential Oil Research* 20: 339–341.

Novák M., Saleminka C. A., Khanb I. 1984. "Biological activity of thealkaloids of *Erythroxylum coca* and *Erythroxylum novogranatense*." *Journal of Ethnopharmacology* 10: 261–274.

Novák P., Krofta K., Matoušek J. 2006. "Chalcone synthase homologums from *Humulus lupulus*: some enzymatic properties and expression." *Biologia Plantarum* 50 (1): 48–54.

Obolskiy D., Pischel I., Feistel B., Glotov N., Heinrich M. 2011. "*Artemisia dracunculus* L. (tarragon): a critical review of its traditional use, chemical composition, pharmacology, and safety." *Journal of Agricultural and Food Chemistry* 59 (21): 11367–11384.

Ochmian I., Skupień K., Grajkowski J., Smolik M., Ostrowska K. 2012. "Chemical composition and physical characteristics of fruits of two cultivars of blue honeysuckle (*Lonicera caerulea* L.) in relation to their degree of maturity and harvest date." *Notulae Botanicae Horti Agrobotanici Cluj-Napoca* 40 (1): 155–162.

Ohsugi M., Basnet P., Kadoka S., Ishii E., Tamura T., Okumura Y., Namba T. 1997. "Antibacterial activity of traditional medicines and an active constituent lupulone from *Humulus lupulus* against *Helicobacter pylori*." *Journal of Traditional Medicines* 14: 186–191.

Okinczyc P., Szumny A., Szperlik J., Kulma A., Franiczek R., Zbikowska B., Krzyzanowska B., Sroka Z. 2018. "Profile of polyphenolic and essential oil composition of polish propolis, black poplar and aspens buds." *Molecules* 23: 1262.

Olecha M., Nowaka R., Peciob L., Łość R., Malmc A., Rzymowskad J., Oleszekb W. 2016. "Multidirectional characterization of chemical composition and health-promoting potential of *Rosa rugosa* hips." *Natural Product Research* 31 (6): 667–671.

Oliveira M. M., Pais M. S. 1990. "Grandular trichomes of *Humulus lupulus* var. Brewer's gold (hops): ultrastructural aspect of peltate trichomes." *Journal of Submicroscopic Cytology and Pathology* 22 (2): 241–248.

Oniga I., Oprean R., Toiu A., Benedec D. 2010. "Chemical composition of the essential oil of *Salvia officinalis* L. from Romania." *Revista medico-chirurgicală a Societăţii de Medici şi Naturalişti din Iaşi* 114 (2): 593–595.

Orav A., Raal A., Arak E. 2008. "Essential oil composition of *Pimpinella anisum* L. fruits from various European countries." *Natural Product Research* 22 (3): 227–232.

Orwa C., Mutua A., Kindt R., Jamnadass R., Anthony S. 2009. *Agroforestree Database: A Tree Reference and Selection Guide*, version 4.0. Nairobi, Kenya: World Agroforestry Centre.

Oukerrou M. A., Tilaoui M., Mouse H. A., Leouifoudi I., Jaafari A., Zyad A. 2017. "Chemical composition and cytotoxic and antibacterial activities of the essential oil of *Aloysia citriodora* Palau grown in Morocco." *Advances in Pharmacological Sciences* 2017: 7801924.

Owokotomo I. A., Ekundayo O., Oguntuase B. J. 2014. "Chemical constituents of the leaf, stem, root and seed essential oils of *Aframomum melegueta* (K. Schum) from South West Nigeria." *International Research Journal of Pure & Applied Chemistry* 4 (4): 395–401.

Özcan M. M., Atalay C. 2006. "Determination of seed and oil properties of some poppy (*Papaver somniferum* L.) varieties." *Grasas y Aceites* 57 (2): 169–174.

Ozcan M. M., Pedro L. G., Figueiredo A. C., Barroso J. G. 2006. "Constituents of the essential oil of sea fennel (*Crithmum maritimum* L.) growing wild in Turkey." *Journal of Medicinal Food* 9 (1): 128–130.

Özden S. 2016. "The economic analysis of the mastic tree (*Pistacia lentiscus* L.) cultivation projects." Paper presented at Colloque international sous le thème "Les espaces forestiers et périforestiers (EFPF) : dynamique et défis," Campus Universitaire Ait Melloul, Université Ibn Zohr, Agadir, Morocco, November 3–5.

Padmanabhan P., Correa-Betanzo J., Paliyath G. 2016. "Berries and Related Fruits." In *The Encyclopedia of Food and Health*, ed. B. Caballero, P. M. Finglas, and F Toldrá, 364–371. Cambridge, Mass.: Academic.

Palmer J., Kaminski C. 2013. *Water: A Comprehensive Guide for Brewers*. Boulder, Colo.: Brewers.

Paraskevopoulou A., Kiosseoglou V. 2016. "Chios Mastic Gum and Its Food Applications." In *Functional Properties of Traditional Foods*, ed. K. Kristbergsson and S. Ötles. New York: Springer.

Parfitt D., Kallsen C., et al. 2008. *Pistachio Cultivars: Pistachio Short Course Production Manual*. Davis: University of California, Davis, Fruit and Nut Research and Information Center.

Patzak J. 2001. "Comparison of RAPD, STS, ISSR and AFLP molecular methods used for assessment of genetic diversity in hop (*Humulus lupulus* L.)." *Euphytica* 121: 9–18.

———. 2002. "Characterization of Czech hop (*Humulus lupulus* L.) genotypes by molecular methods." *Rostlinna Vyroba* 48 (8): 343–350.

———. 2003. "Assessment of somaclonal variability in hop (*Humulus lupulus* L.) in vitro meristem cultures and clones by molecular methods." *Euphytica* 131: 343–350.

Patzak J., Krofta K., Henychová A., Nesvadba V. 2015. "Number and size of lupulin glands, glandular trichomes of hop (*Humulus lupulus* L.), play a key role in contents of bitter acids and polyphenols in hop cone." *International Journal of Food Science and Technology* 50 (8): 1864–1872.

Patzak J., Matoušek J. 2009. "Gene specific molecular markers for hop (*Humulus lupulus* L.)." *Acta Horticulturae* 848: 73–80.

Patzak J., Nesvadba V., Henychova A., Krofta K. 2010. "Assessment of the genetic diversity of wild hops (*Humulus lupulus* L.) in Europe using chemical and molecular analyses." *Biochemical Systematics and Ecology* 38: 136–145.

Paul J. H. A., Seaforth C. E., Tikasingh T. 2011. "*Eryngium foetidum* L.: a review." *Fitoterapia* 82 (3): 302–308.

Pavlovic M., Pavlovic V. 2011. "Model evaluation of quality attributes for hops (*Humulus lupulus* L.)." *Agrociencia* 45: 339–351.

Pavlovica M., Cerenaka A., Pavlovicb V., Rozmanb C., Pazekb K., Bohanecc M. 2010. "Development of DEX-HOP multi-attribute decision model for preliminary hop hybrids assessment." *Computers and Electronics in Agriculture* 75 (1): 181–189.

Peredo E. L., Ángeles Revilla M., Reed B. M., Javornik B., Cires E., Fernández Prieto J. A., Arroyo-Garciá R. 2010. "The influence of European and American wild germplasm in hop (*Humulus lupulus* L.) cultivars." *Genetic Resources and Crop Evolution* 57: 575–586.

Pereire J. A., Oliveira I., Sousa A., Ferreira I. C. F. R., Bento A., Estevinho L. 2008. "Bioactive properties and chemical composition of six walnut (*Juglans regia* L.) cultivars." *Food and Chemical Toxicology* 46: 2103–2111.

Pérez-Ortega G., González-Trujano M. E., Ángeles-López G. E., Brindis F., Vibrans H., Reyes-Chilpa R. 2016. "*Tagetes lucida* Cav.: ethnobotany, phytochemistry and pharmacology of its tranquilizing properties." *Journal of Ethnopharmacology* 181: 221–228.

Peter K. V., ed. 2006. *Handbook of Herbs and Spices*, vol 3. Sawston, UK: Woodhead.

Peter K. V., Tassou C. C., Nirmal Babu K., Nychas G.-J. E., Skandamis P. N., Shylaja M. R., Rodenburg R. 2004. *Handbook of Herbs and Spices*, vol. 2. Sawston, UK: Woodhead.

Petkova N., Vrancheva R., Mihaylova D., Ivanov I., Pavlov A., Denev P. 2015. "Antioxidant activity and fructan content in root extracts from elecampane (*Inula helenium* L.)." *Journal of BioScience and Biotechnology* 4 (1): 101–107.

Pettersson E., Runeberg J. 1961. "The chemistry of the order Cupressales. Heartwood constituents of *Juniperus procera* Hochst. and *Juniperus californica* Carr." *Acta Chemica Scandinavica* 15: 713–720.

Piaru S. P., Mahmud R., Majid A. M. S. A., Ismail S., Man C. N. 2012. "Chemical composition, antioxidant and cytotoxicity activities of the essential oils of *Myristica fragrans* and *Morinda citrifolia*." *Journal of the Science of Food and Agriculture* 92: 593–597.

Pignatti S. 2017. *Flora d'Italia*, vols. 1–4. Bologna, Italy: Edagricole.

Pineau B., Paisley A. G., Jin D., Wohlers M. W., Jia Y., Jaeger S. R., Beatson R. A. 2013. "Sensory characterisation of dried hop cones for specific aroma traits." *Acta Horticulturae* 215–220.

Pirinen H., Dalman P., Karenlampi S., Tammisola J., Kokko H. 1998. "Description of three new arctic bramble cultivars and proposal for cultivar identification." *Agricultural and Food Science in Finland* 7: 455–468.

Piwowarski J. P., Granica S., Kosinski M., Kiss A. K. 2014. "Secondary metabolites from roots of *Geum urbanum* L." *Biochemical Systematics and Ecology* 53: 46–50.

Pliszka B., Waźbińska J., Huszcza-Ciołkowska G. 2009. "Polyphenolic compounds and bioelements in fruits of Eastern teaberry (*Gaultheria procumbens* L.) harvested in different fruit maturity phases." *Journal of Elementology* 14 (2): 341–348.

Plowman T. 1982. "The identification of coca (*Erythroxylum* species): 1860–1910." *Botanical Journal of the Linnean Society* 84: 329–353.

Pluháčková (Kocourková) H., Kocourková B., Ehrenbergerová J., Fojtová J. 2010. "The essential oil content and its composition in hop varieties (*Humulus lupulus* L.)." *Úroda* 58: 99–102.

Pollastrelli Rodrigues B., Gomes da Silva A., Mauri R., Da Silva Oliveira J. T. 2011. "*Cariniana legalis* (Mart.) Kuntze (Lecythidaceae): descrição dendrológica e anatômica." XIII Encontro Latino Americano de Iniciação Científica e IX Encontro Latino Americano de Pós-graduação Idade do Vale do Paraíba, São Paulo, Brasil.

Polley A., Seigner E., Ganal M. W. 1997. "Identification of sex in hop (*Humulus lupulus*) using moluecular markers." *Genome* 40: 357–361.

Popernack J. L., Loeffler-Urazumbetov R., Beard A. 2014. *Antioxidant Activity in Hops (*Humulus lupulus*)*. Duluth, Minn.: College of Saint Scholastica, Department of Chemistry and Biochemistry.

Popescu G. S., Velciov A.-B., Costescu C., Gogoasa I., Gravila C., Petolescu C. 2014. "Chemical characterisation of white (*Morus alba*), and black (*Morus nigra*) mulberry fruits." *Journal of Horticulture, Forestry and Biotechnology* 18 (3): 133–135.

Prabhakara Rao P. G., Galla Narsing R., Jyothirmayi T., Sathyanarayana A. 2015. "Characterisation of seed lipids from *Bixa orellana* and *Trachyspermum copticum*." *Journal of the American Oil Chemists Society* 92: 1483–1490.

Prayaga Murty P., Narasimha Rao G. M. et. al. 2014. "Report on *Hylocereus undatus* (Haw.) Britton & Rose, Andhra Pradesh, India." *Journal of Science* 4 (4): 221–222.

Primetta A. 2014. "Phenolic compounds in the berries of the selected Vaccinium species: the potential for authenticity analyses." *Dissertations in Forestry and* Natural *Sciences* 134. University of Eastern Finland, Kuopio.

Priya Devi S., Thangam M., Ramachadrudu K., Ashok Kumar J., Singh N. P. 2013. "Genetic diversity of kokum (*Garcinia indica*) in goa-tree and fruit characters." *Indian Council of Agricultural Research Technical Bulletin* 33.

Raal A., Orav A., Arai E. 2007. "Composition of the essential oil of *Salvia officinalis* L. from various European countries." *Natural Product Research* 21 (5): 406–411.

Raal A., Orav A., Gretchushnikova T. 2014. "Composition of the essential oil of the *Rhododendron tomentosum* Harmaja from Estonia." *Natural Product Research* 28 (14): 1091–1098.

———. 2014. "Essential oil content and composition in *Tanacetum vulgare* L. herbs growing wild in Estonia." *Journal of Essential Oil Bearing Plants* 17 (4): 670–675.

Ragasa C. Y., Galian R. F., Shen C.-C. 2014. "Chemical constituents of *Annona muricata*." *Der Pharma Chemica* 6 (6): 382–387.

Rahman A.-U., Ahmed D., Asif E., Ahmad S. 1991. "Chemical constituents of *Buxus sempervirens*." *Journal of Natural Products* 54 (1): 79–82.

Rahman A.-U., Ahmed D., Choudhary M. I., Turkoz S., Sener B. 1988. "Chemical constituents of *Buxus sempervirens*." *Planta Medica* 54 (2): 173–174.

Raina V. K., Srivastava S. K., Aggarwal K. K., Syamasundar K. V., Sushil K. 2001. "Essential oil composition of *Syzygium aromaticum* leaf from Little Andaman, India." *Flavour and Fragrance Journal* 16: 334–336.

Rakesh P., Sujit P., Janeshwer Y. 2015. "Isolation and identification of compounds from the leaves extract of *Hibiscus syriacus* L." *Asian Journal of Pharmacy and Technology* 5 (1): 8–12.

Rančič A., Soković M., Vukojević J., Simić A., Marin P., Duletić-Laušević S., Dejan Djoković. 2005. "Chemical composition and antimicrobial activities of essential oils of *Myrrhis odorata* (L.) Scop, *Hypericum perforatum* L. and *Helichrysum arenarium* (L.) Moench." *Journal of Essential Oil Research* 17 (1): 341–345.

Rao N. B., Sita Kumari O. 2014. "Phyto chemical analysis of all spice *Pimenta dioica* leaf extract." *World Journal of Pharmacy and Pharmaceutical Sciences* 4 (1): 1400–1404.

Rao P. P., Rao G. N., Jyothirmayi T., Satyanarayana A., Karuna M. S. L., Prasad R. B. N. 2015. "Characterisation of seed lipids from *Bixa orellana* and *Trachyspermum copticum*." *Journal of the American Oil Chemists' Society* 92 (10): 1483–1490.

Rao P. S., Navinchandra S., Jayaveera K. N. 2012. "An important spice, *Pimenta dioica* (Linn.) Merill: a review." *International Current Pharmaceutical Journal* 1 (8): 221–225.

Reed B. M., Okut N., D'Achino J., Narver L., DeNoma J. 2003. "Cold storage and cryopreservation of hops

(*Humulus* L.) shoot cultures through application of standard protocols." *CryoLetters* 24: 389–396.

Ren X., He T., Chang Y., Zhao Y., Chen X., Bai S., Wang L., Shen M., She G. 2017. "The genus *Alnus*, a comprehensive outline of its chemical constituents and biological activities." *Molecules* 22: 1–37.

Reza V. R. M., Abbas H. 2007. "The essential oil composition of *Levisticum officinalis* from Iran." *Asian Journal of Biochemistry* 2 (2): 161–163.

Riaz G., Chopra R. 2018. "A review on phytochemistry and therapeutic uses of *Hibiscus sabdariffa* L." *Biomedicine & Pharmacotherapy* 102: 575–586.

Ricci D., Epifano F., Fraternale D. 2017. "The essential oil of *Monarda didyma* L. (Lamiaceae) exerts phytotoxic activity in vitro against various weed seeds." *Molecules* 22 (2): 222.

Roborgh R. H. J. 1964. "Classification and botanical description of imported varieties of hops (*Humulus lupulus*) in Nelson, New Zealand." *New Zealand Journal of Botany* 2 (1): 10–18.

Rodriguez-Burruezo A., Prohens J., Raigon M. D., Nuez F. 2009. "Variation for bioactive compounds in ají (*Capsicum baccatum* L.) and rocoto (*C. pubescens* R. & P.) and implications for breeding." *Euphytica* 170: 169–181.

Rodríguez-Reyna I. S., Varela-Rodríguez L., Hernández-Rodríguez P., Ramos-Martínez E., González-Horta C., Talamás-Rohana P., Sánchez-Ramírez B. 2016. "Acute toxicity evaluation of *Rhus trilobata* stems extract in BALB/c mice." *Toxicology Letters* 259: 5193.

Rohloff J., Uleberg E., Nes A., Krogstad T., Nestby R., Martinussen I. 2015. "Nutritional composition of bilberries (*Vaccinium myrtillus* L.) from forest fields in Norway: Effects of geographic origin, climate, fertilization and soil properties." *Journal of Applied Botany and Food Quality* 88: 274–287.

Ronse Decraene L. P., Smets E. F. 1999. "The floral development and anatomy of *Carica papaya* (Caricaceae)." *Canadian Journal of Botany* 77: 582–598.

Rösch M. 2008. "New aspects of agriculture and diet of the early medieval period in Central Europe: waterlogged plant material from sites in South-Western Germany." *Vegetation History and Archaeobotany* 17: 225–238.

Roshan A., Verma N. K., Kumar C. S., Chandra V., Singh D. P., Panday M. K. 2012. "Phytochemical constituent, pharmacological activities and medicinal uses through the millennia of *Glycyrrhiza glabra* Linn: a review." *International Research Journal of Pharmacy* 3 (8): 45–55.

Roslon W., Osinska E., Mazur K., Geszprych A. 2014. "Chemical characteristics of European goldenrod (*Solidago virgaurea* L. subsp. *virgaurea*) from natural sites in Central and Eastern Poland." *Acta Scientiarum Polonorum Hortorum Cultus* 13 (1): 55–65.

Ross K. A., Godfrey D., Fukumoto L. 2015. "The chemical composition, antioxidant activity and α-glucosidase inhibitory activity of water-extractable polysaccharide conjugates from Northern Manitoba lingonberry." *Cogent Food & Agriculture* 1: 1–19.

Routson K. J., Volk G. M., Richards C. M., Smith S., Nabhan G., Wyllie de Echeverria V. 2012. "Genetic variation and distribution of Pacific crabapple." *Journal of the American Society for Horticultural Science* 137 (5): 325–332.

Ruiz-Rodríguez B-M., Morales P., Fernández-Ruiz V., Sánchez-Mata M.-C., Cámara M., Díez-Marqués C., Pardo-de-Santayana M., Molina M., Tardío J. 2011. "Valorization of wild strawberry-tree fruits (*Arbutus unedo* L.) through nutritional assessment and natural production data." *Food Research International* 44 (5): 1244–1253.

Russell W. S. 1846. *Guide to Plymouth, and Recollections of the Pilgrims*. Boston: G. Coolidge.

Salanță L., Tofană M., Socaci S., Mudura E., Pop C., Pop A., Fărcaș A. 2015. "Evaluation and comparison of aroma volatile compounds in hop varieties grown in Romania." *Romanian Biotechnological Letters* 20 (6): 11049–11056.

Salem M. Z. M., Zeidler A., Böhm M., Mervat E. A. M., Ali H. M. 2015. "GC/MS analysis of oil extractives from wood and bark of *Pinus sylvestris*, *Abies alba*, *Picea abies*, and *Larix decidua*." *BioReources* 10 (4): 7725–7737.

Salim N., Abdelwaheb C., Rabah C., Ahcene B. 2009. "African chemical composition of *Opuntia ficus-indica* (L.) fruit." *Journal of Biotechnology* 8 (8): 1623–1624.

Salvesen P. H., Kanz B., Moe D. 2009. "Historical cultivars of *Buxus sempervirens* L. revealed in a preserved 17th century garden by biometry and amplified fragment length polymorphism (AFLP)." *European Journal of Horticultural Science* 74 (3): 130–136.

Sánchez-Mata M. C., Cabrera Loera R. D., Morales P., Fernández-Ruiz V., Cámara M., Díez Marqués C., Pardo-de-Santayana M., Tardío J. 2011. "Wild vegetables of the Mediterranean area as valuable sources of bioactive compounds." *Genetic Resources and Crop Evolution* 59 (3): 431–443.

Sartori A. L. B., De Azevedo Tozzi A. M. G. 2004. "Revisão taxonômica de *Myrocarpus* allemão (Leguminosae, Papilionoideae, Sophoreae)." *Acta Botanica Brasilica* 18 (3): 521–535.

Sauvain M., Rerat C., Moretti C., Saravia E., Arrazola S., Gutierrez E., Lema A. M., Muñoz V. 1997. "A study of the chemical composition of *Erythroxylum coca* var. coca leaves collected in two ecological regions of Bolivia." *Journal of Ethnopharmacology* 56 (3): 179–191.

Scheinvar L. 1981. "Especies, variedades y combinaciones nuevas de Cactaceas del Valle de Mexico." *Phytologia* 49: 313–338.

Seca A. M. L., Silva A. M. S. 2005. "The Chemical Composition of the *Juniperus* Genus (1970-2004)." In *Recent Progress in Medicinal Plants*, vol. 16, *Phytomedicines*, ed. J. N. Govil, V. K. Sing, and R. Bhardwaj. New Delhi: Studium.

Seefelder S., Ehrmaier H., Schweizeier G., Seigner E. 2000. "Male and female genetic linkage map of hops, *Humulus lupulus*." *Plant Breeding* 119: 249–255.

Sefidkon F., Dabiri M., Mirmostafa S. A. 2004. "The composition of *Thymus serpyllum* L. oil." *Journal of Essential Oil Research* 16 (3): 184–185.

Semen E., Hiziroglu S. 2005. "Production, yield and derivatives of volatile oils from eastern redcedar (*Juniperus virginiana* L.)." *American Journal of Environmental Sciences* 1 (2): 133–138.

Serna-Saldivar S. O. 2010. *Cereal Grains: Properties, Processing, and Nutritional Attributes*. Boca Raton, Fla.: CRC.

Setzer W. N. 2016. "Chemical composition of the leaf essential oil of *Lindera benzoin* growing in North Alabama." *American Journal of Essential Oils and Natural Products* 4 (3): 1–3.

Sharopov F. S., Setzer W. N. 2012. "The essential oil of *Salvia sclarea* L. from Tajikistan." *Records of Natural Products* 6 (1): 75–79.

Shehata A. M., Skirvin R. M., Norton M. A. 2008. "Leaf morphology affects horseradish regeneration in vitro." *International Journal of Vegetable Science* 15 (1): 24–27.

Shikov A. N., Poltanov E. A., Dorman H. J. D., Makarov V. G., Tikhonov V. P., Hiltunen R. 2006. "Chemical composition and in vitro antioxidant evaluation of commercial water-soluble willow herb (*Epilobium angustifolium* L.) extracts." *Journal of Agricultural and Food Chemistry* 54: 3617–3624.

Si X.-T., Zhang M.-L., Shi Q.-W., Kiyota H. 2006. "Chemical constituents of the plants in the genus *Achillea*." *Chemistry & Biodiversity* 3: 1163–1180.

Sika K. C., Adoukonou-Sagbadja H., Ahoton L., Saidou A., Ahanchede A., Kefela T., Gachomo E. W., Baba-Moussa L., Kotchoni S. O. 2015. "Genetic characterization of cashew (*Anacardium occidentale* L) cultivars from Benin." *Journal of Horticulture* 2 (1): 7.

Silva S., Costa E. M., Calhau C., Morais R. M., Pintado M. M. E. 2017. "Production of a food grade blueberry extract rich in anthocyanins: selection of solvents, extraction conditions and purification method." *Journal of Food Measurement and Characterization* 11: 1248–1253.

Silvanini A., Torello Marinoni D., Beccaro G. L., Ganino T. 2011. "La caratterizzazione varietale del germoplasma di *Castanea sativa* Mill." Review no. 15. *Italus Hortus* 18 (3): 47–61.

Simpson W. J., Smith A. R. W. 1992. "Factors affecting antibacterial activity of hop compounds and their derivates." *Journal of Applied Bacteriology* 72: 327–334.

Sims C. A., Juliani H. R., Mentreddy S. R., Simon J. E., 2014 "Essential oils in holy basil (*Ocimum tenuiflorum* L.) as influenced by planting dates and harvest times in North Alabama." *Journal of Medicinally Active Plants* 2 (3): 33–41.

Skof S., Cerenak A., Jakse J., Bohanec B., Javornik B. 2012. "Ploidy and sex expression in monoecious hop (*Humulus lupulus*)." *Botany* 99 (3): 355–361.

Skomra U., Bocianowski J., Agacka M. 2013. "Agro-morphological differentiation between European hop (*Humulus lupulus* L.) cultivars in relation to their origin." *Journal of Food Agriculture and Environment* 11 (3–4): 1123–1128.

Skupień K. 2006. "Chemical composition of selected cultivars of highbush blueberry fruit (*Vaccinium corymbosum* L.)." *Folia Horticulturae* 18 (2): 47–56.

Small E. 1978. "A numerical and nomenclatural analysis of morpho-geographic taxa of *Humulus*." *Systematic Botany* 3 (1): 37–76.

Snoussi M., Noumi E., Dehmani A., Flamini G., Aouni M., Alsieni M. Al-sieni A. 2016. "Chemical composition and antimicrobial activities of *Elettaria cardamomum* L. (Manton) essential oil: a high activity against a wide range of food borne and medically important bacteria and fungi." *Journal of Chemical, Biological and Physical Sciences* 6 (1): 248–259.

Snoussi M., Noumi E., Trabelsi N., Flamini G., Papetti A., De Feo V. 2015. "*Mentha spicata* essential oil: chemical composition, antioxidant and antibacterial activities against planktonic and biofilm cultures of Vibrio spp. strains." *Molecules* 20: 14402–14424.

Solberg S. O., Brantestam A. K., Kylin M., Bjørn G. K., Thomsen M. G. 2014. "Genetic variation in Danish and Norwegian germplasm collection of hops." *Biochemical Systematics and Ecology* 52: 53–59.

Solberg S. O., Kylin M., Bjørn M. K., Thomsen M. J., Brantestam A. K. 2013. "A diversity study of the Danish and Norwegian collections of hops (*Humulus lupulus* L.)." In *Pre-Breeding: Fishing in the Gene Pool. Abstracts of Oral Presentations and Posters of the European Plant Genetic Resources Conference 2013, NordGen, SLU, Alnarp*, ed. R. O. Ortiz Rios, 71. Alnarp, Sweden: Swedish University of Agricultural Sciences, NordGen.

Sottile F., Del Signore M. B., Giuggioli N. R., Peano C. 2017. "The potential of the sorb (*Sorbus domestica* L.) as a minor fruit species in the Mediterranean areas: description and quality traits of underutilized accessions." *Progress in Nutrition* 19 (1): 41–48.

Sparrow J. 2015. *Le birre del Belgio III. Degustare e produrre lambic, oud bruin e Flemish red*. Milan: Edizioni LSWR.

Śpiewak R., Dutkiewcz J. 2002. "Occupational airborne and hand dermatitis to hop (*Humulus lupulus* L.) with non-occupational eelapses." *Annals of Agricultural and Environmental Medicine* 9: 249–252.

Srečec S., Čeh B., Savić Ciler T., Ferlež Rus A. 2013. "Empiric mathematical model for predicting the content of alpha-acids in hop (*Humulus lupulus* L.) cv Aurora." *SpringerPlus* 2 (59): 1–5.

Srečec S., Krpan V. Z., Marag S., Špoijaric I., Kvaternjak I., Mršić G. 2011. "Morphogenesis, volume and number of hop (*Humulus lupulus* L.) glandular trichomes, and their influence on alpha-acid accumulation in fresh bracts of hop cones." *Acta Botanica Croatica* 70 (1): 1–8.

Srečec S., Zechner-Krpan V., Petravić-Tominac V., Košir I. J., Čerenak A. 2012. "Importance of medical effects of xanthohumol, hop (*Humulus lupulus* L.) bioflavonoid in restructuring of world hop industry." *Agriculturae Conspectus Scientificus* 77 (2): 61–67.

Srečec S., Zechner-Krpan V., Petravić-Tominac V., Mršić G., Špoljarić I., Marag S. 2010. "ESEM comparative studies of hop (*Humulus lupulus* L.) peltate and bulbous glandular trichomes structure." *Agriculturae Conspectus Scientificus* 75 (3): 145–148.

Stajner N., Jakse J., Kozjak P., Javornik B. 2005. "The isolation and characterisation of microsatellites in hop (*Humulus lupulus* L.)." *Plant Science* 168: 213–221.

Stajner N., Satovic Z., Cerenak A., Javornik B. 2007. "Genetic structure and differentiation in hop (*Humulus lupulus* L.) as inferred from microsatellites." *Euphytica* 161 (1): 301–311.

Standley P. C., Steyermark J. A. 1952. *Flora of Guatemala*. Fieldiana: Botany, vol. 24, pt. 3. Chicago: Chicago Natural History Museum.

Stankovic M. Z., Nikolic N. C., Stajevic L. P., Petrovic S. D., Cakic M. D. 2005. "Hydrodistillation kinetics and essential oil composition from fermented parsley seeds." *Chemical Industry & Chemical Engineering Quarterly* 11 (1): 25–29.

Stanković N., Mihajilov-Krstev T., Zlatković B., Matejić J.,Stankov Jovanović V., Kocić B., Čomić L. 2016. "Comparative study of composition, antioxidant, and antimicrobial activities of essential oils of selected aromatic plants from Balkan Peninsula." *Planta Medica* 82 (7): 650–661.

Steele M. 2012. IPA: *Brewing Techniques, Recipes and the Evolution of India Pale Ale*. Boulder, Colo.: Brewers.

Stewart A. 2013. *The Drunken Botanist: The Plants That Create the World's Great Drinks*. New York: Algonquin.

Stewart C. D., Jones C. D., Setzer W. N. 2014. "Essential oil compositions of *Juniperus virginiana* and *Pinus virginiana*, two important trees in Cherokee traditional medicine." *American Journal of Essential Oils and Natural Products* 2 (2): 17–24.

St-Gelais A., Collin G., Helbig J., Gagnon H. 2018. "Essential oils from the foliage of *Picea sitchensis* from British Columbia." *American Journal of Essential Oils and Natural Products* 6 (3): 19–26.

St-Gelais A., Collin G., Pichette A. 2016. "Aromas from Quebec. V. Essential oils from the fruits and stems of *Heracleum maximum* Bartram and their unsaturated aliphati-cacetates." *Journal of Essential Oil Research* 29 (2): 126–136.

Šuštar-Vozlič J., Javornik B. 1999. "Genetic relationships in cultivars of hop, *Humulus lupulus* L., determined by RAPD analysis." *Plant Breeding* 118: 175–181.

Svoboda P. 1991. "Cultivation of isolated hop tips (*Humulus lupulus* L.) in vitro." *Plant Soil and Environment* 37 (8): 643–648.

Szewczyk K., Zidorn C. 2014. "Ethnobotany, phytochemistry, and bioactivity of the genus *Turnera* (Passifloraceae) with a focus on *Damiana-Turnera diffusa*." *Journal of Ethnopharmacology* 152 (3): 424–443.

Tabanca N., Avonto C., Wang M., Parcher J. F., Ali A., Demirci B., Raman V., Khan I. A. 2013. "Comparative investigation of *Umbellularia californica* and *Laurus nobilis* leaf essential oils and identification of constituents active against *Aedes aegypti*." *Journal of Agricultural and Food Chemistry* 61: 12283–12291.

Taha K. F., Shakour Z. T. A. 2016. "Chemical composition and antibacterial activity of volatile oil of *Sequoia sempervirens* (Lamb.) grown in Egypt." *Medicinal and Aromatic Plants* 5 (3).

Takayama C., Meirade-Faria F., Alvesde Almeida A. C., Dunder R. J., Manzo L. P., Rabelo Socca E. A., Batista L. M., Salvador M. J., Monteiro Souza-Brito A. R., Luiz-Ferreira A. 2016. "Chemical composition of *Rosmarinus officinalis* essential oil and antioxidant action against gastric damage induced by absolute ethanol in the rat." *Asian Pacific Journal of Tropical Biomedicine* 6 (8): 677–681.

Takoi K., Degueil M., Shinkaruk S., Thibon C., Maeda K., Ito K., Bennetau B., Dubourdieu D.,Tominaga T., 2009 "Identification and characteristics of new volatile thiols derived from the hop (*Humulus lupulus* L.) cultivar Nelson Sauvin." *Journal of Agriculture and Food Chemistry* 57 (6): 2493–2502.

Teixeira da Silva J. A. 2017. "Fake peer reviews, fake identities, fake accounts, fake data: beware!" *AME Medical Journal* 2 (3): 28.

Thangaselvabai T., Sudha K. R., Selvakumar T., Balakumbahan R. 2011. "Nutmeg (*Myristica fragrans*

Houtt)—the twin spice—a review." *Agricultural Research Communication Centre* 32 (4): 283–293.

Thavanapong N. 2006. "The essential oil from peel and flower of *Citrus maxima*." Master of Pharmacy thesis. Silpakorn University, Program of Pharmacognosy Graduate School, Bangkok, Thailand.

Thiem B. 2003. "*Rubus chamaemorus* L., a boreal plant rich in biologically active metabolites: a review." *Biology Letters* 40 (1): 3–13.

Thomas A. L., Jett L. 2004. *Performance of Eleven Asparagus Cultivars in Southwest Missouri. Midwest Vegetable Variety Trial Report for 2004.* Purdue University Bulletin No. 2004-B17538. West Lafayette, Ind.: Purdue University.

Tomomatsu A., Itoh T., Wijaya C.H, Nasution Z., Kumendong J., Matsuyama A. 1996. "Chemical constituents of sugar-containing sap and brown sugar from palm in Indonesia." *Japanese Journal of Tropical Agriculture* 40 (4): 175–181.

Toncer O., Karaman S., Diraz E., Sogut T., Kizil S. 2017. "Essential oil composition of *Thymus* × *citriodorus* (Pers.) Schreb. at different harvest stages." *Notulae Botanicae Horti Agrobotanici Cluj-Napoca* 45 (1): 185–189.

Torras J., Grau M. D., Lopez J. F., De Las Heras F. X. C. 2007. "Analysis of essential oils from chemotypes of *Thymus vulgaris* in Catalonia." *Journal of the Science of Food and Agriculture* 87: 2327–2333.

Trojak-Goluch A., Skomra U. 2013. "Artificially induced polyploidization in *Humulus lupulus* L. and its effect on morphological and chemical traits." *Breeding Science* 63: 393–399.

Turland N. J., Wiersema J. H., Barrie F. R., Greuter W., Hawksworth D. L., Herendeen P. S., Knapp S., Kusber W.-H., Li D.-Z., Marhold K., May T. W., McNeill J., Monro A. M., Prado J., Price M. J., Smith G. F. 2018. *International Code of Nomenclature for Algae, Fungi and Plants (Shenzhen Code). (Regnum Vegetabile, 159).* Oberreifenberg (Schmitten), Germany: Koeltz Botanica.

Tyl C., Ismail B. P. 2018. "Compositional evaluation of perennial wheatgrass (*Thinopyrum intermedium*) breeding populations." *International Journal of Food Science and Technology* 54 (3): 660–669.

Usman L. A., Hamid A. A., Muhammad N. O., Olawore N. O., Edewor T. I., Saliu B. K. 2010. "Chemical constituents and anti-inflammatory activity of leaf essential oil of Nigerian grown *Chenopodium album* L." *EXCLI Journal* 9: 181–186.

Usui M., Kakuda Y., Kevan P. G. 1994. "Composition and energy values of wild fruits from the boreal forest of Northern Ontario." *Canadian Journal of Plant Science* 74: 581–587.

Van de Steen J. 2003. *Les Trappistes: les abbaye et leurs bières.* Brussels: Éditions Racine.

Van Opstaele F. 2011. "Hoppy aroma of beer. Characterisation of sensorially differentiated hop aromas and evaluation in brewing practice." PhD dissertation, Arenberg Doctoraatsschool, Groep Wetenschap & Technologie, Faculteit Bio-ingenieurswetenschappen, Katholieke iteit Leuven, Flanders, Belgium.

Van Opstaele F., Goiris K., Syryn E., De Rouck G., Jaskula B., De Clippeleer J., Aerts G., De Cooman L. 2006. "Hop: aroma and bitterness perception." *Cerevisia* 31 (4): 167–188.

Vanhoenacker G., Van Rompaey P., De Keukeleire D., Pat S. 2001. "Chemotaxonomic features associated with flavonoids of cannabinoid-free *Cannabis* (*Cannabis sativa* subsp. *sativa* L.) in relation to hops (*Humulus lupulus* L.)." *Natural Product Letters* 16 (1): 57–63.

Varkey T. K., Mathew J., Baby S. 2014. "Chemical variability of *Citrus maxima* essential oils from South India." *Asian Journal of Chemistry* 26 (8): 2207–2210.

Vasas A., Orbán-Gyapai O., Hohmann J. 2015. "The genus *Rumex*: review of traditional uses, phytochemistry and pharmacology." *Journal of Ethnopharmacology* 175: 198–228.

Vaughan D. A. 1989. "The genus *Oryza* L. Current status of taxonomy." *IRRI Research Paper Series* 138.

Veldkamp J. F. 2013. "Nomenclatural notes on *Boesenbergia* Kuntze (Zingiberaceae)." *Philippine Journal of Science* 142: 215–221.

Verdian-rizi M., Hadjiakhoondi A. 2008. "Essential oil composition of *Laurus nobilis* L. of different growth stages growing in Iran." *Zeitschrift für Naturforschung C* (63c): 785–788.

Verma R. K., Mishra G., Singh P., Jha K. K., Khosa R. L. 2011. "*Alpinia galanga*. An important medicinal plant: a review." *Der Pharmacia Sinica* 2 (1): 142–154.

Verma R. S., Chauhan A., Padalia R. C., Jata S. K., Thulb S., Sundaresan V. 2013. "Phytochemical diversity of *Murraya koenigii* (L.) Spreng. from Western Himalaya." *Chemistry & Biodiversity* 10: 628–641.

Verma R. S., Padalia R. C., Chauhan A. 2011. "Chemical investigation of the volatile components of shade-dried petals of Damask rose (*Rosa damascena* Mill.)." *Archives of Biological Sciences (Belgrade, Serbia)* 63 (4): 1111–1115.

———. 2015. "Chemical composition of root essential oil of *Acorus calamus* L." *National Academy Science Letters* 38 (2): 121–125.

Verma R. S., Rahman L. U., Chanotiya C.S, Verma R. K., Chauhan A., Yadav A., Singh A., Yadav A. K. 2010. "Essential oil composition of *Lavandula angustifolia* Mill.

cultivated in the mid hills of Uttarakhand, India." *Journal of the Serbian Chemical Society* 75 (3): 343–348.

Verma V. K., Siddiqui N. U. 2011. "Bioactive chemical constituents from the plant *Verbena officinalis* Linn." *International Journal of Pharmaceutical Sciences and Research* 3 (4): 108–109.

Vool E., Karp K., Noormets M., Moor U., Starast M. 2009. "The productivity and fruit quality of the arctic bramble (*Rubus arcticus* ssp. *arcticus*) and hybrid arctic bramble (*Rubus arcticus* ssp. *arcticus* × *Rubus arcticus* ssp. *stellatus*)." *Acta Agriculturae Scandinavica, Section B, Soil & Plant Science* 59 (3): 217–224.

Vouillamoz J. F., Carron C.-A., Malnoë P., Baroffio C. A., Carlen C. 2012. "*Rhodiola rosea* 'Mattmark', the first synthetic cultivar is launched in Switzerland." *Acta Horticulturae* 955: 185–190.

Vucic D. M., Petkovic M. R., Rodic-Grabovac B. B., Stefanovic O. D., Vasic S. M., Comic L. R. 2013. "Phenolic content, antibacterial and antioxidant activities of *Erica herbacea* L." *Acta Poloniae Pharmaceutica – Drug Research* 70 (6): 1021–1026.

Vulić J. J., Vračar L. O., Šumić Z. M. 2008. "Chemical characteristics of cultivated elderberry fruit." *Acta Periodica Technologica* 39: 85–90.

Wach D., Gawroński J., Dyduch-Siemińska M., Kaczmarska E., Błażewicz-Woźniak M. 2016. "Phenotypic and genotypic variability of cultivars of highbush blueberry (*Vaccinium corymbosum* L.) grown in the Lublin region." *Acta Scientiarum Polonorum Hortorum Cultus* 15 (6): 305–319.

Wainio W. W., Forbes E. B. 1941. "The chemical composition of forest fruits and nuts from Pennsylvania." *Journal of Agriculture* 62 (10): 627–635.

Wajs-Bonikowska A., Szoka L., Karna E., Wiktorowska-Owczarek A., Sienkiewic M. 2016. "Composition and biological activity of *Picea pungens* and *Picea orientalis* seed and cone essential oils." *Chemistry & Biodiversity* 14 (3).

Wang S., Li P. 2013. "Flavonoids from *Humulus lupulus*." *China Journal of Chinese Materia Medica* 38 (10): 1539–1542.

Weathington M. 2014. "*Fagus grandifolia* 'White Lightning.'" *Horticultural Science* 49 (8): 1086–1087.

Weinoehrl S., Feistel B., Pischel I., Kopp B., Butterweck V. 2011. "Comparative evaluation of two different *Artemisia dracunculus* L. cultivars for blood sugar lowering effects in rats." *Phytotherapy Research* 26 (4): 625–629.

Weiss L. M., Kearns J. K. 2015. "Caffeine and theobromine analysis of *Paullinia yoco*, a vine harvested by Indigenous Peoples of the upper Amazon." *Tropical Resources* 3: 6–15.

Wesolowska A., Jadczak D., Grzeszczuk M. 2011. "Chemical composition of the pepper fruit extract of hot cultivars *Capsicum annuum* L." *Acta Scientiarum Polonorum Hortorum Cultus* 10 (1): 171–184.

White C., Zainasheff J. 2010. *Yeast: The Practical Guide to Beer Fermentation*. Boulder, Colo.: Brewers.

———. 2016. *Gli ingredientti della birra. Il lievito. Guida pratica alla fermentazione della birra*. Milan: Edizioni LSWR.

Whittock, S., Koutoulis, A. 2011. "New hop (*Humulus lupulus* L.) aroma varieties from Australia." Conference of the Scientific Commission of the International Hop Growers' Convention, Lublin, Poland.

Williams T. P. 1977. *Comprehensive Index to the Flora of Guatemala*. Fieldiana, Botany, vol. 24, pt. 13. Chicago: Field Museum of Natural History.

Wojtyniak K., Szymański M., Matławska I. 2013. "*Leonurus cardiaca* L. (Motherwort): a review of its phytochemistry and pharmacology." *Phytotherapy Research* 27: 1115–1120.

Woodske D. 2013. *Hop Variety Handbook: Learn More About Hops . . . Craft Better Beer*. Scotts Valley, Calif.: CreateSpace.

Wyler P., Angeloni L. H. P., Alcarde A. R., da Cruz S. H. 2015. "Effect of oak wood on the quality of beer." *Journal of the Institute of Brewing* 121: 62–69.

Xi-Tao Y., Sang-Hyun L., Wei L., Ya-Nan S., Seo-Young Y., Hae-Dong J., Young-Ho K. 2014. "Evaluation of the antioxidant and anti-osteoporosis activities of chemical constituents of the fruits of *Prunus mume*." *Food Chemistry* 156: 408–415.

Yong-Chun J., Ke Y., Jing Z. 2011. "Chemical composition, and antioxidant and antimicrobial activities of essential oil of *Phyllostachys heterocycla* cv. Pubescens varieties from China." *Molecules* 16: 4318–4327.

Yurry U., Mei-Lan J., Ok-Tae K., Young-Chang K., Seong-Cheol K., Seon-Woo C., Ki-Wha C., Serim K., Chan-Moon C., Yi L. 2016. "Identification of Korean ginseng (*Panax ginseng*) cultivars using simple sequence repeat markers." *Plant Breeding and Biotechnology* 4 (1): 71–78.

Zahri S., Moubarik A., Charrier F., Chaix G., Baillères H., Nepveu G., Charrier B. 2008. "Quantitative assessment of total phenol contents of European oak (*Quercus petraea* and *Quercus robur*) by diffuse reflectance NIR spectroscopy on solid wood surfaces." *Holzforschung* 62: 679–687.

Zanoli P., Zavatti M. 2008. "Pharmacognostic and pharmacological profile of *Humulus lupulus* L." *Journal of Ethnopharmacology* 116: 383–396.

Zatylny A. M., Ziehl W. D., St. Pierre R. G. 2005. "Physicochemical properties of fruit of chokecherry (*Prunus virginiana* L.), highbush cranberry (*Viburnum trilobum* Marsh.), and black currant (*Ribes nigrum* L.) cultivars grown in Saskatchewan." *Canadian Journal of Plant Science* 85 (2): 425–429.

Zawiślak G. 2015. "Comparison of chemical composition of the essential oil from *Marrubium vulgare* L. and *M. incanum* Desr. during the second year of cultivation." *Acta Agrobotanica* 68 (1): 59–62.

Zdunić G., Šavikin K., Pljevljakušić D., Djordjević B. 2016. "Black (*Ribes nigrum* L.) and Red Currant (*Ribes rubrum* L.) Cultivars." In *Nutritional Composition of Fruit Cultivars*, ed. M. S. J. Simmonds and V. R. Preedy, 101–126. Cambridge, Mass.: Academic.

Zebulon O. L., Okoye S. C. 1993. "Chemical analysis of sorrel leaf (*Rumex acetosa*)." *Food Chemistry* 48 (2): 205–206.

Zhao G., Ren Y., Ma H. 2016. "Extraction and characterization of *Raphanus sativus* seed oil obtained by different methods." *Tropical Journal of Pharmaceutical Research* 15 (7): 1381–1385.

Zhao J. 2011. "The extraction of high value chemicals from heather (*Calluna vulgaris*) and bracken (*Pteridium aquilinum*)." PhD dissertation. University of York, Department of Chemistry, York, England, UK.

Zhou Y., Wang S., Lou H., Fan P. 2018. "Chemical constituents of hemp (*Cannabis sativa* L.) seed with potential antineuroinflammatory activity." *Phytochemistry Letters* 23: 57–61.

Zimmer A. R., Leonardi B., Miron D., Schapoval E., De Oliveira J. R., Gosmann G. 2012. "Antioxidant and anti-inflammatory properties of *Capsicum baccatum*: from traditional use to scientific approach." *Journal of Ethnopharmacology* 139: 228–233.

Websites

https://www.absinth.cz
https://www.actaplantarum.org
https://www.allaboutagave.com
https://www.aob.oxfordjournals.org
https://www.areabirra.it
https://www./article/dn12910-ancient-beer-pots-point-to-origins-of-chocolate
https://www.atuttabirra.com
https://www.baladin.it
https://www.bamboobeer.ca
https://www.bangbrewing.com
https://www.beer.wikia.com
https://www.beeradvocate.com
https://www.beerlegends.com
https://www.beer-shop.it
https://www.beverfood.com
https://www.birraaltaquota.it
https://www.birradegliamici.com
https://www.birramia.it
https://www.birramoretti.it
https://www.birraofelia.it
https://www.bjcp.org
https://www.blogvs.it
https://www.bluemoonbrewingcompany.com
https://www.botanical.com
https://www.brasseriepietra.corsica
https://www.brewersfriend.com
https://www.brewtheplanet.it
https://www.brooklynbrewshop.com
https://www.broom02.revolvy.com
https://www.byo.com
https://www.chilibeer.com
https://www.citrusgenomedb.org
https://www.conifers.org
https://www.craftbeer.com
https://www.crea.gov.it/it/comunicati-stampa/il-grano-saraceno-per-la-vita-e-per-lo-sport
https://www.cronachedibirra.it/cultura-birraria/5617/alla-scoperta-della-segale-lingrediente-brassicolo-del-momento
https://www.dodimalto.it
https://www.dogfish.com
https://www.dreher.it
https://www.efloras.org
https://www.erbemedicinali.netsons.org
https://www.fermentobirra.com
https://www.floraitaliae.actaplantarum.org
https://www.floranorthamerica.org
https://www.fondazioneslowfood.com

BIBLIOGRAPHY

https://www.gazzagolosa.gazzetta.it/2016/07/09/birra-saracena/honest-food.net
https://www.giornaledellabirra.it
https://www.gruitale.com
https://www.guidabirreartigianali.it
https://www.hobbybirra.info
https://www.homebrewersassociation.org
https://www.hopworksbeer.com
https://www.hort.purdue.edu
https://www.ibirranti.com
https://www.inbirrerya.com
https://www.indiabiodiversity.org
https://www./it/arca-del-gusto-slow-food/birra-umqombothi
https://www.juniper.orst.edu
https://www.landinstitute.org
https://www.lefthandbrewing.com
https://www.marcadoc.it
https://www.microbirrifici.org
https://www.mixerplanet.com
https://www.mondobirra.org
https://www.mumbeer.be
https://www.nchomebrewing.com
https://www.newbelgium.com
https://www.newscientist.com
https://www./northernbrewers
https://www.nzflora.info
https://www.oaks.of.the.world.free.fr
https://www.omnipollo.com
https://www.oregonfruit.com
https://www.patagoniaprovisions.com
https://www.plantillustrations.org
https://www.plants.usda.gov
https://www.powo.science.kew.org
https://www.prota4u.org
https://www.quercusalbabeer.com
https://www.ratebeer.com
https://www.reflora.jbrj.gov.br
https://www.ricettariomedievale.it
https://www.sangabriel.it
https://www.schlenkerla.de/rauchbier/beschreibungi.html
https://www.shebrewsgoodale.wordpress.com
https://www.tela-botanica.org
https://www.thefullpint.com
https://www.thegarlicfarm.co.uk
https://www.tropical.theferns.info
https://www.ucjeps.berkeley.edu
https://www.unabirralgiorno.blogspot.it
https://www.untappd.com
https://www.usahops.org
https://www.vanherbaryum.yyu.edu.tr
https://www.vecchiaerboristeria.it
https://www.williamsbrosbrew.com
https://www.wood-database.com
https://www.worldagroforestry.org
https://www.worldbeercup.org
https://www.yakimavalleyhops.com

Index of Common Names

achiote tree, 93–94
adzuki bean, 562
afares oak, 436–437
African cucumber, 179–180
African marigold, 520–521
African myrrh, 167
African padauk, 432–433
agave, 46–47
agrimony, 48–49
agua/rosa de Jamaica, 252
ají, 115
Alaska white spruce, 394–395
alder, 51–52
alehoof, 244
Algerian oak, 440–441
all-heal, 559–560
alligator pepper, 43–44
allspice, 398–399
almond, 424–425
almug, 432
alpine heath, 208–209
Alpine strawberry, 230–231
amaranth, 55–57
amargo, 435–436
amburana, 57–58
American beech, 222
American chestnut, 123–124
American crabapple, 321–322
American cranberry, 552–553
American elder, 495
American ginseng, 376–377
American marigold, 520–521
American pepper, 500–501
American plane tree, 414
American silverberry, 202
American wintergreen, 236
Andrew's bottle gentian, 238
anice, 400
anise, 400
anise hyssop, 44–45
anise mint, 44–45
annatto, 93–94
apple may hawthorn, 173
apple mint, 340–341
apple tree, 322–324

apricot, 418–419, *420*
apriplum, 423–424
aprium, 423–424
arabica coffee, 164–165
arctic bramble, 474
arctic raspberry, 474
aricirana, 65–66
Arizona elderberry, 495
arrowleaf elephant ear, 566
artichoke, 190–191
ash, 232–233
asparagus, 79–80
Atlantic white cedar, 128–129
Australian acacia, 34–35
Australian kino, 214–215
Australian red rum, 214–215
Aztec marigold, 520–521
azuki bean, 562

Bahama grass, 191–192
bakeapple, 475–476
baldcypress, 525, 527
bald-cypress, 525, 527
balm, 335–336
balm of heaven, 546–547
balsam, 361–362
balsa wood tree, 361–362
bamboo, 84–85
banana tree, 348–350
Barbary fig, 365
Barbary matrimony vine, 317–318
barley, 15, 254–266, *255*, *267*
bartender's lime, 151–152
basil, 362–363
basil beebalm, 343
basket oak, 445–446
bastard oak, 457
bastard white oak, 457
bay laurel, 302
bee balm, 343–344
bee bread, 95
beetroot, 86, *87*
belladonna, 80–81
Benjamin bush, 311–312
bergamot, 142, 144

Bermuda grass, 191–192
betony, 516–517
bible hyssop, 369
biblical-hyssop, 369
bigarade, 140–141
big huckleberry, 553
big marigold, 520–521
big sagebrush, 74
bilberry, 549–550, 551, 554, *555*, 556
birch, 90–91
bishop's hat, 208
bishop's wort, 516–517
bishopwort, 516–517
bitter ash, 435–436
bitter orange, 140–141
bitter root, 239–240, *241*
bitter wood, 435–436
bitterwort, 239–240, *241*
black acacia, 34–35
blackberry, 206–207, 474–475, 480, 484–485
black birch, 88–89
black crowberry, 206–207
black currant, 464–465
black elder, 496
black henbane, 282, 284
Black Hills spruce, 394–395
black huckleberry, 237, 553
black locust, 467–468
black mint, 522
black mulberry, 346
black mustard, 98–99
black pepper, 407, *408*
black peppercorn, 407, *408*
black-seeded gourd, 181–182
black spruce, 395–396
blackthorn, 428–429
black walnut, 293–294
blackwood, 34–35
blackwood acacia, 34–35
blessed milkthistle, 505
blessed thistle, 127
blood amaranth, 56
bloodwood tree, 248–249
blue agave, 47–48

613

blue-berried honeysuckle, 315–316
blueberry, 550–551
blue giant hyssop, 44–45
blue gum, 216–217
blueleaf huckleberry, 551
blue mistflower, 169
blue spruce, 396–397
bobwood, 361–362
bog bilberry, 557
bog blueberry, 557
bog heather, 209–210
bog-myrtle, 350–351, *352*
bog myrtle, 350–351, *352*
Bolivian bark, 136
borage, 95
borecole, 99
Botany Bay kino, 214–215
bottle gentian, 238
box, 103–104
boxberry, 236
boxwood, 103–104
boysenberry, 485–486
Brazilian arrowroot, 330–331
Brazilian cherry, 218–219
bread wheat, 535–536, *537*
Breckland thyme, 530–531
bridewort, 227
Brindonia tallow tree, 235
broadleaf plantain, 413–414
broadleaved lavender, 303–304
broccoli, 99–100, *101*
broom, 193–194
broomcorn millet, 378
broomtail millet, 378
broom teatree, 309
buckwheat, 220–221
burdock, 68
bur oak, 444–445
burr oak, 444–445
butternut squash, 184–185

cabreuva, 355–356
cabreúva, 354–355
cacao tree, 528–529
cactus pear, 365
cajà, 515–516
cajú, 58–59
Calabrian pine, 401–402
calamansi, 146–147
calamondin, 146–147
California bay, 546–547

California bay laurel, 546–547
California incense-cedar, 108–109
California juniper, 295–296
California mugwort, 71–72
Californian pepper tree, 500–501
Calysaia bark, 136
camerise, 315–316
campeachy wood, 248–249
Canada tea, 236
Canarian oak, 440–441
candyleaf, 517
cannabis, 110–111
canneberge, 552–553
canoe birch, 89–90
cantaloupe, 178–179
canterberry, 236
Cape gooseberry, 392
caraway, 120–121
cardamom, 204
cardus marianus, 505
Caribbean pine, 400–401
Caribbean pitch pine, 400–401
carob tree, 128
carrot, 195–197
cascade bilberry, 551
cascade blueberry, 551
cascara buckthorn, 231
cascara sagrada, 231
cashew tree, 58–59
cassava, 330–331
catmint, 359–360
catnip, 359–360
catsfoot, 244
catswort, 359–360
caucasus oak, 444
Cayenne cherry, 218–219
celery, 64–65
celery top pine, 390–391
century plant, 46–47
Ceylon cinnamon tree, 137–138
chamomile, 129–130, 333–334
checkerberry, 236
checker tree, 511–512
chequers, 511–512
cherry, 419, 421
cherry birch, 88–89
chestnut, 124–125
chestnut oak, 447
chia, 492–493
chili, 112, 114, *117*
Chinabark, 136–137

Chinese boxthorn, 317–318
Chinese gall, 462
Chinese ginger, 94–95
Chinese ginseng, 374, 376
Chinese keys, 94–95
Chinese licorice, 246
Chinese parsley, 170–171
Chinese plum, 426
Chinese sumac, 462
Chinese wolfberry, 317–318
chinkapin oak, 448
chinquapin oak, 448
chittem bark, 231
chokecherry, 429
chopi, 567
chop suey green, 243
Christmas orange, 159–160
chrysanthemum, 132–133
chrysanthemum greens, 243
cigarbox wood, 126
cilantro, 170–171
cinchona bark, 136–137
cinnamon bush, 546–547
citron, 157–158
citrus thyme, 529–530
clary, 494–495
clary sage, 494–495
cleavers, 233–234
clementine, 144–145
closed bottle gentian, 238
closed gentian, 238
cloudberry, 475–476
clove, 519–520
cloves, 519–520
clovetree, 519–520
coca-bush, 213
cocculus, 59–60
cocculus indicus, 59–60
cockle, 314–315
cocoa tree, 10, 528–529
coconut palm, 162–163
coconut tree, 162–163
cocoyam, 566
coffee shrub of Arabia, 164–165
coixseed, 163–164
cold hardy mandarin, 159–160
colewort, 240
Colorado blue spruce, 396–397
common bamboo, 84–85
common bean, 387–388
common beech, 222–224

INDEX OF COMMON NAMES

common cattail, 545
common chamomile, 333–334
common dandelion, 525, *526*
common elderberry, 495
common fig, 224–225, *226*
common flax, 312–313
common guava, 430–431
common hazel, 172–173
common hedgenettle, 516–517
common hop, 268–276, *280*
common horehound, 331–332
common hornbeam, 119–120
common jasmine, 292–293
common juniper, 296, *297*
common lilac, 518–519
common lime, 532–533
common linden, 532–533
common madder, 472
common millet, 378
common myrrh, 167
common plantain, 413–414
common rue, 487
common sage, 493–494
common sea buckthorn, 203
common sundew, 199–200
common sunflower, 249–250
common thyme, 531–532
common verbena, 561
common vervain, 561
common violet, 563
common walnut, 294–295
common wheat, 535–536, *537*
common white horehound, 331–332
common wormwood, 74–75, *76*
coriander, 170–171
corinthian raisins, 563–564
corinth raisins, 563–564
cork oak, 458–459
cork tree, 361–362
corkwood, 361–362
corn, 567–568, *569*
cornelian cherry, 171–172
costmary, 523–524
cotton, 247–248
couch grass, 191–192
countess powder, 136–137
cowberry, 558
cowgrass, 533–534
cow oak, 445–446
cow parsnip, 250–251
crabapple, 320–322, 324–325, 327

crab apple, 326
crackleberry, 237
cranberry, 552–553
creeping Charlie, 244
creeping Jenny, 244
creeping juniper, 298
creeping lemon thyme, 529–530
creeping thyme, 530–531
cross-leaved heath, 209–210
crowberry, 206–207
crown daisy, 243
crowndaisy chrysanthemum, 243
Cryptomeria, 177
Cuban cedar, 126
Cuban pine, 400–401
cubeb, 405–406
cucumber, 180–181
culantro, 212–213
culinary sage, 493–494
cumin, 186–187
cumquat, 153–154
curaçao, 141–142, *143*
currants, 563–564
curry leaf-tree, 348
curry tree, 348
cypress, 525, 527

damask rose, 468
damiana, 544
damiane, 544
dandelion, 525, *526*
danewort, 495
darnel, 314–315
darnel ryegrass, 314–315
date, 388–389
date palm, 388–389
date plum, 198–199
deadly nightshade, 80–81
devil's grass, 191–192
dewberry, 482
dill, 61–62
dog banana, 77–78
dog rose, 469–470
dog's tooth grass, 191–192
dogwood, 460
dollof, 227
dotted beebalm, 345
doub palm, 95–96
Douglas fir, 430
Douglas's sagewort, 71–72
doum palm, 285

down tree, 361–362
downy birch, 92–93
downy oak, 451–452
dragonfruit, 281
dream plant, 71–72
Duke of Argyll's tea plant, 317–318
Duke of Argyll's tea tree, 317–318
Durand oak, 457
durmast oak, 449–450
durum wheat, 538–539
dwarf bilberry, 549–550
dwarf cherry, 421–422
dwarf elder, 495
dwarf ginseng, 377–378
dwarf huckleberry, 549–550
dwarf raspberry, 474
dyer's madder, 472

early blueberry, 554, 556
Eastern European licorice, 246
eastern juniper, 299–300
eastern may hawthorn, 173
Eastern purple coneflower, 201–202
eastern red cedar, 299–300
eastern spicy-wintergreen, 236
eastern teaberry, 236
eastern white pine, 404–405
edible chrysanthemum, 243
Egyptian doum palm, 285
Egyptian oregano, 369
Einkorn wheat, 540–541
elder, 496
elderberry, 496
elecampagne, 289
elephant ear, 566
elfdock, 289
elm, 545–546
elm-leaf blackberry, 483–484
elm-leafed sumac, 462–463
emmer, 536, 538
emmer wheat, 536, 538
English chamomile, 129–130
English lavender, 302–303
Englishman's foot, 413–414
English oak, 452–454
English plantain, 412
English violet, 563
English walnut, 294–295
enkir, 540–541
enzian, 239–240, *241*
epazote, 200–201

estragon, 72–73
eucalyptus kino, 214–215
European beech, 222–224
European black elderberry, 496
European box, 103–104
European elder, 496
European goldenrod, 509–510
European hornbeam, 119–120
European mountain ash, 510
European raspberry, 477–478, *479*
European strawberry, 230–231
European white birch, 92–93

false pepper, 500–501
fat-hen, 130–131
fennel, 44–45, 228
fennel giant hyssop, 44–45
fenugreek, 534
fernleaf yarrow, 37, 39
fern leaf yarrow, 37, 39
fiddlehead, 334–335
field balm, 244
field bean, 387–388
fig, 224–225, *226*
fig-leaf gourd, 181–182
fig opuntia, 365
fingerroot, 94–95
fireweed, 207–208
fitweed, 212–213
flageolet bean, 387–388
flat-leaved vanilla, 560–561
flax, 312–313
florist's daisy, 133
florist's violet, 563
fom wisa, 43–44
foxtail amaranth, 55
fragrant crabapple, 321–322
fragrant giant hyssop, 44–45
freijo, 169–170
French bean, 387–388
fuzzy kiwifruit, 42–43

galangal, 53–54
garapa, 65–66
garden angelica, 62–63
garden asparagus, 79–80
garden bean, 387–388
garden chamomile, 129–130
garden lovage, 311
garden nasturtium, 543–544
garden parsley, 386–387

garden sage, 493–494
garden sorrel, 486–487
garden strawberry, 229–230
garden thyme, 531–532
garden violet, 563
garland chrysanthemum, 243
garland crabapple, 321–322
garlic, 50–51
Garry oak, 442
genepi, 73–74
Georgian oak, 450–451
geranium, 383
German chamomile, 333–334
German liquorice, 246
German thyme, 531–532
gesho, 460
giant eabane, 289
giant fleabane, 290
giant hyssop, 44–45
giant inula, 289, 290
gillover-the-ground, 244
ginger, 570
gingerbread palm, 285
gingko, 242
ginkgo, 242
ginkgo tree, 242
ginseng, 374, 376
glasswort, 490
globe artichoke, 190–191
goji, 317–318
goldenberry, 392
golden root, 502, 504
gooseberry, 466
grains of paradise, 43–44
grand plantain, 413–414
grapefruit, 148–149
grapevine, 564–565
grápia, 65–66
great basil, 362–363
great basin sagebrush, 74
great burdock, 68
greater galangal, 53–54
greater yam, 197–198
great millet, 512–513, *514*
great willowherb, 207–208
great yellow gentian, 239–240, *241*
green bean, 387–388
green cardamom, 204
green pompon horehound, 331–332
grenadille, 380–382
ground apple, 129–130

ground-ivy, 244
ground ivy, 244
guanabana, 63–64
guanábanaa, 63–64
guarana, 382
guava, 430–431
guinea grains, 43–44
guinea pepper, 43–44
gulf-cypress, 525, 527
gum, 214–215
gum myrrh, 167
Guyana arrowroot, 197–198

habanero, 115–116
hairy blue elderberry, 495
hard wheat, 538–539
haricot bean, 387–388
hashish, 110–111, *113*
hawthorn, 174–175
headache tree, 546–547
heather, 105–106, *107*
hedgehog liquorice, 246
hemlock, 168
hemp, 111–112, *113*
henbane, 282, 284
herabol myrrh, 167
herb bennet, 240
herb-of-grace, 487
hibiscus, 253
hibiscus tea, 252
highbush blackberry, 474–475
high-bush blueberry, 550–551
Himalayan goji, 317–318
hog millet, 378
hog plum, 515–516
holy basil, 363–364
Honduran sarsaparilla, 507
Honduran yellow pine, 400–401
Honduras pine, 400–401
honeyberry, 315–316
honey plant, 335–336
honeysuckle, 316–317
hop, 268–276, *280*
horehound, 331–332
horned cucumber, 179–180
horned melon, 179–180
horseheal, 288–289
horsemint, 343–345
horseradish, 69–70
huacatay, 522
hulled wheat, 536, 538

INDEX OF COMMON NAMES

Hungarian chamomile, 333–334
Hungarian licorice, 246
Hungarian oak, 441
hyssop, 286

incense cedar, 108–109
incense-cedar, 108–109
Indian banana, 77–78
Indian celery, 250–251
Indian cress, 543–544
Indian rhubarb, 250–251
intermediate wheatgrass, 205–206
Iowa crabapple, 325
iris, 292
iron tree, 341
ironwood, 100, 102
Italian camomilla, 333–334
Italian chicory, 134, *135*

jackfruit, 75, 77
Jamaica, 252
Jamaican sarsaparilla, 507
Jamaica pepper, 398–399
Japanese apricot, 426
Japanese cedar, 177
Japanese crabapple, 327
Japanese-green, 243
Japanese horseradish, 219–220
Japanese pepper, 567
Japanese persimmon, 198–199
Japanese rose, 470–471
Japanese wineberry, 482
jarrah, 217
jasmine, 292–293
java pepper, 405–406
jequitibá, 118–119
jequitibá branco, 118–119
jequitiba-rosa, 118–119
jequitiba rosa, 118–119
jequitibá vermelho, 118–119
jessamine, 292–293
Jesuitõs bark, 136–137
Jesuitõs powder, 136–137
Jesuit's tea, 200–201
Job's tears, 163–164
Job's-tears, 163–164

Kabylic's oak, 440–441
kaffir lime, 152–153
kahikatea, 194–195
kaki, 198–199

kale, 99
Kamut, 542
Karri, 215
kashfi millet, 378
kauri, 45–46
kauri pine, 45–46
kava, 406–407
kava-kava, 406–407
kawa, 406–407
keg fig, 198–199
key lime, 151–152
Khorasan wheat, 542
kino australiensis, 214–215
kiwano, 179–180
knotberry, 475–476
knoutberry, 475–476
kokum, 235
kokum butter tree, 235
Korean pepper, 567
kumquat, 153–154

labrador tea, 304–305
lady of the meadow, 227
lamb's quarter, 130–131
lamb's tongue, 412
laraha, 141–142, *143*
larch, 301–302
large flowered barrenwort, 208
laurel, 302
laurel blanco, 169–170
lavender giant hyssop, 44–45
Lebanese oregano, 369
legal cariniana, 118–119
lemon balm, 335–336
lemon beebalm, 342
lemon beebrush, 52–53
lemonberry sumac, 463–464
lemon grass, 189–190
lemongrass, 189–190
lemon guava, 430–431
lemon mint, 342
lemon-scented thyme, 529–530
lemon thyme, 529–530
lemon tree, 155–156
lemon verbena, 52–53
lentil, 305, 307
lesser calamint, 160, 162
lesser galangal, 94–95
liche, 313–314
lichee, 313–314
licorice, 246–247

licorice mint, 44–45
liechee, 313–314
lilac, 518–519
limequat, 145–146
lingonberry, 558
linseed, 312–313
lipstick tree, 93–94
liquorice, 246–247
litchi, 313–314
lizhi, 313–314
li zhi, 313–314
locust bean, 128
loganberry, 480–481
logwood, 248–249
lojabark, 136–137
long coriander, 212–213
longleaf pine, 402–403
longspur barrenwort, 208
loquat, 211–212
lovage, 311
love lies bleeding, 55
low-bush salmonberry, 475–476
low chamomile, 129–130
loxa bark, 136–137
lychee, 313–314

macaroni wheat, 538–539
maesil, 426
magrood lime, 152–153
mahaleb cherry, 425–426
maidenhair tree, 242
maidi, 96–97
makrut lime, 152–153
Malabar gourd, 181–182
malanga, 566
mandarin, 158–159, *161*
mandioca, 330–331
mandrake, 327–328
mango, 328–330
manioc, 330–331
manuka, 309
mānuka, 309
manuka myrtle, 309
manure weed, 130–131
maple, 35–36
maracujá, 380–382
maracuya, 380–382
Marian thistle, 505
marigold, 104–105, 521
marionberry, 476–477
marjoram, 368

617

marshmallow, 54–55
marsh rosemary, 305, *306*
marsh samphire, 490
Mary thistle, 505
master of the woods, 234–235
mastic, 409
matrimony vine, 317–318
mauritius papeda, 152–153
mayhaw, 173
may hawthorn, 173
meadow queen, 227
meadowsweet, 227
meadow-wort, 227
meadsweet, 227
mead wort, 227
mede berry, 317–318
Mediterranean milk thistle, 505
melde, 130–131
melegueta pepper, 43–44
mesquite, 417–418
Mexican coriander, 212–213
Mexican cotton, 247–248
Mexican grain amaranth, 56
Mexican lime, 151–152
Mexican marigold, 521
Mexican mint marigold, 521
Mexican tarragon, 521
Mexican-tea, 200–201
meydi, 96–97
milfoil, 37, 39
milk thistle, 505
millet, 378
mint, 336–338
Mirbeck's oak, 440–441
Mongolian oak, 446–447
monks cress, 543–544
moso bamboo, 391–392
mossy-cup oak, 444–445
mother's daisy, 129–130
motherwort, 307–308
mountain ash, 218, 510
mountain chestnut oak, 447
mountain coffee, 164–165
mountain huckleberry, 553
mountain laurel, 546–547
mountain savory, 499–500
mountain sumac, 462–463
mugwort, 74–75, *76*
mulberry, 345–346, *347*
mume, 426
Murray red gum, 214–215

muskmelon, 178–179
musk pumpkin, 184–185
musky gourd, 184–185
muster John Henry, 522
myrrh, 167
myrtle, 357–358
myrtle-leaved orange tree, 147–148
myrtle pepper, 398–399

naartjie, 159–160
narrowleaf plantain, 412
narrow-leaved crabapple, 320
needle juniper, 298–299
nettle, 547–548
New Mexican elderberry, 495
newspice, 398–399
New Zealand Christmas bush, 341
New Zealand Christmas tree, 341
New Zealand pohutukawa, 341
New Zealand tea, 309
New Zealand white pine, 194–195
Nicaragua pine, 400–401
North American beech, 222
northern bilberry, 557
northern red oak, 456
northern white pine, 404–405
Norwegian angelica, 62–63
nosebleed, 37, 39
nutgall tree, 462
nutmeg, 351, 353
nux vomica, 518

oat, 81–82, *83*
oil grass, 189–190
olive tree, 364–365
Omani lime, 151–152
onion, 49–50
opium poppy, 378–380
orange tree, 149–150
oregano, 369–370
Oregon crabapple, 324
Oregon myrtle, 546–547
Oregon pine, 430
Oregon white oak, 442
oriental wheat, 542
ossame, 43–44
ostrich fern, 334–335
Oswego tea, 343–344
ovalleaf huckleberry, 554, 556
overcup oak, 443

ox-eye daisy, 310
oxeye daisy, 310

Pacific crabapple, 324
pale gentian, 239–240, *241*
palmwood, 285
palmyra palm, 95–96
palo santo, 102–103
pamplemousse, 156–157
papaw, 116, 118
papaya, 116, 118
paper birch, 89–90
Paraguayan palo santo, 100, 102
Paraguay tea, 287
parsley, 386–387
partridgeberry, 558
passion fruit, 380–382
pasta wheat, 538–539
pau carga, 118–119
pawpaw, 77–78, 116, 118
pea, 411
peach, 427–428
peanut, 66–67
pear, 434–435
pearl millet, 383–384
pecan, 121–122
pedunculate oak, 452–454
pendant amaranth, 55
pennyroyal, 338
peppadew, 115
pepper, 398–399
peppercorn tree, 500–501
peppernut tree, 546–547
pepper tree, 500–501
pepperwood, 546–547
perforate Saint John's wort, 284
perilla, 384–385
perilla-mint, 384–385
Persian cumin, 120–121
Persian oak, 444
Persian walnut, 294–295
persimmon, 198–199
Peruvian bark, 136–137
Peruvian mastic, 500–501
Peruvian pepper, 500–501
Peruvian peppertree, 500–501
physalis, 392
pimenta, 398–399
pineapple, 60–61
pineapple mint, 340–341
pineapple strawberry, 229–230

INDEX OF COMMON NAMES

pineapple weed, 334
pistachio, 410
pistachio nut, 410
pitanga, 218–219
pitch pine, 400–401
plant species, 572–573
plum, 422–423
plumcot, 423–424
pluot, 423–424
poet's jasmine, 292–293
pohutukawa, 341
pōhutukawa, 341
poison berry, 59–60
poison darnel, 314–315
poison hemlock, 168
poison nut, 518
polished willow, 491–492
pomegranate, 433–434
pomello, 156–157
pomelo, 156–157
pommelo, 156–157
Ponderosa pine, 403–404
pop bean, 387–388
poplar, 416–417
Porsild spruce, 394–395
Portuguese lavender, 303–304
post oak, 457–458
potato, 508–509
prairie crabapple, 325
prickly pear, 365
pride of the meadow, 227
Prince-of-Wales feather, 57
prince's feather, 56
pubescent oak, 451–452
pummelo, 156–157
pumpkin, 182–186
purple amaranth, 56
purple betony, 516–517
purple clover, 533–534
purple coneflower, 201–202
purple yam, 197–198

Quaker buttons, 518
quassia, 435–436
Queen Anne's lace, 195–197
queen of the meadow, 227
Queensland kauri pine, 45–46
quince, 188–189
quinine, 136–137
quinine tree, 136–137
quinoa, 131–132

rabbitbrush, 210–211
radicchio, 134, *135*
radish, 459–460
raspberry, 477–478, *479*
red amaranth, 56
red bush, 78–79
red cedar, 126
red cinchona, 136–137
red clover, 533–534
redcurrant, 465–466
red currant, 465–466
red-cypress, 525, 527
red gum, 214–215
red huckleberry, 556
red juniper, 299–300
red medlar, 317–318
red millet, 378
red raspberry, 477–478, *479*
red river gum, 214–215
red sandalwood, 432
red sanders, 432
red sorrel, 252
red spruce, 397
red willow, 491–492
redwood, 504–505
Rhodiola rosea, 502, 504
rhubarb, 461
ribleaf, 412
ribwort plantain, 412
rice, 371–374, *375*
river red, 214–215
river red eucalyptus, 214–215
rock chestnut oak, 447
rockmelon, 178–179
rock samphire, 175
roiboos, 78–79
Roman chamomile, 129–130
Roman licorice, 246
root, 239–240, *241*
rosebay willowherb, 207–208
rosella, 252
roselle, 252
rose mallow, 253
rosemary, 471–472, *473*
rose of Sharon, 253
roseroot, 502, 504
round-leaved mint, 340–341
round-leaved sundew, 199–200
rowanberry tree, 510
rubber rabbitbrush, 210–211
rue, 487

rugose rose, 470–471
run-away-robin, 244
rutabaga, 97–98
rye, 501–502, *503*

saffron, 176–177
sage, 493–494
sagebrush, 74
Saint-Joseph's-wort, 362–363
Saint Mary's thistle, 505
salad burnet, 497
salmonberry, 482–483
samphire, 175
sarsaparilla, 507
sassafras, 498
satsuma mandarin, 159–160
satsuma orange, 159–160
saunderswood, 432
sawtooth blackberry, 474–475
sawtooth coriander, 212–213
scarlet beebalm, 343–344
scarlet sumac, 462–463
scented mayweed, 333–334
scotch broom, 193–194
scotch thistle, 505
scutch grass, 191–192
seaberry, 203
seafennel, 175
semen strychnos, 518
service tree, 511
sessile oak, 449–450
Seville orange, 140–141
shaddick, 156–157
shaddock, 156–157
shagbark hickory, 122–123
shellbark hickory, 122–123
shiny leaf buckthorn, 460
shiso, 384–385
short-stemmed gentian, 238–239
Sicilian sumac, 462–463
Siebold crabapple, 327
silverberry, 202
silver birch, 90–91
silver fir, 34
sitka rose, 470–471
sitka spruce, 398
skunkbush sumac, 463–464
slash pine, 400–401
slender-leaved elecampagne, 288–289
sloe, 428–429

619

small burnet, 497
small nettle, 548–549
smooth-barked kauri, 45–46
smooth-bark kauri, 45–46
snap, 387–388
snap bean, 387–388
Somali myrrhor, 167
sorghum, 512–513, *514*
sorrel, 252, 486–487
sour cherry, 421–422
sour orange, 140–141
soursop, 63–64
southern blue-gum, 216–217
southern crabapple, 320
southern-cypress, 525, 527
southern marigold, 522
southern white cedar, 128–129
soybean, 245
Spanish cedar, 126
Spanish chestnut, 124–125
Spanish sage, 492–493
spearmint, 339–340
spelt wheat, 541–542
spicebush, 311–312, 546–547
spike lavender, 303–304
spinach, 513, 515
spineless cactus, 365
spotted beebalm, 345
spring heath, 208–209
spruce, 393–394
square-twig blueberry, 553
squawbush, 463–464
star anise, 288
starflower, 95
starfruit, 82, 84
St. Benedict's herb, 240
stemless gentian, 238–239
sticky-willie, 233–234
stinging nettle, 547–548
stinking nightshade, 282, 284
Stinking Roger, 522
St. John's-bread, 128
St. John's-wort, 284
St. John's wort, 284
St. Joseph's rod, 253
St. Lucie cherry, 425–426
strawberry, 229–230
strawberry tree, 67
string bean, 387–388
strychnine tree, 518
sugar beet, 85–86

sugarcane, 487–488, *489*
sugarleaf, 517
sugar maple, 36–37, *38*
sugar palm, 69, 95–96
sumac, 463–464
sumach, 462–463
sumak, 462–463
summak, 462–463
summer haw, 173
summer jasmine, 292–293
summer savory, 499
sunflower, 249–250
suriname cherry, 218–219
swamp chestnut oak, 445–446
swamp cypress, 525, 527
swamp white oak, 438, 440
swede, 97–98
Swedish turnip, 97–98
sweet bay laurel, 302
sweetberry honeysuckle, 315–316
sweet birch, 88–89
sweet cicely, 356–357
sweet crabapple, 321–322
sweet elder, 495
sweet flag, 40, 42
sweetgale, 350–351, *352*
sweet gale, 350–351, *352*
sweetleaf, 517
sweet leaf, 517
sweet mace, 521
sweet orange tree, 149–150
sweet potato, 290–291
sweetscented marigold, 521
sweet violet, 563
sweet wood, 246–247
sweet woodruff, 234–235
swordleaf inula, 288–289
sycamore, 414
Syrian ketmia, 253
Syrian oregano, 369

Tahitian vanilla, 560–561
tailed pepper, 405–406
taioba, 566
tala palm, 95–96
tall blackberry, 474–475
tall huckleberry, 553
tamamoro, 192–193
tamarillo, 192–193
tamarind, 522–523
tangelo, 151

tangerine, 158–160, *161*
Tanner's sumach, 462–463
tansy, 524–525
taro, 165–166
tarragon, 72–73
tart cherry, 421–422
Tasmanian blue-gum, 216–217
tassel flower, 55
tea, 109–110
teak, 527–528
tear grass, 163–164
temple juniper, 298–299
ten-months yam, 197–198
Texas tarragon, 521
thimbleberry, 481
thinleaf huckleberry, 553
Thompson seedless, 563–564
thulasi, 363–364
thyme, 531–532
Tibetan goji, 317–318
tidewater redcypress, 525, 527
tobacco, 360–361
toddy palm, 95–96
tomato, 318–319
Toringo crabapple, 327
tortoise-shell bamboo, 391–392
totara, 415–416
tree of music, 495
tree tomato, 192–193
true cardamom, 204
true cinnamon tree, 137–138
true jasmine, 292–293
true lavender, 302–303
true myrtle, 357–358
trumpet gentian, 238–239
tulasi, 363–364
tulsi, 363–364
tumeric, 187–188
tunhoof, 244
Turkish yenibahar, 398–399
turmeric, 187–188

unshu mikan, 159–160
upland cotton, 247–248
upland sumac, 462–463

vanilla, 560–561
variegated thistle, 505
velvet flower, 55
velvet-leaf elder, 495
vermillion, 432–433

INDEX OF COMMON NAMES

viñátigo, 385–386
vinhatico, 385–386, 415, 571–572

walewort, 495
wasabi, 219–220
watercress, 358–359
watermelon, 139–140
water yam, 197–198
western blueberry, 557
western crabapple, 325
western white spruce, 394–395
western yellow pine, 403–404
West Indian lime, 151–152
West Indian vanilla, 560–561
wheat, 535–536, *537*
whig plant, 129–130
white bergamot, 343
white birch, 89–90
white cedar, 128–129
whitecedar falsecypress, 128–129
white-cypress, 525, 527
white goosefoot, 130–131
white henbane, 281–282, *283*
white horehound, 331–332
white jasmine, 292–293
white man's foot, 413–414
white millet, 378
white mulberry, 345–346, *347*
white mustard, 506
white oak, 437–438, *439*
white sassafras, 498

white spruce, 394–395
white yam, 197–198
whortleberry, 554, *555*, 556–557
wild baby's breath, 234–235
wild bee-balm, 344–345
wild bergamot, 343, 344–345
wild carrot, 195–197
wild celery, 62–63
wild century plant, 46–47
wild chamomile, 333–334
wild elder, 495
wild marigold, 522
wild marjoram, 369
wild rice, 570–571
wild rosemary, 305, *306*
wild service tree, 511–512
wild sesame, 384–385
wild strawberry, 230–231
wild thyme, 530–531
wild triga, 205–206
willow, 490–491
wineberry, 482
wine palm, 95–96
wine raspberry, 482
winged yam, 197–198
winter-flowering heather, 208–209
wintergreen, 236
winter heath, 208–209
winter savory, 499–500
winter squash, 182–185
wiregrass, 191–192

wolfwillow, 202
wood avens, 240
wood betony, 516–517
woodland strawberry, 230–231
woodruff, 234–235
wood violet, 563
wormseed, 200–201
wormwood, 70–71
woundwort, 509–510

xoconostle, 366, *367*

yagar, 96–97
yagcar, 96–97
yam, 197–198, 290–291
yarrow, 37, 39–40, *41*
yeast, 6–7
yellow bark, 136
yellow birch, 88
yellow chestnut oak, 448
yellow gentian, 239–240, *241*
yellow guava, 430–431
yellow lantern chili, 115–116
yellow mustard, 506
yerba mate, 287
yuca, 330–331
yuzu, 154–155

za'atar, 369
zante currants, 563–564
zatar, 369

Library of Congress Cataloging-in-Publication Data
Names: Caruso, Giuseppe, 1965– author, illustrator. | Kosmos (Firm : Reggio Emilia, Italy), translator.
Title: The botany of beer: an illustrated guide to more than 500 plants used in brewing / Giuseppe Caruso; translated by Kosmos SRL
Other titles: Botanica della birra. English Description: New York City : Columbia University Press, [2022] | Includes index.
Identifiers: LCCN 2021042062 (print) | LCCN 2021042063 (ebook) | ISBN 9780231201582 (hardback) | ISBN 9780231554176 (ebook)
Subjects: LCSH: Beer. | Botany. | Brewing.
Classification: LCC TP577 .C343 2022 (print) | LCC TP577 (ebook) | DDC 663/.42—dc23/eng/20211105
LC record available at https://lccn.loc.gov/2021042062
LC ebook record available at https://lccn.loc.gov/2021042063